Volume

	in^3	ft^3	yd^3	m^3	qt	liter	barrel	gal. (U.S.)
1 in^3 =	1	—	—	—	—	0.02	—	—
1 ft^3 =	1,728	1	—	.0283	—	28.3	—	7.480
1 yd^3 =	—	27	1	0.76	—	—	—	—
1 m^3 =	61,020	35.315	1.307	1	—	1,000	—	—
1 quart (qt) =	—	—	—	—	1	0.95	—	0.25
1 liter (l) =	61.02	—	—	0.001	1.06	1	—	0.2642
1 barrel (oil) =	—	—	—	—	168	159.6	1	42
1 gallon (U.S.)	231	0.13	—	—	4	3.785	0.02	1

Energy and Power

1 kilowatt-hour = 3,413 Btus = 860,421 calories
1 Btu = 0.000293 kilowatt-hour = 252 calories = 1,055 joule
1 watt = 3.413 Btu/hr = 14.34 calories/min = 1 joule per second
1 calorie = the amount of heat necessary to raise the temperature of 1 gram (1 cm^3) of water 1 degree Celsius
1 quadrillion Btu = (approximately) 1 exajoule
1 joule = 0.239 calorie = 2.778×10^{-7} kilowatt-hour = force of 1 N applied over 1 meter

Mass and Weight

1 pound = 453.6 grams = 0.4536 kilogram = 16 ounces
1 gram = 0.0353 ounce = 0.0022 pound
1 short ton = 2,000 pounds = 907.2 kilograms
1 long ton = 2,240 pounds = 1,008 kilograms
1 metric ton = 2,205 pounds = 1,000 kilograms
1 kilogram = 2.205 pounds

Temperature

F is degrees Fahrenheit
C is degrees Celsius (centigrade)

$$F = \frac{9}{5}C + 32$$

Fahrenheit		Celsius
32	Freezing of H_2O (Atmospheric Pressure)	0
50	———	10
68	———	20
86	———	30
104	———	40
122	———	50
140	———	60
158	———	70
176	———	80
194	———	90
212	Boiling of H_2O (Atmospheric Pressure)	100

ENVIRONMENTAL GEOLOGY

Eighth Edition

EDWARD A. KELLER
University of California, Santa Barbara

PRENTICE HALL, UPPER SADDLE RIVER, NEW JERSEY 07458

Library of Congress Cataloging-in-Publication Data

Keller, Edward A.
Environmental geology / Edward A. Keller. -- 8th ed.
p. cm.
Includes bibliographical references and index.
ISBN 0-13-022466-9
1. Environmental geology. I. Title.
QE38.K45 2000
304.2--dc21 99-046519

Senior Editor: Patrick Lynch
Editor in Chief: Paul F. Corey
Assistant Vice President of Production & Manufacturing: David W. Riccardi
Executive Managing Editor: Kathleen Schiaparelli
Associate Creative Director: Amy Rosen
Manufacturing Manager: Trudy Pisciotti
Editors in Chief of Development: Carol Trueheart and Ray Mullaney
Marketing Manager: Christine Henry
Art Director: Ann France
Art Manager: Gus Vibal
Art Editor: Karen Branson
Director of Creative Services: Paul Belfanti
Special Projects Manager: Barbara Murray
Cover Designer: Ann France
Interior Designer: Jennifer Bergamini
Photo Research: Beth Boyd
Photo Research Administrator: Melinda Reo
Cover photograph: Paricutín volcano by Tad Nichols
Cover photo caption excerpted from Fisher, R.V., Heiken, G., and Hulen, J.B. 1997. *Volcanoes.* Princeton, NJ: Princeton University Press.
Production Editor: Tim Flem/PublishWare
Page Layout: PublishWare

Printed in the United States of America

10 9 8 7 6 5 4 3 2

ISBN 0-13-022466-9

Prentice-Hall International (UK) Limited, *London*
Prentice-Hall of Australia Pty. Limited, *Sydney*
Prentice-Hall Canada Inc., *Toronto*
Prentice-Hall Hispanoamericana, S.A., *Mexico*
Prentice-Hall of India Private Limited, *New Delhi*
Prentice-Hall of Japan, Inc., *Tokyo*
Prentice-Hall *(Singapore)* Pte, Ltd.
Editora Prentice-Hall do Brasil, Ltda., *Rio de Janeiro*

BRIEF CONTENTS

To Valery Rivera

Through her administrative action, innovative and creative application of educational theory in the classroom, and serving as an inspirational role model, she continues to improve the lives and education of a precious resource—the children of the Americas and of the world.

CONTENTS

PREFACE

Environmental geology is the application of geologic information to the entire spectrum of interactions between people and our physical environment. Those students who become more interested in the subject may go on to become environmental geologists. They will take advanced courses in subjects such as engineering geology and hydrogeology.

Study of environmental geology is facilitated by previous exposure to an introductory physical geology or geography course. However, most students may not have the flexibility in their undergraduate curriculum to take more than a single geology course. As a result, *Environmental Geology* is designed so that students who have had no previous exposure to the geological sciences may comprehend and understand the principles of environmental geology. This requires that a fair amount of physical geology is presented along with discussion of relationships between geology and the environment. One important objective of the book is to present a broad spectrum of subject matter relevant to students studying a wide variety of traditional sciences such as chemistry, biology, geology, physical geography, and physics; liberal arts students studying subjects such as anthropology, economics, environmental studies, human geography, literature, political science, and sociology; and students who may be preparing for professional schools such as engineering, architecture, or planning.

ORGANIZATION

The organization of the eighth edition is similar to the seventh with appropriate expansion of the two water chapters: water supply and management, and water pollution. The continuing emphasis on water results because of the tremendous importance of water resources and their use to all people on earth. Of particular significance is the added material concerning surface and groundwater processes. Since the seventh edition, the subjects of earth system science and air pollution are now stand-alone chapters. The subject of earth system science and global change is of particular importance to environmental geology, and considerable discussion has been added concerning such global environmental problems as potential global warming, stratospheric ozone depletion, and acid rain.

The subject material for *Environmental Geology* is arranged in five parts. Part 1 introduces fundamental principles important in the study of environmental geology. The purpose is to set the philosophical framework for the remainder of the text, as well as to introduce important aspects of geology necessary to undertake a study of applied geology. Emphasis is on geologic process and the study of the earth as a system. This is facilitated through study of earth materials such as minerals, rocks, soils as they relate to processes operating in the solid earth, biosphere, hydrosphere (surface and groundwaters), and atmosphere.

Part 2 addresses the important subject of natural processes (hazards) that continue to make life on earth occasionally difficult for people. These include flooding, landslides, earthquakes, volcanoes, and coastal erosion. Discussion of hazardous natural processes is facilitated through the introduction of numerous case histories that represent the spectrum of hazards and the response of society. New case histories or discussions include the 1993 floods of the Mississippi River and the 1994 earthquake that damaged part of the Los Angeles, California, area.

Part 3 focuses on human interactions with the environment and includes detailed discussions of water resources, water pollution, integrated waste management, and geologic aspects of environmental health. In particular, the material in Chapters 10 and 11 has been completely reorganized and updated in light of important issues related to water as a resource and water pollution. Finally, because groundwater processes are becoming so important in many environmental geology areas, additional discussion is provided so that students might better understand the physical processes related to the occurrence and movement of groundwater.

Part 4 includes the important topics of mineral and energy resources as they pertain to needs of society and environmental issues. Emphasis is placed on the origin and significance of our mineral and energy resources, as well as their availability and environmental impact associated with their development.

Part 5 introduces the subjects of earth system science, global change, air pollution, environmental evaluation, and environmental law. For the eighth edition, the material on earth system science and global change as it relates to global environmental problems has been revised to include a more thorough discussion of stratospheric ozone depletion and global warming. The subjects of environmental impact analysis, land-use planning, and environmental law have been combined into a single chapter that integrates these subjects and highlights the usefulness of environmental geology to society.

SOME SPECIAL FEATURES

The eighth edition of *Environmental Geology* includes a complete reorganization of the chapters:

- The eighth edition presents some quantitative examples in the *Putting Some Numbers On* boxes, which discuss a variety of processes, including: exponential growth; soil mechanics; flood frequency analysis; landslide stability analysis; construction of a beach budget; groundwater movement; and construction of the design response spectra for evaluating the earthquake hazard at a site.
- Each chapter begins with a photograph and an accompanying short explanatory narrative, followed by learning objectives. The purpose of the learning objectives is to inform students about what is important in that chapter.
- Selected *Case History* boxes are separated from the general text so that students might better focus on important issues.
- *A Closer Look* special feature boxes discuss important issues or concepts separately from the general text. For example, *A Closer Look: Wetlands* in Chapter 10 defines wetlands and important environmental processes of wetlands.
- Key terms presented in the text are listed at the end of each chapter to further assist students in their study of the material.
- Study questions are provided for each chapter with the intent of stimulating discussion of important issues related to environmental geology. There are only a few questions for each chapter but the emphasis is on developing critical thinking skills about the text material.
- Useful Web sites are listed in *Appendix A: World Wide Web: An Introductory Statement.*

SPECIAL NOTES TO STUDENTS

I wrote part of the first edition of *Environmental Geology* while I was still a graduate student at Purdue University. As a student I was very interested in trying to understand this new field of environmental geology that a few people were beginning to talk about. I taught a class in environmental geology at Indiana University–Purdue University in Indianapolis, Indiana. At that time, in the early 1970s, we were just becoming conscious of environmental problems and were more concerned with identifying the problems rather than dealing with solutions. Since then, the field of environmental geology has grown substantially, and there are many practicing professional environmental geologists. The focus has also shifted to finding solutions to environmental problems. This requires increased understanding of application of geologic processes to the environment. As a result, we need to know more about how the earth works and how the subject of geology interrelates with other disciplines, such as biology, chemistry, physics, and geography. Some of the most fruitful areas for research for future students will be interdisciplinary studies that combine one or more of the other sciences. Environmental geology, however, is more than just science. Our science may provide potential solutions to environmental problems and outline risk of such processes as earthquakes and flooding. What we decide to do about our environmental problems is more related to our value system. As a result, subjects such as psychology, social anthropology, history, human geography, political science, law and society, and economics come to the forefront. People who will solve environmental problems in the future will continue to be more interdisciplinary in their research approach and their study of the environment. The upshot of all this is that our study of the environment is becoming broader on one hand and more rigorous on the other. In response to this, I have made no attempt to dodge difficult subjects such as hydraulic conductivity, fluid pressure in rocks and soils, moment magnitude of earthquakes, and processes related to stratospheric ozone depletion. I am confident that students remain interested in obtaining the best possible education and expect to be challenged. I have made every attempt to discuss difficult subjects in a way that will enable students to better understand what is happening. However, many of these subjects are difficult, and considerable study is necessary to understand them thoroughly. Finally, more advanced courses in areas such as hydrogeology, geochemistry, and seismology are suggested for those students who wish to pursue these subjects further.

I have learned a great deal in preparing *Environmental Geology*. I hope you enjoy reading the book and hope that some of you may pursue environmental geology as a career. A few of you may become so intensely interested that you will pursue a research and teaching career to better understand how our earth works and how we might better solve environmental problems.

ACKNOWLEDGMENTS

Successful completion of a textbook that includes hundreds of photographs and illustrations as well as case histories would not be possible without the cooperation of many individuals, companies, and agencies. In particular, I greatly appreciate the work of agencies such as the U.S. Geological Survey, which has an extensive environmental program, as well as individual state geological surveys, which have provided a great deal of information and concepts important in the development of the subject of environmental geology. To all of those individuals who are so helpful in this endeavor, I offer my sincere appreciation. Reviews of chapters or the entire book in this or previous editions by Roger J. Bain, Daniel Botkin, Douglas G. Brookings, William Chadwick, Laurence R. Davis, P. Thompson Davis, Thomas L. Davis, John G. Drost, Anne Erdmann, Edward B. Evenson, Richard V. Fisher, Stanley T. Fisher, Robert

B. Furlong, H. G. Goodell, Cal Janes, Donald L. Johnson, Ernest K. Johnson, Ernest Kastning, James Kennett, Björn Kjerfve, Harold L. Krivoy, Gilbert LaFreniere, James R. Lauffer, Gene W. Lene, Hugo Loaiciga, Mel Manalis, Robert Mathews, Richard Mauger, Marc McGinnes, Patricia Miller, Gary L. Millhollen, Robert M. Norris, June A. Oberdorfer, Roderic A. Parnell, Anne D. Pasch, John S. Pomeroy, James Dennis Rash, Charles J. Ritter, Derrick J. Rust, Paul T. Ryberg, William C. Sherwood, Robert Shuster, Mark T. Steward, Samual E. Swanson, Taz Talley, Russel O. Utgard, and William S. Wise are acknowledged and are greatly appreciated.

I am particularly indebted to the staff at Prentice Hall, including Patrick Lynch (Senior Editor), Kathleen Schiaparelli (Executive Managing Editor), Tim Flem (Production Editor), Karen Branson (Art Editor), and Beth Boyd (Photo Researcher).

The Environmental Studies Program and the Department of Geological Sciences at the University of California, Santa Barbara, continue to provide a stimulating environment in which to do research and write. I would like to thank the many people who have readily given their time in thoughtful discussion and help in preparation of many aspects of *Environmental Geology*. In particular, I would like to acknowledge the help of Ellie Dzuro and Molly Trecker.

Edward A. Keller
Santa Barbara, California

About the Author

Edward A. Keller is professor of environmental geology at the University of California, Santa Barbara, where he holds a joint appointment with the Environmental Studies Program and the Department of Geological Sciences. Professor Keller's research focuses on natural hazards, including wildfire, flooding, and earthquakes. He is also evaluating the environmental impacts of ecotourism. Professor Keller received his doctorate from Purdue University in 1973; as a graduate student, he wrote the first five chapters of the first edition of *Environmental Geology*. Professor Keller enjoys teaching and research, and hopes his textbooks help students to appreciate geology and to assist them in becoming better and more informed citizens in our complex world.

PART I

Foundations of Environmental Geology

CHAPTER 1
Philosophy and Fundamental Concepts

CHAPTER 2
Earth Materials and Processes

CHAPTER 3
Soils and Environment

People today may be more concerned with our environment than ever before since the Industrial Revolution. We have come to realize that, while we are generally better off now, rapid increase in world population and development of resources is straining our planet's natural support system. We are concerned about ozone depletion, potential global warming, and pollution of our air, water, and land. We are struggling with how to provide a future for our children—and theirs—that will include a high-quality environment with sufficient space for people as well as the incredible diversity of life with which we share our planet. In short, we are struggling to develop what we term "sustainability."

Chapters 1 through 3 provide the philosophical framework and basic concepts for the remainder of the book. It is hoped that these chapters will whet your interest in learning more about environmental geology. Chapter 1 discusses what environmental geology is, how geologists work, the cultural basis of environmental awareness, environmental ethics, and the fundamental concepts important to the study of environmental geology. Chapter 2 introduces the physical environment through the geologic cycle. The term *cycle* emphasizes that most earth materials, such as air, water, minerals, and rock, although changed physically and chemically and transported from place to place, are constantly being reworked, conserved, and renewed by natural earth processes. Chapter 2 also introduces basic earth science terminology and the engineering properties of earth materials, excluding soil. Chapter 3 discusses soil in terms of its development, classification, engineering properties, and other factors important to land-use planning.

I Philosophy and Fundamental Concepts

Statues at Easter Island. (*Tom Till/Tony Stone Images*)

Polynesian people first reached Easter Island approximately 1500 years ago. The small island is located several thousand kilometers west of South America and has a semi-arid climate. When the Polynesians first arrived, they were greeted by a green island covered with forest. By the sixteenth century, the 7000 people there had established a complex society spread among small villages that raised crops and chickens supplemented by fish from the ocean. The people carved massive 8-m-high statues from volcanic rock that were moved into place at various locations in the island using tree trunks as rollers. Europeans reached Easter Island in the seventeenth century, and the only symbols of the once-vibrant civilization were the statues.

The society evidently collapsed in just a few decades, probably the result of the degradation of the island's limited resource base. As the human population of the island increased, more and more land was cleared for agriculture while remaining trees were used for fuel and for moving the statues into place. The soils beneath the forest cover were protected and held water in the semi-arid environment, and soil nutrients were probably supplied from dust by thousands of miles away that reached the island with the winds. With the clearing of the forest, the soils eroded and the agricultural base of the society was diminished. Loss of the forest also resulted in loss of forest products necessary for building homes and boats, and, as a result, the people were forced to live in caves. They could no longer rely on fish as a source of protein. As population pressure increased, wars between villages became common, as did slavery and even cannibalism, in attempts to survive in an environment depleted of its resource base.

The story of Easter Island is a dark one that vividly points to what can happen when an isolated area is deprived of its resources through human activity. The lesson learned was that limited resources cannot support an ever-growing human population. There is fear today that our planet (an isolated island in space) may be reaching the same threshold that faced

LEARNING OBJECTIVES

In this chapter we discuss what environmental geology is and some aspects of culture and society that are particularly significant to environmental awareness. We also present some basic concepts of environmental science that provide the philosophical framework of this book. After reading this chapter you should be prepared to discuss:

- *What environmental geology is.*
- *What the scientific method is.*
- *The development of environmental ethics and, in particular, the emerging land ethic.*
- *Important factors contributing to the "environmental crisis."*
- *Why environmental problems transcend political and religious systems.*
- *Why the number-one environmental problem is increasing human population.*
- *The concept of sustainability, what a sustainable economy would look like and do, and why sustainability is the environmental objective of the future.*
- *Basic ideas of systems theory, including feedback, growth rates, and changes in systems.*
- *What earth system science is and how it is related to the Gaia hypothesis.*
- *The concept of uniformitarianism and why it is important to environmental geology.*
- *Why land- and water-use planning needs to balance economic considerations and less tangible variables such as aesthetics.*
- *How the cumulative impact of land use gives us a responsibility to future generations.*

Web Resources

Visit the "Environmental Geology" Web site at www.prenhall.com/keller to find additional resources for this chapter, including:

- ▶ Web Destinations
- ▶ On-line Quizzes
- ▶ On-line "Web Essay" Questions
- ▶ Search Engines
- ▶ Regional Updates

the people of Easter Island in the sixteenth century. As we move into the twenty-first century, we are facing limits of resources in a variety of areas, including soils, fresh water, forests, rangelands, and ocean fisheries. A big question from an environmental perspective and for the history of humans on earth is: Will we recognize limits of the earth's resources before it is too late to avoid the collapse of human society on a global scale? Today there are no more frontiers on the earth, and we have a nearly fully integrated global economy. The lesson from Easter Island is clear: develop a sustainable global economy that ensures the survival of our resource base and other living things on earth, or suffer the consequences.*

1.1 Introduction to Environmental Geology

Everything has a beginning and an end. Our earth began about 4.6 billion years ago when a cloud of interstellar gas known as a *solar nebula* collapsed, forming protostars and planetary systems. Life on earth began about 3 billion years ago, and since then multitudes of diverse organisms have emerged, prospered, and died out, leaving only their fossils to mark their place in earth's history. Just a few million years ago, our ancestors set the stage for the present dominance of the human species. As certainly as our sun will die, we too will eventually disappear. Viewed in terms of billions of years, our role in earth's history may be insignificant, but for us living now and for our children and theirs, our impact on the environment is significant indeed.

The place of humanity in the universe is well stated in the *Desiderata* (1): "You are a child of the universe, no less than the trees and the stars; you have a right to be here. And whether or not it is clear to you, no doubt the universe is unfolding as it should." Geologically speaking, we have been here for a very short time. Dinosaurs, for example, ruled the land for more than 100 million years. We do not know how long our own reign will be, but the fossil record suggests that all species eventually become extinct. How will the history of our own species unfold, and who will write it? We can hope that we will leave something more than some fossils marking a brief time in the record when *Homo sapiens* flourished. As we evolve, it is hoped that we will be more environmentally aware and find ways to live in harmony with our planet.

Environmental geology is applied geology. Specifically, it is the use of geologic information to help us solve conflicts in land use, to minimize environmental degradation, and to maximize the beneficial results of using our natural and modified environments. The application of geology to these problems includes the study of:

- ▶ *Natural hazards* (such as floods, landslides, earthquakes, and volcanic activity) in order to minimize loss of life and property.

*Source: Brown, L. R. and Flavin, C. 1999. A new economy for a new century. In *State of the world 1999*, ed. L. Stark, pp. 3–21. New York: W. W. Norton.

- *Landscapes* for site selection, land-use planning, and *environmental impact analysis*.
- *Earth materials* (such as minerals, rocks, and soils) to determine their potential use as resources or waste disposal sites and their effects on human health.
- *Hydrologic processes* of groundwater and surface water to evaluate water resources and water pollution problems.
- *Geologic processes* (such as deposition of sediment on the ocean floor, the formation of mountains, and the movement of water on and below the surface of the earth) to evaluate local, regional, and global change.

Considering the breadth of its applications, we can define environmental geology as the branch of earth science that studies the entire spectrum of human interactions with the physical environment.

The **environment** is the total set of circumstances that surround an individual or a community. It includes all of the physical conditions, such as air, water, gases, and landforms, that affect the growth and development of an individual or a community. It also includes the social and cultural conditions, such as ethics, economics, and aesthetics, that affect individual or communal behavior. Therefore, an introduction to environmental geology must consider not only the earth processes, resources, and structures described by earth scientists, but also the society and culture influencing how we perceive and react to our physical surroundings.

1.2 How Geologists Work: The Scientific Method

It's the thrill of discovery of something that was previously unknown about how the world works that excites geologists and drives them on in continued work. Most scientists are motivated by a basic curiosity of how things work. Creativity and insight that may result in scientific breakthroughs often start with asking the right question. Given that we know so little about how our world works, how do we go about studying it? Most studies start with identification of some problem of interest to the investigator. If little is known about the topic or process being studied, then the first step is to try to qualitatively understand what is going on. This involves making careful observations in the field or perhaps a laboratory. Based upon the observations, the scientist then may develop a question or series of questions about those observations. Next the investigator suggests an answer or several possible answers to the question or questions. These are hypotheses to be tested. The best **hypotheses** can be tested by designing an experiment that involves data collection, organization, and analysis. Following collection and analysis of the data it is interpreted and a conclusion may be drawn. The conclusion is then compared with the hypothesis and the hypothesis may be rejected or tentatively accepted. Often a series of questions or multiple hypotheses are developed and tested. If all hypotheses suggested to answer a particular question are rejected, then a new set of hypotheses must be developed. The above method is sometimes referred to as the **scientific method.** If a hypothesis withstands a sufficient number of experiments to test it, it may be accepted as a theory. A **theory** is a strong scientific statement that the hypothesis behind the theory is likely to be true but has not been conclusively proven. New evidence often disproves existing hypotheses or scientific theory; absolute proof of scientific theory is not possible. Thus, much of the work of science is to develop and test hypotheses, striving to reject current hypotheses and develop better ones.

Geologists often begin their observations in the field or in the laboratory by taking careful notes about what they are observing. In field studies geologists may identify and describe the earth materials present and make maps to show how these materials are distributed at the surface of the earth.

The important variable that distinguishes geology from most of the other sciences is the consideration of time. Geologists are interested in earth history over time periods that are nearly incomprehensible to most people. The geologic time scale highlighting biological evolution is provided on the inside back cover of this book. Notice that humans evolved during the Pleistocene, in the last 1.65 million years, which is less than 0.05 percent of the age of the earth. In answering environmental geology questions we are often most interested in the latest Pleistocene (last 18,000 years) but most interested in the last few thousand or few hundred years of the Holocene, which started approximately 10,000 years ago. Thus in geologic study it is often the task of the geologist to design hypotheses to answer questions integrated through time. For example, we may wish to test the hypothesis that the burning of fossil fuels such as coal and oil is causing global warming. One way to test this would be to show that prior to the Industrial Revolution (when we started burning a lot of coal and later oil), the mean global temperature was significantly less. We might be particularly interested in the last few hundred to few thousand years before temperature measurements were recorded at various spots around the planet as they are today. In order to test the hypothesis that global warming is occurring, the investigator might examine prehistoric earth materials that might provide indicators of global temperature. This might involve studying glacial ice or sediments from the bottom of the ocean or lake. Properly completed, studies can provide information to test the hypothesis that global warming is occurring or to reject the hypothesis.

With our increased understanding of what environmental geology is and how geologists work, we will now consider some of the philosophical underpinnings of studying the environment and introduce fundamental concepts of environmental geology.

1.3 Culture and Environmental Awareness

Environmental awareness involves the entire way of life that we have transmitted from one generation to another. To uncover the roots of our present condition we must look to the past to see how our culture and our social institutions—political, economic, ethical, religious, and aesthetic—affect the way we perceive and respond to our physical environment.

To solve environmental problems such as overpopulation, disposal of hazardous waste, global warming, and resource depletion, we must look to the future. If our social institutions are to contribute to the solutions, fundamental changes in how society works at both the personal and institutional level will be necessary. The magnitude of this adjustment may be comparable to that of the Industrial Revolution, which changed the relationship between people and the environment by producing ever-increasing demands on resources and releasing ever-increasing quantities of toxic waste into the surroundings.

Global environmental concerns are now serious political issues requiring cooperative efforts by both industrial and developing countries. For example, we cannot expect South American nations to better manage the rain forest if the United States and other industrial countries continue to place heavy economic pressure on South America to export tremendous quantities of resources. How can we expect poor, developing societies to respect the environment when wealthier industrial societies remain largely unwilling to do so? Fortunately, public concern for the environment appears to be increasing; if so, we should begin to see real progress in finding political solutions to environmental problems.

1.4 Environmental Ethics

When statesman and conservationist Stewart Udall published *The Quiet Crisis* in 1963 (2), most people were unaware that resource depletion and environmental degradation were problems. Today we are all acutely aware of what Udall called "a crisis of survival." Our emerging environmental consciousness is changing our lifestyles, our institutions, and our ethics.

An ethical approach to the environment is the most recent development in the long history of human ethical evolution, which has included changes in our concept of property rights. In earlier times, human beings were often held as property, and their masters had the unquestioned right to dispose of them as they pleased. Slaveholding societies certainly had codes of ethics, but these codes did not include the modern idea that people cannot be property. Similarly, until very recently, few people in the industrialized world questioned the right of landowners to dispose of land as they saw fit. Only within this century has the relationship between civilization and its physical environment begun to emerge as a relationship involving ethical considerations.

Environmental (including ecological and land) ethics involve limitations on social as well as individual freedom of action in the struggle for existence in our stressed environment. A **land ethic** assumes that we are responsible not only to other individuals and society but also to the total environment, namely the larger community consisting of plants, animals, soil, atmosphere, and so forth. According to this ethic, we are the land's citizens and protectors, not its conquerors. This role change requires us to revere and love our land rather than allow economics to determine all land use (see *A Closer Look: The Land Ethic and the American Experience*).

A land ethic affirms the right of all resources, including plants, animals, and earth materials, to continued existence and, at least in certain locations, continued existence in a natural state. This statement is sometimes interpreted as granting to individual plants and animals the fundamental right to live. That is an unrealistic interpretation, however, for if we are part of the environment we must extract from it the energy necessary to survive. Therefore, although the land ethic assigns animals such as deer, cattle, or chickens the right to survive as species, it does not necessarily assign rights to an individual deer, cow, or chicken. The same argument may be given to justify the use of stream gravel for construction material, or the mining and use of other resources necessary for our well-being. However, unique landscapes with high aesthetic value, such as endangered species, are in need of protection within our ethical framework.

Although widespread ecological awareness is a recent phenomenon, these ideas about the land ethic were expressed by Aldo Leopold nearly 50 years ago (3). Stewart Udall restated our moral responsibility to the land as follows: "Each generation has its own rendezvous with the land, for despite our fee titles and claims of ownership, we are all brief tenants on this planet. By choice or by default, we will carve out a land for our heirs. We can misuse the land and diminish the usefulness of resources, or we can create a world in which physical affluence and spiritual affluence go hand in hand" (2). The resounding message is that humanity is an integral part of the environment. We have a moral obligation to ensure that those who follow us also experience the pleasure of belonging to and cooperating with the entire earth community.

Environmental knowledge is increasing rapidly, and we are more aware than ever of environmental problems at local, regional, and global levels. This makes the field of environmental ethics an emerging subject of intense interest. Important topics in this field include ethical grounds for decision making, environmental racism and social justice, the value of landscape, moral relations with nonhumans, the rights of inanimate objects, and our obligations to future generations (4). If our response to environmental problems is based both on scientific knowledge and on a new perception of our kinship with the earth, we can develop policies that have profound beneficial effects on ourselves and the rest of the natural world.

1.5 The Environmental Crisis

The demands made on diminishing resources by a growing human population, along with the ever-increasing production of human waste, has produced what is popularly referred to as the **environmental crisis.** This crisis in America and the world is a result of overpopulation, urbanization, and industrialization, combined with too little ethical regard for our land and inadequate institutions to cope with environmental stress (5). Today the raid on resources continues on a global scale:

A CLOSER LOOK The Land Ethic and the American Experience*

Arriving in late fall of 1620, after two months on the stormy North Atlantic, 73 men and 29 women from the Mayflower confronted what seemed to them a wild and savage land. The colonists feared the wilderness, and they lacked the skills and knowledge they needed to adapt quickly to their new environment. Nevertheless, they brought three things that assured their eventual success in the New World. First, they brought a new technology. The Pilgrims reportedly did not even have a saw when they landed, but they did have the Iron Age skills necessary to ensure relentless subjugation of the land and its inhabitants. In the long run, the ax, gun, and wheel asserted their supremacy. Second, the colonists carried with them the social blueprints for remaking the New World. They knew how to organize work, use work animals, and sell their surplus to overseas markets. Third, they brought with them a concept of land ownership completely different from that of the Native Americans, whose bonds to the land were religious and based on kinship with nature rather than exclusive possession. The colonists' idea of ownership involved an absolute title to land, regardless of who actually worked the land or how far away the owner was. After the Native Americans were displaced, land use or abuse depended entirely on the attitude of the owner (2).

Today, America, as in its early years, suffers greatly from the **myth of superabundance.** This myth assumes that the land and resources in America are inexhaustible and that management of resources is therefore unnecessary. Stewart Udall writes that the land myth was instrumental in environmental degradation ever since the birth of land policy in the eighteenth century. Even young Thomas Jefferson, who in later life was to become aware of the value of conservation, said there was such a great deal of farmland that it could be wasted as he pleased.

Potentially damaging resource utilization, sometimes called the "raid on resources," in the western United States probably began with the mountain men and their beaver trapping in the 1820s. This was only the beginning; there followed the invention of machines that were capable of large-scale removal of resources and landscape alteration. The inventions included the sawmills that precipitated the destruction of the American forests, and the "Little Giant" hose nozzle that could tear up an entire hillside in the search for California gold. Hydraulic mining was finally outlawed in 1884. However, the effects of these damaging land-use practices are still visible today.

The seeds of conservation were planted in the latter part of the nineteenth century by men such as Secretary of the Interior Carl Schurz and geologist and explorer John Wesley Powell. Their messages concerning conservation of resources and land-use planning, although largely ignored when first introduced, today stand as landmarks in perceptive and innovative conservation.

The historical roots of our landscape heritage are full of lessons. We have learned the hard way that our resources are not infinite and that land and water management is necessary for meaningful existence. This conclusion has become even more significant over the years, as American society continues to urbanize and consume resources at an ever-increasing rate (2).

*Summarized in part from Stewart Udall, *The Quiet Crisis*. New York: Avon Books, pp. 25–27.

- Deforestation and accompanying soil erosion and water and air pollution occur on many continents (Figure 1.1).
- Mining of resources such as metals, coal, and petroleum wherever they occur produces a variety of environmental problems (Figure 1.2).
- Development of both ground and surface water resources results in loss and damage to many environments on a global scale (see *Case History: Aral Sea*).

On a positive note, we have learned a great deal from the environmental crisis. We know a lot about the relationship of environmental degradation to resource utilization; many innovative plans have been and are being developed to lessen or eliminate environmental problems; and we have made real progress, particularly in developed countries, in reducing many environmental problems such as water pollution and management of hazardous waste.

▲ **FIGURE 1.1** Clear-cut timber harvesting exposes soils, compacting them and generally contributing to an increase in soil erosion and other environmental problems. (*Edward A. Keller*)

To consider solutions, we must understand the nature and origins of the crisis. Some writers have suggested that a particular religion or culture or a specific economic or political system is responsible for the way people treat the land. However, such proposals cannot be rigorously defended. It is well known that Communist China and the formerly Communist countries of Eastern Europe have severe environmental problems, as do the capitalist United States and Western Europe (Figure 1.3). Some argue that the Judeo-

▲ **FIGURE 1.2** Large open pit mines such as this one east of Silver City, New Mexico, are necessary if we are to obtain resources. However, they do cause disturbance to the surface of the land and reclamation may be difficult or nearly impossible in some instances. (*Michael Collier*)

▲ **FIGURE 1.3** Foaming pollutants on a stream near Zabrze, Poland. Zabrze is one of the Polish towns that has experienced some of the worst industrial pollution in Eastern Europe. (*Simon Fraser/Science Photo Library/Photo Researchers, Inc.*)

Christian heritage, which views nature as having been created to serve humankind, is responsible for Western attitudes toward the environment (7). However, prehistoric people also exploited and disrupted their land, as have ancient and modern peoples of both Eastern and Western religions. Although the ideals of some cultures may suggest that land is sacred, there is a considerable gap between ethical ideals and actual land-use practices (5,8).

Because environmental degradation apparently transcends both political systems and religious beliefs, we must look further for the primary cause of our present condition. A more satisfactory explanation holds that environmental problems are the natural result of human inventiveness and that they began when people first used tools to better their chances for survival. Each innovation expanded the niche of *Homo sapiens* in a harsh world and assured that everyone who followed had an easier time. A pattern of development emerged in which innovation led to a larger population, which placed greater demands on resources and also demanded more innovation. This spiral has continued to the present, when there are signs that we are on a collision course with our environment.

The situation we seem headed for may be analogous to what happens to deer when, through our artificial management, their numbers exceed the carrying capacity of the land. Deprived of their natural enemies, these "artificial deer" increase in numbers until they have eaten all available food, causing a serious shortage of winter feed and decreased reproduction of food plants for the following spring. Everything from wildflowers to trees is gradually impoverished, and the deer either become dwarfed from malnutrition or starve (3). Are we, with our increasingly artificial environment, doing to ourselves what we have sometimes done to the deer? It is important at this juncture to acknowledge that science must be part of the solution and not part of the problem. The explosion of deer populations occurred because environmental factors controlling population growth were not understood and therefore not applied to managing the herds. Solving environmental problems begins with understanding the science of the problem.

1.6 Fundamental Concepts of Environmental Science

In the rest of this chapter we discuss some fundamental concepts of environmental science in general and environmental geology in particular. The eight concepts presented here do not constitute a list of everything that is important to environmental scientists, and they are not meant to be memorized. However, understanding the general thesis of each concept will help you to understand and evaluate the philosophical and technical material presented in the rest of the text.

CONCEPT ONE: Population Growth

The number-one environmental problem is increase in human population.

The number-one environmental problem is the ever-growing human population. The well-known human ecologist Garrett Hardin has stated that the total environmental impact of

CASE HISTORY Aral Sea

Water diversion for agriculture in what was the southern USSR has nearly eliminated the Aral Sea in a period of only 30 years. A potential tourist vacation spot in 1960, it is now a dying sea surrounded by thousands of square kilometers (km^2) of salt flats, and the change is permanently damaging the economic base of the region.

The area of the Aral Sea in 1960 was about 67,000 km^2. Diversion of the two main rivers that fed the sea has resulted in a drop in surface elevation of more than 20 m and loss of about 28,000 km^2 of surface area (Figure 1.A). Towns that were once fishing centers on the shore of the sea are today about 30 km inland. Loss of the sea's moderating effect is changing the regional climate: The winters are now colder and the summers warmer. Windstorms pick up salty dust and spread it over a vast area, damaging the land and polluting the air.

The lesson of the Aral Sea is how quick regional change from environmental damage can be. Many worry that what people have done to the Aral region is symptomatic of what we are doing on many fronts on a global scale (6).

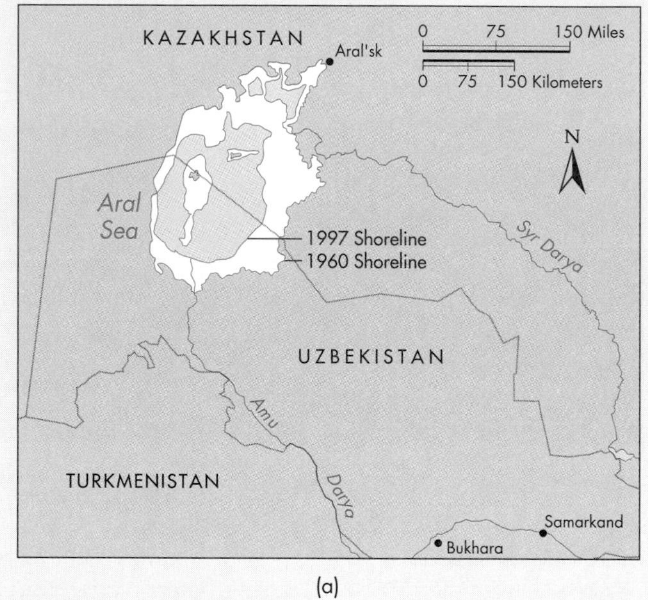

(a)

(b)

▲ **FIGURE 1.A** (a) The Aral Sea is a dying sea, surrounded by thousands of square kilometers of salt flats. (*Courtesy of Philip P. Micklin*) (b) Water diversion for agriculture has nearly eliminated the sea. The two ships shown here are stranded high and dry along the shoreline, which contains extensive salt flats formed as the Aral Sea has evaporated in recent years. (*David Turnley/Black Star*)

population is equal to the product of the impact per person times the population. Therefore, as population increases, the total impact must also increase. As population increases, more resources are needed and, given our present technology, greater environmental disruption results (Figure 1.4).

Overpopulation has been a problem in some areas for at least several hundred years, but it is now apparent that it is becoming a global problem. From 1830 to 1930, the world population doubled from 1 billion to 2 billion people. By 1970, it had nearly doubled again, and by the year 2000, it is expected that there will be about 6.2 billion people on earth. By the middle of the next century there will probably be 10 billion to 15 billion inhabitants on earth! The problem is sometimes called the *population bomb*, because the **exponential growth** of the human population results in the explosive increase shown in Figure 1.5. Exponential growth means that the number of people added to the population each year is not constant; rather, it is a constant percentage of the current population.

▲ **FIGURE 1.4** Korem Camp, Ethiopia, in 1984. Hungry people forced to flee their homes as a result of political/military activity gather in camps such as these. Surrounding lands may be devastated by overgrazing from stock animals, gathering of firewood, and just too many people in a confined area. The result may be famine. (*David Burnett/Contact Press Images*)

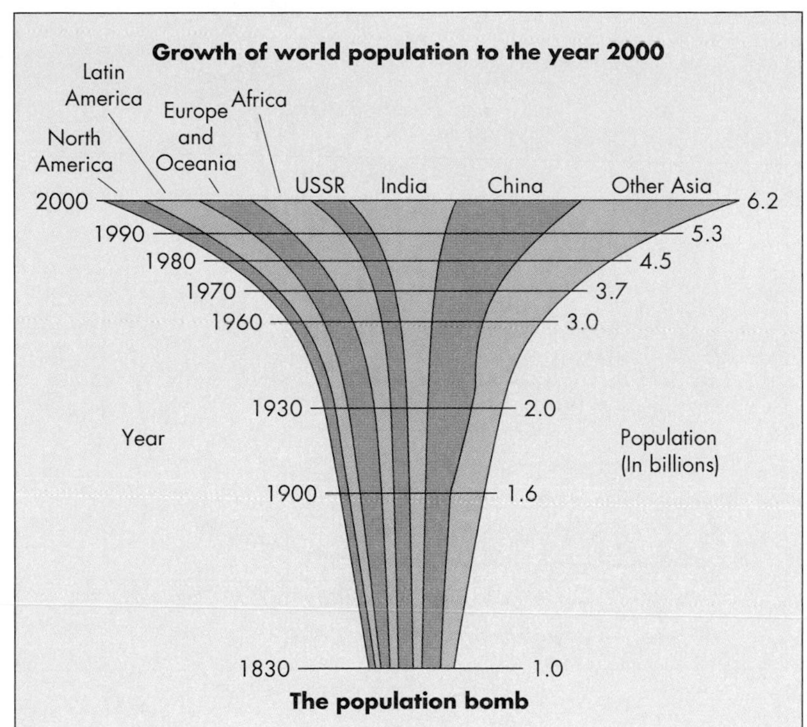

◀ **FIGURE 1.5** The population bomb. (*Modified after* U.S. *Department of State*)

As an extreme example of exponential growth (Figure 1.6a), consider the student who, upon taking a job for one month, requests from the employer a payment of 1 cent for the first day of work, 2 cents for the second day, 4 cents for the third day, and so on. In other words, the payment would double each day. What would be the total? It would take the student eight days to earn a wage of more than $1 per day, and by the eleventh day, earnings would be more than $10 per day. Payment for the sixteenth day of the month would be more than $300, and on the last day of the 31-day month, the student's earnings for that one day would be more than $10 million! This is an extreme case because the constant rate of increase is 100 percent per day, but it shows that exponential growth is a very dynamic process. The human population increases at a much lower rate—1.4 percent annually today—but even this slower exponential growth eventually results in a dramatic increase in numbers (Figure 1.6b). Exponential growth will be discussed further under Concept Three, when we consider systems and change.

Because the world's population is increasing exponentially, many scientists are concerned that it will be impossible to supply resources and a high-quality environment for the billions of people who may be added to the global population in the twenty-first century. Increasing population at local, regional, and global levels compounds nearly all environmental geology problems, including pollution of ground and surface waters; production and management of hazardous waste; and exposure of people and human structures to natural processes (hazards) such as floods, landslides, volcanic eruptions, and earthquakes.

The population problem has no easy answer. In the future we may be able to mass produce enough food from a nearly landless agriculture to support our ever-growing numbers. However, this does not solve the problems of the space available to people and the quality of their lives. Some studies suggest that the present population is already over a comfortable carrying capacity for the planet (the maximum number of people the earth can hold without causing environmental degradation that reduces the ability of the planet to support the population). *The role of education is paramount* in the population problem. As people (particularly women) become more educated, the population growth rate tends to decrease. As the rate of literacy increases, population growth is reduced. Given the variety of cultures, values, and norms in the world today, it appears that our greatest hope for population control is in fact through education.

When resource and other environmental data are combined with population growth data, the conclusion is clear: It is impossible, in the long run, to support exponential population growth with a finite resource base. Therefore, one of the primary goals of environmental work is to ensure that we can defuse the population bomb. Pessimistic scientists believe that population growth will take care of itself through disease and other catastrophes, such as famine. Optimistic scientists hope that we will find better ways to control the population of the world within the limits of our available resources, space, and other environmental needs.

CONCEPT TWO: Sustainability

Sustainability is the environmental objective.

There is little doubt that we are using living environmental resources such as forests, fish, and wildlife faster than they can be naturally replenished. We have extracted minerals,

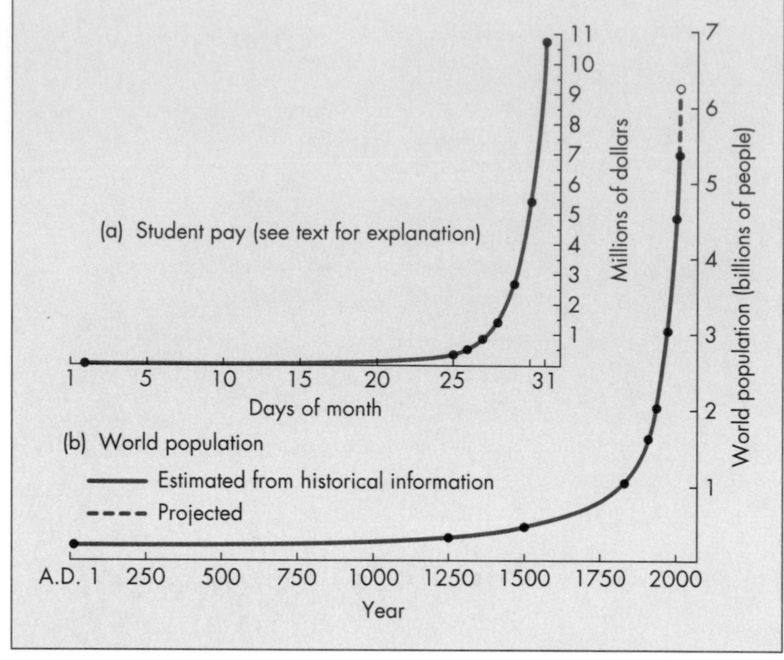

▶ **FIGURE 1.6** Exponential growth. (a) Example of a student's pay, beginning at 1 cent for the first day of work and doubling daily for 31 days. (b) World population. Notice both curves have the characteristic "J" shape, with a slow initial increase followed by a rapid increase. The actual shape of the curve depends on the scale at which the data are plotted. It often looks like the tip of a skateboard. (*Population data from U.S. Department of State*)

oil, and groundwater without sufficient concern for their limits or for the need to recycle them. As a result there are shortages of some resources. We must learn how to sustain our environmental resources so that they continue to provide benefits for people and other living things on the planet. **Sustainability** is something that we are still struggling to define. Some would define it as ensuring that future generations have equal opportunity to the resources that our planet offers. Others would argue that sustainability refers to types of development that are economically viable, do not harm the environment, and are socially just.

The environmental statement of the 1990s was "save our planet." Is earth's very survival really in danger? In the long view of planetary evolution, it seems highly likely that planet earth will survive us. Our sun is likely to last another several billion years at least, and if all humans became extinct in the next few years, life would still flourish on our planet. The changes we have made in the landscape, the atmosphere, and the waters might last for a few hundreds or thousands of years, but they would eventually be cleansed by natural processes. What we are concerned with, as environmentalists, is the quality of the *human* environment on earth.

As a result of thinking about the quality of the environment, we have begun to consider what is sometimes called the **sustainable global economy.** By *economy*, environmentalists mean the careful management and wise use of the planet and its resources, analogous to the management of money and goods considered by economists. By focusing on the concept of a sustainable global economy, we are assuming that under present conditions the global economy is *not* sustainable. In this view, increasing numbers of people have resulted in pollution of the land, air, and water to such an extent that the ecosystems upon which people depend are in danger of collapse.

How might we define a sustainable global economy? Such an economy might have the following attributes (9):

- Populations of humans and other organisms in harmony with the natural support systems such as air, water, and land (including ecosystems).
- An energy policy that does not pollute the atmosphere, cause climatic perturbations such as global warming, or present unacceptable risk (a political or social decision).
- A utilization plan for renewable resources such as water, forests, grasslands, agricultural lands, and fisheries that does not deplete the resources or destroy ecosystems.
- A resource utilization plan for nonrenewable resources that doesn't damage the global environment and provides for a share of our nonrenewable resources to be available to future generations.
- A social, legal, and political system dedicated to a sustainable and socially just global economy, with a democratic mandate to produce such an economy.

Recognizing the fact that population is the environmental problem, we should keep in mind that a sustainable global economy will not be constructed around a completely stable global population. Rather, such an economy requires a balance-of-nature approach, in which the size of the human population fluctuates (within some stable range) as necessary to maintain its desired relationship to other components of the environment.

To achieve a sustainable global economy, it is necessary that (9):

- We develop an effective population-control strategy. This will require considerable education of people in developing countries, since literacy and population growth are inversely related.

- We completely restructure our energy programs. It has been argued that a sustainable global economy is impossible if it is based upon use of fossil fuels. Our new energy plans will have to consider the concept of an integrated energy policy with a much greater emphasis upon renewable energy sources (such as solar and wind). Above all, conservation of energy must have a central place in our management of energy resources.
- We institute economic planning, including development of a tax structure that encourages population control and wise use of resources. Financial aid for developing countries is also necessary.
- We institute social, legal, political, and educational changes that have as their goal the maintenance of a quality local, regional, and global environment. This must be a serious commitment that the peoples of the world cooperate to achieve.

CONCEPT THREE: Systems

Understanding the earth's systems and their changes is critical to solving environmental problems. The earth itself is an open system with respect to energy, but essentially a closed system with respect to materials.

A **system** is any defined part of the universe that we select for study. Examples of systems are a planet, a volcano, an ocean basin, or a river (Figure 1.7). Most systems contain component parts that mutually adjust, so that changes in one part brings about changes in the others.

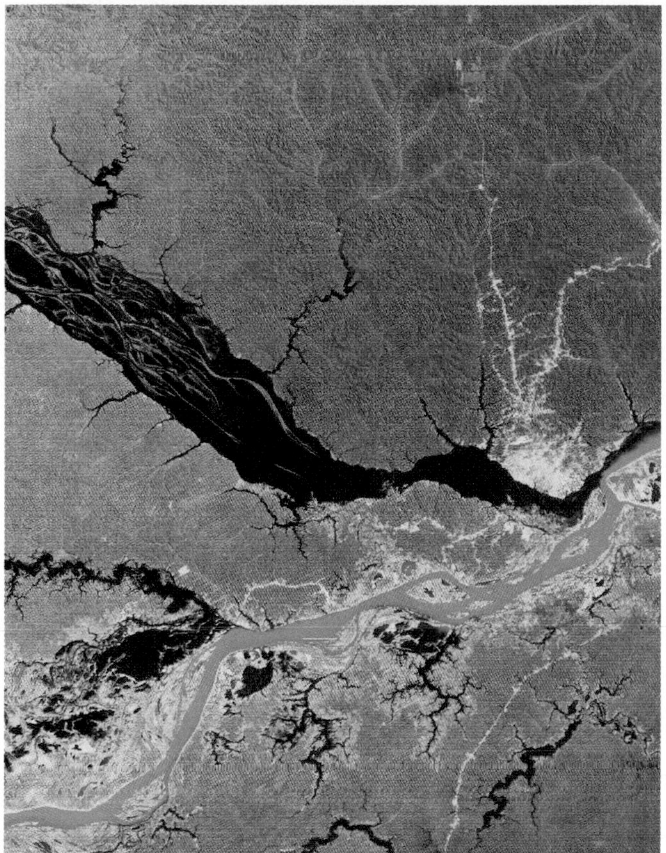

▲ **FIGURE 1.7** Landsat image showing the Amazon River (blue) and its confluence with the Rio Negro (black). The blue water of the Amazon is heavily laden with sediment, whereas the water of the Rio Negro is nearly clear. Note as the two large rivers join, the waters do not mix initially but remain separate for some distance past the confluence. The Rio Negro is in flood stage. The red is the Amazon Rain Forest and the white lines are areas of disturbances, such as roads or other human activity. (*Earth Satellite Corporation/Science Photo Library/Photo Researchers, Inc.*)

The Earth as a System The earth may be considered a system with four parts: the **atmosphere** (air); the **hydrosphere** (water); the **biosphere** (life); and the **lithosphere** (rock, soil). The mutual interaction of these parts is responsible for the surface features of the earth today. Any change in the magnitude or frequency of processes in one part causes changes in the other parts.

This propensity for coordinated change in the various parts of the environment is known as the **principle of environmental unity.** Put more simply, this principle says that everything affects everything else. For example, an increase in the magnitude of the processes that uplift mountains may affect the atmosphere by causing regional changes in precipitation patterns as a new rain shadow is produced. This in turn affects the local hydrosphere as more or less runoff reaches the ocean basins. Biospheric changes as a result of changes in the environment (such as change in types of plants and animals present) also can be expected. Eventually, the steeper slopes will also affect the lithosphere: Increased erosion will change the rate and types of sediments produced and thus the types of rocks created from the sediments.

These interactions among the variables in systems are not random. We can understand them by examining each variable to determine how it interacts with other variables and how it varies spatially over a site, area, or region. The spatial distribution of the oceans affects the amount of sunlight they receive, which affects the amount of evaporation that occurs in each region. These changes in the hydrosphere in turn affect atmospheric conditions by increasing or decreasing the amount of water in the atmosphere. We know that the earth is not static; rather, it is a dynamic, evolving system in which material and energy are constantly changing. Such dynamics suggest that the earth is an **open system,** that is, one that exchanges energy or material with its surroundings. This is certainly true with respect to energy, since the earth receives energy from the sun and radiates energy back into space.

There is also some exchange of matter between earth and its surroundings: Meteors fall to earth, and a small amount of earth material escapes into space as gas. However, most of the earth's material is continuously recycled within the system. When we consider natural cycles such as the water and rock cycles, we can best think of the earth as a **closed system** or, more accurately, a coalition of many closed systems (10). The rain that falls today will eventually return to the atmosphere, and the sediment deposited yesterday will be transformed into solid rock. We will discuss the water and rock cycles in Chapter 2.

Feedback **Feedback** is a system response in which output of the system (something happening) is an input (back into the system) causing change. There are two types of feedback in systems, called *negative* and *positive*. In **negative feedback,** the outcome moderates or decreases the process, inducing the system to approach a steady state. For example, imagine a river that maintains a certain width, volume, and velocity of flow. The river channel and banks are a type of open system called a *steady-state system*; that is, water flows through the system, but there is no net change. If the width, volume, and velocity of water increase (in response to increased regional rainfall or changes caused by urbanization), the increased flow may erode the river banks, causing the channel to widen. This widening allows the larger volume of water to flow more slowly again, so that a new steady state is established.

In **positive feedback,** often called the *vicious cycle*, the outcome of a change amplifies the initiating event, which in turn amplifies the outcome, and so on. For example, off-road vehicle use (Figure 1.8) may set a positive feedback cycle into motion by uprooting plants on a slope. Loss of the plants increases erosion, which causes still more plants to be lost, which further increases erosion. Eventually an area intensively used by off-road vehicles may be nearly denuded of all vegetation and have a very high erosion rate.

Growth Rates Because growth is an important change in many systems, we must understand something about growth rates. Exponential growth, described under Concept One, is particularly significant to the systems with which we are concerned. Consider again Figure 1.6, which illustrates two examples of exponential growth. In each case, the thing being considered (student pay or world population) increases quite slowly at first, then begins to increase more rapidly—and then very rapidly.

▲ **FIGURE 1.8** Off-road vehicle trails on this slope have caused extensive soil erosion. The site is a managed off-road vehicle park in the western Transverse Ranges of California. (*Edward A. Keller*)

There are two important measures of exponential growth:

- The **growth rate** measured as a percentage.
- The **doubling time.**

As already stated, exponential growth is characterized by a constant percent increase; for example, a population might be increasing at a rate of 2 percent per year, so that the number of individuals added annually is greater than the number added the year before. *Doubling time* is the time necessary for the quantity of whatever is being measured to double. A rule of thumb is that doubling time is roughly equal to 70 divided by the growth rate. Using this approximation, we find that a population with a 2 percent annual growth rate would double in about 35 years. Exponential growth assuming constant growth rate is not difficult to calculate and allows for important environmental questions to be addressed (see *Putting Some Numbers On Exponential Growth*).

Many systems in nature display exponential growth some of the time, so it is important that we be able to recognize such growth. In particular, we need to recognize exponential growth with positive feedback, for positive feedback cycles may be very difficult to stop. For example, we are having a very difficult time controlling (stopping) the growth of earth's human population.

Predicting Changes in Systems Some changes in natural systems are predictable, and anyone looking for solutions to environmental problems must be able to recognize systems in which predictable change takes place. Recognizing positive and negative feedback systems and calculating growth rates are important in making predictions concerning resource management. In addition, we need to understand how the input and output of a system affect the supply of a given resource and how all of these factors affect natural cyclical processes. In this discussion of changes in systems, we will begin with some of the methods used for analyzing simple systems or portions of systems, then examine the complex changes that take place on a larger scale.

Input-output analysis is an important method for analyzing change in open systems—that is, in systems with an input and output of materials or energy. Figure 1.9 identifies three types of change in a pool or stock of materials; in each case the net change depends on the relative rates of the input and output. Where the input into the system is equal to the output (Figure 1.9a), a rough steady state is established and no net change occurs. The example shown is a university in which students are brought in as freshmen and graduated four years later at a constant rate. Thus, the pool of university students remains a constant size. Our planet is a roughly steady-state system with respect to energy: Incoming solar radiation is roughly balanced by outgoing radiation from the earth.

In the second type of change, the input into the system is less than the output (Figure 1.9b). Examples include the use of resources such as fossil fuels or groundwater, or the harvest of certain plants or animals. If the input is much less

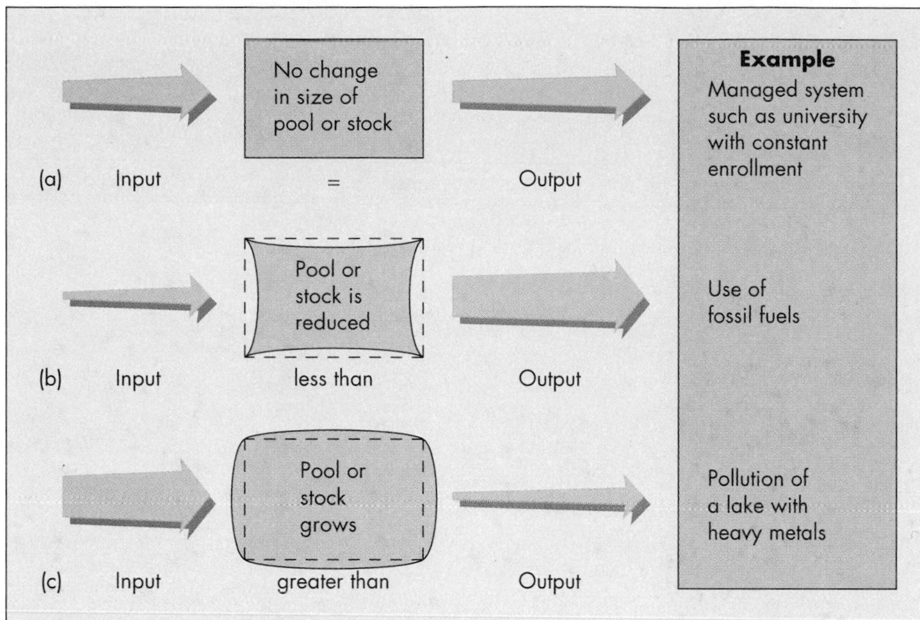

◀ **FIGURE 1.9** Major ways in which a pool or stock of some material may change. (*Modified after* P. R. *Ehrlich*, A. H. *Ehrlich*, *and* J. P. *Holdren*. 1977. Ecoscience: Population, Resources, Environment, *3rd ed.* W. H. *Freeman*)

than the output, then the fuel or water source may be completely used up, or the plants or animals may become extinct. In a system where input exceeds output (Figure 1.9c), the stock of whatever is being measured will increase. Examples are the buildup of heavy metals in lakes or the pollution of soil and water.

By evaluating rates of change or the input and output of a system, we can derive an **average residence time** for a particular material, such as a resource. The residence time is a measure of the time it takes for the total stock or supply of the material to be cycled through a system. To compute the average residence time (assuming constant size of the system and constant rate of transfer), we simply take the total size of the stock and divide it by the average rate of transfer through the system. For example, if a reservoir holds 100 million cubic meters (m^3) of water and both the average input from streams entering the reservoir and the average output over the spillway are 1 m^3/sec, then the average residence time for a cubic meter of water in the reservoir is 100 million seconds, or about 3.2 years. We can also calculate average residence time for systems that vary in size and rates of transfer, but the mathematics is more difficult. It is often possible to compute a residence time for a particular resource and, knowing this, apply the information to developing sound management principles.

Rates of change and input-output analysis help us follow materials through their natural cycles. As more and more demands are made on the earth and its limited resources, it becomes increasingly important for us to understand the magnitude and frequency of the processes that maintain these cycles. For instance, if we hope to manage the water resources of a region, we must know the nature and extent to which natural processes will supply groundwater and surface water. Or, if we are concerned with getting rid of dangerous chemicals in a disposal well, we must know how the procedure will interact with natural cycles to ensure that we or our heirs will not be exposed to hazardous chemicals. This becomes especially critical in dealing with radioactive wastes, which must be contained for periods ranging from several centuries to as long as a quarter of a million years. Therefore, it is exceedingly important to recognize earth cycles and determine the length of time involved in various parts of specific cycles. Table 1.1 lists the residence times of selected earth materials, and Table 1.2 gives the rates of some natural processes.

Complex Systems and Earth System Science So far we have dealt with ways of analyzing small portions of the environment, but ecologists also deal with larger natural systems, including the earth itself. The common expression **"balance of nature"** describes an early model for change in natural systems. According to this model, natural systems untampered with by human activity tend toward some sort of equilibrium. We gave an example of this tendency earlier when we talked about a river channel adapting to an increased volume of water. Although many parts of the environment do tend toward an equilibrium due to negative feedback, it is worthwhile to ask how accurate and how widely applicable the equilibrium model really is. As we look at natural systems for longer periods and in greater detail, we find that equilibrium is seldom obtained or maintained for a very long time. Instead, changes in systems are best described in terms of **complex response, thresholds,** and **disturbance.**

Consider a river system that periodically experiences floods. A large flood can be viewed as a disturbance within the river system that can cause considerable change to both the physical and biological environment. Usually, however, a flood must be of a certain magnitude before change occurs. For example, the banks of a river are often composed of sand, gravel, silt, and clay, bound together by the frictional and cohesive forces of these materials as well as by roots of

Table 1.1 Residence times of some selected materials

Earth Materials	Some Typical Residence Times
Atmosphere circulation	
Water vapor	10 days (lower atmosphere)
Carbon dioxide	5 to 10 days (with the sea)
Aerosol particles	
Stratosphere (upper atmosphere)	Several months to several years
Troposphere (lower atmosphere)	One week to several weeks
Hydrosphere circulation	
Atlantic surface water	10 years
Atlantic deep water	600 years
Pacific surface water	25 years
Pacific deep water	1300 years
Terrestrial groundwater	150 years (above 760 m depth)
Biosphere circulation[a]	
Water	2,000,000 years
Oxygen	2000 years
Carbon dioxide	300 years
Seawater constituents[a]	
Water	44,000 years
All salts	22,000,000 years
Calcium ion	1,200,000 years
Sulfate ion	11,000,000 years
Sodium ion	260,000,000 years
Chloride ion	Infinite

[a]Average time it takes these materials to recycle within the atmosphere and hydrosphere.

Source: *The Earth and Human Affairs* by the National Academy of Sciences. Copyright © 1972 by the National Academy of Sciences (Canfield Press). By permission of Harper & Row, Publishers.

the plants growing on the banks. Erosion of the banks will not occur unless the power of the moving floodwaters exceeds the resistance of the banks.

We may look upon the resistance of the banks as a threshold that, if crossed, will result in change—in this case, bank erosion. Suppose that a high-magnitude storm occurs in the headwater portions of a large river system. The threshold of bank erosion is crossed and erosion occurs, carrying materials from the slopes and river banks to the lower part of the river system. If the lower part of the river is unable to carry all of the sediment delivered, deposition occurs. Thus, in the upper part of the drainage system erosion is the predominant process, whereas deposition predominates downstream. Because of these differences, the eroded channel upstream behaves differently in later storms from the downstream area where deposition occurred. This is an example of complex response, which is characterized by changes in a system brought about by internal processes in the system itself.

Thus, changes in systems are not necessarily related to maintaining a balance or equilibrium. Rather, they are often complex, depending upon disturbance and crossing of thresholds. The lesson to be learned is that disturbance in natural systems is common and probably necessary for the systems to function and provide diversity in the physical and biological environments. That is, a flood may change the shape of a river channel (scouring pools, for example) that produces more physical diversity in habitat for fish and other aquatic life. The more-diverse physical habitat encourages a larger diversity in the plants and animals living in the river. Furthermore, human activity is only one type of disturbance. Events such as hurricanes, floods, wildfire (Figure 1.10), volcanic eruptions, and earthquakes also cause natural systems to change in complex ways as thresholds of change are exceeded.

Earth system science is an emerging field of study that attempts to understand the entire planet in terms of its systems. It asks how components such as the atmosphere,

Table 1.2 Rates of some natural processes

Earth Processes	Some Typical Rates
Erosion	
Average U.S. erosion rate[a]	6.1 cm per 1000 years
Colorado River drainage area	16.5 cm per 1000 years
Mississippi River drainage area	5.1 cm per 1000 years
N. Atlantic drainage area	4.8 cm per 1000 years
Pacific slope (California)	9.1 cm per 1000 years
Sedimentation[b]	
Colorado River	281 million metric tons per year
Mississippi River	431 million metric tons per year
N. Atlantic coast of U.S.	48 million metric tons per year
Pacific slope (California)	76 million metric tons per year
Tectonism	
Seafloor spreading	
N. Atlantic	2.5 cm per year
E. Pacific	7 to 10 cm per year
Faulting	
San Andreas (California)	1 to 5 cm per year
Mountain uplift	
Cajon Pass, San Bernardino Mts. (California)	1 cm per year

[a]Thickness of the layer of surface of the continental United States eroded per 1000 years.
[b]Includes solid particles and dissolved salts.

Source: *The Earth and Human Affairs* by the National Academy of Sciences. Copyright © 1972 by the National Academy of Sciences (Canfield Press). By permission of Harper & Row, Publishers.

(a)

(b)

▲ **FIGURE 1.10** (a) Aerial view of burned forest land from the 1988 Yellowstone National Park fire along the Firehole River. The steam is from geysers and hot springs. Note the sharp line between the burned, brown forest and the unburned, green forest. (*Michael Collier*) (b) Carpets of wildflowers that emerged following the Yellowstone fire are part of the natural recovery. (*Jack W. Dykinga*)

PUTTING SOME NUMBERS ON Exponential Growth

Exponential growth is a powerful process related to positive feedback where the quantity of what is being evaluated (for example, human population increase, consumption of resources such as oil or minerals, or rate of land converted to urban purposes) grow at a fixed rate (a percentage) per year. Exponential growth of human population is shown on Figure 1.6.

Calculating exponential growth is surprisingly easy and involves a rather simple equation:

$$N = N_o e^{kt}$$

where N is the future value of whatever is being evaluated; N_o is present value; e is a constant 2.71828; k is equal to the rate of increase (a decimal representing a percentage); and t is the number of years over which the growth is to be calculated. This equation may be solved utilizing a simple hand calculator, and a number of interesting environmental questions may be asked. For example, assume that we wanted to know what the world population is going to be in the year 2020 given that the population in 2000 is 6.2 billion and the population is growing at a constant rate of 1.36 percent per year (0.0136 as a decimal). Precise figures of human population and growth rates may be obtained from a variety of sources including the U.S. Bureau of Census. Assuming that world population is 6.2 billion in the year 2000 and that the growth rate is 1.36 percent per year, then we can estimate the world population for the year 2020 by applying the equation above:

$$N \text{ (world population in 2020)} = (6.2\times10^9)(e^{(0.0136 \times 20)})$$

$$N = (6.2\times10^9)(e^{0.272}) = (6.2\times10^9)(2.71828^{0.272})$$

N (population projected to year 2020 based upon the above assumptions) = 8.14×10^9, or 8.14 billion persons.

Our equation for exponential growth may also be rearranged to project the time in the future when the earth will reach a certain population. In this case we must assume a beginning population, a population at some time in the future, and the growth rate. Thus, t may be solved by the following equation:

$$t = (1/k)\ ln(N/N_o)$$

where all the terms have been defined and ln is the natural logarithm to the base 2.71828. If we use our previous example and state the question that if the population growth remains constant at 1.36 percent per year and the population in the year 2000 is 6.2 billion people, in what year will it reach 8.14 billion? Substituting in the above equation $t = (1/0.0136)\ ln\ (8.14\times10^9/6.2\times10^9)$. Solving this we see that t is equal to 20 years.

A word of caution concerning the use of exponential growth—it is based upon the assumption of constant growth rate. In trying to put arguments concerning exponential growth in a critical thinking framework, it is important to recognize that assumptions we make are statements accepted as true without proof (11). Rates of growth represented as a percentage may in fact not be constant. As a result, the estimations we make when applying the exponential growth equation based upon a constant rate of increase need to be critically examined in terms of how realistic the constant growth rate is. The longer period of time over which we apply constant rates of growth, the more unlikely our predictions are likely to be. In spite of these shortcomings, analysis of exponential growth is one way to provide insight into predicting future change, and growth or decline of a number or quantity of particular material of interest. The equation to predict decline of a quantity assuming constant rate of reduction as a percentage is:

$$N = N_o e^{-kt}$$

where terms are defined as above.

hydrosphere, biosphere, and lithosphere have formed, evolved, and been maintained; how these components interact with each other and function; and how they will continue to evolve over periods ranging from a decade to a century (12) (see *A Closer Look: The Gaia Hypothesis*). The challenge is to learn to predict changes likely to be important to society, and then to develop management strategies to minimize adverse environmental impacts. For example, study of atmospheric chemistry suggests that our atmosphere has changed. Trace gases such as carbon dioxide have increased by about 100 percent since 1850. Chlorofluorocarbons (CFCs) released at the surface have migrated to the stratosphere, where they react with energy from the sun, causing ozone depletion. The important topics of global change and earth system science will be discussed in Chapter 16 of this book, following topics such as natural hazards, energy resources, and waste management.

CONCEPT FOUR: Limitation of Resources

The earth is the only suitable habitat we have, and its resources are limited.

Concept Four includes two fundamental truths: First, that earth is indeed the only place to live that is now accessible to us; and second, that our resources are limited, and while some resources are renewable, many are not. Therefore, we eventually will need large-scale recycling of many materials, and a large part of our solid and liquid waste-disposal problems could be alleviated if these wastes were recycled. In other words, many things that are now considered pollutants could be considered resources out of place (see Chapter 12).

There are two major views on natural resources. One school holds that finding resources is not so much a problem as is finding ways to use them. In other words, the entire earth, including the ocean and atmosphere, has raw materials that can be made useful if we can develop the necessary ingenuity and skill (13). The basic assumption is that as long as there is freedom to think and innovate, we will be able to produce sufficient energy and locate sufficient resources to meet our needs. Evidence supports this line of reasoning: First, efficient and intelligent use of materials has historically been a successful venture; second, we know more about extracting minerals and fuel than we did in the past and so can find new resources faster and mine lower-grade mineral deposits; and third, recycling of resources can help us meet the needs of the future.

The second school holds that "cornucopian premises" such as the one outlined above are fallacious, because a finite resource base cannot support an exponential increase of people forever. Furthermore, Preston Cloud claims that we are in a resource crisis for a number of reasons (14): first, improvements in medical technology contributing to overpopulation; second, an unrealistic view of the necessity of an ever-increasing gross national product based on obsolescence and waste; third, the finite nature of the earth's accessible minerals; and fourth, increased risk of irreversible damage to the environment as a result of overpopulation, waste, deforestation, burning of fossil fuels, and overuse of many resources including water, energy, soil, minerals, animals and forests (14).

CONCEPT FIVE: Uniformitarianism

The physical processes modifying our landscape today have operated throughout much of geologic time. However, the magnitude and frequency of these processes are subject to natural and artificially induced change.

Understanding the natural processes that are now forming and modifying our landscapes helps us make inferences about a landscape's geologic history. The idea that "the present is the key to the past," called **uniformitarianism,** was first suggested by James Hutton (remember him as the father of geology and the Gaia hypothesis, discussed on page 18) in 1785 and is heralded today as a fundamental concept of the earth sciences. As the name suggests, uniformitarianism holds that the processes we observe today also operated in the past.

Uniformitarianism does not demand or even suggest that the magnitude and frequency of natural processes remain constant with time. Furthermore, we now know that the principle cannot be extended back to include all of geologic time: Processes operating in the oxygen-free environment of the earth's first 2 billion years were quite different from those operating today. However, for as long as the earth has had an atmosphere, oceans, and continents similar to those of today, we can infer that the present processes were operating.

For example, if we know the characteristic landforms associated with today's alpine glaciers, we can infer that valleys with similar landforms were once glaciated even if no glacial ice is present today. Similarly, if we find ancient gravel deposits with all the characteristics of stream gravel on the top of a mountain, uniformitarianism suggests that a stream must have flowed there at one time. We can conclude that what was originally a stream valley has been changed by differential erosion and/or uplift to a mountaintop, a phenomenon known as *inversion of topography*. This and many other geologic phenomena would be difficult to infer without the principle of uniformitarianism.

In making inferences about geologic events, we must consider the effects of human activity on natural earth processes. For example, rivers flood regardless of human activities, but such activities can greatly increase or decrease the magnitude and frequency of flooding. Therefore, to predict the long-range effects of a natural process, we must be able to determine how our future activities will change its magnitude and rate. In this case, the present is the key to the future. We can assume that the same processes will operate but that their magnitudes and rates will vary as the environment adjusts to human activity. We recognize that ephemeral landforms such as beaches and lakes will continue to appear and disappear in response to natural processes and that human influence may be small in comparison.

Effects of human activity may be very pronounced in a local area. One year of erosion at a construction site or human-constructed slope such as grass embankments to a highway bridge (Figure 1.11) may exceed many years of erosion from an equivalent tract of woodland or even agricultural land (16). This erosion results from exposure of the soil following the removal of vegetation. Therefore, to maximize the value of geologic knowledge in land-use planning, we must be able to use our understanding of natural earth processes in both a historical and a predictive mode.

▲ **FIGURE 1.11** Photo of soil erosion from human-constructed slope. (*Edward A. Keller*)

A CLOSER LOOK The Gaia Hypothesis

In 1785 at a meeting of the prestigious Royal Society of Edinburgh, James Hutton, the "father of geology," said he believed that planet earth is a super organism (Figure 1.B). He compared the circulation of earth's water, with its contained sediments and nutrients, to the circulation of blood in an animal. With Hutton's metaphor, the oceans are the heart of the earth and the forest the lungs (15). Two hundred years later James Lovelock (a British scientist and professor) introduced the **Gaia hypothesis,** reviving the idea of a living earth. The hypothesis is named for Gaia, the Greek goddess Mother Earth.

The Gaia hypothesis is best stated as a series of hypotheses:

- Life significantly affects the planetary environment. Very few scientists would disagree with this concept.
- Life affects the environment for the betterment of life. This hypothesis has some support from studies showing that life on earth plays an important role in regulating planetary climate so that it is neither too hot nor too cold for life to survive. For example, it is believed that single-cell plants floating near the surface of the ocean partially control carbon dioxide content of the atmosphere and thereby global climate (15).
- Life deliberately (consciously) controls the global environment. Very few scientists accept this third hypothesis. Systems of positive and negative feedback that operate in the atmosphere, on the surface of the earth, and in the oceans are probably sufficient to explain most of the mechanisms by which life affects the environment. On the other hand, humans are beginning to make decisions concerning the global environment, so the idea that humans can consciously influence the future of earth is not an extreme view. Some people have interpreted this idea as support for the broader Gaia hypothesis.

▲ **FIGURE 1.B** Satellite image of earth centering on the North Atlantic Ocean, North America, and the polar ice sheets. Given this perspective of our planet, it is not difficult to conceive it as a single large system. (*Earth Imaging/Tony Stone Images*)

The real value of the Gaia hypothesis is that it has stimulated a lot of interdisciplinary research to understand how our planet works. As interpreted by most scientists, the hypothesis does not suggest foresight or planning on the part of life but rather that automatic processes of some sort are operating. From a geologic perspective this means that throughout much of earth history life has affected the physical and chemical systems of the earth, just as these systems affect life.

For instance, when environmental geologists examine recent mudflow deposits in an area designated to become a housing development, they must use uniformitarianism to infer where there will be future mudflows, as well as to predict what effects urbanization will have on the magnitude and frequency of future flows.

CONCEPT SIX: Hazardous Earth Processes

There have always been earth processes that are hazardous to people. These natural hazards must be recognized and avoided where possible and their threat to human life and property minimized.

Because the geologic processes we know today were operating long before humans made their appearance, we have always been obligated to contend with processes that make our lives more difficult. Interestingly, *Homo sapiens* appear to be a product of the Ice Age, one of the harshest of all environments.

Early in the history of our species, our struggle with natural earth processes was probably a day-to-day experience. However, our numbers were neither great nor concentrated, so losses from hazardous earth processes were not very significant. As people learned to produce and maintain a constant food supply, the population increased and became more concentrated locally. The concentration of population and resources also increased the impact of periodic earthquakes, floods, and other natural disasters. This trend has continued, so that many people today live in areas likely to be damaged by hazardous earth processes or susceptible to the adverse impact of such processes in adjacent areas.

Earth processes that often cause loss of life or property damage include flooding, earthquakes, volcanic activity, landslides, and mudflows. The magnitude and frequency of these processes depend on such factors as a region's climate, geology, and vegetation. For example, the effects of running water as an erosional or depositional process depend on the intensity of rainfall, the frequency of storms, how much and how fast the rainwater is able to infiltrate rock or soil, the rate of evaporation and transpiration of water back into the atmosphere, the nature and extent of the vegetation, and topography.

We can recognize many natural processes and predict their effects by considering climatic, biologic, and geologic conditions. After earth scientists have identified potentially hazardous processes, they have the obligation to make the information available to planners and decision makers, who can then consider ways of avoiding or minimizing the threat to human life or property.

CONCEPT SEVEN: Geology as a Basic Environmental Science

The fundamental component of every person's environment is the geologic component, and understanding our environment requires a broad-based comprehension and appreciation of the earth sciences and related disciplines.

All geology can be considered environmental. An understanding of our complex environment requires considerable knowledge of such disciplines as **geomorphology,** the study of landforms and surface processes; **petrology,** the study of rocks and minerals; **sedimentology,** the study of environments of deposition of sediments; **tectonics,** the study of processes that produce continents, ocean basins, mountains, and other large structural features; **hydrogeology,** the study of surface and subsurface water; **pedology,** the study of soils; **economic geology,** the application of geology to locating and evaluating mineral materials; and **engineering geology,** the application of geologic information to engineering problems. Beyond this, the serious earth scientist should also be aware of the contributions to environmental research from such areas as biology, conservation, atmospheric science, chemistry, environmental law, architecture, and engineering, as well as physical, cultural, economic, and urban geography.

Environmental geology, then, is the domain of the generalist with a strong interdisciplinary interest. This in no way denies the significant contributions of specialists in various aspects of environmental studies, or the importance of focusing on specific problems or areas of research. It merely suggests that, although our research interests may be specialized, we should always be aware of other disciplines and their contribution to environmental geology. Also, many projects may be studied best by an interdisciplinary team of scientists.

The importance of the interdisciplinary nature of environmental geology becomes apparent when we explore the nature of environmental problems (see *A Closer Look: Geology and Ecosystems*). Most projects are complex, involving many facets that may be generalized into three categories: physical, biological, and of human use and interest. These are essentially the same categories used by Luna B. Leopold (a famous geomorphologist and son of Aldo Leopold, who introduced the concept of the land ethic discussed earlier) to evaluate river valleys, and their extension to other areas of environmental research seems appropriate (17). *Physical factors* include physical geography, geologic processes, hydrologic processes, rock and soil types, and climatology. *Biologic factors* include the nature of plant and animal activity, changes in biologic conditions or processes, and spatial analysis of biologic information. *Human use and interest factors* include land use, economics, aesthetics, interaction between human activity and the physical and biological realms, and environmental law.

Obviously, no one project or research interest will involve all possible factors in each of the three categories, and there is considerable interaction between the categories. Projects such as waste-disposal operations, highway construction, mass transit systems, urban land-use planning, and mining of resources may be concerned with all of the categories. For example, the planning, construction, and operation of a sanitary landfill site (a place where we dispose of urban waste) is concerned with physical factors such as physical location, topography, soil type, and hydrologic conditions; biologic processes such as decay of the organic refuse and contamination of the surrounding biologic realm; and human interests such as good engineering practice and compliance with laws and regulations.

CONCEPT EIGHT: Our Obligation to the Future

The effects of land use tend to be cumulative, and therefore we have an obligation to those who follow us.

Several million years ago, when early hominids roamed the grasslands, marshy deltas, and adjacent forests of ancient Lake Rudolf along the Great Rift Valley system of East Africa, these prehistoric people were completely dependent upon their immediate environment. Their effect on that environment was probably insignificant as they hunted game and were in turn hunted by predators. This relationship between people and the environment probably existed until about 800,000 years ago, when our ancestors developed skill in the use of fire.

Human use of fire brought with it new environmental effects. First, fire was capable of affecting large areas of forest or grasslands. Second, it was a repetitive process capable of damaging the same area at rather frequent intervals. Third, it was a rather selective process, in that certain species were locally exterminated, whereas species that exhibited a resistance to or rapid recovery from fire were favored. Early use of fire for protection and hunting probably had a significant effect on the environment, and as people became more and more dependent on an increasing variety of resources for clothing, lodging, and hunting, they also increased their capacity to observe and test the environment. This early experimentation probably led to planting and primitive agriculture about 7000 B.C.

Emergence of agriculture was the first instance of an artificial land use capable of modifying the natural environment. It also set the stage for the development of a more or less continuously occupied site or cluster of sites that introduced further modification of the environment, such as shelter for living space, primitive latrines, and protective barriers against predators and other people. These early sites probably became the first areas to experience pollution resulting from waste disposal and soil erosion resulting from removal of indigenous vegetation (18). By increasing

A CLOSER LOOK Geology and Ecosystems

An ecosystem is a community of organisms and its nonliving environment in which matter cycles and energy flows. More importantly, sustained life on our planet is a characteristic of ecosystems rather than individual organisms or populations of a single species.

Understanding ecosystems requires us to consider physical, biological, and hydrological processes. For example, if we are interested in studying the forest ecosystem that supports a salmon fishery in the Pacific Northwest (Figure 1.C), we must consider interactions among variables such as:

- Large organic debris (large stems and pieces of woody debris) that produces fish habitat in the stream.
- Supply and transport of stream gravel necessary for fish to spawn.
- Cycling of nutrients within the stream system that provides the food base for the aquatic system.
- Land use in the surrounding forest that may cause changes in supply of water and sediment to the stream system.
- The role of disturbance such as wildfire or high-magnitude storms in maintaining the ecosystem.

Such considerations tend to transcend disciplines such as biology, geology, and hydrologic sciences. Understanding the entire ecosystem requires understanding how each of these components interacts with the others. One of the most fruitful areas for future ecological research may be the interactions of physical and biological processes operating at the ecosystem level.

▲ **FIGURE 1.C** Large redwood stem in Prairie Creek shown here is responsible for the development of a scour pool important in providing fish habitat for the stream ecosystem. Such large debris may reside in the stream channel for centuries. (*Charles A. Mauzy/Tony Stone Images*)

the food supply, agriculture allowed population growth, which in turn necessitated the clearing of additional land (this is an example of positive feedback). This activity certainly influenced an area's ecological balance as some species were domesticated or cultivated and others were removed as pests. It is not surprising that the increase in human population has been paralleled by an increase in the number of extinctions among birds and mammals.

The most significant lesson of this developmental history is that as cities and farms increase, demand for diversification of land use increases, and the effects tend to be cumulative. If this is so, then from an ethical and moral standpoint we need to examine the effects of land use in a historical framework, if only to ensure that our children and their children can survive in the environment they inherit. This is especially critical because it has been determined that at least since cities arose 6000 years ago, and perhaps as far back as 15,000 years ago, the entire surface of the earth has been altered by human activity. In other words, little if any land can be considered original or untouched (10). Furthermore, owing in part to human population increase that demands more land for urban and agricultural purposes, human-induced change to the earth is increasing at a rapid rate. A recent study of human activity and ability to move soil and rock concluded that human activity (agriculture, urbanization, etc.) moves as much or more soil and rock on an annual basis—40 to 45 Gt per year (1 gigaton [Gt] is one billion or 10^9 tons)—than any other earth process, including mountain building (34 Gt/year), river transport of sediment (24 Gt/year), and Pleistocene glaciers (10 Gt/year). This combined with the visual changes to the earth (leveling hills, etc.) suggests that human activity is the most significant process shaping the surface of the planet (19).

The lessons in land use are explicit. Where sound conservation practices were used, there were successful adjustments of population, but where wasteful exploitation of resources was practiced, the results varied from gullied fields to silted-up irrigation reservoirs and canals to the ruin of prosperous cities. Histories of long-populated areas show that soil erosion has frequently destroyed productive land and retarded the progress of civilization. Conservation of our soils must remain a national priority, for as stated by W. C. Lowdermilk, "One generation of people replaces another, but productive soils destroyed by erosion are seldom restorable and never replaceable" (20,21).

The misuse of land has continued in modern times. For an American example, consider Ducktown, Tennessee (22). The land surrounding Ducktown looks more like the Painted Desert of Arizona than the lush vegetation of the Blue Ridge Mountains of the southeastern United States (Figure 1.12). The story starts in 1843 with what was thought to be a gold rush that turned out to be a rush for copper. By 1855, some 30 companies were transporting copper ore by mule over the mountains to a site called "Copper Basin"

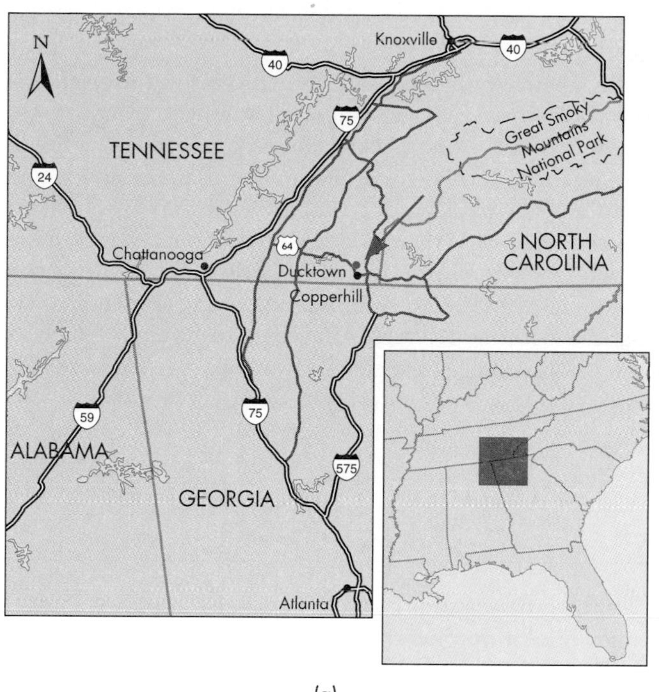

(a)

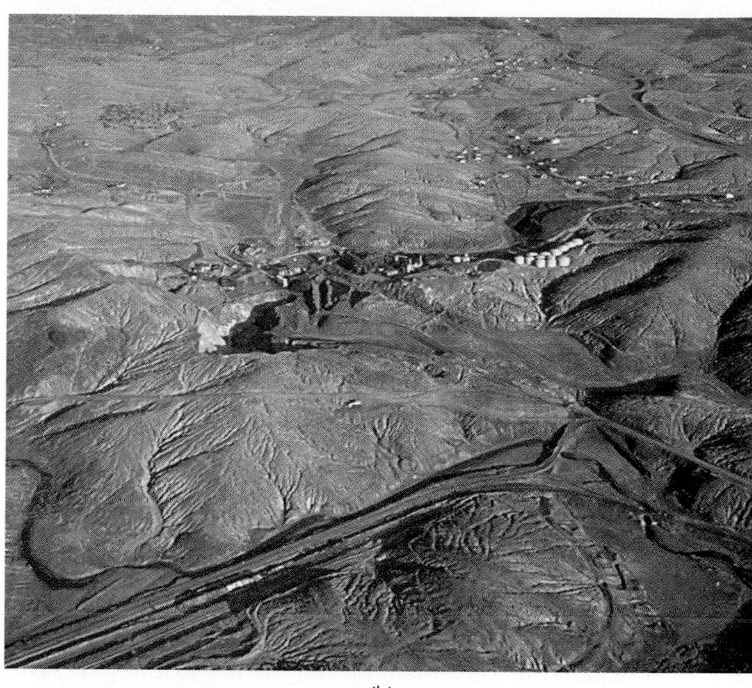

(b)

(c)

▲ **FIGURE 1.12** The lasting effects of land abuse. (a) Location of Ducktown, Tennessee. (b) The human-made desert resulting from mining activities around Ducktown more than 100 years ago. Extensive soil erosion and loss of vegetation has occurred and complete recovery will probably take more than 100 years. (*Kristoff, Emory/NGS Image Sales*) (c) Ducktown area in recent years showing the process of recovery. (*Tennessee Valley Authority*)

and to Ducktown. Huge ovens—open pits 200 m long and 30 m deep—were constructed to separate the copper from zinc, iron, and sulfur. The local hardwood forest was cut to fuel these ovens, and the tree stumps were pulled to be turned into charcoal. Eventually every tree over an area of about 130 km^2 was removed. The ovens produced great clouds of noxious gas that was reportedly so thick that mules wore bells to keep from colliding with people and with each other. The sulfur dioxide gas and particulates produced acid rain and acid dust that killed what vegetation remained. Loss of vegetation led to extensive soil erosion, leaving behind a hard mineralized rock cover resembling a desert. The scarred landscape is so large that it is one of the few human landmarks visible from space.

The devastation resulting from the Ducktown mining activity produced adverse economic and social change. Nevertheless, people in Ducktown in the 1980s remained optimistic. A sign at the entry to the town states "Copper made us famous. Our people made us great." Revegetation started in the 1930s, and by 1970 approximately two-thirds of the area had become covered with some vegetation. However, it will probably take hundreds of years for the land to recover. What is being learned concerning restoration of Copper Basin will provide useful information for other areas in the world where human-made deserts occur, such as the area around the smelters in Sudbury, Ontario. Finally, there is worry for mining areas, particularly in developing countries, where landscape destruction similar to that at Copper Basin is still happening (22).

In summary, meeting our obligations to future generations will not be easy. Both land- and water-use planning will need to balance economic considerations with less tangible considerations such as sustaining our land, water, and scenic resources. The basic point is that there are varying environmental values just as there are varying economic values. Logging old-growth redwood trees may be the most short-term, economic (profitable) use of a forest, but this needs to be balanced with longer-term landscape degradation such as soil erosion, damage to stream ecosystems, loss of a scenic resource, and loss of the forest to future generations.

SUMMARY

Environmental geology is the use of geologic information to better understand and manage our environment. It is applied geology. The important variable that distinguishes geology from the other sciences is the consideration of time. The work of geologists utilizes the testing of hypotheses and the scientific method.

The cultural bases for environmental degradation are ethical, economic, political, and perhaps religious, but environmental problems are not confined to any one political or social system. Our ethical framework appears to be slowly expanding and may eventually include the total environment in a land ethic. This ethic affirms the right of all resources, including plants, animals, and earth materials, to continued existence and, at least in certain locations, continued existence in a natural state. The immediate causes of the environmental crisis are overpopulation, urbanization, and industrialization, which have occurred with too little ethical regard for our land and inadequate institutions to cope with environmental stress. Solving them involves both scientific understanding and the fostering of social, economic, and ethical behavior that allows solutions to be implemented.

Eight fundamental concepts establish a philosophical framework for our investigation of environmental geology: (1) The increasing world population is the number-one environmental problem. (2) Sustainability is the environmental objective. (3) The earth is essentially a closed system with respect to materials, and an understanding of feedback and rates of change in such systems is critical to solving environmental problems. (4) The earth is the only suitable habitat we have, and its resources are limited. (5) Today's physical processes are modifying our landscape and have operated throughout much of geologic time, but the magnitude and frequency of these processes are subject to natural and artificially induced change. (6) Earth processes that are hazardous to people have always existed. These natural hazards must be recognized and avoided where possible, and their threat to human life and property must be minimized. (7) The fundamental component of every person's environment is the geologic one, and understanding our environment requires an understanding of the earth sciences and related disciplines. (8) The effects of land use tend to be cumulative, and we therefore have an obligation to those who follow us.

REFERENCES

1. **Ermann, M.** 1927. *Desiderata.* Terre Haute, IN.
2. **Udall, S. L.** 1963. *The quiet crisis.* New York: Avon Books.
3. **Leopold, A.** 1949. *A Sand County almanac.* New York: Oxford University Press.
4. **Pierce, A., and Van de Veer, D. (eds.)** 1995. *People, penguins and plastic trees: Basic issues of environmental ethics,* 2nd ed., 485 p. Belmont, CA: Wadsworth.
5. **Moncrief, L. W.** 1970. The cultural basis for our environmental crisis. *Science* 170:508–12.
6. **Ellis, W. S.** 1990. A Soviet sea lies dying. *National Geographic* 177(2):73–92.
7. **White, L., Jr.** 1967. The historical roots of our ecological crisis. *Science* 155:1203–7.
8. **Yi-Fu, T.** 1970. Our treatment of the environment in ideal and actuality. *American Scientist* 58:244–49.
9. **Brown, L. R., Flavin, C., and Postel, S.** 1991. *Saving the planet.* New York: W. W. Norton & Co.
10. **National Research Council.** 1971. *The earth and human affairs.* San Francisco: Canfield Press.
11. **McConnell, R. L., and Abel, D. C.** 1999. *Environmental Issues.* Upper Saddle River, NJ: Prentice-Hall.
12. **Earth Systems Science Committee.** 1988. *Earth systems science.* Washington, DC: National Aeronautics and Space Administration.
13. **Holman, E.** 1952. Our inexhaustible resources. *Bulletin of the American Association of Petroleum Geologists* 6:1323–29.

14. **Cloud, P. E., Jr.** 1968. Realities of mineral distribution. In *Man and his physical environment*, ed. G. D. McKenzie and R. O. Utgard, pp. 194–207. Minneapolis: Burgess.
15. **Lovelock, J.** 1988. *The ages of Gaia*. New York: W. W. Norton & Co.
16. **Wolman, M. G., and Schick, A. P.** 1967. Effects of construction on fluvial sediment, urban and suburban areas of Maryland. *Water Resources Research* 3:451–64.
17. **Leopold, L. B.** 1969. Quantitative comparison of some aesthetic factors among rivers. *U.S. Geological Survey Circular* 620.
18. **Nicholson, M.** 1970. Man's use of the earth: Historical background. In *Man's impact on environment*, ed. T. R. Detwyler, pp. 10–21. New York: McGraw-Hill.
19. **Hooke, LeB.** 1994. On the efficiency of humans as geomorphic agents. *GSA Today* 4(9):217, 224–25.
20. **Lowdermilk, W. C.** 1943. Lessons from the Old World of the Americans in land use. *Smithsonian Report for 1943*, pp. 413–28.
21. _______. 1960. The reclamation of a man-made desert. *Scientific American* 202:54-63.
22. **Barnhardt, W.** 1987. The death of Ducktown. *Discover*, October, pp. 35-43.

KEY TERMS

environmental geology (p. 3)
hypothesis (p. 4)
scientific method (p. 4)
land ethic (p. 5)
environmental crisis (p. 5)
myth of superabundance (p. 6)
exponential growth (p. 8)
sustainability (p. 10)
sustainable global economy (p. 10)
system (p. 11)
principle of environmental unity (p. 11)
feedback (p. 12)
input-output analysis (p. 12)
average residence time (p. 13)
earth system science (p. 14)
uniformitarianism (p. 17)
Gaia hypothesis (p. 18)

SOME QUESTIONS TO THINK ABOUT

1. We state in the text that the evolution of ethics is an important environmental trend. Can this statement be rigorously defended? Present an argument that it is good for the environment to have an expanded view of ethics.
2. Stewart Udall writes about the "myth of superabundance" and "the raid on resources." What do you think we have learned in the United States today from our history of how we have treated the environment? Do we still think like the young Thomas Jefferson or have fundamental changes occurred in the relationship between people and the environment?
3. Assuming that there is an environmental crisis today, what possible solutions are available to alleviate the crisis? How will solutions in developing countries differ from those in highly industrialized societies? Will religion or political systems have a bearing on potential solutions? If so, how?
4. It has been argued that we must control human population because otherwise we won't be able to feed everyone. Assuming that we could feed 10 billion to 15 billion people, would we still want to have a smaller population than that? Why?
5. We state that sustainability is the environmental objective. Construct an argument to support this statement. Is the idea of sustainability and building a sustainable economy different in developing and poor countries from those that are affluent with a high standard of living? How and why?
6. The so-called Gaia hypothesis actually consists of three hypotheses. What would need to be done to test each of them? Many scientists would say that one of the hypotheses is fringe science and cannot be tested. Do you agree or disagree, and why? Perhaps the Gaia hypotheses are best considered as only a metaphor. Do you agree?
7. The concept of environmental unity is an important one today. Consider some major development being planned for your region and outline how the principle of environmental unity could help in determining potential environmental impact. In other words, consider a development and then a series of resultant consequences. Some of the impacts may be positive and some may be negative in your estimation.
8. Why is it important in land- and water-use planning to strive for a balance between economic considerations and less tangible variables such as aesthetics? In answering, consider what you have learned about both environmental ethics and sustainability.

2 Earth Materials and Processes

The Grand Canyon of the Colorado River. (*Jeremy Walker/Tony Stone Images*)

The Grand Canyon of the Colorado River at sunrise is an awesome sight. The Colorado River has excavated through sedimentary rocks, which were deposited over a period of several hundred million years as part of the rock cycle and uplifted by active tectonic processes from ancient ocean basins to their present position. Waters of the Colorado River are a product of the global hydrologic cycle, which circulates water from the oceans to the land and back again. Today Colorado River waters are some of the most regulated and managed on earth. A system of dams provides waters for thirsty lands and cities of the southwestern United States. As a result, the once dynamic and biologically rich delta of the Colorado, where it enters the Sea of Cortez in Mexico, has been severely degraded.

2.1 The Geologic Cycle

Throughout much of the 4.6 billion years of earth history, the materials on or near the earth's surface have been created, maintained, and destroyed by numerous physical, chemical, and biological processes. Continuously operating processes produce the earth materials—land, water, and atmospheric—necessary for our survival. Collectively, these processes are referred to as the **geologic cycle,** which is really a group of subcycles (1):

- Tectonic cycle
- Rock cycle
- Hydrologic cycle
- Biogeochemical cycles

The Tectonic Cycle

The term *tectonic* refers to the large-scale geologic processes that deform the earth's crust, producing landforms such as ocean basins, continents, and mountains. Tectonic processes are driven by forces deep within the earth, as shown in Figure 2.1.

Earth's Lithosphere and Crust The outer layers of the earth are shown in Figure 2.2. The outermost layer, called the **lithosphere,** is stronger and more rigid than deeper

LEARNING OBJECTIVES

In this chapter we focus on the geologic materials and processes most important to the study of the environment. If you have previously taken a physical geology course, a simple review of this material focusing on the environmental aspects will be sufficient. If this is your first geology course, you will need to study the material more carefully. Primary learning objectives for the chapter are:

- *To acquire a basic understanding of the geologic cycle and its subcycles (tectonic, rock, hydrologic, and biogeochemical).*
- *To make a modest acquaintance with some of the important mineral and rock types and their environmental significance.*
- *To gain an appreciation of the landforms, deposits, and environmental problems resulting from wind and glacial processes.*

Web Resources

Visit the "Environmental Geology" Web site at www.prenhall.com/keller to find additional resources for this chapter, including:

- Web Destinations
- On-line Quizzes
- On-line "Web Essay" Questions
- Search Engines
- Regional Updates

material. Below the lithosphere lies the **asthenosphere,** a hot and plastically flowing layer of relatively low-strength rock that is present to a depth of about 250 km. Through detailed study of ocean basins and continents, earth scientists have established that the lithosphere averages about 100 km in thickness, ranging from a few kilometers beneath the crests of mid-ocean ridges to 120 km beneath ocean basins to 20 to 400 km beneath continents. The outer part of the lithosphere is the **crust.** Crustal rocks are less dense on the average than the rocks below, but oceanic crust is denser

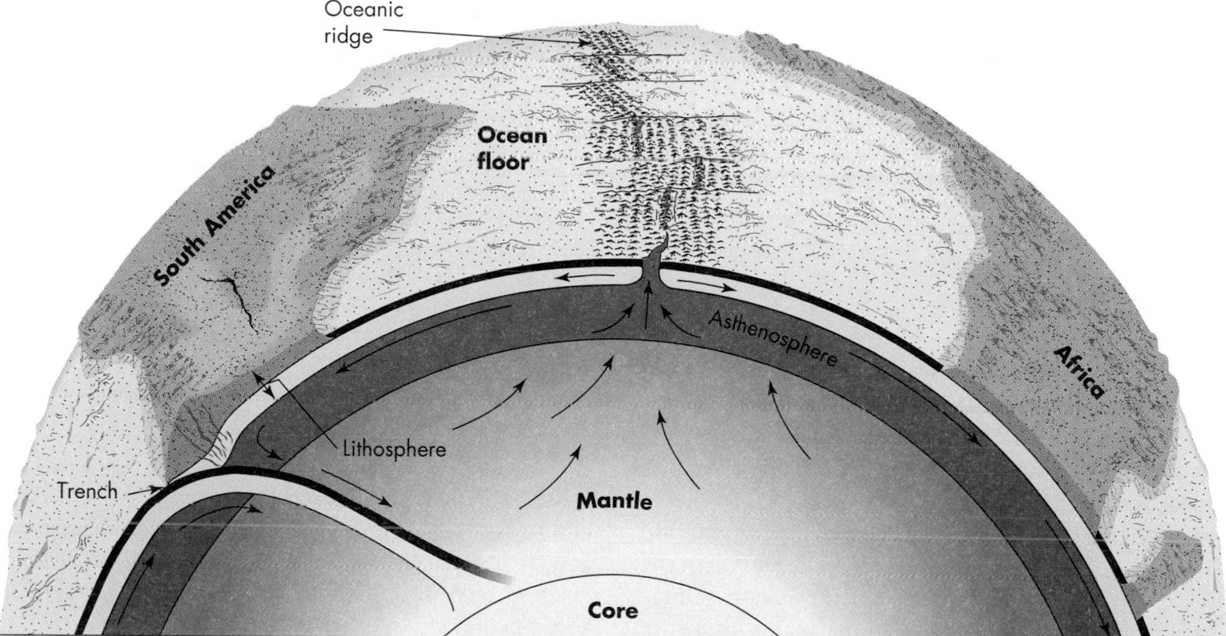

▲ **FIGURE 2.1** Model of plate movement and mantle. (*Modified after* Grand, S. P. 1994. Journal of Geophysical Research, *v.* 99, *pp.* 11591-621.) The outer layer (or lithosphere) is approximately 100 km thick and is stronger and more rigid than the deeper asthenosphere, which is a hot and slowly flowing layer of relatively low-strength rock. The oceanic ridge is a spreading center where plates pull apart, drawing hot, buoyant material into the gap. After these plates cool and become dense, they descend at oceanic trenches (subduction zones), sometimes to depths as deep as the core–mantle boundary, completing the convection system. This process of spreading produces ocean basins, and mountain ranges often form where plates converge at subduction zones. A schematic diagram of earth's layers is shown in Figure 2.2b. (*Modified after Hamblin*, W. K. 1992. Earth's Dynamic Systems, *6th ed.* New York: Macmillan.)

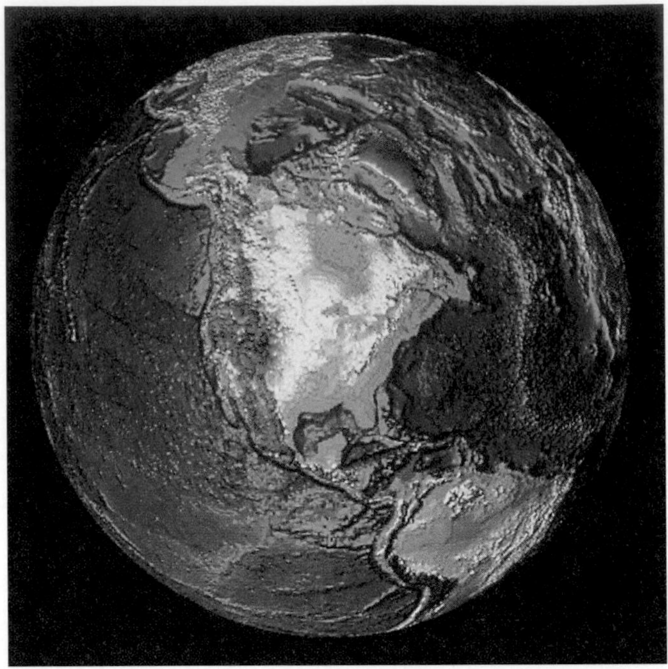

(a)

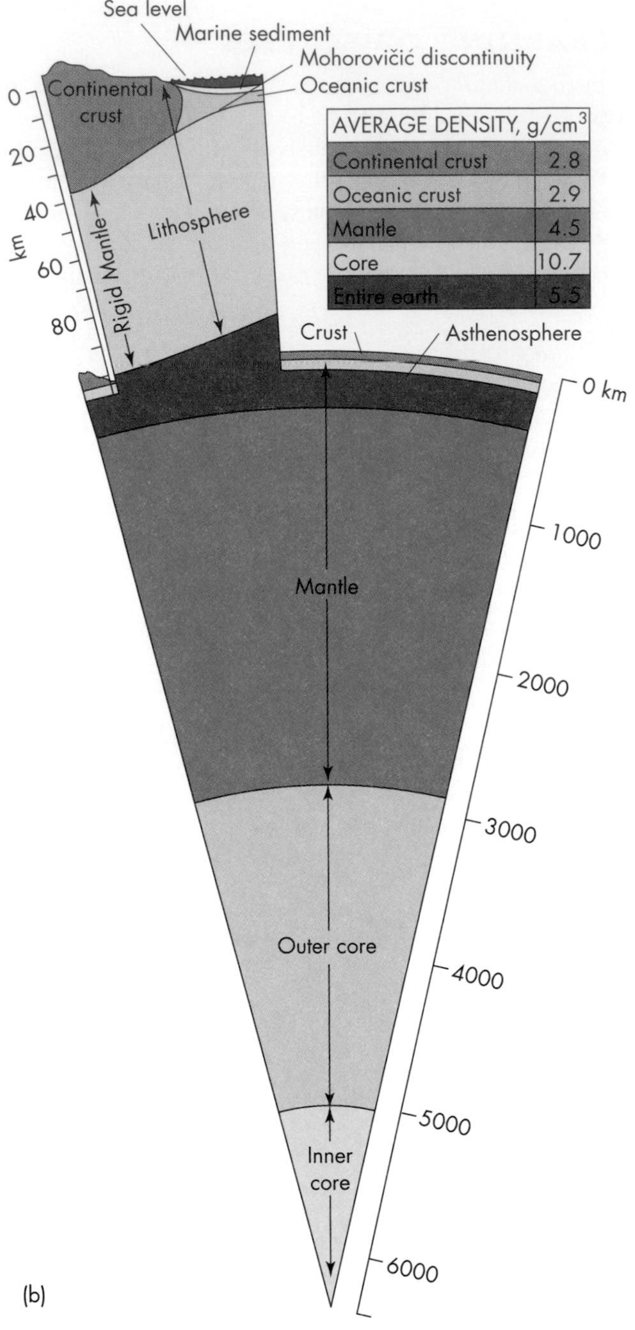

(b)

▲ **FIGURE 2.2** (a) Earth from space. (*National Geophysical Data Center*/NOAA) (b) Idealized diagram showing the internal structure of earth and its layers extending from the center to the surface. Notice that the lithosphere includes the crust and part of the mantle, and the asthenosphere is located entirely within the mantle. Properties of the various layers have been estimated based on interpretation of geophysical data (primarily seismic waves from earthquakes) and examination of rocks thought to have been brought up from below by tectonic processes and meteorites, thought to be pieces of an old earthlike planet. (*From Levin*, H. L. 1986. Contemporary Physical Geology, 2*nd ed*. *Philadelphia: Saunders*)

than continental crust. Oceanic crust is also thinner: The ocean floor has an average crustal thickness of about 7 km, whereas continents are about 30 km thick on the average and up to 70 km thick beneath mountainous regions.

***Movement of the* Lithospheric Plates** Unlike the asthenosphere, which is thought to be more or less continuous, the lithosphere is broken into large pieces called *lithospheric plates* that move relative to one another (Figure 2.3) (2). Processes associated with the origin, movement, and destruction of these plates are collectively known as **plate tectonics.**

A lithospheric plate may include both a continent and part of an ocean basin or an ocean basin alone. Some of the largest plates are the Pacific, North American, South American, Eurasian, African, and Australian plates, but there are many smaller plates that are significant on a regional scale. Figure 2.3a shows some of these small plates, including the Juan de Fuca plate off the Pacific Northwest Coast of the United States and the Cocos plate off Central America. The boundaries between lithospheric plates are geologically active areas where most earthquakes and volcanic activity occur. In fact, plate boundaries are defined as the areas where most seismic activity takes place (Figure 2.3b). It is at these boundaries that plates are formed and destroyed, cycling materials from the interior of the earth to the crust and back again (Figure 2.4). Because of this continuous recycling, tectonic processes are collectively called the **tectonic cycle.**

As the lithospheric plates move over the asthenosphere, they carry the continents with them (3). The idea that continents move is not new; it was first suggested in the early twentieth century by German scientist Alfred Wegener, who presented evidence for *continental drift* based on the shape of continents, particularly across the Atlantic Ocean and similarity in fossils found in South America and Africa. Wegener's hypothesis was not taken seriously, however, because

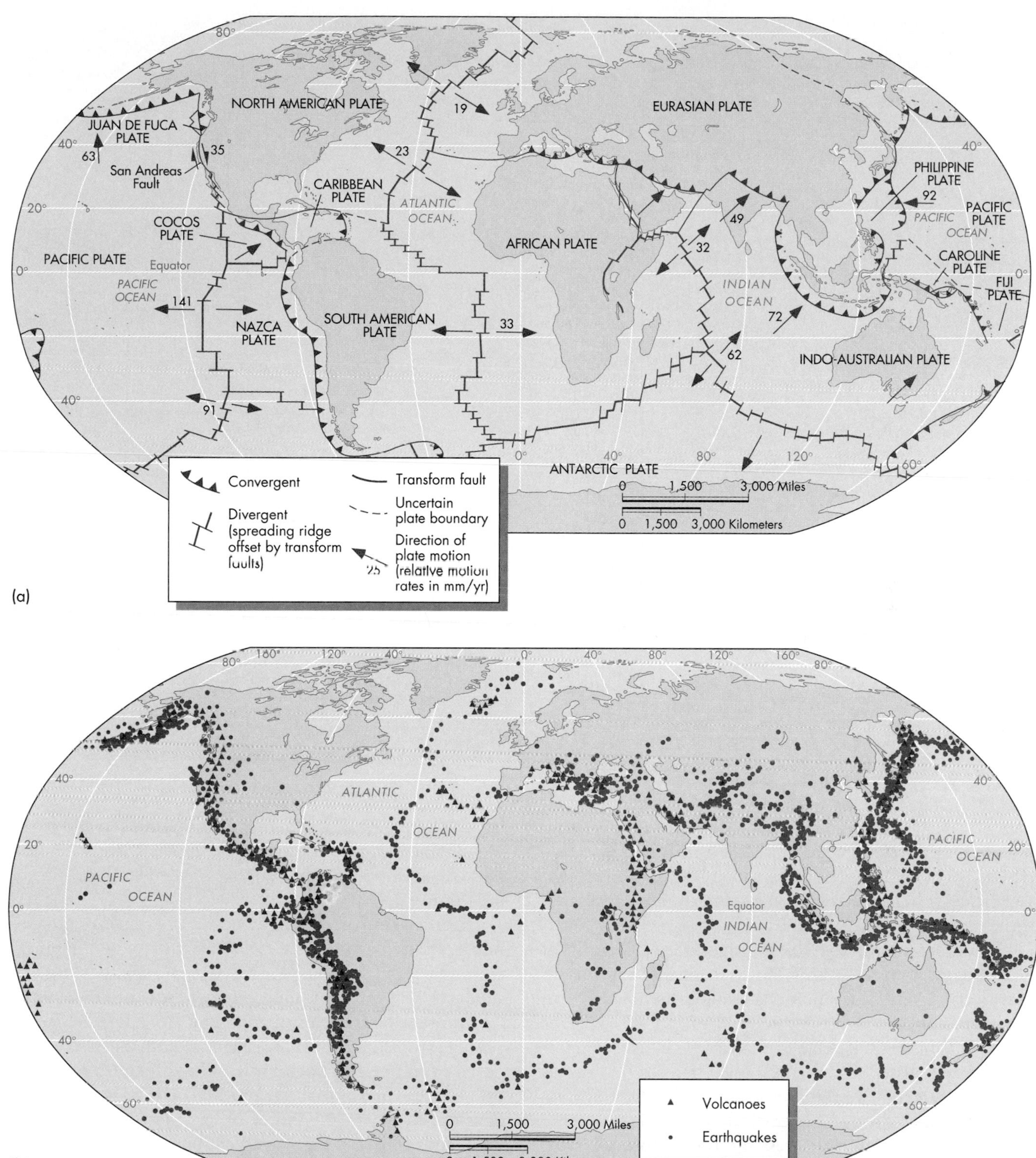

▲ **FIGURE 2.3** (a) Map showing the major tectonic plates, plate boundaries, and direction of plate movement. (*Modified from Christopherson*, R. W. 1994. Geosystems, *2nd ed. Englewood Cliffs*, NJ: *Macmillan*) (b) Map showing location of volcanoes and earthquakes. Notice the correspondence between this map and the plate boundaries. (*Modified after Hamblin*, W. K. 1992. Earth's Dynamic Systems, *6th ed. New York*: *Macmillan*)

the existence of lithospheric plates was not known, and geologists realized that continents could not plow through the denser material of the ocean floor. It was not until the late 1960s, when *seafloor spreading* was discovered, that a plausible mechanism for continental movement was put forward. In the seafloor regions called *spreading ridges*, new crust is continuously added to the edges of lithospheric plates, pushing the plates away from the ridge (Figure 2.4, left). As one

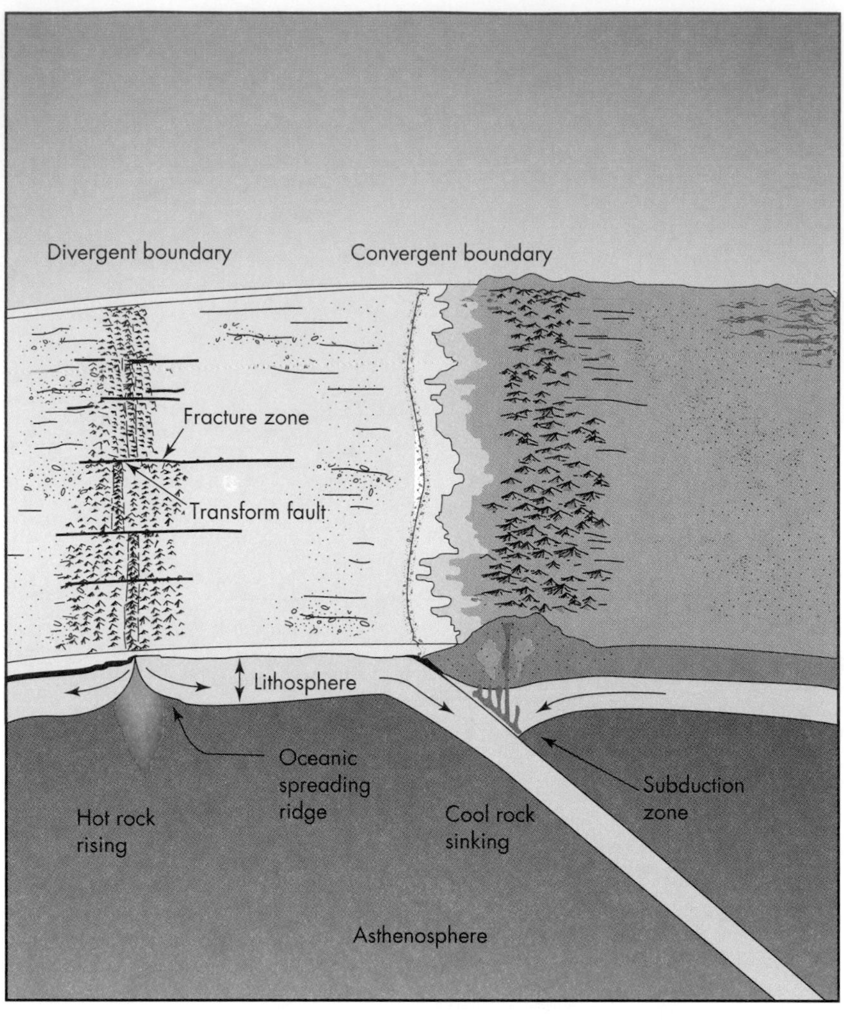

▶ **FIGURE 2.4** Diagram of the model of plate tectonics. New oceanic lithosphere is being produced at the spreading ridge (divergent plate boundary). Elsewhere oceanic lithosphere returns to the interior of earth at a convergent plate boundary (subduction zone). (*Modified from Lutgens, F., and Tarbuck, E.* 1992. Essentials of Geology. *New York: Macmillan*)

edge of a plate is being formed, the opposite edge is being destroyed, as the lithosphere sinks into a *subduction zone* (Figure 2.4, right). Thus, a continent does not move through oceanic crust but is carried along with it by the movement of the plate.

Plate tectonics is to geology what Darwin's origin of species is to biology—a unifying concept that explains an enormous variety of phenomena. Biologists now have an understanding, through the discovery of DNA, of the genetic mechanisms that drive evolutionary change. In geology we are still seeking the exact mechanism that drives plate tectonics, but we think it is most likely *convection*. As rocks are heated deep in the earth they become less dense and rise to spreading ridges. As the rocks move laterally, they cool, eventually becoming dense enough to sink at subduction zones.

Types of Plate Boundaries

As a result of the cycle of plate formation, movement, and destruction, plates have the three types of boundaries shown in Figures 2.3 and 2.4—divergent, convergent, and transform (4). It is important to understand that these boundaries are not narrow cracks but zones that are tens or hundreds of kilometers across. **Divergent boundaries** occur where new lithosphere is being produced and plates are moving away from each other. Typically this occurs at mid-ocean ridges and the process is called seafloor spreading (Figure 2.4).

Convergent boundaries occur where plates are colliding. If an oceanic plate collides with a plate carrying a continent, the higher-density oceanic plate sinks, or subducts, beneath the leading edge of the continental plate. Convergent boundaries of this type are called *subduction zones*. However, if the leading edges of both plates are composed of continental rocks, it is more difficult for subduction to start. In this case a *continental collision boundary* may develop, in which the edges of the plates crumple into mountains. This type of convergent boundary condition has produced some of the highest linear mountain systems on earth, such as the Alpine and Himalayan mountain belts (Figure 2.5).

If you examine Figures 2.3a and 2.4, you will see that a spreading zone is not a single, continuous rift, but a series of short rifts that are offset from one another. **Transform boundaries,** or *transform faults*, occur where the offset segments of two plates slide past one another, as shown in Figure 2.4. Most such boundaries occur in oceanic crust, but some occur on continents. A well-known continental transform boundary is the San Andreas fault in California, where a segment of the Pacific plate is sliding horizontally past a segment of the North American plate (see Figures 2.3a and 2.6).

At some locations, three plates join, and these areas are known as **triple junctions.** Figure 2.6 shows two such junctions: the meeting point of the Juan de Fuca, North American, and Pacific plates on the west coast of North America and the junction of the spreading ridges associated with the Pacific, Cocos, and Nazca plates west of South America.

▲ **FIGURE 2.5** Mountain peaks (the Dolomites) in northern Italy are part of the Alpine mountain system formed from the collision between Africa and Europe. (*Edward A. Keller*)

Rates of Plate Motion Directions of plate motion relative to each other are shown on Figure 2.3a. In general, the rates are about as fast as your fingernails grow, varying from about 2 to 15 cm/yr. The San Andreas fault moves horizontally on the average of about 3.5 cm/yr, so that features such as rock units or streams may be displaced where they cross the fault (Figure 2.7). During the past 4 million years, there has been about 200 km of displacement on the San Andreas fault. Los Angeles, which is on the Pacific plate, is slowly moving toward San Francisco, which is on the North American plate. In about 20 million years the cities will be side by side, and if people are present they will be arguing over which is a suburb of the other. Plates do not necessarily move steadily past each other; they are generally restrained by the friction between them, moving locally or regionally only when the strain becomes so great that they snap apart, causing an earthquake. The horizontal displacement along a transform boundary such as the San Andreas fault may be

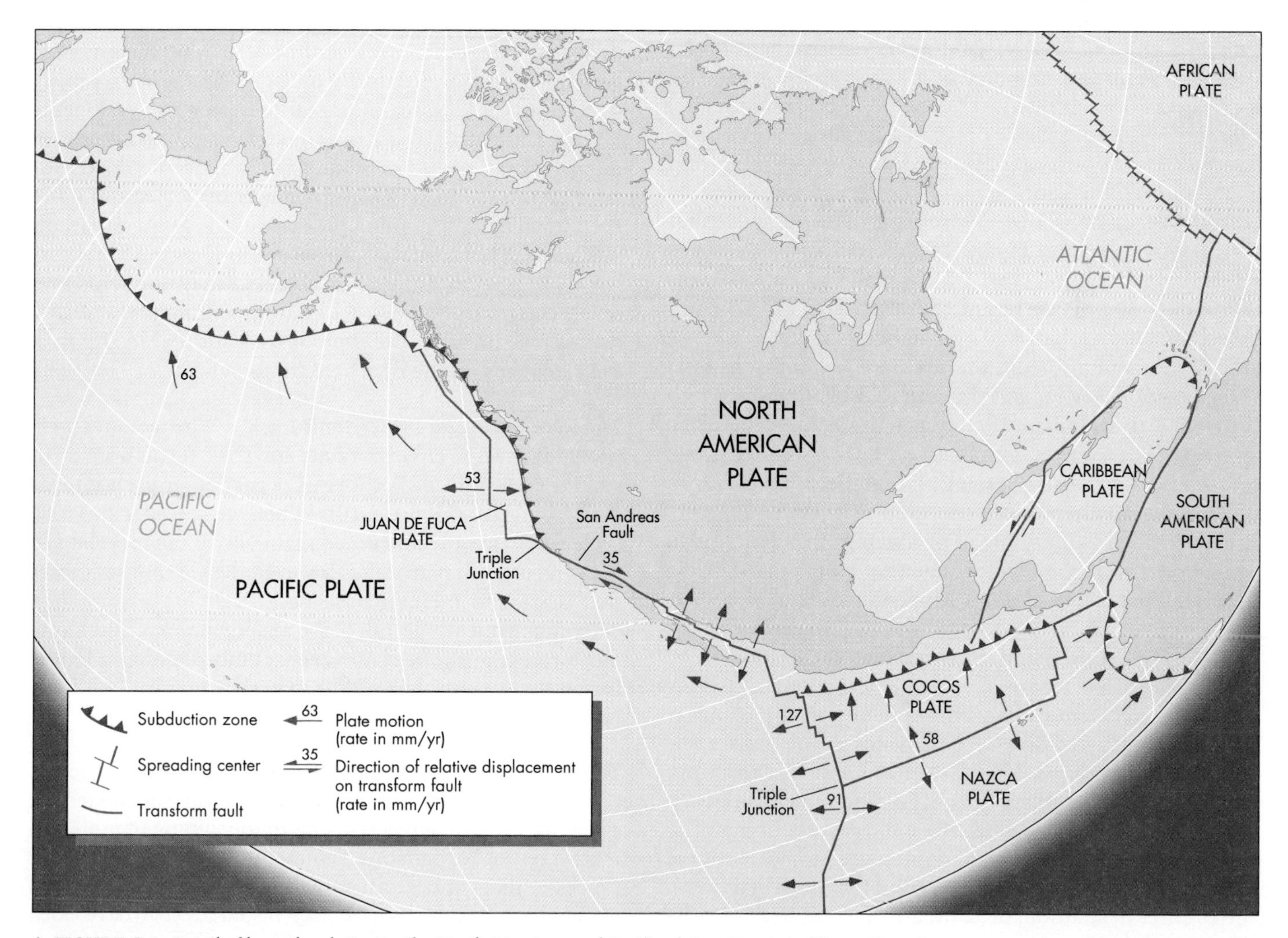

▲ **FIGURE 2.6** Detail of boundary between the North American and Pacific plates. (*Courtesy of Tanya Atwater*)

▲ **FIGURE 2.7** Stream offset along the San Andreas fault in California. The stream bed is the white line. Notice that the stream has a pronounced right bend where it crosses the fault that is running from the right to the left side of the photograph near the bottom. (*Michael Collier*)

as much as several meters during a great earthquake. Fortunately, such an event generally occurs at any given location only once every few hundred years.

Pangaea and the Present Continents Movement of the lithospheric plates is responsible for the present shapes and locations of the continents. There is good evidence that the most recent global episode of continental movement started about 200 million years ago, with the breaking up of a supercontinent called *Pangaea* (a name first proposed by Alfred Wegener). Pangaea itself was formed by earlier continental collisions. Figure 2.8a shows Pangaea as it was 180 million years ago, as well as the present position of the continents and ocean basins (Figure 2.8d). Ocean-floor spreading over the past 200 million years separated Eurasia and North America from the southern landmass, Eurasia from North America, and the southern continents (South America, Africa, India, Antarctica, and Australia) from one another. The Tethys Sea, between Africa and Europe (Figure 2.8a–c) closed, leaving the small remnant known today as the Mediterranean Sea. About 50 million years ago, India collided with China, producing the Himalayan Mountains (the highest mountains in the world) and Tibetan Plateau. That collision is still happening today.

The Tectonic Cycle and Environmental Geology The importance of the tectonic cycle to environmental geology cannot be overstated. Everything living on the planet is affected by plate tectonics. As the plates slowly move a few centimeters per year, so do the continents and ocean basins, producing zones of resources (oil, gas, and minerals), as well as earthquakes and volcanoes. Tectonic processes occurring at plate boundaries largely determine the properties of the rocks and soils upon which we depend for construction and agriculture. In addition, plate motion over millions of years changes or modifies flow patterns in the oceans, influencing global climate as well as regional variation in precipitation. These changes affect the productivity of the land and its desirability as a place to live.

The Rock Cycle

Rocks are aggregates of one or more **minerals,** naturally occurring crystalline substances with defined properties. We will discuss the properties of rocks and minerals in Section 2.2, focusing here on the processes of the **rock cycle,** the largest of the geologic subcycles. The rock cycle is linked to all the other subcycles; it depends on the tectonic cycle for heat and energy, the biogeochemical cycle for materials, and the hydrologic cycle for water, which is used in the processes of weathering, erosion, transportation, deposition, and lithification of sediments.

Although rocks vary greatly in their composition and properties, they can be classified into three general types, or families, according to how they were formed in the rock cycle (Figure 2.9). We may consider this cycle a worldwide rock-recycling process driven by the earth's internal heat, which melts the rocks subducted in the tectonic cycle. *Crystallization* of the molten rock produces **igneous rocks** beneath and on the earth's surface. Rocks at or near the surface break down chemically and physically by **weathering,** forming sediments that are transported by wind, water, and ice to depositional basins such as the ocean. The accumulated layers of sediment eventually undergo *lithification* (conversion to solid rock), forming **sedimentary rocks.** Deeply buried sedimentary rocks may be *metamorphosed* (altered in form) by heat, pressure, or chemically active fluids to produce **metamorphic rocks,** which may be buried still more deeply and melt, beginning the cycle again. Some variations of this idealized sequence are shown in Figure 2.9. For example, an igneous or metamorphic rock may be altered into a new metamorphic rock without undergoing weathering or erosion. In brief, the type of rock formed in the rock cycle depends on the rock's environment.

The Rock Cycle and the Tectonic Cycle The tectonic cycle provides several environments for rock formation, with specific rock-forming processes occurring at each type of plate boundary (Figure 2.10). When we consider the rock cycle alone, we are concerned mainly with the recycling of rock and mineral materials. However, the tectonic processes that drive and maintain the rock cycle are essential in determining the properties of the resulting rocks. Therefore, our interest in the tectonic cycle is more than academic, because it is upon these earth materials that we build our homes, factories, roads, and other structures.

The Rock Cycle and Mineral Resources The rock cycle is responsible for concentrating as well as dispersing materials, a fact that is extremely important to the mining of minerals. If it were not for igneous, sedimentary, and metamorphic processes that concentrate minerals, it would be difficult to extract these resources. We take resources that have been concentrated by a rock-cycle process, transform them through industrial activities, and then return them in a diluted form to the cycle, where they are dispersed by other earth processes (5). Once we have done this, the resource cannot be concentrated again within a useful time frame.

Consider as an example phosphate minerals, which are widely used in fertilizers. Phosphates are mined in a concentrated form, transformed and used in agriculture, then dispersed by ground and surface waters, air currents, and other agents. Eventually, phosphates may become abundant enough to contaminate water, but—with the exception of using phosphate-rich water from wastewater treatment plants for irrigation—it is seldom sufficiently concentrated to be recycled. Similar examples may be cited for many other resources used in paints, solvents, and industrial products.

The Hydrologic Cycle

The **hydrologic cycle** is the movement of water from the oceans to the atmosphere and back again (Figure 2.11). Driven by solar energy, the cycle operates by way of evaporation, precipitation, surface runoff, and subsurface flow. Only a very small amount of the total water in the cycle is active near the earth's surface at any one time. For example, all water in the atmosphere, rivers, and shallow subsurface environment on earth is only about 0.3 percent of the total (more than 97 percent is in the oceans). Nevertheless, this small amount of non-marine water at or near the surface of the earth is tremendously important in facilitating the movement and sorting of chemical elements in solution, sculpturing the landscape, weathering rocks, transporting and depositing sediments, and providing our water resources. Water resources and hydrologic processes are discussed in detail in Chapters 5 and 10.

Biogeochemical Cycles

A **biogeochemical cycle** is the transfer or cycling of an element or elements through the atmosphere (the layer of gases surrounding the earth), lithosphere (earth's rocky outer layer), hydrosphere (oceans, lakes, rivers, and groundwater), and biosphere (the part of the earth where life exists). It follows from this definition that biogeochemical cycles are intimately related to the tectonic, rock, and hydrologic cycles. The tectonic cycle provides water from volcanic processes, as well as the heat and energy required to form and change the earth materials transferred in biogeochemical cycles. The rock and hydrologic cycles are involved in many transfer and storage processes of chemical elements in water, soil, and rock.

Biogeochemical cycles can most easily be described as the transfer of chemical elements through a series of storage compartments or reservoirs (e.g., air, soil, groundwater, vegetation). When a biogeochemical cycle is well understood, the rates of transfer, or *flux*, between all of the compartments is known. However, determining these rates on a global basis is a very difficult task. The amounts of such important elements as carbon, nitrogen, and phosphorus in each compartment and their rates of transfer between compartments are only approximately known.

To better understand biogeochemical cycles we will discuss the **global carbon cycle**, diagrammed in Figure 2.12. Carbon in its pure form is found in some minerals such as graphite and diamonds and in recently discovered exotic molecules composed of 60 carbon atoms each and known as "buckyballs." More important to us, however, are the many compounds carbon forms with other elements. Carbon is the basic building block of life because it readily combines with other carbon atoms and with oxygen and hydrogen to form biological compounds. As a result, the carbon biogeochemical cycle is intimately related to the cycles of oxygen and hydrogen. Some carbon compounds are gases under ordinary conditions. The most important of these gases is carbon dioxide (CO_2), which is found in the atmosphere, soil, and other parts of the environment. Carbon dioxide in the atmosphere or in soil readily dissolves in water, forming a weak acid known as *carbonic acid* (H_2CO_3), which is very important in the weathering of rocks at or near the surface of the earth.

Figure 2.12 shows the amount of carbon in the various storage compartments as well as rates of transfer between compartments on an annual basis (6). Notice that nearly all the carbon on earth is stored in marine sediments and sedimentary rocks. This storage compartment dwarfs all of the others, which together account for only about 0.05 percent of the total. Nevertheless, the carbon stored in the atmosphere and transferred from there to the land and oceans is of particular environmental importance. As the carbon dioxide concentration in the atmosphere increases because of natural change or the burning of fossil fuels, the global climate also changes. (Global warming is discussed in Chapter 16.) Notice an apparent contradiction in Figure 2.12. About 3 billion metric tons per year of carbon are *stored* in the atmosphere from the burning of fossil fuels and land use change, but the amount of carbon *entering* the atmosphere from burning fossil fuels is about 5.4 billion tons per year, with another 1.6 billion tons per year from land use change (including deforestation). Thus, total human input of carbon into the atmosphere is about 7 billion metric tons per year. Where have the other 4 billion tons per year gone? This is an important research question that has not yet been answered. Presumably they have gone into the large storage compartment consisting of shallow ocean water (at least 2 billion metric tons per year), with the remainder in soils, plants on land, or plants in the ocean.

Because the atmosphere is a dynamic environment and rapid transfer of carbon takes place between the atmosphere and the plants on land and in the ocean, the average residence time of carbon in the atmosphere is short. Biogeochemical cycles that do not have a gaseous phase, such as the phosphorus cycle, are much different. The phosphorus cycle is usually very slow, taking hundreds of thousands to millions of years. The transfer of phosphorus from land to the oceans is primarily by way of erosion of phosphate-rich rocks (the phosphorus is often in bones and teeth of fish and other animals in the rock). Phosphorus returns to the land by way of the rock cycle, which produces phosphate-rich sedimentary rock, and the tectonic cycle, which uplifts the rock to where it can again be

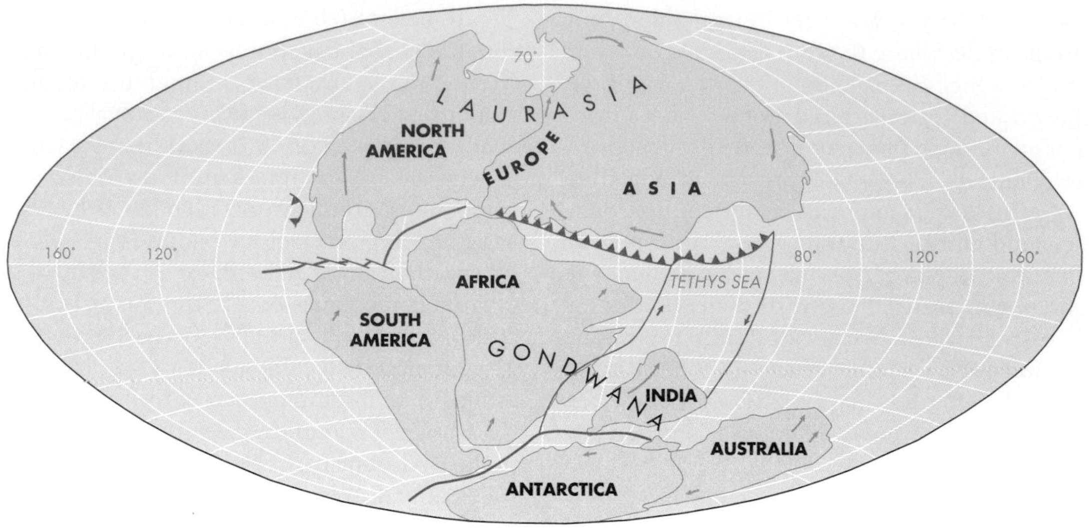

(a) 180 Million Years Ago

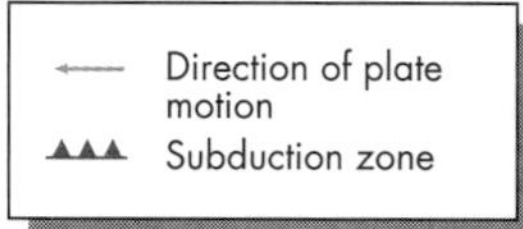

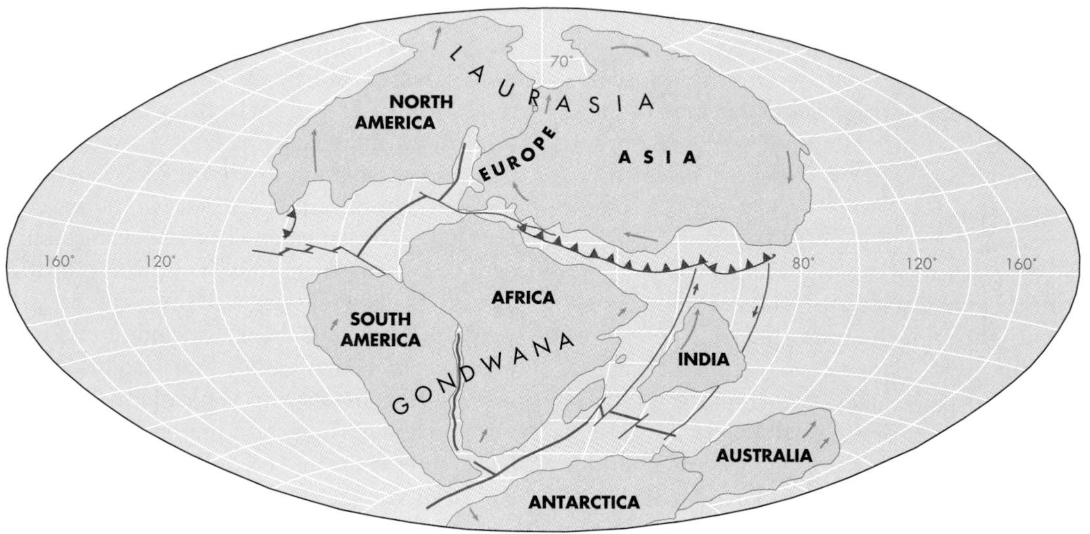

(b) 135 Million Years Ago

▲ **FIGURE 2.8** The supercontinent Pangaea (Laurasia and Gondwana) started to break up approximately 200 million years ago. Shown here is what are thought to have been the positions of the continents at 180 million years ago (a); 135 million years ago (b); 65 million years ago (c); and at present (d). Arrows show directions of plate motion. See text for further explanation of the closing of the Tethys Sea, the collision of India with China, and the formation of mountain ranges. (*From Dietz*, R. S., *and Holden* J. C. 1970. Journal of Geophysical Research 75: 4939-56. *Copyright by the American Geophysical Union. Modifications and block diagrams from Christopherson*, R. W. 1994. Geosystems, *2nd ed. Englewood Cliffs*, NJ: *Macmillan*)

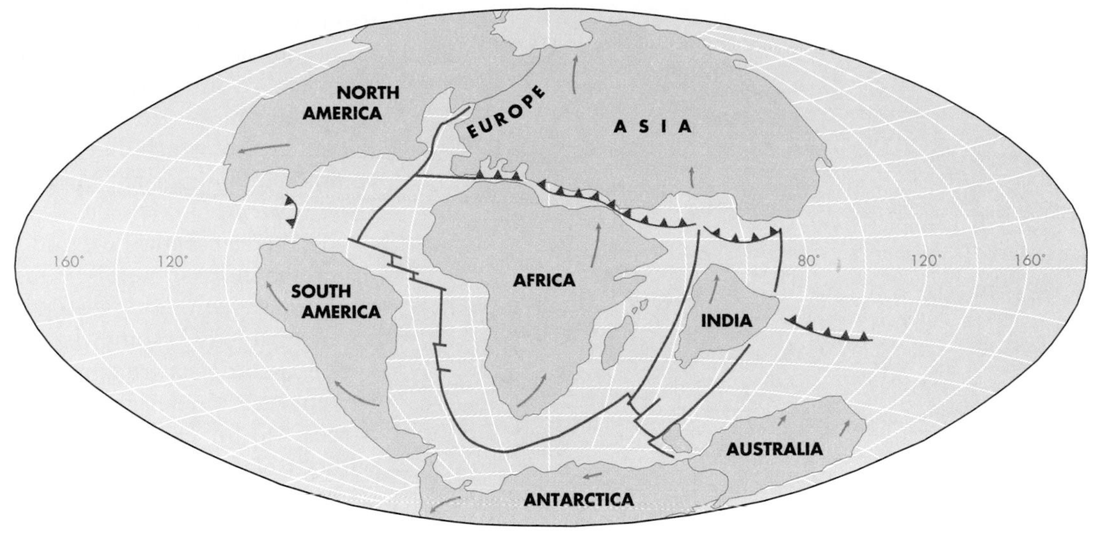

(c) 65 Million Years Ago

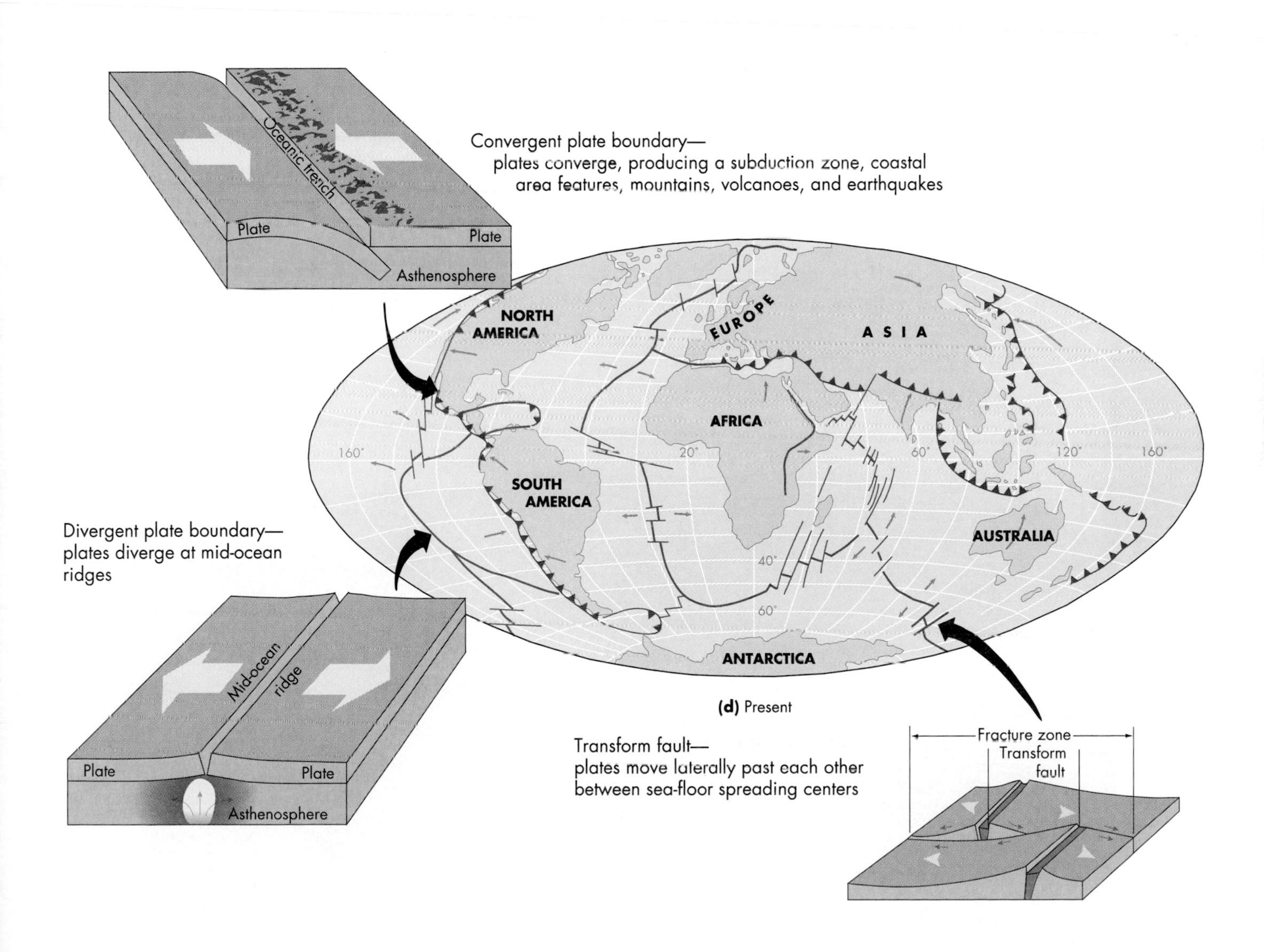

Oceanic trench
Plate
Plate
Asthenosphere

Convergent plate boundary—
plates converge, producing a subduction zone, coastal area features, mountains, volcanoes, and earthquakes

NORTH AMERICA
EUROPE
ASIA
AFRICA
SOUTH AMERICA
AUSTRALIA
ANTARCTICA
160°
20°
60°
120°
160°
40°
60°

Divergent plate boundary—
plates diverge at mid-ocean ridges

Mid-ocean ridge
Plate
Plate
Asthenosphere

(d) Present

Transform fault—
plates move laterally past each other between sea-floor spreading centers

Fracture zone
Transform fault

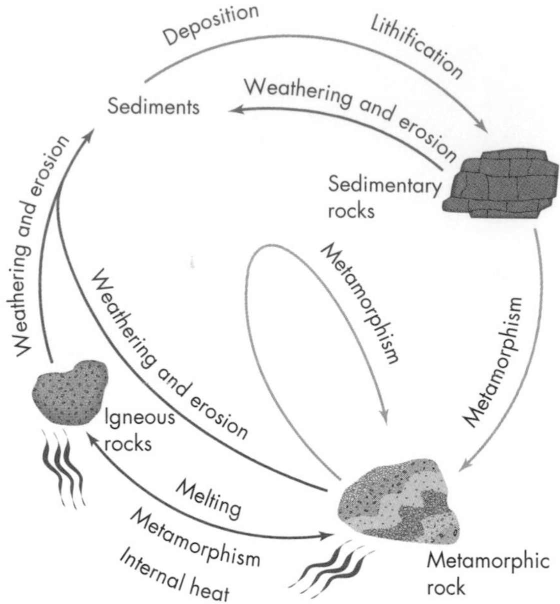

▲ **FIGURE 2.9** The rock cycle.

exposed to erosion. Less frequently, phosphorus is returned to the land as **guano** (bird or bat feces), collected (mined), and used as fertilizer.

2.2 Rocks

Our discussion of the geologic cycle and its subcycles has established that earth materials—rocks and minerals—are constantly being created, changed, and destroyed by dynamic earth processes. We can now begin to focus on the properties and uses of earth materials. In environmental geology we are primarily concerned with rocks, but to understand their properties we need to know something about the minerals of which they are composed. Some of the most common and important minerals are discussed in *A Closer Look: Minerals.*

In the last section we said that a rock is an aggregate of one or more minerals. This is the definition used in traditional geologic investigations. The terminology of applied geology is somewhat different, however, because applied geologists are concerned with properties that affect engineering design. In applied geology, the term *rock* is reserved for earth materials that cannot be removed without blasting; earth materials that can be excavated with normal earth-moving equipment are called *soils.* Thus, a very friable (loosely compacted or poorly cemented) sandstone may be considered a soil, whereas a well-compacted clay may be called a rock. This pragmatic terminology conveys more useful information to planners and designers than the conventional terms. Clay, for example, is generally unconsolidated and easily removed. A contractor might assume that a material described as clay is a soil and bid low for an excavation job, thinking that the material can be removed without blasting. If the clay turns out to be well compacted, however, he may have to blast it, which is much more expensive. This kind of error can be avoided by making a preliminary investigation and forewarning the contractor to consider the clay a rock. (Soils and their properties will be considered in detail in Chapter 3.)

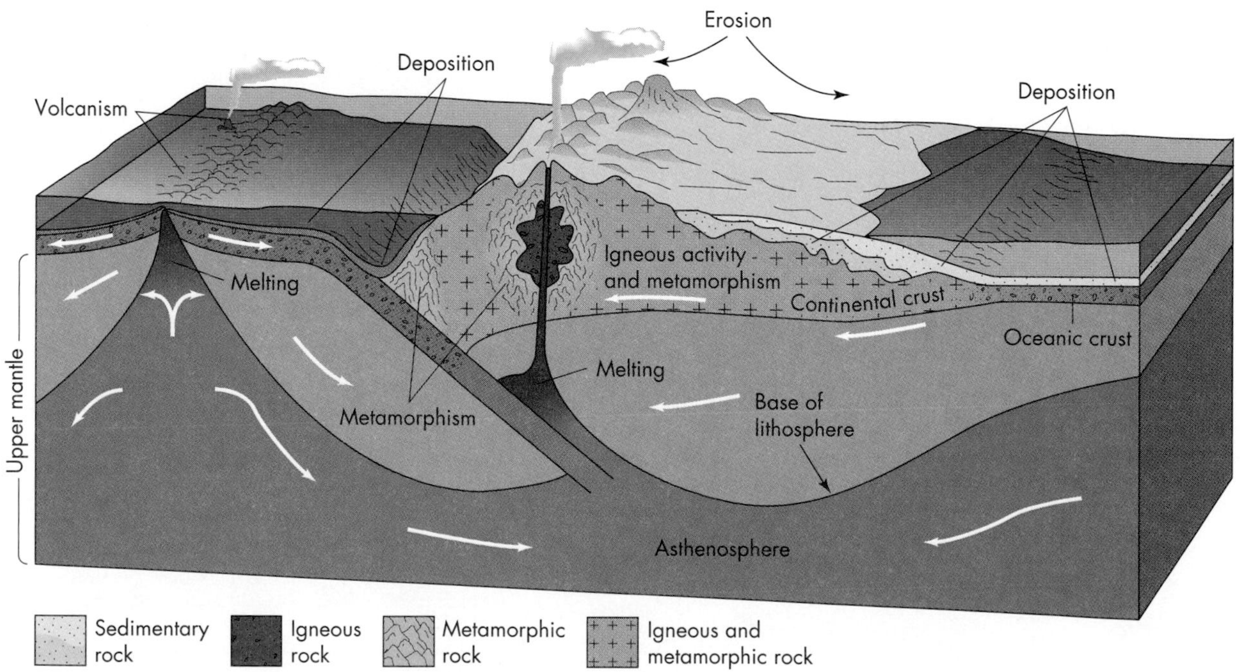

▲ **FIGURE 2.10** Idealized diagram showing some of the environments in which sedimentary, igneous, and metamorphic rocks form. (*From S. Judson, M. E. Kauffman, and L. D. Leet.* 1987. Physical Geology, *7th ed. Englewood Cliffs, NJ: Prentice-Hall*)

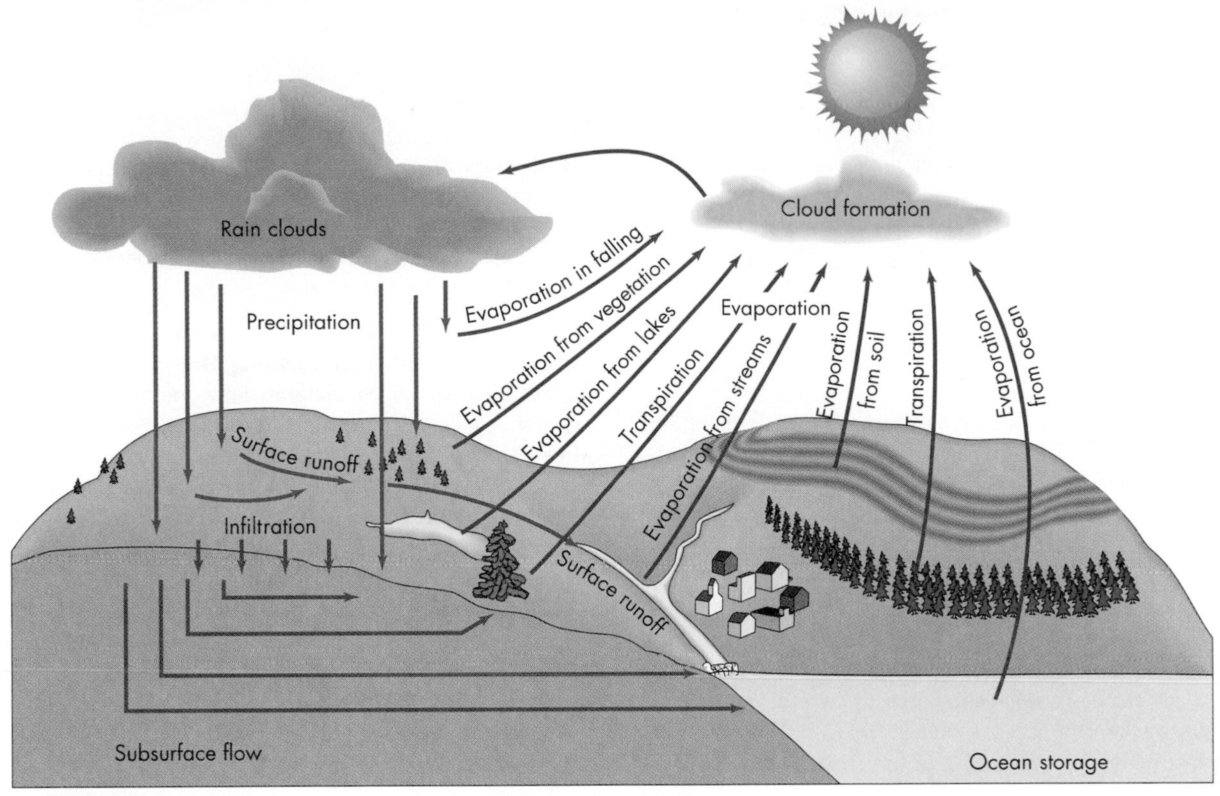

▲ **FIGURE 2.11** Idealized diagram showing the hydrologic cycle illustrating important processes and transfer of water. (*Modified after Council on Environmental Quality and Department of State*, 1980. The Global 2000 Report to the President, *Vol.* 2. *Washington*, DC.)

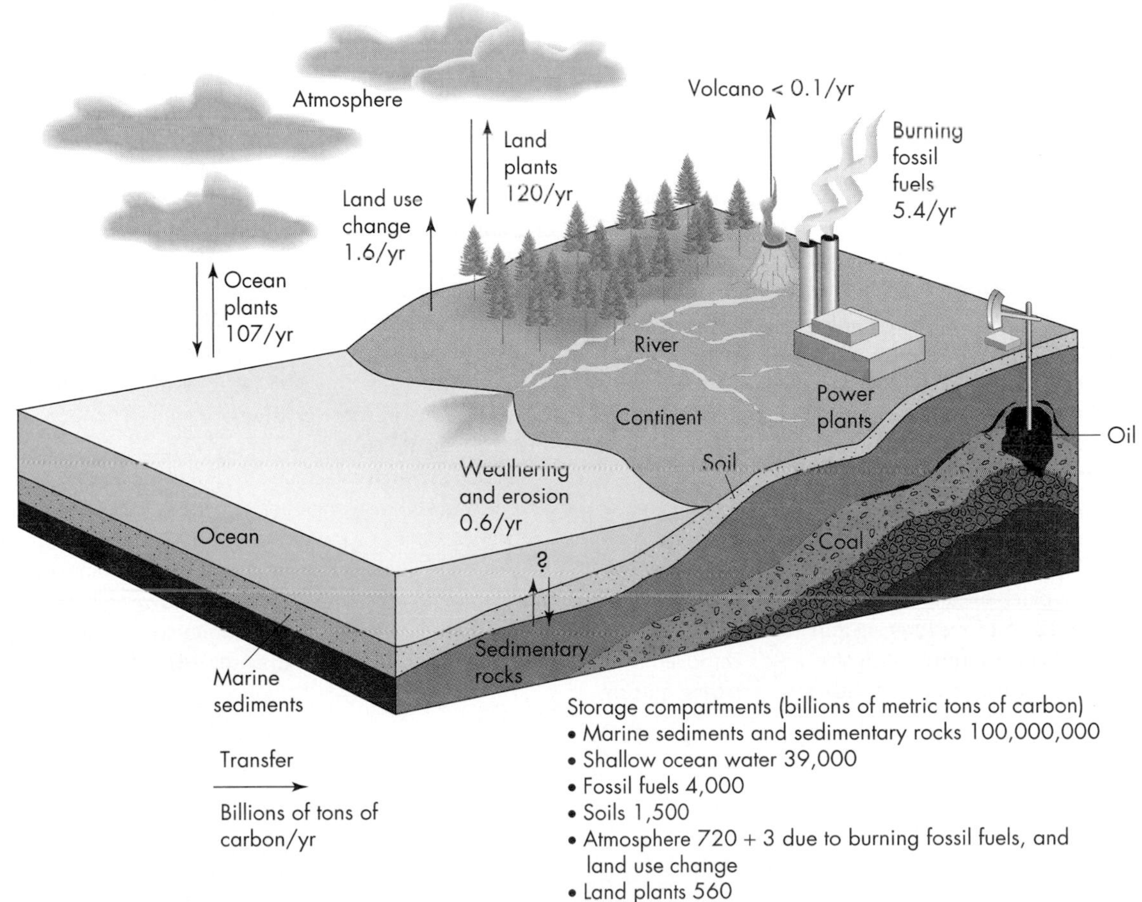

▲ **FIGURE 2.12** Idealized carbon cycle. (*Modified after* G. *Lambert*, 1987. *La gaz carbonique dans l'atmosphere.* La Recherche, 18:782–83.)

A CLOSER LOOK Minerals

A **mineral** is defined as an element or chemical compound that is normally crystalline and has formed as a result of geologic processes. By *crystalline* we mean that the atoms in the mineral are arranged in a regular repeating geometric pattern. Water is not a mineral, but its solid crystalline form ice is. Although there are more than 2000 minerals, we need to know only a few of them to identify most rocks. Some selected minerals are shown in Figure 2.A.

Silicates Nearly 75 percent by weight of the earth's crust is oxygen and silicon. These two elements in combination with a few others (aluminum, iron, calcium, sodium, potassium, and magnesium) account for the chemical composition of minerals that make up about 95 percent of the earth's crust. Minerals that include the elements silicon (Si) and oxygen (O) in their chemical composition are called **silicates;** these are the most abundant of the **rock-forming minerals.** The three most important rock-forming silicate minerals or mineral groups are *quartz, feldspar,* and *ferromagnesian* (Figure 2.A).

Quartz (a form of silicon dioxide, SiO_2) is one of the most abundant minerals in the crust of the earth. It can usually be recognized by its hardness, which is greater than that of glass, and by the characteristic way it fractures—conchoidally (like a clam shell). Pure quartz is white, but some quartz crystals contain impurities that make them rose, purple, black, or another color. Some of these colored quartzes are semiprecious gemstones, such as amethyst. Because it is very resistant to the natural processes that break down minerals, quartz is the common mineral in river and most beach sands.

Feldspars are aluminosilicates, containing silicon, oxygen, and aluminum (Al) in combination with potash (potassium, K), soda (sodium, Na), or lime (calcium, Ca). The most abundant group of rock-forming minerals in the earth's crust, feldspars are commercially important in the ceramics and glass industries. They are generally white, gray, or pink and are fairly hard. Feldspars weather chemically to form clays, which are hydrated aluminosilicates (aluminosilicates combined with water molecules). This process has important environmental implications. All rocks become fractured, and water, which facilitates chemical weathering, enters the fractures. If clay forms along the fractures of a feldspar, it can greatly reduce the strength of the rock and increase the chance of a landslide.

Ferromagnesian minerals are a group of silicates in which the silicon and oxygen combine with iron (Fe) and magnesium (Mg). These are the dark minerals in most rocks (for example, see the black mica *biotite* in Figure 2.A). Because they are not very resistant to weathering and erosional processes, ferromagnesian minerals tend to be altered or removed relatively quickly from their location. They also weather quickly, combining with oxygen to form oxides such as limonite (rust) and with other elements to form clays and soluble salts. These materials, when abundant, may produce weak rocks. Builders must be cautious when evaluating a construction site (for example, for a highway, tunnel, or reservoir) that contains rocks high in ferromagnesians.

Other Important Rock-forming Minerals In addition to silicates, rock-forming minerals important in environmental studies include oxides, carbonates, sulfides, and native (pure) elements. Earth materials containing useful minerals (especially metals) that can be extracted are called **ores.** Iron and aluminum, probably the most important metals in our industrial society, are extracted from ores containing iron and aluminum **oxides.** The most important iron ore is hematite (Fe_2O_3), and the most important aluminum ore is bauxite (Al_2O_3). Magnetite (Fe_3O_4), also an iron oxide but economically less important than hematite, is common in many rocks. It is a natural magnet (lodestone) that attracts and holds iron particles. Where particles of magnetite are abundant, they may produce a black sand in streams or beach deposits.

Environmentally, the most important **carbonate** mineral is *calcite* (calcium carbonate, $CaCO_3$), shown in Figure 2.A. This mineral is the major constituent of limestone and marble, two very important rock types. Weathering of such rocks by water dissolves the calcite, often producing caverns, sinkholes (surface pits), and other unique features. In areas built on limestone, subsurface water in cavern systems may quickly become polluted by urban runoff. In addition, construction of highways, reservoirs, and other engineering structures is a problem where caverns or sinkholes are encountered.

The **sulfide** minerals, such as pyrite, or fool's gold (iron sulfide, FeS_2, Figure 2.A), are sometimes associated with environmental degradation. This occurs most often when roads, tunnels, or mines cut through coal-bearing rocks that contain sulfide minerals. The exposed sulfides oxidize in the presence of water to form compounds such as sulfuric acid, which may enter and pollute streams and other environments. This is a major problem in the coal regions of Appalachia and many other regions of the world where sulfur-rich coal and sulfide minerals are mined (see Chapter 11).

Native elements such as gold, silver, copper (Figure 2.A), and diamonds have long been sought as valuable minerals. They usually occur in rather small accumulations but occasionally are found in sufficient quantities to justify mining. As we continue to mine these minerals in ever lower-grade deposits, the environmental impact will continue to increase (large mines have a larger impact; see Chapter 14).

The Strength of Rocks

The strength of earth materials varies with their composition, texture, and location. Weak rocks, such as those containing many altered ferromagnesian minerals, may creep or nearly flow under certain conditions and be very difficult to tunnel through. On the other hand, granite, a very common igneous rock, is generally a strong rock that needs little or no support. Granite buried deep within the earth, however, may also flow and be deformed. Thus, the strength of a rock can be quite different in different environments.

(a)

(b)

(c)

(d)

(e)

(f)

(g)

▲ **FIGURE 2.A** Some common minerals: (a) Cluster of quartz crystals from Brazil. Some are clear and some are rose quartz. Quartz is a very hard and common rock-forming mineral (*Arnold Fisher/Science Photo Library/Photo Researchers, Inc.*); (b) One of the several varieties of feldspar, the most common rock-forming mineral in the earth's crust (*Stuart Cohen/Comstock*); (c) Yellow and pink clay minerals on the wall of Paint Mines, Colorado. These are only two examples of the many clay minerals that, when present in soils, may exhibit undesirable properties such as low strength, high-water content, poor drainage, and high shrink-swell potential (*Mark D. Phillips/Photo Researchers, Inc.*); (d) The dark mineral in this hand specimen of a rock is the black mica called biotite. It is a common mineral in some granitic rocks as well as some metamorphic rocks (*Barry L. Runk/Grant Heilman Photography, Inc.*); (e) Calcite, the abundant mineral in limestone and marble. Limestone terrain is associated with caverns, sinkholes, subsidence, and potential water pollution and construction problems (*Betty Crowell/Faraway Places*); (f) Pyrite (fool's gold) is iron sulfide, a common mineral associated with coal, which reacts as water and oxygen to form sulfuric acid (*Betty Crowell/Faraway Places*); and (g) Fragment of native copper (*Comstock*).

The strength of an earth material is usually described as the compressive, shear, or tensile stress necessary to break a sample of it (see *A Closer Look: Stress and Strain*).

Table 2.1 shows the compressive strength ranges of some common rocks. However, rocks are neither *homogeneous* (of uniform composition throughout) nor *isotropic* (having properties with the same values along every direction), and both of these conditions must exist before we can assign completely reliable strengths to any material. We therefore assign a safety factor (SF) to ensure that a rock will more likely perform as expected; that is, the rock is loaded only to a fraction of its assumed strength. For example, if we assign a safety factor of 10 to a certain kind of rock, we will load that rock to only 10 percent of its compressive strength. The testing necessary to establish strengths of rocks can be expensive, so it is usually done only for large engineering structures. For small structures, the experienced earth scientist can render a judgment consistent with good engineering practice. However, this does not apply to the evaluation of soils, which may require extensive testing even for small structures (see Chapter 3).

Engineers planning large structures must also be aware of *rock fracture systems*, since the strength of rocks is greatly affected by the frequency and orientation of fractures.

A CLOSER LOOK Stress and Strain

Stress is the force per unit area that exists in a specified plane within rocks or other earth materials. The usual unit of force in science is the Newton (N), defined as the force necessary to produce an acceleration of 1 m/sec^2 to a mass of 1 kg. Stress is measured in N/m^2 or Pascals (Pa), 1 N/m^2 = 1 Pa. Stress cannot be measured directly, but it can be inferred or calculated (7). Stress may be compressive, tensile, or shear, as determined by the directions of the forces acting on the material (Figure 2.B, left). **Strain** is the deformation (shortening, lengthening, or rotating) induced by a stress. Each type of stress produces a corresponding type of strain (Figure 2.B, right).

Some materials under stress deform in an elastic manner, others in a plastic manner. In **elastic deformation** the deformed material returns to its original shape after the stress is removed; examples are compressing a rubber ball and stretching a rubber band. **Plastic deformation** is characterized by permanent strain; that is, the material does not return to its original shape after the stress is removed. Examples of plastic deformation are compressing snow into a snowball and stretching chewing gum.

Materials that rupture before little plastic deformation occurs are **brittle,** while those that rupture after considerable plastic deformation are **ductile.** Glass at room temperature is hard and brittle, but glass heated to a high enough temperature is soft and ductile, deforming plastically before rupturing. That is why glassblowers can make such beautiful objects. Figure 2.C shows the stress and resulting strain experienced by a ductile, deforming substance and a brittle substance (8).

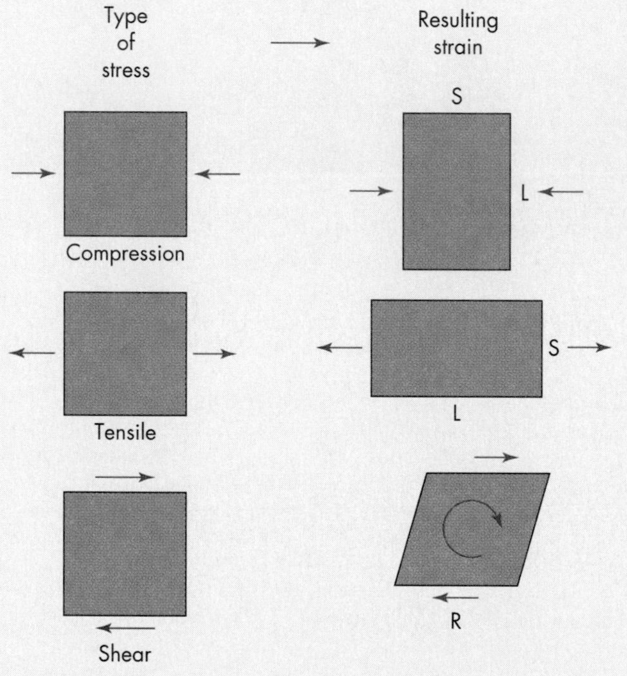

▲ **FIGURE 2.B** Common types of stress and resulting strain. S = shortening; L = lengthening; R = rotating.

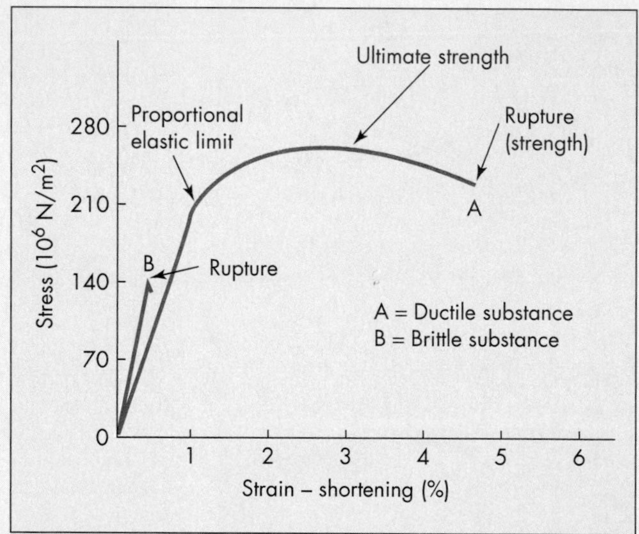

▲ **FIGURE 2.C** Typical stress-strain diagram. (*After* M. P. Billings, 1954. Structural Geology. *New York:* Prentice-Hall. *Reproduced by permission.*)

Furthermore, rocks can be ruptured by movement along *faults*—fractures or fracture systems in which one side moves relative to the other. Fractures in rock can range in size from small hairline fractures to huge fault zones such as the San Andreas fault zone, which scars the California landscape for hundreds of kilometers (Figure 2.13). Even a small active fault near the proposed site of a large masonry dam can be sufficient reason to look for another site (see *Case History: Baldwin Hills Dam*).

The problems and dangers associated with fractures are not confined to active fault systems (those that have moved in the past few thousand years). Once a fracture has developed, it is subject to weathering, which may produce clay minerals. These minerals may be unstable and may become a lubricated surface that facilitates landslides, or the clays may wash out and create open conduits for water to move through the rocks. Therefore, although active faulting is obviously a hazard to the stability of engineered structures, even small, inactive fractures, and faults, must be inspected and carefully evaluated to maximize structural safety.

Types of Rocks

Our discussion of the generalized rock cycle established that there are three rock families: igneous, sedimentary, and metamorphic. Table 2.1 summarizes the characteristics of some common rocks of each type. Rocks are traditionally classified according to their *mineralogy* (mineral composition) and *texture* (size, shape, and arrangement of grains). Texture is more significant in environmental geology, however, be-

▲ **FIGURE 2.13** The San Andreas fault on the Corrizo Plain in California. The fault trace runs from the lower left to upper right diagonally across the photograph, looking like a gigantic plow has been dragged across the landscape. (*James Balog/Tony Stone Images*)

cause, along with *structure* (the various types of fractures), it determines the strength and utility of rock. Granitic rocks, for example, can be given a number of different names (i.e., granite, gabbro) according to their mineralogy, but the engineering properties of all types of granitic rock are nearly the same. This does not mean that mineralogy is unimportant in applied geology—in specific cases it can be extremely significant. However, when we consider the utility of an entire group of rocks, such as intrusive igneous rocks, mineralogy is often less important than texture and structure.

Igneous Rocks Igneous rocks are rocks that have crystallized from **magma,** a mobile mass of hot, quasi-liquid earth material that is probably generated in the upper asthenosphere or the lithosphere (see Figure 2.10). Igneous and metamorphic rocks together comprise more than 90 percent of all rocks in the earth's crust.

Table 2.1 Common rocks (engineering geology terminology and compressive strength)

Type	Texture	Materials	Examples of Approximate Compressive Strength (10^6 N/m^2)
Igneous			
Intrusive			
Granitic[a]	Coarse[b]	Feldspar, quartz	100–280
Ultrabasic	Coarse	Ferromagnesians, ±quartz	Can be very low
Extrusive			
Basaltic[c]	Fine[d]	Feldspar, ±ferromagnesians, ±quartz	50 to greater than 280
Volcanic breccia	Mixed—coarse and fine	Feldspar, ±ferromagnesians, ±quartz	
Welded tuff	Fine volcanic ash	Glass, feldspar, ±quartz	Less than 35
Sedimentary			
Detrital			
Shale	Fine	Clay	2–215
Sandstone	Coarse	Quartz, feldspar, rock fragments	40–110
Conglomerate	Mixed—very coarse and fine	Quartz, feldspar, rock fragments	90
Chemical			
Limestone	Coarse to fine	Calcite, shells, calcareous algae	50–60
Metamorphic		**Parent Material**	
Foliated			
Slate	Fine	Shale or basalt	180
Schist	Coarse	Shale or basalt	15–130
Gneiss	Coarse	Shale, basalt, or granite	160–190
Nonfoliated			
Quartzite	Coarse	Sandstone	150–600
Marble	Coarse	Limestone	100–125

[a]Textural name used by engineers for a group of coarse-grained, intrusive igneous rocks, including granite, diorite, and gabbro.
[b]Individual mineral grains can be seen with naked eye.
[c]Textural name used by engineers for a group of fine-grained, extrusive igneous rocks, including rhyolite, andesite, and basalt.
[d]Individual mineral grains cannot be seen with naked eye.
[e]Some data from References 7,12.

CASE HISTORY Baldwin Hills Dam

In the early afternoon of December 14, 1963, people living below the Baldwin Hills Reservoir in southern California were hastily evacuated from their homes, as water began pouring through a break in the dam. Two hours after the evacuation started, the dam ruptured, claiming five lives and causing $11 million in damage.

The failure of the dam, which was 71 m high and 198 m long, resulted from gradual movement along faults, which fractured the tile drain, foundation, and asphaltic membrane of the dam and reservoir floor. The reservoir was on the edge of an oil field, and withdrawal of oil between 1923 and 1963 caused subsidence at the reservoir of about 1 m. After the reservoir had drained, cracks could be seen in the asphaltic concrete pavement. These cracks extended continuously along the length of the reservoir and were aligned with the deep gash in the dam (Figure 2.D).

When construction of the reservoir began in 1947, it utilized the most advanced knowledge available. The fault, along with several others, was discovered during construction, examined, and judged not dangerous to the stability of the dam. Because of a previous experience with faults (the 1928 failure of the St. Francis Dam), the design of the Baldwin Hills Dam included some provisions for the faults in the foundation material, including construction of a compacted clay reservoir lining. In addition, a series of periodic inspections was initiated to ensure the safety of the operation. Sometime between dam construction and when it failed, the fault moved a few centimeters (there was no earthquake). This caused the clay liner to fail. Water from the reservoir entered cracks along the fault and emerged as leaks that eroded the dam, causing the failure. The failure came quite suddenly. Were it not for fortunate circumstances that gave several hours' warning and enabled officials to act quickly, the loss of life might have been much greater. Cause of the fault movement was controversial, but one hypothesis links the failure to subsidence of the oil field (11,12).

If magma slowly cools and crystallizes well below the surface of the earth, the resulting igneous rock is called **intrusive.** Intrusive igneous rocks tend to be coarse-grained—that is, individual mineral grains can be seen with the naked eye—due to the relatively slow cooling during crystallization. When intrusive igneous rocks such as granite (composed mostly of feldspar and quartz, Table 2.1, Figure 2.14a) are exposed on the surface, we may conclude that erosion has removed the material that originally covered them (Figure 2.15). As the molten magma rises toward the surface, it displaces the rock it intrudes, often breaking off portions of the intruded rock and incorporating them as it crystallizes. These foreign blocks, known as inclusions, are good evidence of forcible intrusion.

Intrusive igneous rocks are generally strong and have a relatively high unconfined compressive strength (9). The strength of a granite depends upon its grain size and degree of fracturing. In general, granites with relatively large grains are stronger than granites with relatively small grains. Fresh, unweathered intrusive rocks, unless extensively fractured, are usually satisfactory for all types of engineering construction and operations (10). If such rocks can be economically quarried, they are often good sources of construction material—crushed stone for concrete aggregate, facing, or veneer on buildings; broken stone for fill material; or riprap (large blocks of rock) to protect slopes against erosion.

Despite the generally favorable physical properties of intrusive rock, it is not safe to assume that all such rocks are satisfactory for their intended purpose. For example, they may not provide a sufficiently stable and safe foundation for very large, heavy structures. This is particularly true for the ultrabasic intrusive igneous rocks that contain weak ferromagnesian minerals (Table 2.1). Before a large, heavy structure is built, engineering geologists must conduct a field evaluation that includes mapping, drilling, and laboratory tests on the foundation rocks. The project may be jeopardized if they identify a fault zone in the proposed site, even though the material is granite or other igneous intrusive rock.

Igneous rocks that crystallize on or just below the surface of the earth are called **extrusive.** They are formed from *lava* (magma that flows from a volcano in a molten state) or from *pyroclastic debris* (magma that has solidified rapidly as it was blown out of a volcano). Extrusive rocks crystallized from lava are called *basaltic rocks* (composed of feldspar ± ferromagnesians ± quartz, Table 2.1, Figure 2.14b) and may be stronger than granite (9). However, if the lava flow is mixed with volcanic *breccia* (angular fragments of broken lava and other material) or with thick vesicular or scoriaceous zones produced as gas escapes from cooling lava, the strength of the resulting rocks may be greatly reduced. Furthermore, lava flows that have cooled and solidified often exhibit extensive *columnar jointing* (Figure 2.16), which is a type of fracturing that forms during cooling and may lower the strength of the rock. Solidified lava flows also may have subterranean voids known as *lava tubes*, which may collapse from the weight of the overlying material or carry large amounts of groundwater, either of which can cause problems during the planning, design, or construction phases of a project.

Pyroclastic debris, or *tephra*, ejected from a volcano produces a variety of extrusive rocks. Debris consisting of rock and glass fragments less than 4 mm in diameter is called *volcanic ash;* when this ash is compacted, cemented, or welded together, it is called **tuff.** Although the strength of a tuff depends upon how well cemented or welded it is, generally it is a soft, weak rock that may have very low compressive strength (see Table 2.1) (9). Some tuff may be altered to a clay known as **bentonite,** an extremely unstable material. When it is wet, bentonite expands to many times its origi-

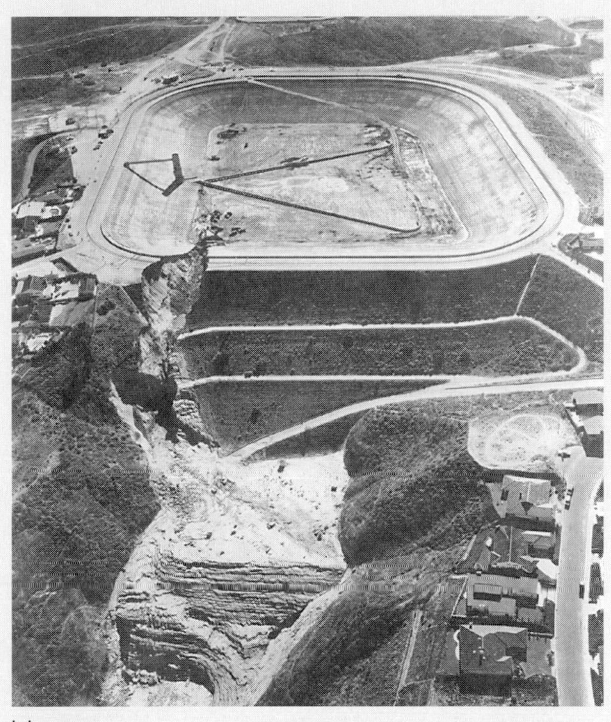

(a)

(b)

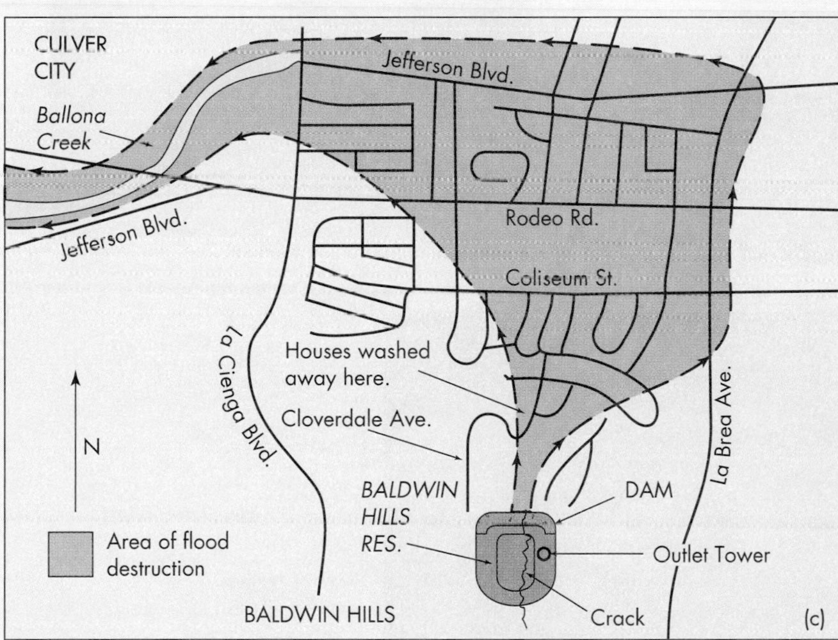

▲ **FIGURE 2.D** Failure of the Baldwin Hills Reservoir, aerial and ground views (a, b), and map showing area flooded (c). ([a] *Courtesy of the California Department of Water Resources.* [b] *Courtesy of Los Angeles Department of Water and Power.* [c] *From Walter E. Jessup, "Baldwin Hills Dam Failure." Reprinted with permission from the February 1964 issue of* Civil Engineering—ASCE, *official monthly publication of the American Society of Civil Engineers, vol. 34, 1964.*)

nal volume. Pyroclastic activity also produces larger fragments that, when mixed with ash and cemented together, form **volcanic breccia,** or **agglomerate.** The strength of this material may be comparable to that of intrusive rocks or be much lower, depending upon how the individual particles are held together. In general, volcanic breccia and agglomerates make poor concrete aggregates because of the large and variable amount of fine materials produced when the rock is crushed.

Because extrusive rocks vary considerably in their composition, it is difficult to make generalizations about their suitability for specific uses; careful field examination is always

▲ **FIGURE 2.14** Some common rock types: (a) A specimen of granite. The pink fine-grained mineral is feldspar, the dark mineral is ferromagnesian, and the very light-colored mineral is quartz. The large crystal in the center is known as a phenocryst, which is a coarse crystal in a finer matrix of other crystals. The phenocryst in this case is feldspar (*Betty Crowell/Faraway Places*); (b) Layered rocks here are basalt flows along the Palouse River in Washington (*Betty Crowell/Faraway Places*); (c) Rocks exposed here in the foreground are sandstone. The parent material was a sand dune that has been cemented and turned to stone, Zion National Park, Utah (*George Hunter/Tony Stone Images*); (d) Rock shown here is a conglomerate found in Death Valley, California. Individual particles that are cemented together are easily seen, and the vertical lines are fractures in the rock (*Betty Crowell/Faraway Places*); (e) Rock exposed here is a highly weathered limestone in Yunnan Province of China. The pinnacles are formed by chemical weathering of the limestone and removal of cover soil material (*Betty Crowell/Faraway Places*); (f) Schist, a foliated metamorphic rock where parallel alignment of mineral grains, in this case mica, produces the foliation and reflects the light (*Marvin B. Winter/Photo Researchers, Inc.*); (g) Gneiss, a foliated metamorphic rock with minerals segregated into white bands of feldspar and dark bands of ferromagnesian minerals. (*John Buitenkant/Photo Researchers, Inc.*)

▲ **FIGURE 2.15** The rock slopes shown here for Yosemite Valley, California, are composed of granite, which is a very strong rock and so can maintain high vertical cliffs. (*Edward A. Keller*)

▲ **FIGURE 2.16** Columnar jointing or fracturing of basalt. These joints or fractures form due to contraction during cooling. This is Devil's Post Pile National Monument, Sierra Nevada, California. (*Courtesy of* A. G. *Sylvester*)

necessary before large structures are built on such rocks (9). Planning, design, and construction of engineering projects on extrusive rocks, especially pyroclastic debris, can be complicated and risky (10). This was tragically emphasized on June 5, 1976, when Teton Dam in Idaho failed, killing 14 people and inflicting approximately $1 billion in property damage. Just before the failure, a whirlpool several meters across was seen in the reservoir, strongly suggesting that a tunnel of free flowing water had developed beneath the dam. The dam had been built upon highly fractured volcanic rocks, and highly erodible wind-deposited clay-silts were used in construction of its interior, or core. The open fractures in the rocks beneath the dam were probably not completely filled with cement slurry (grout) during construction, so water began moving under the foundation as the reservoir filled. When the moving water came in contact with the clay-silt core, it quickly eroded a tunnel through the base of the dam, causing the whirlpool. The dam collapsed just minutes later, and a wall of water up to 20 m high rushed downstream, destroying homes, farms, equipment, animals, and crops along a 160-km stretch of the Teton and Snake rivers.

Sedimentary Rocks Sedimentary rocks (Figure 2.17), which comprise about 75 percent of all rocks exposed at the surface of the earth, form when sediments are transported, deposited, and then lithified by natural cement, compression, or another mechanism. There are two types of sedimentary rocks: *detrital* sedimentary rocks, which form from broken parts of previously existing rocks; and *chemical* sedimentary rocks, which are deposited when chemical or biochemical processes cause solid materials to form from substances dissolved in water.

Detrital sedimentary rocks are classified according to their texture as shale, sandstone, and conglomerate. **Shale** is by far the most abundant, accounting for about 50 percent of all sedimentary rocks. It is the finest textured of the three detrital rock types, being made up of clay- and silt-sized particles. Earlier we defined *clay* in terms of its mineral composition (see *A Closer Look: Minerals*), but engineering geologists also use the term in a textural sense to mean the finest-grained sediment (less than 0.004 mm in diameter). When layers of clay or silt (slightly coarser-grained sediment, 0.076 to 0.004 mm in diameter) are compacted or cemented, they form shale. Shale causes many environmental problems, and its presence is a red flag to the applied-earth scientist.

There are two types of shale: compaction shale and cementation shale. *Compaction shale* is held together primarily by molecular attraction of the fine clay particles. It is a very weak rock and can cause several environmental problems. First, depending upon the type of clay (that is, on its mineralogy), compaction shale may have a high potential

▲ **FIGURE 2.17** The Grand Canyon of the Colorado River exposes a spectacular section of sedimentary rocks. In this view looking toward the south rim at sunset, the top sedimentary unit is the Kaibab Limestone, below which are a series of sandstones, shales, and the famous Red Wall Limestone. (*Jack W. Dykinga*)

to absorb water and swell. Second, depending upon the bonding between its depositional layers, or *bedding planes*, this rock may have a high risk of sliding, even on gentle slopes. Third, compaction shale has a high potential for *slaking*, meaning that contact with water will cause its surface to break away and curl up. Slaking seriously retards bonding between the rock and such structures as the foundations of buildings and dams. Fourth, because of the elastic nature of clays, compaction shale tends to rebound if stress conditions change, making it a very poor foundation material. Finally, compaction shale may have an extremely low compressive strength (see Table 2.1).

Depending upon the degree and type of cementing material, *cemented shale* can be a very stable, strong rock suitable for all engineering purposes. However, the presence of any shale at a building site calls for a close look to determine the type and extent of the rock and its physical properties.

Sandstones and conglomerates (Figure 2.14c,d) are coarse-grained and make up about 25 percent of all sedimentary rock. Sand-sized particles making up sandstone are 0.076 to 2 mm in diameter. Conglomerate contains gravel-sized particles greater than 2 mm in diameter. Depending on the type of cementing material, these rocks may be very strong and stable for engineering purposes. Common cementing materials are silica, calcium carbonate, iron oxide, and clay. Of these, silica is the strongest; calcium carbonate tends to dissolve in weak acid; and clay may be unstable and tend to wash away. It is always advisable to evaluate carefully the strength and stability of cementing materials in the detrital sedimentary rocks.

Chemical sedimentary rocks are classified according to their mineral composition; they include halite (rock salt, $NaCl$), gypsum (hydrous calcium sulfate, $CaSO_4 \bullet 2H_2O$), and **limestone,** which is composed mostly of the mineral calcite (calcium carbonate, $CaCO_3$) (Figure 2.14e). Limestones make up about 25 percent of all sedimentary rocks and are by far the most abundant of the chemical sedimentary rocks. Human use and activity generally do not mix well with limestone. Although this rock may be strong enough to support construction, it weathers easily to form subsurface cavern systems and solution pits. In limestone areas, most of the streams may be diverted to subterranean routes, where they easily become polluted by contaminated runoff entering the groundwater. The mineralogy of limestone is such that little or no natural purification of the polluted water occurs. In addition, construction may be hazardous in areas with abundant caverns and sinkholes.

Metamorphic Rocks Metamorphic rocks (Figure 2.18) are changed rocks. Heat, pressure, and chemically active fluids produced in the tectonic cycle may alter the mineralogy and texture of rocks, in effect producing new rocks. There are two types of metamorphic rocks. In *foliated* metamorphic rocks, the elongated or flat mineral grains have a preferential parallel alignment, producing a banding of light and dark minerals. *Nonfoliated* metamorphic rocks have no preferential alignment or segregation of minerals.

▲ **FIGURE 2.18** Metamorphic rocks exposed at a mountain range scale, Kejser Franz Joseph Fiord, East Greenland. The east-dipping (center) white unit is a marble. This complex exposure is 800 m high and includes several normal (extensional) faults that merge downward to a very large extensional fault known as a *detachment fault*. Metamorphic rock types exposed include gneisses as well as the beautiful white marble. (*Ebbe Hartz, University of Oslo*)

Foliated metamorphic rocks, such as slate, schist, and gneiss, have a variety of physical and chemical properties (Table 2.1), so it is difficult to generalize about the usefulness of such rocks for engineering projects. **Slate** (a fine-grained metamorphic rock formed from metamorphism of a shale or basalt) is generally an excellent foundation material. It was once used for constructing chalkboards for schools and beds for pool tables. It is used for roofing material and decorative stone counters. **Schist** (a coarse-grained metamorphic rock, Figure 2.14f), when composed of soft minerals, is a poor foundation material for large structures. **Gneiss** (a coarse-grained, banded metamorphic rock, Figure 2.14g), is usually a hard, tough rock suitable for most engineering purposes.

Foliation planes of metamorphic rocks are potential planes of weakness. The strength of the rock, its potential to slide, and the movement of water through the rock all vary with the orientation of the foliation. Consider, for example, the construction of roadcuts and dams in terrain where foliated metamorphic rocks are common. In the case of roadcuts (Figure 2.19), the preferred orientation is with the foliation planes dipping away from the roadcut. Foliation planes inclined toward the road result in unstable blocks that might fall or slide toward the road. Also, groundwater will tend to flow down the foliation, causing a drainage problem at the road. For construction of dams (Figure 2.20), the preferred orientation is with nearly vertical foliation planes parallel to the horizontal axis of the structure (9). This position minimizes the chance of leaks and unstable blocks, which are a serious risk when the foliation planes are parallel to the reservoir walls (see *Case History: St. Francis Dam*).

Important nonfoliated metamorphic rocks include **quartzite** and marble. Quartzite, a metamorphosed sandstone (with quartz grains), is a hard, strong rock suitable for

(a)

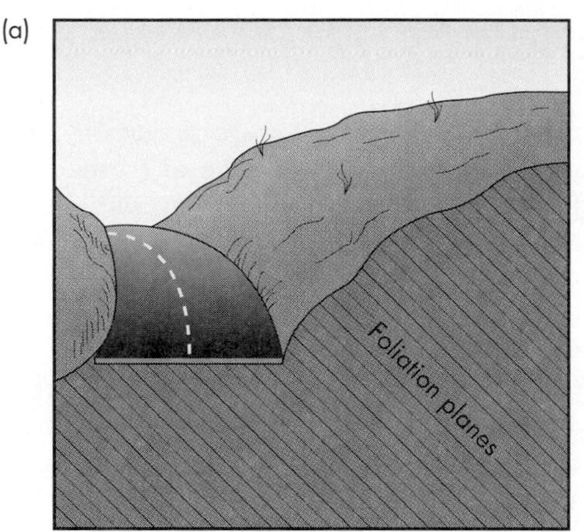

(b)

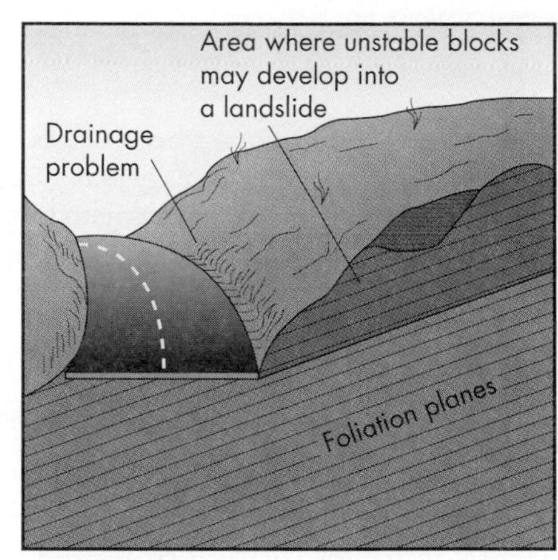

▲ **FIGURE 2.19** Two possible orientations of foliation in metamorphic rock and the effect on highway stability. Where the foliation is inclined away from the road (a), there is less likelihood that unstable blocks will fall on the roadway. Where the foliation is inclined toward the road (b), unstable blocks of rock above the road may produce a landslide hazard.

many engineering purposes. The engineering properties of **marble** (composed of the mineral calcite) are similar to those of its parent rock (limestone), and cavern systems and surface pits resulting from chemical weathering (dissolving) the rock should be expected.

2.3 Surficial Processes: Ice and Wind

Rocks and landforms are shaped not only by tectonic activity originating deep within the earth, but also by processes occurring on the surface. These *surficial processes* include wind-, water-, and ice-related processes, which are responsible for the erosion, transport, and deposition of tremendous quantities of surficial earth materials in the rock cycle. Furthermore, these processes both modify and create a substantial number of landforms in environmentally sensitive areas—coastal, desert, arctic, and subarctic. We have mentioned the importance of water in our brief discussion of the hydrologic cycle, and we will consider its action in more detail in Chapters 5 and 10. In the rest of this chapter we will see how ice and wind affect rocks and landforms, and briefly consider some of the environmental problems associated with wind and ice.

Ice

The cold-climate phenomenon of ice has become an important environmental topic. As more people live and work in the higher latitudes, we will have to learn more about regions of glaciation and permafrost to ensure the best use of these sometimes fragile environments.

Glaciation A **glacier** is a land-bound mass of moving ice. Glaciers that cover large tracts of land are called *continental glaciers*, or *ice sheets;* those confined to mountain valleys at high altitudes or latitudes are called *mountain* (or

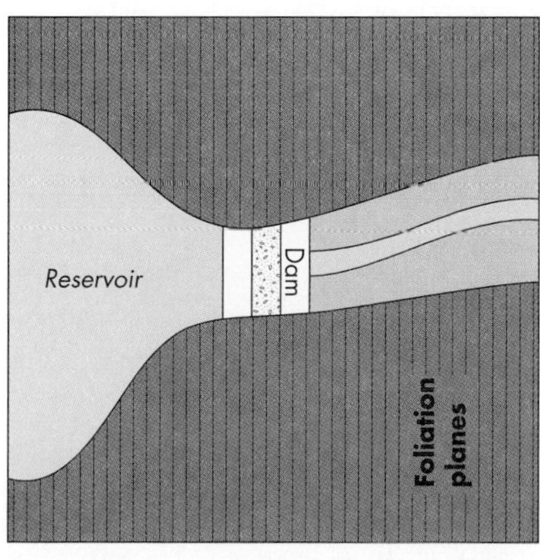

(a)

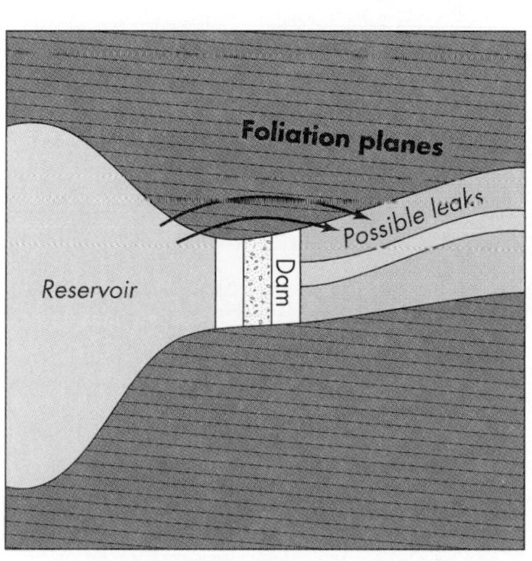

(b)

◀ **FIGURE 2.20** Two possible orientations of foliation in metamorphic rocks at a dam and reservoir site. The most favorable orientation of the foliation is shown in part (a), where the foliation is parallel to the axis of the dam. The least favorable orientation is where the foliation planes are perpendicular to the axis of the dam, as shown in (b).

CASE HISTORY St. Francis Dam

On the night of March 12, 1928, more than 500 lives were lost and $10 million in property damage was done as ravaging floodwaters raced down the San Francisquito Canyon near Saugus, California. The 63-m high St. Francis Dam, with a main section 214 m long and holding 47 million cubic meters (m^3) of water, had failed (Figure 2.E).

Causes of the failure were clearly geologic. The east canyon wall was metamorphic rock (schist) with foliation planes parallel to the wall. Before the failure, both recent and ancient landslides indicated the instability of the rock. The west wall was sedimentary rock with prominent high topographic ridges that suggested the rock was strong and resistant. Under semiarid conditions this was true, but when the rocks became wet they disintegrated. This characteristic was not discovered and tested until after the dam had failed. The contact between the two rock types is a fault with an approximately 1.5-m-thick zone of crushed and altered rock [containing clay and the mineral gypsum which is soluble (that is, capable of dissolving in water)]. The fault was shown on California's 1922 fault map but was either not recognized or ignored. Three processes combined to cause the tragedy: landsliding of the metamorphic rock (the primary cause of failure), disintegration and sliding of the sedimentary rock, and leakage of water along the fault zone that washed out or dissolved the crushed rock. Together, these processes destroyed the bond between the concrete and the rock, and precipitated the failure (9,13). This disaster did more than any previous event to focus public attention on the need for geologic investigation as part of siting reservoirs. Such investigations are now standard procedure.

(a)

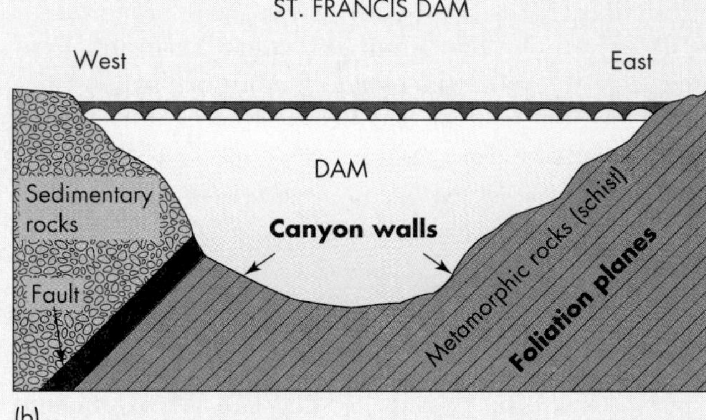

(b)

(c)

▲ **FIGURE 2.E** St. Francis Dam (a) prior to failure; (b) geology along the axis of the dam; and (c) after failure. (*Los Angeles Department of Water and Power*)

alpine) glaciers (Figure 2.21). Only a few thousand years ago, the most recent continental glaciers retreated from the Great Lakes region of the United States. Figure 2.22 shows the maximum extent of these ice sheets. Several times in the last 1.65 million years, during the epoch known as the Pleistocene, or Ice Age (see the geologic time scale on the inside back cover), the ice has advanced southward. Ice sometimes covered as much as 30 percent of the land area of the earth, including the present sites of major cities such as New York and Chicago.

▲ **FIGURE 2.21** Confluence of Muldrow and Traleika glaciers below Mount Denali, Alaska. The dark material on the glaciers is sediment being carried on the top of the ice. The fractures in the white glacier ice are crevasses. (*Michael Collier*)

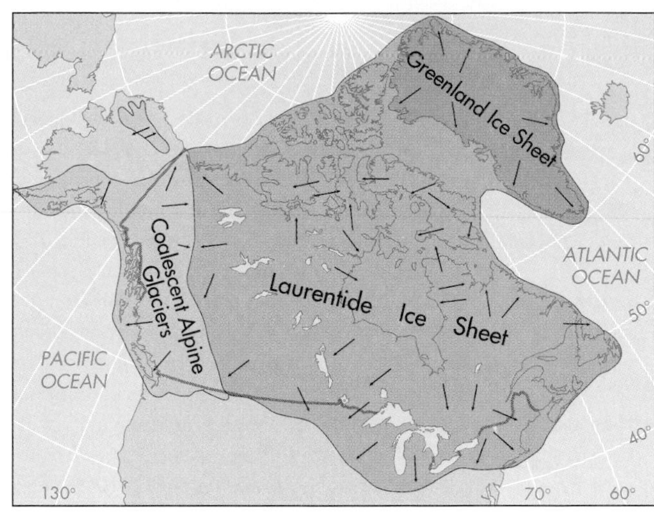

▲ **FIGURE 2.22** Maximum extent of ice sheets during the Pleistocene glaciation. (*From* R. J. *Foster,* General Geology, *4th ed. Columbus*, OH: *Charles* E. *Merrill*, 1983)

Today, glacial ice covers only about 10 percent of the land area of the earth. Most of this ice is located in the Antarctic ice sheet, with lesser amounts in the Greenland ice sheet and in the mountain glaciers of Alaska, southern Norway, the Alps, and the Southern Alps in New Zealand. Scientists are still speculating as to whether the continental ice sheets will advance once again. We may, indeed, still be in the Ice Age.

Glaciers typically move less than 1 m per day, and many move only a few centimeters per day. However, Alaskan glaciers have quite irregular rates of advance and retreat. For example, the Black Rapids Glacier advanced several kilometers down its valley in a period of only five months in 1936 and 1937, then started to retreat. Such a rapid advance, called a *glacial surge*, can radically change a local environment. Scientists gathered valuable information from the recent rapid advance of the Hubbard Glacier (see *Case History: Hubbard Glacier*), which should be applicable to future changes in glaciers in other parts of the world. As more people move into areas that are still partly glaciated, we will have more situations in which human use of the land comes into conflict with glacial processes.

Continental and mountain glaciers have produced a variety of depositional and erosional landforms that are easily seen in the landscape. **Glacial deposits** of earth materials are prominent characteristics of many regions. For example, the flat, nearly featureless *ground moraine*, or *till plains*, of central Indiana are composed of till, material carried and deposited by the continental glaciers that completely buried preglacial river valleys. It is hard to believe that beneath these glacial deposits is a topography formed by running water, much like the hills and valleys of southern Indiana that glaciers never reached. Till is heterogeneous material deposited directly by the ice; it may include, for example, boulders, gravel, sand, silt, and clay particles. In addition to till, glacial deposits include **outwash,** material carried away from glaciers by meltwater in streams. Outwash consists of materials such as silt, sand, and gravel.

Finally, deposition of glacial deposits favors the formation of lakes. Blocks of glacial ice frequently are incorporated with glacial deposits. When a large ice block melts, a depression may form at the surface that fills with water, forming a shallow pond or lake. These lakes may be short-lived and soon fill with sediment and decaying vegetation (water plants). As the plants are buried, they may be transformed to high organic soils or peat deposits. Such deposits are mined for peat from areas glaciated during the Pleistocene in the midwestern United States.

Glaciers are also agents of erosion:

- The Great Lakes in the United States are in part due to glacial erosion of preglacial lowlands.
- Alpine or valley glaciers can convert V-shaped valleys formed by river erosion to U-shaped (Figure 2.23).
- Alpine glaciers can form sharp peaks or "horns" (for example, the Matterhorn, in France) or sharp ridge lines (see Figure 2.5, Italian Alps, the Dolomites).
- Glacial erosion can carve grooves or shallow furrows in rock, showing the direction the glaciers move.
- Glacial erosion near where continental glaciers form, as, for example, in Canada (Figure 2.22), have removed vast quantities of rock debris and transported them south to be deposited as glacial till. This explains the occasional find of gem-quality diamonds in Indiana streams (they came from Canada!).

Because glaciation produces such a variety of landforms, the environmental geology of a recently glaciated area may be complex. The wide variety of earth materials found in an area recently glaciated requires that detailed evaluation of the physical properties of surficial and subsurface materials be conducted before planning, designing, and building structures such as dams, highways, and large buildings. For example, if glacial lake deposits and

CASE HISTORY Hubbard Glacier

In 1971 the Hubbard Glacier in southern Alaska started to advance more rapidly. By the summer of 1986 the glacier was surging at a rate of 30 or more meters per day, slowing to about 6 m per day by September. The 1986 surge blocked Russell Fiord behind an ice dam, forming Russell Lake (Figure 2.F).

The ice advanced so rapidly that it trapped seals, porpoises, and other marine animals in the new lake. People concerned for the porpoises tried to capture and transport them, and some of the seals attempted to walk around the ice dam to the sea. Water rising in Russell Lake precipitated concern that it might spill over into the Situk River, greatly increasing the discharge and possibly damaging the ecosystem there. Located nearby, on glacial deposits that are about 1000 years old, is the village of Yakutat. The residents of the village make a living by fishing in the ocean and in the Situk River, which is well known for its salmon fisheries. They feared that if the lake overflowed into the river, it would bring with it large quantities of silt that would damage the spawning beds. Concerned that their livelihood would be lost, the villagers suggested to local, state, and federal authorities that a canal should be constructed to partially drain the lake and thus maintain the water quality in the river.

On October 8, 1986, the ice dam ruptured, allowing the trapped marine animals to escape and temporarily eliminating the threat of overflow into the Situk River. However, if the glacier continues to advance, it may completely fill Yakutat Bay as it has done in the past, making Russell Lake a more permanent feature. In that case, although the people of Yakutat today are concerned about glacial water entering the river, people in generations to come may be equally alarmed if the glacier retreats, Russell Lake becomes a fiord again, and the discharge to the river is drastically reduced.

peat are present, then differential subsidence of the ground upon loading (building a heavy structure) is a real possibility.

Permafrost Permanently frozen ground, or **permafrost,** is a widespread natural phenomenon in the higher latitudes, underlying about 20 percent of the land area in the world (Figure 2.24). Two main types of permafrost, called *discontinuous* and *continuous*, are defined based on the areal extent of frozen ground. In the *continuous permafrost* areas of Greenland and northernmost Asia and North America, the only ice-free ground is beneath deep lakes or rivers. The presence of the water keeps the ground from freezing. Farther south is the region of *discontinuous permafrost*, characterized in its higher latitudes by scattered islands of thawed ground in a predominantly frozen area. The percentage of unfrozen ground increases at lower latitudes, until finally, at the southern border of the permafrost, all the ground is unfrozen.

Permafrost varies from north to south in thickness as well as in continuity. About 85 percent of Alaska is underlain by continuous or discontinuous permafrost that varies in thickness from about 400 m in the north to less than 0.3 m at the southern margin of the frozen ground (10,14). A cross section of permafrost shows an upper active layer that thaws during the summer, as well as unfrozen layers within the permafrost below the active layer. The thickness of the active layer depends upon such factors as exposure, surface slope, amount of water, and, particularly, the presence or absence of a vegetation cover, which greatly affects the thermal conductivity of the soil. When vegetation is removed, the surface soil is no longer insulated from solar radiation and the permafrost may melt.

Special engineering problems are associated with the design, construction, and maintenance of roads, railroads, airfields, pipelines, and buildings in permafrost areas. The vast range of problems occurring in different earth materi-

(a)

(b)

▲ **FIGURE 2.23** (a) V-shaped valley formed by river erosion, Yellowstone River, Wyoming (*Betty Crowell/Faraway Places*); (b) U-shaped valley scoured by glacier ice, Sierra Nevada, California. (*Michael Collier*)

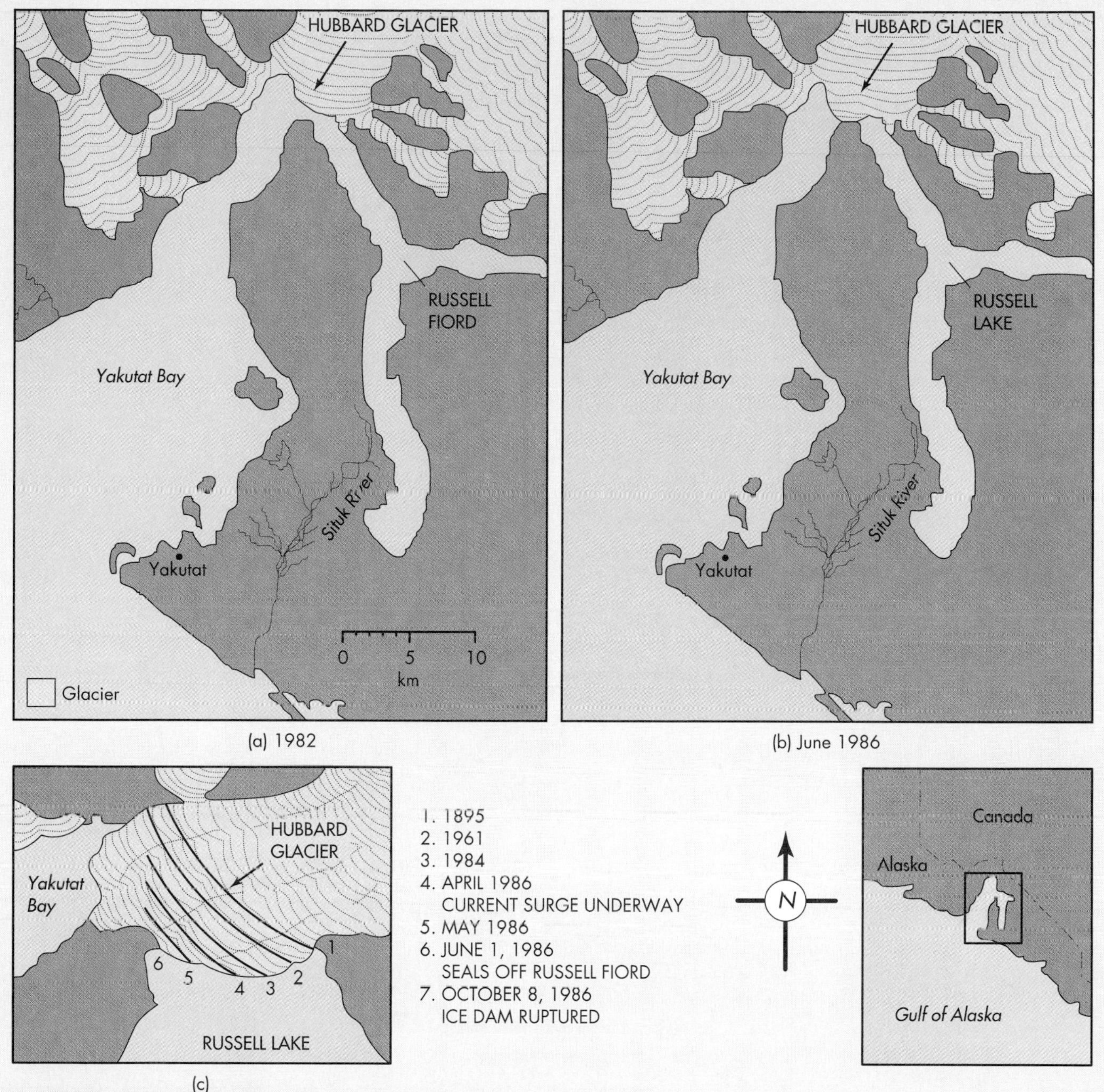

▲ **FIGURE 2.F** The 1986 surge of the Hubbard Glacier. Position of the glacier is shown for (a) 1982, (b) June 1986, and (c) various stages over the past century. (*Data from* U.S. *Forest Service*)

als in these areas is beyond the scope of our discussion. In general, the major concerns are associated with permafrost in silty (fine-grained), poorly drained materials. Because these materials hold a good deal of frozen water, they produce a lot of water on melting. Melting produces unstable materials, resulting in settling, subsidence, landslides, and lateral or downslope flowage of saturated sediment. Heaving (rising) and subsidence of the ground caused by freezing and thawing of the active layer are responsible for many of the engineering problems in the arctic and subarctic regions (10,14).

Presence of permafrost has led to very high maintenance costs and relocation or abandonment of highways, railroads, and other structures (Figure 2.25). Gravel roads may experience severe differential subsidence (sinking at the surface) caused by thawing of permafrost. On the North Slope of Alaska, tractor trails constructed by bulldozing off the vegetation cover over permafrost have caused problems. Small ponds on abandoned trails often form during the first summer following the trail construction, and they continue to grow deeper and wider as the permafrost continues to thaw (14).

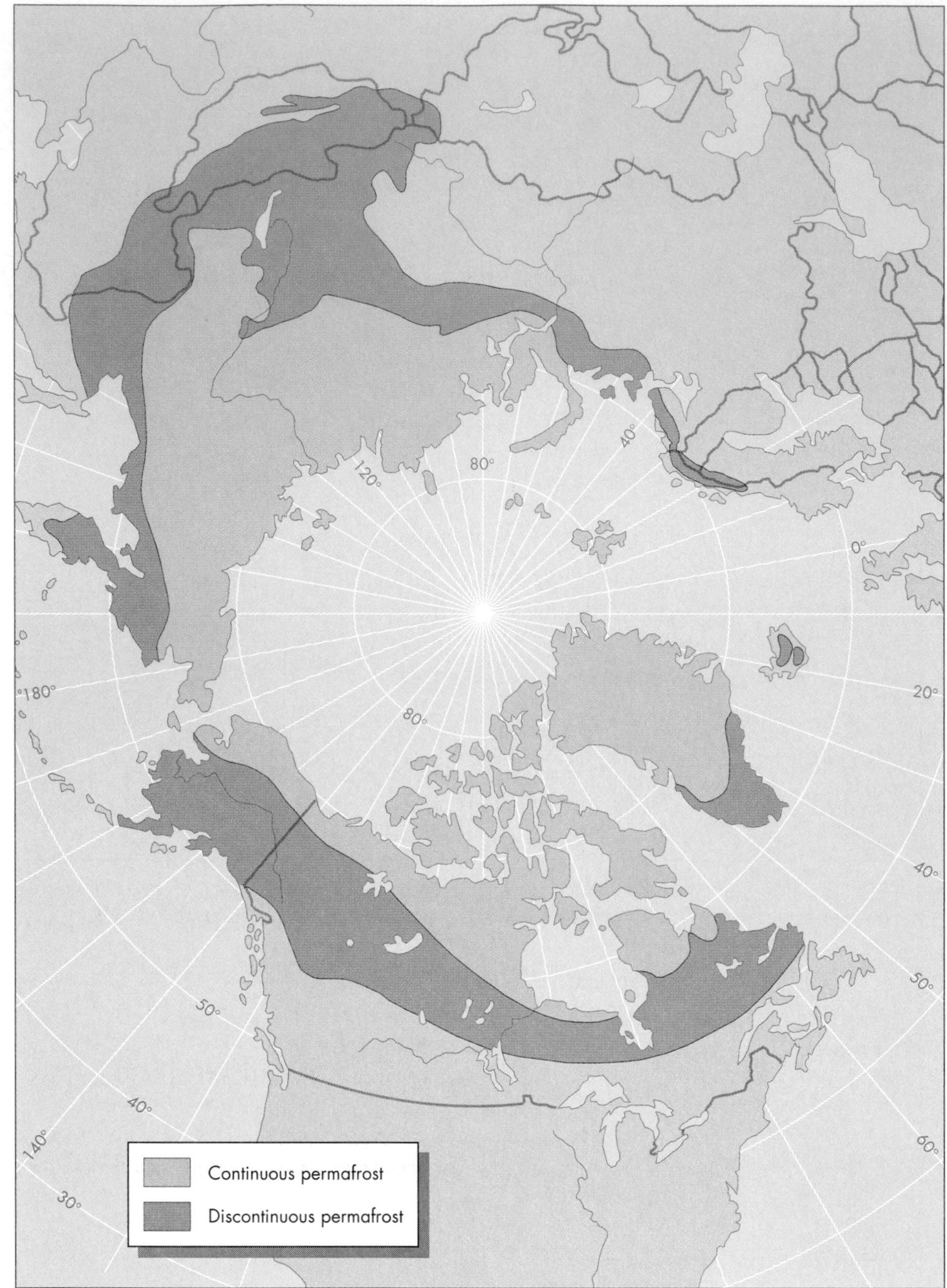

▶ **FIGURE 2.24** Extent of permafrost zones in the Northern Hemisphere. (*From* O. J. *Ferrians*, Jr., R. *Kachadoorian*, *and* G. W. *Greene*. 1969. Permafrost and Related Engineering Problems in Alaska. U.S. *Geological Survey Professional Paper* 678)

Experience has shown that two basic methods of construction can be used on permafrost. The **active method** is used where the permafrost is thin or discontinuous or contains relatively little ice; it works best in coarse-grained, well-drained soils since these are not particularly frost-susceptible. The method consists of thawing the permafrost and, if the thawed material is strong enough, using conventional construction methods. The **passive method** is used where it is impractical to thaw the permafrost. The basic principle is to keep the permafrost frozen and not upset the natural balance of environmental factors. This balance is so sensitive that even the passage of a tracked vehicle that destroys the vegetation cover will upset it, initiating melting that is nearly impossible to reverse. Special design of structures and foundations to minimize melting of the permafrost is the key to the passive method (14).

Wind

Windblown deposits are generally subdivided into two groups: sand deposits (mainly dunes) and **loess,** which is windblown silt. Extensive deposits of windblown sand and silt cover thousands of square kilometers in the United States (Figure 2.26).

Sand Dunes Sand dunes and related deposits are found along the coasts of the Atlantic and Pacific oceans and the Great Lakes. Inland sand is found in areas of Nebraska,

(a)

(b)

▲ **FIGURE 2.25** (a) Damage to a house resulting from subsidence due to thawing of permafrost as a result of poor construction practice in Alaska. (*Steve McCutcheon/Anchorage Museum of History & Art*) (b) Severe differential subsidence of a gravel road near Umiat, Alaska, caused by thawing of ice-wedge polygons in permafrost. (*O. J. Ferrians; courtesy of U.S. Geological Survey*)

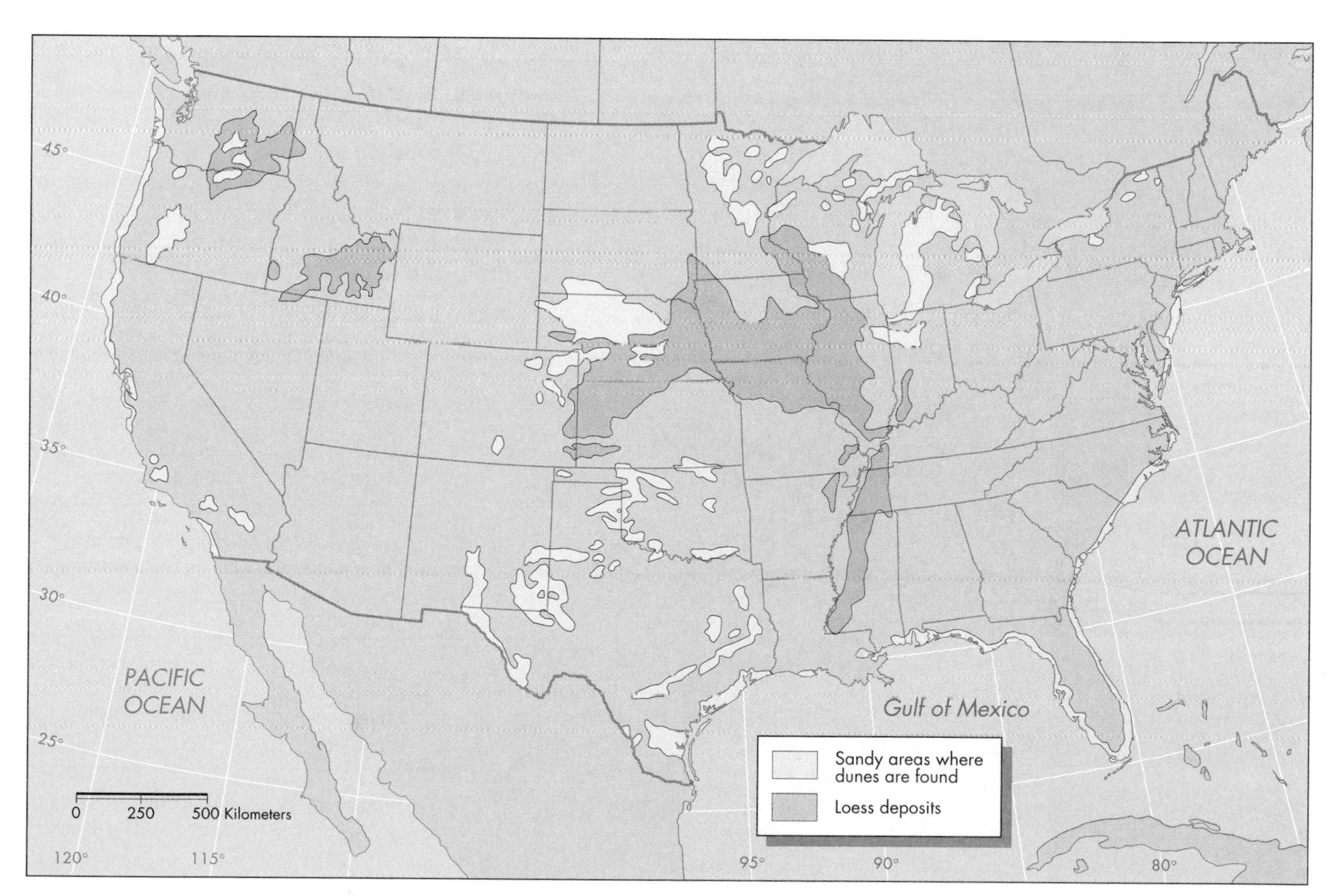

▲ **FIGURE 2.26** Distribution of windblown deposits within the conterminous United States. (*Reproduced, by permission, from Douglas S. Way.* 1973. Terrain Analysis. *Stroudsburg, PA: Dowden, Hutchinson & Ross*)

southern Oregon, southern California, Nevada, and northern Indiana and along large rivers flowing through semiarid regions—for example, the Columbia and Snake rivers in Oregon and Washington. **Sand dunes** are constructed under the influence of wind from sand moving close to the ground. Simple dunes have a gentle up-wind slope and a steep (~33°) down-wind slope called the *slip face*. When the wind blows, sand moves up the gentle slope and down the slip face, causing the form of dune to move in the direction of the wind. The process is known as *dune migration* (Figure 2.27). Dunes have a variety of sizes and shapes (summarized in Figure 2.28) and develop under a variety of conditions. Regardless of where they are located, how they form, or whether they are active or relic (formed in the past but are no longer active) or otherwise stabilized, they tend to cause environmental problems. Migrating sand is particularly troublesome, and stabilization of sand dunes is a major problem in construction and maintenance of highways and railroads that cross sandy areas of deserts (Figure 2.29). Building and maintaining reservoirs in sand-dune terrain is even more troublesome and often extremely expensive. These reservoirs should be constructed only if very high water loss can be tolerated. Canals in sandy areas should be lined to hold water and control erosion (12).

Loess In contrast to sand, which seldom moves more than a meter off the ground, windblown silt and dust can be carried in huge dust clouds thousands of meters in altitude (Figure 2.30). A typical dust storm 500 to 600 km in diameter may carry more than 100 million tons of silt and dust, sufficient to form a pile 30 m high and 3 km in diameter (15). Terrible dust storms in the 1930s probably exceeded even this, perhaps carrying more than 58,000 tons of dust per square kilometer.

Most of the loess in the United States is located in and adjacent to the Mississippi Valley, but some is found in the Pacific Northwest and Idaho. It is derived primarily from glacial outwash, that is, from material deposited by streams carrying glacial meltwater during the Ice Age. Retreat of the glaciers left behind large unvegetated areas adjacent to rivers. These areas were highly susceptible to wind erosion, so silt in the outwash was blown away and redeposited as loess. We know this because loess generally decreases rapidly in thickness with distance from major rivers, including the Mississippi, Missouri, Illinois, and Ohio. Loess deposits may be *primary loess*, which has been essentially unaltered since being deposited by the wind, or *secondary loess*, which has been reworked and transported a short distance by water or intensely weathered in place (9).

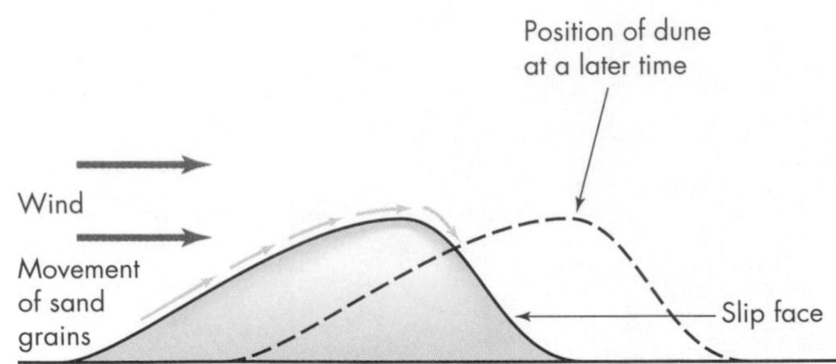

▲ **FIGURE 2.27** Movement of a sand dune. The wind moves the individual sand grains along the surface of the gentle windward side of the dune until they fall down the steeper leeward face. In this manner, the dune slowly migrates in the direction of the wind. (*After Robert J. Foster.* 1983. General Geology, *4th ed. Columbus*, OH: *Charles* E. *Merrill*)

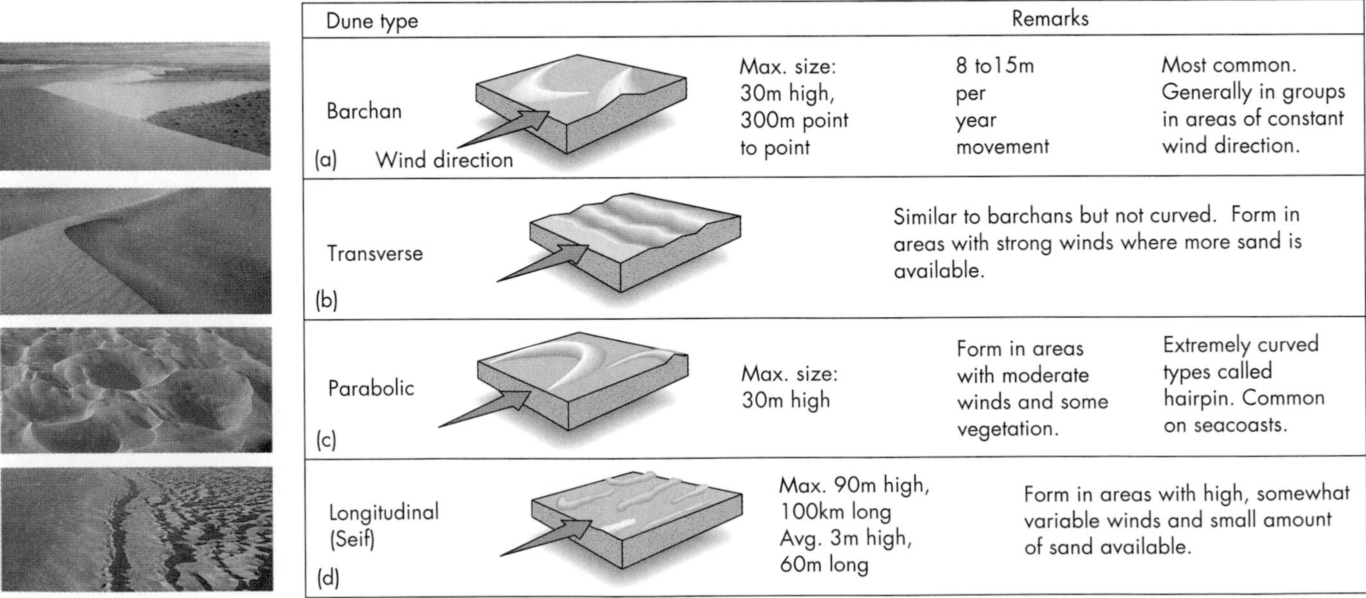

▲ **FIGURE 2.28** Types of sand dunes. (*After* R. J. *Foster.* 1983. General Geology, *4th ed. Columbus*, OH: *Charles* E. *Merrill*) (a) Barchan Dunes, White Sands, New Mexico (*Michael Collier*); (b) Transverse sand dune, Mesquite Flat Sand Dunes, Death Valley, California (*Gary Yeowell/Tony Stone Images*); (c) Parabolic sand dunes in Algodones dune belt southeast of Salton Sea, California (*John* S. *Shelton*); (d) Longitudinal (seif) sand dune along the western edge of the Algodones dune belt, California (*John* S. *Shelton*)

◀ **FIGURE 2.29** People digging out of a sand dune in the Sahara Desert, Africa, after a windstorm. (*Steve McCurry/Magnum Photos, Inc.*)

(a)

(b)

◀ **FIGURE 2.30** (a) Dust storm caused by a cold front at Manteer, Kansas, in 1935. (U.S. *Department of Commerce*) (b) Farmer and sons walking into the face of a dust storm, Oklahoma, 1936. (*Arthur Rothstein/AP/Wide World Photos*)

Although loess is strong enough to form nearly vertical slopes (Figure 2.31), it rapidly consolidates when subjected to a load (such as a building) and wetted. This process, called **hydroconsolidation,** occurs when the clay films or calcium carbonate cement around the silt grains wash away. Loess may therefore be a dangerous foundation material. Settling and cracking of a house reportedly took place overnight when water from a hose was accidentally left on. On the other hand, when loess is properly compacted and remolded, it acquires considerable strength and resistance to erosion, making it a more reliable material to support buildings (9).

▶ **FIGURE 2.31** At this location near the Missouri River, Missouri, brown-colored loess (wind-blown silt) overlies white sedimentary rocks. (*John Buitenkant/Photo Researchers, Inc.*)

SUMMARY

Processes that create, maintain, change, and destroy earth materials (water, sediment, minerals, and rocks) are collectively referred to as the *geologic cycle*. More correctly, the geologic cycle is a group of subcycles, including the tectonic, hydrologic, rock, and biogeochemical cycles. In various activities, people use earth materials found in different parts of these cycles. The materials may originally be found in a concentrated state; but after various human processes (mining, processing, etc.), they are dispersed, contaminated, or polluted in another part of the geologic cycle. Once dispersed or otherwise altered, these materials cannot easily be concentrated or made available for human use in a reasonable time.

Application of geological information to environmental problems requires an understanding of both geological and engineering properties of earth materials. Minerals and rocks behave predictably, but are variable and therefore suited to different land uses. Thus, compaction shales are generally poor foundation materials for large engineering structures, whereas granite with few fractures is satisfactory for most purposes. In addition, the strength of rocks is greatly influenced by active fault systems and by old fractures that have been weathered and altered to unstable minerals.

Geologically recent continental glaciation has produced a variety of different types of earth materials such as glacial till, lake deposits, and water-transported sediment, all of which may be located in a given area. As a result, careful evaluation of surface and subsurface deposits is necessary before planning, designing, and building in recently glaciated areas.

Permafrost underlies about 20 percent of the world's land area, producing a fragile and sensitive environment. Special engineering procedures are necessary to minimize adverse effects from artificial thawing of fine-grained, poorly drained, frozen ground.

Windblown sand, silt, and dust cover many thousands of square kilometers. Major deposits of sand are concentrated along coastal and interior areas, and deposits of windblown silt, or loess, are concentrated along major rivers that carried the meltwater from continental glaciers during the Pleistocene glaciation.

Finally, engineering problems caused by migrating and stabilized sand dunes involve expensive construction and maintenance costs for highways, buildings, and hydraulic structures. Loess deposits, unless properly compacted, make hazardous foundations for engineering purposes because loess may hydroconsolidate and settle when it becomes wet.

REFERENCES

1. **Longwell, C. L., Flint, R. F., and Sanders, J. E.** 1969. *Physical geology*. New York: John Wiley.
2. **Le Pichon, X.** 1968. Sea-floor spreading and continental drift. *Journal of Geophysical Research* 73:3661–97.
3. **Isacks, B., Oliver, J., and Sykes, L.** 1968. Seismology and the new global tectonics. *Journal of Geophysical Research* 73:5855–99.
4. **Dewey, J. F.** 1972. Plate tectonics. *Scientific American* 22:56–68.
5. **Committee on Geological Sciences.** 1972. *The earth and human affairs*. San Francisco: Canfield Press.
6. **Lambert, G.** 1987. La gaz carbonique dans l'atmosphere. *La Recherche* 18, No. 89, pp. 778–787.
7. **Rahn, P. H.** 1986. *Engineering geology*. New York: Elsevier.

8. **Billings, M. P.** 1954. *Structural geology*, 2nd ed. Englewood Cliffs, NJ: Prentice-Hall.
9. **Krynine, D. P., and Judd, W. R.** 1957. *Principles of engineering geology and geotechnics*. New York: McGraw-Hill.
10. **Schultz, J. R., and Cleaves, A. B.** 1955. *Geology in engineering*. New York: John Wiley.
11. **Jessup, W. E.** 1964. Baldwin Hills Dam failure. *Civil Engineering* 34:62–64.
12. **Costa, J. E., and Baker, V. R.** 1981. *Surficial geology*. New York: John Wiley.
13. **Rogers, J. D.** 1992. Reassessment of the St. Francis Dam failure. In *Engineering Geology Practice in Southern California*, eds. R. Proctor and B. Pipkin, pp. 639–66. Association of Engineering Geologists, Special Publication No. 4.
14. **Ferrians, O. J., Jr., Kachadoorian, R., and Greene, G. W.** 1969. *Permafrost and related engineering problems in Alaska*. U.S. Geological Survey Professional Paper 678.
15. **Way, D. S.** 1973. *Terrain analysis*. Stroudsburg, PA: Dowden, Hutchinson & Ross.

KEY TERMS

geologic cycle (p. 24)
plate tectonics (p. 26)
tectonic cycle (p. 26)
mineral (p. 30)
rock cycle (p. 30)
igneous rocks (p. 30)
sedimentary rocks (p. 30)
metamorphic rocks (p. 30)
hydrologic cycle (p. 31)
biogeochemical cycle (p. 31)
global carbon cycle (p. 31)
glacier (p. 45)
permafrost (p. 48)
loess (p. 50)
sand dunes (p. 52)

SOME QUESTIONS TO THINK ABOUT

1. Assume that the supercontinent Pangaea (Figure 2.8) never broke up. Now deduce how earth processes, landforms, and environments might be different from how they are today with the continents spread all over the globe. *Hint*: Think about what the breakup of the continents did in terms of building mountain ranges and producing ocean basins that affect climate and so forth.
2. The carbon cycle (Figure 2.12) is an important biogeochemical cycle. If the burning of fossil fuels were to emit twice as much carbon dioxide into the atmosphere as it does now, what would be the effect on earth processes?
3. You are working as part of a resource-management team for a newly created park in the Mojave Desert of California. The main road through an area of particular public concern has been troubled by migrating sand dunes that, under natural conditions, would eventually cover the road. What steps could you take to keep the road open while protecting the environment?
4. Consider the case history of the Hubbard Glacier that surged in 1986, producing Russell Lake. Assuming that the water was rising in Russell Lake at such a rate that it was certain to spill over into the Situk River, what would be the best course of action? That is, should people take steps to keep the spillover from happening or should nature be allowed to take its course? Discuss the pros and cons of both of these approaches.

3 Soils and Environment

Soil erosion on the Piedmont of the southeastern United States. (*Edward A. Keller*)

Soil erosion is a serious environmental problem on a global scale. Shown here are the red soils of the Piedmont of the southeastern United States at Charlotte, North Carolina, where a series of gullies has developed. The erosion has multiple effects including loss of soil resources as well as degradation of the quality of the water that the sediment enters. Such soil erosion is particularly a problem in urban environments, where vegetation may be removed prior to development. Although we have many safeguards in effect to minimize soil erosion resulting from urbanization, the problem persists in many parts of the United States and is severe in many parts of the world, where protection of soil resources may not be a high priority.

3.1 Introduction to Soils

Soil may be defined in several ways. To soil scientists, soil is solid earth material that has been altered by physical, chemical, and organic processes such that it can support rooted plant life. To engineers, on the other hand, soil is any solid earth material that can be removed without blasting. Both of these definitions are important in environmental geology. Geologists must be aware not only of the different definitions, but also of the different points of view of researchers in various fields concerning soil-producing processes and the role of soils in environmental problems.

Consideration of soils, particularly with reference to land-use limitations, is becoming an important aspect of environmental work:

- In the field of *land-use planning*, land capability (suitability of land for a particular use) is often determined in part by the soils present, especially for such uses as urbanization, timber management, and agriculture.
- Soils are critical when we consider *waste management* because interactions between waste, water, soil, and rock often determine the suitability of a particular site to receive waste.

LEARNING OBJECTIVES

Soils comprise an important part of our environment. Virtually all aspects of the terrestrial environment interact at one level or another with soils. With this in mind, primary learning objectives for this chapter are:

- *To become familiar with soils terminology and the processes responsible for the development of soils.*
- *To understand what soil fertility is, as well as the interactions of water in soil processes.*
- *To gain a modest acquaintance with how we classify soils, particularly for engineering purposes.*
- *To understand the more important engineering properties of soil.*
- *To understand relationships between land use and soils.*
- *To be aware of what sediment pollution is and how it may be minimized.*
- *To have a modest acquaintance with the process of desertification and what drives it.*
- *To be aware of how soils surveys are useful in land-use planning.*

Web Resources

Visit the "Environmental Geology" Web site at www.prenhall.com/keller to find additional resources for this chapter, including:

- ▶ Web Destinations
- ▶ On-line Quizzes
- ▶ On-line "Web Essay" Questions
- ▶ Search Engines
- ▶ Regional Updates

▶ Study of soils helps land-use planners evaluate *natural hazards,* including floods, landslides, and earthquakes. In the case of floods, because floodplain soils differ from upland soils, consideration of soil properties helps delineate natural floodplains. Evaluation of the relative ages of soils on landslide deposits may provide an estimate of the frequency of slides and thus assist in planning to minimize their impact. The study of soils has also been a powerful tool in establishing the chronology of earth materials deformed by faulting, which has led to better calculations of the recurrence intervals of earthquakes at particular sites.

3.2 Soil Profiles

The development of a soil from inorganic and organic materials is a complex process. Intimate interactions of the rock and hydrologic cycles produce the weathered rock materials that are basic ingredients of soils. **Weathering** is the physical and chemical breakdown of rocks and the first step in soil development. Weathered rock is further modified by the activity of soil organisms into soil, which is called either *residual* or *transported,* depending on where and when it has been modified. The more insoluble weathered material may remain essentially in place and be modified to form a residual soil, such as the red soils of the Piedmont in the southeastern United States. If weathered material is transported by water, wind, or glaciers and then modified in its new location, it forms a transported soil, such as the fertile soils formed from glacial deposits in the American Midwest.

A soil can be considered an open system that interacts with other components of the geologic cycle. The characteristics of a particular soil are a function of *climate, topography, parent material* (the rock or alluvium from which the soil is formed), *time* (age of the soil), and *organic processes* (activity of soil organisms). Many of the differences we see in soils are effects of climate and topography, but the type of parent rock, the organic processes, and the length of time the soil-forming processes have operated are also important.

Soil Horizons

Vertical and horizontal movements of the materials in a soil system create a distinct layering, parallel to the surface, collectively called a *soil profile.* The layers are called zones or **soil horizons.** Our discussion of soil profiles will mention only the horizons most commonly present in soils. Additional information is available from detailed soils texts (1,2).

Figure 3.1a shows the common master (prominent) soil horizons. The ***O* horizon** and ***A* horizon** contain highly concentrated organic material; the differences between these two layers reflect the amount of organic material present in each. Generally, the *O* horizon consists entirely of plant litter and other organic material, while the underlying A horizon contains a good deal of both organic and mineral material. Below the *O* or *A* horizon, some soils have an ***E* horizon,** or *zone of leaching,* a light-colored layer that is leached of iron-bearing components. This horizon is light in color because it contains less organic material than the *O* and *A* horizons and little inorganic coloring material such as iron oxides.

The ***B* horizon,** or *zone of accumulation,* underlies the *O, A,* or *E* horizon and consists of a variety of materials translocated downward from overlying horizons. Several types of *B* horizon have been recognized. Probably the most

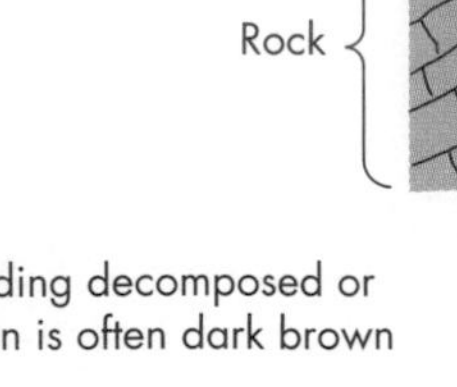

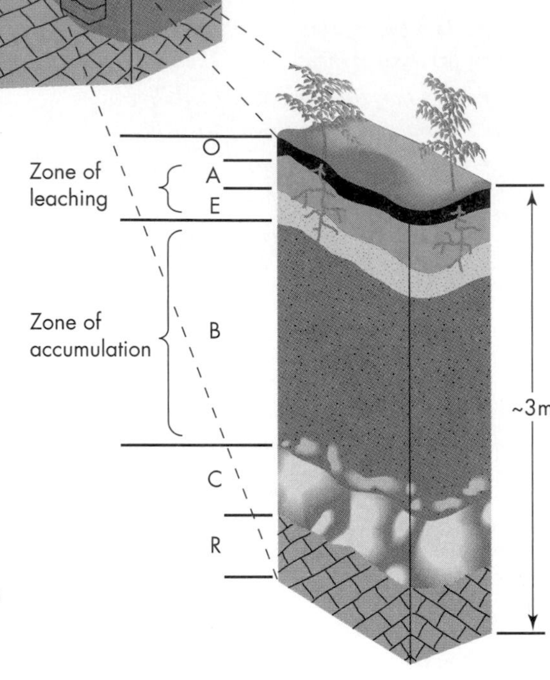

O. Horizon is composed mostly of organic materials, including decomposed or decomposing leaves, twigs, etc. The color of the horizon is often dark brown or black.

A. Horizon is composed of both mineral and organic materials. The color is often light black to brown. Leaching, defined as the process of dissolving, washing, or draining earth materials by percolation of groundwater or other liquids, occurs in the A horizon and moves clay and other material such as iron and calcium to the B horizon.

E. Horizon is composed of light-colored materials resulting from leaching of clay, calcium, magnesium, and iron to lower horizons. The A and E horizons together comprise the zone of leaching.

B. Horizon is enriched in clay, iron oxides, silica, carbonate or other material leached from overlying horizons. This horizon is known as the zone of accumulation.

C. Horizon is composed of partially altered (weathered) parent material; rock as shown here but the material could also be alluvial in nature, such as river gravels in other environments. The horizon may be stained red with iron oxides.

R. Unweathered (unaltered) parent material.

(a)

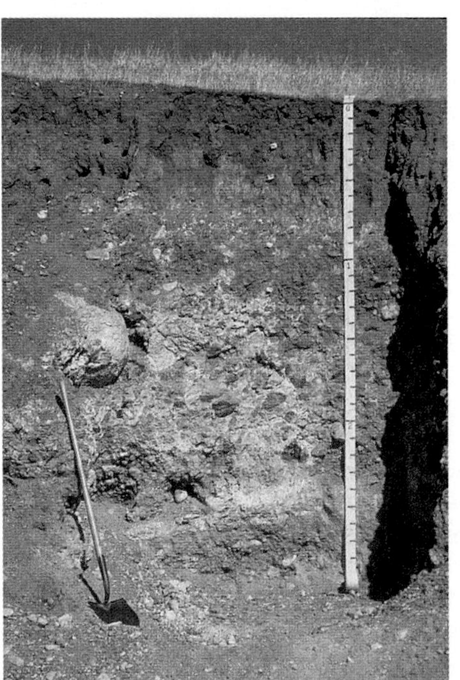

(b)

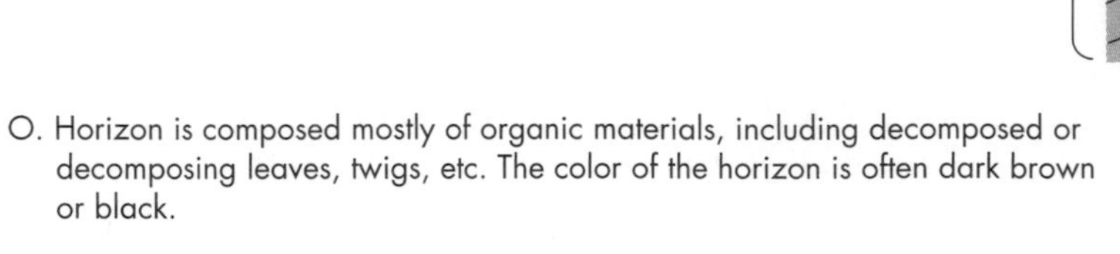

◀ **FIGURE 3.1** Soil profiles. (a) Idealized diagram showing a soil profile with soil horizons. (b) Soil profile showing a black *A* horizon, a light-red *B* horizon, a white *K* horizon rich in calcium carbonate, and a lighter *C* horizon. (*Edward* A. *Keller*)

important type is the *argillic B*, or B_t horizon. A B_t horizon is enriched in clay minerals that have been translocated downward by soil-forming processes. Another type of *B* horizon of interest to environmental geologists is the B_k horizon, characterized by accumulation of calcium carbonate. The carbonate coats individual soil particles in the soils and may fill some pore spaces (the spaces between soil particles), but it does not dominate the morphology (structure) of the horizon. A soil horizon that is so impregnated with calcium carbonate that its morphology is dominated by the carbonate is designated a ***K* horizon** (Figure 3.1b). Carbonate completely fills the pore spaces in *K* horizons, and the carbonate often forms in layers parallel to the surface. The term *caliche* is often used for irregular accumulation or layers of calcium carbonate in soils.

The ***C* horizon** lies directly over the unaltered parent material and consists of parent material partially altered by weathering processes. The ***R* horizon,** or unaltered parent material, is the consolidated bedrock that underlies the soil. However, some of the fractures and other pore spaces in the bedrock may contain clay that has been translocated downward (1).

The term *hardpan* is often used in the literature on soils. A hardpan soil horizon is defined as a hard (compacted) soil horizon. Hardpan is often composed of compacted and/or cemented clay with calcium carbonate, iron oxide, or silica. Hardpan horizons are nearly impermeable and thus restrict the downward movement of soil water.

Soil Color

One of the first things we notice about a soil is its color, or the colors of its horizons. The *O* and *A* horizons tend to be dark because of their abundant organic material. The *E* horizon, if present, may be almost white, owing to the leaching of iron and aluminum oxides. The *B* horizon shows the most dramatic differences in color, varying from yellow-brown to light red-brown to dark red, depending upon the presence of clay minerals and iron oxides. The B_k horizons may be light-colored due to their carbonates, but they are sometimes reddish as a result of iron oxide accumulation. If a true *K* horizon has developed, it may be almost white because of its great abundance of calcium carbonate. Although soil color can be an important diagnostic tool for analyzing a soil profile, one must be cautious about calling a red layer a *B* horizon. The original parent material, if rich in iron, may produce a very red soil even when there has been relatively little **soil profile development.**

Soil color may be an important indicator of how well drained a soil is. Well-drained soils are well aerated (oxidizing conditions), and iron oxidizes to a red color. Poorly drained soils are wet, and iron is reduced rather than oxidized. The color of such a soil is often yellow. This distinction is important because poorly drained soils are associated with environmental problems such as lower slope stability and inability to be utilized as a disposal medium for household sewage systems (septic tank and leach field).

Soil Texture

The texture of a soil depends upon the relative proportions of sand-, silt-, and clay-sized particles (Figure 3.2). *Clay particles* have a diameter of less than 0.004 mm, *silt particles* have diameters ranging from 0.004 to 0.074 mm, and *sand particles* are 0.074 to 2.0 mm in diameter. Earth materials with particles larger than 2.0 mm in diameter are called *gravel*, *cobbles*, or *boulders* depending on the particle size. Note that the sizes of particles given here are for engineering classification and are slightly different from those used by the U.S. Department of Agriculture for soil classification.

Soil texture is commonly identified in the field by estimation, then refined in the laboratory by separating the sand, silt, and clay and determining their proportions. A useful field technique for estimating the size of sand-sized or smaller soil particles is as follows: It is sand if you can see individual grains; silt if you can see the grains with a 10× hand lens; and clay if you cannot see grains with such a hand lens. Another method is to feel the soil: Sand is gritty (crunches between the teeth), silt feels like baking flour, and clay is cohesive. When mixed with water, smeared on the back of the hand, and allowed to dry, clay cannot be dusted off easily, whereas silt or sand can.

Soil Structure

Soil particles often cling together in aggregates, called *peds*, that are classified according to shape into several types. Figure 3.3 shows some of the common structures of peds found

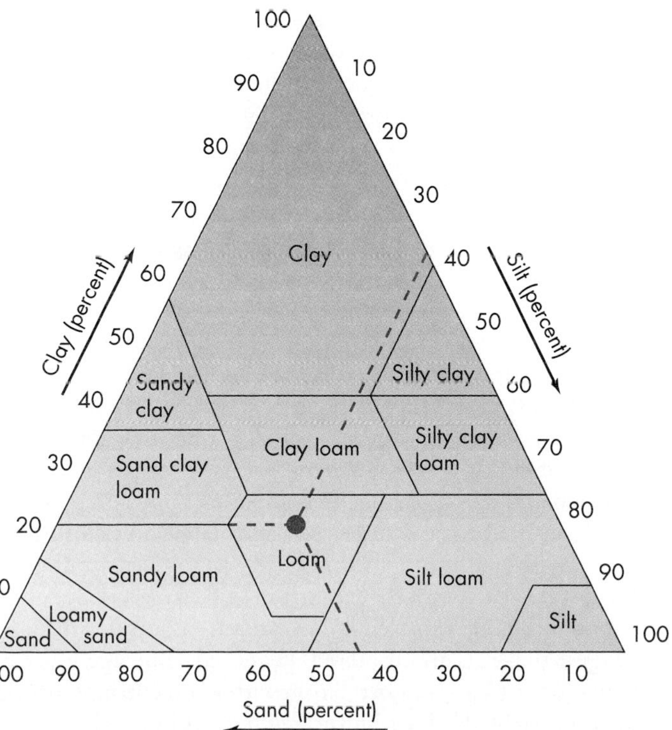

◀ **FIGURE 3.2** Soil textural classes. The classes are defined according to the percentage of clay-, silt-, and sand-sized particles in the soil sample. The point connected by dashed lines represents a soil composed of 40% sand, 40% silt, and 20% clay, which is classified as loam. (U.S. *Department of Agriculture standard textural triangle*)

▶ **FIGURE 3.3** Chart of different soil structures (peds).

Type	Typical size range	Horizon usually found in	Comments
Granular	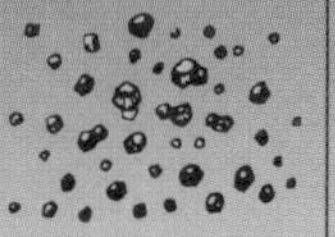 1–10 mm	A	Can also be found in B and C horizons
Blocky	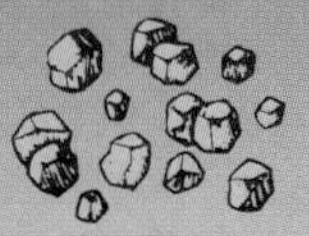 5–50 mm	B_t	Are usually designated as angular or subangular
Prismatic	 10–100 mm	B_t	If columns have rounded tops, structure is called *columnar*
Platy	 1–10 mm	E	May also occur in some B horizons

in soils. The type of structure present is related to soil-forming processes, but some of these processes are poorly understood (1). For example, *granular structure* is fairly common in *A* horizons, whereas *blocky* and *prismatic structures* are most likely to be found in *B* horizons. Soil structure is an important diagnostic tool in helping to evaluate the development and approximate age of soil profiles. In general, as the profile develops with time, structure becomes more complex and may go from granular to blocky to prismatic as the clay content in the *B* horizons increases.

Relative Profile Development

Most environmental geologists will not have occasion to make detailed soil descriptions and analyses of soil data. However, it is important for geologists to recognize differences among weakly developed, moderately developed, and well-developed soils, that is, to recognize their **relative profile development.** These distinctions are useful in preliminary evaluation of soil properties and help determine whether the opinion of a soil scientist is necessary in a particular project:

- *A weakly developed soil profile* is generally characterized by an *A* horizon directly over a *C* horizon (there is no *B* horizon or it is very weakly developed). The *C* horizon may be oxidized. Such soils tend to be only a few hundred years old in most areas, but may be several thousand years old.
- *A moderately developed soil profile* may consist of an *A* horizon overlying an argillic B_t horizon that overlies the *C* horizon. A carbonate B_k horizon may also be present but is not necessary for a soil to be considered moderately developed. These soils have a *B* horizon with translocated changes, a better-developed texture, and redder colors than those that are weakly developed. Moderately developed soils often date from at least the Pleistocene (more than 10,000 years old).
- *A well-developed soil profile* is characterized by redder colors in the B_t horizon, more translocation of clay to the B_t horizon, and stronger structure. A *K* horizon may also be present but is not necessary for a soil to be considered strongly developed. Well-developed soils vary widely in age, with typical ranges between 40,000 and several hundred thousand years and older.

Soil Chronosequences

A **soil chronosequence** is a series of soils arranged from youngest to oldest on the basis of their relative profile development. Such a sequence is valuable in hazards work, because it provides information about the recent history of a landscape, allowing us to evaluate site stability when locating such critical facilities as a waste disposal operation or a large power plant. A chronosequence combined with numerical dating (applying a variety of dating techniques, such as radiocarbon, ^{14}C, to obtain a date in years before the present time of the soil) may provide the data necessary to make such inferential statements as, "There is no evidence of ground rupture due to earthquakes in the last 1000 years," or "The last mudflow was at least 30,000 years ago." It takes a lot of work to establish a chronosequence in soils in a particular area. However, once such a chronosequence is developed and dated, it may be applied to a specific problem.

Consider, for example, the landscape shown in Figure 3.4, an offset alluvial fan along the San Andreas fault in the Indio Hills of southern California. The fan is offset about 0.7 km (700,000 mm). Soil pits excavated in the alluvial fan suggest that it is about 20,000 years old. The age was estimated on the basis of correlation with a soil chronosequence in the nearby Mojave Desert, where numerical dates for

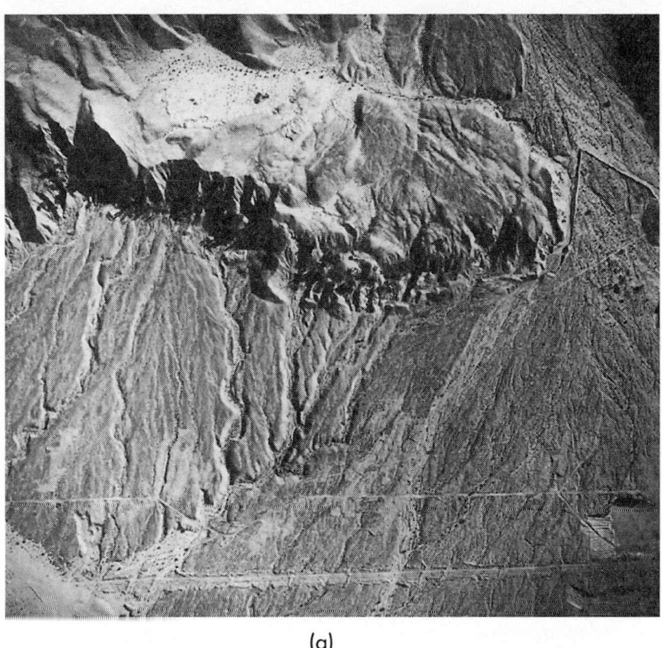

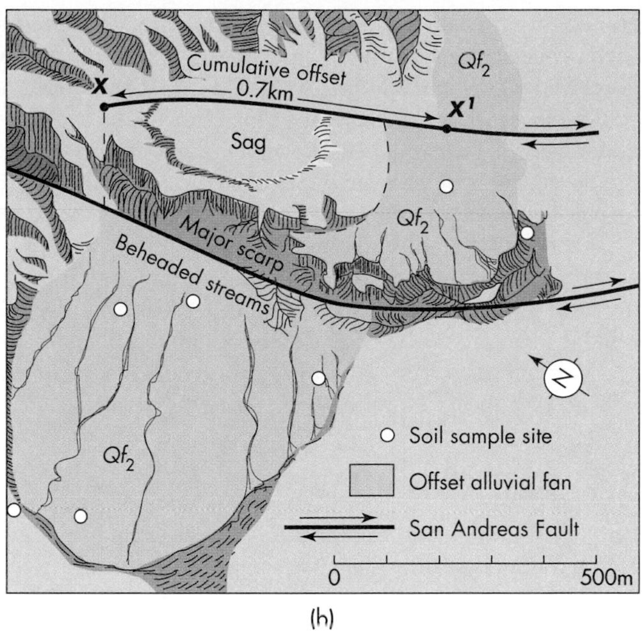

▲ **FIGURE 3.4** Offset alluvial fan along the San Andreas fault near Indio, California. (a) Aerial photograph. (*Woodward-Clyde Consultants*) (b) Sketch map.

similar soils are available. Soil development on the offset alluvial fan allowed the age of the fan to be estimated. This allowed the slip rate (the amount of offset of the fan divided by the age of the fan—that is, 700,000 mm ÷ 20,000 years, which is about 35 mm/yr) for this part of the San Andreas fault to be estimated at 35 mm annually (3). The slip rate for this segment of the fault was not previously known. The rate is significant because it is a necessary ingredient in the eventual estimation of the probability and recurrence interval of large, damaging earthquakes.

3.3 Soil Fertility

A soil may be considered a complex ecosystem. A single cubic meter of soil may contain millions of living things, including small rodents, insects, worms, algae, fungi, and bacteria. These organisms are important in the mixing and aeration of soil particles and in releasing or converting nutrients in soils into forms that are useful for plants (4). **Soil fertility** refers to the capacity of the soils to supply the nutrients (such as nitrogen, phosphorus, and potassium) needed for plant growth when other factors are favorable (5).

Soils developed on some floodplains and glacial deposits contain sufficient nutrients and organic material to be naturally fertile. Other soils, developed on highly leached bedrock or on loose deposits with little organic material, may be nutrient-poor with low fertility. Soils are often manipulated to increase plant yield by applying fertilizers to supply nutrients or materials that improve the soil's texture and moisture retention. Soil fertility can be reduced by soil erosion or leaching that removes nutrients, by interruption of natural processes (such as flooding) that supply nutrients, or by continued use of pesticides that alter or damage soil organisms.

3.4 Water in Soil

If you analyze a block of soil, you will find it is composed of bits of solid mineral and organic matter with pore spaces between them. The pore spaces are filled with gases (mostly air) or liquids (mostly water). If all the pore spaces in a block of soil are completely filled with water, the soil is said to be in a *saturated condition;* otherwise it is said to be *unsaturated.* Soils in swampy areas may be saturated year-round, whereas soils in arid regions may be saturated only occasionally.

The amount of water in a soil, called its *water content* or its *moisture content,* can be very important in determining such engineering properties as the strength of a soil and its potential to shrink and swell. If you have ever built a sand castle at the beach, you know that dry sand is impossible to work with, but that moist sand will stand vertically, producing walls for your castle. Differences between wet and dry soils are also very apparent to anyone who lives in or has visited areas with dirt roads that cross clay-rich soils. When the soil is dry, driving conditions are excellent, but following rainstorms, the same roads become mud pits and nearly impassable (6).

Water in soils may flow laterally or vertically through soil pores, which are the void spaces between grains, or in fractures produced as a result of soil structure. The flow is termed *saturated flow* if all the pores are filled with water and *unsaturated flow* when, as is more common, only part of the pores is filled with water.

In unsaturated flow, movement of water is related to such processes as thinning or thickening of films of water in pores and on the surrounding soil grains (6). The water molecules closest to the surface of a particle are held the tightest, and as the films thicken, the water content increases and the outer layers of water may begin to move. Flow is therefore fastest in the center of pores and slower near the edges.

The study of soil moisture relations and movement of water and other liquids in soils, along with how to monitor movement of liquids, is an important research topic. It is related to many water pollution problems, such as movement of gasoline from leaking underground tanks or migration of liquid pollutants from waste disposal sites.

3.5 Soil Classification

Both terminology and classification of soils are a problem in environmental studies because we are often interested in both soil processes and the human use of soil. A taxonomy (classification system) that includes engineering as well as physical and chemical properties would be most appropriate—but none exists. We must therefore be familiar with two separate systems of soil classification: *soil taxonomy*, used by soil scientists, and the *engineering classification*, which groups soils by material types and engineering properties.

Soil Taxonomy

Soil scientists have developed a comprehensive and systematic classification of soils known as **soil taxonomy,** which emphasizes the physical and chemical properties of the soil profile. This classification is a sixfold hierarchy, with soils grouped into Orders, Suborders, Great Groups, Subgroups, Families, and Series. The eleven Orders (Table 3.1) are mostly based on gross soil morphology (number and types of horizons present), nutrient status, organic content (plant debris, etc.), color (red, yellow, brown, white, etc.), and general climatic considerations (amount of precipitation, average temperature, etc.). With each step down the hierarchy, more information about a specific soil becomes known.

Soil taxonomy is especially useful for agricultural and related land-use purposes. It has been criticized for being too complex and for lacking sufficient textural and engineering information to be of optimal use in site evaluation for engineering purposes. Nevertheless, the serious earth

Table 3.1 General properties of soil order used with soil taxonomy by soil scientists

Order	General Properties
Entisols	No horizon development; many are recent alluvium; synthetic soils are included; are often young soils.
Vertisols	Include swelling clays (greater than 35 percent) that expand and contract with changing moisture content. Generally form in regions with a pronounced wet and dry season.
Inceptisols	One or more of horizons have developed quickly; horizons are often difficult to differentiate; most often found in young but not recent land surfaces, have appreciable accumulation of organic material; most common in humid climates but range from the Arctic to the tropics; native vegetation is most often forest.
Aridisols	Desert soils; soils of dry places; low organic accumulation; have subsoil horizon where gypsum, caliche (calcium carbonate), salt, or other materials may accumulate.
Mollisols	Soils characterized by black, organic-rich *A* horizon (prairie soils); surface horizons are also rich in bases. Commonly found in semiarid or subhumid regions.
Andisols	Soils derived primarily from volcanic materials; relatively rich in chemically active minerals that rapidly take-up important biologic elements such as carbon and phosphorus.
Spodosols	Soils characterized by ash-colored sands over subsoil, accumulations of amorphous iron-aluminum sesquioxides and humus. They are acid soils that commonly form in sandy parent materials. Are found principally under forests in humid regions.
Alfisols	Soils characterized by a brown or gray-brown surface horizon and an argillic (clay-rich) subsoil accumulation with an intermediate to high base saturation (greater than 35 percent as measured by the sum of cations, such as calcium, sodium, magnesium, etc.). Commonly form under forests in humid regions of the midlatitudes.
Ulfisols	Soils characterized by an argillic horizon with low base saturation (less than 35 percent as measured by the sum of cations); often have a red-yellow or reddish-brown color; restricted to humid climates and generally form on older landforms or younger, highly weathered parent materials.
Oxisols	Relatively featureless, often deep soils, leached of bases, hydrated, containing oxides of iron and aluminum (laterite) as well as kaolinite clay. Primarily restricted to tropical and subtropical regions.
Histosols	Organic soils (peat, muck, bog).

Source: After Soil Survey Staff. 1994. *Keys to Soil Taxonomy*, 6th ed. Soil Conservation Service, U.S. Department of Agriculture.

Table 3.2 Unified soil classification system used by engineers

Major Divisions				Group Symbols	Soil Group Names
COARSE-GRAINED SOILS (Over half of material larger than 0.074 mm)	Gravels	Clean Gravels	Less than 5% fines	GW	Well-graded gravel
				GP	Poorly graded gravel
		Dirty Gravels	More than 12% fines	GM	Silty gravel
				GC	Clayey gravel
	Sands	Clean Sands	Less than 5% fines	SW	Well-graded sand
				SP	Poorly graded sand
		Dirty Sands	More than 12% fines	SM	Silty sand
				SC	Clayey sand
FINE-GRAINED SOILS (Over half of material smaller than 0.074 mm)	Silts	Non-plastic		ML	Silt
				MH	Micaceous silt
				OL	Organic silt
	Clays	Plastic		CL	Silty clay
				CH	High plastic clay
				OH	Organic clay
Predominantly Organics				PT	Peat and Muck

scientist must have knowledge of this classification because it is commonly used by soil scientists and Quaternary geologists, those who study earth materials and processes of recent (last 1.65 million years) earth history.

Engineering Classification of Soils

The **unified soil classification system,** widely used in engineering practice, is shown in Table 3.2. Because all natural soils are mixtures of coarse particles (gravel and sand), fine particles (silt and clay), and organic material, the major divisions of this system are *coarse-grained soils, fine-grained soils,* and *organic soils*. Each group is based on the predominant particle size or the abundance of organic material. Coarse soils are those in which more than 50 percent of the particles (by weight) are larger than 0.074 mm in diameter. Fine soils are those with less than 50 percent of the particles greater than 0.074 mm (7). Organic soils have a high organic content and are identified by their black or gray color and sometimes by an odor of hydrogen sulfide, which smells like rotten egg.

3.6 Engineering Properties of Soils

All soils above the **water table** (the surface below which all the pore space in rocks is saturated) have three distinct parts, or phases: *solid material, liquid,* and *gas* (as, for example, air or carbon dioxide). The usefulness of a soil is greatly affected by the variations in the proportions and structure of the three phases. The types of solid materials, the particle sizes, and the water content are some of probably the most significant variables that determine engineering properties (see *Putting Some Numbers On Properties of Soils*). For planners, the most important engineering properties of soils are plasticity, strength, sensitivity, compressibility, erodibility, permeability, corrosion potential, ease of excavation, and shrink-swell potential.

Plasticity, which is related to the water content of a soil, is used to help classify fine-grained soils for engineering purposes. The *liquid limit* (LL) of a soil is the water content (w) above which the soil behaves as a liquid; the *plastic limit* (PL) is the water content below which the soil no longer behaves as a plastic material. The numerical difference between the liquid and plastic limits is the *plasticity index* (PI), the range in moisture content within which a soil behaves as a plastic material. Soils with a very low plasticity index (5%) may cause problems because only a small change in water content can change the soil from a solid to a liquid state. On the other hand, a large PI, one greater than 35 percent, is suggestive of a soil likely to have excessive potential to expand and contract on wetting and drying.

Soil strength is the ability of a soil to resist deformation. It is difficult to generalize about the strength of soils. Numerical averages of the strength of a soil are often misleading, because soils are often composed of mixtures, zones, or layers of materials with different physical and chemical properties.

The strength of a particular soil type is a function of cohesive and frictional forces. **Cohesion** is a measure of the

(continued on page 66)

PUTTING SOME NUMBERS ON Properties of Soil

The subject of **soil mechanics** is the study of soils with the objective of understanding and predicting the behavior of soil for a variety of purposes, including foundations of buildings, use as construction materials, and construction of slopes and embankments for a variety of engineering projects (8). In 1948 Carl Terzagi published a book on soil mechanics that for the first time utilized concepts of physics and mathematics integrated with geology and engineering to put the "art" of soil mechanics on a "firm foundation" (pun intended). Our purpose here is to provide basic information concerning selected properties of soils and the roll of fluid pressure in the strengths of soils.

A phase diagram for soil that consists of solids, water, and gas (mostly air) is shown in Figure 3.A. Notice that the left side of the diagram is in terms of volume and the right side in terms of weight. The symbols are defined as follows: V_m is the volume of the mass (volume of total sample), V_v is the volume of the voids (spaces between grains that may contain soil gases or liquids, such as water), V_a is the volume of air, V_w is the volume of water, V_s is the volume of the solids, W_m is the weight of the mass (weight of

Volume | Weight

			Phase		
V_m	V_v	V_a	Gas	W_a	W_m
		V_w	Water	W_w	
		V_s	Solid	W_s	

V_m = volume of the mass
V_a = volume of gas (air)
V_v = volume of voids
V_w = volume of water
V_s = volume of solids

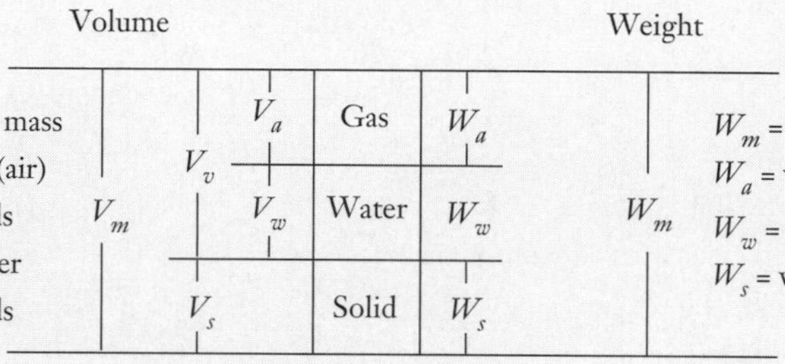

W_m = weight of the mass
W_a = weight of gas (air) = 0
W_w = weight of water
W_s = weight of solids

weight, W, is a force, units, N (Newton, kg m/sec^2); **mass,** m, units, kg;
volume, V, units are m^3

unit weight (γ) = weight/volume; unit weight of mass = γ_m = weight of mass/volume mass = W_m/V_m;
unit weight of solids = γ_s = weight of solids/volume of solids = W_s/V_s;
unit weight of water = γ_w = weight of water/volume of water = W_w/V_w = constant = 1×10^4N/m^3 (or 10^4N/m^3) = 62.4 lb/ft^3

density, ρ = mass/volume (unit mass), units are kg/m^3 or gm/cm^3

Remember γ_w = mass water × acceleration of gravity ÷ volume of water
acceleration of gravity, g; $g = 9.8$ m/sec$^2 \cong 10$ m/sec^2 or 32 ft/sec^2;
$\rho_w = 1$ gm/cm^3 = 10^3 kg/m^3; then $\gamma_w \cong 10^3$ kg/m$^3 \times 10$ m/sec$^2 \cong 10^4$ N/m^3 (62.4 lb/ft^3)

specific gravity, G = unit weight of sample/unit weight of water, G_s is specific gravity of solids, G_m is specific gravity of the mass

$$G_s = \frac{\gamma_s}{\gamma_w} = \frac{W_s}{V_s\gamma_w};\ G_m = \frac{\gamma_m}{\gamma_w} = \frac{W_m}{V_m\gamma_w}, \text{ no units}$$

porosity, $n = \dfrac{V_v}{V_m}$, no units reported as a decimal, reported as a percentage, in calculations as a decimal

void ratio, $e = \dfrac{V_v}{V_s}$, no units reported as a decimal

moisture content (also called water content), $w = \dfrac{W_w}{W_s}$, reported as a percentage, in calculations as a decimal

volumetric water content, $\Theta = \dfrac{V_w}{V_m}$, reported as a percentage, in calculations as a decimal

degree of saturation with respect to water, $S_w = \dfrac{V_w}{V_v}$, reported as a percentage, calculations as a decimal

degree of saturation with respect to air, $S_a = \dfrac{V_a}{V_v}$, reported as a percentage, in calculations as a decimal

$S_w + S_a = 1$

▲ **FIGURE 3.A** Phase diagram for soil consisting of solids, water, and gas (mostly air). Also shown are selected symbols and definitions used for soil mechanics in geotechnical engineering.

total sample), W_a is the weight of air (taken as zero), W_w is the weight of the water, and W_s is the weight of the solids. Given this, we may define a number of soil properties including unit weight (γ), density (ρ), specific gravity (G), porosity (n), void ratio (e), moisture content (w) and the degree of saturation (S_w) as shown on Figure 3.A.

Probably the best way to understand the properties shown is to solve a problem or two. For example, Figure 3.B is a problem where void ratio (e), specific gravity of the solids (G_s), and degree of saturation with respect to water (S_w) are known, and we are asked to compute the water content (w). Water content is a particularly important property of soils because as it varies, so does the strength of the soil as well as other properties, including the unit weight and degree of saturation. Of particular importance to understanding the hydrology of the vadose zone, which is the unsaturated soil above the water table, is the volumetric water content (Θ). As the volumetric water content increases in the vadose zone, the rate of movement of water in the vadose zone increases.

As a second example, assume you have a moist soil with a volume (V_m) of 4.2×10^{-3}m^3 and that the weight of the mass (W_m) is 60 N. After drying, the mass has a weight of 48 N. If the specific gravity of the solids (G_s) is 2.65, compute the degree of saturation with respect to water (S_w), the volumetric water content (Θ), and void ratio (e). It's probably worth mentioning here that there are three volumetric ratios we have introduced: the void ratio (e), porosity (n), and degree of saturation (S_w). The void ratio (e) is expressed as a decimal, while the other two are expressed as a percentage. In theory the void ratio can vary from a minimum of zero to infinity but values of soils composed of sand and gravel vary from about 0.3 to 1.0, whereas for fine-grained soils composed of clay-sized particles, the void ratio may vary from approximately 0.4 to 1.5 (8). Returning now to our problem: The first step is to produce a phase diagram as shown on Figure 3.C. The weight of the mass (W_m) is 60 N; after drying this is reduced to 48 N. Thus the weight of the water (W_w) is 12 N and that value can be placed on the diagram. The weight of the solids (W_s) is the difference between W_m and W_w, which is 48 N. The volume of the mass (V_m) is given, and this is also placed on the diagram. Using the fact that the specific gravity of the solids (G_s) is 2.65, we can calculate the volume of the solids (V_s) as 1.8×10^{-3}m^3. Similarly, because we know the weight of the water, we can use the unit weight of the water, which is a constant (see Figure 3.A, to calculate the volume of the water and place this also on the diagram). Knowing the volume of the water and the volume of the solids, we can calculate the volume of the voids and the volume of air. Given this information, we may then apply our definitions for degree of saturation with respect to water, water content, and void ratio to calculate that S_w = 50 percent, Θ = 29 percent, and e = 1.33.

(continued on next page)

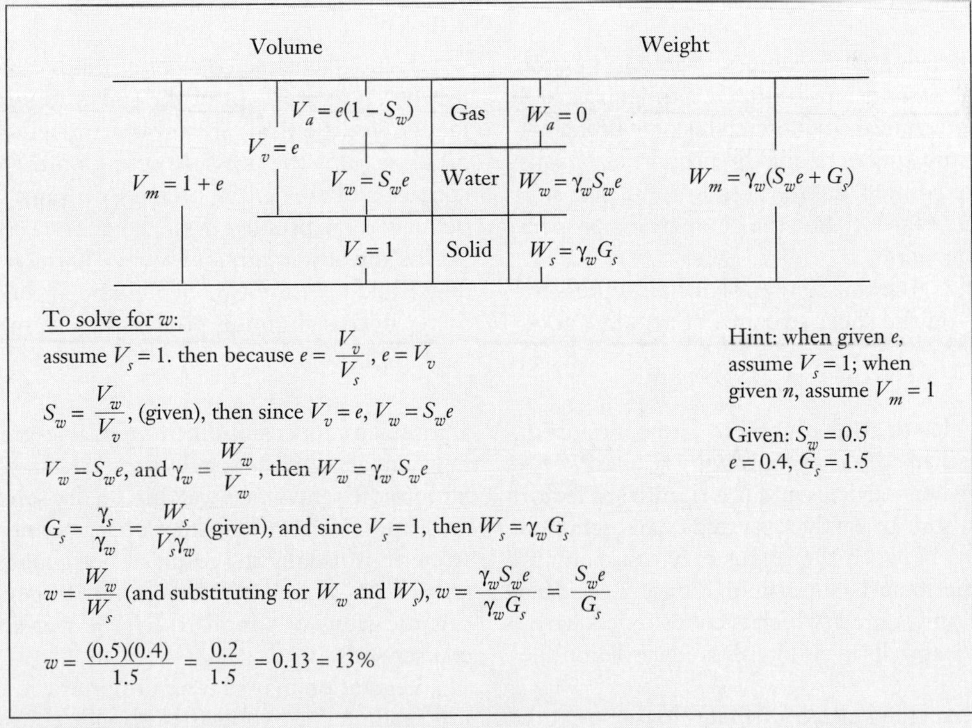

▲ **FIGURE 3.B** Example problem: A soil has a value of $e = 0.40$; $G_s = 1.5$; and $S_w = 0.50$. What is the water content w?

Volume			Weight		
	$V_a = 1.2\times10^{-3}m^3$	Gas	$W_a = 0$		
$V_m = 4.2\text{x}10^{-3}m^3$	$V_w = 1.2\times10^{-3}m^3$	Water	$W_w = 12$ N	$W_m = 60$ N
	$V_s = 1.8\times10^{-3}m^3$	Solid	$W_s = 48$ N		

To solve for S_w, Θ, e:

$$G_s = \frac{W_s}{V_s\gamma_w}\text{, then } V_s = \frac{W_s}{G_s\gamma_w} = \frac{48\text{ N}}{2.65(1\times10^4\text{ N/m}^3)} = \frac{48\text{ N}}{2.65\times10^4\text{ N/m}^3}$$

$$V_s = 1.8\times10^{-3}\text{m}^3$$

$$\gamma_w = \frac{W_w}{V_w}\text{, then } V_w = \frac{W_w}{\gamma_w} = \frac{12\text{ N}}{1\times10^4\text{ N/m}^3}$$

$$V_w = 1.2\times10^{-3}\text{m}^3$$

Then since $V_m = V_a + V_w + V_s$, then $V_a = V_m - V_w - V_s$

$$V_a = 1.2\times10^{-3}\text{m}^3$$

$$\text{Now } S_w = \frac{V_w}{V_v} = \frac{1.2\times10^{-3}\text{m}^3}{2.4\times10^{-3}\text{m}^3} = 0.50 = 50\%$$

$$\Theta = \frac{V_w}{V_m} = \frac{1.2\times10^{-3}\text{m}^3}{4.2\times10^{-3}\text{m}^3} = 0.29 = 29\%$$

$$e = \frac{V_v}{V_s} = \frac{2.4\times10^{-3}\text{m}^3}{1.8\times10^{-3}\text{m}^3} = 1.33$$

▲ **FIGURE 3.C** Example problem: Given $V_m = 4.2\times10^{-3}\text{m}^3$, $W_m = 60$ N, $W_s = 48$ N, and $G_s = 2.65$, calculate S_w, Θ, e.

A large number of soil mechanics problems may be solved using the relationships outlined. Students of soil mechanics or geotechnical engineering become proficient at manipulating the equations for the properties of soils to calculate water content, degree of saturation, porosity, and other properties that define the engineering properties of the soil.

Our next task is to evaluate the role of fluid pressure (water pressure) on the shear strength of a soil. A good way to introduce the concept of fluid pressure and its effect on shear strength of the soil is to introduce the effective and neutral pressures in a soil. The effective pressure ($\bar{p}$) is the pressure acting on the grain-to-grain contact; and the neutral pressure or water pressure (μ_w) is defined as the product of the height of a column of water (h) and the unit weight of water (γ_w), which is $1\times10^4\text{N/m}^3$. The (fluid pressure) water pressure at the bottom of a 2 meter deep swimming pool is $\mu_w = (2\text{ m})(1\times10^4\text{N/m}^3)$,

ability of soil particles to stick together. The cohesion of particles in fine-grained soils is primarily the result of electrostatic forces between particles and is a significant factor in the strength of a soil. In partly saturated coarse-grained soils, moisture films between the grains may cause an apparent cohesion due to surface tension (Figure 3.5). This explains the ability of wet sand (which is cohesionless when dry) to stand in vertical walls in children's sand castles on the beach (9).

Friction between grains also contributes to the strength of a soil. The total frictional force is a function of the density, size, and shape of the soil particles, as well as the weight of overlying particles forcing the grains together. Frictional forces are most significant in coarse-grained soils rich in sand and gravel. Frictional forces explain why you don't sink far into the sand when walking on dry sand on beaches. Because most soils are a mixture of coarse and fine particles, the strength is usually the result of both cohesion and internal friction. Although it is dangerous to generalize, clay and organic-rich soils tend to have lower strengths than do coarser soils.

Vegetation may play an important role in soil strength. For example, tree roots may provide considerable apparent cohesion through the binding characteristics of a continuous root mat or by anchoring individual roots to bedrock beneath thin soils on steep slopes.

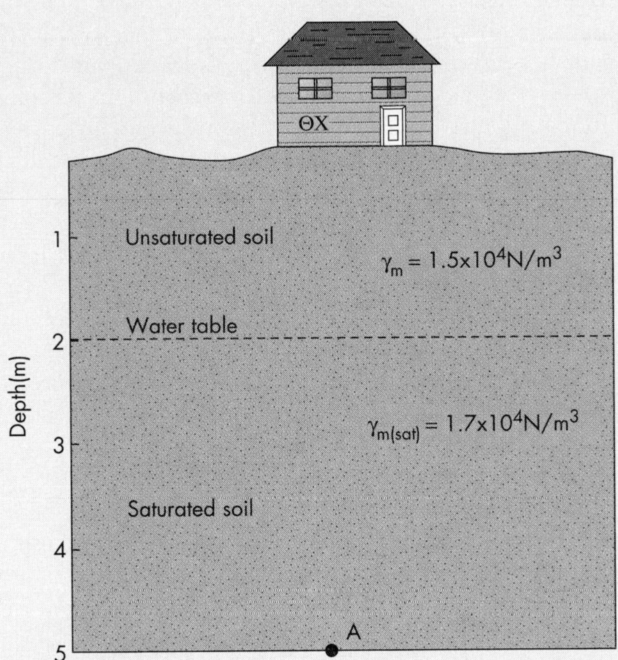

▲ **FIGURE 3.D** Idealized diagram showing a building, water table, and depth to point A where we calculate the effective pressure. See text for calculation.

and $\mu_w = 2\times10^4 N/m^2$. If the pool were filled with water-saturated sand and gravel, the fluid pressure at the bottom would still be $2\times10^4 N/m^2$, as the pore spaces between grains are connected and the height of the continuous water column (although a torturous path around grains) remains 2 m. The total pressure (p) is then equal to the sum of the effective and the neutral pressure ($p = \bar{p} + \mu_w$). This relationship can be rearranged to give the equation $\bar{p} = p - \mu_w$. The quantity $\bar{p}$ (the effective pressure) is a term in the equation that defines the shear strength of a soil:

$$S = C + \bar{p} \tan ø$$

where S is the shear strength; C is cohesion; $\bar{p}$ is the effective pressure; and ø is the angle of internal friction. If we substitute for $\bar{p}$, the equation becomes:

$$S = C + (p - \mu_w) \tan ø$$

Thus we see that the shear strength equation of a soil includes the effective pressure, which can be written as the difference between the total pressure and the water pressure.

An example here will help clear up some of the mystery concerning these relationships. Figure 3.D shows a simple diagram in which the water table is found at a depth of 2.0 m. We are interested in the effective pressure at point A at a depth of 5.0 m. The unit weight of the soil above the water table is given at $1.5\times10^4 N/m^3$, unit weight of the saturated soil below the water table is given at $1.7\times10^4 N/m^3$, and the unit weight of water is a constant $1\times10^4 N/m^3$.

$$\bar{p} \text{ (at point } A) = p - \mu_w$$

$$\bar{p} = (2.0 \text{ m})(1.5\times10^4 N/m^3) + (3.0 \text{ m})(1.7\times10^4 N/m^3) - (3.0 \text{ m})(1\times10^4 N/m^3)$$

$$\bar{p} = (3.0\times10^4 N/m^2) + (5.1\times10^4 N/m^2) - (3.0\times10^4 N/m^2)$$

$$\bar{p} = 5.1\times10^4 N/m^2$$

What would $\bar{p}$ be if the water table were to rise to the surface?

$$\bar{p} = (5.0 \text{ m})(1.7\times10^4 N/m^3) - (5.0 \text{ m})(1\times10^4 N/m^3)$$

$$\bar{p} = (8.5\times10^4 N/m^2) - (5\times10^4 N/m^2)$$

$$\bar{p} = 3.5\times10^4 N/m^2$$

Thus we see by raising the water table the effective pressure at point A has been significantly reduced. Since the effective pressure is part of the equation for the shear strength of the soil, we learn an important principle: If the water table rises, water pressure increases and shear strength is reduced. Conversely if the water table drops, the shear strength increases. The inverse relationship between water pressure and shear strength has important consequences in environmental geology. For example, if we wish to stabilize a landslide we may attempt to drain the potential slide mass, lowering the water table and increasing the shear strength, thus decreasing the likelihood of failure.

Sensitivity measures changes in soil strength resulting from disturbances such as vibrations or excavations. Sand and gravel soils with no clay are the least sensitive. As fine material becomes abundant, soils become more and more sensitive. Some clay soils may lose 75 percent or more of their strength following disturbance (7). Sand with a high water content may liquefy when shaken or vibrated. This process is called **liquefaction.** To observe this, stand on the wet sand of a beach and vibrate your feet. The sand will often liquefy and you will sink in a bit. Liquefaction is discussed further in Chapter 7.

Compressibility is a measure of a soil's tendency to *consolidate*, or decrease in volume. Compressibility is partly a function of the elastic nature of the soil particles and is directly related to settlement of structures. Excessive settlement will crack foundations and walls. Coarse materials, such as gravels and sands, have a low compressibility, and settlement will be considerably less in these materials than in highly compressible fine-grained or organic soils.

Erodibility refers to the ease with which soil materials can be removed by wind or water. Easily eroded materials (soils with a high erosion factor) include unprotected silts, sands, and other loosely consolidated materials. Cohesive soils (more than 20 percent clay) and naturally cemented soils are not easily moved by wind or water erosion and therefore have a low erosion factor.

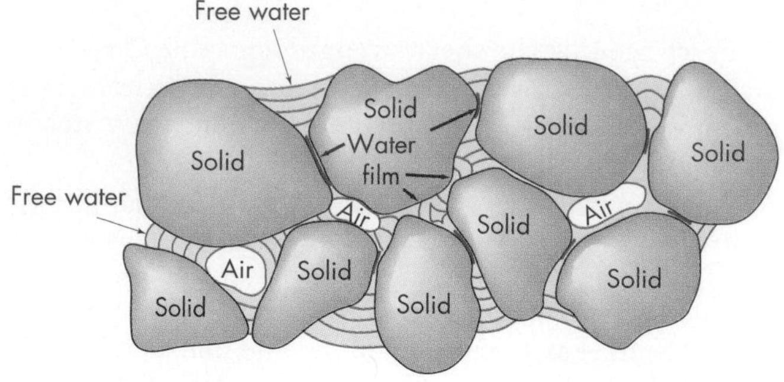

▲ **FIGURE 3.5** Partly saturated soil showing particle-water-air relationships. Particle size is greatly magnified. Attraction between the water and soil particles (surface tension) develops a stress that holds the grains together. This apparent cohesion is destroyed if the soil dries out or becomes completely saturated. (*After* R. *Pestrong*. 1974. Slope Stability. *American Geological Institute*.)

Permeability is a measure of the ease with which water moves through a material. Clean gravels or sands have the highest permeabilities. As fine particles in a mixture of clean gravel and sand increase, permeability decreases. Clays generally have a very low permeability.

Corrosion is a slow weathering or chemical decomposition that proceeds from the surface into the ground. All objects buried in the ground—pipes, cables, anchors, fenceposts—are subject to corrosion. The corrosiveness of a particular soil depends upon the chemistry of both the soil and the buried material and on the amount of water available. It has been observed that the more easily a soil carries an electrical current (low resistivity due to more water in the soil), the higher its **corrosion potential.** Therefore, measurement of soil resistivity is one way to estimate the corrosion hazard (10).

Ease of excavation pertains to the procedures, and hence equipment, required to remove soils during construction. There are three general categories of excavation techniques. *Common excavation* is accomplished with an earth mover, backhoe, or dragline. This equipment essentially removes the soil without having to scrape it first (most soils can be removed by this process). *Rippable excavation* requires breaking the soil up with a special ripping tooth before it can be removed (examples include a tightly compacted or cemented soil). *Blasting* or *rock cutting* is the third, and often most expensive, category (for example, a hard silica-cemented soil, as with concrete, might need to be cut with a jackhammer to be removed).

Shrink-swell potential refers to the tendency of a soil to gain or lose water. Soils that tend to increase or decrease in volume with water content are called **expansive soils.** The swelling is caused by the chemical attraction of water molecules to the submicroscopic flat particles, or plates, of certain clay minerals. The plates are composed primarily of silica, aluminum, and oxygen atoms, and layers of water are added between the plates as the clay expands or swells (Figure 3.6a) (11). Expansive soils often have a high *plasticity index*, reflecting their tendency to take up a lot of water while in the plastic state. *Montmorillonite* is the common clay mineral associated with most expansive soils. With sufficient water, pure montmorillonite may expand up to 15 times its original volume, but fortunately most soils contain limited amounts of this clay, so it is unusual for an expansive soil to swell beyond 25 to 50 percent. However, an increase in volume of more than 3 percent is considered potentially hazardous (11).

Expansive soils in the United States cause significant environmental problems and, as one of our most costly natural hazards, are responsible for more than $3 billion in damages annually to highways, buildings, and other structures. Every year more than 250,000 new houses are constructed on expansive soils. Of these, about 60 percent will experience some minor damage such as cracks in the foundation, walls, or walkways (Figure 3.6b,c), but 10 percent will be seriously damaged—some beyond repair (12,13).

Damages to structures on expansive soils are caused by volume changes in the soil in response to changes in moisture content. Factors affecting the moisture content of an expansive soil include *climate, vegetation, topography, drainage, site control,* and *quality of construction* (12). Regions such as the southwestern United States, which have a pronounced wet season followed by a dry season, thus allowing for a regular shrink-swell sequence, are more likely to experience an expansive soil problem than regions where precipitation is more evenly distributed throughout the year. Vegetation can cause changes in the moisture content of soils. Especially during a dry season, large trees draw and use a lot of local soil moisture, facilitating soil shrinkage (Figure 3.6b). Therefore, in areas with expansive soil, trees should not be planted close to foundations of light structures (such as homes).

Topography and drainage are significant because adverse topographic and drainage conditions cause ponding of water around or near structures, increasing the swelling of expansive clays. However, homeowners and contractors can do a great deal to avoid this problem. Laboratory testing prior to construction can identify soils with a shrink-swell potential. Proper design of subsurface drains, rain gutters, and foundations can minimize damages associated with expansive soils by improving drainage and allowing the foundation to accommodate some shrinking and swelling of the soil (12).

It should be apparent that some soils are more desirable than others for specific uses. Planners concerned with land use will not make soil tests to evaluate engineering properties of soils, but they will be better prepared to design with nature and take advantage of geologic conditions if they understand the basic terminology and principles of earth materials. Our discussion of engineering properties established two general principles. First, because of their low strength, high sensitivity, high compressibility, low permeability, and variable shrink-swell potential, clay soils should be avoided in projects involving heavy structures, structures with minimal allowable settling, or projects needing well-drained soils. Second, soils that have high corrosive potentials or that require other than common excavation

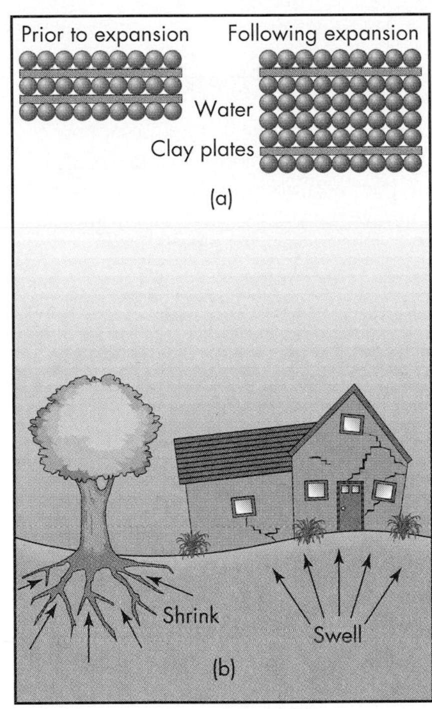

▲ **FIGURE 3.6** Expansive soils. (a) Expansion of a clay (montmorillonite) as layers of water molecules are incorporated between clay plates. (b) Effects of soil shrinking and swelling at a home site. (*After Mathewson and Castleberry*, Expansive Soils: Their Engineering Geology, *Texas* A & M *University*) (c) Cracked wall resulting from expansion of clay soil under the foundation. (*Courtesy of* U.S. *Department of Agriculture*)

should be avoided if possible. If such soils cannot be avoided, extra care, special materials and/or techniques, and higher-than-average initial costs (planning, design, and construction) must be expected. Secondary costs (operation and maintenance) may also be greater. Table 3.3 summarizes engineering properties of soils in terms of the unified soil classification.

3.7 Rates of Soil Erosion

Rates of soil erosion as a volume, mass, or weight of soil eroded from a location in a specified time and area (example, kilograms per year per hectare, kg/yr/ha) vary with engineering properties of the soil, land use, topography, and climate.

There are several approaches to measuring rates of soil erosion. The most direct method is to make actual measurements on slopes over a period of years (at least several) and use these values as representative of what is happening over a wider area and longer time span. This approach is not often taken, however, because data from individual slopes and drainage basins are very difficult to obtain.

A second approach is to use data obtained from resurveying reservoirs to calculate the change in the reservoirs' storage capacity of water (Table 3.4); the depletion of storage capacity (of water) tells us the volume of stored sediment. If we use the data from many reservoirs, we can obtain a sediment yield rate curve for an entire region. Figure 3.7 is an example of such a curve for drainages in the southwestern United States. This approach has merit but is unlikely to be reliable for small drainage basins, where sediment yield can vary dramatically. Note that on Figure 3.7 there is an inverse relationship between sediment yield per unit area and drainage area (upstream area contributing water and sediment, in this case to a reservoir). Smaller drainage areas have a larger sediment yield per unit area because smaller drainage areas tend to be steeper with great erosion potential.

A final approach is to use an equation to calculate rates of sediment eroded from a particular site. Probably the most commonly used equation is the Universal Soil Loss Equation (14). This equation uses data about rainfall runoff, the size and shape of the slope, the soil cover, and erosion-control practices to predict the amount of soil eroded from its original position (15) (see *A Closer Look: Universal Soil Loss Equation*).

3.8 Sediment Pollution

Sediment is probably our greatest pollutant. In many areas, sediment pollution chokes the streams, fills in lakes, reservoirs, ponds, canals, drainage ditches, and harbors, buries vegetation, and generally creates a nuisance that is difficult to remove. Natural pollutional sediment—eroded soil—is truly a resource out of place. It depletes soil at its site of origin (Figure 3.8), reduces the quality of the water it enters, and may deposit sterile materials on productive croplands or other useful land (Figure 3.9) (16). Rates of storage depletion

Table 3.3 Generalized sizes, descriptions, and properties of soils

Soil	Soil Component	Symbol	Grain Size Range and Description	Significant Properties
Coarse-grained components	Boulder	None	Rounded to angular, bulky, hard, rock particle; average diameter more than 25.6 cm.	Boulders and cobbles are very stable components used for fills, ballast, and riprap. Because of size and weight, their occurrence in natural deposits tends to improve the stability of foundations. Angularity of particles increases stability.
	Cobble	None	Rounded to angular, bulky, hard, rock particle; average diameter 6.5–25.6 cm.	
	Gravel	G	Rounded to angular, bulky, hard, rock particles greater than 2 mm in diameter.	Gravel and sand have essentially the same engineering properties, differing mainly in degree. They are easy to compact, little affected by moisture, and not subject to frost action. Gravels are generally more pervious and more stable and resistant to erosion and piping than are sands. The well-graded* sands and gravels are generally less pervious and more stable than those that are poorly graded and of uniform gradation.
	Sand	S	Rounded to angular, bulky, hard, rock particles 0.074–2 mm in diameter.	
Fine-grained components	Silt	M	Particles 0.004–0.074 mm in diameter; slightly plastic or nonplastic regardless of moisture; exhibits little or no strength when air dried.	Silt is inherently unstable, particularly with increased moisture, and has a tendency to become quick when saturated. It is relatively impervious, difficult to compact, highly susceptible to frost heave, easily erodible, and subject to piping and boiling. Bulky grains reduce compressibility, whereas flaky grains (such as mica) increase compressibility, producing an elastic silt.
	Clay	C	Particles smaller than 0.004 mm in diameter; exhibits plastic properties within a certain range of moisture; exhibits considerable strength when air dried.	The distinguishing characteristic of clay is cohesion or cohesive strength, which increases with decreasing moisture. The permeability of clay is very low; it is difficult to compact when wet and impossible to drain by ordinary means; when compacted, is resistant to erosion and piping; not susceptible to frost heave; and subject to expansion and shrinkage with changes in moisture. The properties are influenced not only by the size and shape (flat or platelike), but also by their mineral composition. In general, the montmorillonite clay mineral has the greatest and kaolinite the least adverse effect on the properties.
	Organic matter	O	Organic matter in various sizes and stages of decomposition.	Organic matter present even in moderate amounts increases the compressibility and reduces the stability of the fine-grained components. It may decay, causing voids, or change the properties of a soil by chemical alteration; hence, organic soils are not desirable for engineering purposes.

*The term *well-graded* indicates an even distribution of sizes and is equivalent to *poorly sorted*.

Note: The Unified Soil Classification does not recognize cobbles and boulders with symbols. The size range for these as well as the upper limit for clay are classified according to Wentworth (1922).

Source: After Wagner, "The Use of the Unified Soil Classification System by the Bureau of Reclamation." International Conference on Soil Mechanics and Foundation Engineering, Proceedings [London], 1957.

Table 3.4 Summary of reservoir capacity and storage depletion of the nation's reservoirs

Reservoir Capacity (acre-ft)[a]	Number of Reservoirs	Total Initial Storage Capacity (acre-ft)[a]	Total Storage Depletion		Individual Reservoir Storage Depletion		Average Period of Record
			(acre-ft)[a]	%	Average %/yr	Median %/yr	yr
0–10	161	685	180	26.3	3.41	2.20	11.0
10–100	228	8,199	1,711	20.9	3.17	1.32	14.7
100–1,000	251	97,044	16,224	16.7	1.02	.61	23.6
1,000–10,000	155	488,374	51,096	10.5	.78	.50	20.5
10,000–100,000	99	4,213,330	368,786	8.8	.45	.26	21.4
100,000–1,000,000	56	18,269,832	634,247	3.5	.26	.13	16.9
Over 1,000,000	18	38,161,556	1,338,222	3.5	.16	.10	17.1
Total or average	968	61,239,020	2,410,466	3.9	1.77	.72	18.2[b]

[a] 1 acre-ft = approximately 1234 m^3.

[b] The capacity-weighted period of record for all reservoirs was 16.1 years.

Source: F. E. Dendy, "Sedimentation in the Nation's Reservoirs." *Journal of Soil and Water Conservation* 23, 1968.

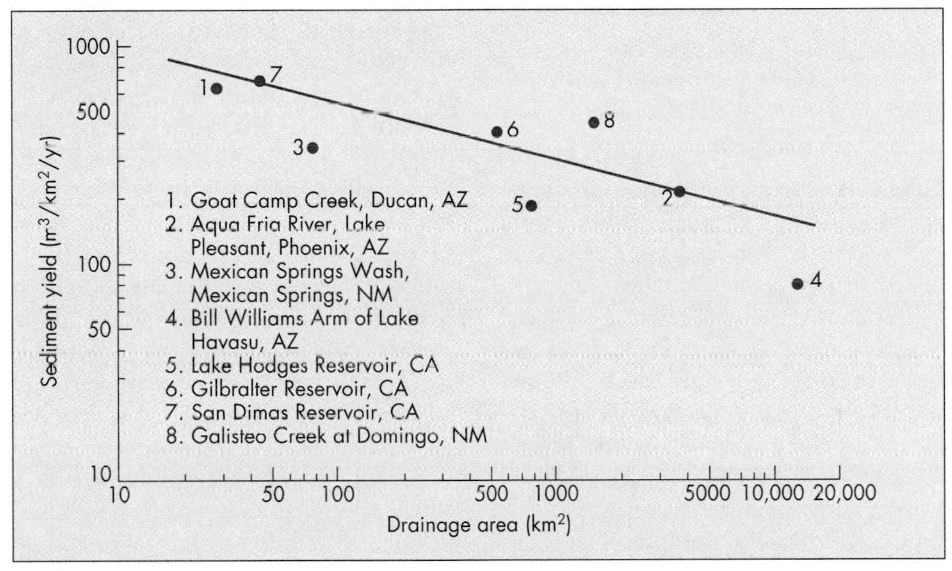

▲ **FIGURE 3.7** Sediment yield for the southwestern United States, based on rates of sediment accumulation in reservoirs. (*Data from* R. I. *Strand*. 1972. *"Present and Prospective Technology for Predicting Sediment Yield and Sources."* Proceedings of the Sediment-Yield Workshop. U.S. *Department of Agriculture Publication* ARS-5-40, 1975, *p.* 13)

A CLOSER LOOK Universal Soil Loss Equation

The Universal Soil Loss Equation is:

$$A = RKLSCP$$

where A = the long-term average annual soil loss for the site being considered

R = the long-term rainfall runoff erosion factor

K = the soil erodibility index

L = the hillslope/length factor

S = the hillslope/gradient factor

C = the soil cover factor

P = the erosion-control practice factor.

The advantage of using this equation is that once the various factors have been determined and multiplied together to produce predicted soil loss, conservation practices may be applied through factors C and P to reduce the soil loss to the desired level. In the case of slopes that are amenable to shaping, factors of K, L, and S may also be manipulated to achieve desired results in terms of sediment loss. This is particularly valuable when evaluating construction sites, such as those for the development of shopping centers, and along corridors such as pipelines and highways. For construction sites, the Universal Soil Loss Equation can be used to predict the impact of sediment loss on local streams and other resources and to develop management strategies for minimizing this impact (14,15).

▲ **FIGURE 3.8** Serious soil erosion and gully formation in central California related to diversion of runoff water. The surface was essentially ungullied several months prior to the photograph. (*Edward A. Keller*)

▲ **FIGURE 3.9** Soil erosion has caused unwanted red sediment at this site in Charlotte, North Carolina. (*Edward A. Keller*)

caused by sediment filling in ponds and reservoirs are shown in Table 3.4. These data suggest that small reservoirs will fill up with sediment in a few decades, whereas large reservoirs will take hundreds of years to fill.

Most natural pollutional sediments consist of rock and mineral fragments ranging in size from sand particles less than 2 mm in diameter, to silt, clay, or very fine colloidal particles. Human-made pollutional sediments include debris resulting from the disposal of industrial, manufacturing, and public wastes. Such sediments include trash directly or indirectly discharged into surface waters. (Litter is not confined to roadsides and parks.) Most of these sediments are very fine-grained and difficult to distinguish from the naturally occurring sediments unless they contain unusual minerals or other particular characteristics. The principal sources of artificially induced pollutional sediments are disruptions of the land surface for construction, farming, deforestation, and channelization works. In short, a great deal of sediment pollution is the result of human use of the environment and civilization's continued change of plans and direction. It cannot be eliminated, only ameliorated.

In this century in the United States, emphasis on soil and water conservation has grown considerably. The first programs emphasized stabilization of areas of excessive wind and water erosion, as well as flood control and irrigation water for arid and semiarid land. As these programs progressed, adverse environmental effects from some of the work, such as accelerated stream erosion and rapid reservoir siltation, were discovered. At the same time, demands increased for expanded use of areas improved for recreation, water supply, and navigation. These new environmental effects and increased demands necessitated research and development to find solutions to new or rediscovered problems. Design changes developed and applied to correct the ero-

CASE HISTORY Reduction of Sediment Pollution, Maryland

A study in Maryland demonstrates that sediment-control measures can reduce sediment pollution in an urbanizing area (17). The suspended sediment transported by the northwest branch of the Anacostia River near Colesville, Maryland, with a drainage area of 54.6 km^2, was measured over a 10-year period. During that time, construction within the basin involved about 3 percent of the area annually. The total urban land area in the basin was about 20 percent at the end of the 10-year study.

Sediment pollution was a problem because the soils are highly susceptible to erosion, and there is sufficient precipitation to ensure their erosion when not protected by a vegetative cover. Most of the sediment is transported during spring and summer rainstorms (17). A sediment-control program was initiated, and the sediment yield was reduced by about 35 percent. The basic sediment-control principles were to tailor development to the natural topography, expose a minimum amount of land, provide protection for exposed soil, minimize surface runoff from critical areas, and trap eroded sediment on the construction site. Specific measures included scheduled grading to minimize the time of soil exposure, mulch protection and temporary vegetation to protect exposed soils, sediment diversion berms, stabilized waterways (channels), and sediment basins. This Maryland study concluded that even further sediment control can be achieved if major grading is scheduled during periods of low erosion potential and if better sediment traps are designed to control runoff during storms (17).

sion and sedimentation problems included contour plowing, changes in farming practices, and construction of small dams and ponds to hold runoff or trap sediments.

Currently, the same basic problems—erosion control and sediment pollution—occupy a significant portion of the public's attention. Sources of the sediment have expanded, however, to include those from land development, highways, mining, and production of new and unusual chemical compounds and products, all in the interest of a better life. Sediment pollution affects rivers, streams, the Great Lakes, and even the oceans, and the problem promises to be with us indefinitely. Solutions will involve sound conservation practices, particularly in urbanizing areas where tremendous quantities of sediment are produced during construction (see *Case History: Reduction of Sediment Pollution, Maryland*). Figure 3.10 shows a typical sediment- and erosion-control plan for a commercial development. The plan calls for diversions to collect runoff and a sediment basin to collect sediment and keep it on the site, thus preventing stream pollution. A generalized cross section of a sediment-control basin is shown in Figure 3.11.

3.9 Land Use and Environmental Problems of Soils

Human activities affect soils by influencing the pattern, amount, and intensity of surface-water runoff, erosion, and sedimentation. The most important of these human influences are the conversion of natural areas to various land uses and manipulations of our surface water.

Figure 3.12 shows the changes in water and sediment processes that might occur as a landscape is modified from natural forest (Figure 3.12a) to various uses. Trees modify the effects of precipitation by intercepting rain as it falls and by returning water to the atmosphere through evapotranspiration (the combined effect of evaporation and transpiration).

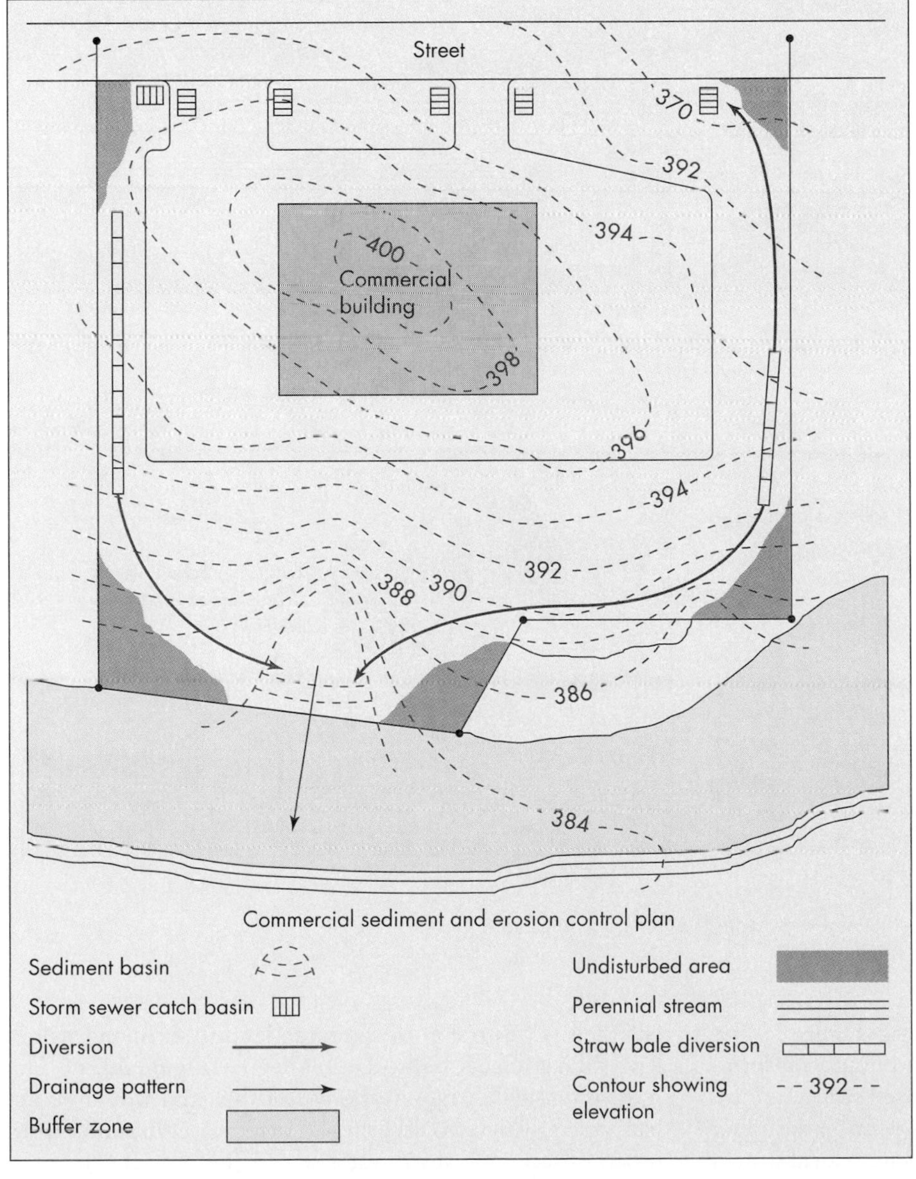

◀ **FIGURE 3.10** Example of a sediment- and erosion-control plan for a commercial development. (*Courtesy of Braxton Williams, Soil Conservation Service*) (384 to 400 ft = 117 to 122 m)

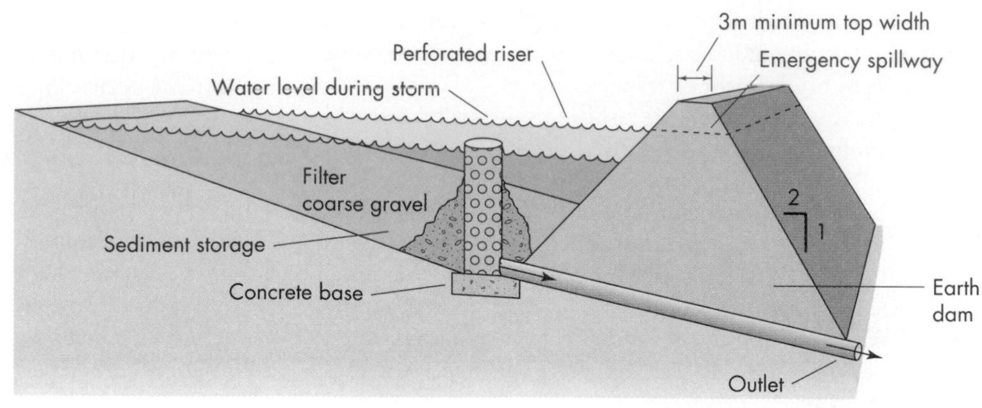

▶ **FIGURE 3.11** Cross section of a sediment basin. Storm water runs into the sediment basin, where the sediment settles out and the water filters through loose gravel into a pipe outlet. Accumulated sediment is periodically removed mechanically. (*After* Erosion and Sediment Control. 1974. *Soil Conservation Service*)

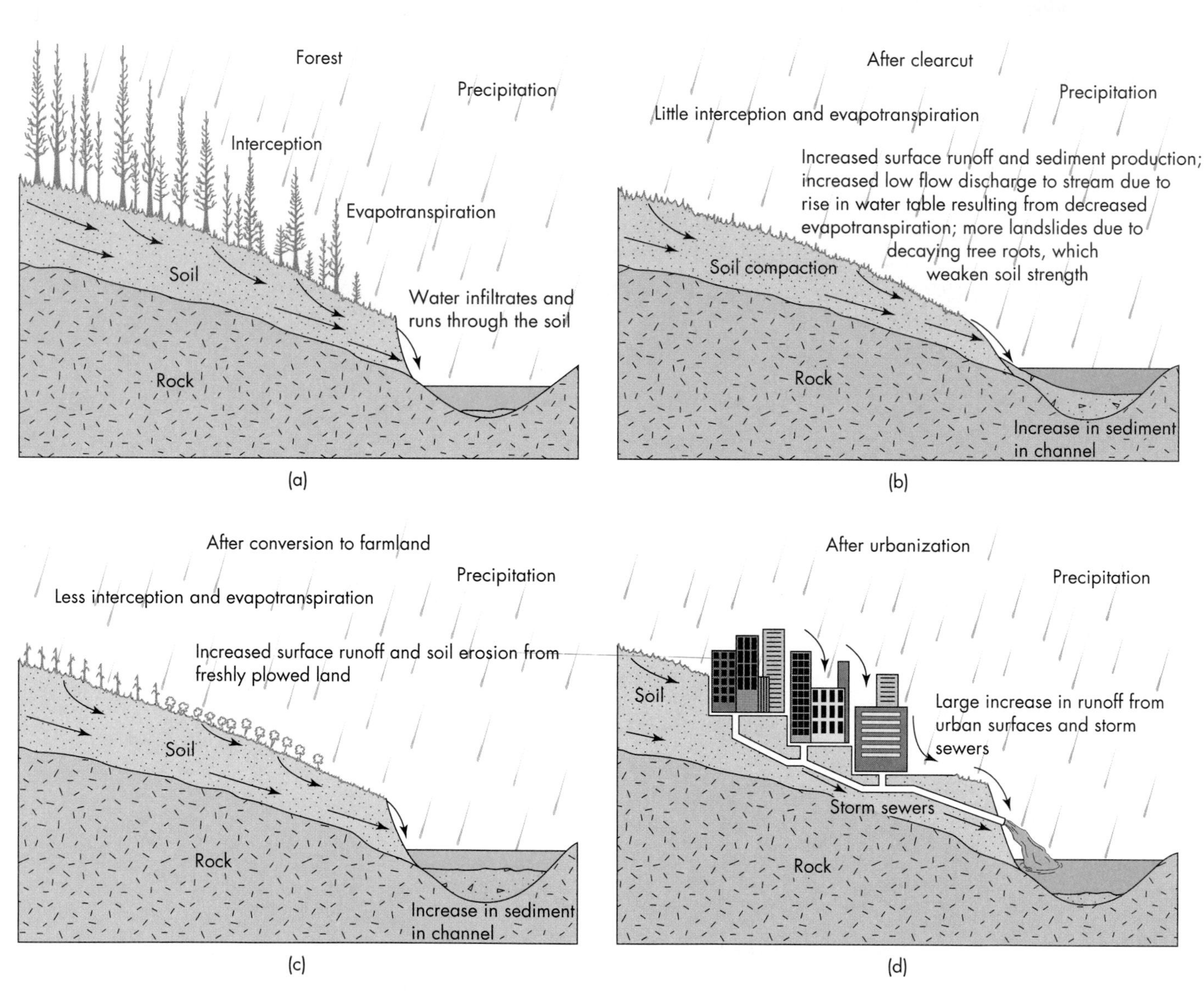

▲ **FIGURE 3.12** Water relationships before and after land-use changes. (a) A natural forested slope. (b) After clearcut. (c) After conversion to farmland. (d) After urbanization.

Following timber harvesting by clear-cutting (Figure 3.12b), one can expect decline in interception and evapotranspiration, resulting in rise in water table, increased surface runoff and production of sediment. As tree roots decay, landsliding may also increase, further delivering sediment to local stream channels. Conversion of the same land to farmland would have similar effects, but to a lesser degree (Figure 3.12c). The most intensive change in the use of this land would be urbanization, which would bring large increases in runoff from urban surfaces and stormwater sewers (Figure 3.12d).

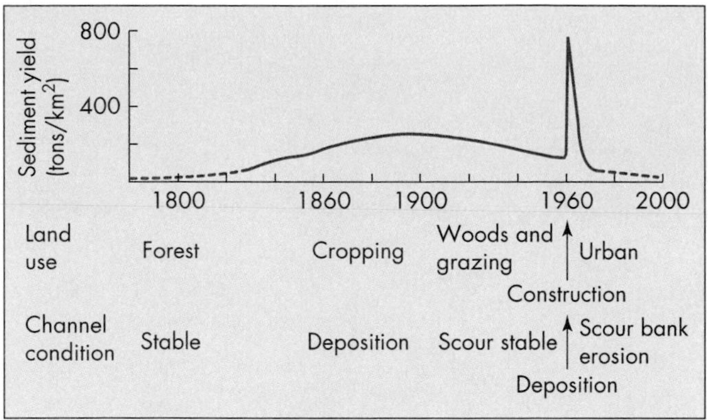

▲ **FIGURE 3.13** Effect of land use on sediment yield and channel condition. The graph shows observed and estimated changes in the Piedmont region of the eastern United States, from before the beginning of extensive farming through a period of construction and urbanization. (*After M. Gordon Wolman.* 1966. Geografiska Annaler 49A)

▲ **FIGURE 3.14** Urbanization and the construction of freeway on-ramps can contribute to soil erosion. This eroding embankment is near the University of California, Santa Barbara, and the community of Isla Vista. (*Edward A. Keller*)

Figure 3.13 summarizes the estimated and observed variation in sediment yield that accompanied changes in land use in the Piedmont region from 1800 to the present. Notice the sharp peak in sediment production during the construction phase of urbanization. These data suggest that the effects of land-use change on the drainage basin, its streams, and its sediment production may be quite dramatic. Streams and naturally forested areas are assumed to be relatively stable, that is, without excessive erosion or deposition. A land-use change that converts forested land for agricultural purposes generally increases runoff and erosion. As a result, streams become muddy and may not be able to transport all the sediment delivered to them. The channels will therefore aggrade (partially fill with sediment), possibly increasing the magnitude and frequency of flooding.

Urbanization

The conversion of agricultural, forested, or rural land to highly urbanized land causes dramatic changes. During the construction phase, there occurs a tremendous increase in sediment production (Figure 3.14), which may be accompanied by a moderate increase in runoff. The response of streams in the area is complex and may include both channel erosion (widening) and deposition, resulting in wide, shallow channels. The combination of increased runoff and shallow channels increases the flood hazard. Following the construction phase, the land is mostly covered with buildings, parking lots, and streets, so the sediment yield drops to a low level. However, the runoff increases further because of the large impervious areas and use of storm sewers, again increasing the magnitude and frequency of flooding. Streams respond to the lower sediment yield and higher runoff by eroding their channels, which makes them deeper.

The process of urbanization directly affects soils in several ways:

- Soil may be scraped off and lost. Once sensitive soils are disturbed, they may have lower strengths when they are remolded.
- Materials may be brought in from outside areas to fill a depression prior to construction, resulting in a much different soil than was previously there.
- Draining soils and pumping them to remove water may cause dessication (drying out) and other changes in soil properties.
- Soils in urban areas are susceptible to soil pollution resulting from deliberate or inadvertent addition of chemicals to soils. This problem is particularly serious if hazardous chemicals have been applied.

Off-Road Vehicles

Urbanization is not the only land use that causes increased soil erosion and hydrologic changes. In recent years, the popularity of off-road vehicles (ORVs) has increased enormously, and demand for recreational areas to pursue this interest has led to serious environmental problems as well as conflicts between users of public lands.

There are now millions of ORVs, many of which are invading the deserts, coastal dunes, and forested mountains of the United States. For example, during the Thanksgiving holiday period in 1998, about 100,000 ORV enthusiasts converged on the Imperial Sand Dunes Recreation Area near Glamis, California, just north of the Mexico–U.S. border. Riding ORVs can be hazardous to your health! Each year at the Imperial site, about 10 people are killed and about 100 suffer serious spinal-cord injuries. Environmental problems associated with ORVs (including snowmobiles) are known, from the shores of North Carolina and New York to sand dunes in Michigan and Indiana to deserts and beaches in the western United States. These problems are not insignificant. A single motorcycle need travel only 8 km to have an impact on 1000 m^2, and a four-wheel-drive vehicle has an impact over the same area by traveling only 2.4 km. In some desert areas, the tracks produce scars that may remain part of the landscape for hundreds of years (18,19).

(a)

(b)

▲ **FIGURE 3.15** Serious erosion problems caused by off-road vehicle use in (a) mountains and (b) coastal dunes. (*Edward A. Keller*)

Major areas of environmental concern are soil erosion, changes in hydrology, and damage to plants and animals. The ORVs cause direct mechanical erosion and facilitate wind and water erosion of materials loosened by their passing (Figure 3.15). Runoff from ORV sites is as much as eight times greater than for adjacent unused areas, and sediment yields are comparable to those found on construction sites in urbanizing areas (19). Hydrologic changes from ORV activity are primarily the result of near-surface soil compaction that reduces the ability of the soil to absorb water. Furthermore, what water is in the soil becomes more tightly held and thus less available to plants and animals. In the Mojave Desert, the tank tracks produced 50 years ago are still visible and the compacted soils have not recovered (19). Compaction also changes the variability of soil temperature. This is especially apparent near the surface, where the soil becomes hotter during the day and colder at night. Animals are killed or displaced and vegetation damaged or destroyed by intensive ORV activity. The damage is a result of a combination of soil erosion, compaction, and changes in temperature and moisture content (19).

There is little doubt that as a management strategy some land must be set aside for ORV use. The question is how much land should be involved, along with how environmental damage can be minimized. Sites should be chosen in closed basins with minimal soil and vegetation variation. The possible effects of airborne removal (removal by wind) must be evaluated carefully, as must the sacrifice of nonrenewable cultural, biological, and geological resources (18). A major problem remains: Intensive ORV use is incompatible with nearly all other land use, and it is very difficult to restrict damages to a specific site. Material removed by mechanical, water, and wind erosion will always have an impact on other areas and activities (18,19).

In recent years the demand for off-road "mountain bikes" has grown dramatically, and this is having a negative impact on the environment. Bicycles have in the past damaged mountain meadows in Yosemite National Park, where they are no longer allowed and are now restricted to bike paths and paved roads. Their intensive use contributes to trail erosion in many areas (Figure 3.16). Users of mountain bikes—who are lobbying to gain entry into even more locations in the national forests, parks, and wilderness areas—state that bicycles cause less erosion than do horses. Although this is true, the problem is that all-terrain bicycles are cheaper and easier to maintain than horses, so many more people may own them. Thus, the cumulative effect of bicycles on trails may be greater than that of horses. Also, hikers and other visitors may not mind seeing animals such as horses in wilderness areas but may be less receptive to bicycles, which are fast and almost silent. This is a sensitive area for managers in the wilderness because many people who enjoy riding all-terrain bicycles come from environmentally concerned groups. As with motorized off-road vehicles, management plans will have to be developed to ensure that overenthusiastic people do not damage sensitive environments.

▲ **FIGURE 3.16** Deep paths (ruts) have been excavated by bicycles in a sandy soil at this California site. (*Edward A. Keller*)

3.10 Soil Pollution

Soil pollution came to the forefront of public attention in 1983 when the town of Times Beach, Missouri, with a population of about 2400, was evacuated and purchased by the federal government, becoming a ghost town (Figure 3.17). The evacuation and purchase occurred after it was discovered that oil sprayed on the town's roads to control dust contained dioxin, a colorless chlorine-containing hydrocarbon compound known to be extremely toxic to mammals. Although it has never been proven that dioxin has killed any person and the dose necessary to cause adverse health effects to people is not well known, dioxin is suspected of being a carcinogen (cancer-causing material).

Soil pollution occurs when materials detrimental to people and other living things are inadvertently or deliberately applied to soils. Many types of materials, including organic chemicals (for example, hydrocarbons or pesticides) and heavy metals (for example, selenium, cadmium, nickel, or lead), may act as soil contaminants. On the other hand, soils, particularly those with clay particles, may selectively attract, absorb, and bind toxins and other materials that otherwise would contaminate the environment. Soils may also contain organisms that break down certain contaminants into less harmful materials. As a result, soils offer opportunities to reduce environmental pollution. However, contaminants in soils and the products of their breakdown by soil and biochemical processes may be toxic to ecosystems and humans if they become concentrated in plants or are transported into the atmosphere or water (2).

Problems arise when soils intended for uses other than waste disposal are contaminated, or when people discover that soils have been contaminated by previous uses. Of particular concern is the building of houses and other structures, such as schools, over sites where soils have been contaminated. Many sites contaminated from old waste disposal facilities or inadvertent dumping of chemicals are now being discovered; some of these are being treated. Treatment of soils to remove contaminants, however, can be a very costly endeavor. In recent years, contamination of soil and water by leaking underground tanks has become a significant environmental concern. Businesses are now adding systems to monitor storage tanks so that leaks may be detected before significant environmental damage occurs. Treatment of contaminated soil can be very expensive and ranges from excavation and disposal or treatment such as incineration to **bioremediation,** which is a technology that utilizes natural or enhanced microbial action in the soil to degrade organic contaminants (such as oil or an organic solvent) into harmless products (such as carbon dioxide and water). Often this is done *in situ* (at the site), not requiring excavation and moving large quantities of contaminated soil (Figure 3.18) (2,20).

▲ **FIGURE 3.17** Examination of soils contaminated by dioxin at Times Beach, Missouri. The town was evacuated due to the dioxin scare. (O. *Franken/Sigma*)

3.11 Desertification

Desertification may be defined as the conversion of land from some productive state to that more resembling a desert. The term was probably first used to describe the advance of the Sahara Desert in Algeria and Tunisia. Driving forces of desertification include, among others, overgrazing, deforestation, adverse soil erosion, poor drainage of irrigated land, and overuse of water supplies.

Considered a major problem today, desertification is most pronounced during times of drought stress. In recent years it has been associated with great human misery, characterized by malnutrition and starvation of millions of people, particularly in India and Africa. The effects of desertification, particularly around highly populated areas, are greatly aggravated by political stress associated with dislocated people concentrated in large encampments. When this situation occurs, the surrounding countryside may be picked clean of all natural vegetation. Livestock overgrazes the available food sources to produce a human-induced desert.

Figure 3.19 shows the degree of desertification hazard worldwide. Although such a map is useful in identifying general areas that may have problems with desertification, it is of little value in identifying particular case histories or in specifying local reasons for desertification. Of special importance in evaluating desertification is consideration of ecological principles and linkages among people, animals, and the physical environment, including long-term climatic cycles that affect hydrologic processes and soil conditions. The process of desertification is not ordinarily characterized by a continuous change in the land along some advancing front. Rather, it is a patchy process that proceeds over a wide area, with local degradation occurring in response to local conditions of water, geology, soil, and land use (21).

Desertification has not received as much attention in North America as in other parts of the world because the effects have not been nearly so severe in terms of damage to human systems. Nevertheless, desertification is significantly reducing the productivity of land in North America, and in some places it is even threatening the ability of the land to support life. Major symptoms of desertification in North America are (21):

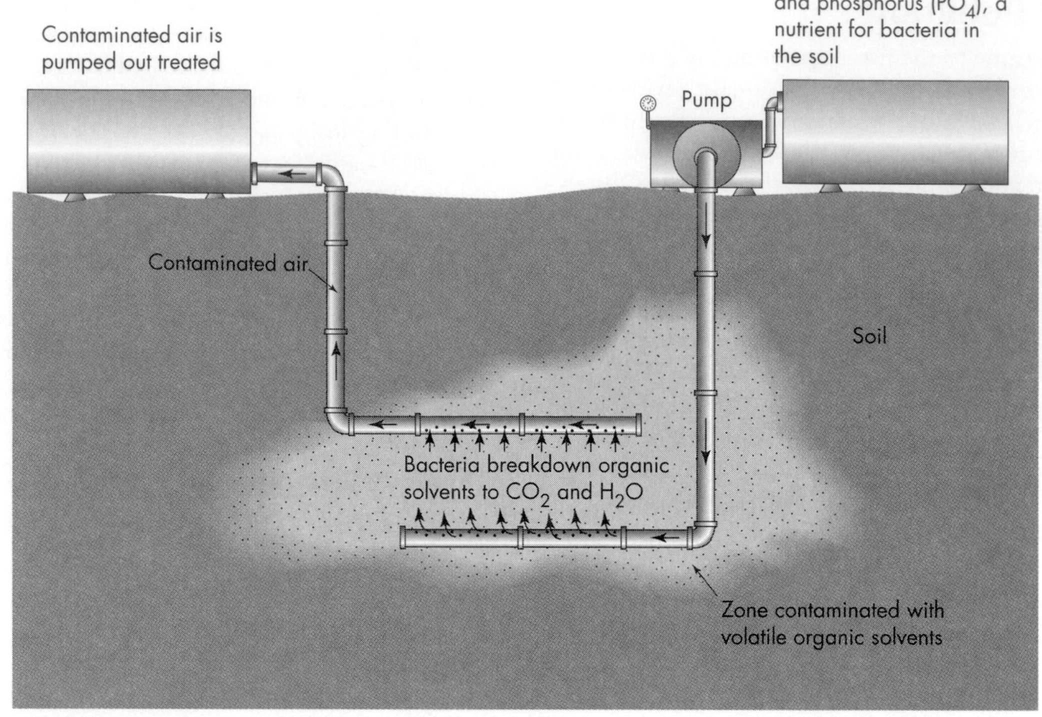

▲ **FIGURE 3.18** Idealized diagram illustrating the process of enhanced bioremediation of soil contaminated by an organic solvent. Methane (CH_4), phosphorus (PO_4), and air (with O_2) are nutrients pumped intermittently into the contaminated area and released from the lower slotted pipe. The upper pipe (also slotted) sucks contaminated air from the soil. The nutrients stimulate the growth of bacteria. The supply of methane (which is a carbon source) is then stopped and the carbon-hungry bacteria go after the inorganic solvents, degrading them to carbon dioxide and water as part of their life cycle. Such a process can greatly reduce the treatment time and cost (5). (*Modified after* Hazen, T. C. 1995. *Savanna River site—a test bed for cleanup technologies.* Environmental Protection, *April, pp.* 10–16)

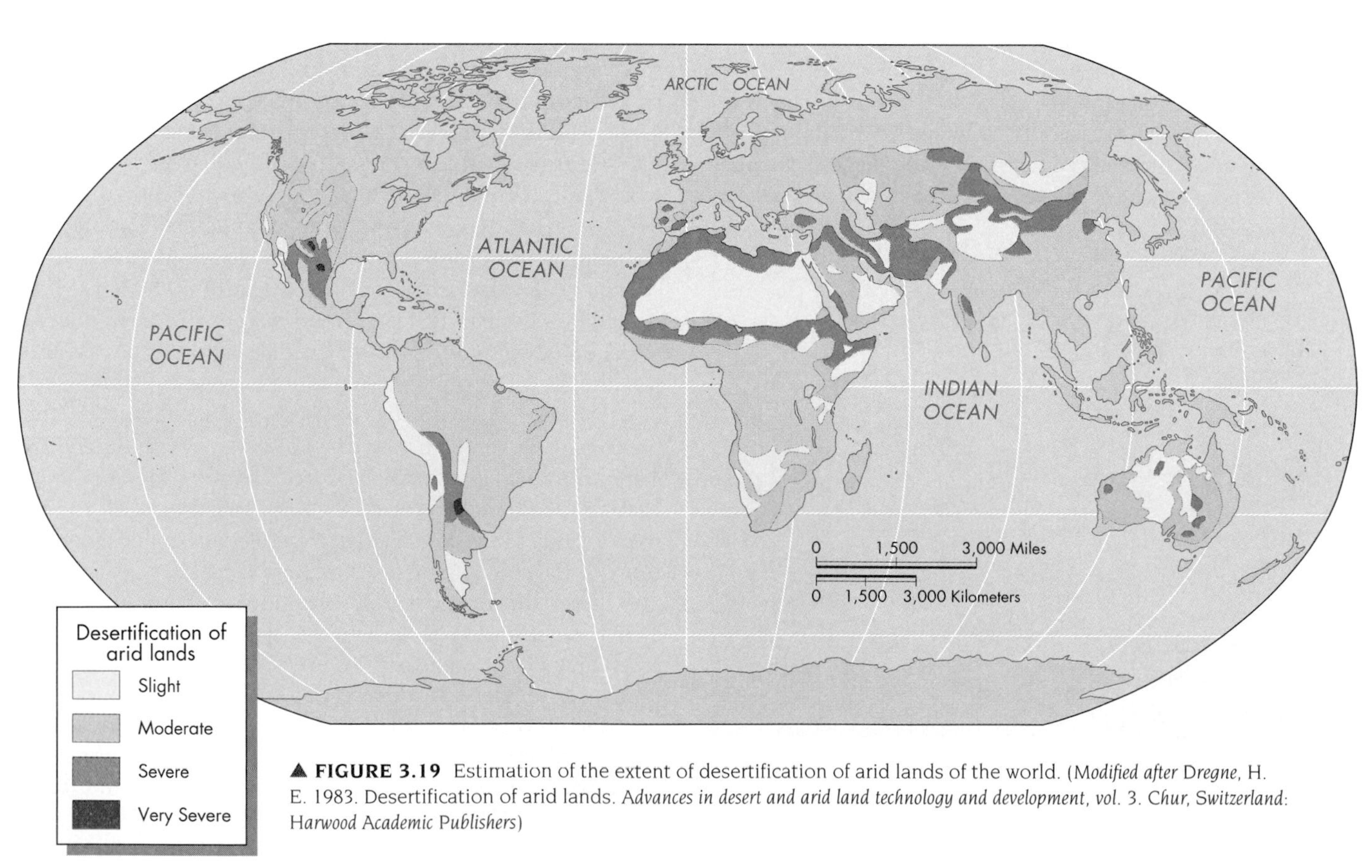

▲ **FIGURE 3.19** Estimation of the extent of desertification of arid lands of the world. (*Modified after* Dregne, H. E. 1983. Desertification of arid lands. *Advances in desert and arid land technology and development, vol.* 3. *Chur, Switzerland:* *Harwood Academic Publishers*)

- Declining groundwater table.
- Salinization of soil and near-surface-soil water.
- Reduction in areal extent of surface water in streams, ponds, and lakes.
- Unnaturally high rates of soil erosion.
- Damage to native vegetation.

An arid area undergoing desertification may have any or all of these symptoms to a lesser or greater extent (see *Case History: Desertification of the San Joaquin Valley*). Furthermore, the symptoms are interrelated: salinization of topsoil, for example, may lead to loss of vegetation, which then leads to accelerated soil erosion.

Prevention, minimization, and reversal of desertification involves the following (21,22):

- Protection and improvement of high-quality land rather than expending great amounts of time and money on poor land
- Application of simple, sound range management techniques to protect the land from overgrazing by livestock
- Application of sound conservation measures to agricultural lands to protect soil resources
- Use of appropriate technology to increase crop production that allows poorer lands to be returned to less-intensive land uses other than intensive agriculture (for example, forestry, wildlife, or grazing)
- Increased land restoration efforts through vegetation management, stabilization of sand dunes, and control of soil erosion

3.12 Soil Surveys and Land-Use Planning

Soils greatly affect the determination of the best use of land, and a soil survey is an important part of planning for nearly all engineering projects. A **soil survey** should include a soil description; soil maps showing the horizontal and vertical extent of soils; and tests to determine grain size, moisture content, shrink-swell potential, and strength. The purpose of the survey is to provide necessary information for identifying potential problem areas before construction.

The information from detailed soil maps can be extremely helpful in land-use planning if used in combination with guidelines to the proper use of soils. Soils can be rated according to their limitations for a specific land use, such as housing, light industry, septic-tank systems, roads, recreation, agriculture, and forestry. Soil characteristics that help determine these limitations include slope, water content, permeability, depth to rock, susceptibility to erosion, shrink-swell potential, bearing strength (ability to support a load, such as a building), and corrosion potential. Table 3.5 shows how some of these characteristics limit the use of land for recreation.

Table 3.5 Soil limitations for buildings in recreational areas

Soil Items Affecting Use	Degree of Soil Limitation[a]		
	None to Slight	**Moderate**	**Severe[b]**
Wetness	Well to moderately well drained soils not subject to ponding or seepage. Over 1.2 meters to seasonal water table.	Well and moderately well drained soils subject to occasional ponding or seepage. Somewhat poorly drained, not subject to ponding. Seasonal water table 0.6–1.2 meters.[c]	Somewhat poorly drained soils subject to ponding. Poorly and very poorly drained soils.
Flooding	Not subject to flooding	Not subject to flooding	Subject to flooding
Slope	0%–8%	8%–15%	15% +
Rockiness[d]	None	Few	Moderate to many
Stoniness[d]	None to few	Moderate	Moderate to many
Depth to hard bedrock	More than 1.5 meters	0.9–1.5 meters[c]	Less than 1 meter

[a]Soil limitations for septic-tank filter fields; hillside slippage, frost heave, piping, loose sand, and low bearing capacity when wet are not included in this rating, but must be considered. Soil ratings for these items have been developed.

[b]Soils rated as having severe soil limitations for individual cottage sites may be best from an aesthetic or use standpoint, but they do require more preparation or maintenance for such use.

[c]These items are limitations only where basements and underground utilities are planned.

[d]Rockiness refers to the abundance of stones or rock outcrops greater than 25 cm in diameter. Stoniness refers to the abundance of stone 8 to 25 cm in diameter.

Source: Reproduced from *Soil Surveys and Land Use Planning* (1966) by permission of the Soil Science Society of America.

CASE HISTORY Desertification of the San Joaquin Valley

California's San Joaquin Valley, one of the most productive agricultural areas in the world, is threatened by the forces of desertification. These forces include poor drainage of irrigated land, overgrazing, overuse of groundwater, cultivation of highly erodible soils, and damage resulting from off-road vehicles (18).

Figure 3.E shows the areas of the San Joaquin Valley with present or potential drainage problems. The drainage problem is primarily confined to the western and southern parts of the valley, where the precipitation is only about 12 cm annually. In many areas of the valley, the soils are becoming more saline, and salty water is rising closer to the surface. Today as many as 16,000 ha are affected by a high salty water table; by the year 2080, as many as 460,000 ha may be unproductive (21).

The foothill area in the southern part of the San Joaquin Valley is particularly vulnerable to erosion, owing to the combined effects of overgrazing, lack of windbreaks in agricultural lands, and off-road-vehicle activities. This was dramatically shown during and following a high-magnitude windstorm on December 20, 1977. Winds that may have locally reached velocities as high as 300 km per hour caused moderate to heavy damage to buildings, crops, automobiles, wildlife, and soil over an area of 2000 km^2. The winds mobilized about 42,000 tons per km^2 of soil over 600 km^2 of grazing lands in a 24-hour period. Five people were killed in automobile accidents caused by impaired visibility, and the windblown soil carried valley-fever spores present in the soil that triggered a significant increase in the incidence of the disease in distant downwind locations.

Along the flanks of the mountains at the southern end of the valley, soil erosion from the windstorm was severe. In some locations, the entire soil was stripped by the wind and even the bedrock was eroded. Animals were literally excavated from their burrows by the wind erosion and killed. Figure 3.F shows a fencepost wind-scoured by the December 1977 storm. The soil loss on this site was approximately 20 cm, but some sites experienced soil losses as high as 60 cm. The wind-scoured land was vulnerable to further erosion, and rainstorms the following February caused accelerated erosion and flooding (24).

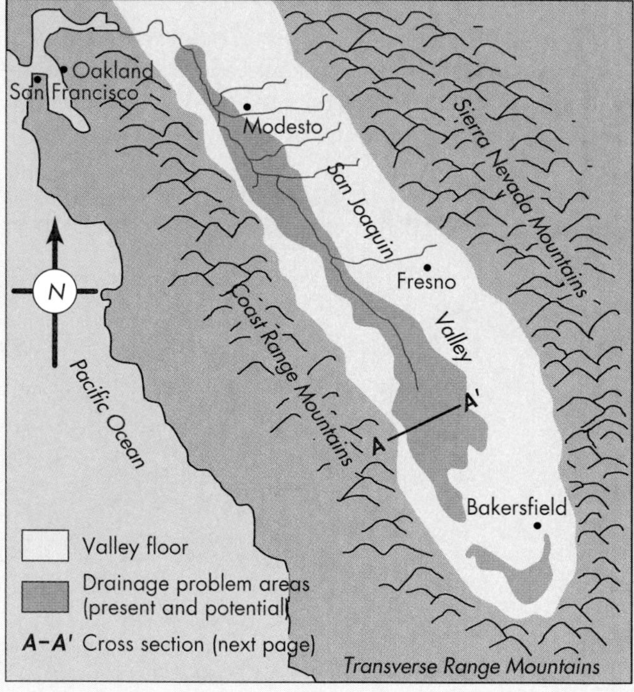

▶ **FIGURE 3.E** Drainage problems in the San Joaquin Valley, California. (right) Areas with present or potential drainage problems. (top, next page) Cross section showing surface landforms and drainage conditions. (*Modified from Sheridan,* Desertification of the United States. 1981. *Council on Environmental Quality*)

Land-use limitations for a specific area can be determined from a detailed soil map and the accompanying description of the soil types. As an example, consider Figure 3.20, which shows the results of a soil evaluation to determine limitations for buildings in a planned recreation area (23). Figure 3.20a is a detailed soil map and brief description of soils for 4.1 km^2 in Aroostook County, Maine. This information and the limitations shown in Table 3.5 were used to produce the limitations map in Figure 3.20b, which can be used as a guide to where buildings could be located.

The standard limitation classes shown in Table 3.5 are defined as follows:

- None to slight—soils are relatively free of limitations that affect the planned use or have limitations that can readily be overcome.
- Moderate—soils have moderate limitations resulting from effects of soil properties. These limitations can normally be overcome with correct planning, careful design, and good construction.

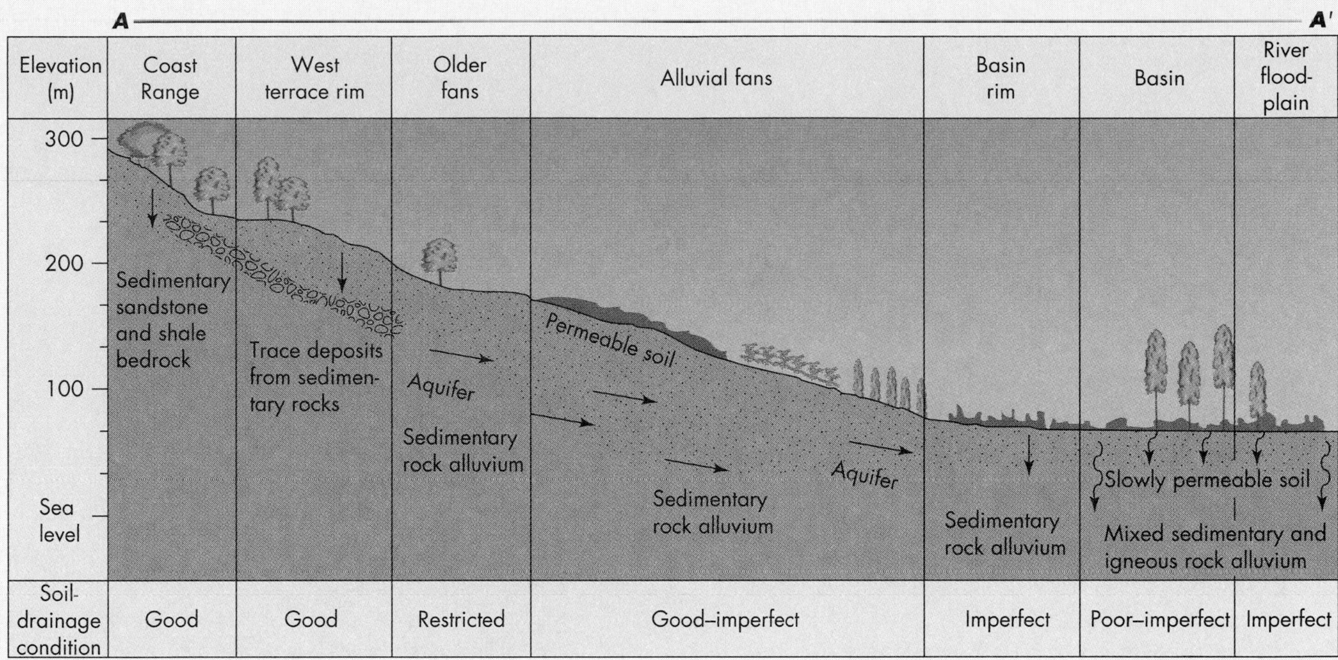

(b)

◀ **FIGURE 3.F** Windstorm damage in the San Joaquin Valley. This fencepost was scoured to a depth of 10.5 cm as a direct result of the December 20, 1977, windstorm. Total soil erosion on slopes in this area was about 20 cm. (*Courtesy of* H. G. Wilshire *and* U.S. *Geological Survey*)

- Severe—soils have severe limitations that make the proposed use doubtful. Building on these soils requires very careful planning and more extensive design and construction measures, possibly including major soil reclamation work (23).

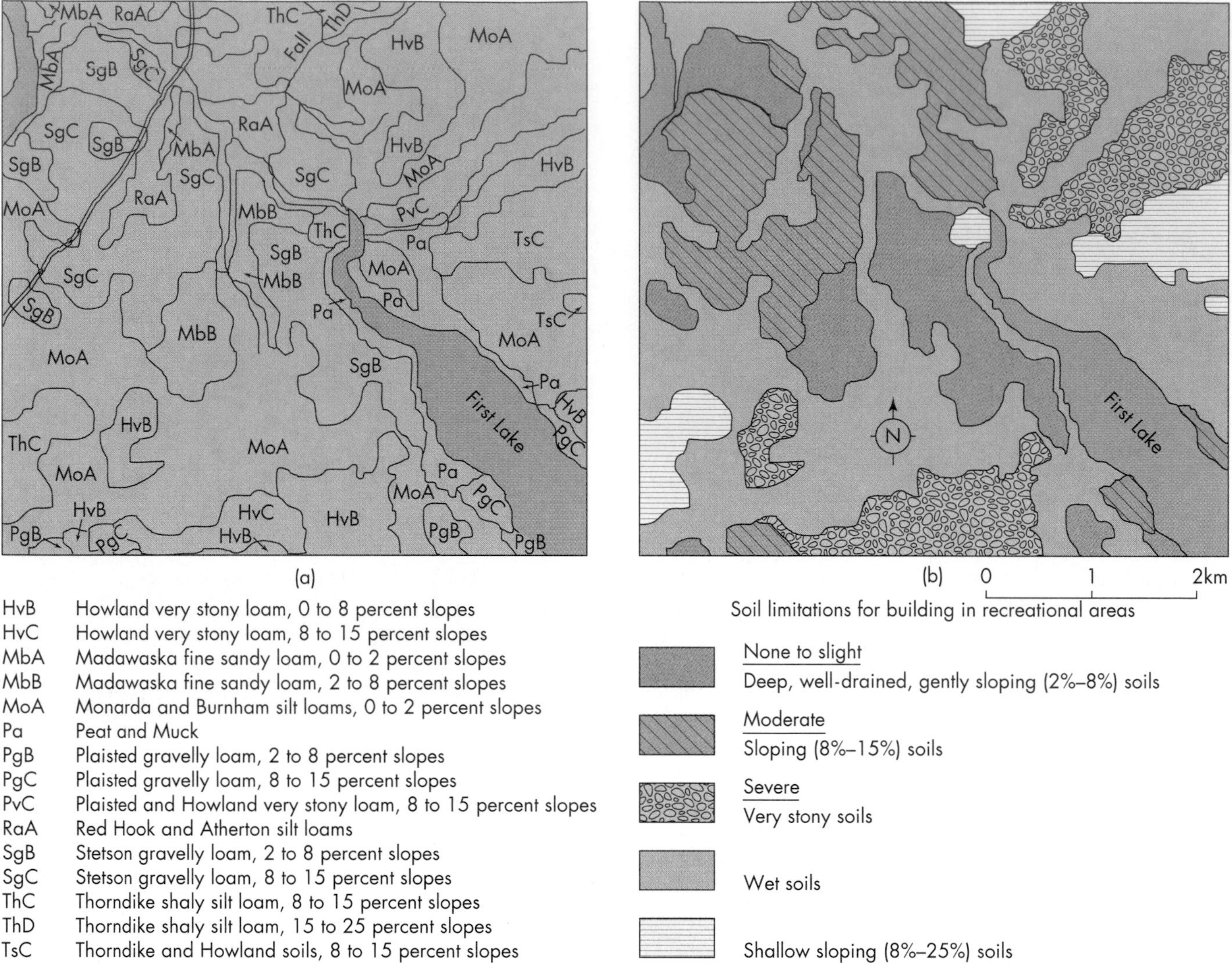

▲ **FIGURE 3.20** Use of soil maps for land-use planning. (a) Detailed soil map of a 4-km^2 tract of land, Aroostook County, Maine. (b) Soil limitations for buildings in recreational areas for the same tract. (*Reproduced from* Soil Surveys and Land Use Planning, 1966, *by permission of the Soil Science Society of America*)

SUMMARY

Engineers define soil as earth material that may be removed without blasting, whereas to a soil scientist, a soil is solid earth material that can support rooted plant life. A basic understanding of soils and their properties is becoming crucial in several areas of environmental geology, including land-use planning, waste disposal, and evaluation of natural hazards such as flooding, landslides, and earthquakes.

Soils result from interactions of the rock and hydrologic cycles. As open systems, they are affected by such variables as climate, topography, parent material, time, and organic activity. Soil-forming processes tend to produce distinctive soil layers, or horizons, defined by the processes that formed them and the type of materials present. Of particular importance are the processes of leaching, oxidation, and accumulation of materials in various soil horizons. Development of the argillic *B* horizon, for example, depends on the translocation of clay minerals from upper to lower horizons. Three important properties of soils are color, texture (particle size), and structure (aggregation of particles).

An important concept in studying soils is relative profile development. Very young soils are weakly developed. Soils older than 10,000 years tend to show moderate development, characterized by stronger development of soil structure, redder soil color, and more translocated clay in the *B* horizon. Strongly developed soils are similar to those of moderate development, but the properties of the *B* soil horizon tend to be better developed. Such soils can range in age from several tens of thousands of years to several hundred thousand years or older. A soil chronosequence is a series of soils arranged from youngest to oldest in terms of relative soil profile development. Establishment of a soil chronosequence in a region is useful in evaluating rates of processes and recurrence of hazardous events such as earthquakes and landslides.

A soil may be considered as a complex ecosystem in which many types of living things convert soil nutrients into forms that plants can use. *Soil fertility* refers to the capacity of the soil to supply nutrients needed for plant growth.

Soil has a solid phase consisting of mineral and organic matter; a gas phase, mostly air; and a liquid phase, mostly water. Water may flow vertically or laterally through the pores (spaces between grains) of a soil. The flow is either saturated (all pore space filled with water) or, more commonly, unsaturated (pore space partially filled with water). The study of soil moisture and how water moves through soils is becoming a very important topic in environmental geology.

Several types of soil classification exist, but none of these integrate both engineering properties and soil processes. Environmental geologists must be aware of both the soil-science classification (soil taxonomy) and the engineering classification (unified soil classification).

Basic understanding of engineering properties of soils is crucial in many environmental problems. These properties include plasticity, soil strength, sensitivity, compressibility, erodibility, permeability, corrosion potential, ease of excavation, and shrink-swell potential. Shrink-swell potential is particularly important because expansive soils in the United States today cause significant environmental problems and is one of our most costly natural hazards.

Rates of soil erosion can be determined by direct observation of soil loss from slopes, by measurement of accumulated sediment in reservoirs, or by calculation (the most common method). The Universal Soil Loss Equation uses variables that affect erosion to predict the amount of soil moved from its original position. These variables can often be manipulated as part of a management strategy for minimizing erosion and sediment pollution from a particular site. Sediment, both natural and human-made, may be one of our greatest pollutants. It reduces water quality and chokes streams, lakes, reservoirs, and harbors. With good conservation practice, sediment pollution can be significantly reduced.

Land use and manipulation of surface water affect the pattern, amount, and intensity of surface water runoff, soil erosion, and sediment pollution. Urbanization often involves loss of soil, change of soil properties, accelerated soil erosion during construction, and pollution of soils. Use of motorized and nonmotorized off-road vehicles causes soil erosion, changes in hydrology, and damage to plants and animals. Soil pollution occurs when hazardous materials are inadvertently or deliberately added to soils. Pollution limits the usefulness of soils or even renders them hazardous to life. Use of new pollution abatement technology such as bioremediation is assisting in solving some soil pollution problems. Desertification, now a major problem, is the conversion of land from a productive state to one more resembling a desert. It is associated with malnutrition and starvation, particularly in parts of Africa and India. Driving forces of desertification include overgrazing, deforestation, adverse soil erosion, poor drainage of irrigated land, overdraft of water supplies, and damage from off-road vehicles.

Soil maps are extremely useful in land-use planning. Soils can be rated according to their limitations for various land uses, and this information can be combined with information from detailed soil maps to produce a simplified map that shows limitations for a specific use.

REFERENCES

1. **Birkland, P. W.** 1984. *Soils and geomorphology.* New York: Oxford University Press.
2. **Brady, N. C., and Weil, R. R.** 1996. *The nature and properties of soils*, 11th ed. Upper Saddle River, NJ: Prentice Hall.
3. **Keller, E. A., Bonkowski, M. S., Korsch, R. J., and Shlemon, R. J.** 1982. Tectonic geomorphology of the San Andreas fault zone in the southern Indio hills, Coachella Valley, California. *Geological Society of America Bulletin* 93:46–56.
4. **Anonymous.** 1979. *Environmentally sound small-scale agricultural projects.* Mt. Rainier, MD: Mohonk Trust, Vita Publications.
5. **Olson, G. W.** 1981. *Soils and the environment.* New York: Chapman and Hall.
6. **Singer, M. J., and Munns, D. N.** 1996. *Soils*, 3rd ed. Upper Saddle River, NJ: Prentice Hall.
7. **Krynine, D. P., and Judd, W. R.** 1957. *Principles of engineering geology and geotechnics.* New York: McGraw-Hill.
8. **West, T. R.** 1995. *Geology applied to engineering.* Upper Saddle River, NJ: Prentice Hall.
9. **Pestrong, R.** 1974. *Slope stability.* New York: McGraw-Hill, American Geological Institute.
10. **Flawn, P. T.** 1970. *Environmental geology.* New York: Harper & Row.
11. **Hart, S. S.** 1974. Potentially swelling soil and rock in the Front Range Urban Corridor. *Environmental Geology* 7. Colorado Geological Survey.
12. **Mathewson, C. C., Castleberry, J. P., II, and Lytton, R. L.** 1975. Analysis and modeling of the performance of home foundations on expansive soils in central Texas. *Bulletin of the Association of Engineering Geologists* 17(4): 275–302.
13. **Jones, D. E., Jr., and Holtz, W. G.** 1973. Expansive soils: The hidden disaster. *Civil Engineering*, August: 49–51.
14. **Wischmeier, W. H., and Meyer, L. D.** 1973. Soil erodibility on construction areas. In *Soil erosion: Causes, mechanisms, prevention and control.* Highway Research Board Special Report 135. Washington, DC, pp. 20–29.
15. **Dunne, T., and Leopold, L. B.** 1978. *Water in environmental planning.* San Francisco: W. H. Freeman.

16. **Robinson, A. R.** 1973. Sediment: Our greatest pollutant? In *Focus on environmental geology*, ed. R. W. Tank, pp. 186–92. New York: Oxford University Press.
17. **Yorke, T. H.** 1975. Effects of sediment control on sediment transport in the northwest branch Anacostia River Basin, Montgomery County, Maryland. *U.S. Geological Survey Journal of Research* 3:487–94.
18. **Wilshire, H. G., and Nakata, J. K.** 1976. Off-road vehicle effects on California's Mojave Desert. *California Geology* 29:123–32.
19. **Wilshire, H. G., et al.** 1977. Impacts and management of off-road vehicles. *Geological Society of America.* Report to the Committee on Environment and Public Policy.
20. **Hazen, T. C.** 1995. Savanna River site—a test bed for cleanup technologies. *Environmental Protection,* April:10–16.
21. **Sheridan, D.** 1981. *Desertification of the United States.* Council on Environmental Quality, Washington, DC.
22. **Dregne, H. E.** 1983. Desertification of arid lands. *Advances in desert and arid land technology and development,* vol. 3. Chur, Switzerland: Harwood Academic Publishers.
23. **Montgomery, P. H., and Edminster, F. C.** 1966. Use of soil surveys in planning for recreation. In *Soil surveys and land use planning,* ed. L. J. Bartelli et al., pp. 104–12. Soil Science Society of America and American Society of Agronomy.
24. **Wilshire, H. G., and Nakata, J. K.** 1981. *Field observations of the December 1977 wind storm, San Joaquin Valley, California.* Geological Society of America Special Paper 86, pp. 233–51.

KEY TERMS

soil (p. 56)
weathering (p. 57)
soil horizons (p. 57)
soil profile development (p. 59)
relative profile development (p. 60)
soil chronosequence (p. 60)
soil fertility (p. 61)
unified soil classification system (p. 63)
soil strength (p. 63)
soil sensitivity (p. 67)
liquefaction (p. 67)
compressibility (p. 67)
erodibility (p. 67)
permeability (p. 68)
corrosion potential (p. 68)
shrink-swell potential (p. 68)
expansive soils (p. 68)
soil pollution (p. 77)
desertification (p. 77)
soil survey (p. 79)

SOME QUESTIONS TO THINK ABOUT

1. How and why could processes such as clear-cut logging (where all trees are cut) and use of off-road vehicles lead to loss of soil fertility?
2. One of your friends who is an environmentalist really likes to ride mountains bikes in steep terrain. She particularly likes racing downhill on ski slopes during the summer months. What are some conflicts she may have in reconciling her sport with potential damage to the environment?
3. Our discussion of desertification in North America mentioned several major symptoms. Consider whether any of these symptoms are present in the region where you live. Is your region undergoing desertification at some level?

PART 2

Hazardous Earth Processes

CHAPTER 4
Natural Hazards: An Overview

CHAPTER 5
Rivers and Flooding

CHAPTER 6
Landslides and Related Phenomena

CHAPTER 7
Earthquakes and Related Phenomena

CHAPTER 8
Volcanic Activity

CHAPTER 9
Coastal Hazards

The earth is a dynamic, evolving system with complex interactions of internal and external processes. *Internal* processes are responsible for moving giant lithospheric plates and continents. Interactions at plate junctions generate internal stress that causes rock deformation, resulting in earthquakes, volcanic activity, and tectonic creep (slow surface movement along a fault zone). These processes, in turn, trigger numerous external events such as landslides, mudflows, and tsunamis (giant sea waves). Other *external* events, such as water flow and the movement of waves, result from external interactions of the hydrosphere, atmosphere, and lithosphere, but even these events are affected by internal processes. For example, tremendous flooding can be caused by hurricane activity, but the nature and extent of the flooding is affected by landforms in the vicinity of the coast, which were produced by the internal workings of the earth. Likewise, the action of running water or waves is affected by the uplifted land surface they erode.

Hazardous earth processes, or *natural hazards*, periodically give rise to *natural disasters*—events that cause loss of human life or property. Most people pay little attention to these processes until disaster strikes. This casual attitude has complex causes: It reflects both the optimism people feel about their lives and homes and a limited understanding of natural phenomena. Part 2 of this book is designed to provide you with a basic understanding of how internal and external earth processes give rise to disasters and how people interact with these hazardous processes. Chapter 4 provides a brief overview of hazardous earth processes and natural disasters. Chapters 5 through 9 discuss specific types of destructive events: river floods, landslides, earthquakes, volcanic eruptions, and coastal processes. We will consider the natural and artificial influences on the magnitude (size or severity) and frequency of each type of disaster, as well as the effects of each type of hazard on human land use and activity.

4 Natural Hazards: An Overview

Earthquake damage, January 1995, Kobe, Japan. (*Shinya Inui/Friday/Sygma Photo News*)

The catastrophic January 1995 earthquake that devastated Kobe, Japan, causing damages in excess of $100 billion and the deaths of more than 5000 people, was a "wake-up call": our cities are vulnerable to natural hazards on a scale we had not previously envisaged. Japan was confident that it was prepared to respond to earthquakes, and yet they apparently were caught off-guard. Emergency relief did not arrive until about 10 hours after the earthquake, and buildings and other structures thought relatively safe failed catastrophically. Not only earthquakes have potential for catastrophe; in 1998 Hurricane Mitch—with winds in excess of 270 kilometers per hour—struck Central America, devastating the landscape as homes, schools, roads, and bridges were blown down or washed away by torrential rains. More than 11,000 people in Nicaragua and Honduras were killed, with at least that many more missing. Perhaps as much as one-third of the total population of Honduras were left homeless! The catastrophic effects of Hurricane Mitch were particularly devastating because much of the landscape has been altered by human use, such as deforestation and urbanization. As human population continues to increase during the twenty-first century, the effects of natural processes, such as hurricanes, floods, landslides, earthquakes and volcanic eruption, will become more damaging to human society simply because more people are at risk.

4.1 Hazards as Natural Processes

During the past two decades, natural disasters, such as earthquakes, floods, cyclones, and hurricanes, have killed several million people on this planet, with an average annual loss of life of about 150,000. The financial loss probably exceeds $20 billion and does not include social losses, such as loss of employment, mental anguish, and reduced productivity.

LEARNING OBJECTIVES

The study of hazardous processes constitutes one of the main activities of environmental geology. Learning objectives for the chapter are:

- *To know the conditions that make some natural earth processes hazardous to people.*
- *To understand how a natural process that gives rise to disasters may also be beneficial to people.*
- *To make a preliminary acquaintance with the various natural processes that constitute hazards to people and property.*
- *To understand the requirements for predicting natural disasters.*
- *To know the basic components of risk assessment for natural hazards.*
- *To become familiar with people's perceptions of and adjustments to hazards.*
- *To be able to discuss the impact and recovery from natural disasters and catastrophes.*
- *To understand why increase in population and changing land use, particularly in developing countries, increases the threat of loss of life and property from natural disasters.*

Web Resources

Visit the "Environmental Geology" Web site at www.prenhall.com/keller to find additional resources for this chapter, including:

- ▶ Web Destinations
- ▶ On-line Quizzes
- ▶ On-line "Web Essay" Questions
- ▶ Search Engines
- ▶ Regional Updates

Two individual disasters—a cyclone accompanied by flooding in Bangladesh in 1970 and an earthquake in China in 1976—each claimed more than 300,000 lives. In 1991 another cyclone struck Bangladesh, claiming another 145,000 lives (Figure 4.1). The 1995 earthquake in Kobe, Japan, claimed more than 5000 lives, destroyed many thousands of buildings, and caused more than $100 billion in property damage (Figure 4.2). These terrible disasters were caused by natural hazards that have always existed—atmospheric disturbance and tectonic movement—but their extent was affected by human population density and land-use patterns.

Natural hazards are basically natural processes. These processes may become hazardous when people live or work in areas where these processes occur, or when land-use changes, such as urbanization or deforestation, increases the occurrence and/or magnitude of processes such as flooding or landsliding. It is the environmental geologist's job to identify potentially hazardous processes and make this information available to planners and decision makers so that they can formulate various alternatives to avoid or minimize the threat to human life or property. However, the naturalness of hazards is a philosophical barrier that we encounter whenever we try to minimize their adverse effects.

The *impact* of a disastrous event is in part a function of its magnitude (amount of energy released) and frequency (recurrence interval), but it is influenced by many other factors,

▲ **FIGURE 4.1** Aftermath of the 1991 cyclone that devastated Bangladesh, killing approximately 145,000 people. (*Bartholomew/Liaison Agency, Inc.*)

▲ **FIGURE 4.2** The earthquake that struck Kobe, Japan, in January 1995 had a devastating effect on the people of the city while causing in excess of $100 billion damage. More than 5,000 people were killed. (*Mike Yamashita/Woodfin Camp and Associates*)

A CLOSER LOOK The Magnitude-Frequency Concept

The **magnitude-frequency concept** is the assertion that there is generally an inverse relationship between the magnitude of an event and its frequency. In other words, the larger the flood, the less frequently such a flood occurs. The concept also includes the idea that much of the work of forming the earth's surface is done by events of moderate magnitude and frequency, rather than by common processes with low magnitude and high frequency or extreme events of high magnitude and low frequency.

As an analogy to the magnitude-frequency concept, consider the work of logging a forest done by resident termites, human loggers, and elephants. The termites are numerous and work quite steadily, but they are so small that they can never do enough work to destroy all the trees. The people are fewer and work less often, but being stronger than termites they can accomplish more work in a given time. Unlike the termites, the people can eventually fell most of the trees (Figure 4.A). The elephants are stronger still and can knock down many trees in a short time, but there are only a few of them and they rarely visit the forest. In the long run the elephants do less work than the people and bring about less change.

In our analogy it is humans who, with a moderate expenditure of energy and time, do the most work and change the forest most drastically. Similarly, natural events with a moderate energy expenditure and moderate frequency are often the most important shapers of the landscape. For example, most of the sediment carried by rivers in regions with a subhumid climate (most of the eastern United States) is transported by flows of moderate magnitude and frequency. However, there are many exceptions. In arid regions, for example, much of the sediment in normally dry channels may be transported by rare high-magnitude flows produced by intense but infrequent rainstorms. Along the barrier-island coasts of the eastern United States, high-magnitude storms often cut inlets that cause major changes in the pattern and flow of sediment.

▲ **FIGURE 4.A** Human beings with our high technology are able to down even the largest trees in our old-growth forests. The lumberjack shown here is working in a national forest in the Pacific Northwest. (*William Campbell/Sygma Photo News*)

including climate, geology, vegetation, population, and land use. In general, the frequency of such an event is inversely related to the magnitude: Small earthquakes, for example, occur more often than do large ones (see *A Closer Look: The Magnitude-Frequency Concept*).

The 1990s were designated by the United Nations as the International Decade for Natural Hazards Reduction. The objective of the continuing UN program is to minimize loss of life and property damage resulting from natural hazards. Reaching this objective will require measures to mitigate both specific physical hazards and the biological hazards that often accompany them. For example, after earthquakes and floods, water may be contaminated by bacteria, increasing the spread of diseases.

Benefits of Natural Hazards

It is ironic that the same natural events that take human life and destroy property also provide us with important benefits. River flooding supplies nutrients to floodplains, as in the case of the Mississippi River or the Nile Delta prior to the building of the Aswan Dam. Flooding also causes erosion on mountain slopes, delivering sediment to beaches from rivers and flushing pollutants from estuaries in the coastal environment. Landslides bring benefits to people when landslide debris form dams, making lakes in mountainous areas. These lakes provide valuable water storage and are an important aesthetic resource.

Volcanic eruptions, while having the potential to produce real catastrophes, provide us with numerous benefits. They often create new land: The Hawaiian Islands, for example, are completely volcanic in origin (Figure 4.3). Nutrient-rich volcanic ash may settle on existing soils and quickly become incorporated in them. Earthquakes, too, provide us with valuable services. When rocks are pulverized during an earthquake, they may form an impervious clay zone known as **fault gouge** along the fault. In many places, fault gouge has formed a groundwater barrier upslope from a fault, producing a natural subsurface dam and a water resource. Earthquakes are also important in mountain building and thus are directly responsible for many of the scenic resources of the western United States.

Death and Damage Caused by Natural Hazards

When we compare the effects of various natural hazards, we find that those that cause the greatest loss of human life are not necessarily the same as those that cause the most extensive property damage. Table 4.1 summarizes selected information about the effects of natural hazards in the United States. The largest number of deaths each year is asso-

(a)

(b)

▲ **FIGURE 4.3** New land being added to the island of Hawaii. The plume of smoke in the central part of the photograph is where hot lava is entering the sea (a). Close-up of an advancing lava front near the smoke plume (b). (*Edward A. Keller*)

ciated with tornadoes and windstorms, although lightning, floods, and hurricanes also take a heavy toll. Loss of life due to earthquakes can vary considerably from one year to the next, as a single great quake can cause tremendous human loss. It is estimated that a great earthquake in a densely populated part of California could inflict $100 billion in damages while killing several thousand people (1). The 1994 Northridge earthquake (large but not great) in the Los Angeles area killed some 60 people and inflicted about $20 to $30 billion in property damage. Property damage from individual hazards is considerable. Floods, landslides, expansive soils, and frost each cause mean annual damages in the United States in excess of $1.5 billion. Surprisingly, expansive soils are one of the most costly hazards, causing more than $3 billion in damages annually.

Table 4.1 Effects of selected hazards in the United States

Hazard	Deaths per Year	Occurrence Influenced by Human Use	Catastrophe Potential[b]
Flood	86	Yes	H
Earthquake[a]	50+?	Yes	H
Landslide	25	Yes	M
Volcano[a]	<1	No	H
Coastal erosion	0	Yes	L
Expansive soils	0	No	L
Hurricane	55	Perhaps	H
Tornado and windstorm	218	Perhaps	H
Lightning	120	Perhaps	L
Drought	0	Perhaps	M
Frost and freeze	0	Yes	L

[a]Estimate based on recent or predicted loss over 150-year period. Actual loss of life and/or property could be much greater.
[b]Catastrophe potential: high (H), medium (M), low (L).

Source: Modified after G. F. White and J. E. Haas. 1975. *Assessment of Research on Natural Hazards*. Cambridge, MA: The MIT Press.

An important aspect of all natural hazards is their potential to produce a **catastrophe,** defined as any situation in which the damages to people, property, or society in general are sufficient that recovery and/or rehabilitation is a long, involved process (2). Table 4.1 gives the catastrophe potential—high, medium, or low—for the hazards considered. The events most likely to produce a catastrophe are floods, hurricanes, tornadoes, earthquakes, volcanic eruptions, and large wildfires. Landslides, because they generally cover a smaller area, have a moderate catastrophe potential. The catastrophe potential of drought is also moderate: Though a drought may cover a wide area, there is usually plenty of warning time before its worst effects are experienced. Hazards with a low catastrophe potential include coastal erosion, frost, lightning, and expansive soils (2).

The effects of natural hazards change with time because of changes in land-use patterns, which influence people to develop on marginal lands; urbanization, which changes the physical properties of earth materials; and increasing population. Damage from most hazards in the United States is increasing, but the number of deaths from many hazards is decreasing because of better prediction, forecasting, and warning of hazards.

4.2 Evaluating Hazards: Disaster Prediction and Risk Assessment

Learning how to predict disasters so we can minimize human loss and property damage is an important endeavor. For each particular hazard we have a certain amount of information—

enough in some cases to forecast events accurately. When there is insufficient information to make accurate predictions, the best we can do is to locate areas where disastrous events have occurred and infer where and when similar future events might take place. If we know both the probability and the possible consequences of an event occurring at a particular location, we can assess the risk the event poses to people and property, even if we cannot accurately predict when it will occur.

Disaster Prediction and Warning

The effects of a specific disaster can be reduced if we can forecast the event and issue a warning. Attempting to do this in a given situation involves most or all of the following elements: identifying the location of a hazard, determining the probability that an event of a given magnitude will occur, observing precursor events, predicting the event, and issuing a warning.

Location For the most part, we know *where* a particular kind of event is likely to occur. On a global scale, the major zones for earthquakes and volcanic eruptions have been delineated by mapping earthquake epicenters and recent volcanic rocks and volcanoes. On a regional scale, we know from past eruptions which areas in the vicinity of certain volcanoes are likely to be threatened by large mudflows or ash in the event of future eruptions. This risk has been delineated for several Cascade volcanoes, including Mt. Rainier, and for specific volcanoes in Japan, Italy, Colombia, and elsewhere. On a local scale, detailed work with soils, rocks, and hydrology may identify slopes that are likely to fail (landslide) or where expansive soils exist. Certainly we can predict where flooding is likely to occur, from the location of the floodplain and such evidence from recent floods as the flood debris and high-water line.

Probability of Occurrence Determining the probability that a particular event will occur in a particular location within a particular time span is an essential part of a hazard prediction. For many rivers we have sufficiently long records of flow to develop probability models that can reasonably predict the average number of floods of a given magnitude that will occur in a decade. Likewise, droughts may be assigned a probability on the basis of past occurrence of rainfall in the region. However, these probabilities are similar to the chances of throwing a particular number on a die or drawing an inside straight in poker (this is the element of chance). Although a 10-year flood may occur on the average only once every 10 years, it is possible for several floods of this magnitude to occur in any one year, just as it is possible to throw two straight sixes with a die.

Precursor Events Many hazardous events are preceded by **precursor events.** For example, the surface of the ground may creep (move slowly) for a long period prior to an actual landslide. Often the rate of creep increases up to the final failure and landslide. Volcanoes sometimes swell or bulge before an eruption, and often a significant increase occurs in local seismic activity in the area surrounding the volcano. Foreshocks or anomalous (unusual) uplift may precede earthquakes. Precursor events help predict when and where an event is likely to happen. Landslide creep or swelling of a volcano may result in a warning being issued and people evacuated from a hazardous area.

Forecasting With some natural processes it is possible to **forecast** accurately when the event will arrive. Flooding of the Mississippi River, which occurs in the spring in response to snowmelt or very large regional storm systems, is fairly predictable, and we can sometimes forecast when the river will reach a particular flood stage. When hurricanes are spotted far out to sea and tracked toward the shore, we can forecast when and where they will likely strike land. Tsunamis, or seismic sea waves, generated by disturbance of ocean waters by earthquakes or submarine volcanoes, may also be forecast. The tsunami warning system has been fairly successful in the Pacific Basin, and in some instances the time of arrival of the waves has been forecast precisely.

Warning After a hazardous event has been predicted or a forecast has been made, the public must be warned. The flow of information leading to the **warning** of a possible disaster such as a large earthquake or flood should move along a path similar to that shown in Figure 4.4. The public does not always welcome such warnings, however, especially when the predicted event does not come to pass. In 1982, when geologists advised that a volcanic eruption near Mammoth Lakes, California, was quite likely, the advisory caused loss of tourist business and apprehension on the part of the residents. The eruption did not occur and the advisory was eventually lifted. In July 1986, a series of earthquakes occurred over a four-day period in the vicinity of Bishop, California, in the eastern Sierra Nevada, beginning with an earthquake of magnitude 3 and culminating in a damaging earthquake of magnitude 6.1. Investigators concluded there was a high probability a larger quake would occur in the vicinity in the near future and issued a warning. Local business owners, who feared the loss of summer tourism, felt that the warning was irresponsible; in fact, the predicted quake never materialized.

Incidents of this kind have led some people to conclude that scientific predictions are worthless and that advisory warnings should not be issued. Part of the problem is poor communication between the investigating scientists and reporters for the media (see *A Closer Look: Scientists, Hazards, and the Media*). Newspaper, television, and radio reports may fail to explain the evidence or the probabilistic nature of disaster prediction, leading the public to expect completely reliable statements as to what will happen. Although scientists are not yet able to predict volcanic eruptions and earthquakes accurately, it would seem that they have a responsibility to publicize their informed judgments. An informed public is better able to act responsibly than an uninformed public, even if the subject makes people un-

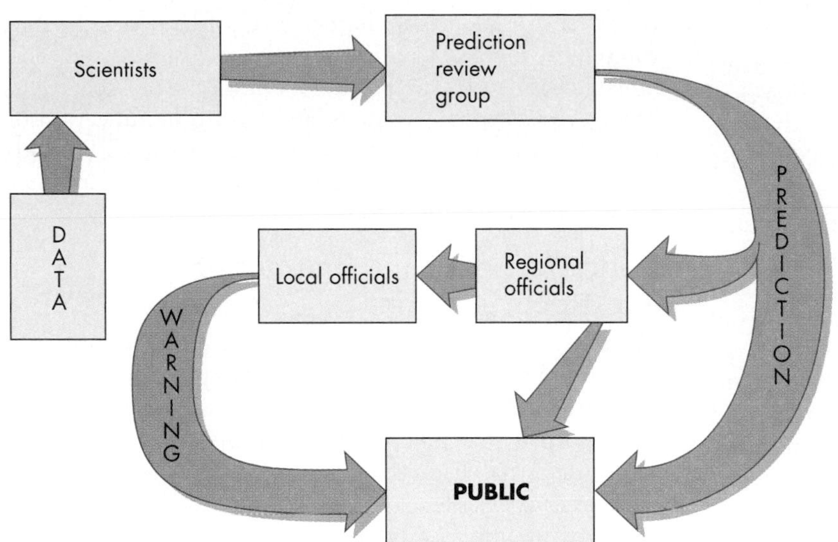

◀ **FIGURE 4.4** Possible flow path for issuance of a prediction or warning for a natural disaster.

comfortable. Ship captains, who depend on weather advisories and warnings of changing conditions, do not suggest that they would be better off not knowing about an impending storm, even though the storm might veer and miss the ship. Just as weather warnings have proved very useful for planning, official warnings of hazards such as earthquakes, landslides, and floods will also be useful to people making decisions about where they live, work, and travel.

Consider once more the prediction of a volcanic eruption in the Mammoth Lake area of California. The seismic data suggested to scientists that molten rock was moving toward the surface. In view of the high probability that the volcano would erupt and the possibility of loss of life if it did, it would have been irresponsible for scientists not to issue an advisory. Although the eruption did not occur, the warning led to the development of evacuation routes and consideration of disaster preparedness. This planning may prove very useful, for it is very likely that a volcanic eruption will occur in the Mammoth Lake area in the future. The most recent event occurred only 600 years ago! As a result of the prediction, the community is better informed than it was before and thus better able to deal with an eruption when it does occur.

A CLOSER LOOK Scientists, Hazards, and the Media

People today learn what is happening in the world by watching television, browsing the Web, listening to the radio, or reading newspapers and magazines. Reporters for the media are generally more interested in the impact of a particular event on people than in its scientific aspects. Even major volcanic eruptions or earthquakes in unpopulated areas may receive little media attention, whereas moderate or even small events in populated areas are reported in great detail. The news media want to sell stories, and what sells is spectacular events that affect people and property (3).

In a perfect world we would like to see good relations between scientists and the news media, but this lofty ideal may be difficult to achieve. Scientists tend to be conservative, critical people, afraid of being misquoted. They may perceive reporters as pushy and aggressive, or as willing to present half-truths while playing up differences of scientific opinion to embellish a story. Reporters, on the other hand, may perceive scientists as uncooperative and aloof, speaking an impenetrable jargon and unappreciative of the deadlines that reporters face (3). These statements about scientists and communicators are obviously stereotypic. Both groups have high ethical and professional standards; nevertheless, communication problems and conflicts of interest often occur.

Because scientists have an obligation to provide the public with information about natural hazards, it is good policy for a research team to pick one spokesperson to talk to the media so that the information is presented as consistently as possible. Suppose, for example, that scientists are studying a swarm of earthquakes near Los Angeles and there is speculation among them about the significance of the swarm. The development of several working hypotheses and future scenarios is the general rule for earth scientists working on a problem. However, when these scientists are dealing with the news media on a topic that concerns people's lives and property, it is better for them to report a consensus than a variety of opinions, or the public may be led to believe that they don't know what they are talking about. Their reports should be conservative evaluations of the evidence at hand, presented with as little jargon as possible. Reporters, for their part, should strive to provide their readers, viewers, or listeners with accurate information that the scientists have verified. Embarrassing scientists by misquoting them will only lead to more mistrust and poor communication between scientists and journalists.

Risk Assessment

Before rational people can discuss and consider adjustments to hazards, they must have a good idea of the risk that they face under various scenarios. The field of risk assessment is a rapidly growing one in the analysis of hazards, and its use and application should probably be expanded.

The **risk** of a particular event is defined as the product of the probability of that event occurring times the consequences should it occur (4). Consequences (damages to people, property, economic activity, public service, and so on) may be expressed in a variety of scales. If, for example, we are considering the risk from earthquake damage to a nuclear reactor, we may evaluate the consequences in terms of radiation released, which then can be related to damages to people and other living things. In any such assessment, it is important to calculate the risks for various possible events—in this example, for earthquakes of various magnitudes. A large event has a lower probability of occurring than does a small one, but its consequences are likely to be greater.

Determining *acceptable risk* is more complicated, for the risk that society or individuals are willing to take depends upon the situation. Driving an automobile is fairly risky, but most of us accept that risk as part of living in a modern world. On the other hand, acceptable risk from a nuclear power plant is very low because we consider almost any risk of radiation poisoning unacceptable. Nuclear power plants are controversial because many people perceive them as high-risk facilities. Even though the probability of an accident due to a geologic hazard, such as an earthquake, may be quite low, the consequences could be high, resulting in a relatively high risk.

A frequent problem of risk analysis is lack of reliable data for analyzing either the probability or the consequences of an event. It can be very difficult to assign probabilities to geologic events, such as earthquakes and volcanic eruptions, because the known chronology of past events is often very inadequate (4). Similarly, it may be very difficult to determine the consequences of an event or series of events. For example, if we are concerned with the consequences of releasing radiation into the environment, we need a lot of information about the local biology, geology, hydrology, and meteorology, all of which may be complex and difficult to analyze. Despite these limitations, risk analysis is a step in the right direction. As we learn more about determining the probability and consequences of a hazardous event, we should be able to provide the more reliable analyses necessary for decision making.

4.3 The Human Response to Hazards

The ways in which we deal with hazards are too often primarily *reactive:* Following a disaster, we engage in search and rescue, fire fighting, and providing emergency food, water, and shelter. There is no denying that these activities reduce loss of life and property and need to be continued. However, the move to a higher level of hazard reduction will require increased efforts to *anticipate* disasters and their impact. Land-use planning to avoid hazardous locations, hazard-resistant construction, and hazard modification or control (such as flood control channels) are some of the adjustments that anticipate future disastrous events and may reduce our vulnerability to them (1).

Reactive Response: Impact of and Recovery from Disasters

The impact of a disaster upon a population may be either direct or indirect. *Direct effects* include people killed, injured, dislocated, or otherwise damaged by a particular event. *Indirect effects* are generally responses to the disaster. They include emotional distress, donation of money or goods, and the paying of taxes levied to finance the recovery. Direct effects have an impact on fewer individuals, whereas indirect effects have an impact on many more people (5,6).

The stages of recovery following a disaster are emergency work, restoration of services and communication lines, and reconstruction. Figure 4.5 shows an idealized model of recovery. This model can be applied to actual recovery activities following events such as the 1994 Northridge earthquake in the Los Angeles area. Restoration following the earthquake began almost immediately (roads were repaired, utilities restored, etc.) in response to an influx of dollars from federal programs, insurance companies, and other sources in the first few weeks and months after the earthquake. The damaged areas moved quickly from the restoration phase to the Reconstruction I stage (which will last beyond the year 2000).

As we move into the Reconstruction II period following the Northridge earthquake, it is important to remember lessons from two past disasters: The 1964 earthquake that struck Anchorage, Alaska, and the flash flood that devastated Rapid City, South Dakota, in 1972. Restoration following the earthquake in Anchorage began almost immediately in response to a tremendous influx of dollars from federal programs, insurance companies, and other sources approximately one month after the earthquake. As a result, reconstruction was a hectic process, with everyone trying to obtain as much of the available funds as possible. In Rapid City, the restoration did not peak until approximately 10 weeks after the flood, and the community took time to carefully think through the best alternatives. As a result, Rapid City today has an entirely different land use on the floodplain, and the flood hazard is much reduced. Conversely, in Anchorage the rapid restoration and reconstruction were accompanied by little land-use planning. Apartments and other buildings were hurriedly constructed across areas that had suffered ground rupture and were simply filled in and regraded. In ignoring the potential benefits of careful land-use planning, Anchorage is vulnerable to the same type of earthquake that struck in 1964. In Rapid City, the floodplain is now a green belt with golf courses and other such activities—a change that has reduced the flood hazard (2,5,6).

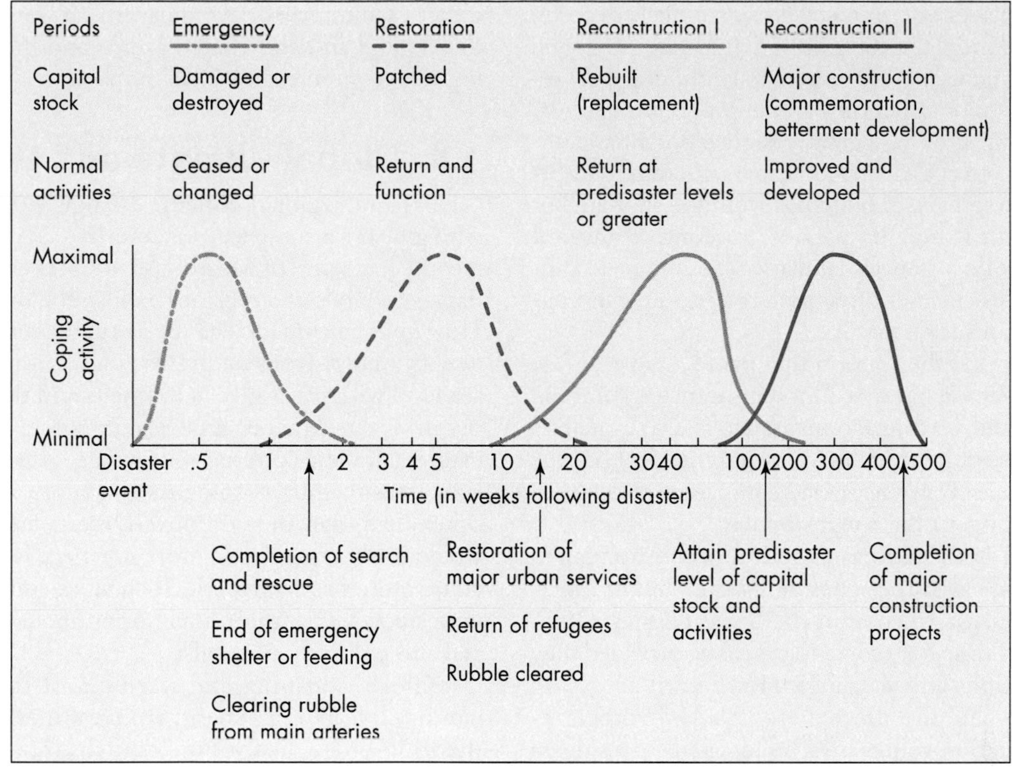

▲ **FIGURE 4.5** Generalized model of recovery following a disaster. (*From Kates and Pijawka*, 1977. *Reconstruction following disaster. In* From rubble to monument: The pace of reconstruction, *eds.* J. E. Haas, R. W. Kates, *and* M. J. Bowden. *Cambridge*, MA: MIT *Press*)

In the Northridge case, the effects of the earthquake on highway overpasses and bridges, buildings, and other structures are being carefully evaluated to determine how improved engineering standards for construction of new structures or strengthening of older structures might be implemented during the Reconstruction II period (see Figure 4.5). Future moderate-to-large earthquakes are certain to occur again in the Los Angeles area. Therefore, we need to continue efforts in the area of earthquake hazard reduction.

Anticipatory Response: Avoiding and Adjusting to Hazards

The options we choose, individually or as a society, for avoiding or minimizing the impacts of disasters depend in part on our hazard perception. A good deal of work has been done in recent years to try to understand how people perceive various natural hazards. This is important because the success of hazard reduction programs depends on the attitudes of the people likely to be affected by the hazard. Although there may be an adequate perception of a hazard at the institutional level, this may not filter down to the general population. This is particularly true for events that occur infrequently; people are more aware of situations such as brush or forest fires (Figure 4.6) that may occur every few years. There may even be institutionalized as well as local ordinances to control damages resulting from these events. For example, homes in some areas of southern California are roofed with shingles that do not burn readily and sometimes even have sprinkler systems, and the lots are often cleared of brush. Such safety measures are often noticeable during the rebuilding phase following a fire.

▲ **FIGURE 4.6** Wildfire in October 1991 devastated this Oakland, California neighborhood. (*Tom Benoit/Tony Stone Images*)

One of the most environmentally sound adjustments to hazards involves **land-use planning.** That is, people can avoid building on floodplains, in areas where there are active landslides, or in places where coastal erosion is likely to occur. In many cities, floodplains have been delineated and zoned for a particular land use. With respect to landslides, legal requirements for soil engineering and engineering geology studies at building sites may greatly reduce potential damages. Damages from coastal erosion can be minimized by requiring adequate setback of buildings from the shoreline or seacliff. Although it may be possible to control physical processes in specific instances, land-use planning to accommodate natural processes is often preferable to a technological fix that may or may not work.

Insurance is another option that people may exercise in dealing with natural hazards. Flood insurance is common in many areas, and earthquake insurance is also available. However, because of large losses following the 1994 Northridge earthquake, several insurance companies announced they would no longer offer the insurance.

Evacuation is an important option or adjustment to the hurricane hazard in the states along the Gulf of Mexico and along the eastern coast of the United States. Often there is sufficient time for people to evacuate, provided they heed the predictions and warnings. However, if people do not react quickly and the affected area is a large urban region, then evacuation routes may be blocked by residents leaving in a last-minute panic. Successful evacuation from volcanic eruptions is mentioned in Chapter 8.

Disaster preparedness is an option that individuals, families, cities, states, or even entire nations can implement. Of particular importance here is training individuals and institutions to handle large numbers of injured people or people attempting to evacuate an area after a warning is issued.

Attempts at *artificial control of natural processes* such as landslides, floods, and lava flows have had mixed success. Even the best-designed artificial structures cannot be expected to always defend against an extreme event. Retaining walls and other structures to defend slopes from failure by landslide have generally been successful when well designed. Even the casual observer has probably noticed the variety of such structures along highways and urban land in hilly areas. Structures to defend slopes have limited impact on the environment and are necessary where construction demands that artificial cuts be excavated or where unstable slopes impinge on human structures. Common methods of flood control are channelization and construction of dams and levees. Unfortunately, flood control projects tend to provide floodplain residents with a false sense of security because no method can be expected to protect people and their property absolutely from high-magnitude floods. We will return to this discussion in Chapter 5.

An option that all too often is chosen is simply bearing the loss caused by a natural disaster. Many people are optimistic about their chances of making it through any sort of disaster and therefore will take little action in their own defense. This is particularly true for those hazards—such as volcanic eruptions and earthquakes—that occur only rarely in a particular area. Regardless of the strategy we choose either to minimize or avoid hazards, it is imperative that we understand and anticipate hazards and their physical, biological, economic, and social impacts.

4.4 Global Climate and Hazards

Global and regional climatic change, possibly associated with global warming (see Chapter 16), may significantly affect the incidence of hazardous natural events, such as storm damage (floods and erosion), landslides, drought, and fires. How might a climatic change affect the magnitude and frequency of disastrous natural events? With global warming, sea level will rise as glacial ice melts and thermally warmed ocean waters expand. As a result, coastal erosion will increase. Climatic patterns will change, causing food production areas to shift as some receive more precipitation and others less than they do now. Deserts and semiarid areas would likely expand, and more northern latitudes could become more productive. Such changes could lead to global population shifts, which might bring about wars or major social and political upheavals.

Global warming and warming of the oceans of the world will feed more energy from warmer ocean water into the atmosphere, likely increasing the frequency and severity of hazardous weather-related processes, including thunderstorms (with tornadoes) and hurricanes. In fact, this may already be happening, because 1998 set a new record for economic losses from weather-related disasters, causing at least \$89 billion in economic losses worldwide. This represents a 48 percent increase over the previous record of \$60 billion set in 1996. In fact, losses from storms, floods, fires, and droughts in 1998, the warmest year on record, are greater than the \$55 billion in losses for the entire decade of the 1980s (Figure 4.7). Global impact on people from weather-related disasters in 1998 was catastrophic, killing approximately 32,000 people while displacing another 300 million people from their homes (7).

4.5 Population Increase, Land-Use Change, and Natural Hazards

Population Increase and Hazardous Events

Population increase is a major environmental problem. As our population continues to increase, putting greater demands on our land and resources, the need for planning to minimize losses from natural disasters also increases. Specifically, an increase in population puts a greater number of people at risk from a natural event and forces more people to settle in hazardous areas, creating additional risks. The risks of both high population density and living in a danger zone are dramatically illustrated by the loss of thousands of lives in Mexico and Colombia in 1985.

Mexico City is the center of the world's most populous urban area. Approximately 23 million people are concentrated in an area of about 2300 km^2, and about

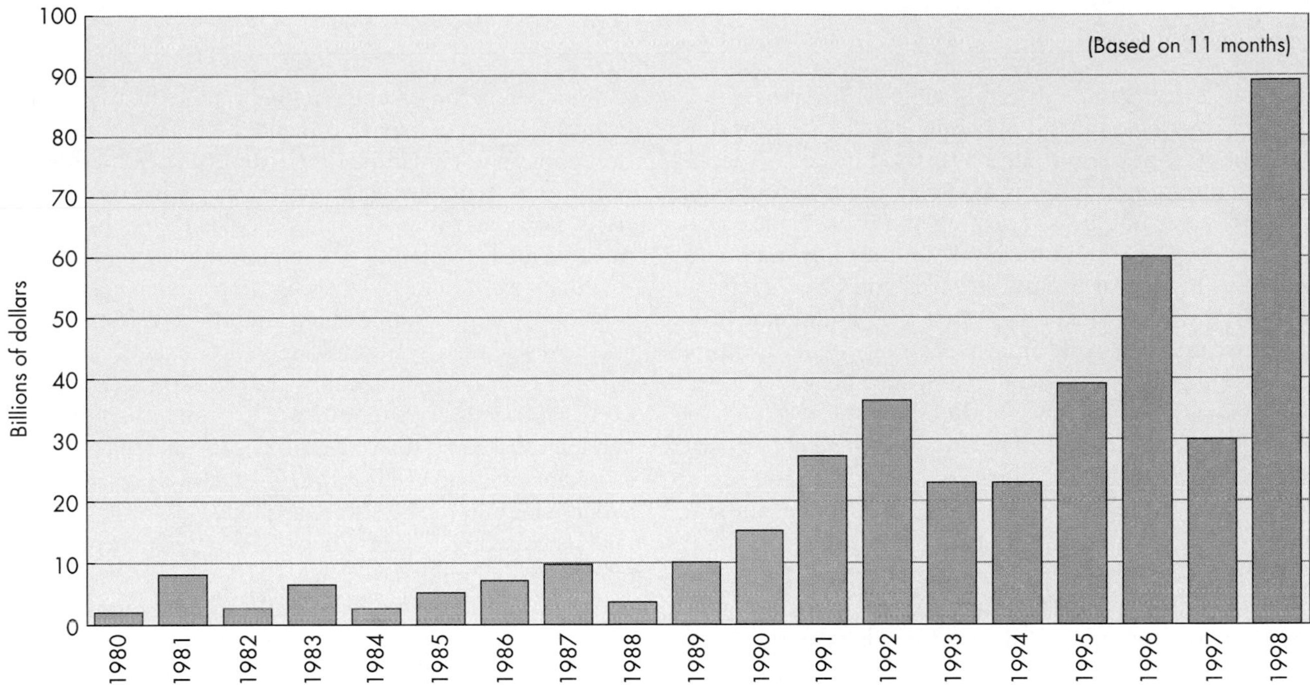

▲ **FIGURE 4.7** Worldwide economic losses from weather-related natural hazards (1980–1998). (*Source: World Watch Institute*, 1998. *Vital Signs Brief* 98-5)

one-third of the families (which average five members) live in a single room. The city is built on ancient lake beds, which accentuate earthquake shaking, and parts of the city have been sinking at the rate of a few centimeters per year, owing in part to groundwater withdrawal. The subsidence has not been uniform, so the buildings tilt and are even more vulnerable to the shaking of earthquakes (8). In September 1985, Mexico endured a magnitude 7.8 earthquake that killed about 10,000 people in Mexico City alone.

When the Colombian volcano Nevado del Ruiz erupted in 1845, a mudflow roared down the east slope of the mountain, killing about 1000 people. Deposits from that event produced rich soils in the Lagunilla River valley, and an agricultural center developed there. The town that the area supported was known as Armero, and by 1985 it had a population of about 22,500. On November 13, 1985, another mudflow associated with a volcanic eruption buried Armero, leaving about 21,000 people dead or missing (see Figure 8.22). A matter of 140 years multiplied the mudflow toll more than 20 times because of population increase. Ironically, the area was decimated by the same event that earlier produced productive soils, stimulating development and population growth (9). The real tragedy is that the mudflow was predicted and evacuation could have saved thousands of lives (3).

Land-Use Change and Hazardous Events

Two of the deadliest catastrophes resulting from natural hazards in 1998 were the flooding of the Yangtze River in China and Hurricane Mitch, which devastated Central America. Hurricane Mitch caused approximately 11,000 deaths, while the floods in the Yangtze River resulted in nearly 4,000 deaths. It has been speculated that damages in Central America and China from these events were particularly severe because of land-use changes that had occurred. For example, Honduras has lost nearly one-half of its forests, and a 11,000-km^2 fire occurred in the region prior to the hurricane. As a result of deforestation and the fire, hillsides that were stripped of vegetation washed away, and with them went farms, homes, roads, and bridges. In central China the story is much the same as the Yangtze River basin has lost about 85 percent of its forest as a result of timber harvesting and conversion of land to agriculture in recent times. As a result of the land-use changes in China, flooding of the Yangtze River is probably much more common than it was previously (7).

The hazardous events that caused catastrophes in 1998 in Central America, China, and other parts of the world may be an early warning sign of things to come. It is apparent that human activities are likely increasing the impacts of natural disasters. In recognition of this, China has banned timber harvesting in the upper Yangtze River basin, unwise floodplain land uses have been prohibited, and several billion dollars have been allocated for reforestation. The lesson being learned is that if we wish to minimize damages from natural hazards in the future, we need to consider land rehabilitation with the goal of achieving sustainable development based on restoration and maintenance of healthy ecosystems (7). It will be difficult, given pressures of human population growth in many parts of the world. This emphasizes the need to control human population growth if we are to solve pressing environmental problems and reach our goal of sustaining our environment.

SUMMARY

Our discussion of natural processes suggests a view of nature as dynamic and changing. This understanding tells us that we cannot view our environment as fixed in time. A landscape without natural hazards would also have less variety; it would be safer but less interesting and probably less aesthetically pleasing. The jury is still out on how much we should try to control natural hazards and how much we should allow them to occur. However, we should remember that disturbance is natural and that management of natural resources must include management for and with disturbances such as fires, storms, and floods.

A fundamental principle of environmental geology is that there have always been earth processes dangerous to people. These become hazards when people live close to the source of danger or modify a natural process or landscape in a way that makes it more dangerous. Natural events that will continue to cause deaths and property damage include flooding, landslides, earthquakes, volcanic activity, wind, expansive soils, drought, fire, and coastal erosion. The frequency of a hazardous event is generally inversely related to its magnitude; its impact on people depends on its frequency and magnitude as well as such diverse factors as climate, geology, vegetation, and human use of the land. The same natural events that create disasters may also bring about benefits, as when river flooding or a volcanic eruption supplies nutrients to soils.

The events causing the greatest number of deaths in the United States are tornadoes and windstorms, lightning, floods, and hurricanes, although a single great earthquake can take a very large toll. Floods, landslides, expansive soils, and frost cause the greatest property damage. Events most likely to produce a catastrophe (a disaster requiring a long, involved recovery) are floods, hurricanes, tornadoes, earthquakes, volcanic eruptions, and fires. Land-use changes, urbanization, and population increase are causing damage from most hazards to increase in the United States, but better prediction and warning are causing deaths from many hazardous processes to decrease.

Some disastrous events can be predicted fairly accurately, including some river floods and the arrival at the coast of hurricanes and tsunamis. Precursor events sometimes give warning of such events as earthquakes and volcanic eruptions. Once an event has been predicted, this information must be made available to planners and decision makers so that they might minimize the threat to human life and property. Of particular significance are how a warning is issued and how scientists communicate with the media and public. For many hazards we cannot determine when a specific event will occur; we can only predict the probability of occurrence, based on the record of past occurrences. The risk associated with an event is the product of the probability of occurrence and the likely consequences.

The impact of a disaster upon a population includes direct effects—people killed, dislocated, or otherwise damaged—and indirect effects—emotional distress, donation of money or goods, and paying taxes to finance recovery. Recovery often has several stages, including emergency work, restoration of services and communication, and reconstruction.

The options that individuals or societies choose for avoiding or adjusting to natural hazards depend in part on hazard perception, which is highest for common events. Options include land-use planning, insurance, evacuation, disaster preparedness, artificial control of natural processes, and bearing the loss. Attempts to control natural processes artificially have had mixed success and usually cannot be expected to defend against extreme events. Regardless of the approach we choose, we must increase our understanding of hazards and do a better job of anticipating them.

As the world's population increases and we continue to modify our environment through changes such as urbanization and deforestation, more people will live on marginal lands and in more hazardous locations. Therefore, as population increases, better planning at all levels will be necessary if we are to minimize losses from natural hazards.

REFERENCES

1. **Advisory Committee on the International Decade for Natural Hazard Reduction.** 1989. *Reducing disaster's toll.* Washington, DC: National Academy Press.
2. **White, G. F., and Haas, J. E.** 1975. *Assessment of research on natural hazards.* Cambridge, MA: The MIT Press.
3. **Peterson, D. W.** 1986. Volcanoes—Tectonic setting and impact on society. In *Studies in geophysics: Active tectonics,* pp. 231–46. Washington, DC: National Academy Press.
4. **Crowe, B. W.** 1986. Volcanic hazard assessment for disposal of high-level radioactive waste. In *Studies in geophysics: Active tectonics,* pp. 247–60. Washington, DC: National Academy Press.
5. **Kates, R. W., and Pijawka, D.** 1977. Reconstruction following disaster. In *From rubble to monument: The pace of reconstruction,* eds. J. E. Haas, R. W. Kates, and M. J. Bowden. Cambridge, MA: The MIT Press.
6. **Costa, J. E., and Baker, V. R.** 1981. *Surficial geology: Building with the earth.* New York: John Wiley.
7. **Abramovitz, J. N., and Dunn, S.** 1998. *Record year for weather-related disasters.* World Watch Institute. Vital Signs Brief 98-5.
8. **Magnuson, E.** 1985. A noise like thunder. *Time* 126(13): 35–43.
9. **Russell, G.** 1985. Colombia's mortal agony. *Time* 126(21): 46–52.

KEY TERMS

magnitude-frequency concept (p. 88)
catastrophe (p. 89)
precursor events (p. 90)
forecasting (p. 90)
warning (p. 90)
risk (p. 92)
disaster preparedness (p. 94)

SOME QUESTIONS TO THINK ABOUT

1. Make a list of all the natural processes that are hazardous to people and property in the region where you live. What adjustments have you and the community in general made to lessen the impacts of these hazards? Could more be done? What? Which alternatives are environmentally preferable?
2. Assume that in the future we will be able to predict with a given probability when and where a large, damaging earthquake will occur, as we do today for hurricanes. If the probability that the earthquake will occur on a given date is quite low, say 10 percent, should the general public be informed of the forecast? Should we wait until a 50 percent confidence or even a 90 percent confidence is assured? Does the length of time between the forecast and the event have any bearing on your answers?
3. Find a friend, and one of you take the role of a scientist and another a news reporter. Assume the news reporter is interviewing the scientist about the nature and extent of hazardous processes in your town. Following the interview, jot down some of your thoughts concerning ways in which scientists communicate with newspeople. Are there any conflicts?
4. Develop a plan for your community to evaluate the risk of flooding. How would you go about determining an acceptable risk?
5. Do you agree or disagree that land-use change and population increase enhances risk from natural processes? Develop a hypothesis and discuss how it might be tested.

5 Rivers and Flooding

Failure of a levee in Monroe County, Illinois, resulting in flooding of farmland during the great floods of the Mississippi River in 1993. (*James A. Finley/AP/Wide World Photos*)

The great Mississippi River floods of 1993 remind us just how vulnerable we are to catastrophic flooding in the United States. Shown above is the failure of a levee in Monroe County, Illinois, which resulted in flooding of farmland. This certainly was not an isolated incident of levee failure, as nearly 80 percent of the private levees along the Mississippi River and its tributaries succumbed to the flood waters. The levees were responsible for providing a false sense of security for the people living and farming behind them. The flood cost more than $10 billion in property damages and has led to an evaluation of how in the future we should deal with flood hazards in the Mississippi River Valley. Some communities will move to higher ground, avoiding the flood hazard altogether. This is the environmentally correct adjustment. In the meantime, the Mississippi River continued to roll along, again flooding some communities in 1995.

5.1 River Processes

For more than 200 years, Americans have lived and worked on floodplains, enticed to do so by the rich alluvial (stream-deposited) soil, abundant water supply, ease of waste disposal, and proximity to the commerce that developed along rivers. Of course, building houses, industry, public buildings, and farms on a floodplain invites disaster, but floodplain residents have refused to recognize the natural floodway of the river for what it is: part of the natural river system. The **floodplain** is the flat surface adjacent to the river channel, periodically inundated by floodwater and in fact produced by the process of flooding (see Figure 5.1d). As a result of not recognizing the floodplain and its relation to the river, flood control and drainage of wetlands (including floodplains) became prime concerns. It is not an oversimplification to say that as the pioneers moved west

LEARNING OBJECTIVES

Flooding is a natural process that becomes a hazard when people choose to live or work on floodplains. The main learning objectives of the chapter are:

- *To gain a general appreciation for river processes.*
- *To understand the nature and extent of the flood hazard and the difference between upstream and downstream floods.*
- *To understand the effects of urbanization on flooding in small drainage basins.*
- *To be aware of the major preventive and adjustment measures for flooding and which ones are environmentally preferable.*
- *To know what potential adverse environmental effects of channelization are and how they might be minimized.*

Web Resources

Visit the "Environmental Geology" Web site at www.prenhall.com/keller to find additional resources for this chapter, including:

- ▶ Web Destinations
- ▶ On-line Quizzes
- ▶ On-line "Web Essay" Questions
- ▶ Search Engines
- ▶ Regional Updates

they had a rather set procedure for modifying the land: First clear the land by cutting and burning the trees, then modify the natural drainage. From that history came two parallel trends: an accelerating program to control floods, matched by an even greater growth of flood damages (1). In this chapter we will consider flooding as a natural aspect of river processes and examine the successes and failures of traditional methods of flood control. We will see that newer approaches, while acknowledging that people will continue to live on floodplains, attempt to work with the natural river processes rather than against them.

Streams and Rivers

Streams and rivers are part of the hydrologic cycle, which transports water by evaporation from the earth's surface (mostly the oceans) to the atmosphere and back again. Some of the water that falls on the land as rain or snow is absorbed; the rest drains, or runs off, following a course determined by the local topography. This **runoff** finds its way to streams, which may merge to form a larger stream or a river. Streams and rivers differ only in size (streams are small rivers), and geologists commonly use the word *stream* for any body of water that flows in a channel. The region (area, in km^2) drained by a single river or river system is called a **drainage basin,** or *watershed* (Figure 5.1a).

A stream's *slope*, or gradient, is its vertical drop per unit of horizontal distance, which can be expressed as meters per kilometer, degrees, or more commonly in hydrology as meters per meter (the units cancel). For example, a slope angle of 0.5° (see Figure 5.1e) is a slope of about 0.009 (m/m) or 9 m per km (9 m in 1000 m). In general, the slope is steepest at higher elevations and is much reduced as the stream approaches its **base level** (the theoretical lowest level to which a river may erode), commonly the ocean (a river may have a temporary base level, such as a lake). The result of rivers flowing downhill to their base level is that they have *longitudinal profiles* (that is, a graph of river elevation vs. distance downstream) like the one shown in Figure 5.1e. This profile is generally concave, like the front of a skateboard. A stream usually has steeper and higher valley sides at high elevations (near its headwaters, where the stream starts; Figure 5.1b) than near its base level (Figure 5.1c). This results because at higher elevations the stream has eroded a deeper valley in the hilly or mountainous terrain often found there. Increased erosion results in part because a steeper channel slope produces higher stream power that can cause a river to transport more sediment and erode its channel deeper than a channel with a low channel slope. Total **stream power** is the product of the volume per unit time of water flowing by a point (discharge), the water surface slope (water surface slope is approximately equivalent to channel slope), and the unit weight of water, which is a constant.

Sediments in Rivers

The total quantity of sediment carried in a river, called its **total load,** includes the bed load, the suspended load, and the dissolved load. The **bed load** moves along the bottom of the channel by bouncing, rolling, or skipping. The bed load of most rivers (usually sand and gravel) is a relatively small component (less than 10%) of the total load. The **suspended load** (usually silt and clay) is carried above the stream bed by the turbulence of the flowing water. It is often the largest part (about 90%) of the total load, and makes rivers look muddy. The **dissolved load** is carried in chemical solution and is derived from chemical weathering of rocks in the drainage basin. It is the dissolved load that may make stream water taste salty (if the dissolved load contains large amounts of sodium and chloride), and may make the stream water hard (if the dissolved load contains high concentrations of calcium and magnesium). The most common constituents of the dissolved load are bicarbonate (HCO_3^-) and sulfate ($SO_4^=$) ions and ions of calcium (Ca^{++}), sodium (Na^+), and magnesium (Mg^+). An ion is an atom or group of

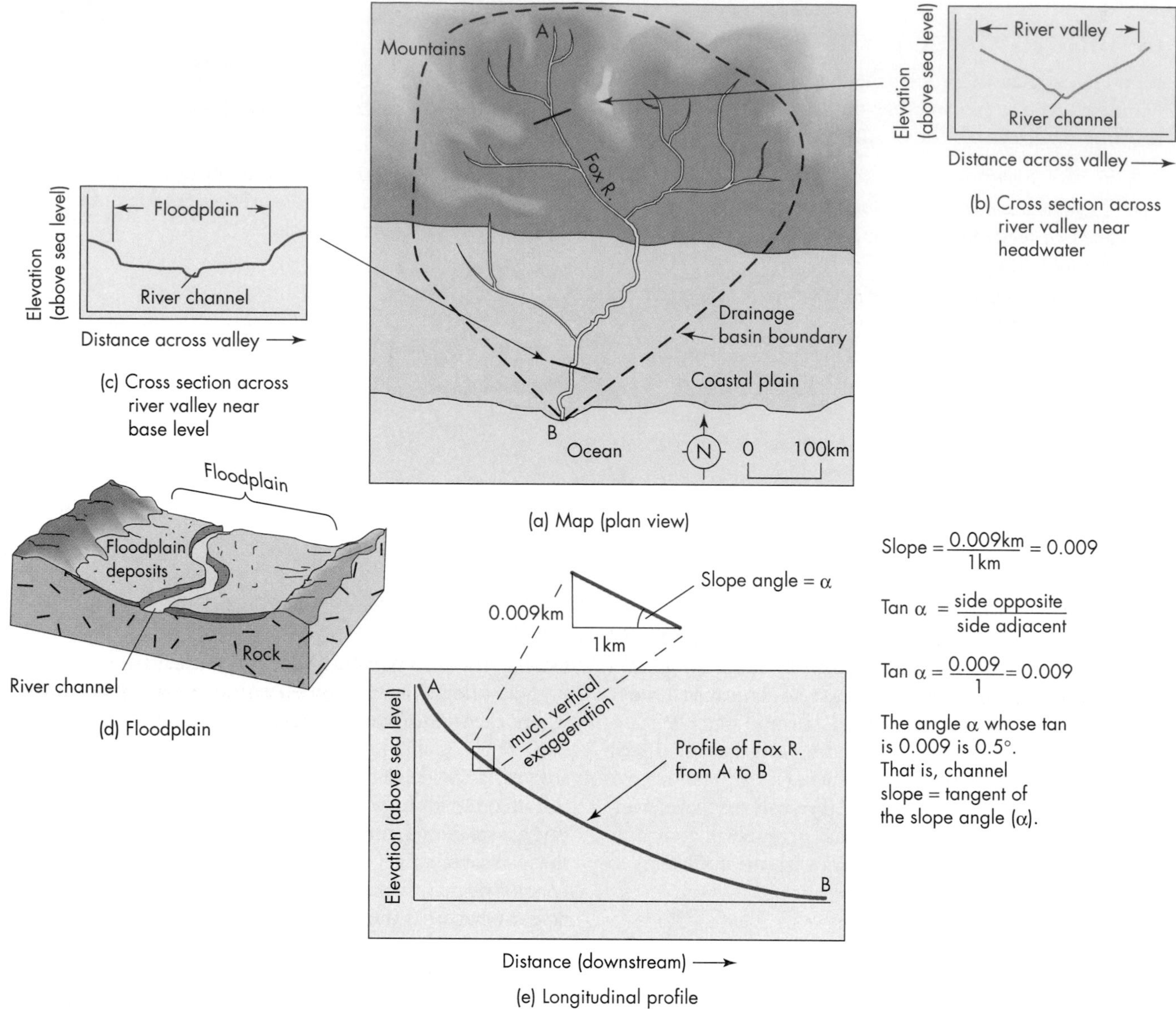

▲ **FIGURE 5.1** Idealized diagram showing drainage basin (a); cross section of valley near headwater (b); near base level (c); floodplain (d); and longitudinal profile (e).

atoms (molecule) with a positive or negative charge resulting from a gain or loss of electrons. Typically, the above five atoms and molecules comprise more than 90 percent of a river's dissolved load. It is the suspended load and the bed load of streams that, when deposited in undesirable locations, produce the sediment pollution discussed in Chapter 3.

River Velocity, Erosion, and Sediment Deposits

Rivers are the basic transportation system of the part of the rock cycle involving erosion and deposition of sediments, and they are a primary erosion agent in the sculpture of our landscape. The velocity of a river varies along its course and affects both erosion and deposition of sediment.

The mean or average water velocity at any point along a river is defined as the ratio of the **discharge** (volume of water passing that point in a given time) to cross-sectional area of flow of the channel. That is, to calculate the average velocity of the water in a river you divide the total discharge by the cross-sectional area of flow. This relationship: $V = Q/A$, or $Q = V \times A$, or $Q = V \times W \times D$, where Q is discharge (m^3 per second, m^3/s, often abbreviated as cms), V is mean velocity of flow (m per second, m/s), A is cross-sectional area of flow (m^2), W is stream width in meters (m), and D is depth of flow in meters (m). The equation $Q = W \times D \times V$ is known as the continuity equation and is one of the most important relationships in understanding flow of water in rivers. We assume that if there are no additions or deletions of flow along a given length of river, then discharge is constant. It follows then that with constant discharge, if the cross-sectional area of flow decreases, then velocity must increase. This explains why a river that flows through a narrow, steep section or reach of channel has a higher velocity of flow than where it spreads out with larger cross-sectional area of flow downstream. A section of river being observed or studied is called a *reach*. A factor that allows the velocity to increase in the narrow reach of channel with reduced cross-sectional area of flow is the steeper channel slope often found there. It has been shown that average velocity of flow of water in a river is proportional to the product of the depth of flow and the slope or gradient of flow. Now we can see why slope is related to stream power (defined earlier as proportional to the product of discharge and slope). We would expect that the power of a river is related to velocity of flow and thus channel slope. If discharge is constant along a reach of river, then stream power is directly proportional to slope. Narrow reaches of a river with steep channel and water slope will have higher stream power than wide reaches with lower slope (assuming again discharge is constant).

In general, a faster-flowing river has the possibility to erode its banks more than a slower-moving one. Furthermore, the faster water flows, the greater the stream power tends to be, and the larger and heavier the sediment particles it can move. The largest and heaviest particles—boulders and gravel—are deposited in river environments at locations where stream power during relatively high flows, when these larger particles are being transported, is relatively low. Sand and silt tend to be deposited at relatively low flows in the lower-gradient, slow-moving reaches where stream power is even lower. Where streams flow from mountains onto plains, they may form fan-shaped deposits known as **alluvial fans** (Figure 5.2). Where a river flows into the ocean or other body of still water, it may deposit sediments that form a **delta,** a triangular land mass extending into the sea (Figure 5.3). The reasons deposition occurs in a specific area of a river channel or on alluvial fans or deltas are complex, related to changes in the river environment beyond the scope of our discussion here.

▲ **FIGURE 5.2** Alluvial fan along the western foot of the Black Mountains, Death Valley. Note the road along the base of the fan. The white materials are salt deposits in Death Valley. (*Michael Collier*)

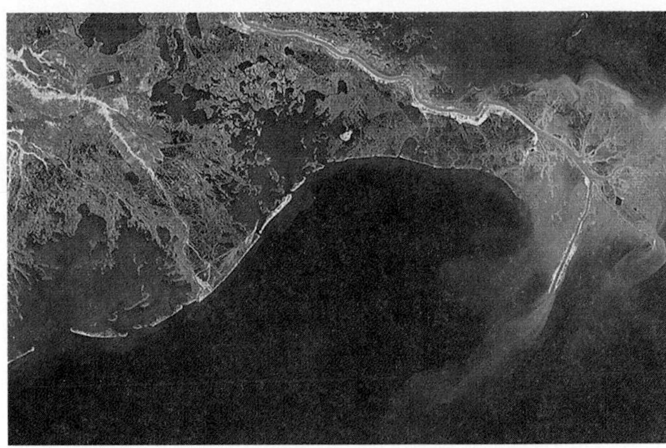

▲ **FIGURE 5.3** Delta of the Mississippi River. In this false color image, red is vegetation and sediment-laden waters are white or light blue, while deeper water with less suspended sediment is a darker blue. The system of distributary channels in the delta in the far right of the photograph looks something like a "bird's foot" and, in fact, the Mississippi River delta is an example of a bird's-foot delta. The distributary channels carry sediment out into the Gulf of Mexico and, because wave action is not strong in the gulf, the river dominates the delta system. (*Landsat image by* U.S. *Geological Survey; courtesy of* John S. *Shelton*)

The largest particle (particle diameter in mm or cm) a river may transport is called its **competency;** while the total load the river carries in a given period of time (units might be kg per second, kg/s) is its **capacity.**

Effect of Land-Use Changes

Streams and rivers are open systems that generally maintain a rough dynamic equilibrium, or steady-state, between the work done (sediment transported by the stream) and the load imposed (sediment delivered to the stream from tributaries and hill slopes) (1). The stream tends to have a slope and cross-sectional shape that provide just the velocity of flow (and stream power) necessary to do the work of moving the sediment load (2). An increase or decrease in the amount of water or sediment the stream receives usually brings about changes in the channel's slope or cross-sectional shape, effectively changing the velocity of the water. The change of velocity may, in turn, increase or decrease the amount of sediment carried in the system. Thus, land-use changes that affect the volume of sediment or of water in a stream may set into motion a series of events resulting in a new dynamic equilibrium.

Consider, for example, a land-use change from forest to agricultural row crops. This change will cause increased soil erosion and an increase in the load supplied to the stream. At first the stream will be unable to transport the entire load and will deposit more sediment, increasing the slope of the channel, which will in turn increase the velocity of water (and also stream power) and allow the stream to move more sediment. The slope (assuming base level is fixed) will continue to increase by deposition in the channel until the velocity (and stream power) increases sufficiently to carry the new load. The notion that deposition of sediment increases channel slope may be counterintuitive to

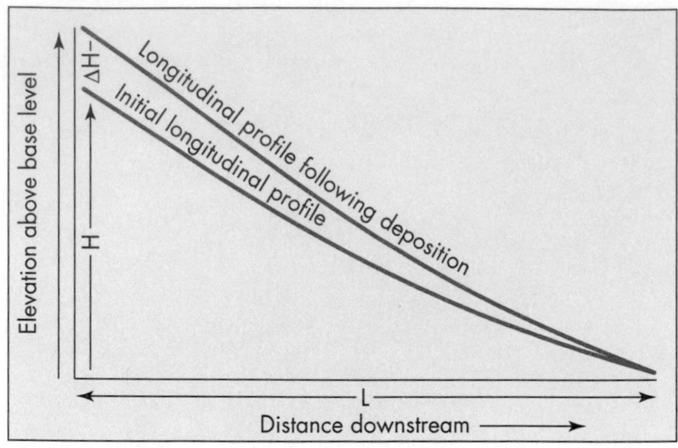

- Initial channel slope = $\frac{H}{L}$
- Slope following deposition = $\frac{H + \Delta H}{L}$
- $\frac{H + \Delta H}{L}$ is greater than $\frac{H}{L}$; so channel slope following deposition is greater than channel slope before deposition
- A similar argument can be constructed to show why channel erosion reduces channel slope.

▲ **FIGURE 5.4** Idealized diagram illustrating that deposition in a stream channel results in an increase in channel gradient.

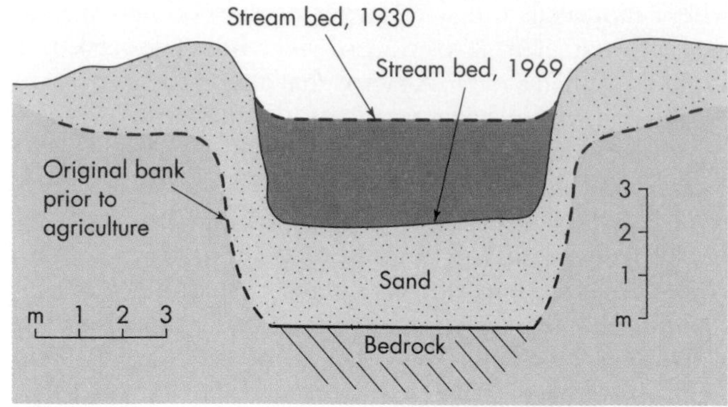

▲ **FIGURE 5.5** Accelerated sedimentation and subsequent erosion resulting from land-use changes (natural forest to agriculture and back to forest) at the Mauldin Millsite on the Piedmont of middle Georgia. (After S. W. Trimble, *"Culturally Accelerated Sedimentation on the Middle Georgia Piedmont," master's thesis* [Athens: University of Georgia, 1969.] Reproduced by *permission.*)

you (see Figure 5.4 for further explanation of this principle). A new dynamic equilibrium may be reached, provided the rate of sediment increase levels out and the channel slope and shape can adjust before another land-use change occurs.

Now suppose the reverse situation occurs—that is, farmland is converted to forest. The sediment load to the stream from the land will decrease (forest lands have lower soil erosion rates than do agricultural lands), less sediment will be deposited in the stream channel, and erosion of the channel will eventually lower the slope, which in turn will lower the velocity of the water. The predominance of erosion over deposition will continue until an equilibrium between the load imposed and work done is achieved again.

The sequence of change just described is occurring in parts of the Piedmont of the southeastern United States. There, land that was forest in the early history of the country was cleared for farming, producing accelerated soil erosion and subsequent deposition of sediment in the stream (Figure 5.5). The land is now reverting to pine forests, and this, in conjunction with soil-conservation measures, has reduced the sediment load delivered to streams. Thus, once-muddy streams choked with sediment are now clearing and eroding their channels. Whether this trend continues depends on future conservation measures and land use.

Consider now the effect of constructing a dam on a stream. Considerable changes will take place both upstream and downstream of the reservoir. Upstream, at the head of the reservoir, the effect will be to slow down the stream, causing deposition of sediment. Downstream, the water coming out below the dam will have little sediment, most of it having been trapped in the reservoir. As a result, the stream may have the capacity to transport additional sediment, and if this happens, erosion will predominate over deposition downstream of the dam. Slope then will decrease until new equilibrium conditions are reached (Figure 5.6). We will return to the topic of dams on rivers in Chapter 10.

Channel Patterns and Floodplain Formation

The configuration of the channel in plan view (as from an airplane) is called the **channel pattern.** The two main channel patterns are braided and meandering and both may be found on the same river. **Braided channels** (Figure 5.7) are characterized by numerous inchannels, gravel bars, and islands that divide and reunite the channel. Formation of the braided channel pattern, as with many other river forms,

▶ **FIGURE 5.6** Upstream deposition and downstream erosion from construction of a dam and a reservoir.

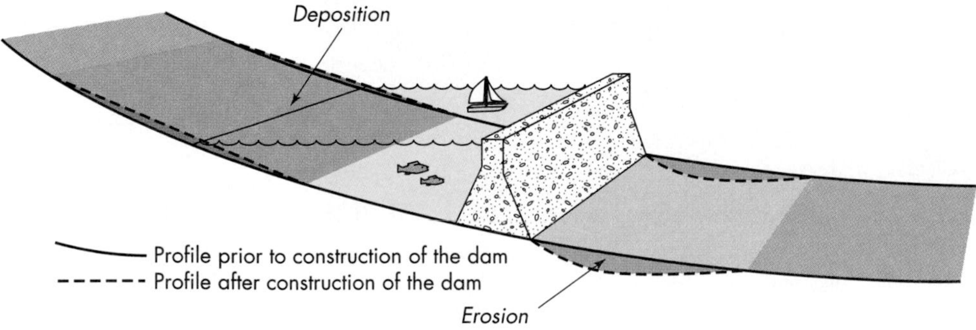

▲ **FIGURE 5.7** The north Saskatchewan River here has a braided channel pattern. Notice the numerous channel bars and islands that subdivide the flow. (*John S. Shelton*)

results from the interaction of flowing water and moving sediment operating with geologic and climatic variables. If the river longitudinal profile is steep and there is an abundance of coarse sediment, the channel pattern is likely to be braided. Braided channels tend to be wide and shallow compared to meandering rivers. Steep slope and coarse sediment favor transport of bed-load material important in the development of gravel bars, which form the "islands" that divide and subdivide the flow. Braided channels tend to be associated with steep glacial meltwater rivers with an abundance of gravel or steep rivers flowing through areas being rapidly uplifted by tectonic processes. Rapid uplift produces steep river gradients and erosive power to produce coarse gravel sediment.

Meandering channels are sinuous, containing individual bends, called **meanders** (Figure 5.8a), that migrate back and forth across the floodplain. On the outside of a bend

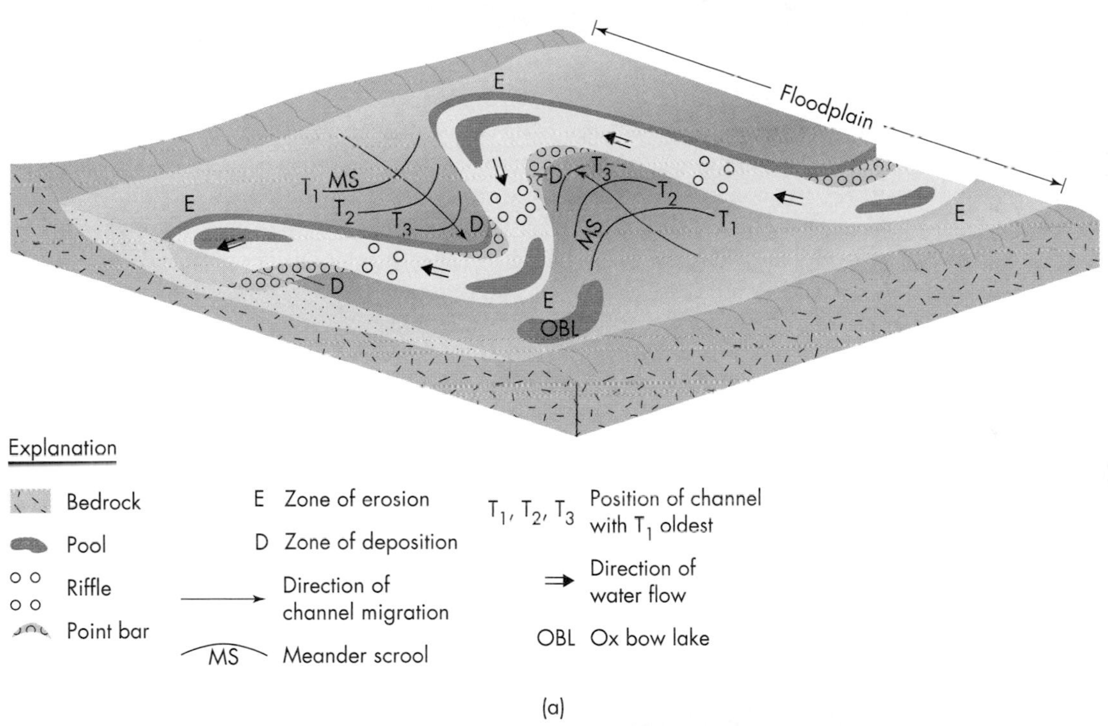

▲ **FIGURE 5.8** Idealized diagram of a meandering stream and important forms and processes (a), and photograph of a meandering stream, Owens River, California (b). (*Edward A. Keller*)

▶ **FIGURE 5.9** Well-developed pool-riffle sequence in Sims Creek near Blowing Rock, North Carolina. A deep pool is apparent in the middle distance, and shallow riffles can be seen in the far distance and the foreground. (*Edward A. Keller*)

the water during high-flow events moves faster, causing more bank erosion; on the inside of a curve, it moves more slowly and sediment is deposited, forming **point bars.** As this differential erosion and sediment deposit continues, meanders migrate laterally, a process that is prominent in constructing and maintaining some floodplains (Figure 5.8b). Overbank deposition during floods causes vertical accretion that is also important in development of floodplains. Much of the sediment transported in rivers is periodically stored by deposition in the channel and on the adjacent floodplain. These areas, collectively called the **riverine environment,** are the natural domain of the river. Lateral migration of bends of rivers and overbank flow combine to produce the floodplain, which is periodically inundated by water and sediment.

Meandering channels often contain a series of regularly spaced pools and riffles (Figure 5.9). Straight channels, a relatively rare channel pattern, may also contain pools and riffles. **Pools** are deep areas produced by scour (erosion) at high flow and characterized at low flow by relatively deep, slow movement of water. Pools are places in which to take a summer swim. **Riffles** are shallow areas produced by depositional processes (fill) at high flow and characterized at low flow by relatively shallow, fast-moving water. We therefore conclude that pools scour (erode) at high flow and fill with sediment at low flow, whereas riffles fill at high flow and scour at low flow (Figure 5.10). Also, notice that while the velocity of a pool is lower than the riffle at low discharge, at high flow the velocity of water through the pool exceeds that of the riffle. This change in velocity may happen in part because the basic shape of a pool and adjacent point bar is like a triangle, whereas that of the riffle is more like a rectangle (Figure 5.11). At low flow the cross-sectional area of flow of a pool may exceed that of an adjacent riffle, and therefore by the continuity equation $Q = A \times V$, the mean velocity of the pool is less than that of the riffle (3). At high flow a change has occurred, and the geometry of the pool and riffle is such that the cross-sectional area of flow of the pool is now less than that of an adjacent riffle. Therefore, by the continuity equation the mean velocity of flow in the pool now exceeds that of the riffle. This is the condition at discharges exceeding the threshold shown on Figure 5.10). It is this pattern of velocity and associated scour and fill that maintains pools and riffles. A pool-riffle sequence is repeated approximately every five to seven times the channel width.

Observation of many streams suggest that those with well-developed pools and riffles tend to have considerable gravel in the streambed and a relatively low slope (less than about 0.015; remember: slope has no units). Streams with finer bed material or steep slopes tend to lack regularly spaced pools and riffles. The reason for this is poorly understood. We do know that pools and riffles have impor-

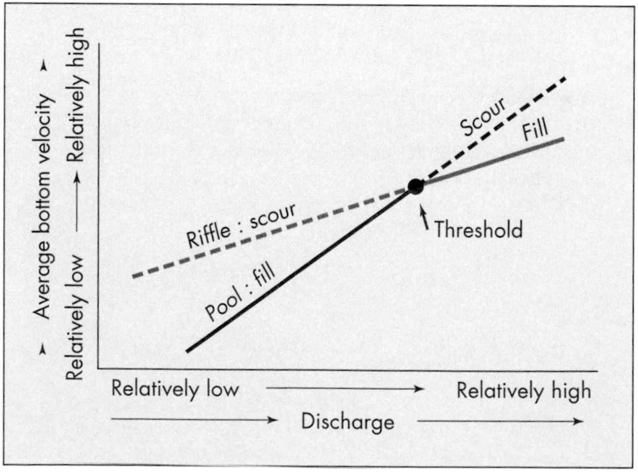

▲ **FIGURE 5.10** Scour-fill pattern characteristic of a pool-riffle sequence. The threshold is a critical discharge at which change in process (scour-to-fill or fill-to-scour) occurs.

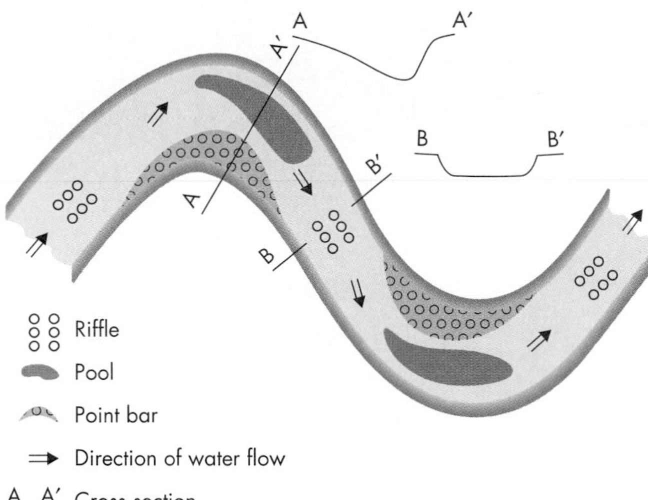

▲ **FIGURE 5.11** Idealized diagram of a stream channel showing form (cross section) of pools and riffles

tant environmental significance. This results in part because the alternation of deep, slow-moving water with shallow, fast-moving water in pools and riffles produces a variable physical environment and increased biologic diversity.

5.2 Flooding

The natural process of overbank flow is termed **flooding.** Most river flooding is a function of the total amount and distribution of precipitation in the drainage basin, the rate at which precipitation infiltrates the rock or soil, and the topography. However, some floods result from rapid melting of ice and snow in the spring or, on rare occasions, from the failure of a dam. Finally, land use can greatly affect flooding in small drainage basins.

The channel discharge (cubic meters per second, m^3/s or cms) at the point where water overflows the channel is called the *flood discharge* and is used as an indication of the *magnitude* of the flood (see *A Closer Look: Magnitude and Frequency of Floods*). The magnitude of a flood may or may not coincide with the extent of property damage. The term **flood stage** frequently connotes that the elevation of the water surface has reached a high-water condition likely to cause damage to personal property. This definition is based on human perception of the event, so the elevation that is considered flood stage depends on human use of the floodplain (4).

The longer flow records are collected, the more accurate the prediction of floods is likely to be. However, designing structures for a 10-year, 25-year, 50-year, or even 100-year flood, or in fact any flow below possible maximum, is a calculated risk because predicting such floods is based on probability. In the long term, a 25-year flood happens on the average of once every 25 years, but two 25-year floods could occur in any given year as could two 100-year floods (5)! As long as we continue to build dams, highways, bridges, homes, and other structures on flood-prone areas, we can expect continued loss of lives and property.

Upstream and Downstream Floods

It is useful to distinguish between upstream and downstream floods (Figure 5.12). **Upstream floods** occur in the upper parts of drainage areas and are generally produced by intense rainfall of short duration over a relatively small area. These floods may not cause severe flooding in the larger streams they join downstream, although they can be quite severe locally. For example, a high-magnitude upstream flood occurred in the Front Range of Colorado in the summer of 1976, when violent flash floods, nourished by a complex system of thunderstorms delivering up to 25 cm of rain, swept through several canyons west of Loveland. This brief local flood killed 139 people and inflicted more than $35 million in damages to highways, roads, bridges, homes, and small businesses. Most of the damage and loss of life occurred in the Big Thompson Canyon, where hundreds of residents, campers, and tourists were caught with little or no warning. Although the storm and flood were rare events (several times the magnitude of the 100-year flood), comparable floods have occurred in the past and others can be expected in the future for similar canyons along the Front Range (8–10).

It is the large **downstream floods,** such as the 1993 Mississippi River floods and the 1997 Red River, North Dakota floods, that usually make television and newspaper headlines. The Mississippi flood is discussed at the end of the chapter. The Red River flood inundated the city of Grand Forks, North Dakota, initiating the evacuation of 50,000 people, causing a fire that burned part of the city center, and inflicting more than $1 billion in damage. Downstream floods cover a wide area and are usually produced by storms of long duration that saturate the soil and produce increased runoff. Flooding on small tributary basins may be limited, but the contribution of increased runoff from thousands of tributary basins may cause a large flood downstream. A flood of this kind is characterized by the downstream migration of an ever-increasing flood wave with large rise and fall of discharge (11). Figure 5.13a shows the 257-km downstream migration of a flood crest on the Chattooga–Savannah River system. It illustrates that a progressively longer time is necessary for the rise and fall of water as the flood wave proceeds downstream, and shows dramatically the tremendous increase in discharge from low-flow conditions to more than 1700 m^3/s in 5 days (12). Figure 5.13b also illustrates the same flood in terms of discharge per unit area, eliminating the effect of downstream increase in discharge. This better illustrates the shape and form (sharpness of peaking) of the flood wave as it moves downstream (12).

A few upstream floods of very high magnitude have been caused directly by structural failure. For example, the most destructive flood in West Virginia's history was caused by the failure of a coal-waste dam on the middle fork of Buffalo Creek (13).

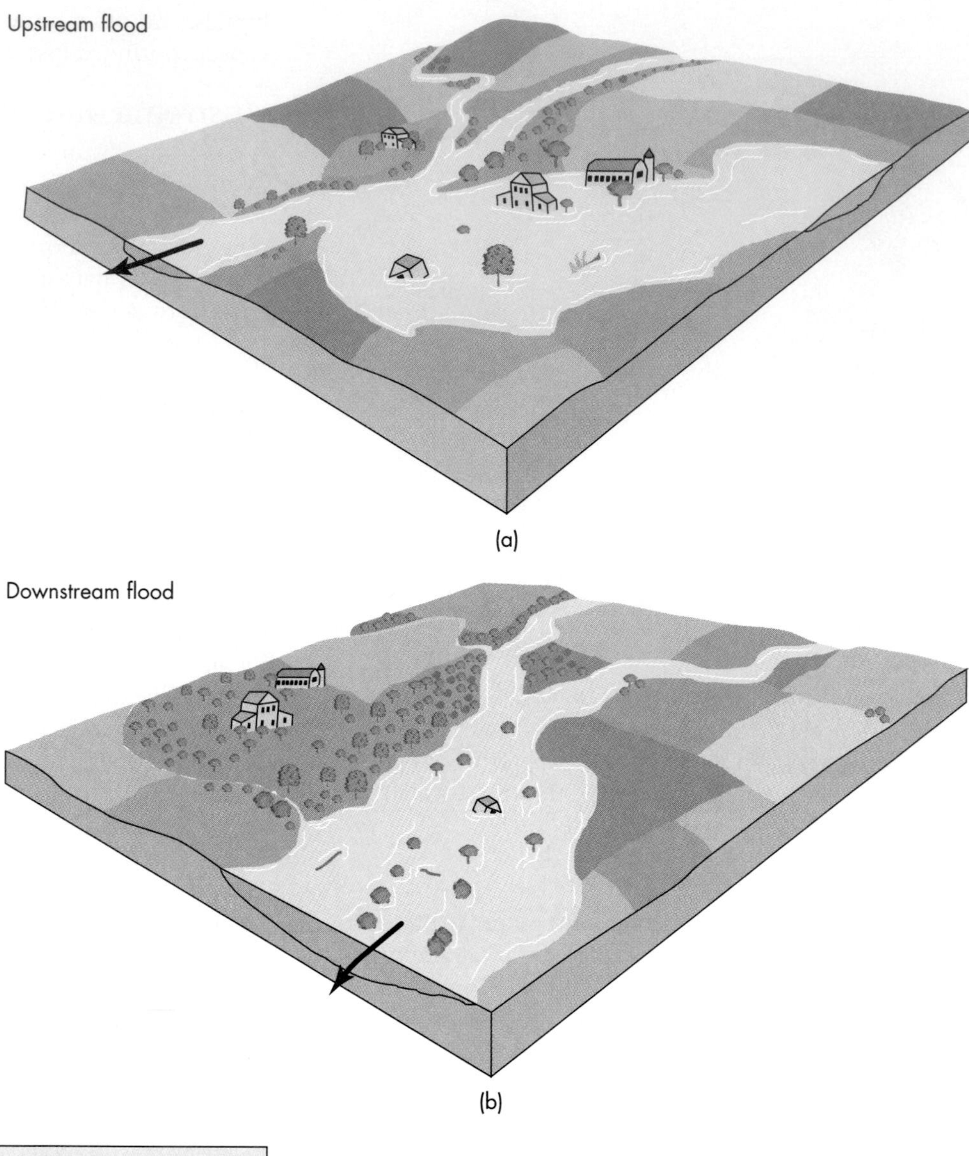

▶ **FIGURE 5.12** Idealized diagram comparing upstream flood (a) to downstream flood (b). Upstream floods generally cover relatively small areas and are caused by intense local storms, whereas downstream floods cover wide areas and are caused by regional storms or spring runoff. (*Modified after* U.S. *Department of Agriculture drawing.*)

Upstream flood

(a)

Downstream flood

(b)

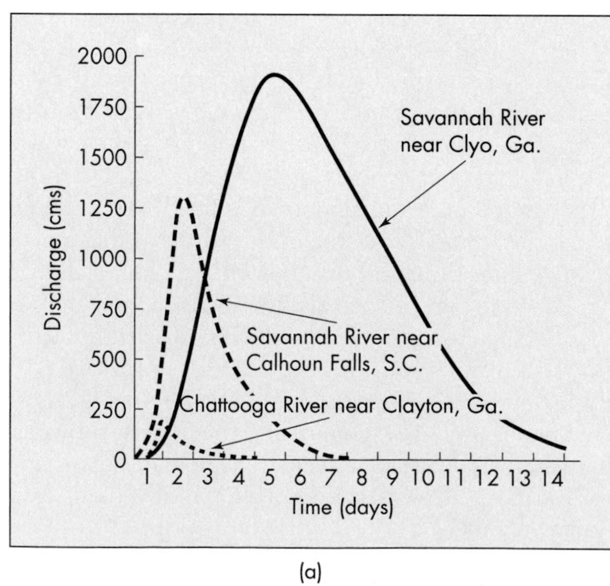

(a)

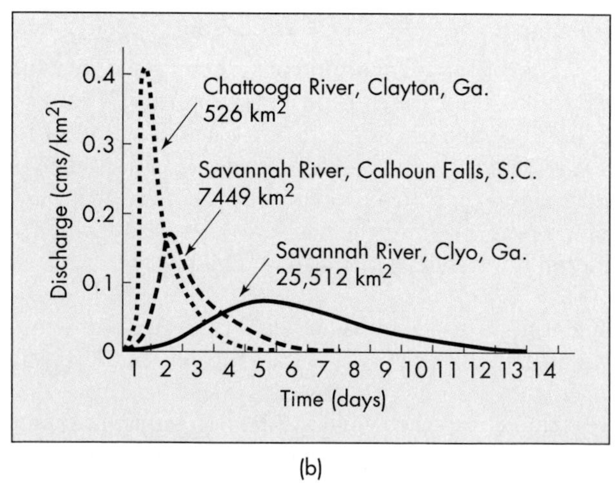

(b)

▲ **FIGURE 5.13** Downstream movement of a flood wave on the Savannah River, South Carolina and Georgia. The distance from Clayton to Clyo is 257 km. (a) Volume of water passing Clayton, Calhoun Falls, and Clyo. (b) Volume of water per unit area at the same points. (*After* William G. Hoyt *and* Walter B. Langbein, Floods [*© copyright* 1955 *by* Princeton University Press], *Fig.* 8, *p.* 39. *Reprinted by permission of* Princeton University Press.)

5.3 Development and Flooding

Human use of land in the urban environment has increased both the magnitude and frequency of floods in small drainage basins of a few square kilometers. The rate of increase is a function of the percentage of the land that is covered with roofs, pavement, and cement (referred to as **impervious cover**) and the percentage of area served by storm sewers. Storm sewers are important because they allow urban runoff from impervious surfaces to reach stream channels quickly. Therefore, impervious cover and storm sewers are collectively a measure of the degree of urbanization. The graph in Figure 5.14 shows that an urban area with 40 percent impervious surface and 40 percent of its area served by storm sewers can expect to have about three times as many overbank flows (floods) as before urbanization. This ratio holds for floods of small and intermediate frequency, but as the size of the drainage basin increases, floods of high magnitude with frequencies of 50 years or so are not much affected by urbanization (Figure 5.15).

Floods are a function of rainfall–runoff relations, and urbanization causes a tremendous number of changes in these relations (see *Case History: Las Vegas, Nevada*). One study showed that urban runoff from the larger storms is nearly five times preurban runoff (5). Estimates of discharge for different recurrence intervals at different degrees of urbanization are shown in Figure 5.16. The estimates dramatically indicate the tremendous increase of runoff with increasing impervious areas and storm sewer coverage.

The increase of runoff with urbanization occurs because less water infiltrates the ground, as suggested by the significant reduction in time between the majority of rainfall and the flood peak **(lag-time)** for urban versus rural conditions (Figure 5.17). Short lag-times, referred to as **flashy discharge,** are characterized by rapid rise and fall of floodwaters. Because little water infiltrates the soil, the low water or dry season flow in urban streams, sustained by groundwater seepage into the channel, is greatly reduced.

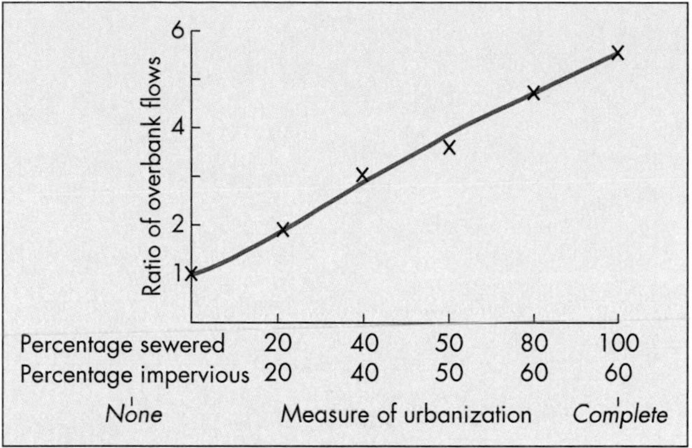

▲ **FIGURE 5.14** Relationship between the ratio of overbank flows (after urbanization to before urbanization) and measure of urbanization. This figure shows that as the degree of urbanization increases, the number of overbank flows per year also increases. (From L. B. *Leopold*, U.S. *Geological Survey Circular* 554, 1968)

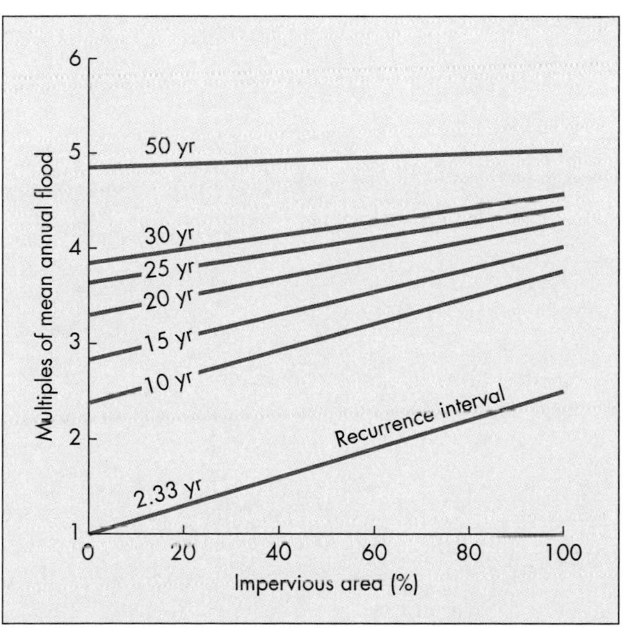

▲ **FIGURE 5.15** Graph showing the variation of flood frequency with percentage of impervious area. The mean annual flood is the average (over a period of years) of the largest flow that occurs each year. The mean annual flood in a natural river basin with no urbanization has a recurrence interval of 2.33 years. Note that the smaller floods with recurrence intervals of just a few years are much more affected by urbanization than are the larger floods. The 50-year flood is little affected by the amount of area that is rendered impervious. (From L. A. Martens, U.S. *Geological Survey Water Supply Paper* 1591C, 1968)

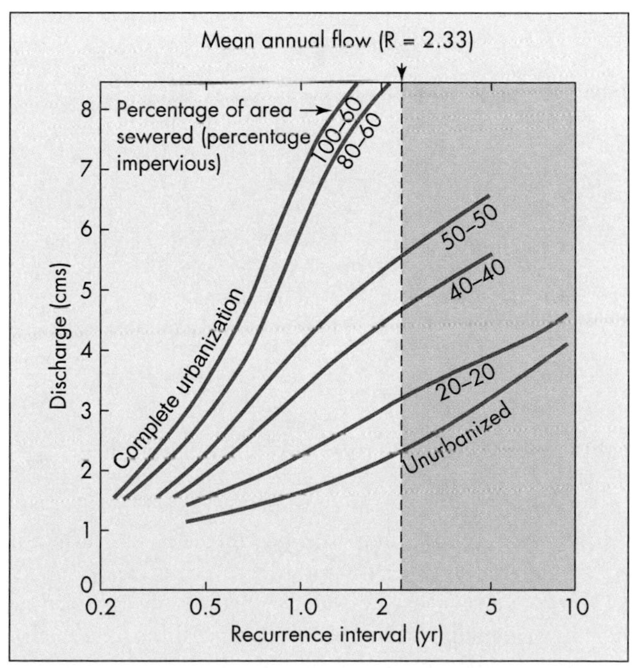

▲ **FIGURE 5.16** Flood frequency curve for a 2.6 km^2 (one-square-mile) basin in various states of urbanization. Note: 100–60 means basin is 100 percent sewered and 60 percent of surface area is impervious. Dashed line shows increase in mean annual flood with increasing urbanization. (*After* L. B. *Leopold*, U.S. *Geological Survey Circular* 554, 1968)

A CLOSER LOOK Magnitude and Frequency of Floods

Flooding is intimately related to the amount and intensity of precipitation and runoff. Catastrophic floods reported on television and in newspapers often are produced by infrequent, large, intense storms. Smaller floods or flows may be produced by less intense storms, which occur more frequently. All flow events that can be measured or estimated from stream-gauging stations (Figure 5.A) can be arranged in order of their magnitude of discharge, generally measured in cubic meters per second (m^3/s). The list of flows so arranged can be plotted on a discharge-frequency curve (Figure 5.B) by deriving the **recurrence interval** R for each flow from the relationship

$$R = (N + 1) \div M$$

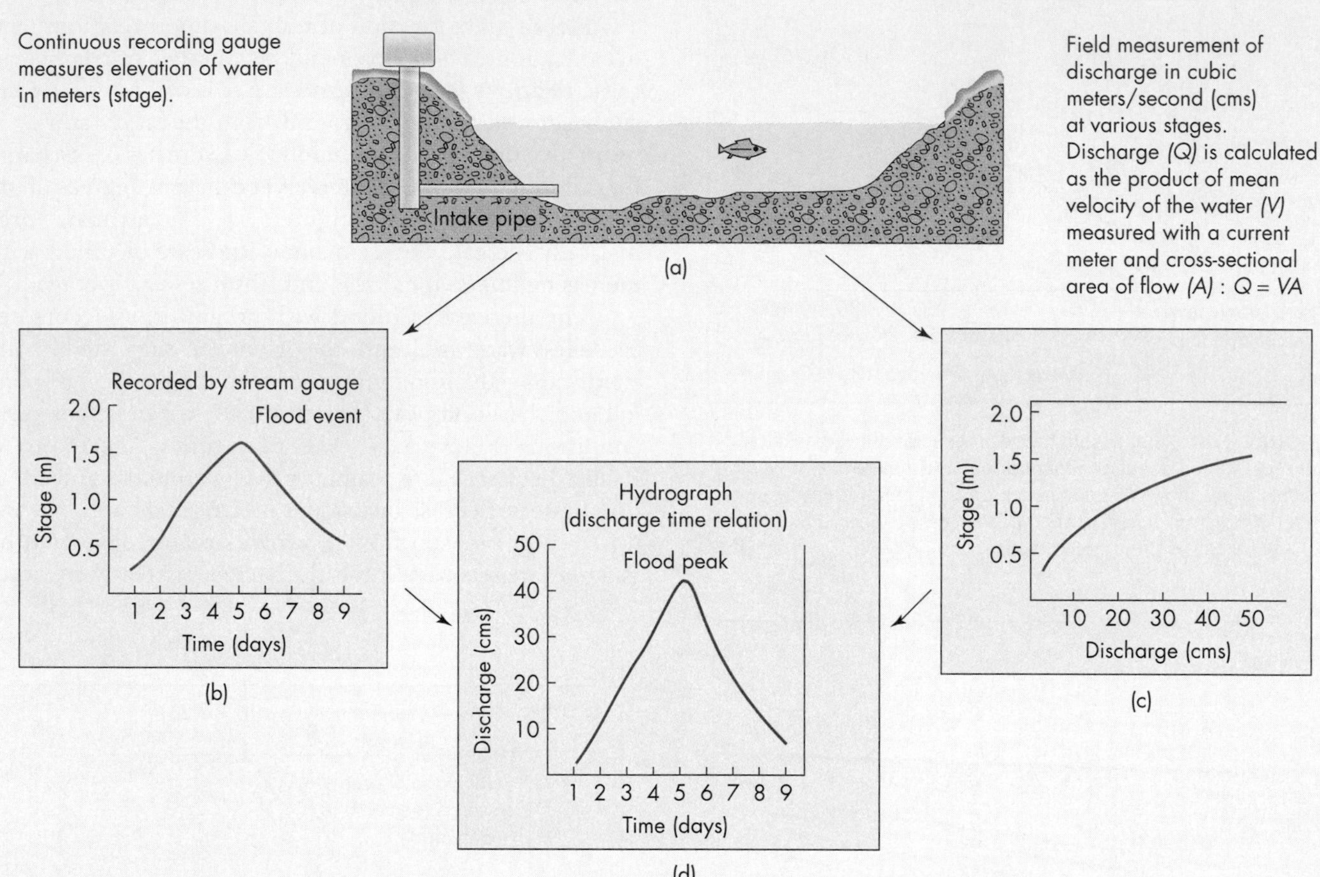

▲ **FIGURE 5.A** Field data (a) consist of a continuous recording of the water level (stage), which is used to produce a stage-time graph (b). Field measurements at various flows also produce a stage-discharge graph (c). Then graphs (b) and (c) are combined to produce the final hydrograph (d).

This effectively concentrates any pollutants present and generally lowers the aesthetic amenities of the stream (7).

Relationships between land use and flooding for small drainage basins may be quite complex. One study concludes that not all types of urbanization increase all runoff and flood events (14). When row crops such as corn and soybeans are replaced by low-density residential development, the predicted runoff and flood peaks for low-magnitude events with recurrence intervals of 2 to 4 years increase, as expected. For events with recurrence intervals exceeding 4 years, however, the predicted runoff and flood peaks for the agricultural land may exceed that for residential development. As row crops are replaced by paved areas and grass, the runoff from the paved areas is greater, but the grass produces less runoff than the agricultural land. Therefore, the effect of the land-use change on runoff and flooding depends on the nature and extent of urbanization and in particular on the proportion of paved and grass-covered areas.

Urbanization is not the only type of development that can increase flooding. Some flash floods have occurred because bridges built across small streams caused temporary debris dams to form (see *Case History: Flash Floods in Eastern Ohio*).

where R is a recurrence interval in years, N is the number of years of record, and M is the rank of the individual flow in the array (6). The highest flow for 9 years of data for the stream shown in Figure 5.B (about 280 m^3/s) has a rank M equal to 1 (7). The recurrence interval of this flood is

$$R = (N + 1) \div M = (9 + 1) \div 1 = 10$$

which means that a flood with a magnitude equal to or exceeding 280 m^3/s can be expected about every 10 years; we call this a 10-year flood. The probability of the 10-year flood happening in any one year is 1 ÷ 10 or 0.1 (10%). The probability of the 100-year flood occurring in any year is 1 ÷ 100 = 0.01 or 1 percent. Studies of many streams and rivers show that channels are formed and maintained by bank-full discharge with a recurrence interval of 1.5 to 2 years (about 30 m^3/s on Figure 5.B). Therefore, we can expect a stream to emerge from its banks and cover part of the floodplain with water and sediment once every year or so.

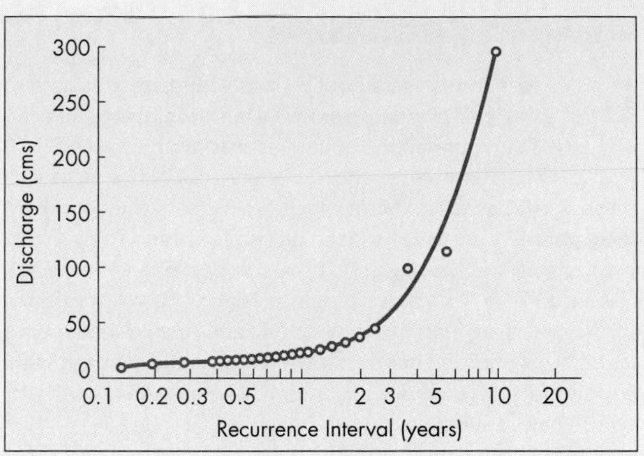

▲ **FIGURE 5.B** Example of a discharge frequency curve. Each circle represents a flow event with recurrence interval plotted on probability paper. (*After* L. B. *Leopold*, U.S. *Geological Survey Circular* 554, 1968)

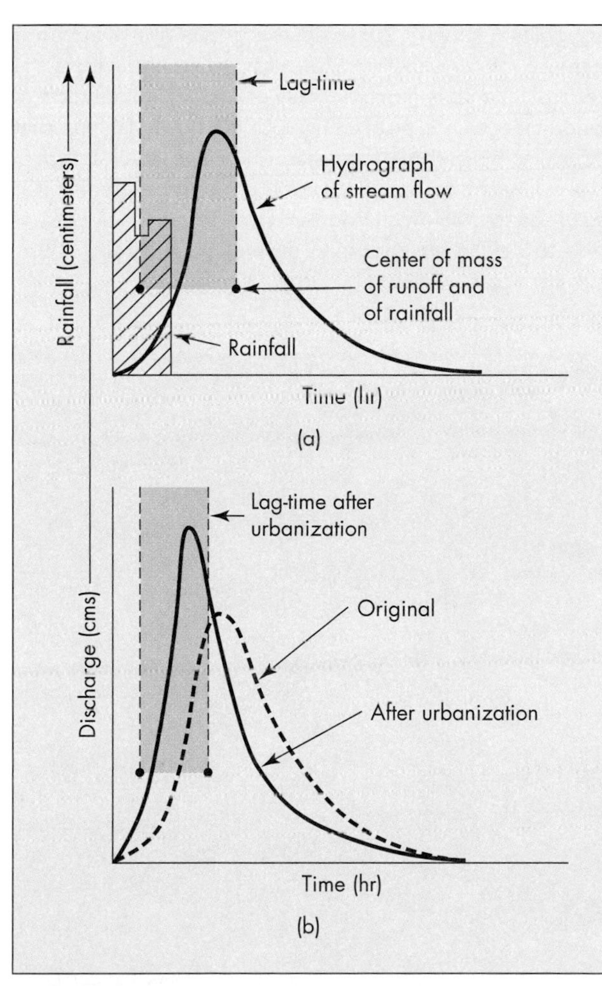

▲ **FIGURE 5.17** Generalized hydrographs. Hydrograph (a) shows the typical lag between the time when most of the rainfall occurs and the time when the stream floods. Hydrograph (b) shows the decrease in lag-time because of urbanization. (*After* L. B. *Leopold*, U.S. *Geological Survey Circular* 554, 1968)

5.4 The Nature and Extent of Flood Hazards

Flooding is one of the most universally experienced natural hazards. In the United States the lives lost to river flooding number about 100 per year, with property damage of about $4 billion annually. The loss of life is low compared to loss in preindustrial societies that lack monitoring and warning systems and effective disaster relief. Although preindustrial societies with dense populations on floodplains lose many more lives to flooding, they have less property damage than do industrial societies (3,4). Table 5.1 lists some severe floods that have occurred in the United States in the past 60 years.

Factors that control the damage caused by floods include:

- Land use on the floodplain.
- Magnitude (depth and velocity of the water and frequency of flooding).
- Rate of rise and duration of flooding.
- Season (for example: crops on floodplain).
- Sediment load deposited.
- Effectiveness of forecasting, warning, and emergency systems.

The effects of flooding may be primary—that is, directly caused by the flood—or secondary, caused by disruption and malfunction of services and systems attributable to the flood (15). Primary effects include injury and loss of life, along with damage caused by swift currents, debris, and sediment to farms, homes, buildings, railroads, bridges, roads, and communication systems. Erosion and deposition of sediment in the rural and urban landscape can also involve a loss of considerable soil and vegetation. Secondary

CASE HISTORY Las Vegas, Nevada

Las Vegas, Nevada, has a flooding problem that dates back to the early 1900s when people began developing the area. The city is surrounded by mountains that drain to alluvial fans with channels known as washes (Figure 5.C). The largest of these is the Las Vegas Wash, which drains from the northwest through Las Vegas to Lake Mead on the Colorado River. Other washes join the Las Vegas Wash in the city; one of these, the Flamingo Wash, has a notorious flood history. Flash floods in the Las Vegas area generally occur in July and August in response to high-magnitude thunderstorm activity, and damage from flooding is severe where major developments have obstructed the natural washes.

Rapid development unfortunately coincided with increased thunderstorm activity from 1975 through the mid-1980s. A flood that came down the Flamingo Wash in July 1975 caused particular concern. At Caesar's Palace (a prominent casino), hundreds of cars in a parking lot, which covered part of the Flamingo Wash, were damaged, many of them moved downstream by floodwaters. Damage to vehicles and from lateral bank erosion along the wash was reported to be as high as $5 million. Following the flood, the Flamingo Wash was routed by way of a tunnel beneath Caesar's Palace. Where the tunnel emerges, the wash flows at the surface again and often down city streets! Flooding again occurred in 1983 and 1984 in the months of July and August. In particular, eight major storms from July through September 1984 resulted in tens of millions of dollars of damage in Clark County, Nevada.

Las Vegas is located in a basin at the foot of alluvial fan surfaces. In many instances, alluvial fans are very porous and surface runoff infiltrates rapidly. In the Las Vegas area, however, the calcium-carbonate-rich soil horizons (K horizon, see Chapter 3) cement alluvial deposits together and retard surface infiltration of water. As a result, the Las Vegas area is more susceptible to flood hazard than otherwise might be expected. Nonetheless, whenever urban development is built directly across active washes, trouble can be expected.

Clark County, Nevada, has developed a plan to address the flood hazard in Las Vegas. It is designed to protect the area from the 100-year peak flow resulting from a 3-hour thunderstorm and includes:

- Construction of channels, pipelines, and culverts to convey floodwaters away from developed areas.
- Construction of storm-water retention basins that will release the flow over a longer period of time.

Because there has been so much development on the washes in Las Vegas, it seems unlikely that any plan will completely alleviate the flood problem. That would require completely routing floodwaters around or safely through all the existing development. But even this precaution would not free the Las Vegas area from flood hazard resulting from the larger storms that will occasionally occur. Implementation of a flood-management plan, along with land-use planning that discourages development on natural washes, will certainly help alleviate the community's flood problems. However, it is a gamble as to how much flood-control measures can be expected to help, given the rapid development that is occurring in Las Vegas.

▶ **FIGURE 5.C** Aerial view of the Las Vegas, Nevada, area. The city is in a natural basin and many structures have been built across natural drainage channels (washes). Arrows show the direction of flow into the urban area and out to Lake Mead. (*Courtesy of Map and Image Library, University of California*)

Table 5.1 Selected river floods in the United States

Year	Month	Location	Lives Lost	Property Damage (Millions of Dollars)
1937	Jan.–Feb.	Ohio and lower Mississippi River basins	137	417.7
1938	March	Southern California	79	24.5
1940	Aug.	Southern Virginia and Carolinas, and eastern Tennessee	40	12.0
1947	May–July	Lower Missouri and middle Mississippi river basins	29	235.0
1951	June–July	Kansas and Missouri	28	923.2
1955	Dec.	West Coast	61	154.5
1963	March	Ohio River basin	26	97.6
1964	June	Montana	31	54.3
1964	Dec.	California and Oregon	40	415.8
1965	June	Sanderson, Texas (flash flood)	26	2.7
1969	Jan.–Feb.	California	60	399.2
1969	Aug.	James River basin, Virginia	154	116.0
1971	Aug.	New Jersey	3	138.5
1972	June	Black Hills, South Dakota (flash flood)	242	163.0
1972	June	Eastern United States	113	3,000.0
1973	March–June	Mississippi River	—	1,200.0
1976	July	Big Thompson River, Colorado (flash flood)	139	35.0
1977	July	Johnstown, Pennsylvania	76	330.0
1977	Sept.	Kansas City, Missouri, and Kansas	25	80.0
1979	April	Mississippi and Alabama	10	500.0
1983	Sept.	Arizona	13	416.0
1986	Winter	Western states, especially California	17	270.0
1990	Jan.–May	Trinity River, Texas	—	1,000.0
1990	June	Eastern Ohio (flash flood)	21	several
1993	June–Aug.	Mississippi River and tributaries	50	>10,000
1997	January	Sierra Nevada, Central Valley, California	23	several hundred
1997	April	Red River, Grand Forks, North Dakota	—	>1,000.0

Sources: NOAA, *Climatological Data, National Summary*, 1970, 1972, 1973, and 1977; U.S. Geological Survey. Updated by author in 1999.

effects can include short-term pollution of rivers, hunger and disease, and displacement of people who have lost their homes. In addition, fires may be caused by short circuits or broken gas mains (15).

5.5 The Response to Flood Hazards

Historically, the response to flooding has been to try to prevent the problem: to control the water by constructing dams, modify the stream by building levees, or even rebuild the entire stream so it will drain the land more efficiently. Every new project has the effect of luring more people to the floodplain in the false hope that the flood hazard is no longer significant. However, we have yet to build a dam or channel capable of controlling the heaviest rainwaters, and when the water finally exceeds the capacity of the structure, flooding can be extensive (16).

There are two general types of response to flood hazards. *Prevention* is the structural approach we have just described. Adjustment measures include *floodplain regulation*, which is assuming greater importance as we begin to recognize the limitations of the structural approach, and flood insurance (15).

Prevention: Physical Barriers

Measures to prevent flooding include construction of physical barriers such as levees (linear mound embankments of compacted earth along the banks of a river, Figure 5.18) and flood walls (usually constructed of concrete as opposed to earthen levees); reservoirs to store water for later release at

CASE HISTORY Flash Floods in Eastern Ohio

On Thursday, June 14, 1990, more than 14 cm of precipitation fell in approximately 3½ hours in some areas of eastern Ohio. Two tributaries of the Ohio River, Wegee and Pipe creeks, generated flash floods near the small town of Shadyside, killing 26 people and leaving 13 missing and presumed dead. The floods were described as walls of water as high as 5 m that rushed through the valley. In all, approximately 70 houses were destroyed and another 40 were damaged. Trailers and houses were seen washing down the creeks, bobbing like corks in the torrent.

The rush of water was apparently related to failure of debris dams that had developed upstream of bridges across the creeks. Runoff from rainfall had washed debris into the channel from sideslopes, and this debris (tree trunks and other material) became lodged against the bridges. When the bridges could no longer contain the weight of the debris, the dams broke loose, sending surges of water downstream. This scenario has been played and replayed in many flash floods around the world. All too often, the supports for bridges are not spaced far enough apart to allow the passage of large debris, which then pile up against the upstream side of the bridge, damming the stream and eventually causing a flood.

▲ **FIGURE 5.18** Levee with a road on top of it protects the bank (left side of photograph) of the lower Mississippi River at this location in Louisiana. (*Comstock, Inc.*)

safe rates; on-site storm-water retention basins (Figure 5.19); channel improvements *(channelization)* to increase channel size and move water off the land quickly; and channel diversions to route floodwaters around areas requiring protection. We will discuss the pros and cons of channel improvements and diversions later in this section.

Unfortunately, the potential benefits of physical barriers are often lost because of increased development on floodplains supposedly protected by these structures. For example, the winters of 1986 and 1997 brought tremendous storms and flooding to the western states, particularly California, Nevada, and Utah. Damages in 1986 exceeded $270 million, and at least 17 people died (17). During one of the storms and floods in 1986, a levee broke on the Yuba River in California, causing more than 20,000 people to flee their homes. An important lesson learned during this flood is that a number of the levees constructed many years ago along

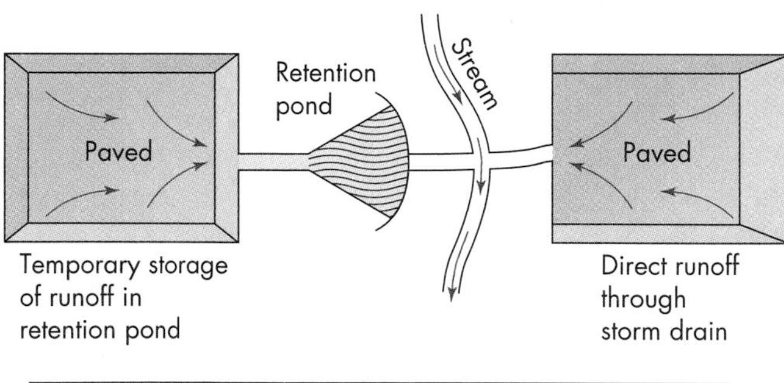

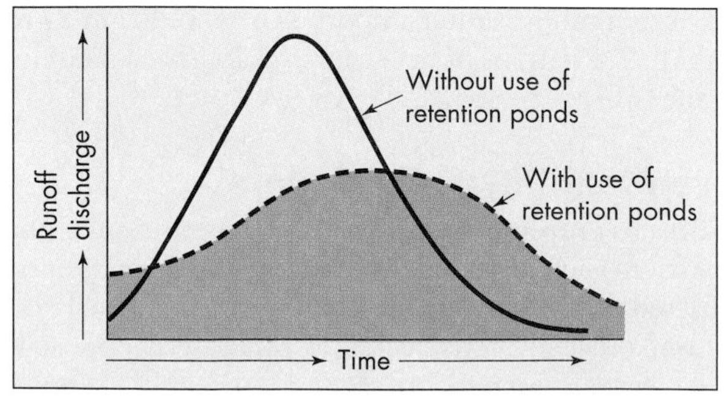

◀ **FIGURE 5.19** (a) Comparison of runoff from a paved area through a storm drain with runoff from a paved area through a temporary storage site (retention pond). Notice that the paved area drained by way of the retention pond produces a lesser peak discharge and therefore is less likely to contribute to flooding of the stream. (*Modified after* U.S. *Geological Survey Professional Paper* 950.) (b) Photograph of a retention pond near Santa Barbara, California. (*Edward A. Keller*)

rivers in California and other states are in poor condition and subject to failure. The 1997 floods damaged campsites and other development in Yosemite National Park. As a result, the park rethought its floodplain management policy—some camping and other facilities were abandoned, and the river is now allowed to "run free."

Some engineering structures designed to prevent flooding have actually increased flooding in the long term (see *Case History: Tucson, Arizona*). What we have learned from all of this is that structural control for floods must go hand in hand with floodplain regulations if the hazard is to be truly minimized (18).

As a final example of using physical barriers to control a river, consider the tremendous undertaking by the U.S. Army Corps of Engineers to keep the lower Mississippi River from shifting its course near New Orleans to a shorter route along the Atchafalaya River, approximately 180 km to the west. The Atchafalaya River via the Red River provides a shorter path to the Gulf of Mexico, and if natural processes were left alone, the Mississippi would shift, leaving Baton Rouge and New Orleans along an abandoned and nearly dry channel. Economic and social costs of such a shift are enormous and so the Corps of Engineers has spent several hundred million dollars in damlike river-control structures to keep the Mississippi where it is. Because the potential new path is much shorter and steeper it may be only a matter of time before the shift takes place, regardless of what structures are built. This view is shared by some geologists but does not reflect the opinion of the Corps of Engineers. The river nearly shifted during the 1973 flood when major damage to one of the control structures occurred. Larger floods are likely, and because the two rivers are close together, the capture of the Mississippi River by a shorter route to the Gulf seems very probable, even if the structures do not fail. That is, overflow and channel cutting could occur at other locations, causing the shift (19).

Adjustment: Floodplain Regulation

From an environmental point of view, *the best approach to minimizing flood damage in urban areas* is **floodplain regulation.** The purpose of floodplain regulation is to obtain the most beneficial use of floodplains while minimizing flood damage and cost of flood protection (20). It is a compromise between indiscriminate use of floodplains, which results in loss of life and tremendous property damage, and complete abandonment of floodplains, which gives up a valuable natural resource.

This is not to say that physical barriers, reservoirs, and channel works are not desirable. In areas developed on floodplains, they will be necessary to protect lives and property. We need to recognize, however, that the floodplain belongs to the river system, and any encroachment that reduces the cross-sectional area of the floodplain increases flooding (Figure 5.20). An ideal solution would be to discontinue floodplain development that necessitates new physical barriers. In other words, the ideal is to "design with nature." Realistically, in most cases, the most effective and practical solution will be a combination of physical barriers

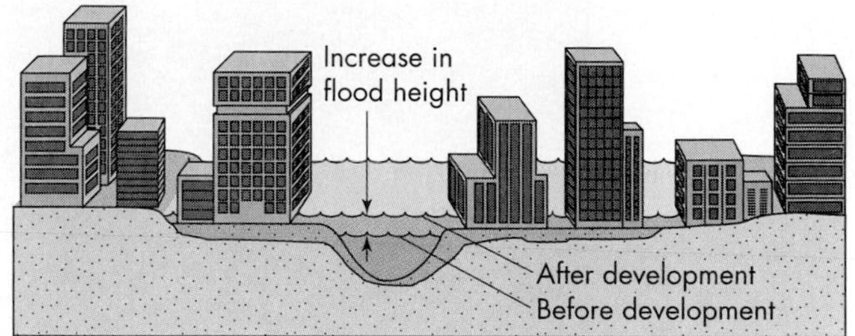

▲ **FIGURE 5.20** Development that encroaches on the floodplain can increase the heights of subsequent floods. (*From Water Resources Council, Regulation of Flood Hazard Areas, vol.* 1, 1971)

and floodplain regulations that will result in less physical modification of the river system. For example, reasonable floodplain zoning in conjunction with a diversion channel project or upstream reservoir may result in a smaller diversion channel or reservoir than would be necessary if no floodplain regulations were used.

Flood-hazard Mapping A preliminary step to floodplain regulation is flood-hazard mapping, which is a means of providing information about the floodplain for land-use planning (23). Flood-hazard maps may delineate past floods or floods of a particular frequency, say, the 100-year flood. They are useful in regulating private development, purchasing land for public use as parks and recreational facilities, and creating guidelines for future land use on floodplains (see *Putting Some Numbers On Flood Hazard Analysis* on page 124).

Developing flood-hazard maps for a particular drainage basin can be difficult and expensive. The maps are generally prepared by analyzing stream-flow data from gauging stations over a period of years. However, flow data are not available in many cases, especially for small streams, so alternative data sources must be used to assess the flood hazard. Methods of upstream flood hazard evaluation may involve estimations of flood peak discharges based on physical properties of the drainage basin. A study of streams in central Texas characterized by periodic intense upstream flooding produced a preliminary empirical model to estimate flood-peak discharges by measuring the *stream magnitude* (number of source streams) of a drainage basin and the *drainage density* (total length of all streams in a basin divided by the area of the basin) (28). In other words, the Texas work produced a statistically valid relationship between measured flood-peak discharge and two measured physical parameters, stream magnitude and drainage density, that can be used to predict floods in basins where hydrologic information is unavailable or insufficient for detailed evaluation.

Flood hazard evaluation for downstream areas may also be accomplished in a general way by direct observation and measurement of physical parameters. For example, extensive flooding of the Mississippi River Valley during the summer of 1993 is clearly shown on images produced from satellite-collected data (Figure 5.21). Floods can also be mapped from aerial photographs taken during flood events, and they

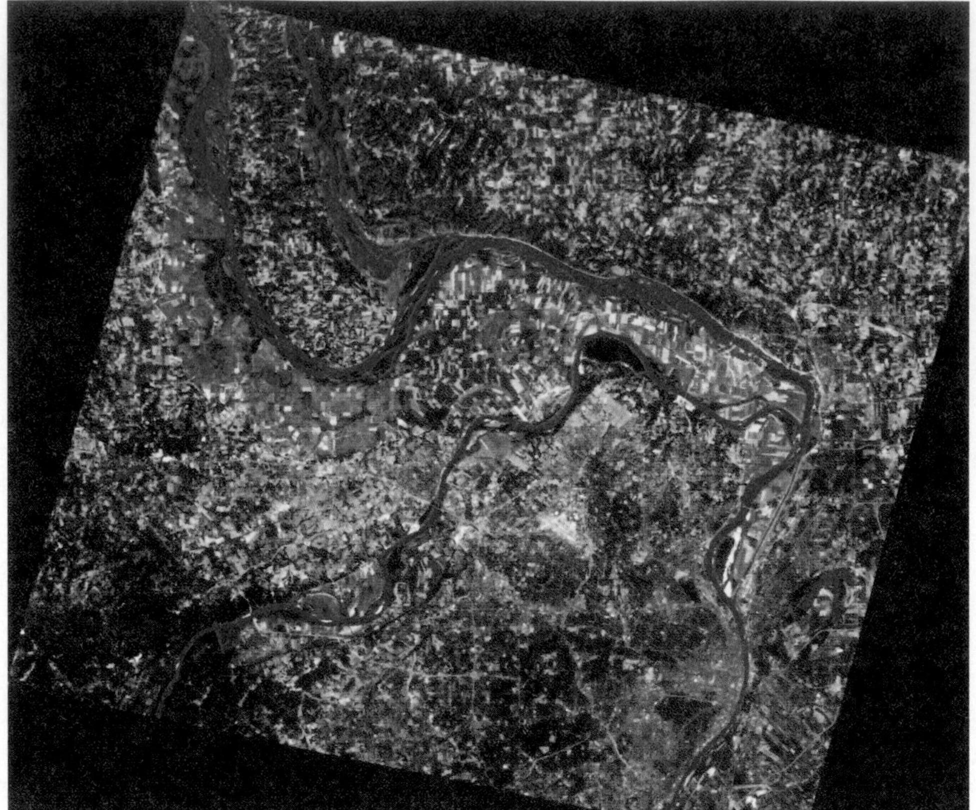

▲ **FIGURE 5.21** This image, which is the synthesis of two images, shows the extent of flooding from the 1993 floods of the Mississippi River. The rivers are shown in 1988 under normal flow conditions in dark blue. The river on the lower left and flowing to the center of the photograph is the Missouri River and it joins the Mississippi River. The smaller river coming in from the top is the Illinois River and the large river that flows from the upper left to the lower right is the Mississippi River. In the lower right-hand corner, the river becomes narrow where it flows by the city of St. Louis, Missouri (orange color area in lower right corner). The river is narrow here because flow is constricted by a series of dikes and flood prevention measures constructed to protect the city. The light blue color is the floodwaters of 1993. Notice the extensive flooding upstream of St. Louis, Missouri. The city with its floodworks is a real "bottleneck" to the flow of water. The town of Alton, Illinois, is the first orange area upstream from St. Louis. This city has a notorious history of flooding. The town is just upstream from the confluence of the Missouri and Mississippi rivers. During the 1993 floods, the width of flooding (the light blue area shown here) from Alton across the flooded area is approximately 8 km. This image was constructed by superposition of images from 1988 and 1993. (*Courtesy of* ITD-SRSC/RSI/SPOT *Image, Copyright* ESA/CNES/*Sygma*)

can be estimated from high-water lines, flood deposits, scour marks, and trapped debris on the floodplain, measured in the field after the water has receded (Figure 5.22).

Careful mapping of soils and vegetation can also help evaluate the downstream flood hazard. Soils on floodplains are often different from upland soils, and, with favorable conditions, certain soils can be correlated with flooding of known frequency. A map based on a study of the Colorado River Valley near Austin, Texas (Figure 5.23), shows a fair correlation between soils and the 100-year flood (28). Vegetation type may facilitate flood-hazard assessment because there is often a rough zonation of vegetation in river valleys that may correlate with flood zones. Some types of trees with shallow roots require an abundant supply of water and benefit from frequent submergence. These trees are often found near the banks of perennial streams that frequently flood. Other species of trees are more restricted to well-drained soils that are not subjected to prolonged or frequent flooding. Although zonation of vegetation is helpful in evaluating flood-prone areas, the cause of the zones is complex and not directly caused by flooding. Therefore, use of vegetation, as with soils, should be combined with other methods of flood-hazard evaluation such as satellite data, aerial photographs, historical records, and floodplain features (28).

The primary advantages of evaluating both upstream and downstream flood hazards from direct observation or properties of the drainage basin and river valley are expediency and cost. The Colorado River study showed that these methods could easily distinguish areas with a frequent flood hazard (1- to 4-year recurrence interval) from areas with an intermediate (10- to 30-year) or infrequent (greater than 100-year) hazard (28). The only disadvantage is that of relative

CASE HISTORY Tucson, Arizona

The September 1983 floods in Arizona killed at least 13 people and inflicted more than $416 million in damages to homes (more than 1300 destroyed or damaged), highways, roads, and bridges. During the flood, extensive bank erosion occurred on the Rillito, a tributary of Santa Cruz River in Tucson (Figure 5.D). The damage clearly suggested that those responsible for flood planning were not prepared for such massive bank erosion. Rather than adopting an overall river-management plan to retard erosion, Tucson had piecemeal bank protection that generated greater channel instability than would have occurred without the structures. That is, although existing bank protection structures protected short reaches of the channel, they enhanced the bank erosion in the unprotected reaches immediately downstream (21). In January 1993, there was major flooding again. Flood peaks were similar to 1983 and serious bank erosion occurred once more, in spite of attempts at additional bank stabilization (soil-cement, i.e., covering the bank with a layer of cement). Once again partial protection only made the problem worse. What was learned in the 1983 flood was apparently ignored and the event was repeated (22).

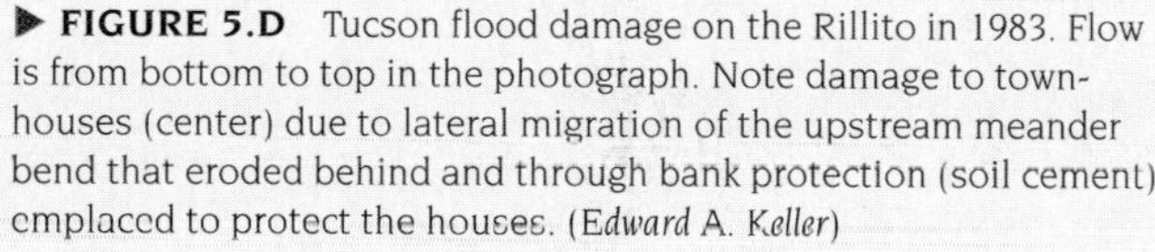

▶ **FIGURE 5.D** Tucson flood damage on the Rillito in 1983. Flow is from bottom to top in the photograph. Note damage to townhouses (center) due to lateral migration of the upstream meander bend that eroded behind and through bank protection (soil cement) emplaced to protect the houses. (*Edward A. Keller*)

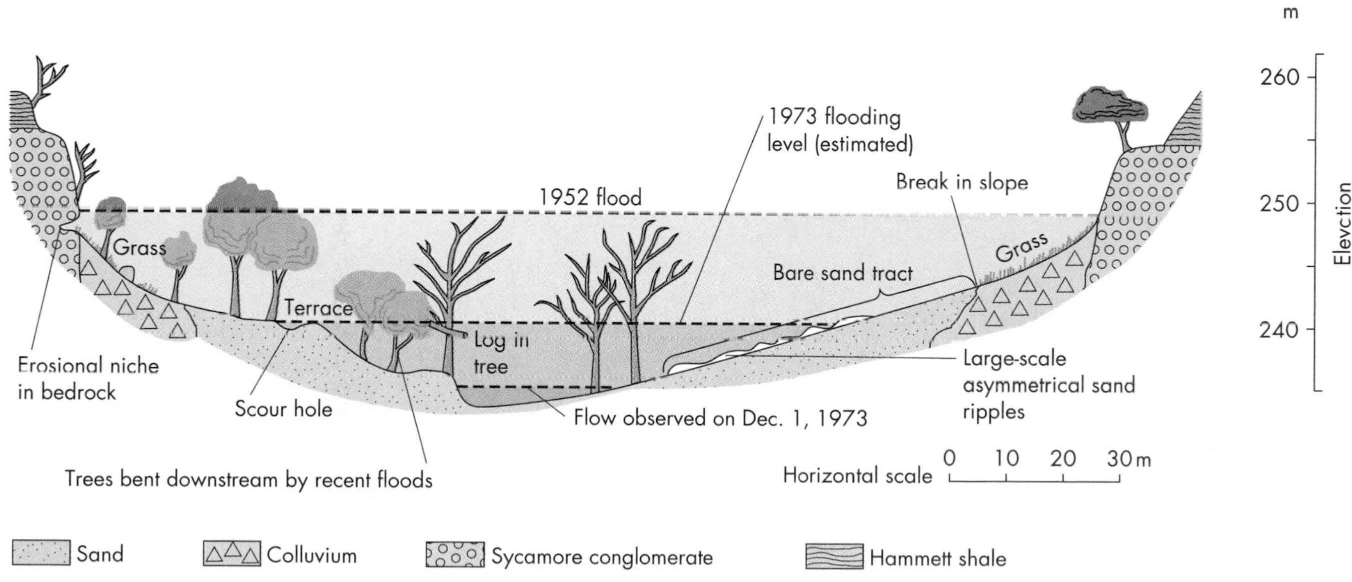

▲ **FIGURE 5.22** Schematic cross section of the Pedernales River Valley, Texas, illustrating floodplain features useful in estimating floods. (*After* V. R. *Baker, "Hydrogeomorphic Methods for the Regional Evaluation of Flood Hazards,"* Environmental Geology, *vol.* 1, *pp.* 261–81, 1976)

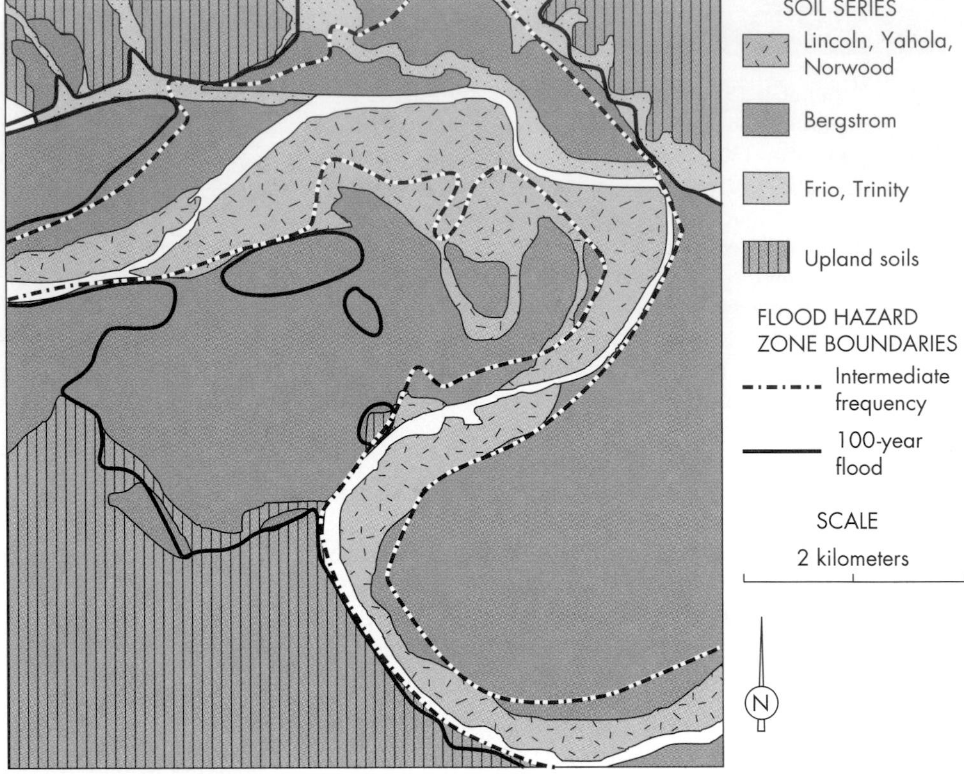

▶ **FIGURE 5.23** Relationship between soils and flooding for a reach of the Colorado River near Austin, Texas. The intermediate frequency floods with a recurrence interval of 10 to 30 years are primarily associated with the Lincoln, Yahola, Norwood, and Bergstrom soils (on the floodplain). The 100-year flood tends to roughly correlate with the lower topographic boundary of the upland soils. (*After* V. R. *Baker, "Hydrogeomorphic Methods for the Regional Evaluation of Flood Hazards,"* Environmental Geology, *vol.* 1, *pp.* 261–81, 1976)

accuracy. Flood-hazard evaluation based on sufficient hydrologic (stream flow) data generally provides more accurate prediction of flood events (revisit *Putting Some Numbers On Flood Hazard Analysis*). In urbanizing areas, the accuracy of flood-hazard mapping based entirely on stream flow data is questionable. A better map may sometimes be produced by assuming projected future urban conditions with an estimated percentage of impervious areas. A theoretical 100-year flood map can then be produced.

Floodplain Zoning Figure 5.24 shows how flood-hazard information is used to designate districts with specific

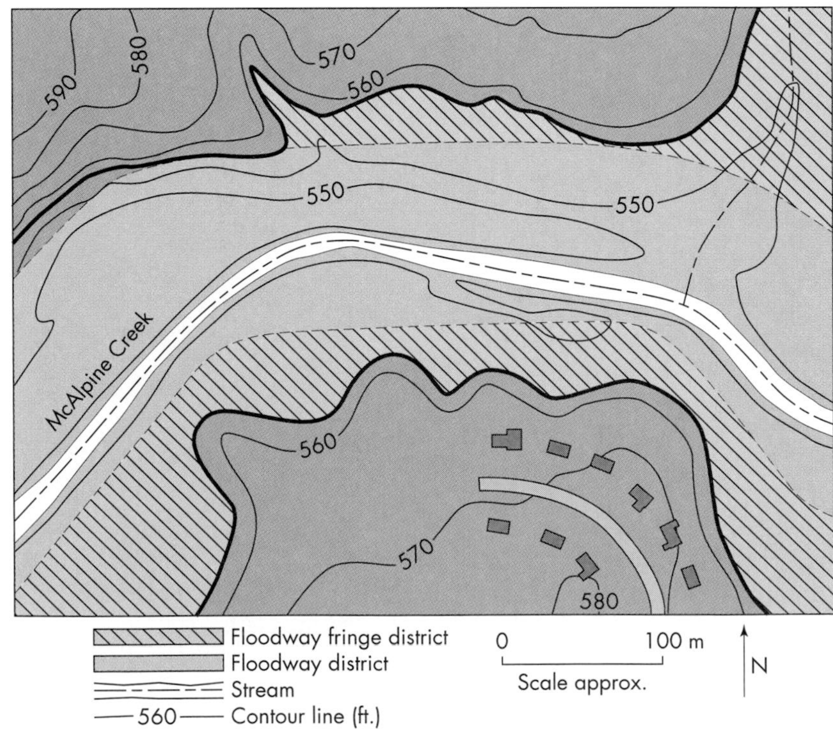

▶ **FIGURE 5.24** Example of a flood hazard map showing the floodway fringe district and the floodway district. (*From County Engineer, Mecklenburg County, North Carolina*) (550 to 580 ft = 168 to 177 m)

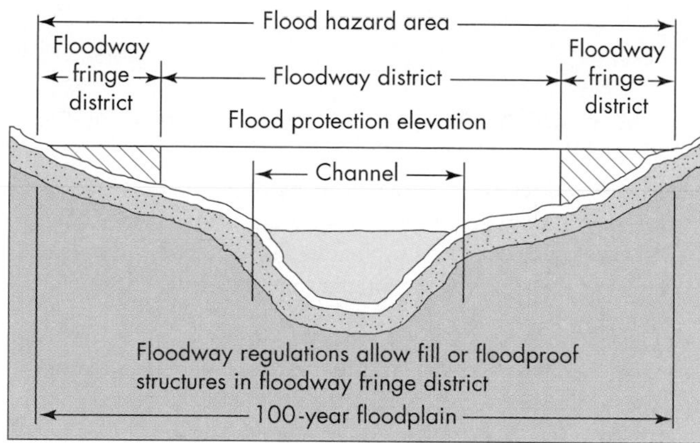

▲ **FIGURE 5.25** Idealized diagram showing a valley cross section and its flood-hazard area, floodway district, and floodway fringe district. (*From* Water Resources Council, Regulation of Flood Hazard Areas, *vol.* 1, 1971)

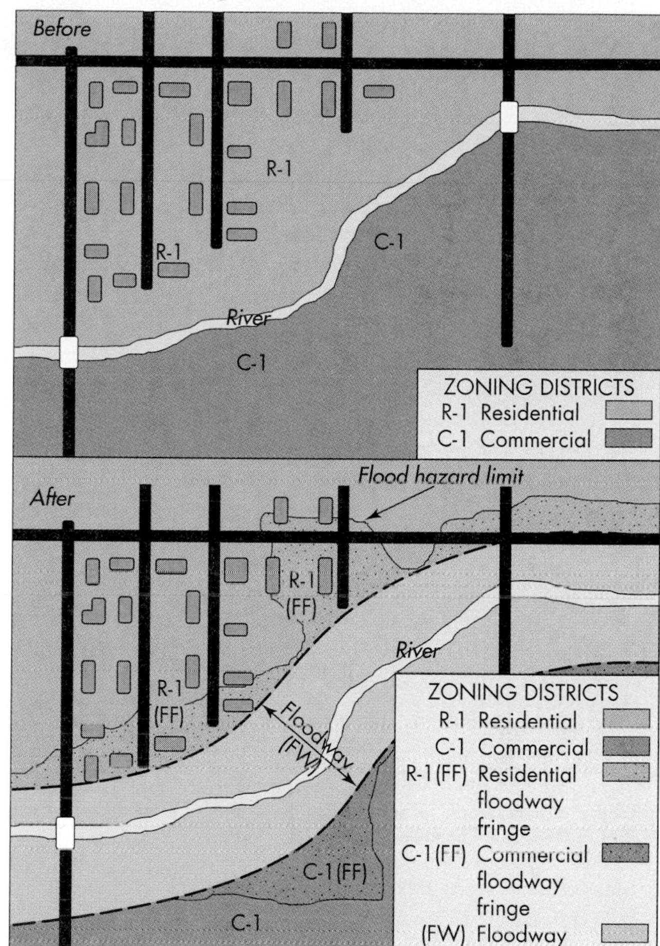

▲ **FIGURE 5.26** Typical zoning map before and after the addition of flood regulations. (*From* Water Resources Council, Regulation of Flood Hazard Areas, *vol.* 1, 1971)

permitted uses in a flood-hazard area. Two districts are mapped in the flood-hazard area, the *floodway district* and the *floodway fringe district*. These districts are shown in an idealized cross section in Figure 5.25. Once the districts are established, planners can use them to establish zoning regulations. Figure 5.26 shows a typical zoning map before and after establishment of floodplain regulations.

The **floodway district** is that portion of the channel and floodplain of a stream designated to provide passage of the 100-year regulatory flood without increasing the elevation of the flood by more than 0.3 m. On this land, permitted uses include farming, pasture, outdoor plant nurseries, wildlife sanctuaries and game farms; loading areas, parking areas, rotary aircraft ports, and similar uses, provided they are farther than 8 m from the stream channel; golf courses, tennis courts, and hiking and riding trails; streets, bridges, overhead utility lines, and storm-drainage facilities; temporary facilities for certain specified activities such as circuses, carnivals, and plays; boat docks, ramps, and piers; and dams, if they are constructed in accordance with approved specifications. Other uses of the floodway district, such as open storage of equipment or material, structures designed for human habitation, or underground storage of fuel or flammable liquids, require special permits.

The **floodway fringe district** consists of the land located between the floodway district and the maximum elevation subject to flooding by the 100-year regulatory flood. Permitted uses in this area include any uses permitted in the floodway district; residential accessory structures, provided they are firmly anchored to prevent flotation; fill material that is graded to a minimum of 1 percent grade and protected against erosion; and structural foundations if firmly anchored to prevent flotation. Aboveground storage or processing of any material that is flammable or explosive or that could cause injury to human, animal, or plant life in times of flooding is prohibited on the floodway fringe district.

The Channelization Controversy

Channelization of streams consists of straightening, deepening, widening, clearing, or lining existing stream channels. Basically, it is an engineering technique, with the objectives of controlling floods, draining wetlands, controlling erosion, and improving navigation (16). Of the four objectives, flood control and drainage improvement are the two most often cited in channelization projects. Thousands of kilometers of streams in the United States have been modified, and thousands more kilometers of channelization are being planned or constructed. Federal agencies alone have carried out several thousand kilometers of channel modification (30). In the past, however, inadequate consideration has been given to the adverse environmental effects of channelization.

Adverse Effects of Channelization Opponents of channelizing natural streams emphasize that the practice is antithetical to the production of fish and wetland wildlife, and that, furthermore, the stream suffers from extensive aesthetic degradation. The argument is as follows:

A CLOSER LOOK History of a River

Philosopher George Santayana observed that "Those who cannot remember the past are condemned to repeat it." Scholars may debate the age-old question of whether or not cycles in human history repeat themselves, but the repetitive nature of natural hazards such as floods is undisputed (29). Better understanding of the historical behavior of a river is therefore valuable in estimating its present and future flood hazards. Consider the Ventura River flood (southern California) of February 1992. With a recurrence interval of approximately 22 years, the flood severely damaged the Ventura Beach RV (recreational vehicle) Resort, which was constructed a few years earlier on an active distributary channel of the delta of the Ventura River (Figure 5.E). Earlier engineering studies suggested that the RV park would not be inundated by flood with a recurrence interval of 100 years. What went wrong?

- It was not recognized that the RV park was constructed on a historically active distributary channel of the Ventura River delta. In fact, early reports did not even mention a delta.
- Engineering models that predict flood inundation are not very good in evaluating distributary channels on river deltas where extensive channel fill and scour as well as lateral movement of the channel is likely.
- Historical documents such as maps dating back to 1855 and more recent aerial photographs that showed the channel were apparently not evaluated. Figure 5.F shows maps rendered from these documents that suggest that the distributary channel was in fact present in 1855 (29).

What went wrong was that the historic behavior of the river was not evaluated as part of the flood-hazard evaluation. If that had been done, it would have been recognized prior to construction of the RV park that a historically active channel was present and the site was unacceptable for development. Nevertheless, necessary permits were issued for development of the park and, in fact, following the flood the park was rebuilt.

Prior to 1992, the distributary channel carried floodwater during 1969, 1978, and 1982. Following the 1992 flood event, the channel again carried floodwaters in the winters of 1993, 1995, and 1998 (repeatedly flooding the RV park). During the 1992 floods, the discharge increased from less than 25 m^3/s to a peak of 1322 m^3/s in only about 4 hours! This is approximately twice as much as the daily high discharge of the Col-

▶ **FIGURE 5.E** Flooding of the Ventura Beach RV Resort in February 1992. The RV park was built directly across a historically active distributary channel of the Ventura River delta. The recurrence interval of this flood is approximately 22 years. A similar flood occurred again in 1995. Notice that U.S. Highway 101 along the Pacific Coast is completely closed by the flood event. (*Mark J. Terrell/AP/Wide World Photos*)

- Drainage of wetlands adversely affects plants and animals by eliminating habitats necessary for the survival of certain species.
- Cutting trees along the stream eliminates shading and cover for fish, while exposing the stream to the sun, which results in damage to plant life and heat-sensitive aquatic organisms.
- Cutting of bottomland (floodplain) hardwood trees eliminates the habitats of many animals and birds and also facilitates erosion and siltation of the stream.
- Straightening and modifying the streambed destroys the diversity of flow patterns, changes peak flow, and destroys feeding and breeding areas for aquatic life.
- Conversion of wetlands from a meandering stream to a straight, open ditch seriously degrades the aesthetic value of a natural area (16). Figure 5.27 summarizes some of the differences between natural streams and those modified by channelization.

It is commonly believed that channelization increases the flood hazard downstream from the modified channel.

orado River through the Grand Canyon in the summer, when it is navigated by river rafters. This is an incredible discharge for a relatively small river with a drainage area of only about 585 km^2. If the flood of 1992 had occurred during nighttime hours, there may have been many deaths rather than the lone fatality that did occur. The river will flood again! A warning system is being developed for the park, but it remains to be seen how effective it will be, given the potential short response time of the river to precipitation. In other words, the park, with or without the RVs and people, is a "sitting duck." This was dramatically illustrated in 1995 and 1998, when winter floods again swept through the park. Although the warning system worked and the park was successfully evacuated, the facility was again severely damaged. There is now a move afoot to purchase the park and restore the land to a more natural delta environment—a good move!

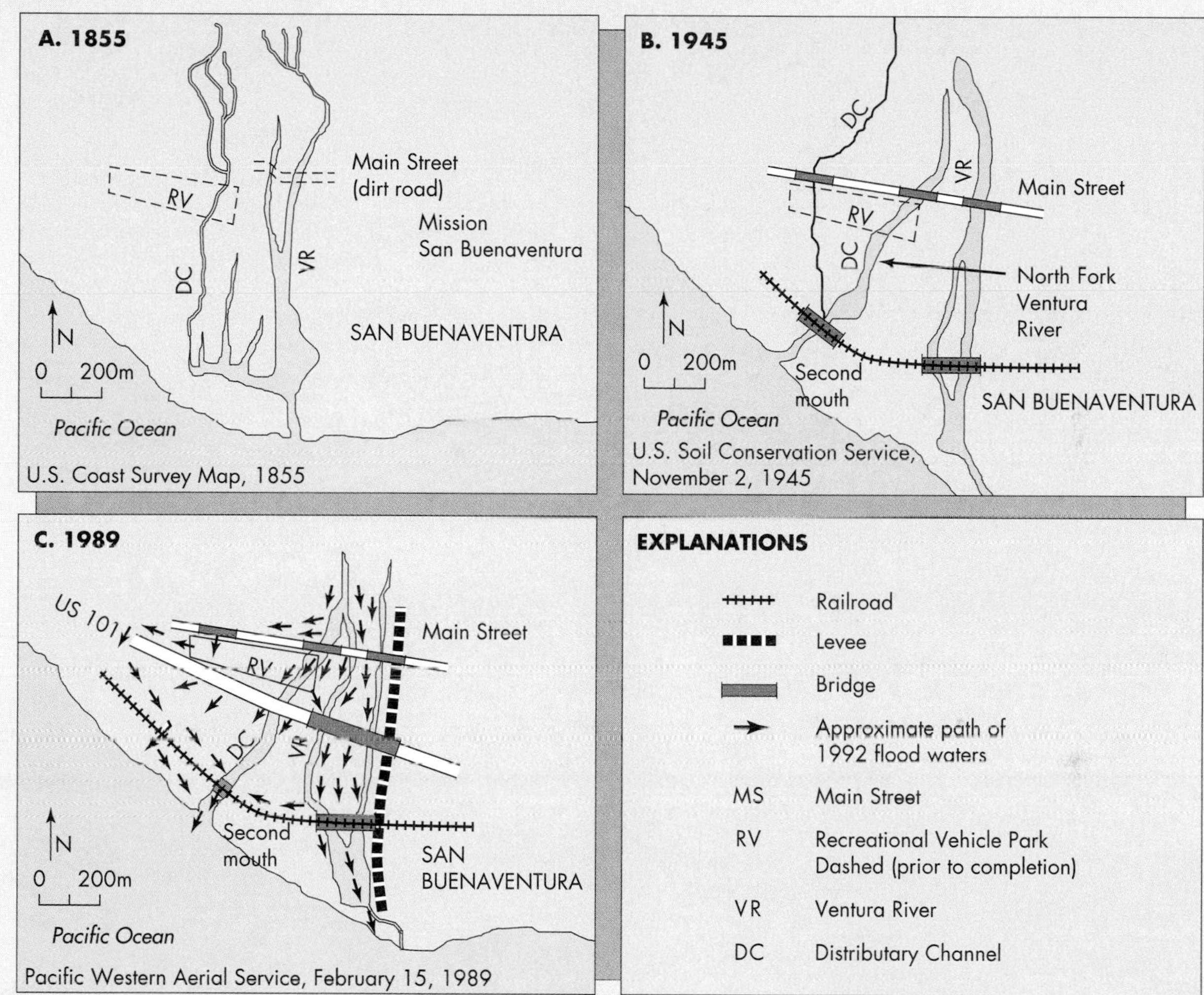

▲ **FIGURE 5.F** Historical maps of the Ventura River delta showing distributary channel and location of a recreational vehicle park. (From E. A. Keller and M. H. Capelli, *Ventura River flood of February 1992: A lesson ignored?* Water Resources Bulletin 25(5):813–31)

Although in many cases this is true, it is far from the entire story. One study concluded that, on the contrary, increases of normal and peak flows from channelization are not particularly significant, and in some cases the flood peaks are actually reduced (30). This results partly because the contribution of runoff from the modified channels tends to be a small fraction of the total basin runoff; peak runoff from channelized streams may not coincide with basinwide runoff, and thus the quicker flow from a modified stream may pass prior to the natural flood flow from the entire basin, thereby reducing the normal aggregated peak flow; and many streams that are modified have gradients so low that no amount of straightening could significantly increase the downstream flow. However, we emphasize that channel modification, especially straightening, can increase flooding directly downstream from the project, and the problem may be compounded if there are a number of projects in the same basin.

Examples of channel-work projects that have adversely affected the environment are well known. For example, from its initiation in 1910, channelization of the Blackwater River in Missouri resulted in channel erosion (enlarge-

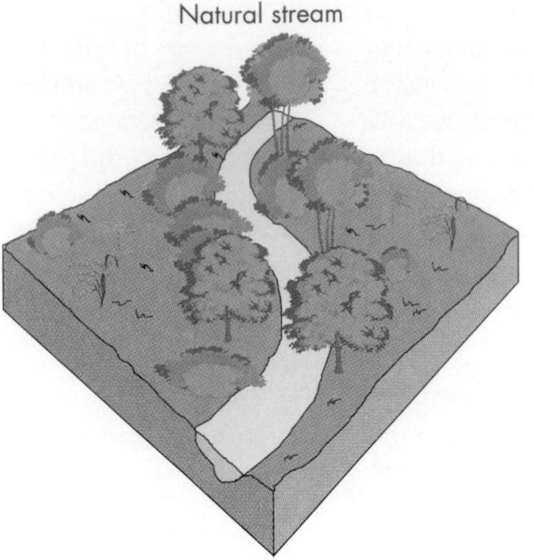

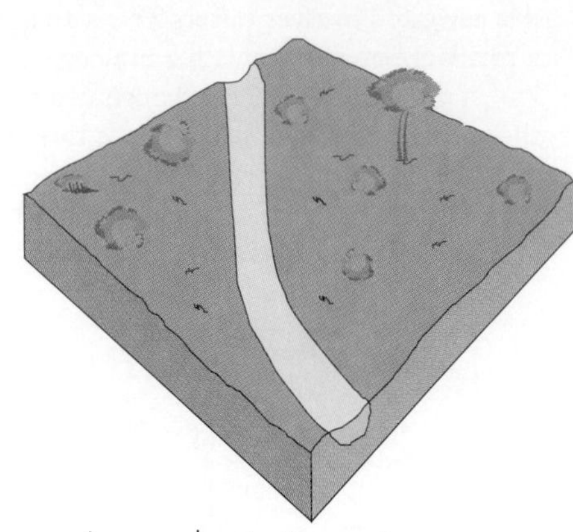

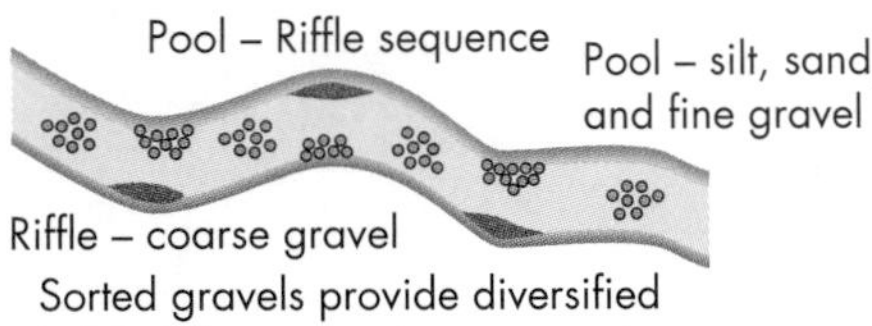

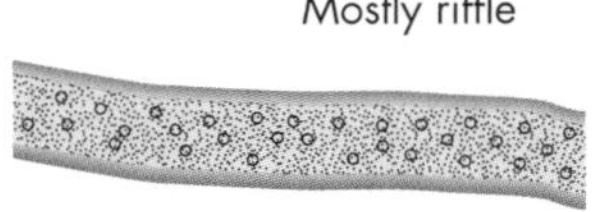

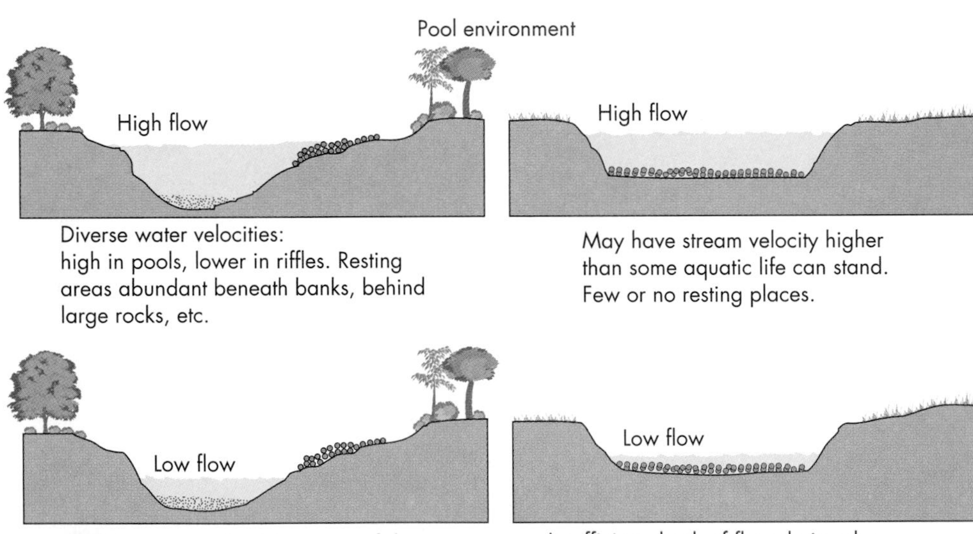

▶ **FIGURE 5.27** A natural stream compared with a channelized stream in terms of general characteristics and pool environments. (*Modified after Corning, Virginia Wildlife, February* 1975)

ment locally exceeded 1000 percent, causing bridges to collapse), reduced biologic productivity, and downstream flooding (31).

Benefits of Channelization and Channel Restoration Not all channelization causes serious environmental degradation; in many cases, drainage projects are beneficial. Benefits are probably best observed in urban areas subject to flooding and in rural areas where previous land use has caused drainage problems. In addition, other examples can be cited in which channel modification has improved navigation or reduced flooding and has not caused environmental disruption.

Many streams in urban areas scarcely resemble natural channels. The process of constructing roads, utilities, and buildings with the associated sediment production is sufficient to disrupt most small streams. **Channel restora-**

tion is often needed to clean urban waste from the channel, allowing the stream to flow freely, to protect the banks by not removing existing trees and, where necessary, planting additional trees and other vegetation. The channel should be made as natural as possible by allowing the stream to meander and, where possible, by providing for variable, low-water flow conditions—fast and shallow (riffle) alternating with slow and deep (pool) (32). Where lateral bank erosion must be absolutely controlled, the outsides of bends may be defended with large stones known as **riprap** (Figure 5.28).

River restoration of the Kissimmee River in Florida may be the most ambitious restoration project ever attempted in the United States. Channelization of the river began in 1962; after 9 years of construction at a cost of $24 million, the meandering river had been turned into an 83-km-long straight ditch. Now, at what will exceed the original cost of channelization, the river may be returned to its original sinuous path. The restoration is being seriously considered because the channelization failed to provide the expected flood protection, degraded valuable wildlife habitat, contributed to water-quality problems associated with land drainage, and caused aesthetic degradation. In Los Angeles, California, a group called "Friends of the River" has suggested that the Los Angeles River be restored. This will be a difficult, if not impossible, task, as most of the riverbed and banks are lined with concrete. Figure 5.29 compares a channel restoration project of Briar Creek, Charlotte, North Carolina, with a concrete-lined channel in southern California.

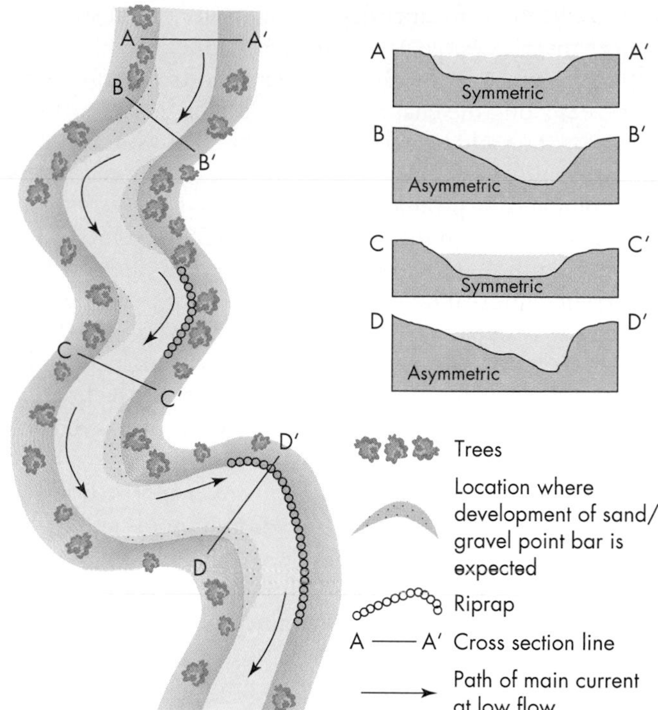

▲ **FIGURE 5.28** Channel-restoration design criteria for urban streams, using a variable cross-channel profile to induce scour and deposition at desired locations. (*Modified after Keller and Hoffman*, 1977, *Journal of Soil and Water Conservation*, *vol.* 32, *no.* 5.)

The Future of Channelization Although considerable controversy and justifiable anxieties surround channelization, it is expected that channel modification will remain a necessity as long as land-use changes, such as urbanization or conversion of wetlands to farmland, continue. Therefore, we must strive to design channels that reduce adverse effects. Effective channelization can be accomplished if project objectives of flood control or drainage improvement are carefully tailored to specific needs (32,33).

If the primary objective is drainage improvement in areas where natural flooding is not a hazard, then there is no need to convert a meandering stream into a straight ditch. Rather, design might involve cleaning the channel and maintaining a sinuous stream. In addition, for gravel-bed streams with a low gradient, a series of relatively deep areas (pools) and shallow areas (riffles) might be constructed by changing cross-sectional shape, asymmetrically to form pools and

(a)

(b)

▲ **FIGURE 5.29** Concrete channel in Los Angeles River system compared to channel restoration in North Carolina. (*Edward A. Keller*)

symmetrically to form riffles, as found in natural streams. Where this is possible, the resulting channel would tend to duplicate nature rather than be alien to it. In addition, cutting trees along the channel bank would be minimized and new growth would be encouraged. This plan would tend to minimize adverse effects by producing a channel that is more biologically productive and aesthetically pleasing (32).

Channel design for flood control is more complicated because natural channels are maintained by flows with a recurrence interval of 1 to 2 years, whereas flood-control projects may need to carry the 25- or even 100-year flood. A 100-year channel cannot be expected to be maintained by a 2-year flood. Such a channel would probably be braided and choked with migrating sandbars. A solution is to construct a pilot channel, a meandering channel designed to be maintained by the 2-year flood and superimposed on the larger flood-control channel. Addition of pools and riffles in the pilot channel, when possible, would help provide fish habitat and better low-flow conditions. The large channel might be vegetated, and the untrained observer might not easily recognize it. This plan will not work in urbanizing areas where sediment production is high and property is not available to be purchased for the pilot channel. However, if sediment is reduced by good conservation practice and the property is available to be purchased, the urban area will benefit from a more aesthetically pleasing and more useful stream.

Perception of Flooding

At the institutional level (government, flood-control agencies, etc.), perception and understanding of flooding is adequate for planning purposes; however, on the individual level, the situation is not as clear. People are tremendously variable in their knowledge of flooding, anticipation of future flooding, and willingness to accept adjustments caused by the hazard.

Progress at the institutional level includes mapping of flood-prone areas (thousands of maps have been prepared), areas with a flash-flood potential downstream from dams, and areas where urbanization is likely to cause problems in the near future. In addition, the federal government has encouraged states and local communities to adopt floodplain management plans (34). Nevertheless, the concept of floodplain management, planning to avoid the flood hazard by not developing on floodplains, or relocating present development to locations off the floodplain needs further consideration and education to be accepted by the general population.

5.6 Mississippi River Flooding, 1973 and 1993

Spring flooding of the Mississippi River in 1973 caused the evacuation of tens of thousands of people as thousands of square kilometers of farmland were inundated throughout the Mississippi River Valley. Fortunately, there were few deaths, but the flooding resulted in approximately $1.2 billion in property damage (18). The 1973 flooding occurred despite a tremendous investment in flood-control dams upstream on the Missouri River. Reservoirs behind these dams inundated some of the most valuable farmland in the Dakotas, and in spite of these structures, the downstream floods near St. Louis were of record-breaking magnitude (35). Impressive as this flood was at the time, it did not compare in magnitude or in the suffering it caused to the flooding that occurred 20 years later.

Flooding of the Mississippi River and its tributaries during the summer of 1993 will likely be remembered as the flood of the century. There was more water than during the 1973 flood, and the recurrence interval exceeded 100 years. The floods lasted from late June to early August, causing 50 human deaths and more than $10 billion in property damages. In all, about 55,000 km^2 of floodplain were inundated, including numerous towns and farmlands (36,37).

The 1993 floods resulted from a major climatic anomaly that covered the upper Midwest and north-central Great Plains, precisely the area that drains into the Mississippi and lower Missouri river systems (38). The trouble started with a wet autumn and a heavy spring snowmelt that saturated the ground in the upper Mississippi River basin. Then, early in June, a high-pressure center became stationary on the East Coast, drawing moist unstable air into the upper Mississippi River drainage basin. This condition kept storm systems in the Midwest from moving east. At the same time, air moving in over the Rocky Mountains initiated unusually heavy rainstorms (38). The summer of 1993 was the wettest on record for Illinois, Iowa, and Minnesota. Cedar Rapids, Iowa, for example, received about 90 cm of rain from April through July—a normal year's rainfall in 4 months (36)! Intense precipitation falling on saturated ground led to a tremendous amount of runoff and unusually high flood peaks during the summer. Floodwaters were high and prolonged, putting tremendous pressure on the levee system that was built in hopes of alleviating flooding.

Prior to construction of the levees, the floodplain of the Mississippi was much wider and contained extensive wetlands. Since the first levees were built in 1718, approximately 60 percent of the wetland in Wisconsin, Illinois, Iowa, Missouri, and Minnesota—all hard hit by the flooding in 1993—have been lost as a result of development and construction of levees. The effect of levees is to constrict the width of the river, which leads to a greater depth of flow and creates a bottleneck that raises the height of floodwaters upstream. In some locations, such as St. Louis, Missouri, levees give way to floodwalls designed to protect the city against high-magnitude floods.

Examination of Figure 5.21, a satellite image from mid-July 1993, shows that the river is relatively narrow at St. Louis, where it is contained by the floodwalls, and very broad upstream near Alton and Portage des Sioux, where extensive flooding occurred. Even so, the rising flood peak came within about 0.6 m of overtopping the floodwalls. Failure of downstream levees partially relieved the pressure and possibly saved St. Louis from flooding. Levee failures (Figure 5.30) were very common during the flood event (39). In fact, something like 80 percent of the private levees along the Mississippi River and its tributaries failed (37). On the other hand, most of the levees built by the federal government survived the flooding and undoubtedly saved lives and property. The problem is that there is not a uniform code for the levees and so some areas' levees are higher or lower than others. Failures occurred as a result of overtopping and breaching, resulting in massive flooding of farmlands and towns (Figure 5.31) (37).

The main lesson learned from the 1993 floods is that construction of levees leads to a false sense of security. It is difficult to design levees to withstand extremely high-magnitude floods for a long period of time. Furthermore, because of loss of wetlands, there is less floodplain space to soak up the floodwaters. The 1993 floods caused such extensive damage and loss of property that many communities along the river are rethinking strategies concerning the flood hazard. Several are considering moving to higher ground! In addition, there is now a FEMA (Federal Emergency Management Agency) program to buy out some floodplain land, with the understanding that homes never be constructed there again. Of course, these adjustments are entirely appropriate. The Mississippi River system flooded riverside communities, including Grafton, Illinois, again in 1995! When will we ever learn?

▲ **FIGURE 5.30** Failure of this levee in Illinois during the 1993 floods of the Mississippi River caused flooding in the town of Valmeyer. (*Comstock, Inc.*)

◀ **FIGURE 5.31** Damage to farmlands during the peak of the 1993 flood of the Mississippi River. (*Comstock, Inc.*)

PUTTING SOME NUMBERS ON Flood Hazard Analysis

We learned earlier in *A Closer Look: Magnitude and Frequency of Floods* that for every flood event we may derive a recurrence interval (R), which provides an estimate of the probable average time period between floods of a particular magnitude. We now return to that theme with the purpose of better understanding how we evaluate the flood hazard and construct flood hazard maps that delineate the land area likely to be inundated by a flood of a particular R, say, the 100-year flood. The 100-year flood may also be represented by Q_{100}, which means the 100-year discharge. The units of discharge commonly used in science are cubic meters per second (cms), but in the United States the units most often encountered are cubic feet per second (cfs). As a result, the data presented here will be in cfs.

Steps in flood hazard analysis are:

- Collect stream flow data from a gauging station or set of gauging stations on a particular river.
- Analyze the stream flow data to estimate magnitude and frequency of flows, and in the case of flood hazards, estimate the discharge from the 100-year flood (Q_{100}) or other discharge of interest to a particular project.
- The 100-year flood might also be estimated from a mathematical model if a stream gauging station is not available or stream gauge data are insufficient.
- Use an appropriate mathematical/computer model to predict the stage (elevation of water surface) expected from the Q_{100} at a variety of topographic cross sections, and construct a flood hazard map showing the area inundated by flood waters.

Probably the best way to understand how flood hazards are estimated is to provide an example. Mission Creek near Santa Barbara, California, has a notorious flood history, including damaging floods of 1995. These floods resulted from a long duration storm that exceeded 8 hours and delivered low- to moderate-intensity rainfall totaling approximately 14 cm (5.5 in.). Mission Creek flooded approximately 500 structures and inflicted about $50 million in property damage. The city of Santa Barbara is built upon an alluvial fan, and the resulting floods were typical of those occurring on alluvial fans (24), consisting of breakouts of waters that spread relatively shallow, fast-moving floodwater down the fan surface, ponding near the ocean. Peak discharge from a stream gauge was estimated by local authorities as exceeding 5000 cfs, with a lower estimate of approximately 3800 cfs. The lower estimate is used here in flood hazard analysis as it was based upon a calculation from field measurements, rather than data from a gauging station, where accumulation of debris may have resulted in an observed stage that was anomalously high. Regardless of which discharge measure is used, the flood was large and caused extensive damage. We begin our analysis of Mission Creek by examining the data on Table 5.A, which shows peak annual flows from 1971 to 1995. This is known as the *annual series*. Notice that in the 1980s and 1990s the years of high discharge are related to El Niño events (see discussion of El Niño in Chapter 16). One way to analyze flood frequency is to arrange the peak annual flows into a series based upon their magnitude (M) where $M = 1$ is assigned the highest flow, which occurred in 1995. The R is then calculated from the equation $R = (N + 1)/M$, where N is the number of years of record. That analysis is shown on Table 5.B. Using this technique we then plot on probability paper the discharge and recur-

Table 5.A Mission Creek at Santa Barbara, California, peak annual flows 1971–1995, arranged per year

Year	Peak flow (cfs)	Comment
1971	360	
1972	1420	
1973	2580	
1974	519	
1975	1130	
1976	353	
1977	569	
1978	2500	El Niño event?[a]
1979	667	
1980	1300	
1981	302	
1982	186	
1983	2300	El Niño year
1984	681	
1985	128	
1986	626	El Niño year
1987	626	El Niño year
1988	139	Drought year
1989	168	Drought year
1990	115	Drought year
1991	468	El Niño year
1992	1130	El Niño year
1993	838	El Niño year
1994	207	El Niño year
1995	3800 (est)	(Highest on record)

[a]El Niño events often bring an increase in precipitation to southern California.

Source: U.S. Geological Survey

Table 5.B Annual peak flow data for Mission Creek at Santa Barbara, CA, 1971–1995, arranged by magnitude *M*, where *M* = 1 is largest event. Also shown are average recurrence intervals *R* (yrs).

Year	Annual Peak Discharge (cm)	Magnitude (*M*)	(*N* + 1) / *M* (*R*, yrs)
1995	3800 (estimate)	1	26.00
1973	2580	2	13.00
1978	2500	3	8.67
1983	2300	4	6.50
1972	1420	5	5.20
1980	1300	6	4.33
1975	1130	7.5[a]	3.47
1992	1130	7.5	3.47
1993	838	9	2.89
1984	681	10	2.60
1979	667	11	2.36
1986	626	12.5	2.08
1987	626	12.5	2.08
1977	569	14	1.86
1974	519	15	1.73
1991	468	16	1.62
1971	360	17	1.53
1976	353	18	1.44
1981	302	19	1.37
1994	207	20	1.30
1982	186	21	1.24
1989	168	22	1.18
1988	139	23	1.13
1985	128	24	1.08
1990	115	25	1.04

[a]Ties are averaged.

Source: U.S. Geological Survey

rence interval for each flow as shown on Figure 5.G. Since we only have 25 years of records, we need to extrapolate the curve to estimate higher magnitude events, such as the Q_{50} or Q_{100}. It is best not to extrapolate the record beyond about two times the length of the record, which means that we could extrapolate to the 50-year flood. However we commonly have to estimate the 100-year flood because this is the so-called "project flood" used in flood hazard analysis. The Water Resources Council (25) in the 1960s recommended using the log-Pearson type III distribution for analyzing flood frequency in flood hazard analysis. This method is thought to produce a better prediction of flood discharge than simple application of the formula used above and shown in Table 5.B. The basic equation used that relates the flood peak and return period with the log-Pearson type III distribution is:

X is equal to $\overline{X} + K\sigma$, where X is the peak discharge at a particular flow frequency of interest (say Q_{100}), $\overline{X}$ is the mean discharge from the set of annual peak flows, K is a frequency factor that increases as R increases, and σ is the standard deviation of the annual peak flows (25). As the name of the distribution implies, logarithms (base 10) are utilized. The method is shown on Table 5.C for the Mission Creek data. We substitute for our general equation above as follows: X is log Q_{100}, $\overline{X}$ is the mean of the logarithmic values of annual peak flows, K is determined from Table 5.D by linear extrapolation, and σ is the standard deviation of the logarithmic values of the annual peak flows. As shown, the last step in the calculation of the

▶ **FIGURE 5.G** Flood frequency of Mission Creek at Santa Barbara, California. Data from Table 5.B.

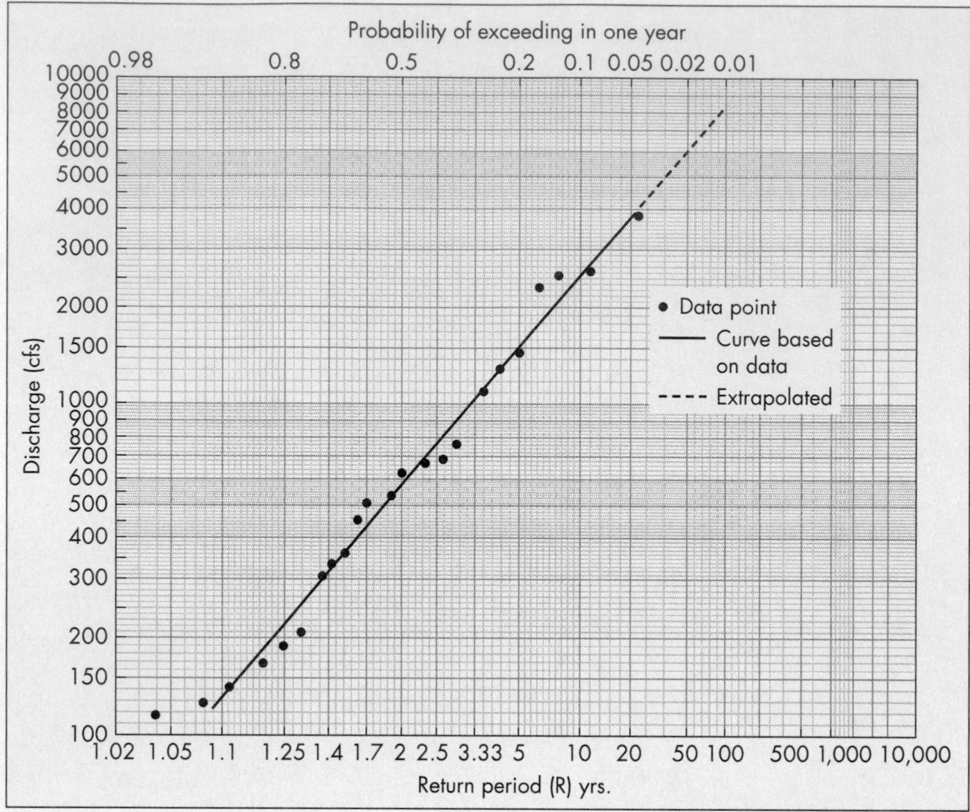

Q_{100} is to take the antilogarithm of the value to produce the desired discharge, in this case Q_{100} (6393 cfs). Notice that there is a different K value for each return period, but to determine K from the table, you must first calculate the value of G from the equation provided. The G value depends upon the actual distribution of the annual peak floods and accounts for the skewness (a measure of the departure of the distribution from one that is symmetrical on both sides of the mean, that is, a departure from the normal distribution, which is roughly "bell-shaped") (26). For our example we have calculated the Q_{100}. Notice that the value of Q_{100} from log-Pearson type III method is less than that from extrapolating the graph on Figure 5.G, where Q_{100} is about 8000 cfs (227 cms). For Q_{50}, the two methods are in closer agreement—4796 cfs (136 cms) for the log-Pearson type III and about 6000 cfs (170 cms) for extrapolation of the graph (Figure 5.G). As an exercise you may wish to calculate other values, say the Q_{10} and Q_{25}. Compare these values with those that you estimate from the graphing of the data shown on Figure 5.G.

The final step in flood hazard analysis is to produce maps that show areas inundated by a particular flood flow (say, Q_{100}). For the system where a channel and floodplain are present, we construct a series of cross sections across the floodplain and channel, and for each of these estimate the stage (elevation of water) that will occur from the Q_{100}. This is commonly done by using a mathematical computer model that solves basic hydrology equations of flow of water in the channel and on the floodplain. A model commonly utilized is the HEC-2 step-backwater model developed by the Hydrologic Engineering Center, Davis, California, in 1990 (27). Data requirements for the program include survey cross sections along the channel and

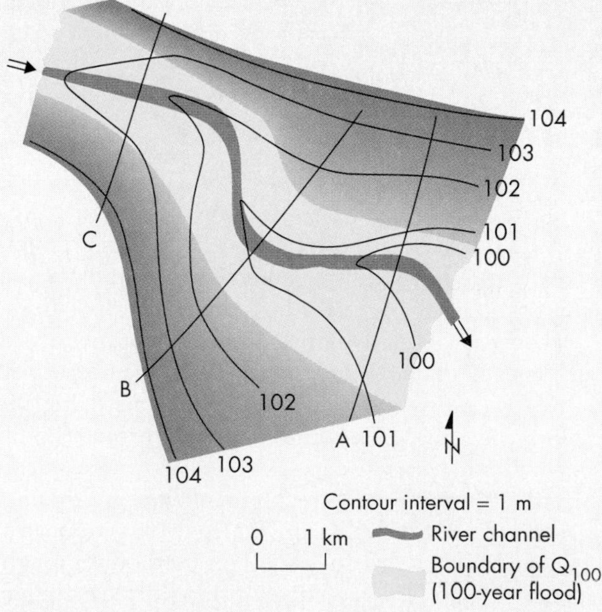

▲ **FIGURE 5.H** Idealized diagram illustrating how flood hazard maps map be produced. See text for explanation.

Table 5.C Mission Creek, Santa Barbara, CA, annual peak flows, 1971–1995. Mission Creek Flood Data Log-Pearson type III distribution. Calculations with logarithms to base 10.

Year	M	Annual Peak Q (cfs)	Q = log peak	$Q - \bar{Q}$	$(Q - \bar{Q})^2$	$(Q - \bar{Q})^3$
1995	1	3800 (estimate)	3.5798	0.8116	0.6587	0.5346
1973	2	2580	3.4116	0.6434	0.4140	0.2664
1978	3	2500	3.3979	0.6297	0.3966	0.2497
1983	4	2300	3.3617	0.5935	0.3523	0.2091
1972	5	1420	3.1523	0.3841	0.1475	0.0567
1980	6	1300[a]	3.1139	0.3457	0.1195	0.0413
1992	7	1300	3.1139	0.3457	0.1195	0.0413
1975	8	1130	3.0531	0.2849	0.0812	0.0231
1993	9	838	2.9232	0.1550	0.0240	0.0037
1984	10	681	2.8331	0.0649	0.0042	0.0003
1979	11	667	2.8241	0.0559	0.0031	0.0002
1986	12.5	626	2.7966	0.0284	0.0008	0.0000
1987	12.5	626	2.7966	0.0284	0.0008	0.0000
1977	14	569	2.7551	−0.0131	0.0002	0.0000
1974	15	519	2.7152	−0.0530	0.0028	−0.0001
1991	16	468	2.6702	−0.0980	0.0096	−0.0009
1971	17	360	2.5563	−0.2119	0.0449	−0.0095
1976	18	353	2.5478	−0.2204	0.0486	−0.0107
1981	19	302	2.4800	−0.2882	0.0831	−0.0239
1994	20	207	2.3160	−0.4522	0.2045	−0.0925
1982	21	186	2.2695	−0.4987	0.2487	−0.1240
1989	22	168	2.2253	−0.5429	0.2947	−0.1600
1988	23	139	2.1430	−0.6252	0.3909	−0.2444
1985	24	128	2.1072	−0.6610	0.4369	−0.2888
1990	25	115	2.0607	−0.7075	0.5006	−0.3541
Sum:			69.2043		4.5876	0.1174

- M = magnitude; $N = 25$ (number of years of record)
- $\bar{Q} = \dfrac{69.2043}{25} = 2.7682$ (is the mean of the logarithmic values of Q)
- $\sigma = \text{standard deviation} = \sqrt{\dfrac{\Sigma(Q - \bar{Q})^2}{N - 1}} = \sqrt{\dfrac{4.586}{24}} = 0.4372$
- $G = \left(\dfrac{n}{(N-1)(N-2)}\right)\left(\dfrac{\Sigma(Q - \bar{Q})^3}{\sigma^3}\right) = \left(\dfrac{25}{(24)(23)}\right)\left(\dfrac{0.1174}{0.4372^3}\right) = 0.064$
- K for Q_{100}, from Table 5.D (for $G = 0.064$) = 2.373 (by linear extrapolation)
- For Q_{100} (100-year flood)
  $\log Q_{100} = \bar{Q} + K\sigma = 2.7682 + (2.373)(0.4372) = 3.8057$
  $Q_{100} = 6393$ cfs (181 cms)

[a]Ties are averaged.

floodplain, initial stage and discharge of water (say, the Q_{100}) and appropriate coefficients, including roughness and whether the channel is expanding or contracting between typical cross sections. An idealized diagram showing the area inundated by the 100-year flood for a hypothetical stretch of river with three cross sections is shown in Figure 5.H. The computer model begins calculations and estimation of flood stage for the downstream cross sections first and then works upstream. Using this method, the area inundated by a flood of a particular frequency may be estimated and mapped. This has been done for thousands of streams and rivers across the United States and is the basic data necessary for evaluating flood hazard and floodplain zonation. This method does have limitations. For example, it doesn't work well for lower Mission Creek because flooding on an alluvial fan may be in multiple distributary channels, and is different from a river with a well-defined floodplain that the HEC-2 model assumes.

Table 5.D Values of *K* for log-Pearson Type III Distribution

	Return Period, Years						
G	2	5	10	25	50	100	200
3.0	−0.396	0.420	1.180	2.278	3.152	4.051	4.970
2.8	−0.384	0.460	1.210	2.275	3.114	3.973	4.847
2.6	−0.368	0.499	1.238	2.267	3.071	3.889	4.718
2.4	−0.351	0.537	1.262	2.256	3.023	3.800	4.584
2.2	−0.330	0.574	1.284	2.240	2.970	3.705	4.444
2.0	−0.307	0.609	1.302	2.219	2.912	3.605	4.298
1.8	−0.282	0.643	1.318	2.193	2.848	3.499	4.147
1.6	−0.254	0.675	1.329	2.163	2.780	3.388	3.990
1.4	−0.225	0.705	1.337	2.128	2.706	3.271	3.828
1.2	−0.195	0.732	1.340	2.087	2.626	3.149	3.661
1.0	−0.164	0.758	1.340	2.043	2.542	3.022	3.489
0.8	−0.132	0.780	1.336	1.993	2.453	2.891	3.312
0.6	−0.099	0.800	1.328	1.939	2.359	2.755	3.132
0.4	−0.066	0.816	1.317	1.880	2.261	2.615	2.949
0.2	−0.033	0.830	1.301	1.818	2.159	2.472	2.763
0.0	0.0	0.842	1.282	1.751	2.054	2.326	2.576
−0.2	0.033	0.850	1.258	1.680	1.945	2.178	2.388
−0.4	0.066	0.855	1.231	1.606	1.834	2.029	2.201
−0.6	0.099	0.857	1.200	1.528	1.720	1.880	2.016
−0.8	0.132	0.856	1.166	1.448	1.606	1.733	1.837
−1.0	0.164	0.852	1.128	1.366	1.492	1.588	1.664
−1.2	0.195	0.844	1.086	1.282	1.379	1.449	1.501
−1.4	0.225	0.832	1.041	1.198	1.270	1.318	1.351
−1.6	0.254	0.817	0.994	1.116	1.166	1.197	1.216
−1.8	0.282	0.799	0.945	1.035	1.069	1.087	1.097
−2.0	0.307	0.777	0.895	0.959	0.980	0.990	0.995
−2.2	0.330	0.752	0.844	0.888	0.900	0.905	0.907
−2.4	0.351	0.725	0.795	0.823	0.830	0.832	0.833
−2.6	0.368	0.696	0.747	0.764	0.768	0.769	0.769
−2.8	0.384	0.666	0.702	0.712	0.714	0.714	0.714
−3.0	0.396	0.636	0.660	0.666	0.666	0.667	0.667

Source: U.S. Soil Conservation Service

SUMMARY

Streams and rivers form a basic transport system of the rock cycle and are a primary erosion agent shaping the landscape. The region drained by a stream system is termed a *drainage basin*. Erosion and deposition of sediments are determined in part by the stream's velocity and stream power at any point, which are determined by the stream's slope, cross-sectional area and shape, and discharge. A river generally maintains a dynamic equilibrium between the work done (sediment transported) and the load imposed (sediment received). A land-use change that affects the amount of water or sediment entering the stream results in a change in the channel's slope and cross-sectional shape and in the velocity of the water.

Sediments deposited by lateral migration of meanders in a stream and by periodic overflow of the stream banks form a floodplain. The magnitude and frequency of flooding are inversely related and are functions of the intensity and distribution of precipitation, the rate of infiltration of water into the soil and rock, and topography. Upstream floods are produced by intense, brief rainfall over a small area. Downstream floods in major rivers are produced by storms of long duration over a large area that saturate the soil, causing increased runoff from thousands of tributary basins. Urbanization has increased flooding in small drainage basins by covering much of the ground with impermeable surfaces such as buildings and roads, increasing the runoff of storm water.

River flooding is the most universally experienced natural hazard. Loss of life is relatively low in developed countries with adequate monitoring and warning systems, but property damage is much greater than in preindustrial societies because floodplains are often extensively developed. Factors that control damage caused by flooding include land use on the floodplain, magnitude and frequency of the flooding, rate of rise and duration of the flooding, the season, the amount of sediment deposited, and the effectiveness of forecasting, warning, and emergency systems.

Environmentally, the best solution to minimizing flood damage is floodplain regulation, but in highly urban areas it will remain necessary to use engineering structures to protect existing development. These include physical barriers such as levees and floodwalls; structures that regulate the release of water, such as reservoirs; and modification of natural channels to accommodate more water (channelization). The realistic solution to minimizing flood damage involves a combination of floodplain regulation and engineering techniques. The inclusion of floodplain regulation is critical because engineered structures tend to encourage further development of floodplains by producing a false sense of security. The first step in floodplain regulation is mapping the flood hazards, which can be difficult and expensive. Planners can then use the maps to zone a flood-prone area for appropriate uses.

Channelization is the straightening, deepening, widening, cleaning, or lining of existing streams. The most commonly cited objectives of channel modification are flood control and drainage improvement. Channelization has often caused environmental degradation, so new projects are closely evaluated. Novel approaches to channel modification that use natural processes are being practiced, and in some cases channelized streams are being restored.

An adequate perception of flood hazards exists at the institutional level; however, on the individual level, more public-awareness programs are needed to help people perceive the hazard of living in flood-prone areas.

Flooding of the Mississippi River Valley in 1993 was of a magnitude that exceeded the 100-year event. The loss of life and extensive damage to crops and property has led some communities to consider moving to higher ground.

REFERENCES

1. **Leopold, L. B., and Maddock, T., Jr.** 1953. *The hydraulic geometry of stream channels and some physiographic implications.* U.S. Geological Survey Professional Paper 252.
2. **Mackin, J. H.** 1948. Concept of the graded river. *Geological Society of America Bulletin* 59:463–512.
3. **Keller, E. A., and Florsheim, J. L.** 1993. Velocity reversal hypothesis: A model approach. *Earth Surface Processes and Landforms* 18:733–48.
4. **Beyer, J. L.** 1974. Global response to natural hazards: Floods. In *Natural hazards*, ed. G. F. White, pp. 265–74. New York: Oxford University Press.
5. **Seaburn, G. E.** 1969. *Effects of urban development on direct runoff to East Meadow Brook, Nassau County, Long Island, New York.* U.S. Geological Survey Professional Paper 627B.
6. **Linsley, R. K., Jr., Kohler, M. A., and Paulhus, J. L.** 1958. *Hydrology for engineers.* New York: McGraw-Hill.
7. **Leopold, L. B.** 1968. *Hydrology for urban land planning.* U.S. Geological Survey Circular 554.
8. **McCain, J. F., Hoxit, L. R., Maddox, R. A., Chappell, C. F., and Caracena, F.** 1979. *Storm and flood of July 31–August 1, 1976, in the Big Thompson River and Cache la Poudre River Basins, Larimer and Weld Counties, Colorado.* U.S. Geological Survey Professional Paper 1115A.
9. **Shroba, R. R., Schmidt, P. W., Crosby, E. J., and Hansen, W. R.** 1979. *Storm and flood of July 31–August 1, 1976, in the Big Thompson River and Cache la Poudre River Basins, Larimer and Weld Counties, Colorado.* U.S. Geological Survey Professional Paper 1115B.

10. **Bradley, W. C., and Mears, A. I.** 1980. Calculations of flows needed to transport coarse fraction of Boulder Creek alluvium at Boulder, Colorado. *Geological Society of America Bulletin*, Part II, 91:1057–90.
11. **Agricultural Research Service.** 1969. *Water intake by soils.* Miscellaneous Publication No. 925.
12. **Strahler, A. N., and Strahler, A. H.** 1973. *Environmental geoscience.* Santa Barbara, CA: Hamilton Publishing.
13. **Davies, W. E., Bailey, J. F., and Kelly, D. B.** 1972. *West Virginia's Buffalo Creek flood: A study of the hydrology and engineering geology.* U.S. Geological Survey Circular 667.
14. **Terstriep, M. L., Voorhees, M. L., and Bender, G. M.** 1976. *Conventional urbanization and its effect on storm runoff.* Illinois State Water Survey Publication.
15. **Office of Emergency Preparedness.** 1972. *Disaster preparedness* 1, 3.
16. **House Report No. 93-530.** 1973. *Stream channelization: What federally financed draglines and bulldozers do to our nation's streams.* Washington, DC: U.S. Government Printing Office.
17. **The angry waters of winter.** 1986. *U.S. News & World Report*, March 3, p. 9.
18. **Rahn, P. H.** 1984. Flood-plain management program in Rapid City, South Dakota. *Geological Society of America Bulletin* 95:838–43.
19. **McPhee, J.** 1989. *The control of nature.* New York: Farrar, Straus & Giroux.
20. **Bue, C. D.** 1967. *Flood information for floodplain planning.* U.S. Geological Survey Circular 539.
21. **Baker, V. R.** 1984. Questions raised by the Tucson flood of 1983. *Proceedings of the 1984 meetings of the American Water Resources Association and the Hydrology Section of the Arizona–Nevada Academy of Science*, pp. 211–19.
22. **Baker, V. R.** 1994. Geologic understanding and the changing environment. *Transactions of the Gulf Coast Association of Geological Societies* XLIV:1–8.
23. **Schaeffer, J. R., Ellis, D. W., and Spieker, A. M.** 1970. *Flood-hazard mapping in metropolitan Chicago.* U.S. Geological Survey Circular 601C.
24. **National Research Council.** 1996. *Alluvial fan flooding.* Washington, DC: National Academy Press.
25. **Hydrologic Committee.** 1967. *A uniform technique for determining flood flow frequencies.* Water Resources Council Bulletin 15.
26. **Davis, J. C.** 1986. *Statistics and data analysis in geology*, 2nd ed. New York: John Wiley & Sons, Inc.
27. **U.S. Army Corps of Engineers.** 1982. *HEC-2 water surface profiles users' manual.* Hydrologic Engineering Center, Davis, California.
28. **Baker, V. R.** 1976. Hydrogeomorphic methods for the regional evaluation of flood hazards. *Environmental Geology* 1:261–81.
29. **Keller, E. A., and Capelli, M. H.** 1992. Ventura River flood of February, 1992: A lesson ignored? *Water Resources Bulletin* 28(5):813–31.
30. **Arthur D. Little, Inc.** 1972. *Channel modifications: An environmental, economic and financial assessment.* Report to the Council on Environmental Quality.
31. **Emerson, J. W.** 1971. Channelization: A case study. *Science* 173:325–26.
32. **Keller, E. A., and Hoffman, E. K.** 1978. Urban streams: Sensual blight or amenity. *Journal of Soil and Water Conservation* 32(5):237–40.
33. **Rosgen, D.** 1996. *Applied river morphology.* Lakewood, CO: Wildland Hydrology.
34. **Edelen, G. W., Jr.** 1981. Hazards from floods. In *Facing geological and hydrologic hazards, earth-science considerations*, ed. W. W. Hays. U.S. Geological Survey Professional Paper 1240-B:39–52.
35. **U.S. Department of Commerce.** 1973. *Climatological data, national summary* 24, no. 13.
36. **Anonymous.** 1993. The flood of 93. *Earth Observation Magazine*, September, pp. 22–23.
37. **Mairson, A.** 1994. The great flood of 93. *National Geographic* 185(1):42–81.
38. **Bell, G. D.** 1993. The great midwestern flood of 1993. *EOS. Transactions of the American Geophysical Union* 74(43):60–61.
39. **Anonymous.** 1993. Flood rebuilding prompts new wetlands debate. *U.S. Water News*, November, p. 10.

KEY TERMS

floodplain (p. 98)
stream power (p. 99)
discharge (p. 100)
competency (p. 100)
capacity (p. 100)
channel pattern (p. 102)
braided channels (p. 102)
meandering (p. 103)
point bar (p. 104)
riverine environment (p. 104)
pool (p. 104)
riffle (p. 104)
upstream floods (p. 105)
downstream floods (p. 105)
recurrence interval (p. 108)
floodplain regulation (p. 113)
floodway district (p. 117)
floodway fringe district (p. 117)
channelization (p. 117)
channel restoration (p. 120)

SOME QUESTIONS TO THINK ABOUT

1. You are a planner working for a community that is expanding into the headwater portion of a drainage basin. You are aware of the effects of urbanization on flooding and wish to make recommendations to avoid some of these effects. Outline a plan of action.
2. You are aware that at the institutional level the perception of flooding is adequate. However, at the individual level the situation is not so clear. How could you develop a plan to communicate the potential of flood hazard to people in your community?
3. You are working for a county flood-control agency that has been channelizing streams for many years. The preferable method has been to use bulldozers to straighten and widen the channel. Recently your agency has been criticized for causing extensive environmental damage. You have developed new plans of channel restoration that you wish to have implemented for a stream maintenance program. Describe what you might do (devise a plan of action) to convince the official in charge of the maintenance program that your ideas will improve the urban stream environment and help reduce the potential of flood hazard.

6 Landslides and Related Phenomena

Reactivated prehistoric landslide destroyed this home in Southern California. (*Edward A. Keller*)

This landslide in the foothills above the community of Santa Barbara, California, is threatening several homes and has the potential to block a nearby stream channel, perhaps causing a flood hazard. The homes built on these slopes were constructed on top of a prehistoric landslide, and the present problem results from reactivation of that slide. In order to minimize the landslide hazard of this area and many others in the United States, we need to carefully recognize and map prehistoric as well as historic landslides to avoid building homes and other structures on unstable lands.

6.1 Introduction to Landslides

Landslides and related phenomena cause substantial damage and loss of life. Each year about 25 people are killed by landslides in the United States, and this number increases to between 100 and 150 if we include collapses of trenches and other excavations. The total annual cost of damages exceeds $1 billion (1).

Landslides and other types of ground failure are natural phenomena that would occur with or without human activity. However, human land use has led to an increase in these events in some situations and a decrease in others. For example, landslides may occur on previously stable hillsides that have been modified for housing development; on the other hand, landslides on naturally sensitive slopes are sometimes averted by means of stabilizing structures or techniques.

In its more restricted sense, the term **landslide** refers to a rapid downslope movement of rock or soil as a more or less coherent mass. It is also used as a comprehensive term for any type of downslope movement of earth materials. Other general terms for downslope movement of earth materials are **slope failure** and *mass wasting*. In this chapter we consider landslides in the restricted sense, as well as the related phenomena of earthflows and mudflows, rockfalls, and snow or debris avalanches. For the sake of convenience, we sometimes refer to all of these as landslides. We will also

LEARNING OBJECTIVES

Landslides constitute a serious natural hazard in many parts of the world, particularly in urban areas. Learning objectives of this chapter are:

- *To gain a basic understanding of the processes operating on slopes and the mechanisms by which slope failures may occur.*
- *To understand the role of driving and resisting forces on slopes and how these are related to determination of slope stability.*
- *To learn how topography, climate, vegetation, water, and time affect slope processes and the incidence of landslides.*
- *To understand how human use of the land has resulted in landslides.*
- *To become familiar with methods of identification, prevention, warning, and correction of landslides.*
- *To gain an appreciation for processes related to land subsidence.*

Web Resources

Visit the "Environmental Geology" Web site at www.prenhall.com/keller to find additional resources for this chapter, including:

- ▶ Web Destinations
- ▶ On-line Quizzes
- ▶ On-line "Web Essay" Questions
- ▶ Search Engines
- ▶ Regional Updates

discuss *subsidence*, a type of ground failure characterized by near vertical deformation (downward sinking) of earth material that often produces circular surface pits but may produce linear or irregular patterns of failure.

6.2 Slope Processes and Slope Stability

Slopes are the most common landforms, and although most slopes appear stable and static, they are dynamic, evolving systems. Material on most slopes is constantly moving down the slope at rates that vary from imperceptible creep of soil and rock to thundering avalanches and rockfalls moving at tremendous velocities. These slope processes are one significant reason that stream valleys are much wider than the stream channel and adjacent floodplain. As with floods, it may not be the largest and least frequent event nor the smallest and most common one that moves the most material down slopes as valleys develop. Rather, events of moderate magnitude and frequency may play the most important role.

To examine slope processes, it is useful to identify slope elements. The slope in Figure 6.1a shows four distinct elements: a **convex slope, or crest**; a nearly vertical **free-face** (cliff); a **debris slope** at approximately 30 to 35°; and a lower

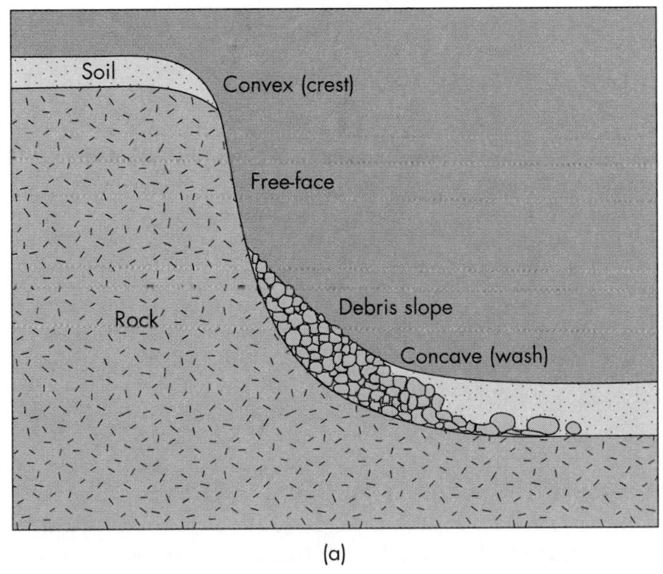

(a)

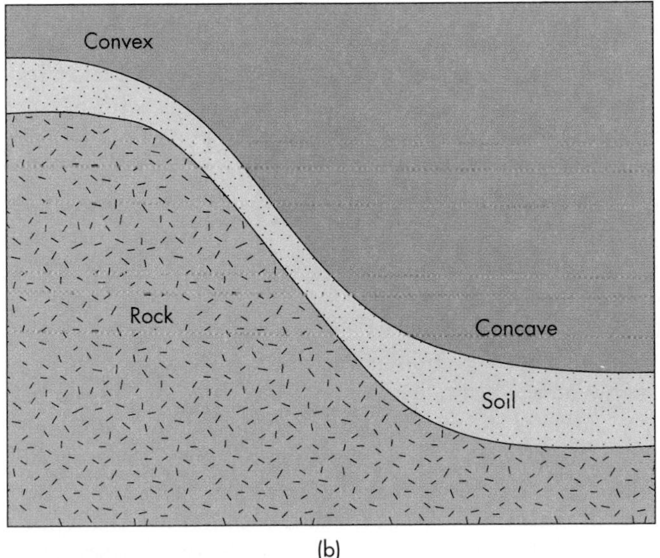

(b)

▲ **FIGURE 6.1** Common slope elements. (a) Slope elements common in semiarid regions or on rocks resistant to weathering and erosion. (b) A convex-concave slope more common to semihumid regions or in areas with relatively soft rocks.

concave slope, or **wash slope.** Note that *wash*, as used by geologists, means an area of loose earth material that has been transported and deposited by water. (In Chapter 5 the word was used to mean a dry stream channel, a usage common in the western United States). All slopes are composed of one or more of these elements, and different slope processes are associated with each element. The convex slope is associated with a slow downslope movement of rock and soil known as *creep*. The free-face is usually associated with processes such as *rockfall*, and the debris slope is where the material from the free-face accumulates. The angle of the debris slope is the **angle of repose,** which is the steepest angle that a given loose material can maintain. The concave slope is produced by processes associated with running water. Steep slopes with a free-face like the one shown in Figure 6.1a commonly occur in semiarid regions and in places where resistant rocks are found.

In subhumid regions and in areas with relatively soft rocks, we find a simpler form of slope, as is also illustrated in Figure 6.1. The elements of this slope are an upper convex slope and a lower concave slope. These gentler (more gradual) slope profiles are often associated with a thick soil cover and often are underlain by rocks of low resistance. Thus, we see that the form of a slope is controlled in part by climate and rock resistance. However, other processes such as stream erosion or wave erosion may produce a prominent free-face, as may tectonic processes such as uplift.

The two types of slope shown in Figure 6.1 are distinguished by the distribution of their soil cover as well as by their shape. On the steeper slope, soil is found at the crest and on the wash slope, but not on the free-face, where weathering is accompanied by rapid erosional removal of the weathered material. On the gentler slope, the soil is thick at the top and bottom portions of the slope and thin in the steeper central portion, where downslope processes are more rapid. Removal of weathered material in the central part of the slope nearly matches its accumulation there.

Earth materials on slopes may fail and move or deform in several ways, which are illustrated in Figure 6.2. **Flowage,** or **flow,** is the downslope movement of unconsolidated material in which the particles move about and mix within the mass. Very slow flowage is called *creep;* extremely rapid flowage is an *avalanche*. Sliding is the downslope movement of a coherent block of earth material. In both flowage and sliding, the moving material is in contact with the slope; **falling,** in contrast, refers to the free fall of earth material, as from the free-face (cliff) of a slope. **Subsidence,** which may occur on slopes or on flat ground, is the sinking of a mass of earth material below the level of the surrounding material.

Landslides are commonly complex combinations of sliding and flowage. As an example, Figure 6.3 shows a failure consisting of an upper slump that is transformed to a flow in the lower part of the slide. (Slumping is a type of sliding, as we will discuss shortly.) Such complex landslides may form when water-saturated earth materials flow from the lower part of the slope, undermining the upper part and causing slumping of blocks of earth materials.

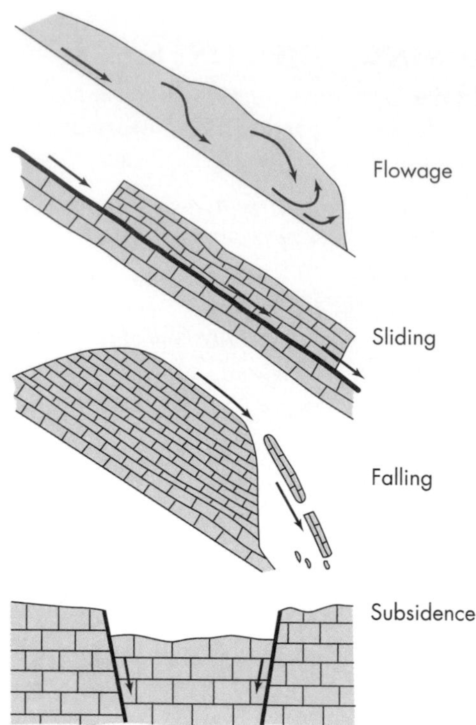

▲ **FIGURE 6.2** Common ways that earth materials fail and move in landslides and related phenomena.

Table 6.1 classifies the common types of downslope movements and reflects the terminology used in this discussion and in other chapters. Important variables in classifying downslope movements are type of movement (slide, fall, flow, or complex movement), slope material type, amount of water present, and rate of movement. In general, the movement is considered rapid if it can be discerned with the naked eye; otherwise it is classified as slow. Actual rates vary from a slow creep of a few millimeters or centimeters per year to very rapid, at 1.5 m/day, to extremely rapid, at several tens of meters per second (2).

Slope Stability

To determine the causes of landslides we must examine slope stability, which can be expressed in terms of the forces acting on slopes. These forces are determined by the interrelationships of the following variables: type of earth materials, slope angle (topography), climate, vegetation, water, and time.

Forces on Slopes The stability of a slope expresses the relationship between **driving forces,** which tend to move earth materials down a slope, and **resisting forces,** which tend to oppose such movement. *The most common driving force is the downslope component of the weight of the slope material,* including anything superimposed on the slope, such as vegetation, fill material, or buildings. *The most common resisting force is the shear strength of the slope material acting along potential slip planes.* Recall from our discussion of soils in Chapter 3 that shear strength is a function of an earth material's cohesion and internal friction. Potential slip planes are geologic planes of weakness in the slope material.

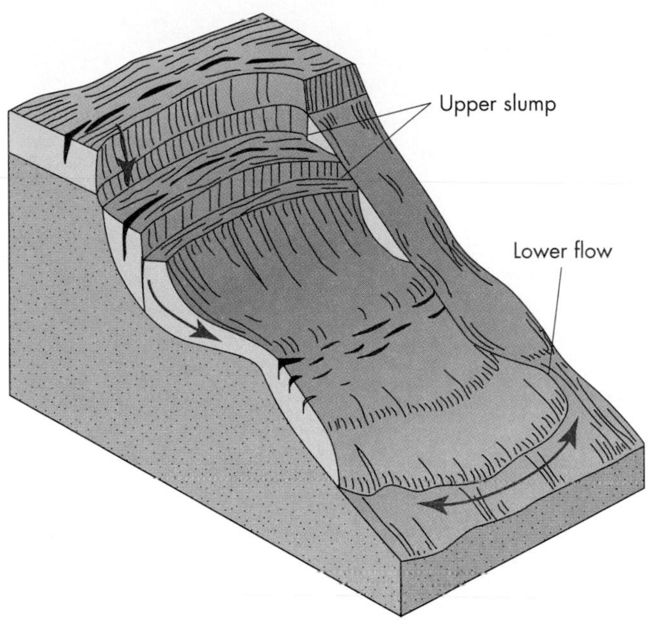

▲ **FIGURE 6.3** Block diagram of a common type of landslide consisting of an upper slump motion and a lower flow. (*After* R. *Pestrong*. 1974. Slope Stability, *American Geological Institute*)

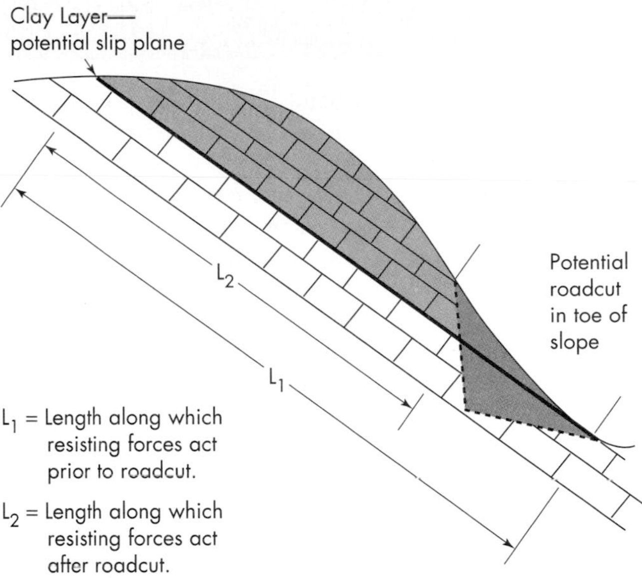

▲ **FIGURE 6.4** Effects on slope stability of a roadcut in the toe of a slope.

Table 6.1 Classification of landslides and other downslope movements

Type of Movement	Materials: Rock	Materials: Soil
Slides (variable water content and rate of movement)	Slump blocks Translation slide	Slump blocks (Rotational slide) Soil slip (planar)
Falls	Rock fall	Soil fall
Slow	Rock creep	Soil creep
↓	Unconsolidated materials (saturated)	
Flows	Earthflow Mudflow Debris flow	
Rapid	Debris avalanche	
Complex	Combinations of slides and flows	

Slope stability is evaluated by computing a **factor of safety** (FS), defined as the *ratio of the resisting forces to the driving forces*. If the factor of safety is greater than 1, the resisting forces exceed the driving forces and the slope is considered stable. If the factor of safety is less than 1, the driving forces exceed the resisting forces and a slope failure can be expected. Driving and resisting forces are not static: As local conditions change, these forces may change, increasing or decreasing the factor of safety. For example, consider construction of a roadcut in the toe of a slope with a potential slip plane (Figure 6.4). The roadcut reduces the driving forces on the slope because some of the slope material has been removed. However, the cut also reduces the resisting forces because the length of the slip plane is reduced, and the resisting force (shear strength) acts along this plane. If you examine Figure 6.4, you can see that only a small portion of the potential slide mass has been removed, while a relatively large portion of the length of the slip plane has been removed. Therefore, the overall effect of the roadcut is to *lower the factor of safety*, because the reduction of the driving force is small compared to the reduction of the resisting force. Factor of safety is commonly computed for natural slopes and slopes constructed as part of site development or highway construction (see *Putting Some Numbers On Landslides*).

The Role of Earth Material Type The material composing a slope affects both the type and the frequency of downslope movement. Slides have two basic patterns of movement, rotational and translational. In **rotational slides,** or **slumps,** the sliding occurs along a curved slip surface (Figure 6.5a). Because the movement follows a curve, slump blocks tend to produce topographic benches (sometimes rotated and tilted in the upslope direction) like those in Figure 6.5a. Slumps are most common on soil slopes, but they also occur on some rock slopes, especially weak rock such as shale. **Translational slides** are planar; that is, they occur along inclined slip planes within a slope. (The slide shown in Figure 6.2 is translational.) Common translation slip planes in rock slopes include fractures in all rock types, bedding planes (Figure 6.5b), clay partings in sedimentary rocks, and foliation planes in metamorphic rocks. In some areas, very shallow slides in soil over rock parallel to the slope, known as **soil slips,** a kind of translation slide, also occur (Figure 6.5c). For soil slips, the slip plane is usually above the bedrock but below the soil,

PUTTING SOME NUMBERS ON Landslides

Analysis of a slope for landslide potential involves determination of resisting and driving forces and the ratio of the two, which is the factor of safety (FS). This may be done for both translational slides or rotational slides. For example, consider the cross section shown on Figure 6.A, which shows a limestone bluff with a potential slip plane composed of clay that is found between bedding planes of the limestone and is inclined at an angle of 30° to horizontal. The potential slip plane is said to "daylight" in the bluff, presenting a potential landslide hazard. Assume that the area above the slip plane in the cross section is 500 m^2, the unit weight of the limestone is 1.6×10^4 N/m^3, shear strength of the clay has been determined from laboratory studies to be 9×10^4 N/m^2, and the length of the slip plane is 50 m. The FS is calculated as the ratio of resisting to driving forces by the equation:

$$FS = \frac{SLT}{W \text{ sine } \Theta}$$

We use what is known as the "unit thickness method," which we use to analyze the resisting and driving forces for a slice (cross section in Figure 6.A) of the bluff, orientated perpendicular to the bluff, which is 1 m thick. The resisting forces are the product of *SLT*, where *S* is the shear strength of the clay, *L* is the length of the slip plane, and *T* is the unit thickness. The driving force is the downslope component of the weight of the slope material above the potential slip plane. This is *W* sine Θ, where *W* is equal to the product of the area above the slip plane *(A)*, the unit weight of the slope material; and the unit thickness *(T)*. Then $W = (500\ m^2)(1\ m)(1.6\times10^4\ N/m^3) = 8\times10^6$ N. *W* is then multiplied by the sine of the angle of the slip plane, and the product (*W* sine Θ) is the downslope component of the weight of the slope materials above the potential slip plane (Figure 6.A).

The factor of safety is calculated as

$$FS = \frac{SLT}{W \text{ sine } \Theta}$$

$$FS = \frac{(9\times10^4\ N/m^2)(50\ m)(1\ m)}{(500\ m^2)(1\ m)(1.6\times10^4\ N/m^3)(0.5)}$$

$$FS = 1.125$$

The conclusion from our analysis, which resulted in a factor of safety of 1.125, is not all that encouraging; generally, a safety factor less than 1.25 is considered as conditionally stable. What could be done to increase the factor of safety to at least 1.25? One possibility would be to remove some of the weight of the limestone above the potential slip plane by reducing the steep slope of the limestone bluff. We can calculate the volume of limestone that would be needed to be removed per unit thickness of the slope using the following equation, setting the factor of safety to 1.25 and solving for *W*.

$$FS = 1.25 = \frac{SLT}{W \text{ sine } \Theta}$$

Then, rearranging:

$$W = \frac{SLT}{1.25 \text{ sine } \Theta}$$

$$W = \frac{(9\times10^4\ N/m^2)(50\ m)(1\ m)}{(1.25)(0.5)}$$

$$W = 7.2\times10^6\ N$$

The weight of the slope material above the slip plane for a unit thickness of our original slope is 8×10^6 N, and therefore $8\times10^6\ N - 7.2\times10^6\ N$ or 8×10^5 N of limestone must be removed per unit thickness. To convert this to a volume of slope (per unit thickness), use the relationship weight of limestone removed = volume (*V*) limestone removed per unit thickness × unit weight of limestone. That is $8\times10^5\ N = (V)(1.6\times10^4\ N/m^3)$. Solving for *V* yields 50 m^3. That is, the original volume of limestone per unit thickness must be reduced by 50 m^3 to a value of 450 m^3 to increase the factor of safety to 1.25. This could be done by removing a uniform thickness of about 1 m of limestone from the top of the slope or a tapered wedge

within a slope material known as **colluvium,** a mixture of weathered rock and other material.

Soil type is a factor in both falls and slides. If a very resistant rock overlies a rock of very low resistance, then rapid undercutting of the resistant rock may cause a slab failure (rock fall) (Figure 6.6).

The strength of the slope materials may greatly influence the magnitude and frequency of landslides and related events. For example, **creep** (the very slow downslope movement of soil and/or rock), **earthflows** (downslope flow of saturated earth materials, may be slow or rapid), slumps, and soil slips are much more common on shale slopes or slopes on weak pyroclastic (volcanic) materials than on slopes on more resistant rock such as well-cemented sandstone, limestone, or granite. In fact, shales are so notorious for landslide activity that what is called **"shale terrain"** is characterized by irregular, hummocky topography produced by a variety of downslope movement processes. For all types of land use, from agricultural to urban, on shale or other weak rock slopes, one must carefully consider the possible landslide hazard prior to development.

The Role of Slope and Topography The *slope angle* (usually called the *slope*) greatly affects the relative magnitude of driving forces on slopes. As the angle of a potential slip plane increases, the driving force also increases; so,

thickening upslope equivalent to 50 m^2 area (of the cross section) per unit thickness of slope (50 m^3).

The above evaluation of the factor of safety is overly simplified, as it assumes that the shear strength is constant along the slip plane, which often is not the case. Our example also doesn't consider fluid pressure in the slide mass (see Chapter 3), which is usually important in landslide analysis. As a result, more detailed evaluation would be necessary, but the principle is the same. That is, resisting and driving forces are calculated and their ratio (FS) is computed. A similar type analysis may be done for rotational or other types of failures, but the mathematics is more complex. Environmental and engineering geologists working on slope stability problems often use computer programs and analyze a number of potential slip planes for both translational and rotational failures as part of slope stability analysis.

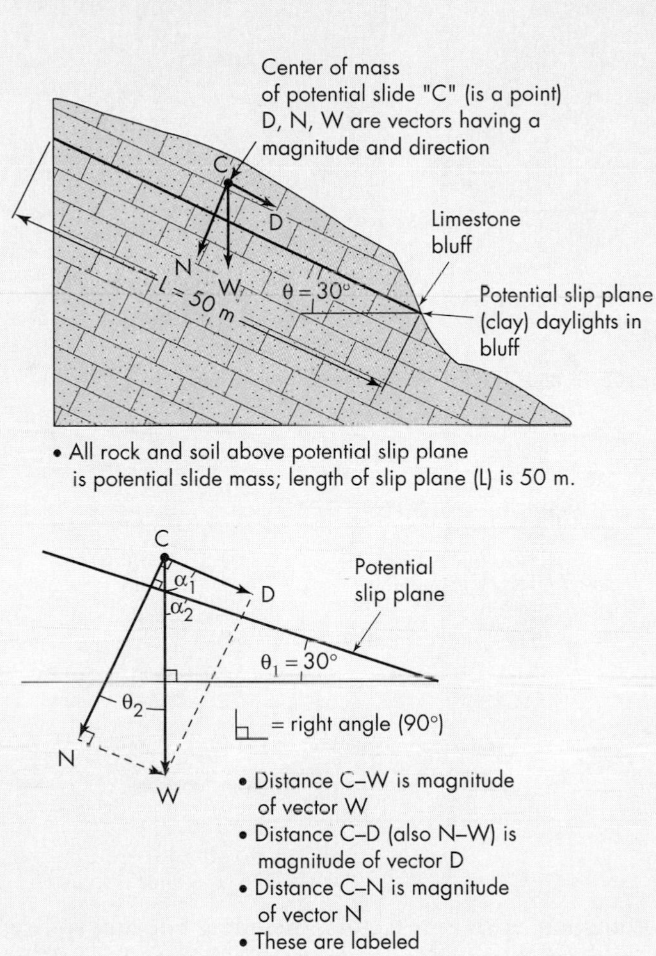

Some trigonometry

$\alpha_1 = \alpha_2$: parallel lines crossed by a line

Show $\theta_1 = \theta_2$ after which $\theta_1 = \theta_2 = \theta$ (slope angle)

$\theta_1 + \alpha_2 = 90°$ right triangle

$\theta_1 + \alpha_1 = 90°$, because $\alpha_1 = \alpha_2$

$\theta_2 + \alpha_1 = 90°$ right triangle

Then $\theta_1 = \theta_2$, follows from $\theta_1 + \alpha_2 = 90°$ & $\theta_2 + \alpha_1 = 90°$

Triangle C W N is a right triangle with hypotenuse W

$$\sin \theta_2 = \frac{\text{side opposite}}{\text{hypotenuse}} = \frac{D}{W}, \text{ or } D = W \sin \theta_2 = W \sin \theta$$

$$\cos \theta_2 = \frac{\text{side adjacent}}{\text{hypotenuse}} = \frac{N}{W}, \text{ or } N = W \cos \theta_2 = W \cos \theta$$

W is the weight of slope materials (potential slide mass) above the slip plane acting down under gravity.

$D = W \sin \theta$ is the downslope component of the weight of the potential slide mass (W); the driving force.

$N = W \cos \theta$ is the normal component (at right angles to the slip plane) of the weight (W) of the potential slide mass. N is part of the shear strength along the slip plane and is thus part of the resisting force.

▲ **FIGURE 6.A** Cross section of a limestone slope (bluff) with a clay layer (potential slip plane) that **daylights.** See text for calculation of factor of safety (FS).

everything else being equal, landslides should be most frequent on steep slopes. A study of landslides that occurred during two rainy seasons in California's San Francisco Bay area established that 75 to 85 percent of landslide activity is closely associated with urban areas on slopes greater than 15 percent, or 8.5° (3). Nationally, the coastal mountains of California and Oregon, the Rocky Mountains, and the Appalachian Mountains have the greatest frequency of landslides. All the types of downslope movement listed in Table 6.1 occur in those locations.

To some extent, the type of landslide activity is also a function of slope and topography. For example, rockfalls and **debris avalanches** (very rapid downslope movement of soil, rock, and organic debris, such as trees) are associated with very steep slopes, and in southern California shallow soil slips are common on steep saturated slopes. These soil slips are often transformed downslope into earthflows or **mudflows** that may be extremely hazardous (Figure 6.7). At the other extreme, earthflows may occur on moderate slopes, and the effect of creep can be observed on slopes of only a few degrees. Another term (you must be tired of new terms by now!) is **debris flow,** which is the downslope flow of relatively coarse material. More than 50 percent of particles in a debris flow are coarser than sand. Debris flows may move very slowly or rapidly depending on conditions. Mudflows and debris flows associated with volcanic processes are discussed in Chapter 8.

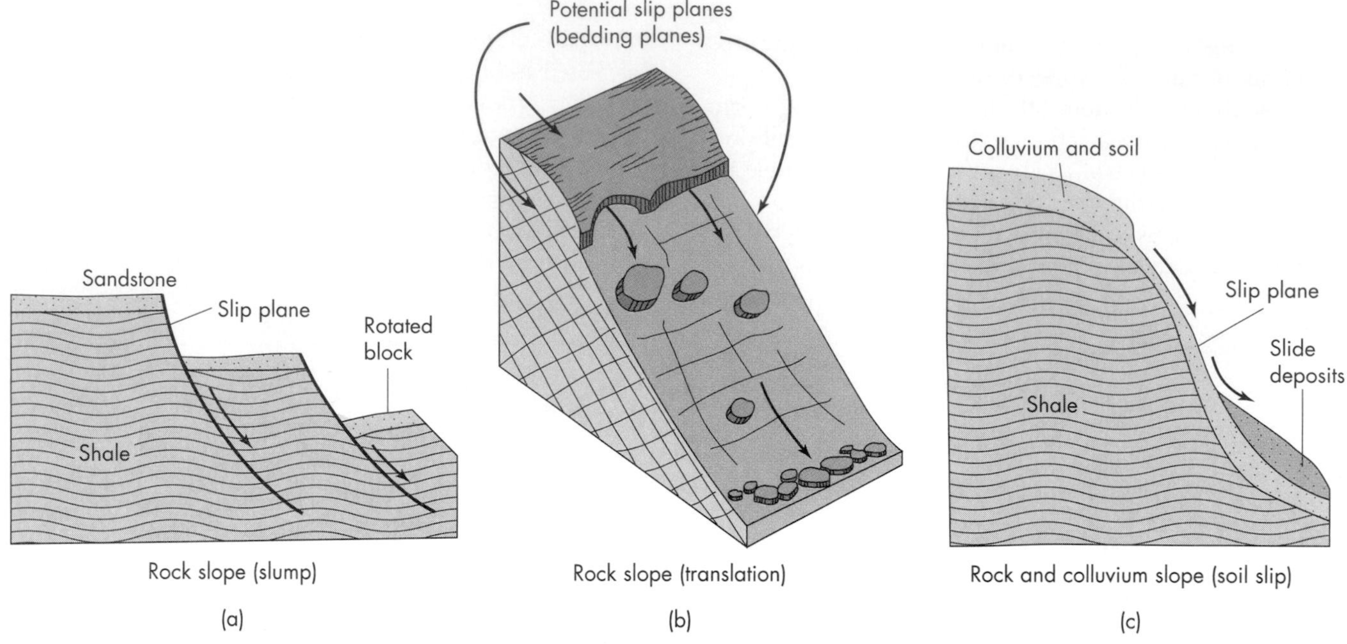

▲ **FIGURE 6.5** Patterns of downslope movement: (a) rotational, (b) translational, and (c) shallow soil slip, also translational.

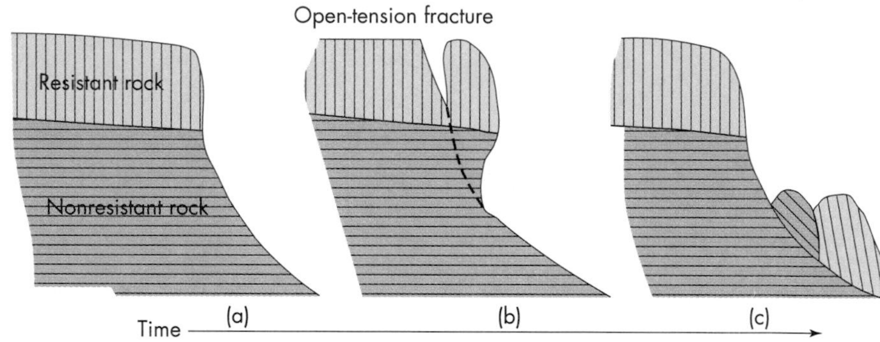

▶ **FIGURE 6.6** Development of a slab slide (type of rock fall). Undercutting of resistant "cap" rock at time (b) causes the development of tension fractures and eventual failure by rock fall at time (c).

The Role of Climate and Vegetation Climate and vegetation can influence the type of landslide or other downslope movement that occurs on a particular slope. Climate controls the nature and extent of precipitation and thus the moisture content of slope materials. For example, both earthflows (which usually occur on slopes) and mudflows or debris flows (which initially may be confined to channels) involve downslope movement of water-saturated earth materials. However, earthflows are common and mudflows relatively rare in humid regions, where a good deal of water infiltrates into slopes, facilitating earthflows. Furthermore, humid regions have many perennial streams (streams that flow throughout the year), which continuously transport the materials delivered from slopes out of the drainage basin. Nevertheless, hazardous debris flows do occur in humid regions in response to high-magnitude storms.

The role of vegetation in landslides and related phenomena is complex because the vegetation in an area is a function of several factors, including climate, soil type, topography, and fire history, each of which also influences what happens on slopes. Vegetation is a significant factor in slope stability for three reasons.

1. Vegetation provides a cover that cushions the impact of rain falling on slopes, facilitating infiltration of water into the slope while retarding grain-by-grain erosion on the surface.
2. Vegetation has root systems that tend to provide an apparent cohesion to the slope materials.
3. Vegetation adds weight to the slope.

Most problems concerning slope stability and vegetation are complicated but related to disturbing, changing, or removing vegetation on or above slopes. Figure 6.8 shows a complex landslide that failed in 1995. The steep slope where the slide occurred is a 40,000-year-old seacliff. At the top of the slope, the topography becomes relatively flat, and an av-

(a)

(b)

▲ **FIGURE 6.7** (a) Shallow soil slip (North Carolina); (b) shallow debris flow (Klamath River, California). Note the long narrow track and debris on the bank of the river. The logging road near bend of the failure may have helped destabilize the slope. (*Edward A. Keller*)

ocado orchard was developed there. It is speculated that irrigation of the orchard increased soil moisture, reducing the stability of the slope by decreasing the resisting forces. (The influence of water on slope stability is discussed below.) There was also a road constructed across the slope that may have contributed to the failure (see Section 6.3). Other complicating factors are that the slide followed heavy precipitation and was a reactivation of an older, smaller slide on a steep slope in an area with many previous natural slides. Thus, the cause and the slide are both complex, involving linkages between weak slope material, topography, water, and human use.

Disturbance or removal of vegetation by logging has also been associated with an increase in landslides. Clearcutting, or removal of all trees, has caused several problems:

- The rate of transpiration (loss of soil water through leaves) is greatly reduced; thus, soil moisture increases and slope stability is reduced.
- In specific instances, infiltration of water into a slope may be increased. This is especially likely if a permeable soil on relatively low-gradient slopes is covered in winter with thick snow pack that slowly melts in the spring.

(a)

(b)

▲ **FIGURE 6.8** Complex landslide consisting of an upper slump block and lower flow at La Conchita, California, in 1995 (a). Close-up of destruction of home in La Conchita. The house was originally three stories (b). (*Edward A. Keller*)

- With the exception of redwood trees, which regenerate after logging, the roots of cut trees decay with time, reducing their strength and thus the apparent cohesion of the soil. This tends to reduce resisting forces within the slope, helping to explain the increased frequency in shallow landslides several years following timber harvesting (4).

In some cases the *presence* of vegetation increases the probability of a landslide, especially for shallow soil slips on steep slopes. One type of soil slip in southern California coastal areas occurs on steep-cut slopes covered with ice plants (Figure 6.9). During especially wet winter months, the shallow-rooted ice plants take up water, adding considerable weight to steep slopes (each leaf stores water and looks like a small canteen!) and increasing the driving forces. By intercepting rainfall, the plants also cause an increase in the infiltration of water into the slope, which decreases the resisting forces. When failure occurs, the plants and several centimeters of roots and soil slide to the base of the slope.

Soil slips on natural steep slopes are a more serious problem in southern California (Figure 6.10). In this case chaparral (dense shrubs or brush) facilitates an increase in water infiltrating into the slope, lowering the safety factor. One study concluded that, in some instances, susceptibility to soil slip may be greater on vegetated slopes than on slopes where vegetation had been recently removed by fire. This should not be interpreted to mean that burning reduces the landslide hazard. Even though soil slips may sometimes be reduced by removal of vegetation, they are not eliminated; in addition, the grain-by-grain erosion caused by rain splash and sheet wash on the surface greatly increases. The eroded sediment tends to fill up the ravines (steep stream valleys) with a meter or more of debris that may be mobilized into debris flows and mudflows during wet winters (5).

The Role of Water Water is almost always directly or indirectly involved with landslides, so its role is particularly important. Much of the chemical weathering of rocks, which slowly reduces their shear strength, is caused by the chemical action of water in contact with soil and rock near the surface of the earth. Natural water (H_2O) is often acidic because it reacts with carbon dioxide (CO_2) in the atmosphere and soil to produce a weak acid—carbonic acid (H_2CO_3). This reaction is $H_2O + CO_2 \rightarrow H_2CO_3$. This chemical weathering is especially significant in areas with limestone, which is very susceptible to weathering and decomposition by carbonic acid.

The ability of water to erode affects the stability of slopes as well. Stream or wave erosion (Figure 6.11) may remove material and steepen a slope, thus reducing the safety factor. This problem is particularly critical if the toe of the slope is an old, inactive landslide that is likely to move again if stability is reduced (Figure 6.12). Therefore, it is important to recognize old landslides along potential roadcuts and other excavations prior to construction and to isolate and correct potential problems.

It has been argued that water has a lubricating effect on individual soil grains and potential slide planes, but this is incorrect: Water is not a lubricant (6). On the contrary, the surface tension of the water surrounding soil grains provides an apparent cohesion. For example, dry sand will form a cone when dumped, whereas moist sand will stand nearly vertical because surface tension (apparent cohesion) in the water holds the grains together. With all its pore spaces filled, saturated sand may flow like mud.

The effects of water on slopes and landslides are quite variable. First, saturation of earth materials causes a rise in pore-water pressure. In general, as the pore-water pressure (the water pressure that develops in the pore spaces between grains, as a result of water filling the pores; see Chapter 3) in slopes increases, the shear strength (resisting force) of the slope decreases and the weight (driving force) increases; thus, the net effect is to lower the factor of safety. This is thought to be a significant factor in the development of soil slips and debris avalanches, as well as other types of landslides. A rise in pore-water pressure is present prior to many landslides, and most landslides are caused by an abnormal increase of the water pressure in the slope-forming materials (6).

(a)

(b)

▲ **FIGURE 6.9** Shallow soil slips on steep slopes covered with shallow-rooted ice plants near Santa Barbara, California. (a) An embankment on a road; (b) beneath a home. (*Edward A. Keller*)

(a)

(b)

▲ **FIGURE 6.10** (a) Shallow soil slips on steep southern California, vegetated slopes. (*Edward A. Keller*) (b) Home in southern California destroyed by debris flow that originated as a soil slip. This 1969 event claimed two lives. (*John Shadle/Los Angeles Department of Building and Safety*)

(a)

(b)

◀ **FIGURE 6.11** (a) Stream bank erosion caused this failure, which damaged a road, San Gabriel Mountains, California. (*Edward A. Keller*) (b) Beachfront home being threatened by a landslide, Cove Beach, Oregon. (*Gary Braasch/Tony Stone Images*)

Soil slips generally occur during heavy rainfall when near-surface temporary or **perched water table** conditions may be present (Figure 6.13). During a rainstorm, the rate of surface infiltration in the unsaturated (vadose) zone of the soil or colluvium exceeds the rate of deep percolation in the rock below the colluvium, and even though some water moves as seepage parallel to the slope, a temporary (perched) water table develops. Failure of the slope occurs when resisting forces are reduced sufficiently—that is, when the factory of safety becomes less than 1. The factor of safety is at a minimum when the perched water table rises to the surface, indicating that the potential slide mass is entirely saturated.

A second way that water may reduce the stability of slopes is by **rapid draw-down,** the rapid lowering of the

(a)

(b)

▲ **FIGURE 6.12** (a) Aerial view of landslide along the Santa Barbara coastal area. Note bulge in coastline; (b) close-up of the slide that destroyed two homes. The slide is a reactivation of an older failure. (*Don Weaver*)

water level in a reservoir or river (at a rate of at least a meter per day). When the water is at a relatively high level, a large amount may enter the banks, a phenomenon called *bank storage*. Then, when the water level suddenly drops, the water stored in the banks is left unsupported. This produces an abnormal distribution of pore-water pressures that reduces the resisting forces; simultaneously, the weight of the stored water increases the driving forces. For this reason, bank failures (slumps) tend to occur along streams after floodwaters have receded.

A third way that water can cause landslides is by contributing to spontaneous **liquefaction** of clay-rich sediment, or **quick clay.** When disturbed, some clays lose their shear strength, behave as a liquid, and flow. The shaking of clay below Anchorage, Alaska, during the 1964 earthquake produced this effect and was extremely destructive. Other examples of slides associated with sensitive clays are found in Quebec, Canada, where several large slides have destroyed numerous homes and killed about 70 people in recent years. The slides occur on river valley slopes in initially solid material that is converted into a liquid mud as the sliding movement begins (7). They are especially interesting because the liquefaction of clays occurs without earthquake shaking. The slides are often initiated by river erosion at the toe of the slope and, although they start in a small area, may develop into large events. Because they often involve the reactivation of an older slide, future problems may be avoided by restricting development. Fortunately, older slides, even though masked from ground view by vegetation, are often visible on aerial photographs.

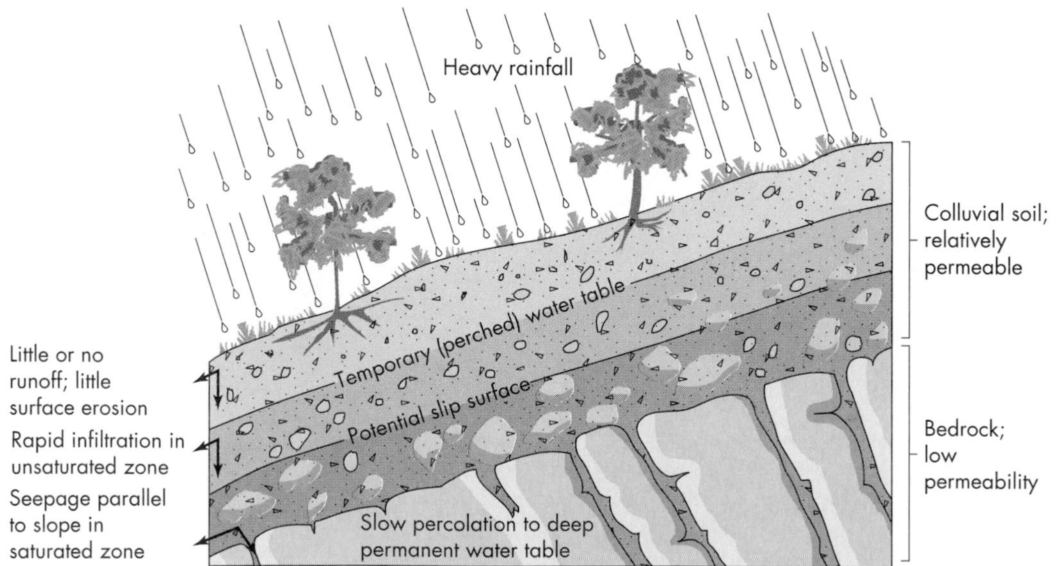

▲ **FIGURE 6.13** Idealized diagram showing development of a perched water table in colluvial material during heavy rainfall and relationship to increased instability of the slope. (*After* R. H. *Campbell*. 1975. U.S. *Geological Survey Professional Paper* 851.)

Seepage of water from artificial sources, such as reservoirs, septic systems, and unlined canals, into adjacent slopes may also affect slope stability by adding weight (the addition of water) to a slope or by removing cementing materials, as was the case in the St. Francis Dam failure (see Chapter 2). Seepage may also cause an increase of the pore-water pressure in adjacent slopes, causing a reduction in the resisting force.

The Role of Time The forces on slopes often change with time. For example, both driving and resisting forces may change seasonally as the moisture content or water-table position alters. These changes are greater in especially wet years, as reflected by the increased frequency of landslides in or following wet years. In other slopes, a continuous reduction in resisting forces may occur with time, perhaps due to weathering, which reduces the cohesion in slope materials, or to a regular increase in water pressure from natural or artificial conditions. A slope that is becoming less stable with time may have an increasing rate of creep until failure occurs (Figure 6.14a).

A slope's factor of safety may also decrease with time because of progressive wetting, which causes disarrangement of the soil particles in the slope, lowering internal friction and thus the strength of the slope materials. Figure 6.14b illustrates this phenomenon as a result of road building and culvert installation that periodically dumps water downslope. This situation emphasizes an important point: We may design slopes that are stable when constructed, but stability has a way of changing with time. Therefore, slope design should include provisions to minimize processes that might progressively weaken slope materials.

Causes of Landslides

The **real causes** of landslides—an increase in driving force or a decrease in resisting force—are often masked by **immediate causes** such as earthquake shocks, vibrations, or a sudden increase in the amount of water entering a slope. The distinction between real and immediate causes is very important when a landslide case is heard in court and an earth scientist is expected to provide a definitive statement concerning the cause of a landslide (8). For example, a translation slide may have as an *immediate cause* heavy rains that saturated the earth material, whereas the *real* cause is the potential to slide upon weak, clay layers. Another example might be the failure of an artificial slope in a housing development, where the immediate cause is an earthquake shock, but the real cause is a poorly designed slope.

Causes of landslides may also be grouped according to whether they are external or internal. **External causes** produce an increase in the shear stress (driving force per unit

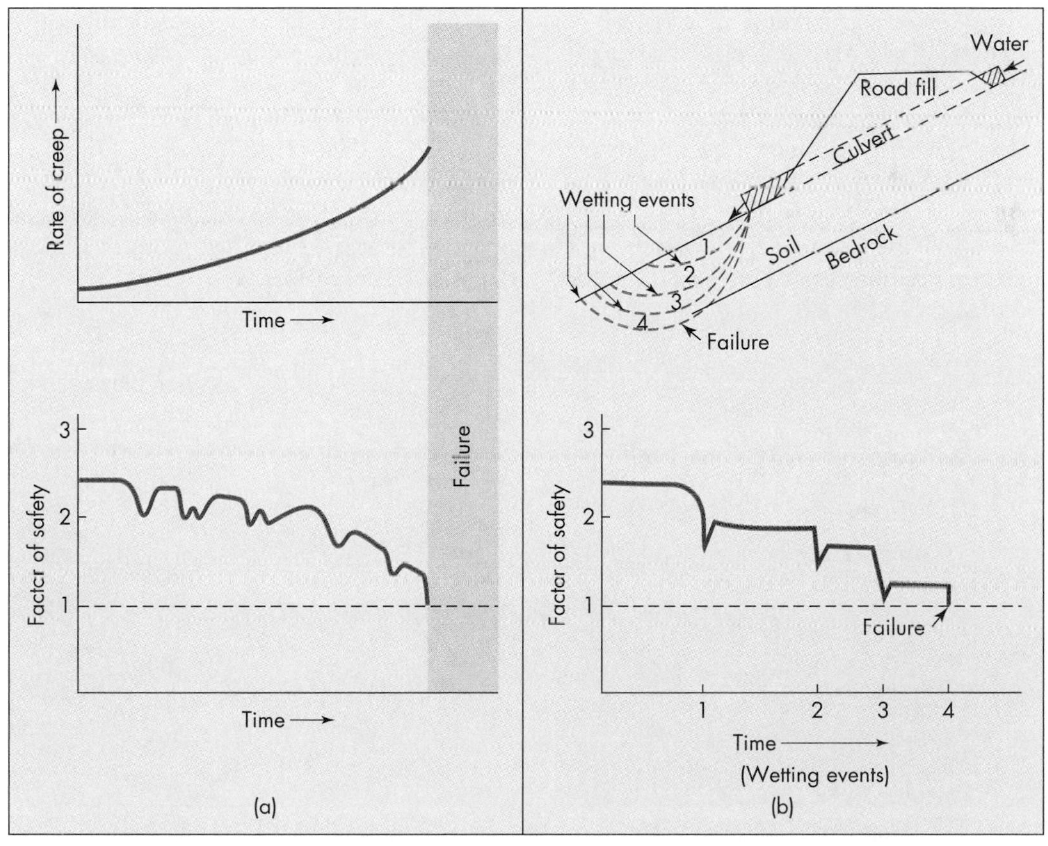

▲ **FIGURE 6.14** Idealized diagrams showing the influence of time on the development of a landslide: progressive creep (a) and progressive wetting (b). Progressive creep is symptomatic of a decreasing safety factor, whereas progressive wetting may cause a reduction in the resisting forces and thus produce a lower safety factor. (*Diagram [b] after* C. S. Yee *and* D. R. Harr. 1977. Environmental Geology, *vol.* 1, *p.* 374)

area) at relatively constant shear strength (resisting force per unit area). Examples of external causes include loading of a slope, steepening of a slope by erosion or excavation, and earthquake shocks. **Internal causes** produce landslides without any recognized external changes and include processes that *reduce the shear strength*. Examples include such changes as increase in pore-water pressure or decrease in cohesion of the slope materials. In addition, some causes of landslides are *intermediate*, having some attributes of both external and internal causes. For example, rapid drawdown involves an increase in the shear stress (caused by weight of water remaining in the slope) accompanied by decrease in shear strength (caused by high pore pressure). Other intermediate causes include spontaneous liquefaction, and subsurface weathering and erosion (6).

6.3 Human Use and Landslides

Effects of human use and interest on the magnitude and frequency of landslides vary from nearly insignificant to very significant. In cases where our activities have little to do with the magnitude and frequency of landslides, we need to learn all we can about where, when, and why they occur, to avoid hazardous areas and to minimize damage. In cases where human use has increased the number and severity of landslides, we need to learn how to recognize, control, and minimize their occurrence wherever possible.

Mixtures of adverse geologic conditions such as weak soil or rock and potential slip planes on steep slopes with torrential rains, heavy snowfall, or seasonably frozen ground (permafrost) will continue to produce landslides, mudflows, and avalanches regardless of human activities. These are natural processes reacting to natural conditions. However, human land-use and settlement patterns (urbanization and deforestation) affect the scope of the disaster (see the discussion in Chapter 4 of landslides resulting from Hurricane Mitch in 1998). Let us compare, for example, the effects of debris avalanches in a sparsely populated and a heavy populated area.

A widespread episode of debris avalanches occurred in August 1969. Remnants of Hurricane Camille, moving eastward from Kentucky and the Appalachian Mountain ridges, mixed with a mass of saturated air to produce thunderstorms of catastrophic proportions. These storms locally produced 71 cm of rain in 8 hours and triggered a great many debris avalanches in central Virginia. The storms claimed 150 lives. The greatest loss of life was the result of flooding, although most people died from broken bones and blunt-force injuries rather than drowning. It is impossible to estimate how many perished as a result of the avalanches, but the amount of debris delivered to channels in floods certainly was significant. The avalanches generally followed preexisting depressions, moved a layer of soil and vegetation 0.3 to 1 m thick, and left a pronounced linear scar of exposed bedrock. The average amount of rock and soil debris moved was 2500 m^3, or about 36,000 metric tons (9).

Although loss of life in Virginia was terrible, it was relatively low because of the sparse population. In 1970 inhabitants of Yungay and Ranrahirca, Peru, were not so fortunate when a debris avalanche triggered by an earthquake roared 3660 m down Mt. Huascaran at a speed in excess of 300 km/hr, killing about 20,000 persons, depositing many meters of mud and boulders, and leaving only scars where the villages had been (Figure 6.15) (10).

Many landslides have been caused by interactions of adverse geologic conditions, excess moisture, and artificial changes in the landscape and slope material. The Vaiont Dam and Reservoir slide of 1963 in Italy is a classic example (see *Case History: Vaiont Dam*). Other examples include landslides associated with timber harvesting and numerous slides in urban areas such as Rio de Janeiro, Brazil; Los Angeles, California; Hamilton County, Ohio; and Allegheny County, Pennsylvania.

(a)

(b)

▲ **FIGURE 6.15** Yungay, Peru, prior to the earthquake and debris avalanche (a); and after the earthquake and debris avalanche (b). The debris avalanche had its origin near the north peak of Nevado Huascaran (the right peak in the photograph). Generated by the earthquake, the debris avalanche moved a distance of approximately 14 km downslope and 4000 m lower in only 3 to 4 minutes (average speed of approximately 320 km/hr). (*Lloyd S. Cluff*)

◀ **FIGURE 6.16** Panoramic view of Rio de Janeiro, Brazil, showing the steep (sugarloaf) hills. Combination of steep slopes, fractured rock, shallow soils, and intense precipitation contribute to the landslide problem, as do human activities such as urbanization, logging, and agriculture. Virtually all of the bare rock slopes were at one time vegetated, and that vegetation has been removed by landsliding and other erosional processes. (*Tony Stone Images*)

Timber Harvesting and Landslides

The possible cause-and-effect relationship between timber harvesting and erosion in northern California, Oregon, and Washington is a controversial topic. Landslides become important in the discussion because there is good reason to believe that landslides, especially shallow soil slips, debris avalanches, and more deeply seated earthflows, are responsible for much of the erosion. In fact, one study in the western Cascade Range of Oregon concluded that shallow slides are the dominant erosion process in the area. Whereas timber-harvesting activities (clear-cutting and road building) over approximately a 20-year observation period on geologically stable land did not greatly increase landslide-related erosion over the same period of time, logging on weak, unstable slopes did increase landslide erosion by several times that which occurred on forested land (11).

The construction of roads in areas to be logged is an especially serious problem because roads may interrupt surface drainage, alter subsurface movement of water, and adversely change the distribution of mass on a slope by cut-and-fill (grading) operations (11). As we learn more about erosional processes in forested areas, we are developing improved management procedures to minimize the adverse effects of timber harvesting. Nevertheless, we are not yet out of the woods with respect to landslide erosion problems associated with timber harvesting.

Urbanization and Landslides

Human use and interest in the landscape are most likely to cause landslides in urban areas where there are high densities of people and supporting structures such as roads, homes, and industries. Examples from Rio de Janeiro, Brazil, and Los Angeles, California, illustrate the situation.

With a population of more than 6 million people, Rio de Janeiro may have more slope-stability problems than any other city its size. The city is noted for the beautiful granite peaks that spectacularly frame the urban area (Figure 6.16). Combinations of steep slopes and fractured rock mantled with surficial deposits contribute to the problem. In earlier times, many such slopes were logged for lumber and fuel and to clear space for agriculture. This early activity was followed by landslides associated with heavy rainfall. More recently, lack of room on flat ground has led to increased urban development on slopes. Vegetation cover has been removed, and roads leading to development sites at progressively higher areas are being built. Excavations have cut the toe of many slopes and severed the soil mantle at critical points. In addition, placing slope fill material below excavation areas has increased the load (driving force) on slopes already unstable before the fill. Because this area periodically experiences tremendous rainstorms, it is easy to see that Rio de Janeiro has a serious problem (13).

In February 1988 an intense rainstorm dumped more than 12 cm of rain on Rio de Janeiro in 4 hours. The storm caused flooding and mudslides that killed about 90 people, leaving some 3000 residents homeless. Restoration costs exceeded $100 million. Many of the landslides were initiated on steep slopes where housing is precarious and control of stormwater runoff nonexistent. It was in these hill-hugging shantytown areas that most of the deaths from mudslides occurred. However, one wing of a three-story nursing home in a more affluent mountainside area was also knocked down by a landslide, killing about 25 patients and members of the staff. If future disasters are to be avoided, Rio de Janeiro is in dire need of measures to control storm runoff and increase slope stability.

Los Angeles, and more generally southern California, has experienced a remarkable frequency of landslides associated with hillside development. Landslides in southern California result from complex physical conditions, in part because of the great local variation in topography, rock and soil types, climate, and vegetation. Interactions between the natural environment and human activity are complex and notoriously unpredictable. For this reason, the area has the sometimes dubious honor of showing the ever-increasing value of urban geology (14). Los Angeles has led the nation

CASE HISTORY Vaiont Dam

The world's worst dam disaster occurred on October 9, 1963, when approximately 2600 lives were lost at the Vaiont Dam in Italy. As reported by George Kiersch (12), the disaster involved the world's highest thin-arch dam (267 m at the crest), yet, strangely, no damage was sustained by the main shell of the dam or the abutments (12). The tragedy was caused by a huge landslide, in which more than 238,000,000 m^3 of rock and other debris moved at speeds of about 95 km/hr down the north face of the mountain above the reservoir, completely filling it with slide material for 1.8 km along the axis of the valley to heights of nearly 152 m above the reservoir level (Figure 6.B). The rapid movement created a tremendous updraft of air and propelled rocks and water up the north side of the valley, higher than 250 m above the reservoir level. The slide and accompanying blasts of air and water and rock produced strong earthquakes recorded many kilometers away. It blew the roof off one man's house well over 250 m above the reservoir and pelted the man with rocks and debris. The filling of the reservoir produced waves of water more than 90 m high that swept over the abutments of the dam. More than 1.5 km downstream, the waves were still more than 70 m high, and everything for kilometers downstream was completely destroyed. The entire event (slide and flood) was over in less than 7 minutes.

The landslide followed a 3-year period of monitoring the rate of creep on the slope, which varied from less than 1 to as many as 30 cm per week, until September 1963, when it increased to 25 cm daily. Finally, on the day before the slide, it was about 100 cm per day. Engineers expected a landslide, but one of much smaller magnitude, and they did not realize until October 8, one day before the slide, that a large area was moving as a uniform, unstable mass. Animals grazing on the slope had sensed danger and moved away on October 1.

The slide was caused by a combination of factors. First, adverse geologic conditions, including weak rocks and limestone with open fractures, sinkholes, and clay partings, which were inclined toward the reservoir, produced unstable blocks (Figure 6.C), and very steep topography created a strong driving (gravitational) force. Second, pore-water pressure was increased in the valley rocks because of the impounded water in the reservoir. Groundwater migration into bank storage raised the pore-water pressure and reduced the resisting forces in the slope. The rate of creep before the slide increased as the water table rose in response to higher reservoir levels. Third, heavy rains from late September until the day of the disaster further increased the weight of the slope materials, raised the pore-water pressure in the rocks, and produced runoff that continued to fill the reservoir even after engineers tried to lower the reservoir level.

It was concluded that the cause of the disaster was an increase in the driving forces accompanied by great decrease in the resisting forces, as rising groundwater in the slope increased the pore-water pressure along planes of weakness in the rock (12).

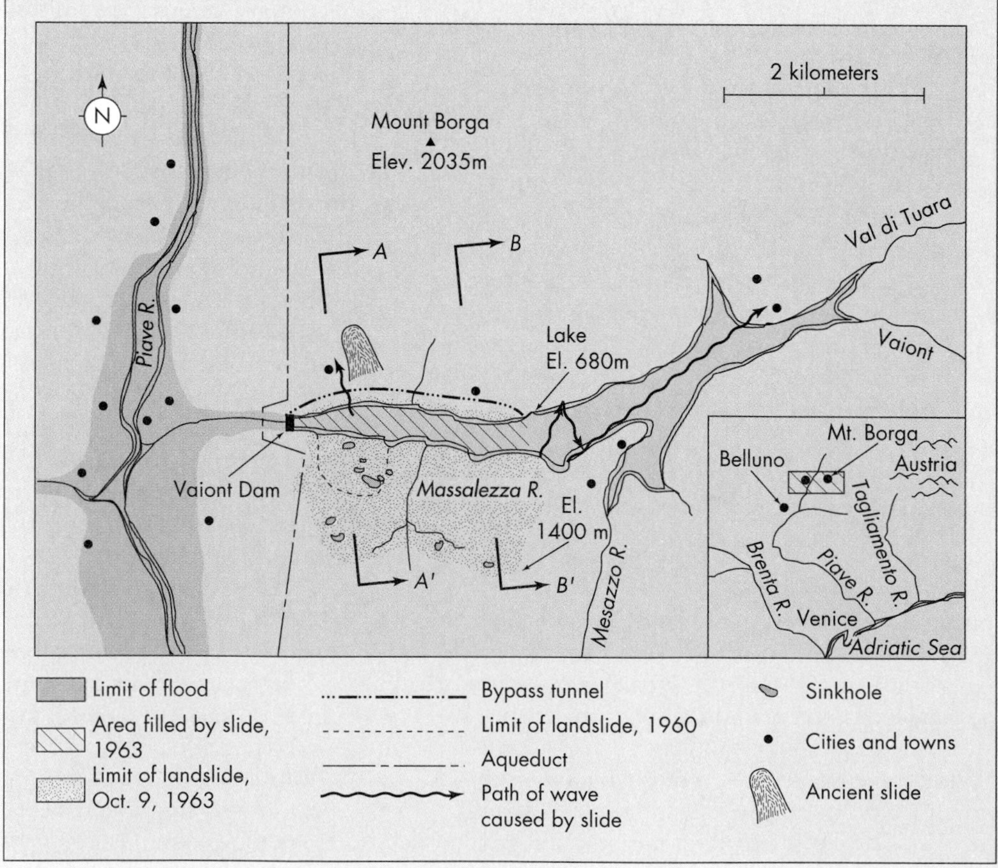

▶ **FIGURE 6.B** Sketch map of the Vaiont Reservoir showing the 1963 landslide that displaced water that overtopped the dam and caused severe flooding and destruction over large areas downstream. A–A′ and B–B′ are section lines shown on Figure 6.C. (*After Kiersch.* 1964. Civil Engineering 34)

◀ **FIGURE 6.B (*continued*)** Photograph of the Vaiont Dam following the landslide. Notice that the concrete dam is still intact but the reservoir above the dam is completely filled (or nearly so) with landslide deposits. (ANSA)

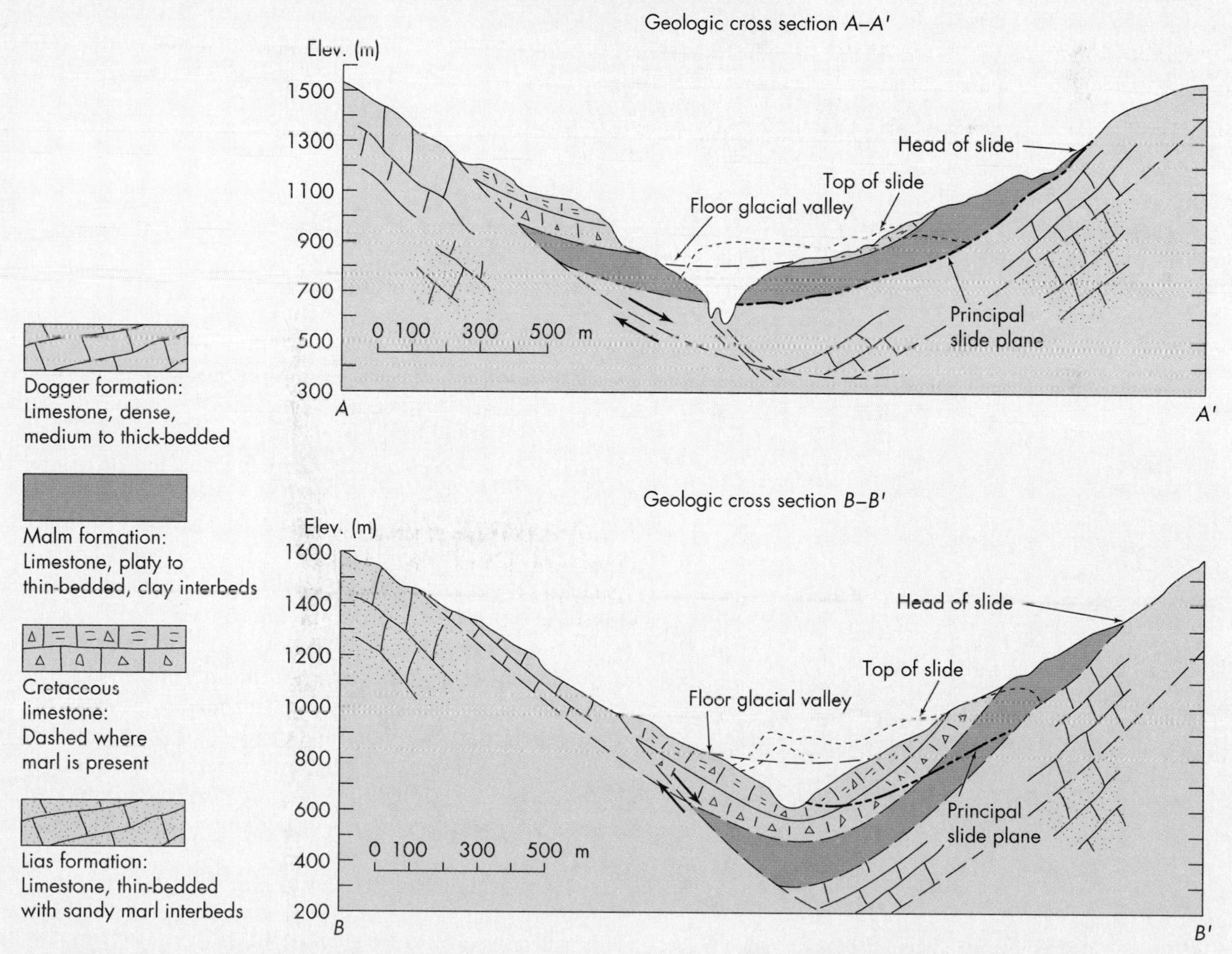

▲ **FIGURE 6.C** Generalized geologic cross sections through the slide area of the Vaiont River Valley. Location of sections are shown on Figure 6.B. (*After Kiersch*. 1964. Civil Engineering 34)

A CLOSER LOOK Determining Landslide Hazard and Risk

The procedure for evaluating the landslide hazard is to first make a landslide inventory, which may be a reconnaissance map showing areas that apparently have experienced slope failure. This may be done by aerial photographic interpretation with field check. At a more detailed level, the landslide inventory may be a map that shows definite landslide deposits in terms of their relative activity as being active, inactive (geologically young), or inactive (geologically old). An example of such a map for part of Santa Clara County, California, is shown on the left in Figure 6.D. Information concerning past landslide activity may then be combined with land-use considerations to develop a slope stability or landslide-hazard map with recommended land uses, as shown on the right in Figure 6.D. The latter map is of most use to planners, whereas the former supplies useful information for the engineering geologist. These maps do not take the place of detailed fieldwork to evaluate a specific site but serve only as a general guideline for land-use planning and more detailed geologic evaluation.

Determining the landslide risk and making landslide-risk maps is more complicated, for this involves probability of occurrence and assessment of potential losses. The **specific risk** *(Rs)* associated with a landslide of a particular magnitude is

$$Rs = E \times H \times V, \text{ where}$$

- *E* is the elements at risk (i.e., value of property and social and economic activity) in the area being considered.
- *H* is the probability that a landslide of specified magnitude will occur (in a given time period).
- *V* is the vulnerability, defined as the proportion of the elements at risk *(E)* affected by the specified landslide. The value of *V* ranges from 0 (no damage) to 1 (complete destruction) (16).

For example, if an urban area has a value of $100 billion and the probability of a large landslide happening in a 10-year period is 1 in a 1000, or 0.001, and the vulnerability is 1 percent (.01), then $Rs = 100 \times 10^9 \times 10^{-3} \times 10^{-2} = 100 \times 10^4 =$ $1 million. Once the risks for various areas are determined, the information may be combined to produce a landslide-risk map. The method outlined here to produce risk maps is a variation of that generalized in Chapter 4 and is applicable to other hazards such as earthquakes, wildfires, and hurricanes.

in developing codes concerning grading (artificial excavation and filling) for development.

Landslides affect 60 percent of the length of seacliffs in southern California (see Figure 6.12), and the retreat of the seacliff is probably controlled by landslides (14). Similar estimates for slopes are not available, but the complex geology and terrain features, as well as evidence from old landslide scars and landslide deposits, suggest that slopes historically have been active. However, human activity has tremendously increased the magnitude and especially the frequency of landslides.

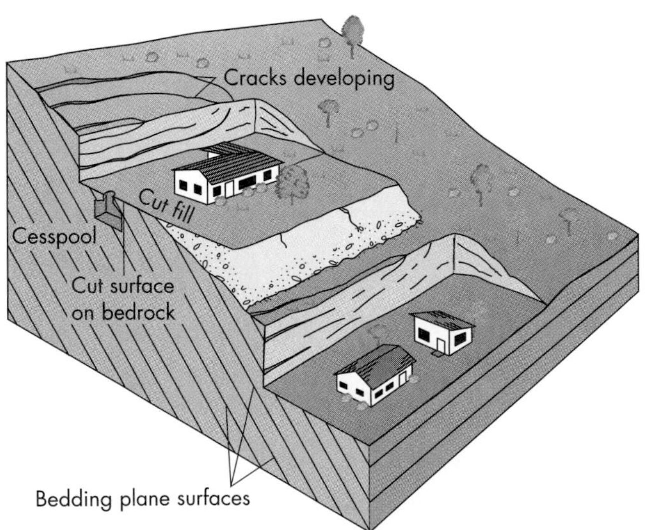

▲ **FIGURE 6.17** Development of artificial translation landslides. Stable slopes may be made unstable by removing support from the bedding plane surfaces. The cracks shown in the upper part of the diagram are an early sign that a landslide is likely to occur soon. (*Reprinted by permission from* F. B. *Leighton*. 1966. *"Landslides and Urban Development,"* Engineering Geology in Southern California [*Whittier, California: Association of Engineering Geologists*])

The grading process in southern California (cutting benches in slopes for home sites, called "pads"—hence the 1960s saying, "come on over to my pad") has been responsible for many landslides. It took natural processes many thousands, if not millions, of years to produce valleys, ridges, and hills. In this century, we have developed the machines to grade them. F. B. Leighton writes: "With modern engineering and grading practices and appropriate financial incentive, no hillside appears too rugged for future development" (14). No earth material can withstand the serious assault of modern technology. Thus, human activity is a geological agent capable of carving the landscape as do glaciers and rivers, but at a tremendously faster pace. We can convert steep hills almost overnight into a series of flat lots and roads, and such conversions have led to numerous artificially induced landslides. As shown in Figure 6.17, oversteepened slopes mixed with increased water from sprinkled lawns or septic systems, as well as the additional weight of fill material and a house, make formerly stable slopes unstable. Any project that steepens or saturates a slope, increases its height, or places an extra load on it may cause a landslide (14).

Landslides on both private and public land in Hamilton County, Ohio, have been a serious problem. The slides occur in glacial deposits (mostly clay, lakebeds, and till) and colluvium and soil developed on shale; the average cost of damage exceeds $5 million per year. Major landslides in Cincinnati have damaged highways and several private structures (1).

Modification of sensitive slopes associated with urbanization in Allegheny County, Pennsylvania, is estimated to

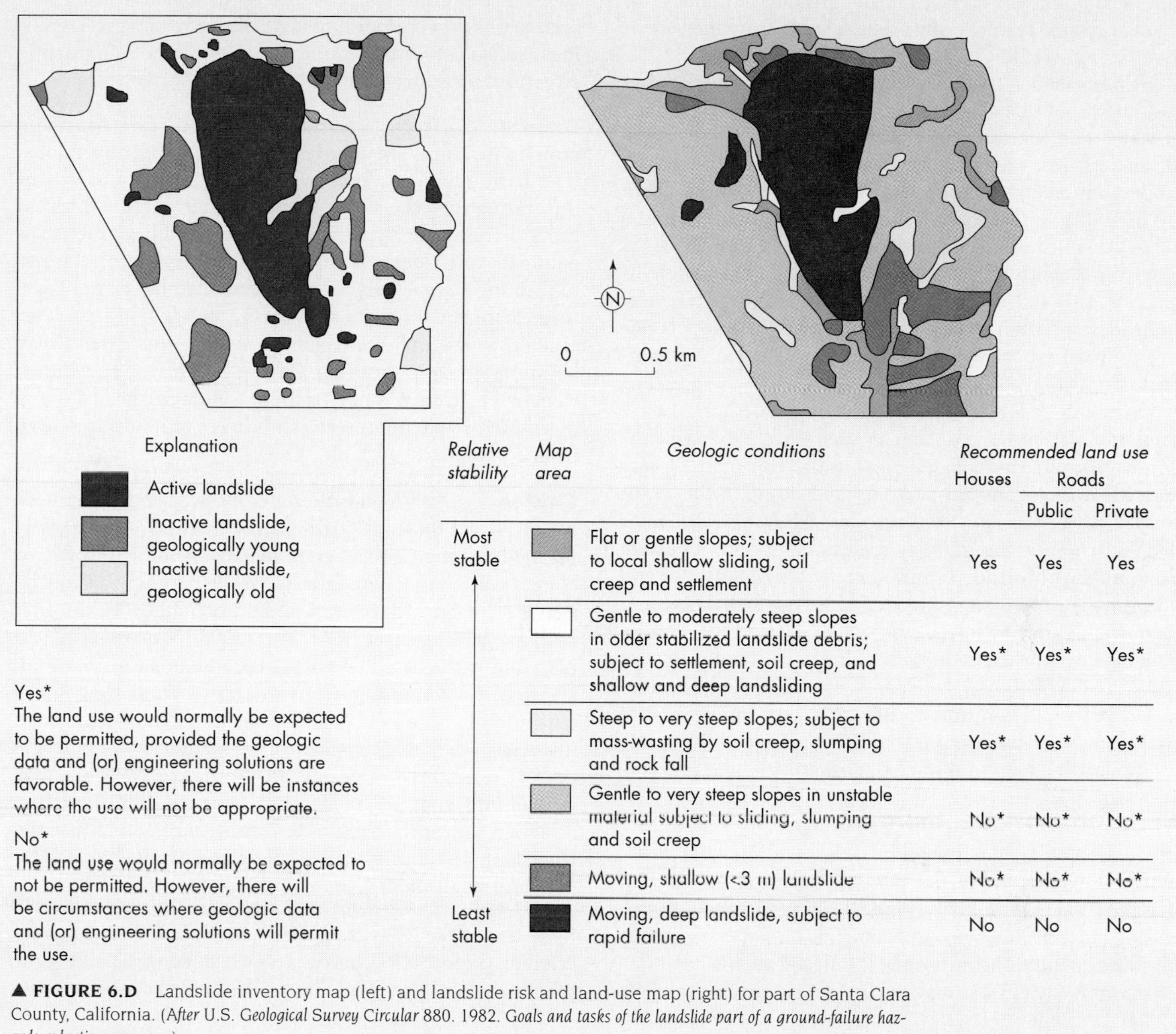

N
0
0.5 km
Explanation
Active landslide
Inactive landslide, geologically young
Inactive landslide, geologically old
Yes*
The land use would normally be expected to be permitted, provided the geologic data and (or) engineering solutions are favorable. However, there will be instances where the use will not be appropriate.
No*
The land use would normally be expected to not be permitted. However, there will be circumstances where geologic data and (or) engineering solutions will permit the use.

Relative stability	Map area	Geologic conditions	Recommended land use: Houses	Roads: Public	Roads: Private
Most stable		Flat or gentle slopes; subject to local shallow sliding, soil creep and settlement	Yes	Yes	Yes
		Gentle to moderately steep slopes in older stabilized landslide debris; subject to settlement, soil creep, and shallow and deep landsliding	Yes*	Yes*	Yes*
		Steep to very steep slopes; subject to mass-wasting by soil creep, slumping and rock fall	Yes*	Yes*	Yes*
		Gentle to very steep slopes in unstable material subject to sliding, slumping and soil creep	No*	No*	No*
		Moving, shallow (<3 m) landslide	No*	No*	No*
Least stable		Moving, deep landslide, subject to rapid failure	No	No	No

▲ **FIGURE 6.D** Landslide inventory map (left) and landslide risk and land-use map (right) for part of Santa Clara County, California. (*After* U.S. *Geological Survey Circular* 880. 1982. *Goals and tasks of the landslide part of a ground-failure hazards reduction program.*)

be responsible for 90 percent of the landslides, which produce an average of about $2 million in damages annually. Most of the landslides are slow-moving, but one rockfall in an adjacent county crushed a bus and killed 22 passengers. Most of the landslides in Allegheny County result from construction activity that loads the top of a slope, cuts into a sensitive location such as the toe of a slope, or alters water conditions on or in a slope (15).

6.4 Minimizing the Landslide Hazard

To minimize the landslide hazard, it is necessary to identify areas in which landslides are likely to occur, design slopes or engineering structures to prevent landslides, to warn people in danger areas of impending slides, and to control slides after they have started moving.

Identification of Potential Landslides

Identifying areas with a high potential for landslides is a first step in developing a plan to avoid landslide hazards. Slide tendency can be recognized by examining geologic conditions in the field and by examining aerial photographs to identify previous slides. This information can then be used to evaluate the risk and produce slope stability maps (see *A Closer Look: Determining Landslide Hazard and Risk*).

The individual homeowner, buyer, or builder can evaluate the landslide hazard on hillside property by looking for specific physical evidence that may indicate a potential or real landslide problem. Signs include cracks in buildings or walls around yards; doors and windows that stick or jam; retaining walls, fences, or posts that are not aligned in a normal way; breakage of underground pipes or other utilities; leaks in swimming pools; tilted trees and utility poles with taut or sagging wires; cracks in the ground; hummocky or steplike ground features; and seeping water from the base or toe of a slope (15).

The presence of one or more of these features is not absolute proof that a landslide is likely. For example, cracks in walls may also be caused by expansive soils or creep. Other features, such as hummocky or steplike ground on moderately steep slopes (greater than 15 percent, or 15-m fall in 100-m horizontal distance), probably do represent a potential landslide hazard that should be evaluated by a geologist. Furthermore, it is advisable not to limit your inspection only to the property in which you are interested; landslides are often larger than individual lots. Inspect adjacent areas, especially those that are upslope and downslope from your property (15).

Grading codes to minimize the landslide hazard have been in effect in the Los Angeles area since 1963. Motivation to institute these codes came in the aftermath of very high loss of lives and property to landsliding in the 1950s and 1960s (see *Case History: Portuguese Bend Landslide*). Since detailed engineering geology studies have been required, the number of hillside homes damaged by landslides and floods has been greatly reduced. Although initial building costs are greater because of the strict codes, they are more than balanced by the reduction of losses in subsequent wet years. And even though landslide disasters during extremely wet years will continue to plague us, the application of geologic and engineering information prior to hillside development can help minimize the hazard.

Prevention of Landslides

Prevention of large, natural landslides is difficult, but common sense and good engineering practice can do much to minimize the hazard. For example, loading the top of slopes, cutting into sensitive slopes, placing fills on slopes, or changing water conditions on slopes should be avoided or done very cautiously (15). Common engineering techniques for landslide prevention include provisions for surface and subsurface drainage, removal of unstable slope materials (grading), construction of retaining walls or other supporting structures, or some combination of these (2,10).

Drainage Control Surface and subsurface **drainage control** measures are usually effective in stabilizing a slope. The basic idea is to keep water from running across or infiltrating into the slope. Surface water may be diverted around the slope by a series of drains. This practice is common for roadcuts (Figure 6.18a). The amount of water infiltrating a slope may also be controlled by covering the slope with an impermeable layer, such as soil-cement, asphalt, or even plastic (Figure 6.18b). Groundwater may be inhibited from entering a slope by excavating a cutoff trench. The trench is filled with gravel or crushed rock and positioned so as to intercept and divert groundwater away from a potentially unstable slope (2).

Grading Although **grading of slopes** for development has increased the landslide hazard in many areas, carefully planned grading can be used to increase slope stability. Two common techniques are reducing the gradient of a slope by a single cut-and-fill operation, and benching. In the first case, material from the upper part of a slope is removed and placed near the base. The overall gradient is thus reduced, and material is removed from an area where it contributes to the driving force and placed at the toe of the slope, where it increases the resisting forces. This method is not practical on very steep, high slopes. As an alternative, the slope may be cut into a series of benches or steps. The benches, designed with surface drains to divert runoff, do reduce the slope and, in addition, are good collection sites for falling rock and small slides (2).

Slope Supports Retaining walls constructed from concrete cribbing (Figure 6.19), gabions (stone-filled wire baskets), or piles (long concrete, steel, or wooden beams driven into the

(a)

(b)

▲ **FIGURE 6.18** (a) Drains on a roadcut to remove surface water from the cut before it infiltrates the slope. (b) Covering a slope with a soil-cement in Greece to reduce infiltration of water and provide strength. (*Edward A. Keller*)

CASE HISTORY Portuguese Bend Landslide

The Portuguese Bend landslide along the southern California (Los Angeles) coast (Figure 6.E) has damaged or destroyed more than 150 homes. The slide is part of an older, larger slide that was reactivated partly by road-building activities and partly by alteration of a delicate subsurface water situation by urban development. The recent movement started in 1956 during construction of a county road that placed approximately 23 m of fill over the upper area of the landslide, increasing the driving forces. During subsequent litigation, the county of Los Angeles was found responsible for the landslide.

Movement of the Portuguese Bend slide was continuous from 1956 to 1978, averaging approximately 0.3 to 1.3 cm per day. The rate accelerated to more than 2.5 cm daily in the late 1970s and early 1980s following several years of above-normal precipitation. Total displacement near the coast has been more than 200 m. The Abalone Cove slide, part of the Portuguese Bend slide complex, began to move during the above-mentioned wet period. The new slide prompted additional geologic investigation, and a landslide-control program was initiated. The program consisted of several dewatering wells installed in 1980 to remove groundwater from the slide mass. By 1985 the slide had apparently been stabilized (17). However, depending on future conditions related to precipitation and groundwater conditions, it may again cause problems.

During a two-decade period of activity, one home on the Portuguese Bend landslide moved about 25 m and constantly shifted in position. Other homes have moved up to 50 m in the same time, and some people living in homes about 1 km from the ocean apparently adjusted to the slow movement and the ever-changing view! Structures remaining in the active slide area had to be adjusted every year or so with hydraulic jacks, and utilities are on the surface. With one exception, no new homes have been constructed since the landslide began to move. The remaining occupants have elected to adjust to the landslide rather than bear total loss of their property. Nevertheless, few geologists would probably choose to live there now.

(a)

(b)

▲ **FIGURE 6.E** (a) Entrance to the Portuguese Bend development (*Edward A. Keller*) and (b) aerial view of the Portuguese Bend landslide. Note kink in the pier near the toe of the landslide. Eventually most of the homes as well as the swim club and pier shown here were destroyed by the slow-moving landslide. (*John S. Shelton*)

▲ **FIGURE 6.19** Retaining wall (concrete cribbing) used to help stabilize a roadcut. (*Edward A. Keller*)

ground) are designed to provide support at the base of a slope. They should be keyed in well below the slope base, backfilled with permeable gravel or crushed rock, and provided with drain holes to reduce the chances of water pressure building up in the slope. A less common method of increasing slope stability involves insertion of heavy bolts (rock bolts) into holes drilled through potentially unstable rocks into stable rocks. This technique was used to secure the slopes at the Glen Canyon Dam on the Colorado River and the Hanson Dam on the Green River in Washington (18).

Preventing landslides can be expensive, but the rewards can be even greater. It has been estimated that the benefit-to-cost ratio for landslide prevention ranges from approximately 10 to 2000. That is, for every dollar spent on landslide prevention, the savings will vary from \$10 to \$2000 (19).

The cost of not preventing a slide is illustrated by a massive landslide in Utah known as the Thistle slide. This slide moved across a canyon in April 1983, creating a natural dam about 60 m high and flooding the community of Thistle, the Denver-Rio Grande railroad and its switchyard, and a major U.S. highway (Figure 6.20) (19). The landslide and resulting flooding caused approximately $200 million in damages.

The Thistle slide involved a reactivation of an older slide, which had been known for many years to be occasionally active in response to high precipitation. Therefore, it could have been recognized that the extremely high amounts of precipitation in 1983 would cause a problem. In fact, a review of the landslide history suggests that the Thistle landslide was recognizable, predictable, and preventable! Analysis of the pertinent data suggests that emplacement of subsurface drains and control of surface runoff would have lowered the water table in the slide mass enough to have prevented failure. Cost of preventing the landslide was estimated to be between $300,000 and $500,000, a small amount compared to the damages (19). Because the benefit-to-cost ratio in landslide prevention is so favorable, it seems prudent to evaluate active and potentially active landslides in areas where considerable damage may be expected and possibly prevented.

▲ **FIGURE 6.20** Thistle landslide, Utah. This landslide, which occurred in 1983, involved the reactivation of an older slide. The landslide blocked the canyon, creating a natural dam, flooding the community of Thistle as well as the Denver-Rio Grande Railroad and a major U.S. highway. (*Michael Collier*)

Landslide Warning Systems

Landslide warning systems do not prevent landslides, but they can provide time to evacuate people and their possessions, and to stop trains or reroute traffic. Surveillance provides the simplest type of warning. Hazardous areas can be visually inspected for apparent changes, and small rockfalls on roads and other areas can be noted for quick removal. Having people monitor the hazard has advantages of reliability and flexibility but becomes disadvantageous during adverse weather and in hazardous locations (20). Other warning methods include electrical systems, tilt meters, and geophones that pick up vibrations from moving rocks. Shallow wells can be monitored to signal when slopes contain a dangerous amount of water. In some regions, monitoring rainfall is useful for detecting when a threshold precipitation has been exceeded and shallow soil slips become much more probable.

Landslide Correction

After movement of a slide has begun, the best way to stop it is to attack the process that started the slide. In most cases, the cause of the slide is an increase in water pressure, and in such cases, an effective drainage program must be initiated. This may include surface drains at the head of the slide to keep additional surface water from infiltrating and subsurface drainpipes or wells to remove water and lower the water pressure. Draining tends to increase the resisting force of the slope material and therefore stabilizes the slope.

The tremendous success of drainage is demonstrated by this description from Karl Terzaghi (6). After a high-magnitude rainstorm, movement on a 30° slope of deeply weathered metamorphic rock was noted. The slide plane was approximately 40 m below the surface, and the slide area was about 150 m wide by 300 m long. The slide was close to a hydroelectric power station, so immediate action was deemed necessary. Fieldwork established that if the water level could be lowered approximately 5 m, then the increase in resisting force would be sufficient to stabilize the slide. Drainage was accomplished by trenches and horizontal drill holes extending into the water-bearing zones of the rock. After drainage, the movement stopped, and even though the next rainy season brought record rainfall, no new movement was observed (6).

6.5 Snow Avalanche

An avalanche is a rapid downslope movement of snow. If abundant rock, soil, and trees are incorporated, it may be much like a debris avalanche. As with landslides, **snow avalanches** are subject to driving and resisting forces on the slope.

Approximately 10,000 snow avalanches occur each year in the mountains of the western United States, and about 1 percent of these cause loss of human life or property damage, killing an average of seven people and inflicting $300,000 in damage (21). Loss of life is increasing, however, as more people venture into mountain areas for recreation during the winter.

Avalanches can occur in dry or wet snow and are of two general types: **loose-snow avalanches,** which occur in cohesionless snow and tend to be relatively small and shallow failures; and **slab avalanches,** which may initially vary from about 100 to 10,000 m^2 in area and 0.1 to 10 m in thickness (21). Large slab avalanches are the most dangerous, releasing tremendous energy by mobilizing up to a million tons of snow and ice and moving downslope at velocities of 5 to 30 m/sec (18 to 100 km per hr) or more. Horizontal thrust (or impact) from such events tends to vary from 5 to 50 tons/m^2, but may in extreme cases exceed 100 tons/m^2. To appreciate the magnitude of this thrust, consider that a thrust of only about 3 tons/m^2 is necessary to collapse a frame house, and 100 tons/m^2 can move reinforced concrete structures (21).

Avalanches are initiated when a mass of snow and ice on a slope fails because of the overload of a large volume of new snow, or when internal changes in a snowpack produce zones of weakness (low shear strength) along which failure occurs. When conditions are very unstable, even the weight of a single skier can start an avalanche. Once started, avalanches tend to follow certain paths, chutes, or tracks (see Figure 6.21a). These are often well channeled; however, unconfined tracks on open slopes also occur. Avalanche tracks often have several branches near the top that coalesce downslope; thus, it is possible for several avalanches to move through the main track in a short period of time as snowpacks in upper branches fail. Failure to recognize this possibility has caused the loss of several lives when workers clearing debris from a first avalanche have been struck by a second (21).

The avalanche hazard can be reduced by avoiding dangerous areas; stabilizing slopes by clearing them with carefully placed explosions; building structures to divert or retard avalanches; and reforesting avalanche paths, since large avalanches are seldom initiated on densely forested slopes (21).

Avalanches are primarily a threat to skiers on high, steep mountain slopes, but they also threaten mountain resorts, villages, railways, highways, and even sections of some cities. For example, Juneau, Alaska, has a significant avalanche hazard. In the last 100 years, a major avalanche chute above Juneau has released snow and ice six times, events that reached the sea. No damaging avalanches have occurred over the last quarter century, however, so an entire subdivision has been constructed across the chute (Figure 6.21b). If another large event occurs, it will destroy about 30 homes, part of a school, and a motel, and eventually roar into the harbor where several hundred boats are docked. It has been estimated that a home in the chute area, with a 40-year life span, has a 96 percent probability of being struck by an avalanche, yet the people who live there have been almost nonchalant about the hazard (22).

(a)

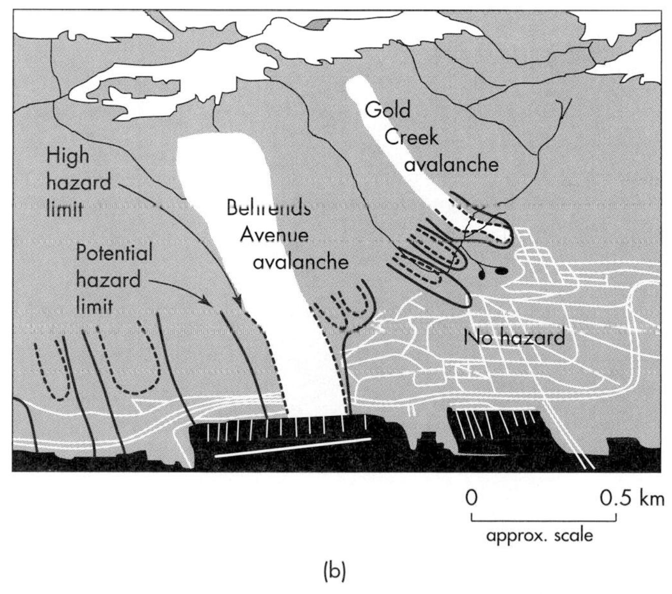

(b)

◀ **FIGURE 6.21** (a) Avalanche chute or track in the Swiss Alps. (*Edward A. Keller*) (b) Map of part of Juneau, Alaska, avalanche hazard. (*After D. Cupp.* 1982. National Geographic 162:290–305)

6.6 Subsidence

Interactions between geologic conditions and human activity have been factors in numerous incidents of **subsidence,** the very slow to rapid sinking or settling of earth materials. Most subsidence is caused by withdrawal of fluids from subsurface reservoirs or by the collapse of surface and near-surface soil and rocks over subterranean voids.

Withdrawals of oil with associated gas and water, of groundwater, and of mixtures of steam and water for geothermal power have caused subsidence (23). In all cases, the general principles are the same. Fluids in earth materials below the earth's surface have a high fluid pressure that tends to support the material above. This is why a large rock at the bottom of a swimming pool seems lighter: Buoyancy produced by the liquid tends to lift the rock. If support or buoyancy is removed from earth materials by pumping out the fluid, the support is reduced, and surface subsidence may result.

The actual subsidence mechanism involves compaction of individual grains of the earth material as the grain-to-grain load increases because of a lowering of fluid pressure. Subsidence of oil fields generally involves considerable reduction of fluid pressure, up to 2.8×10^7 Pa (N/m^2), at great depth (thousands of meters) over a relatively small area, less than 150 km^2. On the other hand, subsidence resulting from withdrawal of groundwater generally involves a relatively low reduction of fluid pressure, often less than 1.4×10^6 Pa, at relatively shallow depths (less than 600 m), over a large area, sometimes many hundreds of square kilometers (23). See Chapter 2 for a discussion of stress (pressure).

Thousands of square kilometers of the central valley of California have subsided as a result of overpumping groundwater (Figure 6.22). More than 5000 km^2 in the Los Banos–Kettleman City area alone have subsided more than 0.3 m, and within this area, one 113-km stretch has subsided an average of more than 3 m, with a maximum of about 9 m (Figure 6.22). As the water was mined, the fluid pressure was reduced and the grains were compacted (Figure 6.23) (24,25); the effect at the surface was subsidence. Similar examples of subsidence caused by overpumping are documented near Phoenix, Arizona; Las Vegas, Nevada; Houston–Galveston, Texas; and Mexico City, Mexico. The subsidence can cause surface fissures (open cracks) to form in sediments, and these fissures can be hundreds of meters long and several meters deep (25).

Sinkholes

Subsidence is also caused by removal of subterranean earth materials (rock) by natural processes. Voids often form within soluble rocks such as limestone and dolomite, and the resulting lack of support for overlying rock may cause it to collapse. The result is the formation of a **sinkhole,** a circular area of subsidence caused by collapse into a subterranean void. Some sinkholes are more than 30 m across and 15 m deep. One near Tampa, Florida, collapsed suddenly in 1973, swallowing part of an orange orchard. What may be the largest sinkhole in the United States formed in 1972 near Montevallo, Alabama (26). A massive hole 120 m wide and 45 m deep, named the "December Giant" by the press, de-

▲ **FIGURE 6.22** (a) Principal areas of land subsidence in California resulting from groundwater withdrawal. (*After* W. B. Bull. 1973. Geological Society of America Bulletin 84. *Reprinted by permission.*) (b) Photograph illustrating the amount of subsidence in the San Joaquin Valley, California. Marks on telephone pole are positions of the ground surface in recent decades. The photo shows approximately 8 m of subsidence. (*Courtesy of Ray Kenny*)

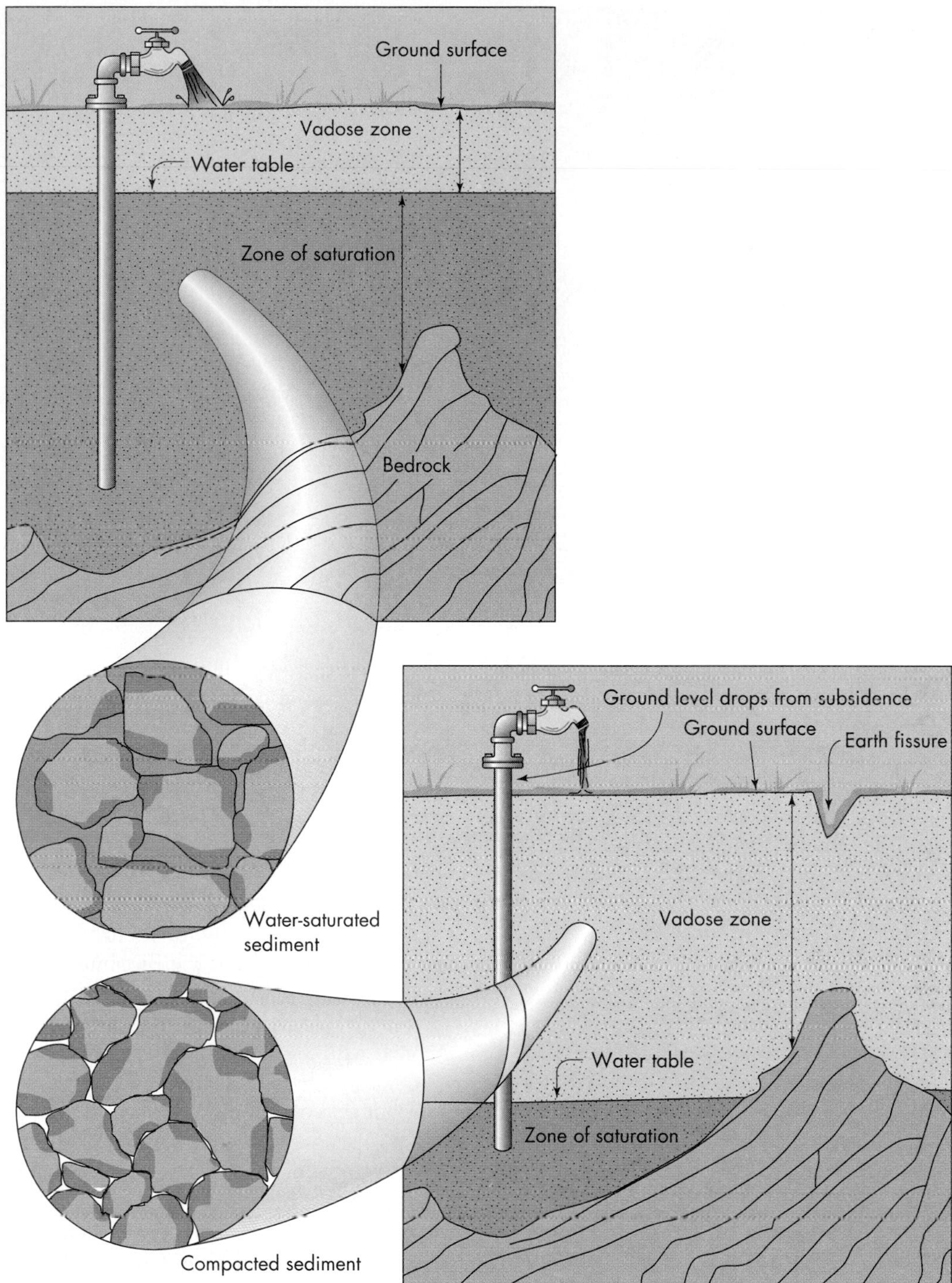

▲ **FIGURE 6.23** Idealized diagram showing how surface subsidence results from pumping groundwater. Pore (empty) spaces between grains collapse following pumping. (*From* R. *Kenny*. 1992. *Fissures*. Earth 2(3):34–41)

veloped suddenly when topsoil and subsurface clay collapsed into an underlying limestone cavern. The collapse was caused by loss of support to the clay and soil over the cavern. A nearby resident reported hearing a roaring noise accompanied by breaking timber and earth tremors that shook his house.

Sinkholes have caused considerable damage to highways, homes, sewage facilities, and other structures. Natural or artificial fluctuations in the water table are probably the trigger mechanism. High water-table conditions favor solutional enlarging of the cavern, and the buoyancy of the water helps support the overburden. Lowering of the water table eliminates some of the buoyant support and facilitates collapse. This was dramatically illustrated on May 8, 1981, in Winter Park, Florida, when a large sinkhole began developing. The sink grew rapidly for 3 days, swallowing part of a community swimming pool, parts of two businesses, several automobiles, and a house (Figure 6.24). Damage caused by the sinkhole exceeded $2 million. Sinkholes form nearly every year in central Florida when the groundwater level is lowest.

▶ **FIGURE 6.24** The Winter Park, Florida, sinkhole that grew rapidly for three days, swallowing part of a community swimming pool as well as several businesses, houses, and automobiles. (*Leif Skoogfors/Woodfin Camp and Associates*)

The Winter Park sinkhole formed during a drought, when groundwater levels were at a record low. Although exact positions cannot be predicted, their occurrence is greater during droughts; in fact, several smaller sinks appeared at about the same time as the Winter Park event.

Sometimes a natural sinkhole has been filled and built upon, with disastrous results. A dramatic example is the Allentown sinkhole (see *Case History: Lehigh Valley, Pennsylvania*). In this case the cumulative effects of not recognizing a sinkhole, filling it with urban debris, and subsequently developing the site were probably responsible for the sudden failure.

Salt Deposits and Subsidence

Serious subsidence events have been associated with mining of salt, coal, and other minerals. Salt is often mined by solution methods: Water is injected through wells into salt deposits, the salt dissolves, and water supersaturated with salt is pumped out. The removal of salt leaves a cavity in the rock and weakens support for the overlying rock, which may lead to large-scale subsidence. In 1970 one event near Detroit produced a subsidence pit 120 m across and 90 m deep. Another near Saltville, Virginia, produced 75 m of subsidence relatively quickly. Two homes went down with the Saltville subsidence. According to local residents, one family moved out the day before the event because a family member had dreamed the mountain was falling. Mining of salt by other methods can also produce subsidence (see *Case History: Lake Peigneur, Louisiana*).

Large sinks associated with bedded salt may also occur without mining. For example, in June 1980 a large depression southwest of Kermit, Texas, known as the "Wink Sink," developed over a period of about 48 hours. At the end of that time, the sinkhole was approximately 110 m across and 34 m deep. The Wink Sink and similar features evidently form by natural processes when groundwater slowly dissolves caverns in bedded salt underlying less soluble rock, such as sandstone. When a cavern reaches a critical size, the overlying rocks can no longer be supported, and collapse occurs. Because this is a natural process and other caverns undoubtedly exist, future sinks probably will develop in the area without warning (28).

Coal Mining and Subsidence

In coal mining, the practice of full recovery (removing all the coal) in subsurface mines has produced subsidence problems. A good example is the Pittsburgh area, where mining has been going on for more than a century. In the early years, companies purchased mining rights permitting removal of the coal with no responsibility for surface damage. Results were not so serious when mining was conducted under farmland, but as recent rapid urbanization has progressed faster than coal can be extracted, problems have resulted. If all the coal is removed, the chance of subsidence and damage to homes is high; however, if about 50 percent of the coal is left, it will usually provide sufficient support. The Bituminous Mine Subsidence and Land Conservation Act of 1966 provided for protection of public health, welfare, and safety by regulating coal mining, but this act will cause hundreds of millions of tons of coal to remain in the ground, attesting to the nature of trade-offs when there is conflict in surface and subsurface human use of the land (29).

Subsidence incidents have also been reported over coal mines that have not been worked for more than 50 years (29). On a January morning in 1973, a few residents of Wales (Britain) were driving over a section of the road that suddenly collapsed into a pit 10 m deep. Their car tottered on the brink while they scrambled to safety. The collapse, which occurred over an air shaft of a lost mine, disrupted some utility service. Other similar subsidence events have happened in the past and are likely to occur in the future.

CASE HISTORY Lehigh Valley, Pennsylvania

On June 23, 1986, a large subsidence pit developed at the site of an unrecognized, filled sinkhole in Lehigh Valley near Allentown, in eastern Pennsylvania. Within a period of only a few minutes, the collapse left a pit approximately 30 m in diameter and 14 m deep. Fortunately, the damage was confined to a street, parking lots, sidewalks, sewer lines, water lines, and utilities. Seventeen residences adjacent to the sinkhole narrowly escaped damage or loss; subsequent stabilization and repair costs were nearly one-half million dollars. Figure 6.F shows the generalized geology of Lehigh Valley. The northern part of the valley is underlain by shale, whereas limestone comprises the southern portion. The valley is bounded by resistant sandstone rocks to the north and resistant Precambrian granitic and gneissic rocks to the south (27).

Photographs from the 1940s to 1969 provide evidence of the sinkhole's history. In the 1940s the sinkhole was delineated by a pond of approximately 65 m in diameter. By 1958 the pond had dried up, the sinkhole was covered by vegetation, and the surrounding area was planted in crops. Ground photographs in 1960 suggest that people were using the sinkhole as a site to dump tree stumps, blocks of asphalt, and other trash. By 1969 there was no surface expression of the sinkhole; it evidently was completely filled and corn was planted over it.

Even though the sinkhole was completely filled with trash and other debris, it still received runoff water that was later increased in volume by urbanization. Sources of water included storm runoff from adjacent apartments and townhouses, as well as streets and parking lots. It is also suspected that an old, leaking water line contributed to runoff into the sinkhole area. In addition, urbanization placed increased demand on local groundwater resources, resulting in the lowering of the water table. Geologists believe that hydrologic conditions contributed to the sudden failure. The increased urban runoff facilitated the loosening or removal of the plug (soil, clay, and trash) that filled the sinkhole, while the lowering of the groundwater table reduced the overlying support, as was the case with the Winter Park sinkhole (27).

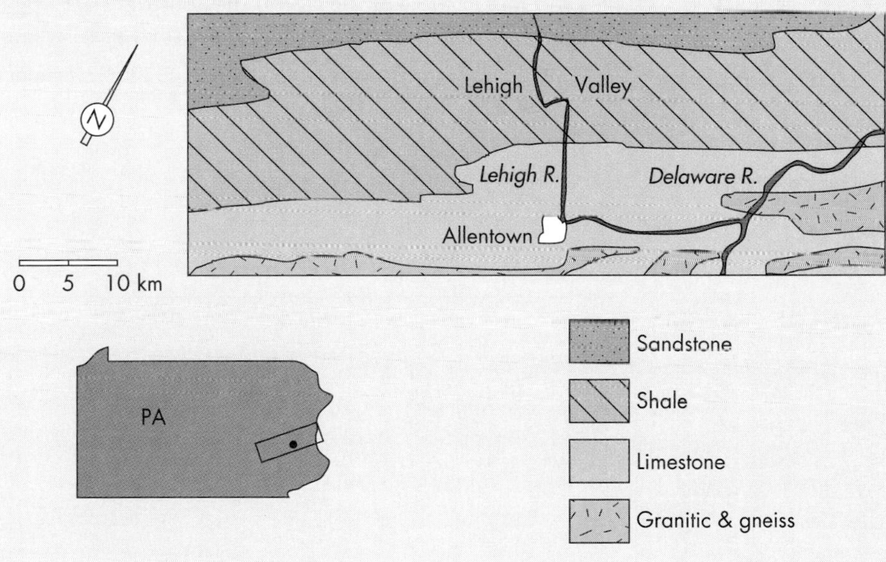

◀ **FIGURE 6.F** Generalized geologic map of the Lehigh Valley in eastern Pennsylvania. [*Modified from* P. H. *Dougherty and* M. *Perlow*, Jr. 1987. Environmental Geology and Water Science 12(2):89–98.]

6.7 Perception of the Landslide Hazard

The common reaction of southern California homeowners to talk of landslides is, "It could happen on other hillsides, but never this one" (14). As with flooding, landslide-hazard maps will probably not prevent people from moving into hazardous areas, and prospective hillside occupants who are initially unaware of the hazards may not be swayed by technical information. The infrequency of large slides tends to reduce awareness of the hazard where evidence of past events is not readily visible. Unfortunately, it often takes catastrophic events such as the recent massive landslide in the Laguna Hills area of California, which claimed numerous expensive homes, to bring the problem to the attention of many people. In the meantime, residents in many parts of the Rocky Mountains, Appalachian Mountains, and other areas continue to build homes in areas subject to future (and even present) landslides.

CASE HISTORY Lake Peigneur, Louisiana

A bizarre example of subsidence associated with a salt mine occurred on November 21, 1980, in southern Louisiana, when shallow Lake Peigneur (with an average depth of 1 m) drained following collapse of the salt mine below. The collapse occurred after an oil-drilling operation apparently punched a hole into an abandoned mine shaft of a still active multimillion-dollar salt mine located about 430 m below the surface of the Jefferson Island salt dome. As the hole enlarged because of water entering the mine—scouring and dissolving pillars of salt—the roof of the mine collapsed, producing a large subsidence pit (Figure 6.G).

The lake drained so fast that 10 barges, a tugboat, and an oil-drilling barge disappeared in a whirlpool of water into the mine, which in places has tunnels as wide as four-lane freeways and 25 m high. (Mining is done with the aid of trucks and bulldozers.) Fortunately, the 50 miners and 7 people on the oil rig escaped. The subsidence also claimed more than 25 ha of Jefferson Island, including historic botanical gardens, greenhouses, and a $500,000 private home. What is left of the gardens is disrupted by large fractures that roughly step the land down to the new edge of the lake. These fractures are tensional in origin and are commonly found on the margins of large subsidence pits.

Lake Peigneur immediately began refilling with water from a canal connecting it to the Gulf of Mexico, and 9 of the barges popped to the surface 2 days later. There was fear at first that even larger subsidence would take place as pillars of salt holding up the roof of the salt dome dissolved. However, the hole was apparently sealed by debris in the form of soil and lake sediment that was pulled into the mine. Approximately 15 million m^3 of water entered the salt dome, and the mine became a total loss. The previously shallow lake now has a large, deep hole in the bottom, which undoubtedly will change the aquatic ecology. In a 1983 out-of-court settlement, the salt mining company reportedly was compensated $30 million by the oil company involved. The owners of the botanical garden and private home apparently were compensated $13 million by the oil company, drilling company, and mining company.

The flooding of the mine raises important questions concerning the structural integrity of salt mines. The federal strategic petroleum reserve program is planning to store 75 million barrels of crude oil in an old salt mine of the Weeks Island salt dome about 19 km from Jefferson Island. On the other hand, while the role of the draining lake in the collapse is very significant, few salt domes have lakes above them. The Jefferson Island subsidence is thus a very rare type of event.

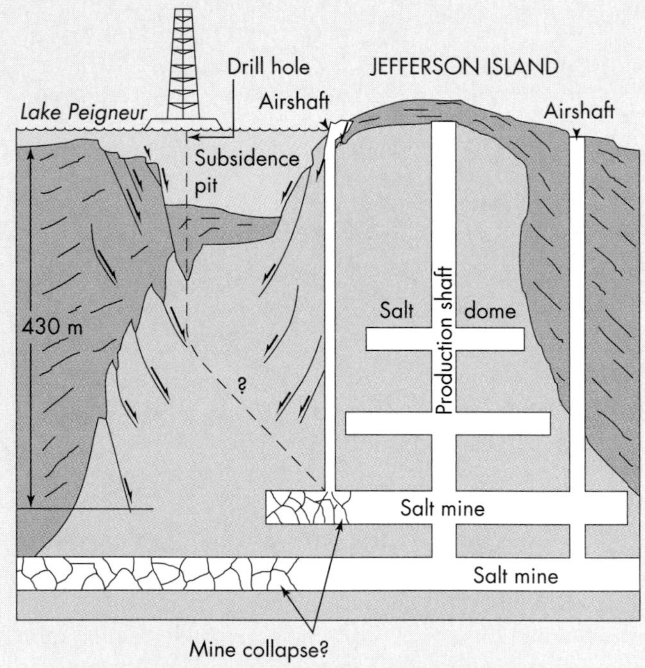

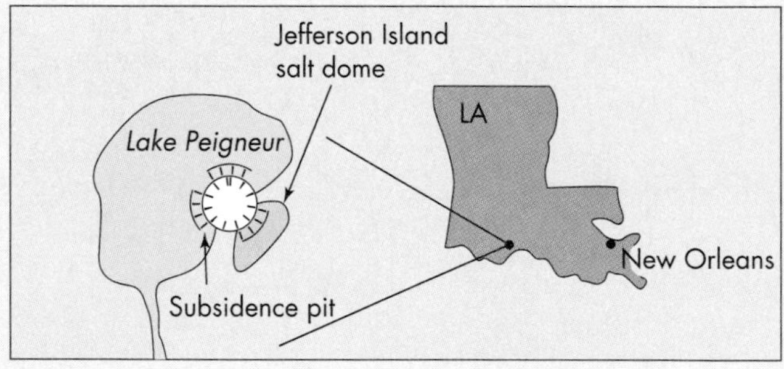

FIGURE 6.G Idealized diagram showing the Jefferson Island Salt Dome collapse that caused a large subsidence pit to form in the bottom of Lake Peigneur, Louisiana.

SUMMARY

Landslides and related phenomena cause substantial damage and loss of life. Although they are natural events, their occurrence can be increased or decreased by human activity.

The most common landforms are slopes—dynamic, evolving systems in which surficial material is constantly moving downslope at rates varying from imperceptible creep to thundering avalanches. All slopes are composed of one or more slope elements, including the crest, free-face, debris slope, and wash slope. The presence of particular slope elements on a specific slope is related to climate and rock type, which affect slope processes. Slope failure may involve *flowage*, *sliding*, or *falling* of earth materials; landslides are often complex combinations of sliding and flowage.

Forces producing landslides are determined by the interactions of several variables: the type of earth material on the slope, topography, climate, vegetation, water, and time. The cause of most landslides can be determined by examining the relations between forces that tend to make earth materials slide *(driving forces)* and forces that tend to oppose movement *(resisting forces)*. The most common driving force is the weight of the slope materials, and the most common resisting force is the shear strength of the slope materials. The *factor of safety* of a slope is the ratio of resisting forces to driving forces; a ratio greater than 1 suggests that the slope is stable. The type of rock or soil on a slope influences both the type and the frequency of landslides.

Water has an especially significant role in producing landslides. Water in streams, lakes, or oceans erodes the toe area of slopes, increasing the driving forces. Excess water increases the weight of the slope materials while raising the water pressure, which in turn decreases the resisting forces in the slope materials. A rise in pore-water pressure occurs before many landslides, and, in fact, most landslides are a result of an abnormal increase in water pressure in the slope-forming materials.

Effects of human use on the magnitude and frequency of landslides vary from insignificant to very significant. Where landslides occur independent of human activity, we need to learn enough about them to avoid development in hazardous areas or to provide protective measures. Where human use has increased the number and severity of landslides, we have to learn how to minimize these occurrences. In some cases dams and reservoirs have increased migration of groundwater into slopes, resulting in slope failure. Logging operations on weak, unstable slopes have increased landslide erosion. Grading of slopes for development has created or increased erosion problems in many urbanized areas of the world.

To minimize landslide hazard, it is necessary to establish identification, prevention, and correction procedures. Monitoring and mapping techniques facilitate identification of hazardous sites. Where identification of potential landslides has been used to establish grading codes, landslide damage has been decreased. Prevention of large natural slides is very difficult, but good engineering practices can do much to minimize the hazard when it cannot be avoided. Engineering techniques for landslide prevention include drainage control, proper grading, and construction of supports, such as retaining walls. Correction of landslides must attack the processes that started the slide; this usually means initiating a drainage program that lowers water pressure in the slope.

Snow avalanches present a serious hazard on snow-covered, steep slopes. Loss of human life because of avalanches is increasing as more people venture into mountain areas for winter recreation.

Withdrawal of fluids such as oil and water and subsurface mining of salt, coal, and other minerals have both caused widespread *subsidence*. In the case of fluid withdrawal, the cause of subsidence is a reduction of fluid pressures that tend to support overlying earth materials. In the case of solid material removal, subsidence may result from loss of support for the overlying material. The latter situation occurs naturally when voids are formed in soluble rock such as limestone and the collapse of overlying earth material produces sinkholes.

Perception of the landslide hazard by most people, unless they have prior experience, is low. Furthermore, hillside residents, like floodplain occupants, are not easily swayed by technical information.

REFERENCES

1. **Fleming, R. W., and Taylor, F. A.** 1980. *Estimating the cost of landslide damage in the United States.* U.S. Geological Survey Circular 832.
2. **Pestrong, R.** 1974. *Slope stability.* New York: McGraw-Hill, American Geological Institute.
3. **Nilsen, T. H., Taylor, F. A., and Dean, R. M.** 1976. *Natural conditions that control landsliding in the San Francisco Bay region.* U.S. Geological Survey Bulletin 1424.
4. **Burroughs, E. R., Jr., and Thomas, B. R.** 1977. *Declining root strength in Douglas fir after felling as a factor in slope stability.* USDA Forest Service Research Paper INT-190.
5. **Campbell, R. H.** 1975. *Soil slips, debris flows, and rainstorms in the Santa Monica Mountains and vicinity, southern California.* U.S. Geological Survey Professional Paper 851.
6. **Terzaghi, K.** 1950. *Mechanisms of landslides.* The Geological Society of America: Application of Geology to Engineering Practice, Berken volume: 83–123.
7. **Leggett, R. F.** 1973. *Cities and geology.* New York: McGraw-Hill.
8. **Krynine, D. P., and Judd, W. R.** 1957. *Principles of engineering geology and geotechnics.* New York: McGraw-Hill.
9. **Williams, G. P., and Guy, H. P.** 1973. *Erosional and depositional aspects of Hurricane Camille in Virginia, 1969.* U.S. Geological Survey Professional Paper 804.
10. **Office of Emergency Preparedness.** 1972. *Disaster preparedness* 1, 3.
11. **Swanson, F. J., and Dryness, C. T.** 1975. Impact of clear-cutting and road construction on soil erosion by landslides in the Western Cascade Range, Oregon. *Geology* 7:393–96.
12. **Kiersch, G. A.** 1964. Vaiont Reservoir disaster. *Civil Engineering* 34:32–39.
13. **Jones, F. O.** 1973. *Landslides of Rio de Janeiro and the Sierra das Araras Escarpment, Brazil.* U.S. Geological Survey Professional Paper 697.

14. **Leighton, F. B.** 1966. Landslides and urban development. In *Engineering geology in southern California*, eds. R. Lung and R. Proctor, pp. 149–97, a special publication of the Los Angeles Section of the Association of Engineering Geology.
15. **Briggs, R. P., Pomeroy, J. S., and Davies, W. E.** 1975. *Landsliding in Allegheny County, Pennsylvania*. U.S. Geological Survey Circular 728.
16. **Jones, D. K. C.** 1992. Landslide hazard assessment in the context of development. In *Geohazards*, eds. G. J. McCall, D. J. Laming, and S. C. Scott, pp. 117–141. New York: Chapman & Hall.
17. **Ehley, P. L.** 1986. The Portuguese Bend landslide: Its mechanics and a plan for its stabilization. In *Landslides and landslide mitigation in southern California*, ed. P. L. Ehley, pp. 181–90, Guidebook for fieldtrip, Cordilleran Section of the Geological Society of America meeting, Los Angeles, California.
18. **Morton, D. M., and Streitz, R.** 1975. Mass movement. In *Man and his physical environment*, eds. G. D. McKenzie and R. O. Utgard, pp. 61–70. Minneapolis: Burgess.
19. **Slosson, J. E., Yoakum, D. E., and Shuiran, G.** 1986. Thistle, Utah, landslide: Could it have been prevented? *Proceedings of the 22nd Symposium on Engineering Geology and Soils Engineering*, pp. 281–303.
20. **Piteau, D. R., and Peckover, F. L.** 1978. Engineering of rock slopes. In *Landslides*, eds. R. Schuster and R. J. Krizek. Transportation Research Board, Special Report 176:192–228.
21. **Perla, R. I., and Martinelli, M., Jr.** 1976. *Avalanche handbook*. U.S. Department of Agriculture, Forest Service, Agriculture Handbook 489.
22. **Cupp, D.** 1982. Battling the juggernaut. *National Geographic* 162:290–305.
23. **Poland, J. F., and Davis, G. H.** 1969. Land subsidence due to withdrawal of fluids. In *Reviews in engineering geology*, eds. D. J. Varnes and G. Kiersch, pp. 187–269, The Geological Society of America.
24. **Bull, W. B.** 1974. Geologic factors affecting compaction of deposits in a landsubsidence area. *Geological Society of America Bulletin* 84:3783–802.
25. **Kenny, R.** 1992. Fissures. *Earth* 1(3):34–41.
26. **Cornell, J., ed.** 1974. *It happened last year—earth events—1973*. New York: Macmillan.
27. **Dougherty, P. H., and Perlow, M., Jr.** 1987. The Macungie sinkhole, Lehigh Valley, Pennsylvania: Cause and repair. *Environmental Geology and Water Science* 12(2):89–98.
28. **Baumgardner, R. W., Gustavson, T. C., and Hoadley, A. D.** 1980. Salt blamed for new sink in W. Texas. *Geotimes* 25(9):16–17.
29. **Vandale, A. E.** 1967. Subsidence: A real or imaginary problem. *Mining Engineering* 19(9):86–88.

KEY TERMS

landslide (p. 132)
driving forces (p. 134)
resisting forces (p. 134)
factor of safety (p. 135)
rotational slide (p. 135)
translational slide (p. 135)
perched water table (p. 141)
rapid draw-down (p. 141)
quick clay (p. 142)
specific risk (p. 148)
drainage control (p. 150)
grading of slopes (p. 150)
snow avalanche (p. 152)
subsidence (p. 154)
sinkhole (p. 154)

SOME QUESTIONS TO THINK ABOUT

1. In this chapter we established that variables such as climate, topography, vegetation, water, and time are important in affecting the nature and occurrence of landslides. Write down as many links as you can between these various processes to discover how they might be interrelated. For example, climate is obviously related to water and vegetation on slopes.
2. Your consulting company is hired by the national park or parks in your region to estimate the future risk from landsliding. Outline a plan of attack of what must be done to achieve this objective.
3. Why do you think that many people are not easily swayed by technical information concerning hazards such as landslides? Assume you have been hired by a community to make the citizens more aware of the landslide hazard in their area, which has a lot of steep topography. Outline a plan of action and defend it.

7 Earthquakes and Related Phenomena

Some of the extensive damage to the town of Golcuk in western Turkey from the magnitude 7.4 earthquake that struck on August 17, 1999. Note that the very old mosque (left) is still standing whereas many modern buildings collapsed. (*Enric Marti/AP/Wide World Photos*)

The magnitude 7.4 earthquake that struck Turkey on August 17,1999, leveled thousands of concrete buildings, left 600,000 people homeless, and killed approximately 15,000 people, with about 15,000 more people missing. Poor and improper construction is thought to be a factor in the collapse of newer buildings due to seismic shaking. Lessons learned from this earthquake will hopefully assist us in our preparation for future urban earthquakes.

7.1 Introduction to Earthquakes

Catastrophic earthquakes are devastating events that can destroy large cities and take thousands of lives in a matter of seconds. A sixteenth-century earthquake in China reportedly claimed 850,000 lives. In our own century, a 1923 earthquake near Tokyo killed 143,000 people, and a 1976 earthquake in China killed several hundred thousand. In 1985 an earthquake originating beneath the Pacific Ocean off Mexico caused 10,000 deaths in Mexico City, several hundred kilometers from the source. Recently, the 1995 earthquake that devastated Kobe, Japan, killed more than 5000 people while causing about $100 billion in property and other damage (Figure 7.1).

Newspaper and television reports of earthquakes usually give their moment magnitude. The moment magnitude (see page 174) is a reflection of the energy released by an earthquake, with each one-unit change (as from 7.0 to 8.0) representing about a thirtyfold increase in energy. An earthquake of magnitude 8.0 or higher is called a *great earthquake*, and a quake with a magnitude between 7.0 and 7.9 is called a *major earthquake*. The events mentioned above were all great earthquakes. But magnitude is only one of the factors determining an earthquake's destructiveness; equally important are human factors, such as population density, land use, and types of buildings. The great San Francisco quake of 1906 caused about 700 deaths and more than $25 million

.ARNING OBJECTIVES

Γhe study of earthquakes is an exciting field with tremendous social consequences. Learning objectives for this chapter are:

- *To understand the relationship of earthquakes to faulting.*
- *To know how we determine the magnitude of an earthquake.*
- *To become familiar with the terminology of earthquake waves.*
- *To know how we estimate seismic risk.*
- *To become familiar with the major effects of earthquakes.*
- *To understand the components of the earthquake cycle.*
- *To be aware of the various ways that we might some day predict earthquakes.*
- *To understand the processes of earthquake-hazard reduction and how people adjust to and perceive the hazard.*

Web Resources

Visit the "Environmental Geology" Web site at www.prenhall.com/keller to find additional resources for this chapter, including:

- ▶ Web Destinations
- ▶ On-line Quizzes
- ▶ On-line "Web Essay" Questions
- ▶ Search Engines
- ▶ Regional Updates

▲ **FIGURE 7.1** This elevated road collapsed as a result of intense seismic shaking associated with the 1995 Kobe, Japan, earthquake. (N. *Hosaka/Liaison Agency, Inc.*)

in property damage. Though the loss of life was enormous, it does not compare with the losses suffered in the Chinese, Japanese, and Mexican earthquakes, for reasons we will discuss in this chapter.

Because many earthquakes are initiated near plate boundaries, there are continuous linear or curvilinear zones in which most seismic activity takes place (Figure 7.2). Most large U.S. earthquakes are in the West, particularly near the North American and Pacific plate boundaries. However, large damaging **intraplate earthquakes** have occurred far from plate boundaries. For example, in 1886 a large intraplate earthquake nearly destroyed Charleston, South Carolina, taking 60 lives. The central Mississippi Valley is subject to infrequent but extremely damaging intraplate earthquakes (see *A Closer Look: Intraplate Earthquakes at New Madrid*).

California, which straddles two lithospheric plates at a location where they move past one another, experiences frequent, damaging earthquakes. The 1989 Loma Prieta earthquake on the San Andreas fault system south of San Francisco killed 62 people and caused $5 billion in property damage. The 1994 Northridge earthquake in the Los Angeles urban area killed 61 people and caused about $30 billion in damages. Neither of these were great earthquakes. It has been estimated that a great earthquake occurring today in a densely populated part of southern California could inflict $100 billion in damage and kill several thousand people (2). Thus, the Northridge quake, as terrible as it was, was not the anticipated "big one." Because earthquakes have the proven potential for producing a catastrophe, much of our research in earthquakes is devoted to understanding earthquake processes. The more we know about the probable location, magnitude, and effects of an earthquake, the better we can estimate the damage that is likely to occur and make plans for minimizing loss of life and property.

7.2 Earthquake Processes

Our discussion of global tectonics established that the earth is a dynamic, evolving system. Earthquakes are a natural consequence of the processes that form the ocean basins, continents, and mountain ranges of the world, and most of them occur along the boundaries of lithospheric plates (Figure 7.3).

Faults and Fault Movement

The process of fault rupture, or **faulting**, can be compared to sliding two rough boards past one another. Friction along the boundary between the boards may temporarily slow their motion, but rough edges break off and motion occurs at various places along the boundary. Similarly, lithospher-

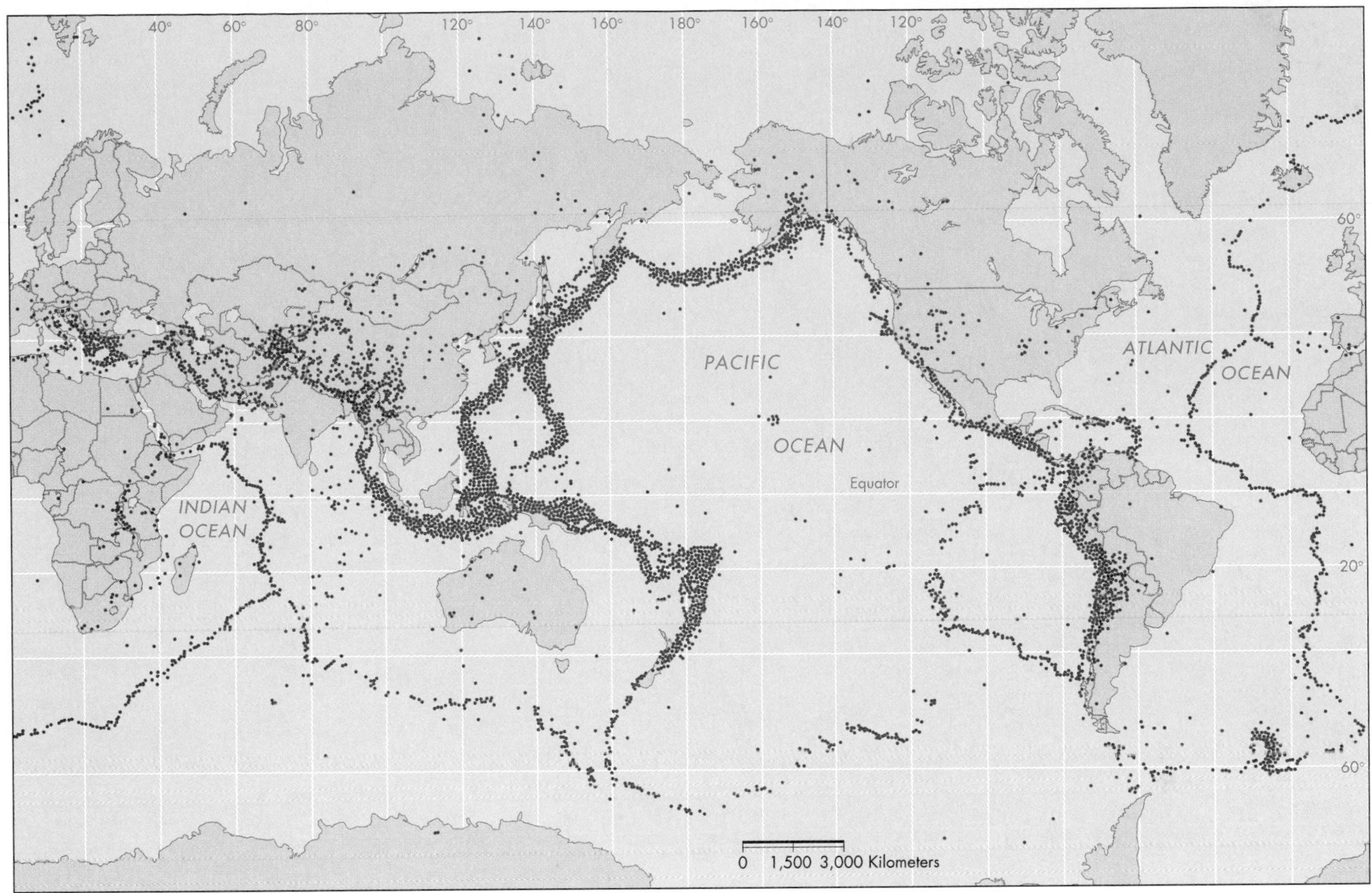

▲ **FIGURE 7.2** Map of global seismicity (1963–1988, Richter magnitude [M] = 5+) delineating plate boundaries and earthquake belts shown in Figure 7.3. (*Courtesy of National Earthquake Information Center*)

ic plates that are moving past one another are slowed by friction along their boundaries. As a result, rocks along the boundary undergo strain (deformation resulting from stress; see Chapter 2). When the stress on the rocks exceeds their strength, the rocks rupture, forming a fault. A **fault** is a *fracture or fracture system along which rocks have been displaced*—that is, one side of the fracture or fracture system has moved relative to the other side. The sudden rupture of the rocks produces earthquake waves, or **seismic waves,** that shake the ground. In other words, the pent-up energy of the strained rocks is released in the form of an earthquake. Faults are therefore **seismic sources,** and identifying them is the first step in evaluating the earthquake (seismic) risk in a given area.

A CLOSER LOOK Intraplate Earthquakes at New Madrid

During the winter of 1811–12, a series of particularly strong earthquakes struck the central Mississippi Valley, nearly destroying the town of New Madrid, Missouri, and killing an unknown number of people. The earthquakes were strong enough to cause church bells to ring in Boston, more than 1600 km away! They produced intense surface deformation, including ground rupture, over a wide area from Memphis, Tennessee, north to the confluence of the Mississippi and Ohio rivers. Forests were flattened, fractures opened so wide that people had to cut down trees to cross them, and the land sank several meters in some areas, causing flooding. It was even reported that the Mississippi River reversed its flow during the shaking (1).

The earthquakes occurred along a seismically active structure known as the *New Madrid Fault Zone,* which is part of the geologic structure known as the *Mississippi Embayment.* The embayment is a deep rift in an area where the lithosphere is weak and has broken repeatedly, even though the plate boundaries are far away. The recurrence interval for large earthquakes along the embayment is estimated at 600 to 700 years (1). Therefore, the possibility of future damage demands that the earthquake hazard be considered in design and construction of such important facilities as power plants and dams.

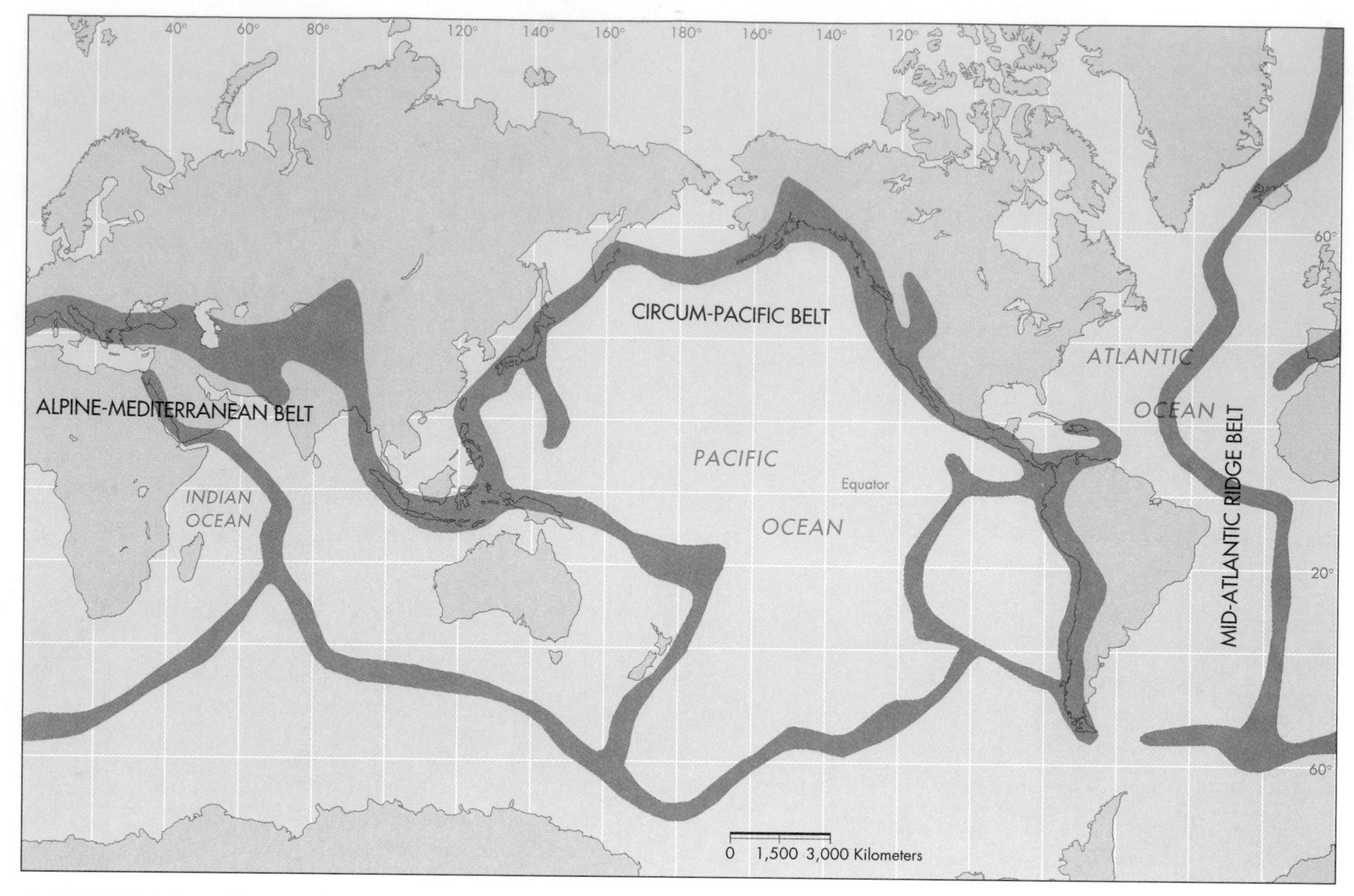

▲ **FIGURE 7.3** Map of the world showing the major earthquake belts as shaded areas. (*Base map from* NOAA)

Fault Types Figure 7.4 shows the major types of faults based upon the sense (direction) of the relative displacement. A **strike-slip fault** is one in which the sides of the fault are displaced horizontally; a strike-slip fault is called **right-lateral** if the right-hand side moves toward you as you sight along the fault line; it is **left-lateral** if the left-hand side moves toward you. A fault with vertical displacement may be a **reverse fault** (a low-angle reverse fault is called a *thrust fault*) or **normal fault** depending on the geometry of the displacement. Reverse and thrust fault displacement is associated with crustal shortening, whereas normal fault displacement is associated with crustal extension.

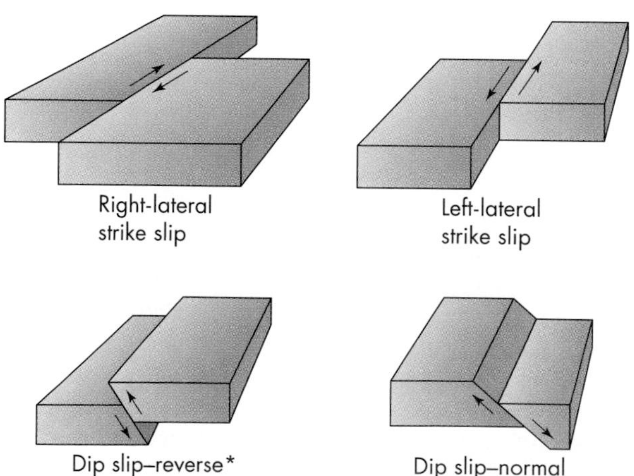

* If dip (inclination) of fault plane is less than 45°, it is called a thrust fault.

▲ **FIGURE 7.4** Types of fault movement based on the sense of motion relative to the fault. (*From* R. L. Wesson, E. J. Helley, K. R. Lajoie, *and* C. M. Wentworth, U.S. *Geological Survey Professional Paper* 941A, 1975.)

The faults shown in Figure 7.4 generally produce surface displacement or rupture. However, there are also **buried faults,** which are usually associated with folded rocks. Displacement and rupture of buried faults do not propagate to the surface even in large earthquakes (see *A Closer Look: The Northridge Earthquake and Buried Faults*).

Fault Zones and Fault Segments Faults almost never occur as a single rupture. Rather, they form **fault zones,** which are a group of related faults roughly parallel to each other in map view. They often partially overlap or form braided patterns. Fault zones vary in width, ranging from a meter or so to several kilometers.

Most long faults or fault zones, such as the San Andreas fault zone, are *segmented*, each segment having an individual history and style of movement. An **earthquake segment** is defined as those parts of a fault zone that have ruptured as a unit during historic and prehistoric earthquakes (3). Rupture during an earthquake generally stops

at the boundaries between two segments; however,
earthquakes may involve several segments of the fault.
ture length from large earthquakes is measured in te
kilometers. *It is the earthquake segment that is most imp*
to evaluation of the seismic hazard.

When the earthquake history of a fault zone
known, the zone may be divided into segments bas
differences in morphology or geometry. However, fr
point of view of earthquake-hazard evaluation, it is
able to define fault segments according to their h
which includes both historic seismic activity and
seismicity (prehistoric seismic activity). **Paleoseism**
is the study of prehistoric earthquakes from the g
environment. The geologic processes that govern **fa**
mentation and generation of earthquakes on individ
ments are the subject of active research.

Fault Activity Most geologists consider a fault
it can be demonstrated to have moved during
10,000 years (Holocene Epoch). The Quaternar
(approximately the past 1.65 million years) is the m
period of geologic time, and most of our land
been produced during that time. Fault displace
has occurred during the Pleistocene Epoch of the Quaternary (about 1.65 million to 10,000 years ago) but not in the Holocene is classified as **potentially active** (Table 7.1).

Faults that have not moved during the past 1.65 million years are generally classified as **inactive.** However, we emphasize that it is often difficult to prove the activity of a fault in the absence of easily measured phenomena such as historical earthquakes. To prove that a fault is active, it may be necessary to determine the past earthquake history or paleoseismicity based on the geologic record. This involves identifying faulted earth materials and determining when the most recent displacement occurred.

dividing that ...
average displacement per event were 1 m and the slip rate 2 mm per year, then the average recurrence interval would be 500 years.

3. **Seismicity:** Using historical earthquakes and averaging the time intervals between events.

Defining slip rate and recurrence interval is easy, and the calculations seem straightforward. However, the concepts of slip rate and recurrence interval are far from simple. Fault slip rates and recurrence intervals tend to change over time, casting suspicion on average rates. For example, it is not uncommon for earthquake events to cluster in time

Table 7.1 Terminology related to fault activity

Era	Period	Epoch	Years Before Present	Fault Activity	
Cenozoic	Quaternary	Historic (Calif.)	200	Active	Potentially active
		Holocene	10,000		
		Pleistocene	1,650,000		
	Tertiary	Pre-Pleistocene	65,000,000	Inactive	
Pre-Cenozoic time			4,500,000,000		
Age of the earth					

Source: After California State Mining and Geology Board Classification, 1973.

...thridge Earthquake and Buried Faults

...ake, which ...geles area on ...d caused about ...ctions of freeways ... structures and more ... Northridge earthquake ...ed events that recently oc-

...d the Northridge earthquake was ...oximately 18 km on a steeply south- ...he rupture quickly propagated upward ... not reach the surface, stopping at a depth ... the same time the rupture progressed lat- ...westward, about 20 km. The geometry of the fault movement is shown in Figure 7.B. The movement produced uplift and folding of part of the Santa Susanna Mountains a few kilometers north of Northridge, elevating them about 38 cm and moving them 21 cm to the northwest (4). Fault displacement near the depth at which the rupture first started (18 km) was about 2 m.

The relationship of a buried reverse fault to folding of rocks is shown in Figure 7.C. Shortening of a sequence of sedimentary rocks has produced folds called *anticlines* and *synclines*. On the left-hand side of Figure 7.C (where the sequence of sedimentary rocks has not been overturned), anticlines are arch-shaped folds and synclines are bowl-shaped. Notice that in the cores of two of the anticlines (middle and left), buried faulting has occurred. It was the rupture of a buried reverse fault like

(a)

(b)

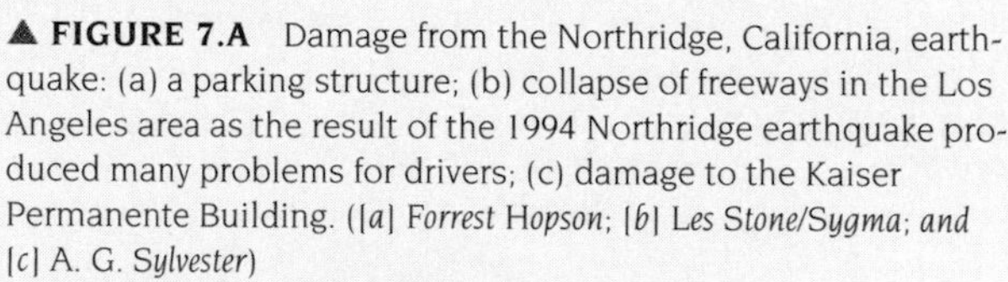

▲ **FIGURE 7.A** Damage from the Northridge, California, earthquake: (a) a parking structure; (b) collapse of freeways in the Los Angeles area as the result of the 1994 Northridge earthquake produced many problems for drivers; (c) damage to the Kaiser Permanente Building. (*[a] Forrest Hopson; [b] Les Stone/Sygma; and [c]* A. G. *Sylvester*)

(c)

and then be separated by long periods of low activity. Thus, both slip rate and recurrence interval will vary according to the time interval for which data is available.

Methods of Estimating Fault Activity Various methods are used to estimate *fault activity*, which includes both historic and prehistoric earthquakes. Paleoseismicity can be estimated by investigating landforms produced or displaced by faulting. Figure 7.5 shows the typical landforms found along active strike-slip faults characterized by predominantly horizontal motion. Features such as offset streams, sag ponds, linear ridges, and fault scarps (steep slopes or cliffs produced by fault movement) may indicate recent faulting. However, care must be exercised

these that produced the Northridge earthquake. Notice that at the surface of the ground, anticlines form ridges and synclines form basins. Faulting during earthquakes causes anticlinal mountains to be uplifted, while subsidence may occur in synclinal valleys.

Until recently we thought it likely that we could map all active faults because the most recent earthquake of each would cause surface rupture. Discovery that some faults are buried and that rupture does not always reach the surface has made it more difficult to evaluate the earthquake hazard in some areas. One way we can study active buried reverse faulting is to evaluate changes in the elevation of anticlinal mountains. Presumably these changes are related to faulting at depth, and the greater the rate of uplift, the greater the presumed earthquake hazard.

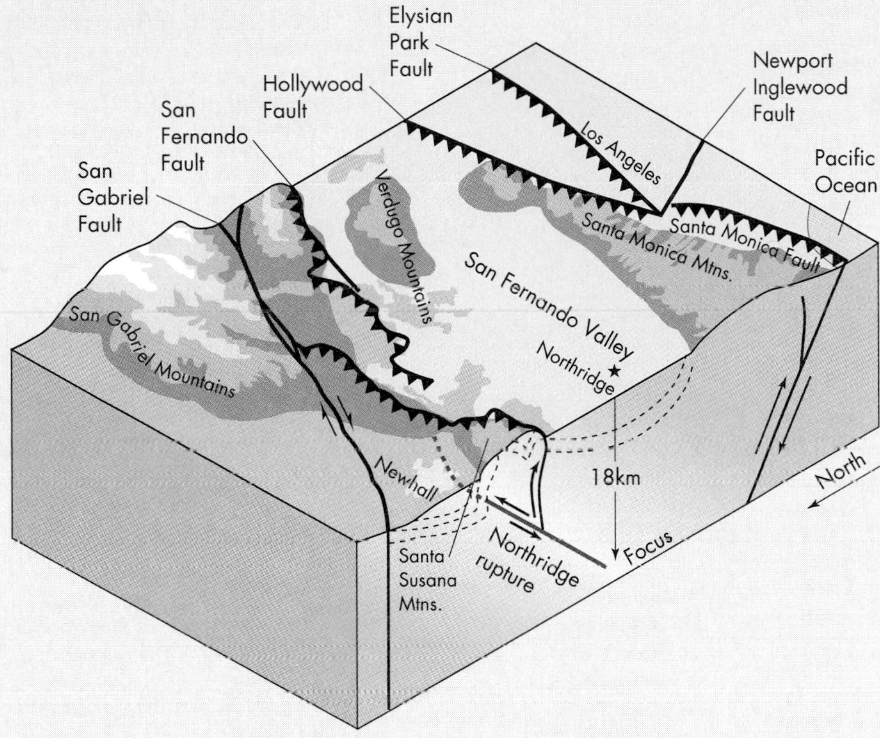

◀ **FIGURE 7.B** Block diagram showing the fault that produced the 1994 Northridge earthquake. (*Drawing courtesy of Pat Williams, Lawrence Livermore Laboratory*)

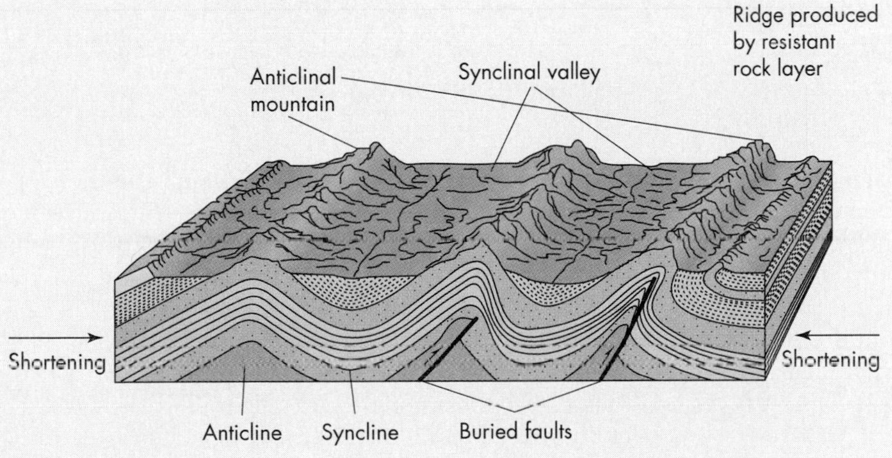

◀ **FIGURE 7.C** Block diagram illustrating several types of common folds and buried reverse faults with possible surface expressions such as anticlinal mountain and synclinal valley. (*Modified after F. Lutgens and E. Tarbuck.* 1992. Essentials of Geology, *4th ed. New York: Macmillan.*)

in interpreting these landforms, because some of them may have a complex origin.

The study of soils can also be useful in estimating the activity of a fault. For example, soils on opposite sides of a fault may be of similar or quite dissimilar age, thus establishing a relative age for the displacement. This method, in conjunction with other data, was used along parts of the Ventura fault in California to study its prehistoric activity (Figure 7.6).

Prehistoric earthquake activity sometimes can be dated radiometrically, providing a numerical date, if suitable materials can be found. For example, radiocarbon dates from faulted sediments have been obtained from the Coyote Creek fault zone and San Andreas fault zone in southern

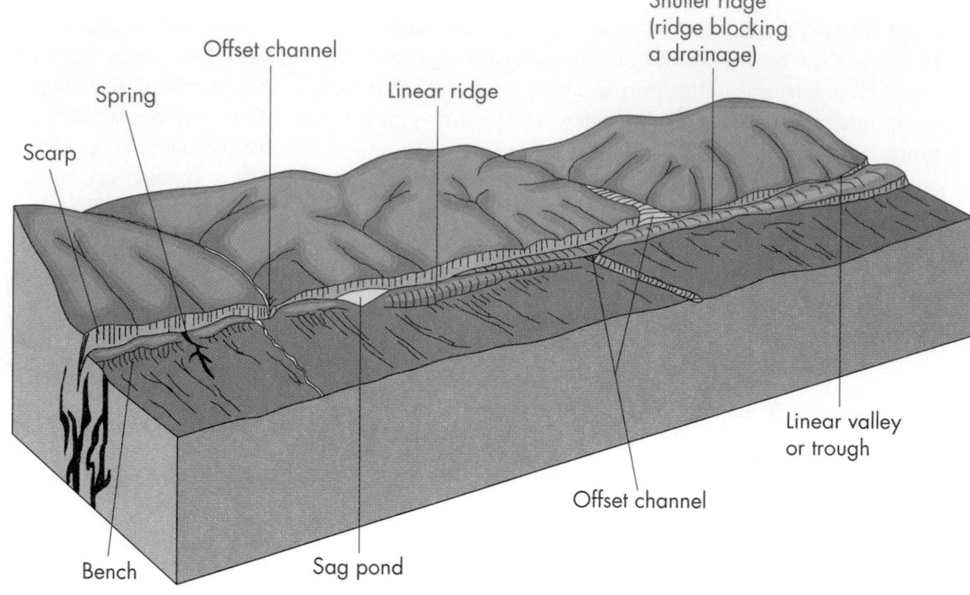

▶ **FIGURE 7.5** Assemblage of fault-related landforms along a large strike-slip fault such as the San Andreas fault. (*After* R. L. Wesson *et al.*, U.S. *Geological Survey Professional Paper* 941A, 1975.)

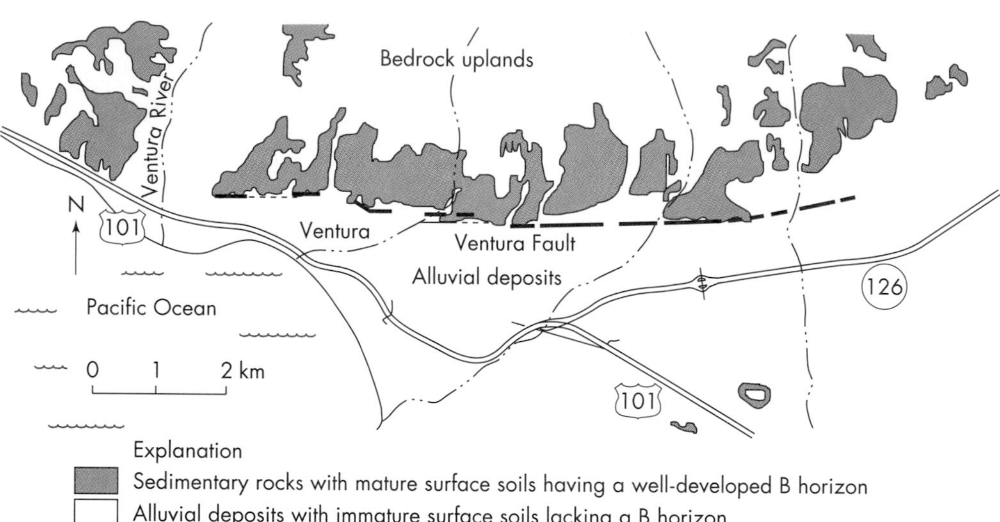

▶ **FIGURE 7.6** Map illustrating the difference in soils on opposite sides of the Ventura fault in southern California. (*After* Wojcicki *et al.*, U.S. *Geological Survey Map* MF-781.)

California. The Coyote Creek fault is characterized by predominant horizontal (strike slip) movement, but there is a vertical component of motion, and a trench across the fault after a 1968 earthquake revealed three older displacements of the flat-lying sediments that could be dated from carbon-14 measurements (Figure 7.7). Data from these measurements suggest that the recurrence interval for earthquakes the size of the 1968 event is about every 200 years (5).

Tectonic Creep Some active faults exhibit **tectonic creep,** that is, gradual displacement not accompanied by felt earthquakes. The process can slowly damage roads, sidewalks, building foundations, and other structures. Tectonic creep has damaged culverts under the football stadium of the University of California at Berkeley, and periodic repairs have been necessary as the cracks developed. Movement of approximately 3.2 cm was measured over a period of 11 years (6). More rapid rates of tectonic creep have been recorded on the Calaveras fault zone, a segment of the San Andreas fault, near Hollister, California. At one location a winery located on the fault is slowly being pulled apart at about 1 cm per year (7). Damages resulting from tectonic creep generally occur along narrow fault zones subject to slow, continuous displacement. However, creep may also be discontinuous and variable in rate. Just because a fault creeps does not mean that damaging earthquakes will not occur. Often the rate of creep on a fault is a relatively small portion of the slip rate, and periodic sudden displacements producing earthquakes can also be expected.

Seismic Waves and Ground Shaking

When a fault ruptures, rocks break apart suddenly and violently, releasing energy in the form of seismic waves. It is the propagation of these waves through the earth (ground) that we perceive as an earthquake, although the term *earthquake* can also refer to the fault rupture that gives rise to the waves.

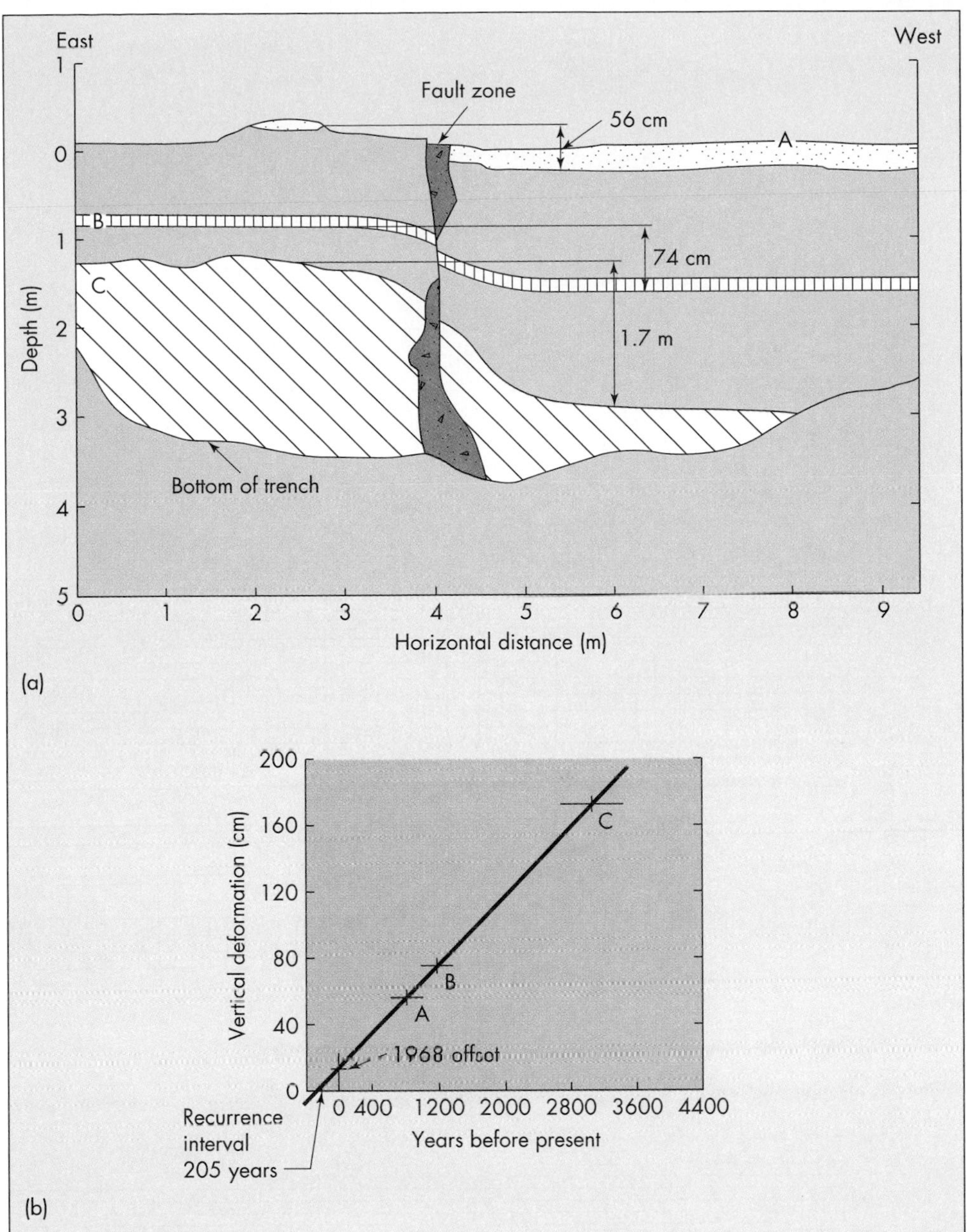

◀ **FIGURE 7.7** Generalized sketch of trench wall showing vertical displacement of flat-lying sediments associated with the Coyote Creek fault, California (a). Older offsets are progressively greater. Graph of vertical deformation against time (b). Derived from radiocarbon dates of the faulted sediments. (*After* M. M. *Clark,* A. *Grantz, and* R. *Meyer.* 1972. U.S. *Geological Survey Professional Paper* 787.)

Focus and Epicenter The point or area within the earth where the earthquake rupture starts is called the **focus** (plural, foci) of the earthquake (8). The **epicenter** is the point on the earth's surface directly above the focus (Figure 7.8a). The location of an earthquake reported by the news media is usually the epicenter, but scientific reporting includes both the location of the epicenter and the depth of the focus.

The foci of earthquakes vary from very shallow (a few kilometers) to almost 700 km below the surface. The deepest earthquakes occur along subduction zones, where slabs of oceanic lithosphere sink to great depths. However, most earthquakes are relatively shallow, with foci of less than about 60 km. In southern California most earthquakes have foci of about 10 to 15 km, although deeper earthquakes have occurred. The shallowness of these earthquakes makes them more destructive than deeper earthquakes of comparable magnitude. A magnitude 7.5 earthquake on strike-slip faults in the Mojave Desert near Landers, California, in 1992 had a focus of less than 10 km. This event caused extensive ground rupture for about 85 km (9) with very local vertical displacement exceeding 2 m and extensive lateral displacement as much as about 6 m (Figure 7.9). If the Landers event had occurred in the Los Angeles Basin, extensive damage and loss of life would have occurred.

Types of Seismic Waves Some of the seismic waves generated by fault rupture travel within the earth and others travel along the surface. Waves traveling *within the earth* are known as **body waves** and are of two types: primary (P) waves and secondary (S) waves (Figure 7.8b). The **P waves,** also called *compressional waves*, are the faster of the two. They

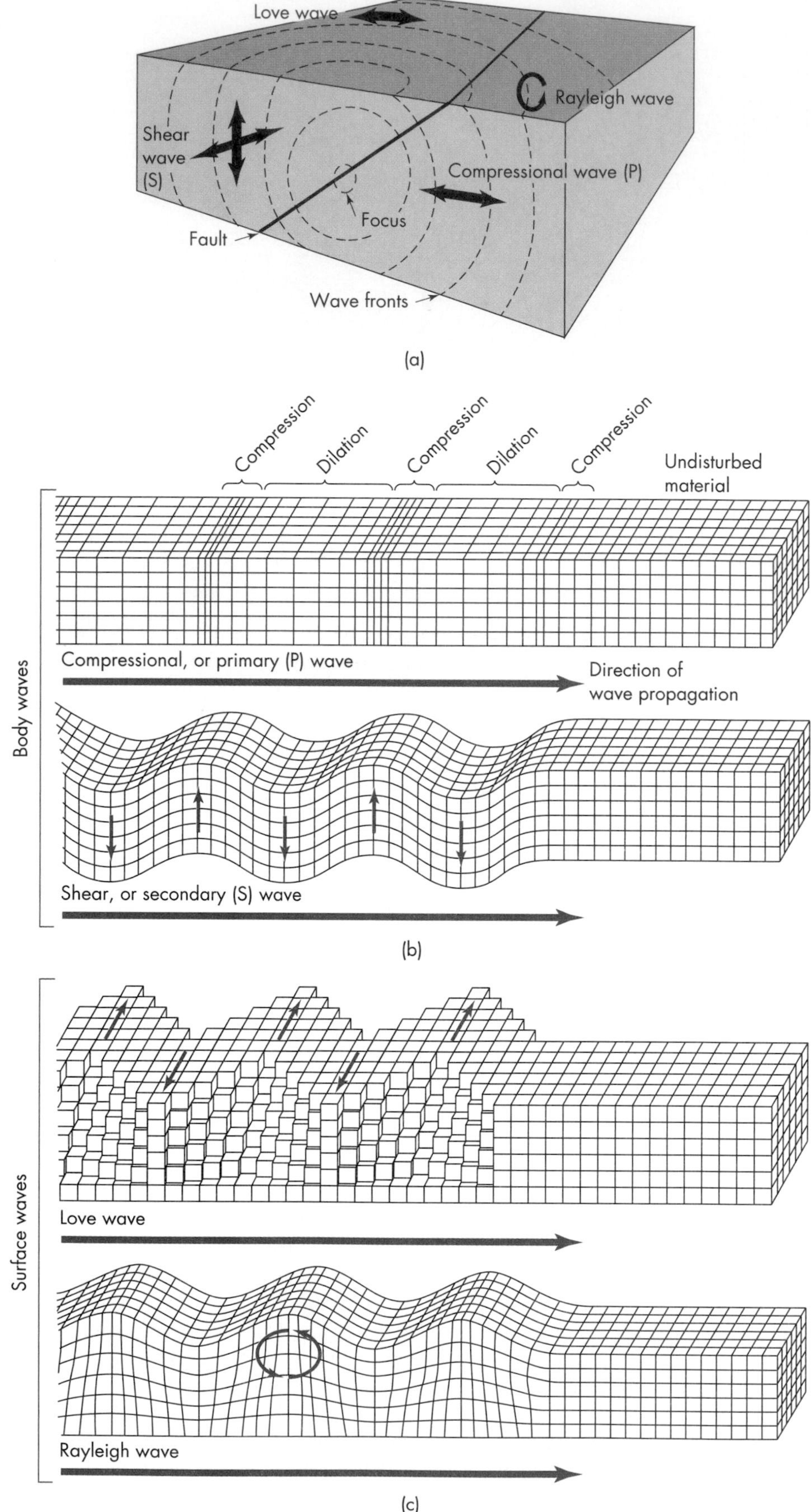

▶ **FIGURE 7.8** (a) Directions of vibration of body waves (P and S) and surface waves (Love and Rayleigh) generated by an earthquake. Also shown are the focus and epicenter of the earthquake event. (b) Propagation of body waves. (c) Propagation of surface waves. (*[a] From* W. W. Hays, 1981, *Hazards from Earthquakes. In* Facing Geologic and Hydrologic Hazards, *ed.* W. W. Hays. U.S. *Geological Survey Professional Paper* 1240-B; *[b] and [c] modified after* B. A. Bolt, 1978, Earthquakes: A Primer. *San Francisco*: W. H. *Freeman*.)

(a)

▲ **FIGURE 7.9** Cracking of ground and damage to building from the magnitude 7.5 Landers earthquake (a). Fence and dirt road offset to the right (b). (*Edward A. Keller*)

(b)

travel through solid, liquid, and gaseous materials in the manner of sound waves, with alternating compression and dilation of the traversed material in the direction of wave propagation. The rate of propagation for P waves through rock types such as granite is approximately 5.5 km/sec. The rate through liquids is much slower; for example, P waves travel about 1.5 km/sec through water. Interestingly, when P waves with frequencies higher than about 15 hertz (cycles per second) are propagated into the atmosphere they are detectable to the human ear. This explains the observation that people sometimes hear an earthquake before they feel the shaking caused by arrival of the slower surface waves (8).

The **S waves,** also called *shear waves,* can travel only through solid materials. Their speed through rocks such as granite is approximately 3 km/sec. The S waves produce an up-and-down motion (sideways shear) at right angles to the direction of wave propagation, similar to the motion produced in a clothesline by pulling it down and letting go. When liquids are subjected to sideways shear, they are unable to spring back, explaining why S waves cannot move through liquids (8).

Waves traveling along the surface of the earth are called **surface waves** (Figure 7.8c). It is these waves that cause much of the earthquake damage to buildings and other structures. Surface waves include **Love waves,** which have a complex *horizontal* ground movement, and **Rayleigh waves,** which have a complex *rolling* motion. Both travel slower than do body waves, but Love waves generally travel faster than do Rayleigh waves.

Because different types of waves and waves of different frequency move at different speeds away from an earthquake source, they become organized into groups of waves traveling at similar velocities. However, near the source of a large earthquake, there is not time for this wave segregation to take place, and the shaking may be severe and complex. Waves traveling through rocks are both reflected and refracted across boundaries between different earth materials and at the surface of the earth, producing amplification that may enhance shaking and damage to buildings and other structures. Furthermore, wave propagation is affected by the initial fault rupture, which may tend to focus earthquake energy in the direction of fault rupture. The complexity of surface shaking is further complicated and accentuated by topography and near-surface soil conditions, which may increase or decrease the amplitude of seismic waves at a particular site (8).

The Effect of Wave Frequency on Shaking Shaking of the ground and of buildings and other structures is greatly affected by the frequency of earthquake waves. The **frequency** of a wave is the number of waves passing a point of reference per second, expressed in hertz (Hz). The **period** is the number of seconds between successive peaks of a wave passing the reference point. The frequency of an earthquake wave is equal to 1 divided by the period (i.e., the reciprocal of the period). Thus, the frequency of a P wave with period of 2 seconds is 1 divided by 2 or 0.5 Hz. Both P and S body waves have a wide range of frequencies, but because of the rapid *attenuation* (diminishing) of higher frequencies as waves propagate away from where they are generated, most body waves have frequencies of 0.5 to 20 hertz. The more complex surface waves have lower frequencies (less than 1 Hz).

Buildings and other structures often have natural vibration frequencies in the same range as earthquake waves. This is unfortunate, because shaking of buildings is facilitated when the frequency of earthquake waves is close to the natural frequency of the building. Low buildings have a higher natural frequency than do taller buildings, so compressional and shear waves tend to cause low buildings to vibrate, while surface waves tend to cause tall buildings to vibrate. Because high-frequency waves attenuate much more quickly than do low-frequency waves, tall buildings may be damaged at long distances from the epicenter of a large earthquake (8,10), whereas low buildings are more sensitive to shaking when they are near the epicenter. This principle was dramatically illustrated in 1985 when a magnitude 8.1 earthquake originating several hundred kilometers away struck Mexico City and damaged or destroyed many tall buildings.

Different earth materials, such as bedrock, alluvium (sand and gravel), and silt and mud, respond differently to seismic shaking. For example, the intensity of shaking of unconsolidated sediments may be much more severe than for bedrock. Figure 7.10 shows how the amplitude of shaking (vertical movement) is greatly increased in such sediments, particularly in silt and clay deposits. This effect is called **material amplification.**

A major lesson from the Mexico City earthquake was that buildings constructed on soils and rocks likely to accentuate and increase a seismic shaking are extremely vulnerable to earthquakes, even if the event is centered several hundred kilometers away. Seismic waves originating offshore initially contained many frequencies, but those that arrived at the city were low-frequency waves of about 0.5 to 1.0 Hz. It is speculated that when these waves struck the lake beds beneath Mexico City, the amplitude of shaking may have increased at the surface by a factor of 4 or 5. Figure 7.11a shows the geology of the city and the location of the worst damage. The intense regular shaking caused buildings to sway back and forth, and eventually many of them collapsed or pancaked as upper stories collapsed onto lower ones (Figure 7.11b). Most of the damage was to buildings with 6 to 16 stories because these buildings had a natural frequency that nearly matched that of the arriving seismic waves (11).

The potential for amplification of surface waves to cause damage was again demonstrated with tragic results during the 1989 magnitude 7.1 Loma Prieta earthquake, which originated south of San Francisco. Figure 7.12 shows the epicenter and the areas that greatly magnified shaking. The collapse of a tiered freeway, which killed 41 people, occurred on a section of roadway constructed on bay fill and mud (Figure 7.13). Where the freeway was constructed on older, stronger alluvium, less shaking occurred and the structure survived. Extensive damage was also recorded in the Marina district of San Francisco (Figure 7.14), primarily in areas constructed on bay fill and mud as well as debris dumped into the bay during the cleanup following the 1906 earthquake (12).

Directivity, which is another amplification effect, results when the amplitude of seismic waves increases in the direction of fault rupture. For example, fault rupture of the 1994 Northridge earthquake was up (north) and to the west (Figure 7.15), resulting in stronger ground motions from seismic shaking to the northwest. The stronger ground motion (peak ground velocity, Figure 7.15) to the northwest is believed to result from the movement (fault rupture during the earthquake) of a block of earth upward and to the northwest (13).

Comparing Earthquakes

There are about a million earthquakes a year that can be felt by people somewhere on earth. However, only a small percentage of these can be felt very far from their source. We can compare earthquakes with one another by calculating the energy they release (magnitude) or by evaluating their impact on people and structures (intensity). It is also useful to compare their ground accelerations, which are a measure of the severity of ground shaking.

Earthquake Magnitude The **Richter magnitude** (M) of an earthquake is a measure of the amount of energy released. It is based on a tracing made by a **seismograph,** an instrument that records earthquake displacement caused by shaking. The essential measurement for determining Richter magnitude, as it was first defined, is the largest amplitude produced during an earthquake by a standard seismograph 100 km from the earthquake epicenter (8). This amplitude is converted to a magnitude on the logarithmic Richter scale. Each integer of the scale represents a tenfold increase in amplitude: A

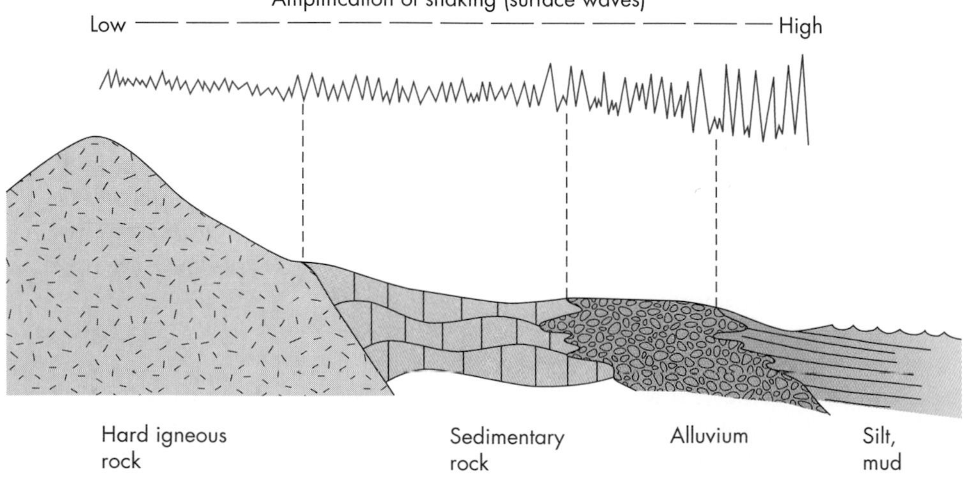

▶ **FIGURE 7.10** Generalized relationship between near-surface earth material and amplification of shaking during a seismic event.

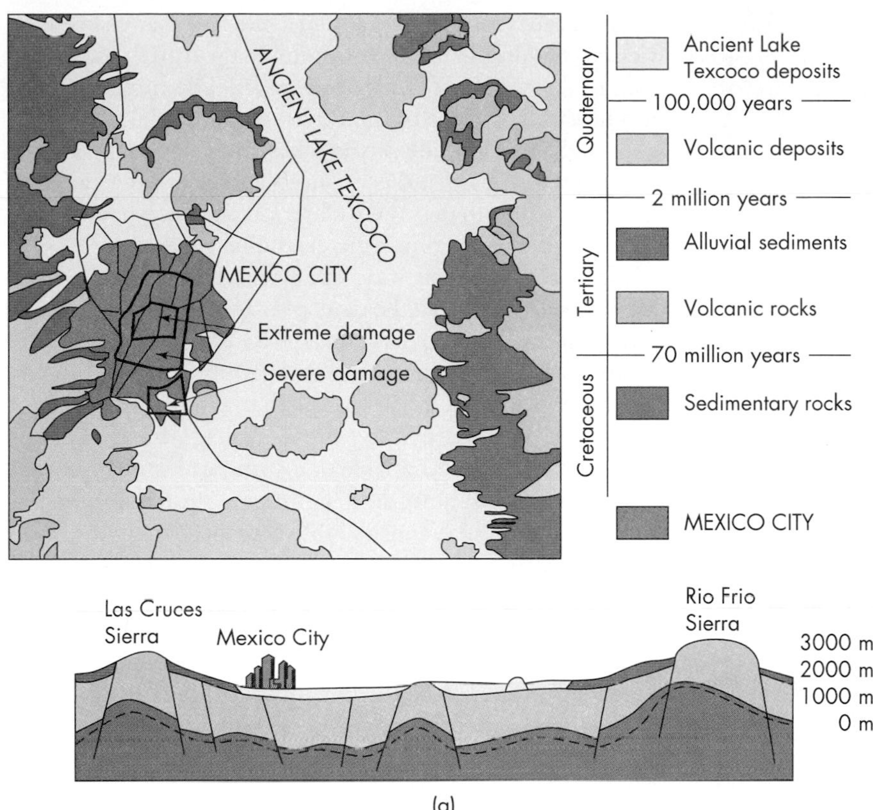

◀ **FIGURE 7.11** Earthquake damage, Mexico City, 1985. (a) Generalized geologic map of Mexico City showing ancient lake deposits where greatest damage occurred. (b) High-rise building, one of many that collapsed. (*Map and photo courtesy of* U.S. *Geological Survey,* T. C. *Hanks and Darrell Herd.*)

magnitude 7 earthquake, for example, produces a displacement on the seismograph recording 10 times larger than does a magnitude 6 earthquake. The energy released in an earthquake is proportional to the magnitude, but the increase in energy from one unit to the next is about thirtyfold: A magnitude 7 earthquake releases about 30 times as much energy as does a magnitude 6 event. About 27,000 (30 × 30 × 30) shocks of magnitude 5 would be required to release as much energy as a single earthquake of magnitude 8.

The Richter magnitude of an earthquake can also be estimated from seismographs at different locations, using graphical solutions to mathematical equations. The maximum amplitude and difference in arrival time of P (primary) and S (secondary) waves from a distant earthquake

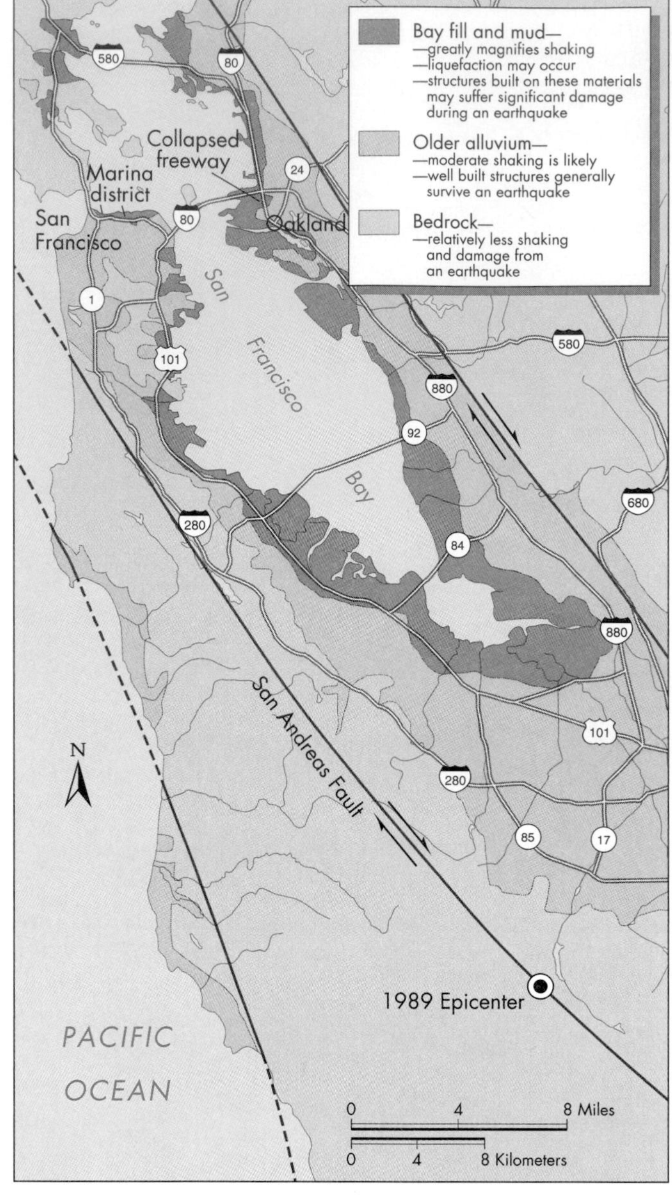

▲ **FIGURE 7.12** San Francisco Bay region, showing San Andreas fault and epicenter of 1989 earthquake, which had a magnitude of 7.1. (*Modified after* T. Hall, *data from* U.S. *Geological Survey*.)

are recorded. Figure 7.16 shows the record of an earthquake with an amplitude of 85 mm and difference in arrival time (S minus P) of 34 seconds. The line connecting the amplitude with the difference in arrival time indicates that the magnitude is 6 and the distance from the epicenter is about 300 km. Records from several seismographs in a region are necessary to locate the epicenter precisely.

The Richter magnitude scale initially was intended as a local magnitude **(M_L)** and the seismic wave used was the largest regardless of type (P, S, or surface). If the magnitude is determined using the largest amplitude of the P waves, the resulting magnitude is termed **M_b** (the *b* stands for body wave), and if the largest amplitude of a surface wave is used, the magnitude is called the **M_s.**

In recent years there has been a move to change from the Richter magnitude to the **moment magnitude (M_w) scale** (see *A Closer Look: Moment Magnitude*). The moment magnitude is usually close to the Richter magnitude, particularly for higher-magnitude earthquakes. However, it has a more sound physical base and is applicable over a wider range of ground motions than is the Richter magnitude, so its use has been encouraged in reporting earthquake statistics.

In a very general way, the Richter magnitude of an earthquake is related to the damage expected. Thus, an event of magnitude 8 or above is considered a **great earthquake,** capable of causing widespread catastrophic damage. In any given year there is a good chance that a magnitude 8 event will occur somewhere in the world. A magnitude 7 event is a **major earthquake,** capable of causing widespread and serious damage. Magnitude 6 signifies a **moderate earthquake** that can cause considerable damage, depending upon factors such as location and surface materials present.

Earthquake Intensity A qualitative way of comparing earthquakes is to use the **Modified Mercalli Scale** (abridged), which has 12 divisions of intensity based on observations concerning the severity of shaking during an earthquake (Table 7.2). Intensity reflects how people perceived and structures responded to the shaking. An *instrumental intensity* can be obtained from data recorded from a dense network of high-quality seismographs. Although a particular earthquake has only one Richter or moment magnitude, different levels of intensity may be assigned to the same earthquake at different locations, depending on proximity to the epicenter and local geologic and engineering features. Figure 7.17 is a map showing the spatial (areal) variability of instrumental intensity for the 1994 magnitude 6.7 Northridge earthquake.

Ground Acceleration During Earthquakes Ground acceleration is the rate of change of the horizontal or vertical velocity of the ground during an earthquake. It is measured in relationship to the acceleration of gravity, which is 9.8 meters per second per second (9.8 $msec^{-2}$). This value is 1 g. An acceleration of 0.5 g would be 4.9 $msec^{-2}$ and 0.1 g is 0.98 $msec^{-2}$. Estimated horizontal and vertical peak accelerations during an earthquake likely to occur in an area are useful information for designing buildings and other structures to withstand seismic shaking.

Although it is possible to estimate ground acceleration at a site based on earthquake magnitude from a distant event, it is difficult to do so for accelerations occurring close to the epicenter. For example, a magnitude 6.0 to 6.9 earthquake can be generally expected to produce accelerations of 0.3 to 0.69. However, during the M = 6.6 San Fernando earthquake, the maximum peak horizontal acceleration recorded at one site was 1.15 g. The 1994 Northridge earthquake produced even higher values of horizontal acceleration in some parts of the Los Angeles urban area. It is the horizontal acceleration that is most likely to cause damage to buildings. Homes constructed of adobe,

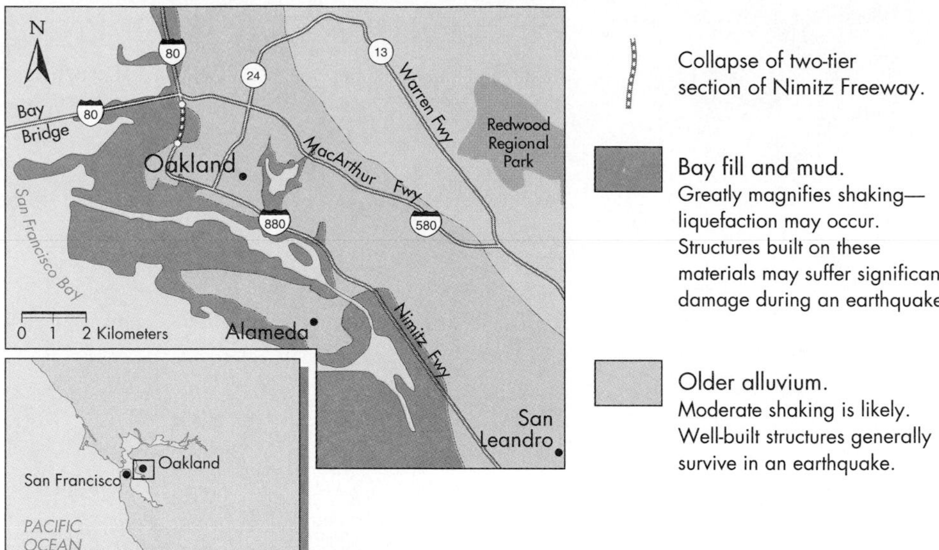

(b)

▲ **FIGURE 7.13** (a) Generalized geologic map of part of the San Francisco Bay showing bay fill and mud and older alluvium. (b) Collapsed freeway. (*Map [a] modified from S. E. Hough et al.*, 26 *April*, 1990. Nature, *Vol.* 344, *No.* 6269, *pp.* 853–855. *Copyright © Macmillan Magazines Ltd.*, 1990. *Used by permission of the author.* [*b*] *Courtesy of Dennis Laduzinsky.*)

▲ **FIGURE 7.14** Damage to buildings in the Marina district of San Francisco resulting from the 1989 earthquake. (*John K. Nakata, courtesy of U.S. Geological Survey*)

which are common in Mexico, South America, and the Middle East, can fail (collapse) under a horizontal acceleration as small as 0.1 g (8).

The Earthquake Cycle

Observations of the 1906 San Francisco earthquake led to a hypothesis known as the **earthquake cycle.** Important features of the hypothesis are related to a drop in elastic strain following an earthquake and reaccumulation of strain prior to the next event.

Elastic Rebound Strain, as we said earlier, is deformation resulting from stress. Elastic strain may be thought of as deformation that is not permanent, provided the stress is released. When the stress is released, the deformed material—a stretched rubber band, for example, or a bent archery bow—returns to its original shape. However, if the stress continues to increase, the deformed material eventually ruptures—the rubber band snaps or the bow breaks—making the deformation permanent (see Chapter 2). When a stretched rubber band or bow breaks, it experiences a *rebound*, in which the broken ends snap back, releasing their pent-up energy. What happens in an earthquake is similar and is called **elastic rebound** (Figure 7.18). Rocks on either side of a fault segment, held together by friction, are stressed by the tectonic forces that pull them in opposite directions. When the deformed rocks finally rupture, the stress is released and the elastic strain decreases quite suddenly as the sides of the fault "snap" into their new (permanently deformed) position. After the earthquake, it takes time for sufficient elastic strain to accumulate again to produce another rupture (14).

Stages of the Earthquake Cycle It is speculated that a typical earthquake cycle has three or four stages. The first is a long period of seismic inactivity following a major earthquake and associated immediate **aftershocks** (earthquakes that occur a few minutes to a few months to a year or so following the main event). This is followed by a second stage characterized

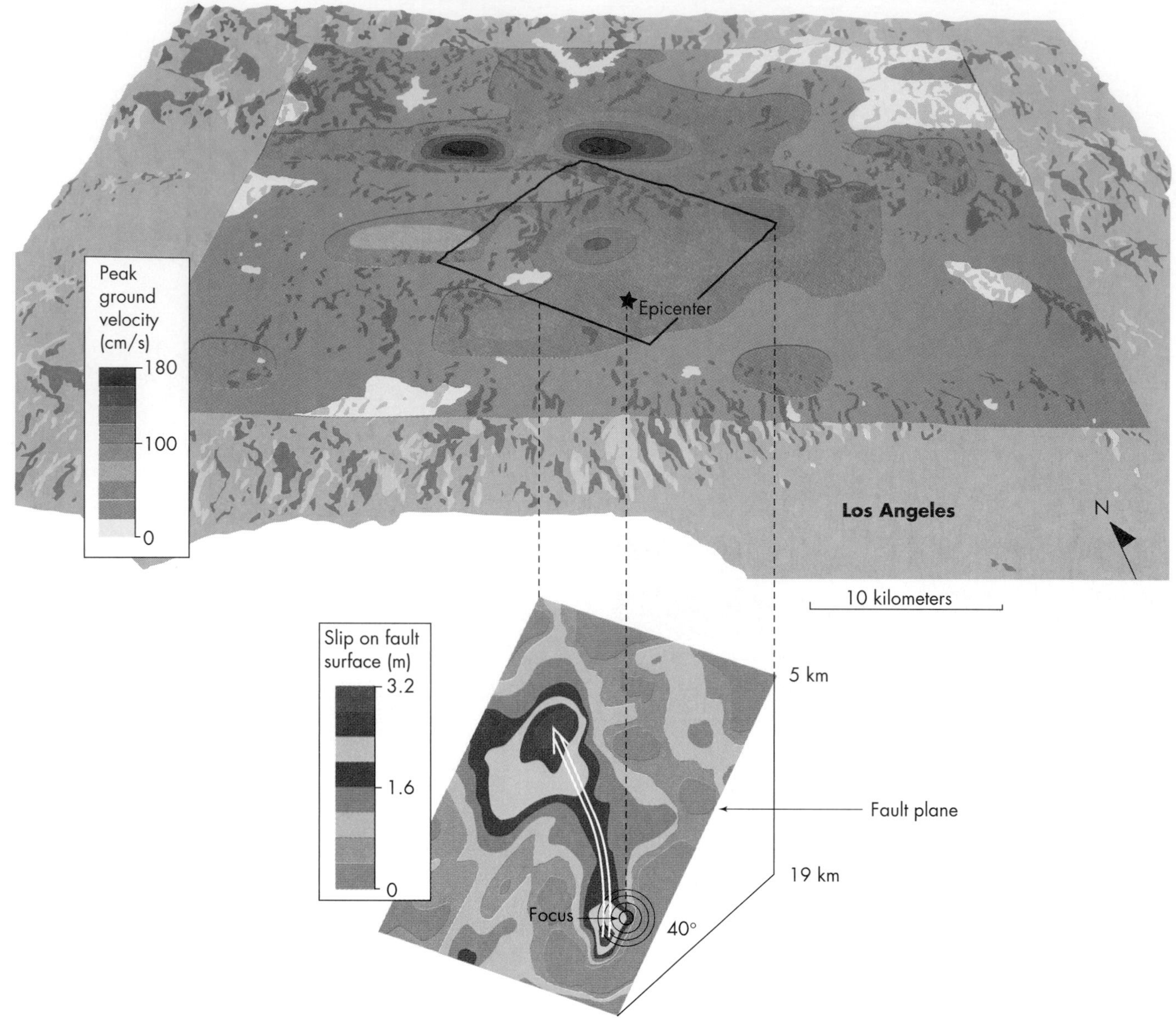

▲ **FIGURE 7.15** Aerial view of the Los Angeles region from the south showing the epicenter of the 1994 Northridge earthquake with peak ground motion in centimeters per second and the fault plane in its subsurface position. The path of the rupture is shown by the large area along with the amount of slip in meters along the fault plane. The area that ruptured along the fault plane is approximately 430 km^2 and the fault plane is dipping at approximately 40° to the south-southwest. Notice that maximum slip and peak ground velocities both occur to the northwest of the epicenter. The fault rupture apparently began at the focus in the southeastern part of the fault plane and proceeded upward and to the northwest as shown by the arrow. (U.S. *Geological Survey*. 1996. U.S.G.S. *response to an urban earthquake, Northridge* 1994. U.S. *Geological Survey Open File Report* 96-263.)

A CLOSER LOOK **Moment Magnitude**

The moment magnitude scale for comparing earthquake energies is based on the seismic moment of earthquakes. Seismic moment is defined as the product of (1) the average amount of slip on the fault that produced the earthquake; (2) the area that actually ruptured; and (3) the shear modulus (resistance of a material to distortion by shear stress; see Figure 2.6) of the rocks that failed (14). Figure 7.D illustrates the important concepts associated with the seismic moment.

The seismic moment of an earthquake can be estimated by examining the records from seismographs, determining the amount and length of rupture, and estimating the shear modulus of rocks involved. The value of the shear modulus is obtained through laboratory testing of rocks. The moment magnitude (M_w) is then determined from the mathematical relationship:

$$M_w = 2/3 \log M_o - 10.7$$

where M_w is the moment magnitude; M_o is the seismic moment, and 10.7 is a constant (14).

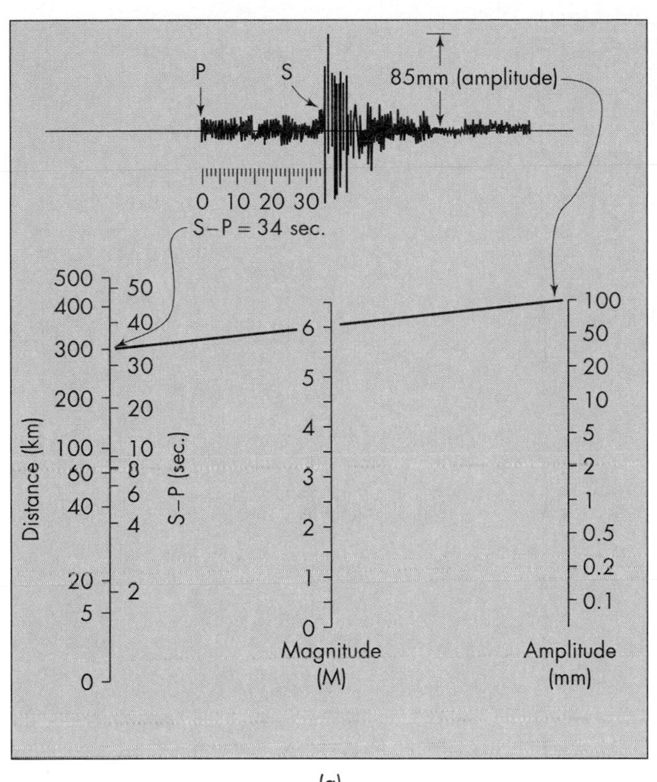

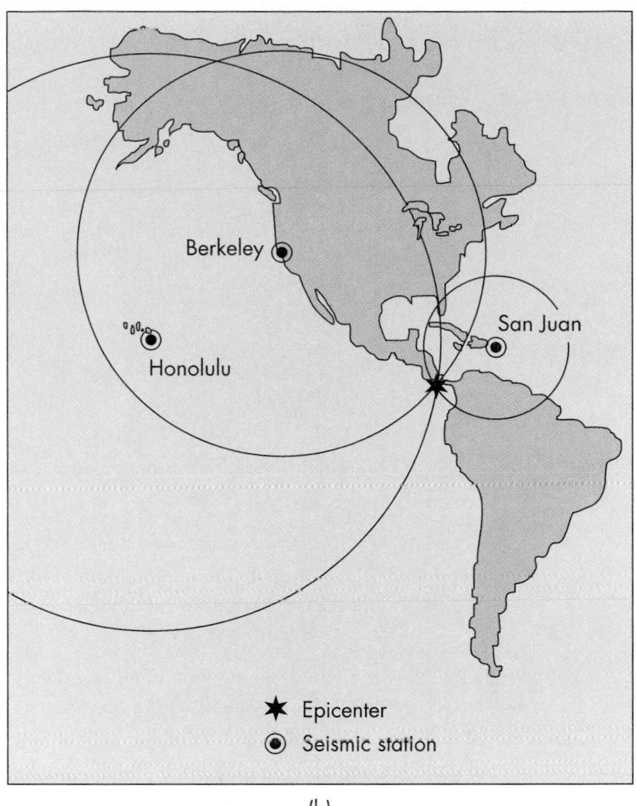

▲ **FIGURE 7.16** (a) Idealized diagram showing the procedure for estimating the magnitude (M) of an earthquake. For our example, the maximum amplitude (85 mm) is measured from the seismic record; the difference in arrival time between the S and P waves (34 seconds) is also taken from the seismic record; and the approximate magnitude of the earthquake as well as distance from the recording station is obtained by placing a straight line between the amplitude in millimeters and difference in arrival time in seconds, as shown on the diagram. Here, the magnitude is 6 and the distance is approximately 300 km. (b) Generalized concept of how the epicenter of an earthquake is located. Distance to event from at least three seismic stations is determined and plotted. The intersection of the arc distances (circle radii) defines the epicenter. For the diagram the epicenter in Central America is located from data supplied by three seismic stations. Accurate location of the epicenter is not always as simple as the above hypothetical example. (*[a] Modified after* B. A. Bolt, 1978, Earthquakes: A Primer. *San Francisco:* W. H. *Freeman.*)

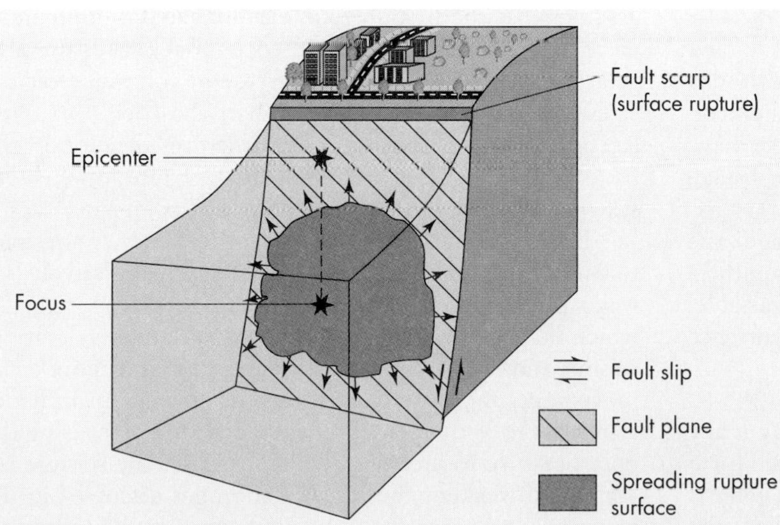

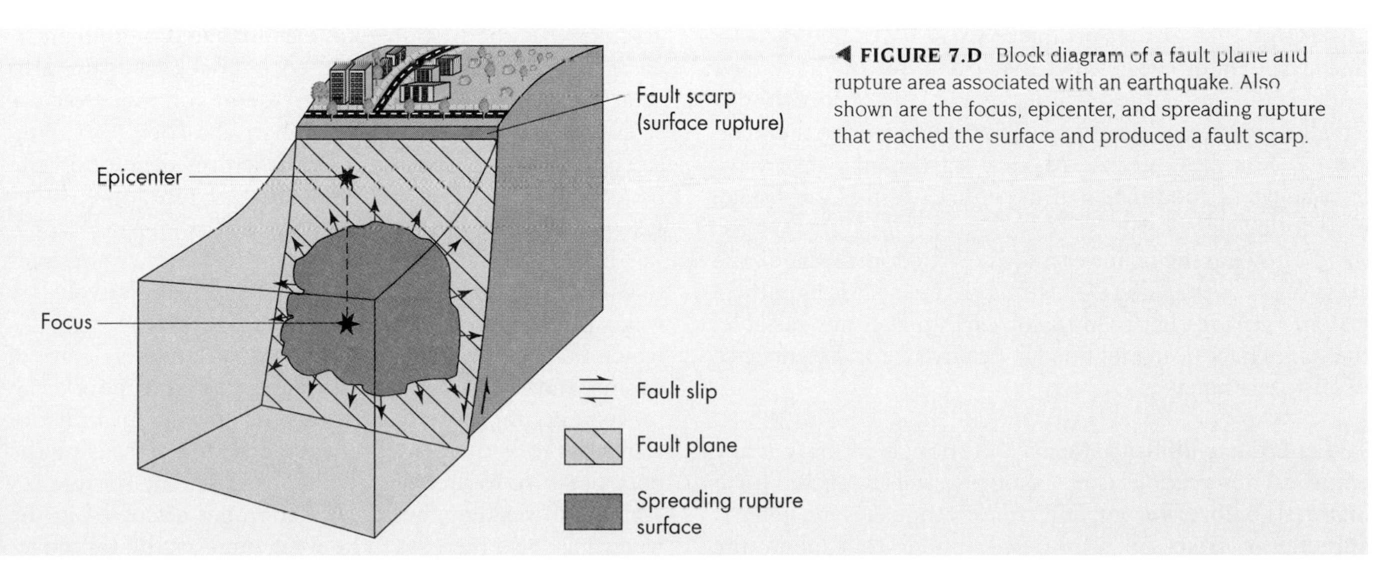

◀ **FIGURE 7.D** Block diagram of a fault plane and rupture area associated with an earthquake. Also shown are the focus, epicenter, and spreading rupture that reached the surface and produced a fault scarp.

Table 7.2 Modified Mercalli Intensity Scale (abridged)

Intensity	Effects
I	Not felt except by a very few under especially favorable circumstances.
II	Felt only by a few persons at rest, especially on upper floors of buildings. Delicately suspended objects may swing.
III	Felt quite noticeably indoors, especially on upper floors of buildings, but many people do not recognize it as an earthquake. Standing motor cars may rock slightly. Vibration like passing of truck. Duration estimated.
IV	During the day, felt indoors by many, outdoors by few. At night some awakened. Dishes, windows, doors disturbed; walls make cracking sound. Sensation like heavy truck striking building; standing motor cars rocked noticeably.
V	Felt by nearly everyone; many awakened. Some dishes, windows, etc., broken; a few instances of cracked plaster; unstable objects overturned. Disturbances of trees, poles, and other tall objects sometimes noticed. Pendulum clocks may stop.
VI	Felt by all; many frightened and run outdoors. Some heavy furniture moved; a few instances of fallen plaster or damaged chimneys. Damage slight.
VII	Everybody runs outdoors. Damage negligible in buildings of good design and construction; slight to moderate in well-built ordinary structures; considerable in poorly built or badly designed structures; some chimneys broken. Noticed by persons driving motor cars.
VIII	Damage slight in specially designed structures; considerable in ordinary substantial buildings, with partial collapse; great in poorly built structures. Panel walls thrown out of frame structures. Fall of chimneys, factory stacks, columns, monuments, walls. Heavy furniture overturned. Sand and mud ejected in small amounts. Changes in well water. Disturbs persons driving motor cars.
IX	Damage considerable in specially designed structures; well-designed frame structures thrown out of plumb; great in substantial buildings, with partial collapse. Buildings shifted off foundations. Ground cracked conspicuously. Underground pipes broken.
X	Some well-built wooden structures destroyed; most masonry and frame structures with foundations destroyed; ground badly cracked. Rails bent. Landslides considerable from river banks and steep slopes. Shifted sand and mud. Water splashed (slopped) over banks.
XI	Few, if any (masonry) structures remain standing. Bridges destroyed. Broad fissures in ground. Underground pipelines completely out of service. Earth slumps and land slips in soft ground. Rails bent greatly.
XII	Damage total. Waves seen on ground surfaces. Lines of sight and level distorted. Objects thrown upward into the air.

Source: From Wood and Neuman, 1931, by U.S. Geological Survey, 1974, *Earthquake Information Bulletin* 6(5), p. 28.

by increased seismicity, as accumulated elastic strain approaches and locally exceeds the strength of the rocks, initiating faulting that produces small earthquakes. The third stage of the cycle, which may occur only hours or days prior to the next large earthquake, consists of **foreshocks** (small to moderate earthquakes occurring before the main event). For example, an $M_L = 6$ earthquake may have foreshocks of about $M_L = 4$. In some cases this third stage may not occur.

Following the major earthquake (the fourth stage), the cycle starts over again (14). Although the cycle is hypothetical and periods between major earthquakes are variable, the stages have been identified in occurrence and recurrence of large earthquakes.

The Dilatancy-diffusion Model Although we have many empirical observations concerning physical changes in earth materials before, during, and after earthquakes, no general agreement exists on a physical model to explain the observations. One model, known as the **dilatancy-diffusion model** (8,15), is illustrated in Figure 7.19. This model assumes that the first stage in earthquake development is an increase of elastic strain in rocks that causes them to **dilate** (undergo an inelastic increase in volume) after the stress on the rock reaches one-half the rock's breaking strength. During dilation, open fractures develop in the rocks, and it is at this stage that the first physical changes take place that might indicate a future earthquake.

Briefly, the model assumes that the dilatancy and fracturing of the rocks are first associated with a relatively low water pressure in the dilated rocks (Figure 7.19, stage 2), which helps to produce lower seismic velocity (velocity of seismic waves through the rocks near a fault), more earth movement, and fewer minor seismic events. An influx of water (stage 3) then enters the open fractures, causing the pore pressure to increase, which increases the seismic velocity and weakens the rocks. Radon gas dissolved in the water may also increase. The weakening facilitates move-

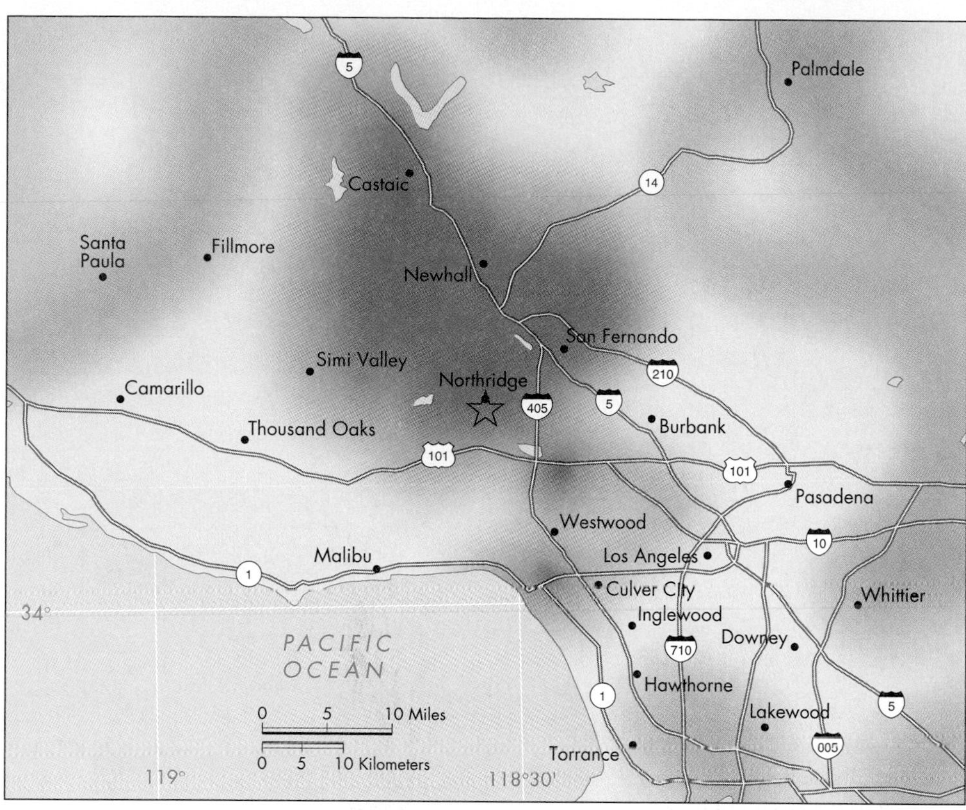

INTENSITY	I	II-III	IV	V	VI	VII	VIII	IX	X+
Shaking	not felt	weak	light	moderate	strong	very strong	severe	violent	extreme
Damage	none	none	none	very light	light	moderate	moderate/hvy	heavy	very hvy
Peak acc (%g)	<0.9	0.9–2.9	2.9–5.3	5.3–9.7	9.7–18	18–33	33–59	59–109	>109
Peak vel (cm/s)	<0.6	0.6–2.2	2.2–4.2	4.2–8	8-15	15–30	30–57	57–110	>110

◀ **FIGURE 7.17** Instrumental intensity map and peak ground acceleration (% g) for the 1994 magnitude 6.7 Northridge, California, earthquake. Map is based upon observations and data from a dense network of high-quality seismographs. Epicenter is a star, and colors match intensity of shaking shown on the table. (*Source:* U.S. *Geological Survey, courtesy of* David Wald.)

ment along the fractures, which is recorded as an earthquake (stage 4). After the movement and release of stress, the rocks resume many of their original characteristics (stage 5) (15).

Considerable controversy surrounds the validity of the dilatancy-diffusion model. One aspect of the model gaining considerable favor is the role of **fluid pressure** (force per unit area exerted by a fluid) in earthquakes. As we learn more about rocks at seismogenic (earthquake generating) depths, it is apparent that much water is present. Deformation of the rocks and a variety of other processes are thought to increase the fluid pressure at depth, and this results in a lowering of the shear strength of the rocks. If the fluid pressure becomes sufficiently high, this can facilitate the occurrence of earthquakes. Empirical data from several environments, including subduction zones and active fold belts, suggest that high fluid pressures are present in many areas where earthquakes occur. Thus, there is increasing interest in the role of fluid flow in fault displacement and the earthquake cycle. The **fault-valve mechanism** (16) hypothesizes that fluid (usually water) pressure rises until failure occurs, thus triggering an earthquake along with upward fluid discharge. This may explain why, following some earthquakes, springs discharge more water and dry streams may start flowing (groundwater enters their channel). Increased flow (discharge) of groundwater may occur for months following a large earthquake.

Earthquakes Caused by Human Activity

Several human activities are known to increase or cause earthquake activity. Damage from these earthquakes is regrettable, but the lessons learned may help to control or stop large catastrophic quakes in the future. Three ways that the actions of people have caused earthquakes are (17):

- Loading earth's crust, as in building a dam and reservoir
- Disposing of waste deep into the ground through disposal wells
- Underground nuclear explosions

Reservoir-induced Seismicity During the first 10 years following the completion of Hoover Dam on the Colorado River in Arizona and Nevada, several hundred local tremors occurred. Most of these were very small, but one had a magnitude of about 5 and two had magnitudes of about 4 (17). An earthquake of about magnitude 6 killed about 200 people in India following construction and filling of a reservoir.

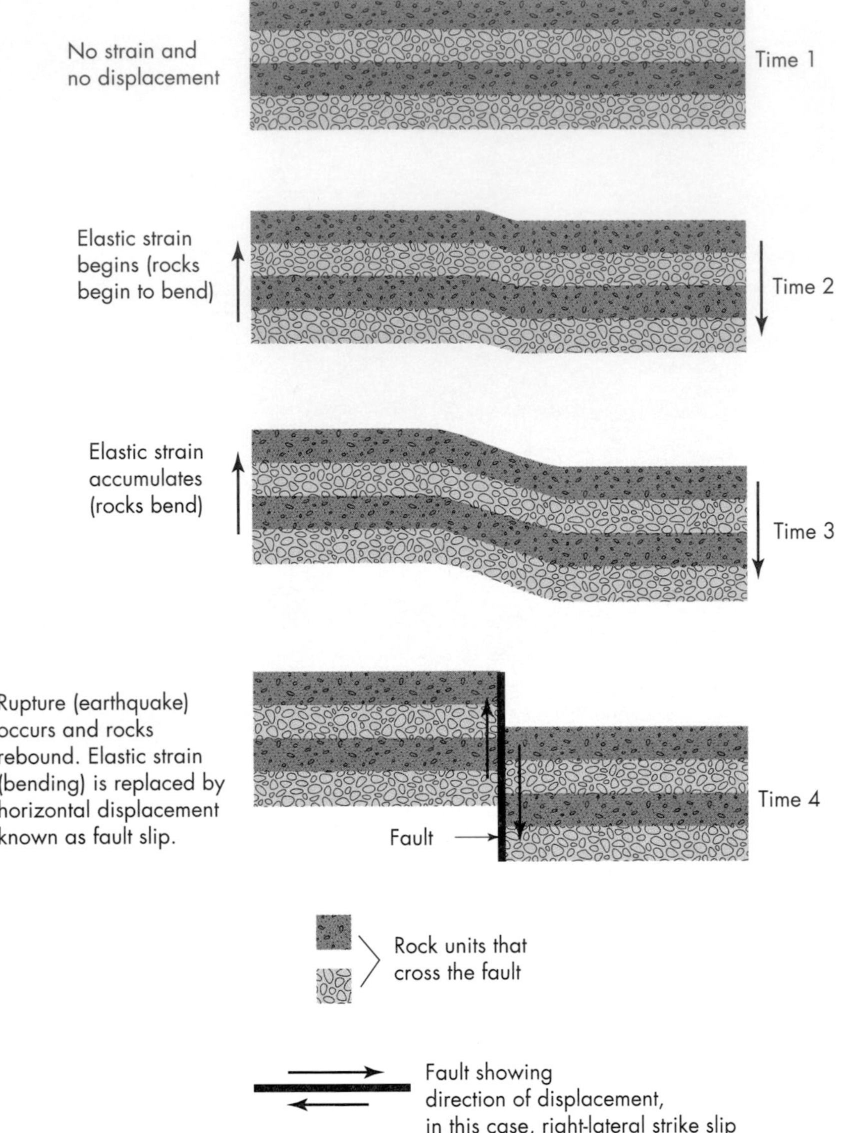

▶ **FIGURE 7.18** Idealized diagram (plan, bird's-eye view) illustrating the earthquake cycle. Time 1 to Time 4 may take hundreds to thousands of years.

Evidently, fracture zones (faults) may be activated by the increased load of water on the land and by increased water pressure in the rocks below the reservoir, resulting in faulting.

Deep Waste Disposal From April 1962 to November 1965, several hundred earthquakes occurred in the Denver, Colorado, area. The largest earthquake was magnitude 4.3 and caused sufficient shaking to knock bottles off shelves in stores. The source of the earthquakes was eventually traced to the Rocky Mountain Arsenal, which was manufacturing materials for chemical warfare. Liquid waste from the manufacturing process was being pumped down a deep disposal well to a depth of about 3600 m. The rock receiving the waste was highly fractured metamorphic rock, and injection of the new liquid increased the fluid pressure, facilitating slippage along fractures. Study of the earthquakc activity revealed a high correlation between the rate of waste injection and the occurrence of earthquakes. When the injection of waste stopped, so did the earthquakes (Figure 7.20) (18). Triggering of earthquakes in Denver by fluid injection of waste was an important occurrence because it directed attention to the fact that earthquakes and fluid pressure are related.

Nuclear Explosions Numerous earthquakes with magnitudes as large as 5.0 to 6.3 have been triggered by underground nuclear explosions at the Nevada nuclear test site (17). Analysis of the aftershocks suggests that the explosions caused some release of natural tectonic strain. This led scientists to speculate as to whether nuclear explosions might be used to prevent large earthquakes by releasing strain before it reached a critical point.

7.3 Effects of Earthquakes

Shaking is not the only cause of death and damage in earthquakes: Catastrophic earthquakes have a wide variety of destructive effects. Effects caused directly by fault

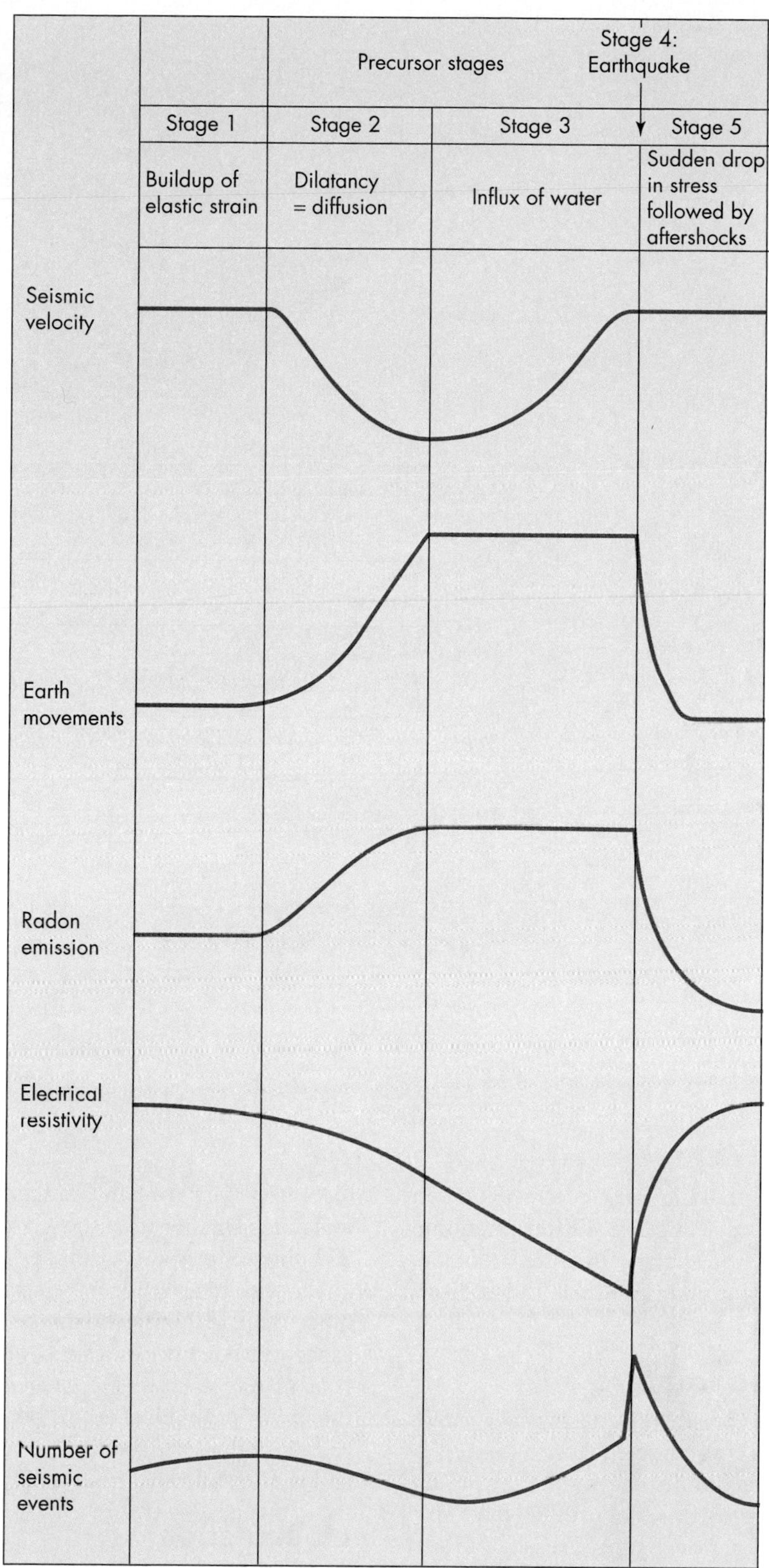

▲ **FIGURE 7.19** Dilatancy-diffusion model to explain the mechanism responsible for earthquakes. The curves show the expected precursory signals. (*Modified from "Earthquake Prediction," Frank Press. Copyright © May 1975 by Scientific American, Inc. All rights reserved.*)

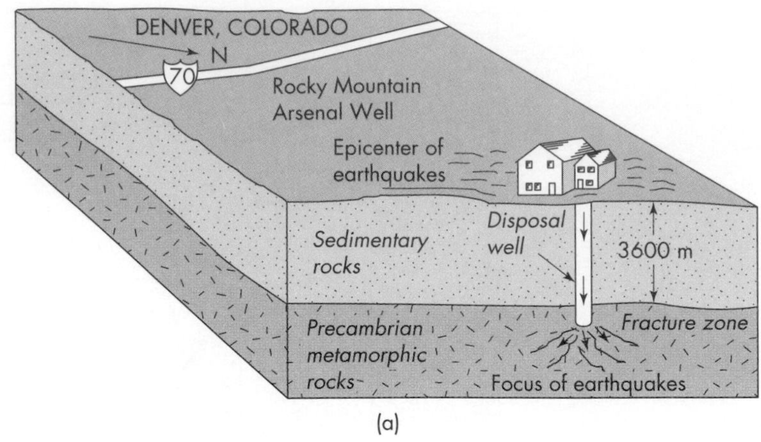

(a)

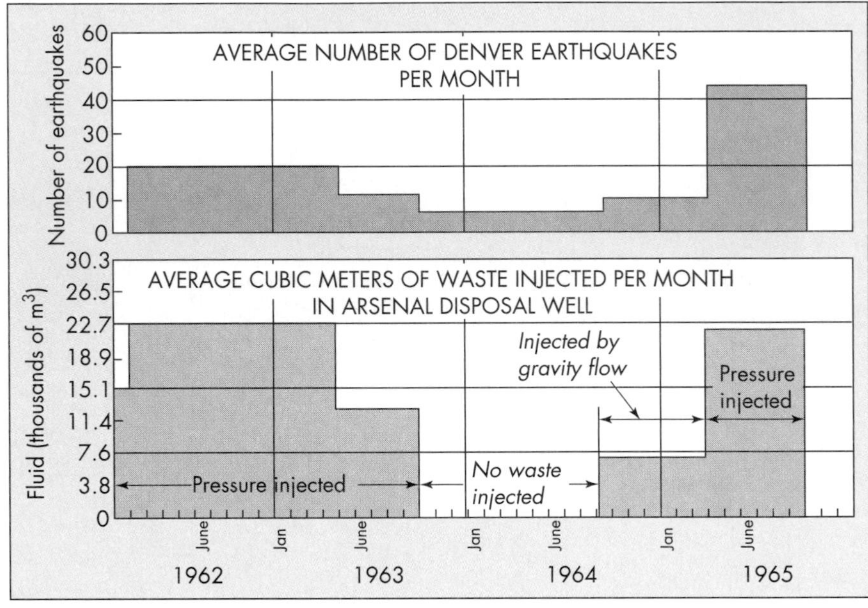

(b)

▶ **FIGURE 7.20** Generalized block diagram showing the Rocky Mountain Arsenal well (a) and graph depicting the relationship between earthquake frequency and rate of injection of liquid waste for five characteristic time periods (b). (*Graph [b] after* D. M. Evans, Geotimes 10, 1966. *Reprinted by permission.*)

movement include ground shaking (and its effects on people and structures) and surface rupture. Other phenomena induced by the faulting and shaking include landslides, fire, liquefaction of the ground, tsunamis, and regional changes in land elevation.

Shaking and Ground Rupture

The immediate effects of a catastrophic earthquake can include violent ground-shaking motion accompanied by widespread surface rupture and displacements. For example, the 1906 San Francisco earthquake produced 6.5 m of horizontal displacement along the San Andreas fault north of San Francisco and a maximum Modified Mercalli Intensity of XI (8). At this intensity, surface accelerations can snap and uproot large trees and knock people to the ground. The shaking may shear or collapse large buildings, bridges, dams, tunnels, pipelines, and other rigid structures (19). The great 1964 Alaskan earthquake ($M_L = 8.25$) caused extensive damage to transportation systems, railroads, airports, and buildings. The 1989 Loma Prieta earthquake (centered south of San Francisco) with a magnitude of 7.1 was much smaller than the Alaska event, yet it caused about $5 billion of property damage. The 1994 Northridge earthquake with magnitude 6.7 caused 61 deaths while inflicting more than $30 billion damage, making it one of the most expensive hazardous events ever in the United States. The Northridge earthquake caused so much damage because there was so much there to be damaged. The Los Angeles region is highly urbanized with high population density, and the seismic shaking was intense.

Liquefaction

Liquefaction is the transformation of water-saturated granular material from a solid to a liquid state. During earthquakes, this may result from an increase in pore-water pressure caused by compaction during intense shaking. Liquefaction of near-surface water-saturated silts and sand causes the materials to lose their shear strength and flow. As a result, buildings may tilt or sink into the liquefied sediments, while tanks or pipelines buried in the ground may rise buoyantly (20).

Landslides

Earthquake shaking often triggers many landslides (a comprehensive term for several types of hillslope failure; see Chapter 6) in hilly and mountainous areas. These can be extremely destructive and cause great loss of life, as demonstrated by the 1970 Peru earthquake. In that event more than 70,000 people died, and of those, 20,000 were killed by a giant landslide that buried towns. Both the 1964 Alaskan earthquake and the 1989 Loma Prieta earthquake caused extensive landslide damage to buildings, roads, and other structures.

Fires

Fire is a major hazard associated with earthquakes. Shaking of the ground and surface displacements can break electrical power and gas lines, thus starting fires. The threat from fire is doubled because fire-fighting equipment may be damaged and essential water mains broken. In individual homes and other buildings, appliances such as gas heaters may be knocked over, causing gas leaks that are ignited. Earthquakes in both Japan and the United States have been accompanied by devastating fires (Figure 7.21). The 1906 San Francisco earthquake has been repeatedly referred to as the "San Francisco fire," and in fact 80 percent of the damage from that event was caused by firestorms that ravaged the city for several days. The 1989 Loma Prieta earthquake also caused large fires in San Francisco, in the city's Marina district. Finally, the earthquake that struck Japan in 1923 killed 143,000 people, 40 percent of whom died in a firestorm that engulfed an open space where they had gathered in an attempt to reach safety (19).

Tsunamis

Tsunamis, or seismic seawaves, can be extremely destructive and present a serious natural hazard. Most of the lives lost in the 1964 Alaskan earthquake were attributed to tsunamis. Fortunately, damaging tsunamis occur infrequently and are usually confined to the Pacific Basin. The frequency of these events in the United States is about one every 8 years (19). Tsunamis originate when ocean water is vertically displaced during large earthquakes, during submarine mass movements, or during submarine volcanic eruption. In open water the waves may travel at speeds as great as 800 km/hr, and the distance between successive crests may exceed 100 km. Wave heights in deep water may be less than 1 m, but when the waves enter shallow coastal waters, they slow to less than 60 km/hr, and their heights may increase to more than 20 m.

▲ **FIGURE 7.21** Fires associated with the 1995 Kobe, Japan, earthquake caused extensive damage to the city. (*Corbis*)

Figure 7.22 shows extensive damage to a small town on the island of Okushiri, Japan, from the July 12, 1993, tsunami produced by an M_s 7.8 earthquake in the Sea of Japan. Vertical run-up (elevation above sea level to which water from the waves reached) varied from 15 m to 30 m (21). There was virtually no warning because the epicenter of the earthquake was very close to the island and the big waves arrived only 2 to 5 minutes after the earthquake. The tsunami killed 120 people and caused $600 million in property damage.

A magnitude M_w 7.1 earthquake on July 17, 1998, caused a large tsunami with wave heights 10 to 15 m. A series of three waves arriving 10 to 20 minutes after the earthquake devastated Sissano Lagoon on the north coast of Papua New Guinea, killing more than 2100 people. The earthquake epicenter was about 50 km offshore. The wave height and run-up on shore was surprisingly high for a subduction zone earthquake of M_w 7.1. The tsunami probably resulted from a combined effect of the earthquake and submarine landslide. The Papua New Guinea event emphasizes the potential devastating damage that can result from unusually large waves produced by a locally generated earthquake (22,23).

Tsunamis can cause catastrophic damage thousands of kilometers from where they are generated. In 1960 an earthquake originating in Chile triggered a tsunami that reached Hawaii 15 hours later, killing 61 people. However, long travel times now allow many tsunamis to be detected in time to warn the coastal communities in their path. Following an earthquake that produces a tsunami, the arrival time of the waves can often be estimated to within 1.5 minutes per hour of travel time. This information has been used to produce tsunami warning systems such as that shown for Hawaii in Figure 7.23.

Consideration is now being given to produce other tsunami warning systems, such as one to warn residents in northwestern California of tsunamis generated by Alaskan or Cascadia subduction-zone earthquakes. Travel time for a tsunami from the Aleutian Islands in Alaska to northern California is about 4 hours, and movement of the tsunami southward can be monitored from changes in water levels at coastal tide gauges. The plan involves placing such gauges on the bottom of the Pacific Ocean, four off Alaska, and three off the northwest coast of California.

The hazard from a tsunami at a particular site on the coast depends in part on local coastal and seafloor topography that may increase or decrease wave height (8). Damage caused by tsunamis is most severe at the water's edge, where boats, harbors and buildings, transportation systems, and

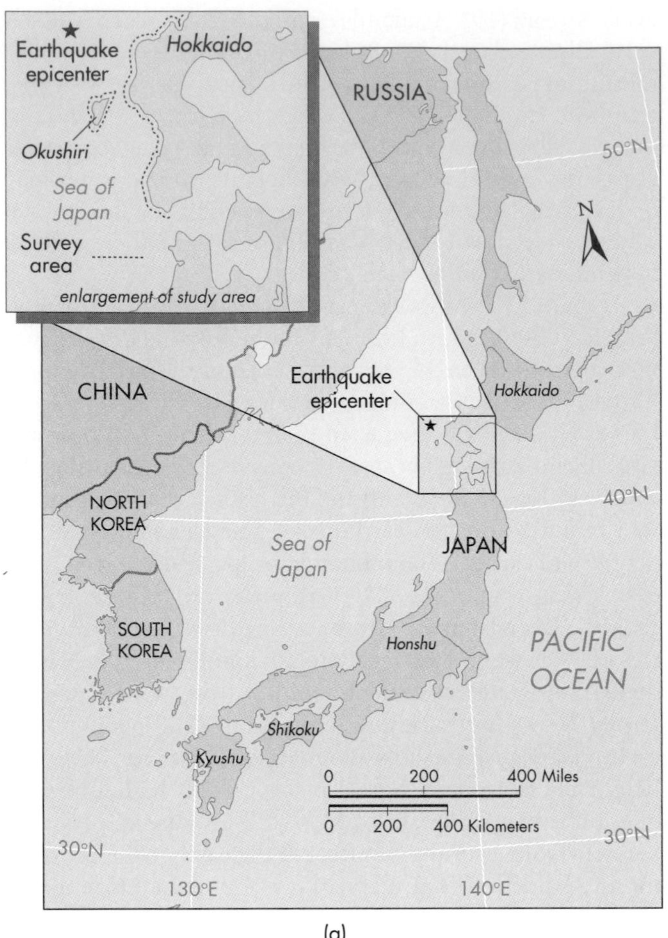

◀ **FIGURE 7.22** (a) Location of the M = 7.8 1993 Hokkaido-Nanse-Oki earthquake. (b) Damage to town on island of Okushira, Japan, as a result of tsunami that struck the island in July 1993. Note the extensive damage to the shoreline area and the fires that are still burning. ([a] *From Hokkaido Tsunami Survey Group 1993. Tsunami devastates Japanese coastal region.* EOS, Transactions of the American Geophysical Union 74(37): 417, 432. [*b*] *Sankei Shimbun/Sygma.*)

utilities may be destroyed. The waves may also cause damage to aquatic and supratidal life in both near- and on-shore environments (19).

Waves caused by landslides may also have considerable effect and cause extensive damage. In 1958 an earthquake induced a landslide into Lituya Bay, Alaska, causing a truly giant wave that produced run-up on land to an elevation of more than 500 m above sea level (24).

Regional Changes in Land Elevation

Vertical deformation, including both uplift and subsidence, is another effect of some large earthquakes. Such deformation can cause regional changes in groundwater levels. The great (magnitude 8.25) 1964 Alaskan earthquake, with a Modified Mercalli Intensity of X–XI (8) caused vertical deformation over an area of more than 250,000 km^2 (25). The deformation included two major zones of warping, each about 500 km long and more than 210 km wide. The uplift, which was as much as 10 m, and the subsidence, which was as much as 2.4 m, triggered effects ranging from severely disturbing coastal marine life to changes in groundwater levels. As a result of subsidence, flooding occurred in some communities, whereas in areas of uplift, canneries and fishermen's homes were displaced above the high-tide line, rendering docks and other facilities inoperable. In 1992 a major earthquake (magnitude 7.1) near Cape Mendocino in northwestern California produced approximately 1 m of uplift at the shoreline, resulting in the deaths of communities of marine organisms exposed by the uplift (26).

7.4 Earthquake Risk and Earthquake Prediction

The great damage and loss of life associated with earthquakes is due in part to the fact that they strike without warning. A great deal of research is being devoted to the problem of anticipating earthquakes. The best we can do at present is to use probabilistic methods to determine the *risk* associated with a particular area or with a particular segment of a fault. Such determinations of risk are a form of **long-term prediction:** We can say that an earthquake of a given magnitude or intensity has a high probability of occurring in a given area or fault segment within a specified number of years. Such prediction may help planners who are considering seismic safety measures or people who are deciding where to live, but it does not help residents of a seismically active area anticipate and prepare for a specific earthquake. **Short-term prediction** specifying the time and place of the earthquake would be much more useful. Such prediction, which depends to a large extent on observation of *precursory phenomena* (changes preceding the event), is still in its infancy but is an area of active investigation.

Estimation of Seismic Risk

The earthquake risk associated with a particular area is shown on **seismic hazard maps** that have been prepared by scientists for the United States. Some of these maps show *relative hazard*, that is, where earthquakes of specified magnitude have occurred. However, a preferable method of as-

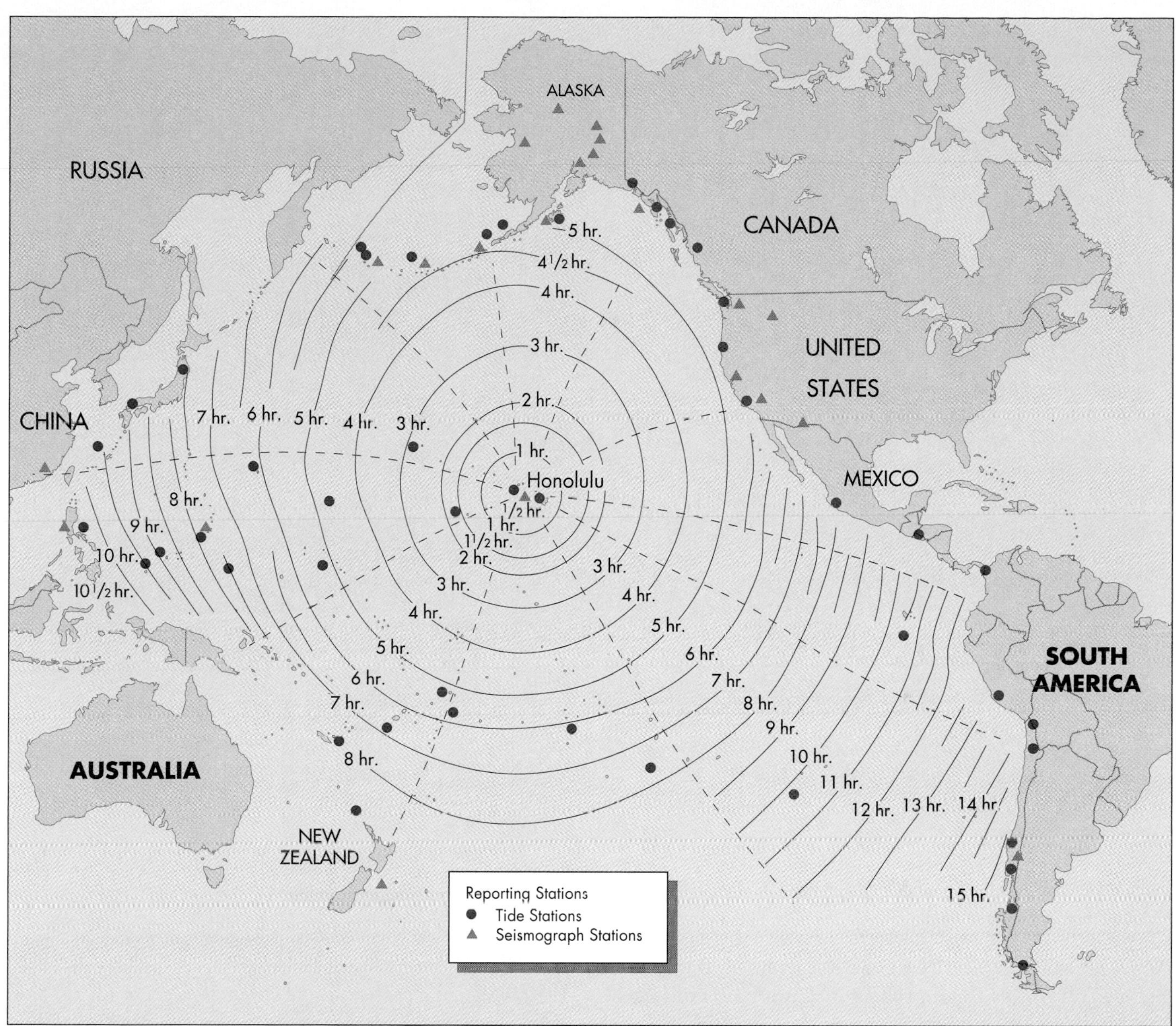

▲ **FIGURE 7.23** Tsunami warning system. Map shows reporting stations and tsunami travel times to Honolulu, Hawaii. (*From* NOAA)

sessing seismic risk is to calculate the *probability* of a particular event or amount of shaking occurring. Figure 7.24, an earthquake-hazard map based upon probability of horizontal ground motion, showing earthquake ground accelerations with a 10 percent probability of being exceeded in a 50-year period. The map is based upon historical seismicity, frequency of earthquakes of various magnitudes, and slip rates on faults. Estimated ground accelerations assume firm rock conditions. Actual hazard of a specific site may vary as a result of material amplification or directivity of seismic shaking. One way of interpreting this map is that its reddest areas represent the regions of greatest seismic hazard, because it is those areas that are likely to experience the greatest seismic shaking. Although regional earthquake-hazard maps are valuable, considerably more data are necessary to evaluate hazardous areas more precisely with the objective of assisting in the development of building codes and determining insurance rates (see *Putting Some Numbers On Earthquake Hazard Evaluation* on page 196).

Conditional Probabilities for Future Earthquakes

Conditional probabilities are estimates of the probability of an earthquake of a given magnitude occurring along a given fault segment within a specified time period. Such estimates are based on a synthesis of historical records and geologic evaluation of prehistoric earthquakes (28) (see *A Closer Look: Twelve Centuries of Earthquakes on the San Andreas Fault in Southern California*). In the late 1980s, geologists in California used these methods to estimate conditional probabilities for major earthquakes along segments of the San

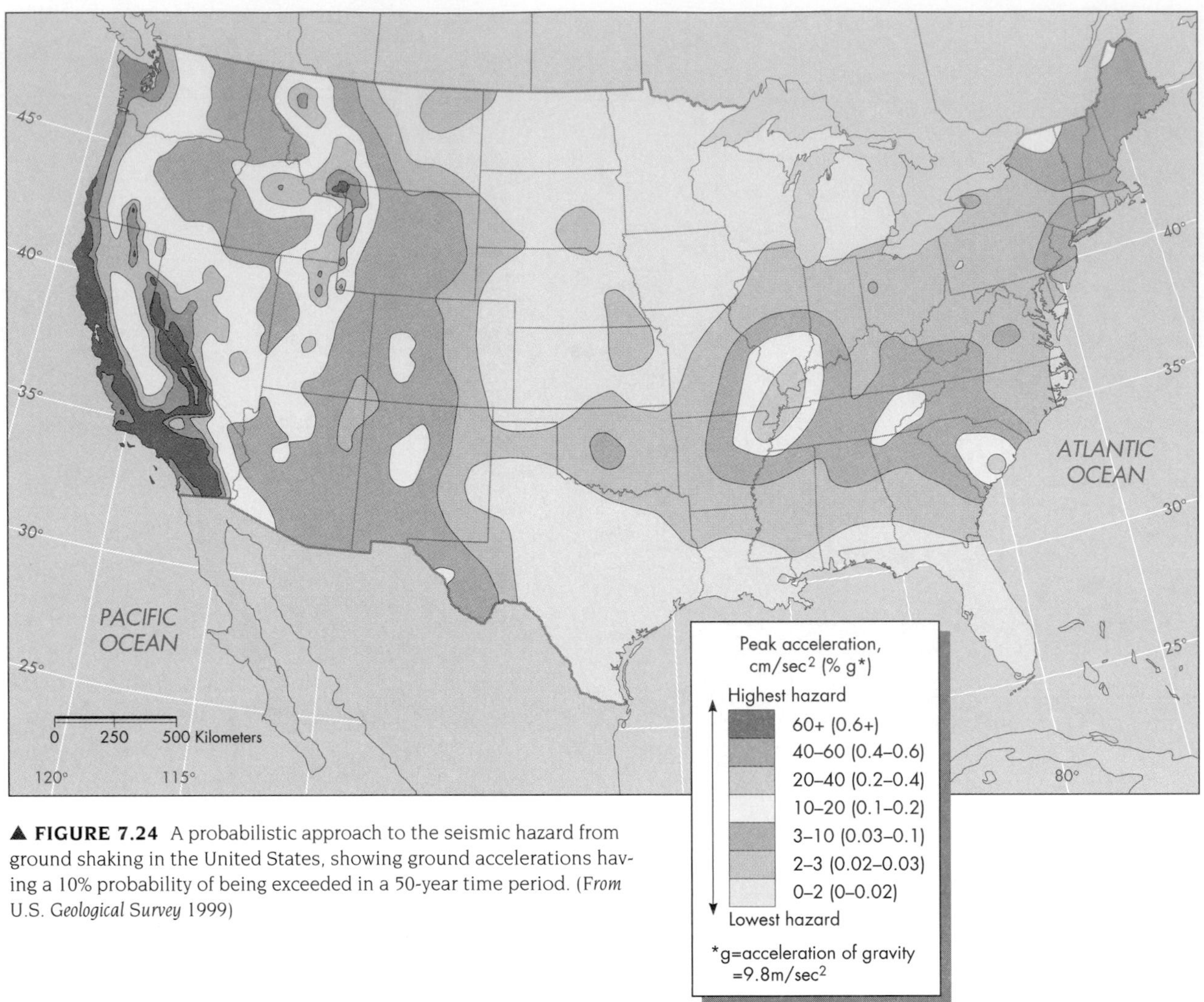

▲ **FIGURE 7.24** A probabilistic approach to the seismic hazard from ground shaking in the United States, showing ground accelerations having a 10% probability of being exceeded in a 50-year time period. (*From* U.S. *Geological Survey* 1999)

Andreas fault for the 30-year period from 1988 to 2018. Figure 7.25 shows these probabilities, with 1.0 equal to a probability of 100 percent.

The conditional probability study of the San Andreas fault suggests that the Parkfield segment, in central California, is almost certain to rupture by the year 2018 (see Figure 7.25). This estimate is based in part on the observation that moderate to large historic earthquakes have occurred on this segment in the years 1857, 1881, 1902, 1922, 1934, and 1966, or on the average every 21 to 22 years. In 1985, before completion of the study, geologists assigned a high probability to the occurrence of a moderate Parkfield event by 1993. This date has now obviously passed, and the expected earthquake did not occur. Nevertheless, because of the many historic earthquakes and high probability for an event in the relatively near future, the Parkfield site remains a natural laboratory in which to study earthquake prediction. Many instruments to study earthquakes have been deployed near Parkfield in anticipation of the event. The southern segment of the fault, in the Coachella Valley, has been assigned a probability of approximately 40 percent to produce a major earthquake by the year 2018.

The conditional probability study, completed in early 1989, assigned a probability of about 30 percent for a major event on the San Andreas fault in the Santa Cruz Mountains. The magnitude 7.1 Loma Prieta earthquake, which occurred in this region on October 17, 1989, has been cited as strong support of the validity of the conditional probability approach. On the other hand, an unforeseen magnitude 7.5 earthquake occurred east of the San Bernardino Mountains in 1992. That event caused ground rupture over nearly 100 km, with maximum right-lateral displacement of about 6 m (30). Because the displacement is almost entirely pure strike slip, the fault that ruptured is probably part of the plate-bounding right-lateral system that includes the San Andreas fault.

Although in retrospect there is evidence for past moderate to large earthquakes on the fault that ruptured in 1992, it was not expected that the next right-lateral earthquake in southern California would be so far to the east of the known San Andreas fault system. There is speculation now that the San Andreas fault, and thus the boundary between the North American and Pacific plates, is either wider than we thought before or in the process of migrating east. Of course, this does not mean that large earthquakes will not

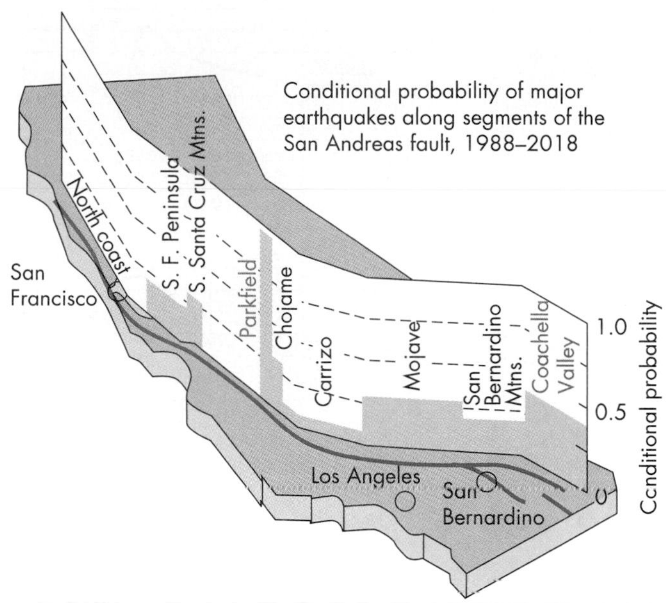

▲ **FIGURE 7.25** Conditional probability of a major earthquake along segments of the San Andreas fault (1988–2018). (*After* T. H. *Heaton et al.* 1989. U.S. *Geological Survey Circular* 1031.)

continue to occur on the main San Andreas fault. The conditional probabilities shown in Figure 7.25 remain our best estimate of the probability of large earthquakes occurring in the next few decades on the San Andreas fault.

Short-Term Prediction

Short-term prediction, or **forecast**, of earthquakes is an area of serious research. Like a weather forecast, an earthquake forecast specifies a relatively short time period in which the event is likely to occur and assigns it a probability of occurring. The Japanese made the first attempts at earthquake prediction with some success, using the frequency of micro-earthquakes, repetitive geodetic level surveys, water-tube tiltmeters, and geomagnetic observations. They found that earthquakes in the areas they studied were nearly always accompanied by swarms of micro-earthquakes that occurred several months before the major shocks. Furthermore, ground tilt correlated strongly with earthquake activity (17).

Chinese scientists made the first successful prediction of a major earthquake in 1975. The Haicheng earthquake of February 4, 1975, had a magnitude of 7.3 and destroyed or damaged about 90 percent of structures in a city of 9000 people. The short-term prediction was based primarily on a series of foreshocks that began 4 days prior to the main event. On February 1 and February 2, there were several shocks with a magnitude less than 1. On February 3, less than 24 hours before the main shock, a foreshock of magnitude 2.4 occurred, and in the next 17 hours, eight shocks with magnitude greater than 3 occurred. Then, as suddenly as it began, the foreshock activity became relatively quiet for 6 hours until the main earthquake occurred (31,32). The lives of thousands of people were saved by the massive evacuation from potentially unsafe housing just before the quake.

Unfortunately, foreshocks do not always precede large earthquakes. In 1976 a catastrophic earthquake, one of the deadliest in recorded history, struck near the mining town of Tanshan, China, killing several hundred thousand people. There were no foreshocks! Earthquake prediction is still a complex problem, and it will probably be many years before dependable short-range prediction is possible. Such predictions most likely will be based upon such factors (precursory phenomena) as:

- Deformation of the ground surface
- Seismic gaps along faults
- Patterns and frequency of earthquakes
- Perhaps, anomalous behavior of animals

Preseismic Uplift and Subsidence Rates of uplift and subsidence, especially when they are rapid or anomalous, may be significant in predicting earthquakes. For example, for more than 10 years before the 1964 earthquake near Niigata, Japan (magnitude 7.5), there was anomalous broad uplift of the earth's crust of several centimeters (Figure 7.26) (15). Similarly, broad and slow uplift of several centimeters occurred over a 5-year period prior to the 1983 (M 7.7) Sea of Japan earthquake. The mechanism responsible for the uplift is thought to be deep, stable fault slip of about 1 m prior to the earthquake (33).

Less well understood but possibly important observations include pre-instrument uplifts of 1 to 2 m that preceded large Japanese earthquakes in 1793, 1802, 1872, and 1927. The uplift was recognized by sudden withdrawals of the sea from the land of as much as several hundred meters in harbors. For example, on the morning of the 1802 earthquake, the sea suddenly withdrew about 300 m from a harbor in response to a preseismic uplift of about 1 m. Four hours later the earthquake struck, destroying many houses and uplifting the land another meter, which caused the sea to withdraw to a greater distance (33).

Seismic Gaps **Seismic gaps** are defined as areas along active fault zones that are capable of producing large earthquakes but have not produced one recently. These areas are thought to store tectonic strain and thus are candidates for future large earthquakes (34). Seismic gaps have been useful in medium-range earthquake prediction. At least 10 large plate-boundary earthquakes have been successfully forecast from seismic gaps since 1965, including one in Alaska, three in Mexico, one in South America, and three in Japan. In the United States, seismic gaps along the San Andreas fault include one near Fort Tejon, California, that last ruptured in 1857 and one along the Coachella Valley, a segment that has not produced a great earthquake for several hundred years. Both gaps are likely candidates to produce a great earthquake in the next few decades (14,34).

As earth scientists examine patterns of seismicity, two ideas are emerging. First, there are sometimes reductions in small or moderate earthquakes prior to a larger event.

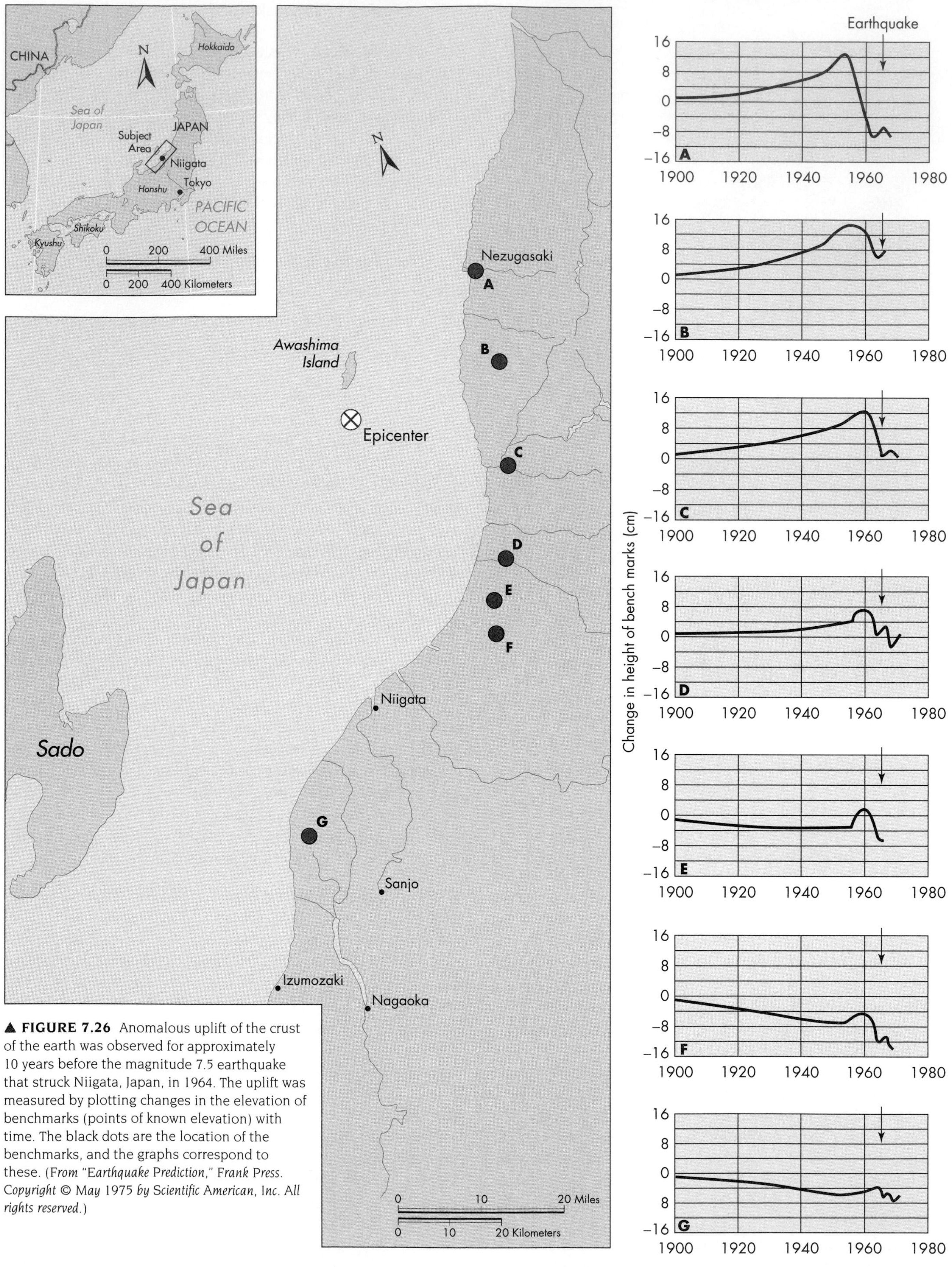

▲ **FIGURE 7.26** Anomalous uplift of the crust of the earth was observed for approximately 10 years before the magnitude 7.5 earthquake that struck Niigata, Japan, in 1964. The uplift was measured by plotting changes in the elevation of benchmarks (points of known elevation) with time. The black dots are the location of the benchmarks, and the graphs correspond to these. (*From "Earthquake Prediction," Frank Press. Copyright © May 1975 by Scientific American, Inc. All rights reserved.*)

For example, prior to the 1978 (M 7.8) Oaxaca, Mexico, earthquake, there was a 10-year period (1963 to 1973) of relatively high seismicity of earthquakes (mostly M 3 to 6.5) followed by a quiescence period of 5 years. Renewed activity beginning 10 months before the M 7.8 event was the basis of a successful prediction (33). Second, small earthquakes may tend to ring an area where a larger event might eventually occur. Such a ring (or donut) was noticed in the 16 months prior to the 1983 M 6.2 event in Coalinga, California. Unfortunately, hindsight is clearer than foresight, and this ring was not noticed or identified until after the event.

Anomalous Animal Behavior Anomalous animal behavior often has been reported prior to large earthquakes. Reports have included unusual barking of dogs, chickens that refuse to lay eggs, horses or cattle that run in circles, rats perched on power lines, and snakes crawling out in the winter and freezing. Anomalous behavior was common before the Haicheng earthquake (31). Although both the significance and the reliability of animal behavior are difficult to evaluate, there has been considerable interest in the topic, and research is ongoing.

Toward Earthquake Prediction

We are still a long way from a working, practical methodology to predict earthquakes reliably. On the other hand, a good deal of information is currently being gathered concerning possible precursor phenomena associated with quakes. Optimistic scientists around the world today believe that eventually we will be able to make consistent long-range forecasts (a few decades to a few thousand years), medium-range predictions (a few years to a few months), and short-range predictions (a few days or hours) for the general locations and magnitudes of large, damaging earthquakes.

Although progress on short-range earthquake prediction has not matched expectations, medium- to long-range forecasting, including hazard evaluation and probabilistic analysis of areas along active faults, has progressed faster than expected (35). The Borah Peak earthquake of October 28, 1983, in central Idaho, has been lauded as a success story for medium-range earthquake-hazard evaluation. Previous evaluation of the Lost River fault suggested that the fault was active (36). The earthquake, which was about magnitude 7, killed two people and caused about $15 million in damage, producing fault scarps up to several meters high and numerous ground fractures along the 36-km rupture zone of the fault. The important fact was that the scarp and faults produced during the earthquake were superimposed on previously existing fault scarps, validating the usefulness of careful mapping of scarps produced from prehistoric earthquakes. The principle is that where the ground has broken before, it may break again!

It is interesting to speculate about a future methodology for predicting earthquakes. Let us assume that we can develop a set of relationships from which an earthquake magnitude and date (or time) of occurrence are predicted. The first step is to identify empirical relationships among precursor events (such as foreshock activity, anomalous tilt, or uplift/subsidence) and probable earthquake magnitude. The precursor events may be measured and evaluated for a specific area or by their intensity, and we would expect to find a relationship with earthquake magnitude. In other words, after observing many earthquakes, we would empirically derive Figure 7.27a. Then, given a set of precursor observations (point A of Figure 7.27a), we could predict the magnitude of a possible earthquake (point B of the same graph).

Let us also assume that a relationship exists between the precursor time interval (time between the start of precursor events and the occurrence of the quake) and the magnitude of an expected earthquake. This relationship is also derived empirically from direct observation of known earthquake activity. Now take the predicted earthquake magnitude from Figure 7.27a, point B, and use this in Figure 7.27b to predict the precursor time interval (point C). Thus, we now know the magnitude, and, if we know when the precursor events started, we can calculate when the expected earthquake should occur. It sounds simple, but it is pure fiction: None of these relationships has been derived with sufficient precision to use

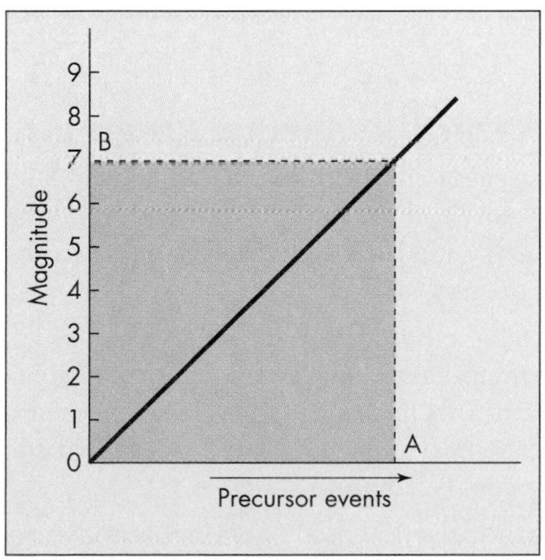

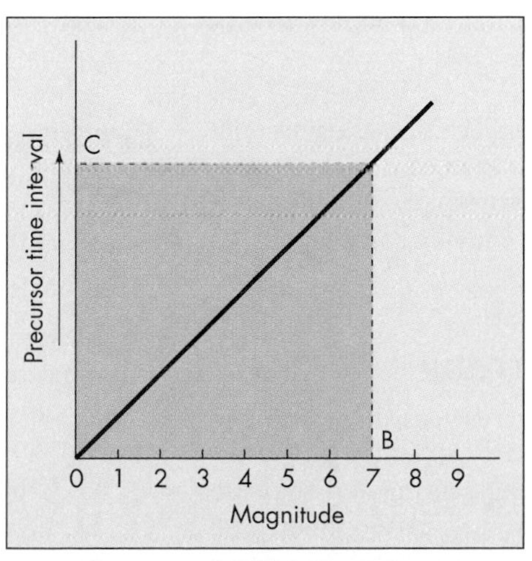

FIGURE 7.27 Hypothetical method of how earthquake magnitude and time of event might be predicted. Graph (a) is used to predict the magnitude based on given precursor events; graph (b) is used to predict the precursor time interval from the earlier predicted magnitude. (See text for full explanation.)

A CLOSER LOOK Twelve Centuries of Earthquakes on the San Andreas Fault in Southern California

Undoubtedly one of the most remarkable sites in the world for paleoseismic studies is Pallett Creek, located approximately 55 km northeast of Los Angeles on the Mojave segment of the San Andreas fault. The site contains evidence for 10 large earthquakes, two of which occurred in historical time (1812 and 1857), and prehistoric events extending back to approximately A.D. 671 (29). High-precision radiocarbon dating methods provide accurate dating of most of the eight prehis-

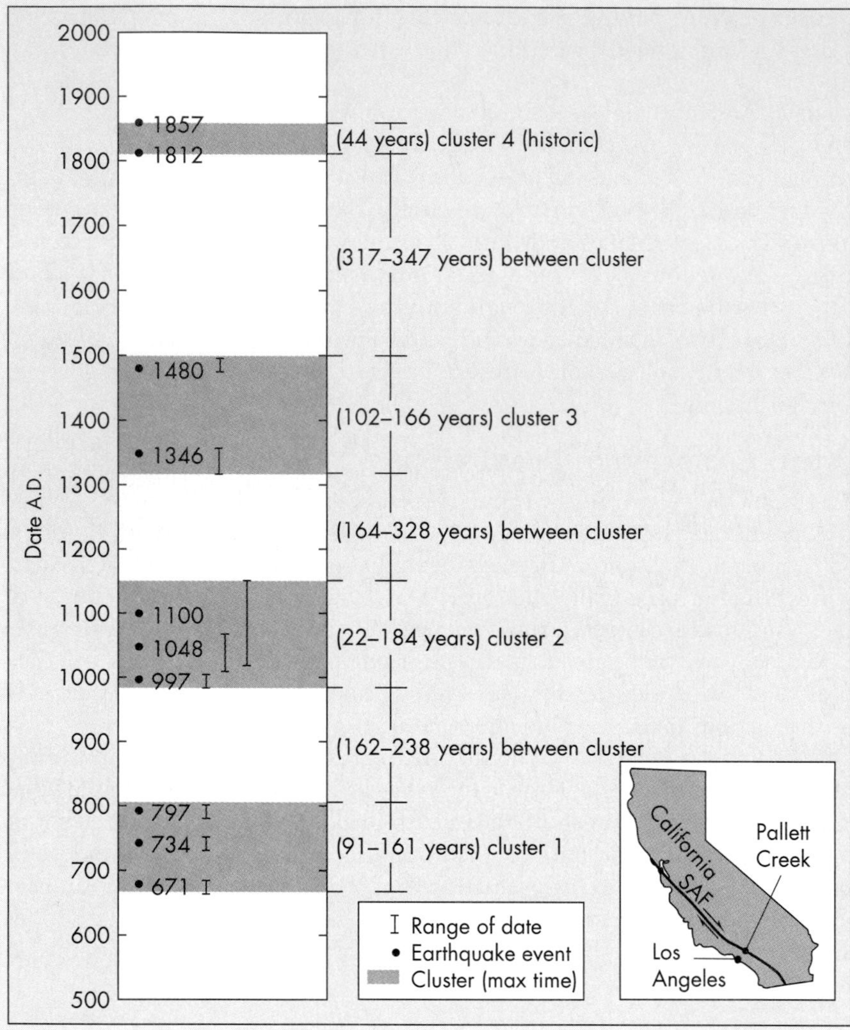

▶ **FIGURE 7.H** Two historical and eight prehistoric earthquakes on the San Andreas fault (Pallett Creek). (*Data from* K. *Sieh,* M. *Stuiver, and* D. *Brillinger.* 1989. *A more precise chronology of earthquakes produced by the San Andreas fault in southern California.* Journal of Geophysical Research 94(B1):603–623.)

in predicting real earthquakes. Furthermore, the potential for errors of measurement with this kind of work is considerable. Nevertheless, it is valuable to explore one model of how earthquake predictions may eventually come about.

7.5 The Response to Earthquake Hazards

Possible responses to the seismic hazards in earthquake-prone areas include development of hazard-reduction programs, careful siting of critical facilities, engineering and land-use adjustments to earthquake activity, and development of a warning system. The extent to which these responses occur depends in part on people's perception of the hazard.

Earthquake Hazard-Reduction Programs

In the United States, geologists at the U.S. Geological Survey, as well as university and other scientists are developing a National Earthquake Hazard-Reduction Program. The major goals of the program (37) are to:

- Develop an understanding of the earthquake source. This involves an understanding of the physical properties and mechanical behavior of faults as well as development of quantitative models of the physics of the earthquake process (see Figure 7.19).
- Determine earthquake potential. This involves characterizing seismically active regions, including determination of the rates of crustal deformation, identification

toric events. When the dates and their accompanying error bars are plotted, it is apparent that the earthquakes are not evenly distributed through time, but tend to cluster (Figure 7.H). Most of the earthquakes within clusters are separated by a few decades, but the time between clusters varies from approximately 160 to 360 years. The earthquakes were identified through careful evaluation of natural exposures and fault trenches at Pallett Creek (29). Figure 7.I shows part of the trench wall and the offset of organic-rich sedimentary layers that correspond to an earthquake that occurred approximately A.D. 1100.

Naturally, the question we wish to ask is when will the next big earthquake occur on the Mojave segment of the San Andreas fault? Studies using conditional probability suggest that probability is approximately 30 percent for the 30-year period from 1988 to the year 2018 (see Figure 7.25). Assuming that the two historic events are a cluster and if clusters can be separated by approximately 160 to 360 years, then the next large earthquake might be expected between the years 2017 and 2207. On the other hand, the data set at Pallett Creek, although the most complete record for any fault in the world, is still a study of small numbers.

▲ **FIGURE 7.I** Part of the trench wall crossing the San Andreas fault at Pallett Creek showing a 30-cm offset. Notice that the strata above the 22-cm scale (central part of photograph) are not offset. The offset sedimentary layers (white beds) and the dark organic layers were produced by an earthquake that occurred about 800 years ago. (*Courtesy of Kerry Sieh.*)

of active faults, determining characteristics of paleoseismicity, calculating long-term probabilistic forecasts, and, finally, developing methods of intermediate- and short-term prediction of earthquakes.

- Predict effects of earthquakes. This includes gathering the data necessary for predicting ground rupture and shaking and for predicting the response of structures that we build in earthquake-prone areas, as well as evaluating the losses associated with the earthquake hazard.
- Apply research results. At this level the program is interested in the transfer of knowledge about earthquake hazards to people, communities, states, and the nation. This knowledge concerns what can be done to better plan for earthquakes and reduce potential losses of life and property.

Earthquakes and Critical Facilities

Critical facilities are those that, if damaged or destroyed, might cause significant-to-catastrophic loss of life, property damage, or disruption of society. In the urban environment, examples include schools, medical facilities, and police and fire stations. Other examples include dams, power plants, and other necessary facilities. There are three aspects of the decision process concerning critical facilities and earthquake hazard (38):

- Evaluation of the hazard
- Evaluation of whether the facility may be designed or modified to accommodate the hazard
- Subjective evaluation of an "acceptable risk"

The first two aspects, which concern making critical facilities safer, have a strong scientific component. However, as no facility can be rendered absolutely safe, we must also be concerned with acceptable risk assessment, which is an issue of public safety rather than of science.

In the case of nuclear power plants, requirements for hazard evaluation reflect a special concern for the siting of these facilities. The evaluation requires estimating fault capability and the maximum credible earthquake that might be expected along faults in the vicinity of the site. A **capable fault,** as defined by the U.S. Nuclear Regulatory Commission, is a fault that has exhibited movement at least once in the last 50,000 years or multiple movements in the last 500,000 years. This is a more conservative criterion and represents a greater safety margin than the 10,000-year criterion commonly used for other facilities. The **maximum credible earthquake** is a believable event based on the tectonic environment, historic earthquakes, and paleoseismicity. It may be used in evaluating the earthquake hazard of all critical facilities.

Estimation of the maximum credible earthquake can sometimes be accomplished by fieldwork to evaluate effects of historical seismicity and prehistoric activity. Length of surface rupture and displacement per event are both related to earthquake magnitude, and equations are available to estimate magnitude if expected surface rupture length or displacement per event is known or assumed (24). Average recurrence of earthquakes of a particular magnitude can also be estimated if a slip rate is known or assumed (Figure 7.28).

Unfortunately, the database concerning fault rupture length and displacement is limited. Therefore, evaluation of the earthquake hazard for many critical facilities will continue to be subjective at best. What geologists do is gather as much pertinent geologic data as possible to ensure that our decisions are based on adequate information. We are presently arguing over what is adequate. Certainly, for a specific fault, we would like to know the length of the fault, style of faulting, total displacement, age of the most recent movement, approximate recurrence interval, and magnitude of the maximum credible earthquake.

In determining the seismic risk to critical facilities, the main problem usually is estimating the activity of a particular fault system. For example, the seismic evaluation for the Auburn Dam site near Sacramento, California, was very controversial. Millions of dollars were spent and hundreds of meters of fault trenches excavated during geological studies, and there is still no agreement as to the hazard associated with the possibly active fault system in the vicinity. The general area was considered to have a relatively low seismic risk until 1975, when a magnitude 5.7 earthquake occurred near Oroville, about 80 km from the Auburn site. This earthquake rekindled concern and initiated a new round of seismic risk evaluation, ultimately leading to the virtual abandonment of the site. The evaluation determined that the faults on and near the site would produce only a very small displacement. However, this was enough to cause major concern because the planned dam could not accommodate even relatively small displacements. Sudden failure of the proposed dam would flood the city of Sacramento, the seat of California government.

The main problem remains: It is often very difficult to absolutely date prehistoric earthquakes. However, we are making progress in understanding earthquakes, identifying active faults, and estimating the maximum credible earthquake for a particular area.

Adjustments to Earthquake Activity

The mechanism of earthquakes is still poorly understood; therefore, such adjustments as warning systems and earthquake prevention are not yet reliable alternatives. There are, however, reliable protective measures we can take:

- **Structural protection,** including the construction of large buildings able to accommodate at least moderate shaking. This has been relatively successful in the United States. The 1988 earthquake in Armenia (M 6.8) was somewhat larger than the 1994 Northridge event in the Los Angeles Basin (M 6.7), but the loss of life and destruction in Armenia was staggering—at least 45,000 killed (compared with 61 in California) and near-total destruction in some towns near the epicenter. Most buildings in Armenia were constructed of unreinforced concrete and instantly crumbled into rubble, crushing or trapping their occupants. The scenario was tragically replayed in May 1995 when an M = 7.5 earthquake struck the town of Neftegorsk on Sakhalin Island, Russia. Two thousand (of a total population of 3000) people died, buried in rubble, when 17 five-story apartment buildings collapsed. The buildings were prefabricated, constructed with unreinforced concrete. This is not to say that the Northridge earthquake did not cause a catastrophe. It certainly did! That event left 25,000 people homeless, caused the collapse of several freeway overpasses, injured about 8000 people, and inflicted many billions of dollars in damages to structures and buildings (Figure 7.A; see *A Closer Look: The Northridge Earthquake and Buried Faults*). However, if most buildings in the Los Angeles Basin had been constructed with unreinforced concrete or adobe, thousands of people would have died.
- **Land-use planning,** including the siting of important structures such as schools, hospitals, and police stations in areas away from active faults or sensitive earth materials likely to accentuate seismic shaking. This will involve study of the response of the ground to seismic shaking on a block-by-block (earthquake zonation, or microzonation) basis because conditions can change quickly when considering response of the ground to

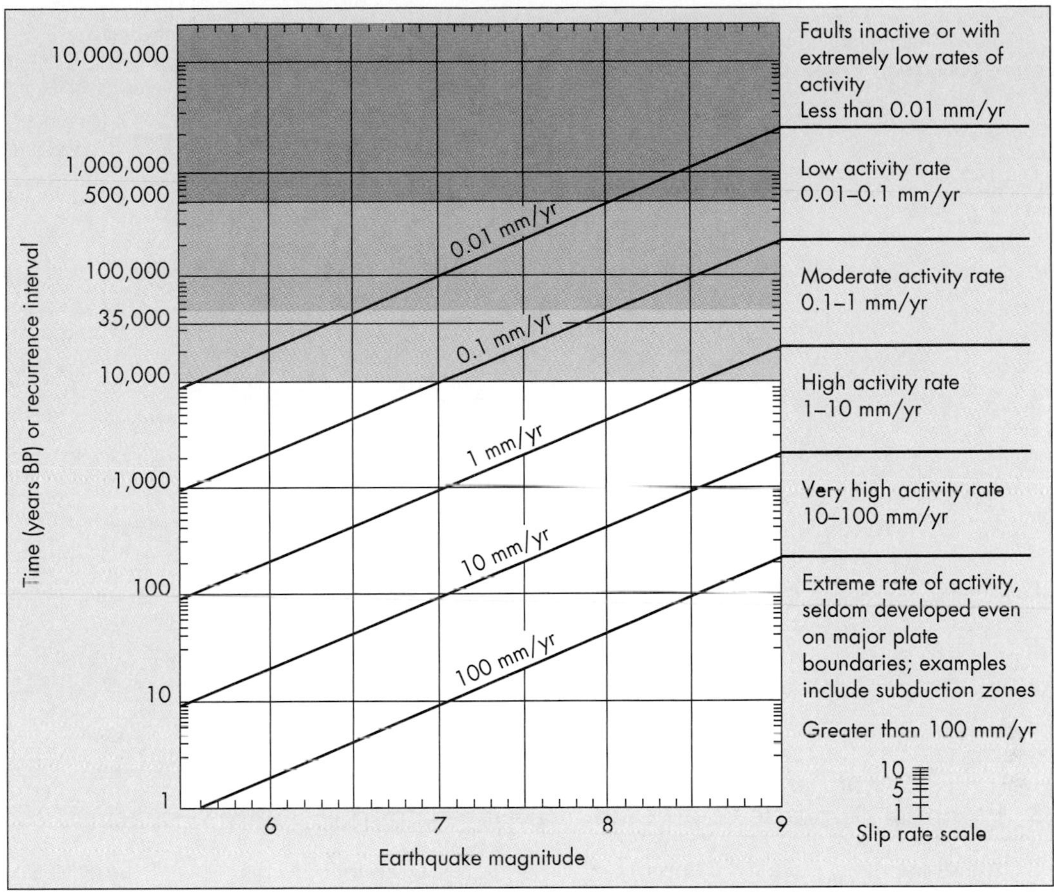

▲ **FIGURE 7.28** Relationships between recurrence interval, slip rate, and earthquake magnitude. (*Modified after B. Slemmons and C. M. Depolo. 1986. Evaluation of Active Faulting and Associated Hazards, in* Active Tectonics. *National Academy Press, pp. 45–63.*)

shaking. For urban areas where property values may be millions to billions of dollars per block, we will need to produce detailed maps of ground response to accomplish microzonation. These maps will assist engineers to design buildings and other structures to better withstand seismic shaking. To do this, microzonation will require a significant investment of time and money, with the first step being development of methods to predict adequately the ground motion from an earthquake at a specific site.

- **Increased insurance and relief measures** to help adjustments following earthquakes. Following the 1994 Northridge quake, total insurance claims were very large and some insurance companies terminated earthquake insurance.

A fourth possible measure is to take little or no action in advance and pay the consequences. This philosophy is not what we espouse, but it is in fact what we often do.

At the same time that we are adjusting to existing hazards, we are increasing the need for adjustment by creating new hazards. When we place any new building or structure on sensitive earth materials or on active faults, we are seeking new problems where there were none. We know from experience that building on unconsolidated deposits such as channel fill or bay fill, which may become unstable (subject to liquefaction) or shake easily during an earthquake, is more hazardous than building on bedrock. We also know that building on active faults is unwise.

We hope eventually to be able to predict earthquakes. The federal plan for issuing prediction and warning is shown in Figure 7.29. The general flow of information is from scientists to a prediction council for verification. A prediction that a damaging earthquake of a specific magnitude will occur at a particular location during a specified time would be issued to state and local officials, who will be responsible for issuing a warning to the public to take defensive active (that has, one hopes, been planned in advance). Potential response to a prediction depends upon lead time (Table 7.3), but even a short time (as little as a few days) would be sufficient to mobilize emergency service, shut down important machinery, and evacuate particularly hazardous areas.

Earthquake Warning Systems

It is technically feasible to develop an earthquake warning system that would provide up to about a 1-minute warning to the Los Angeles area prior to arrival of damaging earthquake waves. This is based on the principle that radio waves

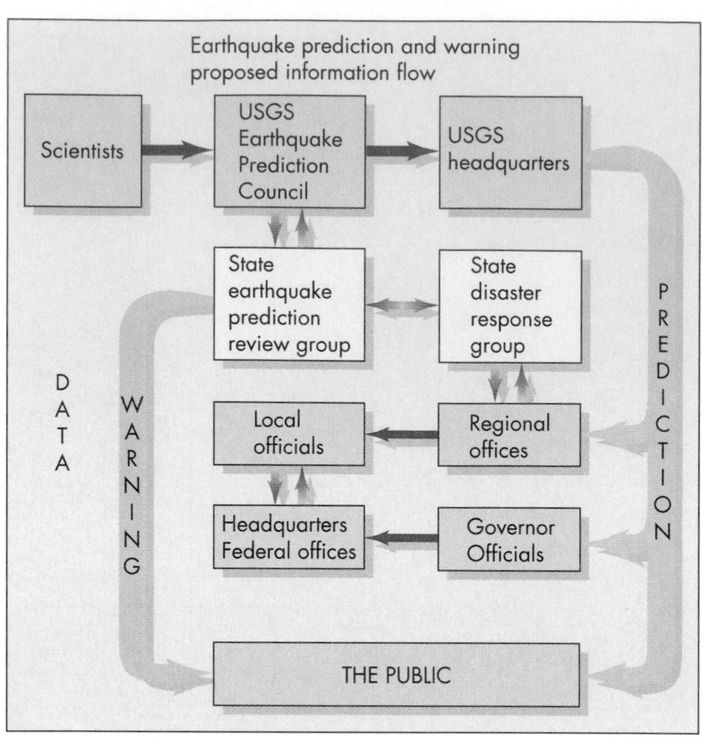

▶ **FIGURE 7.29** A federal plan for issuance of earthquake predictions and warning: the flow of information. (*From* V. E. McKelvey. 1976. Earthquake Prediction—Opportunity to Avert Disaster. U.S. *Geological Survey Circular* 729.)

Table 7.3 Potential response to an earthquake prediction with given lead time

Lead Time	Buildings	Contents	Lifelines	Special Structures
3 days	Evacuate previously identified hazards	Remove selected contents	Deploy emergency materials	Shut down reactors, petroleum products pipelines
30 days	Inspect and identify potential hazards	Selectively harden (brace and strengthen) contents	Shift hospital patients; alter use of facilities	Draw down reservoirs, remove toxic materials
300 days	Selectively reinforce		Develop response capability	Replace hazardous storage
3,000 days		Revise building codes and land-use regulations; enforce condemnation and reinforcement		Remove hazardous dams from service

Source: C. C. Thiel, 1976, *U.S. Geological Survey Circular* 729.

travel much faster than do seismic waves. The Japanese have had a system (using seismometers) for nearly 20 years that provides warning for their high-speed trains, since derailment by an earthquake could result in the loss of hundreds of lives. A system that has been proposed for California involves a sophisticated network of seismometers and transmitters along the San Andreas fault. This system would first sense motion associated with a large quake and then send a warning to the city of Los Angeles, which would relay it to critical facilities, schools, and the general population (Figure 7.30). Depending upon where the earthquake is initiated, the warning time would vary from as little as 15 seconds to as long as 1 minute. This could be enough time for people to shut down machinery and computers and take cover (39). *The earthquake warning system is not a prediction tool*, as it only warns that an earthquake has already occurred.

A potential problem with a warning system is the chance of false alarms. Using the Japanese system for comparison, the number of false alarms would probably be less than 5 percent. However, because the warning time is so short, some people have expressed concern as to whether much evasive action could be taken. There is also concern for li-

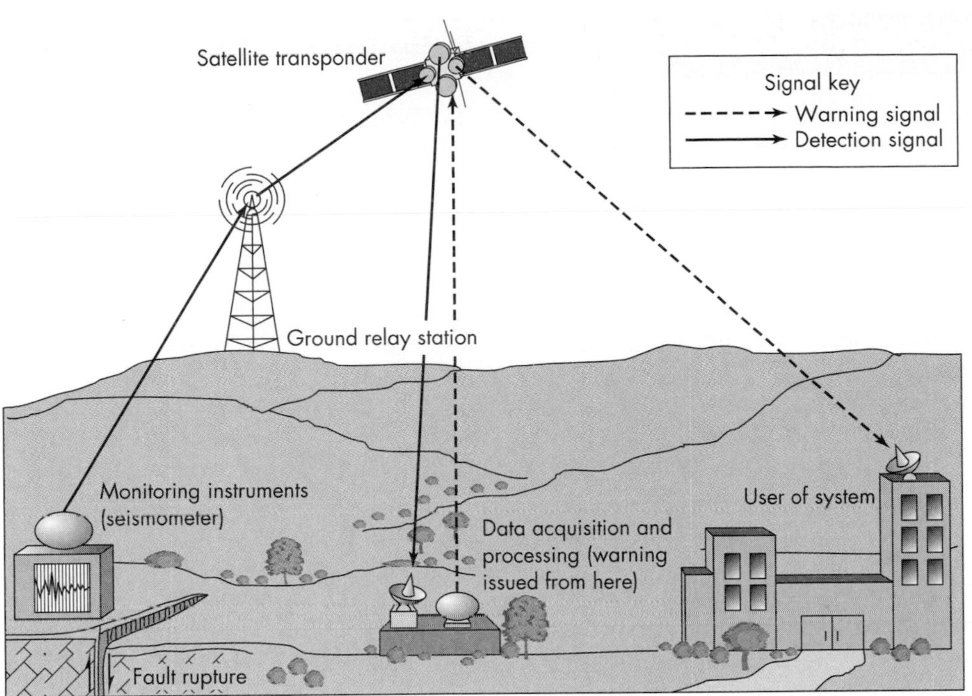

◀ **FIGURE 7.30** Idealized diagram showing an earthquake warning system. (*After* R. *Holden*, R. *Lee, and* M. *Reichle*. 1989. California Division of Mines and Geology Bulletin 101.)

ability issues resulting from false alarms, warning system failures, and damage and suffering resulting from actions taken as the result of early warning.

A recent study of a potential early warning system in California concluded that such a system is technically feasible, but that the potential economic benefits are not sufficient to justify the construction and operating costs. Total expenditure for the construction is estimated to be about $3 million to $6 million, with annual operating costs of $1.5 million to $2.5 million (39). For comparison, damages from the 1994 Northridge earthquake with a magnitude of 6.7 exceeded $30 billion. Larger, more damaging events can be expected. However, it is difficult to place an economic value on all issues related to public safety, especially as they relate to potential injuries or loss of life. Rejection for economic reasons of technology that could warn us of approaching disaster may be unwise.

Perception of the Earthquake Hazard

That terra firma is not so firm in places is disconcerting to people who have experienced even a moderate earthquake. The large number of individuals, especially children, who suffered mental distress following the San Fernando and Northridge earthquakes attests to the emotional and psychological effects. These events were sufficient to influence a number of Los Angeles families who were formerly from the Midwest to move back to the Midwest.

Regardless of their mental distress after earthquakes, most people in areas of potential disaster do not adequately understand the earthquake hazard. In addition, there appears to be a considerable difference between what people say and what they do. Near Tokyo, six new earthquake antidisaster centers and a steel mill are being constructed. Although the centers are to aid victims of earthquakes, the steel mill is being placed near sea level on landfill that is part of a reclamation project for Tokyo Bay. The site is subject to severe earthquake and tsunami hazard (40). The Japanese were caught off-guard by the 1995 Kobe earthquake and the government was criticized for not mounting a quick and effective response. People evidently believed that their buildings and highways were relatively safe compared to those that failed in California a year earlier (Northridge, 1994). The lesson the entire world learned from both Northridge and Kobe was a bitter one. Our modern society is very vulnerable to catastrophic loss from large earthquakes.

People in hazardous areas could remove dangerous structures on old buildings that overhang the streets. They could practice better planning for siting buildings and stop filling in bays and other sensitive areas. But they probably will not (unless ordered by law), because a hazardous event—even a tremendously catastrophic one that appears only once every few generations—is simply not perceived by the general population as a real threat. Yet there are some encouraging signs. California has passed legislation requiring detailed geologic evaluation prior to construction of structures for human use sited on or near active or potentially active faults. However, with the recognition of the buried earthquake threat from faulting that never reaches the surface of the earth, we are recognizing that entire regions such as the Los Angeles Basin and adjacent areas as well are all likely to experience damaging ground shaking from earthquakes in the relatively near future. Minimizing the hazard will require new thinking about the hazard and engineering response to design structures less vulnerable to ground motion from earthquakes. This will require geologic input of data to predict ground motion on a block-by-block scale (microzonation), as discussed earlier.

PUTTING SOME NUMBERS ON Earthquake Hazard Evaluation

The primary purpose of earthquake hazard assessment is to provide a safeguard against loss of life and major structural failures, rather than to limit damage or to maintain functional aspects of society, including roads and utilities (27). The basic philosophy behind the seismic hazard's evaluation is to have what is known as a **"Design Basis Ground Motion"** defined as the ground motion that has a 10 percent chance of being exceeded in a 50-year period either as determined by site-specific seismic hazard analysis or from a hazard map such as that presented earlier (Figure 7.24). The next step then is to evaluate the ground motion at a particular site in which the dynamic effects of the Design Basis Ground Motion likely experienced by buildings are represented by what is known as the **Design Response Spectrum.** The Design Response Spectrum may be calculated from simple equations if the seismic zone factor is known, soil profile type has been identified, and seismic source type is known (27). Seismic source types are listed on Table 7.A in a simple A, B, C format related to the expected maximum moment magnitude and the slip rate as determined from geologic evaluation. In order to produce the Design Response Spectra, we start with the *Seismic Zone Factor Z*, which is shown on Table 7.B and keyed to Figure 7.E. Notice that the state of California, with its considerable seismic hazard, is entirely in zones 3 and 4. The concepts behind the Design Response Spectra are: 1) strong ground motion in earthquakes peak at approximately 2 to 6 Hz (cycles per second), often at about 3 Hz (0.3-second period); and 2) geologists need to evaluate three parameters; PGA (peak ground acceleration, C_a), the spectral acceleration *(S)* at a period of 0.3 seconds, for most one- and two-story buildings, and spectral acceleration *(S)* at 1.0-second periods for buildings taller than two stories.

An example is probably the best way to illustrate how the Design Response Spectra, PGA, and spectral accelerations at desired periods are determined. Assume we

Table 7.A Seismic Source Type[a]

Seismic Source Type	Seismic Source Description	Seismic Source Definition[b]	
		Maximum Moment Magnitude, M_w	Slip Rate, *SR* (mm/yr)
A	Faults that are capable of producing large-magnitude events and that have a high rate of seismic activity	≥ 7.0	≥ 5
B	All faults other than types A and C	≥ 7.0 < 7.0 ≥ 6.5	< 5 > 2 < 2
C	Faults that are not capable of producing large-magnitude earthquakes and that have a relatively low rate of seismic activity	< 6.5	≤ 2

[a]Subduction sources shall be evaluated on a site-specific basis.

[b]Both maximum moment magnitude and slip-rate conditions must be satisfied concurrently when determining the seismic source type.

Source: Uniform Building Code 1997. State of California, Chapter 16.

Table 7.B Seismic Zone Factor *Z*

Zone	1	2A	2B	3	4
Z	0.075	0.15	0.20	0.30	0.40

Note: The zone shall be determined from the seismic zone map in Figure 7.E.

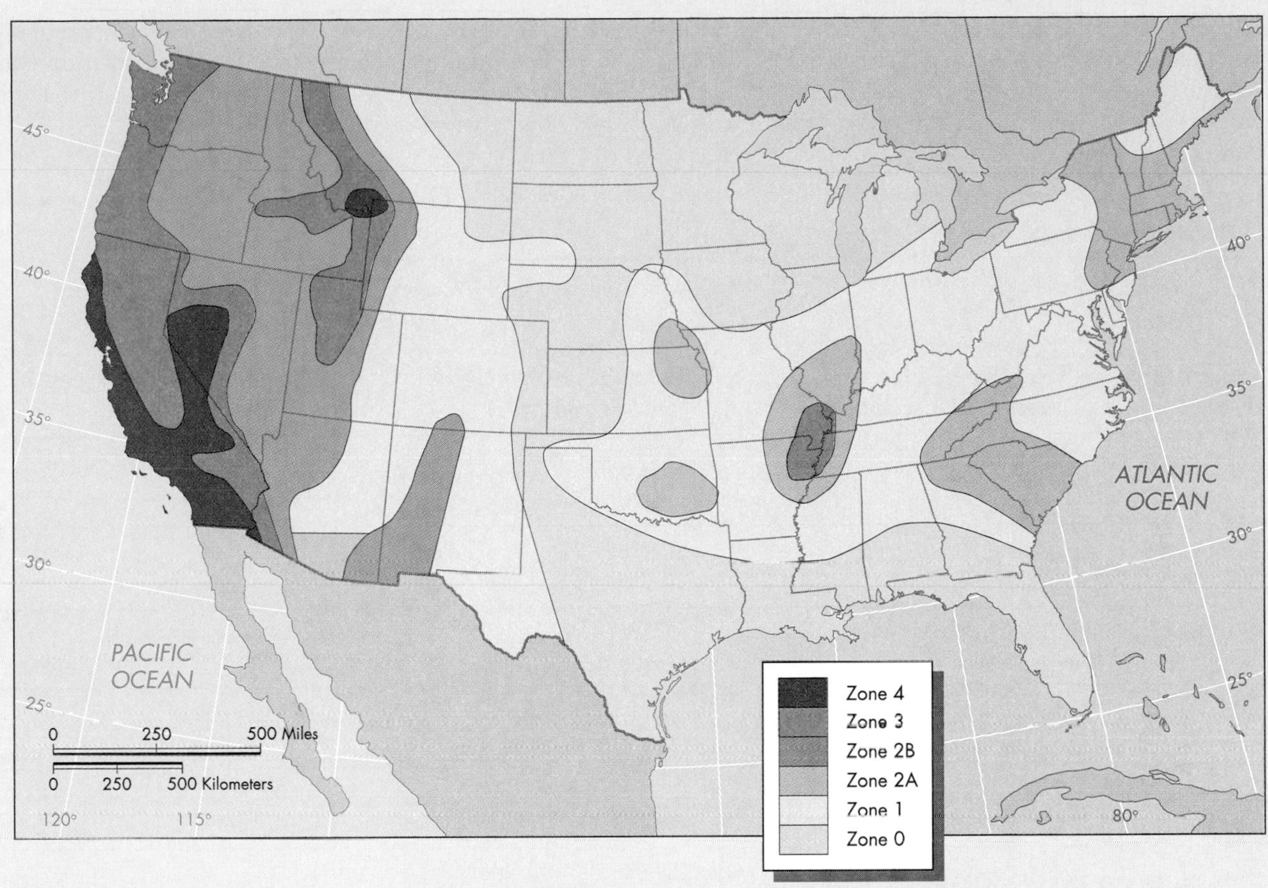

▲ **FIGURE 7.E** Map of seismic zones for the United States. (*From State of California Uniform Building Code*. 1997)

are evaluating a site in southern California near Los Angeles and we have determined that the fault we are interested in evaluating has a maximum moment magnitude (M_w) of 7.2 and a slip rate of 3 mm/yr. We first look to Figure 7.E and see that we are in seismic zone 4. Then, based upon the magnitude and slip rate on Table 7.A, we see we are dealing with a Type B seismic source. We also assume that the site we are interested in evaluating is approximately 17 km away from the fault described above and that the site has a soil profile description of stiff soil (Table 7.C). This means that the average soil properties to a depth of 30 m have relatively low shear wave velocities (the velocity of shear waves from an earthquake traveling through the material). Such material on Table 7.C is designated as S_D. We now have enough information, along with a few more short tables of data, to produce the Design Response Spectra. That information needed is shown on Tables 7.D, 7.E, 7.F, and 7.G. The basic method of deriving the Design Response Spectra is shown on Figure 7.F. Notice that there are control periods defined as T_s and T_o in terms of the two seismic coefficients (C_a, peak ground acceleration, and C_v), as shown on Figure 7.F and Tables 7.D and 7.E. Given all of the above information about the fault and the distance to our site of interest and the soil conditions, we may then construct the Design Response Spectra as shown on Figure 7.G. The value of this spectra is that we can now predict what the horizontal acceleration of the seismic waves for buildings (S at 0.3 sec, and S at 1.0 sec, Figure 7.G) is likely to be for a site on stiff soil. This is the basic information necessary for designing buildings, whether they be one-story single-frame homes or a ten-story high-rise apartment building. Another concept of importance here is that earthquake design for buildings is not the same for every building, and we must design particularly important buildings to withstand even higher ground motions using coefficients that multiply the values obtained in the Design Response Spectra by a particular amount. For example, if we are dealing with hospitals, the *seismic importance factor* is 1.5; we would thus multiply the expected motions by that amount (27).

In summary, the above analysis to produce a Design Response Spectra is a new direction in evaluating seismic hazard. That is, we try to predict the ground motion at a site and from this we can better design our buildings to withstand shaking from earthquakes. What we have learned from recent large earthquakes in southern California area is that horizontal accelerations during earthquakes are almost twice as strong as we thought they were likely to be before careful measurements were made from earthquakes such as the 1994 Northridge earthquake. Other methods of evaluation are available to predict the likelihood of liquefaction during an earthquake given information concerning the type of soil present, the amount of shaking, and other seismic factors.

Table 7.C Soil Profile Types

Soil Profile Type	Soil Profile Name/ Generic Description	Average Soil Properties for Top 100 Feet (30.5 m) of Soil Profile: Shear Wave Velocity, $\bar{v}_s$ feet/second (m/s)	Undrained Shear Strength, $\bar{s}_u$ psf (kPm)
S_A	Hard rock	>5,000 (1,500)	
S_B	Rock	2,500 to 5,000 (760 to 1,500)	
S_C	Very dense soil and soft rock	1,200 to 2,500 (360 to 760)	>2,000 (100)
S_D	Stiff soil profile	600 to 1,200 (180 to 360)	1,000 to 2,000 (50 to 100)
S_E[1]	Soft soil profile	<600 (180)	<1,000 (50)
S_F	Soil requiring site-specific evaluation. See Section 1629.3.1.		

[1]Soil Profile Type S_E also includes any soil profile with more than 100 feet (3048 mm) of soft clay defined as a soil with a plasticity index, $PI > 20$, $w \geq 40$ percent and $s_u < 500$ psf (24 kPa). The plasticity index, PI, and the moisture content, w, shall be determined in accordance with approved national standards.

Source: Uniform Building Code 1997. State of California, Chapter 16.

Table 7.D Seismic Coefficient C_a

Soil Profile Type	Seismic Zone Factor, Z: $Z = 0.075$	$Z = 0.15$	$Z = 0.2$	$Z = 0.3$	$Z = 0.4$
S_A	0.06	0.12	0.16	0.24	$0.32N_a$
S_B	0.08	0.15	0.20	0.30	$0.40N_a$
S_C	0.09	0.18	0.24	0.33	$0.40N_a$
S_D	0.12	0.22	0.28	0.36	$0.44N_a$
S_E	0.19	0.30	0.34	0.36	$0.36N_a$
S_F	See Footnote 1				

[1]Site-specific geotechnical investigation and dynamic site response analysis shall be performed to determine seismic coefficients for Soil Profile Type S_F.

Source: Uniform Building Code 1997. State of California, Chapter 16.

Table 7.E Seismic Coefficient C_v

Soil Profile Type	Seismic Zone Factor, Z				
	Z = 0.075	**Z = 0.15**	**Z = 0.2**	**Z = 0.3**	**Z = 0.4**
S_A	0.06	0.12	0.16	0.24	$0.32N_v$
S_B	0.08	0.15	0.20	0.30	$0.40N_v$
S_C	0.13	0.25	0.32	0.45	$0.56N_v$
S_D	0.18	0.32	0.40	0.54	$0.64N_v$
S_E	0.26	0.50	0.64	0.84	$0.96N_v$
S_F			See Footnote 1		

[1]Site-specific geotechnical investigation and dynamic site response analysis shall be performed to determine seismic coefficients for Soil Profile Type S_F.

Source: Uniform Building Code 1997. State of California, Chapter 16.

Table 7.F Near-Source Factor N_a[1]

Seismic Source Type	Closest Distance to Known Seismic Source[2,3]		
	≤2 km	**5 km**	**≥10 km**
A	1.5	1.2	1.0
B	1.3	1.0	1.0
C	1.0	1.0	1.0

[1]The Near-Source Factor may be based on the linear interpolation of values for distances other than those shown in the table.

[2]The location and type of seismic sources to be used for design shall be established based on approved geotechnical data (e.g., most recent mapping of active faults by the United States Geological Survey or the California Division of Mines and Geology).

[3]The closest distance to seismic source shall be taken as the minimum distance between the site and the area described by the vertical projection of the source on the surface (i.e., surface projection of fault plane). The surface projection need not include portions of the source at depths of 10 km or greater. The largest value of the Near-Source Factor considering all sources shall be used for design.

Source: Uniform Building Code 1997. State of California, Chapter 16.

Table 7.G Near-Source Factor N_v[1]

Seismic Source Type	Closest Distance to Known Seismic Source[2,3]			
	≤2 km	**5 km**	**10 km**	**≥15 km**
A	2.0	1.6	1.2	1.0
B	1.6	1.2	1.0	1.0
C	1.0	1.0	1.0	1.0

[1]The Near-Source Factor may be based on the linear interpolation of values for distances other than those shown in the table.

[2]The location and type of seismic sources to be used for design shall be established based on approved geotechnical data (e.g., most recent mapping of active faults by the United States Geological Survey or the California Division of Mines and Geology).

[3]The closest distance to seismic source shall be taken as the minimum distance between the site and the area described by the vertical projection of the source on the surface (i.e., surface projection of fault plane). The surface projection need not include portions of the source at depths of 10 km or greater. The largest value of the Near-Source Factor considering all sources shall be used for design.

Source: Uniform Building Code 1997. State of California, Chapter 16.

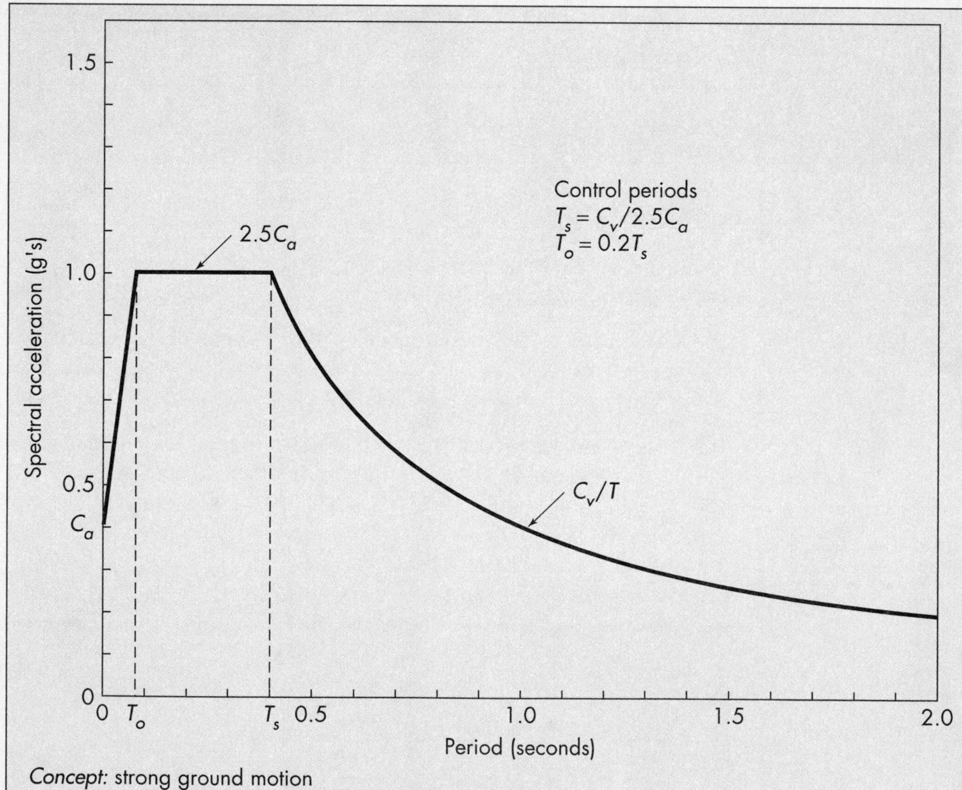

▶ **FIGURE 7.F** Idealized diagram showing the design response spectra. (*From State of California Uniform Building Code*. 1997. *Chapter* 16)

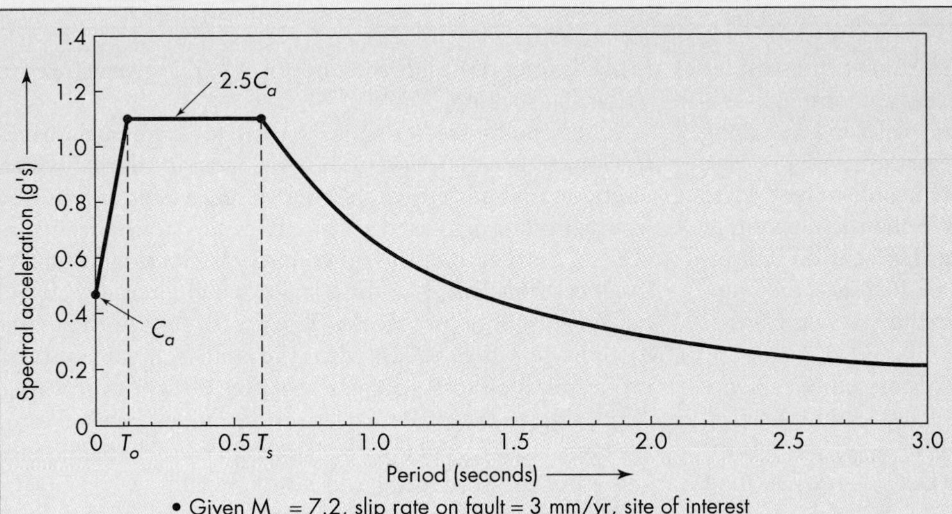

- Given $M_w = 7.2$, slip rate on fault = 3 mm/yr, site of interest is 17 km from fault, near Los Angeles, California.

- From Figure 7.E, we are in seismic zone 4.

- Given, M_w and slip rate, we are dealing with a Type B seismic source (Table 7.A) and the Z factor is 0.40 (Table 7.B).

- C_a from Table 7.D given soil profile type S_D and $Z = 0.4$ is: $C_a = 0.44\ N_a$, near source factor $N_a = 1.0$ (Table 7.F). Then $C_a = 0.44\ \text{m/s}^2$. This is the PGA.

- C_v from Table 7.E = 0.64 N_v, where $N_v = 1$ (Table 7.G), then $C_v = 0.64$ m/s.

- $T_s = C_v/2.5\ C_a = \dfrac{0.64\ \text{m/s}}{(2.5)(0.44\ \text{m/s}^2)} = 0.6$ sec.

- $T_o = 0.2\ T_s = 0.12$ sec.

- S at 0.3 sec = 1.1 g or 10.8 m/s^2 for one- to two-story building (read from graph).

- S at 1.0 sec = 0.65 g or 6.4 m/s^2 (read from graph).

- PGA (peak ground acceleration) = $C_a = 0.44\ \text{m/s}^2$.

◀ **FIGURE 7.G** Example of how to produce the Design Response Spectra, assuming M_w 7.2, slip rate 3 mm/yr at a distance of 17 km from the fault, on stiff soil.

SUMMARY

Large earthquakes rank among nature's most catastrophic and devastating events. Most earthquakes are located in tectonically active areas where lithospheric plates interact along their boundaries, but some intraplate earthquakes do occur.

A *fault* is a fracture or fracture system along which rocks have been displaced. Strain builds up in the rocks on either side of a fault as the sides deform in different directions. When the stress exceeds the strength of the rocks, they rupture, giving rise to seismic waves (earthquake waves) that shake the ground.

Strike-slip faults exhibit horizontal displacement and are either right- or left-lateral. Dip-slip faults exhibit vertical displacement and are either normal or reverse. Some faults are buried and do not rupture the surface, even when their movement causes large earthquakes.

Faults and fault zones have earthquake segments that rupture as a unit. Understanding the activity of earthquake segments is essential to evaluating seismic hazards and requires investigation of the paleoseismicity (prehistoric earthquake activity) of the segments.

A fault is usually considered active if it has moved during the past 10,000 years and potentially active if it has moved during the past 1.65 million years. The slip rate of a fault is its displacement over a specific time interval, and its recurrence interval is the average time between earthquakes; both are subject to change. Methods for determining fault activity include study of landforms and soils and radiometric dating.

Some faults exhibit tectonic creep, a slow displacement not accompanied by felt earthquakes. Although creep is less hazardous than earthquakes, it can cause considerable damage to roads, sidewalks, building foundations, and other structures.

The area within the earth where fault rupture begins is called the *focus* of the earthquake and can be from a few kilometers to almost 700 km deep. The area of the surface directly above the focus is called the *epicenter*. Seismic waves of different kinds travel away from the focus at different rates; much of the damage in earthquakes is caused by surface waves. The severity of shaking of the ground and buildings is affected by the frequency of the seismic waves and by the type of earth material present. Buildings on unconsolidated sediments or landfill, which tend to amplify the shaking, are highly subject to earthquake damage. Seismic shaking may also be amplified in the direction of fault rupture (directivity).

The Richter magnitude of an earthquake is a measure of the amount of energy released. It is determined by the amplitude of the largest horizontal trace recorded on a standard seismograph. A newer scale known as the *moment magnitude*, based on the seismic moment, has been developed. Resulting magnitudes are similar to those of the Richter scale but are more soundly based on physical principles. The intensity of an earthquake is based on the severity of shaking as reported by observers or seismographs and varies with proximity to the epicenter and local geologic and engineering features. Peak ground acceleration during an earthquake is important information for designing structures to withstand shaking.

The hypothesized earthquake cycle for large earthquakes has four stages. A period of seismic inactivity, during which elastic strain builds up in the rocks along a fault, is followed by a period of increased seismicity as the strain locally exceeds the strength of the rocks, initiating local faulting and small earthquakes. The third stage, which does not always occur, consists of foreshocks. The fourth stage is the major earthquake, which occurs when the fault segment ruptures, producing the elastic rebound that generates seismic waves. According to the dilatancy-diffusion model, the stages of the cycle are associated with changes in rock characteristics and water pressure in the rocks along a fault.

Human activity has caused increasing earthquake activity by loading the earth's crust through construction of large reservoirs; by disposing of liquid waste in deep disposal wells, which raises fluid pressures in rocks and facilitates movement along fractures; and by setting off underground nuclear explosions. The accidental damage caused by the first two activities is regrettable, but what we learn from all the ways we have caused earthquakes may eventually help us to control or stop large earthquakes.

Effects of earthquakes include violent ground motion accompanied by fracturing, which may shear or collapse large buildings, bridges, dams, tunnels, and other rigid structures. Other effects include fires, landslides, tsunamis, and regional subsidence and uplift of landmasses, as well as regional changes in groundwater levels. Tsunamis, or seismic seawaves, are produced by earthquakes or other phenomena that displace oceanic water and generate long waves that travel at speeds as great as 800 km/hr. The waves, less than 0.5 m high in deep water, may grow to a height of 15 m or more on reaching the coastline, where they can cause great damage and loss of life. Damaging tsunamis are infrequent, however, and their long travel time allows warnings to be issued.

Prediction of earthquakes is a subject of serious research. To date, long-term and medium-term earthquake prediction based on probabilistic analysis have been much more successful than short-term prediction. Long-term prediction provides important information for land-use planning, developing building codes, and engineering design of critical facilities. Optimistic scientists believe that we will eventually be able to make long-, medium-, and short-range predictions based on previous patterns and frequency of earthquakes, seismic gaps, and monitoring of anomalous uplift and subsidence, ground tilt, micro-earthquake activity, and perhaps anomalous animal behavior.

Reduction of earthquake hazard will be a multifaceted program, including (1) recognition of active faults and earth materials sensitive to shaking, and (2) development of improved ways to predict, control, and adjust to earthquakes (especially designing structures to better withstand shaking). Warning systems and earthquake prevention are not yet reliable alternatives, but more communities are developing emergency plans to respond to a predicted or unexpected catastrophic earthquake. Seismic zoning (including microzonation) and other methods of hazard reduction are being studied.

With the exception of areas that have recently experienced a moderate or larger earthquake, the earthquake hazard in potential disaster areas remains poorly perceived by many inhabitants of the areas. A process that produces even catastrophic damage may not be perceived as a real threat if it strikes the same area only every few generations. However, some communities now have legislation requiring evaluation of proposed construction sites for seismic hazard.

REFERENCES

1. **Hamilton, R. M.** 1980. Quakes along the Mississippi. *Natural History* 89:70–75.
2. **Advisory Committee on the International Decade for Natural Hazard Reduction.** 1989. Reducing disaster's toll. Washington, DC: National Academy Press.
3. **Machette, M. N., Personius, S. F., Nelson, A. R., Schwartz, D. P., and Lunde, W. R.** 1989. Segmentation models and Holocene movement history of the Wasatch Fault Zone. In *Fault segmentation and controls of rupture initiation and termination,* eds. D. P. Schwartz and R. H. Sibson, pp. i–iv. U.S. Geological Survey Open-File Report 89-315.
4. **Davidson, K.** 1994. Learning from Los Angeles. *Earth* 3(5):40–49.
5. **Clark, M. M., Grantz, A., and Meyer, R.** 1972. Holocene activity of the Coyote Creek fault as recorded in the sediments of Lake Cahuilla. In *The Borrego Mountain earthquake of April 9, 1968,* pp. 1112–30. U.S. Geological Survey Professional Paper 787.
6. **Radbruch, D. H., et al.** 1966. *Tectonic creep in the Hayward fault zone, California.* U.S. Geological Survey Circular 525.
7. **Steinbrugge, K. V., and Zacher, E. G.** 1960. Creep on the San Andreas fault. In *Focus on environmental geology,* ed. R. W. Tank, pp. 132–137. New York: Oxford University Press.
8. **Bolt, B. A.** 1993. *Earthquakes.* San Francisco: W. H. Freeman.
9. **Hart, E. W., Bryant, W. A., and Treiman, J. A.** 1993. Surface faulting associated with the June 1992 Landers earthquake, California. *California Geology* (January–February) 46(1):10–16.
10. **Hays, W. W.** 1981. *Facing geologic and hydrologic hazards.* U.S. Geological Survey Professional Paper 1240B.
11. **Jones, R. A.** 1986. New lessons from quake in Mexico. *Los Angeles Times,* September 26.
12. **Hough, S. E., Friberg, P. A., Busby, R., Field, E. F., Jacob, K. H., and Borcherdt, R. D.** 1989. Did mud cause freeway collapse? *EOS, Transactions of the American Geophysical Union* 70, no. 47: 1497, 1504.
13. **U.S. Geological Survey.** 1996. *USGS response to an urban earthquake, Northridge '94.* U. S. Geological Survey Open File Report 96-263.
14. **Hanks, T. C.** 1985. The national earthquake hazards reduction program: Scientific status. *U.S. Geological Survey Bulletin* 1659.
15. **Press, F.** 1975. Earthquake prediction. *Scientific American* 232:14–23.
16. **Sibson, R. H.** 1981. Fluid flow accompanying faulting: field evidence and models in earthquake prediction. In *An International Review, Maurice Ewing Ser.,* vol. 4, eds. D. W. Simpson and P. G. Richards, pp. 593–603. Washington, DC: AGU.
17. **Pakiser, L. C., Eaton, J. P., Healy, J. H., and Raleigh, C. B.** 1969. Earthquake prediction and control. *Science* 166:1467–74.
18. **Evans, D. M.** 1966. Man-made earthquakes in Denver. *Geotimes* 10:11–18.
19. **Office of Emergency Preparedness.** 1972. *Disaster preparedness,* 1,3.
20. **Youd, T. L., Nichols, D. R., Helley, E. J., and Lajoie, K. R.** 1975. Liquefaction potential. In *Studies for seismic zonation of the San Francisco Bay region,* ed. R. D. Borcherdt. U.S. Geological Survey Professional Paper 941A:68–74.
21. **Hokkaido Tsunami Survey Group.** 1993. Tsunami devastates Japanese coastal region. *EOS, Transactions of the American Geophysical Union* 74(37):417–432.
22. **Geist, E. L.** 1998. Source characteristics of the July 17, 1998 Papua New Guinea tsunami. American Geophysical Union, Fall Meeting. *Abstracts,* p. F571.
23. **Davis, H. L.** 1998. A reconstruction of the events of July 17 based upon interviews with survivors. American Geophysical Union, Fall Meeting. *Abstracts,* p. F572.
24. **Slemmons, D. B., and DePolo, C. M.** 1986. Evaluation of active faulting and associated hazards. In *Active tectonics,* pp. 45–62. Washington, DC: National Academy Press.
25. **Hansen, W. R.** 1965. *The Alaskan earthquake, March 27, 1964: Effects on communities.* U.S. Geological Survey Professional Paper 542A.
26. **Oppenheimer, D., Beroza, G., Carver, G., Dengler, L., Eaton, J., Gee, L., Gonzales, F., Jayko, A., Li, W. H., Lisowski, M., Magee, M., Marshall, G., Murray, M., McPherson, R., Romanowicz, B., Sataker, K., Simpson, R., Somerville, P., Stein, R., and Valentine, D.** 1993. The Cape Mendocino, California, earthquakes of April 1992: Subduction at the Triple Junction. *Science* 262:433–438.
27. **State of California Uniform Building Code.** 1997. Chapter 16.
28. **Heaton, T. H., Anderson, D. L., Arabasz, W. J., Buland, R., Ellsworth, W. L., Hartzell, S. H., Lay, T., and Spudich, P.** 1989. *National seismic system science plan.* U.S. Geological Survey Circular 1031.
29. **Sieh, K., Stuiver, M., and Brillinger, D.** 1989. A more precise chronology of earthquakes produced by the San Andreas fault in southern California. *Journal of Geophysical Research* 94(B1):603–623.
30. **Toppozada, T. R.** 1993. The Landers–Big Bear earthquake sequence and its felt effects. *California Geology* (January–February) 46(1):3–9.

31. **Raleigh, B. et al.** 1977. Prediction of the Haicheng earthquake. *Transactions of the American Geophysical Union* 58(5):236–72.
32. **Simons, R. S.** 1977. Earthquake prediction, prevention in San Diego. In *Geologic hazards in San Diego*, ed. A. L. Patrick. San Diego Society of Natural History.
33. **Scholz, C. H.** 1990. *The mechanics of earthquakes and faulting*. New York: Cambridge University Press.
34. **Rikitakr, T.** 1983. *Earthquake forecasting and warning*. London: D. Reidel.
35. **Allen, C. R.** 1983. Earthquake prediction. *Geology* 11:682.
36. **Hait, M. H.** 1978. Holocene faulting, Lost River Range, Idaho. *Geological Society of America Abstracts with Programs* 10(5):217.
37. **Page, R. A., Boore, D. M., Bucknam, R. C., and Thatcher, W. R.** 1992. *Goals, opportunities, and priorities for the USGS Earthquake Hazards Reduction Program*. U.S. Geological Survey Circular 1079.
38. **Cluff, L. S.** 1983. The impact of tectonics on the siting of critical facilities. *EOS* 64(45):860.
39. **Holden, R., Lee, R., and Reichle, M.** 1989. *Technical and economic feasibility of an earthquake warning system in California*. California Division of Mines and Geology Special Publication 101.
40. **Nicholas, T. C., Jr.** 1974. Global summary of human response to natural hazards: Earthquakes. In *Natural hazards*, ed. G. F. White, pp. 274–84. New York: Oxford University Press.

KEY TERMS

fault (p. 163)
fault segmentation (p. 165)
active fault (p. 165)
slip rate (p. 165)
tectonic creep (p. 168)
focus (p. 169)
epicenter (p. 169)
P wave (p. 169)
S wave (p. 171)
surface wave (p. 171)
material amplification (p. 172)
directivity (p. 172)
Richter magnitude (p. 172)
moment magnitude (p. 174)
Modified Mercalli Scale (p. 174)
earthquake cycle (p. 175)
dilantancy-diffusion model (p. 178)
liquefaction (p. 182)
tsunamis (p. 183)
seismic hazard map (p. 184)
seismic gap (p. 187)
capable fault (p. 192)
maximum credible earthquake (p. 192)

SOME QUESTIONS TO THINK ABOUT

1. Assume you are the person in your family, dormitory, or other living unit responsible for developing a plan to minimize the earthquake hazard in the residence. What would be the major components of the plan you develop and how would you go about explaining the plan to others?
2. Assume you are working for the Peace Corps and are in a developing country where most of the homes are built out of unreinforced blocks or bricks. There has not been a large damaging earthquake in the area for several hundred years, but earlier there were several earthquakes that killed thousands of people. How would you present the earthquake hazard to the people living where you are working? What steps might be taken to reduce the hazard?
3. The town you are living in wants to develop a new seismic element as part of its safety planning. One of the concerns is the response of buildings to shaking and in particular the problem of material amplification. Design a research program that could address whether material amplification is likely in your area and how the process could be studied.
4. You live in an area that has a significant earthquake hazard. There is ongoing debate as to whether or not an earthquake warning system should be developed. Some residents are worried that false alarms will cause a lot of problems and others point out that the response time may not be very long. What are your views on this? Do you think it is a responsibility of public officials to finance an earthquake warning system, assuming such a system is feasible? What are potential implications if a warning system is not developed and a large earthquake results in damage that could have been partially avoided with a warning system in place?

8 Volcanic Activity

The village of San Juan Parangaricutiro, Mexico, inundated by a lava flow in 1943. (*Photo by Arno Brehme. Plate 39A from Foshag and Gonzalez-Reyna [1956] and File No. 189-PV-89 in the U.S. National Archives.*)

Several large cities in the world are close to active volcanoes, including Seattle, Washington, and Mexico City, which was dusted by volcanic ash from a nearby volcano in the mid-1990s. Shown above is the village of San Juan Parangaricutiro, which was inundated by a lava flow from the volcano Paricutín in 1943 (also shown, from a different perspective, in the photograph on the cover of this book). The relatively small volcano, shown in the background, is located approximately 150 km from Mexico City. It started in a cornfield and in days grew to more than 100 m high. In two years the volcano attained a height of about 400 m. Today it is dormant. As large urban areas such as Seattle and Mexico City continue to grow, more people are being exposed to potential catastrophic volcanic hazards. The Seattle area, in particular, is in the shadow of Mount Ranier, a large Cascade volcano capable of catastrophic mudflows that could reach outlying areas of the Seattle area, as they did in prehistoric time.

8.1 Introduction to Volcanic Hazards

Fifty to 60 volcanoes erupt each year worldwide. The United States has two or three eruptions annually, most of them in Alaska (1). Eruptions often occur in sparsely populated areas of the world, but when one occurs near a densely populated area, the effects can be catastrophic (2). As the human population grows, more and more people are living on the flanks of active or potentially active volcanoes. In the past 100 years about 100,000 people have been killed by volcanic eruptions—about 28,500 of them in the single decade of the 1980s (1,2). Densely populated countries with many active volcanoes, such as Japan, the Philippines, and Indonesia, are particularly vulnerable (3). The western United States, including Alaska, Hawaii, and the Pacific Northwest, also has many active or potentially active volcanoes, sever-

LEARNING OBJECTIVES

One of the 44 people killed when Mt. Unzen in Japan erupted on June 3, 1991, was Harry Glicken (Figure 8.1). Harry was a brave and dedicated young scientist who, through good fortune, had escaped death in the May 18, 1980, eruption of Mount St. Helens. This chapter on volcanic processes is dedicated to Harry, who loved volcanoes and who was my friend.

Learning objectives for this chapter are:

- *To be familiar with the major types of volcanoes, the rocks they produce, and their plate-tectonic setting.*
- *To know the main effects of volcanic activity, including lava flows, pyroclastic activity, and debris flows/mudflows.*
- *To understand the methods of studying volcanic activity that may result in better prediction of eruptions, including seismic activity, topographic change, emission of gases, and geologic history.*

Web Resources

Visit the "Environmental Geology" Web site at www.prenhall.com/keller to find additional resources for this chapter, including:

- Web Destinations
- On-line Quizzes
- On-line "Web Essay" Questions
- Search Engines
- Regional Updates

▲ **FIGURE 8.1** Harry Glicken at Mount St. Helens observation site in 1980 prior to the eruption that blew the top off the mountain. Harry is looking directly toward the bulge that was developing on the flank of the volcano. When the volcano erupted, the observation post was destroyed and the geologist at the site was killed. *(Courtesy of Harry Glicken)*

al of them located near cities with populations exceeding 350,000 people (Figure 8.2). Table 8.1 lists some historic volcanic events and their effects.

8.2 Volcanism and Volcanoes

Volcanic activity, or volcanism, is directly related to plate tectonics, and most active volcanoes are located near plate junctions. **Magma** (molten rock that includes a small component of dissolved gases) is produced in these regions as spreading or sinking lithospheric plates interact with other earth materials. About 80 percent of all active volcanoes are located in the "ring of fire" that circumscribes the Pacific Ocean, an area corresponding to the Pacific plate (Figure 8.3).

Volcano Types

Each type of volcano has a characteristic style of activity that is partly a result of the viscosity of the magma. *Magma viscosity* is determined mostly by the magma's silica (SiO_2) content, which varies from about 50 to 70 percent, and its temperature. Magma that has emerged from a volcanic vent is called **lava.**

Shield Volcanoes **Shield volcanoes** are by far the largest volcanoes. They are common in the Hawaiian Islands (Figure 8.4) and are also found in Iceland and in some areas of the Pacific Northwest. Although they are shaped like a gentle arch, or shield, they are among the tallest mountains on earth when measured from their bases, often on the ocean floor. Shield volcanoes are characterized by generally nonexplosive eruptions; this is because of the relatively low silica content (about 50%) of the magma. The common rock type of the magma is **basalt,** which is composed mostly of feldspar and ferromagnesian minerals (see Chapter 2 for a review of mineral and rock types).

Shield volcanoes are built up almost entirely from numerous lava flows, but they can also produce a lot of **tephra** (all types of volcanic debris explosively ejected from a volcano), also called *pyroclastic debris*. Accumulation of tephra near a volcanic vent can form features such as the *cinder cone* (tephra cone), shown in Figure 8.5. Accumulation of tephra forms pyroclastic deposits (Greek *pyro*, "fire," and *klastos*, "broken"). **Pyroclastic deposits** may be consolidated to form **pyroclastic rocks.**

The slope of a shield volcano is very gentle near the top (about 3 to 5°), but it increases (to about 10°) on the flanks. This change has to do with the viscosity of the flowing lava. When magma comes out of vents at the top of the volcano, it is quite hot and flows easily, but as it

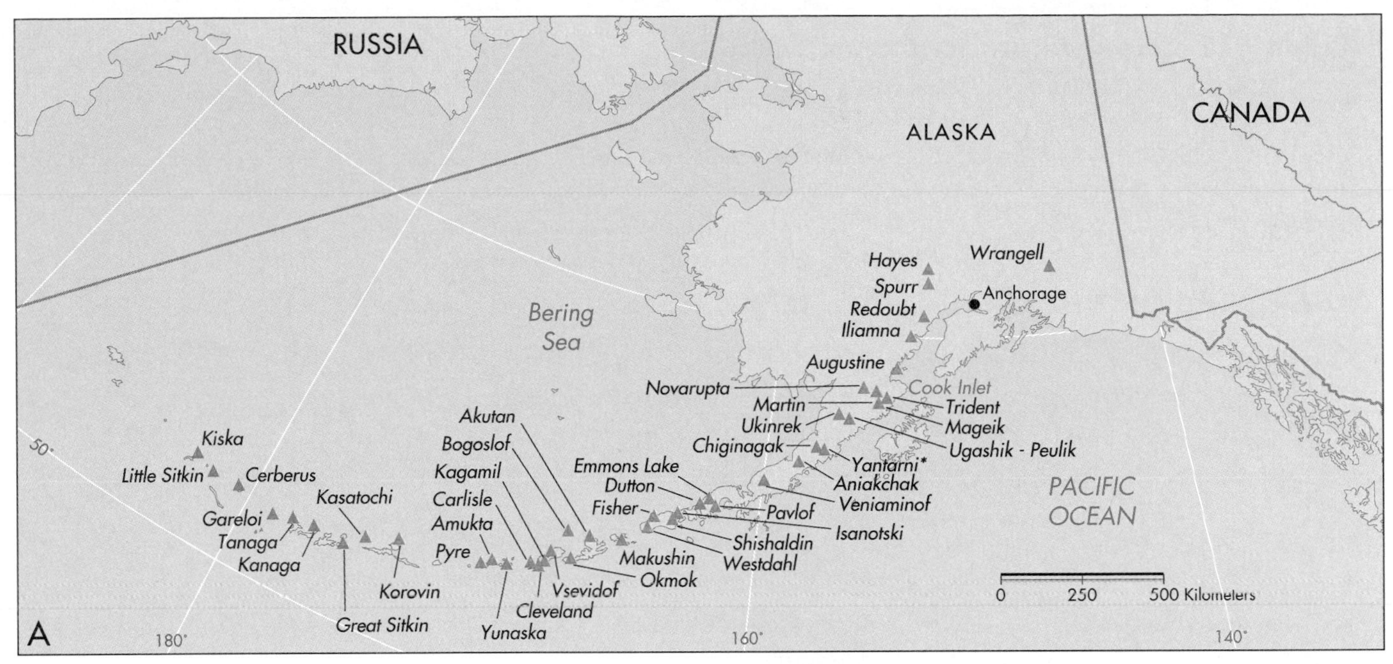

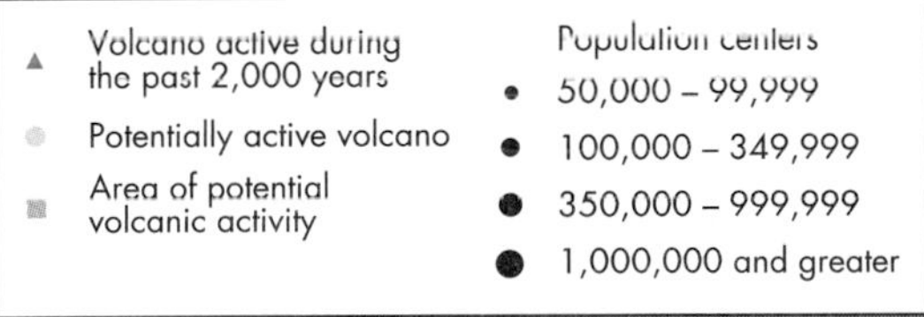

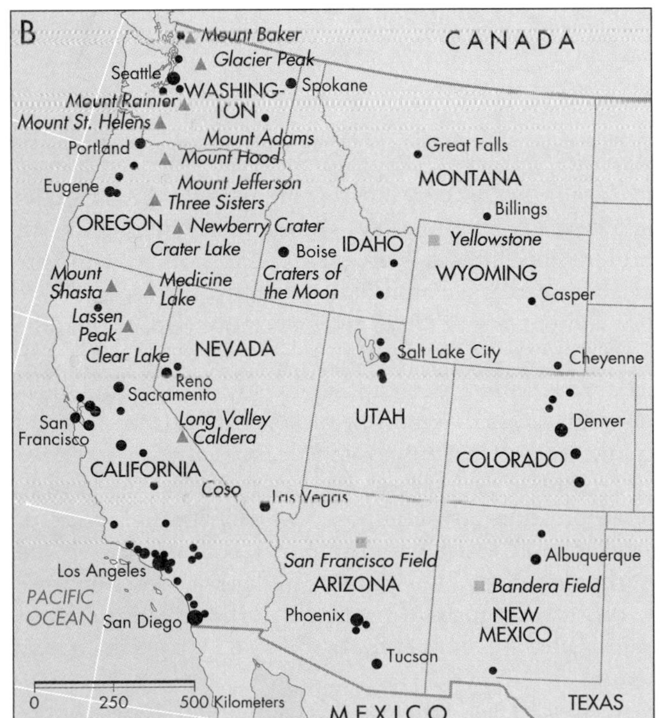

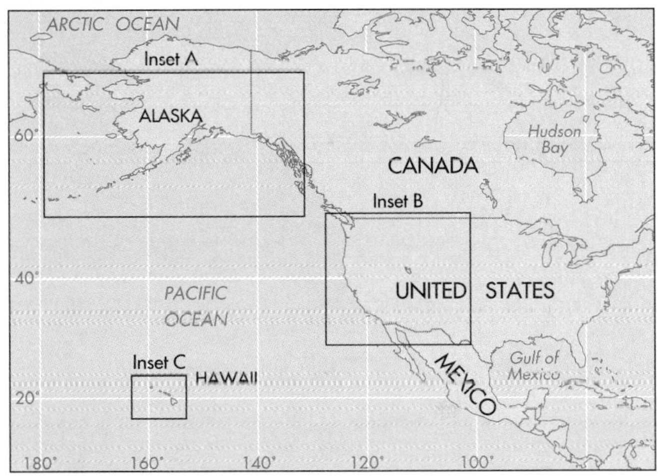

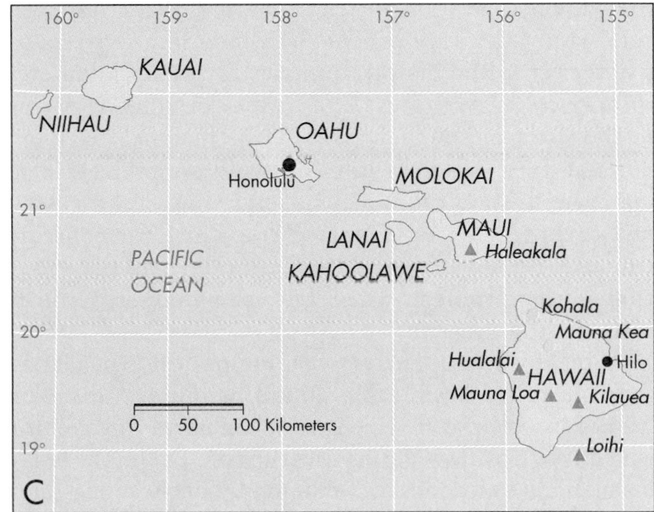

▲ **FIGURE 8.2** Locations of volcanoes in the United States. Index maps show locations of active and potentially active volcanoes and nearby population centers (not all labeled) of the United States. (*Source:* T. L. Wright *and* T. C. Pierson. 1992. *U.S. Geological Survey Circular* 1073.)

Table 8.1 Selected historic volcanic events

Volcano or City	Effect
Vesuvius, Italy, A.D. 79	Destroyed Pompeii and killed 16,000 people. City was buried by volcanic activity and rediscovered in 1595.
Skaptar Jokull, Iceland, 1783	Killed 10,000 people (many died from famine) and most of the island's livestock. Also killed some crops as far away as Scotland.
Tambora, Indonesia, 1815	Global cooling; produced "year without a summer."
Krakatoa, Indonesia, 1883	Tremendous explosion; 36,000 deaths from tsunami.
Mt. Pelée, Martinique, 1902	Ash flow killed 30,000 people in a matter of minutes.
La Soufrière, St. Vincent, 1902	Killed 2000 people and caused the extinction of the Carib Indians.
Mt. Lamington, Papua New Guinea, 1951	Killed 6000 people.
Villarica, Chile, 1963–64	Forced 30,000 people to evacuate their homes.
Mt. Helgafell, Heimaey Island, Iceland, 1973	Forced 5200 people to evacuate their homes.
Mount St. Helens, Washington, USA, 1980	Debris avalanche, lateral blast, and mudflows killed 54 people, destroyed more than 100 homes.
Nevado del Ruiz, Colombia, 1985	Eruption generated mudflows that killed at least 22,000 people.
Mt. Unzen, Japan, 1991	Ash flows and other activity killed 41 people and burned more than 125 homes. More than 10,000 people evacuated.
Mt. Pinatubo, Philippines, 1991	Tremendous explosions, ash flows, and mudflows combined with a typhoon killed more than 300 people; several thousand people evacuated.
Montserrat, Caribbean, 1995	Explosive eruptions, pyroclastic flows, south side of island evacuated, including capital city of Plymouth, several hundred homes destroyed.

Data partially derived from C. Ollier, *Volcanoes* (Cambridge, MA: MIT Press, 1969).

flows down the sides of the volcano, it cools and becomes more viscous, so that it needs a steeper slope for its further downslope transport. However, flowing down the sides is not the only process by which lava moves away from a vent. Magma often moves for many kilometers subsurface in **lava tubes.** These tubes are often very close to the surface, but they insulate the magma, keeping it hot and fluid. After the lava cools and crystallizes, forming rock, the lava tubes may be left behind as long, sinuous cavern systems (Figure 8.6). They form natural conduits for movement of groundwater and may cause engineering problems when they are encountered during construction projects.

Shield volcanoes may have a summit caldera, which is a steep, walled basin often 10 km or more in diameter, formed by collapse, in which a lava lake may form and from which lava may flow during an eruption. Eruptions of lava from shield volcanoes also commonly occur along linear fractures (normal faults) known as rift zones on the flank of a volcano. For example, the Hawaiian shield volcano Kilauea, on Hawaii (the "Big Island"), has experienced rift eruptions from 1983 to 1999 (at this writing) that have added more than 2 km^2 of new land to the island (4).

Composite Volcanoes **Composite volcanoes** are known for their beautiful cone shape (Figure 8.7). Examples in the United States include Mount St. Helens and Mt. Rainier, both in the state of Washington. Composite volcanoes are associated with a magma of intermediate silica content (about 60%), which is more viscous than the lower-silicate magma of shield volcanoes. The common rock type is **andesite,** composed mostly of soda- and lime-rich feldspar and ferromagnesian minerals with small amounts of quartz. Composite volcanoes are characterized by a mixture of explosive activity and lava flows. As a result, these volcanoes are composed of alternating layers of pyroclastic deposits and lava flows and are also called **stratovolcanoes.** They have steep flanks or sides because the **angle of repose** (maximum slope angle for loose material) for many pyroclastic deposits is approximately 30 to 35°.

Do not let the beauty of these mountains fool you. Because of their explosive activity and relatively common occurrence, composite volcanoes are responsible for most of the volcanic hazards that have caused death and destruction throughout history. As the 1980 eruption of Mount St. Helens demonstrated, they can produce gigantic horizontal

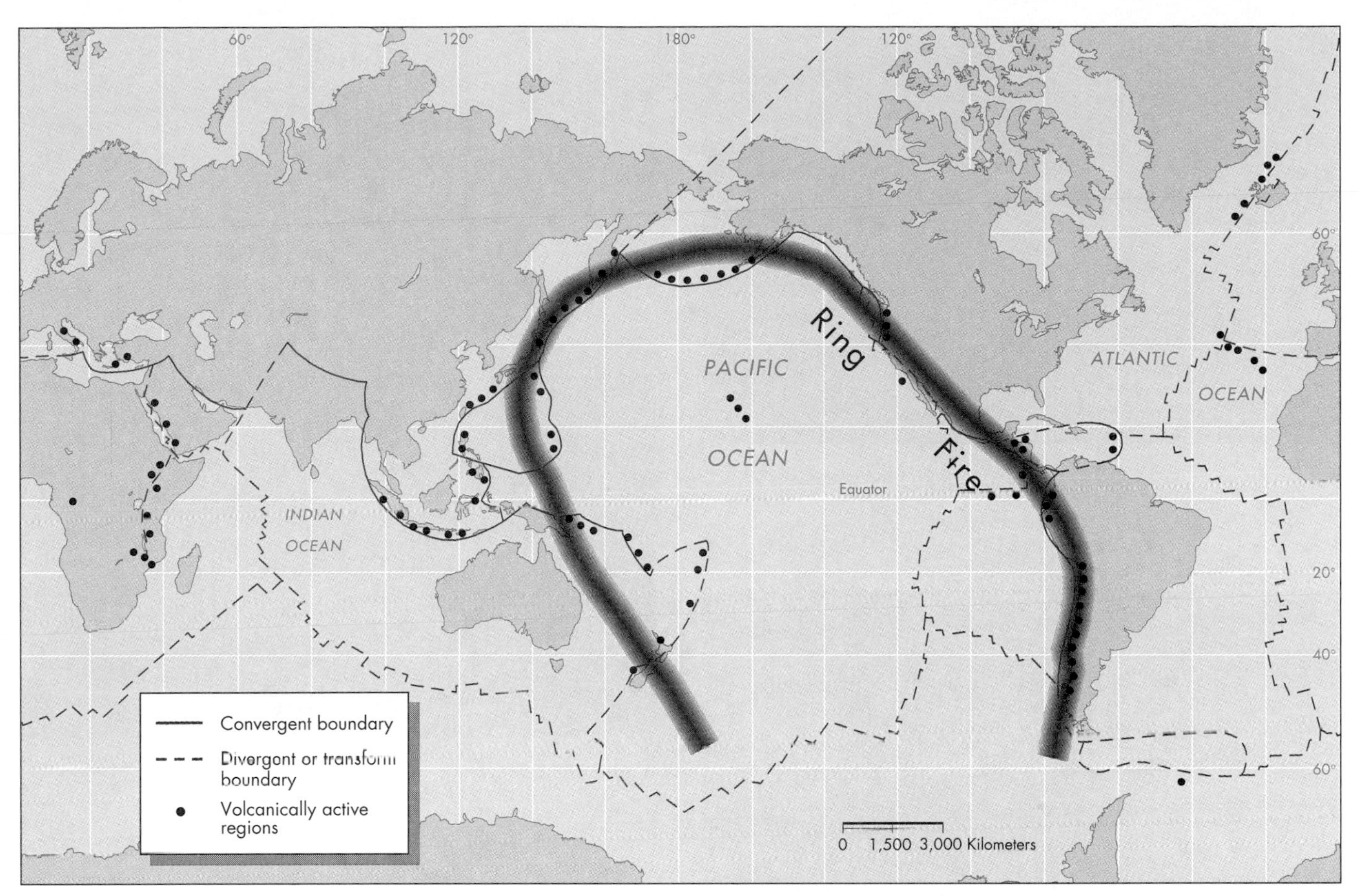

▲ **FIGURE 8.3** The "ring of fire" surrounding the Pacific plate. (*Modified from* J. E. *Costa and* V. R. *Baker.* 1981. Surficial Geology. *Originally published by* John Wiley *& Sons, New York, and republished by* Tech Books, *Fairfax, Virginia.*)

▲ **FIGURE 8.4** Profile of a shield volcano. Notice the profile to the summit is very gently curved like that of a warrior's shield. Mountain shown is Mauna Loa, Hawaii, as viewed from the Hawaiian Volcanic Observatory. (*John* S. *Shelton*)

▲ **FIGURE 8.5** Cinder cone with a small crater at the top near Springerville, Arizona. The black material in the center upper part of the photograph to the upper right-hand corner is a lava flow, which originates from the base of the cinder cone. (*Michael Collier*)

blasts, similar in form to the blast of a shotgun. We should consider such volcanoes armed and dangerous.

Volcanic Domes **Volcanic domes** are characterized by viscous magma with a relatively high silica content (about 70%). The common rock type is **rhyolite,** composed mostly of potassium and soda-rich feldspar, quartz, and minor ferromagnesian minerals. The activity of volcanic domes is mostly explosive, making these volcanoes very dangerous. Mt. Lassen, in northeastern California, is a good example of a volcanic dome. Mt. Lassen's last series of eruptions, from 1914 to 1917, included a tremendous horizontal blast that destroyed a large area (Figure 8.8).

▲ **FIGURE 8.6** Lava tube, Mauna Loa, Hawaii, 1984 eruption. (*Scott Lopez/Wind Cave National Park*)

▲ **FIGURE 8.7** Mt. Fuji, Japan, is a composite volcano with beautiful steep sides. (*Courtesy of James Jenni/Japan National Tourist Organization*)

▲ **FIGURE 8.8** Eruption of Lassen Peak in June 1914. (B. F. *Loomis, courtesy of Loomis Museum Association*)

Volcano Origins

We established earlier that the causes of volcanic activity are directly related to plate tectonics. We now return to that important topic. Understanding the tectonic origins of different types of volcanoes helps explain the chemical differences in their rock types. Figure 8.9 is an idealized diagram showing the relationship of processes at plate boundaries to the volcano types we have described. This diagram illustrates:

- **The occurrence of volcanism at mid-oceanic ridges that produces basaltic rocks** (Figure 8.9). Where these spreading ridge systems occur on land, as, for example, in Iceland, shield volcanoes are formed (Figure 8.10). Basaltic rock, which has a relatively low silica content, wells up directly from the asthenosphere as magma, mixing very little with other materials except oceanic crust, which itself is basaltic.
- **The occurrence of shield volcanoes above hot spots located below the lithospheric plates** (Figure 8.9). For example, the Hawaiian volcanoes are located well within the Pacific plate rather than near a plate boundary. It is currently believed that there is a hot spot below the Pacific plate where magma is generated. Magma moves upward through the plate and produces a volcano on the bottom of the sea that eventually may become an island. Because at the Hawaiian Islands the plate is moving roughly northwest over the stationary hot spot, a chain of volcanoes running northwest to southeast is formed. The "Big Island," Hawaii, is presently near the hot spot and is experiencing active volcanism and growth. Islands to the northwest, such as Molokai and Oahu, have evidently moved off the hot spot as the volcanoes on these islands are no longer active.
- **The occurrence of composite volcanoes associated with andesitic volcanic rocks and subduction zones** (Figure 8.9). These are common volcanoes found around the Pacific Rim. For example, volcanoes in the Cascade Range of Washington, Oregon, and California are related to the Cascadia Subduction Zone (Figure 8.11). Andesitic volcanic rocks are produced at subduction zones, where rising magma mixes with oceanic and continental crust. Because continental crust contains higher silicate content than basaltic magma, this produces rock with intermediate silicate content.
- **Occurrence of caldera-forming eruptions, which may be extremely explosive and violent** (Figure 8.9). These eruptions tend to be associated with rhyolitic rocks, which are produced when magma moves upward and mixes with continental crust. Rhyolitic rocks contain more silicate than do other volcanic rocks because

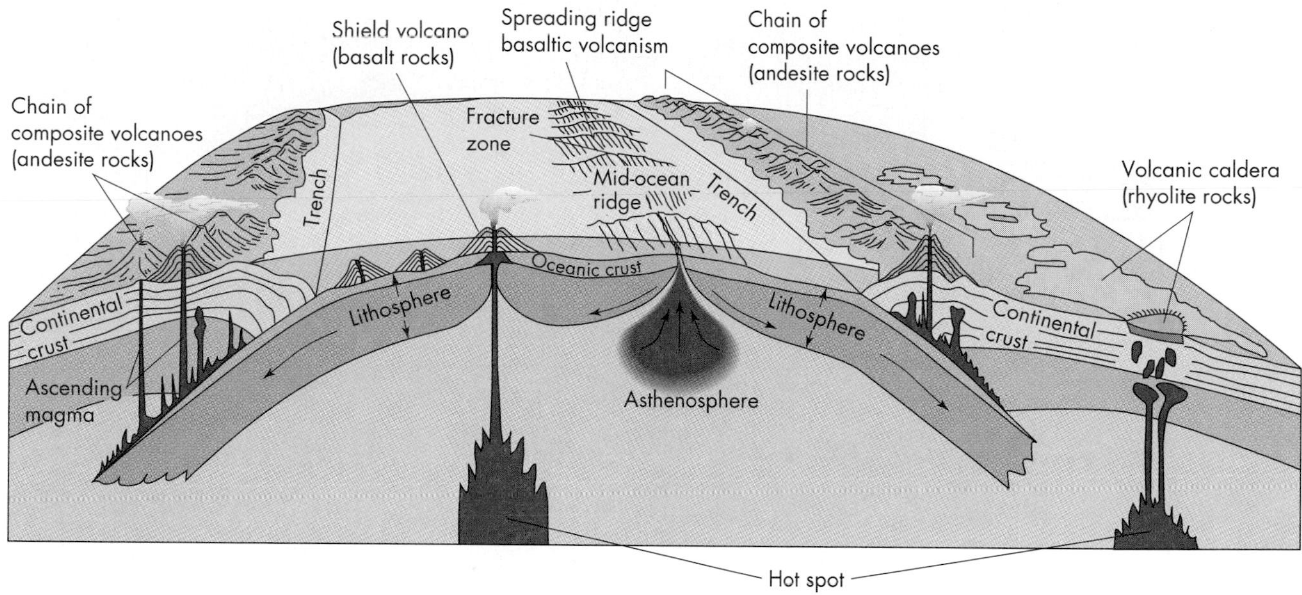

▲ **FIGURE 8.9** Idealized diagram showing plate tectonic processes and their relation to volcanic activity. (*Modified from Skinner and Porter.* 1992. The Dynamic Earth, *2nd ed. New York: John Wiley & Sons, p.* 96, *Figure* 3.32.)

▲ **FIGURE 8.10** Icelandic shield volcano (background) and large fissure (normal fault) associated with spreading of tectonic plates along the Mid-Atlantic Ridge. (*John S. Shelton*)

of the high silicate content of continental crust. Volcanic domes, although not always associated with caldera-forming eruptions, are usually found inland of subduction zones and erupt silica-rich rhyolitic magma.

Our brief discussion does not explain all differences in composition and occurrence of basaltic, andesitic, and rhyolitic rocks. However, it does establish the most basic relationships of plate tectonics to volcanic activity and volcanic rocks.

Volcanic Features

Features that are often associated with volcanoes or volcanic areas include craters, calderas, volcanic vents, geysers, and hot springs.

Craters, Calderas, and Vents The depressions commonly found at the top of a volcano are called **craters.** They form by explosion or collapse and may be flat-floored or funnel-shaped. Craters (which are usually a few kilometers in diameter) are much smaller than **calderas,** which are gigantic, often circular depressions resulting from explosive ejection of magma and subsequent collapse. Calderas may be several tens of kilometers in diameter and contain within them volcanic vents and other features. Volcanic vents are openings through which volcanic materials (lava and pyroclastic debris) are erupted at the surface of the earth. Vents may be roughly circular conduits or elongated fissures through which magma is erupted.

Hot Springs and Geysers Hot springs and geysers are hydrologic features found in some volcanic areas. Groundwater that comes into contact with hot rock becomes heated, and in some cases the heated water discharges at the surface as a **hot spring,** or **thermal spring.** More rarely, the subsurface groundwater system involves circulation and heating patterns that produce periodic release of steam and hot water at the surface, a phenomenon called a **geyser.** World-famous geyser basins or fields are found in Iceland, New Zealand, and Yellowstone National Park in Wyoming (Figure 8.12).

Caldera Eruptions

Giant volcanic depressions, or calderas, are produced by collapse following very rare but extremely violent eruptions. None have occurred anywhere on earth in the last few thousand years, but at least 10 such **caldera eruptions** have occurred in the last million years, three of them in North America. A large caldera-forming eruption may explosively extrude up to 1000 km^3 of pyroclastic debris, mostly ash

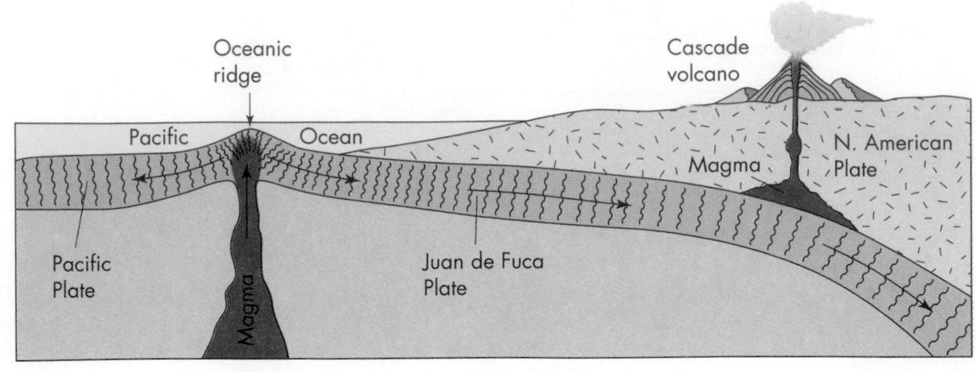

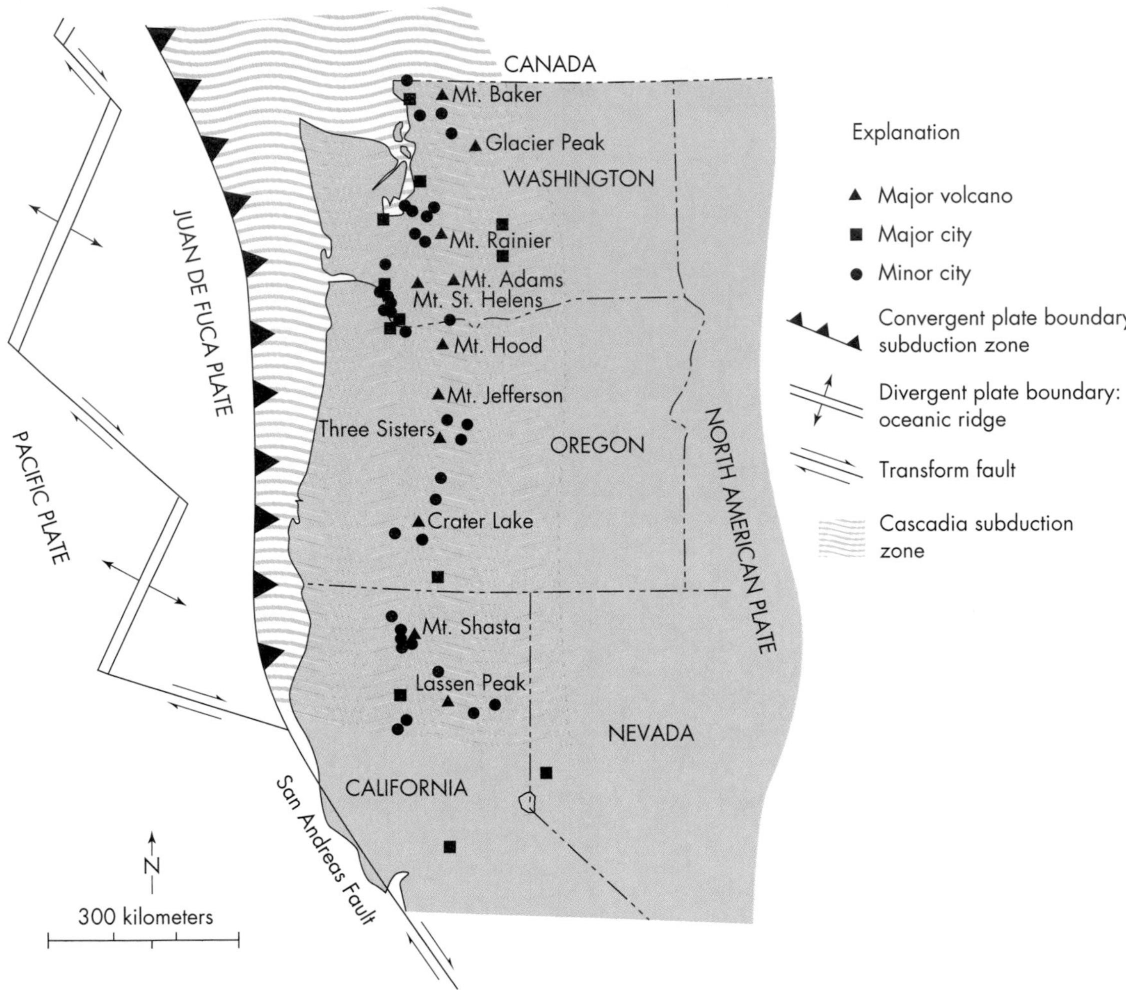

▲ **FIGURE 8.11** Map and plate tectonic setting of the Cascade Range showing major volcanoes and cities in their vicinity. See Figure 3.2 for regional tectonic environment. (*Map modified after Crandall and Waldron.* 1969. Disaster Preparedness. *Office of Emergency Preparedness.*)

(about 1000 times the quantity ejected by the 1980 eruption of Mount St. Helens), producing a crater more than 10 km in diameter and blanketing an area of several tens of thousands of square kilometers with ash. These ash-flow and ash-fall deposits can be 100 m thick near the crater rim and a meter or so thick 100 km away from the source (5). The most recent caldera-forming eruptions in North America occurred about 600,000 years ago at Yellowstone National Park in Wyoming and 700,000 years ago in Long Valley, California. Figure 8.13 shows the area covered by ash in the latter event, which produced the Long Valley Caldera.

The main events in a caldera-producing eruption can be over very quickly, in a few days to a few weeks, but intermittent, lesser-magnitude volcanic activity can linger on for a million years. Thus, the Yellowstone event has left us hot springs and geysers, including Old Faithful, while the Long Valley

▲ **FIGURE 8.12** Eruption of Old Faithful Geyser, Yellowstone National Park, Wyoming. The geyser is named Old Faithful because of its very predictable periodic eruption. (*James Randklev/Tony Stone Images*)

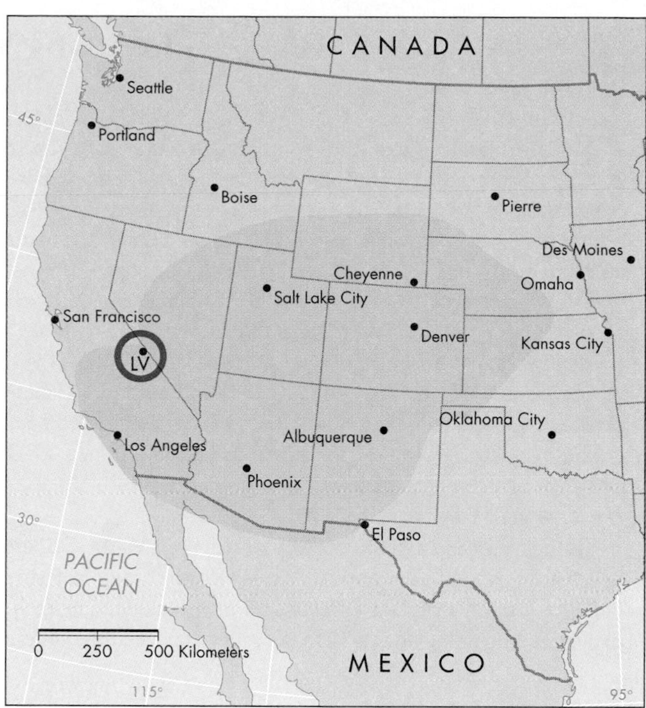

▲ **FIGURE 8.13** Area covered by ash from the Long Valley Caldera eruption approximately 700,000 years ago. Circle near Long Valley (LV) has a radius of 120 km and encloses the area subject to at least 1 m of downwind ash accumulation if a similar eruption were to occur again. Also within this circle, hot pyroclastic flows (ash flows) are likely to occur and, in fact, could extend farther than shown by the circle. (*From C. D. Miller, D. R. Mullineaux, D. R. Crandell, and R. A. Bailey.* 1982. Potential Hazards from Future Volcanic Eruptions in the Long Valley–Mono Lake Area, East-Central California and Southwest Nevada—A Preliminary Assessment. U.S. *Geological Survey Circular* 877.)

event has left us a potential volcanic hazard (see *A Closer Look: The Long Valley Caldera*). Actually, both sites are still capable of producing volcanic activity because magma is still present at variable depths beneath the caldera floors. Both calderas are *resurgent calderas* because their floors have slowly domed upward since the explosive eruptions that formed them.

8.3 Volcanic Hazards

Volcanic hazards include the primary effects of volcanic activity, which are the direct results of the eruption, and secondary effects, which may be caused by the primary effects. Primary effects include lava flows, pyroclastic activity (ash fall, ash flows, and lateral blasts), and the release of gases—mostly steam, but sometimes corrosive or poisonous gases. Secondary effects include debris flows, mudflows, landslides (debris avalanche) floods, and fires (6).

Lava Flows

Lava flows are one of the most familiar products of volcanic activity. They result when magma reaches the surface and overflows the crater or a volcanic vent along the flanks of the volcano. The three major groups of lava have the names of the volcanic rocks: **basaltic,** which is by far the most abundant, **andesitic,** and **rhyolitic.**

Lava flows can be quite fluid and move quite rapidly, or they can be relatively viscous and move slowly. Basaltic lavas, with a silica content of about 50 percent, exhibit a range of velocities. Those with the highest gas content and eruptive temperatures are the fastest moving, with a usual velocity of a meter or so per hour; these lavas have a smooth surface texture when they harden (Figure 8.14). Cooler, less gas-rich basaltic lava flows move at rates of a few meters per day and have a "blocky" texture after hardening (Figure 8.15). With the exception of some flows on steep slopes,

▲ **FIGURE 8.14** Smooth-surface texture lava flow surrounding and destroying a home at Kalapana, Hawaii, on the Big Island of Hawaii in 1990. Such flows destroyed over 100 structures including the National Park Service Visitor Center. These types of flows are often called *ropy lava* because of the surface texture which looks a bit like pieces of rope lying side by side as if coiled. (*Paul Richards/UPI/Corbis*)

A CLOSER LOOK The Long Valley Caldera

About 1980, the rate of uplift in the Long Valley Caldera (Figure 8.A) accelerated from a much lower rate (up to 25 cm in only about 2 years), and this change was accompanied by earthquake swarms. Some of these earthquakes had magnitudes of 5 to 6 and were moving closer to the surface (from an 8-km depth in 1980 to 3.2 km in 1982). This suggested that magma was moving toward the surface, and in doing so, it was breaking rocks and producing earthquakes. Concern over a possible volcanic eruption prompted the U.S. Geological Survey to issue a notice of potential volcanic hazard in 1982. A notice of hazard from continued, severe earthquakes had been issued in 1980. Swarms of earthquakes continued into 1983, and although the notices have been lifted, the future situation remains uncertain (10).

Both a map and a cross section of the Long Valley Caldera and Mammoth Lakes, a popular ski area, are shown in Figure 8.B. The last major volcanic eruption in the caldera since its explosive beginning about 700,000 years ago occurred about 100,000 years ago, but smaller eruptions, including steam explosions and explosive eruptions of rhyolite, occurred as recently as 400 to 500 years ago at the site of the Inyo craters (see Figure 8.B).

Potential future eruptions in the Long Valley area range in magnitude from the unlikely, very large, catastrophic event to more probable smaller eruptions. Geologists now believe that magma could reach the surface. Therefore, monitoring of uplift, tilt, earthquakes, and hot springs activity continues. Figure 8.A shows the potential ash fall and ash flow hazard zones from a possible future Long Valley explosive eruption that could eject about 1 km^3 of material. Citizens of Mammoth Lakes and other nearby communities took the warning seriously, "preparing for the worst (ready to evacuate if necessary) and hoping for the best," while pursuing their business in providing recreation to visiting urbanites (10,11). Given the potentially catastrophic loss of life and property from a major eruption in this area, concern is certainly warranted.

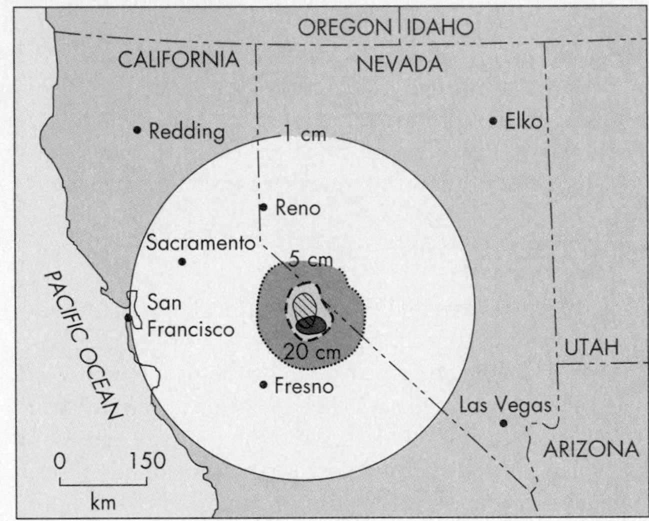

▶ **FIGURE 8.A** Location and potential hazards from a volcanic eruption at the Long Valley Caldera near Mammoth Lakes, California. Red-colored area and diagonal lines show the hazard from flowage events out to a distance of approximately 20 km from recognized potential vents. Lines surrounding the hazard areas represent potential ash thicknesses of 20 cm at 35 km (dashed line), 5 cm at 85 km (dotted line), and 1 cm at 300 km (solid line). These estimates of potential hazards assume an explosive eruption of approximately 1 km^3 from the vicinity of recently active vents. (*From* C. D. *Miller*, D. R. *Mullineaux*, D. R. *Crandell*, *and* R. A. *Bailey*. 1982. Potential Hazards from Future Volcanic Eruptions in the Long Valley–Mono Lake Area, East-Central California and Southwest Nevada—A Preliminary Assessment. U.S. *Geological Survey* Circular 877.)

most lava flows are slow enough that people can easily move out of the way as they approach (7).

Lava flows from Kilauea volcano, Hawaii (Figure 8.14), have been active for the past several years. By May 1990, more than 50 homes in the village of Kalapana were destroyed by lava flows, and by August 1990 lava flowed across part of the famous Kaimu Black Sand Beach and into the ocean. The village of Kalapana as of 1995 has virtually disappeared, and it will be many decades before much of the land is productive again. On the other hand, the eruptions, in concert with beach processes, have produced new black sand beaches. The sand is produced when the molten lava enters the ocean and fragments into sand-sized particles.

Control Methods Several methods, such as bombing, hydraulic chilling, and constructing walls, have been employed to deflect lava flows away from populated or

▲ **FIGURE 8.15** Blocky lava flow engulfing a building during the eruption on the island of Heimaey, Iceland. (*Solarfilma* HF)

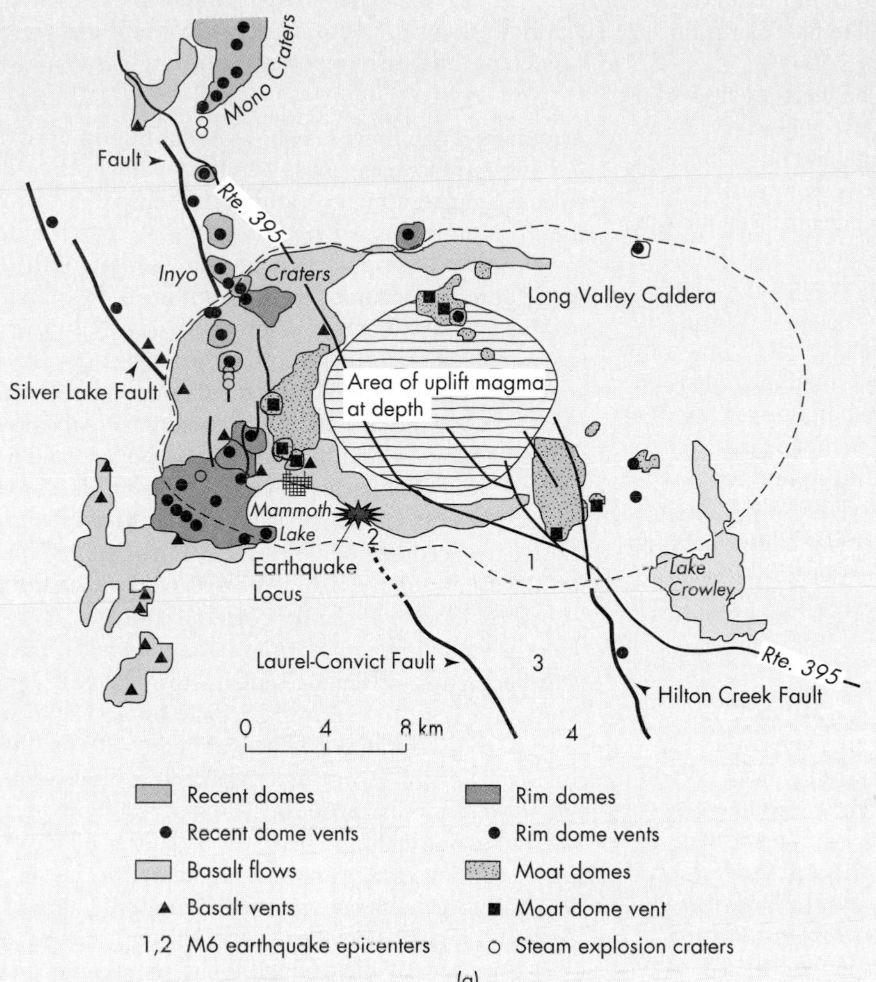

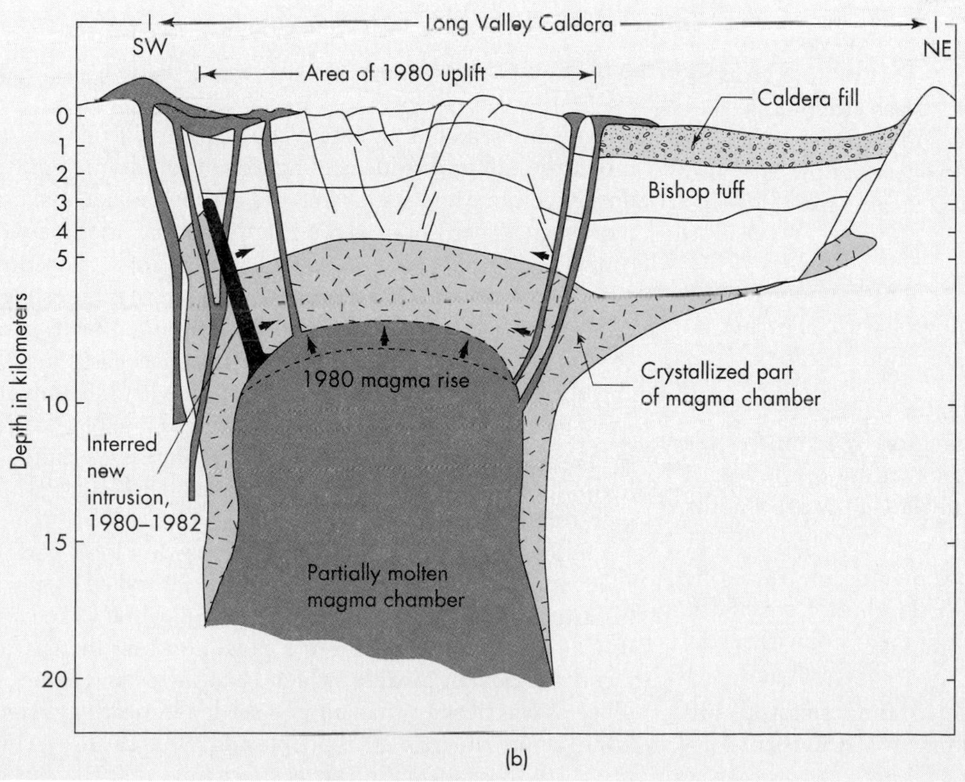

◀ FIGURE 8.B Map (a) and diagram (b) illustrating the volcanic hazard near Mammoth Lakes, California. The map (a) shows the location of past volcanic events, the area of uplift where magma seems to be moving up, and finally the area where earthquake swarms have occurred near Mammoth Lakes. The geologic cross section (b) shows a section (northeast-southwest) through the Long Valley Caldera. Shown are geologic relations inferred for the 1980 magma rise that produced the uplift and swarms of earthquakes. (*From* R. A. *Bailey*. 1983. Mammoth Lakes earthquakes and ground uplift: Precursor to possible volcanic activity? *In* U.S. *Geological Survey Yearbook, Fiscal Year* 1982.)

otherwise high-priority areas. These methods have had mixed success. They cannot be expected to block large flows, and they need further evaluation.

Bombing of the lava flows has been attempted to stop their advance. It has proven most effective against lava flows in which fluid lava is confined to a channel by congealing lava on the margin of the flow. The objective is to partly block the channel, thereby causing the lava to pile up, facilitating an upstream break through which the lava escapes to a less damaging route. The established procedure is successive bombing at higher and higher points as necessary to control the threat. Bombing has some merit for future research, but from our experience so far it seems likely that it will provide little protection in most cases. The results of bombing are unpredictable, and in any event, this method cannot be expected to affect large flows. Poor weather conditions and the abundance of smoke and falling ash can also reduce the effectiveness of bombing (8). Hydraulic chilling of lava flows (cooling the flow with water) has sometimes been successful (see *Case History: Controlling a Lava Flow in Iceland*).

Pyroclastic Hazards

Pyroclastic activity describes explosive volcanism, in which tephra is physically blown from a volcanic vent into the atmosphere. Several types of pyroclastic activity occur. In **volcanic ash eruptions,** or **ash fall,** a tremendous quantity of rock fragments, natural glass fragments, and gas are blown high into the air by explosions from the volcano. **Lateral blasts** are explosions of gas and ash from the side of a volcano. The ejected material travels away from the volcano at tremendous velocity, sometimes exceeding the speed of sound (lateral blasts can be very destructive; see the discussion of Mount St. Helens in this chapter). **Pyroclastic flows,** or **ash flows,** are one of the most lethal aspects of volcanic eruptions. They are avalanches of very hot pyroclastic materials—ash, rock, volcanic glass fragments, and gas—that are blown out of a vent and move very rapidly down the sides of the volcano. Pyroclastic flows are also known as *hot avalanches*, *ignimbrites*, and *nuée ardentes* (French for "glowing cloud").

Ash Fall Volcanic ash eruptions can cover hundreds or even thousands of square kilometers with a carpet of volcanic ash. Ash eruptions create several hazards:

- Vegetation, including crops and trees, may be destroyed.
- Surface water may be contaminated by sediment, resulting in temporary increase in acidity of the water. The increase in acidity generally lasts only a few hours after the eruption ceases.
- Structural damage to buildings may occur, caused by the increased load on roofs. A depth of 1 cm of ash places an extra 2.5 tons of weight on a roof with a surface area of about 140 m^2.
- Health hazards such as irritation of the respiratory system and eyes are caused by contact with the ash and associated caustic fumes (7).
- Engines of jetliners may "flame-out" as melted silica-rich ash forms a thin coating of volcanic glass on fuel injectors and turbine vanes in the engines (9).

Ash Flows Ash flows may be as hot as hundreds of degrees Celsius and move as fast as 100 km/hr down the sides of a volcano, incinerating everything in their path. Fortunately, they seldom occur in populated areas, but they can be catastrophic if a populated area is in the path of the flow. A tragic example occurred in 1902 on the West Indian island of Martinique. On the morning of May 8, a flow of hot, incandescent ash, steam, and other gases *(nuée ardentes)* roared down Mount Pelée and through the town of St. Pierre, killing 30,000 people. A prisoner in jail was one of the two survivors, and he was severely burned and horribly scarred. Reportedly, he spent the rest of his life touring circus sideshows as the "Prisoner of St. Pierre." Flows like these have occurred on volcanoes of the Pacific Northwest in the past and can probably be expected in the future.

Another type of ash flow is a **base surge,** which forms when ascending magma comes in contact with water on or near the earth's surface in a violent steam and ash explosion. Such an eruption occurred in 1911 on an island in Lake Taal, in the Philippines, killing about 1300 inhabitants on the island and lake shore as a tremendous blast swept across the water. A similar event occurred at the same location in 1965, this time claiming perhaps as many as 200 lives. Base-surge eruptions are commonly associated with small volcanoes with bowl-shaped craters like that of Diamond Head, Hawaii. Many extinct volcanoes of this type can be found in Christmas Lake Valley, which is the remains of an ancient lake in south-central Oregon, and in the Tule Lake region of northern California.

Poisonous Gases

Various gases, including water vapor, carbon dioxide, carbon monoxide, sulfur dioxide, and hydrogen sulfide, are emitted during volcanic activity. Water and carbon dioxide make up more than 90 percent of all the emitted gases. Hazardous volcanic gases rarely reach populated areas in toxic concentrations. However, sulfur dioxide can react in the atmosphere to produce acid rain downwind of an eruption. Finally, toxic concentrations of some chemicals emitted as gases may be absorbed by volcanic ash that falls onto the land. Eventually this toxic ash is incorporated into the soil and into plants eaten by people and livestock. Fluorine, for example, is erupted as hydrofluoric acid, which can be absorbed by volcanic ash. It may also be leached into water supplies (2).

Dormant volcanoes may emit gases for long periods following eruptions. This was dramatically and tragically illustrated on the night of August 21, 1986, when Lake Nyos in Cameroon, Africa, vented a poisonous gas, consisting mostly of carbon dioxide, which is colorless and odorless. The gas was heavier than air and settled in nearby villages, killing approximately 2000 people and 3,000 cattle (4) (Figure 8.16) by asphyxiation. It is speculated that the carbon

CASE HISTORY Controlling a Lava Flow in Iceland

The world's most ambitious program to control lava flows was initiated in January 1973 on the Icelandic island of Heimaey, when basaltic lava flows from Mount Helgafell nearly closed the harbor of the island's main town, Vestmannaeyjar, threatening the continued use of the island as Iceland's main fishing port. The situation prompted immediate action. Experiments on the island of Surtsey in 1967 had shown that water could be used to stop the advance of lava, and favorable conditions existed on Heimaey to try it on a large scale. First, the main flows were viscous and slow-moving (advancing at less than a meter per hour), allowing the necessary time to initiate a control program. Second, transportation by sea and the local road system was adequate to move the necessary pumps, pipes, and heavy equipment. Third, water was readily available.

The procedure was to first cool the margin and surface of the flow with numerous fire hoses fed from a 13-cm pipe (Figure 8.C). Next, bulldozers were moved up on the slowly advancing flow, making a track or road on which the plastic pipe was placed. The pipe did not melt as long as water was flowing in it, and small holes in the pipe also helped cool hot spots along various parts of the flow. Each fire hose delivered water at approximately 0.03 to 0.3 m^3 per second to an area about 50 m in back of the edge of the flow. Watering had little effect the first day, but then that part of the flow began to slow down and in some cases stopped.

The program undoubtedly had an important effect on lava flows. It tended to restrict their movement and thus reduced property damage. After the outpouring of lava stopped in June 1973, the harbor was still usable (12). In fact, by fortuitous circumstances, the shape of the harbor was actually improved, and the new rock now provides additional protection from the sea.

(a)

(b)

(c)

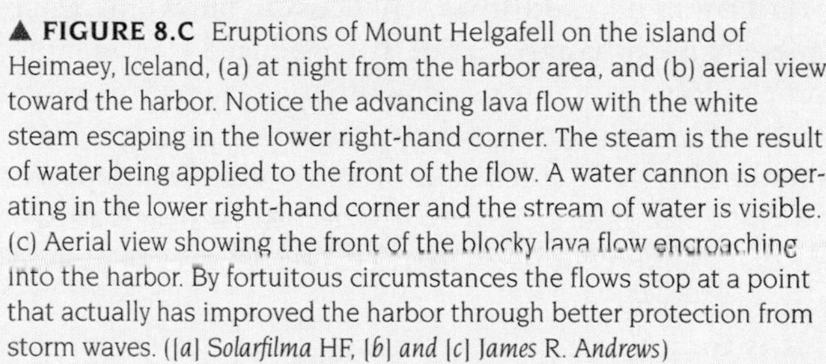

▲ **FIGURE 8.C** Eruptions of Mount Helgafell on the island of Heimaey, Iceland, (a) at night from the harbor area, and (b) aerial view toward the harbor. Notice the advancing lava flow with the white steam escaping in the lower right-hand corner. The steam is the result of water being applied to the front of the flow. A water cannon is operating in the lower right-hand corner and the stream of water is visible. (c) Aerial view showing the front of the blocky lava flow encroaching into the harbor. By fortuitous circumstances the flows stop at a point that actually has improved the harbor through better protection from storm waves. (*[a] Solarfilma HF, [b] and [c] James R. Andrews*)

dioxide and other gases were slowly released into the bottom waters of the lake, which occupies part of the crater of a dormant volcano. As long as such gas is dissolved in the lake water and kept there by the pressure of the overlying water, there is no problem. However, if the lake water is suddenly disturbed by a subaqueous landslide or small earthquake that suddenly brings the water from the bottom of the lake to the surface (as was probably the case with Lake

(a)

(a)

▲ **FIGURE 8.16** (a) Lake Nyos in 1986 released immense volumes of carbon dioxide (T. *Orban/Sygma*); (b) the gas asphyxiated about 2000 people and killed numerous animals (*Peter Turnley/Black Star*)

Nyos), the gases may enter the atmosphere. The Cameroon event is not unique, for in 1984 an underwater landslide in a nearby lake evidently initiated the release of carbon dioxide gas that took the lives of 37 people.

In Japan, volcanoes are monitored to detect releases of poisonous gas, such as hydrogen sulfide. When releases are detected, sirens are sounded to advise people to evacuate to high ground to escape the gas. Scientists speculate that a warning system in Cameroon would not have been effective because the release of gas was very sudden. On the other hand, the event is evidently quite rare and unlikely to occur in other parts of the world with any great regularity (13).

Debris Flows and Mudflows

The most serious secondary effects of volcanic activity are **debris flows** and **mudflows,** collectively known by their Japanese name of **lahar.** Lahars are produced when a large volume of loose volcanic ash and other ejecta becomes saturated and unstable and moves suddenly downslope. The distinction between a debris flow and a mudflow depends upon the dominant size of the particles. In debris flows, more than 50 percent of the particles are coarser than sand (2 mm in diameter).

It is interesting to note that what are probably the largest active landslides on earth are located on the "Big Island" of Hawaii. Presently these are slow moving (about 10 cm/yr) with dimensions of up to 100 km wide, 10 km thick, and 20 km long that extend from a volcanic rift zone (on land) to a submarine terminus. The worry is that some of these slow-moving landslides that contain blocks of rock the size of Manhattan Island might (as they have in the past) become giant, fast-moving submarine debris avalanches, generating huge tsunamis that deposit marine debris hundreds of meters above sea level at nearby islands and cause catastrophic damage around the Pacific Basin. Fortunately, such high-magnitude events apparently happen only every 100,000 years or so (4).

Debris Flows Research completed at several volcanoes suggests that even relatively small eruptions of hot volcanic material may quickly melt large volumes of snow and ice. The copious amounts of meltwater produce floods, which may erode and incorporate sediment such as volcanic ash and other material on the slope of the volcano, forming debris flows. Volcanic debris flows are fast-moving mixtures of sediment (including blocks of rock) and water with the general consistency of wet concrete. Debris flows can travel many kilometers down valleys from the flanks of the volcano where they were produced (7). For example, early in 1990 a pyroclastic flow from the Redoubt volcano in Alaska rapidly melted snow and ice while moving across Drift Glacier. Voluminous amounts of water and sediment produced a debris flow that quickly moved down the valley, with a discharge comparable to that of the Mississippi River at flood stage. Fortunately, the event was in an isolated area, so no lives were lost (2).

Mudflows Gigantic mudflows have originated on the flanks of volcanoes in the Pacific Northwest (6). The paths of two ancient mudflows that originated on Mt. Rainier are shown in Figure 8.17. Deposits of the Osceola mudflow are 5000 years old. This mudflow traveled more than 80 km from the volcano and involved more than 1.9 billion m^3 of debris, equivalent to 13 km^2 of debris piled to a depth of more than 150 m. Deposits of the younger 500-year-old Electron mudflow traveled about 56 km from the volcano and involved in excess of 150 million m^3 of mud.

Hundreds of thousands of people now live within the area covered by these old flows, and there is no guarantee that similar flows will not occur again. Figure 8.18 shows the potential relative risk of lahars for the Mt. Rainier region. Construction of the hazard map is based on a computer model that utilizes topographic data and volumes of lahars to predict areas inundated by flows and degree of hazard (14). Someone in the valley facing the advance of such a flow would describe it as a wall of mud a few meters high, moving at about

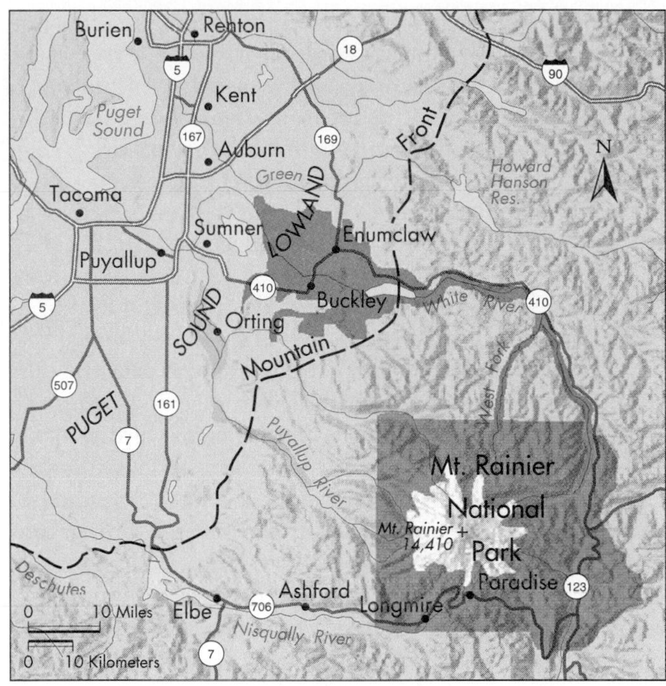

▲ **FIGURE 8.17** Map of Mt. Rainier and vicinity showing the extent of the Osceola mudflow in the White River Valley (colored orange) and the Electron mudflow (colored beige) in the Puyallup River Valley. (*From Crandell and Mullineaux*, U.S. Geological Survey Bulletin 1238, 1967.)

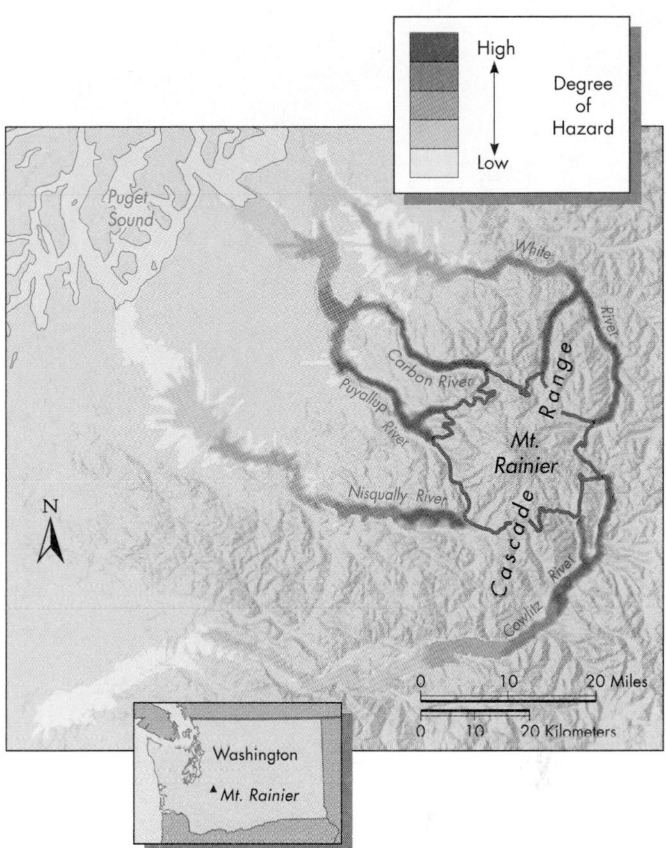

▲ **FIGURE 8.18** Hazard map of the Mt. Rainier region. Showing degree of hazard from a volcanic-related debris flow/mudflow (lahar). Based on a computer model that considers topography of the volcano, and volume of the lahar to predict how far from the volcano a lahar will flow and what area will be inundated. (*Source: Iverson, R. M., Schilling, S. P., and Vallance, J. W. 1998. Objective delineation of lahar-inundation hazard zone.* Geological Society of America Bulletin 110(8):972–984.)

30 km/hr. This observer would first see the flow at a distance of about 1.6 km and would need a vehicle headed in the right direction toward high ground to escape being buried alive (7).

It is instructive to compare the potential hazard of volcanic debris flows and mudflows with that of river flooding. Floods are usually preceded by heavy rains that cause a gradual rise in water level. People in flood-prone areas generally have time to escape, and when the flood recedes, the water and danger are essentially gone. However, catastrophic mudflows can occur with little or no warning, often starting when the volcano is hidden by clouds of smoke. In addition, the debris—rock, mud, and so on, perhaps a few meters thick—remains after the event (7).

Because debris flows and mudflows are confined to valleys, another possible hazard arises when the valley is artificially dammed to produce hydroelectric power. A large debris or mudflow could fill a reservoir, pushing the water over the spillway and causing a severe flood downstream. On the other hand, used wisely, reservoirs might be a safety factor for all except the really large flows. The water level in reservoirs might be drawn down during an upstream volcanic event, and the storage basin behind the dam could be used to contain a possible flow. This is not the intended function of the dam, but it is a fortunate safety mechanism (7).

8.4 Some Case Histories

Mt. Unzen, Japan

Japan has 19 active volcanoes. Nearly 200 years ago Mt. Unzen in southwestern Japan erupted, killing approximately 15,000 people. The volcano awoke from its 200-year dormancy in 1990. It erupted violently in June 1991, and authorities ordered the evacuation of thousands of residents. By the end of 1993, about 0.2 km^3 of lava had erupted and more than 8000 pyroclastic flows had occurred (Figure 8.19), giving Mt. Unzen the dubious honor of being one of the pyroclastic flow centers of the world. This volcano has provided scientists with a crucial natural laboratory for studying such flows (1,3). The 1991 eruption also produced damaging mudflows. A specially designed channel was constructed to contain the mudflows, but, as seen in Figure 8.20, the flows overflowed the channel, burying many homes in mud.

Nevado del Ruiz

The volcano in Colombia known as Nevado del Ruiz erupted on November 13, 1985. The eruption triggered catastrophic mudflows that killed at least 23,000 people while inflicting more than $200 million in property damage. The eruption followed a year of precursory activity, including earthquakes and hot-spring activity. Monitoring of the volcano began in July 1985, and in October a hazard map was completed that correctly identified events that actually occurred on November 13. The report accompanying the map

▲ **FIGURE 8.19** Eruption of Mt. Unzen, Japan, June 1991. The cloud in the mountain area is moving rapidly downslope as an ash flow, and the fire fighter is running for his life. (*Yomiuri Shimbun*/AP/*Wide World Photos*)

gave a 100 percent probability that potentially damaging mudflows would be produced by an eruption, as had been the case with previous eruptions.

The November 13 event included two large explosions. The eruptions produced pyroclastic flows that scoured and melted glacial ice on the mountain, generating the mudflows that raced down river valleys. Figure 8.21 shows the areas impacted by base surges and pyroclastic flows, as well as the location of glacial ice that contributed the water necessary to produce the mudflows. Of particular significance was the mudflow that raced down the River Lagunillas and destroyed part of the town of Armero, where most of the deaths occurred. Figure 8.22 shows the volcano and the town of Armero. Mudflows buried the southern half of the town, sweeping buildings completely off their foundations (15).

The real tragedy of the catastrophe was that the outcome was predicted; in fact, there were several attempts to warn the town and evacuate it. Hazard maps were available and circulated in October, but they were largely ignored. Figure 8.23 shows the hazard map of events expected before the eruption *and* the events that occurred. This graphically illustrates the usefulness of volcanic risk maps (2). Despite these warnings, there was little response, and as a result at least 23,000 people died. Early in 1986 a permanent volcano observatory center was established in Colombia to continue monitoring the Ruiz volcano as well as others in South America. Today, South America should be better prepared to deal with future volcanic eruptions. Had there been better communication lines from civil defense headquarters to local towns, and a fuller appreciation of potential volcanic hazards even 40 km from the volcano, evacuation would have been possible for Armero. It is hoped that the lessons learned from this event will help minimize future loss of life associated with volcanic eruptions and other natural disasters.

Mt. Pinatubo

The June 15–16, 1991, eruptions of Mt. Pinatubo in the Philippines (Figure 8.24) may be the largest of the century. The combined effects of ash fall, debris flows, mudflows, and a typhoon resulted in the deaths of about 350 people. Most fatalities were due to collapse of buildings as volcanic ash accumulated on roofs to thicknesses of 30 cm as far as

(a)

(b)

▲ **FIGURE 8.20** (a) Mudflows from Mt. Unzen in 1991 damaged many homes in Shimabara, Japan. Flows overflowed the channels constructed to contain them. (b) The flows inundated many homes and buildings. (*Michael S. Yamashita*)

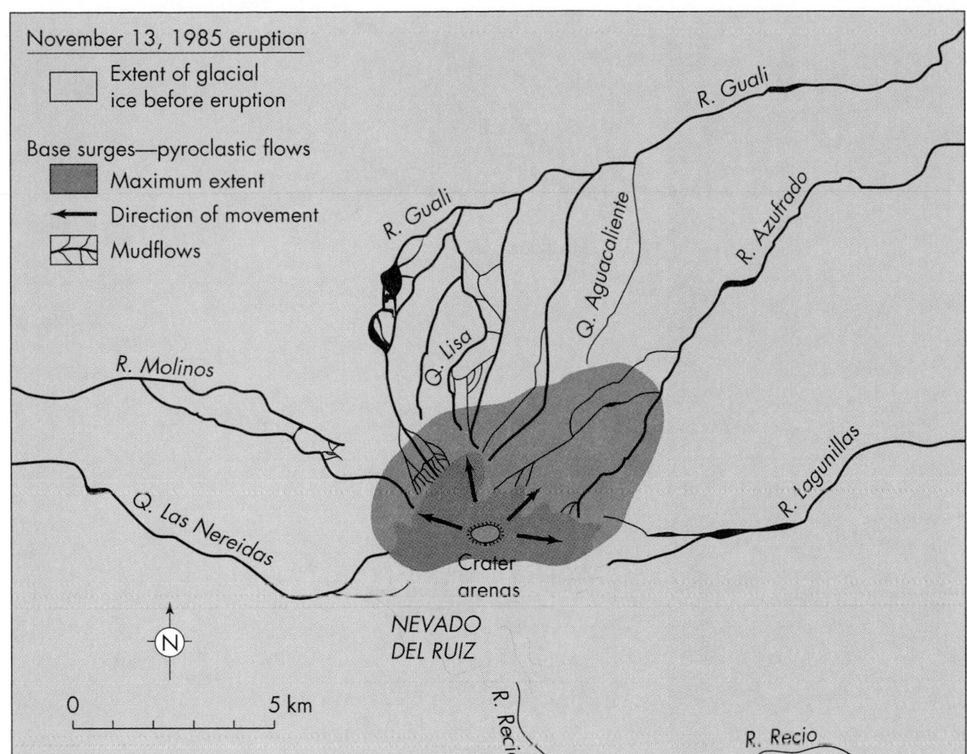

◀ **FIGURE 8.21** Map of the volcano Nevado del Ruiz area showing some of the features associated with the eruption of November 13, 1985. (*Modified after* D. G. *Herd.* 1986. *The Ruiz volcano disaster.* EOS, Transactions of the American Geophysical Union, *May* 13:457–60.)

(a)

(b)

▲ **FIGURE 8.22** In November of 1985 Nevado del Ruiz (a) erupted, generating a mudflow that nearly destroyed the town of Armero, killing 21,000 people (b). ([*a*] *El Espectador/Sygma, and* [*b*] J. *Langevin/Sygma*)

40 km from the volcano (2). Evacuation of 60,000 people from villages and a U.S. military base within a radius of 30 km from the summit saved thousands of lives.

The eruption column from the tremendous explosions at Pinatubo sent a cloud of ash 400 km wide to elevations of 34 km (4). As with similar past events of this magnitude, the aerosol cloud of ash, including sulfur dioxide, remained in the atmosphere, circling the earth for as long as a year. Ash particles and sulfur dioxide scattered incoming sunlight and slightly cooled the global climate during the year following the eruptions (2,16).

Mount St. Helens

The May 18, 1980, eruption of Mount St. Helens in the southwest corner of Washington (see Figure 8.2) exemplifies the many types of volcanic events expected from a Cascade volcano. The eruption, like many natural events, was unique and complex, making generalizations somewhat difficult. Nevertheless, we have learned a great deal from Mount St. Helens, and the entire story is not yet complete.

Mount St. Helens awoke in March 1980, after 120 years of dormancy, with seismic activity and small explosions as

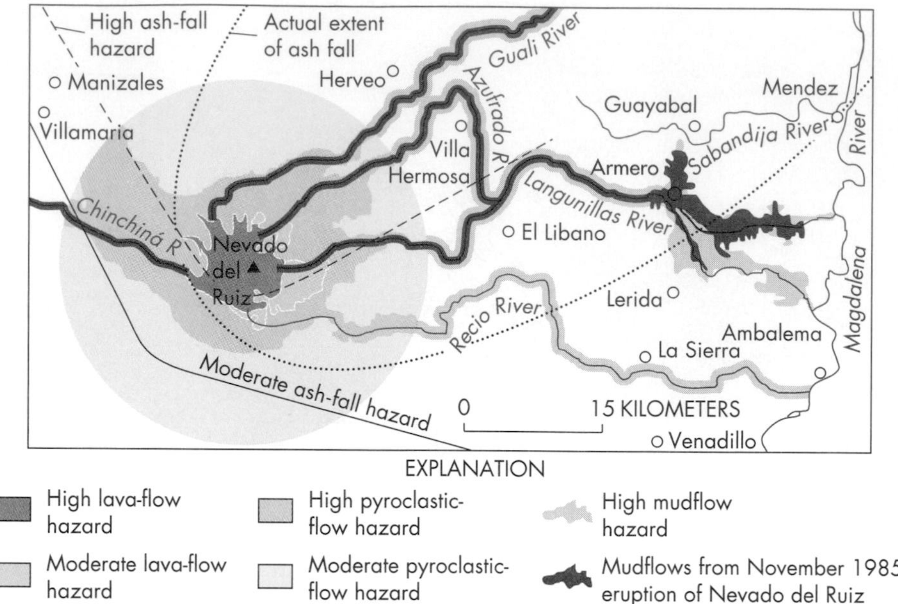

▶ **FIGURE 8.23** Volcanic hazard map produced and circulated 1 month prior to the November 13, 1985, eruption of Nevado del Ruiz and mudflows that buried Armero, Colombia. Shown in red are the actual mudflow deposits. (*Source:* T. L. *Wright and* T. C. *Pierson.* 1992. U.S. *Geological Survey* Circular 1073.)

groundwater came in contact with hot rock. By May 1, a prominent bulge on the northern flank of the mountain could be clearly observed and was growing at a rate of about 1.5 m per day. At 8:32 A.M. on May 18, 1980, a magnitude 5.0 earthquake registered on the volcano triggered a large landslide/debris avalanche (approximately 2.5 km^3), which involved the entire bulge area (Figures 8.25 and 8.26). The avalanche shot down the north flank of the mountain, displacing water in nearby Spirit Lake, struck and overrode a ridge 8 km to the north, then made an abrupt turn and moved for a distance of 18 km down the Toutle River (Figure 8.26a and b).

▲ **FIGURE 8.24** Thousands of people were evacuated, including more than 1000 from U.S. Clark Air Base, during this large ash eruption and explosion of Mt. Pinatubo in the Philippines on June 12, 1991. (*Carlo Cortes/Reuters/Corbis*)

The initial failure of the bulge released internal pressure, and seconds later Mount St. Helens erupted with a lateral blast directly from the area that the bulge had occupied (Figure 8.25c). After the lateral blast, a large vertical cloud rose quickly to an altitude of approximately 19 km (Figure 8.25d). Eruption of the vertical column continued for more than 9 hours, and large volumes of volcanic ash fell on a wide area of Washington, northern Idaho, and western and central Montana (Figure 8.26c). The total amount of volcanic ash ejected was about 1 km^3, and during the 9 hours of eruption a number of ash flows swept down the northern slope of the volcano. The entire northern slope of the volcano, which is the upper part of the north fork of the Toutle River basin, was devastated. Forested slopes were transformed into a gray, hummocky landscape consisting of volcanic ash, rocks, blocks of melting glacial ice, narrow gullies, and hot steaming pits (Figure 8.27) (17).

The first of several mudflows, consisting of a mixture of water, volcanic ash, rock, and organic debris (such as logs), occurred minutes after the start of the eruption. The flows and accompanying floods raced down the valleys of the north and south forks of the Toutle River at estimated speeds of 29 to 55 km/hr, threatening the lives of people camped along the river (see *Case History: Surviving to Tell the Tale*). Water levels in the river reached at least 4 m above flood stage, and nearly all bridges along the river were destroyed. The hot mud quickly raised the temperature of the Toutle River to as high as 38°C. Mud, logs, and boulders were carried 70 km downstream into the Cowlitz River and eventually 28 km farther downstream into the Columbia River. Nearly 40 million m^3 of material was dumped into the Columbia River, reducing the depth of the shipping channel from a normal 12 m to 4.3 m for a distance of 6 km (17).

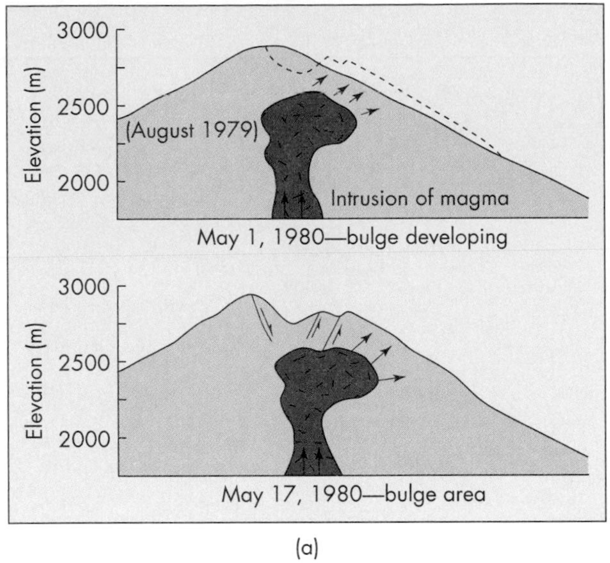

(a)

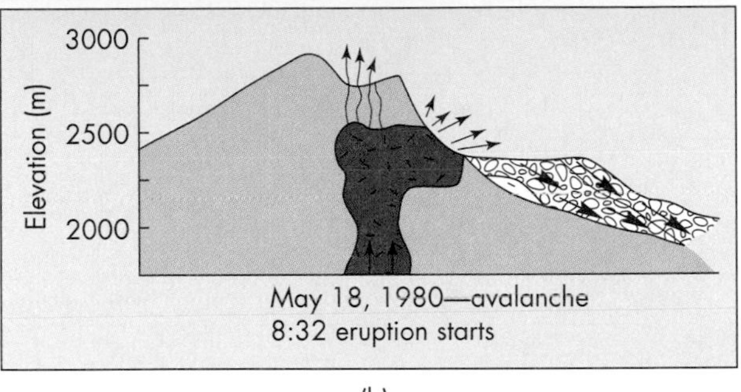

(b)

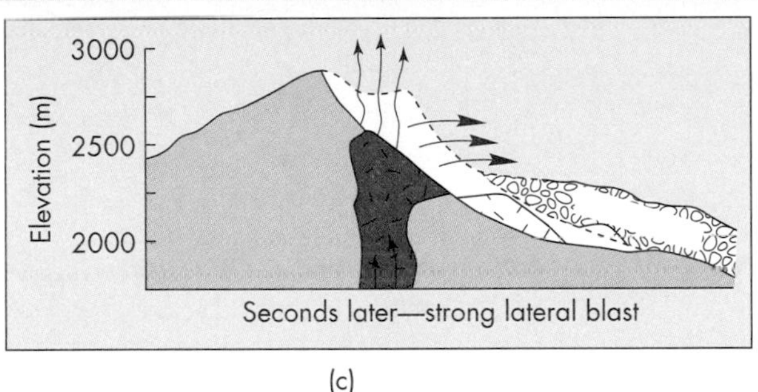

(c)

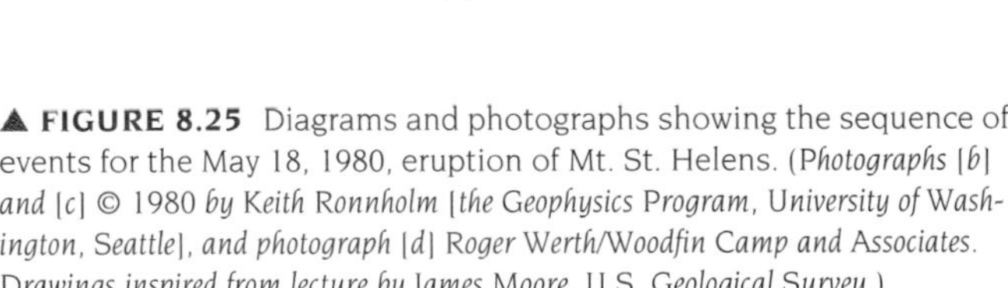

▲ **FIGURE 8.25** Diagrams and photographs showing the sequence of events for the May 18, 1980, eruption of Mt. St. Helens. (*Photographs [b] and [c] © 1980 by Keith Ronnholm [the Geophysics Program, University of Washington, Seattle], and photograph [d] Roger Werth/Woodfin Camp and Associates. Drawings inspired from lecture by James Moore, U.S. Geological Survey.*)

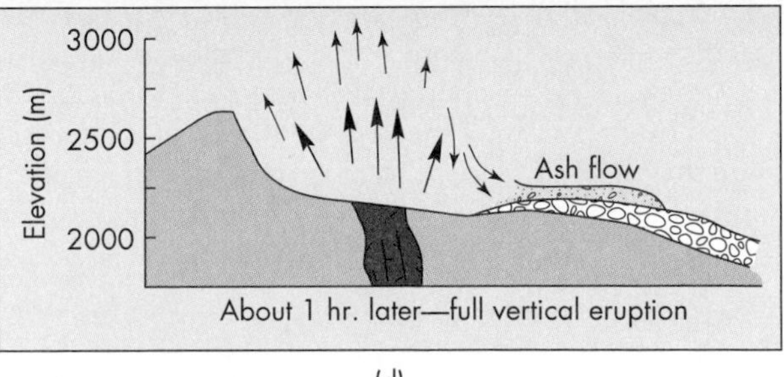

(d)

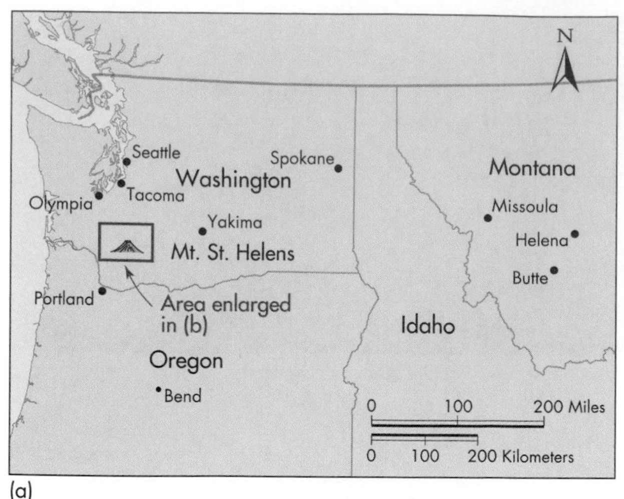

(a)

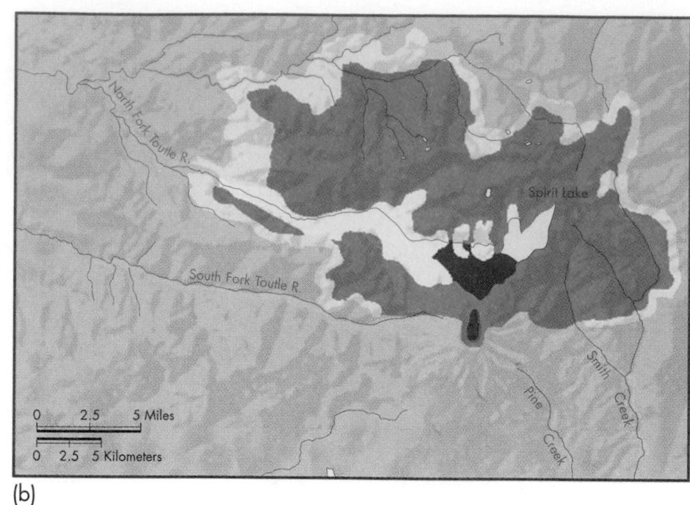

(b)

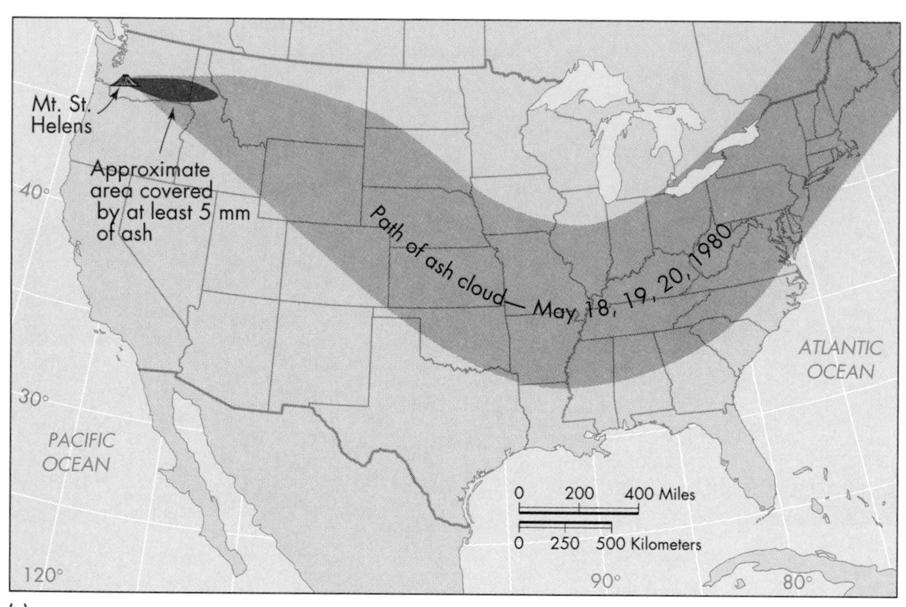

(c)

Debris-avalanche deposits
Down-timber zone
Scorch zone
Pyroclastic flow deposit
Mudflow deposits, scoured areas

▲ **FIGURE 8.26** Mount St. Helens: (a) location; (b) debris-avalanche deposits, tree blowdown, and mudflows associated with the May 18, 1980, eruption; (c) path of the ash cloud from the 1980 eruption. (*Data from various* U.S. *Geological Survey publications*)

CASE HISTORY Surviving to Tell the Tale

On the morning of May 18, 1980, two young people on a fishing trip on the Toutle River were sleeping about 36 km downstream from Spirit Lake. They were awakened by a loud rumbling noise from the river, which was covered by felled trees. They attempted to run to their car, but water from the rising river poured over the road, preventing their escape. A mass of mud then crashed through the forest toward the car, and the couple climbed on top of the roof to escape the mud. They were safe only momentarily, however, as the mud pushed the vehicle over the bank and into the river. Leaping off the roof, they fell into the river, which was by now a rolling mass of mud, logs, collapsed train trestles, and other debris. The water also was increasing in temperature. One of the young people got trapped between logs and disappeared several times beneath the flow but was lucky enough to emerge again. The two were carried downstream for approximately 1.5 km before another family of campers spotted and rescued them.

▲ **FIGURE 8.27** The desolate and barren landscape shown here was produced by the May 18, 1980, eruption of Mount St. Helens. The debris avalanche/debris flow that moved down the Toutle River valley is shown here from the center left of the photograph to the lower right-hand corner. The entire valley is filled full of debris. The surface of the deposit is hummocky, characterized by a scattering of large blocks of volcanic debris. (*John S. Shelton*)

When the volcano could be viewed again following the eruption, its maximum altitude had been reduced by about 450 m, and the original symmetrical mountain was now a huge, steep-walled amphitheater facing northward (Figure 8.28). The debris avalanche, horizontal blast, pyroclastic flows, and mudflows devastated an area of nearly 400 km^2, killing 54 people. More than 100 homes were destroyed by the flooding, and approximately 800 million board feet of timber became flattened by the blast. Total damage was estimated to be nearly $3 billion.

Following the catastrophic eruption of Mount St. Helens, an extensive program was established to monitor volcanic activity, particularly the construction by lava flows of a dome within the crater produced by the May 18, 1980, eruption. During the first 3 years following the May 18 eruption, at least 11 smaller eruptions occurred that contributed to the building of the lava dome to a height of approximately 250 m. In each of these eruptions, lava was extruded near the top of the dome and slowly flowed toward the base. Monitoring the growth of the dome and the deformation of the crater floor, along with such other techniques as monitoring the gases emitted, was useful in predicting eruptions. Each event added a few million cubic meters to the size of the dome (18). By 1995, 15 years following the main eruption, both the mountain and the surrounding area are in many places green once more as life has returned (Figure 8.29). However, the hummocky landscape produced by the landslide deposits is still prominent, a reminder of the catastrophic event of 1980. The area is becoming a tourist attraction visited by about 1 million people in 1993, the year the Coldwater Ridge Visitors Center, located about 11 km northwest of the peak, opened (19).

8.5 Prediction of Volcanic Activity

It is unlikely that we will be able to predict volcanic activity accurately in the near future, but valuable information is being gathered about phenomena that occur prior to eruptions.

(a)

(b)

▲ **FIGURE 8.28** Mount St. Helens (a) before and (b) after the May 18, 1980, eruption. As a result of the eruption, much of the northern side of the volcano was blown away and the altitude of the summit was reduced by approximately 450 m. (*Washington State Tourism Development Division*)

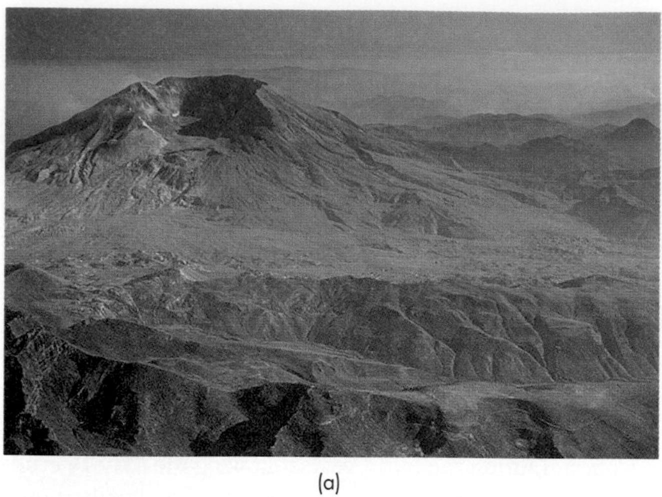
(a)

(b)

▲ **FIGURE 8.29** The eruption of May 18, 1980, of Mount St. Helens produced a barren landscape (a) that is recovering, as illustrated by the flowering lupine 10 years later in July 1990 (b). (*(a) John S. Shelton, and (b) Gary Braasch/Woodfin Camp and Associates*)

One problem is that most prediction techniques require experience with actual eruptions before the mechanism is understood. Thus, we are better able to predict eruptions in the Hawaiian Islands because we have had so much experience there.

Possible methods of predicting volcanic eruptions include:

- Monitoring of seismic activity
- Geophysical observations of thermal and magnetic properties
- Topographic monitoring of tilting or swelling of the volcano
- Monitoring of gas emissions
- Studying the geologic history of a particular volcano or volcanic center (2,16)

Seismic Activity

Our experience with volcanoes such as Mount St. Helens and those on the "Big Island" of Hawaii suggests that earthquakes often provide the earliest warning that a volcanic eruption may occur. In the case of Mount St. Helens, earthquake activity started in mid-March, before the eruption in May. Activity began suddenly with near-continuous shallow seismicity. Unfortunately, there was no additional increase in earthquakes that immediately preceded the May 18 event. In Hawaii, earthquakes have been used to monitor the movement of magma as it approaches the surface. Two months prior to the Mt. Pinatubo eruptions, small steam explosions and earthquakes began (2).

Geophysical Monitoring

Geophysical monitoring of volcanoes is based on the fact that, prior to an eruption, a large volume of magma moves up into some sort of holding reservoir beneath the volcano. The hot material changes the local magnetic, thermal, hydrologic, and geochemical conditions. As the surrounding rocks heat up, the rise in temperature of the surficial rock may be detected by infrared aerial photography. Thus, periodic remote sensing of a volcanic chain may detect new hot points that could indicate possible future volcanic activity. This method was used with some success at Mount St. Helens prior to the main eruption on May 18, 1980.

When older volcanic rocks are heated by new magma, magnetic properties, originally imprinted when the rocks cooled and crystallized, may change. These changes can be detailed by ground or aerial (from an aircraft) monitoring of magnetic properties of the rocks (2,20).

Topographic Monitoring

Monitoring of topographic changes and seismic behavior of volcanoes has been useful in predicting some volcanic eruptions. The Hawaiian volcanoes, especially Kilauea, have supplied most of the data. The summit of Kilauea actually tilts and swells prior to an eruption and subsides during the actual outbreak (Figure 8.30). This movement, in conjunction with earthquake swarms that reflect the moving subsurface magma and announce a coming eruption, was employed to predict a volcanic eruption in the vicinity of the farming community of Kapoho, on the flank of the volcano, 45 km from the summit. As a result, inhabitants were evacuated before the event, in which lava overran and eventually destroyed most of the village (21). Because of the characteristic swelling and earthquake activity before eruptions, scientists expect the Hawaiian volcanoes to continue to be more predictable than others. Monitoring of ground movements, such as tilting, swelling, opening of cracks, or alteration in water level in lakes on or near a volcano, is becoming a useful tool in recognizing changes that might indicate a coming eruption (2).

Monitoring of Volcanic Gases

Monitoring of gases emitted from volcanic areas has the primary objective of recognizing changes in gas geochemistry. In particular, changes in gas composition (relative amounts

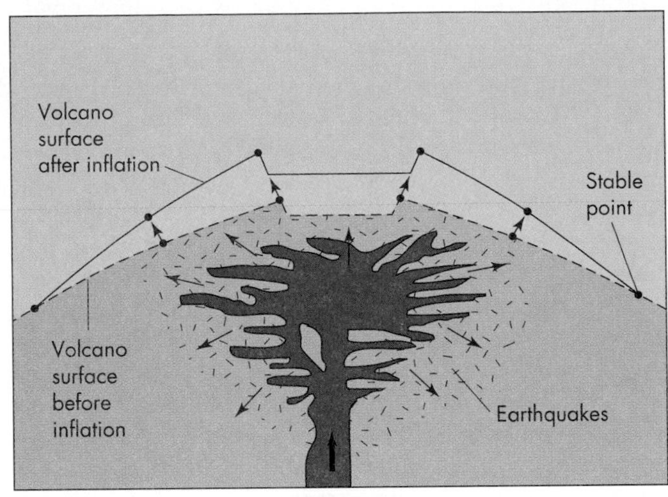

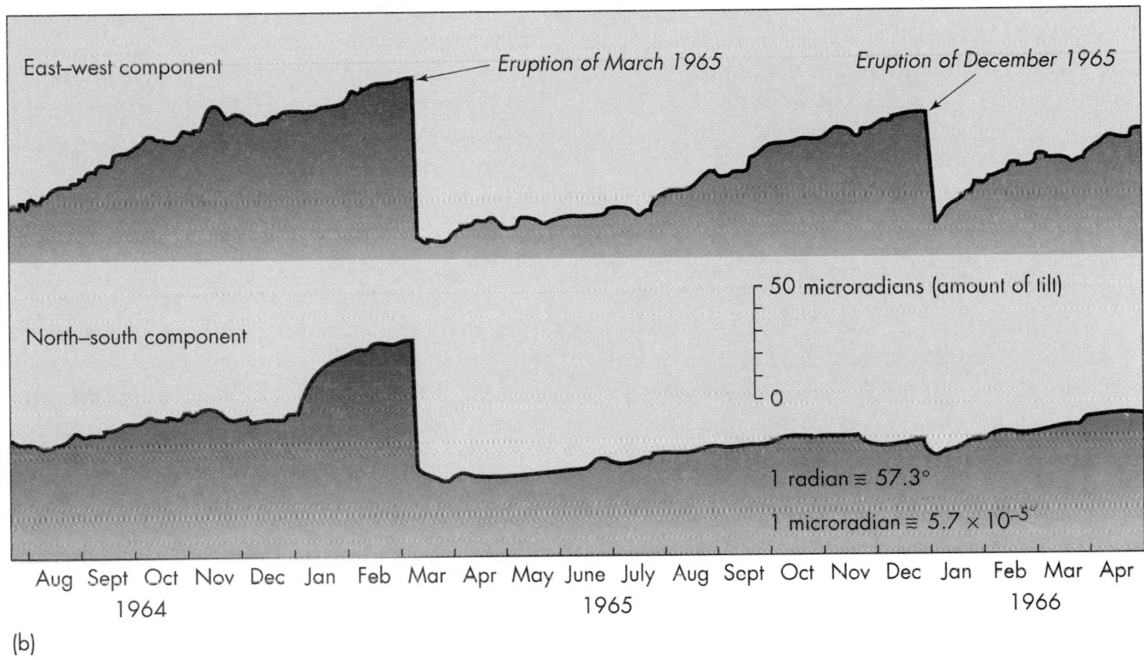

◀ **FIGURE 8.30** (a) Idealized diagram of Kilauea illustrating inflation and surface tilting accompanied by earthquakes as magma moves up. (*Source:* U.S. *Geological Survey Circular* 1073, 1992); and (b) The actual data graph showing east–west component and north–south component of ground tilt recorded from 1964 to 1966 on Kilauea volcano, Hawaii. Notice the slow change in ground tilt before eruption and rapid subsidence during eruption. (*From* R. S. *Fiske and* R. Y. *Koyanagi*, U.S. *Geological Survey Professional Paper* 607.)

of different gases such as steam, carbon dioxide, sulfur dioxide, etc.) or changes in emission rate of sulfur dioxide are thought to be related to changes in subsurface volcanic processes that may indicate movement of magma toward the surface. This technique was useful in studying eruptions at both Mount St. Helens and Mt. Pinatubo. Two weeks prior to the explosive eruptions at Mt. Pinatubo, emissions of sulfur dioxide increased by a factor of about 10 (2).

Geologic History

Understanding the geologic history of a volcano or volcanic system is useful in predicting the types of eruptions likely to occur in the future. The primary tool here is geologic mapping of volcanic rocks and deposits. Attempts are made to date lava flows and pyroclastic activity to determine how recently they have occurred. This information is the primary data necessary to produce maps depicting volcanic hazards at a particular site. Geologic mapping, in conjunction with dating of volcanic deposits at Kilauea, Hawaii, led to the discovery that more than 90 percent of the land surface of the volcano has been covered by lava in only the past 1500 years. The town of Kalapana, destroyed by lava flows in 1990, might never have been built if this information had been known prior to development, because the risk might have been thought too great. The real value of geologic mapping and dating of volcanic events is that it allows development of hazard maps to assist in land-use planning and preparation for future eruptions (2). Such maps are now available for a number of volcanoes around the world.

Progress in Prediction of Eruptions

Although successful short-term predictions are still few and far between, our understanding of volcanic hazards has come a long way. The Mount St. Helens eruption was predicted from a long-range standpoint: Geologists in the 1960s said that it would very likely erupt before the end of the century.

We were also fairly successful in making intermediate-range predictions: Geologists predicted that the increased seismic activity and bulge in the volcano would lead to some sort of eruption in the fairly near future. Hazard zones were established and entry was restricted before the May 18 eruption. Unfortunately, we did not make the very short-range prediction of a few hours or even a few days for the large explosion on May 18.

After the eruption, the area within 30 km of the volcano (45 km to the east and southeast) was closed to entry. This paid off on August 7, when a prediction for an explosive event, based on change in volcanic gas chemistry and earthquake activity, was made just hours prior to the eruption. Thus, as the experience with the volcano increased, chances for successful prediction increased as well. In fact, from June 1980 through December 1982, some 13 eruptions were predicted by monitoring seismicity, deformation of the crater floor, and gas emissions (22). Ending on a positive note, the issuance of a volcanic alert 6 days prior to the main eruption of Mt. Pinatubo allowed for evacuations and is recognized as a successful program that averted a potential disaster (2).

Volcanic Alert or Warning

At what point should the public be alerted or warned that a volcanic eruption may occur? This is an important question being addressed by volcanologists. At present there is no standard code, but one being used with various modifications has been developed by the U.S. Geological Survey. The system is color-coded (by condition) green, yellow, orange, and red, with each color noting increasing concern, as shown in Table 8.2. This table is specifically for Long Valley Caldera in California. Similar systems have been developed for other volcanic areas, including Alaska and the Cascade Mountains of the Pacific Northwest. The hard questions are: When should evacuation begin? and When is it safe for people to return? The color-coded system is a good start, and prior to condition red, evacuation is necessary, but there are gray areas as to when to evacuate between yellow, orange, and red conditions.

8.6 Adjustment to and Perception of the Volcanic Hazard

Apart from the psychological adjustment to losses, the only major human adjustment to volcanic activity is evacuation (23). The small number of adjustments probably reflects the fact that few eruptions occur near populated areas.

The experience of about 5000 Icelanders on the island of Heimaey offers an example of a hardy people's response to a serious destructive hazard. In January 1973 the dormant Mt. Helgafell came alive, and subsequent eruptions nearly buried the town of Vestmannaeyjar in ash and lava flows. The harbor, a major fishing port, was nearly blocked. With the exception of about 300 town officials, fire fighters, and police, the residents were evacuated. They returned 6 months later, in July 1973, to survey the damage and estimate the chances of rebuilding their town and lives. The situation was grim. Ash had drifted up to 4 m thick and covered much of the island and town. Molten lava was still steaming near the volcano. Their first task was to dig out their homes and shops, salvaging what they could. Then they used the same volcanic debris that had buried their town to pave new roads and an airport to allow materials to be moved in and distributed. They also decided to use the volcano to their advantage, and by January 1974, the first home was heated by lava heat.

Because eruptions seldom occur near populated areas, little information is available concerning how people perceive the volcanic hazard. One study of perception evaluated volcanic activity in Hawaii; the study found that a person's age and length of residence near the hazard are significant in that person's knowledge of volcanic activity and possible adjustments (23).

Mount St. Helens provided a valuable example of what types of human difficulties can be expected during and after a high-magnitude physical event that disrupts a large area. The experience should help in devising emergency plans for possible future volcanic eruption or other events such as large earthquakes. For example, the political pressures that arose during the early days of the awakening of Mount St. Helens are quite interesting. On May 17, just a day before the eruption, a caravan of local cabin owners entered the Spirit Lake area at the base of the north flank of the volcano to inspect their properties. They had been requesting admittance for some time, and it was finally granted (they applied political pressure). A second trip was planned for May 18. If the eruption had not taken place on that day, but, say, a week later, one wonders how many people would have returned to the mountain (before the eruption) and been killed.

As we learn more about the magnitude and frequency of natural processes, we are doing a better job of short-range prediction (as, for example, with Mt. Pinatubo and Hawaii), which should help to save lives in the future (2).

Table 8.2 Geologic behavior, color-coded condition, and response: volcanic hazards response plan, Long Valley Caldera, California

Geologic Behavior	Condition	Response
Typical behavior since 1980 includes: *Background:* As many as 10 to 20 small earthquakes with magnitudes (M) less than 3 (M < 3) per day and uplift of the resurgent dome at an average rate of about 1 inch (2.54 cm) per year. *Weak unrest* (likely to occur several times a year): For example, increased number and (or) strength of small earthquakes or a single felt magnitude M > 3 earthquakes. *Moderate unrest* (likely to occur about once a year): For example an M > 4 earthquake or more than 300 earthquakes in a day.	*Green* No immediate risk	*Routine monitoring* plus *Information calls* to U.S. Geological Survey personnel, and town, county, state, and federal agencies regarding locally felt earthquakes and notable changes in other monitored parameters such as ground deformation, fumarole activity, gas emissions, etc.
Intense unrest (may occur about once per decade): For example, a swarm with at least one magnitude 5 earthquake and (or) evidence of magma movement at depth as indicated by an increased rate in ground deformation.	*Yellow* Watch	*Intensified monitoring:* Set up emergency field headquarters at Long Valley Caldera. Initial *Watch* message sent by U.S. Geological Survey to California officials, who promptly inform local authorities. (Includes above information calls.)
Eruption likely within hours or days (may occur every few hundred years): Strong evidence of magma movement at shallow depth.	*Orange* Warning	*Geologic Hazard Warning* issued by U.S. Geological Survey to governors of California and Nevada and others who inform the public. (Includes *Watch* response.)
Eruption under way (may occur every few hundred years).	*Red* Alert	*Sustained on-site monitoring and communication* Maintain intensive monitoring and continuously keep civil authorities informed on progress of eruption and likely future developments.

Notes:

Condition at a given time is keyed to successively more intense levels of geologic unrest, detected by the monitoring network.

Response for a given condition includes the responses specified for all lower levels.

Estimated Recurrence intervals for a given condition are based on the recurrence of unrest episodes in Long Valley Caldera since 1980, the record of magnitude 4 or greater earthquakes in the region since the 1930s, and the geologic record of volcanic eruptions in the region over the last 50,000 years.

Expiration of Watch, Warning, and Alert: This table shows the length of time (in days) a given condition remains in effect after the level of unrest drops below the threshold that initially triggered the condition.

Condition	Expires After	Subsequent Condition*
Watch	14 days	Green (no immediate risk)
Warning (eruption likely)	14 days	Watch
Alert (eruption in progress)	1 day	Warning

*Determined by the level of unrest at the time the previously established condition expires. In the case of the end of an episode of eruptive activity (Alert), a Warning will remain in effect for at least 14 days, depending on the level of ongoing unrest.

Modified from U.S. Geological Survey, 1997.

SUMMARY

Volcanic eruptions often occur in sparsely populated areas, but they have a high potential to produce catastrophes when they happen in populated areas. Volcanic activity is directly related to plate tectonics. Most volcanoes are located at plate junctions, where magma is produced as spreading or sinking lithospheric plates interact with other earth material. The "ring of fire" is a region surrounding most of the Pacific Ocean (Pacific plate) that contains about 80 percent of the world's volcanoes.

Lava is magma that has been extruded from a volcano; the activity of different types of volcanoes is partly determined by the differing silica content and viscosity of their lavas. Shield volcanoes occur at mid-ocean ridges (e.g., in Iceland) and over mid-plate hot spots (e.g., the Hawaiian Islands). Their common rock type is basalt, and they are characterized by relatively nonexplosive lava flows. Composite volcanoes occur at subduction zones, particularly around the Pacific Rim (e.g., in the Pacific Northwest of the United States). They are composed largely of andesite rock and are characterized by explosive eruptions and lava flows. Volcanic domes occur inland of subduction zones (e.g., Mount Lassen). They are composed largely of rhyolite rock and are highly explosive.

Features of volcanoes include vents, craters, and calderas. Other features of volcanic areas are hot springs and geysers. Giant caldera-forming eruptions are violent but rare geologic events. Following their explosive beginning, they often resurge and may present a volcanic hazard for a million years or longer. Recent uplift and earthquakes at the Long Valley Caldera in California are reminders of the potential hazard.

Primary effects of volcanic activity include lava flows, pyroclastic hazards, and occasionally poisonous gases. Construction of walls, bombing, and hydraulic chilling have been used in attempts to control lava flows. These methods have had mixed success and require further evaluation. Pyroclastic hazards include volcanic ash fall, which may cover large areas with a carpet of ash; ash flows, or hot avalanches, which move as fast as 100 km/hr down the side of a volcano; and lateral blasts, which can be very destructive. Secondary effects of volcanic activity include debris flows and mudflows, generated when melting snow and ice or precipitation mix with volcanic ash. These flows can devastate an area many kilometers from the volcano. All of these effects have occurred in the recent history of the Cascade Range of the Pacific Northwest, and there is no reason to believe that they will not occur there in the future.

Sufficient monitoring of geophysical properties, topographic changes, and seismic activity, along with knowledge of the recent geologic history of volcanoes, may eventually result in reliable prediction of volcanic activity. Prediction based on earthquake activity, geophysical changes at a volcano, ground tilting, and volcanic gas emissions is being attempted with some success, particularly in the case of Hawaiian volcanoes. On a worldwide scale, however, it is unlikely that we will be able to predict all volcanic activity accurately in the near future.

Apart from psychological adjustment to losses, the only major human adjustment to volcanic activity is evacuation. Important questions are: When should the public be alerted that a volcanic eruption may occur? When should people be evacuated? When is it safe for people to return?

REFERENCES

1. **Wright, T. L., and Pierson, T. C.** 1992. *Living with volcanoes.* U.S. Geological Survey Circular 1073.
2. **IAVCEE Subcommittee on Decade Volcanoes.** 1994. Research at decade volcanoes aimed at disaster prevention. *EOS, Transactions of the American Geophysical Union* 75(30):340, 350.
3. **Pendick, D.** 1994. Under the volcano. *Earth* 3(3):34–39.
4. **Decker, R., and Decker, B.** 1998. *Volcanoes,* 3rd ed. New York: W. H. Freeman.
5. **Francis, P.** 1983. Giant volcanic calderas. *Scientific American* 248(6):60–70.
6. **Office of Emergency Preparedness.** 1972. *Disaster preparedness* 1, 3.
7. **Crandell, D. R., and Waldron, H. H.** 1969. Volcanic hazards in the Cascade Range. In *Geologic hazards and public problems, conference proceedings,* eds. R. Olsen and M. Wallace, pp. 5–18. Office of Emergency Preparedness Region 7.
8. **Mason, A. C., and Foster, H. L.** 1953. Diversion of lava flows at Oshima, Japan. *American Journal of Science* 251:249–58.
9. **Fisher, R. V., Heiken, G., and Hulen, J. B.** 1997. *Volcanoes.* Princeton, NJ: Princeton University Press.
10. **Bailey, R. A.** 1983. Mammoth Lakes earthquakes and ground uplift: Precursor to possible volcanic activity? *U.S. Geological Survey Yearbook,* 1982:5–13.
11. **Miller, C. D., Mullineaux, D. R., Crandell, D. R., and Bailey, R. A.** 1982. *Potential hazards from future volcanic eruptions in the Long Valley–Mono Lake Area, east-central California and southwest Nevada—A preliminary assessment.* U.S. Geological Survey Circular 877.
12. **Williams, R. S., Jr., and Moore, J. G.** 1973. Iceland chills a lava flow. *Geotimes* 18:14–18.
13. **Sinolowe, J.** 1986. The lake of death. *Time,* September 8:34–37.
14. **Iverson, R. M., Schilling, S. P., and Vallance, J. W.** 1998. Objective delineation of lahar-inundation hazard zone. *Geological Society of American Bulletin* 110:(8):972–84.
15. **Herd, D. G.** 1986. The 1985 Ruiz Volcano disaster. *EOS, Transactions of the American Geophysical Union,* May 13:457–60.

16. **American Geophysical Union.** 1991. Pinatubo cloud measured. *EOS, Transactions of the American Geophysical Union,* 72:29, 305–6.
17. **Hammond, P. E.** 1980. Mt. St. Helens blasts 400 meters off its peak. *Geotimes* 25:14–15.
18. **Brantley, S., and Topinka, L.** 1984. *Earthquake Information Bulletin* 16, no. 2.
19. **Pendick, D.** 1995. Return to Mount St. Helens. *Earth* 4(2):24–33.
20. **Francis, P.** 1976. *Volcanoes.* England: Pelican Books.
21. **Richter, D. H., Eaton, J. P., Murata, K. J., Ault, W. U., and Krivoy, H. L.** 1970. *Chronological narrative of the 1959–60 eruption of Kilauea volcano, Hawaii.* U.S. Geological Survey Professional Paper 537E.
22. **Swanson, D. A., Casadevall, T. J., and Dzurisin, D.** 1983. Predicting eruptions at Mount St. Helens, June 1980 through December 1982. *Science* 221:1369–76.
23. **Murton, B. J., and Shimabukuro, S.** 1974. Human response to volcanic hazard in Puna District, Hawaii. In *Natural hazards,* ed. G. F. White, pp. 151–59. New York: Oxford University Press.

KEY TERMS

magma (p. 206)
shield volcano (p. 206)
basalt (p. 206)
tephra (p. 206)
pyroclastic (p. 206)
lava tube (p. 208)
composite volcano (p. 208)
andesite (p. 208)
volcanic dome (p. 209)
rhyolite (p. 209)
caldera (p. 211)
hot springs and geysers (p. 211)
caldera eruption (p. 211)
lava flow (p. 213)
ash fall (p. 216)
lateral blast (p. 216)
pyroclastic flow (p. 216)
debris flow/mudflow (p. 218)

SOME QUESTIONS TO THINK ABOUT

1. While looking through some old boxes in your grandparents' home, you find a sample of volcanic rock collected by your great-grandfather. No one knows where it was collected. You take it to school and your geology professor tells you that it is a sample of andesite. What might you tell your grandparents about the type of volcano it probably came from, its geologic environment, and the type of volcanic activity that likely produced it?
2. A country in Central America has intentions of developing a moderately large city approximately 30 km from a prominent volcano. Your company has been hired to evaluate the volcanic hazard. Outline a plan of action that would ultimately produce a report discussing the hazards at the site.
3. Using your report of the volcanic-hazard analysis for Question 2, how could you evaluate the specific risk at the site? Consult the discussion of specific risk in Chapter 6 on landslides to help you answer this question.
4. Our discussion of adjustment and perception to the volcanic hazard states that people's perceptions and what they will do in case of an eruption has little to do with proximity to the hazard, home ownership, knowledge of necessary adjustments, and income level. With this in mind, develop a public relations program that could help make people in the planned Central American city (question 2, above) aware of your findings. Keep in mind that the tragedy associated with the eruption of Nevado del Ruiz was in part related to political and economic factors that influenced the fact that the hazard map was not taken seriously. Some people were afraid that the hazard map would result in lower property values in some areas.

9 Coastal Hazards

Coastal erosion along the Mediterranean Coast of southern Spain. (*Edward A. Keller*)

Coastal erosion is a potentially serious problem in many parts of the world, including this Mediterranean coastline in southern Spain not far from the inland city of Granada. Shown here on the beach are large blocks of the seacliff produced in part by wave attack, landslide, and other erosional processes. A number of the buildings are constructed near the top the present seacliff, and a small beach recreation building is at the base. The buildings are vulnerable to erosion in the future! We are at a crossroads today with how to deal with coastal erosion. Defending the coastline is an expensive process and may not be effective. Even more disturbing is the fact that construction of structures to impede coastal erosion may lead to loss of coastal resources, such as beaches, which attract the tourism that provides an economic base to many coastal areas.

9.1 Introduction to Coastal Hazards

Coastal areas are varied in topography, climate, and vegetation, and they are generally dynamic environments. Continental and oceanic processes converge along coasts to produce landscapes that are characteristically capable of rapid change. The impact of hazardous coastal processes is considerable, because many populated areas are located near the coast. This is especially true in the United States, where it is expected that most of the population will eventually be concentrated along the nation's 150,000 km of shoreline, including the Great Lakes. Today, the nation's largest cities lie in the coastal zone, and approximately 75 percent of the population lives in coastal states (1).

The most serious coastal hazards are:

LEARNING OBJECTIVES

In this chapter we focus on one of the most dynamic environments on earth—the coast, where the sea meets the land. Learning objectives are:

- *To understand tropical cyclones and the hazards they produce.*
- *To gain a modest acquaintance with basic terminology of waves, how waves are generated, and what happens when waves enter the shallow waters of coastal zones.*
- *To be able to define a beach in terms of its basic components and the process of littoral transport of sediment.*
- *To be familiar with the major processes related to coastal erosion.*
- *To understand the concepts of the littoral cell, beach budget, and wave climate.*
- *To know the various engineering approaches to shoreline protection, including seawalls, groins, breakwaters, jetties, and beach nourishment.*
- *To gain an appreciation for how human activities affect coastal erosion.*
- *To understand why we are at a crossroads with respect to adjustments to coastal erosion.*

Web Resources

Visit the "Environmental Geology" Web site at www.prenhall.com/keller to find additional resources for this chapter, including:

- Web Destinations
- On-line Quizzes
- On-line "Web Essay" Questions
- Search Engines
- Regional Updates

- Tropical cyclones, which claim many lives and cause enormous amounts of property damage every year.
- Tidal floods caused by combination of a high tide with a storm surge.
- Tsunamis, or seismic seawaves (discussed in Chapter 7), which are particularly hazardous to coastal areas of the Pacific Ocean.
- Coastal erosion, which continues to produce considerable property damage that requires human adjustment.

9.2 Tropical Cyclones

Tropical cyclones are known as **typhoons** in most of the Pacific Ocean and **hurricanes** in the Atlantic. Tropical cyclones have taken hundreds of thousands of lives in a single storm. A tropical cyclone that struck the northern Bay of Bengal in Bangladesh in November 1970 produced a 6-m rise in the sea. Flooding killed approximately 300,000 people, caused $63 million in crop losses, and destroyed 65 percent of the total fishing capacity of the coastal region (2). Another devastating cyclone hit Bangladesh in the spring of 1991, killing more than 100,000 people while inflicting more than $1 billion in damage.

Typhoons and hurricanes are generated as tropical disturbances and dissipate as they move over the land. Wind speeds in these storms are greater than 100 km/yr, and the winds blow in a large spiral around a relatively calm center called the *eye* of the hurricane. Winds of 100 km/hr or greater are generally recorded over an area about 160 km in diameter, while gale-force winds greater than 60 km/hr are experienced over an area about 640 km in diameter.

Most hurricanes form in a belt between 8° north and 15° south of the equator, and the areas most likely to experience cyclones in this zone are those with warm surface-water temperatures. During an average year, about five hurricanes will develop that might threaten the Atlantic and Gulf coasts.

Hurricanes that threaten the East and Gulf coasts of the United States often are generated as thunderstorms in West Africa that move offshore to the west and north; as they move, they gain energy from the warm ocean waters that are drawn into the storm. As the storm grows in intensity, more clouds are formed, a low pressure cell develops, and there is a counterclockwise rotation. The three most likely storm tracks to develop are 1) a storm that heads toward the east coast of Florida, sometimes passing over islands such as Puerto Rico and then, before striking the land, it moves out into the Atlantic to the northeast; 2) a storm that travels over Cuba and into the Gulf of Mexico to strike the Gulf Coast; and 3) a storm that skirts along the East Coast and may strike land from central Florida to New York. Storms that may develop into hurricanes are closely monitored from satellite observation and by flying specially designed aircraft through storms. As a result of better data collection, it is now possible to predict more accurately when and where a hurricane is likely to "hit land," or if it will not hit at all. Accurate prediction of storms and where they will strike land is terribly important to minimize loss of life

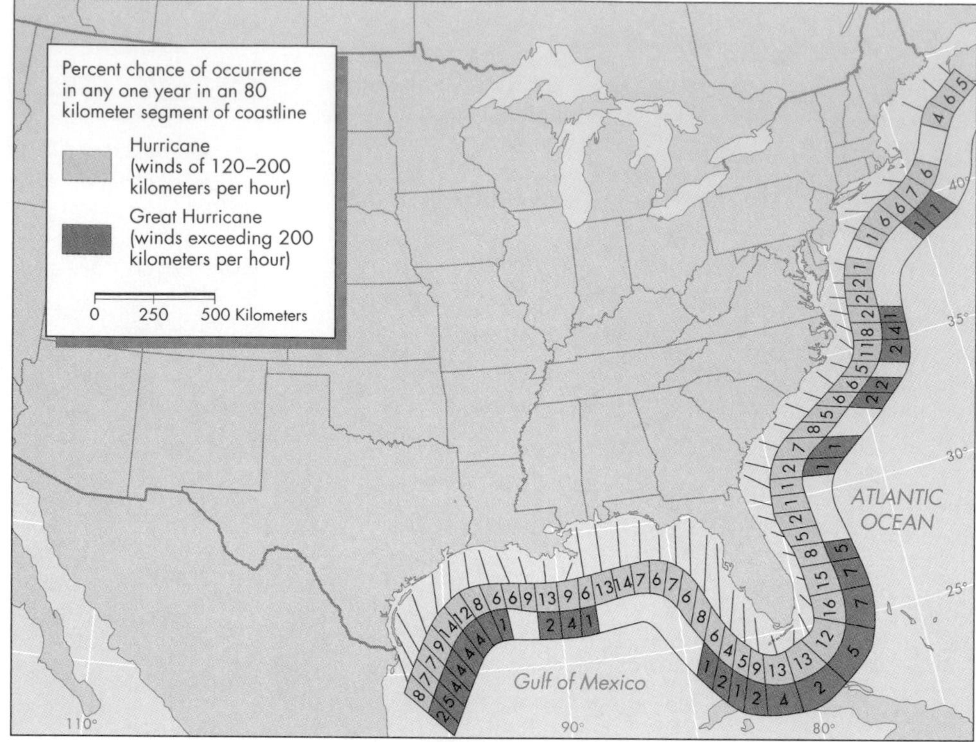

▶ **FIGURE 9.1** Probability that a hurricane will strike a particular 80-km coastal segment in a given year. (*From Council on Environmental Quality*, Environmental Trends, 1981.)

and avoid unnecessary evacuation that may cost many millions of dollars.

The probability that a hurricane will strike a particular 80 km coastal segment of the Atlantic or Gulf coasts in a given year is shown in Figure 9.1. Notice that the probabilities are particularly high in southern Florida and on the Louisiana coastline.

Hazards presented by hurricanes include high winds that may rip shingles and rafters from roofs of buildings, blow over large trees and utility lines, and generally wreak havoc on structures built by people. The process that kills most people and often causes the most damage, however, is flooding resulting from 1) intense precipitation and landward transport of wind-driven waves of ocean waters, and 2) **storm surges.** The most lethal aspect of hurricanes is in fact storm surge, which is a meteorological phenomenon that produces a physical rise in the level of the water surface of the ocean as the storm passes. This happens because storms are associated with lower-than-normal air pressure. Air pressure in a particular location is the weight of the overlying air, measured as a force per unit area (1 atmosphere [atm], average air pressure at sea level is approximately 10^5 N/m^2, or 14.7 lb/in^2). Because lower air pressure exerts less pressure on the ocean, the surface of the water rises up. A storm surge occurring during a hurricane that coincides with high tide may raise the level of the water by as much as 10 m (30 ft) (Figure 9.2). Even smaller storm surges up to about 5 m may move far inland in low-lying coastal areas, inundating homes and buildings while causing fierce coastal erosion at the shoreline (3).

▲ **FIGURE 9.2** Storm surge and high waves produced by Hurricane Gloria on September 27, 1985, in New Jersey. (*Ryan Williams/International Stock Photography, Ltd.*)

A fully developed hurricane is an awesome event. Figure 9.3 shows Hurricane Andrew on August 25, 1992, with winds of up to 300 km/hr moving toward Louisiana after devastating parts of south Florida. Hurricane Andrew was the costliest in U.S. history, with estimated damages in excess of $25 billion. There were 23 human lives lost as a direct result of the storm (most people were evacuated) and 250,000 people were temporarily rendered homeless. Damages to homes and other buildings (Figure 9.4) were extensive: about 25,000 homes were destroyed as entire neighborhoods in Florida were flattened (4). More than 100,000 buildings were damaged, including the National Hurricane Center in Florida, where a radar antenna in a protective dome was torn from the roof (4). Two hurricanes in the summer of 1996 struck the North Carolina coast. The second storm in August inflicted about $3 billion in

◀ **FIGURE 9.3** Hurricane Andrew, August 25, 1992, as shown on a multispectral image. The storm has left south Florida where it did extensive damage and is moving toward Louisiana. (*Hasler, Pierce, Palaniappan, Manyin*/NASA *Goddard Laboratory for Atmospheres*)

(a)

(b)

▲ **FIGURE 9.4** Damage to the coastal area of south Florida from Hurricane Andrew in 1992 resulting from high winds and storm surge. (a) Lighthouse at Cape Florida in Biscayne Bay, Miami, Florida, prior to the arrival of the hurricane (*Wingstock/Comstock*); (b) The same coastline following the hurricane (*Cameron Davidson/Comstock*).

property damage and killed more than 20 people. Despite the increasing population along the Atlantic and Gulf coasts, the loss of lives from hurricanes has decreased significantly because of more effective detection and warning; however, the amount of property damage has greatly increased. Concern is growing that continued population increase in large cities such as Miami and New Orleans, accompanied by unsatisfactory evacuation routes, building codes, and refuge sites, may contribute to a heretofore unexperienced hurricane catastrophe along the Atlantic or Gulf coasts (3).

9.3 Tidal Floods

Hurricanes are not the only coastal storms that inflict damage. For example, storm surges from lesser storms, when combined with a high tide, may produce **tidal floods.** Such an event occurred in Bangor, Maine, on February 2, 1976. The city, located 32 km inland from Penobscot Bay at the confluence of the Kenduskeag Stream and the Penobscot River, was flooded with 3.7 m of water. The flood resulted when a tidal storm surge, caused by strong south-southeasterly winds up to about 100 km/hr off the coast of New England, moved up the funnel-shaped Penobscot Bay and the Penobscot River. When the surge reached Bangor, floodwaters rose very quickly, reaching the maximum depth in less than 15 minutes. Approximately 200 motor vehicles parked in lots along the Kenduskeag Stream were submerged, and when the flood receded about an hour later, damages in the downtown area exceeded $2 million (5). Although this was the first documented tidal flood at Bangor, such flooding has occurred frequently in many parts of the world (see *Case History: The Thames Barrier*).

9.4 Coastal Processes

Waves

Waves that batter the coast are generated by offshore storms. Wind blowing over the water produces a frictional stress along the air–water boundary, and because the air is moving much faster than the water, the moving air transfers some of its energy to the water as waves. The waves, in turn, expend their energy at the shoreline. The size of the waves produced depends upon:

- The *velocity* or speed of the wind. The greater the wind velocity, the larger the waves.
- The *time period or duration* that the wind blows. Storms of longer duration have more time to impart wave energy to the water.
- The *distance*, known as the **fetch,** that the wind blows across the surface of the water. The longer the fetch, the larger the waves are likely to be.

In the area of the storm, the generated ocean waves have a variety of sizes and shapes, but as they move away from their place of origin they become sorted out into groups of similar waves. These groups of waves may travel with very little energy loss for long distances across the ocean to arrive at distant shores.

The basic shape, or *wave form*, of waves moving across deep water is shown on Figure 9.5a. The important parameters are **wave height** (difference in height between wave trough and peak) and **wave length** (distance between successive peaks). The **wave period** (P) is the time in seconds for successive waves to pass a reference point. If you were floating with a life preserver in deep water and could record your motion as waves moved through your area, you would find that you bob up and down and forward and back in a circular orbit, returning to nearly the same place. If you were below the surface with a breathing apparatus, you would still move in circles, but the size of the circle would be smaller. That is, you would move up and down and forward and back in a circular orbit that would remain in the same place while the waves traveled through. This concept is shown on Figure 9.5b. When waves enter shallow water at a depth of less than about one-half their wave length, they "feel bottom." The circular orbits change to become ellipses, and at the bottom the motion may be a very narrow ellipse where the motion is essentially horizontal, that is, forward and back (Figure 9.5c).

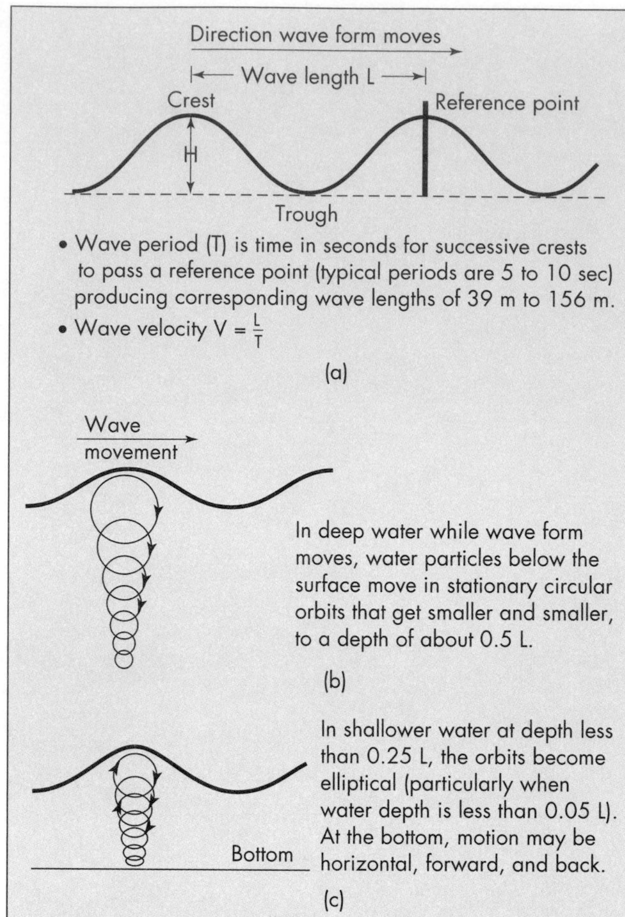

▲ **FIGURE 9.5** (a) Deep-water wave form (water depth is greater than 0.5 L). (b) Motion of water particles associated with wave movement in deep water. (c) Motion of water particles in shallow water at depth less than 0.25 L.

The wave groups generated by storms far at sea are called **swell.** As the swell enters shallower and shallower water, transformations take place that eventually lead to the waves breaking on the shore. With deep-water conditions, there are equations to predict wave height, period, and velocity based upon the fetch, wind velocity, and duration of time that the wind blows over the water. This information has important environmental consequences, because if we can predict the velocity and height of the waves, we can estimate when waves with a particular erosive capability generated by a distant storm will strike the shoreline.

CASE HISTORY The Thames Barrier

In the first century A.D., the Romans founded the city of Londinium on the banks of the River Thames. Until only recently, Londinium, now London, has been at the mercy of tidal floods caused by storm surges. There have been at least seven disastrous floods since the thirteenth century, two of the more recent in 1928 and 1953. It was not a question of *whether* London would flood, but *when* the city would be inundated again. Estimates that a great storm surge could do catastrophic damage (perhaps $6 billion worth), threaten the lives of many people, flood buildings, disrupt government, and close the rail system and underground transit led to construction of the Thames barrier, which was commissioned and tested in 1983. The great barrier, which cost about $750 million, is constructed of ten huge steel gates separated by nine boat-shaped ribs extending over 500 m from bank to bank (Figure 9.A). When not in use, the barrier gates rest on concrete sills in the chalk (limestone) bed of the river and therefore do not present a barrier to navigation. When needed, the gates, powered by electrically driven hydraulics, rotate 90° to stand nearly 15 m above the bed of the river. Along with downstream embankments and smaller barriers, they protect the city of London against even the 1000-year tidal flood (6).

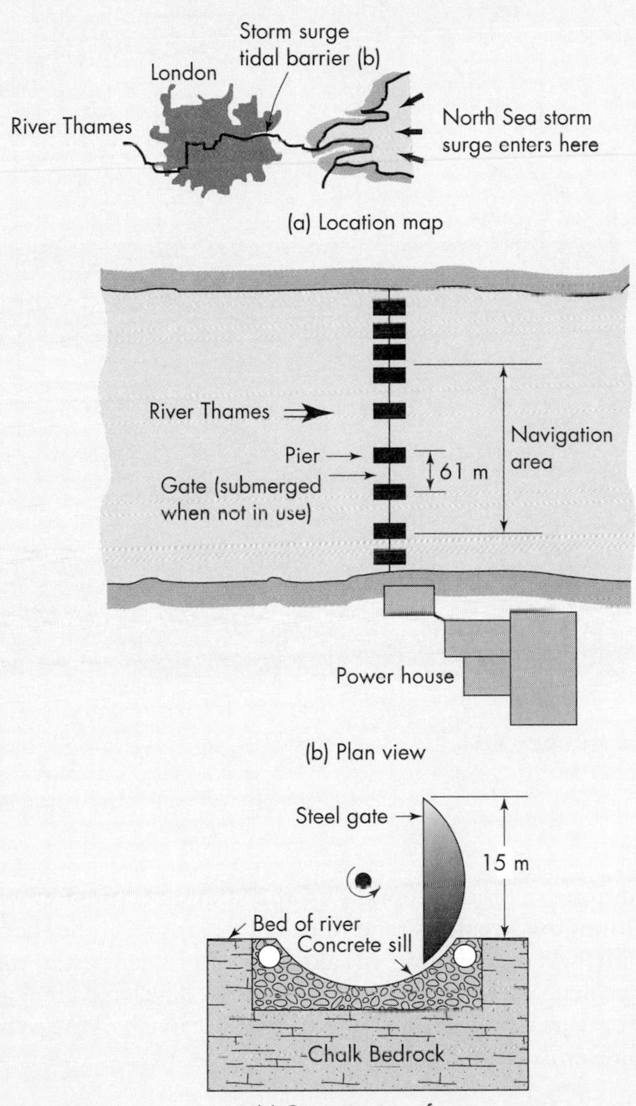

▲ **FIGURE 9.A** Map and diagrams of the storm-surge tidal barrier constructed across the River Thames near London, England. The site of the barrier (a) is just downstream from the main city. A plan view of the river (b) shows the piers and submerged gates across the river where the width is approximately 500 m. The steel gate rests in a concrete sill in the chalk bed of the river (c) and is rotated into place to provide protection from North Sea storm surges as needed. (d) Photograph of Thames River barrier, London, England. (*Adam Woolfitt/Woodfin Camp and Associates*)

We have said that waves expend their energy when they reach the coastline, but just how much energy are we talking about? The amount is surprisingly large. For example, the energy expended on a 400-km length of open coastline by waves with a height of about 1 m (at a particular time) is approximately equivalent to the energy produced by one average-sized nuclear power plant (over the same time) (7). Wave energy is approximately proportional to the square of the wave height. Thus, if wave height increases by a factor of two, the wave energy increases by a factor of four. If we extend the nuclear power plant analogy to waves with a height of 5 m (typical for large storms), then the energy expended, or wave power, along our 400 km of coastline, increases 25 times (5^2) over that of waves with a height of 1 m.

When waves enter the coastal zone and shallow water, wave period remains constant, but wave length and velocity decrease and wave height increases. The waves peak up to a wave height as much as twice their deep-water height. When the water depth is about the same as the wave height, they break, expending their energy on the shoreline. The most dramatic feature of waves entering shallow water is their rapid increase in steepness near the crest. The velocity of the water in the crest eventually exceeds the velocity of the wave form, and the resulting instability causes the wave to break (8).

Although wave heights offshore are relatively constant, the local wave height may increase or decrease when the wave front reaches the nearshore environment due to irregularities in the offshore topography and the shape of the coastline. Figure 9.6 is an idealized diagram showing a rocky point or headland between two relatively straight reaches of coastline. The offshore topography is often similar to that of the shape of the coastline itself. As wave fronts approach the coastline, the shape of the front changes and becomes nearly parallel to the coastline. This occurs because as the waves enter shallow water they slow down first where the water is shallowest, that is, off the rocky point. The result is a bending, or **refraction,** of the wave front. The lines in Figure 9.6 drawn perpendicular to the wave fronts, with arrows pointing toward the shoreline, are known as *wave normals*. Notice that, because of the bending of the wave fronts by refraction, there is a **convergence** of the wave normals at the rocky point and a **divergence** of the wave normals at the beaches. Where wave normals converge, wave height increases, and as a result, wave energy expenditure at the shoreline also increases.

The long-term effect of greater energy expenditure on protruding areas is that wave erosion tends to straighten the shoreline. Of course, protruding rocky points may consist of hard rocks that resist wave erosion. Such points become the more permanent features of the coastline.

The total energy from waves reaching a coastline during a particular time interval may be fairly constant, but there may be considerable local variability of energy expenditure when the waves break on the shoreline. In addition, breaking waves may peak up quickly and plunge or surge, or they may gently spill, depending upon local conditions such as steepness of the shoreline (Figure 9.7). **Plunging breakers** tend to be highly erosive at the shoreline, whereas **spilling breakers** are more gentle and may facilitate deposition of sand on beaches. The large plunging breakers that occur during storms cause much of the coastline erosion we observe.

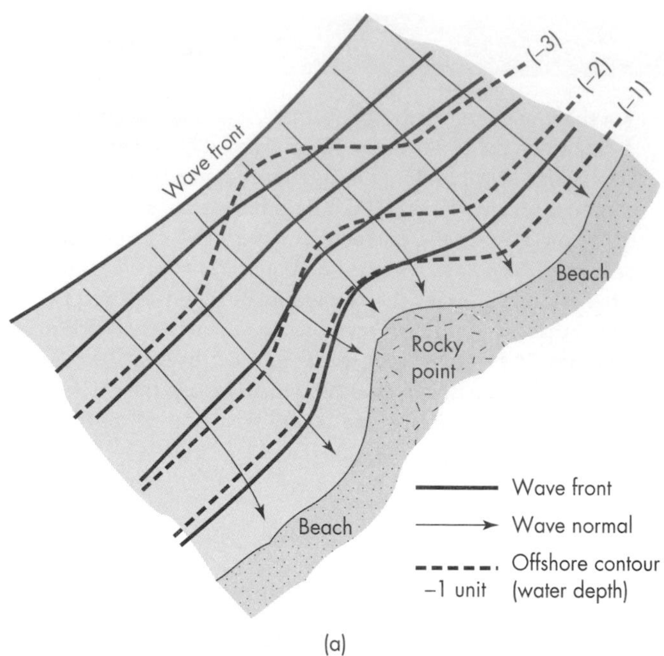

(b)

▲ **FIGURE 9.6** Idealized diagram (a) of the process of wave refraction and concentration of wave energy at headlands. (b) Photograph of large waves striking a rocky headland. (*[b] Douglas Faulkner/Photo Researchers, Inc.*)

Beach Form and Beach Processes

A **beach** is a landform consisting of loose material such as sand or gravel that has accumulated by wave action at the shoreline. Beaches may be composed of a variety of loose material in the shore zone depending on the environment (examples include broken bits of shell and coral, as, for example, beaches on many Pacific islands, and volcanic rock, as, for example, the black sand beaches of Hawaii). Figure

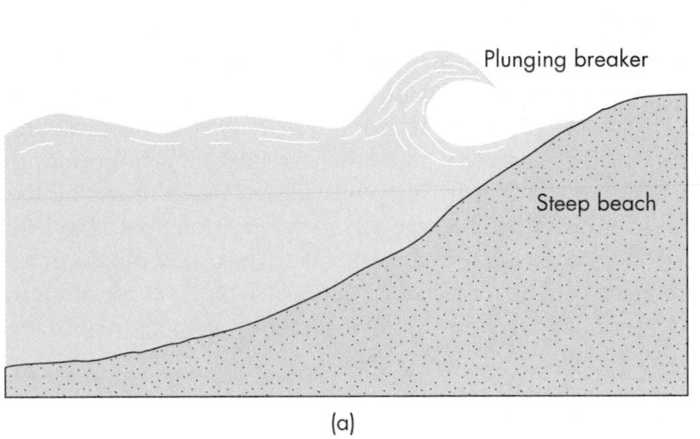

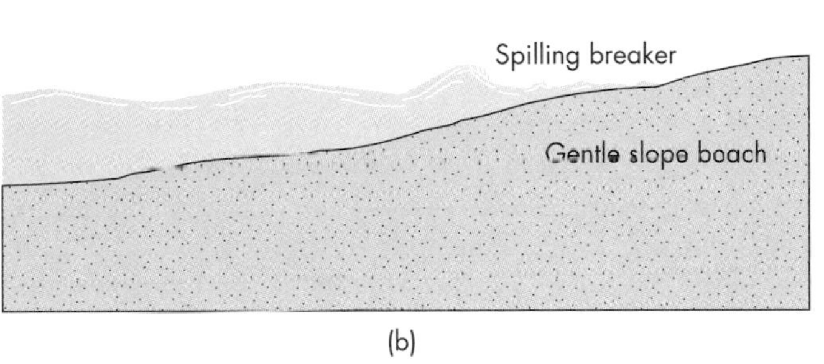

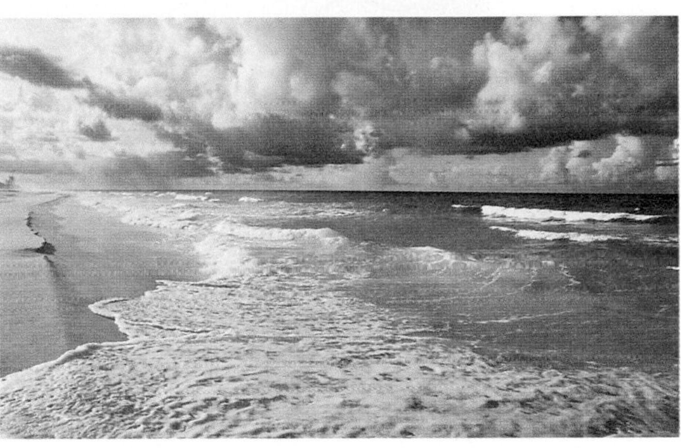

▲ **FIGURE 9.7** Idealized diagram and photos showing (a) plunging breakers and (b) spilling breakers. (*[a] Peter Cade/Tony Stone Images, and [b] Penny Tweedie/Tony Stone Worldwide*)

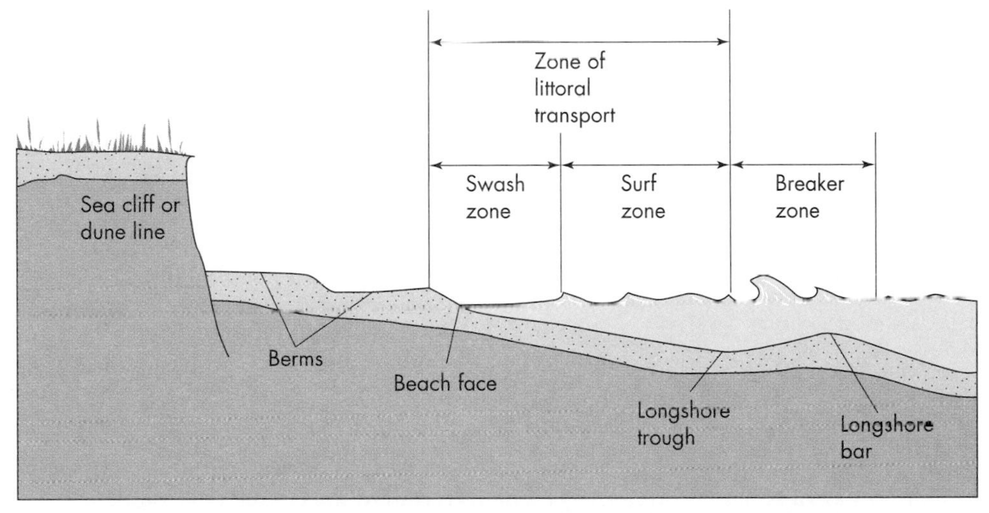

◀ **FIGURE 9.8** Basic terminology for landforms and wave action in the beach and nearshore environment.

9.8 shows the basic terminology of an idealized nearshore environment. The landward extension of the beach terminates at a natural topographic and morphologic change such as a seacliff or dune line. The **berms** are flat backshore areas on beaches (the areas where people sunbathe), formed by deposition of sediment as waves rush up and expend the last of their energy. The **beach face** is the sloping portion of the beach below the berm, and the part of the beach face that is exposed by the uprush and backwash of waves is called the **swash zone.** The **surf zone** is that portion of the seashore environment where turbulent translational waves move toward the shore after the incoming waves break; the **breaker zone** is the area where the incoming waves become unstable, peak, and break. The **longshore trough** and

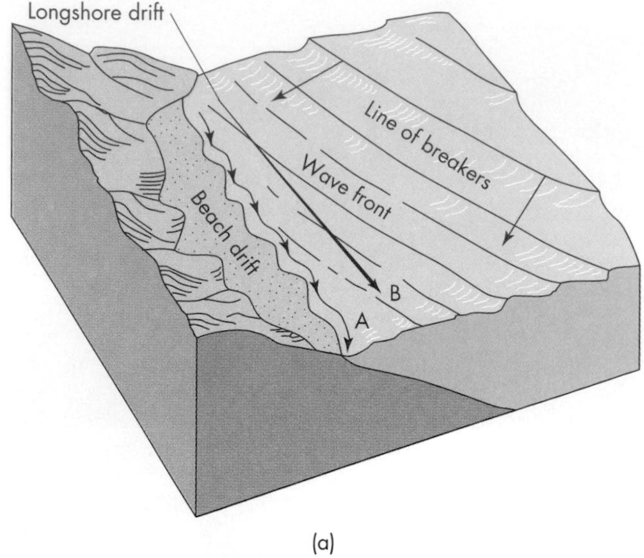

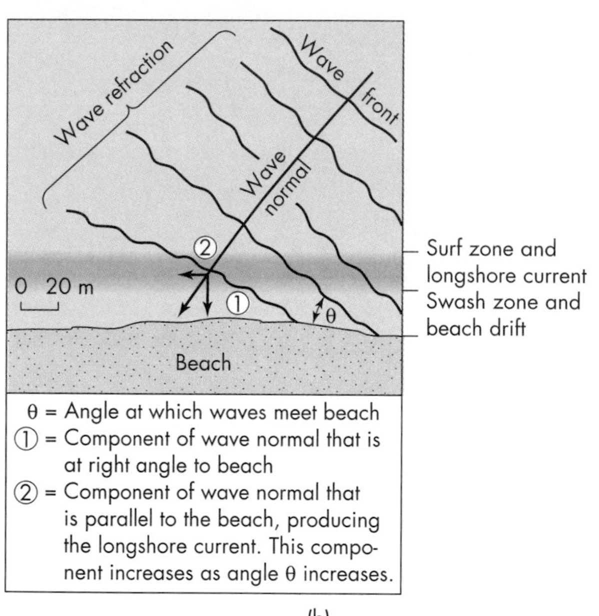

▲ **FIGURE 9.9** (a) Block diagram showing the processes of beach drift and longshore drift, which collectively move sand along the coast (littoral transport). Sediments in the swash zone *(A)* and surf zone *(B)* follow paths shown by the arrows. (b) Two components of wave normals. Component 2 parallel to the beach helps produce the longshore current.

longshore bar are an elongated depression and adjacent ridge of sand produced by wave action. A particular beach, especially if it is wide and gently sloping, may have a series of longshore bars, longshore troughs, and breaker zones (8).

The sand on beaches is not static. Wave action constantly keeps the sand moving in the surf and swash zones, and when waves strike the coast at an angle, the result is **beach drift** and a **longshore drift,** which collectively move beach material along a coast in a process called **littoral transport** (Figure 9.9a). Beach drift is the up-and-back movement of beach material in the swash zone that causes sediment to move along a beach in a zigzag path. The **longshore current** produced by incoming waves striking the coast at an angle occurs in the surf zone. The incoming wave energy (depicted by wave normals in Figure 9.9b) is partitioned into two components—one at right angles and one parallel to the beach (see Figure 9.9b). It is the component parallel to the beach that produces the longshore current. The other component is important in producing beach drift. The movement of sediment by the longshore current is called *longshore drift*. Along both the East and West coasts of the United States, the direction of littoral transport (although it can be quite variable) is most often to the south, and the amount of sediment transported is on the order of 200,000 to 300,000 m^3/yr.

Littoral Cells, Beach Budget, and Wave Climate

The concepts of the littoral cell, beach budget, and wave climate are basic to understanding coastal problems. They are essential in evaluating coastal erosion, which we discuss in the next section.

A **littoral cell** is a segment of coastline that includes an entire cycle of sediment delivery to the coast (usually by rivers but also by coastal erosion and other processes), longshore littoral transport, and eventual loss of sediment from the nearshore environment (9). Figure 9.10 shows five littoral cells in southern California. Each cell involves the erosion, transport, and deposition of more than 200,000 m^3 annually of sand. Furthermore, each cell has an annual **beach budget** of sediment, including sources and losses to the beach environment. Whenever more sediment is transported out of a particular area in a cell than is delivered to that site, erosion results. Negative beach budgets are common, partly because sea level is rising, increasing erosion; dams on rivers store sediments that would otherwise be delivered to coastal areas; and coastal structures interfere with longshore transport of sand.

The ultimate sink for sand in most of the southern California littoral cells is transport through *submarine canyons* to deeper offshore basins. As an example, consider the Santa Barbara littoral cell of Figure 9.10. The sand on the beaches from Point Conception to the sink at Hueneme and Mugu canyons is supplied from streams and rivers that transport sand from the inland mountains to the coastal zone. Both beach drift and longshore drift move the sand to the east and south along the coast. When the sand reaches the heads of the submarine canyons, it is funneled down the canyons, so this part of the shoreline may have no beach. Beaches are again present farther east in the Santa Monica cell, where streams again deliver sand to the coastal environment.

Wave climate is a statistical characterization on an annual basis of wave height, period, and direction, for the purpose of calculating wave energy at a particular site. Figure 9.11 shows three aspects of wave climate (period, direction, and season of arrival) for the coastal area near Oxnard Shores in Ventura County, California. The data for Oxnard Shores suggest that the coast there is vulnerable to northwesterly winter waves that move most of the sediment in the Santa Barbara littoral cell (see Figure 9.10).

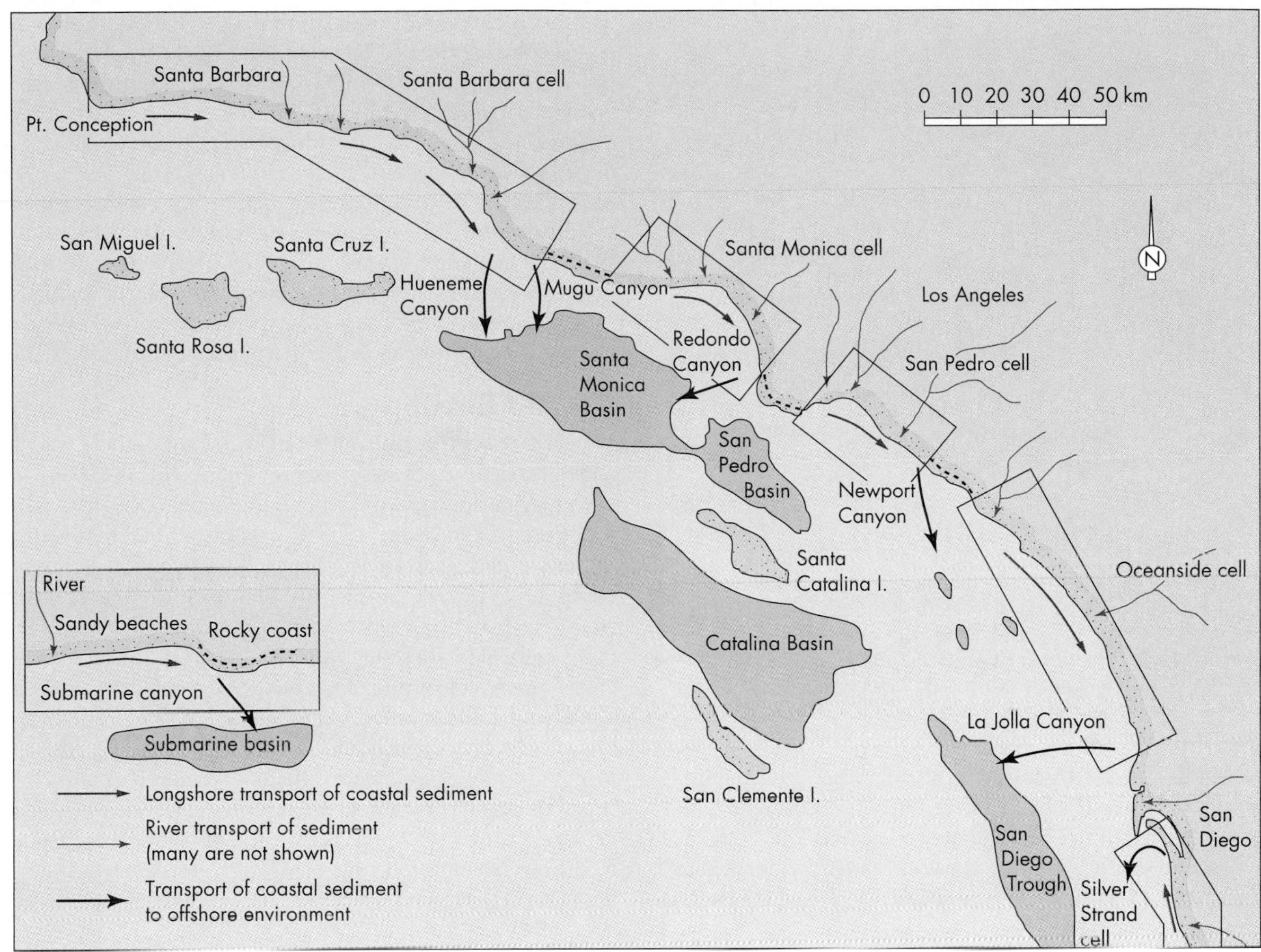

▲ **FIGURE 9.10** Littoral cells in southern California. Shown are five of the major cells, four of which deliver sediment transported along the coast to the submarine basin by way of a submarine canyon. (From D. L. Inman. 1976. Man's impact on the California coastal zone. *State of California, Department of Navigation and Ocean Development.*)

The beach budget and wave climate provide the basic data necessary for formulating and evaluating beach sediment supply and coastal erosion plans (see *Putting Some Numbers On A Beach Budget*). Without this basic information, comprehensive coastal planning is significantly handicapped. Unfortunately, coastal erosion is usually defended against in a piecemeal manner, community by community, rather than from a littoral-cell and beach-budget approach that involves planning for a more extensive length of shoreline. As a result, there is often little coordination of effort, and the problem of beach erosion persists. If communities in the same littoral cell worked together, they would be much more successful in fighting coastal erosion.

9.5 Coastal Erosion

As a result of global rise in sea level and unwise development in the coastal zone, **coastal erosion** is becoming recognized as a serious national and worldwide problem. Compared to other natural hazards, such as earthquakes, tropical cyclones, or floods, erosion of coasts is generally a more continuous, predictable process, and large sums of money are spent in attempts to control it. If extensive development of coastal areas for vacation and recreational living continues, problems of coastal erosion are certain to become more serious.

Erosion Factors

The sand on many beaches is supplied to the coastal areas by rivers that transport it from areas upstream, where it has been produced by weathering of quartz- and feldspar-rich rocks. We have interfered with this material flow of sand from inland areas to the beach by building dams that trap the sand. As a result, some beaches are deprived of sediment and thus erode.

Damming is not the whole story, however. For example, beach erosion along the East Coast of the United States is a result of tropical cyclones and severe storms (known as **Northeasters;** locally, Nor'easters) (7), a rise in sea level, and interference by human use with natural shore processes (10). Tropical cyclones and severe storms can greatly alter a coastline by causing erosion of the beaches and the foreshore

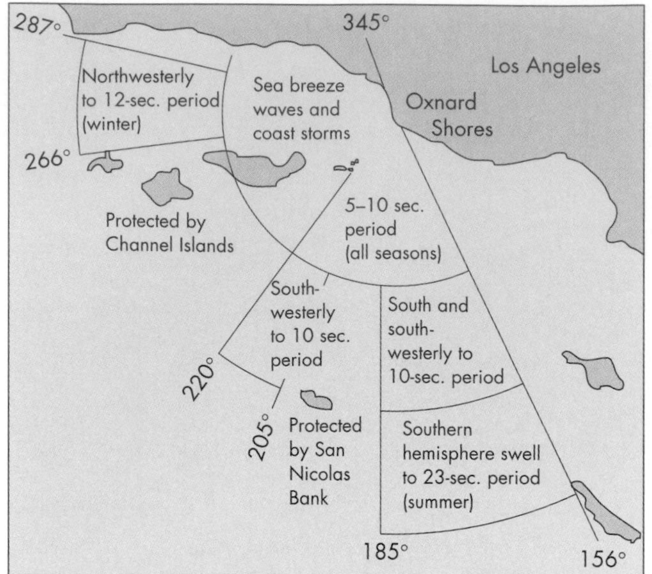

▲ **FIGURE 9.11** Part of the wave climate for Oxnard Shores, west of Los Angeles, California. Shown are the directions and periods of dominant waves that strike the area during particular seasons of the year. (*From* D. L. *Inman.* 1976. Man's impact on the California coastal zone. *State of California, Department of Navigation and Ocean Development.*)

sand-dune system. In addition, storm surges cut channels through the foreshore dunes, and eroded sediment is deposited in back of the beach as *washover deltas* (10).

Recent rise in sea level is *eustatic* (worldwide) and independent of tectonic movement, at the rate of about 2 to 3 mm/yr. Evidence suggests that the rate of rise has increased during the last 60 years as a result of melting of the polar ice caps and thermal expansion of the upper ocean waters, triggered by global warming, which is hypothetically related to increased atmospheric carbon dioxide produced by burning fossil fuels. Sea levels could rise by 700 mm over the next century, ensuring that coastal erosion would become an even greater problem than it is today. There is evidence for submergence of the coast in many areas along the East Coast. Along the Outer Banks of North Carolina (Figure 9.12), the Shackleford Banks that were low-lying forests in 1850 are today salt marshes (11).

Seacliff Erosion

Where a **seacliff** is present along a coastline, there may be added erosion problems, because the seacliff is exposed to both marine and land processes. These processes may work together to erode the cliff at a greater rate than either process could without the other. The problem is further compounded when people interfere with the seacliff environment through inappropriate development.

Figure 9.13 shows a typical southern California seacliff environment at low tide. The rocks of the cliff are steeply inclined and folded shale. A thin veneer of sand and coarser material (pebbles and boulders) mantles the wave-cut platform. A mantle of sand approximately a meter thick covers the beach during the summer, when long gentle waves (spilling breakers) construct a wide berm while protecting the seacliff from wave erosion. During the winter, storm waves (plunging breakers), which have a high potential to erode beaches,

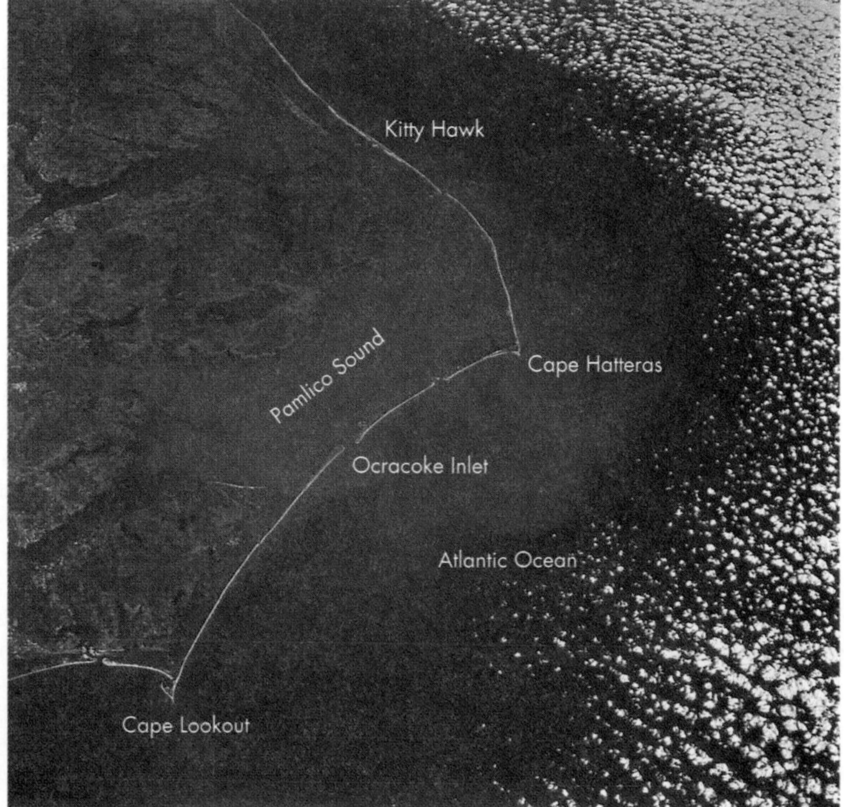

▶ **FIGURE 9.12** The Outer Banks of North Carolina appear in this image from the *Apollo* 9 mission as a thin white ribbon of sand. The Barrier Islands are separated from the mainland by the Pamlico Sound. The brown color in the water is the result of sediment suspended in the water moving within the coastal system. Notice the fan-shaped plume of sediment just seaward of Ocracoke Inlet. The distance from Cape Lookout to Cape Hatteras is approximately 100 km. (*Image courtesy of* NASA)

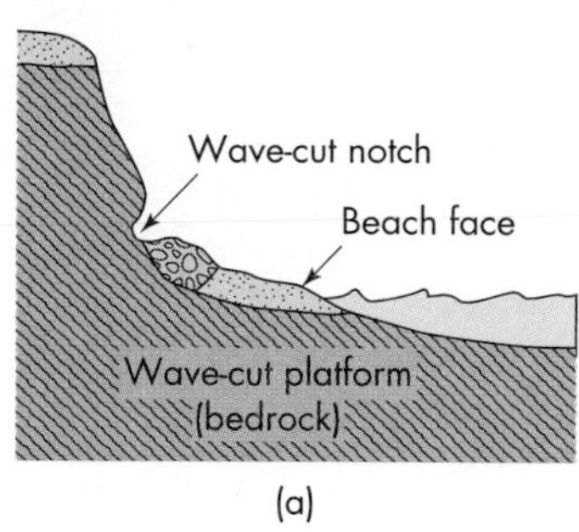

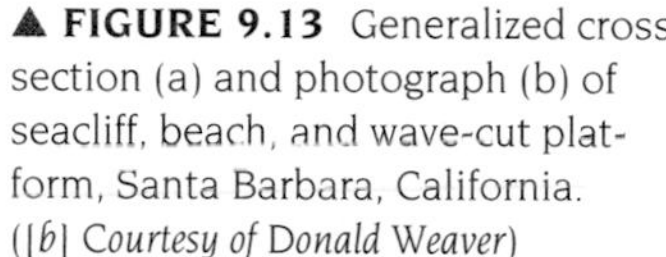

(a)

(b)

▲ **FIGURE 9.13** Generalized cross section (a) and photograph (b) of seacliff, beach, and wave-cut platform, Santa Barbara, California. (*[b] Courtesy of Donald Weaver*)

remove the mantle of sand, exposing the base of the seacliff. Thus, it is not surprising that most erosion of the cliffs in southern California takes place during the winter months.

In addition to wave erosion, processes that attack the seacliff include biological erosion, weathering, rain wash, landslides, and artificially induced erosion (12). Biological processes facilitate and directly cause some erosion of the seacliff; for example, boring mollusks, marine worms, and some sponges can destroy rock. Weathering is significant in weakening the rocks of the seacliff and acts as an aid to erosion: Trees on the top of the seacliff may have roots that penetrate the rock and wedge it apart; salt spray may enter small holes and fractures and, as the water evaporates, the salt crystallizes, exerting pressure on the rock that can weaken it and break off small pieces. Rain wash can cause a considerable amount of seacliff erosion; however, the amount of erosion depends upon the nature and extent of the rainfall and the erodibility of the rocks that make up the seacliff.

A variety of human activities can induce seacliff erosion. Urbanization, for example, results in increased runoff that—if not controlled, carefully collected, and diverted away from the seacliff—can result in serious erosion. Drain pipes that dump urban runoff from streets and homes on the seacliff result in increased erosion. On the other hand, drain pipes that route runoff to the base of the seacliff on the face of the beach result in much less erosion (Figure 9.14).

(a)

(b)

▲ **FIGURE 9.14** The purpose of the pipe in this photograph is to carry surface runoff from the top of the seacliff down to the beach (a). Such pipes can also be used in existing gullies in the seacliff to prevent or at least reduce future erosion of the seacliff (b). (*Edward A. Keller*)

PUTTING SOME NUMBERS ON A Beach Budget

For a given shoreline segment the total volume of sand added to the beach (credits) can be balanced (compared) to losses (debits). This produces the "beach budget" (8).

- If losses are greater than credits, erosion results.
- If losses are less than credits, the beach grows by accretion of sand.
- We need to evaluate a budget over a set period of time, such as 1 year or 10 years.

An example of calculating a beach budget will illustrate the process (see Figure 9.B). Here we determine the budget before and after a dam was constructed, confirming the erosion observed on beaches south of the submarine canyon, where homes are threatened.

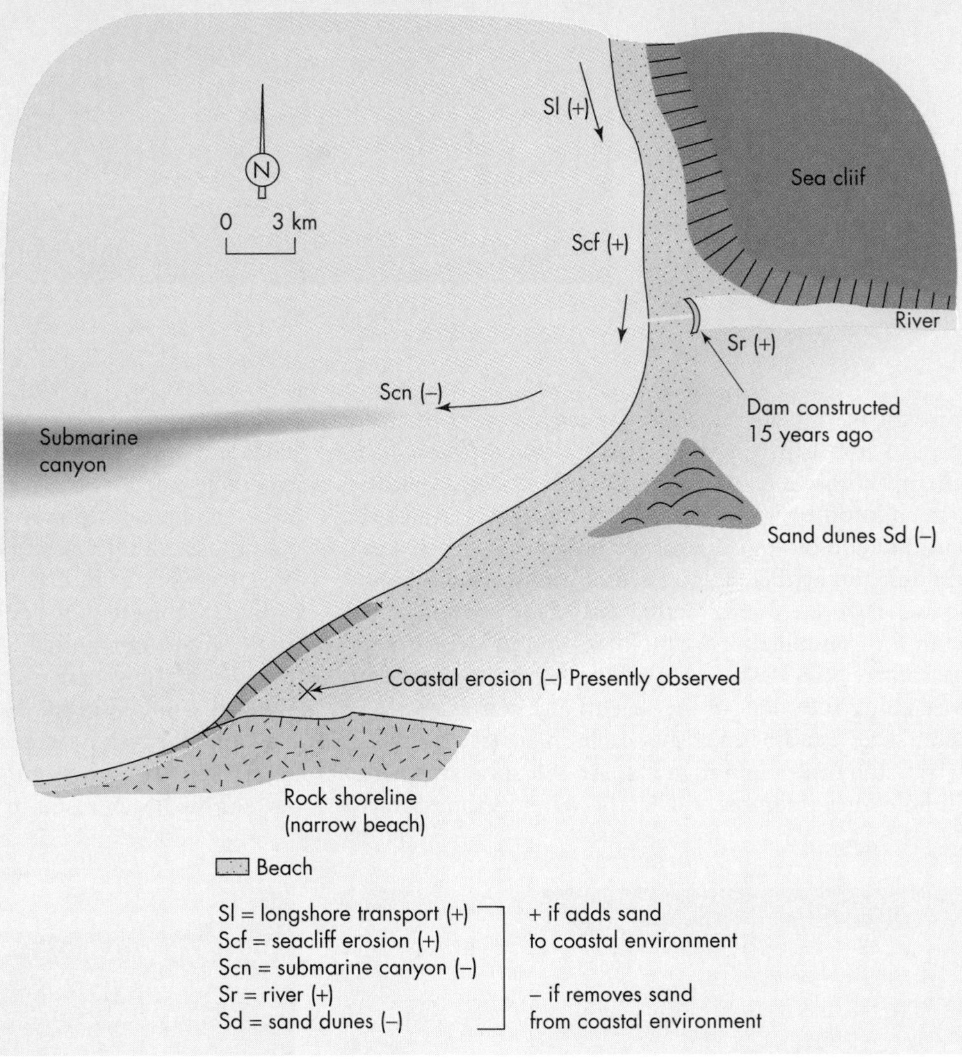

▶ **FIGURE 9.B** Example of a beach budget approach. At present, beach erosion is observed at point *X*. See text for calculation of the beach budget before and following construction of the dam.

Watering lawns and gardens on top of a seacliff also adds a good deal of water to the slope. This water tends to migrate downward through the seacliff toward the base. Where water emerges as small seeps or springs from a seacliff, it effectively reduces the stability of the seacliff, thus facilitating erosion and landslides. Figure 9.13 shows a section of seacliff with a park on top. Recreational use of the seacliff is certainly superior to residential use and should be encouraged. The watering of the large lawn, however, has encouraged the continuous flow of several small seeps along the seacliff. Over a period of years, this may lower the resisting forces of the rocks in the cliff and facilitate an increase in the frequency of landslides in an area where they are already all too common.

Structures such as walls, buildings, swimming pools, and patios may also decrease the stability of the seacliff by increasing the driving forces (Figure 9.15). Strict regulation of development in many areas of the coastal zone now forbids most unsafe construction, but we must continue to live with our past mistakes.

Sources of Sand for Our Beach Budget Example (+ = gain; – = loss):

Littoral transport (+)	South	210,000 m^3/yr
	North	20,000 m^3/yr
	Net	190,000 m/^{3}yr to south
(Scf) Seacliff erosion (+)	Erosion rate is 0.8 m/yr, average height of seacliff is 6 m, total length 3000 m. Assume 60 percent of material eroded remains on beach.	

Scf = (0.8 m/yr)(6 m)(3000 m)(0.6) = 8,640 m^3/yr

(Sr) River source (+) Assume drainage area of 800 km^2

A dam was constructed 15 years ago, reducing the drainage basin delivering sediment to the coast to 100 km^2.

To estimate the sediment delivered to the coast from the river, we can use equations or regional graphs (8). In this case we use Figure 3.C, which presents for the southwestern United States the annual sediment yield for river basins of various size. For example, our basin before the dam has an area contributing sediment of 800 km. From Figure 3.C, this predicts a sediment yield of 330 m^3/km/^{2}yr. Following dam construction, the sediment-contributing area is reduced to 100 km^2 with a yield of 550 m^3/km^2/yr. The increase per unit area results because smaller tributaries below the dam have a higher sediment yield (see Section 3.7, "Rates of Soil Erosion" in Chapter 3).

Assume 30 percent of sediment delivered from the river will remain on the beach; that is, it is sand-sized and larger.

Before dam:

Sr (+) = (330 m^3/km^2/yr)(800 km^2)(0.3) = 79,200 m^3/yr

After dam:

Sr (+) = (550 m^3/km^2/yr)(100 km^2)(0.3)
= 16,500 m^3/yr

(Sd) Loss to sand dunes (–):

Assume dunes advance 0.1 cm/day (0.365 m/yr), the mean height of the dunes is 6 m, and dunes extend 1200 m.

Sd = (0.365 m/yr)(6 m)(1,200 m) ≅ 2,628 m^3/yr.

(Scy) Down submarine canyon (–):

Estimated from offshore observations: 220,000 m^3/yr.

Budget before dam:

Longshore drift (Sl)	+	190,000 m^3/yr
Cliff erosion (Scf)	+	8,640 m^3/yr
River (Sr)	+	79,200 m^3/yr
Sand dunes (Sd)	–	2,628 m^3/yr
Submarine canyon (Scy)	–	220,000 m^3/yr

Budget +55,212 m^3/y; therefore, no coastal beach erosion at point X on Figure 9.B.

Budget after dam:

Sl	+	190,000 m^3/yr
Scf	+	8,640 m^3/yr
Sr	+	16,500 m^3/yr
Sd	–	2,628 m^3/yr
Scy	–	220,000 m^3/yr

Budget –7,488 m^3/yr, erosion is observed at point X on Figure 9.B.

Assume there are several beach homes near point X. What advice would you give them? What would be your response if there were high-rise coastal resorts near point *X*?

The rate of seacliff erosion is variable, and few measurements are available. Along parts of the southern California coast near Santa Barbara, the rate averages 15 to 30 cm/yr, depending on the resistance of the rocks and the height of the seacliff (12). These erosion rates are moderate compared to other parts of the world. Along the Norfolk coast of England, for example, erosion rates in some areas are about 2 m/yr. More remarkable, in the North Sea off Germany, one small island with soft erodible sandstone seacliffs had a perimeter of about 200 km in A.D. 800, and by 1900, the perimeter had been reduced to only about 3 km by seacliff retreat. At that time, a concrete seawall was constructed around the entire island to control the erosion (12).

The main conclusion concerning seacliff erosion and retreat is that it is a natural process that cannot be completely controlled unless large amounts of time and money are invested—and even then there is no guarantee. Therefore, it seems we must learn to live with some erosion. It can be minimized, however, by applying sound conservation practices,

▲ **FIGURE 9.15** Apartment buildings on the edge of the seacliff in the university community of Isla Vista, California. The sign states that the outdoor deck is now open. Unfortunately, the deck is not particularly safe as it is overhanging the cliff by at least 1 m. Notice the exposed cement pillars in the seacliff, originally in place to help support the houses. These decks and apartment buildings are in imminent danger of collapsing into the sea. (*Edward A. Keller*)

such as controlling the water on and in the cliff and not placing homes, walls, large trees, or other loading on top of the cliff, close to the edge.

9.6 Coastal Hazards and Engineering Structures

The engineering structures in the coastal environment, including seawalls, groins, breakwaters, and jetties, are primarily designed to improve navigation or retard erosion. However, because they tend to interfere with the littoral transport of sediment along the beach, these structures all too often cause undesirable deposition and erosion in their vicinity.

Seawalls

Seawalls are structures placed parallel to the coastline to help retard erosion. They may be constructed of concrete, riprap (large stones), wood, or other materials. Seawalls constructed at the base of a seacliff may not be particularly effective because considerable erosion of the seacliff results from subaerial processes on the cliff itself, as well as wave erosion at the base. Use of seawalls has been criticized because seawalls strongly reflect waves, enhancing erosion and producing a narrower beach with less sand. Unless they are carefully designed to complement existing land uses, seawalls generally cause environmental and esthetic degradation (see *Case History: Galveston and Seabright*). For these reasons, design and construction of seawalls must be carefully tailored to specific sites. Some geologists believe seawalls cause sufficient problems that they should be rarely, if ever, used.

Groins

Groins are linear structures placed perpendicular to the shore, usually in groups called **groin fields** (Figure 9.16). The basic idea is that each groin will trap a portion of the sand that is moving in the littoral transport system. A small accumulation of sand will develop updrift of each groin, thus building an irregular but wider beach. The wider beach protects the shoreline from erosion.

The problem is that, although deposition occurs updrift of a groin, erosion tends to occur in the downdrift direction. Thus, a groin or groin field results in a wider, more protected beach in a desired area but may cause a zone of erosion to develop in the adjacent downcoast shoreline. The erosion results primarily as a groin or groin field becomes filled with trapped sediment. Once a groin is filled, sand is transported around its offshore end to continue its journey along the beach. Therefore, erosion may be minimized by artificially filling each groin; known as **beach nourishment,** this requires trucking sand out onto the beach. Thus nourished, the groins will draw less sand from the natural littoral transport system and the downdrift erosion will be reduced (8). Even with beach nourishment and other precautions, however, groins may cause undesirable erosion; therefore, their use should be carefully evaluated.

Breakwaters and Jetties

Breakwaters and jetties provide protection of limited stretches of the shoreline from waves. **Breakwaters** are designed to intercept waves and provide a protected area (harbor) for boat moorings; they may be attached to the beach (Figure 9.17a) or separated (Figure 9.17b). In either case, a breakwater blocks the natural littoral transport of beach sediment, causing the configuration of the coast to change locally as new areas of deposition and erosion develop. In addition to possibly causing serious erosion problems in the downdrift direction, breakwaters act as sand traps that accumulate sand in the updrift direction. Eventually the trapped sand may fill or block the entrance to the harbor as a sand spit or bar develops. As a result, a dredging program, or artificial bypass, is often necessary to keep the harbor open and clear of sediment. The sediment that is removed by dredging should be transported and released on the beach downdrift of the breakwater to rejoin the natural littoral transport system, thus reducing the erosion problem.

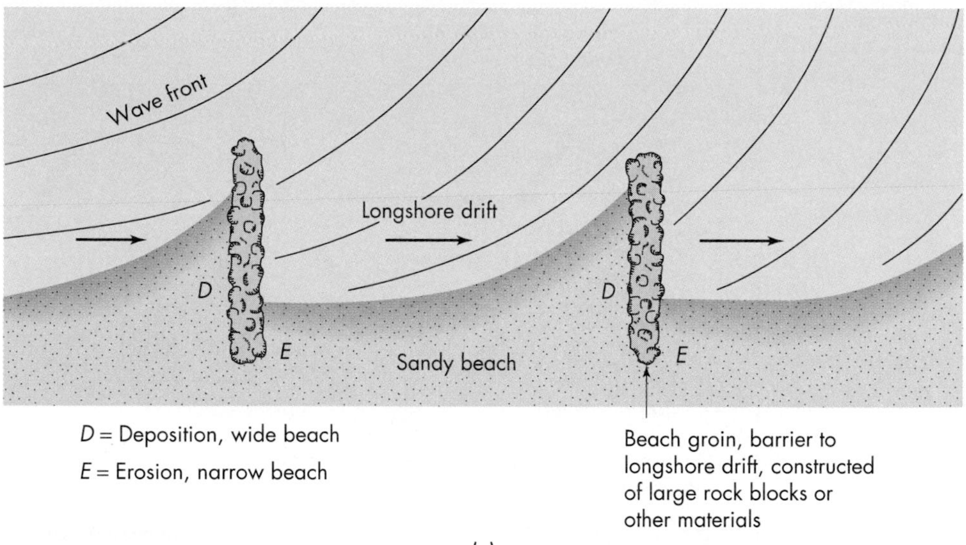

(a)

(b)

(c)

▲ **FIGURE 9.16** Diagram of two beach groins (a) and photos (b) and (c). Deposited sediment builds a wide beach in the updrift direction; in the downdrift direction, the sparsity of sediment for transport can cause erosion to occur. (*Edward A. Keller*)

Jetties are often constructed in pairs at the mouth of a river or inlet to a lagoon, estuary, or bay (Figure 9.17c). They are designed to stabilize the channel, prevent or minimize deposition of sediment in the channel, and generally protect it from large waves (8). Jetties tend to block the littoral transport of beach sediment, thus causing the updrift beach adjacent to the jetty to widen while downdrift beaches erode. The deposition at jetties may eventually fill the channel, making it useless, while downcoast erosion damages coastal development. Mechanically bypassing (dredging) the sediment minimizes but does not eliminate all undesirable deposition and erosion.

There is no way to build a breakwater or jetty on a coast with an active littoral transport system so that it will not interfere with the longshore movement of beach sediment. These structures must therefore be carefully planned, and protective measures must be taken early to eliminate or at least minimize adverse effects. Measures may include installation of a dredging and artificial sediment-bypass system, a beach nourishment (artificial sand deposit) program, seawalls, riprap, or some combination of these (8).

Beach Nourishment

In the discussion above, we introduced the topic of beach nourishment as it pertains to engineering structures in the coastal zone. Beach nourishment can also be an alternative to engineering structures. In its purest form, beach nourishment consists of artificially placing sand on beaches in the hope of constructing a positive beach budget. Beach nourishment is sometimes referred to as the "soft" solution to beach erosion, in contrast to the "hard" solutions, such as constructing groins or seawalls. The basic idea is that the presence of the nourished beach protects coastal property from the attack of waves (8). The procedure has distinct advantages: it is aesthetically preferable to many engineering structures, and it provides a recreation beach as well as some protection from shoreline erosion.

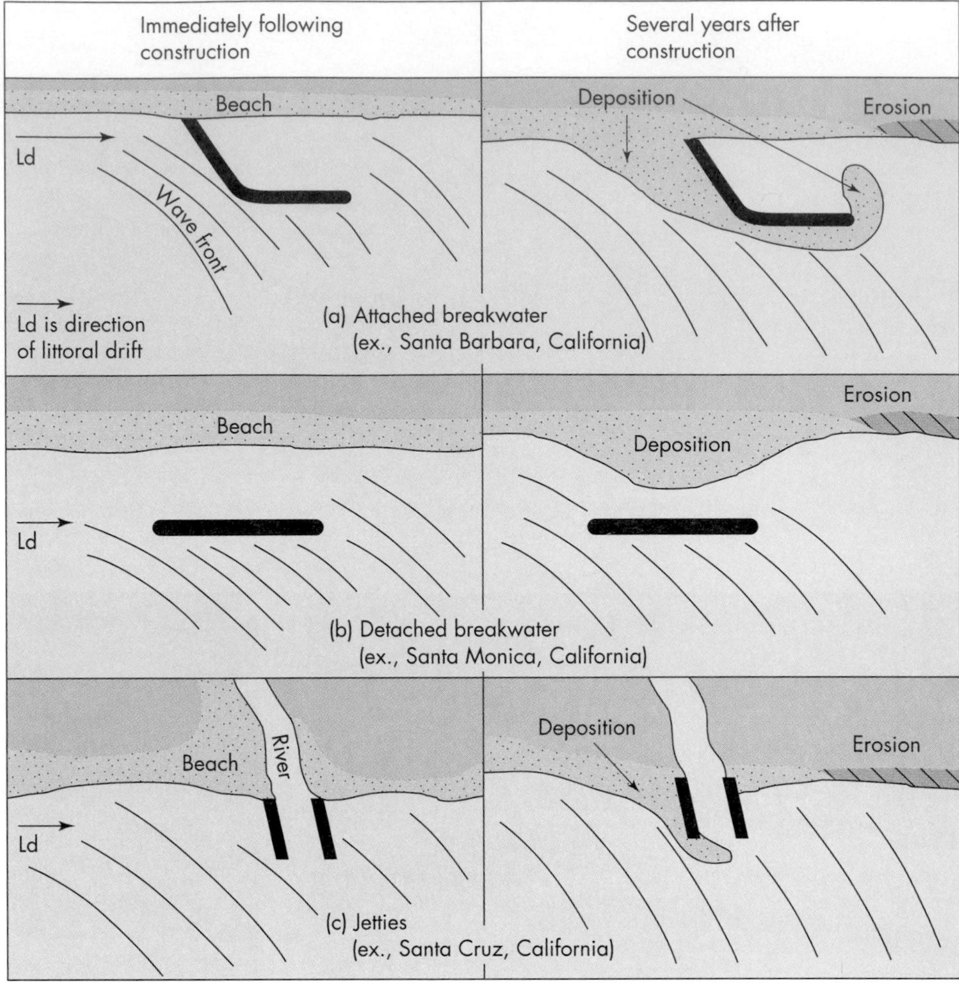

▶ **FIGURE 9.17** Diagrams illustrating the effects of breakwaters and jetties on local patterns of deposition and erosion.

In the mid-1970s the city of Miami Beach, Florida, and the U.S. Army Corps of Engineers began an ambitious beach nourishment program to provide protection from storms and to reverse a serious beach erosion problem that had plagued the area since the 1950s. The natural beach had nearly disappeared by the 1950s, and only small pockets of sand could be found associated with various shoreline protection structures, including seawalls and groins. As the beach disappeared, coastal resort areas, including high-rise hotels, became vulnerable to storm erosion (14). The purpose of the nourishment program for Miami Beach was to produce a positive beach budget and thus a wide beach and to provide additional protection from storm damage. The project cost was about $62 million over 10 years and involved nourishment of about 160,000 m^3 of sand per year to replenish erosion losses. By 1980, about 18 million m^3 of sand had been dredged and pumped from an offshore site onto the beach, producing a 200-m-wide beach (14). Figure 9.18 shows Miami Beach before and after the nourishment. The change is dramatic.

The cross-section design for the project (Figure 9.19) shows the wide berm and frontal dune system that functions as a buffer to wave erosion and storm surge. The Miami project was expanded in the mid- to late 1980s to include dune restoration, which involved establishing native vegetation on the dune shown in Figure 9.19. Public access through the dunes is along special wooden walkways, and other areas of the dunes are protected. The successful Miami Beach nourishment project, which has functioned for more than 20 years and which survived major hurricanes in 1979 and 1992 (8), is certainly preferable to the fragmented erosion-control methods that preceded it.

As of the late 1990s, more than 600 km of coastline in the United States have received some sort of beach nourishment. Not all of this nourishment has had the positive effects reported for Miami Beach. For example, in 1982 Ocean City, New Jersey, nourished a stretch of beach at a cost of more than $5 million. The sand lasted just 2-1/2 months, being eroded out by a series of storms that struck the beach.

On the other hand, the beach sands at Miami may yet be eroded at a much greater cost. Beach nourishment remains controversial, and some consider beach nourishment nothing more than "sacrificial sand" that will eventually be washed away by coastal erosion (13). Nevertheless, beach nourishment has become a preferred method of restoring or even creating recreational beaches and protecting the shoreline from coastal erosion around the world. What is needed

CASE HISTORY Galveston and Seabright

Most of the shoreline in the vicinity of Galveston, Texas, is today protected from erosion by a large system of seawalls, including the largest seawall ever placed on a barrier island. The walls were constructed following a devastating hurricane in 1900 that killed 6000 people. Since then, the seawalls have certainly protected many buildings and property from coastal erosion. The environmental price of the seawalls has been high, however. The walls block access to the coast, have eliminated coastal views, and have caused a narrowing of the beaches (Figure 9.C) (13).

The story at Seabright, New Jersey, has an important lesson: Build a seawall and destroy a beach. The community of Seabright in the late 1800s was a resort in northern New Jersey with a broad sandy beach. Every year, a railroad to the town transported thousands of people to the resort, and the area was a booming tourist attraction. In order to protect the railroad, the residents of Seabright built a large seawall approximately 5 to 6 m high and 8 km long. The wall protected the town and the railroad, but, unfortunately, at the expense of the beach. The beach in front of the seawall has narrowed significantly, and northern Seabright no longer has any beach—only large waves and a large wall (13). Because of case histories such as Seabright, some geologists and coastal engineers are taking a very hard look at the use of seawalls to protect coastal property. Some would go so far as to say that seawalls should never be constructed. This certainly appears to be the case if the community wishes to preserve its beaches.

◀ **FIGURE 9.C** Large seawall protecting Galveston, Texas. Small structure in the foreground and several in the background are beach groins (see Figure 9.16). (*Bob Daemmrich Photo, Inc.*)

▲ **FIGURE 9.18** Miami Beach (a) before and (b) after beach nourishment. (*Courtesy of U.S. Army Corps of Engineers*)

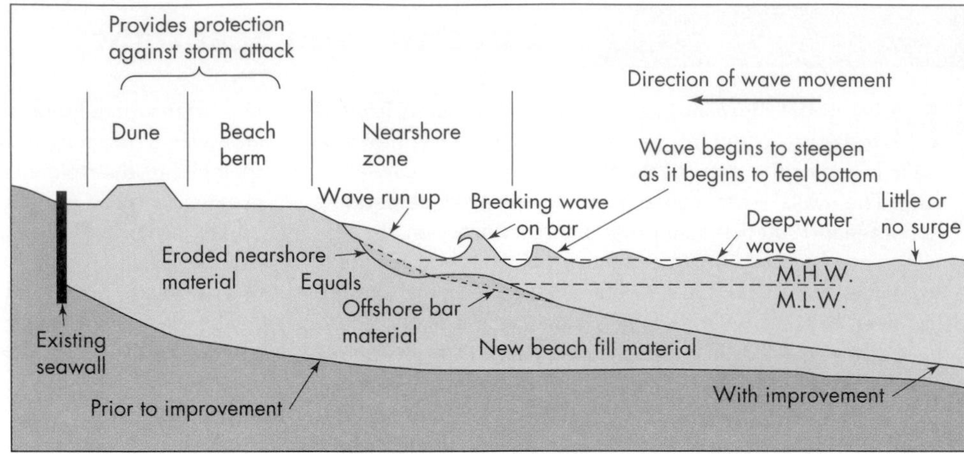

▶ **FIGURE 9.19** Cross section of the Miami Beach nourishment project. The dune and beach berm system provide protection against storm attack. (*From U.S. Army Corps of Engineers*)

are more case histories to document the success or failure of the projects and more public education to inform people of what can be expected from beach nourishment (8).

9.7 Human Activity and Coastal Erosion: Some Examples

Human interference with natural shore processes has caused considerable coastal erosion. Most problems have arisen in areas that are highly populated and developed. For example, artificial barriers often retard the movement of sand, causing beaches to grow in some areas and erode in others, with resulting damage to valuable property.

The Atlantic Coast

The Atlantic Coast from northern Florida to New York is characterized by **barrier islands,** long narrow islands separated from the mainland by a body of water (see Figure 9.12). Many barrier islands have been altered to a lesser or greater extent by human use and interest. Two examples, the barrier island coast of Maryland and the Outer Banks of North Carolina, illustrate the spectrum of human activity.

Demand for the 50 km of Atlantic oceanfront beach in Maryland is very high, and the limited resource is used seasonally by residents of the Washington, D.C., and Baltimore, Maryland, urban centers (Figure 9.20). Since the early 1970s, Ocean City on Fenwick Island has promoted high-rise condominium and hotel development on its waterfront. As a result, the natural frontal dune system of this narrow island has been removed in many locations, resulting in a serious beach erosion problem. More ominous is the almost certain possibility that a future hurricane will cause serious damage to Fenwick Island. The inlet south of Ocean City formed during a hurricane in 1933, and there is no guarantee, despite attempts to stabilize the inlet by coastal engineering, that a new inlet will not form at the present site of Ocean City some time in the future (15).

Across the Ocean City inlet to the south is Assateague Island, which encompasses two-thirds of the Maryland coastline. In contrast to the highly urbanized Fenwick Island, Assateague Island is in a much more natural state. The island is used for passive recreation such as sunbathing, swimming, walking, and wildlife observation. However, both islands are in the same littoral cell and so are connected by a common supply of sand. At least that was the case until 1935, when the jetties at the Ocean City inlet were constructed in an attempt to stabilize the inlet. Since construction of the jetties, coastal erosion in the northern few kilometers of Assateague Island has averaged about 11 m/yr, which is nearly 20 times the long-term rate of shoreline retreat for the Maryland coastline. During this same period, beaches immediately north of the inlet became considerably wider, requiring the lengthening of a recreational pier (16).

Observed changes in the Atlantic coast of Maryland are clearly related to the pattern of longshore drift of sand and human interference. Longshore drift is to the south at an average annual volume of about 150,000 m^3. Construction of the Ocean City inlet jetties interfered with the natural southward flow of sand and diverted it offshore rather than allowing it to continue southward to nourish the beaches on Assateague Island. Starved of sand, the northern portions of the island have experienced serious shoreline erosion during the past 50 years. This example has been cited as the most severe that can be found in the United States of beach erosion associated with engineering structures that block longshore transport of sediment (16). Thus, our fundamental principle involving environmental unity holds: We cannot do one thing only, because everything is connected to everything else.

Another good example of our influence on coastal processes is found in the barrier islands known as the Outer Banks of North Carolina. The southern section near Cape Lookout is nearly unaltered by human activity (Figure 9.21, upper diagram), whereas the northern section near Cape Hatteras has been progressively stabilized and developed, particularly near the beach, which is characterized by a continuous artificial dune system that was built to control erosion (lower diagram). The desire to protect roads and buildings behind the dunes led to the attempt at total stabilization—the idea was to hold everything in place to prevent the sea from washing over the island. In contrast, the

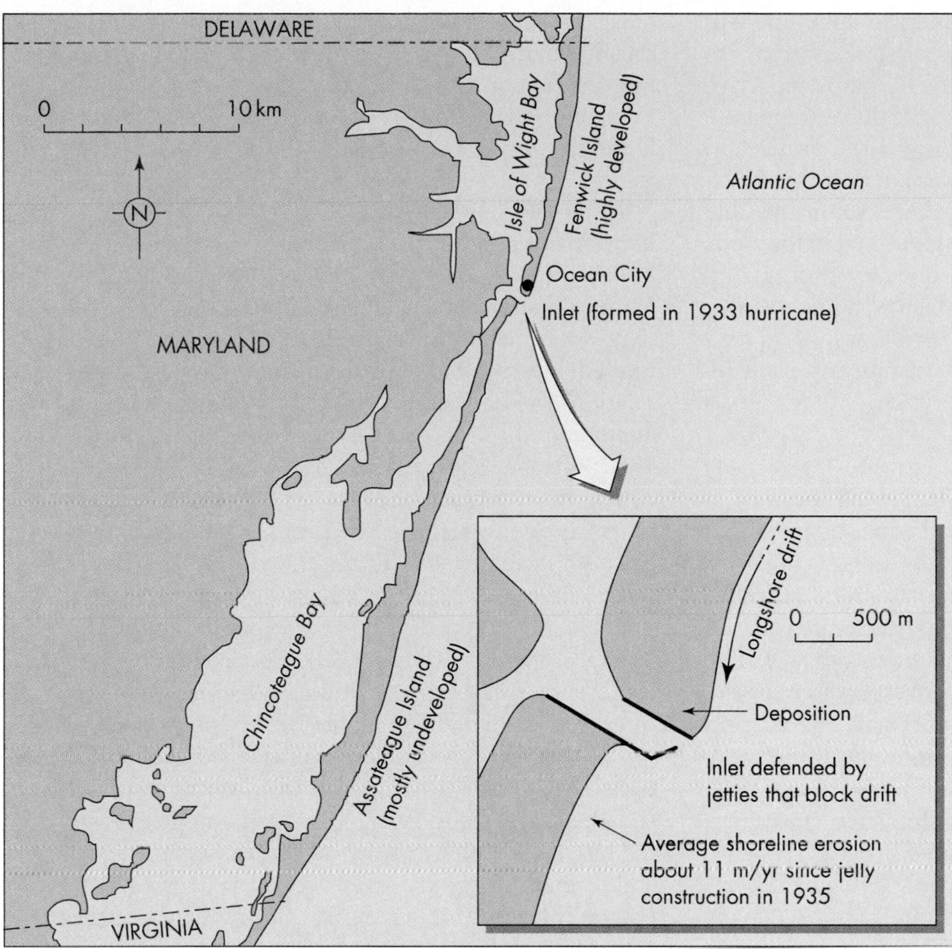

◀ **FIGURE 9.20** The barrier island coast of Maryland. Fenwick Island is experiencing rapid urban development and there is concern for potential hurricane damage. What if a new inlet forms (during a hurricane) at the site of Ocean City? Inset shows details of the Ocean City inlet and effects of jetty construction.

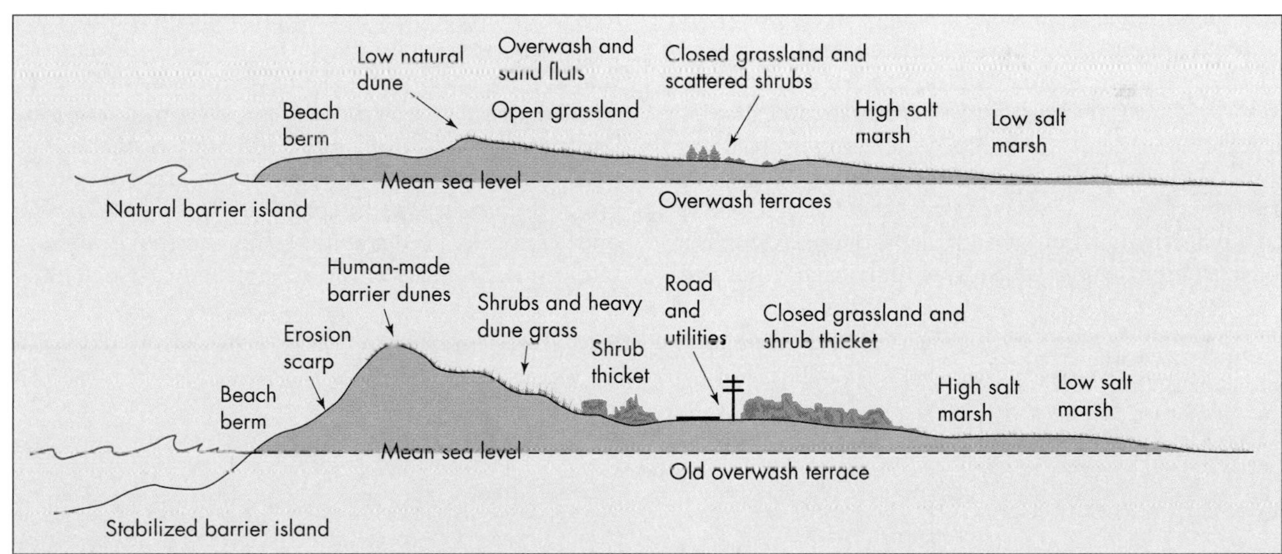

▲ **FIGURE 9.21** Cross sections of the Barrier Islands of North Carolina. The upper diagram is typical of the more natural systems and the lower typical of the artificially stabilized Barrier Islands. (*From* R. Dolan. 1973. Coastal Geomorphology, *ed.* D. R. *Coates* [*Binghamton*, NY: *State University of New York*].)

southern section is characterized by a natural zone of dunes rather than a continuous line. The zone is frequently broken by overwash passes that allow water to flood across the island (10).

Several studies have shown that building the artificial dune line has resulted in erosion and narrowing of the beach and has necessitated spending millions of dollars to stabilize the rapidly eroding artificial dunes (10,17). Even with this effort, the structures and roads behind the dune line will eventually have to be relocated as the shoreline continues to erode and recede. In contrast, where the barrier islands have been left in a natural state, storms cause the dunes to migrate back by natural processes responsible for the origin and maintenance of the barrier island system, rather than wash away. We emphasize, however, that there is considerable controversy as to whether the islands identified in the studies as natural are "typical" barrier islands, for overgrazing with subsequent wind erosion of the frontal dune line may have facilitated the overwash. If this is true, it may be necessary to reevaluate the benefits (dune maintenance and reduced beach erosion) of frequent overwash.

The Gulf Coast

Coastal erosion is also a serious problem along the Gulf of Mexico. One study in the Texas coastal zone suggests that, in specific areas where long-term coastal recession rates can be determined from geologic data, human modification of the coastal zone in the last 100 years has accelerated coastal erosion by 30 to 40 percent over prehistoric rates (18). The human modifications that appear most responsible for the accelerated erosion are construction of coastal engineering structures, subsidence as a result of groundwater withdrawal, and damming of rivers that supply sand to the beaches.

Erosion along the Gulf Coast in Texas has apparently been going on for several thousand years as a natural process, so it is often difficult to determine exactly how much it has increased or decreased due to human interference. Nevertheless, it appears that the natural rate of landward migration of barrier islands and other coastal beach features in many areas has definitely increased in recent years (historic time). For example, Figure 9.22 shows estimated erosion rates for the southwestern Matagorda Peninsula in Texas. The recent (1856 to 1956) erosion rates are 34 percent greater than the prehistoric rates. The situation is quite complex, however, because, as the peninsula has retreated nearly 2 km in the last 1500 years, the nearby Matagorda Island has grown (accreted) by nearly the same amount. Most of the erosion is in response to high-magnitude storms (hurricanes) that produce a net bayward migration of the entire peninsula. The recent situation (post-1965) has been further complicated by construction of a ship channel and accompanying jetties to improve navigation. These structures have produced highly variable erosion rates (18).

The Great Lakes

Erosion is a periodic problem along the coasts of the Great Lakes and has been particularly troublesome along the Lake Michigan shoreline. Damage is most severe during prolonged periods of high lake levels that occur following extended periods of above-normal precipitation. The relationship between precipitation and lake level has been documented by the U.S. Army Corps of Engineers since 1860. The data show that the lake level has fluctuated about 2 m during this time. The lake level has been rising in recent years from a low stage in 1964, and in 1973 it again reached the high level of 1952. In the mid-1980s the lake was at a record high level. During a highwater stage, there is considerable coastal erosion, and many buildings, roads, retaining walls, and other structures are destroyed by wave erosion (Figure 9.23) (19). For example, fall storms in 1985 alone caused an estimated $15 million to $20 million in damages.

During periods of below-average lake level, wide beaches develop that dissipate energy from storm waves and protect the shore. With rising lake-level conditions, however, the beaches become narrow, and storm waves exert considerable energy against coastal areas. Even a small lake-level rise on a gently sloping shore will inundate a surprisingly wide section of beach (19).

Long-term rates of coastal bluff erosion (the equivalent of the seacliff of the ocean shoreline) at many Lake Michigan sites average about 0.4 m/yr (20). Severity of erosion at a particular site depends on such factors as the presence or absence of a frontal dune system (dune-protected bluffs erode at a slower rate); orientation of the coastline (sites exposed to high-energy storm winds erode faster); groundwater seepage (seepage along the base of a coastal bluff causes slope instability, thus increasing the erosion rate); and existence of protective structures (structures may

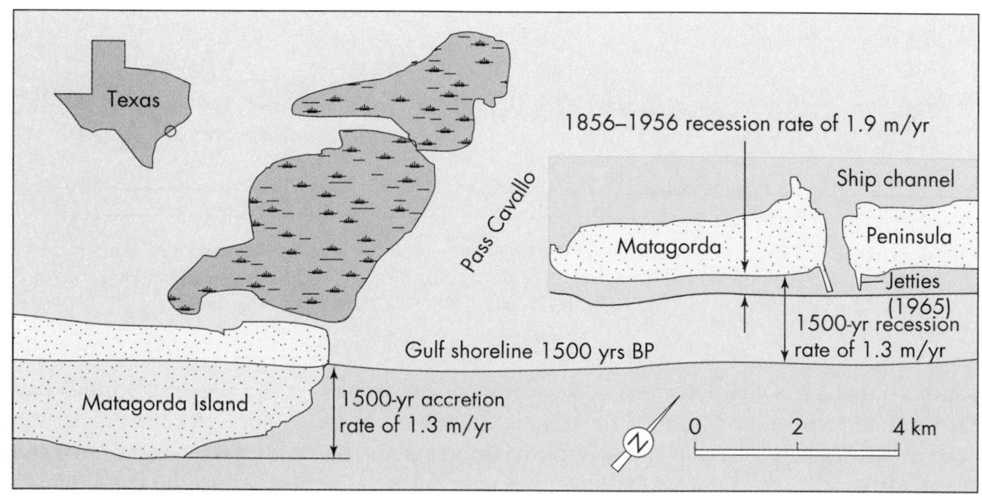

FIGURE 9.22 Estimated rates of coastal erosion for the southwestern Matagorda Peninsula, Texas. (*After Wilkinson and McGowen.* Environmental Geology, *vol.* 1 [*New York: Springer-Verlag,* 1977].)

▲ **FIGURE 9.23** Coastal erosion along the shoreline of Lake Michigan has destroyed this home. (*Steve Leonard/Tony Stone Images*)

be locally beneficial but often accelerate coastal erosion in adjacent areas) (19,20). A few construction options and the relocation alternative are summarized, with respective advantages and disadvantages, in Figure 9.24. In recent years beach nourishment has been attempted for some Great Lakes beaches. The added sands are deliberately sized much coarser than the natural sands that had eroded, which, it is hoped, will reduce the erosion potential.

9.8 Perception of and Adjustment to Coastal Hazards

Perception of Coastal Erosion

Perception of coastal erosion as a natural hazard depends primarily on an individual's past experience, proximity to the coastline, and the probability of suffering property damage. One study of coastal erosion of seacliffs near Bolinas, California, 24 km north of the entrance to San Francisco Bay, established that people living close to the coast in an area likely to experience damage in the near future are generally very well informed and see the erosion as a direct and serious threat (21). People living a few hundred meters from a possible hazard, although aware of the threat, know little about its frequency of occurrence, severity, and predictability. Still farther inland, people are aware that coastal erosion exists but have little perception of the hazard.

Adjustment to Coastal Hazards

Tropical Cyclones People adjust to the tropical cyclone hazard either by doing nothing and bearing the loss or by taking some kind of action to modify potential loss. Bearing the loss is probably the most common individual adjustment. Community adjustments in developed countries include attempts to modify potential loss, such as strengthening the environment with protective structures and land stabilization, and adapting behavior by better land-use zoning, evacuation, and warning (22).

Coastal Erosion Adjustments to coastal erosion fall into one of several categories:

- Beach nourishment that tends to imitate natural processes
- Modification of the wave energy (wave climate) through construction of nearshore structures designed to dissipate wave energy
- Shoreline stabilization through structures such as groins and seawalls
- Land-use change that attempts to avoid the problem.

A preliminary process in any approach to managing coastal erosion is delineation of coastlines subject to erosion (see *A Closer Look: E-Lines and E-Zones*).

We are at a crossroads today with respect to adjustment to coastal erosion. One road leads to ever-increasing coastal defenses in an attempt to control the processes of erosion, and the second path involves learning to live with coastal erosion through flexible environmental planning and wise land use in the coastal zone. In the second path, all structures in the coastal zone (with such exceptions as critical facilities in certain areas) are considered temporary and expendable. Any development in the coastal zone must be in the best interests of the general public rather than a few individuals who develop the oceanfront. Accepting this philosophy requires an appreciation of the following five principles (24):

1. **Coastal erosion is a natural process rather than a natural hazard; erosion problems occur when people build structures in the coastal zone.** The coastal zone is an area where natural processes associated with waves and moving sediment occur. Because such an environment will have a certain amount of natural erosion, the best land uses are those compatible with change. These include recreational activities such as swimming and fishing. When we build in the coastal zone, problems develop.
2. **Any shoreline construction causes change.** The beach environment is dynamic. Any interference with natural processes produces a variety of secondary and tertiary changes, many of which may have adverse consequences. This is particularly true for engineering structures such as groins and seawalls, which affect the storage and flow of sediment along a coastal area.
3. **Stabilization of the coastal zone through engineering structures protects the property of relatively few people at a larger general expense to the public.** Engineering structures along the shoreline are often meant to protect developed property, not the beach itself. With a few exceptions, such as recreational shorelines visited by thousands of people during the tourist season, shoreline protection benefits a relatively small number of property owners. It has been argued that the interests of people who own shoreline property are not compatible with the public interest and that it is unwise to expend large amounts of public funds to protect the property of a few.

SEAWALLS

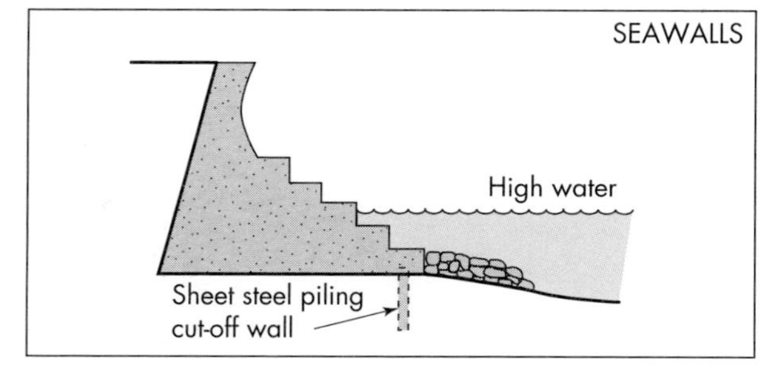

ADVANTAGES
1. Provides protection from wave action and stabilizes the backshore.
2. Low maintenance cost.
3. Readily lends itself to concrete steps to beach.
4. Stabilizes the backshore.

DISADVANTAGES
1. Extremely high first cost.
2. Subject to full wave forces; fail from scour; flanking of foundation.
3. Not easily repaired.
4. Complex design and construction problem. Qualified engineer is essential.
5. Slope design is most important.
6. More subject to catastrophic failure unless positive toe protection is provided.
7. Often cause erosion and loss of the beach.

STONE REVETMENT

ADVANTAGES
1. Most effective structure for absorbing wave energy.
2. Flexible—not weakened by light movements.
3. Natural rough surface reduces wave runup.
4. Lends itself to stage construction.
5. Easily repaired—low maintenance cost.
6. The preferred method of protection when rock is readily available at a low cost.

DISADVANTAGES
1. Heavy equipment required for construction.
2. Subject to flanking and moderate scour.
3. Limits access to beach.
4. Moderately high first cost.
5. Difficult construction where access is limited.

OFFSHORE BREAKWATER

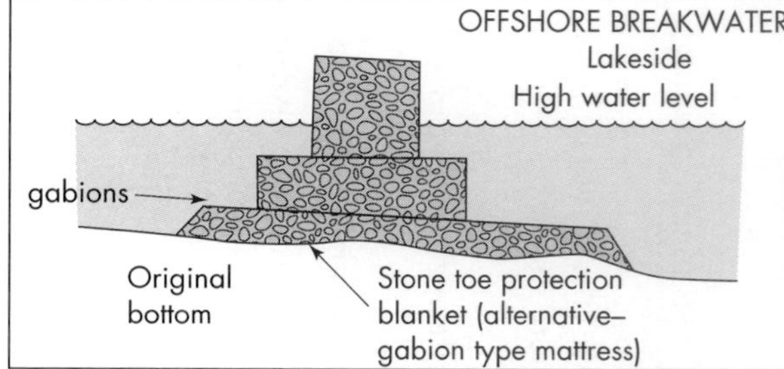

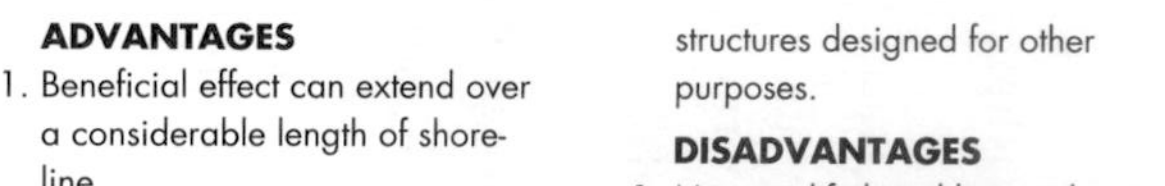

ADVANTAGES
1. Beneficial effect can extend over a considerable length of shoreline.
2. Maintains or enhances recreational value of a beach.
3. Not subject to flanking—can be built in separate reaches.
4. Structure maintenance costs are lower than those of similar structures designed for other purposes.

DISADVANTAGES
1. May modify beachline and cause erosion in downdrift areas.
2. Structure is subject to foundation and scour failures. Floating plant and heavy equipment may be required for construction.

*A gabion is a wire basket filled with rock fragments or coarse gravel.

IMPERMEABLE GROINS

ADVANTAGES
1. Resulting beach protects upland areas and provides recreational benefit.
2. Moderate first cost and low maintenance cost.

DISADVANTAGES
1. Extremely complex coastal engineering design problem. Qualified coastal engineering services are essential. Groins rarely function as intended.
2. Areas downdrift will probably experience rapid erosion.
3. Unsuitable in areas of low littoral drift.
4. Subject to flanking, must be securely tied into bluff.

BROKEN CONCRETE REVETMENT

ADVANTAGES
1. Inexpensive.
2. Easy construction.

DISADVANTAGES
1. Large concrete pieces are difficult to obtain.
2. Large pieces required for underlying filter layer because of large void.
3. Extremely unattractive appearance, unless special care is taken in construction.

RELOCATION

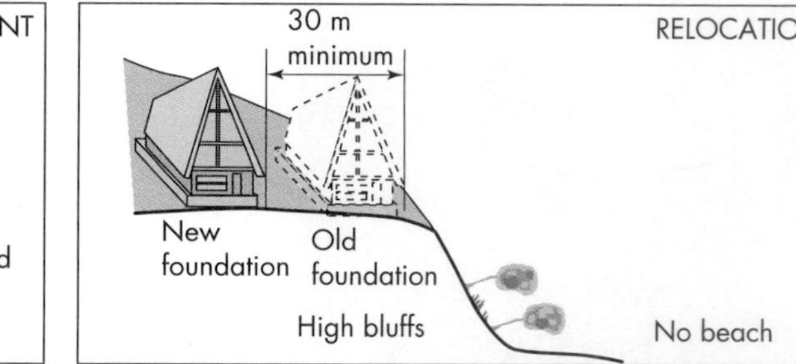

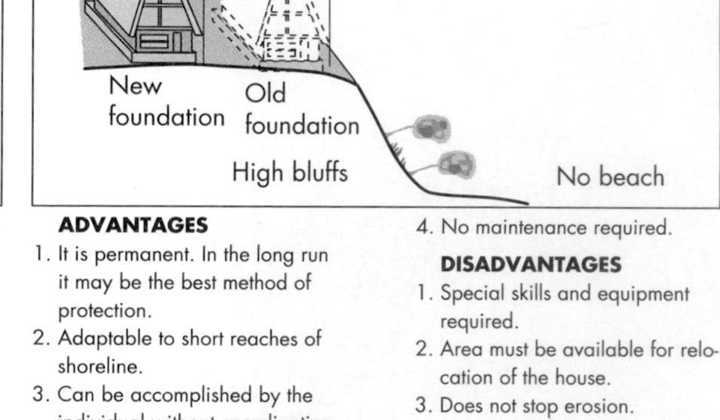

ADVANTAGES
1. It is permanent. In the long run it may be the best method of protection.
2. Adaptable to short reaches of shoreline.
3. Can be accomplished by the individual without coordination through contract with a house mover.
4. No maintenance required.

DISADVANTAGES
1. Special skills and equipment required.
2. Area must be available for relocation of the house.
3. Does not stop erosion.

▲ **FIGURE 9.24** Example of several methods of shoreline protection from wave erosion. (*From* U.S. *Army Corps of Engineers*)

A CLOSER LOOK **E-Lines and E-Zones**

Recently, at the request of the Federal Emergency Management Agency (FEMA), a special committee of the National Research Council (NRC) developed coastal-zone management recommendations (23), some of which are listed below:

- Future erosion rates should be estimated based on historic shoreline change or statistical analysis of the oceanographic environment (waves, wind, sediment supply, etc., that affect coastal erosion).
- E-lines and E-zones based on erosion rates, as shown in Figure 9.D, should be mapped. The E stands for *erosion,* and the E-10 line (for example) is where the coastline is expected to erode to in 10 years. The E-10 zone is considered to be an imminent hazard where no new habitable structures should be allowed. The setback distance depends on the erosion rate. For example, if the rate is 1 m/yr, the E-10 setback is 10 m.
- Movable structures are allowed in the intermediate and long-term hazard zones (E-10 to E-60) (see Figure 9.D).
- Permanent, large structures are allowed at setbacks greater than the E-60 line.
- New structures seaward of the E-60 line (with the exception of those on high bluffs or seacliffs) should be required to be constructed on pilings to withstand erosion associated with a high-magnitude storm with a recurrence interval of 100 years.

NRC recommendations concerning setbacks are considered to be minimum standards for state or local coastal-erosion management programs. A small number of states (including Florida, New Jersey, New York, and North Carolina) use a setback based upon the rate of erosion—most do not. Nevertheless, the concept of E-lines and E-zones has real merit in coastal erosion management, because it is based on rates of erosion-designated setbacks and on allowable construction in the designated zones. This is at the heart of land-use planning to minimize damage from coastal erosion.

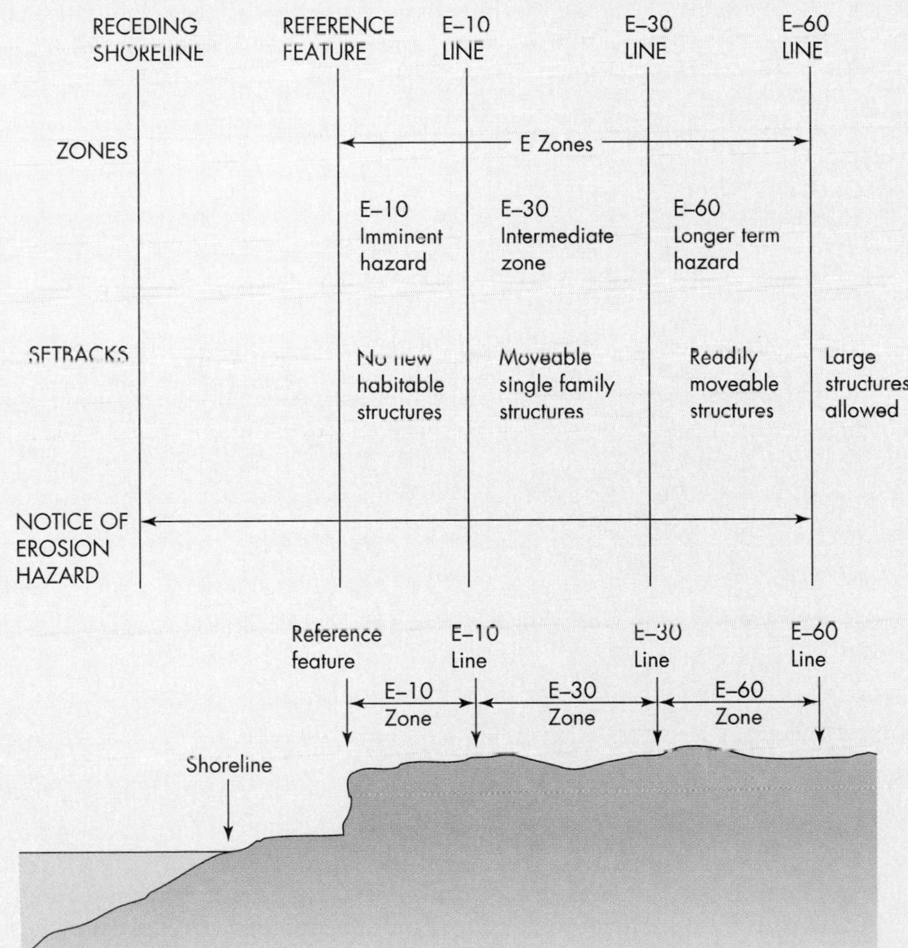

◀ **FIGURE 9.D** Idealized diagram illustrating the concept of the E-lines and E-zones based on the rate of coastal erosion from a reference point such as the seacliff or dune line. The width of the zone depends on the rate of erosion and defines a setback distance. Of course, even with setbacks of 60 years of expected erosion, eventually 60 years down the road structures will be much closer to the shoreline and will become vulnerable to erosion. It is a form of planned obsolescence. (*Source:* National Research Council, 1990. Managing coastal erosion. Washington, DC: National Academy Press.)

4. **Engineering structures designed to protect a beach may eventually destroy it.** Engineering structures often modify the coastal environment to such an extent that it may scarcely resemble a beach. For example, construction of large seawalls causes reflection of waves and turbulence that eventually removes the beach.

5. **Once constructed, shoreline engineering structures produce a trend in coastal development that is difficult, if not impossible, to reverse.** Engineering structures often lead to additional repairs and larger structures, with spiraling costs. In some areas the cost of the structures eventually exceeds the value of the beach property itself. For these and other reasons, several states have recently imposed severe limitations on future engineering construction intended to stabilize the coastline. As sea levels continue to rise and coastal erosion becomes more widespread, nonstructural alternatives to the problem should continue to receive favorable attention because of both financial necessity and the recognition that the amenities of the coastal zone should be kept intact for future generations to enjoy.

▲ **FIGURE 9.25** Cape Hatteras Lighthouse, Outer Banks, North Carolina, was originally constructed approximately 0.5 km from the sea in the late nineteenth century. Before it was moved inland, it was precariously close to being destroyed by coastal erosion. (*Don Smetzer/Tony Stone Images*)

The Cape Hatteras Lighthouse Controversy

In North Carolina a dramatic collision of opinions concerning beach erosion was played out. North Carolina is a leader in the philosophy that beach erosion is a natural process that can be lived with. But erosion threatened the historic Cape Hatteras lighthouse located near Buxton on the Outer Banks (Figure 9.25). When the lighthouse was originally constructed in the late nineteenth century, it was approximately 0.5 km from the sea. Before it was moved inland in the summer of 1999, it was closer to 100 m from the sea, and a major storm could have taken it out. The options were:

- Artificially control coastal erosion at the site, and reverse state policy of yielding to erosion. The U.S. Army Corps of Engineers originally proposed protecting the lighthouse by constructing a $5.6 million seawall around the base (25).
- Do nothing and eventually lose the lighthouse and, thus, an important bit of American history.
- Move the lighthouse inland. This plan was adopted by the National Park Service, which is responsible for the lighthouse. Some local people opposed this plan, fearing the lighthouse would collapse if moved (25).

Relocating the lighthouse reinforced the policy of allowing natural processes to continue and has important implications for future decisions in the coastal zone. During the summer of 1999 the lighthouse was successfully moved at a cost of $12 million to a new location about 0.5 km inland—about the same distance inland the lighthouse was when it was constructed in 1870. Although Hurricane Dennis damaged portions of Hatteras Island in 1999, the lighthouse sustained only superficial damage in its new location.

SUMMARY

The coastal environment is one of the most dynamic areas on earth, and rapid change is a regular occurrence. Migration of people to coastal areas is a continuing trend, and approximately 75 percent of the population in the United States now lives in coastal states.

The most catastrophic coastal hazard is the tropical cyclone hazard. Also called *typhoons* and *hurricanes*, tropical cyclones are violent storms that bring high winds, storm surges (large wind-driven waves), and river flooding. They continue to take thousands of lives and cause billions of dollars in property damage.

The combination of high tides with a storm surge can cause serious tidal flooding of rivers that can inundate upstream communities. The occurrence of such floods in London precipitated the construction of a storm surge (tidal) barrier across the River Thames.

Ocean waves are generated by wind storms at sea and expend their energy on the shoreline. Irregularities in the shoreline account for local differences in wave erosion, which is largely responsible for determining the shape of the coast. Beaches are accumulations of sand or gravel (commonly) deposited at the coast by rivers and shaped by wave action. In actuality, beaches are composed of any loose material such as broken shells or coral, volcanic rock, etc., located in the shore zone. Waves striking a beach at an angle result in littoral transport (the combined effect of beach drift and longshore drift) of the beach sediments. A littoral cell is a segment of coastline that receives sediment from rivers and transports it by littoral transport to a location where it is removed from the coastal environment. The annual beach budget is the balance of sediments gained and lost; negative beach budgets indicate erosion, and are common, partly because of dams that impede the delivery of sediments to beaches. The wave climate is a statistical characterization of wave height, period, and direction. Knowing the beach budget and wave climate of a littoral cell is essential for evaluating coastal erosion in that cell. Using littoral cells as the working unit for coastal planning would ensure greater success than we have at present.

Although coastal erosion causes a relatively small amount of damage compared to other natural hazards, such as river flooding, earthquakes, and tropical cyclones, it is a serious problem along most coasts of the United States, including the shorelines of the Great Lakes. Factors contributing to coastal erosion include river damming, high-magnitude storms, and the worldwide rise in sea level. Seacliffs experience particularly rapid erosion because they are exposed to additional processes, including biological erosion, weathering, rain wash, and human activities, especially those that produce increased runoff of water. Structures on cliffs decrease their stability, contributing to erosion.

Human interference with natural coastal processes, such as the building of seawalls, groins, jetties, breakwaters, artificial dunes, and other structures, is occasionally successful, but in many cases it has caused consi sion. Sand tends to accumulate on the structure and to be eroded on the do problems occur in areas with high popu sparsely populated areas along the Out Carolina are also having trouble with co nourishment (artificial deposition of sand) has had limited success in restoring or widening beaches, but it remains to be seen whether it will be effective in the long term.

Perception of the coastal erosion hazard depends mainly on the individual's experience with and proximity to the hazard.

The most common individual adjustment to tropical cyclones is to do nothing and bear the loss. Community adjustments in developed countries generally attempt to modify the environment by building protective structures designed to lessen potential damage, or to encourage change in people's behavior by better land-use zoning, evacuation, and warning.

Adjustment to coastal erosion in developed areas is often the "technological fix": building seawalls, groins, and other structures or (more recently) beach nourishment. These approaches to stabilizing beaches have had mixed success and may cause additional problems in adjacent areas. Engineering structures are very expensive, require maintenance, and once in place are difficult to remove. The cost of engineering structures may eventually exceed the value of the properties they protect; such structures may even destroy the beaches they were intended to save.

Finally, managing coastal erosion will benefit from careful land-use planning that emphasizes establishment of designated setbacks and allowable construction based upon predicted rates of coastal erosion.

REFERENCES

1. **Coates, D. R., ed.** 1973. *Coastal geomorphology.* Binghamton, NY: State University of New York.
2. **White, A. U.** 1974. Global summary of human response to natural hazards: Tropical cyclones. In *Natural hazards: Local, national, global,* ed. G. F. White, pp. 255–65. New York: Oxford University Press.
3. **Office of Emergency Preparedness.** 1972. *Disaster preparedness* 1, 2.
4. **Lipkin, R.** 1994. Weather's fury. In *Nature on the rampage,* pp. 20–79. Washington, DC: Smithsonian Institution.
5. **Morrill, R. A., Chin, E. H., and Richardson, W. S.** 1979. *Maine coastal storm and flood of February 2, 1976.* U.S. Geological Survey Professional Paper 1087.
6. **Shaw, D. P.** 1983. A barrier to tame the Thames. *Geographical Magazine* 55(3):129–31.
7. **Davis, R. E., and Dolan, R.** 1993. Nor'easters. *American Scientist* 81:428–39.
8. **Komar, P. D.** 1998. *Beach processes and sedimentation.* 2nd ed. Upper Saddle River, NJ: Prentice Hall.
9. **Inman, D. L.** 1976. *Man's impact on the California coastal zone.* Sacramento, CA: State of California Department of Navigation and Ocean Development.
10. **El-Ashry, M. T.** 1971. Causes of recent increased erosion along United States shorelines. *Geological Society of America Bulletin* 82:2033–38.
11. **Godfrey, P. M., and Godfrey, M. M.** 1973. Comparison of ecological and geomorphic interactions between altered and unaltered barrier island systems in North Carolina. In *Coastal geomorphology,* ed. D. R. Coates, pp. 239–58. Binghamton, NY: State University of New York.
12. **Norris, R. M.** 1977. Erosion of sea cliffs. In *Geologic hazards in San Diego,* eds. P. L. Abbott and J. K. Victoris. San Diego: Society of Natural History.
13. **Flanagan, R.** 1993. Beaches on the brink. *Earth* 2(6):24–33.
14. **Carter, R. W. G., and Oxford, J. D.** 1982. When hurricanes sweep Miami Beach. *Geographical Magazine* 54(8):442–48.

. Department of Commerce. 1978. *State of Maryland oastal management program and final environmental impact statement.* Washington, DC: Author.

.6. **Leatherman, S. P.** 1984. Shoreline evolution of North Assateague Island, Maryland. *Shore and Beach,* July:3–10.

17. **Dolan, R.** 1973. Barrier islands: Natural and controlled. In *Coastal geomorphology,* ed. D. R. Coates, pp. 263–78. Binghamton, NY: State University of New York.

18. **Wilkinson, B. H., and McGowen, J. H.** 1977. Geologic approaches to the determination of long-term coastal recession rates, Matagorda Peninsula, Texas. *Environmental Geology* 1:359–65.

19. **Larsen, J. I.** 1973. *Geology for planning in Lake County, Illinois.* Illinois State Geological Survey Circular 481.

20. **Buckler, W. R., and Winters, H. A.** 1983. Lake Michigan bluff recession. *Annals of the Association of American Geographers* 73(1):89–110.

21. **Rowntree, R. A.** 1974. Coastal erosion: The meaning of a natural hazard in the cultural and ecological context. In *Natural hazards: Local, national, global,* ed. G. F. White, pp. 70–79. New York: Oxford University Press.

22. **Baumann, D. D., and Sims, J. H.** 1974. Human response to the hurricane. In *Natural hazards: Local, national, global,* ed. G. F. White, pp. 25–30. New York: Oxford University Press.

23. **National Research Council.** 1990. *Managing coastal erosion.* Washington, DC: National Academy Press.

24. **Neal, W. J., Blakeney, W. C., Jr., Pilkey, D. H., Jr., and Pilkey, O. H., Sr.** 1984. *Living with the South Carolina shore.* Durham, NC: Duke University Press.

25. **McDonald, K. A.** 1993. A geology professor's fervent battle with coastal developers and residents. *The Chronicle of Higher Education* 40(7):A8–89, A12.

KEY TERMS

tropical cyclone (p. 233)
hurricane (p. 233)
tidal flood (p. 236)
wave height (p. 236)
wave length (p. 236)
wave period (p. 236)
plunging breaker (p. 238)
spilling breaker (p. 238)
beach (p. 238)
berms (p. 239)
swash zone (p. 239)
surf zone (p. 239)
breaker zone (p. 239)
longshore trough and bar (p. 239)
littoral transport (p. 240)
longshore current (p. 240)
littoral cell (p. 240)
beach budget (p. 240)
wave climate (p. 240)
coastal erosion (p. 241)
seacliff (p. 242)
seawall (p. 246)
groins (p. 246)
beach nourishment (p. 246)
breakwater (p. 246)
jetties (p. 247)
barrier island (p. 250)

SOME QUESTIONS TO THINK ABOUT

1. There is concern that if global warming is occurring, the ocean waters will also warm and the number of violent storms such as hurricanes will increase. What are the possible consequences of this to human society and how might adverse effects be mitigated?
2. Do you believe that human activity has increased the coastal-erosion problem? Outline a research program that could test this hypothesis.
3. Do you agree or disagree with this statement: All structures in the coastal zone (with the exceptions of critical facilities) are considered temporary and expendable, and, as a result, any development in the coastal zone must be in the best interest of the general public rather than the few who developed the oceanfront.
4. How could you test the following hypotheses: (1) engineering structures designed to protect the beach may eventually destroy it; and (2) once constructed, shoreline engineering structures produce a trend in coastal development that is difficult, if not impossible, to reverse?

PART 3

Human Interaction with the Environment

CHAPTER 10
Water: Process, Supply, and Use

CHAPTER 11
Water Pollution and Treatment

CHAPTER 12
Waste Management

CHAPTER 13
The Geologic Aspects of Environmental Health

In this latter part of the twentieth century, Americans are a metropolitan people. Today, nearly all Americans live in or near urban areas that have populations greater than 50,000. It is expected that by the year 2000, 85 percent of the population will be urban residents. Therefore, urban population growth is now a basic feature of life in the United States.

Chapters 10 through 13 consider complex relationships between people and their environment. Because these relationships are often most stressed in urban areas, special attention is given to the problems stemming from human activities and inactivities that create our special urban landscape. Chapter 10 reviews several relationships between hydrology and human use, emphasizing water processes, supply, use, and management. Chapter 11 discusses the important subject of water pollution.

Chapter 12 is concerned with integrated waste management and the treatment and disposal of wastes—sanitary landfills, ocean dumping, septic systems, and radioactive and hazardous chemical wastes. Recent advances in environmental health involving the fields of geology, geography, and toxicology are discussed in Chapter 13. This subject promises to become more significant as our understanding of the interrelationships of earth processes, earth materials, and health problems increases.

10 Water: Process, Supply, and Use

Las Vegas, Nevada, a desert city, is a human-made oasis. (*Edward A. Keller*)

Availability of water is necessary for desert cities to flourish. Egyptian dynasties prospered when water from the Nile River was abundant and suffered famine during droughts. Shown here is a more modern pyramid and monument in one of America's newest playgrounds, also located near a mighty river, which in this case is the Colorado. The pyramid is a luxury hotel, and the city of Las Vegas, one of the most rapidly growing cities in the country, is growing, thanks in part to an abundance of nearby water that allows the development of casinos, water parks, and extravagant shows and outdoor fountains and gardens.

A major question facing people in many parts of the world today is how long will apparent abundant water supplies last as population continues to increase into the new millennium. It is feared that scarcity of fresh water that is safe from disease is a greatly underestimated resource issue that will face the world in coming decades. On a global basis, 70 percent of the world's fresh water that is derived from groundwater and surface water sources is used for agriculture, while another 20 percent is used for industry and 10 percent for residences. The two most populous countries in the world are China and India, and in these countries groundwater resources used to produce food are being used and degraded rapidly. Groundwaters are commonly mined, and levels of groundwater are receding in many locations at the rate of a meter or so per year. As water resources diminish, harvests of crops nourished by irrigation will diminish, perhaps producing food shortages. Undoubtedly, as we go into the twenty-first century, demand for water will increase and competition for limited water resources will likely become apparent.*

*Source: Brown, L. R., and Flavin, C. 1999. A new economy for a new century. In *State of the world 1999*, ed. L. Stark, pp. 3–21. New York: W. W. Norton.

LEARNING OBJECTIVES

Water is one of our most basic and important resources. Ensuring that we maintain an adequate safe supply of water is one of our most important environmental objectives. The lack of a pollution-free and disease-free water supply constitutes a continued serious environmental problem for billions of people in many regions of the world. In this chapter we will consider the topics of hydrology, water supply and use, water management, and water and ecosystems. Learning objectives of this chapter are:

- *To gain a modest appreciation for the global water resource.*
- *To understand why water is a unique fluid in our environment.*
- *To know the major storage compartments for water in the water cycle.*
- *To become familiar with the main factors that control surface runoff and sediment yield.*
- *To understand the basics of groundwater geology, including movement of groundwater and Darcy's law.*
- *To gain a modest acquaintance with the water budget of the United States.*
- *To understand the main types of water use.*
- *To be familiar with some of the major trends in water uses during the past 40 years.*
- *To be able to discuss some of the ways we can conserve our water resources.*
- *To be able to discuss some of the major principles associated with water management.*
- *To understand some of the environmental consequences of water resources development, including construction of dams and canals.*
- *To know the criteria for identifying a wetlands and to understand the environmental significance of wetlands and wetland loss.*

Web Resources

Visit the "Environmental Geology" Web site at www.prenhall.com/keller to find additional resources for this chapter, including:

- ▶ Web Destinations
- ▶ On-line Quizzes
- ▶ On-line "Web Essay" Questions
- ▶ Search Engines
- ▶ Regional Updates

10.1 Water: A Brief Global Perspective

The global water cycle involves the movement of water from one of the earth's storage compartments to another. In its simplest form (diagrammed in Figure 10.1), the water cycle can be viewed as water moving from the oceans to the atmosphere, falling from the atmosphere as rain, and then returning to the oceans as surface runoff and subsurface flow or to the atmosphere by evaporation. The annual cyclic nature of this global movement of water is illustrated in Figure 10.1. Note that:

1. Where the annual volume of water transferred from the ocean (from evaporation) to the land (47,000 km^3) is balanced by the same volume returning by river and groundwater flow to the ocean (1).
2. The 505,000 km^3 per year of water evaporated from the oceans of the world is balanced by the sum of the water that falls as precipitation in the ocean (458,000 km^3) and the 47,000 km^3 of water that is transferred from the atmosphere to the land.
3. Evaporation of water from the land is 72,000 km^3 per year, and the sum of this and that transferred from the

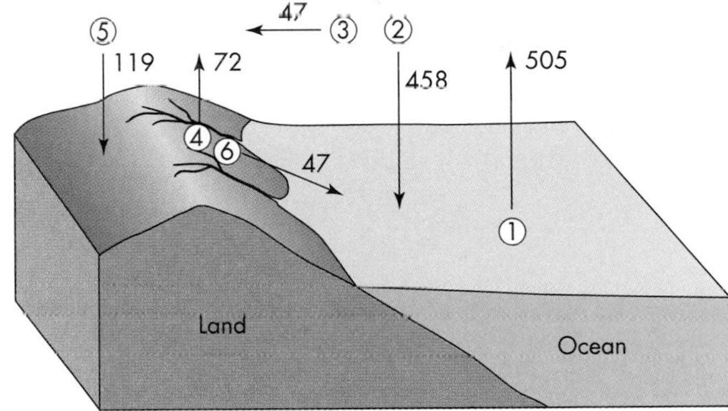

Annual flow of water on earth in thousands of km^3
① Evaporation from oceans
② Precipitation to oceans
③ Transfer of water from atmosphere to land
④ Evaporation (from land) to atmosphere
⑤ Precipitation to land
⑥ Runoff of surface water and groundwater from land to oceans

▲ **FIGURE 10.1** Movement of water in the global water cycle. (*Data from P. H. Gleick. 1993. An introduction to global fresh water issues. In* Water in Crisis, *ed. P. H. Gleick, 1993, pp. 3–12. New York: Oxford University Press.*)

atmosphere (47,000 km^3 per year) is 119,000 km^3 per year that falls as precipitation on the land.

4. The 119,000 km^3 per year that falls as precipitation on the land, 60 percent (72,000 km^3 per year) evaporates, and 40 percent (47,000 km^3 per year) returns to the oceans as surface or groundwater runoff.

But the water that returns is changed because it carries with it sediment (gravel, sand, silt, clay) eroded from the land. The return flow also carries many chemicals (most of which are natural) but also includes many human-made and human-induced compounds, such as organic waste, nutrients, and thousands of chemicals used in our agricultural, industrial, and urban processes. In this chapter we will be particularly concerned with surface runoff and subsurface flow as they relate to human use.

On a global scale, water abundance is not a problem—the problem is water's availability in the right place at the right time in the right form. Water is a heterogeneous resource that can be found in liquid, solid, or gaseous form at a number of locations at or near the earth's surface. Depending upon the specific location of water, the residence time may vary from a few days to many thousands of years (Table 10.1). Furthermore, more than 99 percent of the earth's water is unavailable or unsuitable for beneficial human use because of its salinity (seawater) or its form and location (ice caps and glaciers). Thus, the water for which all the people on earth compete is much less than 1 percent of the total.

As the world's population and the industrial production of many goods increase, the use of water will also accelerate. Today, world per capita use of water is about 700 m^3/yr, and the total human use of water is 3850 km^3/yr. The per capita use in the United States is about 1850 m^3/yr, or more than 2.5 times the world per capita use. It is estimated that by the year 2000, total world use of water (with a decrease in per capita use due to better conservation) will nevertheless increase to 6000 km^3/yr—a significant fraction of the naturally available fresh water.

The total average annual water yield (runoff) from the earth's rivers and groundwater is approximately 47,000 km^3 (Table 10.2), but its distribution is far from uniform. Some runoff occurs in almost uninhabited regions, such as Antarctica, which produces 2310 km^3, or about 5 percent of the earth's total runoff. South America, which includes the largely uninhabited Amazon Basin, provides 12,200 km^3, or about one-fourth of the total runoff. The total runoff from North America is about two-thirds of that for South America, or 8180 km^3. Unfortunately, much of the North American runoff occurs in sparsely settled or uninhabited regions, particularly in the northern parts of Canada and Alaska.

Compared with other resources, water is used in tremendous quantities. In recent years the total amount of water by volume used on the earth annually has been approximately 1000 times the world's total production of minerals, including petroleum, coal, metal ores, and nonmetals (2). Because of its great abundance, water is generally a very inexpensive resource. But because the quantity and the quality of water available at any particular time are highly variable, statistical statements about the cost of water on a global basis are not particularly useful. Shortages of water have occurred and will continue to occur with increasing frequency, leading to serious economic disruption and human suffering (3).

The U.S. Water Resources Council has estimated that water use in the United States by the year 2020 may exceed surface water resources by 13 percent. As early as 1965, 100 million people in the United States used water that had already been used once before, and by the end of the century most of us will be using recycled water. How can we manage our water supply, use, and treatment to maintain adequate supplies?

10.2 Water as a Unique Liquid

To understand water in terms of supply, use, pollution, and management, we first need a modest acquaintance with some of water's characteristics. Water is a unique liquid;

Table 10.1 The world's water supply (selected examples)

Location	Surface Area (km^2)	Water Volume (km^3)	Percentage of Total Water	Water: Estimated Average Residence Time
Oceans	361,000,000	1,230,000,000	97.2	Thousands of years
Atmosphere	510,000,000	12,700	0.001	9 days
Rivers and streams	—	1,200	0.0001	2 weeks
Groundwater: shallow, to depth of 0.8 km	130,000,000	4,000,000	0.31	Hundreds to many thousands of years
Lakes (fresh water)	855,000	123,000	0.009	Tens of years
Ice caps and glaciers	28,200,000	28,600,000	2.15	Up to tens of thousands of years and longer

Source: Data from U.S. Geological Survey.

Table 10.2 Water budgets for the continents

Continent	Precipitation mm/yr (km^3)		Evaporation mm/yr (km^3)		Runoff k^3/yr
North America	756	(18,300)	418	(10,000)	8,180
South America	1,600	(28,400)	910	(16,200)	12,200
Europe	790	(8,290)	507	(5,320)	2,970
Asia	740	(32,200)	416	(18,100)	14,100
Africa	740	(22,300)	587	(17,700)	4,600
Australia and Oceania	791	(7,080)	511	(4,570)	2,510
Antarctica	165	(2,310)	0	(0)	2,310
Earth (entire land area)	800	(119,000)	485	(72,000)	47,000*

*Surface runoff is 44,800; groundwater runoff is 2,200.

Source: Data from I.A. Shiklomanov, 1993. World fresh water resources. In *Water in Crisis*, ed. P. H. Gleick, 1993, pp. 3–12. New York: Oxford University Press.)

without it, life as we know it would be impossible. Every water molecule contains two atoms of hydrogen and one of oxygen. The chemical bonds that hold the molecule together are *covalent*, meaning that each hydrogen atom shares its single electron with the oxygen atom, and the oxygen atom shares its outermost electrons with the hydrogen atom. Although the molecule is electrically neutral (having no net positive or negative charge), the hydrogen end of the molecule is more positively charged, and the oxygen end is more negatively charged, because the electrons, which are negatively charged, are somewhat closer to the oxygen than to the hydrogen. A molecule with one end more negative and the other more positive is called *dipolar*.

The fact that water is dipolar accounts for many of its important properties and for how it reacts in the environment. For example, water molecules are attracted to each other (more positive ends to more negative ends), so they produce thin films, or layers of water molecules, between and around particles important in the movement of water in the unsaturated (vadose) zone above the groundwater table. This process is one of *cohesion*. Water molecules may also be attracted to solid surfaces *(adhesion)*; in particular, the more negative (oxygen) ends of the water molecule are attracted to positive ions such as sodium, calcium, magnesium, and potassium. Because clay particles tend to have a negative charge, they attract the more positive (hydrogen) end of water molecules and so become hydrated. Finally, the dipolar nature of the water molecule is responsible for producing surface tension: Water molecules are more attracted to each other than they are to molecules of air. Surface tension is extremely important in many physical and biological processes involving water moving through small openings and pore spaces (4).

Water is often referred to as the universal solvent. Its ability to dissolve a wide variety of substances (from simple salts to minerals and rocks) makes it an essential and major component of living matter. Water is particularly important in the chemical weathering of rocks and minerals that, along with physical and biochemical processes, initiates soil formation.

Among common substances, water is the only one with a solid form lighter than its liquid form, which explains why ice floats. If ice were heavier than liquid water, it would sink. Although this would be safer for ships traveling in the vicinity of icebergs, properties of the biosphere would be much different from what they are. Rivers, lakes, and the ocean would freeze from the bottom up.

Another important feature of water is its *triple point*, the temperature and pressure at which its three phases—solid (ice), liquid (water), and gas (water vapor)—can exist together. The triple point of water occurs naturally at or near the surface of the earth. This has important implications for transfer of water from the ocean to the atmosphere and biosphere via the **water cycle.** The world would be a much different place if water couldn't evaporate from the oceans to the atmosphere at near-surface conditions (the water cycle would stop). The triple point for some substances, on the other hand, can only be achieved on earth under laboratory conditions.

Water has a tremendous moderating effect on the environment because of its high *specific heat*. Specific heat is defined as the amount of heat (measured in calories) required to raise the temperature of one gram (g) of a substance one Celsius degree. The specific heat of water is 1.0 calorie/g, as compared to the specific heats of most other solvents, which are about 0.5 calories/g. Thus, compared to other common liquids, water has the greatest capacity to absorb and store heat. This storage of heat helps moderate the environment, particularly near large bodies of water.

10.3 Surface Runoff and Sediment Yield

Surface runoff has important effects on erosion and the transport of materials. Water moves materials either in a dissolved state or as suspended particles, and surface water

(a)

(b)

▲ **FIGURE 10.2** (a) Raindrop falling in a cornfield causes soil particles to be lifted into the air, initiating the erosion process; (b) surface runoff often causes the formation of small gulleys such as those shown there. (*[a] Runk-Schoenberger/Grant Heilman Photography, Inc.; [b] courtesy of* U.S. *Department of Agriculture*)

can dislodge soil and rock particles on impact (Figure 10.2). The number and size of the suspended particles moved by surface waters depend in part on the volume and depth of the water and the velocity of flow. The faster a stream or river flows, the larger the particles it can move and the more material is transported. Therefore, the factors that affect runoff also affect sediment erosion, transport, and deposition.

The flow of water on land is divided by watersheds. A **watershed,** or **drainage basin** (Figure 10.3), is an area of ground in which any drop of water falling anywhere in it will leave in the same stream or river. (This definition assumes that the drop is not consumed by the biosphere, evaporated, stored, or transported out of the watershed by subsurface flow.) Large drainage basins can be subdivided into smaller ones. For example, the Mississippi River drainage basin drains about 40 percent of the United States but contains many subbasins such as the Ohio, Missouri, and many others. Drainage basins such as the Ohio may be further divided into smaller basins. Figure 10.3 shows two drainage basins (A and B) that are side by side. Two drops of rain separated by only centimeters along the boundary of a major continental divide may end up a few weeks later in different oceans thousands of kilometers apart. We may also think of a drainage basin as the land area that contributes its runoff to a specific drainage net, the set of channels that makes up a drainage basin. Thus, the drainage basin refers to an area of land, whereas the drainage net refers to the actual river and stream channels in the drainage basin.

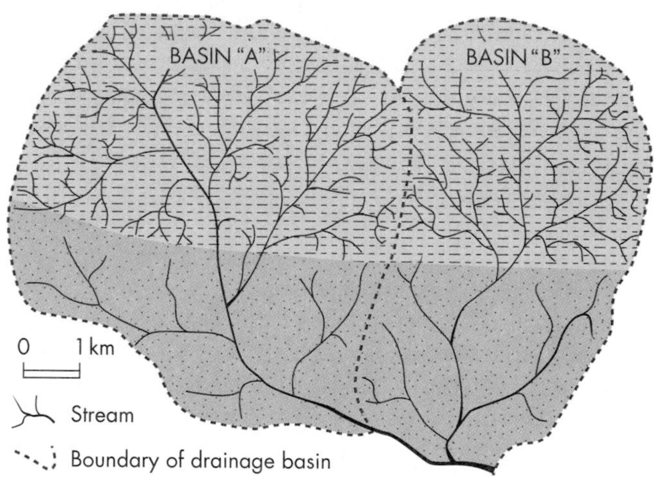

▲ **FIGURE 10.3** Two drainage basins. Water falling on one side of the central boundary will drain into Basin A; on the other side, water will drain into Basin B. In this case, the streams from both basins eventually converge.

Factors Affecting Runoff

The amount of surface-water runoff and the amount of sediment carried by the runoff vary significantly from one drainage basin and river to another. The variation results from geologic, physiographic, climatic, biologic, and land-use characteristics of a particular drainage basin and variations of these factors with time. Even the most casual observer can see the difference in the amount of sediment carried by the same river in flood state and at low flow, since floodwaters are usually more muddy.

Geologic Factors The principal geologic factors affecting surface-water runoff and sedimentation include rock and soil type, mineralogy, degree of weathering, and structural characteristics of the soil and rock. Fine-grained, dense, clay soils and exposed rock types with few fractures generally allow little water to move downward and become part of the subsurface flows. The runoff from precipitation falling on such materials is comparatively rapid. Conversely, sandy and gravelly soils, well-fractured rocks, and soluble rocks absorb a larger amount of precipitation and have less surface runoff. These principles are illustrated in Figure 10.3. The upper parts of basins A and B are underlain by shale, and the lower parts are underlain by sandstone. Because the shale has a greater potential to produce runoff than the more porous sandstone, the *drainage density* (length of channel per unit area) is much greater in the shale areas than in the sandstone areas.

Physiographic Factors Physiographic factors that affect runoff and sediment transport include shape of the drainage basin, relief and slope characteristics, and the orientation of the stream basins to prevailing storms.

The *shape of the drainage basin* is greatly affected by the geologic conditions. For example, drainage may develop along weak, crushed rock associated with fracture zones, producing a long, narrow drainage basin. One principal effect of basin shape on runoff and sedimentation is its role in governing the rate at which water is supplied to the main stream. Basins that are lengthy and narrow and have a long main channel with many short tributaries receive flow from the tributaries much more rapidly than do basins that have a shorter main channel with long, sinuous tributaries. Rivers in drainage basins that experience rapid rise in flow during or after precipitation are said to be "flashy" and can produce flash floods.

The factors of *relief* and *slope* are interrelated: The greater the relief (the difference in elevation between the highest and lowest points of a drainage basin or a river or any landform of interest), the more likely the stream is to have a steep gradient and a high percentage of steep, sloping land adjacent to the channel. Relief and slope are important because they affect not only the velocity of water in a stream but also the rate at which water infiltrates the soil or rock and the rate of overland flow, both of which affect the rate at which surface and subsurface runoff enters a stream.

Orientation of the stream basin to prevailing storms influences the rate of flow, the peak flow, the duration of surface runoff, and the amount of transpiration and evaporation losses. The latter is a factor because basin orientation affects the amount of heat received from the sun as well as exposure to prevailing winds.

Climatic Factors Climatic factors affecting runoff and sediment transport include the type of precipitation that occurs, the intensity of the precipitation, the duration of precipitation with respect to the total annual climatic variation, and the types of storms (whether cyclonic or thunderstorm). In general, discharge of large volumes of water and sediment is associated with infrequent high-magnitude storms that occur on steep, unstable topography underlain by soil and rocks with a high erosion potential.

Biologic Factors Vegetation, animals, and soil organisms all influence runoff and sediment yield. *Vegetation* is capable of affecting stream flow in several ways:

- Vegetation may decrease runoff by increasing the amount of rainfall intercepted and removed by evaporation. Rainfall that is intercepted by vegetation also falls to the ground more gently and is more likely to infiltrate the soil. Experimental clear-cutting of forested watersheds has been shown to increase the stream flow due to decreased evapotranspiration (water used by the trees and released to the atmosphere) following timber harvesting (5).
- Decrease or loss of vegetation due to climatic change, wildfire, or land use will increase runoff and production of sediment. Figure 10.4 shows the response of a small drainage basin following a wildfire in southern California. Figure 10.4a depicts the channel shortly following the fire, which occurred in the summer. These are assumed to be the conditions (as the channel was) before the fire as no storms or other events occurred. Following a moderate rainstorm and runoff event, the entire channel filled with fine gravel derived from the burned slopes (Figure 10.4b). Following another moderate rainstorm and flow, the sediment in the channel was transported out of the system and the channel looked much as it did after the fire before the first storm (Figure 10.4c). What happened? The fire removed vegetation on slopes, and loose material (sediment) that had accumulated on the slopes—but held there by the vegetation before the fire—moved downslope toward the stream channel. This process of dry transport of loose material is called *dry ravel*. When the first rains fell on the burned slopes, runoff was high and a voluminous amount of sediment moved down hillslopes to the stream channel. The stream flow was not sufficient to transport all the sediment, and so much of it was deposited in the channel (Figure 10.4b). Importantly, much of the sediment from the hillslope was removed by the first storm, so when the next storm struck, there was much less sediment carried from hillslopes to the stream. Runoff of water from the burned hillslopes produced lots of stream flow, which scoured the material earlier deposited in the channel. Thus, the effect of the wildfire was to cause a major flushing of sediment from burned slopes out of the drainage basin. This is a common response following wildfire. Less commonly, large debris flows may be produced if intense precipitation of sufficient duration falls on burned slopes containing abundant coarse debris (sediment) (6).
- Streamside vegetation increases the resistance to flow, which slows down the passage of floodwater.
- Streamside vegetation retards stream-bank erosion because its roots bind and hold soil particles in place.
- In forested watersheds, large organic debris (stems and pieces of woody debris) may profoundly affect

(a)

(b)

(c)

▲ **FIGURE 10.4** (a) A small stream channel in southern California shortly after a wildfire that burned vegetation in the drainage basin. Some trees near the channel survived the fire. (b) The scene after the first winter storm. Note the voluminous amount of sediment deposited. (c) After the second winter storm, which scoured the channel. Note the channel looks much as it did following the fire. See text for further explanation. (*Edward A. Keller*)

stream-channel form and process. In steep mountain watersheds, many of the pool environments important for fish habitat may be produced by large organic debris.

Animals affect streams by removing vegetation or burrowing. Large grazing mammals can damage streamside environments, causing bank-erosion problems. Animals burrowing through flood-control levees can start erosion problems that eventually lead to failure of the levees.

Soil organisms alter the physical structure of the soil, which sometimes results in greater percolation of water into the soil, reducing runoff and erosion. Plant roots and burrowing animals can produce macropores (large openings) in soil that can greatly increase the rate at which water moves through soil. Soils with a high organic content tend to be relatively cohesive—they reduce surface erosion and tend to hold water tenaciously—compared to sandy soils, which have low cohesion, high porosity, and high permeability.

Runoff Paths

We have seen that runoff is quite variable and depends upon geologic, physiographic, climatic, and biologic conditions. Under natural conditions with continuous forest cover, the direct surface runoff shown in Figure 10.2 is unusual because trees and lower vegetation intercept the precipitation. In such cases water can easily infiltrate the soil on hillslopes, and runoff is by way of **throughflow,** which is a shallow subsurface flow above the groundwater table (Figure 10.5a). An exception may occur near streams and in hillslope depressions if the groundwater table rises to the surface. Such saturated areas can produce surface runoff even in humid climates with good vegetation cover (point 3a, Figure 10.5a). Areas that are saturated tend to expand and contract, being larger during the time of spring snowmelt than in the late summer or fall, when precipitation is less. In disturbed areas, areas with sparse vegetation cover, semiarid lands, tropical and subtropical areas with clay-rich soil that retards surface infiltration of water, and areas with such land uses as row crops or urbanization, **overland flow** is produced because the intensity (rate) of precipitation is greater than the rate at which water infiltrates the ground (Figure 10.5b).

Thus, we can identify three major paths by which water on slopes can be transported from hillslopes to the stream environment and exported from the drainage basin: overland flow, throughflow, and **groundwater flow** (Figure 10.5a). Groundwater flow is discussed in detail in the next section. Understanding potential paths of runoff for a particular site or area is critical in evaluating hydrologic impacts of projects involving land-use changes. Loss of vegetation and soil compaction during urbanization, for example, will produce more overland flow (point 3b, Figure 10.5b), as will land-use change from forest to row crops.

Sediment Yield

Variations in the natural **sediment yield** (volume or mass of sediment per unit time) for relatively small river basins are listed on Table 10.3. The amount of sediment carried by rivers as part of their work within the rock cycle varies with

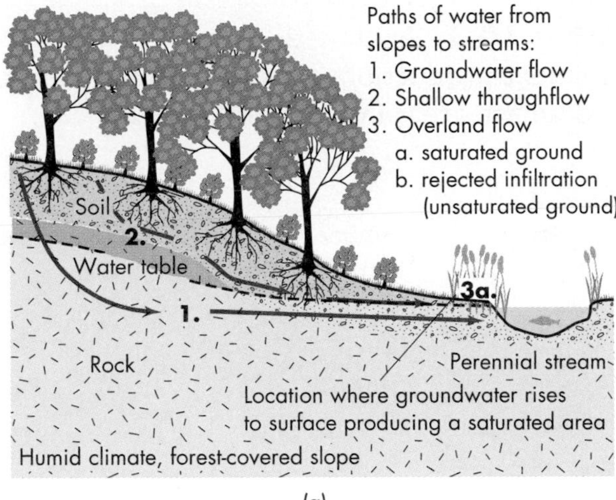

(a)

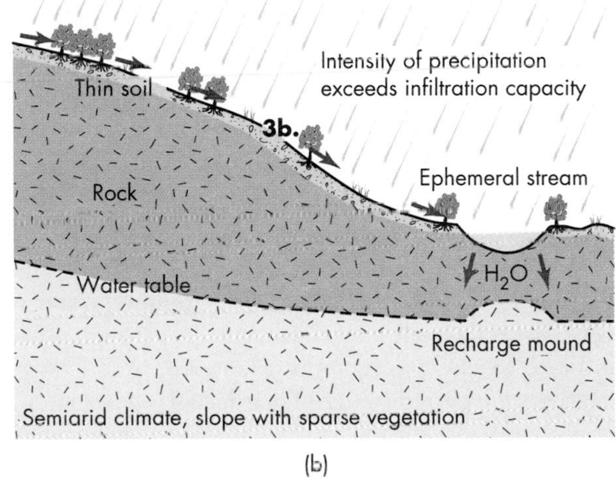

(b)

▲ **FIGURE 10.5** Paths of water from slopes to streams: (a) on a humid, forest covered slope, and (b) in a semiarid climate on a slope with sparse vegetation.

geologic, climatic, topographic, physical, vegetative, and other conditions (recall our earlier discussion of wildfire). Hence, some rivers are consistently and noticeably different in their clarity and appearance, as can be inferred from Table 10.4. Although this table reflects varying degrees of human influence, it demonstrates the sizable variations of sediment load per unit area in various parts of the world. For instance, on the average, the Lo River of China carries nearly 200 times more suspended load than does the Nile River of Egypt. In the United States, the Mississippi is not as "muddy" as the Missouri and the Colorado rivers.

The general relationship between size of drainage basin and sediment load suggests that, as basin size increases, the sediment yield per unit area decreases (Table 10.5). This relationship results from the increase in probability of sediment storage and deposition with increased basin size, the fact that smaller basins tend to be steeper (which increases the energy available for erosion and transport of sediment), and the decreased probability of total basin coverage by a single storm event with increased basin size.

Table 10.3 Estimated ranges in sediment yields from drainage areas of 260 km^2 or less

	Estimated Sediment Yield (metric tons/km^2/yr)[a]		
Region	**High**	**Low**	**Average**
North Atlantic	4,240	110	880
South Atlantic-Gulf	6,480	350	2,800
Great Lakes	2,800	40	350
Ohio	7,391	560	2,780
Tennessee	5,460	1,610	2,450
Upper Mississippi	13,660	40	2,800
Lower Mississippi	28,760	5,460	18,220
Souris-Red-Rainy	1,650	40	175
Missouri	23,470	40	5,250
Arkansas-White-Red	25,760	910	7,710
Texas-Gulf	8,140	320	6,310
Rio Grande	11,700	530	4,550
Upper Colorado	11,700	530	6,310
Lower Colorado	5,670	530	2,100
Great Basin	6,240	350	1,400
Columbia-North Pacific	3,850	120	1,400
California	19,510	280	4,550

[a]The range in high to low values reflects different years with different discharges and ability to erode and transport sediment.

Source: *The Nation's Water Resources*. Water Resources Council, 1968.

10.4 Groundwater

The major source of groundwater is precipitation that infiltrates the surface to enter and move through the top of the **vadose zone** (Figure 10.6). The vadose zone includes all earth material above the water table (for example, soil, alluvium, or rock). Water that infiltrates from the surface may move downward through the vadose zone, which is seldom saturated. Until recently, the vadose zone was called the *unsaturated zone*, but we now know that some saturated areas may exist there at times as water moves through. The vadose zone has special significance because potential pollutants infiltrating at the surface must percolate through the vadose zone before they enter the saturated zone below the water table. Thus, in environmental subsurface monitoring, the vadose zone is an area of early warning for potential pollution to groundwater resources.

Water that percolates through the vadose zone may enter the groundwater system, or **zone of saturation,** where saturated flow occurs. The upper surface of this zone is the **water table.** The **capillary fringe** just above the water table is a belt of variable thickness where water is drawn up by capillary action, which is due both to the attractive force between water and the surfaces of earth materials, and to surface tension (attraction of water molecules to each other).

Table 10.4 Some major rivers of the world ranked by sediment yield per unit area

River	Drainage Basin (10^3 km^2)	Sediment Load per Year (tons/km^2)
Amazon	5,776	63
Mississippi	3,222	97
Nile	2,978	37
Yangtze	1,942	257
Missouri	1,370	159
Indus	969	449
Ganges	956	1,518
Mekong	795	214
Yellow	673	2,804
Brahmaputra	666	1,090
Colorado	637	212
Irrawaddy	430	695
Red	119	1,092
Kosi	62	2,774
Ching	57	7,158
Lo	26	7,308

Source: Data from Holman, 1968.

Table 10.5 Arithmetic average of sediment-production rates for various groups of drainage areas in the United States

Watershed-Size Range (km^2)	Number of Measurements	Average Annual Sediment-Production Rate (m^3/km^2)
Under 25	650	1,810.3
25–250	205	762.2
250–2,500	123	481.2
Over 2,500	118	238.2

Note: Data illustrate that, as the size of a drainage basin (watershed) increases, the sediment production per unit area decreases.

Source: From *Handbook of Applied Hydrology*, by Ven Te Chow. Copyright © 1964 by Mc-Graw Hill. Used with permission of McGraw-Hill Book Company.

In addition to precipitation, other sources of groundwater include water that infiltrates from surface waters, including lakes and rivers, artificial recharge (surface water deliberately injected into the groundwater system), stormwater retention or recharge ponds, agricultural irrigation, and wastewater treatment systems, such as cesspools and septic tank drain fields.

Movement of water into the zone of saturation and through earth materials is an integral part of both the hydrologic cycle and the rock cycle. For example, water may dissolve minerals from materials it moves through and deposit them elsewhere as cementing material, producing sedimentary rocks. Groundwater may transport sediment, heat, gases, and microorganisms. What actually occurs varies with the chemical and physical characteristics of the water, soil,

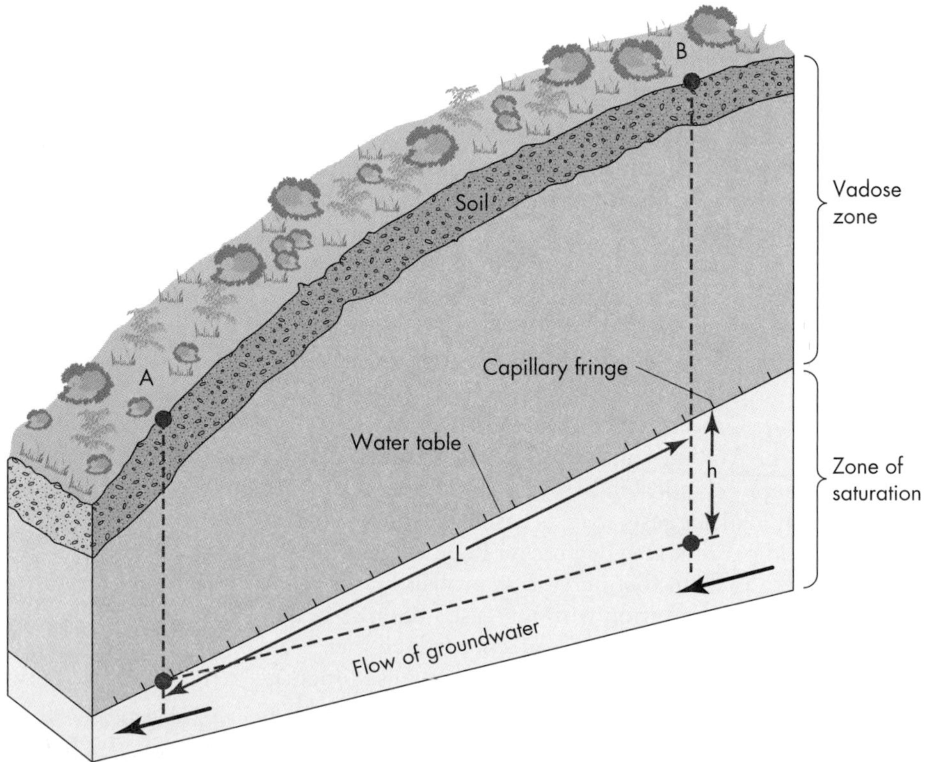

▶ **FIGURE 10.6** Generalized diagram showing zones of groundwater, capillary fringe, and water table.

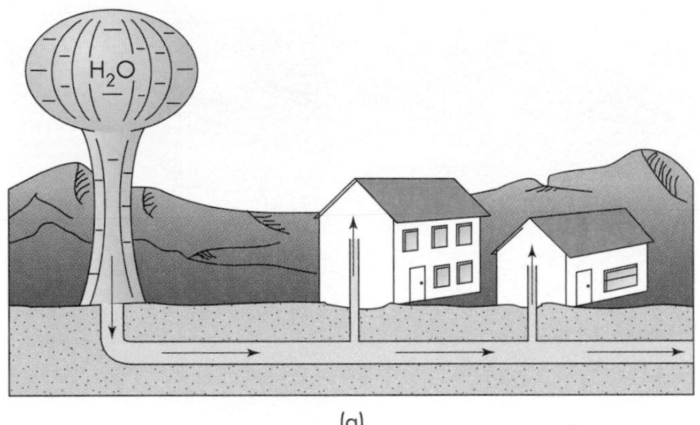

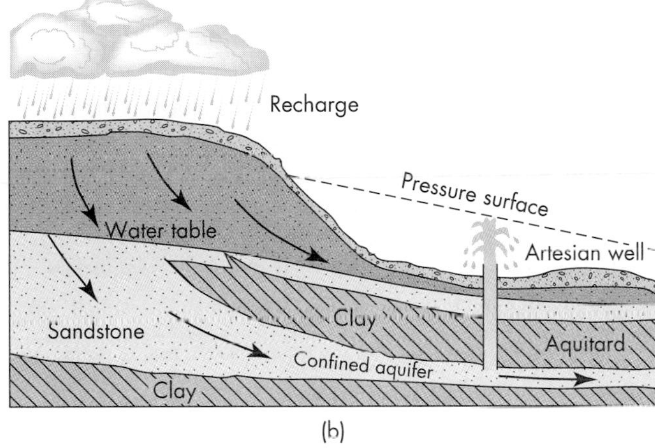

▲ **FIGURE 10.7** Development of an artesian well system. In (a), water rises in homes due to pressure created by water level in the tower. If friction in pipes is small, there will be little drop in pressure. As shown in (b), the pressure surface in natural systems declines away from the source because of friction in the flow system, but water may still rise above the surface of the ground if an impervious layer such as clay is present to cap the groundwater.

▲ **FIGURE 10.8** Discharge of groundwater from Fern spring at the southern end of Yosemite Valley, California. This spring emerges like many others at the base of a hillslope. The width of the small stream emerging from the spring pool in a short cascade or falls is about 2 m. (*Edward A. Keller*)

and rock as the water infiltrates through the biologic and soil-horizon environments above the water table and moves through the groundwater system below the water table.

Aquifers

A zone of earth material capable of supplying groundwater at a useful rate from a well is called an **aquifer.** Gravel, sand, soils, and fractured sandstone, as well as granite and metamorphic rocks with high porosity due to connected open fractures, are good aquifers if groundwater is present. A zone of earth material that will hold water but not transmit it fast enough to be pumped from a well is called an **aquiclude** or **aquitard.** Aquitards often form a *confining layer* through which little water moves. Clay soils, shale, and igneous or metamorphic rocks with little interconnected porosity and/or fractures are likely to form aquitards.

An aquifer is called an **unconfined aquifer** if there is no confining layer restricting the upper surface of the zone of saturation at the water table. If a confining layer is present, the aquifer is called a **confined aquifer,** and the water beneath it may be under pressure, forming **artesian** conditions. These conditions are analogous in their effect to a water tower that produces water pressure for homes (Figure 10.7a). Water in artesian systems tends to rise to about the height of the **recharge zone** (the zone where precipitation infiltrates the surface to move down to the groundwater system), creating an **artesian well** (Figure 10.7b).

In a more general sense **groundwater recharge** is any process that adds water to the aquifer and can be natural infiltration or human-induced as, for example, leakage and infiltration from a broken water line. **Groundwater discharge** is any process that removes groundwater from an aquifer. Included is natural discharge from a **spring** that is present where water flowing in an aquifer intersects the surface of the earth. Spring discharge can form the beginning of a stream or river (Figure 10.8). Groundwater discharge also occurs when water is pumped from a well. Both confined and unconfined aquifers may be found in the same area (Figure 10.9).

When water is pumped from a well, a **cone of depression** forms in the water table or artesian pressure surface (Figure 10.10). A large cone of depression can alter the direction in which groundwater moves within an area. Overpumping of an aquifer causes the water level to lower continuously with time, which necessitates lowering the pump settings or drilling deeper wells. These adjustments are often costly, and they may or may not work, depending

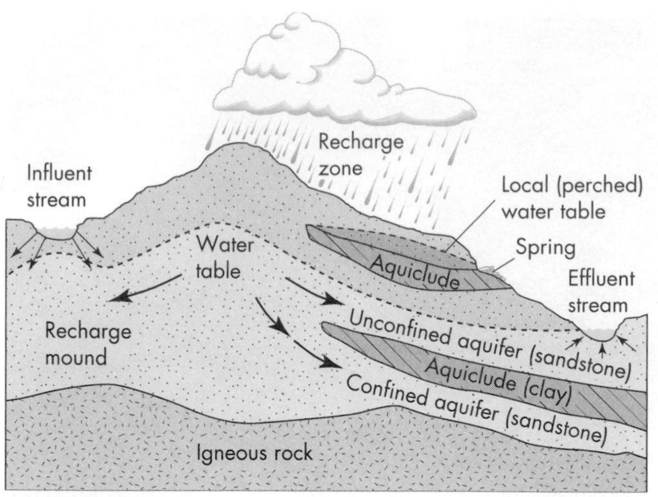

▲ FIGURE 10.9 An unconfined aquifer, a local (perched) water table, and influent and effluent streams.

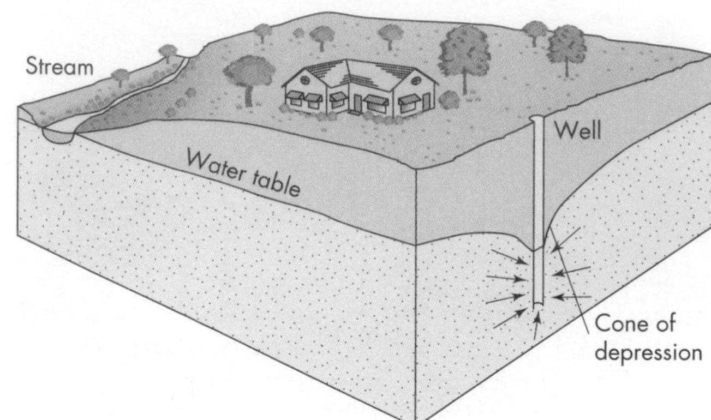

▲ FIGURE 10.10 Cone of depression in water table resulting from pumping water from a well.

on the hydrologic conditions. For instance, continued deepening to correct for overpumping of wells that tap igneous and metamorphic rocks is limited. Water from these wells is pumped from open fracture systems that tend to close or diminish in number and size with increasing depth. Also, the quality of groundwater may be degraded if it is extracted from deeper water containing more dissolved minerals.

Groundwater Movement

Both the rate and the direction of groundwater movement depend upon the gradient of the water table and the properties of the materials present. The **hydraulic gradient** in the simplest cases for an unconfined aquifer is approximately the slope of the water table (see Figure 10.6). The ability of particular material to allow water to move through it is called its **hydraulic conductivity,** which is expressed in units such as meters per day. Expressing the relationship of hydraulic gradient and hydraulic conductivity to groundwater flow quantitatively allows us to solve many problems involving groundwater (see *Putting Some Numbers On Groundwater Flow*).

The hydraulic conductivity of an earth material is a function of both the properties of the material (such as particle diameter, size of pores, and how interconnected the pore spaces are) and the properties of the fluid moving through it (such as viscosity and density). The percentage of void (empty) space in soil or rock is called its **porosity** and depends on the nature and extent of its primary (intergranular) and secondary (fracture) openings. Table 10.6 shows the porosity and hydraulic conductivity of some common earth materials. Notice that some of the most porous materials, such as clay, have a very low hydraulic conductivity. Although clay has a great deal of pore space because of its small, flat particles, the individual openings are very small and hold water tenaciously.

The term *permeability* is also used as a measure of the ability of an earth material to transmit fluid, but only in terms of the properties of that material (not the properties of the fluid). In talking about groundwater, we will use both hydraulic conductivity and permeability to describe hydraulic properties of earth materials; for example, we may say that gravel and sands have high permeabilities compared to silt and clay. However, the term *hydraulic conductivity* is preferred because it is expressed in units that are easily understood and it is commonly used in hydrogeology today.

Interactions Between Surface Water and Groundwater

Interactions between surface water and groundwater are so interrelated that we need to consider both as part of the same resource (Figure 10.11) (7). Nearly all natural surface water environments such as rivers, lakes, and wetlands, as well as human-constructed water environments such as

Table 10.6 Porosity and hydraulic conductivity of selected earth materials

Material	Porosity (%)	Hydraulic Conductivity[a] (m/day)
Unconsolidated		
Clay	45	0.041
Sand	35	32.8
Gravel	25	205.0
Gravel and sand	20	82.0
Rock		
Sandstone	15	28.7
Dense limestone or shale	5	0.041
Granite	1	0.0041

[a]In older works, may be called coefficient of permeability.

Source: Modified after Linsley, Kohler, and Paulhus, *Hydrology for Engineers* (New York: McGraw-Hill, 1958. Copyright © 1958 by McGraw-Hill Book Company. Used by permission of McGraw-Hill Book Company.)

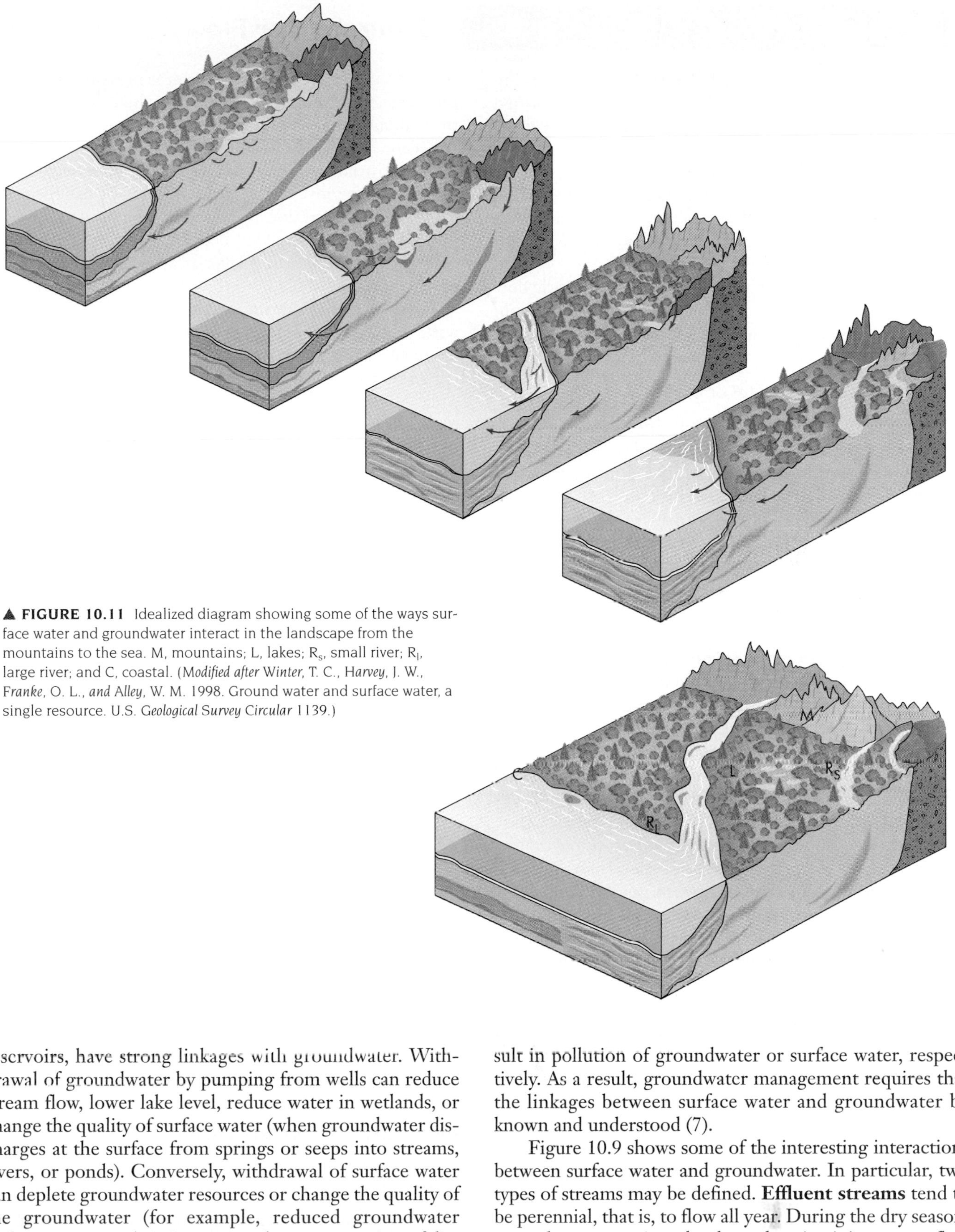

▲ **FIGURE 10.11** Idealized diagram showing some of the ways surface water and groundwater interact in the landscape from the mountains to the sea. M, mountains; L, lakes; R_S, small river; R_l, large river; and C, coastal. (*Modified after* Winter, T. C., *Harvey*, J. W., *Franke*, O. L., *and Alley*, W. M. 1998. Ground water and surface water, a single resource. U.S. *Geological Survey Circular* 1139.)

reservoirs, have strong linkages with groundwater. Withdrawal of groundwater by pumping from wells can reduce stream flow, lower lake level, reduce water in wetlands, or change the quality of surface water (when groundwater discharges at the surface from springs or seeps into streams, rivers, or ponds). Conversely, withdrawal of surface water can deplete groundwater resources or change the quality of the groundwater (for example, reduced groundwater recharge may result in increasing the concentration of dissolved constituents in the groundwater that otherwise would be diluted by mixing with infiltrated surface water). Finally, pollution of either surface water or groundwater can result in pollution of groundwater or surface water, respectively. As a result, groundwater management requires that the linkages between surface water and groundwater be known and understood (7).

Figure 10.9 shows some of the interesting interactions between surface water and groundwater. In particular, two types of streams may be defined. **Effluent streams** tend to be perennial, that is, to flow all year. During the dry season, groundwater seeps into the channel, maintaining stream flow (Figure 10.9, right). **Influent streams** are everywhere along their channel above the groundwater table and only flow in direct response to precipitation. Water from influent streams

PUTTING SOME NUMBERS ON Groundwater Flow

In 1856 an engineer named Henry Darcy was working on the water supply for Dijon, France. He performed a series of important experiments that demonstrated that the discharge *(Q)* of groundwater may be defined as the product of the cross-sectional area of flow *(A)*, the hydraulic gradient *(I)*, and the hydraulic conductivity *(K)*. Thus,

$$Q = KIA$$

The unit on each side of the equation is a volumetric flow rate (such as cubic meters per day), and this relationship is known as **Darcy's law**. The quantity $Q/A = KI$ is the **Darcy flux** *(v)*. We may say that

$$v = Q/A \text{ or } Q = vA$$

Although v has the units of a velocity, the Darcy flux is only an apparent velocity. To determine the actual velocity of groundwater in an aquifer *(vx)* we must remember that the water moves through pore spaces, so its velocity is affected by the porosity of the aquifer material. If we let n represent the porosity, then the actual cross-sectional area of flow is An, and it follows from $Q = vA$ that

$$vx = Q/An = v/n \text{ or } vx = KI/n.$$

The actual velocity vx is about three times the Darcy flux (assuming an average value of $n = 0.33$).

The driving force for groundwater flow is called the **fluid potential** or **hydraulic head,** which at the point of measurement is the sum of the elevation of the water (elevation head) and the ratio of the fluid pressure to the unit weight of water (pressure head). For our simple example in Figure 10.6, the pressure head at both points A and B is atmospheric (defined as 0); thus, the hydraulic heads at A and B are their respective elevations. The difference in hydraulic heads between points A and B *(h)* divided by the flow length *(L)* gives us the hydraulic gradient *(I)*. The condition shown for Figure 10.6 is an unconfined aquifer. If a confining layer is present, then the fluid pressure must be considered in the calculation of the hydraulic gradient. However, Darcy's law still applies.

Groundwater always moves from an area of higher hydraulic head to an area of lower hydraulic head and may therefore move down, laterally, or upward, depending upon local conditions. The water in Figure 10.7 flows upward at the artesian well because the hydraulic head below the clay confining layer is greater than the hydraulic head above it.

Darcy's law has many important applications to groundwater problems. For example, consider an area underlain by sedimentary rocks with a semiarid climate. The area is dissected by a river system in a valley approximately 4 km wide. Alluvial deposits in the valley form an aquifer, and two wells have been drilled approximately 1 km apart in the down-valley direction (Figure 10.A, part a). A cross-valley section between the wells (Figure 10.A, part b) shows that the saturated zone is 25 m thick, consists of sand and gravel, and has a hydraulic conductivity of 100 m/day (1.2×10^{-3} m/sec). Porosity *(n)* of the aquifer materials is 30 percent (0.3). A down-valley section is shown in Figure 10.A, part c. The wells are separated by 1000 m and the elevation of the water in wells 1 and 2 are, respectively, 98 and 97 m. Two questions we might ask concerning the conditions shown in Figure 10.A are:

1. What is the discharge Q (m^3/sec or gallons per day) of water moving through the aquifer in the down-valley direction?
2. What is the travel time *(T)* of the groundwater between wells 1 and 2? This question is particularly interesting from an environmental standpoint if a water pollution event is detected at well 1 and we want to know when the pollution will reach well 2.

Answering these two questions requires us to apply Darcy's law to the situation outlined above. To answer the first question, which asks how much water is moving

moves down through the vadose zone to the water table, forming a recharge mound (Figure 10.9, left). Influent streams may be intermittent or ephemeral in that they flow only part of the year.

From an environmental standpoint, influent streams are particularly important because water pollution in the stream may move downward through the stream bed and eventually pollute the groundwater below. Dry river beds are particularly likely to experience this type of problem. For example, the Mojave River in southern California is dry almost all of the time in the vicinity of Barstow. Solvents introduced into the dry river bed as part of a large cleaning operation for equipment have infiltrated down through the vadose zone to contaminate and threaten groundwater that is used by several communities for municipal purposes, including drinking.

Perceptions About Groundwater

People's perceptions about groundwater affect the way they view our water resource:

- People tend to assume that water is available when, where, and in the amounts they want. We turn on a faucet and expect water—it is somebody else's responsibility to see that we have it.

through the aquifer, recall that $Q = KIA$. We will solve for Q. The hydraulic gradient, as illustrated in Figure 10.6, is the ratio of the difference in elevation of the water between the two wells to the length of the groundwater flow between the wells. The difference in elevation of the groundwater table between the wells is 1 m and the flow length is 1000 m. Thus, the hydraulic gradient (I) is 0.001 (1×10^{-3}). The hydraulic conductivity is given as 1.2×10^{-3} m/sec. The cross-sectional area of the aquifer (A) is 25 m × 4000 m, or 100,000 m^2 $(1\times10^5$ m$^2)$. Multiplying these numbers, we find that Q is equal to 0.12 m^3/sec. This is equivalent to 10,368 m^3/day, which is approximately 2.7 million gallons per day. Of course, all of this water could not be pumped from the aquifer. Pump tests of the wells would be necessary to determine how much of the 2.7 million gallons per day could be pumped without depleting the resource.

Turning now to the second question, which concerns the travel time of the water from one well to the other, we again apply Darcy's law. In this case we calculate the Darcy flux (v), which is

$$v = Q/A = KI$$

Remember that the Darcy flux is only an apparent velocity and does not reflect the fact that the actual movement of the groundwater is through the pore spaces between the grains of sand and gravel in the aquifer. The actual velocity **(vx)** is the ratio of the product of KI to the porosity.

$$vx = KI/n$$

$$= (1.2\times10^{-3} \text{ m/sec})\,(1\times10^{-3})/0.3$$

$$= 4.0\times10^{-6} \text{ m/sec}$$

Travel time (T) then is the ratio of the length of flow (L) to the velocity of the water moving through the pore spaces (vx). This follows from the fact that distance L is the product of velocity vx and time T $(L = vxT)$. Thus, $T =$ 1000 m/4.0×10^{-6} m/sec $= 2.5\times10^8$ sec. This is approximately 7.9 years.

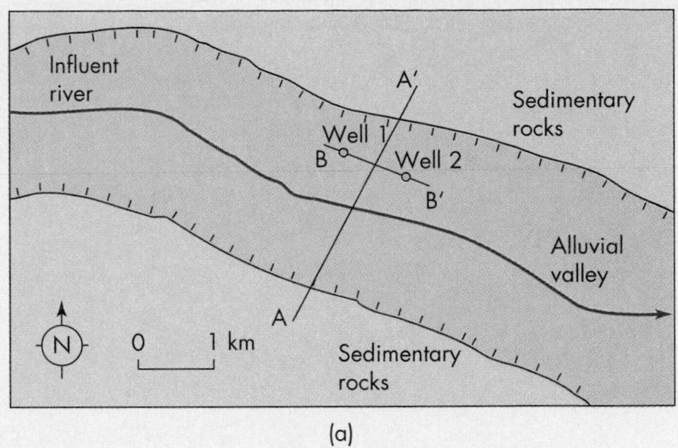

(a)

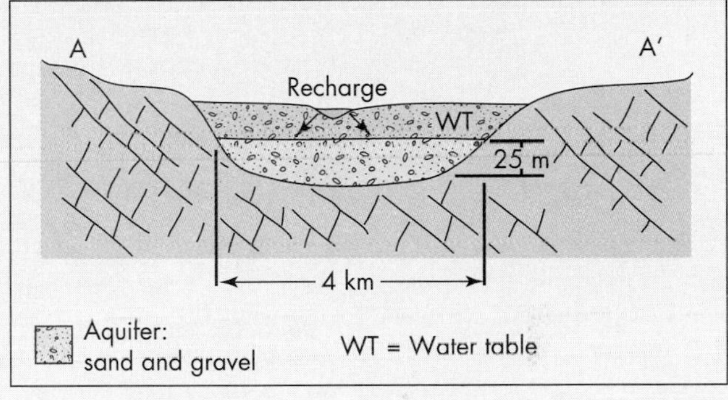

(b)

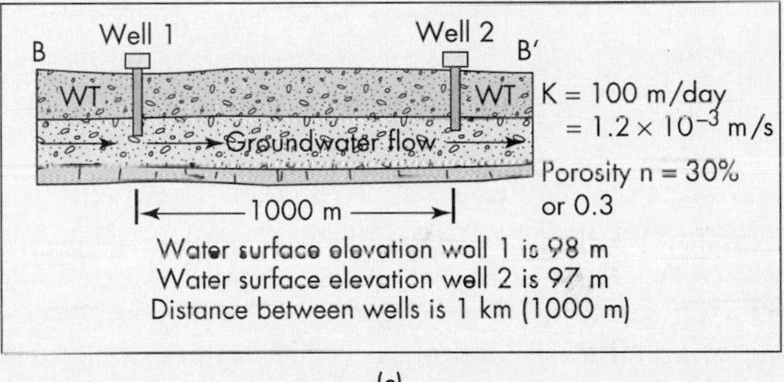

(c)

▲ **FIGURE 10.A** Hypothetical map of an alluvial valley (a); cross-valley profile (b); and profile down-valley (c) showing groundwater conditions.

- Because groundwater is out of sight, it is out of mind or mysterious.
- Groundwater is not as easily measured quantitatively as surface water. Therefore, precise quantitative values of groundwater reserves are not available, and we rely on estimates of the probable reserves.

10.5 Water Supply

The water supply at any place on the land surface depends upon several factors in the hydrologic cycle, including the rates of precipitation, evaporation, stream flow, and subsurface flow. The various uses of water by people also significantly affect water supply. In this section we will focus on the U.S. water supply as an example of the problems occurring in many parts of the world.

The Water Budget

A concept useful in understanding water supply is the **water budget**—the inputs, outputs, and storage of water in a system. The water budget for the conterminous United States is shown in Figure 10.12. The amount of water vapor passing over the United States daily is equivalent to approximately 152,000 million m^3 or 40,000 billion gallons (bg) of

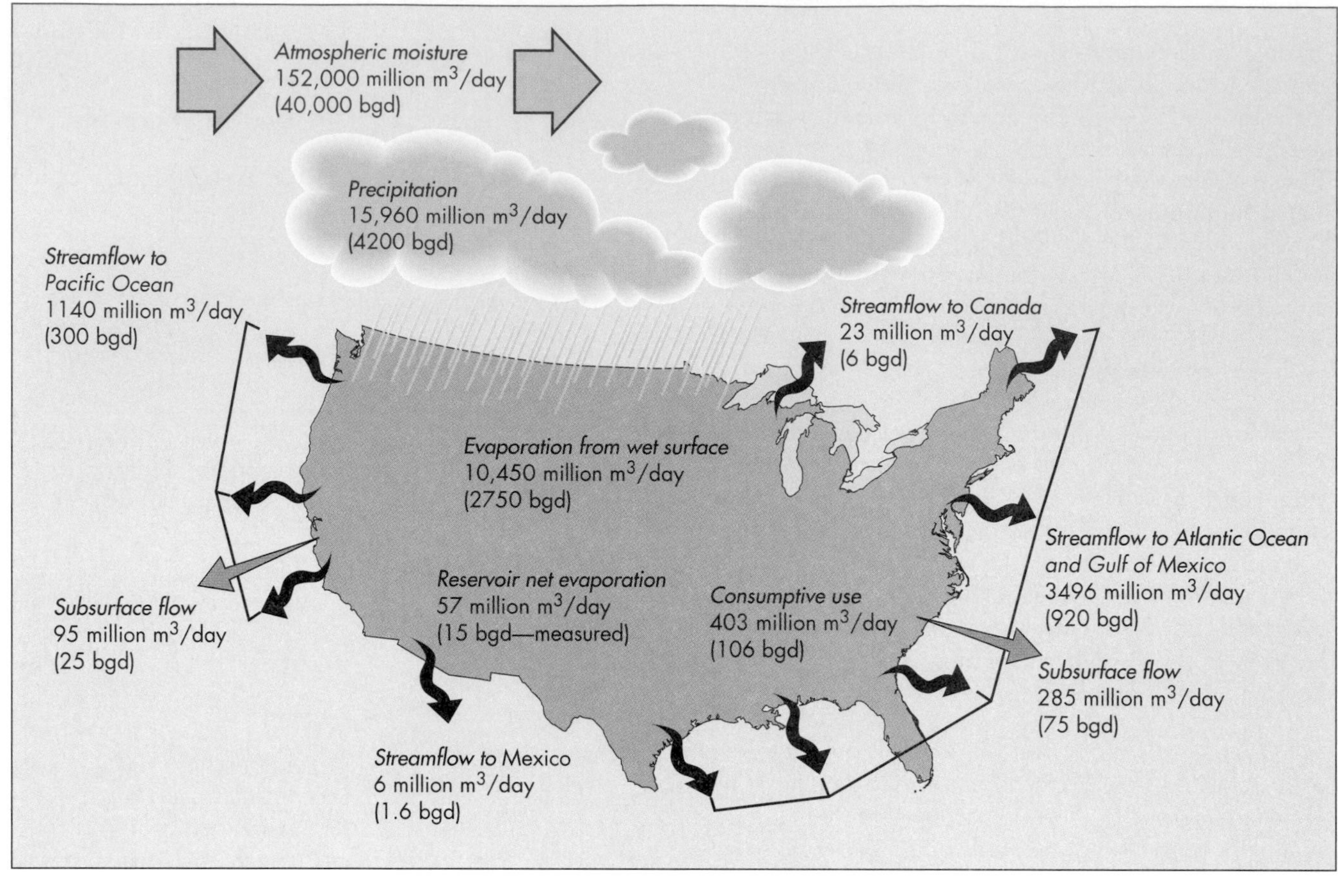

▲ **FIGURE 10.12** Water budget for the conterminous United States. (*From* U.S. *Water Resources Council*. 1978. The Nation's Water Resources, 1975–2000.)

liquid water. Of this amount approximately 10 percent falls as precipitation in the form of rain, snow, hail, or sleet. Approximately two-thirds of the precipitation evaporates quickly or is transpired by vegetation. The remaining one-third, or about 5510 million m^3 (1450 bg) per day, enters the surface or groundwater storage systems, flows to the oceans or across the nation's boundaries, is used by consumption, or evaporates from reservoirs. Unfortunately, owing to natural variations in precipitation that cause either floods or droughts, only a portion of this water can be developed for intensive uses. Thus, only about 2565 million m^3 (675 bg) per day are considered to be available 95 percent of the time (3).

On a regional scale, it is critical to consider annual precipitation and runoff patterns in order to develop water budgets. Potential problems with water supply can be predicted in areas where average precipitation and runoff are relatively low, such as in the southwestern and Great Plains regions of the United States as well as in some of the intermontane valleys in the Rocky Mountain area. The theoretical upper limit of surface water supplies is the *mean annual runoff*, assuming it could be successfully stored. Unfortunately, storage of all the runoff is not possible because of evaporative losses from large reservoirs, the limited number of suitable sites for reservoirs, and need for other water uses such as river transportation and wildlife. As a result, shortages in water supply are bound to occur in areas with low precipitation and runoff. Strong conservation practices are necessary to ensure an adequate supply (3).

Because of the large annual variations in stream flow, even areas with high precipitation and runoff may periodically suffer from droughts. For example, the dry years of 1961, 1966, and 1999 in the northeastern United States, and 1976–1977 and 1985 to 1990 in parts of the western United States, produced serious water shortages. Fortunately, in the more humid eastern United States, stream flow tends to vary less than in other regions, and drought is less likely (3). Nevertheless, the summers of 1986 and 1999 brought droughts in the southeastern and northeastern United States respectively, causing billions of dollars in damage.

The Groundwater Supply

Nearly half the population of the United States uses groundwater as a primary source of drinking water. Fortunately, the total amount of groundwater available in the United States is enormous, accounting for approximately 20 percent of all water withdrawn for consumptive uses. Within the conterminous United States the amount of groundwater within 0.8 km of the land surface is estimated to be between 125,000 and 224,000 km^3. To put this in perspective, the lower estimate is about equal to the total discharge of the Mississippi River during the last 200 years. Unfortunately, owing to the cost of pumping and exploration, much less than the total quantity of groundwater is available (3).

Protecting groundwater resources is an environmental problem of particular public concern because so many people derive their domestic water supplies from groundwater. The residence time for groundwater in aquifers is often measured in hundreds to thousands of years; therefore, once aquifers are damaged by pollutants, it may be difficult or impossible to reclaim them for continued use. Aquifers are also very important because approximately 30 percent of the stream flow in the United States is supplied by groundwater that emerges as springs or other seepages along the stream channel. This phenomenon, known as **base flow,** is responsible for the low flow or dry-season flow of most perennial streams. Therefore, maintaining high-quality groundwater is important in maintaining good-quality stream flow.

In many parts of the country, groundwater withdrawal from wells exceeds natural inflow. In such cases, water is being mined and can be considered a nonrenewable resource. Groundwater overdraft is a serious problem in the Texas–Oklahoma–High Plains area; in California, Arizona, Nevada, New Mexico; and in isolated areas of Louisiana, Mississippi, Arkansas, and the South Atlantic–Gulf Coast region (Figure 10.13a). In the Texas-Oklahoma-High Plains area alone, the overdraft amount is approximately equal to the natural flow of the Colorado River (Figure 10.13b) (3). In this area lies the Ogallala aquifer, which is composed of water-bearing sands and gravel that underlie an area of about 400,000 km^2 from South Dakota into Texas. Although the aquifer holds a tremendous amount of groundwater, it is being used in some areas at a rate that is up to 20 times that of natural recharge by infiltration of precipitation. The water level in many parts of the aquifer has declined in recent years, and eventually a significant portion of land now being irrigated may return to dry farming if the resource is used up.

To date, only about 5 percent of the total groundwater resource has been depleted, but water levels have declined as much as 30 to 60 m in parts of Kansas, Oklahoma, New Mexico, and Texas. As the water table becomes lower, yields from wells decrease and energy costs to pump the water increase. The most severe problems in the High Plains and the Ogallala aquifer today are in those locations where irrigation has been going on the longest—that is, since the 1940s.

In many areas, pumping of groundwater has forever changed the character of the land. For example, rivers in the Tucson, Arizona, area, prior to lowering of the water table through pumping, were perennial, with healthy populations of trout, beaver, and other animals. Today the native riparian trees have died (the water table is below their roots) and the rivers are dry much of the year. Ironically, these processes also increased the flood hazard in Tucson, which currently gets its entire water supply from groundwater sources. Loss of riparian trees and the root strength they provided to stream banks render the channels much more vulnerable to lateral bank erosion. During the 1983 and 1993 floods in Tucson (see Chapter 5), this became very apparent as roads, bridges, and buildings were damaged by the shifting channels. Tree-lined channels are much more stable, but riparian trees need a groundwater table sufficiently close to the surface for healthy growth. Unfortunately, mining of groundwater in the Tucson area has precluded restoration of trees.

Desalination

Desalination of seawater, which contains about 3.5 percent salt (about 40 kilograms [kg] per cubic meter), is an expensive form of water treatment practiced at several hundred plants around the world. The salt content must be reduced to about 0.05 percent for the water to be drinkable and pass water-quality standards. Large desalination plants produce 20,000 to 30,000 m^3 of water per day at a cost of about ten times that paid for traditional water supplies in the United States. Desalinated water has a "place value," which means that the price increases quickly with the transport distance and elevation increase from the plant at sea level. Because the various processes that actually remove the salt require energy, the cost of the water is tied to ever-increasing energy costs. For these reasons, desalination will remain an expensive process that will be used only when alternative water sources are not available. Because of an increasing population and inadequate water supply, accented by recent droughts, some U.S. communities such as Santa Barbara, California, have constructed desalination plants as an emergency measure for future droughts.

Middle Eastern countries in particular will continue to use desalination. In many arid regions, including the Middle East, there are brackish ground and surface waters with a salinity of about 0.5 percent (one seventh that of seawater). Obviously, desalination of this water is less expensive, and plants may be located at inland sites.

10.6 Water Use

To discuss water use, we must distinguish between instream and offstream uses. **Offstream uses** remove or divert water from its source. Examples include water for irrigation, livestock, thermoelectric power generation, industrial processes, and public supply. **Consumptive use** is an offstream use in which water does not return to the stream or groundwater resource immediately after use. This is the water that evaporates, is incorporated into crops or products, or is consumed by animals and humans (3,8). **Instream use** relates to the water that is used but not withdrawn from its source. Examples include use of river water for navigation, hydroelectric power generation, fish and wildlife habitats, and recreation. In general, consumptive use is much less than offstream use, which is much less than instream use. For example, in the United States in 1995, consumptive use was about 100 billion gallons (3.8×10^8 m^3) per day; offstream use was about 400 billion gallons (1.5×10^9 m^3) per day, and instream use (for hydroelectric power generation) was about 3,000 billion gallons (1.1×10^{10} m^3) per day (8).

(a)

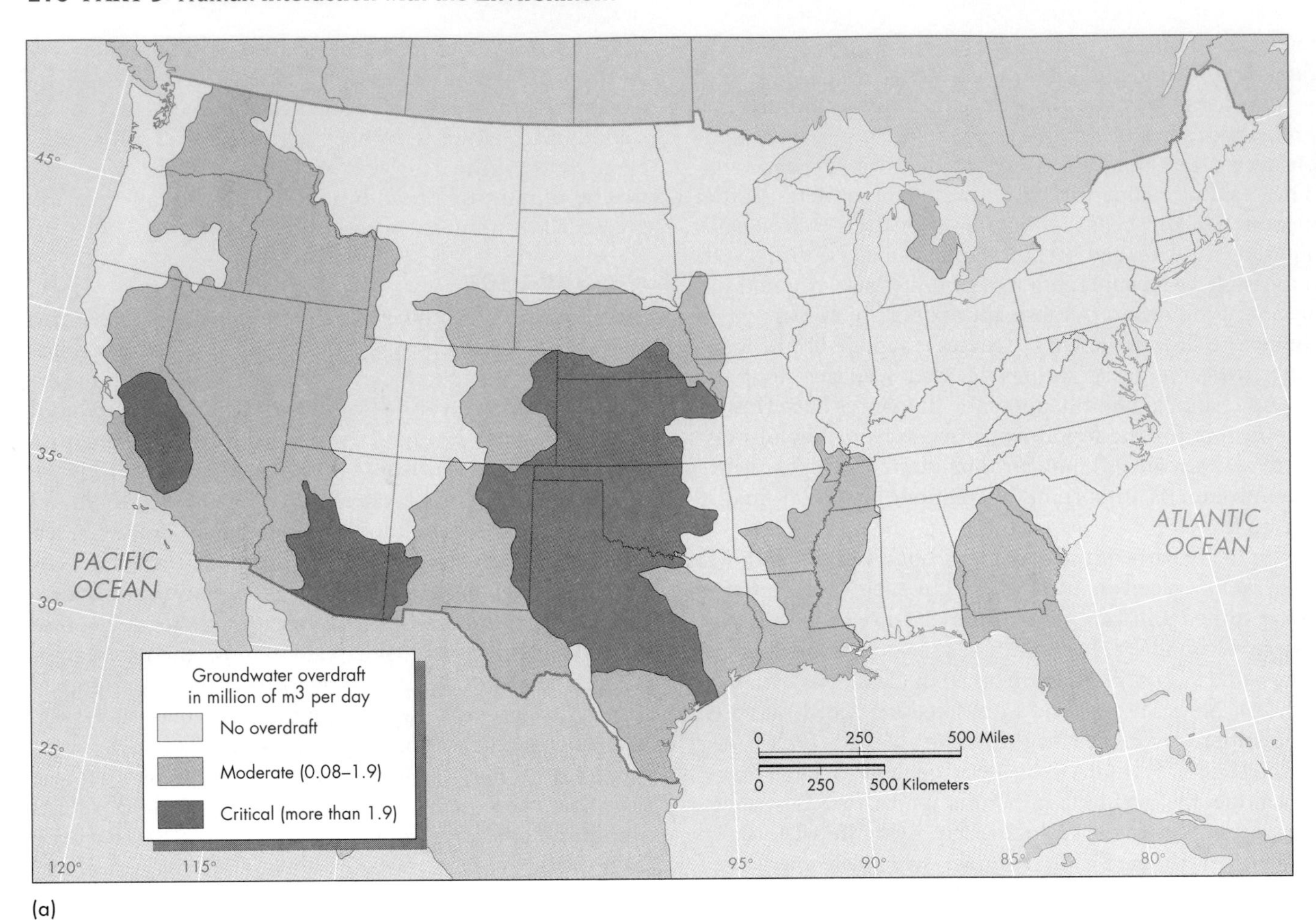

▲ **FIGURE 10.13** (a) Groundwater overdraft for the conterminous United States. (b) A detail of water-level changes in the Texas–Oklahoma–High Plains area. (*Source*: U.S. *Geological Survey*)

Multiple instream uses of rivers and streams usually create controversy because each use requires different conditions to prevent damage or detrimental effects. Fish and wildlife require certain water levels and flow rates for maximum biological productivity, and these levels and rates may differ from the requirements for hydroelectric power generation, which requires large fluctuations in discharges to match power needs. Similarly, both of these may conflict with requirements for shipping and boating. The discharge necessary to move the sediment load in a river may require yet another pattern of flow. Figure 10.14 diagrams the seasonal patterns of discharge for some of these uses.

A major problem concerns how much water may be removed from a stream or river and transported to another location without damaging the stream system. This is a problem in the Pacific Northwest, where certain fish, including the steelhead trout and salmon, are on the decline partly because people have induced alterations in land use (for example, timber harvesting) and stream flows (building dams that block seasonal migration of fish and change downstream hydrology) that have degraded fish habitats.

Important concepts associated with water use are illustrated in Figure 10.15. Surface and groundwater sources are moved to the users, often by way of a public supplier, which

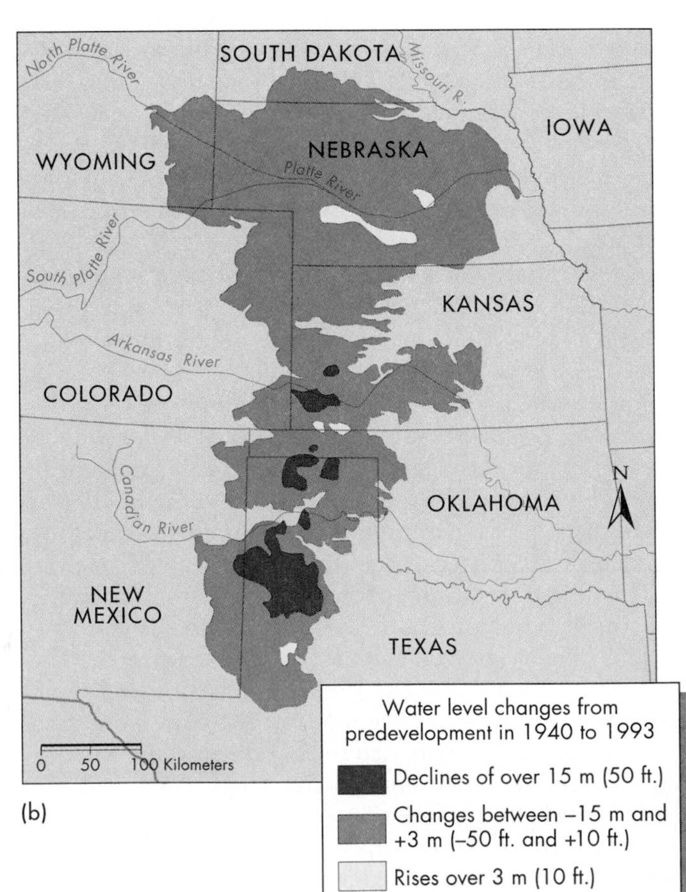

(b)

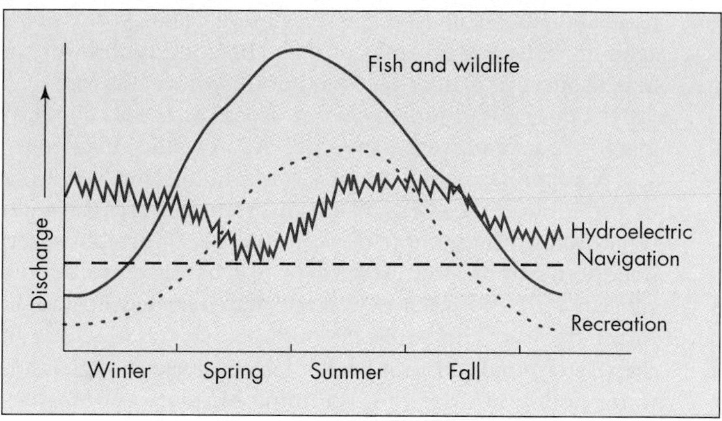

◀ **FIGURE 10.14** Diagram of instream water uses and the varying discharges for each use. Discharge is the amount of water passing by a particular location and is measured in cubic meters per second (cms).

may be a local or regional water district. Knowing the volumes of water moved from point to point allows quantities such as conveyance losses and consumptive use to be calculated (8).

Movement of Water to People

In our modern civilization, water is often moved vast distances from areas with abundant rainfall to areas of high usage. In California, demands are being made on northern rivers for reservoir systems supplying the cities in the southern part of the state. Two-thirds of California's runoff occurs north of San Francisco, where there is a surplus of water, while two-thirds of the water use occurs south of San Francisco, where there is a deficit. In recent years, canals constructed by the California Water Project and the Central Valley Project have moved tremendous amounts of water from the northern to the southern part of the state, adversely affecting ecosystems (especially fisheries) in some northern California rivers through diversion of waters.

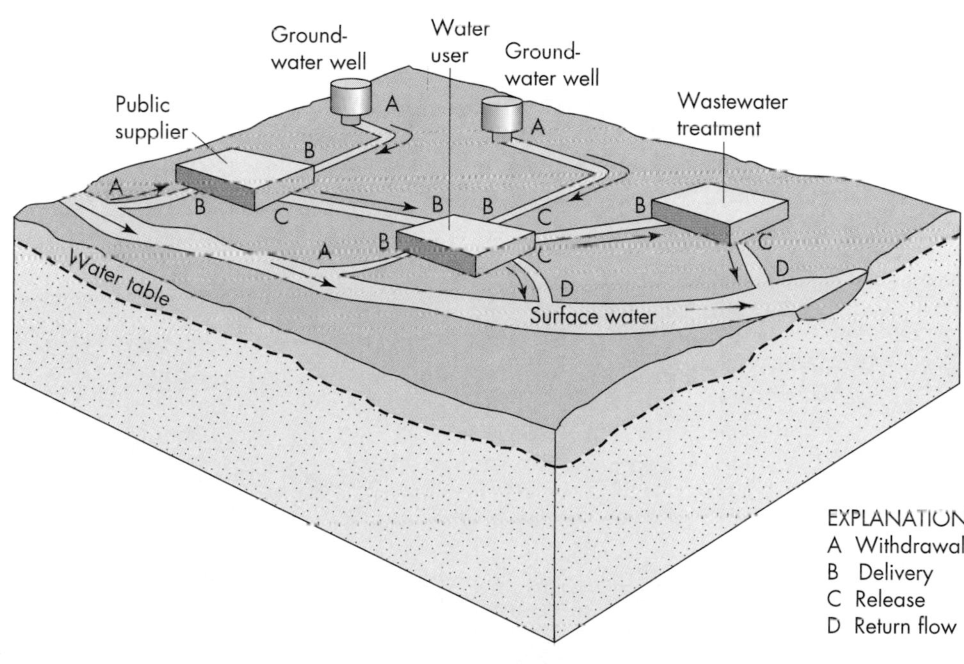

1. Withdrawal—The quantity of water diverted or withdrawn from surface or groundwater (A in sketch).
2. Delivery/release—The quantity of water delivered at the point of use (B) and the quantity released after use (C).
3. Conveyance loss—The quantity of water that is lost in transit, for example, from point of withdrawal to point of delivery (A–B), or from point of release to point of return (C–D).
4. Consumptive use—That part of water withdrawn that is evaporated, transpired, or incorporated into products or crops. In some instances, consumptive use will be the difference between the volume of water delivered and the volume released (B–C).
5. Return flow—The quantity of water that is discharged to a surface or groundwater source (D) after release from the point of use and thus becomes available for further use.

◀ **FIGURE 10.15** Important concepts associated with water use. (*Source:* W. B. Solley, R. R. Pierce, and H. A. Perlman, 1993. Estimated use of water in the United States in 1990. U.S. *Geological Survey Circular* 1081.)

The major water-diversion projects in California are shown in Figure 10.16. Of particular interest is the long-standing dispute between the city of Los Angeles and the people in Owens Valley on the eastern side of the Sierra Nevada. Los Angeles suffered a drought near the end of the nineteenth century and, after looking for a potential additional water supply, settled on the Owens Valley. By various means, some of which were controversial, to say the least (some have contended the water was stolen), the city purchased most of the water rights and constructed the Los Angeles Owens River Aqueduct, completed in 1913. Since that time groundwater has also been pumped and taken from Owens Valley via the aqueduct. As a result of the tremendous exportation of surface water and groundwater, Owens Valley, which before water exportation to Los Angeles contained wetlands and lakes, has suffered from desertification (the production of a more desertlike environment), producing "Owens Dry Lake," perhaps the single largest point source of hazardous alkaline dust in the U.S. Recently, Los Angeles has agreed to reduce water exports to attempt to control the production of dust. Protests that were more violent in the early 1900s are now court battles; only recently have both parties come closer to a settlement that will include limits on the amount of water taken by Los Angeles and projects to halt environmental degradation.

Many large cities in the world must seek water from areas increasingly farther away. For example, New York City has imported water from nearby areas for more than a century. Water use and supply in New York City represent a repeating pattern. Originally, local groundwater, streams, and the Hudson River itself were used. However, water needs exceeded local supply, so in 1842 the first large dam was built more than 48 km north of the city. As the city expanded rapidly from Manhattan to Long Island, water needs again increased. The sandy aquifers of Long Island were at first a source of drinking water, but this water was removed faster than rainfall replenished it. Local cesspools contaminated the groundwater, and salty ocean water intruded.

A larger dam was built at Croton in upstate New York in 1900, but further expansion of the population brought repetition of the same pattern: initial use of groundwater; pollution, salinification, and exhaustion of this resource; and subsequent building of new, larger dams farther upstate in forested areas. The boroughs of Brooklyn and Queens, on the western end of Long Island, have experienced groundwater pollution since the beginning of the twentieth century, and they import upstate water. Eastern counties of Long Island (Nassau and Suffolk) do not import water and, of necessity, have enacted strict regulations to protect and conserve their groundwater supply. Nevertheless, they also are experiencing problems of pollution, salinification, and exhaustion of the resource.

It is important to recognize that New York City and Los Angeles are not unique. Many urban areas are having problems with their water supply as a growing population demands more water, which is becoming harder to obtain. One would think that eventually the cost of obtaining water from long distances would place an upper limit on growth, and to some extent this may be true, but the price of water is often kept artificially low through a variety of government programs. People in urban environments could do much more through increased water conservation to alleviate or reduce the problems related to water supply, but shortages have not yet become sufficiently acute. Nevertheless, urban water districts are developing strategies to

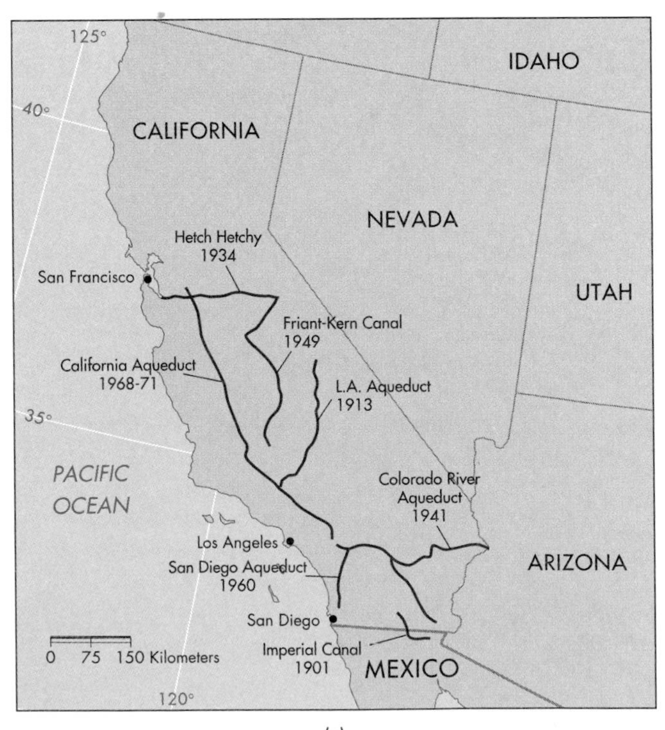

(a)

(b)

◀ **FIGURE 10.16** (a) California aqueducts and irrigation canals. (b) View of the California aqueduct in the San Joaquin Valley. (*Allan Pitcairn/Grant Heilman Photography, Inc.*)

encourage conservation. These include water prices that increase with water use and rebates for installing water-conserving fixtures such as low-flow flush toilets and low-flow shower heads. Manufacturers are also now producing washing machines and other appliances that use less water or have low water-use settings.

As greater quantities of water are needed for cities and agriculture, conflicts will increase and intensive argument will center on instream water use. An important, fruitful area of research is more careful evaluation of what flows are necessary to maintain a natural river system.

Trends in Water Use

Trends in water withdrawals for various uses in the United States provide insight that is both interesting and necessary for managing our water resources. Figure 10.17a shows trends in fresh ground- and surface water withdrawals from 1950 through 1995. These data suggest:

- Surface water withdrawals far exceed groundwater withdrawals.
- Water withdrawals increased until 1980, and since then has decreased and leveled off. The population of the United States was about 151 million in 1950 and has continued to increase, reaching 267 million in 1995. Thus, during the period when water withdrawals decreased and leveled off, population was increasing. This suggests better water management and conservation during the past 15 years (8).

Figure 10.17b shows trends in water withdrawals by water-use category from 1960 through 1995. These data show that:

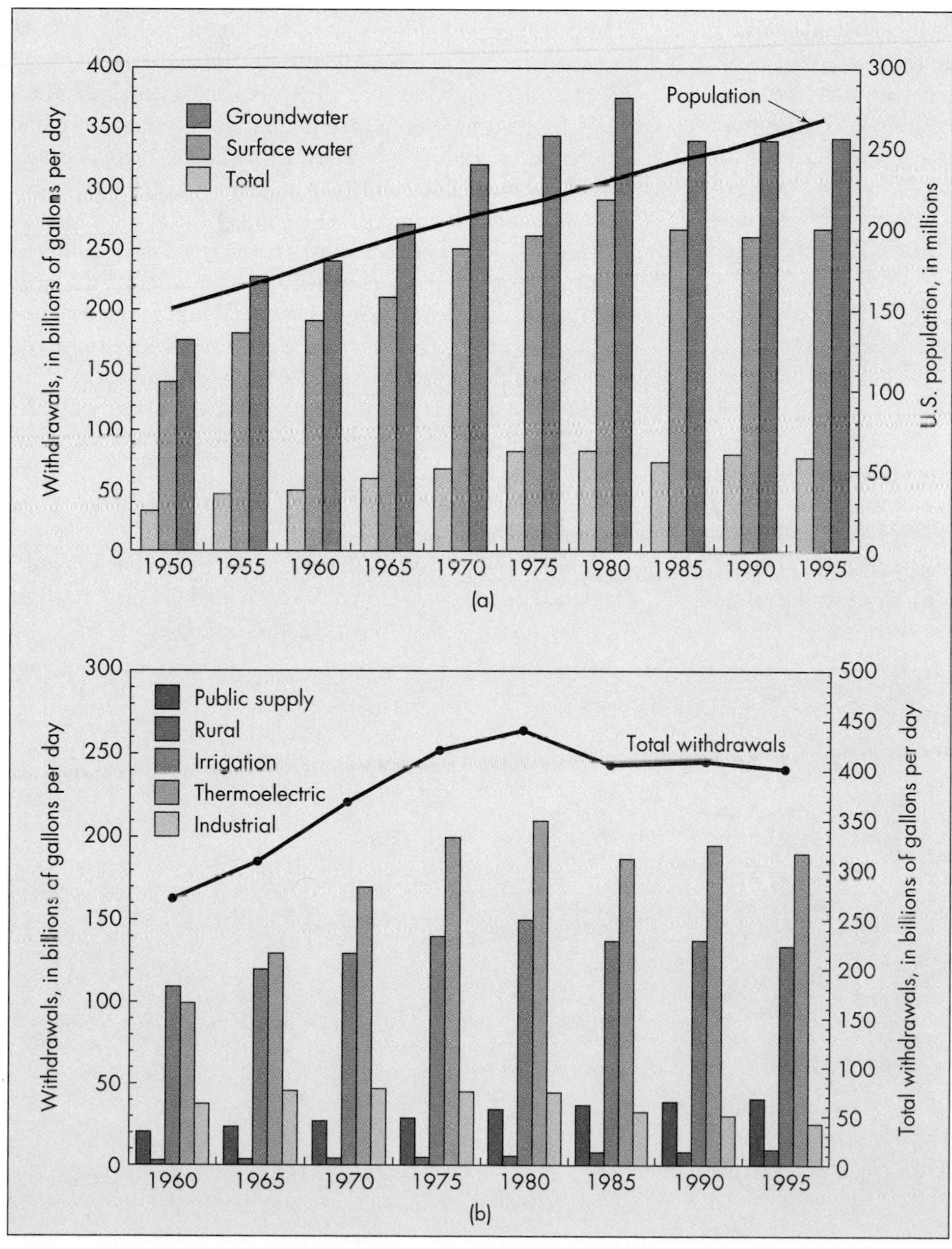

◀ **FIGURE 10.17** Trends in withdrawal of fresh groundwater and surface waters (1950–1995) (a); and withdrawal of both fresh and saline water use by category (1960–1995) (b). (*Source: Solley,* W. B., *Pierce,* R. R., *and Perlman,* H. A. 1998. Estimated use of water in the United States in 1995. U.S. *Geological Survey Circular* 1200.)

- Irrigation needs and the thermoelectric industry are the big users of fresh water.
- Use of water by the public (urban and rural sectors) has increased through the period, a trend presumably related to the increase in population of the country.
- The use of water by agriculture for irrigation leveled off in 1980 and has decreased slightly since then. This presumably is related to efforts in water conservation.
- Water used for the thermoelectric power increased dramatically from 1960 to 1980, as numerous power plants came on-line, and has since decreased somewhat due to more efficient use of water.
- Since 1980, industry has used significantly less fresh water. This is due, in part, to new technologies that require less water as well as improved plant efficiencies and increased water recycling.

There are encouraging signs that the general public is more aware of our water resources and the need to conserve them. As a result, in many states water demands have been reduced. Another encouraging sign is that use of reclaimed wastewater is now much more common, increasing from about 200 million gallons per day in 1955 to 1000 million gallons daily in 1995. This is about 1 percent of the consumptive use of water and 0.25 percent of offstream water use in the United States today. More significantly, the trend seems to be accelerating: From 1990 to 1995 the use of reclaimed wastewater increased by about 36 percent (8).

Water Conservation

What can be done to use water more efficiently and reduce withdrawal and consumption? Improved agricultural irrigation could reduce withdrawals by between 20 and 30 percent. Such improvements in **water conservation** include lined and covered canals that reduce seepage and evaporation; computer monitoring and scheduling of water releases from canals; a more integrated use of surface waters and groundwaters; night irrigation; improved irrigation systems (sprinklers, drip irrigation); and better land preparation for water application.

Domestic use of water accounts for only about 6 percent of the total national withdrawals. However, because this use is concentrated, it poses major local problems. Withdrawal of water for domestic use may be substantially reduced at a relatively small cost with more efficient bathroom and sink fixtures, night irrigation, and drip irrigation systems for domestic plants.

How people perceive the water supply is important in determining how much water is used. For example, people in Tucson, Arizona, perceive the area as a desert (which it is) and cultivate many native plants such as cactus in their yards and gardens (Figure 10.18). Tucson's water supply is from groundwater, which is being mined (used faster than it is being naturally replenished); the water use is about 605 liters (160 gallons) per person per day. Not far away the people of Phoenix, Arizona, use about 983 liters (260 gallons) of water per person per day. Parts of Phoenix use as much as 3780 liters (1000 gallons) per person per day to water large lawns, mulberry trees, and high hedges! Phoenix has been accused of having an "oasis mentality" concerning water use.

Water rates also make a difference. People in Tucson pay about 75 percent more for water than do people in Phoenix, where the water supply is drawn from the Salt River rather than from groundwater. Water rates in Tucson are structured to encourage conservation, and some industries consider water as a cost-control measure (9). The message here is that, because water in the southwestern United States and other locations will be in short supply in the future, we could all do with a little of Tucson's "desert mentality," particularly such large urban areas as Los Angeles and San Diego.

Water removal for steam generation of electricity could be reduced as much as 25 to 30 percent by using cooling

FIGURE 10.18 Home in Tucson, Arizona, with native vegetation and rocks as ground cover. This type of landscaping minimizes water use. (*Edward A. Keller*)

towers designed to use less or no water. Manufacturing and industry might curb water withdrawals by increasing in-plant treatment and recycling of water or by developing new equipment and processes that require less water. Because the field of water conservation is changing so rapidly, it is expected that a number of innovations will reduce the total withdrawals of water for various purposes, even though consumption will continue to increase (3).

10.7 Water Management

Management of water resources is a complex issue that will become more difficult in coming years as the demand for water increases. While this will be especially true in the southwestern United States and other arid and semiarid parts of the world, New York and Atlanta, among other U.S. cities, also face future water-supply problems. Options open to people who want to minimize potential water-supply problems include locating alternative supplies, managing existing supplies better, or controlling growth.

The Future of Water Management

Cities in need of water are beginning to treat water like a commodity that can be bought and sold on the open market, like oil or gas. If cities are willing to pay for water and are allowed to avoid current water regulation, then allocation and pricing as they are now known will change. If the cost rises enough, "new water" from a variety of sources may become available. For example, irrigation districts (water managers for an agricultural area) may contract with cities to supply water to urban areas. They could do this without any less water being available for crops by using conservation measures to minimize present water loss through evaporation and seepage from unlined canals. Currently, most irrigation districts do not have the capital to finance expensive conservation methods, but money paid by cities for water could finance such projects. It seems apparent that water will become much more expensive in the future and, if the price is right, many innovative programs are possible. Serious consideration is being given to ideas as original as towing icebergs (which are composed of frozen fresh water) to coastal areas where fresh water is needed.

Luna Leopold (a leader in the study of rivers and water resources) has suggested that a new philosophy of **water management** is needed—one based on geologic, geographic, and climatic factors as well as on the traditional economic, social, and political factors. He argues that the management of water resources cannot be successful as long as it is naively perceived primarily from an economic and political standpoint. However, this is how water use is approached. The term *water use* is appropriate because we seldom really "manage" water (10). The essence of Leopold's water-management philosophy is summarized in this section.

Surface water and groundwater are both subject to natural flux with time. In wet years, surface water is plentiful, and the near-surface groundwater resources are replenished. During these years, we hope that our flood-control structures, bridges, and storm drains will withstand the excess water. Each of these structures is designed to withstand a particular flow (for example, the 20-year flood), which, if exceeded, may cause damage or flooding.

All in all, Leopold concluded we are much better prepared to handle floods than water deficiencies. During dry years, which must be expected even though they may not be accurately predicted, we should have specific strategies to minimize hardships. For instance, subsurface waters in various locations in the western United States are either too deep to be economically extracted or have marginal water quality. These waters may be isolated from the present hydrologic cycle and therefore may not be subject to natural recharge. Such water might be used when the need is great, but this will be possible only if plans are in place for drilling the wells and connecting them to existing water lines when the need arises. Another possible emergency plan might involve the treatment of wastewater. Reuse of water on a regular basis might be too expensive or objectionable for other reasons, but advance planning to reuse treated water during emergencies might be wise (10).

When dealing with groundwater that is naturally replenished in wet years, we should develop plans to use surface water when it is available and not be afraid to use groundwater during dry years. In other words, groundwater could be pumped out at a rate exceeding the replenishment rate in dry years, but it would be replenished during wet years by both natural and artificial recharge (pumping excess surface water into the ground). This water-management plan recognizes that excesses and deficiencies in water are natural and can be planned for.

A Managed River: The Colorado

No discussion of water resources and water management would be complete without a mention of the Colorado River Basin and the controversy that surrounds the use of its water. People have been using the water of the Colorado River for about 800 years. Early Native Americans in the basin had a highly civilized culture with a sophisticated water-distribution system. Many of their early canals were later cleared of debris and used by settlers in the 1860s (11). Given this early history, it is somewhat surprising to learn that the Colorado was not completely explored until 1869, when John Wesley Powell, who later became director of the U.S. Geological Survey, navigated wooden boats through the Grand Canyon.

Although the waters of the Colorado River Basin are distributed by canals and aqueducts to many millions of urban residents, and to agricultural areas such as the Imperial Valley in California, the basin itself, with an area of approximately 632,000 km^2, is only sparsely populated. Yuma, Arizona, with approximately 42,000 people, is the largest city on the river, and within the basin only the cities of Las Vegas, Phoenix, and Tucson have more than 50,000 inhabitants. Nevertheless, only about 20 percent of the total population of the basin is rural. Vast areas of the basin have extremely low densities of people, and in some areas measuring several thousand square kilometers there are no permanent residents (11).

Rod Nash, writing about the wilderness values of the river, states that at the confluence of the Green and Colorado rivers, it is 80 km to the nearest video game and you are in the heart of a national park (12).

The headwaters of the Colorado River are in the Wind River Mountains of Wyoming, and in its 2300-km journey to the sea the river flows through or abuts seven states—Wyoming, Colorado, Utah, New Mexico, Nevada, Arizona, and California—and Mexico (Figure 10.19). Although the drainage basin is very large, encompassing much of the southwestern United States, the annual flow is only about 3 percent of that of the Mississippi River and less than a tenth of that of the Columbia. Therefore, for its size the Colorado River has only a modest flow, and yet it has become one of the most regulated, controversial, and disputed bodies of water in the world. Conflicts that have gone on for decades extend far beyond the Colorado River Basin itself to involve large urban centers and developing agricultural areas of California, Colorado, New Mexico, and Arizona. The need for water in these semiarid areas has resulted in overuse of limited supplies and deterioration of water quality. Interstate agreements, court settlements, and international pacts have periodically eased or intensified tensions among people who use the waters along the river. The legacy of laws and court decisions, along with changing water-use patterns, continues to influence the lives and livelihood of millions of people in both Mexico and the United States (13).

Waters of the Colorado River have been appropriated among the various users, including the seven states and the Republic of Mexico. This appropriation has occurred through many years of negotiation, international treaty, interstate agreements, contracts, federal legislation, and court decisions. As a whole, this body of regulation is known as the "Law of the River." Two of the more important early documents in this law were the Colorado River Compact of 1922, which divided water rights in terms of an upper and lower basin (see Figure 10.19), and the treaty with Mexico in 1944, which promised an annual delivery of 1.85 km^3 (1.5 million acre-feet [1 acre-foot is the volume of water covering 1 acre to a depth of 1 ft], or 325,829 gallons) of Colorado River water to Mexico. More recent was a 1963 U.S. Supreme Court decision involving Arizona and California. Arizona refused to sign the 1922 compact and had a long conflict with California concerning appropriation of water. The Court decided that southern California must relinquish approximately 0.74 km^3 (600,000 acre-feet) of Colorado River water when the Central Arizona Project is completed. Finally, in 1974 the Colorado River Basin Salinity Control Act was approved by Congress. The act authorized procedures to control adverse salinity of the Colorado River water, including construction of desalination plants to improve water quality.

Management of the Colorado River Basin and its waters has been frustrating in part because the basin is characterized by inherent instabilities (11). For example, in 1922 when the Colorado River Compact was worked out, the hypothesis was that the virgin flow of the river was approximately 20 km^3 (16.2 million acre-feet) per year. That annual flow is now believed to average closer to 16.6 km^3 (13.5 million acre-feet) annually (14). Even these numbers are misleading, however, because of the tremendous hydrologic

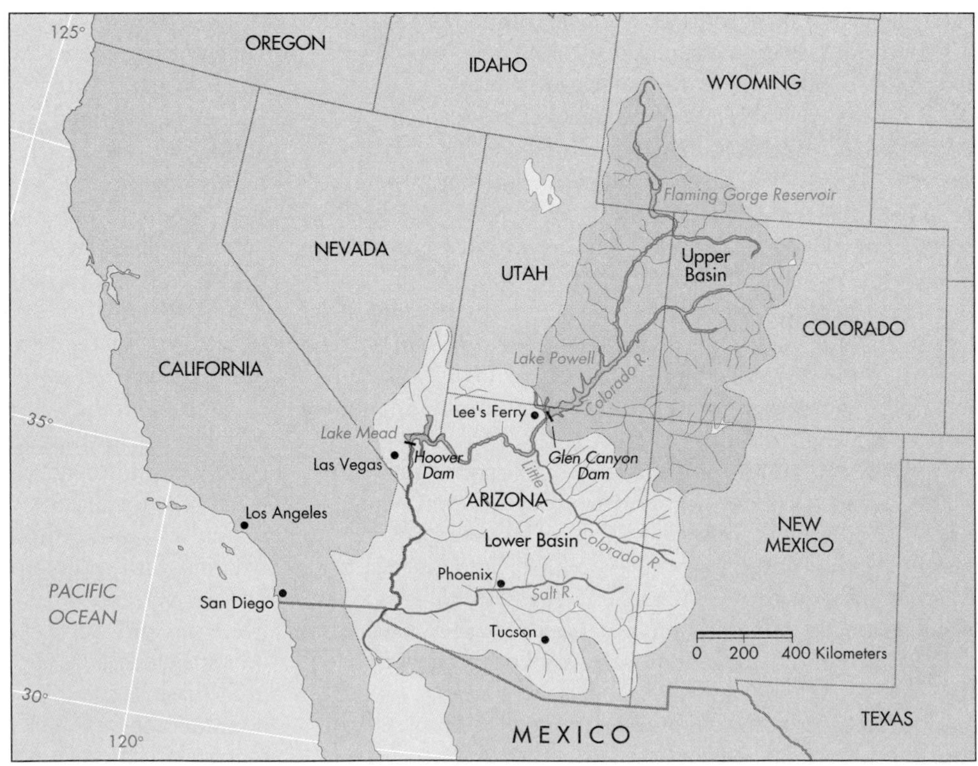

▶ **FIGURE 10.19** The Colorado River Basin.

instability within the basin. Floodwaters in the Colorado River may come from snowmelt floods, long-term winter precipitation events, or short-term summer thunderstorms; thus, the total water available on a year-to-year basis is tremendously variable. Table 10.7 shows the legal water entitlements for the Colorado River Basin. Notice that the actual distribution of water adds up to 14,500 million acre-feet per year, which is greater than the annual flow. This distribution can be obtained because the Colorado River is one of the most regulated rivers in the world. Figure 10.20 shows a profile of the river and some of the major dams and reservoirs. The 19 high dams on the river can store approximately 86.3 km^3 (70 million acre-feet) of water. Of this, approximately 80 percent is stored in two reservoirs, behind Hoover and Glen Canyon dams. This storage, if managed very efficiently, represents a buffer of several years' water supply. However, if a severe drought of several years' duration should occur, delivery of water could become very difficult. The Colorado River was one of the nation's first major rivers to have its entire flow fully appropriated. Balancing the future water needs of various users will continue to be a difficult and frustrating problem.

Construction of dams, reservoirs, and diversions on the Colorado River has generally been viewed as a successful venture from the viewpoint of supplying water. However, this has not always been the case. For example, the present Salton Sea in the Imperial Valley formed in 1905 and 1906 when virtually the entire Colorado River was unintentionally diverted into the southern Imperial Valley (Salton Basin). At that time the Colorado River was completely undammed, and "control works" (structures constructed to control the flow of the Colorado River) located in Mexican territory failed because of flooding in 1905 and 1906. By the time the river was controlled in 1907, the present Salton Sea had formed and was at a level higher than present. Water in the Salton Sea today is maintained through inflow from irrigation waters used to leach salts out of agricultural lands. Should this inflow stop or be reduced, the Salton Sea would soon dry up, owing to high evaporation rates there. Because the lake has become an important recreation (sport fishing) area, its future is controversial. If the lake waters become much saltier than they are now, the ecosystem and present fishery would be significantly damaged. However, the present lake is not unique to the Salton Basin. Other earlier lakes in the Imperial Valley present during recent geologic history have also dried up.

Although water supply is the primary problem in the Colorado River Basin, the problem of how to manage water quality is also significant. Although heavy metals and radioactive materials have become concentrated in the basin's waters and reservoirs, salt is causing the most problems. A salinity of 550 ppm (parts per million) is the upper limit set for human consumption, and more than 750 ppm may damage agriculture. The natural salinity of the Colorado River in the headwaters is only 50 ppm. As the river flows toward the sea, tributaries flow over exposed salt beds, and salt springs add salt to the river, so under natural conditions the salinity of the Lower Colorado is probably in the range of 250 to 380 ppm. However, upstream irrigation and evaporation have increased the salinity of the Lower Colorado River to an average of 1500 ppm, and at times the salinity reaches 2700 ppm. The quality of the water is so poor that Mexican farmers have allowed it to pass their fields rather than damage their crops and soils. The United States and Mexico agreed in 1973 that the United States would deliver water to Mexico with a salinity of no more than 115 ppm greater than the salinity at the Imperial Dam, a short distance upstream from the border. The salinity there is approximately 800 ppm. To achieve this goal, a large desalination plant near the border, costing several hundred million dollars in capital expenses and more than $10 million a year to operate, is necessary. This is a tremendous investment in a structural effort to control the salinity of the river water. Only time will tell how effective it will be (11).

Issues of water and basin management in the Colorado River are complex, but they firmly illustrate some of the major

Table 10.7 Legal and actual distribution of Colorado River water

State	Legal Entitlements (million ac ft per yr)	Actual Distribution (million ac ft per yr)
California	4.400[a]	4.400[f]
Arizona	3.800[a]	2.050[f]
Nevada	0.300[a]	0.300
Lower Basin	**8.500[b]**	**6.750**
Colorado	3.881[c]	2.406
Utah	1.725[c]	1.070
Wyoming	1.050[c]	0.651
New Mexico	0.844[c]	0.523
Upper Basin	**7.500[b]**	**4.650**
Mexico	1.500[d]	1.500
Total	**17.500**	**14.500**

[a]1928 Boulder Canyon Project Act

[b]1922 Colorado River Compact

[c]1948 Upper Colorado River Basin Compact

[d]1944 Mexico-U.S. Treaty

[e]Includes losses to evaporation of 0.6 million ac ft per year in the Upper Basin and 0.9 million ac ft per year in the Lower Basin, and 0.9 million ac ft per year inflow to Lower Basin from local streams.

[f]Agreement at time of Central Arizona Project authorization by Congress Upper Basin amounts agreed to by states as percentages.

Source: W. L. Graf, 1985. *The Colorado River,* Resource Publications in Geography, Association of American Geographers.

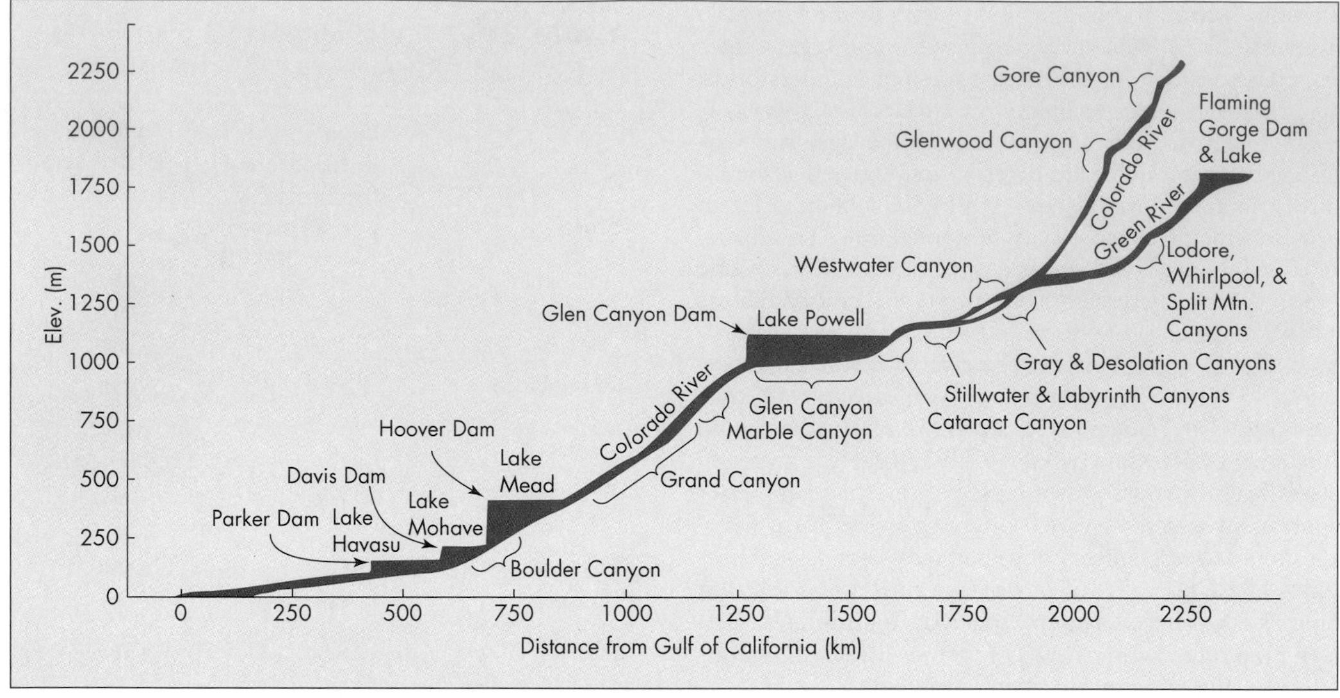

▲ **FIGURE 10.20** Longitudinal profiles of the Colorado and Green rivers, showing the major dams, reservoirs, and canyons. (*From* W. L. *Graf.* 1985. The Colorado River. *Association of American Geographers.*)

problems likely to face other parts of the arid Southwest in coming years: How are we to appropriate water? How can we best control water quality? Answers to these questions are not simple; what we have learned so far from our experiences with the Colorado River should help in future planning.

10.8 Dams, Reservoirs, and Canals

Our discussion of water supply established that many agricultural and urban areas require water delivered from nearby—and, in some cases, not so nearby—sources. To accomplish this a system of water storage and routing by way of canals and aqueducts from reservoirs is needed. The parties interested in water and water development range from government agencies to local water boards and conservation groups. A good deal of controversy often surrounds water development, and the day of developing large projects in the United States without careful environmental review has passed. The resolution of development issues now involves input from a variety of groups that may have very different needs and concerns. These range from agricultural groups who see water development as critical for their livelihood to those whose primary concerns are with wildlife and wilderness preservation. It is a positive sign that the various parties on water issues are now at least able to meet and communicate their needs and concerns.

Dams and Reservoirs

Dams and their accompanying reservoirs are generally designed to be multifunction structures. That is, those who propose the construction of dams and reservoirs point out that reservoirs may be used for activities such as recreation in addition to providing flood control and assuring a more stable water supply. It is important to recognize that reconciling these various uses at a given site is often difficult. For instance, water demands for agriculture might be high during the summer, resulting in a drawdown of the reservoir and the production of extensive mud flats. Those interested in recreation find the low water level and the mud flats to be aesthetically degrading, and these effects of high water demand may also interfere with wildlife (particularly fish) by damaging or limiting spawning opportunities. Finally, as we saw in Chapter 5, dams and reservoirs tend to instill a false sense of security to those living below the water-retention structures, because dams cannot fully protect us against great floods.

There is little doubt that we may need some additional dams and reservoirs if our present practices of water use are continued. Some existing structures may also need to be heightened. As more people flock to urban areas, water demands there are going to increase and additional water storage will be required. This is particularly true in the more arid parts of the country, including the southern California belt extending eastward into Arizona, where populations are growing rapidly.

Conflicts over construction of additional dams and reservoirs are bound to occur. Water developers may view a canyon dam site as a resource for water storage, whereas other people view it as a wilderness area and recreation site for future generations. The conflict is particularly pointed because good dam sites are often sites of high-quality scenic landscape. Unless water-use patterns change in agricultural and urban areas, however, additional water-supply facilities will be a high priority for rapidly growing urban areas, perhaps taking precedence over aesthetic and environmental concerns.

Whenever a dam and reservoir are constructed on a river system, that system is changed forever (see *Case History: The Grand Canyon*). The flow of water and sediment is changed, as are the physical and biological habitats and land uses below the dam. As a result of ecological damage to rivers below dams, a few dams have been removed and several others from Washington State to Florida will likely be removed in coming years. This represents an important shift in our view of the river and its environmental significance. As dams and their reservoirs age and become less useful, or hazardous, their removal may become both an environmental and economic alternative to expensive rebuilding or repairs (15).

Canals

Water from upstream reservoirs may be routed to downstream needs by way of natural watercourses or by canals and aqueducts.

Canals, whether lined or unlined, are often attractive nuisances to people and animals. Where they flow through urban areas, drownings are an ever present threat. When they are unlined, canals may lose a good deal of water to the subsurface flow system. Although it may be argued that this is a form of artificial groundwater recharge, it may be an inefficient one because canals may cross areas with little potential for groundwater development or areas of poor groundwater quality. In these cases, water seeping from unlined canals is essentially lost water.

The construction of canal systems, especially in developing countries, has led to serious environmental problems. For example, when the High Dam at Aswan, Egypt, was completed in 1964, canals were needed to convey the water to agricultural sites. The canals became infested with snails that carry the dreaded disease schistosomiasis (snail fever). This disease has always been a problem in Egypt, but the swift currents of Nile River floodwaters flushed out the snails each year. The tremendous expanse of waters in irrigation canals now provides happy homes for these creatures. The disease is debilitating and so prevalent in parts of Egypt that virtually the entire population of some areas is affected by it. The Egyptian canals are also a home for mosquitoes, some of which carry malaria.

Reservoirs and canal systems are being planned in a variety of environments around the world today. Environmental concern (and laws) in the United States ensures that important environmental review will take place. This is not always true in many developing countries, where attention to environmental concerns is not as high a priority as, for instance, the production of food. In such areas, construction of large and lengthy canals may considerably alter land use and the biologic environment by producing new and different water systems and barriers to migration of wildlife. This is not to say that water development in these countries should not take place, but it does emphasize the need for environmental concern at the ecosystem level when planning and developing water resources. At the very least, we can give developing countries information about our successes and failures in planning water projects, so that they may benefit from our experience. Water development and environmental concern are not necessarily incompatible. However, trade-offs must be made if a quality environment is to be preserved.

10.9 Water and Ecosystems

The major ecosystems of the world have evolved in response to physical conditions that include, among others, climate, nutrient input, soils, and hydrology. Changes in these factors affect ecosystems; in particular, changes induced by humans may have far-reaching consequences. Throughout the world today, with few exceptions, people are degrading natural ecosystems on a regional and global scale. Hydrologic conditions, particularly surface water processes and quality, along with interactions with groundwater, are becoming limiting factors for the existence of many ecosystems. This is particularly true for wetlands (20) (see *A Closer Look: Wetlands*).

Development of water resources often has an extensive impact on ecoystems. Construction of large dams, for example, can permanently change not only rivers but also the bodies of water they supply. Recall our Case History of the Grand Canyon, and see *A Closer Look: The Three Rivers Gorges, China*, in Chapter 15.

SUMMARY

The global water cycle involves the movement, storage, and transfer of water from one part of the cycle to another. The movement of water on land—that is, surface runoff and subsurface flow—is the part of the cycle of most direct concern to people. Globally, water is one of our most abundant renewable resources. However, more than 99 percent of the earth's water is unavailable or unsuitable for human use because of its location or its salinity. Water is used in tremendous quantities compared to other resources, and ensuring an adequate quantity and quality of water will be an increasing problem.

Water's unique properties make it indispensable to life as we know it. Many of these properties arise from the dipolarity (unequal charge distribution) of its molecules. As the

CASE HISTORY The Grand Canyon

The Grand Canyon of the Colorado River (Figure 10.B) provides a good example of a river's adjustment to the impact of a large dam. In 1963 the Glen Canyon Dam was built upstream from the Grand Canyon. Construction of the dam drastically altered the pattern of flow and channel process downstream: From a hydrologic viewpoint, the Colorado River was tamed. Before the Glen Canyon Dam, the river reached a maximum flow in May or June during the spring snowmelt, then flow receded during the remainder of the year, except for occasional flash floods caused by upstream rainstorms. During periods of high discharge, the river had a tremendous capacity to transport sediment (mostly sand and silt) and vigorously scoured the channel. The high floods also moved large boulders off the rapids, which formed because of shallowing of the river where it flows over alluvial fan or debris flow deposits delivered from tributary canyons to the main channel. As the summer low flow approached, the stream was able to carry less sediment, so deposition along the channel formed large bars and terraces, known as **beaches** to people who rafted the river.

After the dam was built, the mean annual flood (the average of the highest flow each year) was reduced from approximately 2500 cubic meters per second (cms) to 800 cms, and the 10-year flood was reduced from about 3500 cms to 860 cms. On the other hand, the dam did control the flow of water to such an extent that the median discharge actually increased from about 210 cms to 350 cms. The flow is highly unstable, however, because of fluctuating needs to generate power, and the level of the river may vary by as much as 5 m per day, with a mean daily high discharge of about 570 cms and a daily low of 130 cms. The dam also greatly affected the sediment load of the Colorado River through the Grand Canyon: The median suspended sediment concentration was reduced by a factor of about 200 immediately downstream from the dam. A lesser reduction in sediment load occurred farther downstream because tributary channels continued to add sediment to the channel (16).

▲ **FIGURE 10.B** The Colorado River in the Grand Canyon. The sandbar in the lower left corner is being used by river rafters whose numbers have begun to impact the canyon environment. As a result, the number of people allowed to raft through the canyon is restricted. (*Larry Minden/Minden Pictures*)

The change in hydrology of the Colorado River in the Grand Canyon has greatly changed the river's morphology. The rapids may be becoming more dangerous because large floods no longer occur—flows that had previously moved some of the large boulders farther downstream. In addition, some of the large sandbars (beaches), which are valuable

"universal solvent," water is an essential component of all organisms and is important in determining the composition of soils. Water forms thin films around soil particles, and these films are important in the movement of water above the groundwater table.

The flow of water on land is divided by drainage basins, or watersheds. Surface-water runoff and sediment yield vary greatly from one drainage basin to another and are influenced by geographic, physiographic, climatic, and biologic factors. The three major paths by which water on slopes reaches a stream channel and is exported from the drainage basin are overland flow (surface flow), throughflow (shallow subsurface flow), and groundwater flow (flow of water below the water table). Understanding these paths is critical to understanding how land-use change may influence runoff and sediment production.

Groundwater occurs in a *zone of saturation* below the water table. Its major source is precipitation that infiltrates the *recharge zone* on the land surface and moves down through the *vadose zone*, which is seldom saturated. An *aquifer* is a zone of earth material capable of supplying water at a useful rate from a well. The presence of a *confining layer* above an aquifer may raise water in the aquifer to the surface. Both the direction and the rate of ground-

wildlife habitat, are eroding because the river is deficient in sediment below the dam and is picking up more sediment and thus causing erosion.

Changes in the river flow (mainly deleting the high flows) have also resulted in vegetational shifts. Before the dam was built, three nearly parallel belts of vegetation were present on the slopes above the river. Adjacent to the river and on sandbars grew ephemeral plants, which were scoured out by yearly spring floods. Above the high-water line were clumps of thorned trees (mesquite and catclaw acacia) mixed with cactus and Apache plume. Higher yet could be found a belt of widely spaced brittle brush and barrel cactus (17). Closing the dam in 1963 tamed the spring floods for 20 years, and plants not formerly found in the canyon, including tamarisk (salt cedar) and indigenous willow, became established in a new belt along the river banks.

In June 1983 a record snowmelt in the Rocky Mountains forced the release of about 2500 cms, which is about three times that normally released and about the same as an average spring flood prior to the dam. The resulting flood scoured the river bed and banks, releasing stored sediment that replenished the sediment on sandbars and scoured out or broke off some of the tamarisk and willow stands. The effect of the large release of water was beneficial to the river environment and emphasizes the importance of the larger events (floods) in maintaining the system in a more natural state. Perhaps management of rivers below some dams should call for periodic release of large flows to help cleanse the system. As an experiment, or "test flood," between March 26 and April 2 of 1996, 1274 cms of water was released from the dam in order to redistribute the sand supply. The floods resulted in the formation of 55 new beaches and added sand to 75 percent of the existing beaches. It also helped rejuvenate marshes and backwaters, which are important habitats to native fish and some endangered species. The experimental flood was hailed as a success (18), but a significant part of the new sand deposits were subsequently eroded away (19).

The 1996 test flood remobilized sand, scouring it from the channel bottom and banks of the Colorado River below Glen Canyon Dam, depositing it on sandbars (beaches). However, little new sand was added to the river system from tributaries to the Colorado River, as they were not in flood flow during the test flood. The sand was mined from the river bed below the dam, and as such is a limited, nonrenewable source that can't supply sand to sandbars on a sustainable basis. A new, creative idea has recently been suggested (19). The plan is to use the sand delivered to the Grand Canyon by the Little Colorado River (a relatively large river with drainage area of 67,340 km^2) (Figure 10.19) that joins the Colorado River in the canyon downstream from Lee's Ferry. In 1993 a flood on the Little Colorado River delivered a large volume of sand to the Colorado River in the Grand Canyon, and prominent beaches were produced. Unfortunately, a year later the beaches were nearly eroded away by the flow of the Colorado River. The problem was that the beaches were not deposited high enough above the bed of the Colorado and so were vulnerable to erosion from normal post-dam flows. The idea suggested in the new study is to time the releases of flood flows from Glen Canyon Dam with sand-rich spring floods of the Little Colorado River. The resulting combined flood of the two rivers would be larger, and the new sand from the Little Colorado would be deposited higher above the channel bed and less likely to be removed by lower flows of the Colorado. Evaluation of the hydrology of the Little Colorado River suggests the opportunity to replenish sand on the beaches occurs, on average, once in 8 years. The proposed plan would restore or recreate river flow and sediment transport conditions to be more as they were prior to the construction of Glen Canyon Dam, conditions that formed and maintained the natural ecosystems of the canyon (19).

One final impact of the Glen Canyon Dam is the increase in the number of people rafting through the Grand Canyon. Although rafting is now limited to 15,000 people annually, the long-range impact on canyon resources is bound to be appreciable. Prior to 1950, fewer than 100 explorers and river runners had made the trip through the canyon. We must concede that the Colorado River of the 1970s and 1980s is a changed river. Despite the 1983 and 1996 floods that pushed back some of the changes, river restoration efforts cannot be expected to return it to what it was before construction of the dam (16,17,19). On the other hand, better management of the flows and sediment transport will improve and better maintain river ecosystems.

water movement depend on the *hydraulic gradient* (in the simplest case, approximately the slope of the water table) and the *hydraulic conductivity* of the earth material through which the water moves. This relationship is expressed quantitatively by *Darcy's law*, which has many important applications to groundwater problems. Interactions between surface water and groundwater are important environmentally because pollution in surface water may eventually contaminate the groundwater.

To evaluate a region's water supply, a water budget is developed to define the natural variability and availability of water. Water supply is limited, even in areas of high precipitation and runoff, by our inability to store all runoff and by the large annual variation in stream flows. In many areas groundwater is being mined (withdrawal exceeds natural replenishment), and in some areas this has permanently changed the character of the land. Desalination of seawater will continue to be used where other water sources are unavailable, but large-scale desalination is not likely because of the high costs of energy and transportation involved.

Water uses are categorized as offstream (including consumptive) and instream. Multiple instream uses—hydroelectric power, recreation, and fish and wildlife habitats—often have conflicting requirements; how water resources should

R LOOK Wetlands

nds (21) refers to landscape features such 'and dominated by trees or shrubs), marsh- frequently or continuously inundated by (a wetland that accumulates peat deposits), prairie potholes (small marshlike ponds), and vernal pools (shallow depressions that occasionally hold water). Some of these features are shown in Figure 10.C. The common feature and operational definition of wetlands is that they are inundated by water or the land is saturated to a depth of a few centimeters for at least a few days most years. The major components used to determine the presence or absence of wetlands are hydrology (wetness), type of vegetation, and type of soil. Of the three, hydrology is often the most difficult to define because some wetlands are only wet for a very short period each year. However, the presence of water, even for short periods on a regular basis, does give rise to characteristic wetland soils and specially adapted vegetation. Recognition of soils and vegetation greatly assists in identifying the wetland itself in many cases (22, 23).

Wetlands and their associated ecosystems have many important environmental features:

- Coastal wetlands such as salt marshes provide a buffer for inland areas from coastal erosion associated with storms and high waves.
- Wetlands are one of nature's natural filters. Plants in wetlands may effectively trap sediment and toxins.
- Freshwater wetlands are a natural sponge. During floods, they store water, helping to reduce downstream flooding. The stored water is slowly released following the flood, nourishing low flows of river systems.
- Wetlands are often highly productive lands where many nutrients and chemicals are naturally cycled while providing habitat for a wide variety of wildlife and plants.
- Freshwater wetlands are often areas of groundwater recharge to aquifers. Some of them—a spring-fed marsh, for example—are points of groundwater discharge.

Although most coastal marshes are now protected in the United States, freshwater wetlands are still threatened in many areas. It is estimated that 1 percent of the nation's total wetlands is lost every 2 years. Freshwater wetlands account for nearly all of this loss. In just the past 200 years about one-half of the wetlands in the United States, including about 90 percent of the freshwater wetlands, have disappeared as a result of being drained for agricultural purposes or filled for urban or industrial development.

Because so many wetlands have been damaged or destroyed, there is a growing effort to restore wetlands. Unfortunately, restoration is not usually an easy task, for wetlands are a result of complex hydrologic conditions that may be difficult to restore if the water has been depleted or is being used for other purposes. Ongoing research is carefully documenting the hydrology of wetlands as well as the movement of sediment and nutrients. As more information is gathered concerning how wetlands work, restoration is likely to be more successful.

be partitioned to meet the various uses is a controversial subject. Water is often transported long distances by canals, from areas of abundant rainfall to areas of high use, sometimes with severe adverse effects on ecosystems. Trends in water use during the last few decades are encouraging: Total withdrawals of water have been reduced and leveled off somewhat as the U.S. population has increased. This suggests water conservation has improved and more water is being reclaimed. During the next several decades, consumptive use of water will increase because of greater demands from a growing population and industry. However, the total water withdrawn from streams and groundwater in the United States may decrease slightly because of greater awareness of the need for conservation by individuals and industries.

Water-resource management needs a new philosophy that considers geologic, geographic, and climatic factors and utilizes creative alternatives. The Colorado River is one of the most heavily regulated rivers in the world, with all of its flow apportioned among a large number of users; it therefore provides many lessons for future water management. Damming of the river has helped ensure delivery of the allocated water and allowed irrigation of formerly dry areas, but it has brought about significant ecosystem changes. In addition, upstream irrigation has greatly increased the salinity of the water downstream.

Construction of dams, reservoirs, and canal systems has caused significant environmental and health problems, especially in developing countries. For example, the Aswan Dam, along with its associated canal system in Egypt, has brought an increase in diseases carried by snails and mosquitoes. As nations continue to develop their water resources, they will need to plan at the ecosystem level to try to avoid these problems.

Water is an integral part of ecosystems, and its increasing use by people is a major contributor to the degradation of ecosystems. Loss or damage of wetlands is an area of particular environmental concern in the United States because significant portions of these ecosystems have already been lost, including 90 percent of the freshwater wetlands.

(a)

(b)

(c)

◀ **FIGURE 10.C** Several types of wetlands: (a) Chesapeake Bay salt marsh (*Comstock*); (b) freshwater cypress swamp in North Carolina (*Carr Clifton/Minden Pictures*); and (c) prairie potholes in the Dakotas (*Jim Brandenburg/Minden Pictures*).

REFERENCES

1. **Gleick, P. H.** 1993. An introduction to global fresh water issues. In *Water in crisis*, ed. P. H. Gleick, pp. 3–12. New York: Oxford University Press.
2. **Council on Environmental Quality and the Department of State.** 1980. *The global 2000 report to the president: Entering the twenty-first century*, vol. 2. Washington, DC: Council on Environmental Quality.
3. **Water Resources Council.** 1978. *The nation's water resources, 1975–2000*, vol. 1.
4. **Singer, M. J., and Munns, D. M.** 1978. *Soils: An introduction*. New York: Macmillan.
5. **Likens, G. E., Bormann, F. H., Pierce, R. S., Eaton, J. S., and Johnson, N. M.** 1977. *The biogeochemistry of a forested ecosystem*. New York: Springer-Verlag.
6. **Florsheim, J. L., Keller, E. A., and Best, D. W.** 1991. Fluvial sediment transport in response to moderate storm flows following chaparral wildfire, Ventura County, southern California. *Geological Society of America Bulletin* 103:504–11.
7. **Winter, T. C., Harvey, J. W., Franke, O. L., and Alley, W. M.** 1998. *Ground water and surface water. A single resource*. U.S. Geological Survey Circular 1139.

8. **Solley, W. B., Pierce, R. R., and Perlman, H. A.** 1998. *Estimated use of water in the United States in 1995.* U.S. Geological Survey Circular 1200.
9. **Alexander, G.** 1984. Making do with less. *National Wildlife* (Special Report), February–March, 11–13.
10. **Leopold, L. B.** 1977. A reverence for rivers. *Geology* 5:429–30.
11. **Graf, W. L.** 1985. *The Colorado River.* Association of American Geographers.
12. **Nash, R.** 1986. Wilderness values and the Colorado River. In *New courses for the Colorado River,* eds. G. D. Weatherford and F. L. Brown, pp. 201–14. Albuquerque: University of New Mexico Press.
13. **Hundley, N., Jr.** 1986. The West against itself: The Colorado River—An institutional history. In *New courses for the Colorado River,* eds. G. D. Weatherford and F. L. Brown, pp. 9–49. Albuquerque: University of New Mexico Press.
14. **Ballard, S. C., Michael, D. D., Chartook, M. A., Clines, M. R., Dunn, C. E., Hock, C. M., Miller, G. D., Parker, L. B., Penn, D. A., and Tauxe, G. W.** 1982. *Water and western energy: Impacts, issues, and choices.* Boulder, CO: Westview Press.
15. **Joseph, P.** 1998. The battle of the dams. *Smithsonian* 29:48–61.
16. **Dolan, R., Howard, A., and Gallenson, A.** 1974. Man's impact on the Colorado River and the Grand Canyon. *American Scientist* 62:392–401.
17. **Lavender, D.** 1984. Great news from the Grand Canyon. *Arizona Highways Magazine,* January, 33–38.
18. **Hecht, J.** 1996. Grand Canyon flood a roaring success, *New Scientist* 151:8.
19. **Lucchitta, I., and Leopold, L.B.** 1999. Floods and sandbars in the Grand Canyon. *Geology Today* 9:1–7.
20. **Covich, A. P.** 1993. Water and ecosystems. In *Water in crisis,* ed. P. H. Gleick, pp. 40–55. New York: Oxford University Press.
21. **Gleick, P. H., Ed.** 1993. *Water in crisis,* Table F.1, p. 288. New York: Oxford University Press.
22. **Levinson, M.** 1984. Nurseries of life. *National Wildlife* (Special Report), February–March, 8–21.
23. **Holloway, M.** 1991. High and dry. *Scientific American* 265(6): 16–20.

KEY TERMS

water cycle (p. 263)
watershed (drainage basin) (p. 264)
throughflow (p. 266)
overland flow (p. 266)
groundwater flow (p. 266)
sediment yield (p. 266)
vadose zone (p. 267)
water table (p. 267)
capillary fringe (p. 267)
aquifer (p. 269)
aquitard (p. 269)
artesian (p. 269)
cone of depression (p. 269)
hydraulic gradient (p. 270)
hydraulic conductivity (p. 270)
porosity (p. 270)
effluent stream (p. 271)
influent stream (p. 271)
Darcy's law (p. 272)
Darcy flux (p. 272)
fluid potential (hydraulic head) (p. 272)
vx (p. 273)
water budget (p. 273)
base flow (p. 275)
desalination (p. 275)
offstream use (p. 275)
consumptive use (p. 275)
instream use (p. 275)
water conservation (p. 280)
water management (p. 281)
wetlands (p. 288)

SOME QUESTIONS TO THINK ABOUT

1. You have been hired by a consulting company to evaluate the water resources of the region in which you live. Your first task is to develop a rough water budget. How would you go about doing this? What sorts of data would you need? How could the data be used to evaluate your water-resource situation?
2. You are working for a planning agency trying to come to grips with potential water use for a moderately sized river basin of about 5000 km^2 that discharges into the ocean. People interested in environmental quality and wilderness want to see adequate river flows to maintain healthy ecosystems along the river, whereas agricultural and urban interests see the flow as a potential source of water to irrigate crops and provide basic water supply. Finally, the river is navigable, and certain users are interested in seeing that there are adequate flows for using the river as a transportation route. After examining the idealized diagram shown in Figure 10.14, you are fairly certain that conflicts of interest will arise in the use of the water in the river. Outline what these conflicts are likely to be and what steps could possibly be taken to help in mediation or conflict resolution concerning the water resources of the river.
3. Find out what (if any) management principles are being used for the water resources of your community. How could some of the suggestions put forth by Luna Leopold be applied to your specific water-management needs? Pay particular attention to those times when water shortages might occur.
4. What sort of wetlands are found in your region? Outline a plan to inventory the wetlands and make an assessment of how much of the resource has been lost or damaged. Is wetlands restoration possible in your region, and what would you need to do to make it successful?

11 Water Pollution and Treatment

Hanauma Bay, Oahu, Hawaii. (*Edward A. Keller*)

Hanauma Bay Beach Park is located only a few kilometers west of the city of Honolulu on the island of Oahu, Hawaii. The bay, the result of collapse or marine erosion of a volcanic peak, is the site of a fringing coral reef that is now a marine sanctuary visited by tourists from around the world.

Hanauma Bay and its coral reef are in poor ecological condition, as are many coral reefs of the world's coastline as a result of a variety of human activities, including global warming of oceanic water, intensive coastal development, and pollution of nearshore waters. Coral reefs are particularly vulnerable to water pollution because they thrive best in clear coastal waters that are naturally relatively nutrient-poor. When raw sewage or even treated wastewater or agricultural runoff enters nearshore environments, nutrient levels may rise, causing problems for coral reefs as other organisms that prey on coral experience a population explosion in response to the increased nutrient load. Coastal development and agriculture may also greatly increase the amount of sediment entering coastal waters that may partially cover coral reefs, blocking sunlight and weakening the coral so that they become vulnerable to disease. It is estimated that 60 percent of the world's coral reefs are today being threatened by human activities.

We once thought that the oceans of the world could never be polluted because they are so vast. Today we are learning that this is far from the case. Beaches in southern California are sometimes closed as a result of pollutants entering the coastal environment from streams and other sources. Raw sewage and chemicals from cities around the world, particularly in developing countries, are being indiscriminately dumped into rivers that enter our lakes and oceans in growing quantities as human population increases. In southern California, beach closures are blamed on everything from seagulls to dogs on beaches to seals, when we ought to be looking more closely at processes of urban runoff and waste disposal into streams and rivers that make their way to the coastal environment.

LEARNING OBJECTIVES

One of the most serious environmental problems for billions of people on earth is the continuous or sporadic lack of a pollution- and disease-free water supply for personal consumption. With this in mind, learning objectives for the chapter are:

- *To be able to define water pollution and discuss some of the common water pollutants.*
- *To gain an appreciation for selected water pollution problems, including cultural eutrophication and acid mine drainage.*
- *To know the difference between point and nonpoint sources of water pollution.*
- *To understand processes by which groundwater may become polluted and how polluted water may be treated.*
- *To become familiar with some of the important issues related to water quality standards.*
- *To understand the principles of wastewater treatment associated with septic-tank sewage disposal and wastewater treatment plants.*
- *To gain an appreciation for processes related to wastewater renovation and wastewater treatment involving resource recovery.*
- *To become familiar with the basic doctrine of law associated with surface water and groundwater as well as some of the major federal water legislation important in protecting water resources.*

Web Resources

Visit the "Environmental Geology" Web site at www.prenhall.com/keller to find additional resources for this chapter, including:

- Web Destinations
- On-line Quizzes
- On-line "Web Essay" Questions
- Search Engines
- Regional Updates

11.1 An Overview of Water Pollution

Water pollution refers to degradation of water quality as measured by biological, chemical, or physical criteria. This degradation is judged according to the intended use of the water, departure from the norm, and public health or ecological impacts. From a public health or ecological point of view, a pollutant is any substance in which an identifiable excess is known to be harmful to desirable living organisms. Thus, excessive amounts of heavy metals, certain radioactive isotopes, phosphorus, nitrogen, sodium, and other useful (even necessary) elements, as well as certain pathogenic bacteria and viruses, are all pollutants. In some instances, a material may be considered a pollutant to a particular segment of the population although not harmful to other segments. For example, excessive sodium as a salt is not generally harmful, but it is to some people on diets restricting intake of salt for medical purposes. Table 11.1 lists some common sources of groundwater pollution.

Problems related to water pollution are extremely variable. Of particular significance are the residence times and reservoir sizes of water in the various parts of the hydraulic cycle, because these factors are related to pollution potential. For example, water in rivers has a short average residence time of about 2 weeks. Therefore, a one-time pollution event (one that does not involve the pollutant's attaching itself to sediment on the riverbed, which would result in a much longer residence time) will be relatively short-lived because the water will soon leave the river environment. On the other hand, the same pollutant may enter a lake or an ocean, where residence times are longer and pollutants more difficult to deal with. Many circumstances, such as sewage spills or pollutant-carrying truck or train crashes, can produce one-time pollution events. News of such events is frequent these days. However, pollution is more likely to result from chronic processes that discharge pollutants directly into rivers (Figure 11.1). Groundwaters, unlike river waters, have long res-

Table 11.1 Common sources of groundwater pollution and/or contamination

Leaks from storage tanks and pipes
Leaks from waste disposal sites such as landfills
Seepage from septic systems and cesspools
Accidental spills and seepage (train or truck accidents, for example)
Seepage from agricultural activities such as feedlots
Intrusion of salt water into coastal aquifers
Leaching and seepage from mine spoil piles and tailings
Seepage from spray irrigation
Improper operation of injection wells
Seepage of acid water from mines
Seepage of irrigation return flow
Infiltration of urban, industrial, and agricultural runoff

▲ FIGURE 11.1 An example of severe water pollution producing a health hazard: This ditch carries sewage and toxic waste to the Rio Grande in Mexico. (*Jim Richardson/Richardson Photography*)

idence times (from hundreds to thousands of years). Therefore, natural removal of pollutants from groundwater is a very slow process, and correction is very costly and difficult.

11.2 Selected Water Pollutants

Many different materials may pollute surface water or groundwater. Our discussion here will focus on oxygen-demanding waste; pathogenic organisms; nutrients; oil; hazardous chemicals; heavy metals; radioactive materials; and sediment.

Oxygen-Demanding Waste

Dead organic matter in streams decays; that is, it is consumed by bacteria, which require oxygen. If there is enough bacterial activity, the oxygen in the water can be reduced to levels so low that fish and other organisms die. A stream without oxygen is a dead stream for fish and many organisms we value. The amount of oxygen used for bacterial decomposition is the **biochemical oxygen demand (BOD)**, a commonly used measure in water quality management. The BOD is measured as milligrams per liter of oxygen consumed over 5 days at 20°C. A high BOD indicates a high level of decaying organic matter in the water.

Dead organic matter in streams and rivers comes from natural sources (for example, dead leaves from a forest) as well as from agriculture and urban sewage. Approximately 33 percent of all BOD in streams results from agricultural activities, but urban areas, particularly those with sewer systems that combine sewage and storm-water runoff, may add considerable BOD to streams during floods, when sewers entering treatment plants can be overloaded and overflow into streams, producing pollution events.

The Council on Environmental Quality defines the threshold for water pollution as a dissolved oxygen content of less than 5 mg per liter (mg/l) of water. The diagram in Figure 11.2 illustrates the effect of BOD on dissolved oxygen content in a stream when raw sewage is introduced as a result of an accidental spill. Three zones are recognized. The *pollution zone* has a high BOD and a reduced dissolved oxygen content as initial decomposition of the waste begins. In the *active decomposition zone*, the dissolved oxygen content is at a minimum owing to biochemical decomposition as the organic waste is transported downstream. In the *recovery zone*, the dissolved oxygen increases and the BOD is reduced because most oxygen-demanding organic waste from the input of sewage has decomposed, and natural stream processes replenish the water with dissolved oxygen. All streams have some capability to degrade organic waste after it enters the stream. Problems result when the stream is overloaded with biochemical oxygen-demanding waste, overpowering the stream's natural cleansing function.

Pathogenic Organisms

Pathogenic (disease-causing) microorganisms are important biological pollutants. Among the major waterborne human diseases are cholera, typhoid infections, hepatitis, and dysentery. Because it is often difficult to monitor the pathogens directly, we use the count of human **fecal coliform bacteria** as a common measure of biological pollution and a standard measure of microbial pollution. These common and usually harmless bacteria are normal constituents of human intestines and are found in all human waste.

All forms of fecal coliform bacteria are not harmless! *E. coli*, a type of fecal coliform bacteria, has been responsible for human illness and deaths in the 1990s. Outbreaks of disease, apparently caused by *E. coli*, occurred from people eating contaminated meat at a popular fast-food restaurant in 1993, and in 1998 *E. coli* apparently contaminated the water in a Georgia water park and a town's water supply in Wyoming, causing illness and one death.

In the past, epidemics of waterborne diseases have killed thousands of people in U.S. cities. Such epidemics have been largely eliminated by separating sewage water and drinking water and treating drinking water prior to consumption. Unfortunately, this is not the case worldwide, and every year several billion people (particularly in poor countries) are exposed to waterborne diseases. For example, as recently as the early 1990s epidemics of cholera occurred in South America. Outbreaks of waterborne diseases are always a threat, even in developed countries.

Perhaps the largest known outbreak of a waterborne disease in the United States took place in 1993. In that year, approximately 400,000 cases of cryptosporidiosis occurred in Milwaukee, Wisconsin. The disease, which causes flu-like

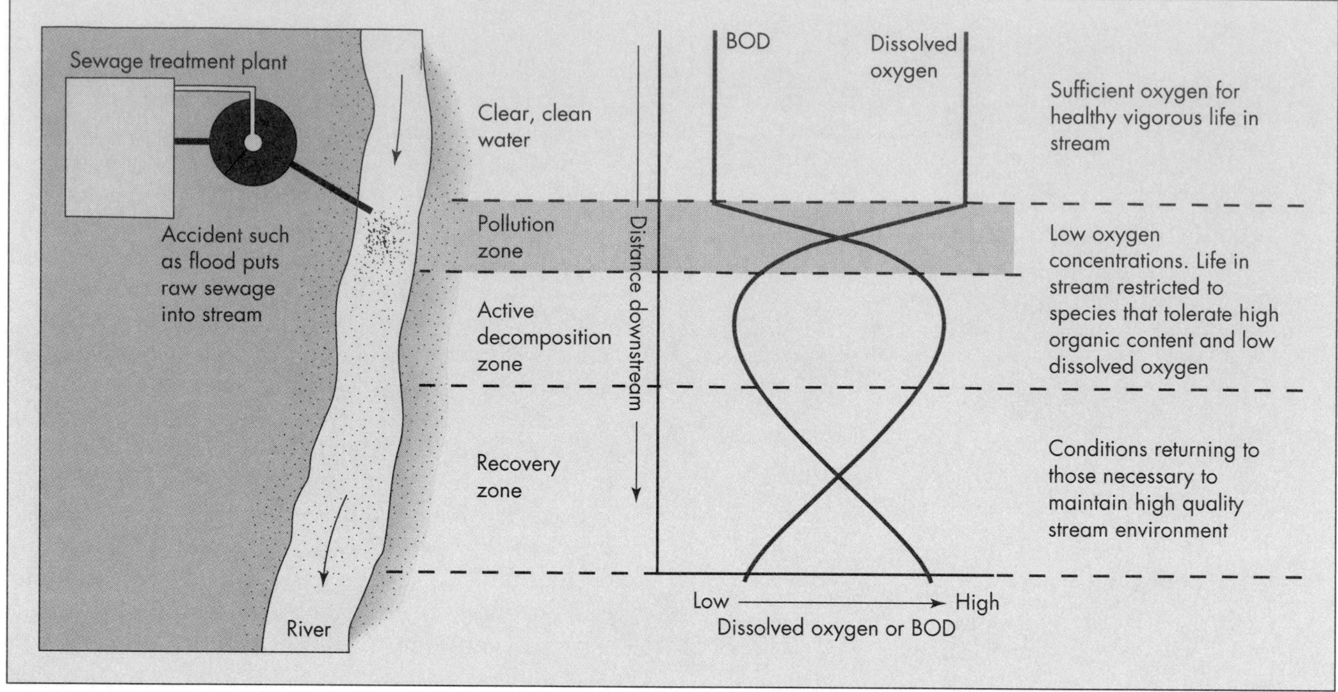

▲ **FIGURE 11.2** Relationship between dissolved oxygen and biochemical oxygen demand (BOD) for a stream following the input of sewage.

symptoms, is carried by a microorganism (a parasite) and can be fatal to people with a depressed immune system, such as AIDS or cancer patients. The parasite is resistant to chlorination, and people in Milwaukee were advised to boil their water during the epidemic. The outbreak was a wake-up call concerning water quality and disease, because many other cities utilizing surface water supplies are just as vulnerable as Milwaukee (1).

The threat of an outbreak of a waterborne disease is accentuated following disasters such as earthquakes, floods, and hurricanes, because these events may damage sewer lines or cause them to overflow, resulting in contamination of water supplies. Following the 1994 Northridge earthquake, people in the San Fernando Valley of the Los Angeles Basin were advised to purify municipal water by boiling.

Nutrients

Nutrients released by human activity may lead to water pollution. Two important nutrients that can cause problems are phosphorus and nitrogen, both of which are released from a variety of materials including fertilizers, detergents, and the products of sewage treatment plants. The concentration of phosphorus and nitrogen in streams is related to land use, as shown in Figure 11.3. Forested land has the lowest concentrations, whereas the highest concentrations are found in agricultural areas—sites of such sources as fertilized farm fields and feedlots (2). Urban areas can also add a lot of phosphorus and nitrogen to local waters, particularly where wastewater treatment plants discharge treated waters into rivers, lakes, or the ocean. These plants are effective in reducing organic pollutants and pathogens, but without advanced treatment nutrients pass through the system.

High concentrations of nitrogen and phosphorus in water often result in the process known as **cultural eutrophication.** Eutrophication (from the Greek for "well-fed") is characterized by rapid increase in the abundance of plant life, particularly algae. In freshwater ponds and lakes, blooms of algae form thick mats that sometimes nearly cover the surface of the water, blocking sunlight to plants below, which eventually die. In addition, as the algae decompose, the oxygen content of the water decreases, and fish and aquatic animals may die.

In the marine environment, nutrients in nearshore waters may cause blooms of seaweed (marine algae) that can become a nuisance when the seaweed is torn loose and piles up on beaches. Algae may also damage or kill coral in tropical areas. For example, the island of Maui in the Hawaiian Islands has a cultural eutrophication problem resulting from nutrients entering the nearshore environment from waste disposal practices and agricultural runoff. The inhabitants of the island may be killing the goose that lays the golden egg. Beaches in some areas become fouled with algae that washes up on the shore, where it rots, smells bad, and provides a home for irritating insects, eventually driving away tourists (Figure 11.4). In the water, algae that covers coral may damage or kill it.

Oil

Oil discharged into surface water, usually the ocean, has caused major pollution problems. The largest oil discharges have usually involved oil-tanker accidents at sea (see *Case History: Exxon Valdez*).

Military activity has become another source of pollution of the marine environment by oil. The huge oil spill in the Persian Gulf during the 1991 war released an unknown

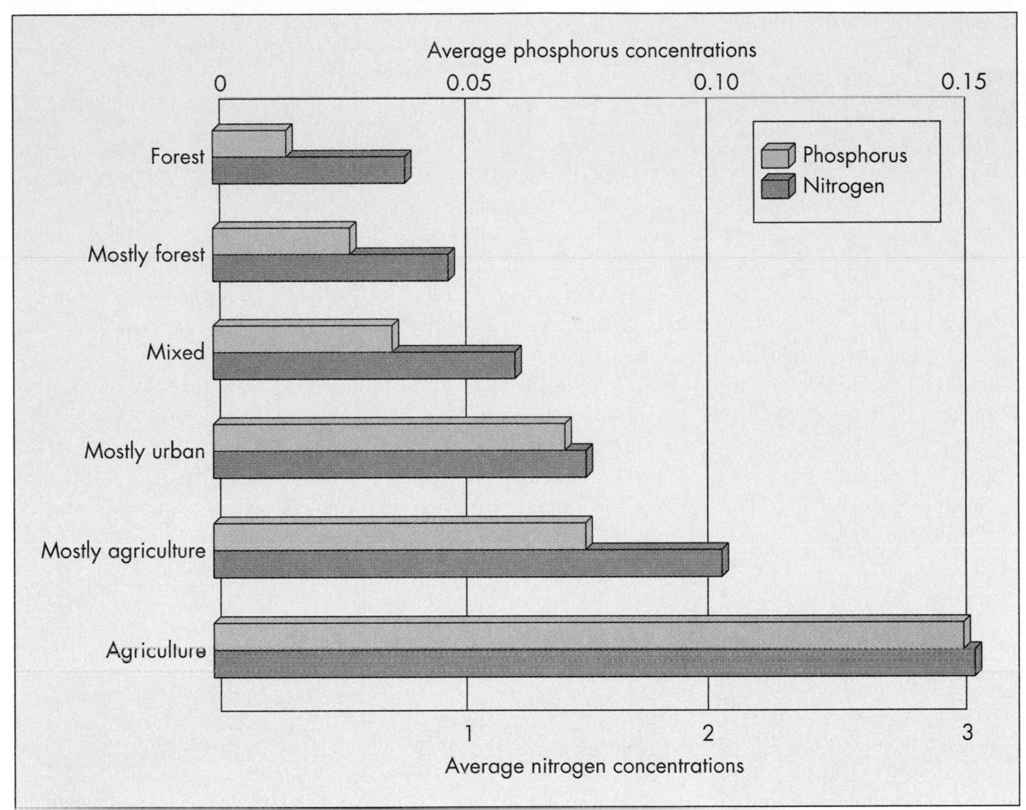

FIGURE 11.3 Relationship between land use and average nitrogen and phosphorus concentration in streams (in milligrams per liter). (*From Council on Environmental Quality*, 1978)

(a)

(b)

(c)

FIGURE 11.4 Oceanfront condominium on the island of Maui, Hawaii. Note brown line along edge of beach, which is an accumulation of marine algae (locally called *seaweed*) (a). On the beach itself, the algae piles up sometimes to a depth of about 0.5 m and is avoided by people using the beach (b). Condominium complexes often have small wastewater treatment plants, such as the one shown here, that provide primary and secondary treatment. Following this treatment, the water is injected underground at a relatively shallow depth. The treatment does not remove nutrients such as phosphorus and nitrogen that apparently encourage the accelerated growth of marine algae in the nearshore environment (c). (*Edward A. Keller*)

CASE HISTORY Exxon Valdez

Just after midnight on March 24, 1989, the oil tanker *Exxon Valdez* ran aground on Bligh Reef, 40 km south of Valdez, Alaska, in Prince William Sound. Crude oil delivered to Valdez via the Trans-Alaskan Pipeline poured out of the ruptured tanks of the vessel at a rate of approximately 20,000 barrels per hour (Figure 11.A). The *Exxon Valdez* was loaded with 1.2 million barrels of North Slope crude, and of this, more than 250,000 barrels (11 million gallons) gushed from the hold of the 300-m tanker. The oil remaining in the *Exxon Valdez* was loaded into another tanker (3).

Any hope of containment of the oil slick was lost as winds began blowing a few days after the accident, spreading the oil. Of the 11 million gallons of spilled oil, 50 percent was deposited on the shoreline, 20 percent evaporated, and only 14 percent was collected by waste recovery (4).

The oil spilled into what is considered one of the most pristine and ecologically rich marine environments of the world (3, 5), and the accident is now known as the worst oil spill in the history of the United States. Following the spill, the oil spread over a very large area. Figure 11.B, left side, shows the extent of oil sheens, tar balls, and mousse that was suspected to have come from the *Exxon Valdez* as of August 10, 1989. Mousse is a thick, weathered patch of oil with the consistency of a soft pudding that often washes up on beaches. Figure 11.B, right side, compares the area of the spill as of August 10 to the eastern seaboard of the United States. Notice that it would extend from Massachusetts to North Carolina! Many species of fish, birds, and marine mammals are present in Prince William Sound, and the long-term impact of the oil spill on the environment is difficult to ascertain.

(a)

(b)

▲ **FIGURE 11.A** Oil spill from the *Exxon Valdez* in Alaska, 1989. (a) Aerial view of oil being offloaded from the leaking tanker *Exxon Valdez* on the left to the smaller *Exxon Baton Rouge* on the right. Floating oil is clearly visible on the water (*Michelle Barns/Liaison Agency, Inc.*); (b) attempting to clean oil from the coastal environment by scrubbing and spraying with hot water. (*I. L. Atlan/Sygma*)

volume of oil into a fragile environment. It may be the world's largest spill, and it is certainly the largest deliberate spill.

Oil spills on land can also lead to serious environmental problems if pipelines rupture, as happened in the fall of 1994 in northern Russia. That event allowed an unknown but very large volume (estimates range from 4 million to 80 million gallons) of crude oil to pollute land and water resources. This brings up an important point: Pipelines 25 to 30 years old are more vulnerable to fatigue and corrosion than are new pipelines. Periodic monitoring of aging systems and repair or replacement of worn-out pipelines should be a priority in minimizing the occurrence of leaks.

Toxic Substances

Many substances that enter surface- and groundwater are toxic to organisms. We will mention three general categories of toxic substances here and discuss them in detail in Chapters 12 and 13.

Hazardous chemicals are synthetic organic and inorganic compounds that are toxic to humans and other living things. When these materials are accidentally introduced into surface or subsurface waters, serious pollution may result. The complex problem of hazardous chemicals and their management is discussed in Chapter 12 ("Waste Management").

Heavy metals such as lead, mercury, zinc, and cadmium are dangerous pollutants and are often deposited with natural sediment in the bottoms of stream channels. If these metals are deposited on floodplains, they may become incorporated into plants, including food crops, and animals. If they are dissolved and the water is withdrawn for agriculture or human use, heavy-metal poisoning can result. Heavy metals are discussed in detail in Chapter 13 ("The Geologic Aspects of Environmental Health").

Radioactive materials in water may be dangerous pollutants. Of particular concern are possible effects to people, other animals, and plants of long-term exposure to low

Soon after the accident, the governor of Alaska declared Prince William Sound a disaster area and applied for federal assistance. Cleanup work on the coastline posed enormous problems to workers attempting the project. Photographs and videotapes of the work suggest an almost futile attempt to clean individual pebbles on beaches. The spill completely disrupted the lives of those who work in the vicinity of Prince William Sound, and only after the passage of time will its impacts be fully understood. Certainly the short-term effects were very significant indeed. These included the death of 100,000–645,000 seabirds, 28 percent of the sea otters, and 13 percent of the harbor seals in the Sound (4,5), as well as disruption of the commercial fisheries, sport fisheries, and tourism. Interruption of the flow of North Slope crude resulted in an almost immediate increase in the price of oil to the lower 48 states. Lessons learned from the *Exxon Valdez* spill have resulted in better management strategies for both the shipment of crude oil and the emergency plans to minimize environmental degradation.

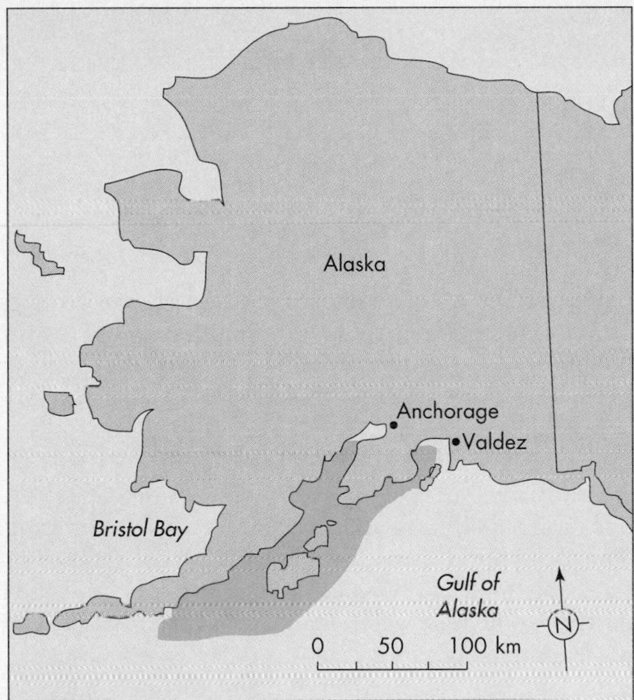

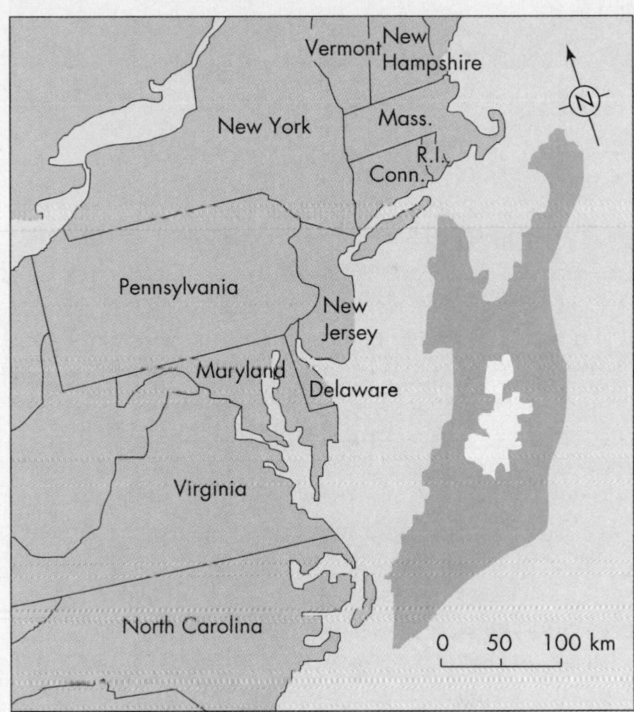

▲ **FIGURE 11.B** (left) Extent of Alaskan oil spill of 1989. (right) Area of 1989 Alaskan oil spill compared to the eastern coast of the United States. (*From* Alaska Fish and Game, *July–August* 1989)

doses of radioactivity. Chapters 12 and 13 discuss radiation in terms of waste disposal and environmental effects.

Sediments

Sediment consists of rock and mineral fragments ranging in size from sand particles less than 2 mm in diameter to silt, clay, and even finer colloidal particles. Our greatest water pollutant by volume, sediment, is a resource out of place. It depletes a land resource (soil), reduces the quality of the water resource it enters, and may deposit sterile materials on productive croplands or other useful land. Sediment pollution is discussed in detail in Chapter 3.

Thermal Pollution

Thermal pollution is the artificial heating of waters, primarily by hot-water emission from industrial operations and power plants. Heated water causes several problems. Even water only a few degrees warmer than the surrounding water holds less oxygen. Warmer water favors different species than does cooler water and may increase growth rates of undesirable organisms, including certain water plants and fish. On the other hand, the warm water may attract and allow better survival of certain desirable fish species, particularly during the winter.

11.3 Surface-Water Pollution and Treatment

Pollution of surface waters occurs when too much of an undesirable or harmful substance flows into a body of water, exceeding the natural ability of that ecosystem to utilize or remove the pollutant or convert it to a harmless form. Water pollutants are emitted from point (localized) sources or nonpoint (diffuse) sources.

▲ **FIGURE 11.5** Pipe discharging partially treated effluent from the Climax Molybdenum Mine in Colorado. (*Jim Richardson/Richardson Photography*)

▲ **FIGURE 11.6** Sediment is being removed here by heavy equipment (background) following the 1995 flood in Goleta, California. The sediment came from a nearby stream that overflowed its bank and deposited all this sediment at a new car dealership. (*Rafael Maldonado/Santa Barbara News-Press*)

Point Sources of Surface-Water Pollution

Point sources are discrete and confined, such as pipes that empty into streams or rivers from industrial or municipal sites (Figure 11.5). In general, point-source pollutants from industries are controlled through on-site treatment or disposal and are regulated by permit. Municipal point sources are also regulated by permit. In older cities in the northeastern and Great Lakes areas of the United States, most point sources are outflows from combined sewer systems that carry both storm-water flow and municipal waste. During heavy rains, urban storm runoff may exceed the capacity of the sewer system, causing it to back up and overflow, delivering pollutants to nearby surface waters.

An important principle of avoiding water pollution is that water from different sources should not be mixed; it should remain separated according to the intended use. For example, agricultural runoff containing pollutants such as nitrates and pesticides should be kept from entering water intended for urban consumption. This is a primary problem for large water delivery systems (as, for example, in California) that supply several different users requiring different water quality.

Nonpoint Sources of Surface-Water Pollution

Nonpoint sources are diffused and intermittent; they are influenced by such factors as land use, climate, hydrology, topography, native vegetation, and geology. Pollution from nonpoint sources, or *polluted runoff,* is difficult to control. Common urban nonpoint sources include urban runoff from streets or fields containing all sorts of pollutants, including heavy metals, chemicals, and sediment (Figure 11.6). When you wash your car in your driveway and the detergent and oil on the surface run down a storm drain that enters a stream, you are contributing to polluted runoff. Polluted runoff is also produced when rainwater washes insecticides from the plants in your garden, then runs off into a stream or infiltrates the surface to contaminate groundwater. Similarly, rain and runoff from factories and storage yards are a source of nonpoint pollution (6). Rural sources of nonpoint pollution are generally associated with agriculture, forestry, or mining (see *A Closer Look: Acid Mine Drainage*).

Reduction of Surface-Water Pollution

A serious attempt is being made in the United States to reduce water pollution and thereby increase water quality. The assumption is that people have a basic right to have water safe to drink, swim in, and use in agriculture and industry. At one time water quality near major urban centers was considerably worse than it is today; in one instance, in 1969, the Cuyahoga River flowing through Cleveland, Ohio, was inadvertently set on fire. The fire was an environmental shock to the city and state, which responded by passing laws to reduce discharge of pollutants into the river. Today the river is much cleaner and is being used for recreational purposes (7).

In recent years the number of success stories, including the Cuyahoga, has been very encouraging. Perhaps the best-known case is the Detroit River. In the 1950s and the early 1960s the Detroit River was considered dead, having been an open dump for sewage, chemicals, garbage, and urban trash. Tons of phosphorus were discharged daily into the river, and a film of oil up to 0.5 cm thick was often present. Aquatic life was damaged considerably, and thousands of ducks and fish were killed. Although today the Detroit River is not a pristine stream, considerable improvement has resulted from industrial and municipal pollution control. Oil and grease emissions were reduced by 82 percent, and phosphorus and sewage discharges have also been greatly diminished. Fish once again are found in the Detroit River, and the shoreline is usually clean. Other success stories include New York's Hudson River (see *Case History: Cleaning Up the Hudson*), New Hampshire's Pemigewasset River, North Carolina's French Broad River, and the Sa-

A CLOSER LOOK Acid Mine Drainage

Acid mine drainage does not refer to an acid mine, but to acidic water that drains from mines. Specifically, **acid mine drainage** is water with a high concentration of sulfuric acid (H_2SO_4) that drains from some mining areas to pollute surface-water resources. The acid is produced by a simple weathering reaction: When sulfide minerals associated with coal or a metal (zinc, lead, or copper) come into contact with oxygen-rich water near the surface, the sulfide mineral oxidizes. For example, pyrite (FeS_2) is a common sulfide often associated with coal, and when pyrite oxidizes in the presence of water, sulfuric acid is formed. The sources for the water may be surface water that infiltrates into mines or shallow groundwater that moves through mines. Similarly, surface and shallow groundwaters that come into contact with mining waste (tailings) also may react with sulfide minerals found there to form acid-rich waters.

When waters with a high concentration of sulfuric acid migrate away from a mining area, they may pollute surface and groundwater resources. If the acid-rich water runs into a natural stream or lake, significant ecological damage can result, because the acid water is extremely toxic to plants and animals in aquatic ecosystems. Acid mine drainage is a significant water pollution problem in many areas of the United States, including parts of Wyoming, Illinois, Indiana, Kentucky, Tennessee, Missouri, Kansas, Oklahoma, West Virginia, Maryland, Pennsylvania, Ohio, and Colorado. The total impact associated with acid mine drainage is very significant because thousands of kilometers of streams have been polluted (Figure 11.C).

The Tar Creek area of Oklahoma was at one time designated as the nation's worst hazardous waste site by the U.S. Environmental Protection Agency (EPA). The creeks in the area were severely polluted by acid-rich water from abandoned mines of the Tri-State Mining District of Arkansas, Oklahoma, and Missouri. Sulfide deposits containing both lead and zinc were first mined in the late nineteenth century, and mining ended in some areas in the 1960s. During operation of the mines, subsurface areas were kept dry by pumping out groundwater that was constantly seeping in. After mining ceased, the groundwater table naturally rose again, and some mines flooded and overflowed into nearby streams, polluting them.

▲ **FIGURE 11.C** Water seeping from this Colorado mine is an example of acid mine drainage. The water is also contaminated by heavy metals. (*Tim Haske/Profiles West/Index Stock Imagery, Inc.*)

vannah River in the southeastern United States. These examples are evidence that water pollution abatement has positive results (8).

An innovative system that uses naturally occurring earth materials to purify water for public consumption is reported from a Michigan community on the Lake Michigan shore. The city of Ludington, with a summer population exceeding 10,000, uses sands and gravels below the lake bottom to prefilter and thus treat the lake water for municipal use. A system of lateral intakes is buried in sand and gravel 4 to 5 m below the lake bottom, where the water depth is at least 5 m. The water is pumped out for municipal use, and in some cases the only additional treatment is chlorination. We will say more about the ability of rock and soil to filter out impurities in the discussion of groundwater pollution.

11.4 Groundwater Pollution and Treatment

Approximately one-half of all people in the United States today depend on groundwater as their source of drinking water. We are therefore concerned about the introduction into aquifers of chemical elements, compounds, and microorganisms that do not occur there naturally. The hazard presented by a particular groundwater pollutant depends upon several factors, including the volume of pollutant discharged, the concentration or toxicity of the pollutant in the environment, and the degree of exposure to people or other organisms (12).

Most of us have long believed that groundwater is pure and safe to drink, so many of us find it alarming to learn that it may be easily polluted by any one of several sources (see Table 11.1). In addition, the pollutants, even the very toxic ones, may be difficult to recognize. One of the best-known examples of groundwater pollution is the Love Canal near Niagara Falls, New York, where burial of chemical waste has caused serious water pollution and health problems, which we will discuss in Chapter 12.

Unfortunately, Love Canal is not an isolated case. Hazardous chemicals have been found or are suspected to be in groundwater supplies in nearly all parts of the world, developed and developing nations alike. Developed industrial countries produce thousands of chemicals; many of these, particularly pesticides, are exported to developing countries, where they protect crops that eventually are imported by the same industrial countries, completing a circle.

CASE HISTORY Cleaning Up the Hudson

The Hudson River assessment and cleanup of PCBs (polychlorinated biphenyls) is a good example of people's determination to clean up our rivers. The PCBs, which have a chemical structure similar to DDT and dioxin, were used mainly in electrical capacitors and transformers; discharge of the chemicals from two outfalls on the Hudson River started about 1950 and terminated in 1977. Approximately 295,000 kg of PCBs are believed to be present in Hudson River sediments. Concentrations in the sediment are as high as 1500 parts per million (ppm) near the outfalls, compared to *less* than 10 ppm several hundred kilometers downstream at New York City (9, 10). An important source of PCBs in the New York metropolitan area has been sewage effluent and urban runoff. These have delivered up to about half of the PCB load to the Hudson–New York harbor in recent years (11). The U.S. Food and Drug Administration (FDA) permits less than 2.5 ppm PCBs in dairy products, whereas the New York State limit for drinking water is 0.1 part per billion (ppb).

It is known that PCBs are carcinogenic and can cause disturbances of the liver, nervous system, blood, and immune response system in humans. Furthermore, PCBs are nearly indestructible in the natural environment and become concentrated in the higher rungs of the food chain—thus the concern! Water samples in the 240-km tidal reach of the Hudson River have yielded average PCB concentrations ranging from 0.1 to 0.4 ppb, but PCBs are concentrated to much higher levels in some fish. As a result, fishermen on the lower Hudson have suffered a significant economic impact from the contamination because nearly all commercial fishing was banned, and sport fishing was greatly reduced (9,10).

Cleanup of the Hudson River has considered two alternatives (10,11).

- Dredging "hot spots" where concentrations of PCBs are greater than 50 ppm. It is anticipated that dreging would reduce the time necessary for the river to clean itself up by natural processes, such as sediment transport to the ocean; burial of the most contaminated sediments by river processes; and biogeochemical dechlorination by organisms in the sediment of the river bed.
- No action. This alternative would allow natural processes to clean up the PCBs. This assumes that sources of input of the chemicals have been greatly reduced.

The no-action alternative, perhaps by default because there has been continued postponement of removal by dredging in the upper Hudson River, is what is happening. Natural cleansing is occurring and the concentrations of PCBs on sediment particles transported downstream in the river were several times lower in the mid-1980s compared to the mid-1970s. Half-life response times in the Hudson River system (that is, the time for the concentration of PCB on the sediment to be reduced by one-half) is about 3.5 years. Inputs of PCBs have been greatly reduced due to EPA restrictions on manufacture of the chemicals (11).

For example, Costa Rica imports several pesticides, including DDT, aldrin, endrin, and chlordane, that are banned or heavily restricted in the United States. Thus, these chemicals are polluting the surface and groundwater of Costa Rica and other places where they are still being used, and residual concentrations of some of them are returned to us on crops we import (13).

In the United States today, the problem of groundwater pollution is becoming more apparent as testing of water becomes more common. For example, Atlantic City, New Jersey, and Miami, Florida, are two eastern cities threatened by polluted groundwater that is slowly migrating toward municipal wells. It is estimated that 75 percent of the 175,000 known waste disposal sites in the country may be producing **plumes** (body of earth material above and/or below the water table contaminated by a water pollutant) of hazardous chemicals that are migrating into and polluting groundwater resources. Because many of the chemicals are toxic or suspected carcinogens, it appears we have been conducting a large-scale experiment on the effects of chronic low-level exposure of people to potentially harmful chemicals. Unfortunately, the final results of the experiment will not be known for many years (14). Preliminary results suggest we had better act now before a hidden time bomb of health problems explodes.

Comparison of Groundwater and Surface-Water Pollution

Differences in the physical, geologic, and biologic environments make the problems associated with groundwater pollution significantly different from those of surface-water pollution. In the case of surface-water pollution, the rapidity of the flow results in rapid dilution and dispersion of pollutants, and the availability of oxygen and sunlight contributes to their rapid degradation. The situation is markedly different for groundwater, where the opportunity for dilution and dispersion of pollutants is limited and opportunity for bacterial degradation of pollutants is generally confined to the soil a few meters or so below the surface. The channels through which groundwater moves are often very small and variable, so the rate of movement is quite slow, except in some large solution channels within limestones. Furthermore, the lack of oxygen in groundwater kills the aerobic (oxygen-requiring) microorganisms that help degrade pollutants but may provide a happy home for anaerobic varieties that live in oxygen-deficient environments.

The often long residence time for groundwaters (hundreds to thousands of years) reflects the deep, insulated type of storage that aquifers provide. Not all groundwater takes hundreds of years to rejoin the other, more rapidly moving parts of the hydrologic cycle, but most of it is well below the influence of transpiration by plants and evaporation into the atmosphere. Where it is not that deep, it is most susceptible to evapotranspiration, discharge to streams, and use or abuse by humans. The latter is of increasing concern because of the potential long-term damage to this resource, the high cost of cleaning up polluted groundwater, and the increasing need to use it as per capita water use increases.

Exchanges Between Groundwater and Its Surroundings

The soil, sediment, and rocks through which groundwater passes may act as natural filters. The water may actually exchange materials with the soil and rock. Under the right conditions, this filtering system cleanses the water, trapping and biodegrading disease-causing microorganisms and particulates that contain toxic compounds. However, if the soil or rock surface is already highly contaminated or contains naturally occurring toxic elements such as arsenic, the natural exchange processes may make the water toxic.

An interesting example with serious environmental implications comes from the western San Joaquin Valley in California, where selenium, a very toxic heavy metal present in the soil, is released by application of irrigation waters. Subsurface drainage of the selenium-rich water from fields has entered the surface waters and has caused birth defects in waterfowl. The extent of the general problem related to agriculture drainage water is only now being learned. Selenium is also toxic to people. Like many trace metals, selenium has a dual character: It is necessary for life processes at trace concentrations, but it is toxic at some higher concentrations. The environmental impact of the selenium problem is discussed in Chapter 18.

Groundwaters moving through rock and soil dissolve a mixture of minerals and some gases that can be nuisances to some human uses. Some examples are iron as ferrous hydroxide, which colors the water brown and leaves a brown discoloration on laundry and plumbing fixtures; calcium, which creates in part the so-called hardness of water; and hydrogen sulfide, which produces a "rotten egg" odor.

The ability of most soils and rocks to filter out solids, including pollution solids, by physical means is well recognized. This ability varies with different sizes, shapes, and arrangements of filtering particles, as evidenced in the use of selected sands and other materials in water filtration plants. Also known, but perhaps not so generally, is the ability of clays and other selected minerals to capture and exchange some elements and compounds when they are dissociated in solutions as positively or negatively charged elements or compounds. Such exchanges, along with sorption and precipitation processes, are important in the capture of pollutants. These processes, however, have definable units of capacity and are reversible. They also can be overlooked easily in designing facilities to correct pollution problems by relying on soils and rocks of the geologic environment for treatment. This oversight, which can result in possible groundwater pollution, is especially significant in land application of wastewaters.

Salt Water Intrusion

Aquifer pollution is not solely the result of disposal of wastes on the land surface or in the ground. Overpumping or mining of groundwater so that inferior waters migrate from adjacent aquifers or the sea can also cause contamination problems. Hence, human use of public or private water supplies can accidentally result in aquifer pollution. Intrusion of salt water into freshwater supplies has caused problems in coastal areas of New York, Florida, and California, among other areas (including many islands) (see *Case History: The Threatened Groundwater of Long Island*).

Figure 11.7 illustrates the general principle of saltwater intrusion. The groundwater table generally is inclined toward the ocean, while a wedge of salt water is inclined toward the land. Thus, with no confining layers, salt water near the coast may be encountered at depth. Because fresh water is slightly less dense than salt water (1.000 compared to 1.025 g/cm^3), a column of fresh water 41 cm high is needed to balance 40 cm of salt water. A more general relationship is that the depth to salt water below sea level is 40 times the height (H in Figure 11.7) of the water table above sea level. When wells are drilled, a cone of depression develops in the freshwater table, which may allow intrusion of salt water as the interface between fresh and salt water rises (forming a core of ascension) in response to the loss of fresh-water mass.

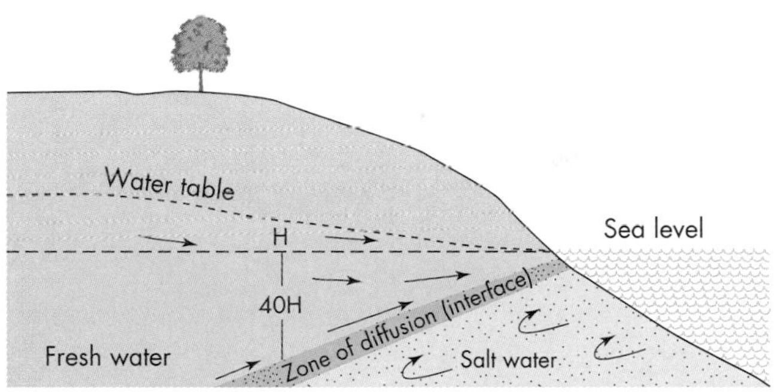

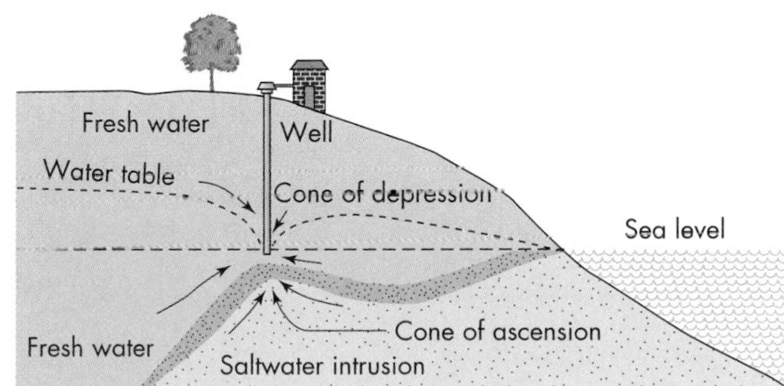

▲ **FIGURE 11.7** How saltwater intrusion might occur: The upper drawing shows the groundwater system near the coast under natural conditions, and the lower drawing shows a well with both a cone of depression and a cone of ascension. If pumping is intensive, the cone of ascension may be drawn upward, delivering salt water to the well. The H and 40 H represent the height of the freshwater table above sea level and the depth of salt water below sea level, respectively.

CASE HISTORY The Threatened Groundwater of Long Island

Long Island, New York, provides a good example of an area with groundwater problems. Two counties on the island, Nassau and Suffolk, with a population of several million people, are entirely dependent on groundwater for their water supply. In terms of the total volume of water and number of people who use it, the groundwater resource for Long Island is one of the world's largest, but it is threatened by two major problems, intrusion of salt water and shallow-aquifer contamination, particularly in Nassau County (15).

Figure 11.D shows the general movement of groundwater under natural conditions for Nassau County. Salty groundwater is restricted from inland migration by the large wedge of

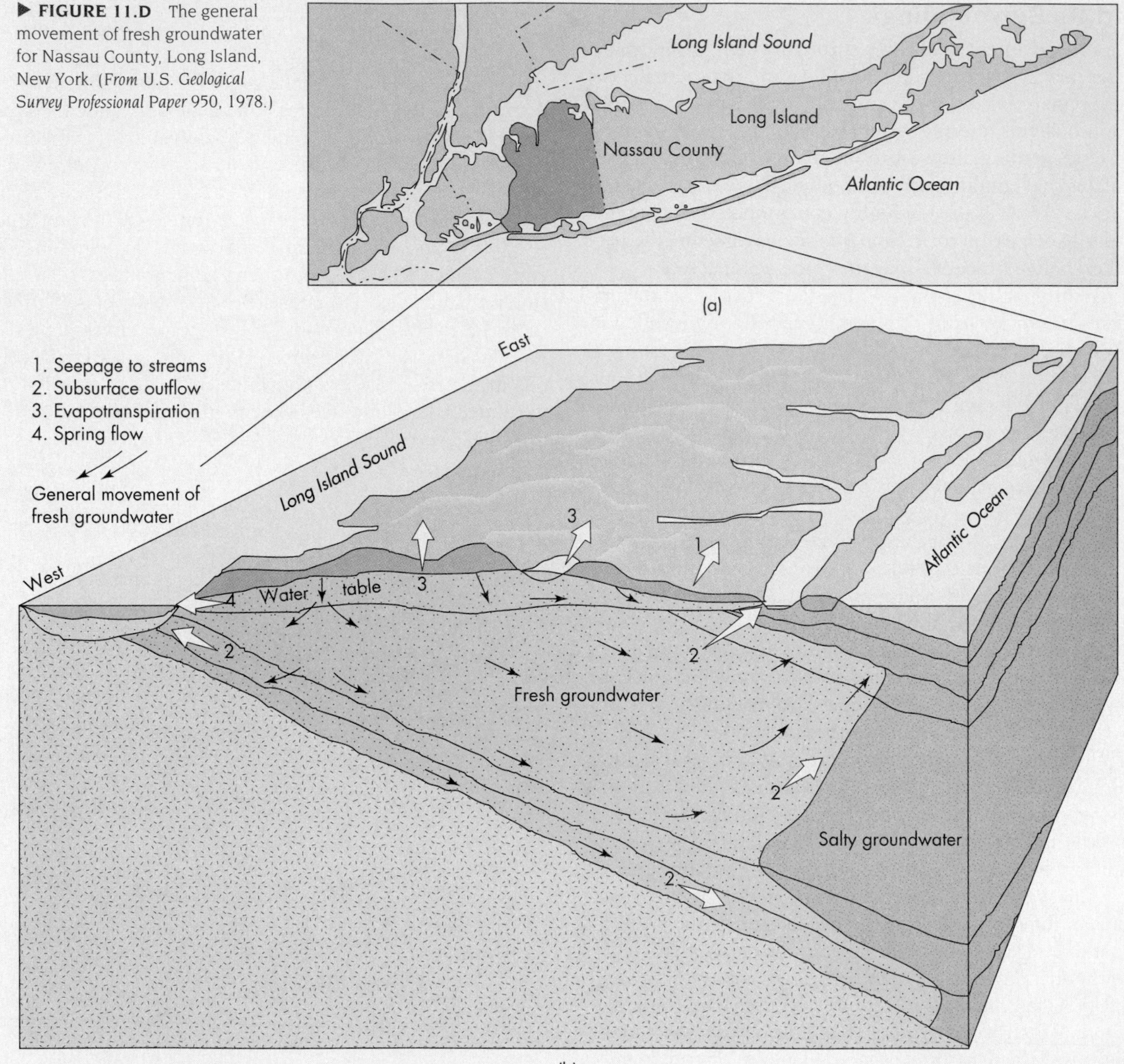

▶ **FIGURE 11.D** The general movement of fresh groundwater for Nassau County, Long Island, New York. (From U.S. *Geological Survey Professional Paper* 950, 1978.)

Groundwater Treatment

In view of the difficulty of detecting groundwater pollution, the long-term residency of groundwater, the degradation of the polluted aquifer, and the difficulty and expense of aquifer recovery, a strong argument can be made that no wastes or possible pollutants should be allowed to enter any part of the groundwater system. This is an impossible dream. Rather, the response to groundwater pollution must be to learn more about how natural processes treat wastes, so that when soil and rocks cannot treat, store, or recycle wastes, we can develop processes to make the pollutants treatable, storable, or recyclable.

Correction of aquifer and vadose zone contamination is not impossible, though it may be a complex and expensive

fresh water moving beneath the island. Groundwater resources of Nassau County are in two main aquifers, as shown in Figures 11.D and 11.E. The upper aquifer is composed of young glacial deposits that yield large amounts of water at depths of less than 30 m. Below the glacial deposits are older marine sedimentary rocks consisting of interbedded sands, clays, and silts (the Magothy Aquifer). Most of the fresh water in Nassau County is pumped from this lower aquifer, from sandy beds at depths below 30 m. Most of the water-bearing sands are confined by overlying clay and silt beds of low permeability. That is, the aquifer is composed of alternating and discontinuous layers of sand and finer-grained silt and clay. Because of the confining layers in the aquifer the water is under artesian pressure, which causes it to rise to within 15 m of the surface in wells.

Despite the huge quantities of water in Nassau County's groundwater system, intensive pumping in recent years has caused water levels to decline as much as 15 m in some areas. As groundwater is removed near coastal areas, the subsurface outflow to the ocean decreases, allowing salt water to migrate inland. Saltwater intrusion in the deep aquifer has occurred in Nassau County. The mechanism of salt intrusion is more complex than the idealized mechanism shown in Figure 11.7. As fresh water from sandy beds in the deep aquifer is intensely pumped, salt water is drawn inland as a series of na wedges. Although the problem of saltwater intrusion is n yet widespread, the saltwater front is being carefully monitored as part of a comprehensive management program.

The most serious groundwater problem on Long Island is shallow-aquifer pollution associated with urbanization. Sources of pollution in Nassau County include urban runoff, household sewage from cesspools and septic tanks, salt used to deice highways, and industrial and solid waste. These pollutants enter surface waters and then migrate downward, especially in areas of intensive pumping and declining groundwater levels. Figure 11.E shows the extent of high concentration of dissolved nitrate in deep groundwater zones. The greatest concentrations are located beneath densely populated urban zones, where water levels have dramatically declined and where nitrates from such sources as cesspools, septic tanks, and fertilizers are routinely introduced into the hydrologic environment (15). Landfills, sites where urban waste are buried, have been of particular concern because urban waste (garbage) often contains many pollutants. When landfills are located on sandy (permeable) soil over a shallow aquifer, groundwater pollution is inevitable. As a result, most landfills on Long Island have been closed.

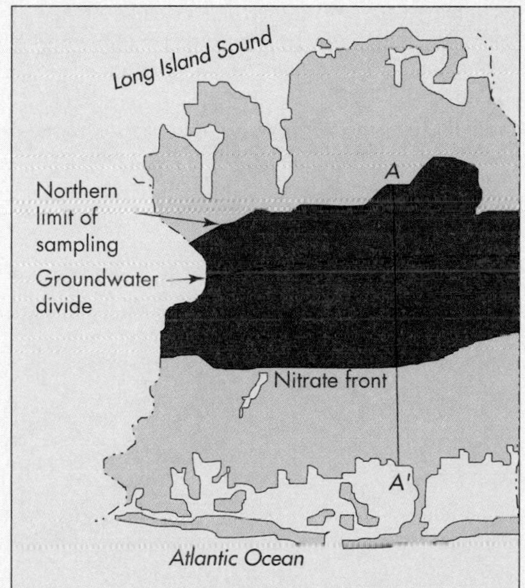

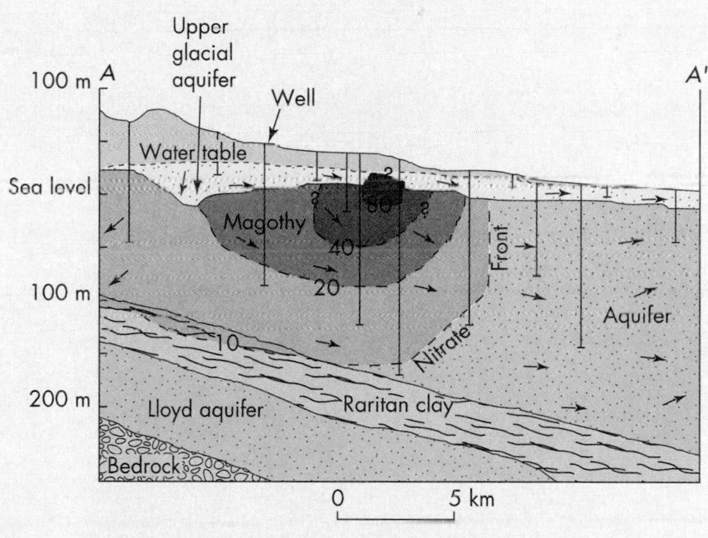

▲ **FIGURE 11.E** Extent of high concentration of dissolved nitrate in groundwater zone, Nassau County, Long Island, New York. The greatest concentrations are located beneath densely populated urban zones, where water levels have dramatically declined and nitrates are more abundant because of urban waste disposal and horticulture practices. Contours shown in milligrams per liter of dissolved nitrate. (From *U.S. Geological Survey Professional Paper* 950, 1978)

problem requiring careful evaluation and treatment. Important steps involved in correcting a groundwater pollution problem are:

- **Characterizing the geology.** This is particularly important because features such as more permeable buried channels, soil macropores, and geologic structures such as fractured, folded, and faulted rocks may be the dominant factors controlling the direction of groundwater flow.
- **Characterizing the hydrology.** Factors such as depth to groundwater, direction of flow, and rate of flow must be determined. Characterizing the hydrology

tifying relationships between surface water processes affecting the site.

:ontaminants and their transport ıinants are identified through careful gathering of samples. Some conta- oline, are floaters; that is, most of the gasoline will be found on top of the water table because it is lighter than water. However, some components in gasoline are soluble in water, so there will also be a dissolved phase below the water table. On the other hand, contaminants such as trichloroethylene (TCE), which is a dry-cleaning solvent that is heavier than water, will sink rather than float. Other pollutants, such as some salts, are very soluble in water and will move with the general flow of the groundwater environment.

- **Initiating the treatment process.** Table 11.2 briefly outlines some of the methods for treating groundwater and vadose zone water. The specific treatment selected depends upon variables such as type of contaminant, method of transport, and characteristics of the local environment, such as depth to water table and geologic characteristics.

As an example of groundwater treatment, consider Figure 11.8. Figure 11.8a shows a site with a service station and underground gasoline tank that is leaking. Most of the contaminant is floating on top of the water table, but some is also dissolved and moves with the groundwater. The direction of the migrating vapor phase is away from the leakage plume. Figure 11.8b shows the same location after a system consisting of dewatering wells and a vapor extractor well has been installed. The dewatering wells lower the groundwater table locally, and the vapor extraction well, which uses a vacuum pump, collects the contaminant in a vapor phase, where it may be treated.

Underground gasoline tanks that leak are a very common phenomena in today's urban environment. In recent years regulation of underground tanks has been tightened. It

Table 11.2 Methods of treating groundwater and vadose zone water

Extraction Wells	Vapor Extraction	Bioremediation	Permeable Treatment Bed
Pump out contaminated water and treat it by filtration, oxidation, or air stripping (volatilization of contaminant in an air column), or by biological processes.	Uses vapor extraction well and then treatment.	Injection of nutrients and oxygen to encourage growth of organisms that degrade the contaminant in the groundwater.	Provides contact treatment as contaminated water plume moves through a treatment bed in the path of groundwater movement. Encourages neutralization of the contaminant by chemical, physical, or biological processes.

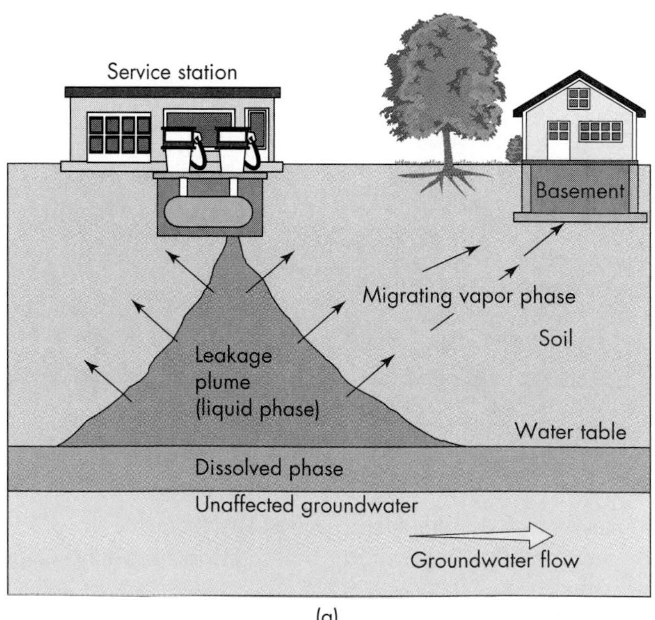

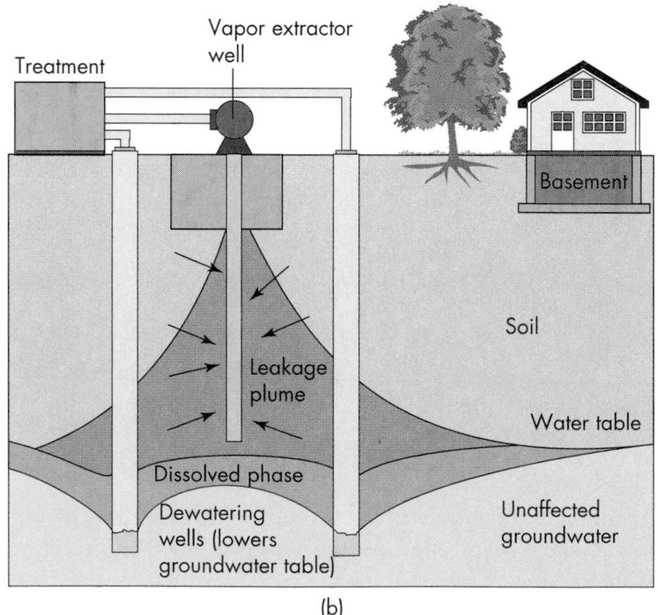

▲ **FIGURE 11.8** Idealized diagram of a leaking buried tank (a) and possible remediation method (b). See text for explanation. (*Courtesy of University of California–Santa Barbara Vadose Zone Laboratory and David Springer.*)

is not uncommon to see drill rigs testing gasoline station sites and to notice later that the tanks have been excavated and treatment for leaking gasoline has begun. In a particular pollution case it may be difficult to show where a contaminant such as gasoline has come from. At many intersections there may be two or more gasoline stations that have had a series of buried tanks over a long period of time. Litigation over responsibilities concerning groundwater pollution from leaking underground tanks may be difficult to resolve.

11.5 Water Quality Standards

A question people commonly ask is: How safe is our water supply? Americans are used to believing that their drinking water is of high quality, some of the best in the world. For the most part this is true, but in recent years we have gained an ability to detect specific contaminants in parts per billion (ppb) of water, or in some cases even parts per trillion (ppt). The question that then arises is: How dangerous might some of these chemicals be? You may think that such small amounts of contaminants cannot possibly be dangerous, but as the U.S. Environmental Protection Agency (EPA) reminds us, a single microscopic virus can cause a disease. Physicians are able to delineate fairly clearly what diseases are caused by particular "bugs," but we are less sure about the effects of long-term exposure to very small amounts of chemicals.

In response to this concern, Congress has mandated the EPA to establish minimum national drinking water standards for a variety of chemicals and other materials. In 1986 Congress expanded the Safe Drinking Water Act of 1974 to include 83 contaminants, for 26 of which the EPA had already set **Maximum Contaminant Levels (MCLs).** Among other regulations, the new legislation banned the use of lead in the installation or repair of water systems used for drinking water. Health effects associated with lead toxicity are very well known. At high concentrations, lead causes damage to the nervous system and the kidneys and is particularly toxic to infants and pregnant women (16).

The EPA was also required by the 1986 amendment to issue **Maximum Contaminant Level Goals (MCLGs)** along with an MCL. The MCLGs, which were recognized as an unenforceable health goal, were set at the maximum level at which a particular contaminant was not expected to cause adverse health effects over a lifetime of exposure. By law MCLs must be set as close to the MCLG as economics and technology allow (16).

The EPA has set standards for a number of contaminants that might possibly be found in our drinking water. However, only two substances for which these standards have been set are thought to pose an immediate health threat when standards are exceeded. These are (16):

- **Coliform bacteria**—because they may indicate that the water is contaminated by harmful disease-causing organisms.
- **Nitrate**—because contamination above the standard is an immediate threat to young children. In youngsters under a year old, high levels of nitrate may react with their blood to produce an anemic condition known as "blue baby."

Table 11.3 is an abbreviated list enumerating some of the contaminants covered by the National Primary Drinking Water Standards. A complete list can be obtained from the U.S. Environmental Protection Agency. The purpose of the standards and regulations concerning drinking water are (16):

- To ensure that our water supply is treated to remove harmful contaminants.
- To regularly test and monitor the quality of our water supply.
- To provide information to citizens so that they are better informed concerning the quality and testing of their water supply.

How are we doing in the effort to reduce water pollution and improve water quality in the United States? Regulation of toxic chemicals in our water supply has only been going on for a few decades, but there has been progress. Figure 11.9 shows trends in water quality from 1970 to 1989, based on data collected from the U.S. Fish and Wildlife Service monitoring stations. These data suggest that concentrations of selected toxic metals and toxic organic compounds (in fish tissue) have been significantly reduced. Organic compounds such as DDT and PCBs that have been regulated the longest show the greatest decrease. On the other hand, new organic chemicals found in herbicides are present in some areas in concentrations that exceed established health limits (1).

11.6 Wastewater Treatment

Water that is used for municipal and industrial purposes is often degraded during use by a variety of contaminants, including oxygen-demanding materials, bacteria, nutrients, salts, suspended solids, and a variety of other chemicals. In the United States our laws dictate that these contaminated waters must be treated before they are released back into the environment. The annual cost for such treatment is approximately $20 billion, and will increase during the next decade. Because so much money is involved, wastewater treatment is big business. In rural areas the conventional method of treatment is by way of septic-tank disposal systems. In larger communities, wastewaters are generally collected and centralized in water-treatment plants that collect the wastewater from a sewer system.

In many parts of the country, water resources are being stressed, and as a result innovative systems are being developed to reclaim wastewaters so that they can be used for such purposes as irrigating fields, parks, or golf courses, rather than being discharged into the nearest body of water. New technologies are also being developed for treating wastewaters not as a waste but as a resource to be used. Those developing the new technologies say that sewage

Table 11.3 National Primary Drinking Water Standards: Some examples

Contaminant	Maximum Contaminant Level (mg/l)	Comments/Problems
Inorganics		
Arsenic	0.05	Highly toxic
Cadmium	0.01	Kidney
Lead	0.015[a]	Highly toxic
Mercury	0.002	Kidney, nervous system
Selenium	0.01	Nervous system
Asbestos	7 MFL[b]	Benign tumors
Fluoride	4	Skeletal damage
Organic chemicals		
Pesticides		
Endrin	0.0002	Nervous system, kidney
Lindane	0.004	Nervous system, kidney, liver
Methoxychlor	0.1	Nervous system, kidney, liver
Herbicides		
2,4D	0.07	Liver, kidney, nervous system
Silvex	0.05	Nervous system, liver, kidney
Volatile organic chemicals		
Benzene	0.005	Cancer
Carbon tetrachloride	0.005	Possible cancer
Trichloroethylene	0.005	Probable cancer
Vinyl chloride	0.002	Cancer risk
Microbiological organisms		
Fecal coliform bacteria	1 cell/100 ml	Indicator—disease-causing organisms

[a]The action level for lead related to treatment of water to reduce lead to the safe level. There is no MCL for lead.

[b]Million fibers per liter with fiber length >10 microns.

Source: U.S. Environmental Protection Agency.

treatment sites should not have to be hidden from the public. Rather, we should come to expect sewage to be reclaimed at small cost while producing flowers and shrubs in a parklike setting (17).

Septic-Tank Sewage Disposal

Population movement in the United States continues to be from rural to urban, or urbanizing, areas. Although the most satisfactory method of sewage disposal is through municipal sewers and sewage-treatment facilities, construction of an adequate sewage system often has not kept pace with growth. As a result, the individual **septic-tank** disposal system continues to be an important method of sewage disposal. There are more than 22 million systems in operation, and about a half-million new systems are added each year. As a result, septic systems are used by about 30 percent of the people in the United States (18). Not all land, however, is suitable for installation of a septic-tank disposal system, so evaluation of individual sites is necessary and often required by law before a permit can be issued.

The basic parts of a septic-tank disposal system are shown in Figure 11.10. The sewer line from the house or small business leads to an underground septic tank in the yard. Solid organic matter settles to the bottom of the tank, where it is digested and liquefied by bacterial action. The clarified liquid discharges into the drain field, a system of piping through which it seeps into the surrounding soil. As the water moves through the soil, it is further treated and purified by natural processes of filtering and oxidation.

Geologic factors affecting the suitability of a septic-tank disposal system include type of soil, depth to the water table, depth to bedrock, and topography. These variables are generally listed, with soil descriptions associated with a soil survey of a county or other area. Soil surveys are published by the Soil Conservation Service and are extremely valuable in interpreting possible land use, such as suitability for a septic system. However, the reliability of a soils map for predicting limitations of soils is limited to an area larger than a few thousand square meters, and soil types can change within a few meters, so it is often necessary to have an on-

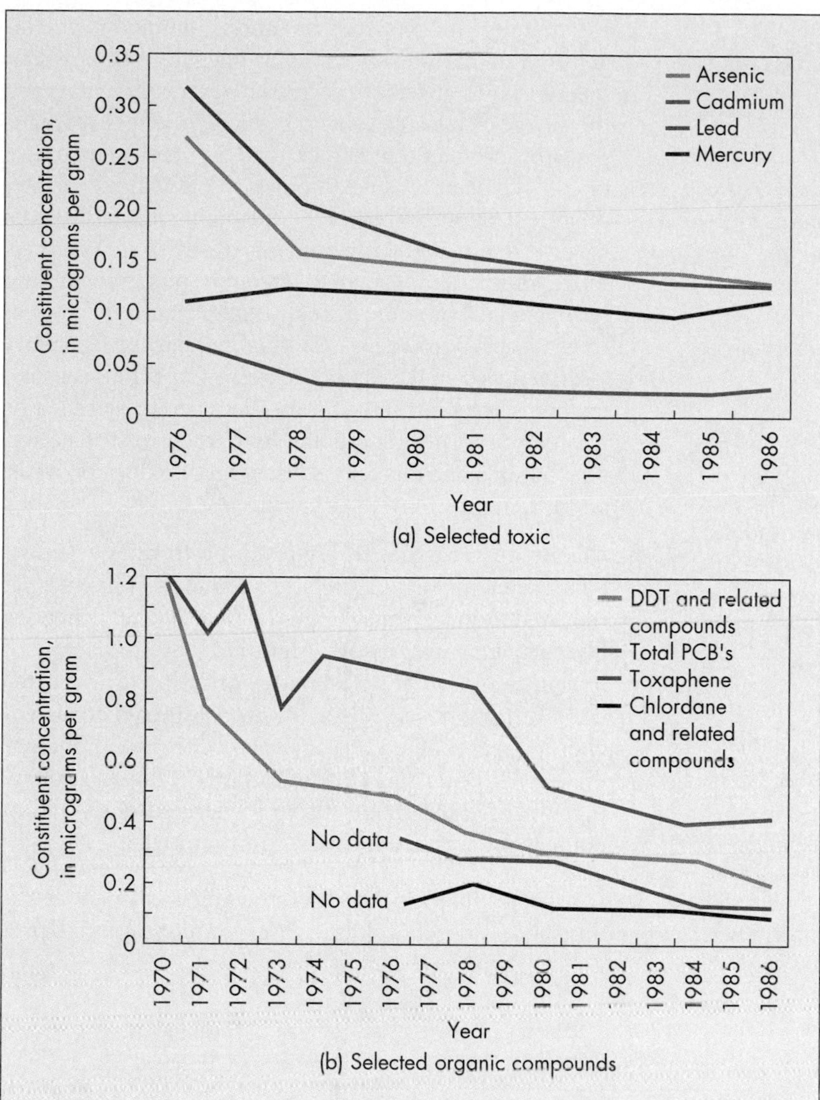

◀ **FIGURE 11.9** Trends from 1970 to 1989 in the concentration of selected toxic metals (a) and toxic organic chemicals (b) in fish tissue, measured at monitoring stations by U.S. Fish and Wildlife Service. (*Source*: R. A. Smith, 1994. *Water quality and health.* Geotimes 39(1):14–21.)

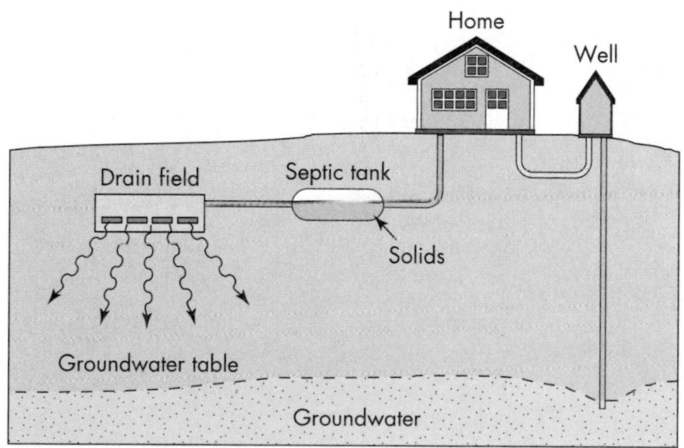

▲ **FIGURE 11.10** Idealized diagram showing septic-tank sewage disposal system.

site evaluation by a soil scientist or soils engineer. To calculate the size of the absorption field needed, it is necessary to know the rate at which water moves through the soil, which is best determined by a percolation test.

Sewage absorption fields may fail for several reasons. The most common cause is poor soil drainage, which allows the effluent to rise to the surface in wet weather. Poor drainage can be expected in areas with clay or compacted soils with low hydraulic conductivity, and in areas that have a high water table, rock with low hydraulic conductivity near the surface, or frequent flooding.

When septic systems fail, waste materials often surface above the drainage field, producing a potential health hazard. This sort of failure is easy to see. Unfortunately, what is happening beneath the ground is not so easy to see, and if extensive leaching of waste occurs, then groundwater resources may be polluted. Of particular concern are septic systems that serve small commercial and industrial activities. These tend to cause more severe problems of groundwater pollution than do septic systems for homes because of the potentially hazardous nature of waste disposed by these activities. Possible contaminants include nutrients such as nitrates; heavy metals such as zinc, copper, and lead; and synthetic organic chemicals such as benzene, carbon tetrachloride, and vinyl chloride. In recent years the EPA has identified a number of commercial and industrial septic systems that have caused sufficient water pollution that cleanup has been necessary (18).

Wastewater Treatment Plants

The main purpose of wastewater treatment is to reduce the amount of suspended solids, bacteria, and oxygen-demanding materials in wastewater. In addition, new techniques are being developed to remove nutrients and harmful dissolved inorganic materials that may be present.

Existing wastewater treatment generally has two or three stages (Figure 11.11):

- **Primary treatment.** This includes screening, which removes the grit (sand, stones, and other large particles); and sedimentation, in which much of the remaining particulate matter settles out to form a mudlike sediment called *sludge*. The sludge is piped to the digester, and the partially clarified wastewater goes on to the secondary stage of treatment. Primary treatment removes 30 to 40 percent of the pollutants from the wastewater (19).
- **Secondary treatment.** The most common secondary treatment is known as *activated sludge*. Wastewater from primary treatment enters the aeration tank, where air is pumped in and aerobic (oxygen-requiring) bacteria break down much of the organic matter remaining in the liquid. This process takes several hours, after which the wastewater is then pumped to the final sedimentation tank, where more sludge settles out and is pumped to the digester. The digester provides an oxygen-poor environment in which anaerobic bacteria digest organic matter in the sludge. This anaerobic digestion produces methane gas, a by-product that can be used as a fuel to help heat or cool the plant or run equipment. Following secondary treatment, about 90 percent of the pollutants in the waste have been removed. However, this treatment does not remove all nutrients, such as nitrogen, phosphorus, and heavy metals, or some human-made chemicals, such as solvents and pesticides (19). The final part of secondary treatment is disinfection of the wastewater. This is usually done with chlorine, but sometimes ozone is used. The treated wastewater is usually discharged to surface waters, but in some places it is discharged to disposal wells, as, for example, in Maui, Hawaii.
- **Advanced treatment.** This is done to remove nutrients, heavy metals, or specific chemicals. This may be required if higher-quality treated wastewater is needed for particular uses as, for example, wildlife habitat or irrigation of golf courses, parks, or crops. The treated wastewater for such uses is often referred to as **reclaimed water.** Methods of advanced treatment include use of chemicals, sand filters, or carbon filters. Following advanced treatment, up to 95 percent of the pollutants in the wastewater have been removed.

A troublesome aspect of wastewater treatment is the handling and disposal of sludge. The amount of sludge pro-

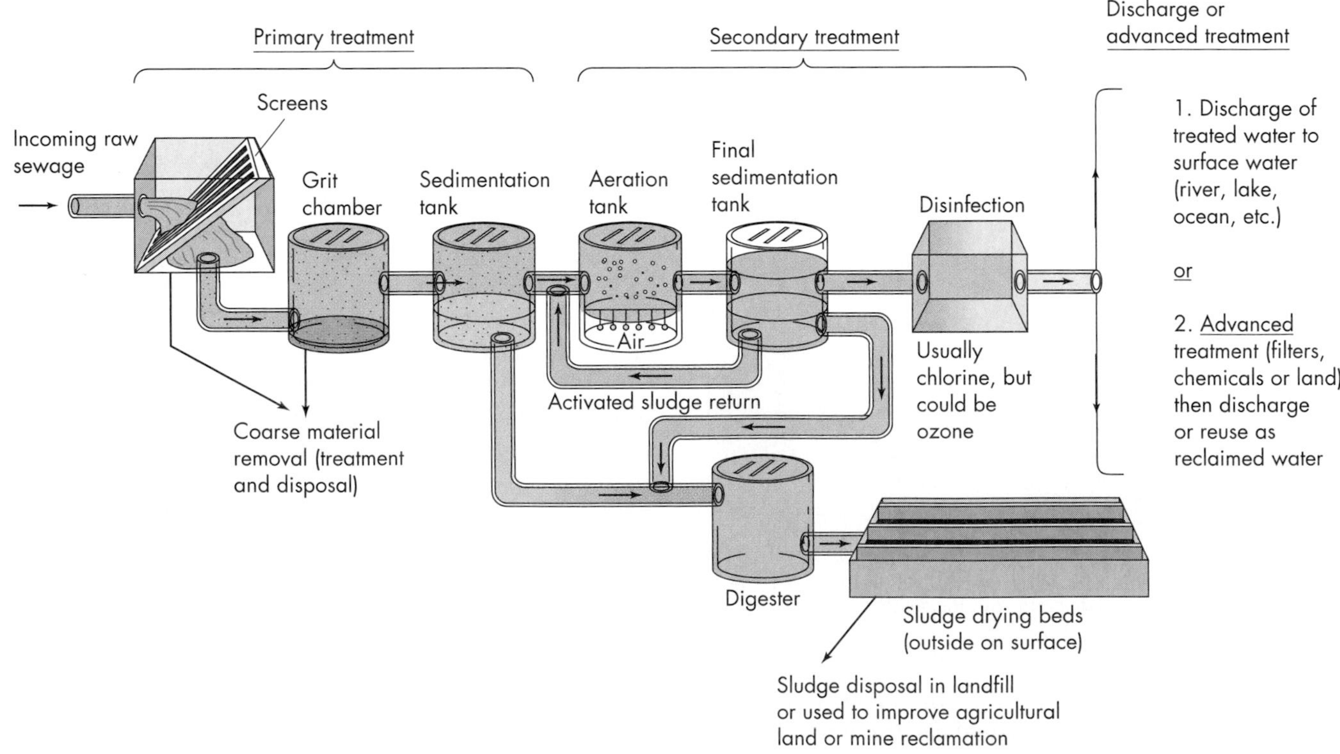

▲ **FIGURE 11.11** Idealized diagram showing activated sludge sewage treatment with (or without) advanced treatment.

duced in the treatment process is conservatively estimated at about 54 to 112 grams per person per day, and sludge disposal accounts for 25 to 50 percent of the capital and operating cost of a treatment plant. Sludge handling and disposal have four main objectives (20):

- To convert the organic matter to a relatively stable form
- To reduce the volume of sludge by removing liquid
- To destroy or control harmful organisms
- To produce by-products whose use or sale reduces the cost of processing the sludge

Final disposal of sludge is accomplished by incineration, burying it in a landfill, using it for soil reclamation, or dumping it in the ocean. From an environmental standpoint, the best use of sludge is to improve soil texture and fertility in areas disturbed by activity, such as strip mining and poor soil conservation.

Although it is unlikely that all the tremendous quantities of sludge from large metropolitan areas can ever be used for beneficial purposes, many smaller towns and many industries, institutions, and agricultural activities can take advantage of municipal and animal wastes by converting them into resources.

Wastewater Renovation

The process of recycling liquid waste, called the **wastewater renovation and conservation cycle,** is shown schematically in Figure 11.12. The major processes in the cycle are return of the treated wastewater to crops by sprinkler or other irrigation system; renovation, or natural purification by slow percolation of wastewater through soil to eventually recharge the groundwater resource with clean water; and reuse (conservation) of the water by pumping it out of the ground for municipal, industrial, institutional, or agricultural purposes (21). Of course, not all aspects of the cycle are equally applicable to a particular wastewater problem. Wastewater renovation from cattle feedlots differs considerably from renovation of water from industrial or municipal sites. But the general principle of renovation is valid, and the processes are similar in theory.

The return and renovation processes are crucial to wastewater recycling, and soil and rock type, topography, climate, and vegetation play significant roles. Particularly important factors are the ability of the soil to safely assimilate waste, the ability of the selected vegetation to use the nutrients, and knowledge of how much wastewater can be applied (21).

Wastewater is recycled on a large scale near Muskegon and Whitehall, Michigan. Raw sewage from homes and industries is transported by sewers to the treatment plant, where it receives primary and secondary treatment. The wastewaters are then chlorinated and pumped into a piping network that transports the effluent to a series of spray irrigation rigs. After the wastewater trickles down through the soil, it is collected in a network of tile drains and transported to the Muskegon River for final disposal. This last step is an indirect advanced treatment, using the natural environment as a filter.

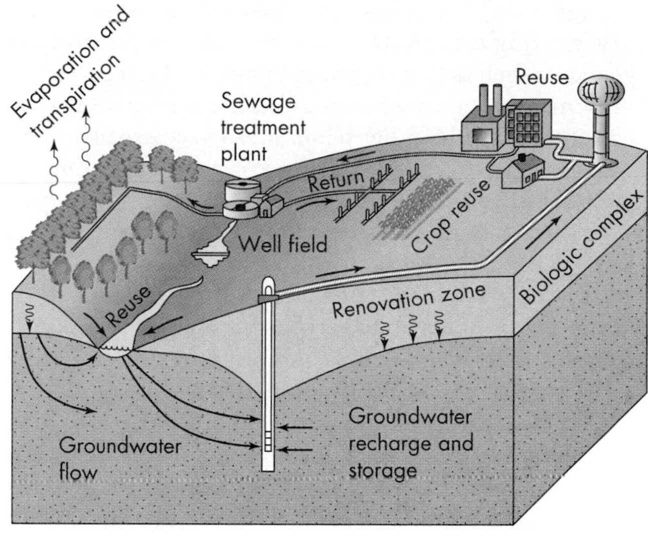

▲ **FIGURE 11.12** The wastewater renovation and conservation cycle. *(From R. R. Parizek and E. A. Myers. 1968. American Resources Administration.)*

The Michigan project is controversial because of concern for possible pollution of surface water and groundwater, as well as problems associated with an elevated groundwater table. However, it provides a possible alternative to direct (at a treatment plant) advanced treatment, and experience gained from this project has been valuable in evaluating other possible sites for recycling wastewater. The system effectively removes most of the potential pollutants, as well as heavy metals and viruses.

In Clayton County, Georgia, just south of Atlanta, a large water renovation and conservation cycle project was recently initiated. The project handles up to 760,000 m^3 (20 million gallons) of wastewater per day, which is applied to a 972-ha (2400-acre) pine forest. Trees will be harvested on a 20-year rotation. The forest is part of the watershed that supplies water to the area; therefore, wastewater is recycled to become part of the drinking-water supply (22).

Wastewater Treatment as Resource Recovery

At the beginning of this section, we said that we hoped people would some day look at raw sewage as a resource and that treatment plants could be constructed in a parklike setting. Pioneering work in this area has been done in Arcata, California, located on Humboldt Bay. For secondary and advanced treatment of wastewater, this community has constructed oxidation ponds that form part of a large wetland system in the bay. Water drawn from the oxidation ponds has also been used to rear Pacific salmon fingerlings. Thus, the wastewater treatment scheme utilizes wetlands as part of the treatment and can produce a resource—in this case Pacific salmon—that are released into the ocean.

The United States has developed a tremendous capacity for treating wastewater. The treatment is primarily in large plants such as that shown in Figure 11.11. This sort of treatment is certainly effective and has a relatively good track record. On the other hand, it is a very expensive endeavor and failures of the system are certainly not unheard of, particularly when the systems are stressed from external factors such as high rates of input of raw sewage during floods. Chlorination in the final stages of secondary treatment is effective in killing pathogens, but it produces toxic chlorine compounds as by-products, some of which are known to cause cancer. Finally, following treatment, a large amount of sludge must be disposed of (17).

We must constantly ask our technology the following question: Is there a better, more economic, and environmentally preferred method? For wastewater treatment, that question cannot yet be answered. However, experiments are being conducted to test the hypothesis that a more environmentally preferable resource recovery system of wastewater treatment might be possible. By *resource recovery* we mean that the process of treatment would produce resources such as methane gas (which can be used as a fuel) and production of ornamental plants or other plants that have commercial value. Figure 11.13 is an idealized diagram of a pilot plant that illustrates treatment starting with screening and filtration. The next and often final steps are anaerobic treatment followed by a process known as *nutrient-film treatment* (either outdoors or in a greenhouse). In nutrient-film treatment, nutrient-rich wastewater flows in a thin film over the inclined surfaces of plant beds. This constitutes the secondary treatment. Additional advanced treatment can be obtained by further biological purification utilizing other plants. The anaerobic bacteria that perform the primary wastewater treatment also produce methane gas.

Results of pilot studies suggest resource-recovery plants produce a relatively small amount of sludge and that the treated wastewater is of high quality (meets secondary treatment standards). Some of the problems that this new technology face in being more widely used are (17):

- We have a tremendous investment in more traditional wastewater treatment plants and are familiar with them.
- There is a general lack of economic incentives to provide for new technologies.
- There is a general lack of personnel capable of designing, building, and operating these systems. However, as more universities are developing true environmental engineering programs, this problem may be rectified.

11.7 Water Law and Federal Legislation

Water is absolutely necessary to life and to all aspects of human use of the land. As a result, water resources are probably the most legislated and discussed commodity in the area of envi-

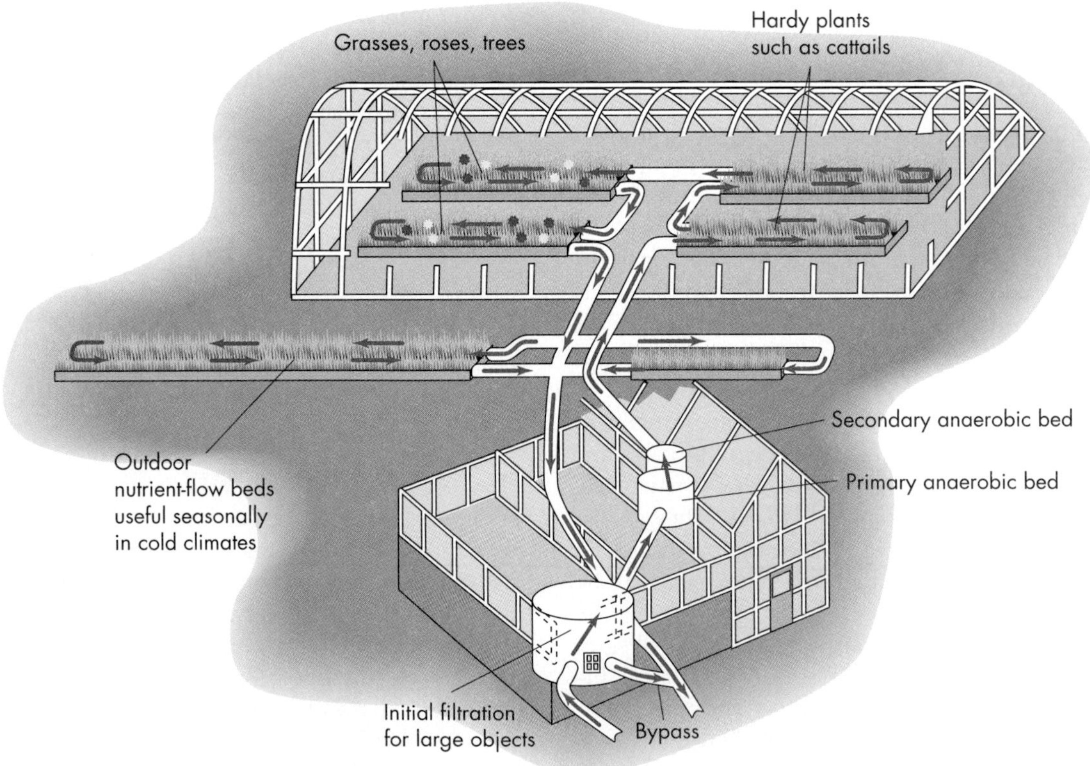

▲ **FIGURE 11.13** Idealized diagram illustrating process of resource-recovery wastewater treatment. The recovery is of two types: methane (energy) from the anaerobic beds; and ornamental plants, flowers, etc. (*After* W. J. Jewell. 1994. *Resource-recovery wastewater treatment*. American Scientist 82[4]:366–75.)

ronmental law. In the United States today, an elaborate framework of state water laws surrounds the use of surface waters and groundwaters. In addition, the federal government has attempted to regulate water quality by means of legislation.

Surface-Water Law

American states generally fall into one of two camps with respect to surface-water rights: those that apply the **Riparian Doctrine,** and those that apply the **Prior Appropriation Doctrine** (23).

The **Riparian Doctrine** was the prevailing water law in most states prior to 1850 and is still used in most of the eastern half of the United States. Riparian rights to water are restricted for the most part to owners of the land adjoining a stream or body of standing water. It is important to keep in mind that a water right is not a legal title to the water—it is simply the legal right to use water in a manner dictated by law (23). Therefore, under the Riparian Doctrine, the right to use water is considered real property, but the water itself does not belong to the property owner. Riparian water rights are considered property that enters into the value of the land and may be transferred, sold, or granted to other people (24). Under riparian rights, landowners have the right to make reasonable use of water on their land, provided the water is returned to its natural stream before it leaves the property. A property owner also has the right to receive the full flow of the stream undiminished in quantity and quality but is not entitled to make withdrawals of water that infringe upon the rights of other riparian owners (25).

The **Prior Appropriation Doctrine** in water law holds that prior usage is a significant factor. That is, the first person to divert and use water from a surface water supply has the primary water right and this may be passed to successive owners. Furthermore, the right to use water is separate from other property rights (23). Appropriation water law is common in the western part of the United States, and generally states with the least abundant water supply must manage their water most closely. For example, the state of Arizona, with an average precipitation of less than 38 cm/yr, must manage water very closely indeed. As a result, the Arizona state constitution states that riparian water rights are not authorized and declares all water subject to appropriation. Preferred uses are domestic, municipal, and irrigation (24).

Comparison of the two doctrines suggest that management of water resources is considerably more effective when the principles of appropriation are applied. Because riparian law requires judicial decision, it is therefore subject to possible variations and interpretations in different courts. As a result, property owners are never sure of their position. The riparian system also tends to encourage nonuse of water and thus is counterproductive in times of shortage. On the other hand, states with appropriation systems have the power to make and enforce regulations based on sound hydrologic principles that are more likely to lead to effective management of water resources (24).

Court decisions concerning water use and the environment have also been involved with the government's obligation to protect our common heritage, including ecosystems. This is known as the *Public Trust Doctrine* (see *Case History: Mono Lake and the Public Trust Doctrine*).

Groundwater Law

In the United States, laws governing groundwater use go back to the right of absolute ownership of the water beneath a particular person's land. This doctrine is known as the **English Rule,** or the **Absolute Ownership Doctrine.** Under this doctrine landowners could pump at will and take as much water as they wished, even though that water was shared in a common groundwater aquifer with adjacent landowners. This sort of arrangement works pretty well in a region with a wet, humid climate, such as England or the eastern United States, where there is usually plenty of water. Even in the eastern United States, however, problems may arise during drought conditions.

In the western United States, where water is a much more scarce resource, it became apparent early on that absolute ownership led to major problems, and legal modifications were made that limited property owners' rights to groundwater. Under one such modification, known as the **American Rule,** or the **Reasonable Use Doctrine,** the amount of groundwater withdrawn is based upon the reasonable and beneficial purposes the water is used for on the land above the aquifer. Establishing what is reasonable use may be difficult, however. Problems also arise from the fact that the doctrine is applied through a system of laws regulating the issuing of pumping permits (23,26). California has developed what is known as the **Correlative Rights Doctrine** as a reasonable alternative to the idea of absolute ownership of the groundwater. This doctrine recognizes a landowner's right to use water beneath the land, but it limits these rights by making provisions for other landowners whose property overlies a common groundwater source. All of the landowners have equal or correlative rights to a reasonable amount of groundwater when that water is applied to beneficial use of the land over the groundwater basin (aquifer) (26).

Establishing correlative rights is not an easy endeavor. It requires determination of the availability of water on an annual basis to determine the **safe yield** of the aquifer. If the total withdrawal of water by pumping is less than the average annual recharge of the aquifer, then the excess water can be apportioned to users. On the other hand, if an overdraft exists (groundwater withdrawal exceeds recharge), then the water rights are apportioned among all users, with the total withdrawals set at the average annual recharge for the aquifer (23).

Most of the states in the western United States have also adopted the Prior Appropriation Doctrine, cited earlier with respect to surface water. As noted previously, this doctrine states that the first user of groundwater has a right to continue that use, provided the water is put to a beneficial use without waste. These rights are then superior to rights of people who at a later time appropriate water. In those states that utilize the Prior Appropriation

CASE HISTORY Mono Lake and the Public Trust Doctrine

Mono Lake, located in the Mono basin at the foot of the Sierra Nevada east of Yosemite National Park in California (Figure 11.F), is the focus of a recent controversy centering around the very existence of the lake. From the lake's watershed, approximately 100,000 acre-feet of water per year are diverted south to the city of Los Angeles. The water is diverted from streams before entering the lake. Mono Lake measures approximately 21 by 13 km, with an average depth of about 17 m, making it the largest lake by volume contained entirely within the state of California. It is fed by a number of streams from the Sierra Nevada and some groundwater flow as well. During the last million years, geologic events associated with uplift of the Sierra Nevada, volcanic activity, and glaciation have left the lake with no natural outlet. Mono Lake is therefore salty, having a salinity approximately three times that of seawater.

Mark Twain visited Mono Lake in the 1860s and had little good to say about it except that the alkaline waters made laundry work easy. In *Roughing It*, Twain wrote, "Half a dozen little mountain brooks flow into Mono Lake but not a stream of any kind flows out of it. What it does with its surplus water is a dark and bloody mystery" (27). What happens to the water, of course, is that it evaporates. In fact, approximately 22 cm/yr of water evaporates from the surface of Mono Lake. Under natural conditions this loss is matched from streams that feed the lake system (27).

Mono Lake and its basin have a long and interesting history dating back at least to 1853, when Native Americans living in Yosemite were pursued by the military to the shores of the lake. About that time, gold was discovered in the area, initiating a small gold rush that lasted until approximately 1889. In 1913 the city of Los Angeles considered importing water into the growing urban area, and by 1930 funds had been approved for the construction of dams, reservoirs, and a tunnel to divert water from the eastern Sierra and Mono Lake area. In 1941 diversion of water from Mono basin streams began in earnest, and by 1981 the lake level had dropped (by evaporation) approximately 15 m. This decreased the volume of the lake by approximately one-half, which increased the salinity by 100 percent.

Brine shrimp grow in great abundance in the lake and provide the major food source for migrating birds. If the salinity were to become too high, the brine shrimp would die and the birds would have no food during a crucial stage in their migration. More significantly, lowering of the lake formed a land bridge to several volcanic islands in the lake that are major breeding grounds for California gulls. In 1979, after the land bridge had formed, coyotes entered the nesting area and chased off all 34,000 nesting birds. In addition, the lowering of the lake level exposed nearly 9000 ha of highly alkaline lake bed. During windy periods, alkali dust may rise into the atmosphere several hundred meters and be transported both around and out of the basin, causing air pollution (27). Extremely wet years in 1983 and 1984 caused the lake level to rise a bit, but it was still much lower than the 1941 level. Figure 11.F shows the 1980 situation with inflow and diversion of waters (27).

People interested in the preservation of Mono Lake and its ecosystem would like to see the lake level stabilized approximately 3 m above that necessary to support the healthy ecosystem. They advocate a wet year/dry year plan that would limit diversion to the dry years when the city of Los Angeles really needs the water. They further advocate a statewide program to conserve urban and agricultural water.

No one disagrees with the advocacy of water conservation. The city of Los Angeles, however, which receives approximately 17 percent of its water supply from the Mono basin, would like to see diversions continue at a rate greater than that advocated by those who want to see the lake preserved. The people in favor of continued diversion point out that the project produces a good deal of energy (approximately 300 million kilowatt hours per year, which saves approximately half a million barrels of oil annually). They would like to see the diversions continued and the lake level eventually stabilized at about 15 m below the 1981 level. One of their arguments is that the city of Los Angeles has invested more than $100 million in the area since the 1930s and really needs the water.

The Mono Lake story is an important one in environmental law because in 1983 the California Supreme Court reaffirmed the public interest in protecting natural resources through what is known as the Public Trust Doctrine. The 1983 decision states that it is the duty of the state to protect the people's common heritage, including streams, lakes, marshlands, and tidelands. In essence, the court decided that public trust obligates the state of California to protect lakes such as Mono as much as possible, even if this means reexamining past water allocations (27). In effect, the city of Los Angeles is forced to reduce diversion of water to such an extent that the Mono Lake ecosystem remains healthy.

Doctrine, the water rights are generally managed through a permit procedure supervised by a state government official (26).

In sum, a variety of doctrines and laws govern use of groundwater. One issue that constantly comes forward is what constitutes "beneficial use." To some people, beneficial use might be ensuring sufficient water for a river system to support a healthy ecosystem and aesthetic values. To others, beneficial uses may be limited to activities such as agriculture or public water supply. In still other cases, people might argue that recreational use of water is a beneficial use. These arguments come up because when groundwater is withdrawn from an area, it often affects the surface water supplies as well. When pumping lowers the water table below the bed of a perennial stream, the flow may cease and the riparian vegetation die, damaging the ecosystem. Downstream users of the surface water supply would also be denied their water. Because surface and groundwaters are so interrelated, the laws governing water use are complex and sometimes difficult to apply to specific situations.

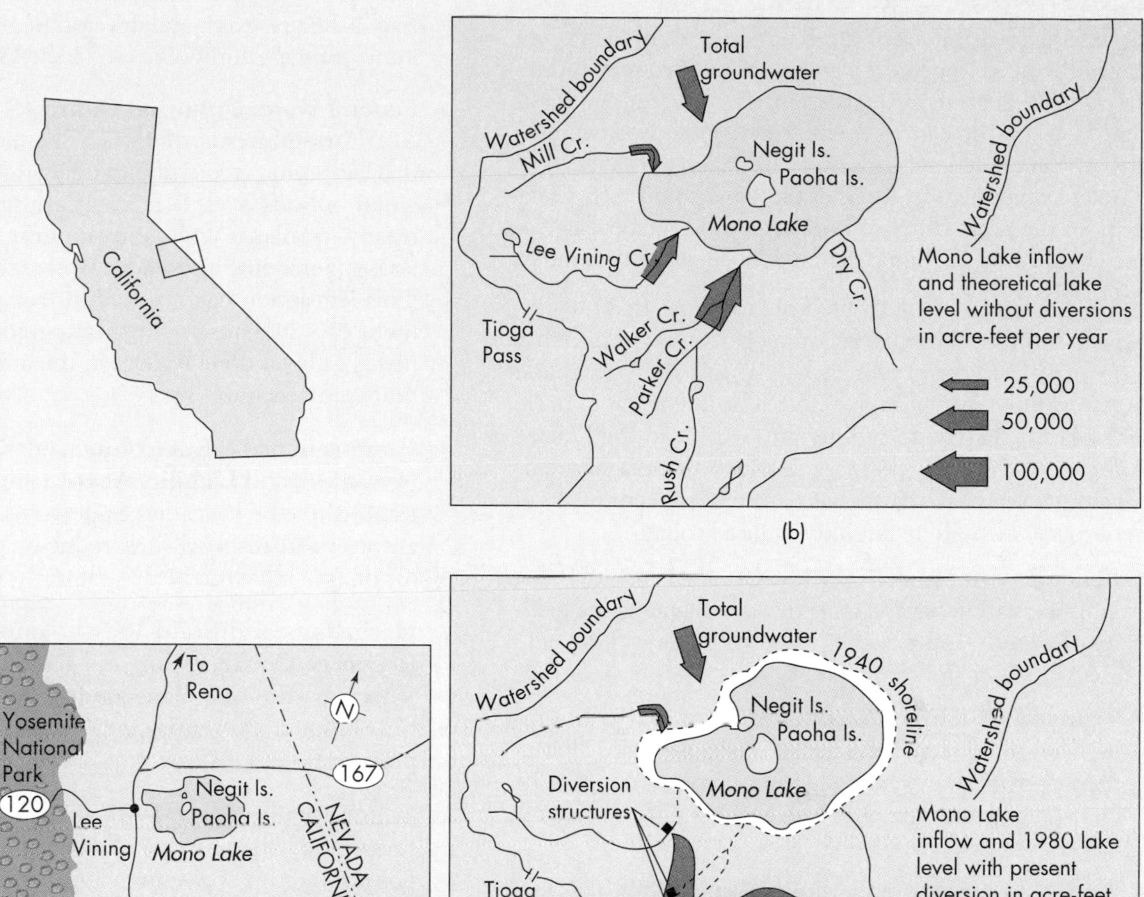

(d)

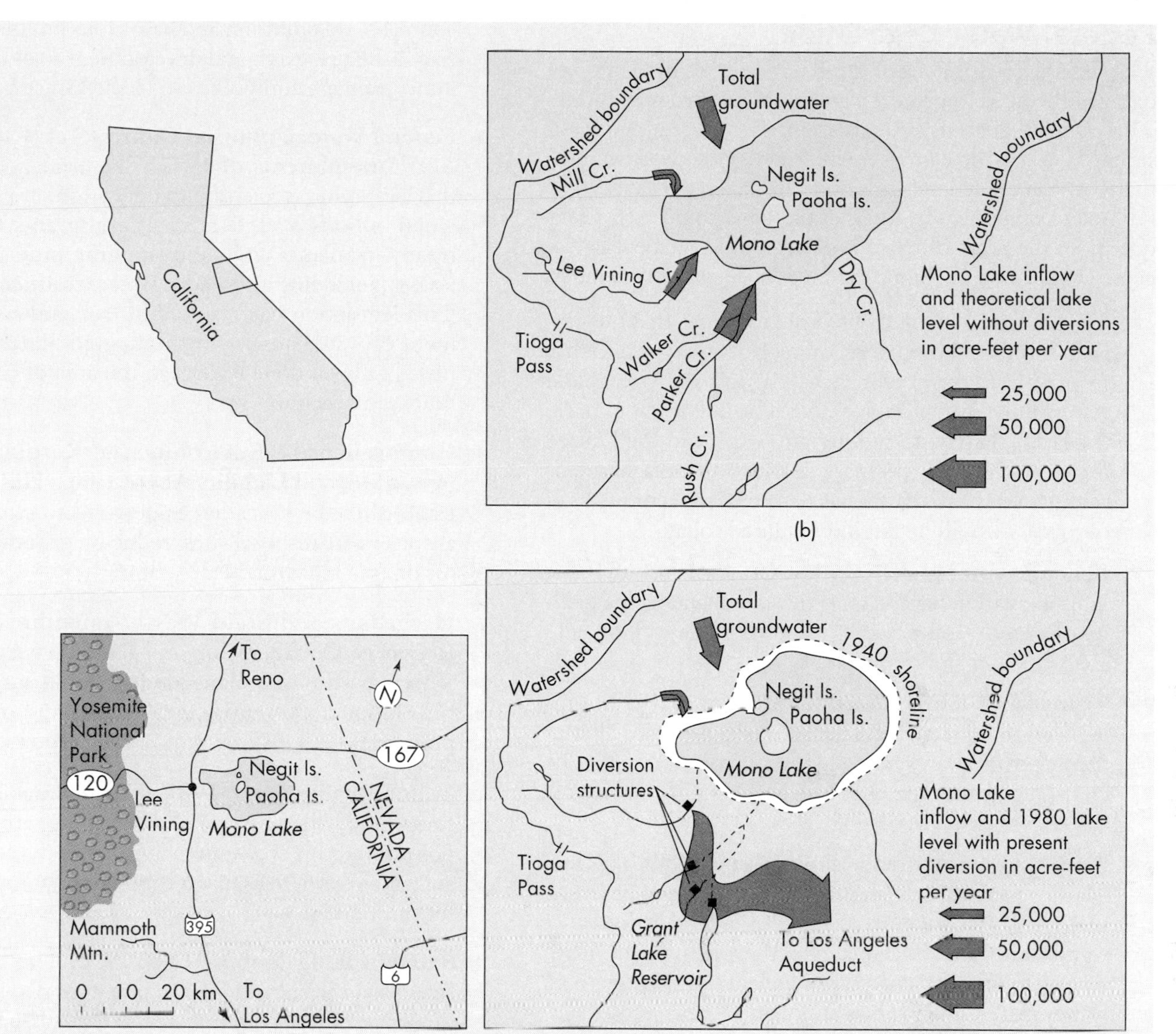

▲ FIGURE 11.F (a) Location of Mono Lake. (b) Situation without water diversion. (c) With water diversion. (d) Aerial view of Mono Lake in California showing the white ring or shoreline exposed from the lake's being at a lower level. (*Peter Essick/Aurora & Quanta Productions*)

Federal Water Legislation

The federal government of the United States has long recognized the need to protect water resources from pollution. The ultimate goal of the legislation is to protect our water supply as well as the natural environment. Following is a summary of selected federal legislation in the area of water pollution and water quality. Legislation that also protects water quality in the areas of hazardous waste management is discussed in Chapter 12. Significant federal legislation includes:

- **The Refuse Act of 1899.** This act states that it is unlawful to throw, discharge, or deposit any type of refuse from any source except that flowing from streets and sewers into any navigable water. The intent was to make it against the law to pollute any stream in the United States. Of course, the part dealing with discharge of sewage water into rivers has had to change, or many of our rivers would be terribly polluted today.
- **Federal Water and Pollution Control Act of 1956.** This legislation has the objective of enhancing the quality of water resources and preventing, controlling, and abating water pollution.
- **Fish and Wildlife Coordination Act of 1958.** This legislation state that water resource projects such as dams, power plants, and flood-control works must be coordinated with the U.S. Fish and Wildlife Service for the conservation of wildlife.
- **National Environmental Policy Act of 1969.** This legislation is extremely significant in requiring environmental impact statements prior to federal actions that significantly affect the quality of the environment. With respect to water resources, this includes dams, reservoirs, power plants, and bridges, among other projects.
- **Water Quality Improvement Act of 1970.** This legislation expanded the power of the 1956 act through control of oil pollution and hazardous pollutants. It also established research and development to eliminate acid mine drainage and pollution in the Great Lakes.
- **Federal Water Pollution Control Act (Clean Water Act) Amendments of 1972.** The primary purpose of this legislation is to clean up the nation's waters. It provided billions of dollars in federal grants for sewage treatment plants while encouraging innovative technology, including alternative water treatment methods. This legislation has resulted in tremendous improvement of water quality in the United States, although little has been done to date in the area of encouraging innovative technology.
- **Comprehensive Environmental Response, Compensation, and Liability Act of 1980.** This legislation establishes the so-called Superfund to clean up hazardous waste disposal sites, reducing groundwater pollution (see Chapter 12).
- **Hazardous and Solid Waste Amendments to the Resource Conservation and Recovery Act of 1984.** This legislation regulates underground storage tanks, thus reducing potential for gasoline and other liquid pollutants to damage groundwater resources.
- **Water Quality Act of 1987.** This act establishes as national policy the control of nonpoint sources of water pollution. This was important in the development of state management plans to control nonpoint water pollution sources.
- **Safe Drinking Water Act of 1996.** This act emphasizes sound science and development of risk-based water quality standards, and provided for consumer awareness of water quality and assistance obtaining improvement in the water system infrastructure.

SUMMARY

Water pollution is the degradation of water quality as measured by physical, chemical, or biological criteria. These criteria take into consideration the intended use for the water, departure from the norm, effects on public health, and ecological impacts.

The major water pollutants are oxygen-demanding waste, measured by biochemical oxygen demand (BOD); pathogens, measured by the fecal coliform bacteria count; nutrients that lead to eutrophication, in which overgrowth of algae deprives water of oxygen and sunlight; oil; toxic substances, including synthetic organic and inorganic compounds, heavy metals, and radioactive materials; heat; and sediment.

Surface-water pollutants have either point or nonpoint sources. Point sources include pipes that empty industrial and municipal wastes into streams, and combined sewer systems that carry both waste and storm-water flow in older cities. Nonpoint sources, or polluted runoff, are more difficult to control than point sources. Nonpoint sources include urban, agricultural, forestry, and mining runoff carrying a wide variety of pollutants. *Acid mine drainage* refers to water with a high concentration of sulfuric acid that drains from some coal or metal-mining areas, causing surface-water and groundwater pollution in many parts of the United States.

Since the 1960s there has been a serious attempt to reduce surface-water pollution and improve water quality in the United States. Although the program has been quite successful, water quality is still substandard in some areas.

In the case of surface water, pollution processes are slowed by dilution and dispersion of pollutants and degradation of pollutants in the presence of sunlight and oxygen. In the case of groundwater, the depth, slow flow, and long residency time of the water limit the opportunities for these natural controls to operate. On the other hand, many soils and rocks act as filters, exchanging certain elements and compounds with groundwater. In moving through an aquifer, groundwater may improve in quality, but it may also be rendered unsuitable for human use by natural or artificial contaminants. Pollution of an aquifer can result from disposal of wastes on the land surface or in the ground. It can also result from overpumping of groundwater in coastal areas, leading to intrusion of salt water into freshwater aquifers. Because we cannot prevent all pollutants from entering groundwater, and reversal of aquifer and vadose zone contamination is complex and expensive, we must find ways to assist the natural processes that limit groundwater pollution.

Development of water quality standards in the United States has been mandated by federal legislation and involves setting of Maximum Contaminant Levels (MCLs) for contaminants that might be found in our drinking water. The major purposes of the standards are to ensure that our water supply is treated to remove harmful contaminants and that water quality is regularly tested and monitored. Monitoring of toxic metals and organic compounds in fish indicates that the levels of toxins in water have been significantly reduced, particularly for toxins that have been regulated the longest.

Wastewater treatment facilities include septic-tank sewage disposal systems and wastewater treatment plants. Septic-tank systems, utilized by homes and small commercial and industrial activities, are very common in the United States today. Failure of these systems may cause significant pollution to groundwater resources. Wastewater treatment plants collect and process water from municipal sewage systems. Primary and secondary treatment by wastewater plants removes up to 90 percent of the pollutants in the wastewater. These include oxygen-demanding materials, bacteria, and suspended solids. Advanced wastewater treatment may be utilized to remove heavy metals and nutrients so that water can be reclaimed for other uses, including wildlife habitat or application to farm fields, parks, and golf courses. The use of reclaimed water is growing fast in the United States, particularly in areas where water shortages are most likely to occur. Considerable research is ongoing to develop methods of wastewater treatment that involve resource recovery. Typically, such treatment involves use of the biologic environment as part of the treatment process.

Water law for surface water and groundwater resources is complex and varies from one U.S. state to another. In some cases the right to use water is based upon living adjacent to a water resource or over a groundwater basin, whereas in other cases water resources are appropriated and regulated by regional and state agencies. The federal government has a long history of enacting laws in an attempt to control water pollution. As a result, we have some of the highest water quality standards in the world and are attempting to control and abate water pollution problems.

REFERENCES

1. **Smith, R. A.** 1994. Water quality and health. *Geotimes* 39(1):19–21.
2. **Water Resources Council.** 1978. *The nation's water resources, 1975–2000*, vol. 1. Washington, DC: U.S. Government Printing Office.
3. **Special oil spill issue.** 1989. *Alaska Fish and Game* 21, July–August.
4. **Rice, S. D., ed.** 1996. *Proceedings of the Exxon Valdez Oil Spill Symposium*. Anchorage, AK: American Fisheries Society.
5. **Hollway, M.** 1991. Soiled shores. *Scientific American* 265:102–106.
6. **Parfit, M.** 1993. Troubled waters run deep. *National Geographic* 184(5A):78–89.
7. **Miller, G. Tyler, Jr.** 1994. *Living in the environment*. Belmont, CA: Wadsworth.
8. **Council on Environmental Quality.** 1979. *Environmental quality*. Washington, DC: U.S. Government Printing Office.
9. **Anonymous.** 1982. U.S. Geological Survey activities, fiscal year 1982. *U.S. Geological Survey Circular* 875:90–3.
10. **Geiser, K., and Waneck, G.** 1983. PCBs and Warren County. *Science for the People*.
11. **Bopp, R. F., and Simpson, H. J.** 1989. *Contamination of the Hudson River: The sediment record in contaminated marine sediments—assessment and remediation*. Washington, DC: National Academy Press.
12. **Pye, U. I., and Patrick, R.** 1983. Ground water contamination in the United States. *Science* 221:713–18.
13. **Weir, D., and Schapico, M.** 1980. The circle of poison. *The Nation*, November 15.
14. **Carey, J.** 1984. Is it safe to drink? *National Wildlife*, February–March:19–21.
15. **Foxworthy, G. L.** 1978. Nassau County, Long Island, New York—Water problems in humid country. In *Nature to be commanded*, eds. G. D. Robinson and A. M. Spieker, pp. 55–68. U.S. Geological Survey Professional Paper 950.
16. **Environmental Protection Agency.** 1991. *Is your drinking water safe?* EPA 570-9-91-0005.
17. **Jewell, W. J.** 1994. Resource-recovery wastewater treatment. *American Scientist* 82(4):366–75.
18. **Bedient, P. B., Rifai, H. S., and Newell, C. J.** 1994. *Groundwater contamination*. Englewood Cliffs, NJ: Prentice Hall.
19. **Leeden, F., Troise, F. L., and Todd, D. K.** 1990. *The water encyclopedia*, 2nd ed. Chelsea, MI: Lewis Publishers.
20. **American Chemical Society.** 1969. *Clean our environment: The chemical basis for action*. Washington, DC: U.S. Government Printing Office.

21. **Parizek, R. R., and Myers, E. A.** 1968. Recharge of ground water from renovated sewage effluent by spray irrigation. *Proceedings of the Fourth American Water Resources Conference*, pp. 425–43.
22. **Bastian, R. K., and Benforado, J.** 1983. Waste treatment: Doing what comes naturally. *Technology Review*, February–March:59–66.
23. **Fetter, C. W.** 1994. *Applied hydrogeology*. New York: Macmillan.
24. **Legal Approach to Water Rights.** 1972. In *Water quality in a stressed environment*, ed. W. A. Pettyjohn, pp. 255–76. Minneapolis: Burgess.
25. **Research and Documentation Corp.** 1970. Private remedies for water pollution. *Environmental Law:* 47–49.
26. **Smith, Z. A.** 1989. *Groundwater in the West*. New York: Academic Press.
27. **Mono Lake Committee.** 1985. *Mono Lake: Endangered oasis*. Lee Vining, CA: author.

KEY TERMS

water pollution (p. 292)
biochemical oxygen demand (BOD) (p. 293)
fecal coliform bacteria (p. 293)
cultural eutrophication (p. 294)
point sources (p. 298)
nonpoint sources (p. 298)
acid mine drainage (p. 299)
Maximum Contaminant Levels (MCLs) (p. 305)
Maximum Contaminant Level Goals (MCLGs) (p. 305)
septic tank (p. 306)
primary treatment (p. 308)
secondary treatment (p. 308)
advanced treatment (p. 308)
reclaimed water (p. 308)
wastewater renovation and conservation cycle (p. 309)

SOME QUESTIONS TO THINK ABOUT

1. The island of Maui, one of the Hawaiian Islands, has a strong tourist industry. Near some of the urban areas, the beaches are occasionally spoiled by accumulation of decaying algae (seaweed) that may smell so bad that it drives people from the beaches. The algae evidently increase (bloom) in the shallow waters offshore in response to input of nutrients from urban wastewater and/or agricultural runoff. Urban wastewaters are treated to secondary standards in a series of small units for a particular development and, in some cases, for larger communities. These waters are injected into the ground near the ocean. How could you develop a research plan to try to determine if the eutrophication that is taking place is the result of the injection of urban wastewater or agricultural runoff? How might each of these pollution sources be controlled to preserve the water quality in the nearshore marine environment and eliminate the algae blooms?
2. For your community, develop an inventory of point and nonpoint sources of water pollution. Carefully consider how each of these might be eliminated or minimized as part of a pollution abatement strategy.
3. Visit a wastewater treatment plant. What are the processes utilized at the plant, and could the concept of resource-recovery or wastewater renovation cycle be utilized? What would be the advantages and disadvantages of using a biologic system, such as plants, as part of the wastewater treatment procedures?
4. How safe do you think your water supply is? If you drink bottled water, how safe is it? Upon what are you basing your answers? What do you need to know to give informed answers?

12 Waste Management

Sign protesting a hazardous waste disposal site in California. (*Edward A. Keller*)

How a society manages its waste is a measure of its commitment to the environment. Shown here is a sign to protest the Casmalia Toxic Dump in California. Although the disposal site, like many others, has been closed for environmental reasons, the legacy from hazardous waste continues on a global scale. For example, it is estimated that approximately 25 percent of the people in Russia live in areas where the concentrations of toxic pollutants exceed standards by ten times. As recently as 1991 a survey of 100 nations reported that 90 percent of those countries that responded stated that uncontrolled dumping of industrial hazardous waste was a problem and two-thirds of those reporting stated that hazardous chemical waste was disposed of in uncontrolled sites. It is believed that the casual treatment of hazardous waste and municipal solid waste (particularly in developing countries) has and will continue to adversely affect the overall environment and the health of people and have costly economic consequences in many locations of the world.*

12.1 Concepts of Waste Management: An Overview

People in the United States and throughout the world are facing a tremendous solid-waste disposal problem, particularly in growing urban areas. The problem boils down to the simple fact that urban areas are producing too much waste and there is far too little space for disposal. About half the cities in the United States are estimated to be running out of

*Source: Gardner, G., and Sampat, P. 1999. Forging a sustainable materials economy. In *State of the world 1999*, ed. L. Stark, pp. 41–59. New York: W. W. Norton.

LEARNING OBJECTIVES

Development of management strategies to deal with our waste problems is an important environmental concern. Learning objectives for this chapter are:

- *To gain an appreciation for the evolution of concepts of waste management from "dilute and disperse" to integrated waste management, to materials management with the visionary goal of zero production of waste.*
- *To understand the principles behind "reduce, recycle, and reuse."*
- *To know the various alternatives for solid-waste disposal.*
- *To understand important processes related to sanitary landfills, including generation of leachate, site selection, design, monitoring, and federal legislation.*
- *To understand the principles of hazardous chemical-waste management in terms of what responsible management is, alternative management strategies, and federal legislation pertaining to hazardous waste.*
- *To gain a basic understanding of the strategies and policies associated with radioactive waste management.*
- *To understand some of the problems and potential solutions associated with ocean dumping.*

Web Resources

Visit the "Environmental Geology" Web site at www.prenhall.com/keller to find additional resources for this chapter, including:

- ▶ Web Destinations
- ▶ On-line Quizzes
- ▶ On-line "Web Essay" Questions
- ▶ Search Engines
- ▶ Regional Updates

landfill space (see *Case History: The Fresh Kills Landfill*). Cost is another limiting factor—expenditures for landfill disposal have skyrocketed in recent years (1).

All types of societies produce waste, but industrialization and urbanization have caused an ever-increasing effluence that has greatly compounded the problem of waste management. Although tremendous quantities of liquid and solid waste from municipal, industrial, and agricultural sources are being collected and recycled, treated, or disposed of, new and innovative programs remain necessary if we are to keep ahead of what might be called a waste crisis. Disposal or treatment of liquid and solid waste by federal, state, and municipal agencies costs billions of dollars every year. In fact, it is one of the most costly environmental expenditures of governments, accounting for the majority of total environmental expenditures (2).

A possible solution to the solid-waste problem would be to develop new disposal facilities. Unfortunately, no one wants to live near a waste-disposal site, be it a sanitary landfill for municipal waste, an incinerator facility that can reduce the volume of waste by 75 percent, or a disposal operation for hazardous chemical materials. This obviously creates serious siting problems even if the local geographic, geologic, and hydrologic environment is favorable. The siting problem also involves issues of *social justice*. Waste-management facilities are all too frequently located in areas where the people are of low social and economic status or belong to a minority ethnic group or race. Investigation of the issues involved in siting waste facilities to which many people object based on perceived environmental problems is part of an emerging field known as *environmental justice* (3). The consensus seems to be that people have little confidence in the ability of government or industry to preserve and protect public health as it relates to waste disposal (1).

The waste-disposal industry in the United States, which represents a $20 billion sector of the economy, is accustomed to the relatively simple system of collection of waste and landfill disposal (1). The rise in public consciousness concerning environmental problems and solutions is forcing the disposal industry to explore new solid-waste management systems. What has emerged is the concept known as **integrated waste management (IWM),** a complex set of management alternatives including source reduction, recycling, composting, landfill, and incineration (1).

Earlier Views

During the first century of the Industrial Revolution, the volume of waste produced was relatively small and the concept of "dilute and disperse" was adequate. Factories were located near rivers because the water provided easy transport of materials by boat, ease of communication, sufficient water for processing and cooling, and easy disposal of waste into the river. With few factories and sparse population, "dilute and disperse" seemed to remove the waste from the environment (4).

Unfortunately, as industrial and urban areas expanded, the concept of "dilute and disperse" became inadequate, and a new concept known as "concentrate and contain" became popular. It is now apparent, however, that containment was and is not always achieved. Containers, natural or artificial, may leak or break and allow waste to escape. As a result, another concept developed, known as "resource recovery." This philosophy holds that wastes may be converted to useful

CASE HISTORY The Fresh Kills Landfill

The Fresh Kills landfill is located on the western shore of Staten Island (Figure 12.A). The landscape is a mixture of wetlands (salt marsh), woodlands, and grasslands. The landfill opened in 1948 and is the only landfill operating in the city of New York. The landfill has an area of approximately 7500 ha and at its peak in 1986 received more than 21,000 tons of waste per day. During the 1990s the amount of waste the facility accepted began to slow down as the city eliminated commercial deliveries and the people of New York began to recycle returnable bottles, plastic containers, and newspapers. Currently Fresh Kills receives between 12,000 and 14,000 tons of waste per day, at a cost of approximately $44 per ton (5).

The Fresh Kills landfill is one of the largest such facilities in the world, but space for the waste generated by the city of New York is running out. Today less than about 2000 ha of the facility are actually used for landfilling. As a result the city is negotiating to begin trucking waste out of New York City at a cost of $50 to $70 per ton as the landfill enters its final closure stages. The cost of closing the landfill will be more than $1 billion, including 30 years of monitoring after closure in approximately December 2001. The city of New York has an ambitious plan to transform one of the world's largest landfills into an environmentally sound and aesthetically pleasing natural area. As a result of the disposal activities, the elevation of parts of the landfill, when closed, will be approximately 80 m, a sizable hill for coastal New York! As part of the closure plan, a slurry wall (underground concrete barrier) containment system designed to prevent the migration of untreated leachate (noxious, polluted liquid produced when water infiltrates through waste material) outside of the landfill has been constructed. A subsurface leachate collection system inside the containment wall has more than 30 wells to collect leachate for treatment. A leachate treatment center will be utilized to neutralize more than one million gallons (3800 m^3) of leachate per day. At the site, storm-water runoff is diverted to retention ponds and about 150 wells collect 283,000 m^3 (10 million ft^3) of methane daily from two sections of the landfill. The gas produced from the wells is purified and sold to Union Gas in Brooklyn (5).

The primary goal is ensure that following closure of the landfill, the Fresh Kills site will be environmentally safe. A secondary goal is to transform the site into an area that is aesthetically pleasing. It has been determined that maintaining a "lawn" of grass the size of the landfill site would be too costly, at more than $20 million over the 30-year monitoring period. Therefore, the city Sanitation Department has developed a series of test plots and experiments to reestablish native woodland communities at the site. The trees require minimal care, and the plan is to let nature take over. Thousands of shrubs and trees have been planted and to date the trees have grown moderately well and the shrubs have done great. A side benefit of the replanting is that many birds have come to perch in the trees, further dispersing seeds and therefore adding new species of plants to the site. Initially it was feared that replanting of the landfill site with trees would produce roots that might reach down and break the clay cap used to contain waterborne pollutants. To date this has not been a problem. The land reclamation plan of the site, if successful, will result in open space that is guaranteed to remain undeveloped in the future. The reclamation of the site to a mixture of marsh, woodland, and grasslands as it was before the landfill is a positive action that is being attempted in other landfills around the United States facing closure. Nevertheless, preservation of the original land 150 years ago as a natural reserve would have been environmentally preferred. Today, a large preserve would be a treasure, as is Central Park in New York City (6). The lesson learned from Fresh Kills, in spite of the positive aspects of land reclamation, is that waste management facilities are a tremendous financial burden to society. Furthermore, we have failed in the past 50 years to move from a throw-away waste-disposal-oriented society to sustain natural resources. We are now beginning to move in a new direction, toward a wasteless society, which is a real and pressing necessary goal.

▲ **FIGURE 12.A** Aerial view of a small part of the Fresh Kills landfill in New York. (*Comstock*)

materials, in which case they are no longer wastes but resources. However, even with our state-of-the-art technology, large volumes of waste cannot be economically converted or are essentially indestructible. Therefore, we still have waste-disposal problems (4).

Modern Trends: Integrated Waste Management

There is a growing awareness that many of our waste-management programs simply involve moving waste from one site to another and not really properly disposing of it. For example, waste from urban areas may be placed in landfills, but eventually these may cause further problems from the production of methane gas, as mentioned above (which is a resource if managed properly), or noxious liquids that leak from the site to contaminate the surrounding areas. Disposal sites are also capable of producing significant air pollution. It is safe to assume that waste management is going to be a public concern for a long time. Of particular importance will be the development of new methods of waste management that will not endanger the public health or cause a nuisance.

Integrated waste management (IWM) emerged in the 1980s as a set of management alternatives, including resource reduction, recycling, reuse, compositing, landfill, and incineration (1). **Reduce, recycling, and reuse** are the three Rs of IWM, and it is believed that the primary objective of recycling could reduce the weight of urban refuge disposed in landfills by approximately 50 percent.

The recycling option of IWM, which has been seriously pursued for nearly two decades, has been responsible for the generation of entire systems of waste management that have produced tens of thousands of jobs while reducing the amount of urban waste from homes in the United States sent to landfills from 90 percent in the 1980s to about 65 percent today. In fact, many firms have combined waste reduction with recycling to reduce by 50 percent to 90 percent of the waste they deliver to landfills. In spite of this obvious success, IWM is being criticized for not effectively advancing policies to prevent waste production by over-emphasizing recycling. In the long term, waste management policies that rely on recycling cannot be successful. As human population continues to increase, in one doubling of the population, we will be where we are today with respect to waste disposal, if we depend upon landfilling for 50% of the waste we produce. That is, if we continue on with today's management of waste, in approximately 50 to 70 years, when the U.S. population has doubled again, we will be producing the same volume of waste sent to landfills that we do today, given a 50 percent rate of recycling. Clearly, emphasizing recycling is not a sustainable sollution to our waste problem. With this in mind, the concept of IWM needs to be rethought and expanded to include what is termed *materials management* (7).

12.2 Materials Management

Materials management is part of IWM, but it provides a new goal. That goal is "zero production of waste," so that what is now thought of as waste will be a resource! This is a visionary goal, requiring more sustainable use of materials combined with resource conservation. It is believed that materials management as an extension of IWM can be established by (7):

- Eliminating subsidies for extraction of virgin materials such as timber, minerals, and oil.
- Establishing "green building" incentives that use recycled materials and products in new construction.
- Establishing financial penalties for production of those products that do not meet the objectives of material management practices.
- Providing financial incentives for those industrial practices and products that benefit the environment by enhancing sustainability, such as encouraging products that reduce waste production and use recycled materials.
- Providing incentives for the production of new jobs in the technology of materials management and practice of reducing, recycling, and reusing of resources. This is the essence of materials management and sustainable resource utilization.

The concept of materials management for "zero waste" is part of what is known as **industrial ecology.** The idea is to produce urban and industrial systems that model natural ecosystems, where waste from one part of the system is a resource for another part.

With this introduction to modern trends and integrated waste management, it is advantageous to break the management treatment and disposal of waste into several categories: solid-waste disposal; hazardous chemical-waste management; radioactive waste management; and ocean disposal.

12.3 Solid-Waste Disposal

Disposal of solid waste is primarily an urban problem. In the United States alone, urban areas produce about 640 million kg of solid waste each day, an amount that could cover more than 1.6 km^2 of land to a depth of 3 m (8). Figure 12.1 summarizes major sources and types of solid waste, and Table 12.1 lists the generalized composition of solid waste at a disposal site in 1986 and projected for the year 2000. We emphasize that this is only an average composition, and considerable variation can be expected because of differences in such factors as land use, economic base, industrial activity, climate, and season of the year. It is no surprise that paper is by far the most abundant solid waste.

In some areas, infectious wastes from hospitals and clinics can create problems if they are not properly sterilized before disposal. Some hospitals have facilities to incinerate such wastes. In large urban areas, huge quantities of toxic materials may also end up at disposal sites. Urban landfills are now being considered hazardous waste sites that will require costly monitoring and cleanup.

The common methods of solid-waste disposal, summarized from a U.S. Geological Survey report, include onsite disposal, composting, incineration, open dumps, and sanitary landfills (9).

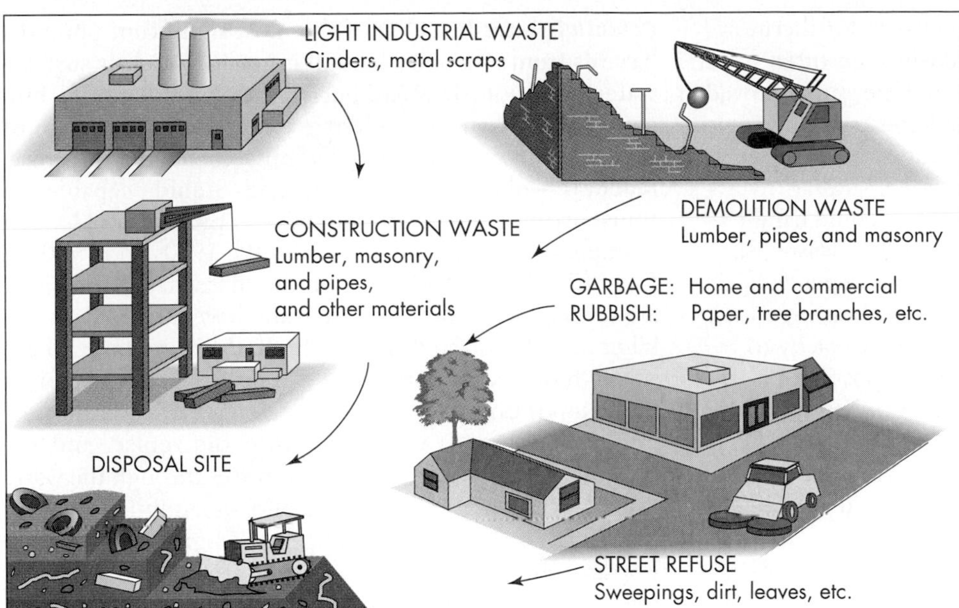

◀ **FIGURE 12.1** Types of materials or refuse commonly transported to a disposal site.

Table 12.1 Generalized composition of urban solid waste (by weight) for 1986 and projected for the year 2000

Material	1986 (%)	2000 (%)
Paper	36	39
Yard waste	20	19
Plastics	7	9
Metals	9	9
Food waste	9	7
Glass	8	7
Wood	4	4
Other	7	6

Source: A. M. Ujihara and M. Gough, "Managing Ash from Municipal Waste Incinerators," in *Resource for the Future* (Center for Risk Management: 1989).

On-site Disposal

By far the most common on-site disposal method in urban areas is the mechanical grinding of kitchen food waste. Garbage disposal devices are installed in the wastewater pipe system from a kitchen sink, and the garbage is ground and flushed into the sewer system. This effectively reduces the amount of handling and quickly removes food waste, but final disposal is transferred to the sewage treatment plant where solids such as sewage sludge still must be disposed of (9).

Hazardous liquid chemicals may be inadvertently or deliberately disposed of in sewers, requiring treatment plants to handle toxic materials. Illegal dumping in urban sewers has only recently been identified as a potential major problem.

Composting

Composting is a biochemical process in which organic materials decompose to a humuslike material. It is rapid, partial decomposition of moist, solid, organic waste by aerobic organisms. The process is generally carried out in the controlled environment of mechanical digesters (2). Although composting is not common in the United States, it is popular in Europe and Asia, where intense farming creates a demand for the compost (4). A major drawback of composting is the necessity to separate the organic material from the other waste. Therefore, it is economically advantageous only when organic material is collected separately (10). Nevertheless, composting is considered part of integrated waste management, and its importance is expected to grow.

Incineration

Incineration is the reduction of combustible waste to inert residue by burning at high temperatures (900 to 1000°C). These temperatures are sufficient to consume all combustible material, leaving behind only ash and noncombustibles. Incineration ideally reduces the volume of waste that must be disposed of by 75 to 95 percent (9). However, because of maintenance and waste-supply problems, the actual reduction of waste by incineration is closer to 50 percent. As we have already mentioned, this is about the same savings that can be gained from waste reduction and recycling (8). The advantages of incinerating urban waste are twofold:

- Incineration can effectively convert a large volume of combustible waste to a much smaller volume of ash to be disposed of at a landfill.
- Combustible waste can be used to supplement other fuels in generating electrical power.

Burning urban waste is certainly not a clean process. The burning produces air pollution and toxic ash that must be disposed of at landfills. Smokestacks from incinerators may emit nitrogen and sulfur oxides, which are precursors of acid rain, as well as carbon monoxide and heavy metals such as lead, cadmium, and mercury. The smokestacks can be fitted with devices to trap some of the pollutants, but the process of pollution abatement is expensive. Furthermore, the incinerators themselves are expensive and often need government subsidies to be established. One study showed that an investment of $8 billion could construct incinerators capable of burning about 25 percent of the solid waste generated in the United States, whereas a similar investment in recycling and composting facilities could handle as much as 75 percent of the nation's solid urban waste (8).

The economic viability of incinerators depends upon revenue from the sale of energy produced by burning waste. As a result, incinerators need to run at near capacity to remain profitable. With the increase in composting and recycling, the economics are far from certain, because those processes compete directly with incineration. However, it is safe to say that waste reduction and recycling can reduce the volume of waste that must be disposed of at a landfill at least as much as incineration can (8).

Open Dumps

Open dumps are the oldest and most common way of disposing of solid waste. In many cases, open dumps are located wherever land is available, without regard to safety, health hazards, and aesthetic degradation. The waste is often piled as high as equipment allows. In some instances, the refuse is ignited and allowed to burn; in others, it is periodically leveled and compacted (9). In addition to being unsightly, open dumps generally create a health hazard by breeding pests, polluting the air, and often contaminating groundwater and surface water. In the United States, open dumps have given way to the better-planned and managed sanitary landfills, but they are still common in many poor countries of the world.

Sanitary Landfills

A **sanitary landfill** as defined by the American Society of Civil Engineering is a method of solid-waste disposal that functions without creating a nuisance or hazard to public health or safety. Engineering principles are used to confine the waste to the smallest practical area, reduce it to the smallest practical volume, and cover it with a layer of compacted soil or specially designed tarps at the end of each day of operation, or more frequently if necessary. It is this covering of the waste that makes the sanitary landfill sanitary. The cover effectively denies continued access to the waste by insects, rodents, and other animals. It also isolates the refuse from the air, thus minimizing the amount of surface water entering into and gas escaping from the wastes (10).

The sanitary landfill as we know it today emerged in the late 1930s. Two types are used: area landfill on relatively flat sites and depression landfill in natural or artificial gullies or pits. Normally, refuse is deposited, compacted, and covered at the end of each day. The finishing cover (cap) is at least 50 cm of compacted soil (clay) designed to minimize infiltration of surface water (9). Compaction and subsidence can be expected for years following completion of a landfill. Therefore, any subsequent development that cannot accommodate potential subsidence should be avoided.

Potential Hazards One of the most significant potential hazards from a sanitary landfill is groundwater or surface-water pollution. If waste buried in a landfill comes into contact with water percolating down from the surface or with groundwater moving laterally through the refuse, **leachate**—obnoxious, mineralized liquid capable of transporting bacterial pollutants—is produced (11). For example, two landfills dating from the 1930s and 1940s in Long Island, New York, have produced leachate plumes that are several hundred meters wide and have migrated several kilometers from the disposal site. Both the nature and the strength of leachate produced at a disposal site depend on the composition of the waste, the length of time that the infiltrated water is in contact with the refuse, and the amount of water that infiltrates or moves through the waste (9). The concentration of pollutants in landfill leachate is much higher than in raw sewage or slaughterhouse waste. Fortunately, the amount of leachate produced from urban waste disposal is much less than the amount of raw sewage.

Another possible hazard from landfills is uncontrolled production and escape of methane gas, which is generated as organic wastes decompose. For example, gas generated in an Ohio landfill migrated several hundred meters through a sandy soil to a housing area, where one home exploded and several others had to be evacuated. Properly managed, methane gas (if not polluted with toxic materials) is a resource. At new and expanded landfills, methane is often confined by barriers made of plastic liner and clay and collected in specially constructed wells. The technology for managing methane is advancing, and landfills across the country are now producing methane and selling the gas as one way to help reduce costs associated with waste management.

Site Selection Factors controlling the feasibility of sanitary landfills include:

- Topographic relief
- Location of the groundwater table
- Amount of precipitation
- Type of soil and rock
- Location of the disposal zone in the surface-water and groundwater flow system

The best sites are those in which natural conditions ensure reasonable safety in disposal of solid waste. This means that there is little (or acceptable) pollution of ground- or surface waters, and that conditions are safe because of climatic, hydrologic, geologic, or human-induced conditions or combinations of these (12).

The best sites for landfills are in arid regions. Disposal conditions are relatively safe there because in a dry environment, regardless of whether the burial material is permeable or impermeable, little or no leachate is produced. On the other hand, some leachate will always be produced in a humid environment, so an acceptable level of leachate production must be established to determine the most favorable sites. What is acceptable varies with local water use, local regulations, and the ability of the natural hydrologic system to disperse, dilute, and otherwise degrade the leachate to a harmless state.

The most desirable site in a humid climate is one in which the waste is buried above the water table in clay and silt soils of low hydraulic conductivity. Any leachate produced will remain in the vicinity of the site, where it will be degraded by natural filtering and by exchange of some ions between the clay and the leachate. This holds even if the water table is fairly high, as it often is in humid areas, provided material with low hydraulic conductivity is present (13). For example, if the refuse is buried over a fractured-rock aquifer, as shown in Figure 12.2, the potential for serious pollution is low because the leachate is partly degraded by natural filtering as it moves down to the water table. Furthermore, the dispersion of contaminants is confined to the fracture zones (2). However, if the water table were higher or if the cover material were thinner with a moderate to high hydraulic conductivity, then widespread groundwater pollution of the fractured-rock aquifer might result.

If a landfill site is characterized by an inclined limestone-rock aquifer overlain by sand and gravel with high hydraulic conductivity (Figure 12.3), considerable contamination of the groundwater could result. Leachate moves quickly through the sand and gravel soil and enters the limestone, where open fractures or cavities may transport the pollutants with little degradation other than dispersion and dilution. Of course, if the inclined rock is all shale, with low hydraulic conductivity, little pollution will result.

The following general guidelines (13) should be followed in site selection for sanitary landfills:

- Limestone or highly fractured rock quarries and most sand and gravel pits make poor landfill sites because these earth materials are good aquifers.
- Swampy areas, unless properly drained to prevent disposal into standing water, make poor sites.

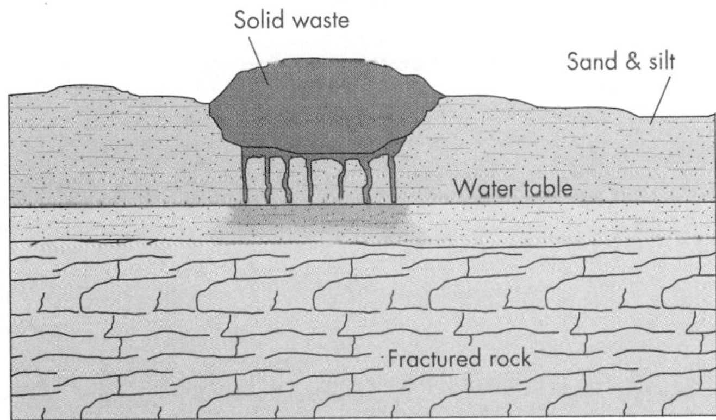

▲ **FIGURE 12.2** Waste-disposal site where the refuse is buried above the water table over a fractured rock aquifer. Potential for serious pollution is low because leachate is partially degraded by natural filtering as it moves down to the water table. (*After* W. J. *Schneider.* 1970. U.S. *Geological Survey Circular* 601F.)

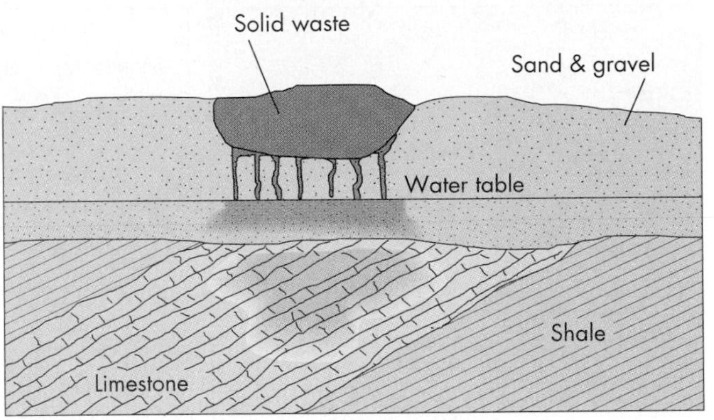

▲ **FIGURE 12.3** Solid-waste disposal site where the waste is buried above the water table in permeable material with high hydraulic conductivity in which leachate can migrate down to fractured bedrock (limestone). The potential for groundwater pollution may be high because of the many open and connected fractures in the rock. (*After* W. J. *Schneider.* 1970. U.S. *Geological Survey Circular* 601F.)

- Floodplains likely to be periodically inundated by surface water should not be considered as acceptable sites for refuse disposal.
- Areas in close proximity to the coast, where trash (transported by wind or surface water) or leachate in ground- or surface water may pollute beaches and coastal marine waters, are undesirable sites.
- Any material with high hydraulic conductivity and with a high water table is probably an unfavorable site.
- In rough topography, the best sites are near the heads of gullies where surface water is at a minimum.
- Clay pits, if kept dry, may provide satisfactory sites.
- Flat areas are favorable sites, provided an adequate layer of material with low hydraulic conductivity, such as clay and silt, is present above any aquifer.

We emphasize that, although these guidelines are useful, they do not preclude the need for a hydrogeological investigation that includes drilling to obtain samples, permeability testing to determine hydraulic conductivity, and other tests to predict the movement of leachate from the buried refuse (10).

Design of Sanitary Landfills Design of modern sanitary landfills is complex and employs the multiple-barrier approach. Barriers include a compacted clay liner, leachate collection systems, and a compacted clay cap. Figure 12.4 is an idealized diagram showing these features, and Figure 12.5 shows such a landfill being constructed. Depending upon local site conditions, landfills may also have additional synthetic liners made of plastics or other materials and a system to collect natural gas that might accumulate. Finally, sanitary landfills must have a system of monitoring wells and other devices to evaluate potential for groundwater pollution. The subject of monitoring is an important one, and we will now address that issue in greater detail.

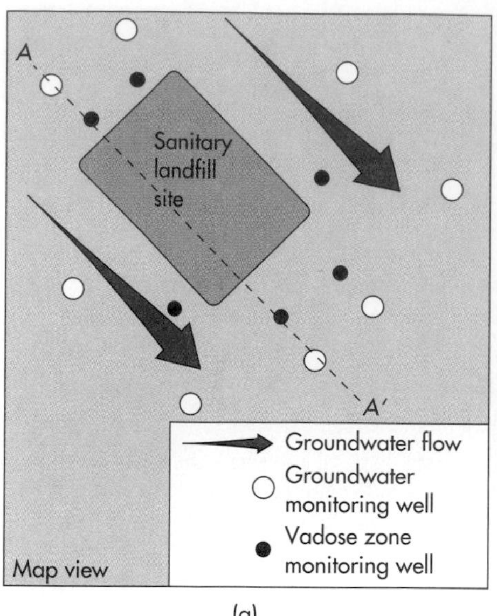

(a)

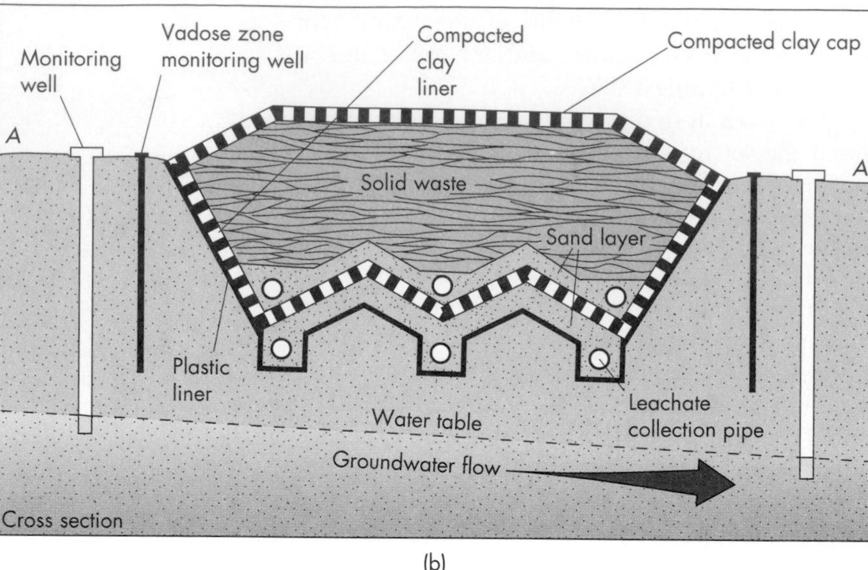

(b)

▲ **FIGURE 12.4** Idealized diagrams showing map view (a) and cross section (b) of a landfill with a double liner of clay and plastic and a leachate collection system.

Monitoring Sanitary Landfills Once a site is chosen for a sanitary landfill, **monitoring** the movement of groundwater should begin before filling commences. After the operation starts, continued monitoring of the movement of leachate and gases should be continued as long as there is any possibility of pollution. This is particularly important after the site is completely filled and the permanent cover material is in place, because a certain amount of settlement always occurs after a landfill is completed. If small depressions form as a result of settlement, surface water may collect, infiltrate the fill material, and produce leachate. Therefore, monitoring and proper maintenance of an abandoned landfill will reduce its pollution potential (10).

▲ **FIGURE 12.5** Rock Creek municipal landfill, Calaveras County, California, under construction. The light brown slope in the central part of the photograph is a compacted clay liner. The sinuous ditch is part of the leachate collection system, and the square pond in the upper part of the photograph is the leachate evaporation pond under construction. (*Courtesy of John Kramer*)

Hazardous waste pollutants from a solid-waste disposal site can enter the environment (14) by as many as seven paths (Figure 12.6):

- Gases in the soil and fill, such as methane, ammonia, hydrogen sulfide, and nitrogen, may volatilize and enter the atmosphere.
- Heavy metals such as lead, chromium, and iron are retained in the soil.
- Soluble substances, such as chloride, nitrate, and sulfate, readily pass through fill and soil to the groundwater system.
- If there is surface runoff, the runoff may pick up leachate and transport it into the surface-water network.
- Some crops and cover plants growing in the disposal area may selectively take up heavy metals and other toxic substances to be passed up the food chain as people and animals ingest them.

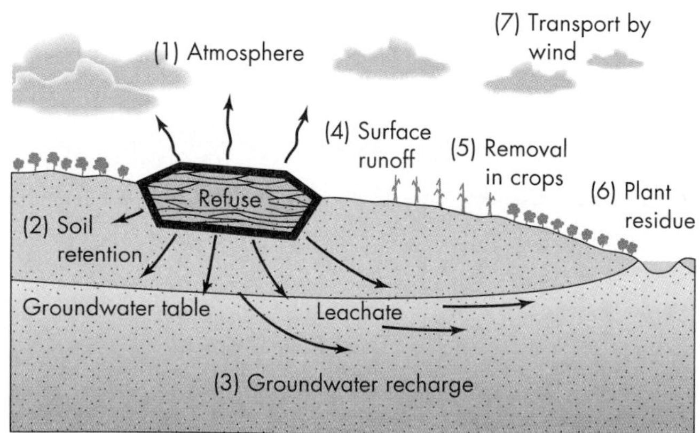

▲ **FIGURE 12.6** Several ways that hazardous waste pollutants from a solid-waste disposal site may enter the environment.

- If the plant residue left in the field contains toxic substances, it will return these materials to the environment through soil-forming and runoff processes.
- Paper, plastics, and other undesirable waste may be transported off-site by wind.

A thorough monitoring program would consider the seven possible paths by which pollutants enter the environment. Potential atmospheric pollution by gas from landfills is a growing concern, and a thorough monitoring program would include periodic analysis of air samples to detect toxic gas before it becomes a serious problem. Many landfills have no surface runoff; therefore, monitoring of on-site surface water is not necessary. However, if surface runoff does occur, thorough monitoring is required, and monitoring of nearby down-gradient streams, rivers, and lakes is necessary. Monitoring of soil and plants should include periodic chemical analysis at prescribed sampling locations.

If permeable water-bearing zones exist in the soil or bedrock below a sanitary landfill, monitoring wells (see Figure 12.4) are needed for frequent sampling of groundwater quality and monitoring of the movement of any leachate that has entered the groundwater (14). Even if the landfill is in relatively impermeable soil overlying dense permeable rock, minimal monitoring of groundwater quality through monitoring wells is still needed. In this case, leachate and groundwater movement may be less than 30 cm/yr. Water in the unsaturated (vadose) zone above the water table must also be monitored to identify potential pollution problems before they contaminate groundwater resources, where correction is very expensive. Waste transported off-site by wind is monitored, collected as necessary, and disposed of.

Sanitary Landfills and Federal Legislation Federal legislation regulates new landfills strictly. The intent of the legislation is to strengthen and standardize the design, operation, and monitoring of sanitary landfills. Those landfills that are unable to comply with the regulations might be shut down. Specific regulations include:

- Landfills may not be sited in certain areas, including floodplains, wetlands, unstable land, and earthquake fault zones. They may not be sited near airports because birds attracted to landfill sites present a hazard to aircraft.
- Landfill construction must include liners and a leachate collection system.
- Operators of landfills must monitor groundwater for specific toxic chemicals.
- Operators of landfills must meet financial assurance criteria. This may be met through posting bonds or insurance to ensure that monitoring of the landfill continues for 30 years after closure.

Under the law, states may opt to obtain approval of a solid-waste-management plan from the Environmental Protection Agency (EPA). A state that opts not to seek approval of its own plan must comply rigidly with the federal standards. Those states with EPA approval are allowed more flexibility. For example, alternative materials for the daily cover over the waste are allowed, as are different groundwater protection standards, documentation of groundwater monitoring, and financial assurance mechanisms. Furthermore, under certain circumstances, expansion of landfills in wetlands and fault zones may be allowed. Given the additional flexibility, it would appear advantageous for states to develop waste-management plans for their landfill facilities and have them approved by the EPA.

12.4 Hazardous Waste Management

The creation of new chemical compounds has proliferated tremendously in recent years. In the United States alone, approximately 1000 new chemicals are marketed annually and about 50,000 chemicals are currently on the market. Although many of these chemicals have been beneficial to people, several tens of thousands of them are classified as definitely or potentially hazardous to people's health (Table 12.2).

The United States is currently generating more than 150 million metric tons of **hazardous waste** each year. In the recent past, as much as half of the total volume of wastes was being indiscriminately dumped (15). This is now illegal, and we do not know how much illegal dumping is going

Table 12.2 Examples of products we use and potentially hazardous waste they generate

Products We Use	Potential Hazardous Waste
Plastics	Organic chlorine compounds
Pesticides	Organic chlorine compounds, organic phosphate compounds
Medicines	Organic solvents and residues, heavy metals (mercury and zinc, for example)
Paints	Heavy metals, pigments, solvents, organic residues
Oil, gasoline, and other petroleum products	Oil, phenols and other organic compounds, heavy metals, ammonia salts, acids, caustics
Metals	Heavy metals, fluorides, cyanides, acid and alkaline cleaners, solvents, pigments, abrasives, plating salts, oils, phenols
Leather	Heavy metals, organic solvents
Textiles	Heavy metals, dyes, organic chlorine compounds, solvents

Source: U.S. Environmental Protection Agency. SW-826, 1980.

on—certainly there is some, particularly in urban sewer systems. Past uncontrolled dumping of chemical waste has polluted soil and groundwater resources in several ways (Figure 12.7).

- Barrels in which chemical waste is stored, either on the surface or buried at a disposal site, eventually corroded and leaked, polluting the surface, soil, and groundwater.
- Liquid chemical wastes dumped in unlined lagoons (shallow ponds for collection of wastes) percolated through the soil and rock and eventually reached the groundwater table.
- Liquid chemical waste has been illegally dumped in deserted fields or along dirt roads.

Old abandoned hazardous landfills and other sites for the disposal of chemical waste have caused serious problems that have been very difficult to correct. A site near Elizabeth, New Jersey, provides an example of the casual dumping of chemicals that was once so widespread. At that site, the remains of about 50,000 charred drums were left standing next to a brick and steel building once owned by a now-

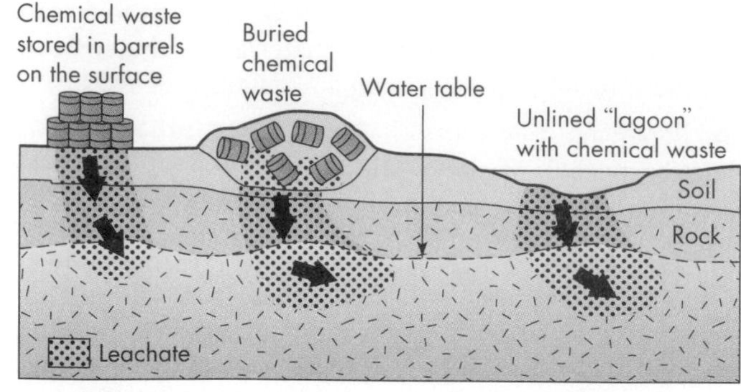

▲ **FIGURE 12.7** Ways that uncontrolled dumping of chemical waste may pollute soil and/or groundwater.

CASE HISTORY Love Canal

In 1976, in a residential area near Niagara Falls, New York, trees and gardens began to die. Children found the rubber on their tennis shoes and on their bicycle tires disintegrating. Dogs sniffing in a landfill area developed sores that would not heal. Puddles of toxic, noxious substances began to ooze to the soil surface; a swimming pool popped its foundations and was found to be floating on a bath of chemicals.

A study revealed that the residential area had been built on the site of a chemical dump. The area was excavated in 1892 by William T. Love as part of a canal between the upper and lower reaches of the Niagara River. The idea was to produce inexpensive hydroelectric power for a new urban-industrial center. When that plan failed, because alternating current was discovered and industry could be located far from the source of power, the canal was unused (except for recreation such as swimming and ice skating) for decades. It seemed a convenient place to dump wastes. From the 1940s to the 1950s, more than 80 different substances from a chemical company were dumped there. More than 20,000 tons of chemical waste, along with urban waste from the city of Niagara Falls, was disposed of in the canal (17). Finally, in 1953, the company dumping the chemicals donated the land to the city of Niagara Falls for one dollar. Eventually, several hundred homes adjacent to an elementary school were built near the site (Figure 12.B). Heavy rainfall and snowfall during the winter of 1976–77 set off the events that made Love Canal a household word.

A study of the site identified a number of substances present there—including benzene, dioxin, dichlorethylene, and chloroform—that were suspected of being carcinogens. Although officials readily admitted that very little was known about the impact of these chemicals and others at the site, there was grave concern for the people living in the area. During the next few years there were allegedly higher-than-average rates of miscarriages, blood and liver abnormalities, birth defects, and chromosome damage. However, a study by the New York State health authorities suggested that no chemically caused health effects had been absolutely established (18–20).

The cleanup of the Love Canal is an important demonstration of state-of-the-art technology in hazardous waste treatment. The objective is to contain and stop the migration of wastes through the groundwater flow system and to remove and treat dioxin-contaminated soil and sediment from stream beds and storm sewers (17). The method being used to minimize further production of contaminated water is to cover the dump site and adjacent contaminated area with a 1-m-thick layer of compacted clay and a polyethylene plastic cover to reduce infiltration of surface water. Lateral movement of water is inhibited from entering or escaping the site by specially designed perforated-tile drain pipe. These procedures will greatly reduce subsurface seepage of water through the site, and the water that does seep out will be collected and treated (18–20).

The homes adjacent to Love Canal were abandoned and bought by the government. Approximately 200 of the homes had to be destroyed. During the 1980s, approximately $175 million was spent for cleanup and relocation at Love Canal. The EPA now considers some of the area clean, and some of the remaining homes were scheduled to be sold in the early 1990s. Because the price of the homes is approximately 20 percent below the market of other areas in Niagara Falls, they are expected to sell. Sales contracts for four homes were approved in late 1990, this despite the reputation of the area and the adverse publicity it attracted. In early 1995 the maintenance and operation of the area was transferred from New York State to a consulting company, which will continue long-term sampling and monitoring (17,21).

What went wrong in Love Canal to produce a suburban ghost town? How can we avoid such disasters in the future? The real tragedy of Love Canal is that it is probably not an isolated incident. There are many hidden "Love Canals" across the country, "time bombs" waiting to explode (18,19).

bankrupt chemical corporation. The drums and other containers, which were stacked four high in places, had been corroding for nearly 10 years. Many of them had been improperly labeled or burned so badly that the nature of the chemicals could not be determined from outside markings. Leaking barrels had allowed unknown quantities of waste to seep into an adjacent stream that eventually flows into the Hudson River.

The New Jersey site was so polluted that cleanup efforts were very difficult. Identification of some of the materials at the site showed that there were two containers of nitroglycerine, numerous barrels of biological agents, cylinders of phosgene and gaseous phosphorus (which are extremely volatile and ignite when exposed to air), as well as a variety of heavy metals, pesticides, and solvents, some of which are very toxic. It took months of work with a large crew of people to remove most of the material from the New Jersey site. Unfortunately, it is difficult to know if all the waste has been removed; additional material may be buried at other sites that are more difficult to locate (16). There are many such stories of terrible problems resulting from chemical waste disposal, but the best known comes from the Love Canal near Niagara Falls, New York (see *Case History: Love Canal*).

Responsible Management

In 1976 the U.S. government moved to begin the management of hazardous waste with the passage of the Resource Conservation and Recovery Act (RCRA), which is intended to provide for "cradle-to-grave" control of hazardous waste. At the heart of the act is the identification of hazardous wastes and their life cycles. Regulations call for stringent record keeping and reporting to verify that wastes do not present a public nuisance or a public health problem. The act also identifies hazardous wastes in terms of several categories:

▲ **FIGURE 12.B** This is an aerial infrared photograph of the Love Canal area in New York. Healthy vegetation is bright red. This portion of Love Canal runs from the upper left corner to the lower right. It appears as a scar on the landscape. Buried chemical waste seeped to the surface to cause numerous environmental problems and concern here. The site became a household name for toxic waste. (*New York State Department of Environmental Conservation*)

- Materials that are highly toxic to people and other living things
- Wastes that may explode or ignite when exposed to air
- Wastes that are extremely corrosive
- Wastes that are otherwise unstable

Recognizing that a great number of waste disposal sites presented hazards, Congress in 1980 passed the Comprehensive Environmental Response Compensation and Liability Act (CERCLA), which established a revolving fund (popularly called the *Superfund*) to clean up several hundred of the worst abandoned hazardous chemical-waste-disposal sites known to exist around the country. The EPA developed a list of Superfund sites (National Priorities List). Figure 12.8 summarizes environmental impact statistics and lists some of the pollutants encountered at Superfund sites.

Although the Superfund has experienced significant management problems and is far behind schedule, a number of sites have been treated. Unfortunately, the funds available are not sufficient to pay for decontamination of all the targeted areas. That would cost many times more, perhaps as much as $100 billion. Furthermore, because of concern that the present technology is not sufficiently advanced to treat all the abandoned waste-disposal sites, the strategy may be simply to confine the waste to those areas until better disposal methods are developed. It seems apparent that the danger of abandoned disposal sites is likely to persist for some time to come.

The federal legislation also changed the way the real estate industry did business. The act has tough liability provisions, and property owners could be liable for costly cleanup of hazardous waste found on their property (even if they did not cause the problem). Banks and other lending institutions could be liable for release of hazardous materials on their property by their tenants. In 1986 the Superfund Amendment and Reauthorization Act (SARA) provided a possible defense for real estate purchasers against liability provided they completed an environmental audit prior to purchase. The audit is a study of past land use at the site (determined by analyzing old maps and aerial photographs, and may involve drilling and sampling of soil and groundwater) to determine if pollutants are present. Such audits now are done on a routine basis prior to purchase of property for development.

The SARA legislation required that certain industries report all releases of hazardous materials into the environment. The list of companies releasing such substances became public and was known as the "Toxic 500 list." Unwanted publicity to companies on the list is thought to have resulted in better and safer handling of hazardous waste by firms that formerly were identified as polluters of the environment. No owner wants his or her company to be the No. 1 (or even the twenty-fifth or hundredth) most serious polluter among U.S. firms (22).

Management of hazardous chemical waste includes several options: recycling, on-site processing to recover byproducts with commercial value, microbial breakdown, chemical stabilization, high-temperature decomposition, incineration, and disposal by secure landfill or deep-well injection. A number of technological advances have been made in the field of toxic waste management, and as land disposal becomes more and more expensive, the trend toward onsite treatment that has recently started is likely to continue. However, on-site treatment will not eliminate all hazardous chemical waste; disposal will remain necessary. Table 12.3 compares hazardous waste reduction technology in terms of treatment and disposal. Notice that all of the technologies available will cause some environmental disruption. No one simple solution exists for all waste-management issues.

Secure Landfill

The basic idea of the **secure landfill** is to confine the waste to a particular location, control the leachate that drains from the waste, collect and treat the leachate, and detect possible leaks. Figure 12.9 demonstrates these procedures. A dike and liners (made of clay and impervious material such as plastic) confine the waste, and a system of internal drains concentrates the leachate in a collection basin from which it is pumped out and transported to a wastewater treatment

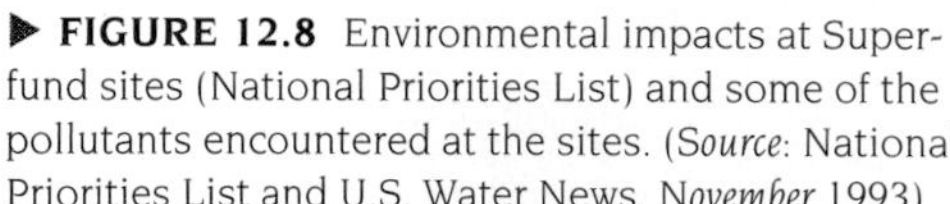

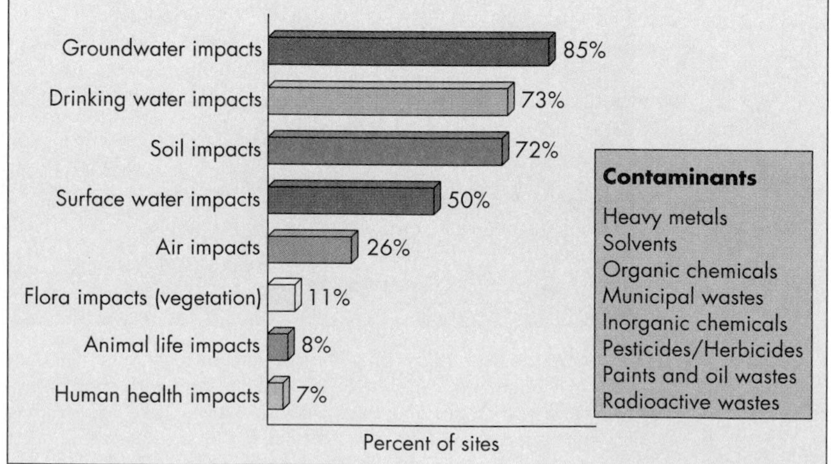

▶ **FIGURE 12.8** Environmental impacts at Superfund sites (National Priorities List) and some of the pollutants encountered at the sites. (*Source*: National Priorities List and U.S. Water News, *November* 1993)

Table 12.3 Comparison of hazardous waste reduction technologies

	Disposal		Treatment		
	Landfills and Impoundments	Injection Wells	Incineration and Other Thermal Destruction	Emerging High-Temperature Decomposition[a]	Chemical Stabilization
Effectiveness: How well it contains or destroys hazardous characteristics	Low for volatiles, questionable for liquids; based on lab and field tests	High, based on theory, but limited field data available	High, based on field data, except little data on specific constituents	Very high; commercial scale tests	High for many metals; based on lab tests
Reliability issues	Siting, construction, and operation Uncertainties: long-term integrity of cells and cover, linear life less than life of toxic waste	Site history and geology; well depth, construction, and operation	Monitoring uncertainties with respect to high degree of DRE: surrogate measures, PICs, incinerability[c]	Limited experience Mobile units, on-site treatment avoids hauling risks Operational simplicity	Some inorganics still soluble Uncertain leachate test, surrogate for weathering
Environment media most affected	Surface and ground water	Surface and ground water	Air	Air	Groundwater
Least compatible wastes[b]	Liner reactive, highly toxic, mobile, persistent, and bioaccumulative	Reactive; corrosive; highly toxic, mobile, and persistent	Highly toxic and refractory organics, high heavy metals concentration	Some inorganics	Organics
Relative costs: Low, Moderate, High	L-M	L	M-H	M-H	M
Resource recovery potential	None	None	Energy and some acids	Energy and some metals	Possible building material

[a]Molten salt, high-temperature fluid well, and plasma arc treatments.
[b]Wastes for which this method may be less effective for reducing exposure, relative to other technologies.
[c]DRE = destruction and removal efficiency. PIC = product of incomplete combustion.

Source: Council on Environmental Quality 1983.

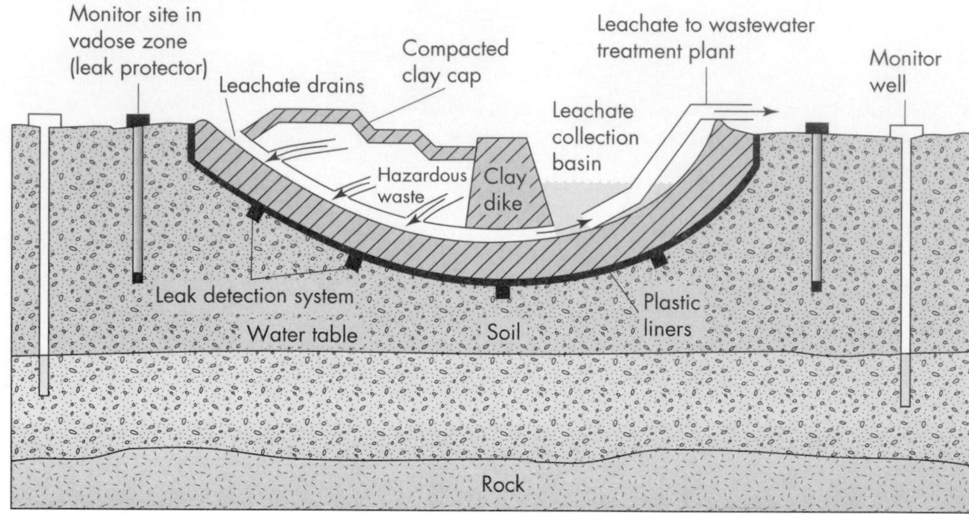

▶ **FIGURE 12.9** A secure landfill for hazardous chemical waste. The impervious liners and systems of drains are an integral part of the system to ensure that leachate does not escape from the disposal site. Monitoring in the vadose zone is important and involves periodic collection of soil water with a suction device.

plant. Designs of new facilities today must include multiple barriers consisting of several impermeable layers and filters as well as impervious covers. The function of impervious liners is to ensure that the leachate does not contaminate soil and, in particular, groundwater resources. However, this type of waste-disposal procedure must have several monitoring wells to alert personnel if and when leachate migrates out of the system possibly to contaminate nearby water resources.

It has been argued that there is no such thing as a really secure landfill, implying that all landfills leak to some extent. This is probably true; impervious plastic liners, filters, and clay layers can fail, even with several backups, and drains can become clogged, causing overflow. Yet landfills that are carefully sited and engineered can minimize problems. Preferable sites are those with good natural barriers, such as thick clay-silt deposits, an arid climate, or a deep water table that minimizes migration of leachate. Nevertheless, land disposal should be used only for specific chemicals suitable for the method.

Land Application

Application of waste materials to the surface-soil horizon is referred to as **land application,** *land spreading,* or *land farming.* Land application may be a desirable treatment method for certain biodegradable industrial wastes, including petroleum oily waste and certain organic chemical-plant waste. A good indicator of the usefulness of land application for disposal of a particular waste is the waste's **biopersistence** (the measure of how long a material remains in the biosphere). The greater or longer the biopersistence, the less suitable the waste is for land-application procedures. Land application is not an effective treatment or disposal method for inorganic substances, such as salts and heavy metals (23).

Land application of biodegradable waste works because, when such materials are added to the soil, they are attacked by microorganisms (bacteria, molds, yeast, and other organisms) that decompose the waste material. The soil may thus be thought of as a microbial farm that constantly recycles matter by breaking it down into more fundamental forms useful to other living things in the soil. Because the upper soil zone contains the largest microbial populations, land application is restricted to the uppermost 15 to 20 cm of the soil profile (23). As with other types of land-disposal technology, the vadose zone and groundwater near the site must be carefully monitored to ensure the disposal system is working as planned and not polluting water resources.

Surface Impoundment

Excavations and natural topographic depressions have been used to hold hazardous liquid waste. These **surface impoundments** are primarily formed of soil or other surficial materials, but they may be lined with manufactured materials such as plastic. The impoundment is designed to hold the waste; examples include aeration pits and lagoons at hazardous-waste facilities. Surface impoundments have been criticized because they are especially prone to seepage, resulting in pollution of soil and groundwaters. Evaporation from surface impoundments can also produce an air-pollution problem. For these reasons, hazardous-waste facilities have been prohibited from receiving noncontainerized liquid waste.

Deep-well Disposal

Another method of hazardous-waste disposal is by injection into deep wells. The term *deep* refers to rock (not soil) that is below and completely isolated from all freshwater aquifers, thereby assuring that injection of waste will not contaminate or pollute existing or potential water supplies. This generally means that the waste is injected into a permeable rock layer several hundred to several thousand meters below the surface in geologic basins confined above by relatively impervious, fracture-resistant rock, such as shale or salt deposits (4).

Deep-well injection of oil-field brine (salt water) has been important in controlling water pollution in oil fields for many years, and huge quantities of liquid waste (brine) pumped up with oil have been injected back into the rock.

Today, several billion liters per day are pumped into subsurface rocks (24). In recent years, the technique has been used more commonly for permanent storage of industrial waste deep underground. A typical well is about 700 m deep, and wastes are pumped into a 60-m-thick zone at a rate of about 400 liters per minute (25).

Deep-well disposal of industrial wastes should not be viewed as a quick and easy solution to industrial-waste problems (26). Even where geologic conditions are favorable for deep-well disposal, natural restrictions include the limited number of suitable sites and the limited space within these sites for disposal of waste. Possible injection zones in porous rock are usually already filled with natural fluids, mostly brackish or briny water. Therefore, to pump in waste, some of the natural fluid must be displaced by compression (even slight compression of the natural fluids in a large volume of permeable rock can provide considerable storage space) and by slight expansion of the reservoir rock as the waste is being injected (25).

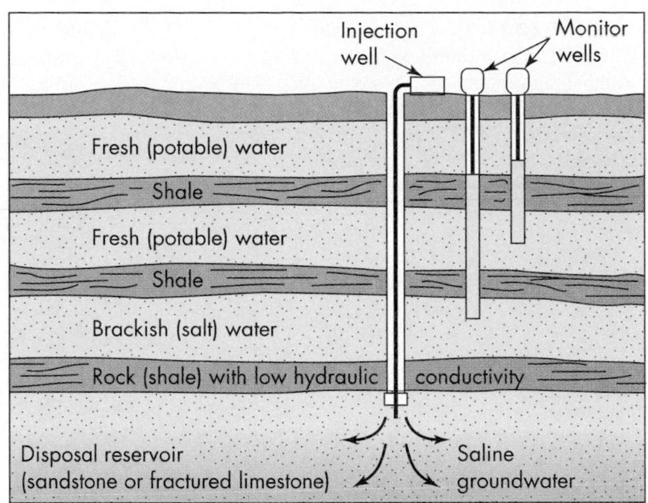

▲ **FIGURE 12.10** A deep-well injection system. The disposal reservoir is a sandstone or fractured limestone capped by impermeable rock and isolated from all fresh water. Monitor wells are a safety precaution to ensure that there is no undesirable migration of the liquid waste into freshwater aquifers above the injection zone.

Problems with Deep-well Disposal Several problems associated with disposal of liquid waste in deep wells have been reported (25,26). Perhaps the best known are the earthquakes that were caused by injecting waste from the Rocky Mountain Arsenal near Denver, Colorado (see Chapter 7). These earthquakes occurred between 1962 and 1965. The injection zone was fractured gneiss at a depth of 3.6 km, and the increased fluid pressure evidently initiated movement along the fractures. This is not a unique case. Similar initiations of earthquakes have been reported in oil fields in western Colorado, Texas, and Utah (25). Similar activation of faults in southern California caused by injection of fluids into the Inglewood oil field for secondary recovery is thought to have contributed to the failure of the Baldwin Hills Reservoir (see Chapter 2).

Feasibility and General Site Considerations The feasibility of deep-well injection as the best solution to a disposal problem depends on four factors: the geologic and engineering suitability of the proposed site; the volume and the physical and chemical properties of the waste; economics; and legal considerations (27). The geologic considerations for disposal wells are twofold (27):

- The injection zone must have sufficient porosity, thickness, hydraulic conductivity, and size to ensure safe injection. Sandstone and fractured limestone are the commonly used reservoir rocks (26).
- The injection zone must be below the level of freshwater circulation and confined by a relatively impermeable rock with low hydraulic conductivity, such as shale or salt, as shown in Figure 12.10.

Optimal use of limited underground storage space is achieved if deep-well injection is used only when more satisfactory methods of waste disposal are not available and when the volume of injected wastes is minimized by good waste management (27). Optimal use includes thorough evaluation of the physical and chemical properties of the waste to ensure that it will not adversely affect the ability of the rock reservoir to accept it. Adverse effects can be minimized in some instances by preinjection treatment of the waste to enhance compatibility with the reservoir rock and the natural fluids present there. It is also advisable to take advantage of natural buffers. For example, if the waste is acidic, use of limestone as a reservoir rock may be advantageous because the acid waste tends to increase the reservoir's permeability and hydraulic conductivity by chemically attacking and enlarging natural fractures, thus allowing more waste to be injected. Care must be taken, however, because certain acids, such as sulfuric acid, may react to plug the porosity of limestone.

Construction and operating costs often determine the feasibility of deep-well injection as the preferred method of waste treatment. Important geologic factors pertaining to construction costs are the depth of the well and the ease with which drilling proceeds. Operating costs depend on the hydraulic conductivity, porosity, and thickness of the injection zone, and the fluid pressure in the reservoir. All of these are important in determining the rate at which the reservoir will accept liquid waste (27).

Consideration of legal aspects of deep-well disposal suggests that existing laws, regulations, and policies are probably adequate. Injection well use in the United States is regulated by federal laws, and certain hazardous wastes are prohibited from being injected underground (22).

Monitoring Disposal Wells An essential part of any disposal system is monitoring. It is very important to know exactly where the wastes are going, how stable they are, and how fast they are migrating. It is especially important in deep-well disposal where toxic or otherwise hazardous materials are involved.

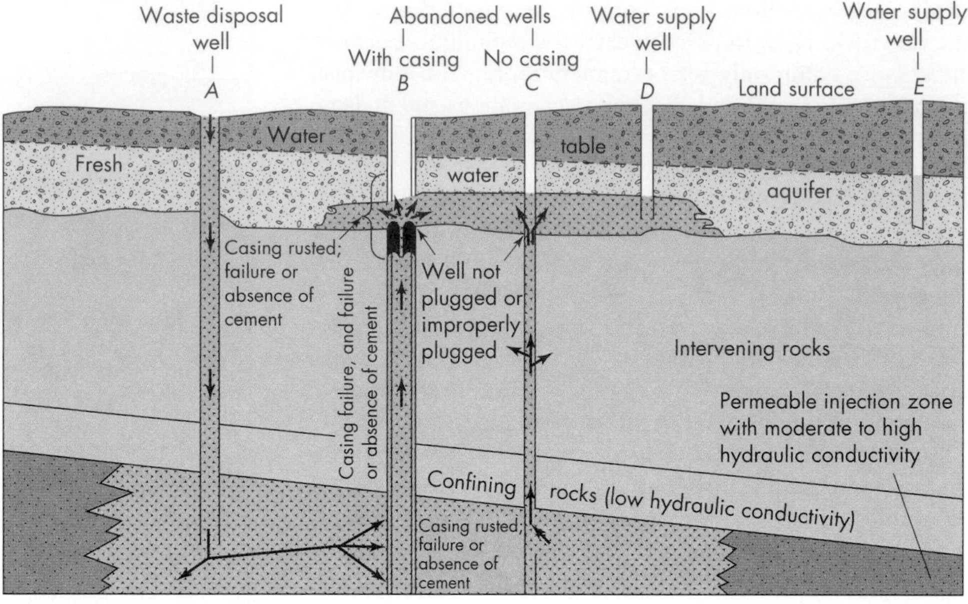

▶ **FIGURE 12.11** How liquid waste might enter a freshwater aquifer through abandoned wells. This diagram illustrates why the location of all abandoned wells should be known, and it emphasizes the necessity of monitoring wells. (*After Irwin and Morton*. 1965. U.S. *Geological Survey Circular* 630.)

Effective monitoring requires that the geology be precisely defined and mapped before initiating the disposal program. Especially important is locating all freshwater-bearing zones and old or abandoned oil or gas wells that might allow the waste to migrate up to freshwater aquifers or to the surface (Figure 12.11). A system of deep monitoring wells drilled into the disposal reservoir in the vicinity of the well can monitor the movement of waste, and shallow monitoring wells drilled into freshwater zones can monitor the water quality to identify quickly any upward migration of the waste.

Incineration of Hazardous Chemical Waste

Hazardous waste may be destroyed through high-temperature incineration. Incineration is considered to be a waste treatment rather than a disposal method because the hazardous waste is not disposed of directly; rather, it undergoes a treatment (incineration) that produces an ash residue to be disposed of in a landfill (23). The technology used in incineration and other high-temperature decomposition or destruction is changing rapidly. Figure 12.12 diagrams one type of high-temperature incineration system that may be used to burn toxic waste. Waste—as liquid, solid, or sludge—enters the rotating combustion chamber, where it is rolled and burned. Ash from this burning process is collected in a water tank, and the remaining gaseous materials move into a secondary combustion chamber, where the process is repeated. Remaining gas and particulates move through a scrubber system that eliminates particulates and acid-forming components. Carbon dioxide, water, and air then are emitted from the stack. As shown in Figure 12.12, ash particulates and wastewater are produced at various parts

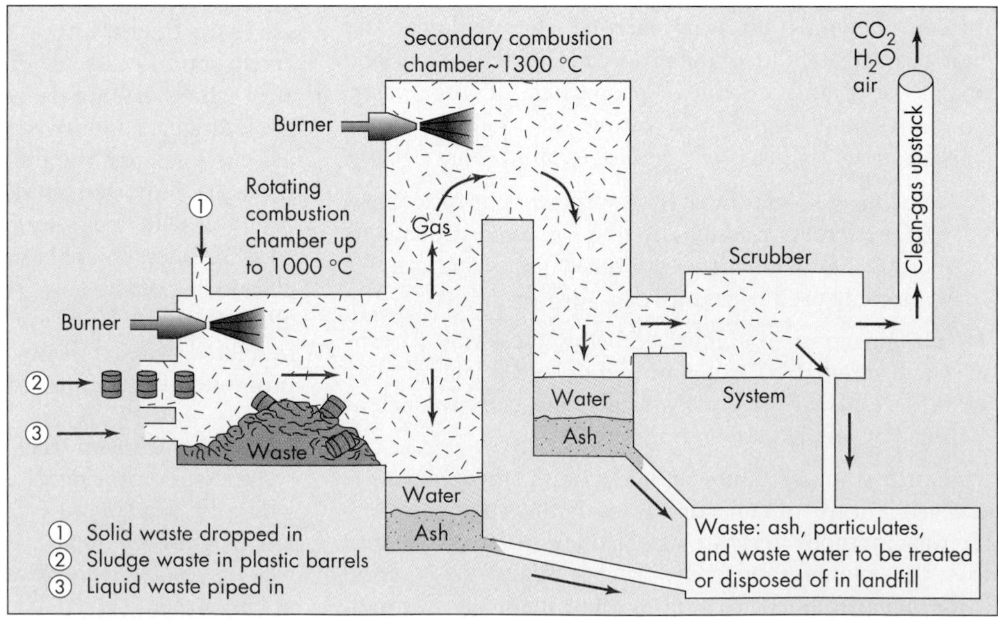

▶ **FIGURE 12.12** High-temperature incinerator system to burn toxic waste.

of the incineration process; these must be either treated or disposed of in a landfill.

More advanced types of incineration and thermal decomposition of waste are being developed. One of these utilizes a molten salt bed that should be useful in destroying certain organic materials. Other incineration techniques include liquid-injection incineration on land or sea, fluidized-bed systems, and multiple-hearth furnaces. Which incineration method is used for a particular waste depends upon the nature and composition of the waste and the temperature necessary to destroy the hazardous components. For example, the generalized incineration system shown in Figure 12.12 could be used to destroy PCBs.

Alternatives to Land Disposal

Direct land disposal of hazardous waste is often not the best initial alternative. Even with extensive safeguards and state-of-the-art designs, land-disposal alternatives cannot guarantee that the waste is contained and will not cause environmental disruption in the future. This holds true for all land-disposal facilities, including landfill, surface impoundments, land application, and injection wells. Pollution of air, land, surface water, and groundwater may result from failure of a land-disposal site to contain hazardous waste. Groundwater pollution is perhaps the most significant result of such failure, because it provides a convenient route for pollutants to reach humans and other living things. Figure 12.13 shows some of the paths that pollutants may take from land-disposal sites to contaminate the environment. These paths include:

- Improper landfill procedures that eventually produce leakage and runoff to surface water or groundwater
- Seepage, runoff, or air emissions from unlined lagoons
- Percolation and seepage resulting from surface application of waste to soils
- Leaks in pipes or other equipment associated with deep-well injection
- Leaks from buried drums, tanks, or other containers

It has been argued that alternatives to land disposal are not being utilized to their full potential. That is, the volume of waste could be reduced and the remaining waste could be recycled or treated prior to land disposal of the residues of the treatment processes. The philosophy of handling hazardous chemical waste should be multifaceted and should include such processes as source reduction, recycling and resource recovery, and treatment; in other words, materials management (28).

Source Reduction Source reduction has the objective of reducing the amount of hazardous waste generated by manufacturing or other processes. For example, changes in the chemical processes, equipment, raw materials, or maintenance measures employed may be successfully utilized to reduce either the amount or the toxicity of the hazardous waste produced (28).

Recycling and Resource Recovery Hazardous chemical waste may contain materials that can be successfully recovered for future use. For example, acids and solvents collect contaminants when they are used in manufacturing processes. These acids and solvents can be processed to remove the contaminants and can then be reused in the same or different manufacturing processes (28).

Treatment Hazardous chemical waste can be treated by a variety of processes to change the physical or chemical composition of the waste in such a way as to reduce its toxic or hazardous characteristics. Examples include neutralizing acids, precipitation of heavy metals, and oxidation to break up hazardous chemical compounds. Incineration, as we have pointed out, is also a type of waste treatment.

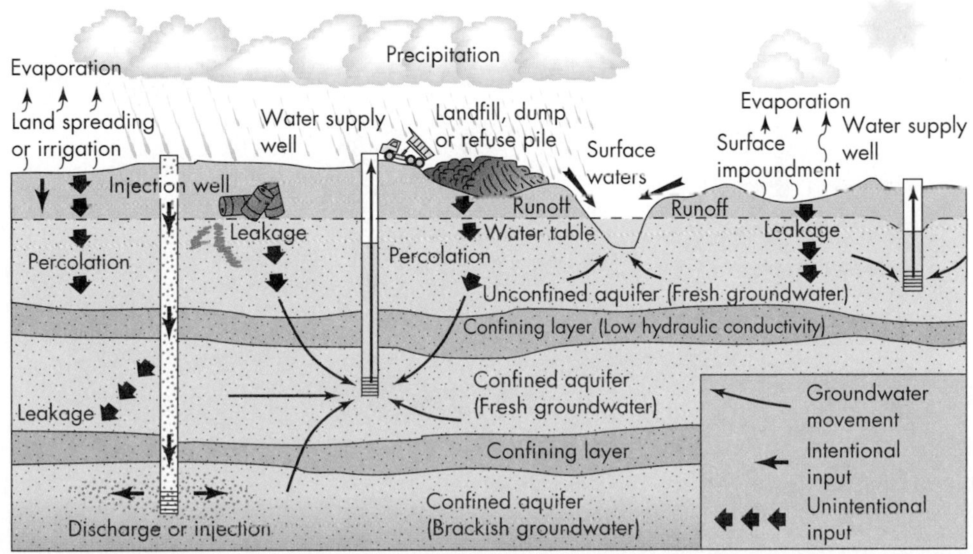

FIGURE 12.13 Examples of how land disposal/treatment methods of managing hazardous wastes may contaminate the environment. (*Modified after* C. B. Cox. 1985. The buried threat. No. 115-5. *Sacramento, CA: California Senate Office of Research*)

The advantages of source reduction, recycling, and treatment include:

- The waste that must be later disposed of is reduced to a much smaller volume, which produces less stress on the dwindling number of acceptable landfill sites.
- Treatment of wastes may make them less toxic and therefore less likely to cause problems in landfills.
- Useful chemicals may be reclaimed and reused.

12.5 Radioactive Waste Management

Radioactive wastes are by-products that must be expected as electricity is produced from nuclear reactors or weapons are manufactured from plutonium. Considering waste-disposal procedures, radioactive waste may be grouped into two general categories: low-level waste and high-level waste. In addition, the tailings (materials that are removed by mining activity but are not processed and remain at the site) from uranium mines and mills must also be considered very hazardous (Figure 12.14). In the western United States, more than 20 million tons of abandoned tailings will produce radiation for at least 100,000 years.

Disposal of Low-level Radioactive Wastes

Low-level radioactive wastes are materials containing only small amounts of radioactive substances. These low-level wastes include a wide variety of items such as residuals or solutions from chemical processing; solid or liquid plant waste, sludges, and acids; and slightly contaminated equipment, tools, plastic, glass, wood, fabric, and other materials (29). Prior to disposal, liquid low-level radioactive waste is solidified or packaged with material capable of absorbing at least twice the volume of liquid present (30).

Radioactive decay of low-level waste does not generate a great deal of heat, and a rule of thumb is that the material must be isolated from the environment for about 500 years to ensure that the level of radioactivity does not produce a hazard. In the United States the philosophy for management of low-level radioactive waste has been "dilute and disperse." Experience suggests that low-level radioactive waste can be buried safely in carefully controlled and monitored near-surface burial areas in which the hydrologic and geologic conditions severely limit the migration of radioactivity (29). Such waste has been buried at 15 main sites in states including Washington, Nevada, New Mexico, Missouri, Illinois, Ohio, Tennessee, Kentucky, South Carolina, and New York.

Several of the burial sites for low-level radioactive waste have not provided adequate protection of the environment. This failure has been due at least in part to a poor understanding of the local hydrologic and geologic environment (30). For example, a study of the Oak Ridge National Laboratory in Tennessee suggests that the water table is less than 7 m below the ground surface in places. The investigation identified migration of radioactive materials from one of the burial sites and concluded that containment of the waste is difficult because of the short residence time of water in the vadose zone. In other words, leachate generated from the disposal sites does not take long to infiltrate the vadose zone and percolate down to the groundwater (30). On the other hand, the depth to the water table at the low-level radioactive waste-disposal facility near Beatty, Nevada, is about 100 m. That site has apparently successfully contained radioactive

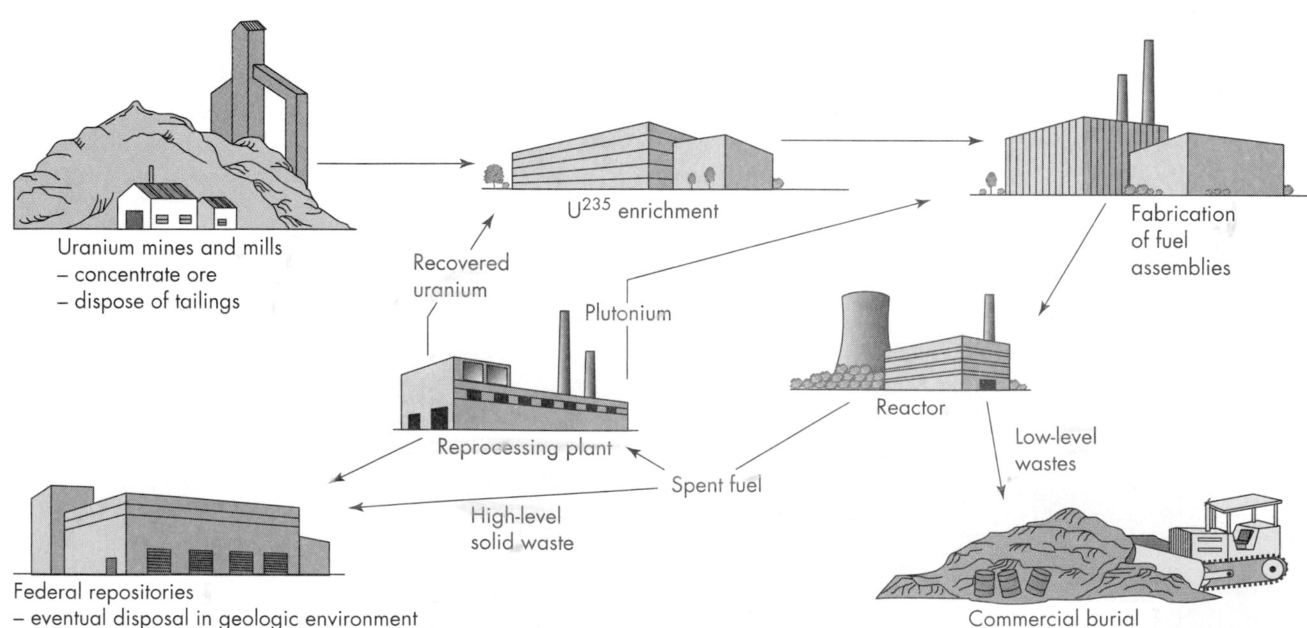

▲ **FIGURE 12.14** The nuclear fuel cycle. The United States does not now reprocess spent fuel. Disposal of tailings, which because of their large volume can be more toxic than high-level radioactive wastes, has been treated casually.

waste, partly because it takes a long time for any leachate generated to enter the groundwater environment (30).

Both the hydrology and the geology of a particular disposal site play major roles in containing radioactive materials at a low-level radioactive-waste facility. Environmental hazards associated with low-level radioactive waste can persist for 500 years, and unfortunately most investigations of sites receiving waste have been ongoing for only about 20 years. This time span is obviously inadequate to estimate properly the success of particular sites. Nevertheless, several criteria for disposal sites have been delineated in an attempt to create a multiple-barrier approach to disposal. Figure 12.15 shows an idealized disposal site and some of the natural factors that provide the multiple barriers likely to help confine the waste material (30). It is emphasized that no one site is likely to have all of these natural factors, so site selection must carefully consider options to minimize migration of radioactive waste from the burial area.

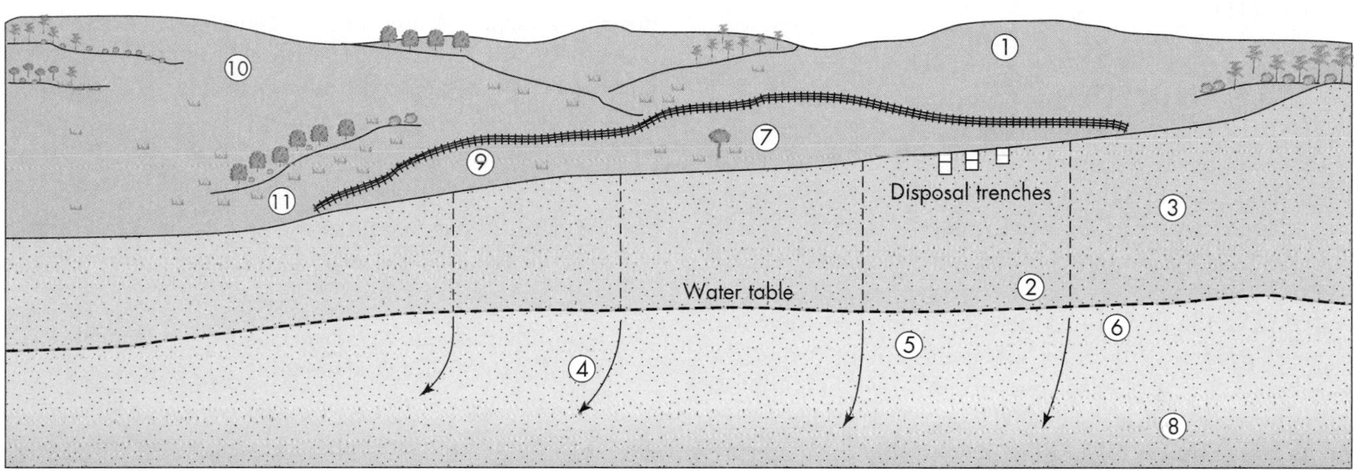

Explanation

1. Low rainfall:	Means less surface infiltration and thus less production of leachate.
2. Deep water table:	In general, the deeper the water table, the longer the time necessary for leachate to reach the groundwater.
3. Modest soil hydraulic conductivity:	High hydraulic conductivity leads to rapid transport rates of fluids and very low hydraulic conductivity may lead to surface ponding; therefore, moderate hydraulic conductivity is deemed more desirable.
4. Slow moving groundwater:	Provides for or facilitates longer residency time and thus time of travel of contaminants from the disposal site to other locations.
5. High adsorption and ion exchange rates:	Facilitates adhesion of radioactive molecules to the surface of other earth materials, resulting in removal of contaminants.
6. Homogeneous geology:	Makes it easier to predict likely movement of contaminants. Complex geology consisting of faults, folds, and other structures provides discontinuities along which waste may preferentially migrate.
7. Topography and soils that minimize erosion:	Reduces the likelihood of exposure of waste by erosional processes.
8. Absence of exploitable resources:	Waste must be isolated for hundreds of years, and so obviously we do not want to pick a site with needed resources.
9. Absence of surface-water bodies:	Reduces likelihood of pollution of water bodies.
10. Low probability for faulting or volcanic activity:	Faulting or volcanic activity at or near a site obviously might jeopardize containment of low-level radioactive waste.
11. Adequate buffer zone:	Recognizes that containment is not always possible and buffer zones provide additional time for radioactive decay to take place and thus are an added safety precaution.

▲ **FIGURE 12.15** Idealized diagram illustrating natural factors associated with suitability for burial of low-level radioactive waste. (*After* J. N. *Fischer.* 1986. U.S. *Geological Survey Circular* 973.)

Disposal of High-level Radioactive Wastes

High-level radioactive wastes are produced as fuel assemblages in nuclear reactors become contaminated with large quantities of fission products. This spent fuel must periodically be removed and reprocessed or disposed of. Fuel assemblies will probably not be reprocessed in the near future in the United States (reprocessing is more expensive than mining and processing new uranium); therefore, the present waste-management problems involve removal, transport, storage, and eventual disposal of spent fuel assemblies (31).

The Scope of the Disposal Problem Hazardous radioactive materials produced from nuclear reactors include fission products such as krypton-85 (half-life of 10 years), strontium-90 (half-life of 28 years), and cesium-137 (half-life of 30 years). The half-life is the time required for the radioactivity to be reduced to one-half its original level. Generally, at least 10 half-lives (and preferably more) are required before a material is no longer considered a health hazard. Therefore, a mixture of the fission products mentioned above would require hundreds of years of confinement from the biosphere. Reactors also produce a small amount of plutonium-239 (half-life of 24,000 years). Because plutonium and its fission products must be isolated from the biological environment for a quarter of a million years or more, their permanent disposal is a geologic problem.

High-level radioactive waste is extremely toxic, and a sense of urgency surrounds its disposal as the total volume of spent fuel assemblies slowly accumulates. But there is also conservative optimism that the waste-disposal problem will be solved. It has been projected that without a disposal program, 40,000 metric tons of spent fuel elements from commercial reactors will be in storage at U.S. reactor sites by the year 2000, awaiting disposal or eventual reprocessing to recover plutonium and unfissioned uranium (31). With reprocessing, solid high-level radioactive waste would occupy only several thousand cubic meters, a volume that would not even cover a football field to a depth of 1 m (32).

Production of plutonium for nuclear weapons also generates high-level radioactive waste. By the year 2000 there will be about 8000 metric tons of this solidified high-level waste being stored at U.S. Department of Energy repositories at Hanford, Washington; Savannah River, Georgia; and Idaho Falls, Idaho. Serious problems have occurred with liquid radioactive waste buried in underground tanks. Sixteen leaks involving 1330 m^3 were located at Hanford from 1958 to 1973. An incident in 1973 involved a leak of 437 m^3 of low-temperature waste. Since then, various improvements, including stronger, double-shelled storage tanks, reduction of the volume of liquid waste stored through the solidification program, and increased reserve capacity, have been made. It is hoped these changes will reduce the chance of future incidents (33).

Storage of high-level radioactive waste is at best a temporary solution that allows the federal government to meet its commitments for accepting waste. Regardless of how safe any "storage" program is, it requires continuous surveillance and periodic repair or replacement of tanks or vaults. Therefore, it is desirable to develop more permanent "disposal" methods in which retrieval may be possible but is not absolutely necessary.

Disposal in the Geologic Environment There is fair agreement that the geologic environment can provide the most certain safe containment of high-level radioactive waste. Because disposal of this waste is a necessity, the federal government is actively pursuing and developing possible alternative methods. Although such concepts as disposal into polar ice caps or into sediment in deep ocean basins have been explored, stable bedrock offers the most promise.

A comprehensive geologic disposal development program should have a number of objectives (32):

- To identify sites that meet the broad criteria of tectonic stability and slow movement of groundwater with long flow paths to the surface
- To conduct intensive subsurface exploration of possible sites to positively determine geologic and hydrologic characteristics
- To predict the future behavior of potential sites on the basis of their present geologic and hydrologic characteristics and possible changes in such variables as climate, groundwater flow, erosion, and tectonics
- To evaluate the risk associated with various predicted changes
- To make a political decision as to whether the risks are acceptable to society

The Nuclear Waste Policy Act of 1982 initiated a comprehensive federal–state, high-level nuclear waste disposal program. The Department of Energy was responsible for investigating several potential sites, and the act originally called for the President to recommend a site by 1987. In December 1987, Congress amended the act to specify that only the Yucca Mountain site in southern Nevada would be evaluated to determine if high-level radioactive waste could be disposed of there. Some scientists and others believe that the site was chosen not so much for its geology (although the rock type at the site does have several favorable qualities for disposal) as because it is an existing nuclear reservation and therefore might draw minimal social and political opposition (33,34).

The rock at the Yucca Mountain site is densely compacted tuff (naturally welded volcanic ash). Fortunately, the site is located in an extremely dry region. Precipitation is about 15 cm/yr and most of this runs off or evaporates. Hydrologists have estimated that less than 5 percent of the precipitation infiltrates the surface and eventually reaches the water table, which is several hundred meters below the surface. The depth to the potential repository is about 300 m below the mountain's surface and, as a result, the repository could be constructed about 200 m above the water table in the vadose zone (35).

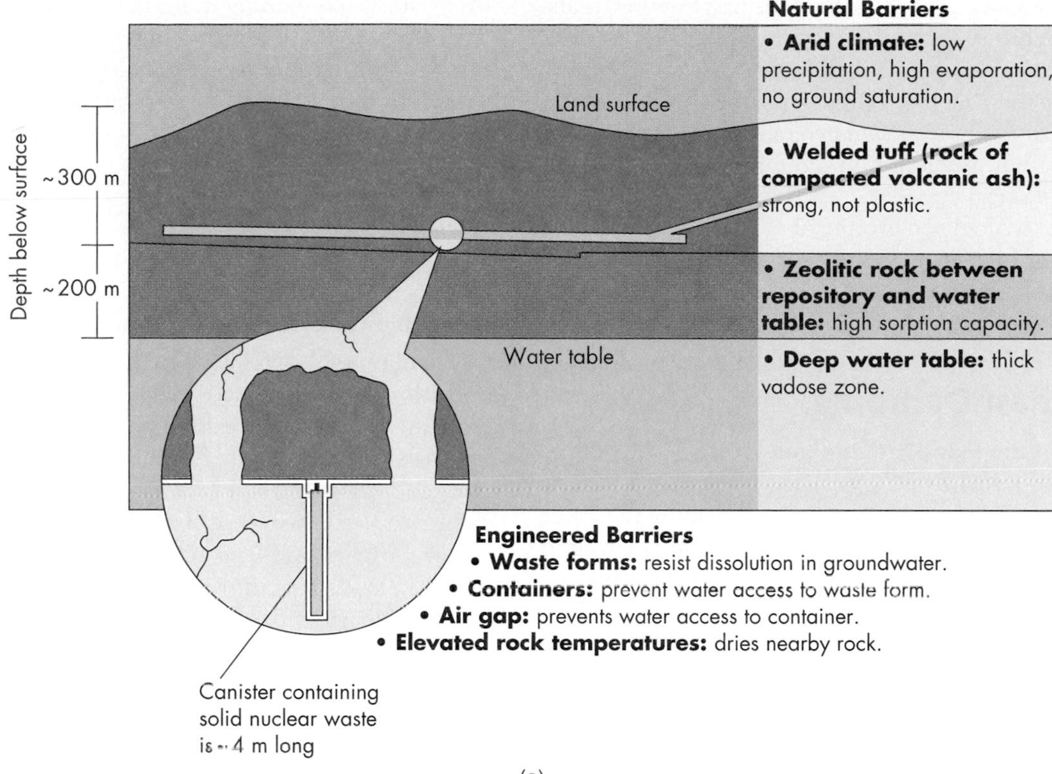

▲ **FIGURE 12.16** (a) Idealized diagram of proposed Yucca Mountain repository. Listed are natural and engineered barriers. The waste would be retrievable for 50 years. (*Source: Lawrence Livermore Laboratory*. 1988. LLL-TB-92.) (b) Aerial view of the Yucca Mountain Site that is being investigated for the disposal of high-level nuclear waste. The site is on the boundary of the Nevada Test Site where the United States has conducted underground nuclear tests. (*Mark Marten/U.S. Department of Energy/Photo Researchers, Inc.*)

During the 1990s the Department of Energy and the U.S. Geological Survey completed extensive scientific evaluation of the Yucca Mountain site. These studies will help determine how well the geologic and hydrologic setting can isolate high-level nuclear waste from the environment. The site is attractive because there are several natural barriers present (Figure 12.16):

- An arid climate, which greatly restricts downward movement of water
- A strong rock (welded tuff that has a high sorption capacity for radioactive material)
- A thick vadose zone (deep water table)

Additional engineering and construction is expected to provide additional barriers to the waste escaping the storage canisters.

If the studies indicate that the Yucca Mountain site could safely isolate radioactive waste, then the Department of Energy will apply to the U.S. Nuclear Regulatory Commission for a license to construct a disposal facility. If opposition from the state of Nevada is worked out and environmental and safety factors are met, the repository might be ready for acceptance of high-level radioactive waste by the year 2010.

Long-term Safety A major problem with the disposal of high-level radioactive waste remains: How credible are long-range geologic predictions, that is, predictions of conditions thousands to millions of years in the future (32)? There is no easy answer to this question because geologic processes vary over both time and space. Climates change over long periods of time, as do areas of erosion, deposition, and groundwater activity. For example, large earthquakes hundreds or even thousands of kilometers from a site may permanently alter groundwater levels. The seismic record for the western

United States is known only for about the past hundred years, so estimates of future earthquake activity are tenuous at best. The result is that geologists can evaluate the relative stability of the geologic past, but they cannot guarantee future stability. Therefore, decision makers (not geologists) need to evaluate the uncertainty of prediction in light of pressing political, economic, and social concerns (32). These problems do not mean that the geologic environment is not suitable for safe containment of high-level radioactive waste, but care must be taken to ensure that the best possible decisions are made on this very critical and controversial issue.

12.6 Ocean Dumping

The oceans of the world, 361 million km^2 of water, cover more than 70 percent of the planet. They play a part in maintaining the world's environment by providing the water necessary to maintain the hydrologic cycle, contributing to the maintenance of the oxygen–carbon dioxide balance in the atmosphere, and affecting global climate. In addition, the oceans are very valuable to people, providing such necessities as foods and minerals.

It seems reasonable that such an important resource as the ocean would receive preferential treatment, yet all too often this is not the case. In much of the developing world, untreated sewage is disposed of in the oceans of the world, as are many industrial and agricultural wastes. The types of wastes that have been dumped in the oceans off the United States include (36):

- **Dredge spoils**—solid materials such as sand, silt, clay, rock, and pollutants deposited from industrial and municipal discharges—removed from the bottom of bodies of water, generally to improve navigation
- **Industrial waste**—acids, refinery wastes, paper mill wastes, pesticide wastes, assorted liquid wastes, and sewage sludge
- **Construction and demolition debris**—cinder block, plaster, excavation dirt, stone, tile, and other materials
- **Solid waste**—refuse, garbage, or trash; explosives
- **Radioactive waste**

In the United States federal law now prohibits ocean dumping of radiological, chemical, and biological warfare agents and any high-level radioactive waste. Furthermore, it provides for regulation of all other waste disposal in the oceans off the United States by the Environmental Protection Agency or, in the case of dredge spoil, by the U.S. Army Corps of Engineers. In addition to materials prohibited by law from being dumped, the EPA has prohibited:

- Material whose effect on marine ecosystems cannot be determined
- Persistent inert materials that float or remain suspended, unless they are processed to ensure that they sink and remain on the bottom
- Material containing more than trace concentrations of mercury and mercury compounds, cadmium and cadmium compounds, organohalogen compounds (organic compounds of chlorine, fluorine, and iodine), and compounds that may form from such substances in the oceanic environment
- Crude oil, fuel oil, heavy diesel oil, lubricating oils, and hydraulic fluids

Ocean Dumping and Pollution

Ocean dumping contributes to the larger problem of **ocean pollution,** which has seriously damaged the marine environment and caused a health hazard to people in some areas. Shellfish have been found to contain pathogens such as polio virus and hepatitis, and at least 20 percent of the nation's commercial shellfish beds have been closed because of pollution. Beaches and bays have been closed to recreational uses. Lifeless zones in the marine environment have been created. Heavy kills of fish and other organisms have occurred, and profound changes in marine ecosystems have taken place (36). Some of the world's major ecosystems, including estuaries, salt marshes, mangrove swamps, beaches, rocky intertidal areas, and coral reefs are threatened by ocean pollution. Ocean pollution is also impacting people and society directly as pollution and contaminated marine organisms may transmit toxic elements and disease to people (37).

Major impacts of marine pollution on oceanic and coastal life (36) include:

- Killing or retarding growth, vitality, and reproductivity of marine organisms by toxic pollutants
- Reduction of the dissolved oxygen necessary for marine life because of increased oxygen demand from organic decomposition of wastes
- Biostimulation by nutrient-rich waste (usually nitrogen and phosphorus), causing excessive blooms of algae (cultural eutrophication; see Chapter 11) in shallow waters of estuaries, bays, and parts of the continental shelf, resulting in depletion of oxygen and subsequent killing of algae that may wash up and pollute coastal areas (as, for example, beaches)
- Habitat change caused by waste-disposal practices that subtly or drastically change entire marine ecosystems

Major impacts on people and society caused by marine pollution include:

- A public health hazard caused by marine organisms transmitting toxic elements and disease to people
- Loss of visual and other amenities as beaches and harbors become polluted by solid waste, oil, and other materials
- Economic loss—the loss of shellfish from pollution in the United States amounts to many millions of dollars annually. In addition, a great deal of money is being

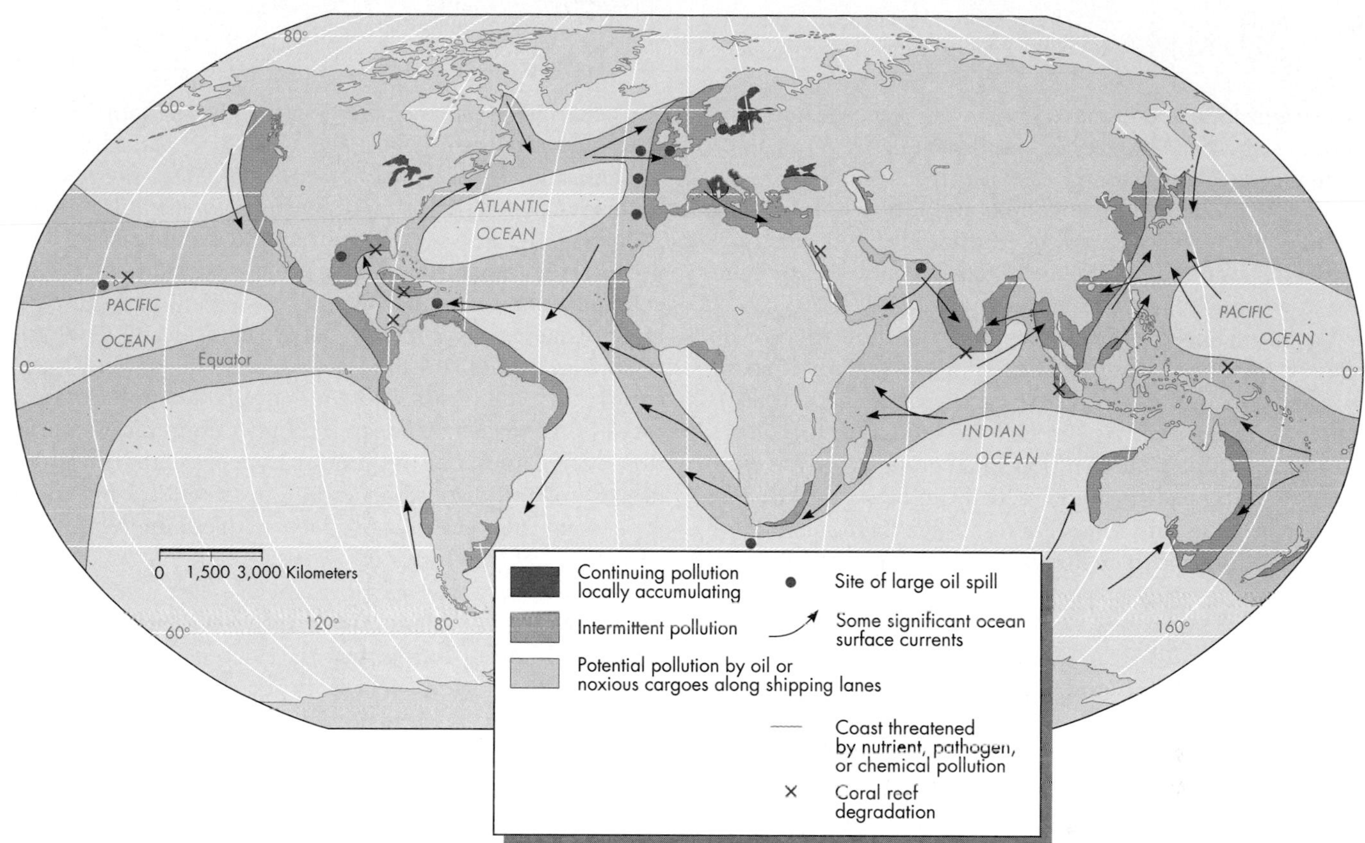

▲ **FIGURE 12.17** Sites of ocean and coastline pollution throughout the world. Impacts are variable, ranging from periodic closure of some beaches in southern California and reduced water quality in Florida Bay due to intermittent pollution to more serious, continuous pollution of areas in the Mediterranean Sea, parts of the Baltic and North seas, and parts of the Great Lakes. (*Adapted from Council on Environmental Quality* 1981. Environmental Trends. *With data from* McGinn, A. P. 1999. *Safeguarding the health of oceans*. World Watch, *Paper* 145. *Washington*, DC: *Worldwatch Institute*, *pp*. 22–23.)

spent cleaning solid waste, liquid waste, and other pollutants in coastal areas (36)

- Closure of beaches to recreational activities such as swimming and surfing

Ocean Dumping: The Conflict

It is unfortunate that the people interested in ocean dumping and those interested in harvesting marine resources such as fish and shellfish both prefer to do their jobs near the shore—the former because of convenience and transportation costs and the latter because of the richness of nearshore fisheries. Figure 12.17 shows the locations in the world's oceans, large lakes, and coastlines that are continually polluted or experience intermittent pollution, have degraded coral reefs or are considered threatened. These areas often coincide with nearshore productive areas of the marine environment and constitute a global problem (36–38).

The solution to the problem of ocean dumping is: first, recognizing that the marine environment is a limited resource, and second, developing economically feasible and environmentally safe alternatives to ocean dumping and pollution. We need to manage the oceans and coastal waters to support sustainable use of resources. Untreated wastewater and chemical wastes need to be treated, and agricultural and urban runoff controlled. Finally, we need to develop better scientific understanding and public awareness at local to international levels of the problems our oceans and coasts are facing as a result of human processes that cause pollution. The United States (and eight other countries) could start by ratifying of the 1982 United Nations Law of the Sea, which, by 1998, 130 other nations had ratified. The Law of the Sea establishes a broad management framework for the oceans and the common heritage of its resources within an environmental context of protecting and conserving marine resources (38).

SUMMARY

Industrialization and urbanization have produced enormous amounts of waste and greatly compounded the problem of waste management. Around many large cities, space for new landfills is becoming hard to find, and few people wish to live near any waste-disposal operation. We are headed toward a disposal crisis if the new methods and ideas of integrated waste management are not acted upon soon.

Waste-management practices since the Industrial Revolution have moved from "dilute and disperse," to "concentrate and contain," to integrated waste management (IWM), which includes alternatives such as reducing, recycling, reusing, landfilling, incineration, and composting. The goal of many of these alternatives, which can be summarized as "reduce, recycle, and reuse," is to reduce the total amount of waste that needs to be disposed of in landfills or incinerators. More recently a new concept of materials management is emerging as part of IWM. The ultimate objection is "zero waste"—all waste will become resources.

The most common method for disposal of urban waste in the United States today is the sanitary landfill, in which the waste deposited each day is covered with a layer of compacted soil. Potential hazards from sanitary landfills are pollution of groundwater by leachate (polluted water) from the site and uncontrolled production of methane. However, if methane is contained, it is a useful by-product of landfill operations. Under arid conditions, buried waste produces little leachate; the most suitable sites for landfills in humid regions are where waste can be buried well above the water table in clay and silt soil of low hydraulic conductivity. Modern sanitary landfills have multiple barriers to prevent leachate from infiltrating the vadose zone and systems of monitoring the vadose zone and groundwater through monitoring wells. The siting and operation of sanitary landfills is regulated by federal laws.

Hazardous waste management may be the most serious environmental problem in the United States. Hundreds or thousands of uncontrolled disposal sites may be time bombs that eventually will cause serious public health problems. Because we will continue to produce hazardous wastes, it is imperative that safe disposal methods be developed and used. Land-disposal options for the management of these wastes include secure landfills, in which the waste is confined and the leachate controlled; land application, in which suitable biodegradable materials are spread on the surface; deep-well injection; and incineration, with disposal of the residue by secure landfill. Alternatives to land disposal include source reduction, on-site processing to recover by-products with commercial value, and chemical stabilization.

Radioactive-waste management presents a serious and ever-increasing problem. Apparently, low-level radioactive waste can be safely buried near the surface if the burial sites are carefully selected and monitored. High-level radioactive wastes from nuclear power plants and weapons-production facilities remain hazardous for thousands of years; those currently being stored will eventually have to be permanently disposed of. The most promising method appears to be disposal in a carefully and continuously monitored area in stable bedrock. The site most extensively studied is Yucca Mountain, Nevada.

Indiscriminate ocean dumping is a significant source of marine pollution. Federal law now prohibits ocean dumping of certain dangerous materials and regulates all waste disposal in U.S. ocean waters. Alternatives to ocean dumping of materials such as polluted dredge spoils and other potentially hazardous materials are being developed, but in many cases such alternatives are not yet practical or economically feasible.

REFERENCES

1. **Relis, P., and Dominski, A.** 1987. *Beyond the crisis: Integrated waste management.* Santa Barbara, CA: Community Environmental Council.
2. **Council on Environmental Quality.** 1973. *Environmental quality—1973.* Washington, DC: U.S. Government Printing Office.
3. **Bullard, R. D.** 1990. *Dumping in Dixie: Race, class and environmental quality.* Boulder, CO: Westview Press.
4. **Galley, J. E.** 1968. Economic and industrial potential of geologic basins and reservoir strata. In *Subsurface disposal in geologic basins: A study of reservoir strata,* ed. J. E. Galley, pp. 1–19. American Association of Petroleum Geologists Memoir 10.
5. **Wright, A. G.** 1998. Big Apple is dumping its last active landfill. *ENR* 240(7):28–31.
6. **Young, W.** 1995. A dump no more. *American Forests,* Autumn, 59–62.
7. **Relis, P., and Levenson, H.** 1998. *Discarding solid waste as we know it: Managing materials in the 21st century.* Santa Barbara, CA: Community Environmental Council.
8. **Young, J. E.** 1991. Reducing waste, saving materials. In *State of the world,* ed. L. R. Brown, pp. 39–55. New York: World Watch Institute, W. W. Norton & Co.
9. **Schneider, W. J.** 1970. *Hydraulic implications of solid-waste disposal.* U.S. Geological Survey Circular 601F.
10. **Turk, L. J.** 1970. Disposal of solid wastes—Acceptable practice or geological nightmare? In *Environmental geology,* pp. 1–42. Washington, DC: American Geological Institute.
11. **Hughes, G. M.** 1972. Hydrologic considerations in the siting and design of landfills. *Environmental Geology Notes, No. 51.* Illinois State Geological Survey.
12. **Bergstrom, R. E.** 1968. Disposal of wastes: Scientific and administrative considerations. *Environmental Geology Notes, No. 20.* Illinois State Geological Survey.

13. **Cartwright, K., and Sherman, F. B.** 1969. Evaluating sanitary landfill sites in Illinois. *Environmental Geology Notes, No. 27*. Illinois State Geological Survey.
14. **Walker, W. H.** 1974. Monitoring toxic chemical pollution from land-disposal sites in humid regions. *Ground Water* 12:213–18.
15. **Environmental Protection Agency.** 1980. *Everybody's problem: Hazardous waste.* SW-826. Washington, DC: U.S. Government Printing Office.
16. **Magnuson, E.** 1980. The poisoning of America. *Time* 116(12):58–69.
17. **New York State Department of Environmental Conservation.** 1994. *Remedial chronology: The Love Canal hazardous waste site.* Albany, NY: New York State.
18. **Elliot, J.** 1980. Lessons from Love Canal. *Journal of the American Medical Association* 240:2033–34, 2040.
19. **Kufs, C., and Twedwell, C.** 1980. Cleaning up hazardous landfills. *Geotimes* 25:18–19.
20. **Albeson, P. H.** 1983. Waste management. *Science* 220:1003.
21. Return to Love Canal. 1990. *Time* 135(22):27.
22. **Bedient, P. B., Rifai, H. S., and Newell, C. J.** 1994. *Ground water contamination.* Englewood Cliffs, NJ: Prentice Hall.
23. **Huddleston, R. L.** 1979. Solid-waste disposal: Landfarming. *Chemical Engineering* 86(5):119–24.
24. **McKenzie, G. D., and Pettyjohn, W. A.** 1975. Subsurface waste management. In *Man and his physical environment*, eds. G. D. McKenzie and R. O. Utgard, pp. 150–56. Minneapolis: Burgess.
25. **Piper, A. M.** 1970. *Disposal of liquid wastes by injection underground: Neither myth nor millennium.* U.S. Geological Survey Circular 631.
26. **Committee of Geological Sciences.** 1972. *The earth and human affairs.* San Francisco: Canfield Press.
27. **Warner, D. L.** 1968. Subsurface disposal of liquid industrial wastes by deep-well injection. In *Subsurface disposal in geologic basins: A study of reservoir strata*, ed. J. E. Galley, pp. 11–20. American Association of Petroleum Geologists Memoir 10.
28. **Cecelia, C.** 1985. *The buried threat.* No. 115-5. Sacramento: California Senate Office of Research.
29. **Office of Industry Relations.** 1974. *The nuclear industry, 1974.* Washington, DC: U.S. Government Printing Office.
30. **Fischer, J. N.** 1986. *Hydrologic factors in the selection of shallow land burial for the disposal of low-level radioactive waste.* U.S. Geological Survey Circular 973.
31. **U.S. Department of Energy.** 1992. DOE's Yucca Mountain Studies. DOE/RW-0345P. Washington, DC: U.S. Government Printing Office.
32. **Bredehoeft, J. D., England, A. W., Stewart, D. B., Trask, J. J., and Winograd, I. J.** 1978. *Geologic disposal of high-level radioactive wastes—Earth science perspectives.* U.S. Geological Survey Circular 779.
33. **Hunt, C. B.** 1983. How safe are nuclear waste sites? *Geotimes* 28(7):21–22.
34. **Heiken, G.** 1979. Pyroclastic flow deposits. *American Scientist* 67:564–71.
35. **U.S. Department of Energy.** 1990. *Yucca Mountain project: Technical status report.* DE90015030. Washington, DC: U.S. Government Printing Office.
36. **Council on Environmental Quality.** 1970. *Ocean dumping: A national policy.* Washington, DC: U.S. Government Printing Office.
37. **Lenssen, N.** 1989. The ocean blues. *World Watch*, July–August, pp. 26–35.
38. **McGinn, A. P.** 1999. Safeguarding the health of oceans. *World Watch*, Paper 145.

KEY TERMS

integrated waste management (IWM) (p. 318)
reduce, recycle, and reuse (p. 320)
materials management (p. 320)
industrial ecology (p. 320)
composting (p. 321)
incineration (p. 321)
sanitary landfill (p. 322)
leachate (p. 322)
monitoring (p. 324)
hazardous waste (p. 325)
secure landfill (p. 328)
land application (p. 330)
biopersistence (p. 330)
surface impoundments (p. 330)
deep-well disposal (p. 331)
low-level radioactive waste (p. 334)
high-level radioactive waste (p. 336)
ocean pollution (p. 338)

SOME QUESTIONS TO THINK ABOUT

1. Complete an audit of your personal waste production and disposal where you live. How much are you presently recycling and how much do you estimate you might recycle at the high end? If everyone in your neighborhood did this, what would be the impact on the local waste situation?
2. Defend or criticize the statement that management strategies consisting of recycling and incineration compete with one another and therefore we should emphasize one of these based upon environmental considerations.
3. For the region in which you live, identify potential hazardous wastes that are produced by homes, businesses, and industry or agriculture. How are these wastes currently being treated and what could be done to develop a better management strategy if there are problems?
4. Some people argue that we should not dispose of high-level radioactive waste deep in the geologic environment where such waste is not retrievable. One alternative would be to store the waste at several different sites on the surface for a period of perhaps as long as 100 years. The idea is that during that time we will find better ways to deal with the radioactive waste. Develop arguments for and against these ideas.

13 The Geologic Aspects of Environmental Health

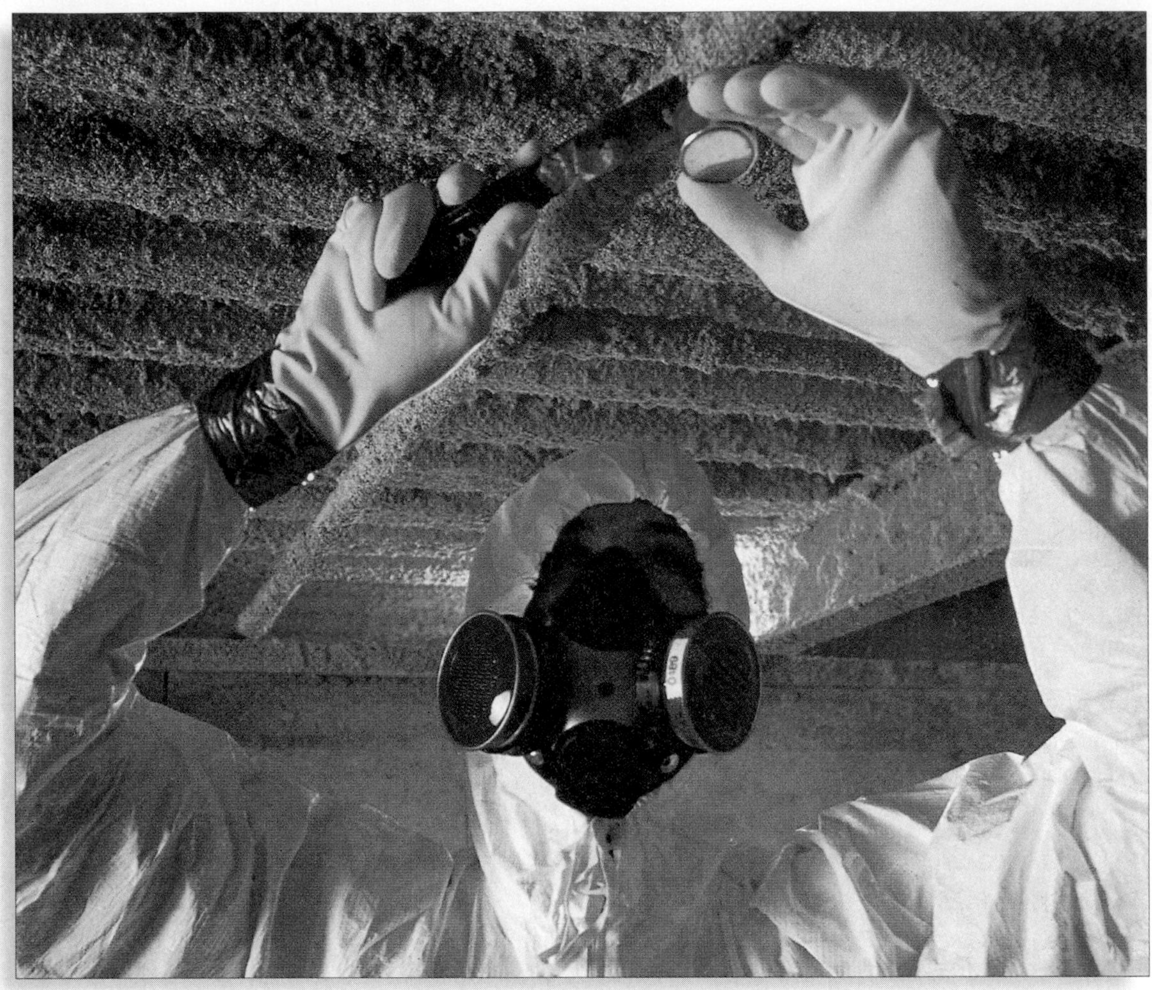

The woman shown here in protective clothing is inspecting asbestos. (*Rene Sheret/Tony Stone Worldwide*)

The woman in the photograph is wearing protective clothing while inspecting asbestos in a building. Industrial and agricultural processes use many materials from the earth and manufacture others for a variety of uses that have been perceived as beneficial to people. In many cases this assumption is correct, but in many others the chemicals and minerals we are producing have adverse environmental consequences to people and ecosystems around the world. Herbicides and pesticides have protected our crops but at a cost, as some of them have been shown to have adverse environmental consequences. For example, DDT was useful in eradicating mosquitoes, which carry malaria, one of the most debilitating diseases in the world. However, DDT is a long-lived organic chemical that works its way up the food chain; it caused environmental problems, particularly to birds, in which the chemical caused a thinning and softening of eggshells, resulting in the death of chicks. Similarly, asbestos, which is mined from fibrous silicate minerals, has many industrial uses related to overheating of machinery and fire prevention. However, workers exposed to certain types of asbestos contracted a lung disease known as asbestosis as well as lung cancers. What is apparent is that the health of people is intimately linked to our environment and exposure to the chemicals we produce and minerals we use.

LEARNING OBJECTIVES

Environmental health is an important field of inquiry because better understanding of relationships between our environment and incidence of disease will help produce strategies to improve our overall health and well-being. Learning objectives for this chapter are:

- *To be able to define disease from an environmental perspective and state the general factors associated with health and disease.*
- *To understand some of the geologic factors related to environmental health.*
- *To be familiar with the concept of dose dependency for trace elements and toxins in the environment.*
- *To gain a modest acquaintance of relationships between chronic disease and the geologic environment.*
- *To know what asbestos is and why there is controversy concerning its potential to produce health problems.*
- *To know what nuclear radiation is in terms of types of radiation and their potential health effects.*
- *To be able to discuss the radon gas problem in homes, schools, and other buildings.*
- *To know the major steps in the processes of risk assessment and risk management for exposure to toxic materials in the environment.*

Web Resources

Visit the "Environmental Geology" Web site at www.prenhall.com/keller to find additional resources for this chapter, including:

- ▶ Web Destinations
- ▶ On-line Quizzes
- ▶ On-line "Web Essay" Questions
- ▶ Search Engines
- ▶ Regional Updates

13.1 Introduction to Environmental Health

As a member of the biological community, the human species has carved a niche in the biosphere—a niche highly dependent on complex interrelations between the biosphere, atmosphere, hydrosphere, and lithosphere. Yet we are only beginning to inquire into and gain a basic understanding of the total range of environmental factors affecting our health and well-being. As we continue our exploration of the geologic cycle—from minute quantities of elements in soil, rocks, and water to regional patterns of climate, geology, and topography—we are making important discoveries about how these factors may influence the death rate and the incidence of certain diseases. In the United States alone, the death rate varies significantly from one area to another (1), and some of the variability is the result of the local, physical, biological, and chemical environment in which we live.

Disease has been described as an imbalance resulting from a poor adjustment between an individual and the environment (2). Disease seldom has a single cause. The geologist's contribution to our understanding of its causes is to help isolate aspects of the geologic environment that may influence the incidence of disease. This tremendously complex task requires sound scientific inquiry coupled with interdisciplinary research with physicians and other scientists. Although the picture is now rather vague, the possible rewards of the emerging field of medical geology are exciting and may eventually play a significant role in environmental health.

To study the geologic aspects of environmental health, one must also consider cultural and climatic factors associated with patterns of disease and death rates. This procedure helps to isolate the geologic influence.

Environmentally Transmitted Infectious Diseases

Environmentally transmitted infectious diseases, spread from interactions of individuals with food, water, air, or soil, constitute one of the oldest health problems that humans face. Those diseases that may be controlled by manipulating the environment, such as improving sanitation and treating water, that are classified as environmental health concerns. Although there is great concern for the toxins and carcinogens we are producing in industrial society today, infectious diseases that are transmitted through the environment constitute the greatest mortality in developing countries. In the United States infectious environmentally transmitted disease has certainly not been eliminated, and each year there are thousands of cases of waterborne illness and food poisoning. They may be spread by: vectors such as mosquitoes or fleas; contact with contaminated food, water, or soil; or transmitted through the ventilation system of a building. Specific examples of environmentally transmitted diseases include Legionnaires' disease, Giardia (chronic diarrhea), salmonella (food poisoning), malaria, plague, dengue fever, Lyme disease, and cryptosporidosis (3).

Cultural Factors

The cultural aspects of a society reflect the sum total of concepts and techniques that people within the society have developed to survive in their environment. These aspects influence the occurrence of disease by creating links or barriers between people and the causes of illness. The nature and extent of the links depend on such factors as local customs and degree of industrialization. Members of societies who live more directly off the land and water are plagued by different health problems than are members of urban societies. Industrial societies have nearly eliminated such diseases as cholera, typhoid, hookworm, and dysentery, but they are more likely to suffer from lung cancer and other diseases related to air, soil, and water pollution.

As an example of a relationship between culture and disease, consider recent construction practices in the United States. We build houses on cement slabs, which may develop cracks, and insulate these houses tightly for energy conservation. These practices have made our homes more susceptible to indoor air pollution, such as radon gas, which is linked to lung cancer. Another example is the high incidence of stomach cancer in Japan (4). The Japanese prefer rice that is polished and powdered; unfortunately, the powder contains asbestos as an impurity. Asbestos, a fibrous mineral, is either a true carcinogen (cancer-causing material) or takes a passive role as a carrier of trace-metal carcinogens.

The incidence of lead poisoning also suggests a cultural, political, and economic influence on patterns of disease. The effects of lead poisoning can include anemia, mental retardation, and palsy. Lead is found in some moonshine whiskey and has resulted in lead poisoning among adults and even unborn or nursing infants whose mothers drank it (5). Some researchers have suggested that widespread lead poisoning was one of the reasons for the fall of the Roman Empire. It has been estimated that the Romans produced about 55,000 metric tons of lead annually for 400 years. Lead was used in pots in which grape juice was processed into a syrup, in cups from which the Romans drank wine, and in cosmetics and medicines. The ruling class also had water piped into their homes through lead pipes. Historians argue that gradual lead poisoning among the upper class resulted in their eventual demise through widespread stillbirths, deformities, and brain damage. The high lead content found in the bones of ancient Romans lends support to this hypothesis (5). Further support comes from a study of ice cores from Greenland glaciers; these cores show that, for the period from 500 B.C. to A.D. 300, lead concentrations in the ice are about four times normal. This suggests that mining and smelting of the metal during the time of the Roman Empire polluted the atmosphere in the Northern Hemisphere with lead (6). With the elimination of lead in gasoline and paints, emissions of new lead have all but ceased in the United States. Nevertheless, lead poisoning of children remains a potentially serious problem in inner-city areas due to children's exposure to lead stored in the soil and old paint of buildings (7).

Climatic Factors

Climatic factors such as temperature, humidity, and amount of precipitation are sometimes intimately related to disease patterns. Two of the worst climate-related diseases, schistosomiasis and malaria, are found in the tropics. These diseases are related to climate because the disease vectors, snails and mosquitoes, respectively, are in part climatically controlled. Schistosomiasis, called *snail fever*, is a significant cause of death among children and saps the energy of millions of people worldwide. Its effects have tremendous socioeconomic consequences, and some researchers consider it the world's most important disease (2).

Assumed relations between culture or climate and the incidence of disease must be viewed with some skepticism because there is seldom a simple answer to environmental health problems. For example, if schistosomiasis were controlled only climatically, then all areas with an appropriate climate, such as much of the Amazon River Basin, would have the disease. Fortunately, this is not the case, and in some instances the reason is geologic. Conditions in the Amazon Basin are nearly optimal for the disease, yet it occurs only in two very limited areas, primarily because there is not enough calcium in the water of most of the region to support schistosomiasis-carrying snails. In some parts of the Amazon Basin, the acidity of the water in the presence of copper and other heavy metals may be responsible for the absence of the snails in an otherwise suitable environment for schistosomiasis (1).

From this introduction to environmental health, we can see some of the complex relations between disease patterns and environment. With this in mind, we will now focus on the geologic aspects of health problems. In this chapter we consider specific geologic factors important to health, effects of trace elements on health, and the significance of the geologic environment to the incidence of heart disease and cancer, which are the leading causes of death in the United States. The chapter concludes with a discussion of radiation and radon gas, including the geologic factors involved and the potentially serious threat to human health.

13.2 Some Geologic Factors of Environmental Health

The soil in which we cultivate plants for food, the rock on which we build our homes and industries, the water we drink, and the air we breathe all influence our chances of developing serious health problems. On the other hand, these same factors can also influence our chances of living a longer, more productive life. Surprisingly, many people still believe that soil, water, or air in a "natural," "pure," or "virgin" state must be "good" and that if human activities have changed or modified them, they have become "contaminated," or "polluted," and therefore "bad." This simple dichotomy is by no means the entire story (8).

Relationships between geology and health are significant research and discussion topics. Although few definite cause-and-effect relationships have been isolated, we are learning more all the time about the subtle ways in which the geologic environment affects general health. Treating the various aspects of medical geology at even the introductory level requires discussion of the natural distributions of elements in the earth's crust and the ways in which natural and artificial processes concentrate or disperse those elements.

Natural Abundances of Elements

A rough inverse relationship exists between the atomic number of an element and its abundance: lighter (lower-numbered) elements are encountered more frequently than most heavier ones. In general, this relation holds for both the lithosphere and the biosphere. Table 13.1, the periodic table of the elements, identifies some of the more abundant elements in the earth's crust and some environmentally important trace elements. Table 13.2 shows the most abundant elements in the rocks of the continental crust. Note that the elements that make up more than 99 percent of these rocks by weight are among the first 26 elements of the periodic table. Table 13.3 shows the distribution of the more abundant elements in the average adult human body. More than 99 percent of the body by weight is composed of the first 20 elements of the periodic table.

Living tissue is composed primarily of 11 elements, the so-called **bulk elements.** These are hydrogen, sodium, magnesium, potassium, calcium, carbon, nitrogen, oxygen, phosphorus, sulfur, and chlorine. For those species that have hemoglobin, iron is added to the list. In addition to the bulk elements, living tissue requires minute quantities of several other elements to function properly. These **trace elements** (or *trace-element metals*) help regulate the dynamic processes of life. Trace elements that have been studied and shown essential for nutrition include fluorine, chromium, manganese, cobalt, copper, zinc, selenium, molybdenum, and iodine. This list is not complete, and it would not be surprising to learn that many more trace elements are essential or at least active in life processes (9,10). Other elements, including nickel, arsenic, aluminum, and barium, accumulate as tissues age and are known as **age elements.** The physiological consequences of the accumulation of some elements in living tissue are known in some cases but completely unknown or poorly understood in others (10).

Concentration and Dispersion of Chemical Substances

The movement of elements and compounds along various paths through the lithosphere, hydrosphere, atmosphere, and biosphere makes up the biogeochemical cycles. Natural processes, such as the release of gas by volcanic activity or the weathering of rock and rock debris, release chemical materials into the environment. In addition, human use may result in the release of materials and substances that lead to pollution or contamination of the environment. In general, but with many exceptions, concentrations of trace elements tend to increase from rock to soil and/or water to plants and animals. Figure 13.1 shows some of the paths that trace elements may take to become concentrated in the human body, possibly causing health problems.

Once released by natural or artificial processes, elements and other substances are cycled and recycled by geochemical and rock-forming processes that may change their concentration. Thus, the concentration of a particular chemical element or compound may be quite different in igneous rock than in sedimentary rock formed from that igneous rock's weathered products. Whether the concentration has increased or decreased depends on the nature of the biogeochemical and rock-forming processes. Table 13.4 lists the concentrations of selected elements in igneous and sedimentary rocks. Although this information is not detailed, it is useful because it indicates a change in the relative abundance of elements produced by rock-forming and biological processes, as, for example, the approximately tenfold increase in selenium from the original weathering of igneous rocks to the formation of shale. With the exception of coal and phosphorites (rock that is rich in calcium phosphate), other types of sedimentary rocks do not show a similar increase in selenium. This example and others suggest that biogeochemical and rock-forming processes, such as weathering, leaching, accretion, and deposition, effectively sort, concentrate, and disperse elements and other substances throughout the environment. Human processes are also responsible for the concentration of chemicals, and as human population increases, so do potential problems (see *A Closer Look: Nitrogen and People*).

Weathering Weathering is the physical and chemical breakdown of rock material and a major process in the formation of soil. Regardless of whether the parent material for soil is bedrock or rock debris transported and deposited by running water, wind, or ice, weathering is a natural process that frees trace elements to be used by the biosphere in life processes.

The artificial counterpart of weathering is pollution or contamination, which also releases trace elements into the environment. For example, lead is released into the environment when lead additives in gasoline are emitted through exhaust systems; mercury, cadmium, nickel, zinc, and other metals are released into the atmosphere and water through industrial and mining operations.

Reduction in lead pollution is becoming an environmental success story. It is no longer used as an additive in gasoline in the United States and many other parts of the world. As a result, the concentration of lead in the atmosphere has dropped dramatically (more than 90%), and this will have positive environmental health effects for millions of people formerly exposed to lead emitted from vehicles burning leaded gasoline.

Leaching Leaching, accretion, deposition, biologic activity, and other processes may concentrate or disperse elements after they are released by natural and artificial processes. Leaching of soils is the natural removal of soluble material (in solution) from the upper to lower soil horizons. Material that leaches out of soil may enter the groundwater system and be dispersed or diluted. If the material is sufficiently abundant or toxic or otherwise harmful, it also may pollute the groundwater. Leaching from soils is most prevalent in warm, humid climates where the soil may be nutrient-poor because the nutrients are removed. Furthermore, trace elements that remain may be concentrated at undesirable levels.

Table 13.1 Periodic table of the elements showing elements that are relatively abundant in the earth's crust and some environmentally important trace elements

Key: Atomic number; * Element relatively abundant in earth's crust; Element symbol; ** Environmentally important trace elements; L Major importance to life; Element name

1 H Hydrogen L																	2 He Helium
3 ** Li Lithium	4 Be Beryllium											5 B Boron	6 C Carbon L	7 N Nitrogen L	8 * O Oxygen L	9 ** F Fluorine	10 Ne Neon
11 * Na Sodium L	12 * Mg Magnesium L											13 * Al Aluminum	14 * ** Si Silicon	15 P Phosphorus L	16 S Sulfur L	17 Cl Chlorine L	18 Ar Argon
19 * K Potassium L	20 * Ca Calcium L	21 Sc Scandium	22 Ti Titanium	23 ** V Vanadium	24 ** Cr Chromium	25 ** Mn Manganese	26 * ** Fe Iron L	27 ** Co Cobalt L	28 ** Ni Nickel	29 ** Cu Copper L	30 ** Zn Zinc L	31 Ga Gallium	32 Ge Germanium	33 ** As Arsenic	34 ** Se Selenium	35 Br Bromine	36 Kr Krypton
37 Rb Rubidium	38 Sr Strontium	39 Y Yttrium	40 Zr Zirconium	41 Nb Niobium	42 ** Mo Molybdenum L	43 Tc Technetium	44 Ru Ruthenium	45 Rh Rhodium	46 Pd Palladium	47 Ag Silver	48 ** Cd Cadmium L	49 In Indium	50 ** Sn Tin	51 Sb Antimony	52 Te Tellurium	53 ** I Iodine	54 Xe Xenon
55 Cs Cesium	56 Ba Barium	57 La Lanthanum	72 Hf Hafnium	73 Ta Tantalum	74 W Wolfram	75 Re Rhenium	76 Os Osmium	77 Ir Iridium	78 Pt Platinum	79 Au Gold	80 ** Hg Mercury	81 Tl Thallium	82 ** Pb Lead	83 Bi Bismuth	84 ** Po Polonium	85 At Astatine	86 ** Rn Radon
87 Fr Francium	88 ** Ra Radium	89 Ac Actinium															

58 Ce Cerium	59 Pr Praseodymium	60 Nd Neodymium	61 Pm Promethium	62 Sm Samarium	63 Eu Europium	64 Gd Gadolinium	65 Tb Terbium	66 Dy Dysprosium	67 Ho Holmium	68 Er Erbium	69 Tm Thulium	70 Yb Ytterbium	71 Lu Lutetium
90 Th Thorium	91 Pa Protactinium	92 ** U Uranium	93 Np Neptunium	94 ** Pu Plutonium	95 Am Americium	96 Cm Curium	97 Bk Berkelium	98 Cf Californium	99 Es Einsteinium	100 Fm Fermium	101 Md Mendelevium	102 No Nobelium	103 Lw Lawrencium

Table 13.2 The relative abundance of the most common elements in the rocks of the earth's crust

Atomic No.	Element		Weight (%)
8	O	Oxygen	46.40
14	Si	Silicon	28.15
13	Al	Aluminum	8.23
26	Fe	Iron	5.63
20	Ca	Calcium	4.15
11	Na	Sodium	2.36
12	Mg	Magnesium	2.33
19	K	Potassium	2.09
	Total		99.34

Table 13.3 Distribution of the more abundant elements in the adult human body

Atomic No.	Element		Weight (%)
8	O	Oxygen	65.00
6	C	Carbon	18.00
1	H	Hydrogen	10.00
7	N	Nitrogen	3.00
20	Ca	Calcium	1.50
15	P	Phosphorus	1.00
16	S	Sulfur	0.25
19	K	Potassium	0.20
11	Na	Sodium	0.15
17	Cl	Chlorine	0.15
12	Mg	Magnesium	0.05
	Total		99.30

Accumulation The term *accumulation in soils* refers to processes that cause or increase retention of material in soil. Examples include salts that may accumulate on the surface and the upper zones of soils through evaporation processes and materials that have been removed by leaching from the *A* horizon and accumulate in the *B* horizon. An example of the latter is found in semiarid regions where accumulation of calcium carbonate (caliche) is found in the *B* horizon of some soils. (See Chapter 3 for a discussion of soil horizons.)

Deposition Deposition of earth materials gives rise to two environmentally important problems. First, heavy metals and some other materials cause biologic disruptions when they are deposited in streams, lakes, and oceans. For example, mercury may become attached (deposited) to suspended sediments and bottom sediments, leading to higher toxic concentrations of mercury in aqueous environments. This may cause biological disruption because plants rooted in the sediment and animals feeding in the sediment take up some of the mercury, which is then passed on at higher and higher concentrations through the food chain (for example, to birds and fish). This process is known as *biomagnification* and can result in pollutants such as mercury reaching concentrations that can cause health problems to people and other animals that eat the contaminated fish.

Second, a deficiency of needed trace elements occurs in some areas because the elements were not originally deposited along with other sediments moved by water, ice, and wind. This second problem is less well understood than the first because it may involve interactions of other processes, such as leaching, with deposition. Erosion and deposition by wind are particularly susceptible to selective removal. One

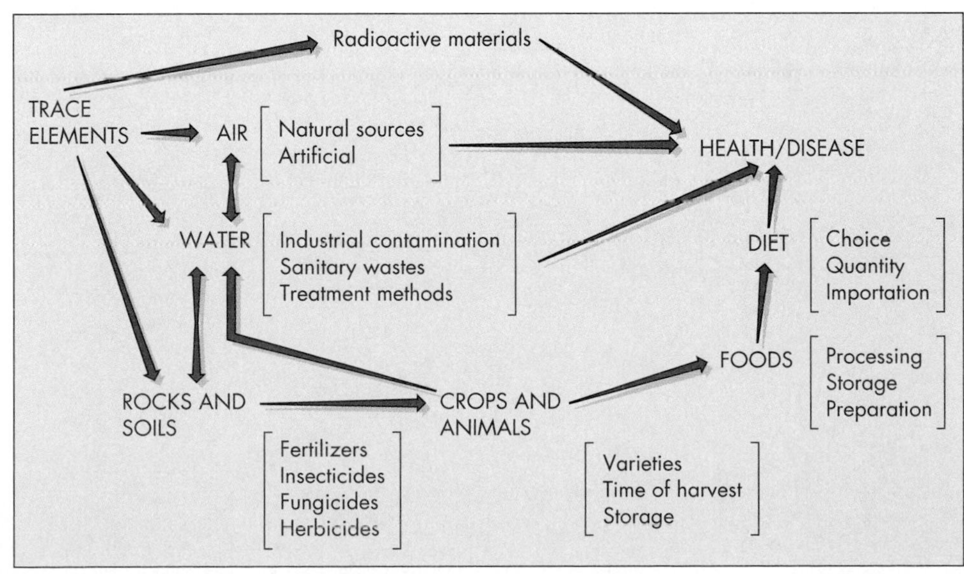

◀ **FIGURE 13.1** A schematic drawing showing mechanisms by which trace elements may find their way to humans and animals, thus influencing the quality of health or producing disease. (*After* K. E. *Beeson*. 1974. Geochemistry and the environment, *vol.* 1. *Reproduced with permission of the* National Academy *of Sciences.*)

Table 13.4 Concentrations of some elements in various natural materials

Type of Material	Concentrations by Elements (ppm)									
	Cadmium	Chromium	Copper	Fluorine	Iodine	Lead	Lithium	Molybdenum	Selenium	Zinc
Ultramafic igneous[a,b]	0–0.02 0.05(?)	1000–3400 1800	2–100 15	—	0.06–0.3 0.1	— 1	— 0.5	— 0.3	— 0.05	— 40
Basaltic igneous[a,b]	0.006–0.6 0.2	40–600 220	30–160 90	20–1060 360	— 0.5	2–18 6	3–50 20	0.9–7 1.5	— 0.05	48–240 110
Granitic igneous[a,b]	0.003–0.18 0.15	2–90 20	4–30 15	20–2700 870	— 0.5	6–30 18	10–120 35	1–6 1.4	— 0.05	5–140 40
Shales and clay[a,b,c]	0–11 1.4	30–590 120	18–120 50	10–7600 800	212–380 5(?)	16–50 20	4–400 80	— 2.6	— 0.6	18–180 90
Black shales (high C)[d]	0.3–8.4 1.0	26–1000 100	20–200 70	—	—	7–150 20	—	1–300 10	—	34–1500 100(?)
Deep-sea clays[a,b]	0.1–1 0.5	— 90	— 250	— 1300	11–50 35(?)	— 80	— 57	— 27	— 0.17	— 165
Limestones[a,b,c,e]	— 0.05	— 10	— 4	0–1200 220	0.4–29 5	— 9	5–10 7	— 0.4	— 0.08	— 20
Sandstones[a,b,c]	— 0.05	— 35	— 2	10–880 180	— 1.7	1–31 7	7–90 30	— 0.2	— 0.05	2–41 16
Phosphorites[f]	0–170 30	30–3000 300	10–100 30	24,000–41,500 31,000	—	10–30 10	—	3–300 30	1–100 18	20–300 50
Coals (ash)[a]	— 2	10–1000 20	2–40 15	40–480 80	1–11 4	2–50 15	2–300 50	0.2–16 5	0.4–3.9 2[g]	7–108 50

Note: The upper figure is the range usually reported; the lower figure, the average.

[a]Turekian and Wedepohl (1961).

[b]Parker (1967).

[c]Becker et al. (1972).

[d]Vine, J. D., and Torutelot, E. B., *Econ. Geol.* 65, 223 (1970).

[e]Wedepohl (1970).

[f]Gulbrandsen, R. A., *Geochim. Cosmochim. Acta* 30, 769 (1966).

[g]U.S. Geological Survey (1972).

Source: M. Fleischer, U.S. Geological Survey. Reproduced with permission of the National Academy of Science.

A CLOSER LOOK Nitrogen and People

The tremendous growth in human population of the earth from about two billion in the 1950s to a point exceeding six billion by the year 2000 is largely the result of modern medicine and sanitation coupled with intensive agriculture that has provided food for six billion people. The introduction of massive amounts of synthetic nitrogen fertilizers during the same period from the 1950s to the end of the twentieth century saw an increase in consumption of those fertilizers from about 10 million tons per year to nearly 80 million tons per year (11). During that period, the growth rate for human population has averaged about 2 percent growth per year while consumption of nitrogen fertilizer has experienced a growth rate of closer to 5 percent per year. It is now suggested that the tremendous growth of use of nitrogen fertilizers has introduced a global unplanned geochemical experiment with potentially dangerous environmental consequences. This results because overapplication of nitrogen fertilizer often causes significant environmental problems to society today including: surface and groundwater pollution in both urban and agricultural areas (see Chapter 11). In some areas there is now sufficient deposition of nitrogen compounds from the atmosphere to soils that adding nitrogen fertilizer is no longer necessary. However, this certainly is not a planned application, and excess of nitrogen may cause unexpected environmental problems. In fact, environmental problems related to synthetic nitrogen fertilizers were not anticipated when we began producing massive amounts of these materials in the mid-twentieth century. This serves to emphasize the principle of environmental unity: you can't do just one thing, as everything affects everything else. A major problem remains: the population of the earth is expected to double again in the coming century, perhaps as early as about the year 2040. Feeding of an additional six billion people will require either massive changes in the way we feed people, or, more likely, a massive increase in productivity of crops used to feed animals and people. Even if we are successful in curbing population growth in coming decades, we will still have at least several billion additional people to feed, and this may require a substantial increase in the amount of nitrogen fertilizers to increase crop yields. This will likely cause further pollution of surface- and groundwaters in major agricultural areas, leading to an increase in the incidence of disease caused by nitrogen pollution and toxicity. How might this be avoided? Potential strategies to reduce the increase in use of nitrogen fertilizers include:

- Major changes in people's eating habits, moving toward vegetarian diets. This would greatly reduce the amount of land necessary to feed people because on average it takes approximately 3 to 4 units of protein from feed crops to produce 1 unit of protein from meat (11). However, we can't simply use rangelands that are now used to raise beef to grow crops. This is because rangelands, often with steeper slopes and thinner soils with limited rainfall and high erosion potential, cannot support sustainable crop production.
- Develop much more efficient ways of applying fertilizers to crops so that they are available in the right amounts at the right time, thus eliminating excess application. This is likely the most promising alternative for sustainable intensive agriculture.

In summary, parallel increase of human population with use of fertilizers suggests that our intensive agricultural practices have in part allowed for the tremendous increase in people on earth. It appears we may have developed a habit of using, and chemical dependency upon, synthetic nitrogen to feed the people on earth (11). Because of the environmental problems associated with use of massive amounts of nitrogen fertilizer, and that tremendous amounts of energy (oil) are required for their production (oil that may not be as available in the twenty-first century as it has been in the twentieth century; see Chapter 15) it appears we are going to have to give priority to breaking or lessening the dependency upon these chemicals. This is proving to be a difficult but hopefully not insurmountable task.

author attributes the lack of lead, iron, copper, cobalt, and other materials in the sandhills of Nebraska to the fact that these metals occur in grains that are smaller and heavier than quartz grains and therefore were not moved along with the quartz grains that formed the hills (3).

13.3 Trace Elements and Health

Every element has a broad spectrum of possible effects on a particular plant or animal. For example, selenium is toxic in seleniferous areas (areas high in selenium), has no observable effect in most areas, and is beneficial to animal production in some livestock-raising areas. The apparent contradiction is resolved when we recognize that the first case is one of oversupply of selenium; the second represents a balanced state; and the third case is one of deficiency, which in some cases is rectified by supplementing the animals' food supply with the element (12). In this section we will consider some of the biological effects of *trace elements*, or *trace substances*. These terms refer to elements or compounds that are needed in minute amounts but may be toxic in larger quantities.

Dose Dependency and Effective Dose

It was recognized many years ago that the effects of a certain trace element on a particular organism depend on the dose or concentration of the element. This **dose dependency** can be represented by a **dose-response curve,** as shown in Figure 13.2a (12,13). When various concentrations of an element present in a biological system are plotted against effects on the organism, three things are apparent. First, although large concentrations may be toxic, injurious, or even lethal (*D-E-F* in Figure 13.2a), trace concentrations may be beneficial or even necessary for life (*A-B*). Second,

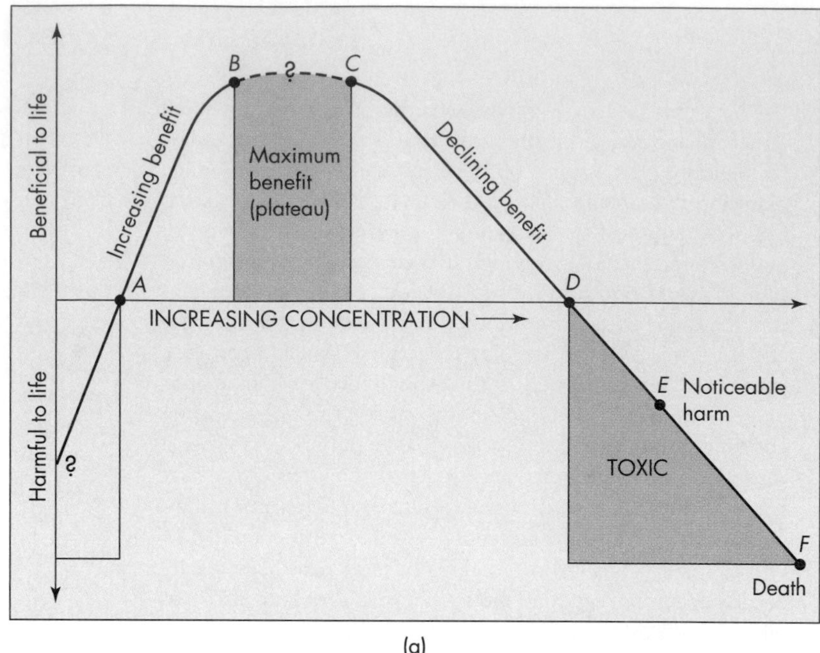

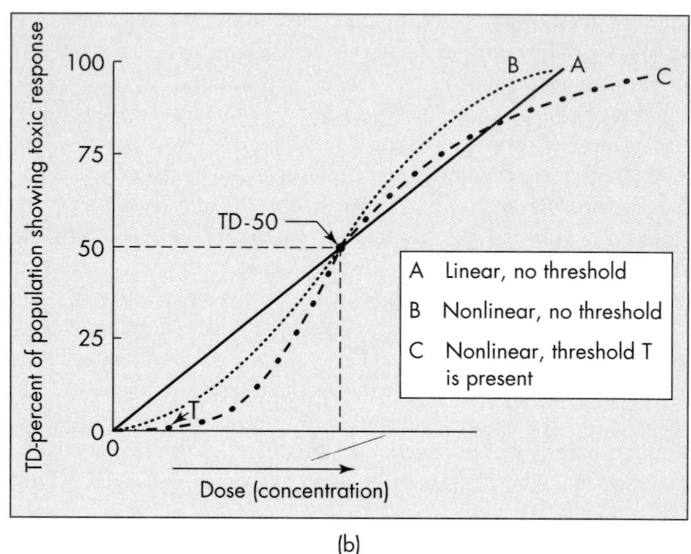

▶ **FIGURE 13.2** (a) Generalized dose-response curve; and (b) generalized toxic dose-response curve. See text for explanation.

the dose-response curve has (*B-C* in Figure 13.2a), a plateau of optimal concentration and maximum benefit to life. Third, there are two threshold concentrations at which harmful effects to life begin. One of these thresholds is point *A* (Figure 13.2a), below which decreasing concentrations cause increasing harm; the other is point *D*, above which increasing concentrations cause increasing harm (12,13).

Points *A*, *B*, *C*, *D*, *E*, and *F* in Figure 13.2a are all significant threshold concentrations. Unfortunately, points *E* and *F* are known only for a few substances (as they affect humans and a few other species), and the really important point *D* is all but unknown (13). The width of the maximum-benefit plateau (points *B* and *C*) for an organism depends on the organism's particular physiological equilibrium (13). In other words, the different phases of activity, whether beneficial, harmful, or lethal, may differ widely both quantitatively and qualitatively for different substances and are therefore completely observable only under special conditions (12).

The term **toxin** refers to materials that are poisonous to people or other living things. The study of toxins and their environmental effects, as well as economic and legal ramifications of toxins in the environment, is known as **toxicology.** Toxicology studies often use the concept of the **effective dose (ED),** which measures the effects of a substance on a population rather than an individual organism. Such studies can produce any of the three dose-response curves shown in Figure 13.2b, depending on the toxin and the species studied. Each of these curves shows the percentage of the population displaying a particular response at increasing doses or concentrations of a particular toxin. Point TD-50 is the concentration **(toxic dose)** at which 50 percent of the population experiences the response, which may be a particular symptom, the onset of a disease, or even death. If the response we are considering is death, the effective dose is called the **lethal dose (LD).** The LD-50 is the concentration (dose) at which 50 percent of the population dies.

The three curves in Figure 13.2b might represent the number of people who develop a rash in response to three different toxins. Curve *A* is a linear curve that suggests a direct relationship between the dose of the toxin and the percentage of population showing the response. In this case, doubling the initial dose will cause twice as many people to develop rashes. Curve *B* is a nonlinear response in which the fastest increase in the percentage of people showing a response occurs between about 25 and 75 percent of the population. Curve *C* is a nonlinear response that shows a threshold effect (point *T* on the curve). That is, as the dose or concentration increases, there is no significant response until a threshold concentration is reached, after which the percent of population showing a response increases.

Imbalances of Selected Trace Elements

A complete discussion of the geology and environmental effects of all trace elements would be a textbook in itself. Our objective here is to discuss representative examples to show the possible effects of imbalances (too much or too little) of trace elements. For this purpose, we have selected fluorine, iodine, zinc, and selenium. In addition, it is valuable to relate trace-element problems to human use of the land, which we will do with examples of mining activity.

Fluorine Fluorine is an important trace element that forms fluoride compounds, or fluorides. The compound calcium fluoride helps prevent tooth decay by facilitating the growth of larger, more-resistant crystals of apatite (calcium phosphate) in teeth. The same processes occur in bones, where fluoride assists in the development of more perfect bone structure that is less likely to fail with old age. Fluorine is fairly abundant in rocks (see Table 13.3) and in some soils and water. Most of the fluorine in soils and water is derived from the parent rock, but it can also be added by volcanic activity, which deposits fluorine-rich volcanic ash on the land. Industrial activity and application of fertilizers have also, on a limited basis, contributed locally to an increase in the concentration of fluorine in soils and water.

Relationships between the concentration of fluorides and health indicate a specific dose-response curve, as shown in Figure 13.3. The optimum fluoride concentration (point *B*) for the reduction of dental caries (tooth decay) is about 1 part per million (ppm). Fluoride levels greater than 1.5 ppm do not significantly decrease the incidence of caries (Figure 13.4), but they do increase the occurrence and severity of mottling (discoloration of teeth) (14). In concentrations of about 4 to 6 ppm, fluoride may help prevent calcification of the abdominal aorta and may reduce the prevalence of osteoporosis, a disease characterized by loss of bone mass and collapsed vertebrae (15). Figure 13.5 compares bone density in people in areas with high-fluoride and low-fluoride water. (The letters *N.S.* in this figure indicate those differences that are not statistically significant. This means for these *N.S.* cases we are not confident that observed differences are real.) Point *C* on the dose-response curve in Figure 13.3 was determined from the study of fluoride benefits in osteoporosis. Point *E* was determined from studies showing that fluoride concentrations of 8 to 20 ppm are associated with excessive bone formation in the periosteum (dense, fibrous, outer layer of bone) and calcification of ligaments that usually do not calcify (9). The positions of points *A*, *D*, and *F* are not known precisely, but fluoride in massive doses is the main ingredient of some rodent poisons.

Iodine Thyroid diseases are probably the best-known example of the relationship between geology and disease. The thyroid gland, located at the base of the neck, requires iodine for normal function. Lack of iodine causes goiter, a tumorous condition involving enlargement of the thyroid

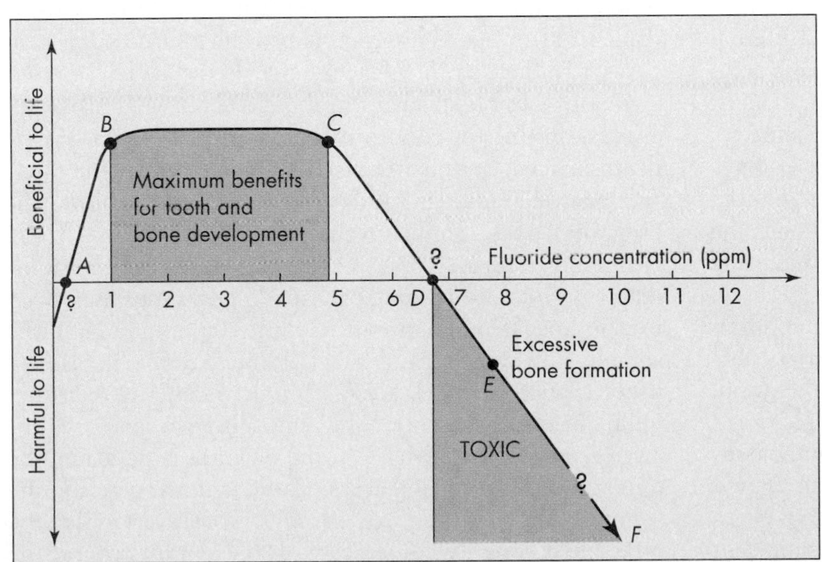

FIGURE 13.3 Dose-response curve for fluoride.

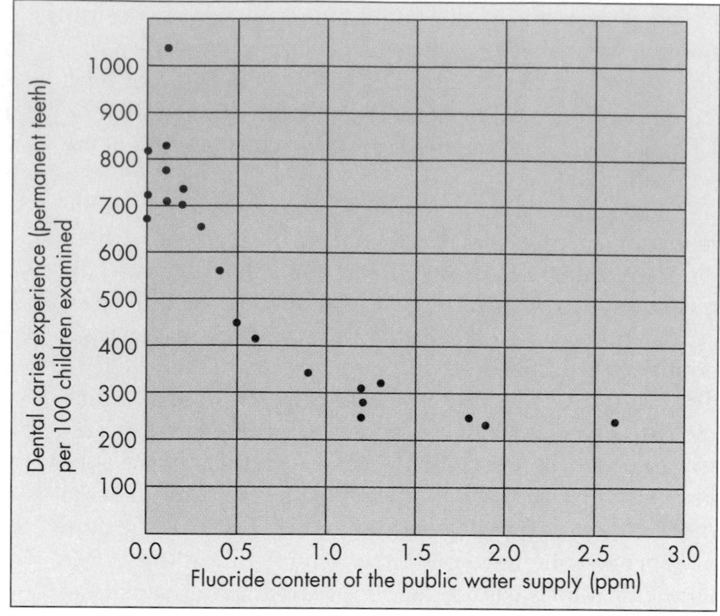

▶ **FIGURE 13.4** Relationship between incidence of dental caries (in permanent teeth) observed in 7257 selected 12- to 14-year-old school children in 21 cities of four states, and the fluoride content of public water supply. (*Reprinted from* F. J. Maier, Water quality and treatment, *3rd ed., by permission of the Association. Copyright 1971 by the American Water Works Association, Inc., 6666 West Quincy Ave., Denver, Colorado 80235. Used with permission of* McGraw-Hill Book Company.)

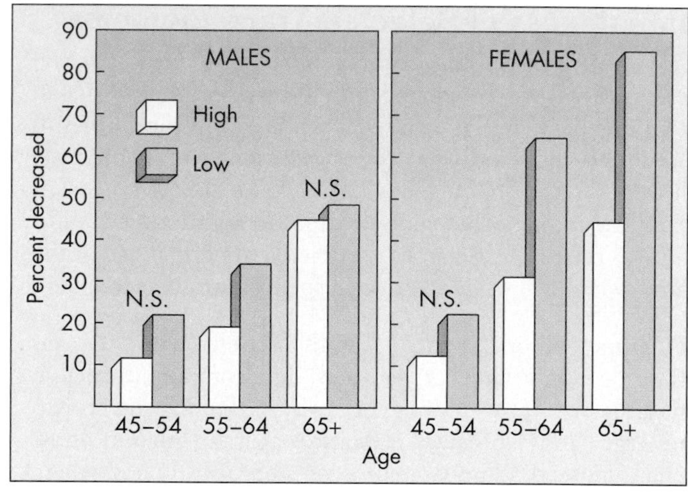

▶ **FIGURE 13.5** Percentage of decreased bone density in subjects from a high-fluoride area compared to those from a low-fluoride area. (*Reprinted from the* Journal of the American Medical Association, *October 31, 1966, vol. 198. Copyright 1966, American Medical Association.*)

gland (16). Furthermore, a child born to a mother who had iodine deficiency during pregnancy may suffer from cretinism, characterized by stunted growth and mental disability. At one time cretinism was fairly common in areas of Mexico and Switzerland that had a high goiter rate (17).

The incidence of goiter is clearly related to deficiencies of iodine, as shown by the relationship between iodine-deficient areas and the occurrence of goiter in the United States (Figure 13.6). Use of iodized salt is now common in the goiter belt, and it has been found that all adolescent and many adult goiters slowly decrease in size when iodized salt is used. In only four years (1924 to 1928), the use of iodized salt in Michigan reduced the incidence of goiter from 38.6 to 9 percent (16). Goiter remains endemic in some parts of the world. For example, in Sri Lanka, nearly 10 million people are at risk, and rates of the disease as high as 44 percent of the population have been observed in some areas. As in the United States, the main cause is iodine deficiency, but there is also a climatic factor, and the goiter belt in Sri Lanka is coincident with a wet climate zone that reduces the natural availability of iodine. The incidence of goiter is higher among low-income rural groups. A program to provide iodized salt is necessary to reduce the incidence of the disease (18).

Considerable speculation abounds over what processes are responsible for concentrating iodine in or removing it from surficial earth materials. The most popular hypothesis is that the iodine in soil has been released by the weathering of rocks. Some of this iodine enters rivers and eventually the sea, so that the oceans have become great iodine reservoirs, containing perhaps 25 percent of the earth's total iodine (19). Because most iodine compounds are quite soluble, however, it is unlikely that much iodine is residual in rocks after long weathering. A more likely theory is that iodine in the ocean enters the atmosphere as gas (or is absorbed onto dust particles in the air) and is deposited by rain and snow onto land areas, where it accumulates in the soil. Another possibility arises from the observation that the goiter belt roughly coincides with the region around the

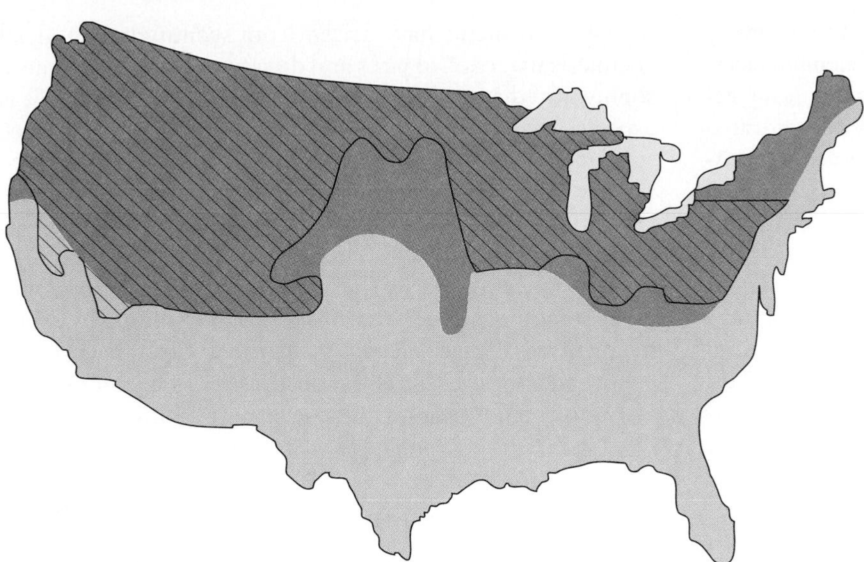

◀ **FIGURE 13.6** Map of the United States showing relationship between goiter occurrence and iodine deficiencies. (*From Armed Forces Institute of Pathology*)

Great Lakes that was glaciated only a few thousand years ago. It can be argued that soils in these areas are low in iodine because glaciation removed and destroyed the older, iodine-rich soils, and the postglaciation soils have not yet had time to accumulate large concentrations of iodine from the atmosphere. (19). This is unlikely, however, considering the solubility of iodine compounds.

Biologic processes significantly affect the amount of iodine available to plants and animals in an area. Plants affect the iodine content in soils by absorbing iodine into their living tissue and by retaining iodine in the organic-rich upper soil horizon (humus). Several analyses have shown that iodine tends to be concentrated in the upper soil horizons. Therefore, biological processes that cycle iodine in soils and plants might be more significant in determining availability of iodine than the amount of iodine in the local bedrock (19).

Zinc Zinc is a trace element necessary to plants, animals, and people. Although zinc is a heavy metal that in excessive amounts has been associated with disease, its role is known primarily from zinc-deficiency studies. Zinc deficiencies are present in 32 states and have resulted in a variety of plant diseases that cause low yields, poor seed development, and even total crop loss (20). Zinc deficiency in plants is related primarily to three soil conditions: low content of zinc, unavailability of zinc present in the soil, and poor soil management. The amount of zinc is low in areas that are highly leached, as in many parts of the Coastal Plain of North Carolina, Georgia, and Florida. Zinc content is also low in the sandhills of northern Nebraska, where wind transport of the sand that formed the hills failed to move the heavy metals along with the lighter quartz grains (20).

Zinc is recognized as essential to all animals and people, especially during early stages of development and growth. Although required concentrations are small, even slight deficiencies can cause loss of fertility, delayed healing, and disorders of bones, joints, and skin. Zinc deficiencies in humans may be associated with lung cancer, some kinds of chronic arterial disease, and other chronic diseases (20). However, zinc deficiencies have been observed mostly in patients, and the interrelationships between the deficiencies and the diseases are poorly understood. Do these people have the disease partially because they suffer from a zinc deficiency, or do they have a zinc deficiency because they have the disease? If the former is true, then zinc therapy, with its known beneficial effects on tissue repair, may be useful in treating some chronic diseases (20).

Adding zinc supplements to the soil may help prevent retardation of plant and animal growth. Care must be taken in adding zinc, however, because if the supplementary zinc is not of high quality, too much cadmium (an element often associated with zinc) may be inadvertently released into the environment. Cadmium has been associated with bone disease, heart disease, and cancer.

Selenium In high concentrations, selenium may be the most toxic element in the environment. It is a good example of why concern is increasing over the need for controlling health-related trace elements. Selenium is required in the diet of animals at a concentration of 0.04 ppm, beneficial to 0.1 ppm, and toxic above 4 ppm. Selenium is of concern to biologists because of its toxicity, even though greater damages to livestock have resulted from selenium deficiency than from selenium toxicity (21,22). Selenium toxicity has recently been discovered in the San Joaquin Valley in central California, where it is threatening a large agricultural area. (The environmental impact of selenium in this region is discussed in Chapter 18.)

The primary source of selenium is volcanic activity. It has been estimated that throughout the history of the earth, volcanoes have released about 0.1 g of selenium for every square centimeter of the earth's surface (21). Selenium ejected from volcanoes is in a particulate form, so it is easily removed from volcanic gas by rain and is usually concentrated near the volcano. This explains why the average concentration of selenium in the crust of the earth is about 0.05 ppm, whereas the soils of Hawaii, which are derived from volcanic material, contain 6 to 15 ppm (21).

Selenium in soils varies from about 0.1 ppm in deficient areas to as much as 1200 ppm in organic-rich soils in toxic areas. Of more concern, however, is the amount of selenium that enters plants. Hawaii, with its high selenium content, does not produce toxic seleniferous plants, but South Dakota and Kansas do, although their soils contain less than 1 ppm selenium (normal expected concentrations resulting from atmospheric fallout and weathering of rocks). This puzzle is resolved by examining the availability of selenium. In acid soils (as in Hawaii), the selenium forms insoluble compounds and is not available to plants, whereas in alkaline soils, selenium may be oxidized to a compound that is extremely soluble in water and thus readily available to plants. Forage crops grown for animals in soil containing soluble selenium are toxic, whereas forage crops grown in soils containing insoluble selenium have selenium deficiencies (21).

Selenium tends to be somewhat concentrated in living organisms. Some plants, called *selenium accumulators*, may contain more than 2000 ppm selenium, whereas other plants in the same area may have less than 10 ppm total selenium (22). Selenium levels in human blood range from 0.1 to 0.34 ppm, which is about a thousand times that found in river water and several thousand times that found in seawater. Marine fish, on the other hand, contain selenium in amounts of about 2 ppm, which is many thousand times that found in seawater (21). Because selenium accumulates in organic material, it may be concentrated in organic sediment and soil. Therefore, fossil fuels such as coal, which developed from organic material, also contain selenium. It has been estimated that the annual release of selenium by combustion of coal and oil in the United States is about 4000 tons (21).

Little is known about selenium deficiency and toxicity in humans. However, selenium deficiency is known to be related to muscular dystrophy in cows and sheep, and one study concluded that it may also be a causative factor for the disease in humans (23). A few cases of selenium poisoning have been reported in people who live on food grown in highly seleniferous soils or who drink seleniferous water in undeveloped countries. In economically developed countries, where interregional shipments of food are standard, there is probably little problem with selenium (22).

Human Use and Imbalances of Trace Elements

Agricultural, industrial, and mining activities have all been responsible for releasing potentially hazardous and toxic materials into the environment. This risk is part of the price we pay for our lifestyle. For example, unexpected and unanticipated problems have arisen from seemingly beneficial chemicals that control pests and disease. As we become more sophisticated at anticipating and correcting problems, it is expected that environmental disruption from release of trace substances will be reduced.

Geologically significant examples of unexpected problems with trace substances are found in mining processes. Ironically, we spend time, energy, and money to extract resources concentrated by the geological cycle, but in doing so we sometimes concentrate and release potentially harmful quantities of trace elements into the environment. A case history illustrating this is the occurrence of a serious bone disease related to mining zinc, lead, and cadmium in Japan (see *Case History: Mining and Toxic Trace Elements*).

13.4 Chronic Disease and Geologic Environment

Health can be defined as an organism's state of adjustment to its own internal and external environment. Observation over many years has suggested that some regional and local variations in human chronic diseases such as cancer and heart disease are related to the geologic environment. Although evidence continues to accumulate, the nature of these associations remains to be discovered. There are two reasons for the lack of conclusive results. First, hypotheses about the relationship between the geologic environment and disease have not been specific enough to be tested adequately. Basic research and field verification need to be better coordinated. Second, many methodological difficulties remain in obtaining reliable and comparable data for medical-geological studies (24). Thus, we know much less about geologic influences on chronic disease than about the contribution of other environmental factors such as climate.

Although our evaluation of geologic contributions to disease remains an educated guess, the benefit to humankind of learning more about these relationships is obvious. The geographic variations in the incidence of heart disease in the United States may be related to the geologic environment, and it has been estimated that two-thirds of the cancerous tumors in the Western Hemisphere result in part from environmental causes, although genetic factors are also important (24).

Heart Disease and the Geochemical Environment

The term *heart disease* here includes coronary heart disease (CHD) and cardiovascular disease (CVD). Variations of heart disease mortality have generally shown interesting relationships with the chemistry of drinking water, and in particular with the hardness of drinking water. Hardness is a function of the amount of calcium and magnesium that is dissolved in water. Higher concentrations of these substances produce higher hardness. Water with low concentrations of these substances is termed *soft*. Studies in Japan, England, Wales, Sweden, and the United States all conclude

CASE HISTORY Mining and Toxic Trace Elements

A serious chronic disease known as *itaiitai* has claimed many lives in Japan's Zintsu River Basin. This extremely painful disease (*itaiitai* means "ouch, ouch") attacks bones, causing them to become so thin and brittle that they break easily. The disease broke out near the end of World War II, when the Japanese industrial complex was damaged and sound industrial-waste disposal practices were largely ignored. Mining operations for zinc, lead, and cadmium dumped mining wastes into the rivers, and farmers used the contaminated water downstream for domestic and agricultural purposes. The cause of the disease was unknown for years, but in 1960 bones and tissues of victims were examined and found to contain large concentrations of zinc, lead, and cadmium (31).

Measurement of heavy-metal concentrations in the Zintsu River Basin showed that although the water samples generally contained less than 1 ppm cadmium and 50 ppm zinc, these metals are selectively concentrated in the sediment and even more highly concentrated in plants (another example of biomagnification). One set of data for five samples shows an average of 6 ppm cadmium in polluted soils. In plant roots, this average increased to 1250 ppm, and in the harvested rice it was 125 ppm. Subsequent experiments showed that rats fed a diet containing 100 ppm cadmium lost about 3 percent of their total bone tissue, and rats fed a diet containing 30 ppm cadmium, 300 zinc, 150 lead, and 150 copper lost an equivalent of about 33 percent of their total bone tissue (31).

Although measurements of heavy-metal concentrations in the water, soil, and plants of the Zintsu River Basin produce somewhat variable results, the general tendency is clear. Scientists are fairly certain that heavy metals, especially cadmium, in concentrations of a few parts per million in the soil and rice produce itaiitai disease (32).

that communities with relatively soft water have a higher rate of heart disease than do communities with harder water.

Perhaps the first report of a relationship between water chemistry and cardiovascular disease came from Japan, where a prevalent cause of death is stroke (apoplexy), a sudden loss of body functions caused by the rupture or blockage of a blood vessel in the brain. The geographic variation of the disease in Japan is related to the ratio of sulfate to bicarbonate in river water. Figure 13.7 shows that areas in Japan with high death rates due to stroke correspond in general to areas where the sulfate-to-bicarbonate ratio is relatively high (25).

To relate this to hardness of the water takes a little explanation. Bicarbonate ions (HCO_3^-) are in part the result of chemical weathering of carbonate-rich rocks such as limestone, composed largely of the mineral calcite ($CaCO_3$), or calcium-magnesium carbonate, composed of the mineral dolomite [$CaMg(CO_3)_2$]. Thus, water high in bicarbonate often also contains high concentrations of calcium and/or magnesium and is relatively hard; a low ratio of sulfate to bicarbonate would indicate relatively high bicarbonate concentrations and hard water. Conversely, a high ratio would indicate relatively soft water.

The abundance of sulfate, especially in the northeastern part of Japan, evidently stems from the sulfur-rich volcanic rock found there. Rivers in the area have relatively soft water. In contrast, rivers in Japan that flow through sedimentary rocks are low in sulfate and high in bicarbonate (relatively hard water), like most river water in the world.

A general inverse relationship between hard water and death rates from heart disease is also present in the United States (26,27). For example, a study in Ohio suggests that sulfate ($SO_4^=$) and bicarbonate (HCO_3^-) concentrations may influence the incidence of heart disease (28). The Ohio study found that counties with sulfate-rich drinking water derived from coal-bearing rocks in the southeastern part of the state tend to have a higher death rate due to heart attack than do counties with low-sulfate, high-bicarbonate drinking water

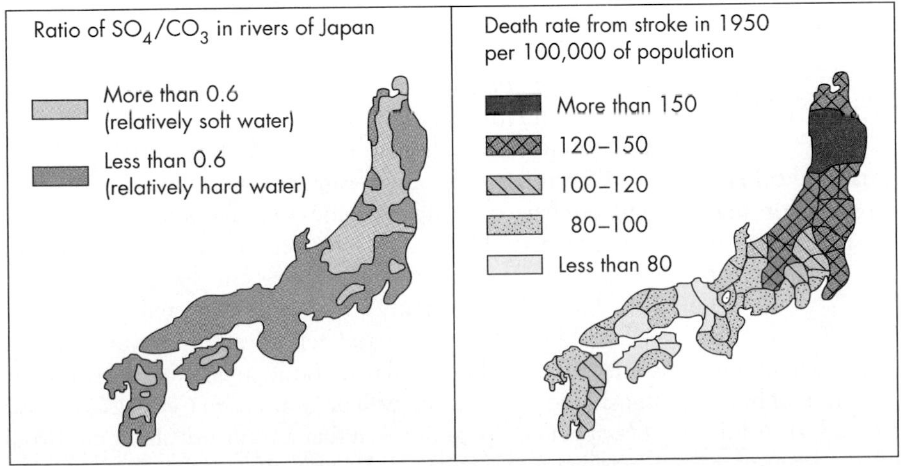

◀ **FIGURE 13.7** Maps of Japan comparing the ratio of SO_4 to CO_3 in rivers to the death rate from stroke in 1950. (*Data from* T. *Kobayashi*, 1957)

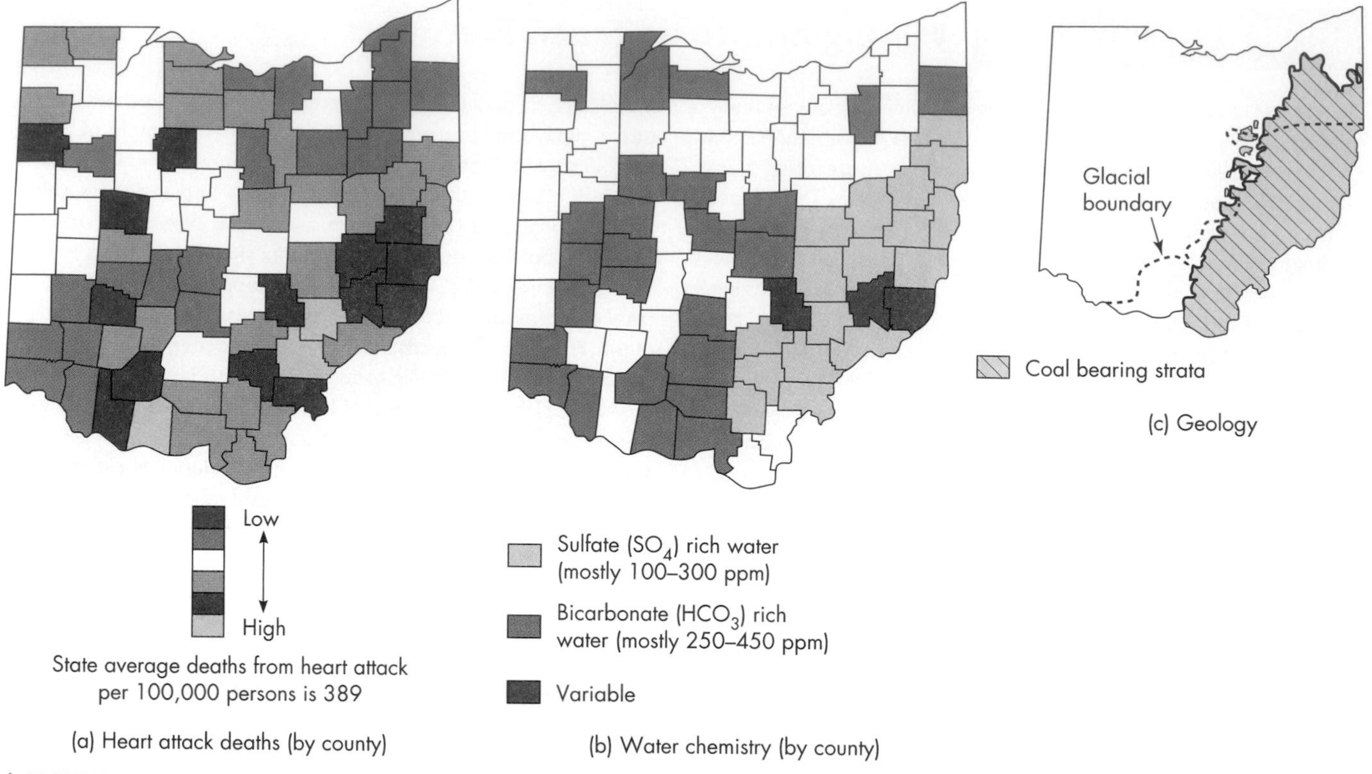

▲ **FIGURE 13.8** Occurrence of heart-attack death by county and distribution of sulfate-rich and bicarbonate-rich surface water by county in Ohio. (*After* R. J. *Bain*. 1979. Geology 7:7–10.)

derived from young glacial deposits, as shown in Figure 13.8. We emphasize, however, that this generally negative correlation is not conclusive. A study in Indiana found a small positive correlation between heart disease and hardness of water, suggesting that other variables may exert considerable influence on rates of heart disease (29).

The correlations we have discussed do not necessarily show a cause-and-effect relationship between the geochemical environment and heart disease. If there is a cause-and-effect relationship, we do not know its nature, but there are several possibilities:

- Soft water is acidic and may, through corrosion of pipes, release into the water trace elements that cause heart disease.
- Some other characteristics of soft water may contribute directly to heart disease.
- Some substances (probably trace elements) dissolved in hard water may help prevent heart disease.

Of course, some combination of these factors (and others) is also possible and is consistent with our observation that a disease may have several causes. Additional research is needed to prove the benefit of hard water and perhaps treat soft water to reduce heart disease.

A study of Georgia soils suggests that certain trace elements found in soils—manganese, chromium, vanadium, and copper—play a beneficial role in the prevention of heart disease (see *Case History: Heart Disease in Georgia*). Additional information is needed to increase our understanding of how other trace elements affect the heart and circulatory system. Cadmium, fluorine, and selenium, among others, require further study. With respect to cadmium, we know that individuals who die from hypertensive complications generally have greater concentrations of cadmium or higher ratios of cadmium to zinc in their kidneys than do individuals who die from other diseases. Surprisingly, however, workers who are exposed to cadmium dust and accumulate the element in their lungs do not have abnormal rates of hypertension (30).

Cancer and the Geochemical Environment

Cancer tends to be related to environmental conditions. As with heart disease, however, relationships between geochemical environment and cancer have not been proved. The causes of the various types of cancers are undoubtedly complex and involve many variables, some of which may be the presence or absence of certain earth materials.

Carcinogenic (cancer-causing) substances in the environment have two origins: some occur naturally in earth materials such as soil and water, whereas others are released into the environment by human use. In recent years, attention has been focused on the many known and suspected carcinogens released by human industrial activities. This awareness has sometimes resulted in alarm about substances whose relationship to cancer is not well demonstrated (see *A Closer Look: Asbestos*). This does not mean that all concern about industrial carcinogens is misplaced. Recent information suggests that

CASE HISTORY Heart Disease in Georgia

An interesting study of patterns of heart disease mortality and their possible relation to the geologic environment was conducted in Georgia by the U.S. Geological Survey (33, 34). The difference between the highest and lowest rates of death caused by heart disease among the 159 counties in Georgia is nearly as great as can be found between any two counties in the United States. Nine counties in northern Georgia with low rates and nine counties in central and south-central Georgia with high rates were selected for geochemical analysis.

The locations of the selected counties are shown in Figure 13.A. In the nine counties with low rates of death from heart disease, the range is from 560 to 682 deaths per 100,000 population for males 35 to 74 years of age during a period of 10 years. In the nine counties with high mortality rates, the range is from 1151 to 1446 deaths per 100,000 population for the same age group and time period (33). Mortality rates for the south-central part of the state are thus about twice as high as for the northern section.

The nine counties with high heart-disease rates are primarily in southern Georgia on the Atlantic Coastal Plain. The landscape characteristically has low relief, sluggish drainage, and swamps. In general, sandy soils overlie Cenozoic marine sedimentary rocks that have undergone intensive weathering. Small-scale agricultural activity is prevalent where relief and drainage are sufficient.

The nine counties with low heart-disease rates are located in northern Georgia in portions of the Appalachian region known as the Piedmont, the Blue Ridge, and the Ridge and Valley region. In the Piedmont, which extends from the Blue Ridge Mountains to the Coastal Plain, rocks are a mixture of Precambrian and Paleozoic metamorphic rocks and intrusive igneous rocks of varying ages. Soils are generally mixtures of sand, silt, and clay (loam) with clay subsoils. The topography of the Blue Ridge is mountainous, with narrow valleys and turbulent streams. The rocks include metamorphic and sedimentary rocks of Precambrian to early Cambrian age and Paleozoic igneous rocks. Soils are acid with relatively high organic content; the high relief is not favorable to agricultural activity. The Ridge and Valley region is characterized by linear ridges and parallel valleys oriented northeast to southwest; the rocks are folded and faulted sedimentary rock of Paleozoic age. Soils vary from rich loam in valleys with limestone bedrock to less fertile clay soils developed on shales and relatively infertile soils derived from sandstone bedrock (33). In the Ridge and Valley region of nearby Virginia, one valley on limestone is named "Rich Valley" and a neighboring valley on shale is named "Poor Valley." Geology makes all the difference!

A detailed geochemical investigation of the soils and plants of the low-death-rate area in northern Georgia and high-death-rate area in south-central Georgia concluded that the geochemical variations in the soil reflected differences in the bedrock (33,34). Thirty elements were analyzed, and significant concentrations of aluminum, barium, calcium, chromium, copper, iron, potassium, magnesium, manganese, niobium, phosphorus, titanium, and vanadium were found. There were differences between the high-death-rate and low-death-rate counties in the concentrations of some of these elements, probably reflecting the fact that soils in the high-death-rate counties of the Coastal Plain derive from unconsolidated sediments (sands and clays) that were previously weathered and leached during deposition. Zirconium was the only element that occurred in significantly larger concentrations in soils of the high-death-rate counties (34). However, manganese, chromium, vanadium, and copper, which are known to have beneficial effects on heart disease, are more highly concentrated in the low-death-rate counties. The low death rate in these areas may therefore be a result of an abundance of beneficial trace elements in the soils rather than low concentrations of harmful elements (34).

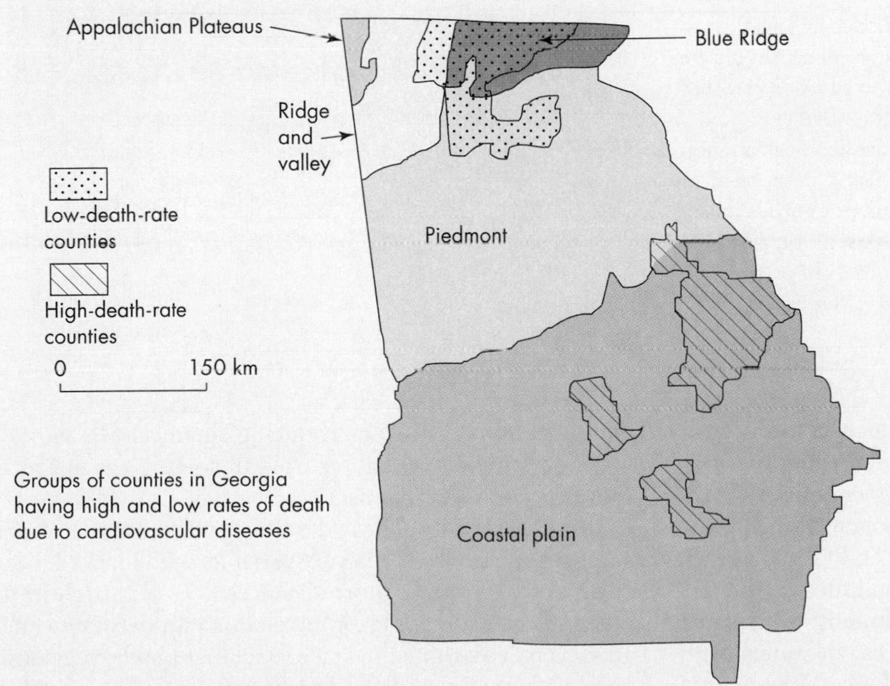

FIGURE 13.A Map showing physiographic regions and Georgia counties having high and low death rates from cardiovascular disease. (*After* H. T. *Shacklette*, H. I. *Sauer*, *and* A. T. *Miesch*. 1970. U.S. *Geological Survey Professional Paper* 574C.)

A CLOSER LOOK Asbestos

The term *asbestos* refers to small, elongated mineral fragments called *fibers,* particularly fragments of the ferromagnesian silicate minerals known as *amphiboles* or of the rock *serpentine*. There have been many industrial uses of asbestos, mostly related to fire prevention and overheating of materials. Asbestos is used in brake linings and as an insulation for a variety of purposes.

The downside of asbestos use is that extensive contact with some varieties of asbestos by workers in industry has led to cases of a lung disease known as *asbestosis,* thought to be caused by the inhalation of the fibers. Concern has also been raised that exposure to asbestos can cause lung cancer. The move to reduce the presence of asbestos or to ban it outright has led in recent years to the removal of asbestos from old structures (particularly schools and other public buildings) in the United States.

Asbestos is not the only mineral that produces fibers—crushing almost any rock is likely to produce some small, elongated mineral fragments. Laboratory experiments in which animals are exposed to a wide variety of mineral fibers that become embedded in lung tissue suggest that tumors may develop as a result of the exposure. These experiments have been criticized because they involve high doses of fibers placed directly on lung-lining tissues. Applying the results of these experiments to humans requires extrapolation to the much lower levels of exposure that people face (35,36).

Research on cancer has designated quartz as a probable carcinogenic material to humans; as a result, any natural material in the United States with more than one-tenth of 1 percent of free silica (this includes quartz) must display hazard-warning signs. Thus, trucks transporting crushed stones theoretically need to carry warnings, and a truck in Delaware transporting crushed stones was actually issued a violation for not displaying such signs! The author reporting the incident made the statement (facetiously) that beaches composed of silicate sand might present a public health hazard, so warnings should be posted. Much misunderstanding, in the same author's view, has resulted from well-meaning efforts to extrapolate from toxicity studies to the environment without understanding common natural characteristics of minerals (36). Sand composed of quartz contains hard and rounded particles, not fibrous dust.

Returning to our discussion of asbestos, all types of asbestos are not equally hazardous. The mineral chrysotile (Figure 13.B) is known as white asbestos and is the variety most commonly utilized in the United States. It has been estimated that approximately 95 percent of the asbestos now in place in the United States is white asbestos. Environmental health studies of miners in Canada, where most of this asbestos comes from, suggest that exposure to white asbestos is not very harmful. In fact, no study has linked lung disease with nonoccupational exposure to white asbestos. On the other hand, blue asbestos, which comes from the mineral crocidolite, is known to cause lung disease (35).

Use of asbestos in buildings in the United States has become a health issue. A great deal of fear is associated with nonoccupational exposure to asbestos, particularly exposure of children, and tremendous amounts of money have been spent to remove white asbestos from schools, public buildings, and homes. It would seem from our discussion of white asbestos that the risk has been overstated and that much of the removal of white asbestos may have been unnecessary. Nevertheless, additional research should be devoted to evaluating health risks associated with the various types of asbestos and rocks that produce small, thin, fibrous fragments.

▲ **FIGURE 13.B** Shown here is a sample of the asbestos mineral chrysotile or white asbestos. The sample shows the fluffy, fibrous nature of this commercial grade asbestos. The sample is from New Jersey. The scale along the bottom is in millimeters. (*Photograph courtesy of* H. *Catherine W. Skinner, Yale University. From* Asbestos and other fibers: Mineralogy, crystal chemistry and health effects *by* H. *Catherine W. Skinner, Malcolm Ross, and Clifford Frondel.* 1988. *New York: Oxford University Press.*)

cancer-causing substances (in variable concentrations) may be found in much of our drinking water. Certainly water polluted with industrial and urban waste containing toxic chemicals, some of them possible carcinogens, is being released into our surficial and groundwater water supplies. The Mississippi River, particularly, has pollution problems. Ironically, present methods of water treatment, which have greatly reduced the historic serious threat of waterborne diseases, may contribute to these problems: When combined with chlorine, some industrial waste produces low concentrations of cancer-producing chemicals. In addition, water-treatment procedures used in some areas fail to remove certain carcinogens.

In northern Iran, near the Caspian Sea, rates of esophageal cancer are tremendously variable, with marked differences occurring over short distances (37). Research in this area indicates a complex interrelationship between cancer and several environmental factors: climate, soils, vegetation, and agricultural practices. The highest correlation between the esophageal cancer rate and an environmental factor is

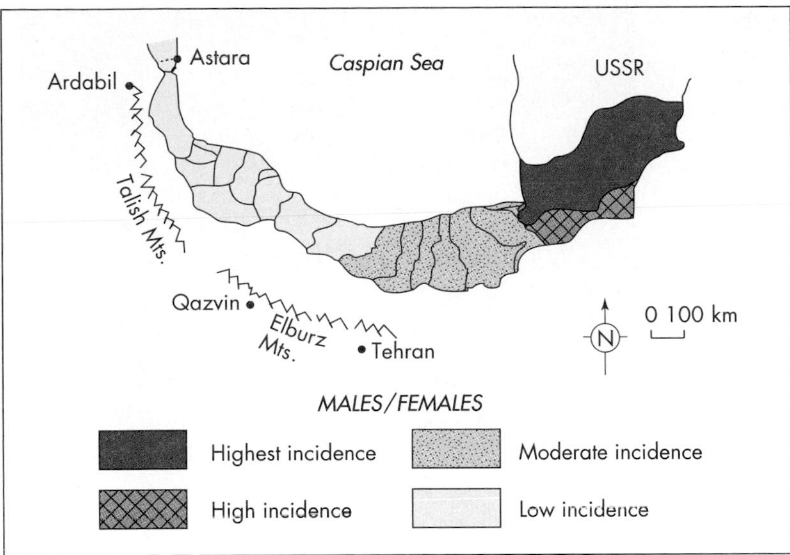

(a) Age-standardized incidence rates of esophageal cancer in the Caspian littoral of Iran

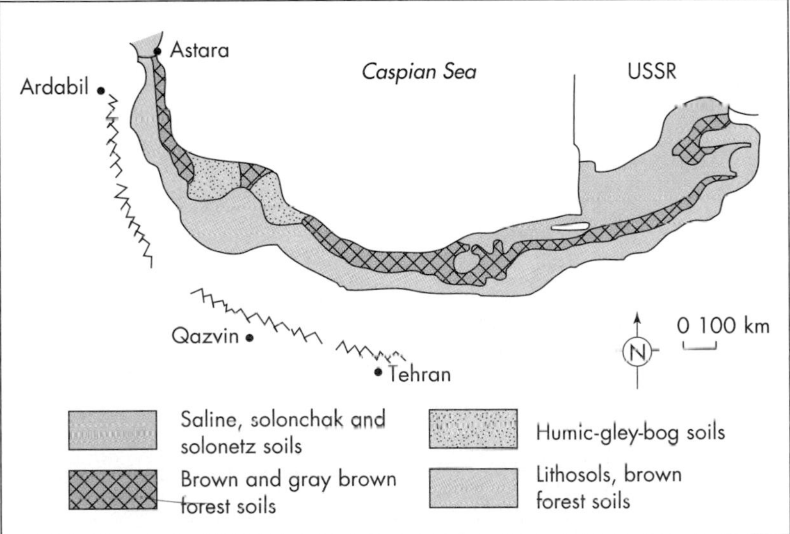

(b) Types of soil in the Caspian littoral of Iran

◀ **FIGURE 13.9** Maps of the esophageal cancer belt in Asia comparing age and standardized incidence rates of the cancer (upper) with soil types (lower). (*From Kmet and Mahboubi,* Science, *vol.* 175, *pp.* 846–53, *February* 1972. *Copyright* © 1972, *American Association for the Advancement of Science.*)

for soil types (Figure 13.9). The highest rates of the disease are associated with saline soils in the dry eastern part of the area; the lowest incidence is in the rain belt in the western section. Rainfall increases east to west by a factor of 4, and the soils become more and more leached of salt. Vegetation and agriculture systematically change from grazing on sparse cover of drought-tolerant plants to lush forests and dry farming of rice, fruit, and tea. Paralleling this change from east to west is the continuously lessening incidence of esophageal cancer (37).

Examples of relations between environment and cancer reinforce the importance of recognizing multiple causes of chronic disease. Regardless of the numerous contradictions and exceptions that can be found, we have sufficient data to be certain that there is no single cause for cancer. We must therefore adopt a multifactorial approach, and variations in the environment are certainly among the significant factors.

13.5 Radioactivity and Radon Gas

Radon, a naturally occurring radioactive gas, is an important environmental concern that arises from natural processes and is not related to any public, industrial, or government activities. Before we consider the radon question, which may emerge as the most significant geology-based environmental health concern in the United States today, we must consider the nature of radioactivity and the health risks posed by exposure to radioactive emissions (radioactive radiation).

The Nature of Radioactivity

All atoms of the same element have the same atomic number (number of protons in the nucleus). Isotopes are atoms of an element that have different numbers of neutrons and therefore different atomic mass numbers (the number of protons plus neutrons in the nucleus). For example, two isotopes of uranium are $^{235}U_{92}$ and $^{238}U_{92}$. The atoms of both of these isotopes have an atomic number of 92, but their atomic mass numbers are 235 and 238, respectively. These isotopes may be written as uranium-235 and uranium-238, or U-235 and U-238.

A radioactive isotope, or **radioisotope,** spontaneously undergoes nuclear decay, meaning that it undergoes a nuclear change while emitting one or more forms of radioactive radiation. The three major kinds of radiation emitted during radioactive decay are called **alpha particles** (α), **beta particles** (β), and **gamma radiation** (γ). Each radioisotope has its own characteristic emissions; some isotopes emit only one type of radiation, and some emit a mixture.

Alpha particles consist of two protons and two neutrons, making them much more massive than other types of radioactive emission. Because alpha decay (emission of an alpha particle) changes the number of protons in the nucleus, the isotope is changed into an isotope of a different element. For example, a radon-222 atom, which has 86 protons, emits an alpha particle and is thereby transformed into a polonium-218 atom, which has 84 protons. Because of their great mass, alpha particles are the slowest-moving (lowest energy) radioactive emissions. They travel the shortest distances (about 5 to 8 cm in air) and penetrate solid matter less deeply than do beta or gamma emissions.

Beta particles are energetic electrons and have a small mass compared to alpha particles. Beta decay occurs when one of the neutrons in the nucleus of the isotope spontaneously changes (38). Note that the electron emitted is a product of the transformation; there are no electrons in the nucleus before beta decay occurs.

In gamma radiation, a type of energy called a *gamma ray* is emitted from the isotope, but the number of protons and neutrons in the nucleus is unchanged. Gamma rays are nuclear rays similar to medical X rays. Gamma rays are emitted by the nucleus and have the highest energy of all radioactive emissions and travel faster and farther and penetrate deeper than do alpha or beta particles.

An important characteristic of a radioisotope is its **half-life,** which is the time required for one-half of a given amount of the isotope to decay to another form. Every radioisotope has a unique characteristic half-life. Radon-222, for example, has a relatively short half-life of 3.8 days. Carbon-14, a radioactive isotope of carbon, has a half-life of 5570 years; uranium-235 has a half-life of 700 million years; and uranium-238 has a half-life of 4.5 billion years.

Some radioisotopes, particularly those of very heavy elements, undergo a series of radioactive decay steps, finally reaching a stable (nonradioactive) isotope. Figure 13.10 shows the decay chain from uranium-238 through radium-226, radon-222, and polonium-218 to the stable isotope lead-206. The two most important facts about each transformation are the type of radiation emitted and the half-life of the isotope that is transformed; this information is given for some of the transformations shown in Figure 13.10. The decay from one radioisotope to another is often stated in terms of parent and daughter atoms. For example, the parent radon-222, a gas with a half-life of 3.8 days, decays by alpha emission to its daughter, polonium-218, a solid with a half-life of 187 seconds.

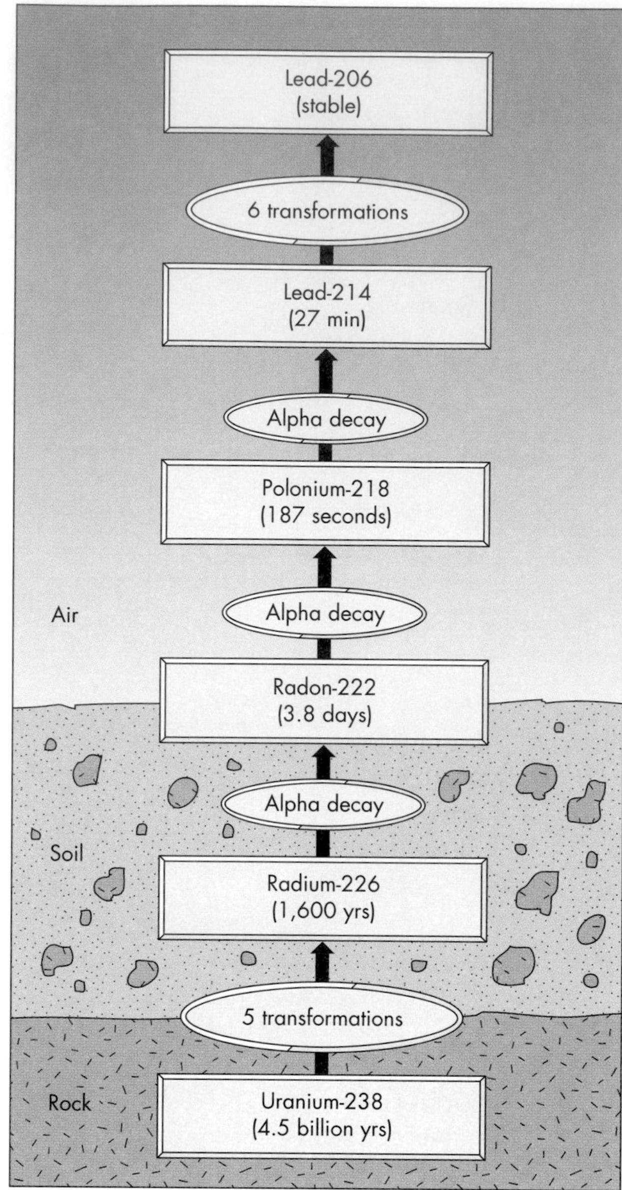

▲ **FIGURE 13.10** Simplified radioactive decay chain from uranium-238 to lead-206. Not all isotopes are shown. Half-lives and type of decay are shown for some isotopes.

Radiation Units

The units used to measure radioactivity depend on whether we are interested in the process of radioactive decay or in the effects of the radiation dose absorbed by living organisms. In either case, the units measure the energy of radioactive radiation.

Measuring Radioactive Decay A commonly used unit for measuring radioactive decay is called the *curie* in honor of Marie and Pierre Curie, who isolated the first known radioactive elements, polonium and radium, in the 1890s.

Marie Curie coined the term *radioactivity*, and she named polonium for her homeland, Poland. The harmful effects of radiation were not known at that time, and both Marie Curie and her daughter, who also worked with radioactive elements, died of leukemia, which is believed to be radiation induced (38).

A curie (Ci) is the amount of radioactivity present in 1 g of radium-226, which undergoes about 37 billion nuclear decays each second. Radium-226 has a half-life of 1622 years; at the end of that time the initial gram of radium will be reduced to 0.5 g, the nuclear decays will be reduced to about 18.5 billion per second, and the amount of radioactivity present will be 0.5 Ci (39). In the International System of measurements, the unit commonly used for radioactive decay is the *becquerel* (Bq) (named for Marie Curie's teacher, physicist Henri Becquerel), which corresponds to one radioactive decay per second (thus $1 \text{ Ci} = 3.7 \times 10^{10} \text{ Bq}$).

The becquerel is a useful unit for working with a radioactive gas, which contains far fewer atoms than the same volume of a solid. If we were working with radon-222, for example, we could state its radioactivity in *becquerels per cubic meter* or in *picocuries per liter* (pCi/l). A picocurie is a trillionth (10^{-12}) of a curie, so picocuries per liter are a measure of the number of decays per second in a liter ($1 \text{ pCi/l} = 37 \text{ Bq/m}^3$).

Measuring the Radiation Dose The dose of radiation received from a radioactive emitter is commonly expressed in rads and rems. A *rad* is the unit of the Radiation Absorbed Dose, and a *rem* is the unit of the *equivalent dose*, which will be explained shortly. The corresponding units of the International System are grays and sieverts: one gray (absorbed dose) is equal to 100 rads, and one sievert (equivalent dose) is equal to 100 rem (38). Both systems of units are commonly used.

Rads and grays measure the energy retained by living tissue that has been exposed to radiation. However, because different types of radiation have different penetration, they cause different degrees of damage to living tissue. The rad or gray is therefore multiplied by a factor known as the *relative biological effectiveness to produce the equivalent dose* (or effective equivalent dose), which is measured in rems or sieverts. The millirem (mrem) and millisievert (mSv)—one one-thousandth of a rem or a sievert—are used to measure small doses of radioactivity (39–41). The units commonly used to measure X rays or gamma rays are the *roentgen* or, in the International System of units, *coulombs per kilogram* (both are related to ion charge per unit mass).

The Health Danger from Radioactive Radiation

How dangerous a radioisotope is to health depends on several factors: the kind of radiation it emits, the nature of the exposure (that is, whether the isotope is outside or inside the body), the half-life of the isotope, and its physical state and chemical activity. Radon-222 has low-energy emissions and a short half-life, but it is dangerous because it is a gas. If the gas is inhaled, its solid daughter products emit their damaging radiations within the body.

Type of Radiation We have already noted that alpha particles travel only about 5 to 8 cm in air before they are stopped by collisions with air particles. In human tissue, which is much denser than air, alpha particles can travel only about 0.005 to 0.008 cm, so they must originate very close to a cell or tissue to damage it (38). Alpha radiation is most dangerous to health when the alpha-emitting isotope is inhaled or ingested: All the emissions are stopped by the body's tissues within a very short distance, so the body absorbs all of the damaging energy. On the other hand, when an alpha-emitting isotope is stored in a container, it is relatively harmless.

Because of their very high energy and deep penetration, gamma rays can be dangerous outside or inside the body. However, if the emitter is ingested, some of the radiation passes to outside the body, reducing the potential damage. Protection from an external source of gamma rays requires thick shielding. Beta radiation is intermediate in its effects, although most beta radiation is absorbed by the body when a beta-emitter is ingested. Beta particles from an external source can be stopped by moderate shielding.

Radiation Dose The natural background radiation (that radiation we are naturally exposed to on or near the surface of earth) received by Americans averages about 1.5 mSv per person per year. The range, however, is about 1.0 to 2.5 mSv (39). In general, the highest levels of radiation are in the mountain states and the lowest level is in Florida. The differences are due primarily to elevation and geology. More cosmic radiation is received at higher elevations because of the thinner atmosphere, and granitic rocks that often contain radioactive minerals are more common in mountainous areas. Florida, with a basic limestone geology and low elevation, has a relatively low level of background radiation (41). Nevertheless, in parts of Florida where phosphate deposits occur, the background radiation is often above average owing to high uranium concentrations in the phosphate-containing rocks.

There is no general agreement about the precise amounts of radiation we receive from low-level background sources. However, major sources of background radiation include potassium-40 and carbon-14, which are present in our bodies and probably deliver between 0.2 and 0.25 mSv per person per year. Potassium is an important electrolyte in our blood, and one isotope of potassium (K-40) is radioactive with a very long half-life. Although potassium-40 makes up only a very small percentage of the total potassium in our bodies, it is present in all of us, so we are all slightly radioactive. Thus, if you share your life with another person, you are exposed to a little bit more radiation (a small hazard compared to the many benefits of companionship). Cosmic rays deliver between 0.35 and about 1.5 mSv per person per year, depending upon elevation, and radioactive materials in rocks and soils deliver on the average about 0.35 mSv. However, the amount delivered from rocks and soils may be much larger in some areas where radon gas seeps into homes.

Anthropogenic (human-created) sources of low-level radiation include X rays for medical and dental purposes, which may deliver an average of 0.7–0.8 mSv per person per year; nuclear weapons testing and nuclear power plants that may be responsible for approximately 0.04 mSv; and burning of fossil fuels such as coal, oil, and natural gas, which may add another 0.03 mSv (39,41).

Occupation and life-style can also affect a person's annual dose of radiation. Every time you fly at high altitudes, you receive an additional dose. If you work at a nuclear power plant or conventional coal-fired plant or in a number of industrial positions, you may be exposed to additional low-level radiation. The amount of radiation received by workers is closely monitored at obvious sites such as nuclear power plants and laboratories that produce X rays; personnel must wear badges that show the total dose of radiation received. Figure 13.11 shows some of the sources of radiation to which we are commonly exposed. Notice that the exposure to radon gas at its high end (that is, in homes with the highest concentrations of radon) can be about equivalent to the exposure experienced by evacuees from Chernobyl in the year of the Russian nuclear power plant accident (42).

A major question concerning radiation exposure is: When does the exposure or dose become a hazard to health? There are no easy answers to this question. We know a good deal about the effects of high doses of radiation on people but much less about continuous exposure to low-level radiation. Most of our information about the effects of high doses comes from studies of atomic-bomb survivors in Japan, of people exposed to high levels of radiation in uranium mines, of workers painting watch dials with luminous paint containing radium, and of people treated with radiation therapy for disease (43). Workers who are exposed to high levels of radiation in mines have been shown to suffer a significantly higher rate of lung cancer than the general population. Mortality studies suggest that there is a delay of 10 to 25 years between the time of exposure and the onset of disease.

The adverse effects of high doses of radiation have been determined from the kinds of studies just described. A dose of about 5000 mSv (5 Sv) is considered lethal to all people exposed to it. A 1000- to 2000-mSv dose causes health problems, including vomiting, fatigue, increased rate of natural abortion of pregnancies of less than 2 months' duration, and temporary sterility in males, and at 500 mSv physiological damage is recorded. The maximal allowable dose of radiation per year for workers in industry is 50 mSv, which is approximately 30 times the average whole-body radiation received by people from natural background sources (39, 41). The maximum permissible annual dose for the general public in the United States is 5 mSv, or about three times the average annual background dose (39).

Although most scientists agree that radiation can cause cancer, there is debate about the relationship between low-level radiation exposure and cancer mortality. Some scientists believe that the relationship is linear, so that any increase in radiation produces an additional hazard. Others believe that the body is able to handle and recover from very low levels of radiation, with health effects beginning to appear beyond some threshold dose. The verdict is still out on this subject, but it seems prudent to take a conservative stand and accept that there may be a linear relationship. That is, any increase in radiation is likely to be accompanied by an increase in adverse health effects.

Radon Gas

Radon is a naturally occurring radioactive gas that is colorless, odorless, and tasteless. Radioactive decay of uranium-238 produces radium-226, which in turn decays to radon-222 (see Figure 13.10). Thus, uranium-bearing rocks are the source of the radon gas that contaminates many homes in the United States.

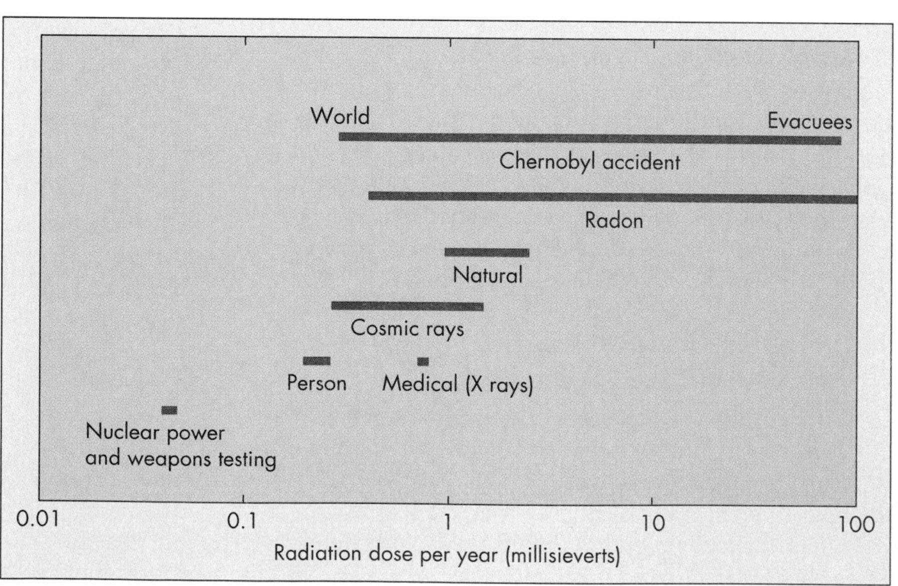

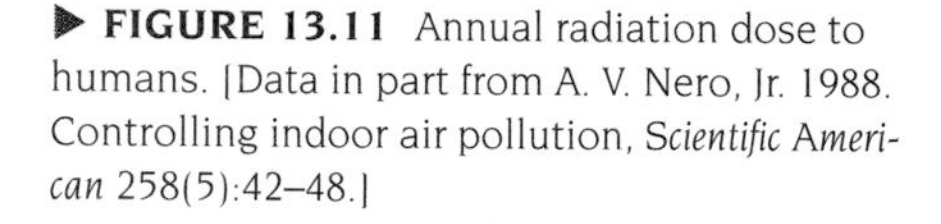

▶ **FIGURE 13.11** Annual radiation dose to humans. [Data in part from A. V. Nero, Jr. 1988. Controlling indoor air pollution, *Scientific American* 258(5):42–48.]

Radon gas has had an interesting history ever since its discovery in 1900 by German chemist Ernest Dorn. In the early 1900s radon water became a popular health fad. One doctor reported in a medical journal that radon had absolutely no toxic effects and in fact was accepted harmoniously by the human system much as sunlight is accepted by plants. In 1916 the American Medical Association actually refused to endorse several radon emanators (a device for dissolving radon in water) because too little radon was produced. During this period, many products containing radium, and thus radon, were on the market, including chocolate candies, bread, and toothpaste. As recently as 1953 a contraceptive jelly that contained radium was marketed in the United States. Today the cell-killing properties of radon daughter products are widely used for the treatment of cancer (38).

Why Radon Gas Is Dangerous People are worried about radon gas because exposure to elevated concentrations of radon is associated with increased risk of lung cancer, particularly for smokers. The risk is thought to increase as the level of radon concentration, the length of exposure to radon, and the amount of smoking increase (44,45).

In recent years approximately 140,000 lung cancer deaths have occurred annually in the United States. Although smoking is the most important factor associated with lung cancer, the EPA estimates that 7000 to 30,000 of these deaths are related to exposure to radon gas. One estimate is that exposure to both tobacco and radon is approximately 10 times as hazardous as exposure to either of these pollutants by itself. Although few direct studies have linked radon gas exposure in homes to increased incidence of lung cancer, estimates of such linkage primarily come from studies of individuals who have experienced high exposure to radiation through such activities as mining uranium.

The health risk from radon gas is related primarily to its daughters, particularly polonium-218, which is a particle and adheres to dust. This dust may be inhaled into the lungs, where alpha radiation can occur as polonium-218 decays to lead-214 with a half-life of approximately 3 minutes (see Figure 13.10). Inhaled aerosols that carry radon daughters such as polonium-218 are trapped in the lungs by sticky mucus on the surface of the bronchial airways of the lungs, and basal cells located just beneath the mucus may be damaged by alpha radiation. It is known that radiation can cause breakage of strands of DNA in cells, and when both strands of DNA are broken, the damage may be permanent, resulting in mutations. The mutations probably do not cause cancer by themselves but are linked to the process of initiation of the disease, which may later be promoted through exposure to chemicals such as those found in cigarette smoke (38). The EPA has radon-risk evaluation charts, one of which is shown in Figure 13.12. The chart relates lifetime exposure to radon gas in picocuries per liter to the estimated risk and estimated number of lung cancer deaths (45).

A recent study suggested radon as a causative factor in inducing myeloid leukemia, melanoma, cancer of the kidney, and some childhood cancers (46). The study is based on statistical analysis of radon concentration in homes and incidence of cancer. It is controversial, however, in part because the database consists of cancer rates and average radon concentrations for several countries rather than evaluations of radon concentrations in homes of cancer victims. The approach, while valid, needs to be tied closely to individuals and their specific environment before more skeptical observers are convinced that radon causes cancers other than lung cancer. The connection of radon to melanoma (a deadly form of skin cancer) is hypothetically related to cell damage to the skin by radioactive decay of radon and its daughter products, which might somehow cluster on the skin. The leukemia is hypothetically related to the assumption that radon is soluble in blood and fat cells, resulting in accumulation of radon in the fatty cells of bone marrow (46). These hypotheses are interesting and have important implications that should be addressed in future research.

In terms of the implied risk, it is known that large concentrations of radon are hazardous, and it has not been shown that small concentrations are harmless. The EPA has set 4.0 pCi/l as the radioactivity level beyond which radon gas is considered to be a hazard. The average outdoor concentration of radon gas is approximately 0.2 pCi/l, and the average indoor level about 1.0 pCi/l. The radioactivity level of 4.0 pCi/l is only an estimate of a concentration target that indoor levels should be reduced to. If you have never smoked, the comparable risk at 4.0 pCi/l is about the same as the risk of drowning; if you are a smoker, the risk is about 100 times the risk of dying in an airplane crash (see Figure 13.12).

The risk of radon exposure can also be estimated according to the number of people who might contract lung cancer, as shown on Figure 13.12. These estimations have been difficult to verify. One U.S. study of nonsmoking women reports not finding a significant positive relationship between radon exposure and lung cancer (47). Another study in Sweden (48) did report a statistically significant positive relation between exposure to radon gas in homes and lung cancer. The latter concluded that residential exposure to radon gas is an important cause of lung cancer in the general population of Sweden (48). Therefore, we can expect that, depending on cumulative exposure, radon will also be a cause of lung cancer in the rest of the world's population. The Swedish study involved both men and women and confirmed that smoking has a multiplicative effect with respect to exposure to radon and incidence of lung cancer.

If risks from radon gas are even close to EPA estimates, then the hazard is comparable to deaths from automobile accidents and hundreds of times higher than risks presented by outdoor pollutants possibly present in water and air. Pollutants in the outdoor environment are generally regulated to reduce the risk of premature death and disease to less than 0.001 percent, or 1 in 100,000. Assignment of risk from indoor pollutants such as organic chemicals suggests a risk of cancer of approximately 0.1 percent. These are small compared to the risk associated with radon. For example, people who live in homes for about 20 years with an average

▶ **FIGURE 13.12** Estimated risk associated with radon. These estimates are calculated as long-term exposure risks for someone living to age 70 and spending about 75% of the time in the home with a designated level of radon. (From U.S. *Environmental Protection Agency*. 1986. A citizen's guide to radon, OPA-86-004.)

RADON RISK IF YOU SMOKE

Radon level	*If 1,000 people who smoked were exposed to this level over a lifetime...*	*The risk of cancer from radon exposure compares to...*	*WHAT TO DO: Stop smoking and...*
20 pCi/l	About 135 people could get lung cancer	← 100 times the risk of drowning	Fix your home
10 pCi/l	About 71 people could get lung cancer	← 100 times the risk of dying in a home fire	Fix your home
8 pCi/l	About 57 people could get lung cancer		Fix your home
4 pCi/l	About 29 people could get lung cancer	← 100 times the risk of dying in an airplane crash	Fix your home
2 pCi/l	About 15 people could get lung cancer	← 2 times the risk of dying in a car crash	Consider fixing between 2 and 4 pCi/l
1.3 pCi/l	About 9 people could get lung cancer	(Average indoor radon level)	(Reducing radon levels below 2 pCi/l is difficult)
0.4 pCi/l	About 3 people could get lung cancer	(Average outdoor radon level)	

Note: If you are a former smoker, your risk may be lower.

RADON RISK IF YOU'VE NEVER SMOKED

Radon level	*If 1,000 people who never smoked were exposed to this level over a lifetime...*	*The risk of cancer from radon exposure compares to...*	*WHAT TO DO:*
20 pCi/l	About 8 people could get lung cancer	← The risk of being killed in a violent crime	Fix your home
10 pCi/l	About 4 people could get lung cancer		Fix your home
8 pCi/l	About 3 people could get lung cancer	← 10 times the risk of dying in an airplane crash	Fix your home
4 pCi/l	About 2 people could get lung cancer	← The risk of drowning	Fix your home
2 pCi/l	About 1 person could get lung cancer	← The risk of dying in a home fire	Consider fixing between 2 and 4 pCi/l
1.3 pCi/l	Less than 1 person could get lung cancer	(Average indoor radon level)	(Reducing radon levels below 2 pCi/l is difficult)
0.4 pCi/l	Less than 1 person could get lung cancer	(Average outdoor radon level)	

Note: If you are a former smoker, your risk may be higher.

concentration of radon of about 25 pCi/l face a 2 to 3 percent chance of contracting lung cancer (42).

The Geology of Radon Gas Both rock type and geologic structure are important in determining how much radon is likely to reach the surface of the earth. Uranium-238 concentrations in rocks and soil can vary greatly. Some rock types, such as sandstone, generally contain less than 1 part per million (ppm) U-238; others, such as some dark shales and some granites, may contain more than 3 ppm U-238. The actual amount of radon that reaches the surface of the earth is related to the concentration of uranium in the rock and soil as well as the efficiency of the transfer processes from the rock or soil to soil-water and soil-gas.

Some regions of the United States contain bedrock with an above-average natural concentration of uranium. An area

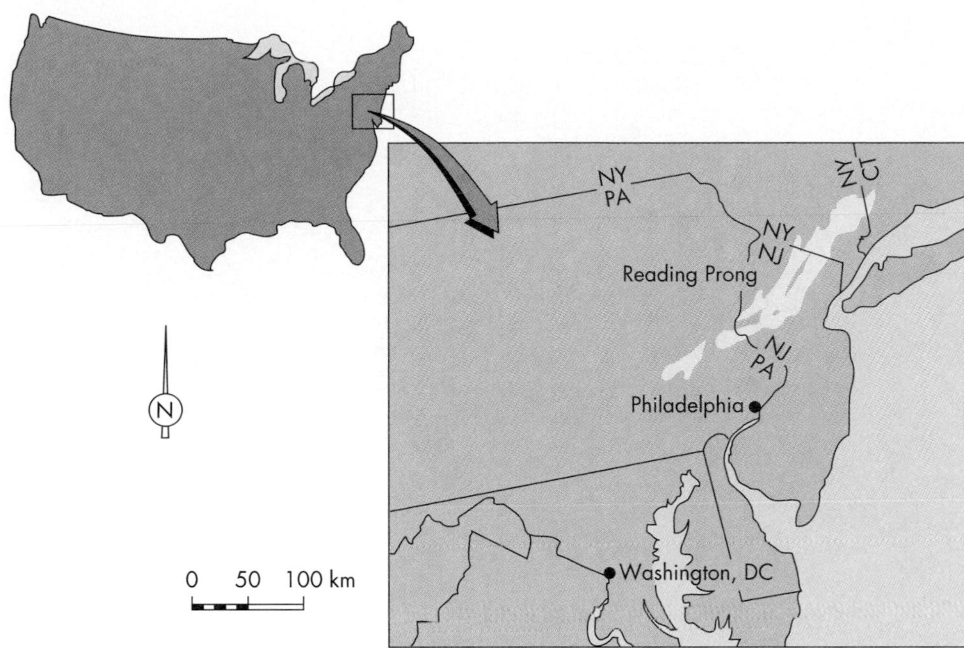

◀ **FIGURE 13.13** The Reading Prong area in the eastern United States where high levels of indoor radon were first discovered in the early 1980s. (*Modified after* U.S. *Geological Survey*. 1986. U.S.G.S. Yearbook.)

in Pennsylvania, New Jersey, and New York famous for elevated concentrations of radon gas is the Reading Prong (Figure 13.13). This region contains a large number of homes with elevated concentrations of radon gas (38). Similarly, two areas in Florida have been identified that have elevated concentrations of uranium in phosphate-rich rocks. Many other states, including Illinois, New Mexico, South Dakota, North Dakota, and Washington, have identified areas with elevated indoor radon concentrations. A dark shale in the Santa Barbara, California, area has been identified as a significant producer of radon gas (Figure 13.14).

Geologic structures such as fracture zones and faults are commonly enriched with uranium and can produce elevated concentrations of radon gas in the overlying soils (49). One study in southern Virginia found that shear zones in granite rock are associated with as much as a tenfold increase in the concentration of radon in overlying soils. Furthermore, one of the highest indoor levels in the United States is identified over fracture zones in Boyertown, Pennsylvania (49). In Santa Barbara, California, one of the highest concentrations of in-house radon measured to date occurs at the crest of an active fold (anticline) involving sand beds that probably overlie organic-rich shale at shallow depth. It is hypothesized that open tension fractures at the crest of the fold allow the radon gas to diffuse upward to the surface.

The amount of radon gas that escapes from bedrock and soil particles is greatly influenced by water content. Relatively high moisture content favors the trapping of radon gas in the spaces between soil and fractures in rocks: Water greatly reduces the distance that radon can travel as a result of radioactive decay from its parent material such as radium-226. Movement of radon gas from fractures in rock and pore spaces in soil is facilitated by relatively low moisture content. That is, the process of diffusion of radon gas is much greater in unsaturated material. Thus, the moisture content affects the amount of radon that may reach the surface of the earth in two opposing ways:

1. High moisture content increases the chances of radon entering and residing in the pore space in rock fractures or between soil grains.
2. High moisture content tends to inhibit radon from moving through the pore spaces in rocks and soils.

Consequently, there is an optimal water content that will result in the maximum amount of radon that actually reaches the surface of the earth. In some soils this optimal water content is approximately 20 to 30 percent (38) (recall from Chapter 3 that water content is the ratio of the weight of the water to the weight of the solids in a sample, expressed as a decimal).

How Radon Gas Enters Homes Three major pathways have been identified by which radon gas enters homes (Figure 13.15). These are:

- Gas that migrates up from soil and rocks into basements and other parts of homes
- Groundwater pumped from wells
- Construction materials such as building blocks made of substances that emit radon gas (43)

Most of the early interest concerning radon gas in homes was related to the use of building materials containing high concentrations of radium. For example, between 1929 and 1975, materials used to manufacture concrete in Sweden contained relatively high concentrations of radium. Similarly, in the 1960s in the United States it was discovered that some homes and other buildings in the Grand Junction area of Colorado were constructed with materials contaminated by uranium mine waste. Some homes in Florida

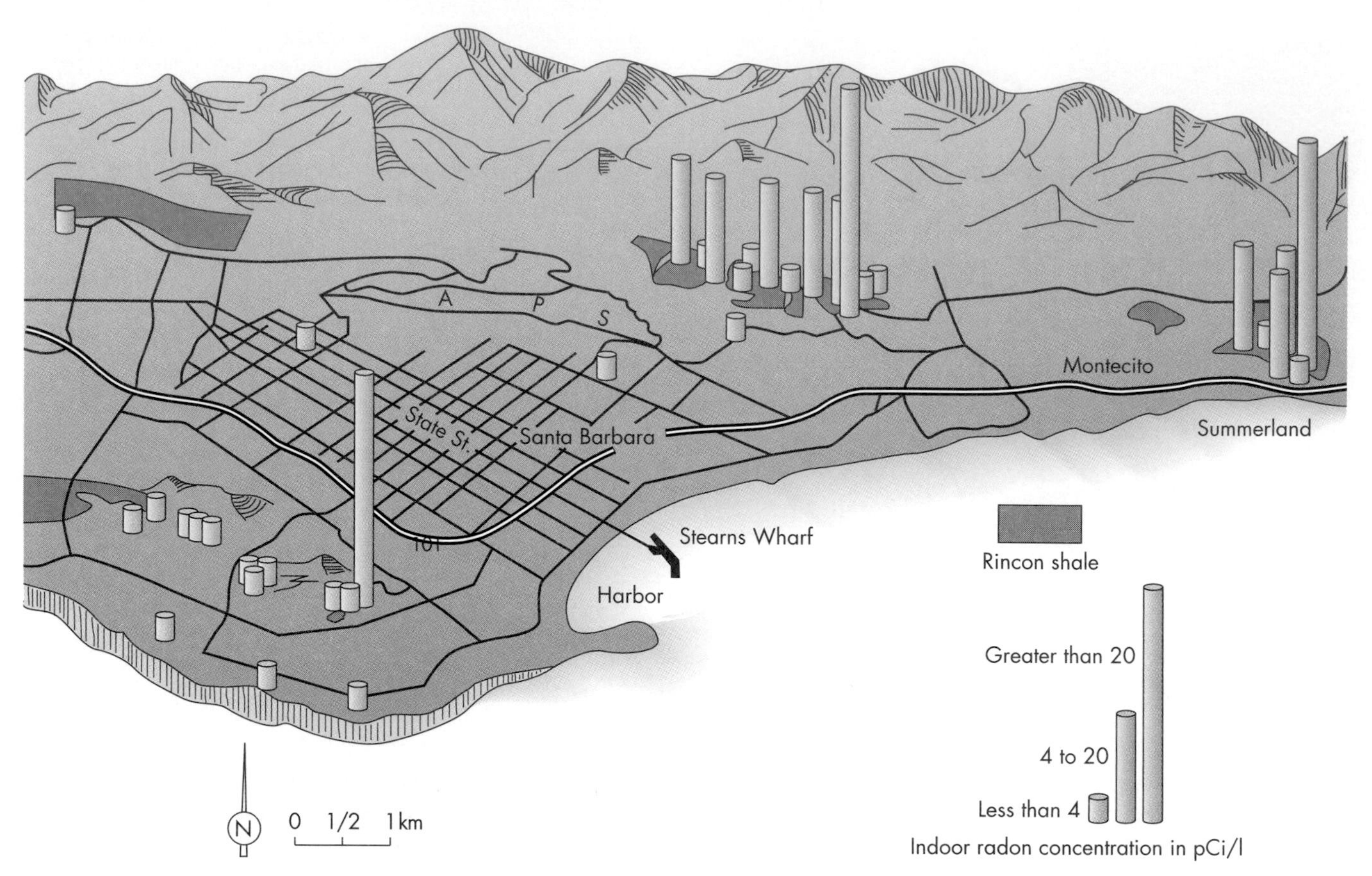

▲ **FIGURE 13.14** Santa Barbara, California. This seaside community has an emerging radon problem related to houses built on Rincon shale.

were discovered to have relatively high concentrations of radon where phosphate mining waste was used as fill dirt. Finally, some homes in New Jersey were built at a landfill site where wastes from a radium processing plant were earlier deposited (50).

In December 1984, scientists discovered that radon gas from natural sources may enter the home and possibly present a serious health hazard. The story of Stanley Watras has been told and retold many times until it is now part of radon legend (38). Nevertheless, it is worth repeating here.

Stanley Watras, of Boyertown, Pennsylvania, had a job as technical advisor in the Limerick Nuclear Power Station. At the entrance to the plant, radiation detectors ensure that no radioactive materials leave the facility. The reactor at the power station had not yet been turned on when Watras, in December 1984, on his way into the plant, set off the alarms! Testing of his clothing suggested that the contamination could not have come from the plant but must have originated where he lived. Power company officials checking Watras's home were astounded to find that the radiation level of the indoor air was 3200 pCi/l, 800 times higher than the 4 pCi/l considered a threshold for hazard by the U.S. Environmental Protection Agency. Scientists were very surprised because until then they did not believe that radon that naturally seeped into homes could be hazardous (38, 48,51). The Watras home held the record for indoor radon concentration until the latter part of the 1980s, when a

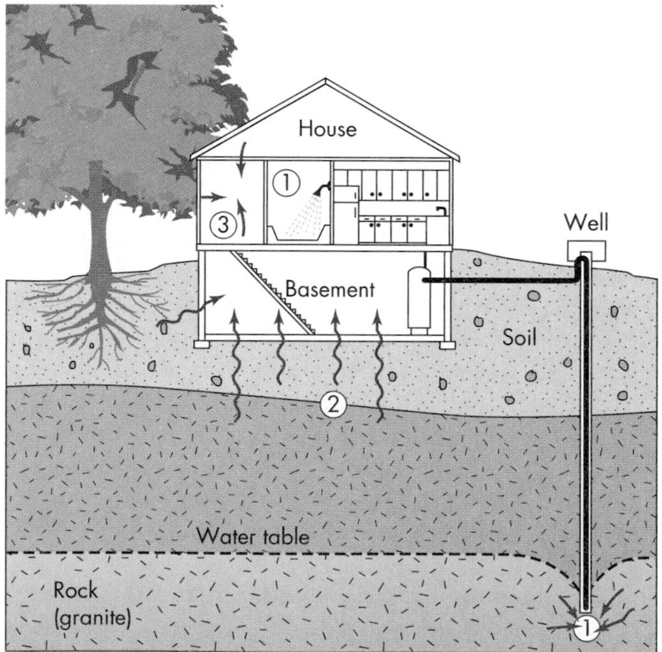

▲ **FIGURE 13.15** How radon may enter homes. (1) Radon in groundwater enters well and goes to house where it is used for water supply, washing dishes, taking showers, and other purposes. (2) Radon gas in rock and soil migrates into basement through cracks in foundation and pores in construction. (3) Radon gas is emitted from construction materials used in building the house. (*Source*: U.S. *Environmental Protection Agency*)

house in Whispering Hills, New Jersey, was discovered with a radiation level of 3500 pCi/l (51).

The occurrence of radon in homes is directly related to the fact that radon is a gas that can move through the small openings in soils and rocks. Radon migrates up with the soil gases and seeps through concrete floors and foundations, floor drains, or small cracks and pores in block walls. Radon enters homes for the same reason that smoke goes up a chimney, in a process known as the *chimney* (or *stack*) *effect:* Houses are generally warmer than the surrounding soil and rock, so as the gases and air rise, radon is drawn into the house. Wind can also be a factor because it increases air flows into and out of buildings. Radon can also enter homes through the water supply, particularly if the home is supplied by a private well, although this contribution is usually much smaller than the gas that seeps up through house foundations. Radon and its daughter products are released into the air when the water is used for daily functions such as showering, dish washing, and clothes washing. In general, approximately 10,000 pCi/l of radon in water will produce about 1 pCi/l of radon in indoor air.

Water pumped from underground sources for municipal supplies may also contain radon gas. However, because most of the water in these systems is stored for a few days and treated for purification, there is a delay between the time the water is pumped and when it is actually used. The storage time allows for the radon to decay away (38). As a result, public and municipal water supplies generally have a relatively low radon content.

Measurement of Radon Gas in Homes Measurement of radon concentration in homes has several inherent difficulties. The concentration of radon may be extremely variable in any one house, depending upon the season of the year and other poorly understood factors related to radon emission from the soil gas. In many parts of the country, higher levels of radon may be expected in the winter than in the summer, when windows are left open for ventilation. Within the ground, emissions of radon can vary with barometric pressure, water content, and other variables that may be difficult to ascertain.

Another factor concerning indoor radiation is that while one house may have a relatively high concentration, others in the vicinity may have much lower levels. Although the geology in a general way can be used as a diagnostic tool to identify areas that might have high concentrations of radon, only in-house testing can verify this. Even slight differences in the geology—as, for example, locations of major fractures—can be quite significant. Fortunately, testing for radon is easy and not very expensive.

Scope and Perception of the Problem Since 1985, awareness of the radon gas problem in the United States has increased. Nevertheless, people have a hard time focusing on a problem that they cannot see, smell, hear, or touch. The problem is further compounded by the fact that some property owners seem to be more concerned about potential loss of property values than the health hazard. There is a considerable hiatus between people's perception of the radon problem and the potential size of the problem itself. This is true even in states such as New Jersey, where the problem has been known about for several years. One survey of homeowners concluded that only about 5 percent of them either had or were planning a radon test; about 10 percent believed that radon might be a problem in their home. The actual percentage of homes expected to have a radon problem in that part of New Jersey is approximately 30 percent (38).

There is no solid estimate of the number of homes in the United States that may have elevated concentrations of radon gas. Looking at the tests that have been done, approximately one home in 12 has an indoor radon level above 4 pCi/l. If this rate is a good average, then approximately 7 million homes in the United States would have elevated rates, and millions more will need to be tested. On the other hand, many tests are taken in areas where radon is already suspected, which may bias the estimate. It should be mentioned that high radon concentrations are not only found in homes but also in other buildings. In a recent survey of more than 100 schools, the EPA found that in more than half of them the radon concentration was above the action level of 4 pCi/l, and one school in Tennessee had a level of 136 pCi/l (44). This information is important because some health officials believe that children are more susceptible to radon damage than are adults because of the small size of their lungs and the fact that they are still developing.

Reducing Concentrations of Radon Gas in Homes A successful program to reduce or limit a potential hazard from radon gas in homes centers upon three major strategies.

- Points of entry of radon may be located and sealed.
- Ventilation of the home may be improved by keeping more windows open or using fans.
- Construction methods that provide a venting system may be installed (52,53).

Probably the simplest method of reducing the radon concentration is to increase the ventilation of the home. Sometimes this is sufficient to solve the problem. Sealing cracks in the foundation can also reduce radon entry, particularly if it is done in specific sites where the radon is actually entering the house. Detectors are available to help identify such locations. Unfortunately, this method often fails to solve the problem, and new cracks may form. Various construction options are open to the home owner, including venting systems in a basement or crawl space. If the house is built on a slab, then subslab ventilation systems can be installed.

There is good news and bad news concerning radon gas. The bad news is that many parts of the United States and other parts of the world have relatively high emissions of radon gas from soil and rocks, and the radiation is producing a hazard in homes. The good news is that in most cases the problem can be fixed relatively easily. Even if a ventilation system is required, it costs only a few thousand dollars, generally a small percentage of the value of the home.

Future research may show that health risks from radon gas exposure are not as great as predicted by the EPA. Some scientists are skeptical that the dose response for radon is a straight line (Figure 13.2b, curve *A*) from low to high concentrations, and they therefore believe that the hazard level of 4 pCi/l may be set too low (54). No matter who is right, it is important for people to become informed now concerning this potential problem. People have difficulty taking appropriate action to deal with the possibility of radon gas in their homes, and in too many cases they seem to be more concerned with property values than their health. When people see the whole picture and know that the problem can be ameliorated, the fear will be reduced.

13.6 Risk Assessment in Toxicology

Risk assessment is the process of determining potential adverse environmental health effects following exposure to a toxic material, whether it be a naturally occurring heavy metal or human-made organic chemicals. The steps in assessing risk (55) are:

- **Hazard identification.** This step determines whether exposure to a particular material is likely to cause environmental health problems. It might consist of investigating populations of people who have been exposed. For example, in identifying the hazard associated with radon gas, investigators have studied miners exposed to radon while mining uranium as well as exposure in homes. Another approach to hazard identification might be conducting experiments at the molecular level to determine how a particular material interacts with living cells.
- **Dose-response assessment.** The purpose of this step is to identify relationships between the dose of a toxic material and its adverse environmental health effects. This is often a controversial procedure because the predicted responses at low doses are often extrapolated from known effects of higher doses, or vice versa. There also may be discussion concerning whether or not thresholds are present or absent. Finally, the dose-response assessment often relies on statistical analysis of data obtained from observations and measurements that are subject to experimental or measurement errors.
- **Exposure assessment.** The objective of exposure assessment is to estimate the duration, frequency, and intensity of exposure to a particular toxin. It is assumed that the hazard associated with exposure is directly proportional to the population that is actually exposed as well as to the intensity, duration, or frequency of exposure. Exposure assessment is also a controversial process because measuring the exposure and concentration of toxic materials at extremely small concentration (parts per million or parts per billion) is difficult.
- **Risk characterization.** This is the final step in the process of risk assessment. It involves taking into consideration the hazard that has been identified, the dose-response assessment, and the evaluation of the exposure as outlined above.

Assessment of the risk from exposure to toxins in the environment allows us to develop action plans for minimizing the risk. This is the process of **risk management,** which integrates risk assessment with the legal, social, political, economic, and technical issues involved in action plans. Both risk-assessment and risk-management plans require making sound scientific judgments based on all the existing information (56).

SUMMARY

The death rate and the incidence of specific diseases vary from one area to another, and some of the variability has environmental causes. These causes are often quite complex, and a particular disease seldom has a one-cause, one-effect relationship.

Cultural factors can affect the geographic pattern of a particular disease. Examples include stomach cancer in Japan and lead poisoning in numerous areas. Climate—including temperature, humidity, and precipitation—is an important factor in determining disease patterns. Schistosomiasis and malaria are the best-known examples of diseases related to climate. Several diseases are also associated with a particular climate, but cause-and-effect relations are not clear.

The natural distribution of matter in the universe is such that the lighter elements are more abundant. The first 26 elements of the periodic table account by weight for the vast majority of the lithosphere and biosphere. Movements of elements along numerous paths through the lithosphere, hydrosphere, biosphere, and atmosphere are termed *biogeochemical cycles.* Along these paths, natural processes such as weathering and volcanic activity, combined with human pollution, release many types of substances into the environment. These substances are then concentrated or dispersed by other natural processes such as leaching, accretion, deposition, and biologic activity.

Every element has an entire spectrum of possible effects on plants and animals. Trace concentrations of a particular element may be beneficial or essential to a particular kind of plant or animal, in which case the element is called a *trace element,* whereas higher concentrations may be toxic and still higher concentrations lethal. The graph of a substance's concentration against its effects on an organism is known as a *dose-response curve*. The concentration of a toxin that produces a given response in 50 percent of a population is called the *effective dose* (ED-50). If the response is toxic, the *toxic dose* (TD-50) is used, and if the response is death, the *lethal dose* (LD-50). Interesting examples of environmentally important elements include iodine, fluorine, zinc, and selenium. Agricultural, industrial, and mining activities have released hazardous and toxic materials into the environment. These materials have been associated with bone disease in humans, metabolic disorders in cattle, and other biological problems.

Relationships between chronic disease and geologic environment are complex and difficult to analyze. A correlation between a given environmental factor and the occurrence of a disease is not proof of a cause-and-effect relationship; furthermore, it is likely that most chronic diseases have multiple causes. Nevertheless, considerable evidence suggests that the geochemical environment is a significant factor in the incidence of some serious chronic health problems. Heart disease rates are apparently related to water chemistry, such as total hardness and concentration of sulfate or bicarbonate ions; one study suggests that heart disease is also related to deficiencies of trace elements. Cancer rates in specific areas appear to be related to the organic content of the soil and the abundance of certain trace elements; in other areas, cancer may be related to saline soils. Asbestos is associated with increased incidence of lung disease in asbestos workers, and it is suspected of causing lung cancer. However, the type of asbestos commonly used in the United States is not thought to produce a serious health hazard, and extensive removal of asbestos from buildings may not be necessary.

Radioisotopes undergo radioactive decay, in which they become transformed into a different element and produce radiation. The major kinds of radioactive radiation are alpha particles, beta particles, and gamma rays, each with a different effect and toxicity. The health hazard from a specific radioisotope depends on several factors, including the kind of radiation it emits, its half-life, and its physical state or chemical activity. Isotopes with low-energy emissions and short half-lives are generally less dangerous than are those with high-energy emissions and long half-lives. However, a low-energy, short half-lived isotope such as radon, which readily enters a gaseous phase or is absorbed in water, becomes dangerous when concentrated and inhaled or ingested. Although we know the hazardous and lethal doses of high-level radiation, we are not sure what exposure to low-level radiation constitutes a hazard. There are two hypotheses: that a direct linear relationship exists between exposure and hazard, and that there is some threshold of radiation beyond which damage occurs.

Radon gas in homes is a serious environmental health problem. Radon is a radioactive daughter product of radium, which is derived in turn from the uranium in rocks. Exposure to radon gas increases the risk of lung cancer, especially for smokers; up to 30,000 lung cancer deaths annually are thought to be related to the indoor radon gas problem. The damage is caused by the solid daughter products of radon, particularly polonium-218, which remain in the lungs after radon gas is inhaled. Factors that control the concentration of indoor radon include geology, radon concentration in the soil and rock, moisture content of soil, type of house construction, and season of the year. Most exposure is from radon that is produced in the underlying rock and enters the atmosphere as gas, although some is from radon dissolved in water from private wells and some derives from materials formerly used in home construction. Fortunately, inexpensive technology is available to reduce or remove the radon problem.

Risk assessment for determining potential adverse environmental health effects associated with toxic materials includes four steps: identifying the hazard; assessing the dose-response relationship; assessing the exposure; and characterizing the risk. Risk assessment is an essential first step of risk management, which develops action plans to minimize health problems related to a particular environmental toxin.

REFERENCES

1. **Sauer, H. I., and Brand, F. R.** 1971. Geographic patterns in the risk of dying. In *Environmental geochemistry in health*, eds. H. L. Cannon and H. C. Hopps, pp. 131–50. Geological Society of America Memoir 123.
2. **Hopps, H. C.** 1971. Geographic pathology and the medical implications of environmental geochemistry. In *Environmental geochemistry in health*, eds. H. L. Cannon and H. C. Hopps, pp. 1–11. Geological Society of America Memoir 123.
3. **Blumenthal, D. S., and Ruttenber, A. J.** 1995. *Introduction to environmental health*, 2nd ed. New York: Springer Publishing Company.
4. **Young, K.** 1975. *Geology: The paradox of earth and man.* Boston: Houghton Mifflin.
5. **Bylinsky, G.** 1972. Metallic menaces. In *Man, health and environment*, ed. B. Hafen, pp. 174–85. Minneapolis: Burgess.
6. **Hong, S., Candelone, J.-P., Patterson, C. C., and Boutron, C. F.** 1994. Greenland ice evidence of hemispheric lead pollution two millennia ago by Greek and Roman civilizations. *Science* 265:1841–43.
7. **Needleman, H. L., Riess, J. A., Tobin, M. J., Biesecker, G. E., and Greenhouse, J. B.** 1996. Bone lead levels and delinquent behavior. *Journal of the American Medical Association* 275:363–369.
8. **Warren, H. V., and Delavault, R. E.** 1967. A geologist looks at pollution: Mineral variety. *Western Mines* 40:23–32.
9. **Furst, A.** 1971. Trace elements related to specific chronic diseases: Cancer. In *Environmental geochemistry in health*, eds. H. L. Cannon and H. C. Hopps, pp. 109–30. Geological Society of America Memoir 123.

10. **Cargo, D. N., and Mallory, B. F.** 1974. *Man and his geologic environment.* Reading, MA: Addison-Wesley.
11. **Smil, V.** 1997. Global populations and the nitrogen cycle. *Scientific American,* July, 76–81.
12. **Mertz, W.** 1968. Problems in trace element research. In *Trace substances in environmental health—II,* ed. D. D. Hemphill, pp. 163–69. Columbia: University of Missouri Press.
13. **Horne, R. A.** 1972. Biological effects of chemical agents. *Science* 177:1152–53.
14. **Maier, F. J.** 1971. Fluorides in water. In *Water quality and treatment,* 3rd ed., pp. 397–440. New York: McGraw-Hill.
15. **Bernstein, D. S., Sadowsky, N., Hegsted, D. M., Guri, C. D., and Stare, F. J.** 1966. Prevalence of osteoporosis in high- and low-fluoride areas in North Dakota. *The Journal of the American Medical Association* 198:499–504.
16. **Gilbert, F. A.** 1947. *Mineral nutrition and the balance of life.* Norman: University of Oklahoma Press.
17. **Spencer, J. M.** 1970. Geologic influence on regional health problems. *The Texas Journal of Science* 21:459–69.
18. **Dissanayake, C. B., and Chandrajith, R. L. R.** 1996. Iodine in the environment and endemic goiter in Sri Lanka. In *Environmental geochemistry and health,* eds. J. D. Appleton, R. Fuge, and G. J. H. McCall, pp. 213–21. Geological Society Special Publication 113. Oxford, England: Geological Society of London.
19. **Shacklette, H. T., and Cuthbert, M. E.** 1967. Iodine content of plant groups as influenced by variations in rock and soil type. In *Relations of geology and trace elements to nutrition,* eds. H. L. Cannon and D. F. Davidson, pp. 31–45. Geological Society of America Special Paper 90.
20. **Pories, W. J., Strain, W. H., and Rob, C. G.** 1971. Zinc deficiency in delayed health and chronic disease. In *Environmental geochemistry in health and disease,* eds. H. L. Cannon and H. C. Hopps, pp. 73–95. Geological Society of America Memoir 123.
21. **Lakin, H. W.** 1973. Selenium in our environment. In *Trace elements in the environment,* ed. E. L. Kothny, pp. 96–111. Advances in Chemistry Series 123, American Chemical Society.
22. **Allaway, W. H.** 1969. Control of environmental levels of selenium. In *Trace substances in environmental health—II,* ed. D. D. Hemphill, pp. 181–206. Columbia: University of Missouri Press.
23. **Orndahl, G., Ridby, A., and Selin, E.** 1982. Myotonic dystrophy and selenium. *Acta Med. Scand.* 211:493–99.
24. **Armstrong, R. W.** 1971. Medical geography and its geologic substrate. In *Environmental geochemistry in health and disease,* eds. H. L. Cannon and H. C. Hopps, pp. 211–19. Geological Society of America Memoir 123.
25. **Kobayashi, J.** 1957. On geographical relationship between the chemical nature of river water and death-rate of apoplexy. *Berichte des Ohara Institute fur landwirtschaftliche biologie* 11:12–21.
26. **Winton, E. F., and McCabe, L. J.** 1970. Studies relating to water mineralization and health. *Journal of American Water-Works Association* 62:26–30.
27. **Schroeder, H. A.** 1966. Municipal drinking water and cardiovascular death-rates. *Journal of the American Medical Association* 195:125–29.
28. **Bain, R. J.** 1979. Heart disease and geologic setting in Ohio. *Geology* 7:7–10.
29. **Klusman, R. W., and Sauer, H. I.** 1975. *Some possible relationships of water and soil chemistry to cardiovascular diseases in Indiana.* Geological Society of America Special Paper 155.
30. **Schroeder, H. A.** 1965. Cadmium as a factor in hypertension. *Journal of Chronic Disease* 18:647–56.
31. **Pettyjohn, W. A.** 1972. No thing is without poison. In *Man and his physical environment,* eds. G. D. McKenzie and R. O. Utgard, pp. 109–10. Minneapolis: Burgess.
32. **Takahisa, H.** 1971. Discussion. In *Environmental geochemistry in health and disease,* eds. H. L. Cannon and H. C. Hopps, pp. 221–22. Geological Society of America Memoir 123.
33. **Shacklette, H. T., Sauer, H. I., and Miesch, A. T.** 1970. *Geochemical environments and cardiovascular mortality rates in Georgia.* U.S. Geological Survey Professional Paper 574C.
34. **Shacklette, H. T., Sauer, H. I., and Miesch, A. T.** 1972. Distribution of trace elements in the occurrence of heart disease in Georgia. *Geological Society of America Bulletin* 83:1077–82.
35. **Ross, M.** 1990. Hazards associated with asbestos minerals. In *Proceedings of a U.S. Geological Survey Workshop on Environmental Geochemistry,* ed. B. R. Doe, pp. 175–76. U.S. Geological Survey Circular 1033.
36. **Skinner, H. C. W., and Ross, M.** 1994. Minerals and cancer. *Geotimes* 39(1):13–15.
37. **Kmet, J., and Mahboubi, E.** 1972. Esophageal cancer in the Caspian littoral of Iran. *Science* 175:846–53.
38. **Brenner, D. J.** 1989. *Radon: Risk and remedy.* New York: W. H. Freeman.
39. **Ehrlich, P. R., Ehrlich, A. H., and Holdren, J. P.** 1970. *Ecoscience: Population, resources, environment.* San Francisco: W. H. Freeman.
40. **Waldbott, G. L.** 1978. *Health effects of environmental pollutants,* 2nd ed. St. Louis: C. V. Mosby.
41. **Van Koevering, T. E., and Sell, N. J.** 1986. *Energy: A conceptual approach.* Englewood Cliffs, NJ: Prentice-Hall.
42. **Nero, A. V., Jr.** 1988. Controlling indoor air pollution. *Scientific American* 258(5):42–48.
43. **University of Maine and Maine Department of Human Services.** 1983. Radon in water and air. *Resource Highlights,* February.
44. **U.S. Environmental Protection Agency.** 1989. *Radon measurements in schools.* Office of Radiation Programs. EPA 520/1 89-010.
45. **U.S. Environmental Protection Agency.** 1986. *A citizen's guide to radon.* OPA-86-004.
46. **Henshaw, D. L., Eatough, J. P., and Richardson, R. B.** 1990. Radon as a causative factor in induction of myeloid leukaemia and other cancers. *Lancet* 335(8696):1008–12.
47. **Alavanja, M. C., Brownson, R. C., Lubin, J. H., Berger, E., Chang, J. C., and Boice, J. D., Jr.** 1994. Residential radon exposure and lung cancer among non-smoking women. *Journal of National Cancer Institute* 80(24):1829–37.
48. **Pershagen, G., Akerblom, G., Axelson, O., Clavensjo, B., Damber, L., Desai, G., Enflo, A., Lagarde, F., Mellander, H., Svartengren, M., and Swedjemark, G. A.** 1994. Residential radon exposure and lung cancer in Sweden. *New England Journal of Medicine* 330(3):159–64.
49. **Gates, A. E., and Gundersen, L. C. S.** 1989. Role of ductile shearing in the concentration of radon in the Brookneal Zone, Virginia. *Geology* 17:391–94.
50. **Hurlburt, S.** 1989. Radon: A real killer or just an unsolved mystery? *Water Well Journal,* June, 34–41.

51. **Egginton, J.** 1989. Menace of Whispering Hills. *Audubon*, January, 28–35.
52. **U.S. Environmental Protection Agency.** 1986. *Radon reduction techniques for detached houses.* EPA 625/5-86-019.
53. **U.S. Environmental Protection Agency.** 1988. *Radon-resistant residential new construction.* EPA 600/8-88/087.
54. **Store, R.** 1993. Radon risk up in the air. *Science* 261:1515.
55. **U.S. Environmental Protection Agency.** 1989. *Glossary of terms related to health exposure and risk assessment.* EPA 450/3-88-016.
56. **Roberts, L.** 1991. Dioxin risks revisited. *Science* 251:624–26.

KEY TERMS

disease (p. 343)
bulk element (p. 345)
age elements (p. 345)
dose dependency (p. 349)
dose-response curve (p. 349)
toxin (p. 350)
toxicology (p. 350)
effective dose (ED) (p. 350)
toxic dose (TD) (p. 350)
lethal dose (LD) (p. 351)
carcinogenic (p. 356)
radioisotope (p. 360)
alpha particles (α) (p. 360)
beta particles (β) (p. 360)
gamma radiation (γ) (p. 360)
half-life (p. 360)
radon gas (p. 363)
risk assessment (p. 368)
risk management (p. 368)

SOME QUESTIONS TO THINK ABOUT

1. Some investigators believe that chlorination of our public water supplies produces chlorine by-products that can cause cancer. It has been reported that as much as 20 percent of colon and other bowel cancers may be related to exposure to these chemicals. Discuss how you might go about testing this hypothesis.
2. Consider the inverse relationship between hardness of water and incidence of heart disease presented in this chapter. Four hypotheses explaining the relationship were stated, including that hardness of water has nothing to do with heart disease. Develop a strategy that can be used for testing each of the hypotheses.
3. Parents in a local PTA are concerned that the asbestos used in pipes, insulation, and ceiling tiles in their children's school presents a health hazard. They come to you for advice on what to do. What would you tell them?
4. Your consulting company has been hired by a school district to determine if there is a radon gas problem at any of their 20 schools. Outline a risk-assessment plan, and indicate how you would report the results and make recommendations for the following scenarios: no indication that radon gas above 4 pCi/l is present; radon levels are above 4 pCi/l but only moderately so in a few of the schools; and radon gas is generally moderately above 4 pCi/l for most of the schools, and the high value is 38 pCi/l.

PART 4

Minerals, Energy, and Environment

CHAPTER 14
Mineral Resources and Environment

CHAPTER 15
Energy and Environment

A fundamental concept of environmental geology is that this earth is our only suitable habitat and its resources are limited. Some resources such as oil, gas, and minerals are recycled so slowly in the geologic cycle that they are essentially nonrenewable. Other resources such as timber, water, air,[1] and food are renewable, but only as long as environmental conditions remain favorable for their natural or planned reproduction. Careless use of air, water, vegetation, and, to a lesser extent, soils may render these resources less renewable than we would wish.

Because resources are limited, important questions arise. How long will a particular resource last? How much short- or long-term environmental deterioration are we willing to concede to ensure that resources are developed in a particular area? How can we minimize environmental disruption associated with mineral and energy resource utilization? How can we make the best use of available resources? These questions have no easy answers. We are now struggling with the first question by seeking better ways to estimate the quality and quantity of resources. It is extremely hard to address the second, third, and fourth questions without a satisfactory answer to the first, but determining how long a particular resource will last is complicated—availability will change as our technological skills and discovery techniques become more sophisticated.

Our present resource problem (or crisis) is related to a people problem—too many people. The earth and its resources are finite, while the population continues to grow. When resource data are combined with population data, the conclusion is clear—it is impossible in the long run to match exponential population growth with exponential production of useful materials dependent upon a finite resource base. Considering that many other countries also aspire to affluence, while the world population increases, the availability of necessary resources becomes questionable. This section of the text will explore geologic and environmental aspects of mineral and energy resources from this perspective.

Mineral resources, discussed in Chapter 14, include a wide variety of earth materials of value to people. Included in these resources are metallic and nonmetallic minerals, as well as sand, gravel, crushed rock, and ornamental rock (dimension stone) that have commercial value. Some earth materials, such as those used for construction material (sand and gravel, etc.), are low-value resources and have primarily a "place value." On the other hand, materials such as diamonds, copper, gold, and aluminum are high-value resources. These materials are extracted wherever they are found and transported around the world to numerous markets regardless of distances.[2]

The subjects of mineral resources and energy resources overlap considerably, as resources found in rocks such as oil and coal are used primarily as sources of energy. For organizational purposes, we will discuss all resources associated with energy production separately. In Chapter 15 we explore some selected geologic and environmental aspects of such well-known energy resources as coal, petroleum, and nuclear sources, as well as such potentially important sources as oil shale, tar sands, and geothermal resources. Finally, we will discuss alternative energy sources such as hydropower, wind, and solar power.

[1]A discussion of the air environment, including air as a renewable resource; the urban air and air pollution; the urban microclimate; and global climate change, is presented in Chapter 16.

[2]Flawn, P. T. 1970. *Environmental geology*. New York: Harper & Row.

14 Mineral Resources and Environment

Aerial view of phosphate mining and processing facilities in central Florida. (C. *Davidson/Comstock*)

Mining of mineral resources requires physical removal of materials at or near the surface of the land. As a result, the landscape is disturbed and unwanted materials may be exposed at the surface. Phosphate mining in areas such as central Florida are no exception. Unless carefully managed, waste from mining and processing facilities can cause environmental degradation affecting human habitat and health as well as ecosystems.

14.1 Minerals and Human Use

Modern society depends on the availability of mineral resources (1,2). Consider the mineral products found in a typical American home (Table 14.1). Specifically, consider your breakfast this morning. You probably drank from a glass made primarily of sand, ate food from dishes made from clay, flavored your food with salt mined from the earth, ate fruit grown with the aid of fertilizers such as potassium carbonate (potash) and phosphorus, and used utensils made from stainless steel, which comes from processing iron ore and other minerals. If you read a magazine or newspaper while eating, the paper was probably made using clay fillers. If the phone rang and you answered, you were using more than 40 minerals used in the telephone. When you went to school or work, you may have turned on a computer or other equipment made largely of minerals.

Minerals are so important to people that, other things being equal, one's standard of living increases with the increased availability of minerals in useful forms. Furthermore, the availability of mineral resources is one measure of the wealth of a society. Those who have been successful in the location, extraction, or importation and use of minerals have grown and prospered. Without mineral resources, modern technological civilization as we know it would not be possible.

LEARNING OBJECTIVES

Our modern society absolutely depends upon the availability of mineral resources. As world population increases, we are facing a resource crisis, and there is fear that the earth may have reached its capacity to absorb environmental degradation related to mineral extraction, processing, and use. With this in mind, learning objectives for the chapter are:

- *To gain an understanding of the relationship between human population and resource utilization.*
- *To understand why minerals are so important to modern society.*
- *To know the difference between a resource and a reserve and why that difference is important.*
- *To be aware of some of the factors controlling the availability of mineral resources.*
- *To be familiar with the geologic processes responsible for producing our mineral resources.*
- *To gain an understanding of the environmental impact of mineral development.*
- *To be aware of the potential benefits of biotechnology to environmental cleanup associated with mineral extraction and production.*
- *To understand the economic and environmental role of recycling of mineral resources.*

Web Resources

Visit the "Environmental Geology" Web site at www.prenhall.com/keller to find additional resources for this chapter, including:

- Web Destinations
- On-line Quizzes
- On-line "Web Essay" Questions
- Search Engines
- Regional Updates

The important role of nonfuel minerals in the U.S. economy is shown in Figure 14.1. The data on this diagram suggest that (3):

- Processed materials from minerals have an annual value of several hundred billion dollars, which is about 5 percent of the U.S. gross domestic product.
- Value of recycled metal and other mineral scrap is about 30 percent of the value of domestic mineral raw materials. This is a significant contribution.

Minerals can be considered our nonrenewable heritage from the geologic past. Although new deposits are forming from present earth processes, these processes are too slow to be of use to us today. Mineral deposits tend to occupy a small area and to be hidden. Deposits must therefore be discovered, and, unfortunately, most of the easy-to-find deposits have already been exploited. If civilization (with our science and technology) were to vanish, our descendants would have a much harder time discovering minerals for technological advance than we and our ancestors did. Unlike biological resources, minerals cannot be managed to produce a sustained yield. Recycling and conservation will help, but eventually the supply will be exhausted.

Resources and Reserves

Mineral resources can be defined broadly as elements, compounds, minerals, or rocks that are concentrated in a form that can be extracted to obtain a usable commodity (4). This definition is unsatisfactory from a practical viewpoint, however, because a resource will not normally be extracted unless extraction can be accomplished at a profit. A more pragmatic definition is that a **resource** is a concentration of a naturally occurring material (solid, liquid, or gas) in or on the crust of the earth in such a form that economical extraction is *currently or potentially feasible* (5). A **reserve** is that portion of a resource that is identified and *currently available*, that is, from which usable materials can be legally and economically extracted at the time of evaluation. The distinction between resources and reserves, therefore, is based on current geologic, economic, and legal factors (Figure 14.2). Resources include (5):

1. Materials that are identified and legally and economically available (reserves)
2. Materials that are identified but legally or economically unavailable (subeconomic resources)
3. Undiscovered materials (hypothetical or speculative resources)

The main point about resources and reserves is that *all resource categories in Figure 14.2 are not reserves!* An analogy from a student's personal finances will help clarify this point. A student's reserves are the liquid assets, such as money in the pocket or bank, whereas the student's resources include the total income the student can expect to earn during his or her lifetime. This distinction is often critical to the student in school because resources are "frozen" assets or next year's income and cannot be used to pay this month's bills (4).

Regardless of potential problems, it is important for planning to estimate future resources. A simple periodical

Table 14.1 A few of the mineral products in a typical American home

Building materials	Sand, gravel, stone, brick (clay), cement, steel, aluminum, asphalt, glass
Plumbing and wiring materials	Iron and steel, copper, brass, lead, cement, asbestos, glass, tile, plastic
Insulating materials	Rock, wool, fiberglass, gypsum (plaster and wallboard)
Paint and wallpaper	Mineral pigments (such as iron, zinc, and titanium) and fillers (such as talc and asbestos)
Plastic floor tiles, other plastics	Mineral fillers and pigments, petroleum products
Appliances	Iron, copper, and many rare metals
Furniture	Synthetic fibers made from minerals (principally coal and petroleum products); steel springs; wood finished with rottenstone polish and mineral varnish
Clothing	Natural fibers grown with mineral fertilizers; synthetic fibers made from minerals (principally coal and petroleum products)
Food	Grown with mineral fertilizers; processed and packaged by machines made of metals
Drugs and cosmetics	Mineral chemicals
Other items	Windows, screens, lightbulbs, porcelain fixtures, china, utensils, jewelry: all made from mineral products

Source: U.S. Geological Survey Professional Paper 940, 1975

listing of the total amount of material available or likely to become available is misleading when used for planning purposes; what is required is a continual reassessment of all components of a total resource by considering new technology, the probability of geologic discovery, and shifts in economic and political conditions (5). A method for doing this, developed by the U.S. Geological Survey and the U.S. Bureau of Mines, involves listing or graphing mineral resources in terms of the resource classification shown in Figure 14.2. Data for an *identified resource* such as a precious metal (for example, gold) or building materials (for example, sand and gravel) can be tabulated as follows:

- **Measure-identified resources** are those that are well known and measured and for which the total tonnage or grade is well established.
- **Indicated-identified resources** are not so well known and measured and therefore cannot be outlined completely by tonnage or grade. Total tonnage or grade can be estimated, but not as well as for measured-identified resources.
- **Inferred-identified resources** have quantitative estimates based on broad geologic knowledge of the deposit. Total tonnage or grade can only be crudely estimated.

The category into which a particular identified resource fits is a function of available geologic information. Obtaining this information involves testing, drilling, and mapping, all of which become more expensive with greater depth.

The example of silver will illustrate some important points about resources and reserves. The earth's crust (to a depth of 1 km) contains almost 2 million million (2×10^{12}) metric tons of silver—this is the earth's crustal resource of silver—an amount much larger than the annual world use, which is approximately 10,000 metric tons. If this silver existed as pure metal concentrated into one large mine, it would represent a supply sufficient for several hundred million years at current levels of use. Most of this silver, however, exists in extremely low concentrations—too low to be extracted economically with current technology. The known reserve of silver, reflecting the amount we could obtain immediately with known techniques, is about 200,000 metric tons, or a 20-year supply at current use levels.

The problem with silver, as with all mineral resources, is not with its total abundance but with its concentration and relative ease of extraction. When an atom of silver is used, it is not destroyed, but it is dispersed and may become unavailable. In theory all mineral resources could be recycled, given enough energy, but this is not possible in practice. Consider lead, which is mined and was for many years used in gasoline. This lead is now scattered along highways across the world and deposited in low concentration in forests, fields, and salt marshes close to these highways. Recovery of this lead is for all practical purposes impossible.

Availability and Use of Mineral Resources

The availability of a mineral in a certain form, in a certain concentration, and in a certain total amount at that concentration is determined by the earth's history; it is a geologic issue that we will consider in the next section. What a mineral resource is and when it becomes limited are technological and social questions that we will consider here.

Types of Mineral Resources Some mineral resources are necessary for life. An example is salt (sodium chloride)—primitive peoples traveled long distances to obtain salt when it was not locally available. Other mineral resources are desired for their beauty, and many more are necessary for maintaining a certain level of technology.

The earth's mineral resources can be divided into several broad categories based on our use:

- Elements for metal production and technology, which can be classified according to their abundance. The

U.S. natural resources:

Rocks, minerals, water, air, etc.

Extractive industries:

Including mining, quarrying, etc.

Domestic mineral raw materials:

Iron ore, copper ore, sand, gravel, stone, etc.

Value: $32 Billion

Mineral processing industries:

Including smelting, refining.

Processed materials of mineral origin:

Steel, aluminum, copper, other metals, brick, glass, cement, inorganic chemicals, fertilizers, etc.

Value: $310 Billion

U.S. economy:

Gross domestic product:

$5,950 Billion

Domestic reclaimed metals and mineral materials:

Scrap iron, aluminum, glass, etc.

Value of old scrap: $10 Billion

Reclaiming and recycling industries:

Scrap dealers, etc.

U.S. scrap and waste materials:

Iron and steel, aluminum, copper, brass, bronze, lead, zinc, glass.

Foreign natural resources

Foreign mineral operations of U.S. firms and/or foreign firms

Imports into U.S. of mineral raw materials:

Copper ore, iron ore, bauxite, etc.

Value: $2 Billion

Imports into U.S. of processed materials of mineral origin:

Steel, aluminum, inorganic chemicals, etc.

Value: $41 Billion

Exports from U.S. of mineral raw materials and processed materials of mineral origin:

Metals inorganic chemicals, fertilizers, etc.

Value: $37 Billion

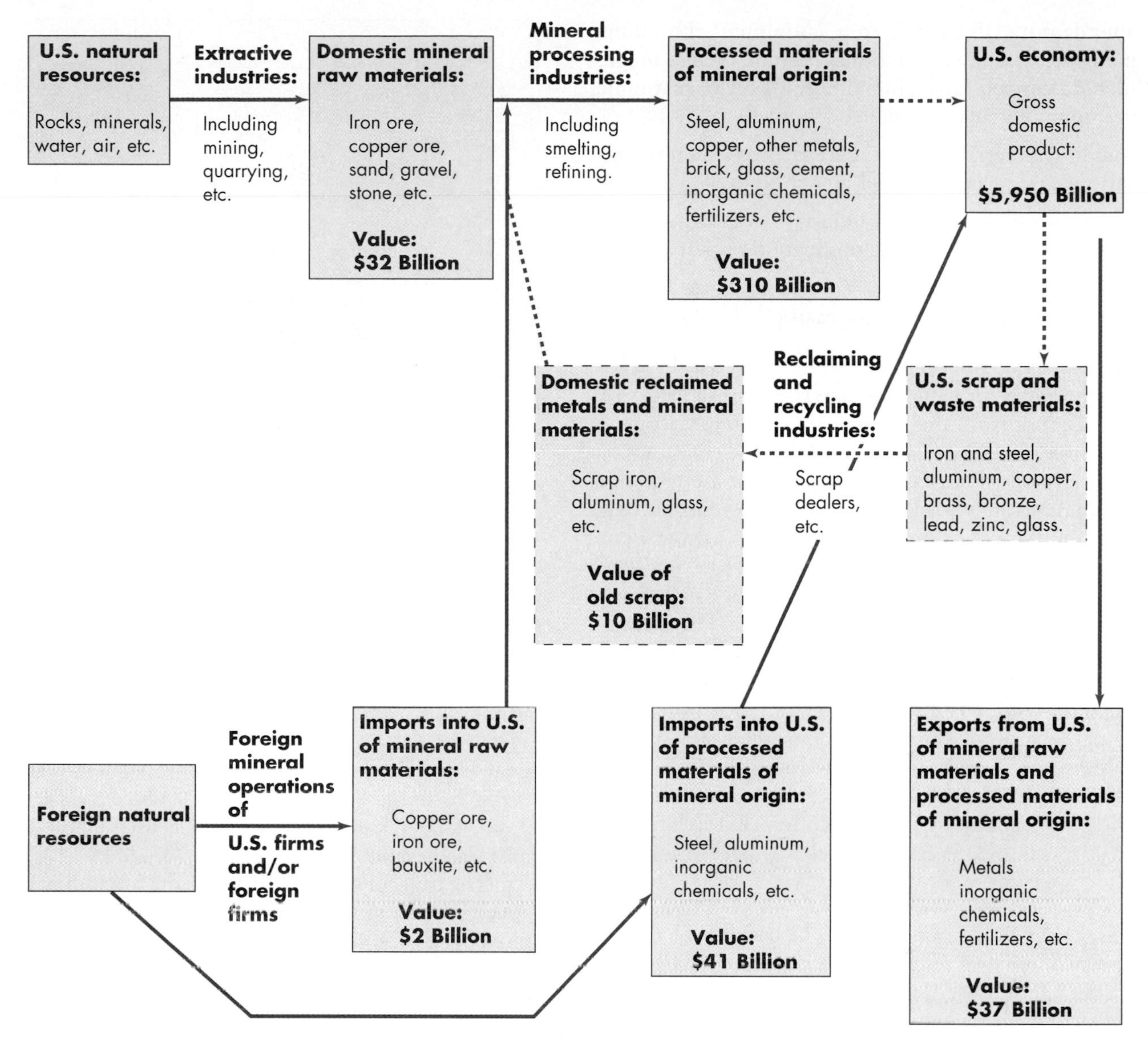

▲ **FIGURE 14.1** The role of nonfuel minerals in the U.S. economy. (*Data from* U.S. *Department of the* Interior, *Bureau of* Mines. 1993. Mineral commodity summaries 1993. I 28.148:993.)

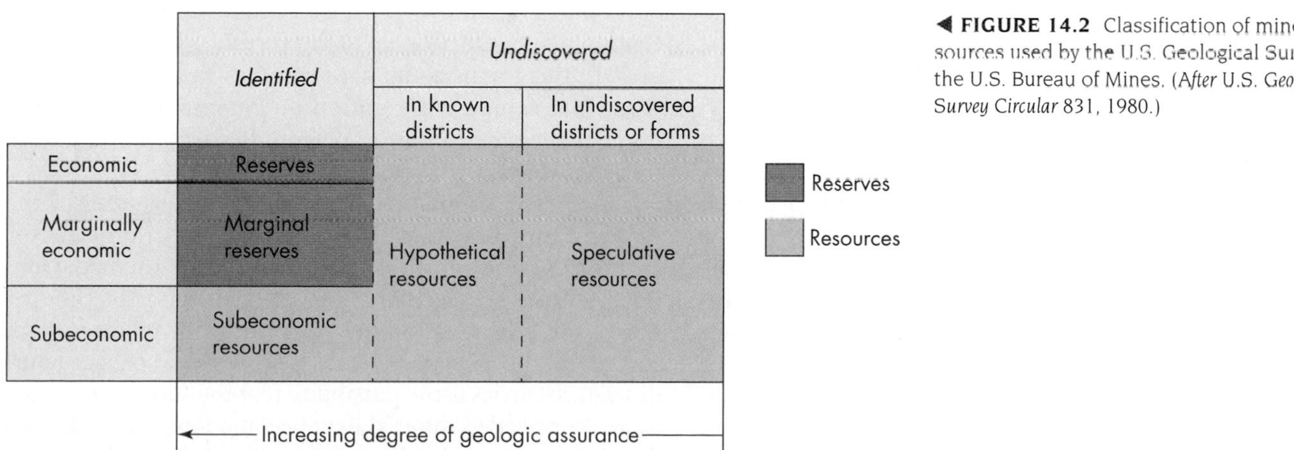

◀ **FIGURE 14.2** Classification of mineral resources used by the U.S. Geological Survey and the U.S. Bureau of Mines. (*After* U.S. *Geological Survey Circular* 831, 1980.)

abundant metals include iron, aluminum, chromium, manganese, titanium, and magnesium. Scarce metals include copper, lead, zinc, tin, gold, silver, platinum, uranium, mercury, and molybdenum.

- Building materials such as aggregate for concrete, clay for tile, and volcanic ash for cinder block
- Minerals for the chemical industry—for example, the many minerals used in the production of petrochemicals
- Minerals for agriculture—for example, fertilizers

When we think of mineral resources, we usually think of the metals used in structural materials, but in fact (with the exception of iron) the predominant mineral resources are not of this type. Consider the annual world consumption of a few selected elements. Sodium and iron are used at a rate of approximately 0.1 billion to 1 billion tons per year. Nitrogen, sulfur, potassium, and calcium are used at a rate of approximately 10 million to 100 million tons per year. These four elements are used primarily as soil conditioners or fertilizers. Zinc, copper, aluminum, and lead have annual world consumption rates of about 3 million to 10 million tons, whereas gold and silver have annual consumption rates of 10,000 tons or less. Of the metallic minerals, iron makes up 95 percent of all the metals consumed, and nickel, chromium, cobalt, and manganese are used mainly in alloys of iron (as in stainless steel). Therefore, we can conclude that the nonmetallics, with the exception of iron, are consumed at much greater rates than elements used for their metallic properties.

Responses to Limited Availability The basic problem with availability of mineral resources is not actual exhaustion or extinction, but the cost of maintaining an adequate reserve, or stock, within an economy through mining and recycling. At some point, the costs of mining exceed the worth of the material. When the availability of a particular mineral becomes a limitation, several solutions are possible:

- Find more sources
- Find a substitute
- Recycle what has already been obtained
- Use less and make more efficient use of what we have
- Do without

Which choice or combination of choices is made depends on social, economic, and environmental factors.

We can use a particular mineral resource in several ways: rapid consumption, consumption with conservation, or consumption and conservation with recycling. Which option is selected depends in part on economic, political, and social criteria. Figure 14.3 shows the hypothetical depletion curves corresponding to these three options. Historically, resources have been consumed rapidly, with the exception of precious metals. However, as more resources become limited, increased conservation and recycling are expected. Certainly the trend toward recycling is well established for such metals as copper, lead, and aluminum.

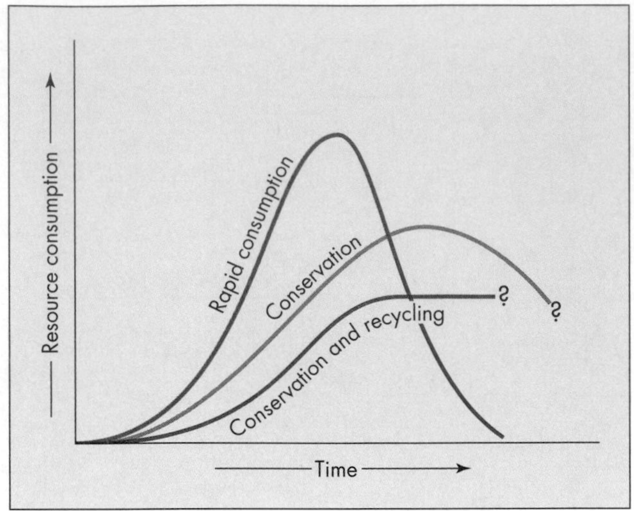

▲ **FIGURE 14.3** Diagram of three hypothetical depletion curves for the use of mineral resources.

As both the world population and the desire for a higher standard of living increase, the demand for mineral resources expands at a faster and faster rate. The more developed countries in the world, with only 16 percent of the earth's population, consume a disproportionate share of mineral resources. For example, 70 percent of the aluminum, copper, and nickel extracted is used by the United States, Japan, and Western Europe (6). About 10,000 kg (10 metric tons) of new mineral material (excluding energy resources) are required each year for each person in the United States (Table 14.2). As less-developed countries become more affluent and use more resources, world per capita mineral consumption is expected to increase. If the world per capita consumption rate of iron, copper, and lead were to rise to the U.S. level, production of these metals would have to increase to several times the present rate. As such an increase in production is very unlikely, affluent countries will have to find substitutes for some minerals or use a smaller proportion of the world annual production. With the exception of construction materials (crushed stone, sand, and gravel), this seems to be happening in the United States, where per capita consumption of aluminum, copper, and lead has decreased about 12 percent from the mid-1970s to the late 1990s.

Domestic supplies of many mineral resources in the United States and other affluent nations are insufficient for current use and must be supplemented by imports from other nations. Table 14.3 shows the deficiency of U.S. reserves for selected nonfuel minerals and major foreign sources for the needed minerals. Of particular concern to industrial countries is the possibility that the supply of a much desired or needed mineral may become interrupted by political, economic, or military instability of the supplying nation. Today the United States, along with many other nations, depends on a steady supply of imports to meet the

Table 14.2 U.S. consumption, 1997, of selected materials along with value and uses

Material	Total Consumption (millions of metric tons)	Per Capita Consumption (kg/person)*	Total 1997 Value (million $)	Some Selected Uses; Percentages Are of Total Consumption
Crushed stone	1,430	5,343	8,070	Unspecified uses, 43%; of the rest, construction aggregates, 83%; chemical and metallurgical, 14%
Sand and gravel (1996 construction)	953	3,557	4,000 (1996)	Concrete aggregates, 43%; 41% unspecified uses; of the rest, road base and covering, 23%, fill, 12%
Iron and steel	114	426	73,000	Automotive, 13%; construction, 14%; containers, 4%
Cement	96	359	6,000	Ready-mix concrete, 70%; concrete products, 10%
Salt	37	138	965	Chemical industries, 45%; highway deicing, 30%; food and agricultural, 7%
Phosphate rock	44	164	1,080	Fertilizers, 93%
Clay	42	156	2,140	Example: kaolinite clay: paper, 56%; fiberglass, 7%; paint, 5%
Aluminum	6.7	25	5,900	Transportation, 32%; packaging, 26%; building, 16%; electrical, 8%
Copper	2.9	11	4,600	Construction, 43%; electrical, 24%; industrial machinery and equipment, 12%
Lead	1.6	6	440	Batteries, fuel tanks, solder, seals, bearings, 71%
Gold	0.00014	0.0005	3,400	Jewelry/arts, 70%; industrial, 23%; dental, 7%

*U.S. population in 1997 was 267,636,000 persons.

Source: Data from U.S. Bureau of Mines. *Mineral commodity summaries*, 1998, 1999.

mineral demand of industries. Of course, the fact that a mineral is imported into a country does not mean that it does not exist in quantities that could be mined within the country. Rather, it suggests that there are economic, political, or environmental reasons that make it easier, more practical, or more desirable to import the material.

14.2 Geology of Mineral Resources

The geology of mineral resources is intimately related to the entire geologic cycle, and nearly all aspects and processes of the cycle are involved to a lesser or greater extent in producing local concentrations of useful materials.

Local Concentrations of Metals

The term **ore** is sometimes used for those useful metallic minerals that can be mined at a profit, and locations where ore is found are anomalously high concentrations of these minerals. The concentration of metal necessary for a particular mineral to be classified as an ore varies with technology, economics, and politics. Before smelting (extraction of metal by heating) was invented, the only metal ores were those in which the metals appeared in their pure form; gold, for example, was originally obtained as a pure, or native, metal. Now gold mines extend deep beneath the surface, and the recovery process involves reducing tons of rock to ounces of gold. Although the rock contains only a minute amount of gold, we consider it a gold ore because we can extract the gold profitably.

The **concentration factor** of a metal is the ratio of its necessary concentration for profitable mining (that is, of its concentration in ore) to its average concentration in the earth's crust. Table 14.4 lists some metallic elements and their average concentrations, concentrations in ore, and concentration factors. Aluminum has an average concentration of about 8 percent in the earth's crust and needs to be found at concentrations of about 35 percent to be mined economically, giving it a concentration factor of about 4. Mercury, on the other hand, has an average concentration of only a tiny fraction of 1 percent and must have a concentration factor of about 10,000 to be mined economically. Nevertheless, mercury ores are common in certain regions, where they and other metallic ores are deposited (see *A Closer Look: Plate Tectonics and Minerals*). The percentage of a metal in ore (and thus the concentration factor) is subject to change as the demand for the metal changes.

Table 14.3 U.S. reliance on imports (by percentage) on selected nonfuel mineral resources for 1992

Mineral	U.S. Reliance	Major Sources (1988–1991)
Arsenic	100%	Chile, France, Mexico
Bauxite and aluminum	100%	Australia, Guinea, Jamaica, Brazil
Columbium (niobium)	100%	Brazil, Canada, Fed. Rep. of Germany
Graphite	100%	Mexico, China, Brazil, Madagascar
Manganese	100%	Rep. of South Africa, France, Gabon
Mica (sheet)	100%	India, Belgium, Brazil, Japan
Strontium (celestite)	100%	Mexico, Spain, Fed. Rep. of Germany
Thallium	100%	Belgium, Japan, United Kingdom (U.K.)
Gemstones (natural and synthetic)	98%	Israel, Belgium, India, U.K.
Asbestos	95%	Canada, Rep. of South Africa
Platinum-group metals	94%	Rep. of South Africa, U.K., U.S.S.R.
Diamonds (industrial stones)	92%	Ireland, Zaire, U.K., Rep. of South Africa
Fluorspar	87%	Mexico, Rep. of South Africa, China, Canada
Tantalum	87%	Fed. Rep. of Germany, Thailand, Australia
Tungsten	85%	China, Bolivia, Fed. Rep. of Germany, Peru
Cobalt	76%	Zaire, Zambia, Canada, Norway
Chromium	74%	Rep. of South Africa, Turkey, Zimbabwe, Yugoslavia
Tin	73%	Brazil, Bolivia, China, Indonesia, Malaysia
Potash	67%	Canada, Israel, U.S.S.R., Fed. Rep. of Germany
Nickel	64%	Canada, Norway, Australia, Dominican Republic
Antimony	58%	China, Mexico, Rep. of South Africa, Hong Kong
Iodine	52%	Japan, Chile
Peat	50%	Canada, Norway, Austria, Ireland
Cadmium	49%	Canada, Mexico, Australia, Fed. Rep. of Germany
Selenium	47%	Canada, U.K., Belgium-Luxembourg, Japan
Barite	44%	China, India, Mexico, Morocco
Silicon	36%	Brazil, Canada, Venezuela, Norway
Gypsum	35%	Canada, Mexico, Spain
Zinc	34%	Canada, Mexico, Peru, Spain
Pumice and pumicite	32%	Greece, Mexico, Ecuador
Magnesium compounds	23%	China, Canada, Greece, Mexico
Sulfur	18%	Canada, Mexico
Nitrogen	15%	Canada, U.S.S.R., Trinidad and Tobago, Mexico
Mica (scrap and flake)	14%	Canada, India
Vermiculite	14%	Rep. of South Africa, China, Brazil
Salt	13%	Canada, Mexico, Bahamas, Chile
Iron and steel	12%	European Community, Japan, Canada, Rep. of Korea
Iron ore	12%	Canada, Brazil, Venezuela, Mauritania
Lead	8%	Canada, Mexico, Peru, Belgium
Cement	7%	Canada, Mexico, Japan, Spain
Perlite	3%	Greece
Sodium sulfate	1%	Canada, Mexico

Source: U.S. Bureau of Mines. 1993. *Mineral commodity summaries, 1993.*

The geology of economically useful deposits of mineral and rock materials is as diversified and complex as are the processes responsible for their formation or accumulation in the natural environment. Most deposits, however, can be related to various parts of the rock cycle within the influence of the tectonic, geochemical, and hydrologic cycles. The genesis of mineral resources with commercial value can be subdivided into several categories:

Table 14.4 Approximate concentration factors of selected metals necessary before mining is economically feasible

Metal	Natural Concentration (Percent)	Percent in Ore	Approximate Concentration Factor
Gold	0.0000004	0.001	2,500
Mercury	0.00001	0.1	10,500
Lead	0.0015	4	2,500
Copper	0.005	0.4 to 0.8	80 to 160
Iron	5	20 to 69	4 to 14
Aluminum	8	35	4

Source: Data from U.S. Geological Survey Professional Paper 820, 1973

- **Igneous processes,** including crystal settling, late magmatic process, and hydrothermal replacement
- **Metamorphic processes** associated with contact or regional metamorphism
- **Sedimentary processes,** including accumulation in oceanic, lake, stream, wind, and glacial environments
- **Biological processes**
- **Weathering processes,** such as soil formations and *in situ* (in-place) concentrations of insoluble minerals in weathered rock debris

Table 14.5 lists examples of ore deposits from each of these categories.

Table 14.5 Examples of different types of mineral resources

Type	Example
Igneous	
Disseminated	Diamonds—South Africa
Crystal settling	Chromite—Stillwater, Montana
Late magmatic	Magnetite—Adirondack Mountains, New York
Pegmatite	Beryl and lithium—Black Hills, South Dakota
Hydrothermal	Copper—Butte, Montana
Metamorphic	
Contact metamorphism	Lead and silver—Leadville, Colorado
Regional metamorphism	Asbestos—Quebec, Canada
Sedimentary	
Evaporite (lake or ocean)	Potassium—Carlsbad, New Mexico
Placer (stream)	Gold—Sierra Nevada foothills, California
Glacial	Sand and gravel—northern Indiana
Deep-ocean	Manganese oxide nodules—central and southern Pacific Ocean
Biological	Phosphorus—Florida
Weathering	
Residual soil	Bauxite—Arkansas
Secondary enrichment	Copper—Utah

Source: Modified from Robert J. Foster. 1983. *General geology,* 4th ed. Columbus, OH: Charles E. Merrill.

Igneous Processes

Most ore deposits caused by igneous processes result from an enrichment process that concentrates an economically desirable ore of metals such as copper, nickel, or gold. In some cases, however, an entire igneous rock mass contains disseminated crystals that can be recovered economically. Perhaps the best-known example is the occurrence of diamond crystals, found in a coarse-grained igneous rock called **kimberlite,** which characteristically occurs as a pipe-shaped body of rock that decreases in diameter with depth (Figure 14.4). Almost the entire kimberlite pipe is the ore deposit, and the diamond crystals are disseminated throughout the rock (6).

Diamonds, which are composed of carbon, form at very high temperatures and pressures (7), perhaps at depths as great as 150 km—well below the crust of the earth and into the mantle. Some kimberlite pipes in South Africa are believed to be as old as 2 billion years. Near the surface, diamonds are not stable over geologic time and will eventually change to graphite (the mineral in "lead" pencils). Do not sell your diamonds yet! The transformation won't happen at surface temperature and pressure, and as a result diamonds are metastable, remaining beautiful and mysterious for periods of time of interest to humans. The fact that the kimberlite pipes are so old suggests that they must be intruded (moved upward) from deep diamond-forming depth to near the surface relatively quickly. If this were not the case, the diamonds would have been transformed to graphite.

Crystal Settling More concentrated ore deposits can result from igneous processes called *crystal settling* that segregate crystals formed earlier from those formed later. For example, as magma cools, heavy minerals that crystallize early may slowly sink or settle toward the lower part of the magma chamber, where they form concentrated layers. Deposits of chromite (ore of chromium) have formed by this process (Figure 14.5).

A CLOSER LOOK Plate Tectonics and Minerals

In a broadbrush approach to the geology of mineral resources, tectonic plate boundaries are related to the origin of such ore deposits as iron, gold, copper, and mercury (Figure 14.A). The basic ideas relate to processes operating at the diverging and converging plate boundaries (see Chapter 2).

The origin of metallic ore deposits at divergent plate boundaries is related to the migration (movement) of ocean water. Cold, dense ocean water moves down through numerous fractures in the basaltic rocks at oceanic ridges and is heated by contact with or heat from nearby molten rock (magma). The warm water is lighter and more chemically active and rises up (convects) through the fractured rocks, leaching out metals. The metals are carried in solution and deposited (precipitated) as metallic sulfides (8). Quite a few hot-water vents with sulfide deposits have been discovered along oceanic ridges, and undoubtedly numerous others will be located.

The origin of metallic ore deposits at convergent plate boundaries is hypothesized to be the result of the partial melting of seawater-saturated rocks of the oceanic lithosphere in a subduction zone. The high heat and pressure that cause the melting also facilitate the release and movement of metals from the partially molten rock. These metals, which originated in the rock, become concentrated and ascend as more fluid components of the magma. The metal-rich fluids are eventually released (or escape) from the magma, and the metals are deposited in a host rock (8).

Perhaps the best example of metallic deposits at subduction zones is the global occurrence of known mercury deposits (Figure 14.B). All the belts of productive deposits of mercury are associated with volcanic systems and are located near convergent plate boundaries. It has been suggested that the mercury, originally found in oceanic sediments of the crust, is distilled out of the downward-plunging plate and emplaced at a higher level above the subduction zone (4). The significant point from the economic view is that convergent plate junctions characterized by volcanism and tectonic activities are likely places to find mercury. A similar argument can be made for other ore deposits, but there is danger in oversimplification as many deposits are not directly associated with plate boundaries (Figure 14.C).

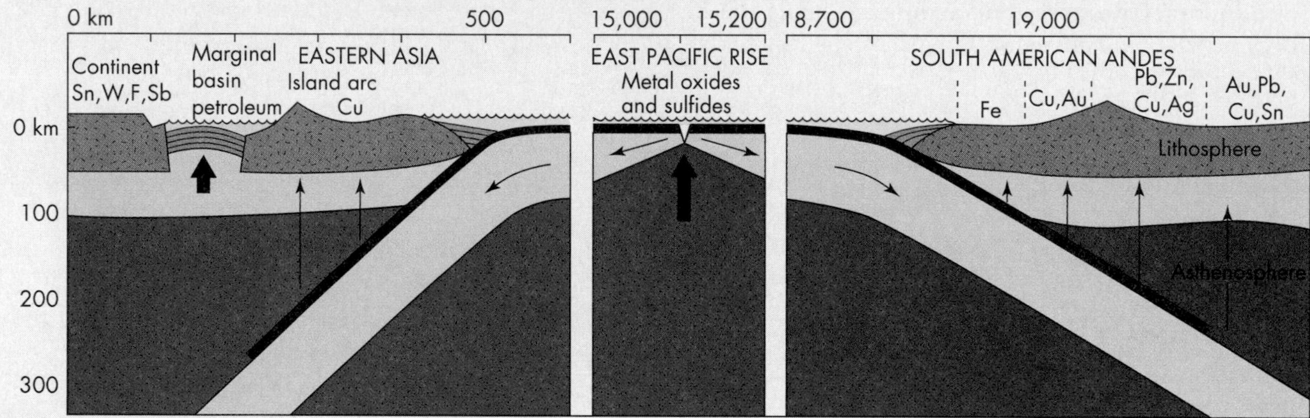

▲ **FIGURE 14.A** Diagram of the relationship between the East Pacific Rise (divergent plate boundary), Pacific margins (convergent plate boundaries), and metallic ore deposits. (*After* NOAA, California Geology 30(5), 1977.)

Late Magmatic Processes and Hydrothermal Replacement Late magmatic processes occur after most of the magma has crystallized, and rare and heavy metalliferous materials in water- and gas-rich solutions remain. This late-stage metallic solution may be squeezed into fractures or settle into interstices (empty spaces) between earlier-formed crystals. Other late-stage solutions form coarse-grained igneous rock known as **pegmatite,** which is rich in feldspar, mica, and quartz, as well as certain rare minerals. Pegmatites have been extensively mined for feldspar, mica, spodumene (lithium mineral), and clay that forms from weathered feldspar.

Hydrothermal (hot-water) mineral deposits are a common type of ore deposit. They originate from late-stage magmatic processes and give rise to a variety of mineralization, including gold, silver, copper, mercury, lead, zinc, and other metals, as well as many nonmetallic minerals. The hydrothermal solutions that form ore deposits are mineralizing fluids that migrate through a host rock, crystallizing as veins or small dikes (Figure 14.6). The mineral material is either produced directly from the igneous parent rock or altered by metamorphic processes as magmatic solutions intrude into the surrounding rock. (Alteration by metamorphic processes, called *contact metamorphism,* is discussed under Metamorphic Processes on page 385) Many hydrothermal deposits cannot be traced to a parent igneous rock mass, however, and their origin remains unknown. It is speculated that circulating groundwater, heated and enriched with minerals after contact with deeply buried magma, might be responsible for some of these deposits (6,9).

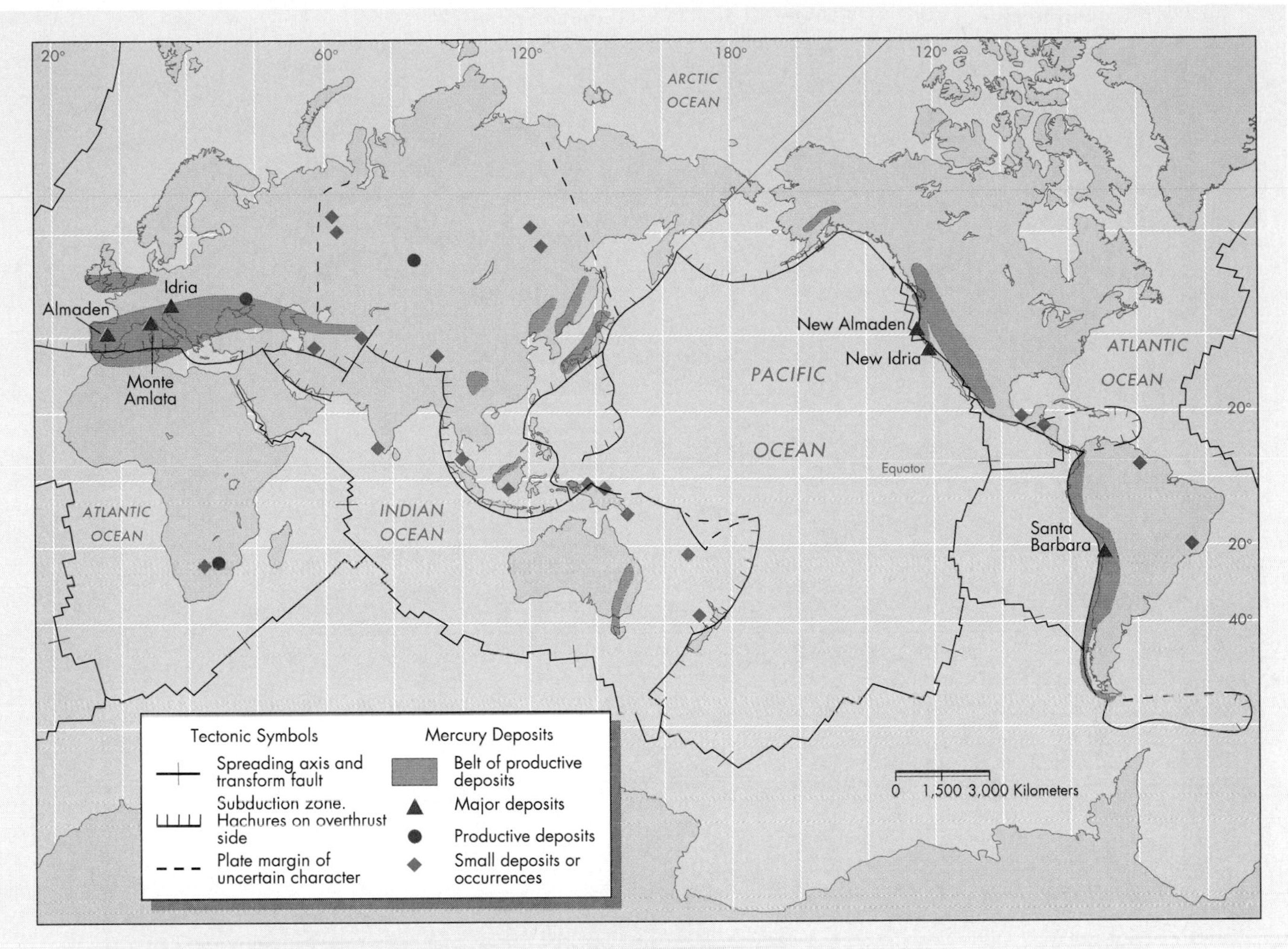

▲ **FIGURE 14.B** Relationship between mercury deposits and recently active subduction zones. (*From* D. A. Brobst *and* W. P. Pratt, *eds.* 1973. U.S. *Geological Survey Professional Paper* 820.)

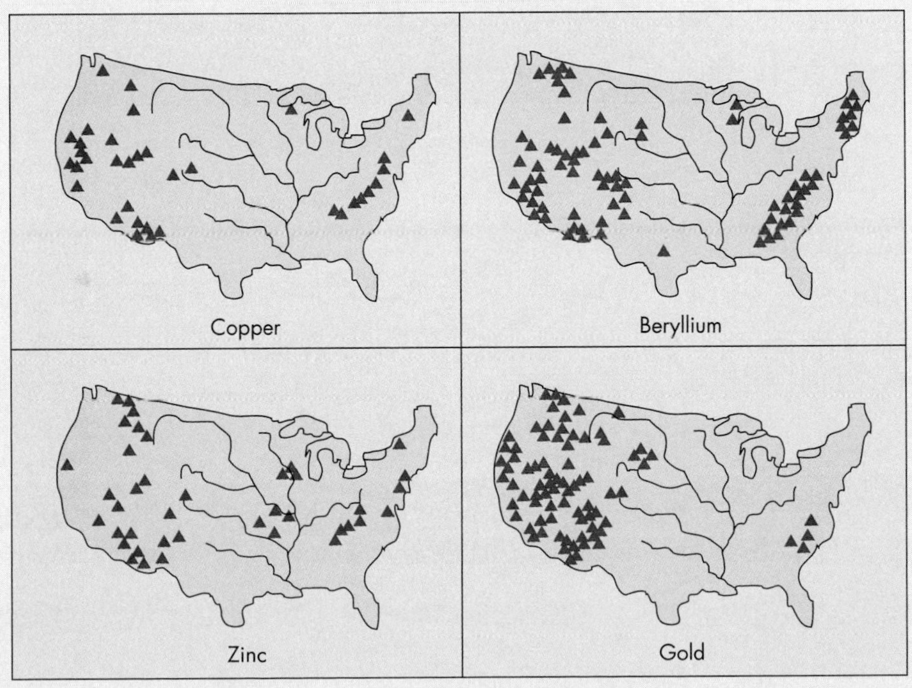

◀ **FIGURE 14.C** Occurrence of economic deposits of some selected metals in the United States. (*Data from mineral resource maps of the* U.S. *Geological Survey*)

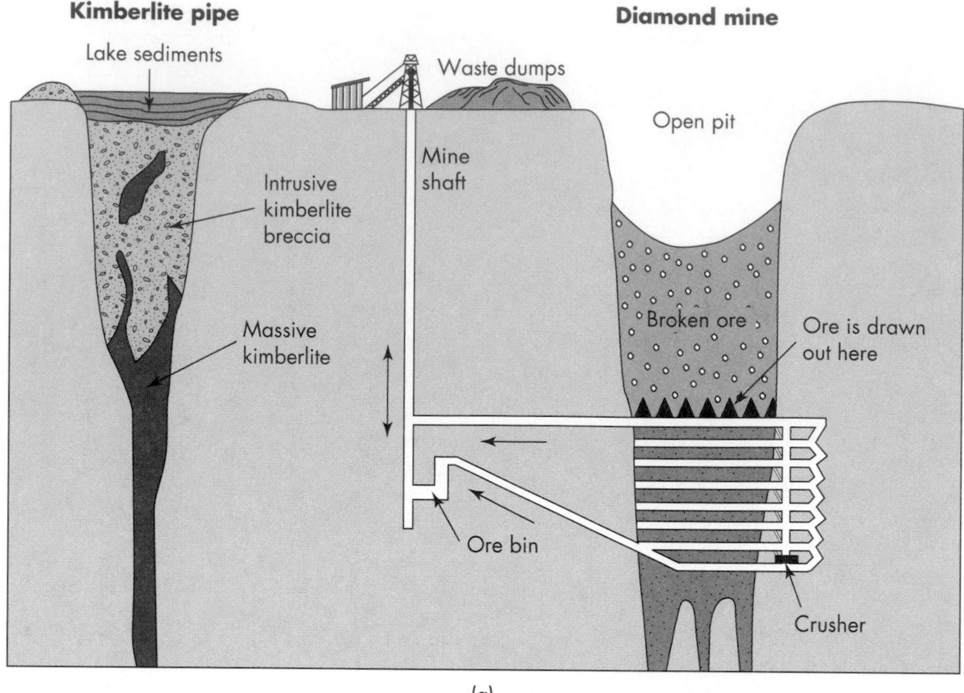

▲ **FIGURE 14.4** (a) Idealized diagram showing a typical South African diamond pipe and mine. Diamonds are scattered throughout the cylindrical body of igneous rock known as kimberlite. (*In* S. E. *Kesler.* 1994. Mineral resources, economics, and the environment. *New York: Macmillan Publishing Co.*) (b) Aerial view of Diamond Mine, Kimberly, South Africa. This is one of the largest hand-dug excavations in the world. (*Helen Thompson/Animals/Animals/ Earth Scenes*)

▶ **FIGURE 14.5** How chromite layers might form. The chromite crystallizes early, and the heavy crystals sink to the bottom and accumulate in layers. (*From Foster.* 1983. General geology, *4th ed. Columbus,* OH: *Charles* E. *Merrill.*)

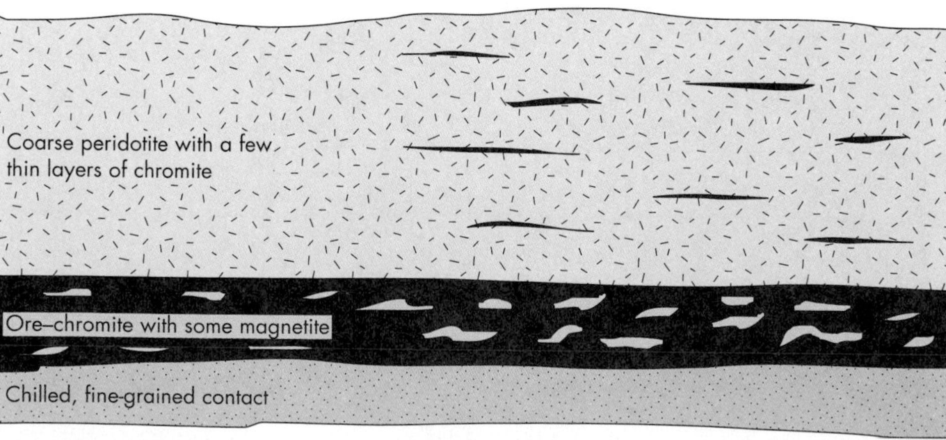

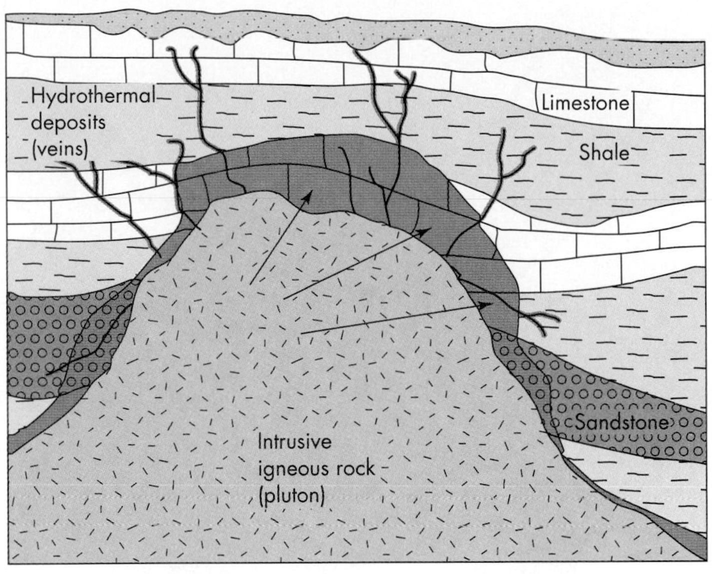

FIGURE 14.6 How hydrothermal and contact metamorphic ore deposits might form.

Contact metamorphic zone where mineral deposits may be present. Notice the zone is wider in the limestone rock than in the sandstone or shale. This results because limestone is chemically more active under contact metamorphism.

Two types of hydrothermal deposits can be recognized: cavity-filling and replacement. *Cavity-filling deposits* are formed when hydrothermal solutions migrate along openings in rocks (such as fracture systems, pore spaces, or bedding planes) and precipitate (crystallize) ore minerals. *Replacement deposits* form as hydrothermal solutions react with the host rock, forming a zone in which ore minerals precipitate from the mineralizing fluids and replace part of the host rock. Although replacement deposits are believed to dominate at higher temperatures and pressures than cavity-filling deposits, both may be found in close association as one grades into the other; that is, the filling of an open fracture by precipitation from hydrothermal solutions may occur simultaneously with replacement of the rock that lines the fracture (6,9).

Hydrothermal replacement processes are significant because, excluding some iron and nonmetallic deposits, they have produced some of the world's largest and most important mineral deposits. Some of these deposits result from a massive, nearly complete replacement of host rock with ore minerals that terminate abruptly; others form thin replacement zones along fissures; and still others form disseminated replacement deposits that may involve huge amounts of relatively low-grade ore (9).

The actual sequence of geologic events leading to the development of a hydrothermal ore deposit is usually complex. Consider, for example, the tremendous, disseminated copper deposits of northern Chile. The actual mineralization is thought to be related to igneous activity, faulting, and folding that occurred 60 million to 70 million years ago. The ore deposit is an elongated, tabular mass along a highly sheared (fractured) zone separating two types of granitic rock. The concentration of copper results from a number of factors:

- A source igneous rock supplied the copper.
- The fissure zone tapped the copper supply and facilitated movement of the mineralizing fluids.
- The host rock was altered and fractured, preparing it for deposition and replacement processes that produced the ore.
- The copper was leached and redeposited again by meteoric water, which further concentrated the ore (10).

Metamorphic Processes

Contact Metamorphism Ore deposits are often found along the contact between igneous rocks and the surrounding rocks they intrude. This area is characterized by **contact metamorphism,** caused by the heat, pressure, and chemically active fluids of the cooling magma interacting with the surrounding rock, called *country*, or *host rock*. The width of the contact metamorphic zone varies with the type of country rock, as shown in Figure 14.6. The zone is usually thickest in limestone because limestone is more reactive: The release of carbon dioxide (CO_2) increases the mobility of reactants. The zone is generally thinnest for shale because the fine-grained texture retards the movement of hot, chemically active solutions, and the zone is intermediate for sandstone. As we have already mentioned, some of the mineral deposits that form in contact areas originate from the magmatic fluids and some from reactions of these fluids with the country rock.

Regional Metamorphism Metamorphism can also result from regional increase of temperature and pressure associated with deep burial of rocks or tectonic activity. This **regional metamorphism** can change the mineralogy and texture of the preexisting rocks, producing ore deposits of asbestos, talc, graphite, and other valuable nonmetallic deposits (9).

Metamorphism has been suggested as a possible origin of some hydrothermal fluids. It is a particularly likely cause in high-temperature, high-pressure zones where fluids might be produced and forced out into the surrounding rocks to form replacement or cavity-filling deposits. For example, the native copper found along the top of ancient basalt flows in the Michigan copper district was apparently produced by metamorphism and alteration of the basalt, which released the copper and other materials that produced the deposits (10).

Our discussion of igneous and metamorphic processes has focused primarily on ore deposits. However, igneous and metamorphic processes are also responsible for producing a good deal of stone used in the construction industry. Granite, basalt, marble (metamorphosed limestone), slate (metamorphosed shale), and quartzite (metamorphosed sandstone), along with other rocks, are quarried to produce crushed rock and dimension stone in the United States. Stone is used in many aspects of construction work; but many people are surprised to learn that, in total value, with the exception of iron and steel, the stone industry is the largest nonfuel mineral industry in the United States (Table 14.2) (3).

Sedimentary Processes

Sedimentary processes are often significant in concentrating economically valuable materials in sufficient amounts for extraction. As sediments are transported, wind and running water help segregate the sediment by size, shape, and density. Thus, the best sand or sand and gravel deposits for construction purposes are those in which the finer materials have been removed by water or wind. Sand dunes, beach deposits, and deposits in stream channels are good examples.

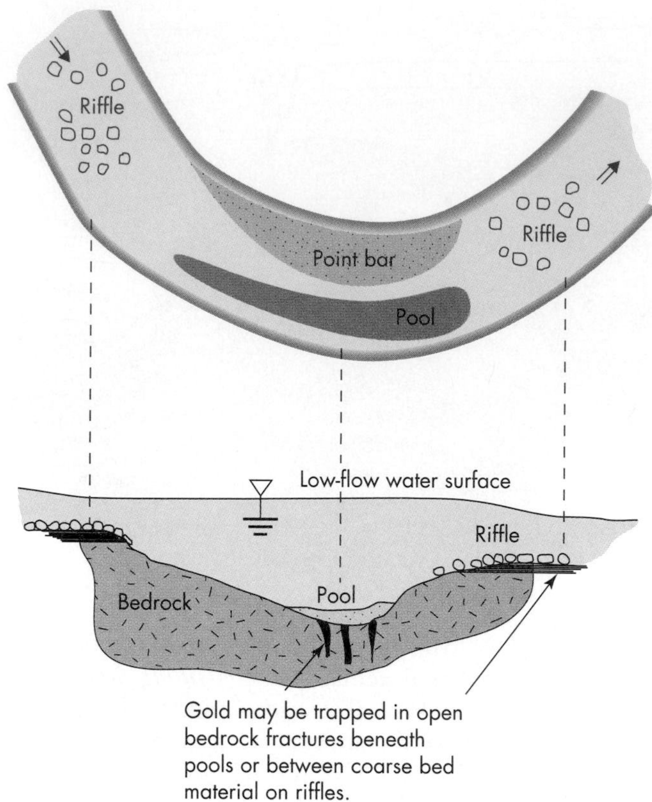

▲ **FIGURE 14.7** Diagram of a stream channel and bottom profile showing areas where placer deposits of gold are likely to occur.

Sand and Gravel The U.S. sand and gravel industry amounts to about $3 billion per year, and, by volume mined (about 850 million tons in 1993), it is one of the largest nonfuel mineral industries in the United States. Currently, most sand and gravel are obtained from river channels and water-worked glacial deposits. The United States now produces more sand and gravel than it needs, but demand is increasing. Environmental restrictions on extraction are causing sand and gravel operations to move away from areas with high population density, and shortages of sand and gravel are expected to increase as zoning and land development restrict locations where they may be extracted (3). Extraction from river channels and active floodplains can cause degradation to the river environment, and objections to river extraction operations are becoming more common.

Placer Deposits Stream processes transport and sort all types of materials according to size and density. Therefore, if the bedrock in a river basin contains heavy metals such as gold, streams draining the basin may concentrate heavy metals to form **placer deposits** (ore formed by deposit of sediments) in areas where there is reduced turbulence or velocity of flow, such as between particles on riffles, in open crevices or fractures at the bottoms of pools, or at the inside curves of bends (Figure 14.7). Placer mining of gold—known as a "poor man's method" because a miner needed only a shovel, a pan, and a strong back to work the streamside claim—helped to stimulate settlement of California, Alaska, and other areas of the United States. Furthermore, the gold in California attracted miners who acquired the expertise necessary to locate and develop other resources in the western conterminous United States and Alaska. Placer deposits of gold and diamonds have also been concentrated by coastal processes, primarily wave action. Beach sands and nearshore deposits are mined in Africa and other places.

Evaporite Deposits Rivers and streams that empty into the oceans and lakes carry tremendous quantities of dissolved material derived from the weathering of rocks. From time to time, geologically speaking, a shallow marine basin may be isolated by tectonic activity (uplift) that restricts circulation and facilitates evaporation. In other cases, climatic variations during the ice ages produced large inland lakes with no outlets, which eventually dried up. In either case, as evaporation progresses, the dissolved materials precipitate, forming a wide variety of compounds, minerals, and rocks called **evaporite** deposits that have important commercial value.

Most evaporite deposits can be grouped into one of three types: *marine evaporites* (solids)—potassium and sodium salts, calcium carbonate, gypsum, and anhydrite; *nonmarine evaporites* (solids)—sodium and calcium carbonate, sulfate, borate, nitrate, and limited iodine and strontium compounds; and *brines* (liquids derived from wells, thermal springs, inland salt lakes, and seawaters)—bromine, iodine, calcium chloride, and magnesium. Heavy metals (such as copper, lead, and zinc) associated with brines and sediments in the Red Sea, Salton Sea, and other areas are important re-

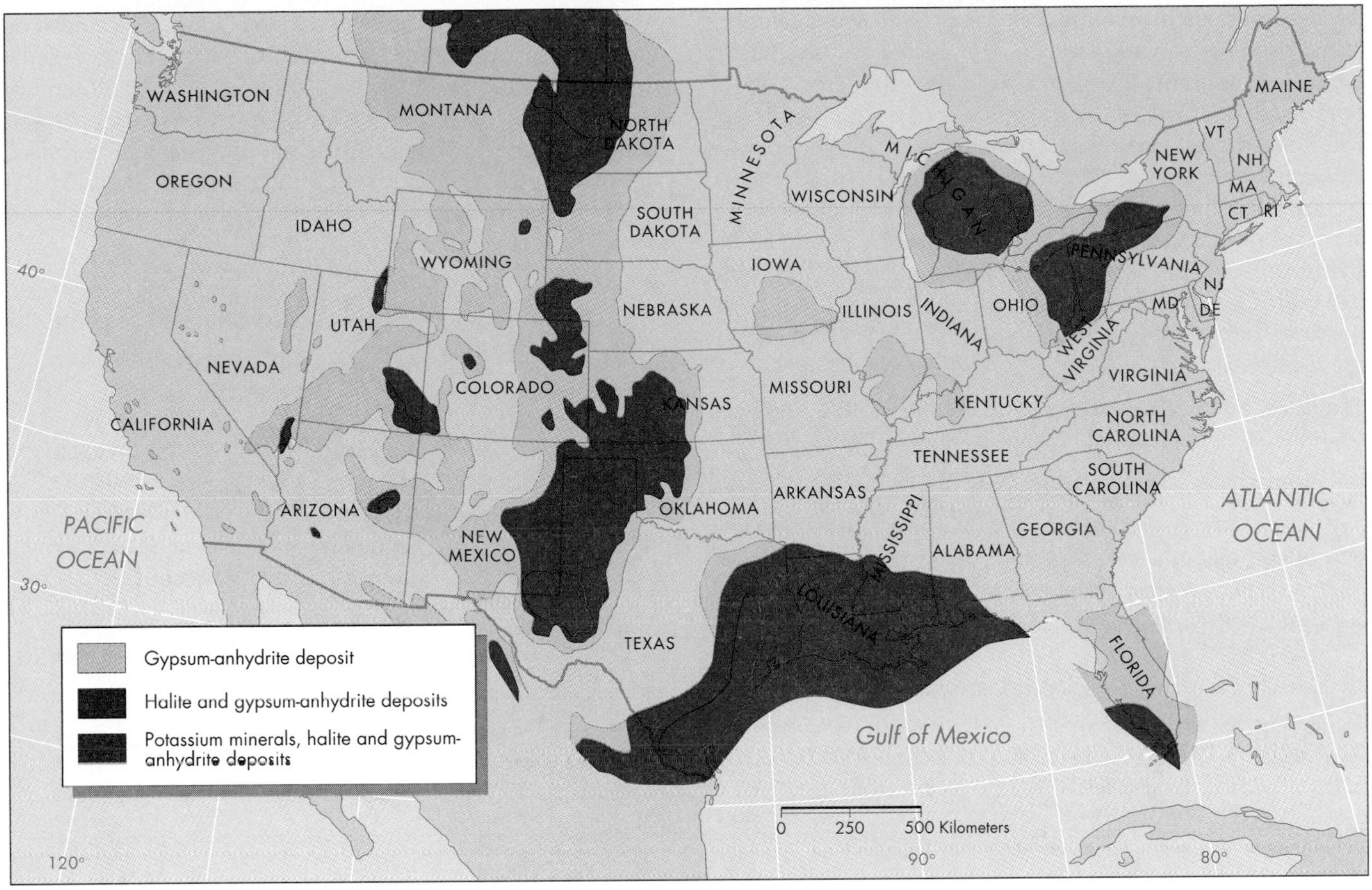

▲ **FIGURE 14.8** Marine evaporite deposits of the United States. (*After* D. A. *Brobst and* W. P. *Pratt, eds.* 1973. U.S. *Geological Survey Professional Paper* 820.)

sources that may be exploited in the future. Extensive marine evaporite deposits exist in the United States (Figure 14.8). The major deposits are halite (common salt, NaCl), gypsum ($CaSO_4 \bullet 2H_2O$), anhydrite ($CaSO_4$), and interbedded limestone ($CaCO_3$). Limestone, gypsum, and anhydrite are present in nearly all marine evaporite basins, and halite and potassium minerals are found in a few. Evaporite materials are widely used in industry and agriculture (11).

Marine evaporites can form stratified deposits that may extend for hundreds of kilometers with a thickness of several thousand meters. The evaporites represent the product of evaporation of seawater in isolated shallow basins with restricted circulation. Within many marine evaporite basins, the different deposits are arranged in broad zones that reflect changes in salinity and other factors controlling the precipitation of evaporites; that is, different materials may be precipitated at the same time in different parts of the evaporite basin. Halite, for example, is precipitated in areas where the brine is more saline, and gypsum where it is less saline (Figure 14.9). Economic deposits of potassium evaporite minerals are relatively rare but may form from highly concentrated brines.

Nonmarine evaporite deposits form by evaporation of lakes in a closed basin. Tectonic activity such as faulting can produce an isolated basin with internal drainage and no outlet. However, to maintain a favorable environment for evaporite mineral precipitation, the tectonic activity must continue to uplift barriers across possible outlets or lower the basin floor faster than sediment can raise it. Even under these conditions, economic deposits of evaporites will not form unless sufficient dissolved salts have washed into the basin by surface runoff from surrounding highlands. Finally, even if

▲ **FIGURE 14.9** White salt deposits forming in Death Valley, California. The salt is deposited as the water evaporates. (*Willard Clay/Tony Stone Worldwide*)

all favorable environmental criteria are present, including an isolated basin with sufficient runoff and dissolved salts, valuable nonmarine evaporites such as sodium carbonate or borate will not form unless the geology of the highlands surrounding the basin is also favorable and yields runoff with sufficient quantities of the desired material in solution (11).

Some evaporite beds are compressed by overlying rocks and mobilized, then pierce or intrude the overlying rocks. Intrusions of salt, called **salt domes,** are quite common in the Gulf Coast of the United States and are also found in northwestern Germany, Iran, and other areas. Salt domes in the Gulf Coast are economically important because:

- They are a good source for nearly pure salt. Some have extensive deposits of elemental sulfur (Figure 14.10).
- Some have oil reserves on their flanks.

Salt domes are also environmentally important as possible permanent disposal sites for radioactive waste, although because salt domes tend to be mobile, their suitability as disposal sites for hazardous wastes must be seriously questioned.

Evaporites from brine resources of the United States are substantial (Table 14.6), assuring that no shortage is likely for a considerable period of time. But many evaporites will continue to have a *place value* because transportation of these mineral commodities increases their price, so continued discoveries of high-grade deposits closer to where they will be consumed remains an important goal (11).

Biological Processes

Organisms are able to form many kinds of minerals, such as the various calcium and magnesium carbonate minerals in shells and calcium phosphate in bones. Some of these minerals cannot be formed inorganically in the biosphere. Thirty-one different biologically produced minerals have been identified. Minerals of biological origin contribute significantly to sedimentary deposits (12).

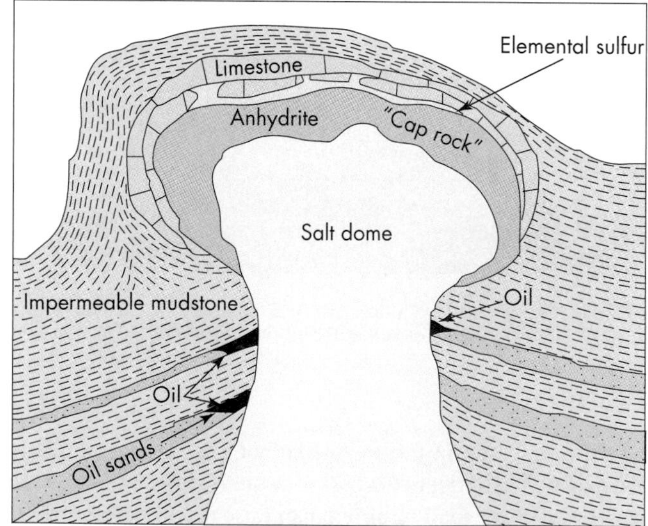

▲ **FIGURE 14.10** A cross section through a typical salt dome of the type found in the Gulf Coast of the United States.

An interesting example of mineral deposits produced by biological processes are phosphates associated with sedimentary marine deposition. Phosphorus-rich sedimentary rocks are fairly common in some of the western states as well as in Tennessee, North Carolina, and Florida. The common phosphorus-bearing mineral in these rocks is apatite, a calcium phosphate associated with bones and teeth. Fish and

Table 14.6 Evaporite and brine resources of the United States expressed in years of supply at current rates of domestic consumption

Commodity	Identified Resources[a] (Reserves[b] and Subeconomic Deposits)	Undiscovered Resources (Hypothetical[c] and Speculative[d] Resources)
Potassium compound	100 years	Virtually inexhaustible
Salt	1000+ years	Unlimited
Gypsum and anhydrite	500+ years	Virtually inexhaustible
Sodium carbonate	6000 years	5000 years
Sodium sulfate	700 years	2000 years
Borates	300 years	1000 years
Nitrates	Unlimited (air)	Unlimited (air)
Strontium	500 years	2000 years
Bromine	Unlimited (seawater)	Unlimited (seawater)
Iodine	100 years	500 years
Calcium chloride	100+ years	1000+ years
Magnesium	Unlimited (seawater)	Unlimited (seawater)

[a]Identified resources: Specific, identified mineral deposits that may or may not be evaluated as to extent and grade, and whose contained minerals may or may not be profitably recovered with existing technology and economic conditions.

[b]Reserves: Identified deposits from which minerals can be extracted profitably with existing technology and under present economic conditions.

[c]Hypothetical resources: Undiscovered mineral deposits, whether of recoverable or subeconomic grade, that are geologically predictable as existing in known districts.

[d]Speculative resources: Undiscovered mineral deposits, whether of recoverable or subeconomic grade, that may exist in unknown districts or in unrecognized or unconventional form.

Source: G. I. Smith et al. 1973. U.S. Geological Survey Professional Paper 820.

▲ **FIGURE 14.11** Large open-pit phosphate mine in Florida. Piles of mining waste with standing water dominate the landscape shown here. Some land has been reclaimed for use as pasture in the upper part of the photograph. (*William Felger/Grant Heilman Photography, Inc.*)

other marine organisms extract the phosphate from seawater to form apatite, and the mineral deposit results from sedimentary accumulations of the phosphate-rich fish bones and teeth. The richest phosphate mine in the world, known as "Bone Valley," is located about 40 km east of Tampa, Florida (Figure 14.11). The deposit is marine sedimentary rocks composed in part of fossils of marine animals that lived 10 million to 15 million years ago, when Bone Valley was the bottom of a shallow sea. That deposit has supplied as much as one-third of the world's phosphate production.

Another important source of phosphorus is guano (bird feces), which accumulates where there are large colonies of nesting sea birds and a climate arid enough for the guano to dry to a rocklike mass. Thus, the formation of one of the major sources of phosphorus depends upon unique biological and geographical conditions.

Weathering Processes

Weathering is responsible for concentrating some materials to the point that they can be extracted at a profit. Weathering processes can produce residual ore deposits in the weathered material and provide secondary enrichment of low-grade ore.

Residual Ore Deposits Intensive weathering of rocks and soils can produce residual deposits of the less soluble materials, which may have economic value. For example, intensive weathering of some rocks forms a type of soil known as *laterite* (a residual soil derived from aluminum- and iron-rich igneous rocks). The weathering processes concentrate relatively insoluble hydrated oxides of aluminum and iron, while more soluble elements such as silica, calcium, and sodium are selectively removed by soil and biological processes. If sufficiently concentrated, residual aluminum oxide forms an aluminum ore known as **bauxite** (Figure 14.12). Important nickel and cobalt deposits are also found in laterite soils developed from ferromagnesian-rich igneous rocks (13).

Insoluble ore deposits such as native gold are generally residual, meaning that unless they are removed by erosion, they accumulate in weathered rock and soil. Accumulation of the insoluble ore minerals is favored where the parent rock is a relatively soluble material such as limestone (Figure 14.13). Care must be taken in evaluating a residual weathered rock or soil deposit because the near-surface concentration may be a much higher grade than ore in the parent, unweathered rocks (14).

Secondary Enrichment Weathering is also involved in *secondary enrichment* processes to produce sulfide ore deposits from low-grade primary ore. Near the surface, primary ore containing such minerals as iron, copper, and silver sulfides is in contact with slightly acid soil water in an oxygen-rich environment. As the sulfides are oxidized, they are dissolved, forming solutions rich in sulfuric acid and in silver and copper sulfate; these solutions migrate downward, producing a leached zone devoid of ore minerals. Figure 14.14 shows a primary ore that has already undergone oxidation and leaching. Below the leached zone, oxidation continues, as the sulfate solutions continue to move toward the groundwater table. Below the water table, if oxygen is no longer available, the solutions are deposited as sulfides, increasing the metal content of the primary ore as much as tenfold. In this way, low-grade primary ore is rendered more valuable, and high-grade primary ore is made even more attractive (9,10).

The presence of a residual iron oxide cap at the surface indicates the possibility of an enriched ore below, but is not always conclusive. Of particular importance to formation of

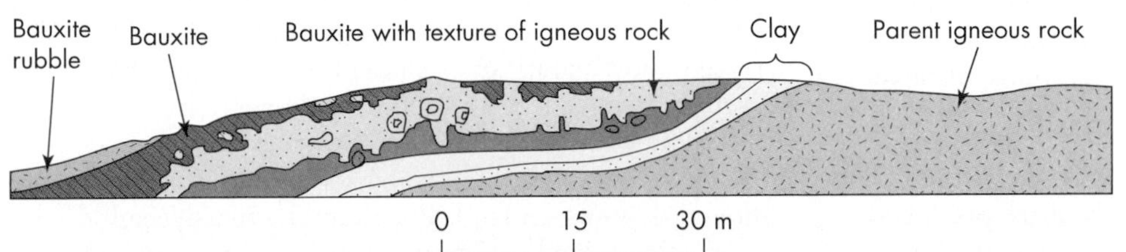

◀ **FIGURE 14.12** Cross section of the Pruden bauxite mine, Arkansas. The bauxite was formed by intensive weathering of the aluminum-rich igneous rocks. (*After G. Mackenzie, Jr., et al. 1958. U.S. Geological Survey Professional Paper 299.*)

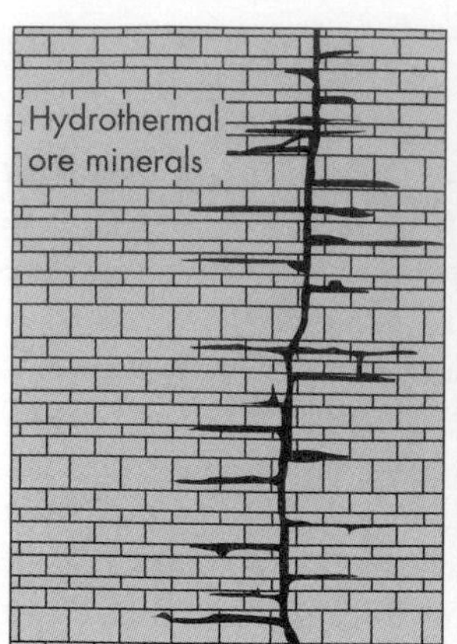

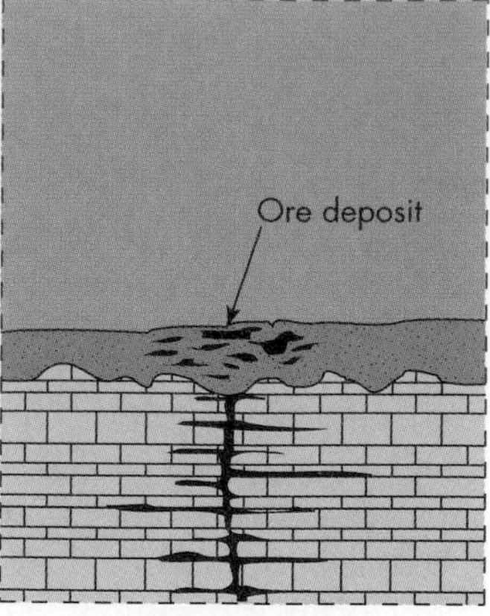

▲ **FIGURE 14.13** How an ore deposit of insoluble minerals might form by weathering and formation of a residual soil. As the limestone that contained the deposit weathered, the ore minerals became concentrated in the residual soil. (*From Foster.* 1983. General geology, *4th ed. Columbus*, OH: *Charles* E. *Merrill*.)

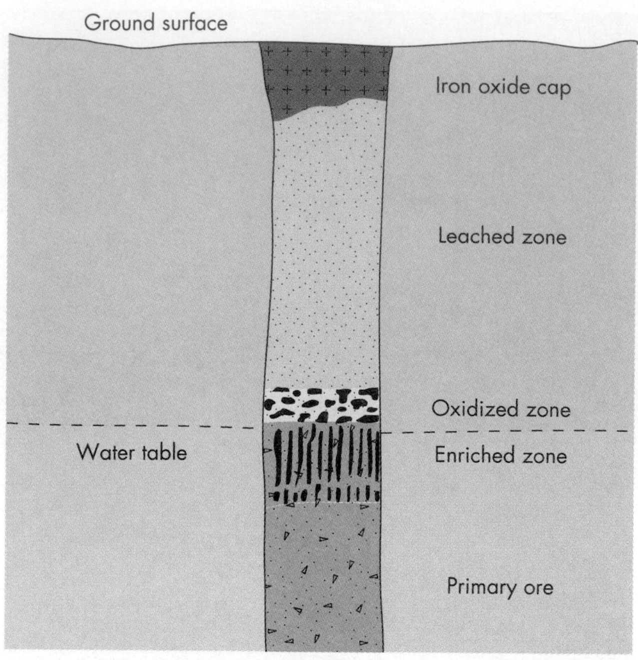

▲ **FIGURE 14.14** Typical zones that form during secondary enrichment processes. Sulfide ore minerals in the primary ore vein are oxidized and altered and then are leached from the oxidized zone and redeposited in the enriched zone. The iron oxide cap is generally a reddish color and may be helpful in locating ore deposits that have been enriched. (*From Foster.* 1983. General geology, *4th ed. Columbus*, OH: *Charles* E. *Merrill*.)

a zone of secondary enrichment is the presence in the primary ore of iron sulfide (for example, pyrite). Without it, secondary enrichment seldom takes place, because iron sulfide in the presence of oxygen and water forms sulfuric acid, which is a necessary solvent. Another factor favoring development of a secondary-enrichment ore deposit is the primary ore being sufficiently permeable to allow water and solutions to migrate freely downward. Given a primary ore that meets these criteria, the reddish iron oxide cap probably does indicate that secondary enrichment has taken place (9).

Several disseminated copper deposits have become economically successful because of secondary enrichment, which concentrates dispersed metals. For example, secondary enrichment of a disseminated copper deposit at Miami, Arizona, increased the grade of the ore from less than 1 percent copper in the primary ore to as much as 5 percent in some localized zones of enrichment (10).

Minerals from the Sea

Mineral resources in seawater or on the bottom of the ocean are vast and, in some cases, such as magnesium, nearly unlimited. In the United States, magnesium was first extracted from seawater in 1940. By 1972, one company in Texas produced 80 percent of our domestic magnesium, using seawater as its raw material source. In 1992 three companies in Texas, Utah, and Washington extracted magnesium, respectively, from seawater, lake brines, and dolomite (mineral composed of calcium and magnesium carbonate).

The deep-ocean floor may eventually be the site of a next mineral rush. Identified deposits include massive sulfide deposits associated with hydrothermal vents, manganese oxide nodules, and cobalt-enriched manganese crusts.

Sulfide Deposits Massive sulfide deposits containing zinc, copper, iron, and trace amounts of silver are produced at divergent plate boundaries (oceanic ridges) by the forces of plate tectonics. Pressure created by several thousand meters of water at ridges forces cold seawater deep into numerous rock fractures, where it is heated by upwelling magma to temperatures as high as 350 °C. The pressure of the heated water produces vents known as *black smokers*, from which the hot, dark-colored, mineral-rich water emerges as hot springs (Figure 14.15). Circulating seawater leaches the surrounding rocks, removing metals that are deposited when the mineral-rich water is ejected into the cold sea. Sulfide minerals precipitate near the vents, forming massive towerlike formations rich in metals. The hot vents are of particular biologic significance because they support a unique assemblage of animals, including giant clams, tube worms, and white crabs. Ecosystems including these animals base their existence on sulfide compounds extruded from black smokers, existing through a process called *chemosynthesis* as opposed to photosynthesis, which supports all other known ecosystems on earth.

The extent of sulfide mineral deposits along oceanic ridges is poorly known, and although leases to some possible deposits are being considered, it seems unlikely that such deposits will be extracted at a profit in the near future. Certainly potential environmental degradation, such as decreased water quality and sediment pollution, will have to be carefully evaluated prior to any mining activity.

Study of the formation of massive sulfide deposits at oceanic ridges is helping geologists understand some of the mineral deposits on land. For example, massive sulfide deposits being mined in Cyprus are believed to have formed at an oceanic ridge and to have been later uplifted to the surface.

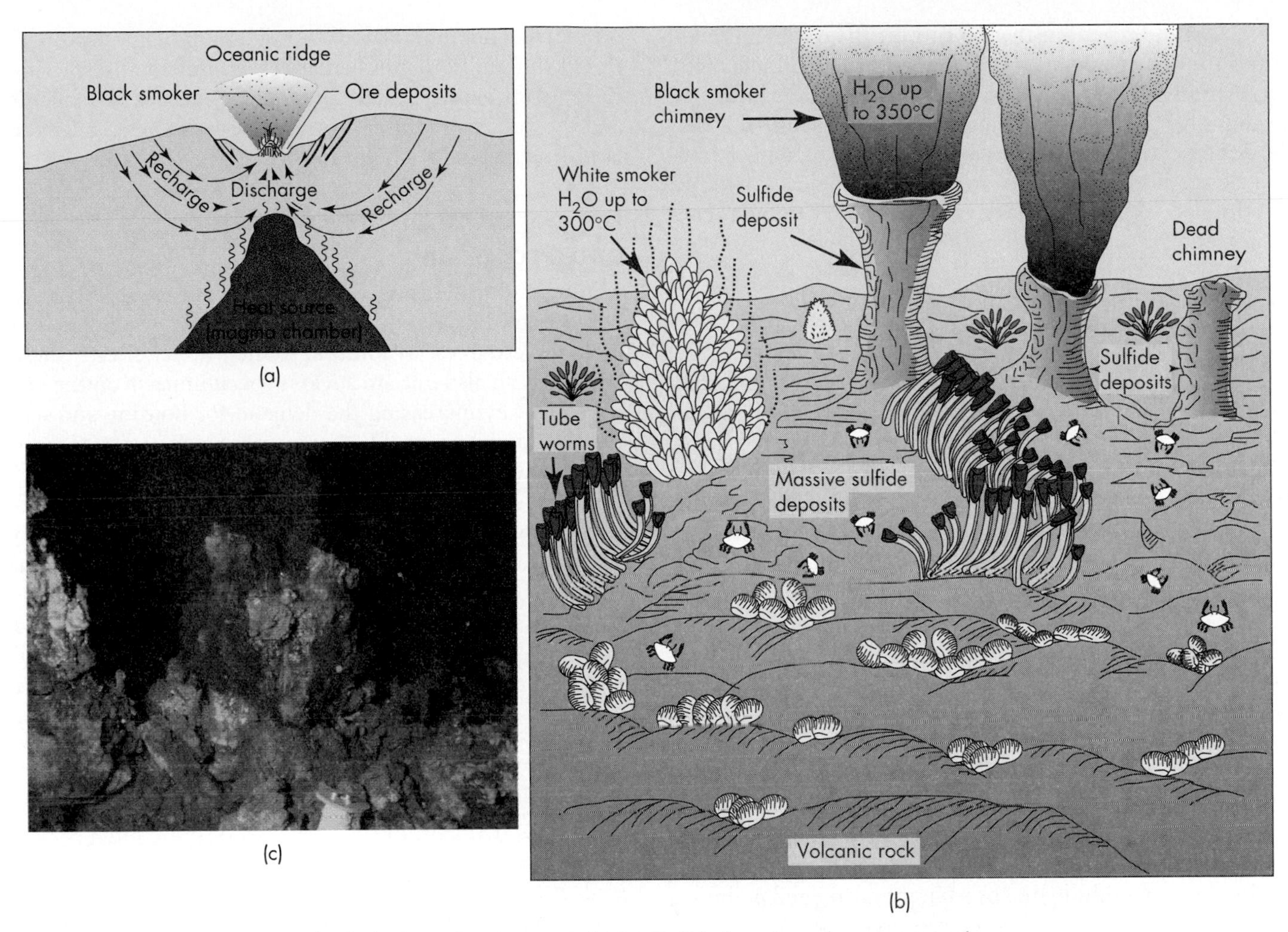

▲ **FIGURE 14.15** (a) Oceanic ridge hydrothermal environment; (b) detail of black smokers where massive sulfide deposits form; (c) photograph of black smoker. (R. *Haymon*)

Manganese Oxide Nodules **Manganese oxide nodules** (Figure 14.16) cover vast areas of the deep-ocean floor. They contain manganese (24%) and iron (14%), with secondary copper (1%), nickel (1%), and cobalt (0.25%). Nodules are found in the Atlantic Ocean off Florida, but the richest and most extensive accumulations occur in large areas of the northeastern, central, and southern Pacific, where they cover 20 to 50 percent of the ocean floor (15).

Manganese oxide nodules are usually discrete, but are welded together locally to form a continuous pavement. Although they are occasionally found buried in sediment, nodules are usually surficial deposits on the seabed. Their size varies from a few millimeters to a few tens of centimeters in diameter (many are marble to baseball sized). Composed primarily of concentric layers of manganese and iron oxides mixed with a variety of other materials, each nodule formed around a nucleus of a broken nodule, a fragment of volcanic rock, or sometimes a fossil. The estimated rate of nodular growth is 1 to 4 mm per million years. The nodules are most abundant in those parts of the ocean where sediment accumulation is at a minimum, generally at depths of 5 to 7 km (8,15).

The origin of the nodules is not well understood; presumably they might form in several ways. The most probable theory is that they form from material weathered from the continents and transported by rivers to the oceans, where ocean currents carry the material to the deposition site in the deep-ocean basins. The minerals from which the nodules form may also derive from submarine volcanism, or may be released during physical and biochemical

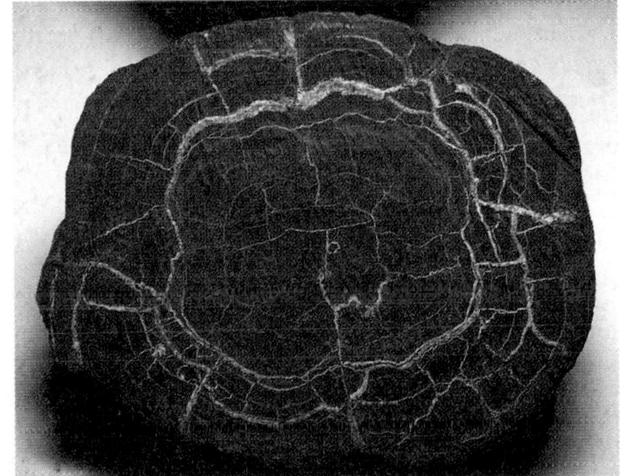

▲ **FIGURE 14.16** Manganese oxide nodule from the ocean bottom. Shown is a cross section through the nodule. Notice the metallic deposits and ring structure. The diameter of the nodule is approximately 7 cm. (*William E. Ferguson*)

processes and reactions that occur near the water–sediment interface during and after deposition of the sediments (15).

Mining of manganese oxide nodules involves lifting the nodules off the bottom and up to the mining ship; this may be done by suction or scraper equipment (6). Although mining of the nodules appears to be technologically feasible, production would be expensive compared to mining manganese on land. In addition, there are uncertainties concerning ownership of the nodules, and nodule mining would cause significant damage to the seafloor and local water quality, raising environmental concerns (6).

***Cobalt-enriched* Manganese Crusts** Oceanic crusts rich in cobalt and manganese are present in the mid- and southwest Pacific, on flanks of seamounts, volcanic ridges, and islands. Cobalt content varies with water depth; the maximum concentration of about 2.5 percent is found at water depths of 1 to 2.5 km. Thickness of the crust averages about 2 cm. The processes of formation are not well understood. Both the nature and the extent of the crusts, which also contain nickel, platinum, copper, and molybdenum, are being studied by U.S. Geological Survey scientists (16).

14.3 Environmental Impact of Mineral Development

Many scientists and others fear that the world is in a resource crisis. Population increase is placing more demands on mineral resources at a time when the planet may be close to the limit of its ability to absorb mineral-related pollution of air, water, land, and biologic resources (6). With this in mind, we will discuss environmental impacts related to nonenergy mineral development. Development of minerals as energy sources is discussed in the next chapter.

The impact of mineral exploitation on the environment depends upon such factors as mining procedures, local hydrologic conditions, climate, rock types, size of operation, topography, and many more interrelated factors. Furthermore, the impact varies with the stage of development of the resource. The exploration and testing stages involve considerably less impact than do the extraction (mining) and processing stages.

Impact of Mineral Exploration and Testing

Exploration and testing activities for mineral deposits vary from collecting and analyzing data gathered by remote-sensing from airplanes or satellites to fieldwork involving surface mapping, drilling, and gathering of geophysical data. Generally, exploration has a minimal impact on the environment, provided care is taken in sensitive areas such as some arid lands, marshlands, and areas underlain by permafrost. Some arid lands are covered by a thin layer of pebbles over fine silt several centimeters thick. The layer of pebbles, called *desert pavement*, protects the finer material from wind erosion. When the pavement is disturbed by road building or other activity, the fine silts may be eroded by wind and/or water, changing physical, chemical, and biological properties of the soil in the immediate environment and scarring the land for many years. In other areas, such as marshlands and the northern tundra, wet organic-rich soils render the land sensitive to even light traffic.

Impact of Mineral Extraction and Processing

Mining and processing of mineral resources are likely to have a considerable adverse impact on land, water, air, and biologic resources. In addition to their direct effects, these activities can also initiate adverse social impacts on the environment by increasing the demand for housing and services in mining areas. These effects are part of the price we pay for the benefits of mineral consumption. It is unrealistic to expect that we can mine our resources without affecting some aspect of the local environment, but we must attempt to hold environmental degradation to a minimum. This can be very difficult, because the demand for minerals continues to increase while deposits of highly concentrated minerals decrease. To provide even more material, we will need increasingly larger operations to mine ever-poorer grades of ore. By the year 2000 the cumulative land use for mining on earth is expected to be 0.2 percent of the land area, or about 300,000 km^2.

Surface mines and quarries today cover less than 0.3 percent, or 29,000 km^2, of the total area of the United States. For comparison, U.S. wilderness, wildlife, and national park lands have a land area of about 500,000 km^2. However, environmental degradation tends to extend beyond the excavation and surface plant areas of both surface and subsurface mines. Large mining operations change the topography by removing material in some areas and dumping waste in others. At best, these actions produce severe aesthetic degradation; often they produce severe environmental degradation as well. The impact of a single mining operation is a local phenomenon, but numerous local occurrences eventually constitute a larger problem.

***Waste from* Mines** In the United States about 60 percent of the land used for mining is disturbed for extraction of minerals, and the remaining 40 percent is used for disposal of mining wastes (mostly overburden, which is the rock removed to get to the ore). This is an enormous waste-disposal problem, representing 40 percent of all the solid waste generated in the country (6). During the past 100 years or so, an estimated 50 billion tons of mining waste have accumulated in the United States, and annual production of waste is now 1 billion to 2 billion tons (17).

***Types of* Mining *and Their* Impact** One of the major practical issues in mining is whether surface or subsurface mines should be developed in a particular area. Surface mining is cheaper but has more direct environmental effects. The trend in recent years has been away from subsurface mining and toward large, open-pit (surface) mines such as the Bingham Canyon Copper Mine in Utah (Figure 14.17).

▲ **FIGURE 14.17** Bingham Copper Mine near Salt Lake City, Utah. Notice the large volume of mine waste. (*Michael Collier*)

This mine is one of the world's largest artificial excavations, covering nearly 8 km^2 to a maximum depth of nearly 800 m.

Sometimes leaching is used as a mining technique. For example, some gold deposits contain such finely disseminated gold that extraction by conventional methods is not profitable. For some of these deposits a process known as *heap leaching* with a dilute cyanide solution is used. The cyanide solution is applied by sprinklers to a heap of crushed gold ore, and as it seeps through the ore it dissolves the gold. The gold-bearing solutions are collected in a plastic-lined pond and treated to recover the gold. Because cyanide is extremely toxic, the mining process must be carefully controlled and monitored. The process has the potential, should an accident occur, to create a serious groundwater pollution problem. Research is also ongoing to develop *in situ* cyanide leaching to eliminate the need for removing ore from the ground. Control and monitoring of the hazardous leaching solution is a difficult problem (18).

Air and Water Pollution Both extraction and processing operations have adverse effects on air and water quality. Dust from mineral mines may affect air resources even though care is often taken to reduce dust by sprinkling water on roads and other dust-producing areas. Smelting (separation of metal from ore by heat) has released enormous quantities of pollutants into the atmosphere, including sulfur dioxide, a major constituent of acid rain and snow. (The problem of acid precipitation is discussed in Chapter 15.)

Water resources are particularly vulnerable to degradation even if drainage is controlled and sediment pollution reduced. Surface drainage is often altered at mine sites, and runoff from precipitation (rain or snow) may infiltrate waste material, leaching out trace elements and minerals. Trace elements leached from mining wastes and concentrated in water, soil, or plants may be toxic, causing diseases in humans and other animals who drink the water, eat the plants, or use the soil. These potentially harmful trace elements include cadmium, cobalt, copper, lead, molybdenum, zinc, and others. The white streaks in Figure 14.18 are mineral

▲ **FIGURE 14.18** A zinc mine in Colorado. The white streaks are mineral deposits apparently leached from tailings. (*Edward A. Keller*)

deposits apparently leached from tailings from a zinc mine in Colorado. Similar-looking deposits may cover rocks in rivers for many kilometers downstream from some mining areas. Specially constructed ponds to collect polluted runoff from mines can help, but they cannot be expected to eliminate all problems.

Groundwater may also be polluted by mining operations when waste comes into contact with slow-moving subsurface waters. Surface water infiltration or groundwater movement through mining waste piles causes leaching of sulfide minerals that may pollute groundwater. The polluted groundwater can eventually seep into streams to pollute surface water. Groundwater problems are particularly troublesome because reclamation of polluted groundwater is very difficult and expensive.

Even abandoned mines can cause serious problems. For example, subsurface mining for lead and zinc in the Tri-State Area (Kansas, Missouri, and Oklahoma), which started in the late nineteenth century and ceased in some areas in the 1960s, has triggered serious water-pollution problems in the 1980s and 1990s. The mines, extending to depths of 100 m below the water table, were kept dry by pumping when the mines were in production. However, since mining stopped, some have flooded and started to overflow into nearby creeks. The water is very acidic because sulfide minerals in the mine react with oxygen and groundwater to form sulfuric acid, a problem known as *acid mine drainage* (see Chapter 11). The problem was so severe in the Tar Creek area of Oklahoma that in 1982 the Environmental Protection Agency designated it as the nation's foremost hazardous-waste site. Acid mine drainage is also a widespread problem in many of the eastern coal fields.

Impact on the Biological Environment Physical changes in the land, soil, water, and air associated with mining directly and indirectly affect the biological environment. Direct impacts include deaths of plants or animals caused by mining activity or contact with toxic soil or water from mines. Indirect impacts include changes in nutrient cycling, total biomass, species diversity, and ecosystem stability due to alterations in groundwater or surface water availability or quality. Periodic or accidental discharge of low-grade pollutants through failure of barriers, ponds, or water diversions or through breach of barriers during floods, earthquakes, or volcanic eruptions also damages local ecological systems.

Social Impact The social impact of large-scale mining results from a rapid influx of workers into areas unprepared for growth. Stress is placed on local services, including water supplies, sewage and solid-waste disposal systems, schools, and rental housing. Land use shifts from open range, forest, and agriculture to urban patterns. The stress on nearby recreation and wilderness areas, some of which may be in a fragile ecological balance, is also increased. Construction activity and urbanization affect local streams through sediment pollution, reduced water quality, and increased runoff. Air quality is diminished as a result of more vehicles, dust from construction, and generation of power.

Adverse social effects may result when miners are displaced by mine closures or automation, because towns surrounding large mines come to depend on the income of employed miners. In the old American West, mine closures produced the well-known "ghost towns." Today, the price of coal and other minerals directly affects the lives of many small towns, especially in the Appalachian Mountain region of the United States, where closures of coal mines are taking their toll. These mine closings result partly from lower prices for coal and partly from rising mining costs.

One of the reasons for the rising cost of mining is increased environmental regulation of the mining industry. Of course, regulations have also helped make mining safer and have facilitated land reclamation. Some miners, however, believe the regulations are not flexible enough, and there is some truth to their arguments. For example, some areas might be reclaimed for use as farmland following mining if the original hills have been leveled, but environmental regulations often require that the land be restored to its original hilly state, even though hills make inferior farmland.

Minimizing the Impact of Mineral Development

Technologically developed countries are making a good deal of progress in reversing the environmental damage done by mineral mining in the past and in minimizing the effects of new extraction and processing operations. Environmental laws regulate emissions and waste disposal and mandate restoration measures following mining operations. In addition, innovative technologies, particularly biotechnology, are providing less environmentally damaging ways of mining.

Environmental Regulation Most of the serious environmental degradation associated with mining in more developed countries is a relic of past practices that are now forbidden or restricted by environmental laws. For example, 100 years or so of smelting nickel ore in the Sudbury, Ontario, area produced a region of barren land of about 100 km^2. Another 350 km^2 or more of land was damaged (Figure 14.19). Extensive harm to land, water, and biologic resources resulted from air pollutants from the smelters that caused deposition of mercury, arsenic, and cadmium, among other metals, in the vicinity. Effluent from the smelters also contained tremendous quantities of sulfur dioxide and was therefore a major source of acid precipitation (6).

Smelters in the United States today must adhere to air quality emission standards of the Clean Air Act. As a result, U.S. smelters recover almost all of the sulfur dioxide from their emissions. Canada has enacted similar clean-air legislation. Smelters in the Sudbury area have reduced their emissions of sulfur dioxide by about 50 percent, and additional reductions are slated for the future. As a result of the decrease in the pollutants from the smelters, there has been some natural recovery. This has been augmented by planting of trees and adding lime to lakes to help neutralize acids.

▲ **FIGURE 14.19** Barren land near Lake St. Charles, Sudbury, Ontario, one of the largest sources of acid rain in North America resulting from emissions from smelters. Vegetation is killed by acid rain and deposition of toxic heavy metals. Tall stacks from the smelters are just barely visible on the horizon. (*Bill Brooks/Masterfile Corporation*)

These restoration measures, along with natural recovery, have resulted in revegetation of about 40 percent of the barren ground that surrounded the smelters (6).

Following mining activities, land reclamation is necessary if the mining has had detrimental effects and if the land is to be used for other purposes. Reclamation of land used for mining is required by law today, and approximately 50 percent of the land utilized by the mining industry in the United States has been reclaimed. Methods of mine reclamation will be discussed in Chapter 15, where we consider the impact of coal mining on the environment.

Biotechnology Several biological processes used for metal extraction and processing are likely to have important economic and environmental consequences. **Biotechnology,** using processes such as biooxidation, bioleaching, biosorption, and genetic engineering of microbes, has enormous potential for both extracting metals and minimizing environmental degradation (see *Case History: Homestake Mine, South Dakota*). Biotechnology is still in its infancy, and its potential uses are just beginning to be realized by the mining and metals industries.

Another promising biotechnology is *bioassisted leaching*, or *bioleaching*, which uses microorganisms to recover metals. In this technique, bacteria oxidize crushed gold ore in a tank, releasing finely disseminated gold that can then be treated by cyanide leaching. A commercial plant in Nevada was constructed to produce 50,000 troy ounces of gold annually by bioassisted leaching. This method is an attractive alternative from both an economic and environmental viewpoint to the cyanide-leaching process for gold extraction described earlier (19).

Biotechnology developed and tested by the U.S. Bureau of Mines is being used to treat acid mine drainage. Engineered (constructed) wetlands (Figure 14.20) at several hundred sites have utilized acid-tolerant plants to remove metals and neutralize acid by biological activity. Both oxidizing and sulfate-reducing bacteria also have an important role in the wetlands. Research is ongoing to develop new improved wetland designs that require little maintenance (19).

The Impact of Mineral Development: Summary and Conclusions

Considering that the demand for mineral resources will increase, the logical approach to environmental degradation is to minimize both on-site and off-site problems by controlling sediment, water, and air pollution through sound engineering and conservation practices. Although these actions will raise the cost of mineral commodities and hence the price of all items produced from these materials, they will yield other returns of equal or higher value to future generations. We must realize, however, that even the most careful measures to control environmental disruption associated with mining will occasionally fail.

The most-developed countries have relatively high pollution-abatement standards for the mineral industry, and environmental pollution is less likely to occur in these countries now than it has in the past. This is not necessarily true for less-developed countries attempting to exploit more of their mineral resources. These countries may not have the technology or the economic resources to ensure that their mining and mineral-processing activities do not cause serious pollution. The more-developed countries that have devised pollution-abatement strategies have the obligation to transfer this technology to less-developed nations to help protect their local and our global environment.

14.4 Recycling of Mineral Resources

A diagram of the cycle of mineral resources (Figure 14.21) reveals that many components of the cycle are connected to waste disposal. In fact, the major environmental impacts of mineral resource utilization are related to waste products.

CASE HISTORY Homestake Mine, South Dakota

The Homestake gold mine in South Dakota provides an interesting example of the recent use of biotechnology to clean up the environment degraded by mining activity. The objective of the Homestake study is to test the use of bacterial biooxidation to convert contaminants in water to substances that are environmentally safe (20).

The mining operation at Homestake discharges water from the gold mine to a nearby trout stream, and the untreated wastewater contains cyanide in concentrations harmful to the trout. The treatment process developed at the Homestake mine uses bacteria that have a natural capacity to oxidize the cyanide to harmless nitrates (19). The bacteria were collected from mine tailing ponds and cultured to allow biological activity at higher cyanide concentrations. They were then colonized on special rotating surfaces through which the contaminated water flowed before being discharged to the stream. The bacteria also extracted precious metals from the wastewater that could be recovered by further processing (19). The system at Homestake reduced the level of cyanide in the wastewater from about 10 ppm to less than 0.2 ppm, which is below the level required by water quality standards for discharge into the trout stream. Because the process of reducing the cyanide produced excess ammonia in the water, a secondary bacteria treatment was designed that converts the ammonia to nitrate compounds, so that the discharged water now meets stream-water quality criteria (19).

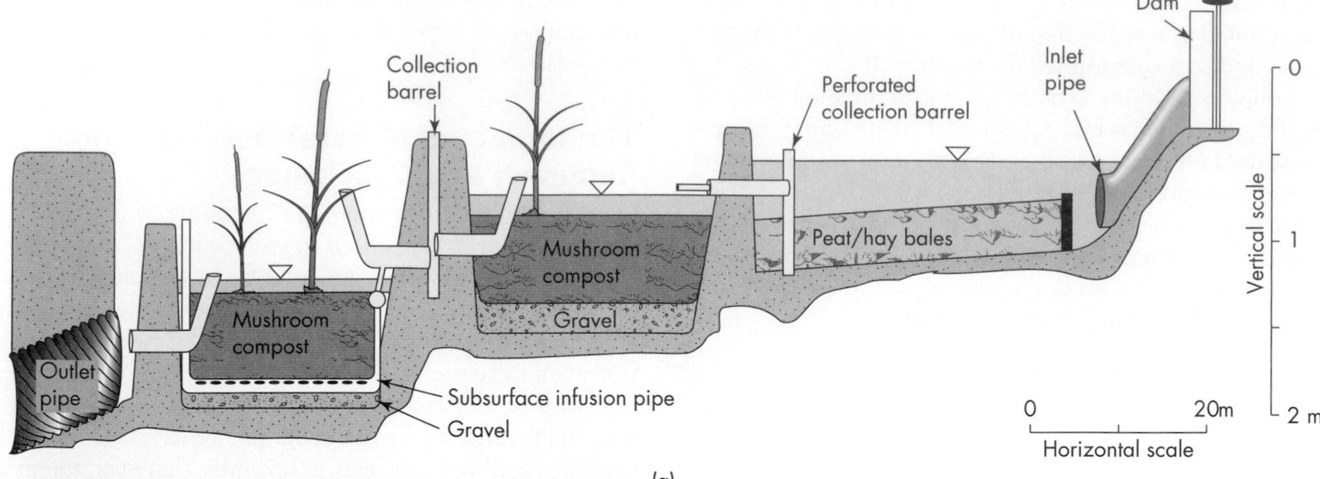

▲ **FIGURE 14.20** Idealized diagram (a) of wetlands constructed to use biotechnology for environmental cleanup of wastewater from mines. The plan calls for several shallow ponds lined with compost, topsoil, or crushed limestone. The plants are cattails. Bacteria live in the substrata of the ponds. (b) Photograph of artificially constructed wetlands. (*Source: Bureau of Mines, Pittsburgh Research Center*)

Wastes produce pollution that may be toxic to humans, are dangerous to natural ecosystems and the biosphere, and are aesthetically undesirable. They may attack and degrade other resources such as air, water, soil, and living things. Wastes also deplete nonrenewable mineral resources with no offsetting benefits for human society. Recycling of resources is one way to reduce these wastes.

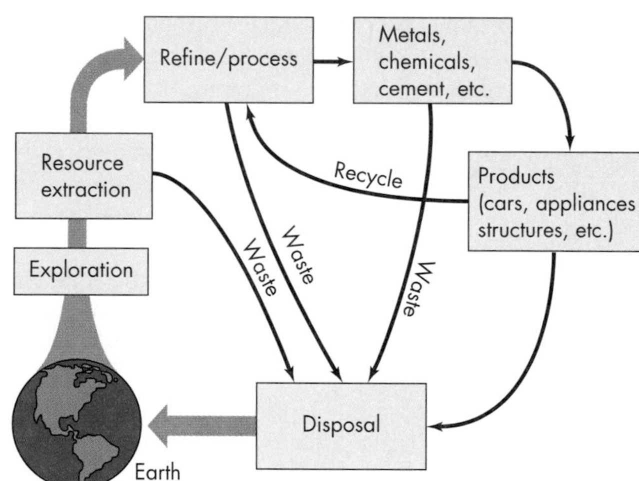

▲ **FIGURE 14.21** Simplified flow chart of the resource cycle.

Recycling Scrap Metal

The practice of recycling metals is not new. Metals such as iron, aluminum, copper, and lead have been recycled for many years. Of the millions of motor vehicles discarded annually, nearly all are dismantled by auto wreckers and scrap processors for metals to be recycled (21). Recycling metals from discarded autos is a sound conservation practice, considering that 90 percent by weight of the average discarded vehicle is metal.

Recycling of metals in 1997 was a $22 billion business in the United States. About 90 percent of all secondary (recycled) metal is iron (including steel). Aluminum is second, followed by copper, lead, and zinc (22). The reasons so much iron is recycled are (22):

- The market is huge, allowing for a large scrap collection and processing industry.
- Not to recycle would be an enormous economic burden.
- Not to recycle would have a profound environmental impact (73 million tons would have to be disposed of).

Figure 14.22 shows the approximate amounts of aluminum, copper, and lead recycled in the United States as a percentage of consumption for 1960, 1998, and 1997. The amount of recycled aluminum, copper, and lead has increased significantly. Today about 65 percent of the lead consumed is recycled, and the corresponding value for aluminum is 40 percent.

Urban Ore

Materials (especially metals) that end up in landfills and other waste-management facilities are sometimes designated as **urban ore** because of the useful materials they may contain

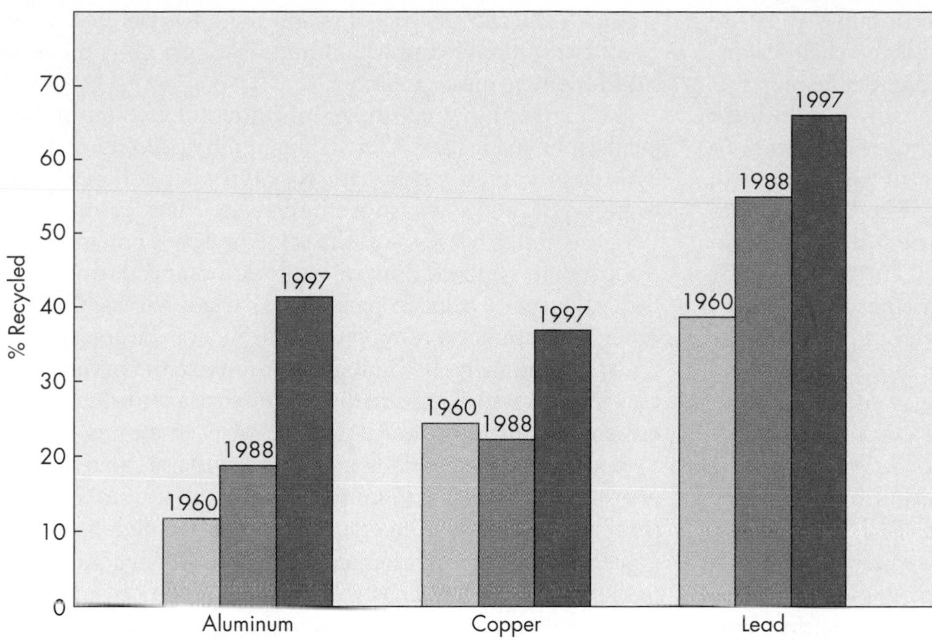

◀ **FIGURE 14.22** Approximate amount of recycled metals as a percentage of U.S. consumption in 1960 and 1992. U.S. consumption for 1992 was: aluminum, 5.5 million metric tons; copper, 2.3 million metric tons; and lead, 1.2 million metric tons. (*Data from* U.S. *Department of the Interior, Bureau of Mines*. 1993. Mineral commodity summaries 1993. I 28.148:993.)

(23). The concept of "urban ore" originated when it was discovered that ash from the incineration of sewage sludge in Palo Alto, California, contained large concentrations of gold (30 ppm), silver (660 ppm), copper (8000 ppm), and phosphorus (6.6%). Each metric ton of the ash contained approximately 1 troy ounce of gold and 20 ounces of silver. The gold was concentrated above natural abundance by a factor of 7500, making the "deposit" double the average grade that is mined today. Silver in the ash had a concentration factor of 9400, similar to that of rich ore deposits in Idaho, and copper had a concentration factor of 145, similar to that of a common ore grade. Commercial phosphorus deposits vary from 2 to 16 percent, so the ash with 6.6 percent phosphorus had the potential of a high-value resource (23).

The ash in the Palo Alto dump represented a silver and gold deposit with a value of about $10 million, and gold and silver worth approximately $2 million were being concentrated and delivered to the dump each year. The sources of the metals in the Palo Alto sewage were the large electronics industry and the photographic industry located in the area. Gold in significant amounts has been found in the sewage of only one other city, and silver is usually present in much smaller concentrations than at Palo Alto. Thus, Palo Alto's unique urban ore presented an unusual opportunity to study and develop methods to recycle valuable materials concentrated in urban waste (23). The city employed a private company to extract the gold and silver. By the early 1990s Palo Alto's industrial companies treated their wastewater to recover the gold and silver. As a result, the mines of Palo Alto are now shut down.

Not all sewage sludge is urban ore. Sludge that contains high concentrations of heavy metals such as cadmium is a toxic material and precludes the application of the sludge for uses such as land reclamation. More efficient pretreatment of industrial wastewater and strict regulations are necessary to avoid production of toxic sewage sludge from urban areas.

Recycling may be one way to delay or partially alleviate a possible resource crisis caused by the convergence of a rapidly rising population and a finite resource base. However, as discussed in Chapter 12, the problem of integrated waste management is complex, and before recycling can become more widespread, improved technology and more economic incentives are needed. Nevertheless, the trends are clearly in place, and the volume of resources recycled will continue to grow.

SUMMARY

Availability of mineral resources is one measure of the wealth of a society. Modern technological civilization as we know it would not be possible without exploitation of these resources. However, we cannot maintain exponential population growth on a finite resource base. Because of the slowness of geological processes, mineral deposits must be considered nonrenewable resources.

A mineral resource is a concentration of a naturally occurring earth material in a form that makes extraction currently or potentially feasible. A mineral reserve is that portion of a mineral resource that is currently available, legally and economically, for extraction. It is important to remember that not all resources are reserves. Unless discovered and captured, resources cannot be used to solve present shortages.

Mineral resources can be classified according to their use as metals, building materials, minerals for the chemical industry, and minerals for agriculture. Nonmetallic minerals are consumed at much greater rates than any metal except iron. When a mineral becomes scarce, the choices are to find more sources, recycle what has already been obtained, find a substitute, use less, or do without. The United States and other affluent nations have insufficient domestic supplies of many mineral resources for current use and must supplement them by imports from other countries. As these countries industrialize and develop, such imports may be more difficult to obtain, and affluent nations may have to find substitutes for some minerals or use a smaller portion of the world's annual production. These adjustments are occurring in the United States, where there has been a 12 percent reduction in per capita consumption of aluminum, copper, and lead over the past 20 years.

The geology of mineral resources is complex and intimately related to various aspects of the geologic cycle. Mineral resources are generally extracted from ores, the name given to naturally occurring anomalously high concentrations of earth materials that can be profitably extracted. To be classified as an ore, a mineral-bearing deposit must have a specific concentration factor, which is the ratio of the mineral's concentration in ore to its average concentration in the earth's crust. The concentration factor of a given mineral reflects both geologic and economic circumstances. Many metallic ores are produced by dynamic earth processes that occur at junctions of lithospheric plates.

Mineral resources for construction and industrial uses are concentrated in the geologic environment by igneous or magmatic processes, such as crystal settling or hydrothermal activity; by metamorphic processes, such as contact and regional metamorphism; by sedimentary processes, including oceanic and lake processes, running water, wind, and moving ice; by biological processes, such as concentration of minerals in bones and teeth; and by weathering processes, including residual ore deposits of insoluble minerals and secondary enrichment of ores. The ocean, which is already the chief source of magnesium, has rich resources of other exploitable important minerals, including metallic sulfides, manganese, and cobalt.

The environmental impact of mineral exploitation depends upon many factors, including mining procedures, local hydrologic conditions, climate, rock types, size of operation, topography, and many more interrelated factors. In addition, the impact varies with the stage of development of the resource. In general, mineral exploration and testing does little damage, except in particularly fragile areas. On the other hand, mineral mining can have major adverse effects on land, water, air, and biological resources; the mining can also initiate social impacts on the environment due to increasing demand for housing and services in mining areas.

Because the demand for mineral resources is going to increase, we must strive to minimize both on-site and off-site problems caused by mineral development through good engineering and conservation practices. The recent application of biotechnology to metal extraction and pollution reduction shows real promise. Environmental degradation associated with mining and mineral processing in the more-developed countries has been much reduced in recent years owing to development of pollution-abatement strategies and legislation to mandate improved pollution-control measures and land reclamation. Such technologies and regulations are not necessarily present in less-developed countries that are striving to develop their mineral resources. It is the responsibility of more highly industrialized nations to transfer technology so that environmental degradation related to mining activities is minimized at the local, regional, and global levels.

Recycling of mineral resources is one way to delay or partly alleviate a crisis caused by the convergence of a rapidly rising population and a limited resource base. Metals used in large quantities by industry have long been recycled in the United States, with iron representing about 90 percent of recycled metal. Recycling a wide variety of materials is not an easy task, and innovative refinements in recycling methods will be necessary to ensure that the recycling trend continues.

REFERENCES

1. **U.S. Department of Interior, Bureau of Mines.** 1991. *Minerals in 1991.* I 28.156/3:991.
2. **Barsotti, A. F.** 1992. Wake up and smell the coffee. *Minerals today,* U.S. Department of Interior, Bureau of Mines, October, pp. 12–17.
3. **U.S. Department of Interior, Bureau of Mines.** 1993. *Mineral commodity summaries, 1993.* I 28.148:993.
4. **Brobst, D. A., Pratt, W. P, and McKelvey, V. E.** 1973. *Summary of United States mineral resources.* U.S. Geological Survey Circular 682.
5. **U.S. Geological Survey.** 1975. *Mineral resource perspectives 1975.* U.S. Geological Survey Professional Paper 940.
6. **Kesler, S. F.** 1994. *Mineral resources, economics and the environment.* New York: Macmillan.
7. **Meyer, H. O. A.** 1985. Genesis of diamond: A mantle saga. *American Mineralogist* 70:344–55.
8. **NOAA.** 1977. Earth's crustal plate boundaries: Energy and mineral resources. *California Geology* 30(5):108–109.
9. **Bateman, A. M.** 1950. *Economic ore deposits,* 2nd ed. New York: John Wiley.
10. **Park, C. F., Jr., and MacDiarmid, R. A.** 1970. *Ore deposits,* 2nd ed. San Francisco: W. H. Freeman.
11. **Smith, G. I., Jones, C. L., Culbertson, W. C., Erickson, G. E., and Dyni, J. R.** 1973. Evaporites and brines. In *United States mineral resources,* eds. D. A. Brobst and W. P. Pratt, pp. 197–216. U.S. Geological Survey Professional Paper 820.
12. **Lowenstam, H. A.** 1981. Minerals formed by organisms. *Science* 211:1126–30.

13. **Cornwall, H. R.** 1973. Nickel. In *United States mineral resources*, eds. D. A. Brobst and W. P. Pratt, pp. 437–42. U.S. Geological Survey Professional Paper 820.
14. **Foster, R. J.** 1983. *General geology*, 4th ed. Columbus, OH: Charles E. Merrill.
15. **Van, N., Dorr, J., Crittenden, M. D., and Worl, R. G.** 1973. Manganese. In *United States mineral resources*, eds. D. A. Brobst and W. P. Pratt, pp. 385–99. U.S. Geological Survey Professional Paper 820.
16. **McGregor, B. A., and Lockwood, M.** (no date). *Mapping and research in the exclusive economic zone.* U.S. Geological Survey and NOAA.
17. **U.S. Department of the Interior, Bureau of Mines.** 1991. *Research 92. Biotechnology—Using nature to clean up wastes.* I 28.115:992, pp. 16–21.
18. **Silva, M. A.** 1988. Cyanide heap leaching in California. *California Geology* 41(7):147–56.
19. **Jeffers, T. H.** 1991. Using microorganisms to recover metals. *Minerals today,* U.S. Department of Interior, Bureau of Mines, June, pp. 14–18.
20. **Haynes, B. W.** 1990. Environmental technology research. *Minerals today,* U.S. Bureau of Mines, May, pp. 13–17.
21. **Davis, F. F.** 1972. Urban ore. *California Geology* 25(5): 99–112.
22. **Staff, Division of Mineral Commodities.** 1994. *Recycled metals in the United States.* U.S. Department of the Interior, Bureau of Mines. Special Publication I 28.151: M56.
23. **Gulbrandsen, R. A., Rait, N., Dries, D. J., Baedecker, P. A., and Childress, A.** 1978. *Gold, silver, and other resources in the ash of incinerated sewage sludge at Palo Alto, California—A preliminary report.* U.S. Geological Survey Circular 784.

KEY TERMS

resource (p. 375)
reserve (p. 375)
ore (p. 379)
concentration factor (p. 379)
kimberlite (p. 381)
pegmatite (p. 382)
hydrothermal (p. 382)
contact metamorphism (p. 385)
regional metamorphism (p. 385)
placer deposits (p. 386)
evaporite (p. 386)
salt domes (p. 388)
bauxite (p. 389)
manganese oxide nodules (p. 391)
biotechnology (p. 395)
urban ore (p. 396)

SOME QUESTIONS TO THINK ABOUT

1. Make an inventory of the mineral and mineral products in your home or workplace. Which are most important in terms of the total volume involved, and which are used most frequently? Based upon your inventory and examination of Table 14.3, how reliant are you on foreign sources for the mineral resources you commonly use?
2. Today, technological changes are coming from all directions. There is electronic mail (e-mail) that may lead us to a paperless society. We are now hearing about e-money, a technology that allows us to pay our bills via personal computers. Some journals and newspapers we subscribe to come over the Internet, and we may store electronically what we really want to keep. What is the impact of this technology be on resource utilization? To answer this question, start by constructing a list of the resources from mineral sources necessary to support our present mostly paper-driven society. Then develop a plan to evaluate the impact on natural resources of the transformation to a paperless society.
3. Because we know that mineral resources are finite, there are two ways to look at our present and future use of these resources. One view is that we are headed toward a mineral crisis as the world's population increases. The other is that the presence of more people heightens the possibility for innovations to increase, and we will therefore find ways to adjust our use of minerals to an expanding population. How could these two hypotheses be tested?
4. Biotechnology and genetic engineering are potential tools for cleaning up the environment. We saw some examples of biotechnology in this chapter: Bacteria can be cultured to neutralize acids in effluent from mines, and artificial wetlands can be constructed in which biological processes purify water polluted by mineral processing and mining. What do you think of this technology and how might it be transferred to the mineral industry in the United States and other countries?

15 Energy and Environment

Offshore oil platform. (*Kim Steele/PhotoDisc, Inc.*)

People of the industrialized world today are hooked on oil as a primary source of energy. We have searched the land and oceans of the world, drilling deep wells to extract oil that is a resource (gift) from our planet's geologic past. The so-called "Oil Age" began with the first automobiles in the late 1800s. Oil also fueled the agricultural revolution through the production of synthetic fertilizers that have allowed us to feed the ever-expanding human population. Although in the past there have been warnings that we are running out of oil or that it is becoming scarce, there has seemed to be sufficient amounts. However, it is believed that depletion of oil will follow after a peak in discovery, and that discovery peaked in the 1960s. Today it is believed we have extracted approximately one-half of all the oil that is likely found within the entire planet. Shortages are expected to develop, perhaps as soon as 20 years or so following peak global production of oil. It is ironic that shortage will likely follow closely peak global production—but that's what is being suggested.

15.1 Energy Supply and Energy Demand

Americans encountered the effects of post-World-War-II energy shortages for the first time in the 1970s, including increases in the prices of energy and products produced from petroleum. Nevertheless, to many people energy still seems

LEARNING OBJECTIVES

Our discussion of mineral resources established that resources are not infinite, and that a finite resource base cannot support an exponential increase in population. The same applies to energy derived from mineral resources. Learning objectives of this chapter are:

- *To become familiar with the general patterns of energy consumption in the United States in terms of energy sources.*
- *To know the types and distribution of the major fossil fuels and the environmental impact associated with their development.*
- *To gain a modest acquaintance with nuclear energy and the important environmental issues associated with it.*
- *To know what geothermal energy is, how it is produced, and its future as an energy source.*
- *To know the main types of renewable energy and the environmental significance of each.*
- *To become familiar with important issues related to energy policy, particularly the difference between hard path and soft path and what constitutes sustainable energy development.*

Web Resources

Visit the "Environmental Geology" Web site at www.prenhall.com/keller to find additional resources for this chapter, including:

- ▶ Web Destinations
- ▶ On-line Quizzes
- ▶ On-line "Web Essay" Questions
- ▶ Search Engines
- ▶ Regional Updates

unlimited, especially given the oil glut of the 1980s and mid-1990s. The United States continues to consume a disproportionate share of the total energy produced in the world: With only 5 percent of the world's population, the United States accounts for about 25 percent of the world's total energy consumption.

Examination of Figure 15.1a reveals that nearly 90 percent of the energy consumed in the United States today is produced from coal, natural gas, and petroleum (oil), which are sometimes called the *fossil fuels* because of their organic origin. As we saw in the last chapter, these fuels must be considered nonrenewable resources. The remaining 10 percent of energy consumed comes from hydropower and, more recently, nuclear power. We still have huge reserves of coal, but major new sources of natural gas and petroleum are becoming scarce. In fact, we import approximately the same amount of oil that we produce, making us very vulnerable to changing world conditions that affect the available supply of crude oil. Few new large hydropower plants can be expected, and planning and construction of new nuclear power plants have become uncertain for a variety of reasons. On the brighter side, alternative energy sources such as solar power for homes, farms, and offices are becoming economically more feasible and thus more common.

Figure 15.1a provides some interesting information about changes in energy consumption in the United States. From 1950 through 1974, there was a sharp increase in energy consumption, but following the shortages of the mid-1970s the rate of increase declined dramatically. The basic unit for energy expenditure is the joule. One joule is defined as a force of one Newton applied over a distance of 1 m. From 1950 through 1975, total U.S energy consumption increased from 30 to approximately 70 exajoules. (One exajoule is equal to 10^{18} joules and approximately equal to one quad, or 10^{15} Btu.) The units for power are time rate of energy or joules per second. One joule per second is a watt. A large power plant produces about 1000 megawatts, which is one billion watts. Electrical energy is commonly sold by kilowatt-hours. This unit is 1000 watts applied over one hour (3600 seconds). Thus, one kilowatt-hour is 3,600,000 joules or 3.6 megajoules. Since 1980 energy consumption has in creased by only about 15 exajoules. This suggests that energy conservation policies, such as requiring new automobiles to be more fuel-efficient and buildings to be better insulated, have been at least partially successful. Nevertheless, at the end of the twentieth century, we remain hooked on the fossil fuels. Further progress may depend on our realizing that quality of life is not directly related to amount of energy consumed.

Projections of energy supply and demand are difficult at best because the technical, economic, political, and social assumptions that underlie such projections are constantly changing. It is clear, however, that we must continue to research, develop, and evaluate potential energy sources and conservation practices to ensure sufficient energy to maintain our industrial society and a quality environment. Of particular importance will be energy uses with applications below 100 °C because a large portion of the total energy consumption (for uses below 300 °C) in the United States is for space heating and water heating (Figure 15.1b). With these ideas in mind, we will cautiously explore some selected geologic and environmental aspects of such well-known energy resources as coal, petroleum, and nuclear sources, as well as potentially important sources such as oil

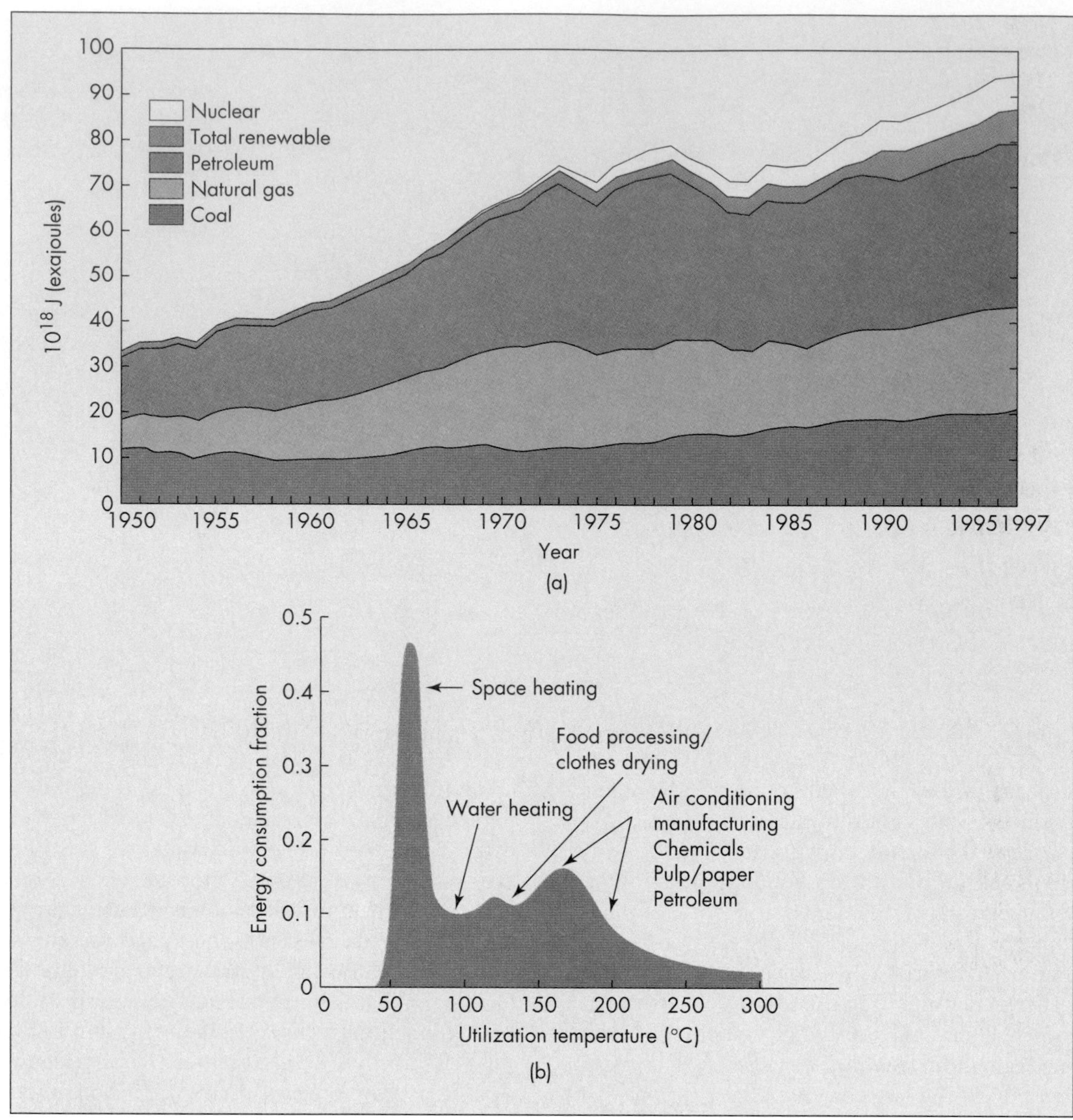

▲ **FIGURE 15.1** (a) Energy consumption for the United States from 1949 to 1997. (b) Spectra of energy use below 300 °C in the United States. (*Data for [a] from Energy Information Administration*, 1994. *Part [b] from* Los Alamos *Scientific Laboratory* [L.A.S.L. 78-24], 1978.)

shale, tar sands, and geothermal resources. We will also discuss renewable energy sources such as hydropower, wind, and solar power.

15.2 Fossil Fuels

The origin of fossil fuels—coal, oil, and gas—is intimately related to the geologic cycle. These fuels are essentially stored solar energy in the form of organic material that has escaped total destruction by oxidation. In the United States today we are dependent upon the fossil fuels for nearly all of the energy we consume. This is changing—but slowly.

The environmental disruption associated with exploration and development of fossil fuels must be weighed against the benefits gained from the energy, but this is not an either-or proposition. Good conservation practices combined with pollution control and reclamation can help minimize the environmental disruption associated with fossil fuels.

Coal

Coal is one of the major fossil fuels. Burning coal accounts for about 20 percent of the total U.S. energy consumption. As we shall discuss, the environmental costs of coal consumption are significant.

Geology of Coal Like other fossil fuels, coal is made up of organic materials that have escaped oxidation in the carbon cycle. Coal is essentially the altered residue of plants that flourished in ancient freshwater or brackish-water swamps, typically found in estuaries, coastal lagoons, and low-lying coastal plains or deltas (1).

(a) Coal swamp forms.

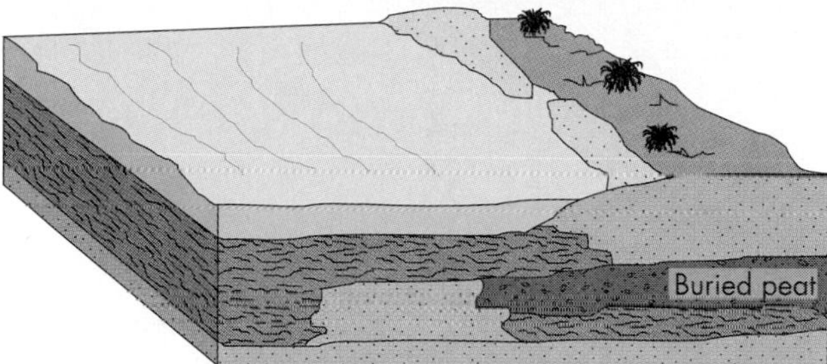

(b) Rise in sea level buries swamp in sediment.

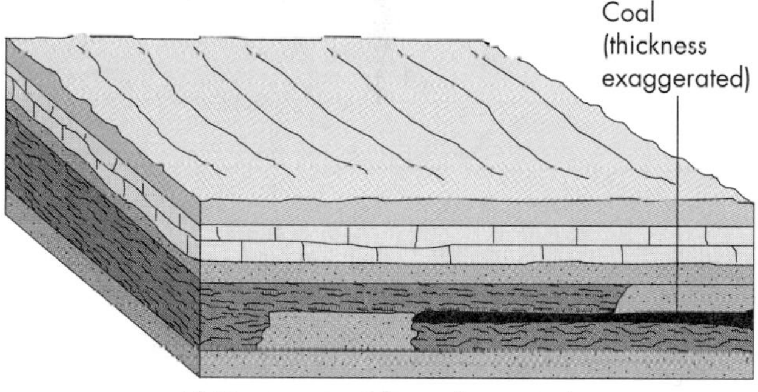

(c) Compression of peat forms coal.

▲ **FIGURE 15.2** The processes by which buried plant debris (peat) is transformed into coal. Considerable lengths of geologic time must elapse before the transformation is complete.

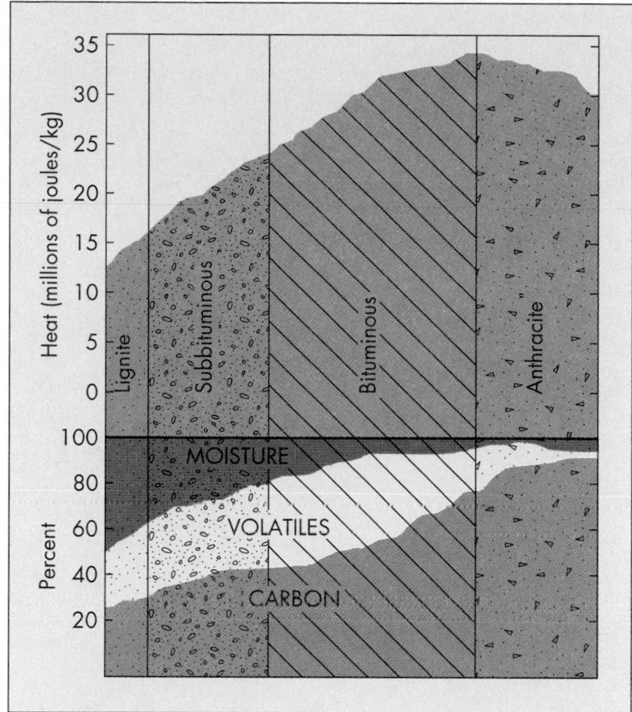

▲ **FIGURE 15.3** Generalized classification of different types of coal based upon their relative content (in percentage) of moisture, volatiles, and carbon. The heat values of the different types of coal are also shown. (*After* D. A. *Brobst and* W. P. *Pratt, eds.* 1973. U.S. *Geological Survey Professional Paper* 820.)

Coal-forming processes (Figure 15.2) start in swamps rich in plant life, where water-saturated soils exclude much of the oxygen normally present in soil. The plants partially decompose in this oxygen-deficient environment and accumulate slowly to form a thick layer of peat. The swamps and accumulations of peat may then be inundated by a prolonged slow rise of sea level (a relative rise, as the land may be sinking) and covered by sediments such as sand, silt, clay, and carbonate-rich material. As more and more sediment is deposited, water and organic gases (volatiles) are squeezed out, and the percentage of carbon increases in the compressed peat. As this process continues, the peat is eventually transformed to coal. Because there are often several layers of coal in the same area, scientists believe the sea level may have alternately risen and fallen, allowing development and then drowning of coal swamps.

Classification and Distribution of Coal Coal is commonly classified according to rank and sulfur content. The rank is generally based on the percentage of carbon and heat value on combustion. Percent carbon increases from lignite to subbituminous to bituminous to anthracite, as shown in Figure 15.3. This figure also shows that heat content is maximum in bituminous coal, which has relatively few volatiles (oxygen, hydrogen, and nitrogen) and low moisture content compared to subbituminous coal. Heat content is minimum in lignite, which has a high moisture content. The distribution of the common coals (bituminous, subbituminous, and lignite) in the conterminous United States is shown in Figure 15.4a. The distribution of world coal reserves, which amount to about 1000 billion metric tons, is shown in Figure 15.4b. The United States has about 25 percent of these reserves. The annual world consumption of about 4 billion tons suggests that at the present rate of consumption, known reserves are sufficient for about 250 years (2). However, if coal consumption increases, as is likely in countries undergoing intensive industrialization, such as China, reserves may be depleted much sooner (3).

The *sulfur content* of coal may be generally classified as low (zero to 1%), medium (1.1 to 3%), or high (greater than 3%). With all other factors equal, the use of low-sulfur coal as a fuel for power plants causes the least air pollution (emission of sulfur dioxide, SO_2). Most coal in the United States is of the low-sulfur variety (Table 15.1); however, by far the most common low-sulfur coal is a relatively low-grade, subbituminous variety found west of the Mississippi River (4). To avoid air pollution, thermal electric power plants on the highly populated East Coast of the United States will have

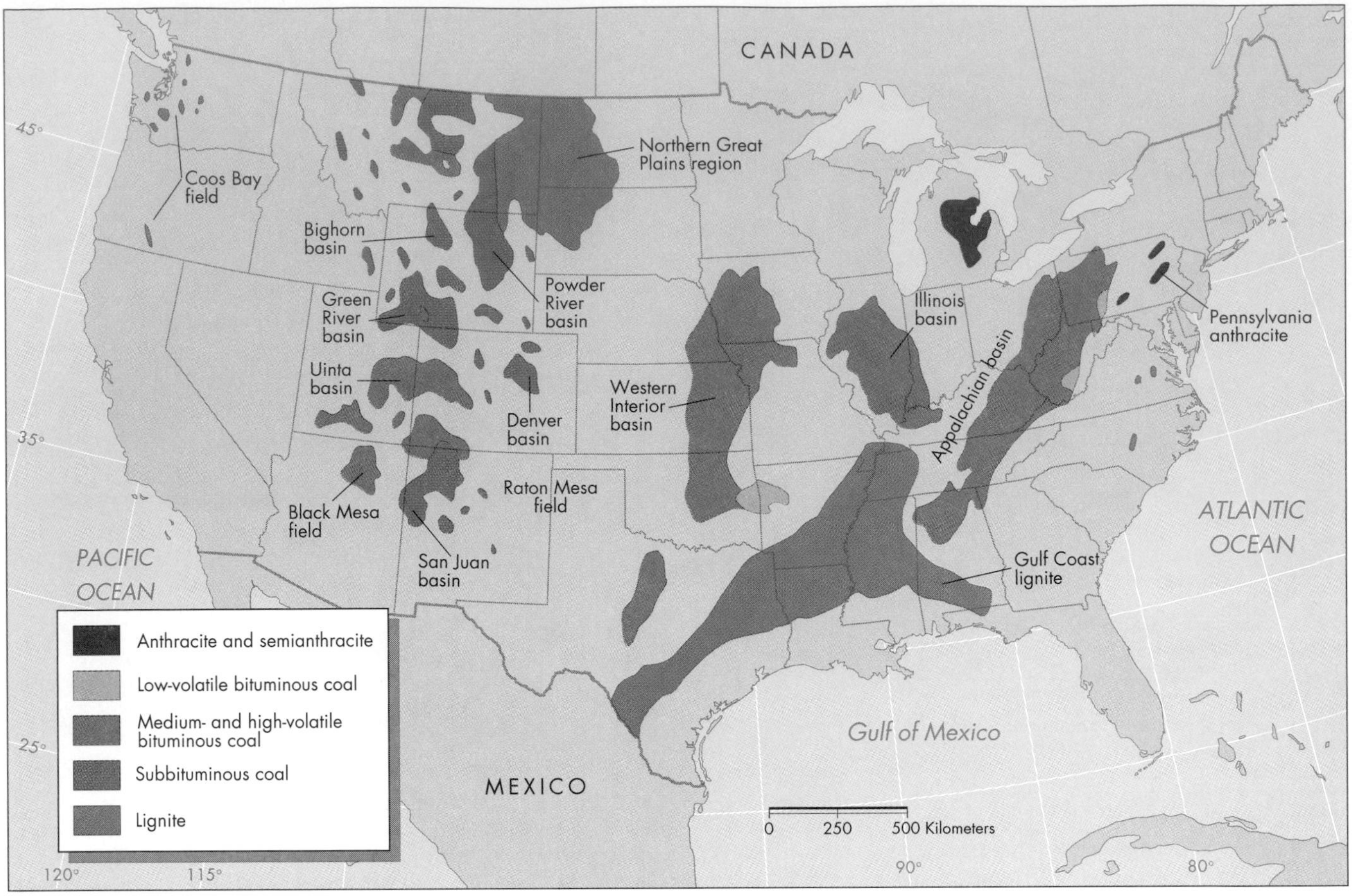

(a)

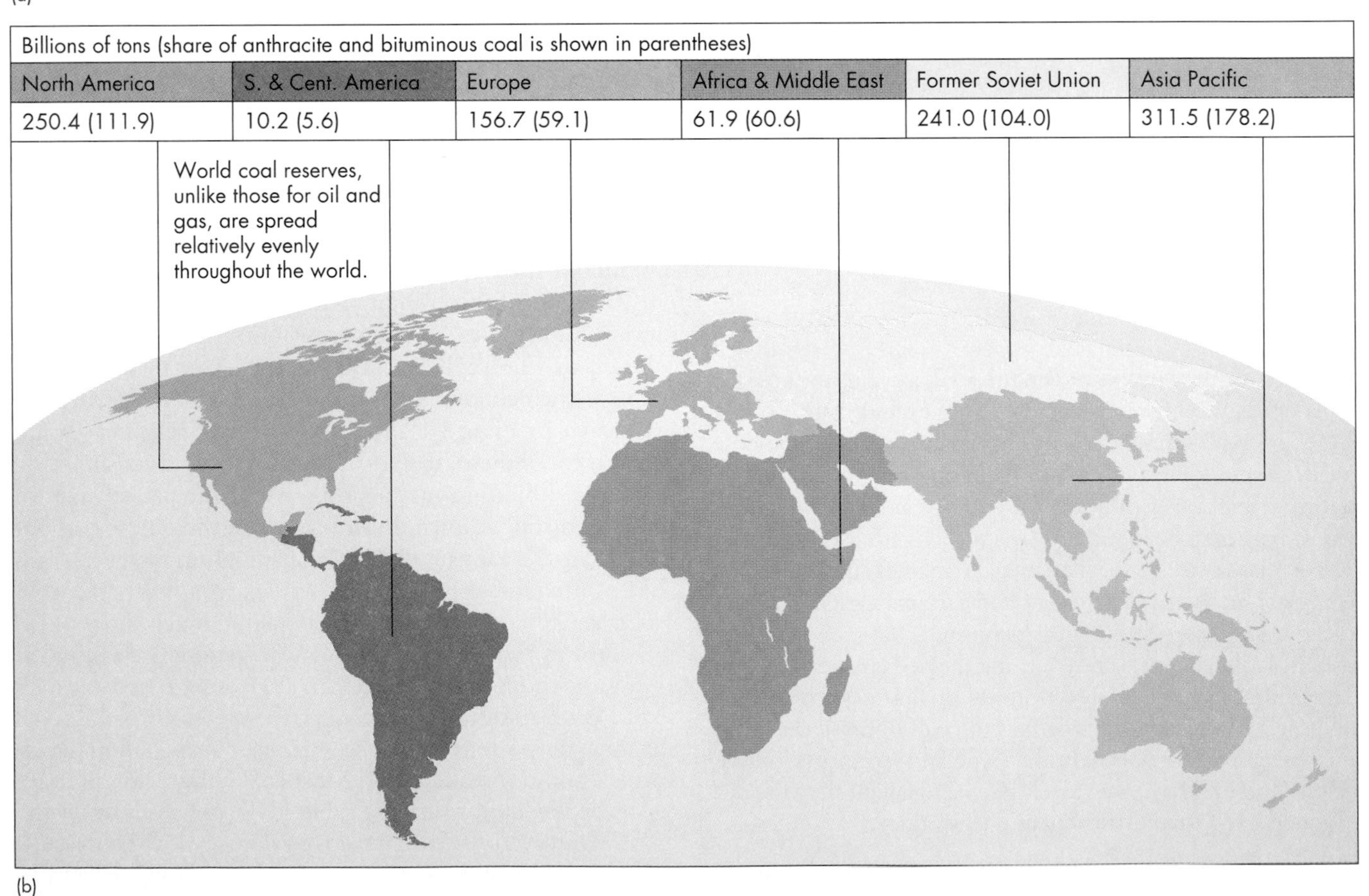

Billions of tons (share of anthracite and bituminous coal is shown in parentheses)					
North America	S. & Cent. America	Europe	Africa & Middle East	Former Soviet Union	Asia Pacific
250.4 (111.9)	10.2 (5.6)	156.7 (59.1)	61.9 (60.6)	241.0 (104.0)	311.5 (178.2)

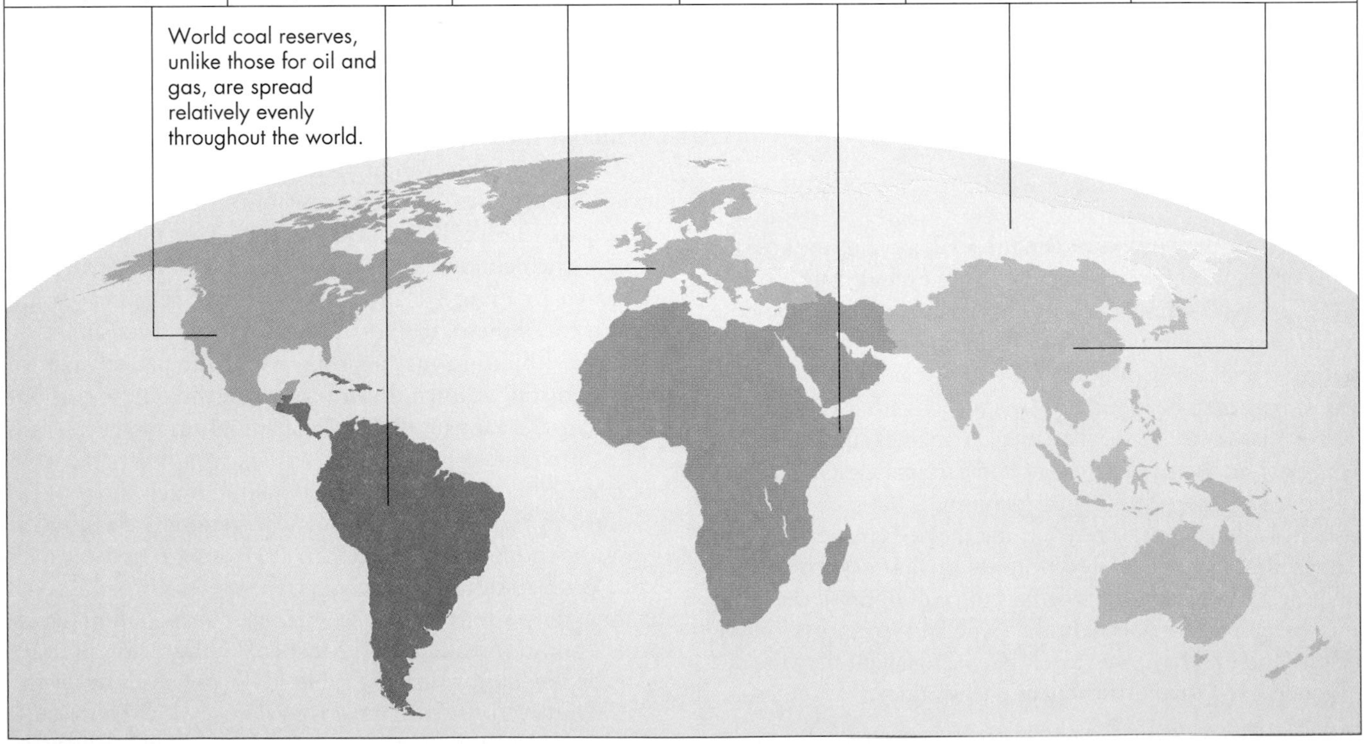

World coal reserves, unlike those for oil and gas, are spread relatively evenly throughout the world.

(b)

▲ **FIGURE 15.4** Coal areas of the conterminous United States (a); and world proven reserves (b). (*Sources:* [*a*] S. *Carbini and* S. P. *Schweinfurth.* 1986. U.S. *Geological Survey Circular* 979; *and* [*b*] *British Petroleum.* 1997. BP *Statistical Review of World Energy.*)

Table 15.1 Distribution of U.S. coal resources according to their rank and sulfur content

	Sulfur Content (Percent)		
Rank	Low 0–1	Medium 1.1–3.0	High 3+
Anthracite	97.1	2.9	—
Bituminous coal	29.8	26.8	43.4
Subbituminous coal	99.6	0.4	—
Lignite	90.7	9.3	—
All ranks	65.0	15.0	20.0

Source: U.S. Bureau of Mines Circular 8312, 1966

▲ **FIGURE 15.5** Acid-rich water draining from an exposed coal seam is polluting this stream in Ohio. (*Kent & Donna Dannen/Photo Researchers, Inc.*)

to continue to treat local coal to lower its sulfur content before burning or capture the sulfur after burning by a process such as scrubbing (discussed in Chapter 17). These treatments increase the cost of thermal electric power, but they may be more economical than shipping low-sulfur coal long distances.

Impact of Coal Mining Much of the coal mining in the United States is still done underground, but **strip mining** (open-pit), which started in the late nineteenth century, has steadily increased, whereas production from underground mines has stabilized. Strip mining is in many cases technologically and economically more advantageous than underground mining. The increased demand for coal will lead to more and larger strip mines to extract the estimated 40 billion metric tons of coal reserves that are now accessible to surface mining techniques. In addition, approximately another 90 billion metric tons of coal within 50 m of the surface is potentially available for stripping if need demands.

The impact of large strip mines varies by region depending on topography, climate, and, most importantly, reclamation practices. In humid areas with abundant rainfall, mine drainage of acid water is a serious problem (Figure 15.5). Surface water infiltrates the spoil banks (material left after the coal or other minerals are removed), where it reacts with sulfide minerals such as pyrite (FeS_2) to produce sulfuric acid. The sulfuric acid then runs into and pollutes streams and groundwater resources. Although acid water also drains from underground mines and roadcuts and areas where coal and pyrite are abundant, the problem is magnified when large areas of disturbed material remain exposed to surface waters. Acid drainage can be minimized through proper use of water-diversion practices that collect surface runoff and groundwater before they enter the mined area and divert them around the potentially polluting materials. This practice reduces erosion, pollution, and water-treatment cost (5).

In arid and semiarid regions, water problems associated with mining are not as pronounced as in wetter regions, but the land may be more sensitive to mining activities such as exploration and road building. In some arid areas, the land is so sensitive that even tire tracks across the land survive for years. Soils are often thin, water is scarce, and reclamation work is difficult.

Common methods of strip mining include **area mining**, which is practiced on relatively flat areas (Figure 15.6), and **contour mining**, which is used in hilly terrain (Figure

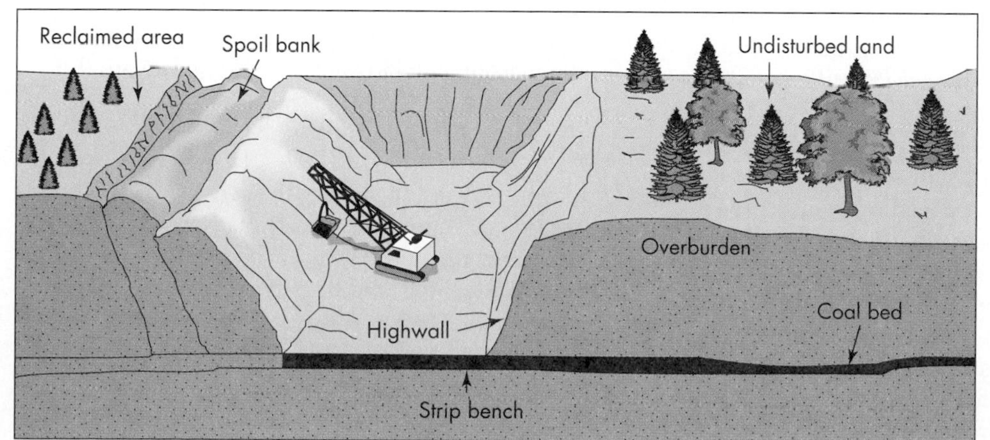

◀ **FIGURE 15.6** Diagram of area strip mining.

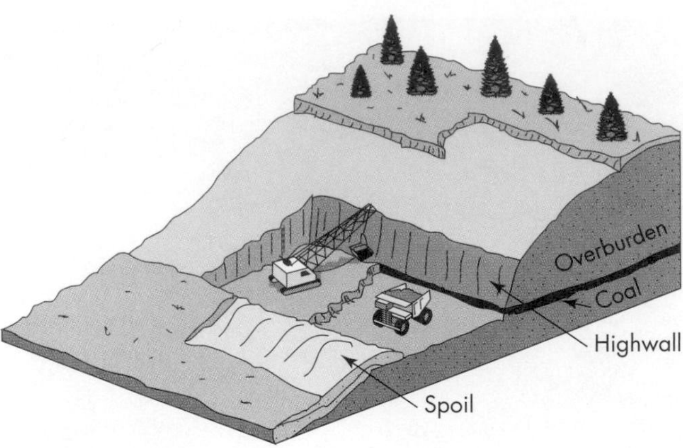

▲ **FIGURE 15.7** Diagram of contour strip mining.

15.7). The width of the cut in contour mining depends on the ease of excavation and topography. As the width increases, more and more overburden must be removed, which increases the cost. Topography thus limits the total width of the cut. The objective is to confine the cut to a given elevation (contour) and work around the hill while maintaining that elevation. Both methods of strip mining severely disturb the landscape by removing indigenous vegetation, overburden, and desired minerals. Unless the mined land is reclaimed, unsightly conical piles or ridges of waste (spoil) remain as a source material to pollute groundwater and surface water. Area mining is used only on relatively flat ground where the potential velocity and erosion of runoff are both low. Therefore, the potential to pollute streams by siltation is less than for contour mining. Potential groundwater pollution is greater for area mining, however, because the increased area allows more precipitation to infiltrate and slowly migrate through the spoil piles (5).

All methods of strip mining have the potential to pollute or destroy scenic, water, biologic, or other land resources, but sound reclamation practices can minimize the damage. The best method is to segregate the overburden to eventually replace the topsoil that is removed. This method has been used rather widely in the coal fields of the eastern United States, and experience has shown that it is a successful way to control water pollution when combined with regrading and revegetation (5).

Federal guidelines now govern strip mining of coal in the United States. They require, basically, that mined land be restored to support its premining use (see *Case History: A Tale of Two Mines*). Restoration includes disposing of wastes, contouring the land, and replanting vegetation. The hope is that after reclamation the mined land will appear and function as it did prior to extraction of the coal; however, as previously mentioned, this task will be difficult and probably not completely successful. The new regulations also prohibit strip mining on prime agricultural land and give farmers and ranchers the opportunity to restrict or veto mining on their land, even if they do not own the mineral rights.

Underground mining of coal and other resources has also caused considerable environmental degradation from

▲ **FIGURE 15.8** An underground fire in an abandoned coal mine in Pennsylvania has melted the snow near this vent. Note the smoke escaping to the atmosphere. (*National Institute for Occupational Safety & Health*)

mine drainage of acid water and has produced serious hazards, such as subsidence over mines or fires in mines (Figure 15.8). In the past, waste material (spoil) from underground mines has been piled on the surface. This produces aesthetic degradation as well as sediment and chemical pollution of water resources resulting from exposure of the mine spoil to surface water and groundwater, producing acid mine drainage, as discussed earlier with strip mining (see also Chapter 11).

Future Use and Environmental Impacts of Coal Limited resources of oil and natural gas are increasing the demand for coal. The crunch on oil and gas supplies is still years away, but when it does come, it may put pressure on the coal industry to open more and larger mines in both the eastern and the western coal beds of the United States. This could have significant environmental impacts for several reasons (6):

- More and more land will be strip-mined and will thus require careful restoration.
- Unlike oil and gas, burned coal leaves ash (5 to 20% of the original amount of the coal) that must be collected and disposed of. Some ash can be used for landfill or other purposes, but about 85 percent is presently useless.
- Handling of tremendous quantities of coal through all stages—mining, processing, shipping, combustion, and final disposal of ash—will have potentially adverse environmental effects, including air pollution; release of

trace elements likely to cause serious health problems into the water, soil, and air; and aesthetic degradation at mines, power plants, and other facilities associated with the coal industry.

The transport of large amounts of coal or of energy derived from coal from production areas to large population centers is a significant environmental issue. Coal can be converted on site to electricity, synthetic oil, or synthetic gas, all of which are relatively easy to transport, but with few exceptions these alternatives present problems. Methods of transporting large volumes of coal for long distances include ships, freight trains, and coal-slurry pipelines. Trains have the advantage of a relatively low cost for new capital expenses and thus will continue to be used. In the United States, trains transport about 75 percent of coal shipments, with barges and ships accounting for another 10 percent (8).

Coal-slurry pipelines, designed to use water to transport pulverized coal, have recently been used in Arizona and Ohio (8). These pipelines are relatively short (437 km and 173 km, respectively) compared to pipelines that would be necessary to carry western coal to markets in the eastern United States. A major problem with water transport of coal is that water is scarce in the western United States (see Chapter 10) and coal-slurry pipelines require a lot of water. For example, a pipeline to transport 30 million metric tons of coal per year would require about 20 million m^3 of water annually. This volume of water can meet the water supply needs of a city of 85,000 people or irrigate 40 km^2 of farmland.

Environmental problems associated with coal, while significant enough to cause concern, are not necessarily insurmountable, and careful planning could minimize them. As world trade increases, new mines with cleaner coal will also become more available. For example, coal mines being developed in Borneo are mining a low grade but very low-sulfur coal that has been termed "solid natural gas" because it is so pure a source of energy. These clean coal deposits in Borneo are up to 66 m thick with only a few meters of overburden (8). Regardless of sources and quality of coal, there may be few alternatives to mining tremendous quantities of coal to feed thermoelectric power plants and to provide oil and gas by gasification and liquefaction processes in the future. An important objective is to find ways to use coal that minimize environmental disruption.

Hydrocarbons: Oil and Gas

Oil (petroleum or crude oil) and **natural gas** are *hydrocarbons*, made up of carbon, hydrogen, and oxygen. Natural energy gas is mostly methane, CH_4 (usually it is more than 80% of the energy gases present at a location). Other natural energy gases include *ethane*, *propane*, *butane*, and *hydrogen* (9). Like coal, the hydrocarbons are, for the most part, fossil fuels in that they form from organic material that has escaped complete decomposition after burial. Oil and natural gas are found in concentrated deposits, which have been heavily mined (obtained from wells), and also in less readily available form in earth materials known as *oil shales* and *tar sands*.

Geology of Oil and Gas Deposits Next to water, oil is the most abundant fluid in the earth's crust, yet the processes that form it are only partly understood. Most earth scientists accept that oil and natural gas are derived from organic materials that are buried with marine or lake sediments. Favorable environments where organic debris might escape oxidation include nearshore areas characterized by rapid deposition that quickly buries organic material or deeper-water areas characterized by a deficiency in oxygen at the bottom that promotes an anaerobic decomposition. Beyond this, the locations where oil and gas form are generally classified as subsiding, depositional basins in which older sediment is continuously buried by younger sediments, thus progressively subjecting the older, more deeply buried material to higher temperatures and pressures (10).

The major source material for oil and natural gas is a fine-grained, organic-rich sediment that is buried to a depth of 1 to 3 km and subjected to heat and pressure that physically compress the source rock. The elevated temperature and pressure, along with other processes, start the chemical transformation of organic debris into oil and gas (Figure 15.9). First to form in an oxygen-deprived environment is biogenic gas from action of microorganisms consuming buried organic matter and producing methane as a by-product. They consume hydrogen and carbon in the reaction $4H_2 + CO_2 \rightarrow CH_4 + 2H_2O$. That is, the microorganisms combine hydrogen with carbon dioxide to form methane and water. About 20 percent of the energy gas discovered in the earth is biogenic gas (9). As the depth of burial, temperature, and pressure increases, the porosity of the source rock is reduced because of compaction, and the increased temperature kills microorganisms in the rocks; at depths of about 3 to 6 km, thermogenic gas and oil form. Thermogenic gas and oil involve the transformation of organic material to oil and gas by physical and chemical reactions associated with pressure and temperature rather than organisms that produce biogenic gas.

Following their formation, the thermogenic hydrocarbons begin an upward migration to a lower-pressure environment with increased porosity. This *primary migration* through the source rock merges into *secondary migration* as the hydrocarbons move more freely into and through coarse-grained, more permeable rocks (Figure 15.9). These porous, permeable rocks (such as sandstone or fractured limestone) are called *reservoir rocks*. If the path is clear to the surface, the oil and gas migrate and escape there, and this may explain why most oil and gas is found in geologically young rocks (less than 100 million years old)—hydrocarbons in older rocks have had a longer time in which to reach the surface and leak out (10). However, the lower amounts of hydrocarbon in very old rocks might also be explained by tectonic processes that uplift rocks containing oil and gas, exposing them to erosion.

At still-deeper depths below about 6 km, only thermogenic gas is formed, both from oil that formed earlier at shallower depth and from other organic material remaining in the rock. The entire process from initial deposition

CASE HISTORY A Tale of Two Mines

Trapper Mine

Trapper Mine on the western slope of the Rocky Mountains in northern Colorado is a good example of a new generation of large coal strip mines. The main operation is designed to minimize environmental degradation during mining and to enable reclaiming the land for dry land farming and grazing of livestock and big game without artificial application of water. The mine will produce 68 million metric tons of coal over a 35-year period, to be delivered to an 800-megawatt power plant adjacent to the mine. To meet this commitment, approximately 20 to 24 km^2 of land will have to be strip-mined.

Four coal seams, varying from about 1 to 4 m thick and separated by various depths of overburden, will be mined. The depth of the overburden varies from zero to about 50 m. The steps in the actual mining are:

- Vegetation and topsoil are removed with dozers and scrapers, and the soil is stockpiled for reuse.
- Overburden along a cut up to 1.6 km long and 53 m wide is removed with a 23-m^3 dragline bucket (Figure 15.A).
- Exposed coal beds are drilled and blasted to fracture the coal, which is removed with a backhoe and loaded into trucks (Figure 15.B).
- The cut is filled, topsoil is replaced, and the land is either planted in a crop or returned to rangeland (Figure 15.C).

At the Trapper Mine the land is reclaimed without artificially applying water. Precipitation (mostly snow) is about 35 cm annually, which is sufficient to reestablish vegetation provided there is adequate topsoil to help hold the soil water. This factor emphasizes that reclamation is site specific; what works at one location or region may not apply to other areas.

Water and air quality are closely monitored at the Trapper Mine. Surface water is diverted around mine pits, and groundwater is intercepted while pits are open. Settling basins constructed downslope from pits trap suspended solids before discharging water into local streams. Air quality at the mine could be degraded by dust produced from blasting, hauling, and grading of the coal. However, dust is minimized by regularly watering or otherwise treating roads and other dust-producing surfaces.

Reclamation of mined land at the Trapper Mine has been very successful during the first years of operation. Although the environmental protection techniques increase the cost of the coal by as much as 50 percent, the payoff will come in the long-range productivity of the land after termination of mining. Some argue that the Trapper Mine is unique in that a fortuitous combination of geology, hydrology, and topography allow for successful reclamation; on the other hand, the mine shows that with careful site selection and planning, strip mining is not incompatible with other land uses.

Star Fire Mine in Eastern Kentucky

Coal has been one of the driving factors of the economy in eastern Kentucky for a long time. Eventually, however, the coal will be mined out, so there must be a shift to other uses of the land. The Star Fire Mine is in the midst of very interesting experiments concerning future land use and mine reclamation.

Essentially, the mining company is involved with long-term land-use planning, so that land being mined now will be devel-

▲ **FIGURE 15.A** Overburden being removed at the Trapper Mine, Colorado. The large drag line shown here has a 23-m^3 bucket. (*Edward A. Keller*)

▲ **FIGURE 15.B** Large backhoe at the Trapper Mine, Colorado, removing the coal, which is then loaded in large trucks and delivered to a power plant just off the mining site. (*Edward A. Keller*)

to deep burial often takes millions of years. At its conclusion about one-half of the original organic material deposited in the basin is still in the rocks and not converted to oil or gas. This organic material, when subjected to increased heat and pressure is eventually transformed to solid carbon such as the mineral graphite used in pencils—which I am using to edit this manuscript (9).

When oil and gas are impeded in their upward migration by a relatively impervious barrier, they accumulate in the reservoir rocks. If the barrier, or *cap rock*, has a favorable

▲ **FIGURE 15.C** Reclaimed land at the Trapper Mine, Colorado. The site in the foreground has just had the soil replaced following mining, whereas the vegetated sites have been entirely reclaimed. (*Edward A. Keller*)

oped for other uses in the future. Valleys in this part of the world tend to be narrow with steep sides, with flatlands for development relatively uncommon. Capitalizing on a variance of the 1977 Surface Mining Control and Reclamation Act, which in general requires restoration to approximate original contours following mining, Star Fire Mine lands are being heavily modified. The variance is allowed, provided the land that is reclaimed has an equal or better economic or public use following reclamation (7).

At Star Fire Mine (Figure 15.D) the process of mining is known as *mountain top removal/valley fill.* The method involves filling the valleys with materials excavated from the surrounding mountains. As a result, the landscape following mining is much different from what was originally there—in fact, the highest part of the land may be where the valleys were. Mining at Star Fire is projected to continue until about the year 2010, and by that time more than 20 km^2 of gently rolling land will have been produced. This will constitute the single largest parcel of potentially developable land in all of eastern Kentucky (7). Furthermore, the land does not have a flood hazard, which is unusual for flatlands in eastern Kentucky.

It remains to be seen whether the experiment at Star Fire Mine will be successful. The mining company owns the mineral rights and the land outright, which provides some of the incentive for producing land with a potential for development. On the other hand, the company is learning to work with huge quantities of fill. Valleys will be filled with more than 100 m of material excavated from adjacent hills. The project has started with the development of a moderate-sized lake on top of about 80 m of unconsolidated mine spoil. Although the lake provides an important aquatic habitat, its construction is viewed as a short-term goal compared to the long-term land-use planning that is being attempted. Ongoing research is attempting to determine whether water lines, sewers, roads, and other structures necessary for development may be safely placed on the massive volumes of mine fill. The importance of the Star Fire Mine Reclamation project is that it is challenging previously believed limitations on land development following mining (7).

(a)

(b)

◀ **FIGURE 15.D** (a) Mining coal at the Star Fire Mine, Kentucky; and (b) grasslands and pond produced as part of mine reclamation. (*Courtesy of Cyprus Mountain Coal Corporation*)

geometry (structure), such as a dome or an anticline, the oil and gas will be trapped in their upward movement at the crest of the dome or anticline below the cap rock (10). Figure 15.10 shows an anticlinal trap and two other possible traps caused by faulting or by an unconformity (buried erosion surface). These are not the only possible types of traps. Any rock that has a relatively high porosity and permeability and that is connected to a source rock containing hydrocarbons may become a reservoir, provided that the upward migration of the oil and gas is impeded by a cap

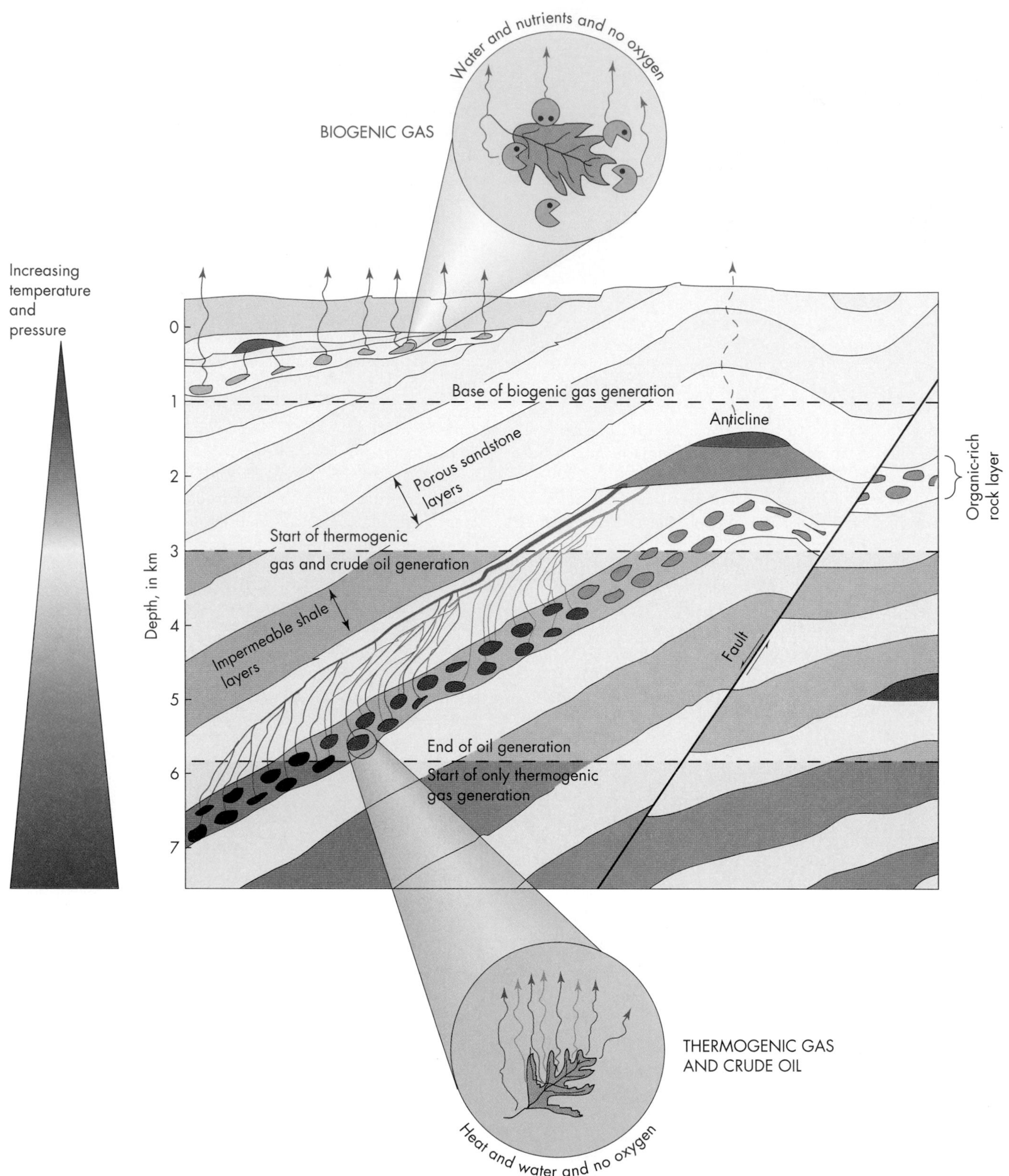

▲ **FIGURE 15.9** Idealized diagram showing depth of formation of biogenic gas and thermogenic gas and oil. See text for full explanation. (*From* P. J. McCabe, D. L. Gautier, M. D. Lewan, *and* C. Turner. 1993. The future of energy gases. U.S. *Geological Survey Circular* 1115.)

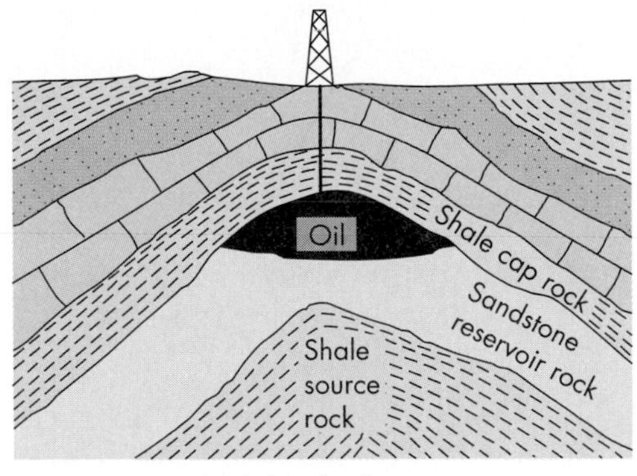

(a) Anticlinal trap

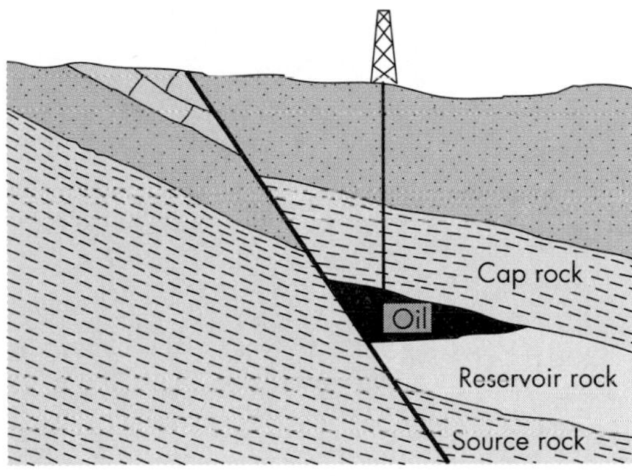

(b) Fault trap

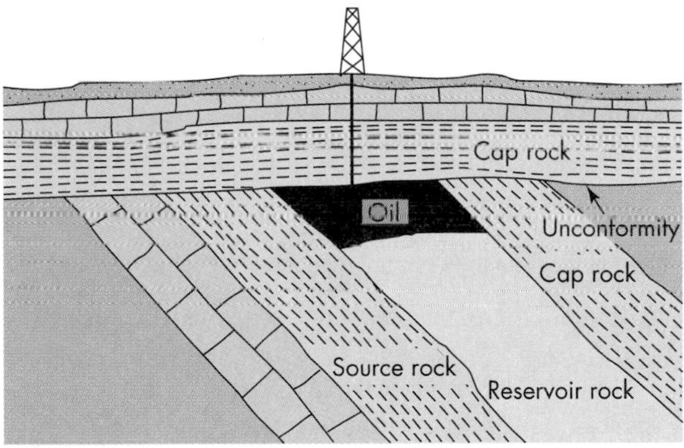

(c) Unconformity (stratigraphic) trap

▲ **FIGURE 15.10** Types of oil traps: (a) anticlinal, (b) fault, and (c) unconformity.

rock so oriented that the hydrocarbons are entrapped at a central high point (10).

Petroleum Production Production wells in an oil field recover petroleum through primary or enhanced recovery methods. *Primary recovery* uses natural reservoir pressure to move the oil to the well, but normally this pumping delivers no more than 25 percent of the total petroleum in the reservoir. To increase the recovery rate to 50 to 60 percent or more, *enhanced recovery* (secondary or tertiary recovery) methods are necessary. The enhancement manipulates reservoir pressure by careful injection of natural gas, water, steam, and/or chemicals into the reservoir, pushing petroleum to wells where it can be lifted to the surface by means of the familiar "horse head" bobbing pumps, submersible pumps, or other lift methods.

Petroleum production always brings to the surface a variable amount of salty water (brine) along with the oil. After separating the oil and water, the latter must be disposed of because it is toxic to the surface environment. Disposal can be accomplished by injection as part of enhanced (secondary) recovery, evaporation in lined open pits, or deep-well disposal outside the field.

Distribution and Amount of Oil and Gas The distribution of oil and gas in space and geologic time is rather complex, but in general, three principles apply:

- First, commercial oil and natural gas are produced almost exclusively from sedimentary rocks deposited during the last 500 million years of the earth's history (10).
- Second, although there are many oil fields in the world, approximately 85 percent of the total production plus reserves occurs in less than 5 percent of the producing fields; 65 percent occurs in about 1 percent of the fields; and, as remarkable as it seems, 15 percent of the world's known oil reserves can be found in two accumulations in the Middle East (10).
- Third, the geographic distribution of the world's giant oil and gas fields (Figure 15.11) shows that most are located near tectonic belts (plate junctions) that are known to have been active in the last 60 million to 70 million years.

It is difficult to assess the petroleum and gas reserves of the United States, much less those of the entire world. However, Figure 15.12 shows a recent estimate of the world's reserves of crude oil and natural gas. Most of the oil reserves (65%) are in the Middle East, and more than 70 percent of the natural gas reserves are in the former Soviet Union and the Middle East.

Impact of Exploration and Development The environmental impact of exploration and development of oil and gas varies from negligible—for remote sensing techniques in exploration—to significant, unavoidable impact for projects such as the Trans-Alaska Pipeline. The impact of exploration for oil and gas can include constructing roads, exploratory drilling, and building a supply line (for camps, airfields, etc.) to remote areas. These activities, except in sensitive regions such as some semiarid-to-arid environments and some permafrost areas, generally cause few adverse effects to the landscape and resources compared to development and consumption activities.

Development of oil and gas fields involves *drilling* wells on land or beneath the sea; *disposing* of wastewater brought to the surface with the petroleum; *transporting* the oil by tankers, pipelines, or other methods to refineries; and *converting* the crude oil into useful products. All along the way, there is a well-documented potential for environmental disruption from problems associated with wastewater disposal,

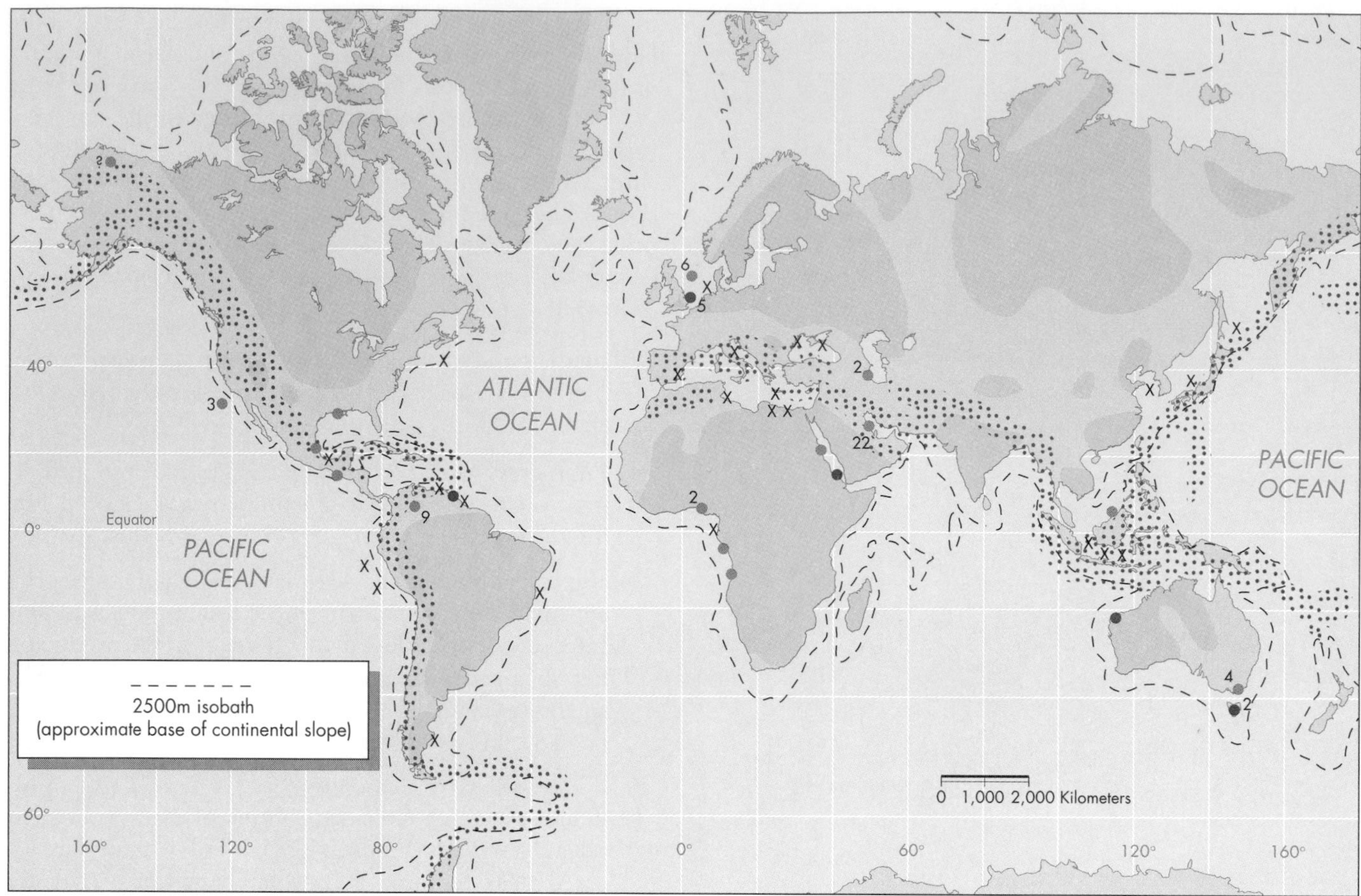

▲ **FIGURE 15.11** Giant oil and gas fields of the world relative to generalized tectonic belts. Active tectonic areas are shown by the stippled pattern, and regions not subjected to tectonic activity for the last 500 million years are shaded. Giant oil fields are denoted by the green dots; giant gas fields by the red dots; and general areas of discovery either of less than giant size or under field development are indicated by an X. Numbers indicate how many fields are in a particular location. (*After* U.S. *Geological Survey* Circular 694, *as adapted from* C. L. *Drake*, American Association of Petroleum Geologists Bulletin, *vol.* 56, *no.* 2, 1972, *with permission.*)

accidental oil spills, leaking pipes in oil fields, shipwrecks of tankers, air pollution at refineries, and other impacts. Serious oil spills have affected the coastlines of Europe and America, spoiling beaches, estuaries, and harbors, killing marine life and birds, polluting groundwater and surface water, and causing economic problems for shorelines that depend on tourist trade. We need additional research and legislation to minimize these occurrences (see Chapter 11).

A familiar and serious impact associated with oil is air pollution, which is produced in urban areas when fossil fuels are burned to produce energy for electricity, heat, and automobiles. The adverse effects of smog on vegetation and human health are well documented (see Chapter 17).

Oil Shales and Tar Sands Recovering petroleum from surface or near-surface oil shale and tar sands involves use of well-established exploration techniques. Numerous deposits are known—what we need are reliable techniques of developing the oil shale resources that will cause a minimum of environmental disruption.

Oil shale is a fine-grained sedimentary rock containing organic matter (kerogen). On heating (destructive distillation), oil shale yields significant amounts of hydrocarbons that are otherwise insoluble in ordinary petroleum solvents (11). The best-known oil shales in the United States are those in the Green River Formation, which is about 50 million years old and underlies approximately 44,000 km^2 of Colorado, Utah, and Wyoming (Figure 15.13). The Green River Formation consists of oil shale interbedded with variable amounts of sandstone, siltstone, claystone, and compacted volcanic ash (tuff).

As with other fossil fuels, the origin of oil shale involves the deposition and only partial decomposition of organic debris. Favorable environments for oil shale are lakes, stagnant streams or lagoons in the vicinity of organic-rich swamps, and marine basins (12). The large, shallow, ancient lakes of Wyoming, Utah, and Colorado, in which the organic material accumulated, varied from fresh to alkaline water. The lake basins subsided slowly and irregularly, causing the center of deposition of the oil shale to shift slowly during the millions of years the deposition continued. For millions of years after the sediment was buried, little tectonic activity disturbed the sediments. More recently, uplift and local tilting of the rocks have exposed some of the oil shale to erosion (11).

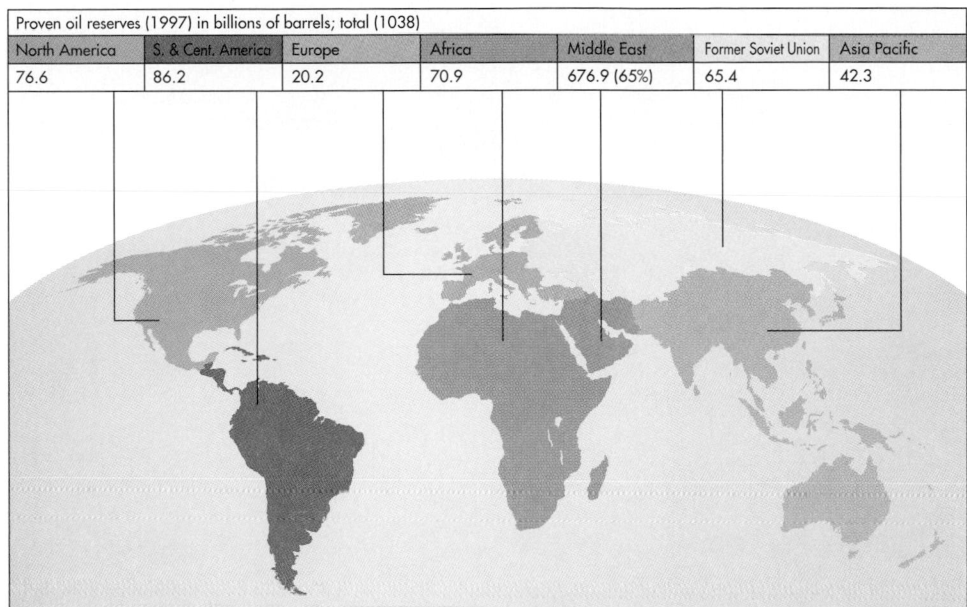

(a)

Proven natural gas (1997) in trillions of cubic meters; total (145)

North America	S. & Cent. America	Europe	Africa	Middle East	Former Soviet Union	Asia Pacific
8.4	6.3	5.6	9.9	48.9 (34%)	56.7 (39%)	9.1

(b)

◀ **FIGURE 15.12** Estimated worldwide reserves of (a) crude oil and (b) natural gas. (*British Petroleum Co.* 1997. BP *Statistical Review of World Energy.*)

Many countries have deposits of oil shale, but there is a tremendous range in thickness of the rock, quality of the oil, and areal extent (13). Identified oil shale resources of the world's land areas are estimated to contain about 3 trillion barrels of oil, but evaluation of the grade of oil is incomplete. Oil shale resources in the United States amount to about 2 trillion barrels of oil, or two-thirds of the total identified in the world; of this, 90 percent, or 1.8 trillion barrels, is located in the Green River Formation. The environmental impact of developing oil shale resources will vary according to the recovery technique. Surface and subsurface mining as well as in-place *(in situ)* techniques have been considered. Mining of oil shale using today's technology has proven too expensive to date. Nevertheless, if oil prices rise sufficiently, oil shale mining will likely be considered again. Oil shale is down but not out!

Tar sands are rocks that are impregnated with tar oil, asphalt, or other petroleum materials and from which recovery of petroleum products by usual methods such as oil wells is not commercially possible. The term *tar sand* is somewhat confusing because it includes several rock types, such as shale and limestone, as well as unconsolidated or consolidated sandstone. The one thing all these rocks have in common is that they contain a variety of semiliquid, semisolid, and solid petroleum products. Some of these products ooze from the rock outcrops, whereas others are difficult to remove even with boiling water (14).

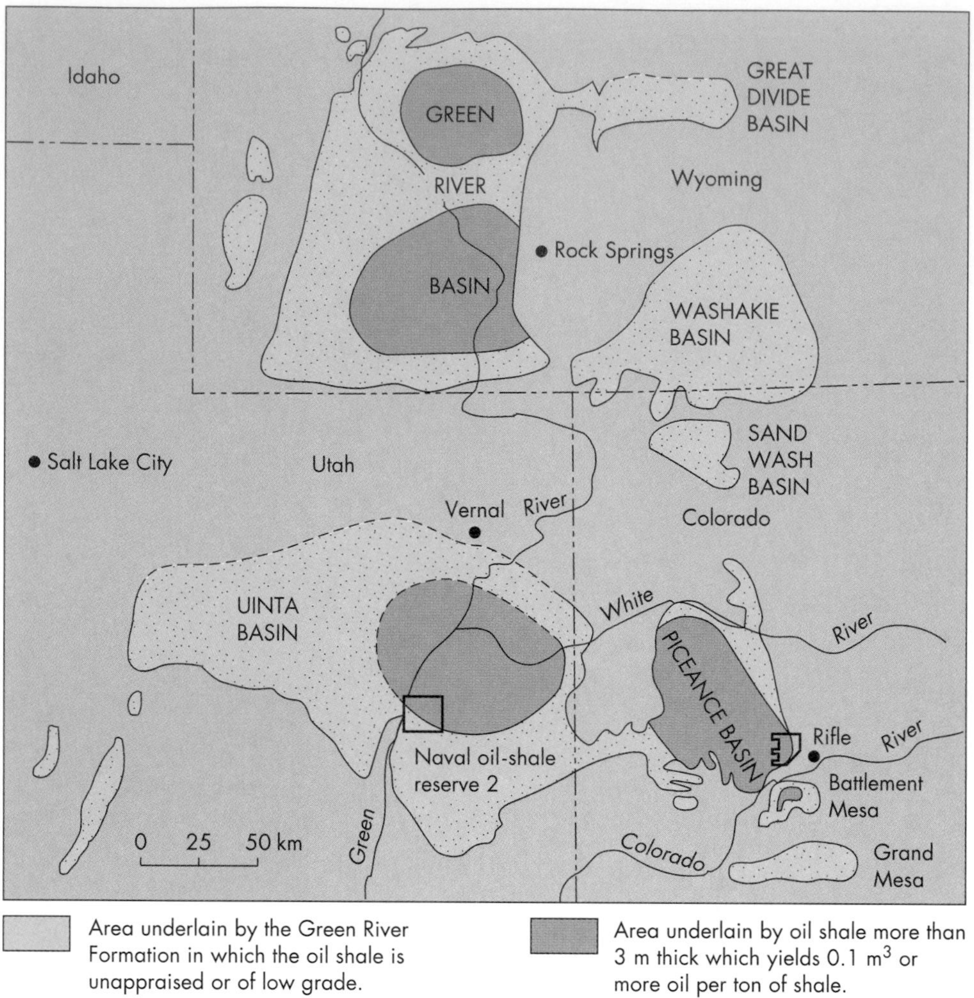

FIGURE 15.13 Distribution of oil shale in the Green River Formation of Colorado, Utah, and Wyoming. (*After* D. C. *Duncan and* V. E. *Swanson*. 1965. U.S. *Geological Survey Circular* 523.)

Oil in tar sands is nearly the same as the heavier oil pumped from wells. The only real difference is that tar sand oil is much more viscous and therefore more difficult to recover. A possible conclusion concerning the geology of tar sands is that they form essentially the same way that the more fluid oil forms, but much more of the volatiles and accompanying liquids in the reservoir rocks have escaped, leaving the more viscous materials behind.

Large accumulations of tar sands have been identified. For example, the Athabasca Tar Sands of Alberta, Canada, cover an area of approximately 78,000 km^2 and contain an estimated reserve of 2 trillion barrels of oil that might be recovered (8,15). In addition, smaller tar sand resources are known in Utah (1.8 billion barrels), Venezuela (900 million barrels), California (100 million barrels), Texas, and Wyoming (8,14).

The Athabasca Tar Sands in Alberta are now yielding about 500,000 barrels of synthetic crude oil per day from two large open-pit (strip) mines (16). The indigenous vegetation is water-saturated plants (muskeg swamp) of decayed and decaying vegetation that can be removed easily only in the winter when it is frozen. The process of recovering oil from tar sands is relatively easy once the sand has been removed from the mine; it involves washing the viscous oil out of the sand with hot water. The oil then flows to the surface and is removed (14). Only about 10 percent of the total oil from the tar sands can be economically recovered by open-pit mining, however, so obtaining additional oil will probably require recovery of the oil in place, that is, without removal of the tar sands by surface or subsurface mining (16). It is hoped that in-place recovery techniques will disturb the surface only minimally.

15.3 Fossil Fuels and Acid Rain

Acid rain is thought to be a regional-to-global environmental problem related to burning of fossil fuels (mostly coal, but also gasoline). Acid rain refers to both wet and dry acid deposition, and although *acid deposition* is a more correct term, *acid rain* is more commonly used. Acid rain is defined as precipitation in which the pH is below 5.6. The pH is a numerical value of the relative concentration of the hydrogen ion (H^+) of a solution used to describe its acidity. A solution with pH of 7 represents a neutral solution (one that is neither an acid nor a base). Natural rainfall is slightly acidic with a pH of about 5.6 because water in the atmosphere combines with carbon dioxide to produce a weak carbonic acid ($H_2O + CO_2 \rightarrow H_2CO_3$). The pH scale (Figure 15.14) is negative loga-

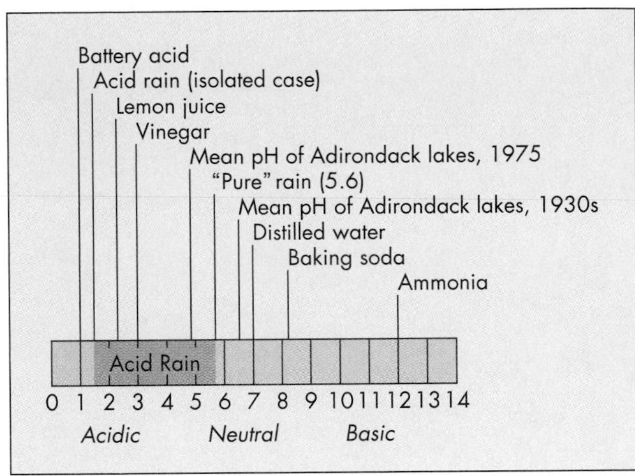

▲ **FIGURE 15.14** The pH scale. (*Modified after* U.S. Environmental Protection Agency, 1980)

rithmic—a pH value of 3 is ten times more acidic than a pH value of 4, and 100 times more acidic than a pH value of 5. Rainfall in Wheeling, West Virginia, was once measured at a pH value of 1.5, nearly as acidic as stomach acid, and pH values as low as 3 have been recorded in other locations.

In the United States today burning fossil fuels releases about 20 million tons each of nitrogen oxide and sulfur dioxide into the atmosphere annually. Following emission, these oxides are transformed to sulfate ($SO_4^{=}$) or nitrate (NO_3^-) particles, which may combine with water vapor to eventually form sulfuric (H_2SO_4) and nitric (HNO_3) acids. These acids may travel long distances with prevailing winds and be deposited as acid rain (Figure 15.15).

The acid rain problem we are most familiar with results from sulfur dioxide, which is primarily emitted from burning coal in power plants that produce electricity in the eastern United States.

Environmental Effects of Acid Rain

The geology and climate patterns, as well as types of vegetation and composition of the soil, all affect potential impacts of acid rain. Figure 15.16 shows areas in the United States that are sensitive to acid rain, and identification of these areas is based on some of the above factors. Areas that are particularly sensitive are those in which the bedrock or soils or water cannot buffer the acid input. Such areas include terrain dominated by granitic rocks as well as those in which soils have little buffering action. By *buffering*, we mean the ability of a material to neutralize acids. Such materials (called *buffers*) include calcium carbonate ($CaCO_3$), the mineral calcite, which is present in many types of soils and rock (limestone). The calcium carbonate reacts with the hydrogen (H^+) ions in the acidic water, removing it by forming bicarbonate ions (HCO_3^-), neutralizing the acid.

The major environmental effects of acid rain include:

- **Damage to vegetation,** especially forest resources such as evergreen trees in Germany and red spruce trees in Vermont. The damage probably results because soils may be damaged from a fertility standpoint, either because nutrients are leached out by the acid or because the acid releases into the soil elements that are toxic to plants.
- **Damage to lake ecosystems.** Acid rain may damage lake ecosystems by interfering with the natural cycling

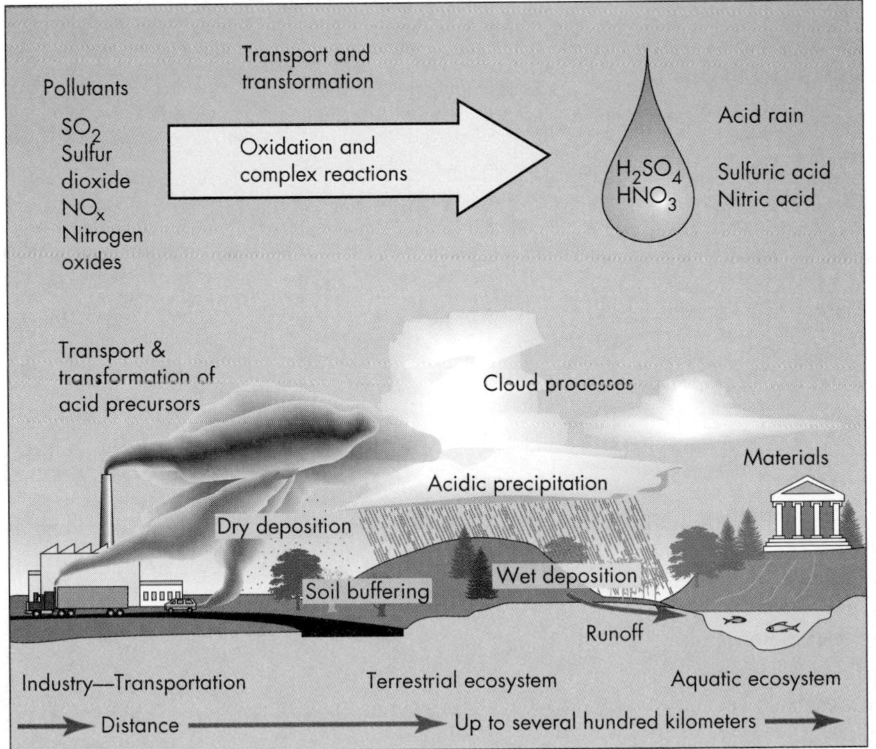

◀ **FIGURE 15.15** Paths and processes associated with acid rain. (*Modified after* Albritton, D. L., as presented in Miller, J. M.)

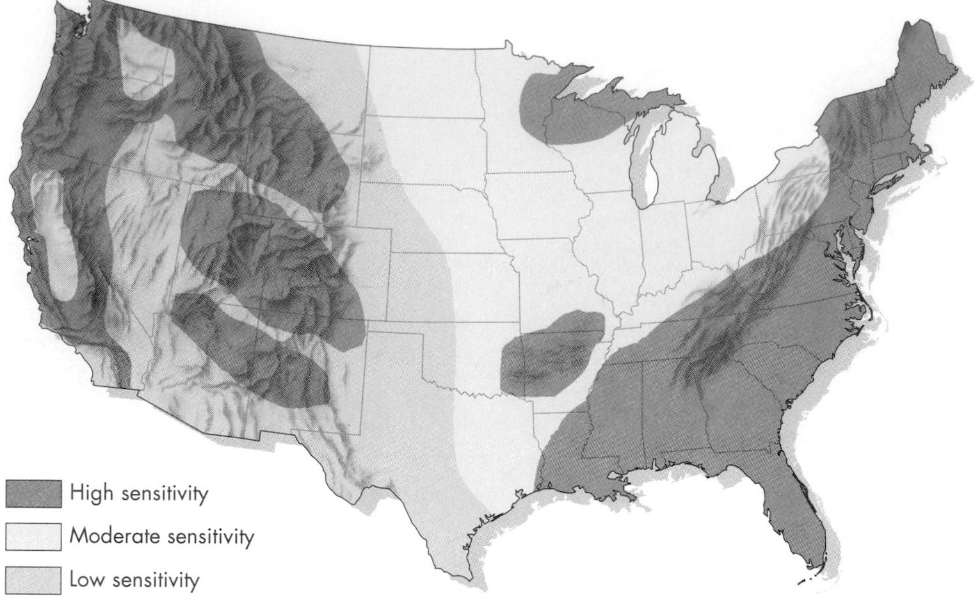

▶ **FIGURE 15.16** Areas in the United States sensitive to acid rain. (*From* U.S. *Environmental Protection Agency*, 1980)

of nutrients and other chemical elements necessary for life by keeping them in solution so that they leave the lake rather than be cycled in the system. As a result, aquatic plants may not grow, and animals that feed on these plants have little to eat. This degradation is passed up the food chain to the fish and other larger animals. Examples of adverse effects of acid rain on lake systems have been described in Canada and Scandinavia.

▶ **Damage to human structures.** Acid rain damages building materials, plastics, cement, masonry, galvanized steel, and several types of rocks, especially limestone, sandstone, and marble. In cities around the world, irreplaceable statues along with buildings have been significantly damaged, resulting in losses that reach billions of dollars per year (17,18) (Figure 15.17).

▲ **FIGURE 15.17** Air pollution and acid rain is damaging buildings and statues in many urban regions on earth. Shown here is the Acropolis in Athens, Greece. Statues here have been damaged to such an extent that originals have been placed inside buildings in specially constructed glass containers. (*Peter Christopher/Masterfile Corporation*)

A Solution to the Acid Rain Problem

A solution to lake acidification is rehabilitation by periodic addition of a buffer material such as lime (calcium carbonate). Although this has been done in several areas, including New York State, Sweden, and Canada, the solution is not satisfactory over a long period and is expensive. The only practical long-term solution to the acid rain problem is to reduce the emissions of the chemicals that cause the problem. From an environmental viewpoint, the best way to do this is to practice strong energy conservation, which would result in lower emissions, or to treat the coal before, during, and after burning to intercept the sulfur dioxide before it is released into the environment. Reduction of nitrogen oxide is more difficult as it is primarily related to the burning of gasoline in automobiles. Nevertheless, control strategies do exist, and it is encouraging that it is a national and international goal to reduce these pollutants and thus the acid rain problem. For example, Phase I of the clean Air Act Amendments of 1990 was implemented in 1995, and resulted in significant decreases in the emission of sulfur dioxide in the Ohio River Valley and mid-Atlantic states of eastern United States. These reductions resulted in the rainfall being less acidic (higher pH) than previous years (1983–94) in these areas (19).

15.4 Oil in the Twenty-first Century

Recent estimates of proven oil and gas reserves in the world suggest that, at present production rates, the oil and natural gas will last only a few decades (40 to 60 years) (2). However, the important question with respect to crude oil is not how long it is likely to last at present production rates but when will we reach peak production? This is an important question because following peak production, less oil will be available, leading to shortages and price shocks if we are not prepared. Perhaps of most concern is that world oil production peak is likely to occur about the year 2020, within the lifetimes of most people living today (16). This is soon and in fact much sooner than generally expected by most people and governments today. As a result there may be little time left to adjust to potential changes in life-style and economies in a post-petroleum era (16). As Walter Youngquist, an expert on energy, has eloquently stated, we are very fortunate today to be living in a brief, bright period of human history that was made possible by our inheritance of a 500-million-year period of oil-forming processes (16). Of course, we will never entirely run out of crude oil and probably will be producing significant amounts in the next several hundred years. The problem is that the people of the world depend upon oil for nearly 40 percent of their energy, and significant shortages will cause major problems (20).

What is the evidence that we are heading toward a potential crisis with respect to availability of crude oil?

- We are now close to having consumed approximately 50 percent of the total crude oil available from traditional oil fields in the earth (16).
- With proven reserves of about 1.04 trillion barrels (2), it is optimistically projected that approximately 2 and 3 trillion barrels of crude oil may ultimately be recovered from our remaining oil resources. World consumption today is about 26 billion barrels per year (71 million barrels per day) but we are using it fast. Presently, for every four barrels of oil we consume, we are finding only one barrel (20).
- Forecasts that predict a decline in production of oil are based upon the estimated amount of oil that may ultimately be recoverable (2 to 3 trillion barrels), along with projections on new discoveries and future rates of consumption. As a result, it has been estimated that the peak in world crude oil production will occur about the year 2020, at a rate of about 90 million barrels per day (20).
- In the United States it is predicted that the production of oil as we know it today will cease by about the year 2090, and that world production of oil will be nearly exhausted by 2100 (20).

What is the appropriate response to the above argument that production rates of oil will fall early in the twenty-first century? The first step should be a major educational program to inform people and governments of the potential depletion and shortages of crude oil. Presently we are operating under ignorance or denial in the face of a potentially serious situation. Education and planning are of paramount importance if we wish to avoid future military confrontation (we have already had one oil war) and avoid food shortages as a result of less oil being used to produce synthetic fertilizers on which our agricultural industries have become dependent.

Before significant shortages of oil occur, we need to develop alternatives such as large-scale gasification and liquefaction of our tremendous coal reserves, extract oil and gas from oil shale, perhaps rely more on atomic energy, and certainly rely more on solar energy. These changes will strongly affect our present petroleum-based society, but there appears to be no insurmountable problem if we implement meaningful short- and long-range plans to phase out oil and natural gas and phase in alternative energy sources.

Phasing in alternative energy sources will require time, research, exploration, and development, so it is crucial that we begin the task now. We are living in an interesting, and some would say potentially perilous, time from an energy standpoint. What we do now (through the next several decades) with respect to education linked to energy planning and policy will significantly impact human society in the next century.

15.5 Nuclear Energy: Fission and Fusion

Energy from Fission

The first controlled nuclear fission was demonstrated in 1942, leading the way to the use of uranium in explosives and as a heat source to provide steam for generation of electricity. Fission of 1 kg of uranium oxide releases approximately the same amount of energy as the burning of 16 metric tons of coal.

Chain Reactions Nuclear **fission** is the splitting of atomic nuclei by neutron bombardment (Figure 15.18). Fission of a uranium (U-235) nucleus releases three neutrons, fission fragments (nuclei of radioactive elements lighter than uranium), and energy in the form of heat. The released neutrons strike other U-235 atoms, releasing more neutrons, fission products, and heat. The process continues in a chain reaction—as more and more uranium is split, it releases ever more neutrons. An uncontrolled chain reaction—the kind used in nuclear weapons—leads quickly to an explosion. Sustained, or stable, nuclear reactions in reactors are used to provide heat for the generation of electricity.

Three types of uranium are found in a naturally occurring uranium sample: U-238, which accounts for approximately 99.3 percent of natural uranium; U-235, which

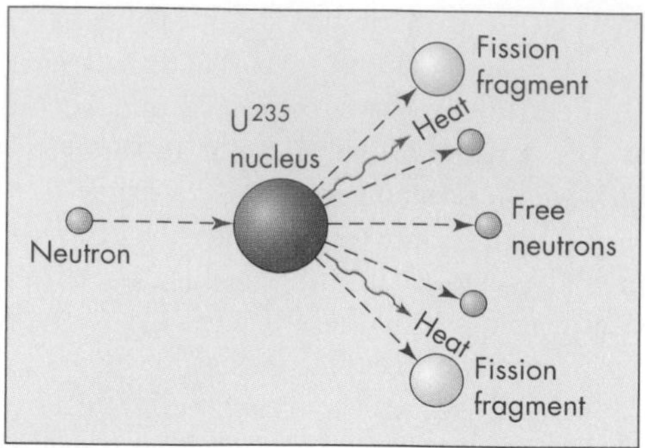

▲ **FIGURE 15.18** Fission of U-235. A neutron strikes the U-235 nucleus, producing fission fragments and free neutrons and releasing heat. The released neutrons may then each strike another U-235 atom, releasing more neutrons, fission fragments, and energy. As the process continues, a chain reaction develops.

makes up about 0.7 percent; and U-234, which makes up about 0.005 percent. Uranium-235 is the only naturally occurring fissionable material and is therefore essential to the production of nuclear energy. Naturally occurring uranium is processed to increase the amount of U-235 from 0.7 percent to about 3 percent before it is used in a reactor. The processed fuel is called *enriched uranium*. Uranium-238 is not naturally fissionable, but it is "fertile material" because upon bombardment by neutrons it is converted to plutonium-239, which is fissionable (21).

Reactor Design and Operation Most reactors today consume more fissionable material than they produce and are therefore known as **burner reactors.** The reactor itself (Figure 15.19) is part of the nuclear steam-supply system,

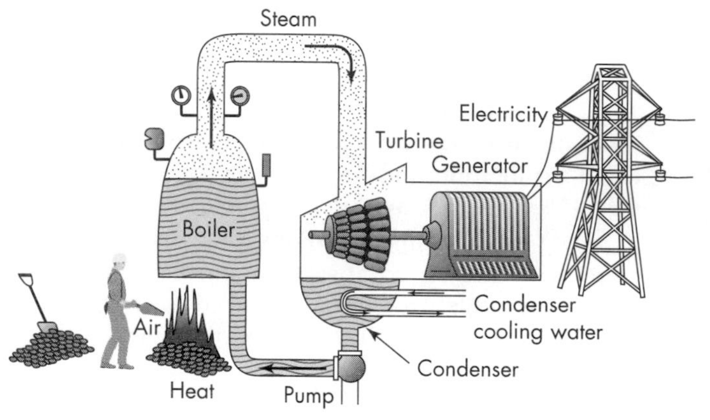

(a)

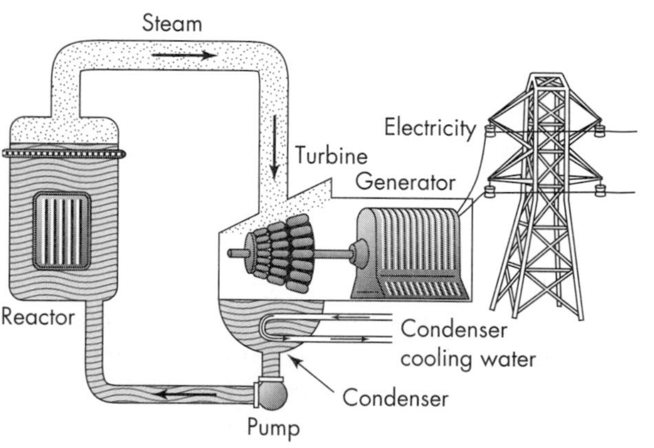

(b)

▲ **FIGURE 15.19** Diagrams: Comparison of (a) fossil-fuel power plant and (b) nuclear power plant with a boiling-water reactor. Notice that the nuclear reactor has exactly the same function as the boiler in the fossil-fuel power plant. (*Reprinted, by permission, from* Nuclear power and the environment, *American Nuclear Society*, 1973) Photographs: (a) This fossil-fuel power plant, Skytell Bridge, Tampa Bay, Florida, burns coal. Components include the storage of coal in the lower right-hand corner, the power plant itself in the center, cooling water leaving the power plant on the left, and the series of electric power lines leading away from the power plant. (*Comstock*) (b) Diablo Canyon Nuclear Power Plant near San Luis Obispo, California. Reactors are in the dome-shaped buildings and cooling water is escaping to the ocean. The siting of this power plant has been and remains very controversial because of its close proximity to faults capable of producing earthquakes that might damage the facility. (*Comstock*)

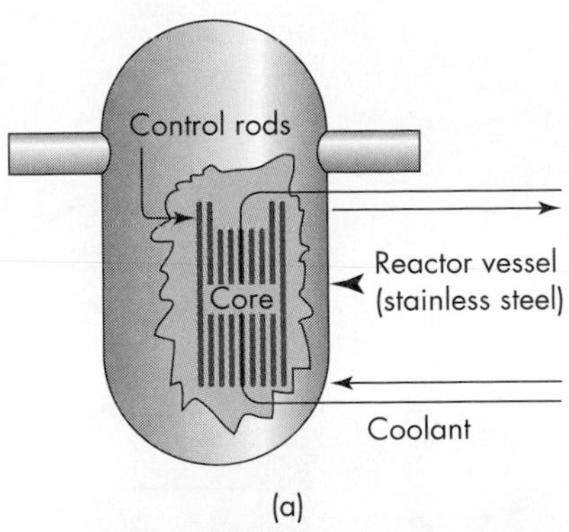

(b)

▲ **FIGURE 15.20** (a) Diagram of the main components of a nuclear reactor. (b) Part of the interior of a nuclear reactor is shown here. The site is the Flamanville Nuclear Power Station, France. The core of the reactor is in the blue water pool in the center. The crane above is used to carry fuel rods to change the fuel before the reactor is operated. The power plant is a pressurized water reactor. Nuclear power accounts for approximately 75 percent of France's electricity. (*Catherine Pouedras/Science Photo Library/Photo Researchers, Inc.*)

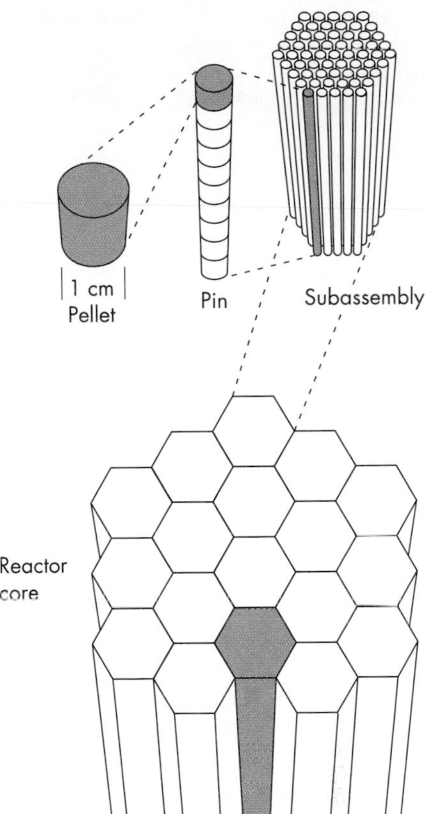

▲ **FIGURE 15.21** Fuel pellets of enriched uranium are placed in hollow tubes forming fuel pins that are bundled into fuel subassemblies, which are placed in the reactor as part of the core. (*Modified after* Energy Research and Development Administration. 1976. ERDA-76-107.)

which produces the steam to run the turbine generators that produce electricity (22).

The main components of the reactor shown in Figure 15.20 are the core, control rods, coolant, and reactor vessel. The *core* of the reactor, where the chain reaction takes place, is contained in a heavy stainless steel *reactor vessel*. (For extra safety and security, the entire reactor is contained in a reinforced concrete building called a *containment structure*.) *Fuel pins*, which consist of enriched uranium pellets in hollow tubes less than 1 cm in diameter, are packed together (40,000 or more in a reactor) into fuel subassemblies in the core (Figure 15.21). A stable fission chain reaction is maintained by controlling fuel concentration and the number of released neutrons that are available to cause fission. A minimum fuel concentration is necessary to keep the chain reaction *critical* (self-sustaining), and control of the number of neutrons is necessary to regulate the reaction rate. The *control rods* contain materials that capture neutrons, preventing them from bombarding other nuclei. When the rods are pulled out of the core, the chain reaction speeds up; when they are inserted into the core, the reaction slows down (23).

Pumps circulate a coolant (usually water) through the reactor, extracting the heat produced by fission. When the coolant is water, it not only removes heat but also acts as a moderator, slowing down the neutrons and facilitating efficient fission of uranium-235 (23). Figure 15.22 shows how a type of reactor called a *pressurized water reactor* (PWR) uses the heat generated in the reactor core to make steam. A pressurized water reactor is one type of *light water reactor*, so called because it uses ordinary water (light water) as a coolant. Water circulating in the primary *coolant loop* (a closed system) is heated by the reactor core; it then transfers its heat to a *steam generator* (or *heat exchanger*), turning the water in the *secondary loop* to steam. The steam is then used to drive the turbine that turns the generator (22). Using a heat exchanger and secondary loop to carry heat away from the reactor allows water in the primary loop, which is radioactive, to remain isolated within the containment structure.

A second type of light water reactor in use is the *boiling water reactor* (BWR), shown in Figure 15.23. A boiling water

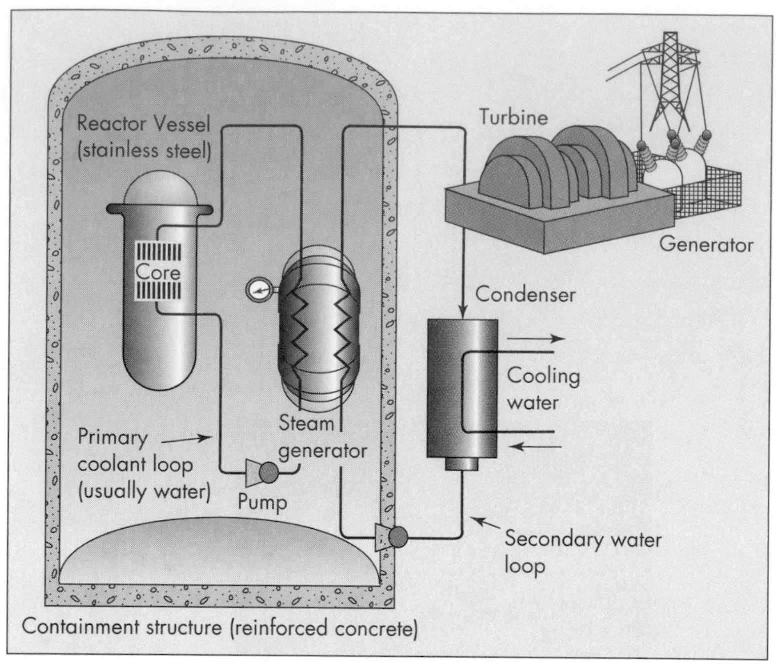

▶ **FIGURE 15.22** Pressurized water reactor (PWR) showing the three main parts of the nuclear steam supply system: reactor, primary coolant loop, and steam generator (heat exchanger). (*Modified after Energy Research and Development Administration*. 1976. ERDA-76-107.)

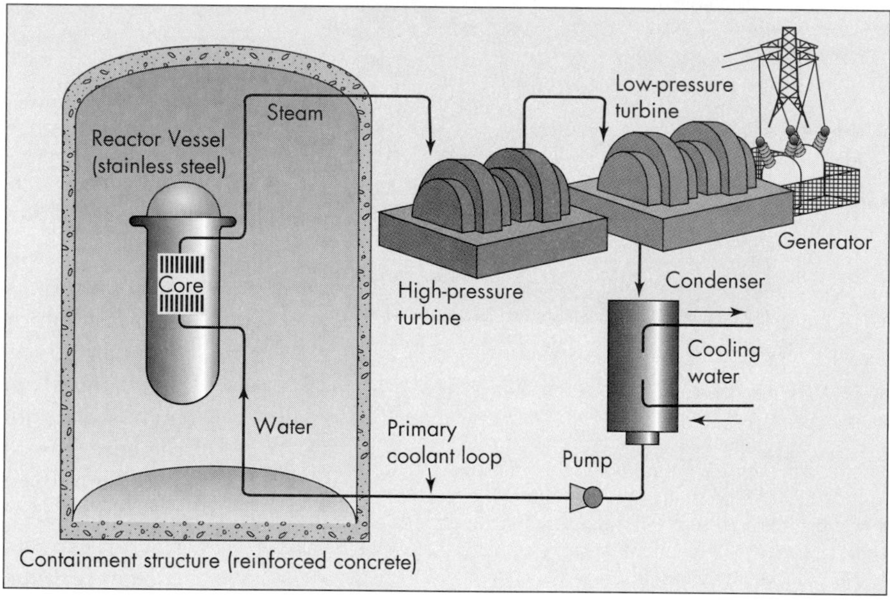

▶ **FIGURE 15.23** Boiling water reactor (BWR) showing the two main parts of the nuclear steam supply system: reactor and primary coolant loop. (*Modified after Energy Research and Development Administration*. 1976. ERDA-76-107.)

reactor is a *direct cycle system* because there is no heat exchanger—the primary coolant loop goes from the reactor core directly to a turbine. Because the steam that turns the turbine comes directly from the reactor core, great care must be taken to keep hazardous radiation from leaking into the turbine system, which is outside the main containment structure. The turbines must be heavily shielded, increasing construction and maintenance costs (22).

To conserve U-235, reactors that produce more fissionable material (nuclear fuel) than they use have been developed. These **breeder reactors,** using a fuel core of fissionable material (plutonium-239) surrounded by a blanket of fertile material (U-238), produce (or *breed*) additional plutonium-239 from the U-238. The transformation from U-238 to Pu-239 occurs in the breeder reactor at the same time that the plutonium nuclei in the core are undergoing fission, providing heat that produces steam to generate electricity.

Geology and Distribution of Uranium The natural concentration of uranium in the earth's crust is about 2 parts per million (ppm). Uranium originates in magma and is concentrated to about 4 ppm in granitic rock, where it is found in a variety of minerals. Some uranium is also found with late-stage igneous rocks such as pegmatites. To be mined at a profit, uranium must have a concentration factor of 400 to 2500 times the natural concentration (21).

Fortunately, uranium forms a large number of minerals, many of which can be found in high-grade deposits being mined today. Three types of deposits have produced most of the uranium in the last few years: sandstone impregnated with uranium minerals, veins of uranium-bearing materials localized in rock fractures, and placer deposits in river or delta deposits (now coarse-grained sedimentary rock) more than 2.2 billion years old (24).

Uranium in sandstone is often found in thin lenses interbedded with mudstone. It is hypothesized that the uranium in these deposits was derived by leaching from volcanic glass associated with the sedimentary rocks or from granitic rocks exposed along the margins of the sedimentary basins. Supposedly, the uranium was transported by groundwater and then precipitated into the pore spaces of the sandstone under reducing (oxygen-deficient) conditions (21).

Uranium in ancient river or delta sediment was deposited before the occurrence of abundant free oxygen in the atmosphere. Therefore, it is believed, the uranium mineral was deposited as stream-rounded particles along with gold, pyrite, and other typical placer material (24). Although sandstone ores have been the main source of uranium in this country, vein deposits are a significant source in Australia, Canada, France, and Africa, and ancient placer deposits are being mined in Africa and Canada (24). The amount of uranium from the known deposits is sufficient to last into the early twenty-first century.

***Risks Associated with* Fission Reactors** Nuclear energy and the possibly adverse effects associated with it have been subjects of vigorous debate. The debate is healthy because we should continue to examine the consequences of nuclear power very carefully.

Nuclear fission uses and produces radioactive isotopes. Various amounts of radiation are released into the environment at every step of the nuclear cycle: mining and processing of uranium, controlled fission in reactors, reprocessing of spent nuclear fuel, and final disposal of the radioactive wastes. Serious hazards are associated with the transporting and disposing of nuclear material (see Chapter 12), as well as with supplying other nations with reactors. Furthermore, because the plutonium produced by nuclear reactors can be used to make nuclear weapons, terrorist activity and the possibility of irresponsible actions by governments add a risk that is present in no other form of energy production.

An uncontrolled chain reaction—a nuclear explosion—cannot occur in a nuclear reactor because the fissionable material is not concentrated enough. However, unwanted chemical reactions in a reactor can produce explosions that release radioactive substances into the environment. Although the chance of a disastrous accident is estimated to be very low, it increases with every reactor put into operation. Major accidents have already occurred, including the disastrous accident at Chernobyl in 1986 (see *Case History: Two Reactor Accidents*).

As a result of the Chernobyl disaster, risk analysis in nuclear power is now a real-life experience rather than a computer simulation. We are more acutely aware now that as long as people build nuclear power plants and manage them, there will be the possibility of accidents—it is as much a problem of human nature as anything else. Although the probability of a serious accident is very small at a particular site, the consequences of the event may be great enough to constitute an unacceptable risk. Perhaps we may need to conclude that we are not ready as a people to handle nuclear power. However, there is a move to build safer nuclear reactors. One of the ideas is to design smaller modular units that would be much less susceptible to serious accidents. Smaller modules at a particular site have another advantage: If one or more of them are down for maintenance, the rest will still produce power.

***The Future of* Energy *from* Fission** In 1982 the demand for electric power decreased for the first time in many years. It will increase in the future, however, as may the demand for nuclear power, which produced about 20 percent of our electricity in 1993. By the year 2000, it was expected that heat from nuclear reactors would have the capacity to produce about 22 percent of the electrical power in the United States. About 110 reactors are now in operation (more than 80% of which are in the eastern United States); many more would be needed to realize the projected energy from uranium. Because of increased costs, environmental considerations, and other factors, the projected generating capacity is too high and has been revised to 20 percent of electricity produced.

Nuclear fission may indeed be the answer to our energy problems and perhaps someday will provide unlimited cheap energy. But along with nuclear power comes the responsibility of ensuring that nuclear power is used for people, not against them, and that future generations will inherit a quality environment free from worry about hazardous nuclear waste. The full impact of what began in 1942 is still to be determined.

Energy from Fusion

The long-range energy strategy for the United States is to develop and provide sufficient sustainable energy resources to meet energy demand in the twenty-first century and beyond. Energy sources capable of meeting long-term sustainable needs are those that are for all practical purposes inexhaustible—as, for example, solar, hydrogen, and fusion reactors. Our discussion here will focus on fusion.

In contrast to nuclear fission, which involves splitting the nuclei of heavy atoms, nuclear **fusion** involves combining the nuclei of light elements to form heavier ones, releasing energy in the process. Fusion reactions are the source of energy in our sun and other stars. Figure 15.24 shows the fusion of two hydrogen isotopes (atoms with different numbers of neutrons), a reaction that could be used to produce energy in a fusion reactor. The two isotopes, deuterium (D) and tritium (T), are injected into a reactor chamber where necessary conditions (temperature, time, and density) for fusion are maintained. Products of the D-T fusion include helium, carrying 20 percent of the energy released, and neutrons, carrying 80 percent of the energy released (Figure 15.24) (25).

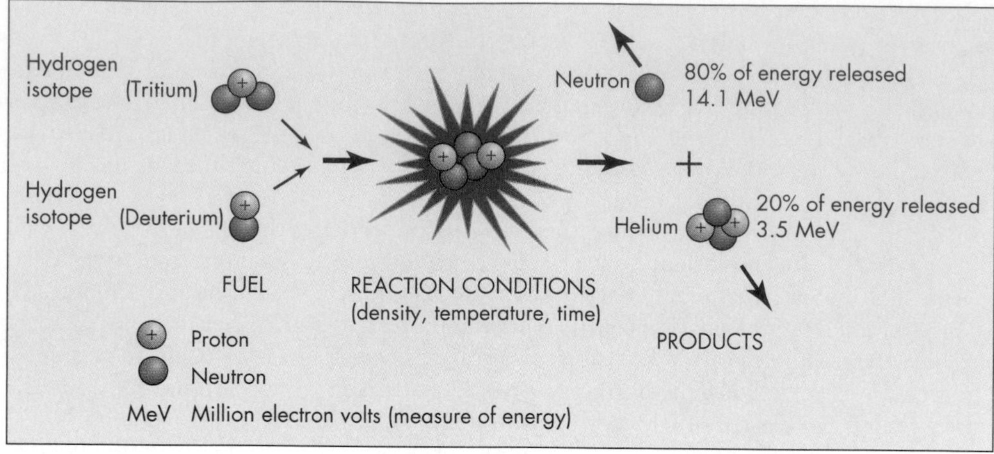

▶ **FIGURE 15.24** Deuterium-tritium (D-T) fusion reaction. (*Modified from* U.S. *Department of Energy.* 1980. DOE/ER-0059.)

Energy from fusion has a variety of applications, including heating and cooling buildings and producing synthetic fuels, but producing electricity is probably the most important. It is expected (but not proven) that fusion power plants will be economically competitive with other sources of electric energy. From an environmental view, fusion certainly appears attractive: Land-use and transportation impacts are small compared to fossil fuel or fission energy sources. Fusion produces no fission products and little radioactive waste, and it has a much lower risk of potential hazard resulting from an accident (26). On the other hand, fusion power plants will probably use materials that are harmful to humans, such as lithium, which is toxic when inhaled or ingested. Other potential hazards are the strong magnetic fields and microwaves used in confining and heating plasma (an electrically neutral material composed of ions, positively charged nuclei and negatively charged electrons), and short-lived radiation emitted from the reactor vessel (25).

15.6 Geothermal Energy

The use of **geothermal energy**—natural heat from the earth's interior—is an exciting application of geologic knowledge and engineering technology. The idea of harnessing the earth's internal heat is not new: Geothermal power was developed in Italy using dry steam in 1904 and is now used to generate electricity at numerous sites around the world and a few in the western United States and Hawaii. At many other sites geothermal energy not hot enough to produce electrical power is used to heat buildings or for industrial purposes. Existing geothermal facilities utilize only a small portion of the total energy that might eventually be tapped from the earth's reservoir of internal heat; the geothermal resource is vast. If only 1 percent of the geothermal energy in the upper 10 km of the earth's crust could be captured, this would amount to 500 times the total global oil and gas resource (29).

Geology of Geothermal Energy

Natural heat production within the earth is only partly understood. We do know that some areas have a higher flow of heat from below than others, and that for the most part these locations are associated with the tectonic cycle. Oceanic ridge systems (divergent plate boundaries) and convergent plate boundaries, where mountains are being uplifted and volcanic island arcs are forming, are areas where this natural heat flow from the earth is anomalously high. The coincidence of geothermal power plants with areas of known active volcanism is no accident.

The increase in temperature with depth below the earth's surface, measured in degrees per kilometer, is called the *geothermal gradient.* In general, the steeper the gradient, the greater the heat flow to the surface. A steep geothermal gradient indicates that hot rock is closer to the surface than usual. A moderate gradient of 30° to 45 °C per kilometer is found over vast regions of the western United States (Basin and Range area) and especially the Battle Mountain region (Figure 15.25a), which is therefore (with some exceptions) a good prospect for geothermal exploration. Areas with a steep geothermal gradient generally have a relatively high heat flow (rate at which geothermal heat is escaping at the surface of the earth). Figure 15.25b shows heat flow (in greater detail than Figure 15.25a) for the western United States, along with locations of existing geothermal power plants. Total power production is about 3000 megawatts, more than 90 percent of which is in California. About two-thirds of the total is in The Geysers facility in northern California (*G*, in Figure 15.25b). A typical commercial geothermal well will produce between 5 and 8 megawatts of electrical power (26).

Steep geothermal gradients in the United States are concentrated in western states with recent or ongoing tectonic and volcanic activity. The great width of the belt of moderate-to-steep geothermal gradient is something of an anomaly, and for years geologists have vigorously debated the origin of the rock structure and tectonic activity of the western region. The data used to produce the maps in Figure 15.25 are not sufficient to accurately define the limits of the regions (30), so the map has limited value for locating specific sites where geothermal energy resources could be developed.

Within the region of relatively high heat flow, commercial development of geothermal power is most likely where a heat source such as convecting magma is relatively near the

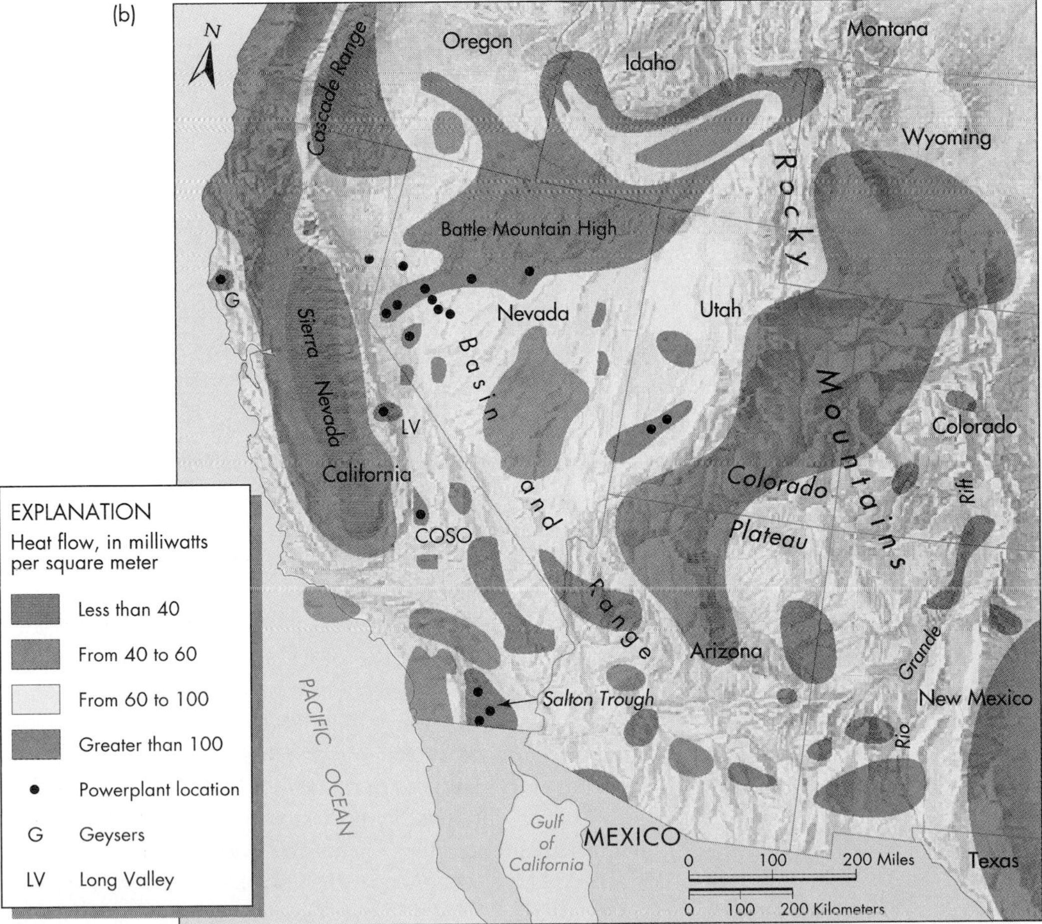

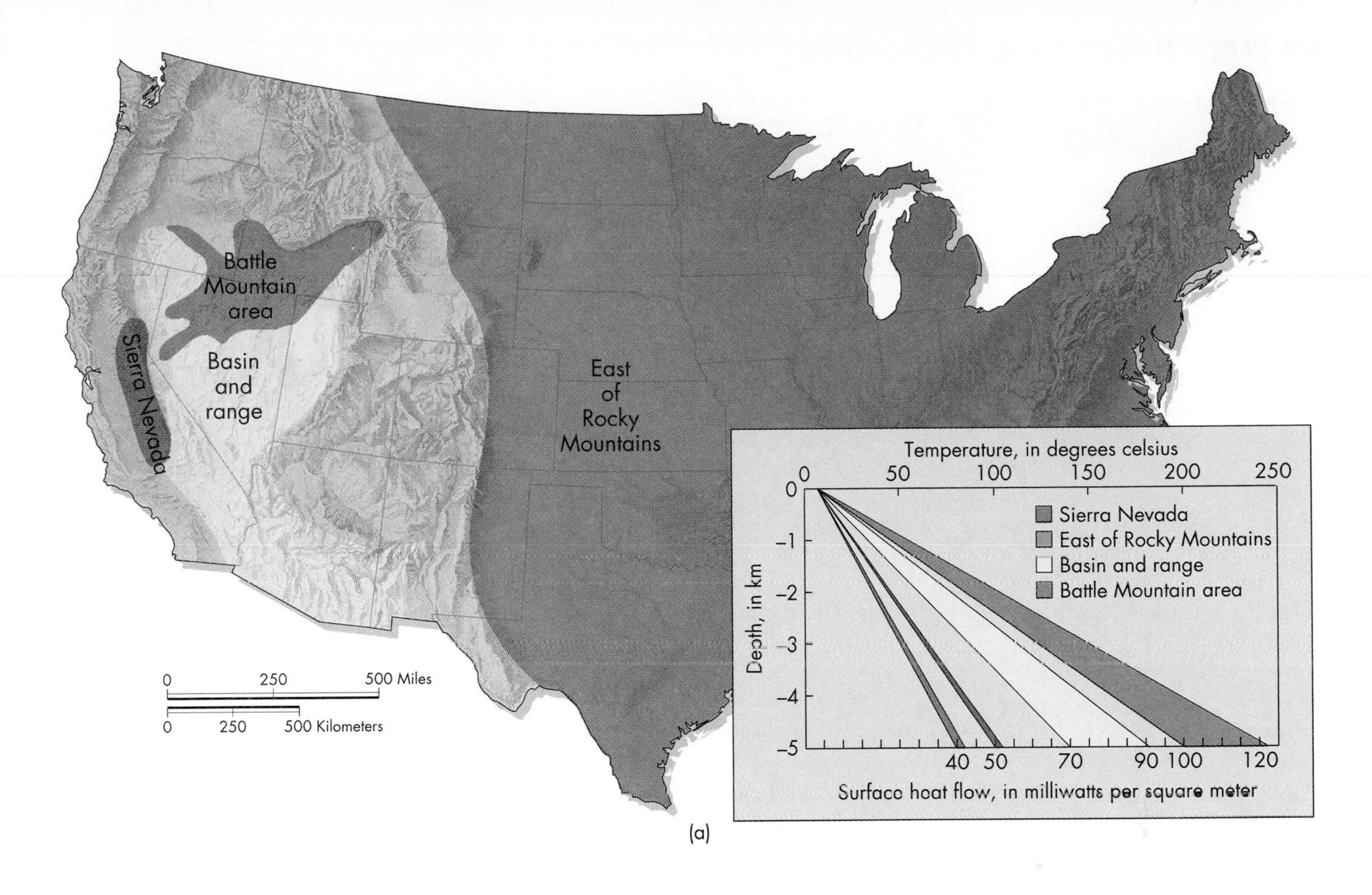

▲ **FIGURE 15.25** Geothermal gradients and generalized heat flow (a) in the United States; more detailed map (b) of heat flow for the western United States and location of geothermal power plants. (*From* W. A. *Duffield,* J. H. *Sass, and* M. L. *Sorey.* 1994. Tapping the earth's natural heat. U.S. *Geological Survey Circular* 1125.)

CASE HISTORY Two Reactor Accidents

Three Mile Island

On March 28, 1979, malfunctions in the nuclear plant at the Three Mile Island nuclear power plant near Harrisburg, Pennsylvania, resulted in radiation release within one of the nuclear facilities and release of radioisotopes into the environment. The release into the environment was at a low level per person exposed; exposure from the plume emitted into the atmosphere has been estimated at 100 mrem, which is low in terms of the amount of radiation required to cause acute toxic effects (see Chapter 13). However, radiation levels were much higher near the site of the accident. On the third day after the accident, 1200 mrem/hr were measured at ground level near the site.

The Three Mile Island incident made clear that there are many problems with the way our society has dealt with nuclear power. Historically, nuclear power has been relatively safe, and the state of Pennsylvania was somewhat unprepared to deal with the accident. For example, there was no state bureau for radiation health, and the Pennsylvania Department of Health did not have a single book on radiation medicine (the medical library had been dismantled 2 years before for budgetary reasons). One of the major impacts of the incident was fear, yet there was no state office of mental health or any authority to allow anyone from the Pennsylvania Department of Health to sit in on briefing sessions to discuss the accident and what should be done.

Because the long-term chronic effects of exposure to low levels of radiation are not well understood, the effects of the Three Mile Island exposure—although apparently small—are difficult to estimate. This case shows that society needs to improve its ability to handle crises that could arise from sudden releases of pollutants from present-day technology. It also shows our lack of preparedness and apparent readiness to treat a nuclear power plant as an acceptable risk (27).

Chernobyl

The preparedness problem was still more dramatically illustrated by the events that began unfolding on the morning of Monday, April 28, 1986. Workers at a nuclear power plant in Sweden, frantically searching for the source of high levels of radiation near the plant, concluded that it was not their installation that was leaking radiation—rather, the radioactivity was coming from their Soviet neighbors by way of prevailing winds. Confronted, the Soviets announced late on Monday that there had been an accident at their nuclear power plant at Chernobyl (Figure 15.E). This was the first notice to the world that the worst accident in the history of nuclear power generation had occurred.

It is speculated that the system that supplies cooling waters for the reactor failed, causing the temperature in the reactor core to rise to more than 3000 °C and melting the uranium fuel. Explosions occurred that removed the top of the building over the reactor, and the graphite surrounding the fuel rods—used to moderate the nuclear reactions in the core—ignited. The fires produced a cloud of radioactive particles that rose high into the atmosphere, then drifted north and west.

Radiation killed about 30 people at and near the plant site in the days following the accident, and millions of others were exposed to potentially harmful radiation as the radioactive cloud drifted first over parts of the Soviet Union, then north to Scandinavia, and then to eastern Europe. During the next 25 to 30 years, an increase in cancers will document the impact on humans from the accident. About 115,000 people in the 30-km exclusion zone around the accident site were evacuated and about 24,000 people received an average radiation dose of about 430 mSv (about 285 times natural background radiation in the United States). One estimate is that the radiation exposure to the 24,000 people will have caused in excess of 100 cases of leukemia by the year 1998 (28).

One of the avoidable tragedies of Chernobyl was the way the Soviet government handled the accident. Delays in warning people in their own country and outside resulted in unnecessary exposure to radiation. The Soviets have been accused of not giving attention to reactor safety and of using outdated equipment; the Chernobyl reactor was of a type that is not used in the United States because of its inadequate safety features. Nevertheless, people are now wondering if such an event could happen elsewhere. With about 400 reactors producing power in the world today, the answer has to be yes. Many countries such as Japan, Britain, and France, which depend heavily on nuclear power plants for generating electric power, are rethinking the risks of siting the plants near densely populated areas.

surface (3 to 10 km) and in thermal contact with circulating groundwater. Likely sites for exploration are areas with naturally occurring hot springs and geysers that reflect near-surface hot spots; areas of recent volcanic activity, particularly those characterized by high-silica magma, because they are more likely to have stored heat near the surface, accessible to drilling (30); and localized hot spots with little or no near-surface expression that are discovered by direct and indirect (geophysical) subsurface exploration. The best location is not necessarily that with the most prominent surface expression (geyser and hot-spring activity) because geothermal systems with a high rate of upflow may not be as well insulated as systems with little or no leakage (31). There is likely to be more stored heat at shallow depths in a well-insulated geothermal reservoir.

Based on geologic criteria, several geothermal systems can be defined: hydrothermal convection systems, hot igneous systems, geopressured systems, and normal (cold)

(a)

(b)

◀ **FIGURE 15.E** (a) Color radar image of the area surrounding the Chernobyl Nuclear Power Plant, Ukraine. The arrow points to the site of the four nuclear reactors. The large blue body of water in the center of the photograph is the power station's 12-km-long cooling pond. The town of Chernobyl is the round whitish area in the lower right area of the image. One of the reactors overheated and suffered a meltdown and explosion during the 1986 accident. An exclusion zone with a radius of 30 km around the site includes the town of Chernobyl. This image was taken in 1994. (NASA/*Science Photo Library/Photo Researchers, Inc.*) (b) Damage to Reactor #4. People shown here are measuring the radiation level. (APN/SIPA Press)

groundwater systems (32,33). Each system has a different origin and different potential as an energy source.

Hydrothermal Convection Systems Hydrothermal convection systems are characterized by a permeable layer in which a variable amount of hot water circulates. They are of two basic types: *vapor-dominated systems* and *hot-water systems*. Vapor-dominated hydrothermal convection systems are geothermal reservoirs in which both water and steam are present at depth (Figure 15.26). Near the surface, where pressure is lower, the water flashes to superheated steam, which can be tapped and piped directly into turbines to produce electricity. These systems characteristically have a slow recharge of groundwater, meaning that the hot rocks boil off more water than can be replaced in the same amount of time by natural recharge or by injection of water from a condenser following power generation (32). Vapor-dominated systems are not very common. Only three have

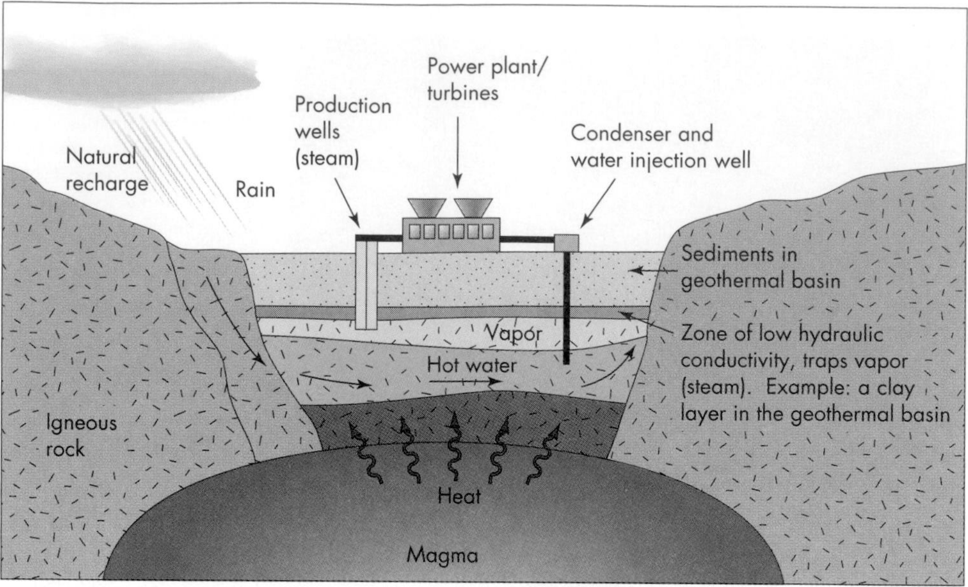

▶ **FIGURE 15.26** Idealized diagram of a vapor-dominated geothermal system.

been identified in the United States: The Geysers, 145 km north of San Francisco, California; Mt. Lassen National Park, California; and Yellowstone National Park, Wyoming. The national parks are not available for energy development. This has been a much-debated issue. Yellowstone contains about 10,000 hot springs and geysers (geothermal features that include eruptions of hot water and steam). Legislation passed by Congress now protects Yellowstone's hot springs and geysers. What is being debated is exactly what this protection entails. At the very least it requires an adequate buffer zone to ensure that energy development outside the park does not damage Yellowstone's geothermal features (29). Steam from The Geysers power plant in California has been producing electric energy for years (Figure 15.27).

In the United States, hot-water hydrothermal convection systems are about 20 times more common than vapor-dominated systems. Hot-water systems, with surface temperatures greater than 150 °C, have a zone of circulating hot water (but no steam) that when tapped moves up to a zone of reduced pressure, yielding a mixture of steam and water at the surface. The water must be removed from the steam before the steam can be used to drive the turbine (33). One problem with this system is disposal of the water. However, as ideally shown in Figure 15.28 the water can be injected back into the reservoir to be reheated.

▲ **FIGURE 15.27** Aerial view of The Geysers power plant north of San Francisco, California. The entire facility is the world's largest geothermal electricity development. (*Courtesy of Pacific Gas and Electricity* [PG&E])

Hot Igneous Systems Hot igneous systems consist of hot rock that is not in thermal contact with groundwater. Such systems may consist of molten magma at temperatures of 650° to about 1200 °C (depending on the type of magma) or of a large quantity of hot, dry rock. Hot igneous systems contain more stored heat per unit volume than any of the other geothermal systems; however, because they lack circulating hot water, innovative methods to use the heat will have to be developed (32). Some of these geothermal reservoirs have hot, dry rocks accessible to drilling, in which case the rocks might first be drilled and then fractured with explosives or hydrofracturing techniques. Then cold water might be injected into the rock at one location and pumped out at elevated temperatures at another location to recover the heat. The induced circulation of water would effectively mine the heat of the dry rock system (34).

A pilot project by the Los Alamos Scientific Laboratory in the Jemez Mountains of New Mexico generated about 4 megawatts of electricity from a hot igneous geothermal system for a period of 4 months, using an innovative nearly closed system of circulating water. Injected water was heated at a depth of about 3000 m to about 150 °C and pumped to the surface. The water was then piped through a heat exchanger, causing rapid expansion of freon that turned a turbine to generate electricity (35,36). The project was put on hold in 1993 for lack of funding.

Geopressured Systems Geopressured systems exist where the normal heat flow from the earth is trapped by impermeable clay layers that act as an effective insulator. The favorable environment is characterized by rapid

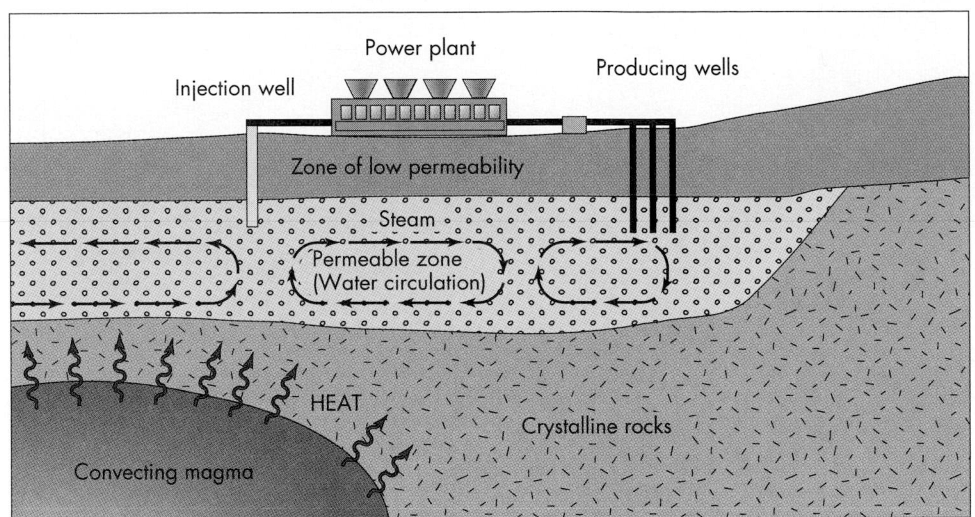

◀ **FIGURE 15.28** A hot-water geothermal system. At the power plant, the steam will be separated from the water and used to generate electrical power. The water will be injected back into the geothermal system by a disposal well. (*Courtesy of Pacific Gas and Electricity [PG&E]*)

deposition of sediment and regional subsidence. Deeply buried water trapped in the sediments develops considerable fluid pressure and is heated (29,33).

Perhaps the best-known regions where geopressurized systems develop are in the Gulf Coast of the United States, where temperatures of 150° to 273 °C at depths of 4 to 7 km have been identified. These systems have the potential to produce large quantities of electricity for three reasons:

- They contain hot-water thermal energy that could be extracted.
- They contain mechanical energy from the high-pressure water that could be used to turn hydraulic turbines to produce electricity.
- As an added asset, the waters contain considerable amounts of dissolved methane gas, up to 1 m^3 per barrel of water, that could be extracted (29,33,34).

The energy potential from the heat and methane gas far exceeds that of the high-pressure water (37).

Groundwater Systems The idea of using groundwater at normal shallow underground temperatures is a relatively new one. At a depth of about 100 m, groundwater typically has a temperature of about 13 °C. This is cold if you want to use it to take a bath but warm compared to winter air temperature in the eastern United States. Compared to summer temperatures, it is cool. Devices that transfer heat between groundwater and the air in a building can exploit these temperature differences to heat buildings in the winter and cool them in the summer. Although initially expensive owing to the drilling of wells, geothermal systems using constant (normal) temperature groundwater are in service in several midwestern and eastern U.S. locations. As energy costs rise, such systems will become more attractive.

Environmental Impact of Geothermal Energy Development

The adverse environmental impact of intensive geothermal energy development may be less than for development of other energy sources, but it is nevertheless considerable. Geothermal energy is developed at a particular site, and environmental problems include on-site noise, gas emissions, and industrial scars. Fortunately, development of geothermal energy does not require the extensive transportation of raw materials or refining typical of the fossil fuels. Geothermal plants produce less than 1 percent of the nitrous oxides and 5 percent of the carbon dioxide produced by coal-burning power plants producing comparable amounts of power (29). Finally, geothermal energy does not produce the atmospheric particulate pollutants associated with burning fossil fuels nor does it produce any radioactive waste.

Except for vapor-dominated systems, geothermal development produces considerable thermal pollution from hot wastewaters, which can be saline or highly corrosive. The plan is to dispose of these waters by reinjecting them into the geothermal reservoir, but that kind of disposal has problems: Injecting fluids may activate fracture systems in the rocks and cause earthquakes. In addition, the original withdrawal of fluids may compact the reservoir, causing surface subsidence. It is also feared that subsidence could occur because, as the heat in the system is extracted, the cooling rocks will contract (33).

The Future of Geothermal Energy

The energy potential of geothermal sources is vast. Figure 15.29 compares the worldwide resource base for most of the nonrenewable energy sources—note that the scale is logarithmic. (Annual world energy consumption is approximately 320 exajoules, an amount too small to appear in this figure.) The hot, dry rock resource base alone is several hundred times that of the fossil fuels, although it is much less than the potential of fusion, should that source ever come on-line (35).

Geothermal energy appears to have a future. Over a 30-year period, the estimated yield from this vast resource far exceeds that of hundreds of modern nuclear power plants. The resource is both identified and recoverable at this time (disregarding cost), and it is expected that many more systems with presently recoverable energy are yet to

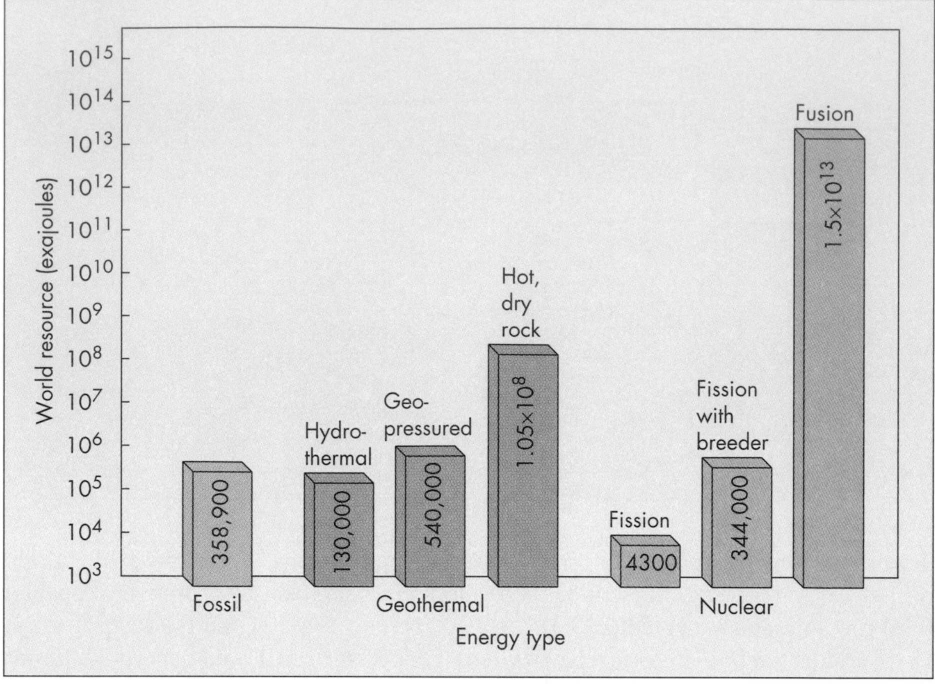

▶ **FIGURE 15.29** World energy resource base. Notice the large geothermal resource compared to fossil fuels. (*Modeled after* J. W. *Tester,* D. W. *Brown, and* R. M. *Potter.* 1989. Hot dry rock geothermal energy. *Los Alamos National Laboratory,* LA-11514-MS.)

be discovered (32). Furthermore, geothermal energy in the form of hot water (not steam) using normal heat flow may be a future source of energy to heat homes and other buildings along the Atlantic Coast, where alternative energy sources are scarce.

At present, geothermal energy supplies only a small fraction of 1 percent of the electrical energy produced in the United States. With the exception of the unusual vapor-dominated systems such as The Geysers in California, it is more expensive to produce electricity from geothermal reservoirs than from fossil fuels. Commercial development of the energy will not proceed rapidly until the economics are equalized. Even if this happens, the total output from geothermal sources is not likely to exceed a few percent—10 percent at most—of electrical output in the near future. This is true even for California, where geothermal energy has been produced and where expanding facilities are likely (34). Nevertheless, the growth in power produced from geothermal sources has increased dramatically in the past 35 years and continued (although slow) growth can be expected.

15.7 Renewable Energy Sources

The transition from fossil fuels to renewable energy sources has begun. In the United States today, oil and natural gas still supply approximately 65 percent of our energy needs, but as oil and gas become scarce and/or more costly, we will slowly change to new energy sources. The three major sources of energy likely to be used in the near future are *fossil fuels,* such as coal, oil shale, and tar sands; *nuclear energy* (fission today and perhaps fusion sometime in the future); and *renewable energy,* broadly defined to include direct solar energy, hydropower, biomass, and wind. Geothermal energy might also be added to the list of renewable sources. However, it may be better to consider geothermal energy as a sustainable resource. That is, for geothermal energy to be renewable it must not be extracted faster than it is naturally replenished. Development of a strategy to withdraw the energy at the rate it is replenished is an example of sustainable energy production (29). An important point concerning renewable energy sources is that in many parts of the world these may be the only energy sources indigenous to a given region.

We clearly see the necessity for seriously considering renewable energy sources when we examine the environmental impact associated with other possible sources. The abundant fossil fuels such as coal are often damaging to the environment throughout the fuel cycle, from mining to processing to consumption. Fossil fuels also carry the threat of global climate modification through increased discharge of carbon dioxide, particulates, and other materials. Nuclear energy, while imposing no threat of climate modification, is associated with serious problems such as waste disposal, accidents, and weapons proliferation. Nuclear energy also releases waste heat into the environment through on-site cooling processes and through transportation and use of the electricity it produces. Finally, we must recognize that all fossil fuels and nuclear energy from fission are ultimately exhaustible (38).

Renewable energy sources are replenished quickly enough to maintain a constant supply if they are not overused, so with good management we may regard them as sustainable. Because their energy is derived directly or indirectly from the sun, they can be considered forms of solar energy. Thus, broadly defined, **solar energy** includes several energy sources, as shown in Figure 15.30. Energy derived directly from the sun's heat is called *direct solar energy;* it can be used for heating homes or generating small amounts of

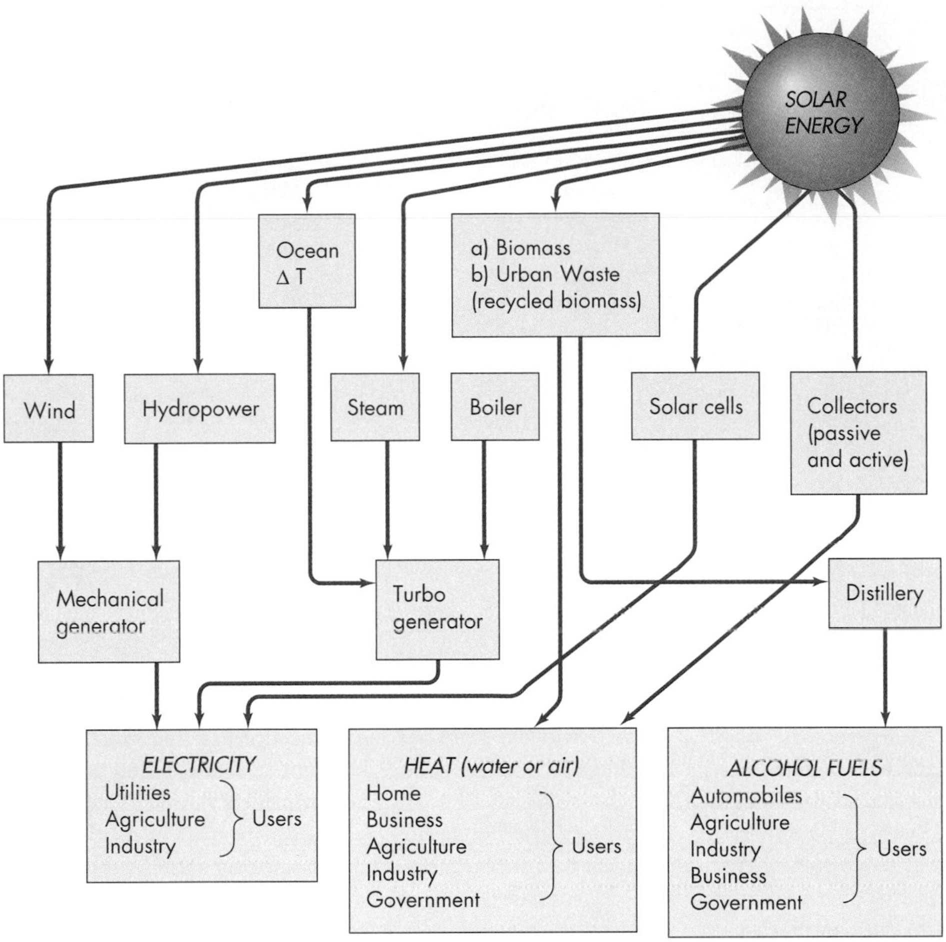

SOLAR ENERGY
Ocean ΔT
a) Biomass
b) Urban Waste (recycled biomass)
Wind
Hydropower
Steam
Boiler
Solar cells
Collectors (passive and active)
Mechanical generator
Turbo generator
Distillery
ELECTRICITY
Utilities
Agriculture
Industry
Users
HEAT (water or air)
Home
Business
Agriculture
Industry
Government
Users
ALCOHOL FUELS
Automobiles
Agriculture
Industry
Business
Government
Users

◀ **FIGURE 15.30** Routes of the various types of renewable solar energy.

electricity. Energy from wind, water, or biomass, which derive their energy from the sun, is called *indirect solar energy* and is used mainly for generating electric power.

In addition to sustainability, the advantages of renewable energy sources are that they generally cause minimal environmental degradation and, with the exception of burning biomass or urban waste, they do not add carbon dioxide to the atmosphere. A disadvantage of these sources is that (except perhaps for hydropower, biomass, and ocean thermal conversion) they are intermittent and localized. In addition, some of these sources, such as solar cells (photovoltaics), which produce electricity directly from solar energy, are considerably more expensive than fossil or nuclear energy sources. Nevertheless, with the possible exception of nuclear fusion, some form of solar energy appears to be the only long-term energy alternative at this time.

Direct Solar Energy

The use of direct solar energy is not new. Nearly 2500 years ago the Greeks designed homes to capture sunlight, and ancient cliff-dwelling Native Americans in the southwestern United States used the asymmetry of valleys and differential exposure to sunlight for natural winter heating and summer shading. Amory Lovins points out that throughout history there has been a steady evolution in solar architecture and technology, but progress has been periodically interrupted by the influx of apparently plentiful, cheap fuels such as new forests or large deposits of oil, natural gas, coal, and uranium. Earlier energy crises, from shortages of wood during the Greek and Roman eras to coal strikes in the United States at the end of the nineteenth century, led to the rediscovery of earlier knowledge concerning practical applications of solar energy. These lessons should not be neglected in the current energy situation (39).

The total amount of solar energy that reaches the earth's surface is tremendous. On a global scale, 2 weeks' worth of solar energy is roughly equivalent to the energy stored in all known reserves of coal, oil, and natural gas on earth. In the United States, on the average, 13 percent of the solar energy entering the atmosphere arrives at the ground, an amount equivalent to approximately 177 watts per m^2/hr. The actual amount at a particular site is quite variable, however, depending upon the time of year and the cloud cover (40).

Two questions often asked about direct solar energy are: Where can it be used and how much energy will it save? These questions are not easy to answer. From an availability standpoint, Figure 15.31 shows in a general way the estimated year-round availability of solar energy in the United States. But this view of solar energy availability is analogous to trying to paint a picture with a paint roller. Locations for solar energy are site-specific, and detailed observation in the field is necessary to evaluate the solar energy potential in a given area.

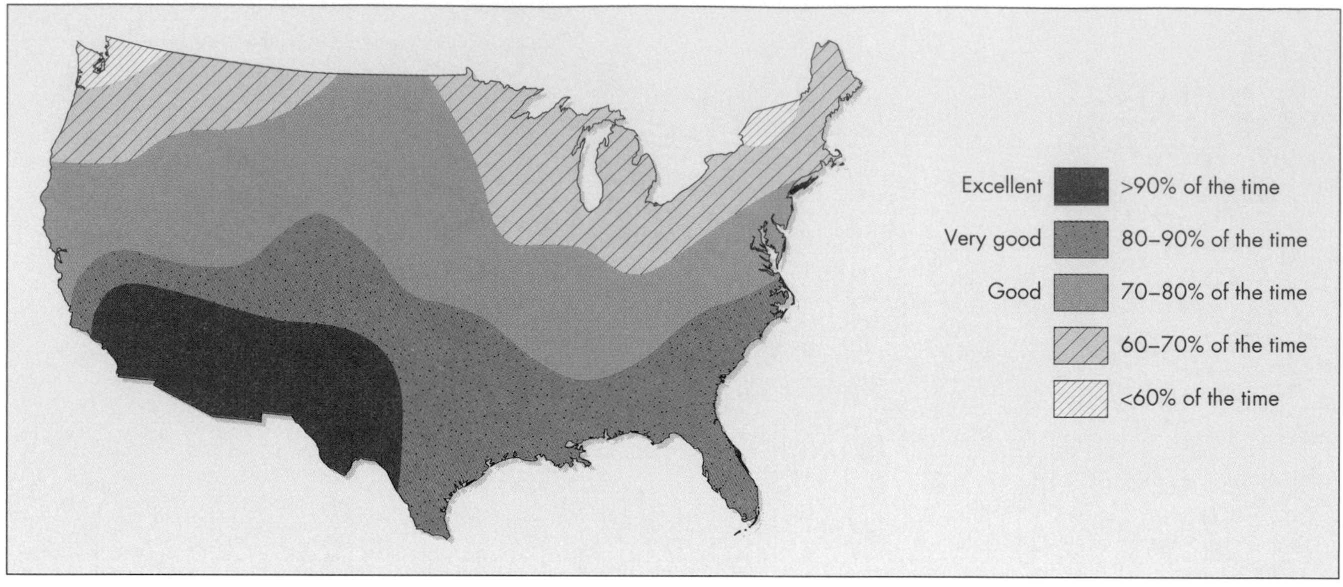

▲ **FIGURE 15.31** Estimated year-round usability of solar energy for the conterminous United States. (*Modified after* *National Wildlife Federation*, 1978)

Active and Passive Systems Solar energy can be used directly through either passive or active solar energy systems. *Passive systems* often involve architectural design, without implementation of mechanical power, to enhance and take advantage of natural changes in solar energy that occur throughout the year. Many homes and other buildings in the southwestern United States and other parts of the country now use passive systems for at least part of their energy needs. Two examples of passive systems are (1) use of a wall especially designed to absorb solar energy during the day—at night the stored heat in the wall radiates to warm a room; and (2) constructing overhangs on buildings that allow low-angle winter sunlight to penetrate and warm a room while blocking entry of higher-angle, more intense summer sunlight (helping cool the room).

Active solar energy systems require mechanical power, usually pumps and other apparatus to circulate air, water, or other fluids from solar collectors to a heat sink, where the heat is stored until used. Solar collectors are usually flat panels consisting of a glass cover plate over a black background upon which water is circulated through tubes. Shortwave solar radiation enters the glass and is absorbed by the black background. As longer-wave radiation is emitted from the black material, the water in the circulating tubes is heated, typically to temperatures of 38° to 93 °C (100° to 200 °F) (Figure 15.32) (40). In the United States, the number of solar energy systems that use collectors such as these is rapidly growing. Typical uses of active systems are for heating water for buildings and swimming pools.

Solar Cell Another potentially important aspect of direct solar energy involves solar cells, or **photovoltaics,** which convert sunlight directly into electricity. These cells are expensive, costing about five times more than electricity produced from traditional fossil-fuel sources (41). On the other hand, the field of solar technology is changing quickly, and low-cost solar cells may become more widely available in the future. The annual commercial world electrical production of electricity from photovoltaics is about 60 megawatts, spread about equally among the United States, the European Community, and Japan. Most photovoltaic systems are used in remote areas, where they provide electricity for communication, refrigeration, research, equipment, and lighting, among other uses (Figure 15.33). Evaluation of photovoltaic power stations to produce commercial electricity by utility companies is ongoing (41).

Solar Energy Power Plants During the 1980s, solar energy power plants of several varieties were developed. One of the early varieties was a "power tower" that worked by collecting solar energy as heat and delivering the energy in the form of steam to turbines to produce electric power (Figure 15.34). An experimental 10-megawatt power tower is located in California's Mojave Desert, near Barstow. The tower is approximately 100 m high and is surrounded by approximately 2000 mirror modules, each with a reflective area of about 40 m^2. The mirrors are adjusted continually as the earth rotates to reflect as much sunlight into the tower as possible. Other experimental power towers are being considered in Japan and some other countries. Continued research into the technology is worthwhile and should continue.

The most successful solar power experiment to date is also located in the Mojave Desert. The project consists of eight "solar farms," each consisting of a power plant surrounded by hundreds of solar collectors (Figure 15.35). Called the LUZ solar electric generating system, as of 1990 it was generating 275 megawatts of electricity for the southern California area. This is more than 90 percent of the world's solar-generated electricity that is connected to a util-

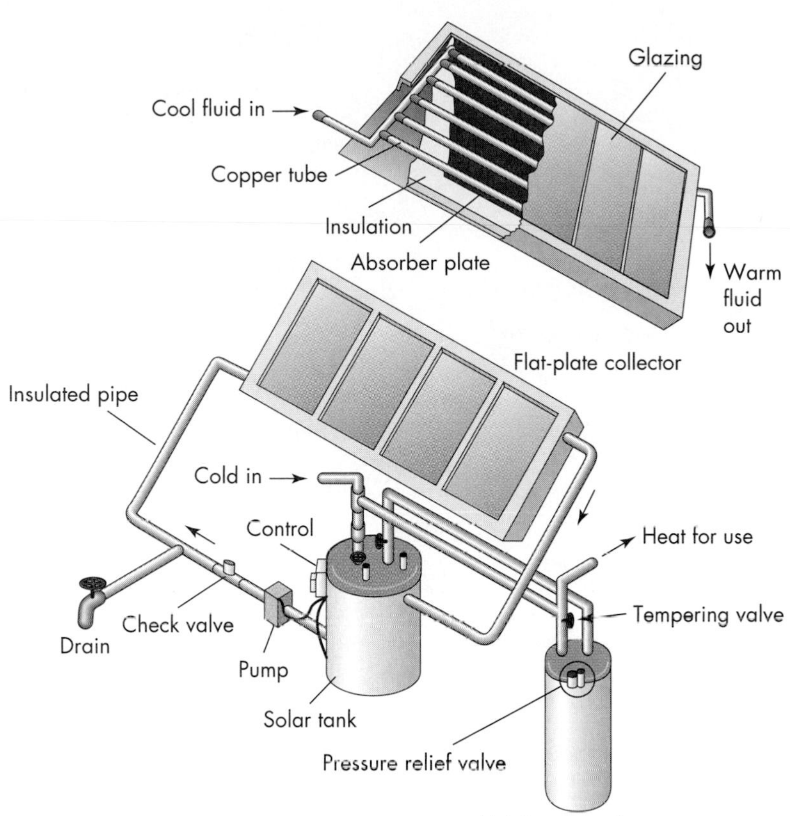

◀ **FIGURE 15.32** Details of a flat-plate solar collector. (*After Farallones Institute.* 1979. The integral urban house: Self-reliant living in the city. *San Francisco: Sierra Club Books.*)

ity system (42). The system works by utilizing solar collectors (mirrors) to heat a synthetic oil that flows through heat exchangers to drive steam turbine generators. Basic components of the system are shown in Figure 15.36. The LUZ system has a natural-gas burner backup system that ensures uninterrupted power generation during periods of peak demand or on cloudy days. The modular nature of the system allows construction of individual solar farms in a period as short as a year. The cost of the energy produced is close to the average cost of production of electricity from other sources and is less expensive than electricity produced from new nuclear power plants (42).

In summary, the LUZ system is a combination of solar technology and conventional power generation. The system looks very promising and further improvements in efficiency are probable. At present the LUZ system uses natural gas to generate approximately 25 percent of the power it produces. If the price of fossil fuels increases in the future, the profits from the combined system will increase substantially (42). One would think that a successful system such as LUZ could not fail, yet in late 1991 the company filed for bankruptcy and was required to reorganize to continue operating. Some people believe LUZ failed not because of technical problems, but as a result of the lack of a coherent energy policy in the United States. Insufficient government support to the young solar energy industry and regulations that limited the amount of natural gas that could be used contributed to LUZ's failure (43).

▲ **FIGURE 15.33** Photovoltaic cells being used at this remote site in the Granite Mountains of the Mojave Desert, California, to power a meteorology research station. (*Edward* A. *Keller*)

Ocean Thermal Energy Conversion A last example of direct use of solar energy involves using part of the natural oceanic environment as a gigantic solar collector. The surface temperature of ocean water in the tropics is often about 28 °C (82 °F). At the bottom of the ocean, however, at a depth as shallow as about 600 m, the temperature of the water may be 2° to 6 °C (36° to 43 °F). Low-efficiency systems have been designed to exploit this temperature differential and produce electricity. Experiments in September 1994, as part

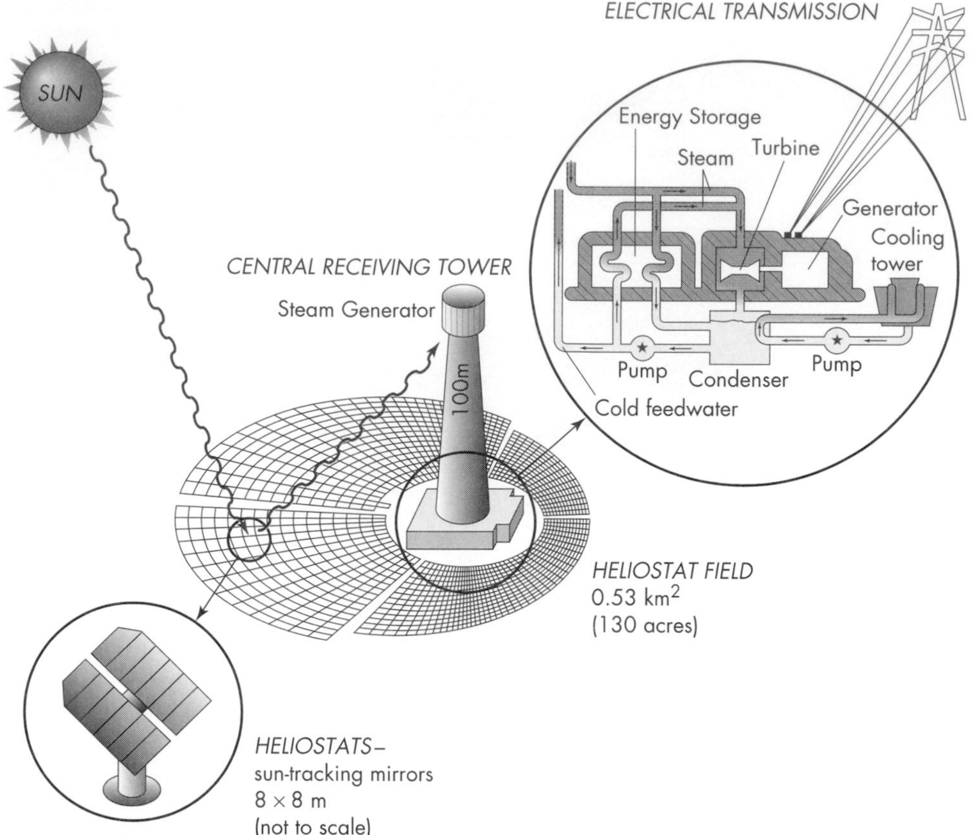

▶ **FIGURE 15.34** Idealized diagram showing how a solar power tower works. (*Modified after drawing by Southern California Edison Co.*)

▲ **FIGURE 15.35** The LUZ solar farm, in the Mojave Desert of California. Numbers show the parts of the power plant: (1) solar collector assembly; (2) natural gas boiler; (3) turbine generator; (4) steam generator and solar super heaters; (5) control building; (6) cooling tower; and (7) connection facilities to the electrical grid system in southern California. (*Courtesy of* LUZ)

of a 5-year program at the Kailua-Kona, Hawaii, test plant, using an open system approach generated 255 kilowatts while netting 104 kilowatts. The open system withdraws warm seawater from the ocean. The warm water is run through a vacuum chamber (evaporator) that causes the water to flash to steam at a temperature of about 22 °C. The steam drives a turbine, producing electricity. Salt water that does not evaporate is discharged to the ocean. Cold ocean water pumped into the plant is used in a heat exchanger to condense the steam, producing fresh water as a by-product. Large amounts of incoming warm ocean water are necessary to run the system because less than 0.5 percent of the water is turned to steam.

Closed systems have also been used at the Hawaii station to produce electricity. These use the warm ocean water in a heat exchanger to vaporize pressurized liquid ammonia. The ammonia vapor propels a turbine that produces electricity. Cold ocean water is used in a condenser to return the ammonia to a pressurized liquid again. No fresh water is produced with the closed system (44). Whether large-scale ocean thermal plants are ever built will depend primarily upon whether suitable locations are discovered close to potential markets and whether the plants are economically feasible (40). Answers to these questions are uncertain, so the future of ocean thermal plants remains speculative.

Experiments with ocean thermal energy conversion on the big island of Hawaii have proved very interesting. The laboratory at Keahole Point on the Kona coast is ideally located because the seafloor slopes off very steeply to deep, cold waters. Although only a small amount of energy has actually been produced, a payoff may be aquaculture. The cold, nutrient-rich water is being experimented with by several companies to grow seafoods such as abalone, salmon, oysters, and kelp. Several of these may be grown together in large polyculture tanks or ponds. The laboratory is also using the cold water for air conditioning, resulting in a significant saving on its electric bill.

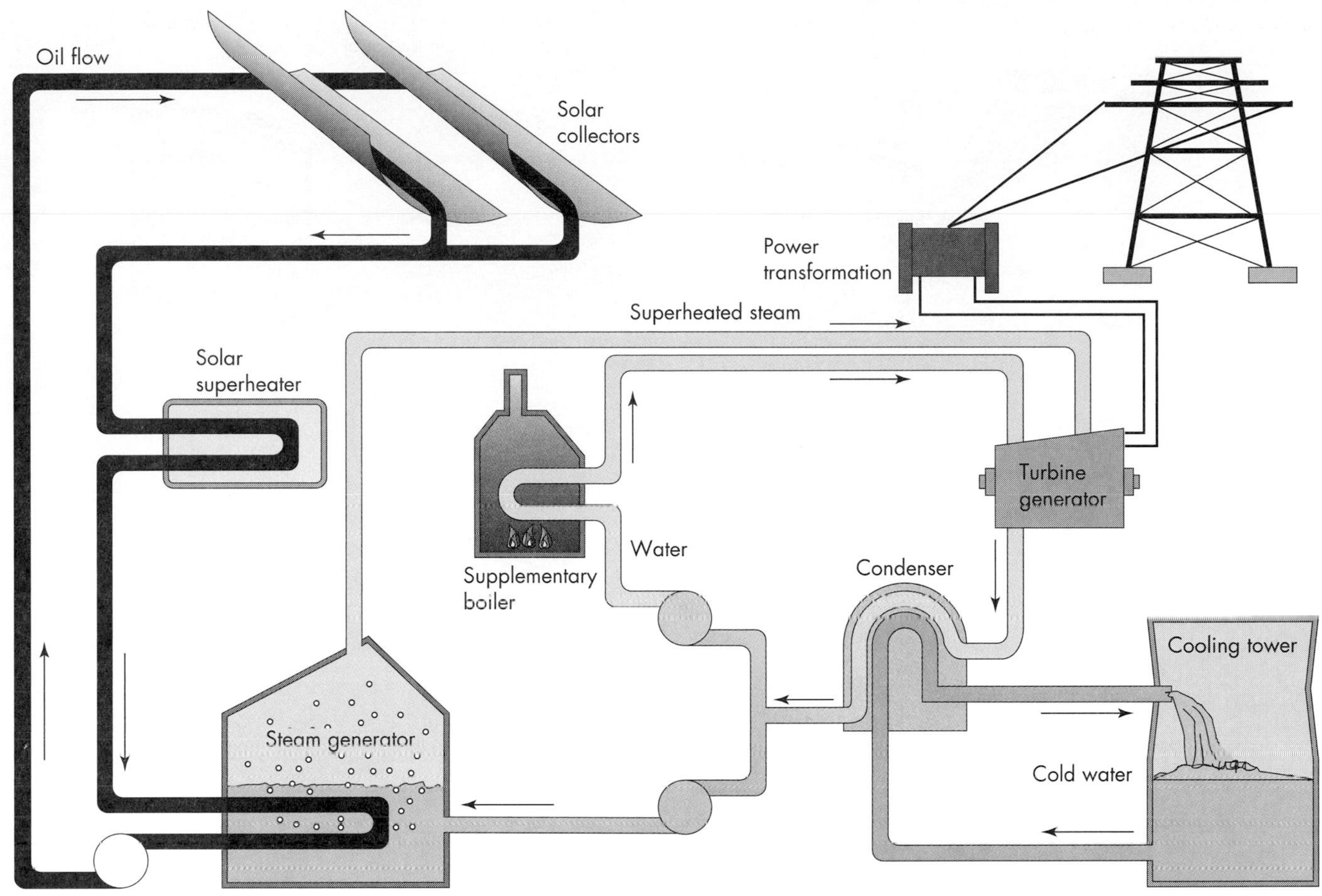

▲ **FIGURE 15.36** Idealized diagram showing how the LUZ solar electric generating system works. (*Courtesy of* LUZ)

Water Power

The use of **water power** as a source of energy has been a successful venture, at least since the time of the Roman Empire. Waterwheels were turning in western Europe in the seventeenth century, harnessing the energy of moving water and converting it to mechanical energy. During the eighteenth and nineteenth centuries, large waterwheels provided the energy to power grain mills, sawmills, and other machinery in the United States.

Hydroelectric Power Today, hydroelectric power plants provide about 15 percent of the total electricity produced in the United States. Although the total amount of electrical power produced by running water will increase somewhat in the coming years, the percentage may be reduced as other energy sources such as nuclear, direct solar, and geothermal increase at a more rapid rate.

Most of the acceptable sites for large dams to produce hydropower are probably already being utilized. On the other hand, small-scale hydropower systems may be much more common in the future. These are systems designed for individual homes, farms, or small industries. They will typically have power outputs of less than 100 kilowatts. Termed *micro-hydropower systems* (45), they represent one of the world's oldest and most common energy sources, and numerous sites in many areas have potential for producing small scale electrical power today. This is particularly true in mountainous areas, where potential energy from stream water is most available. Micro-hydropower development is by its nature very site-specific, depending upon local regulations, economic situation, and hydrologic limitations. Of particular importance is the fact that hydropower can be used to generate either electrical power or mechanical power to run machinery. Such plants may help cut the high cost of importing energy and help small operations become more independent of local utility providers (45).

An interesting aspect of hydropower is "pump storage" (Figure 15.37), the objective of which is to make better use of the total electrical energy from different sources through energy management. The basic idea is that, during times when demand for power is low, electricity from oil, coal, or nuclear plants may be used to pump water up to a storage site or reservoir (high pool). Then, when demand for electricity is high, the stored water flows back down to the low pool through generators to supplement the power supply. It is important to keep in mind that pump storage systems are not as efficient as are conventional hydroelectric plants. The fact is that about three units of energy from an oil, gas, or nuclear power plant are needed to produce two delivered units of energy from a pump storage facility. The advantage

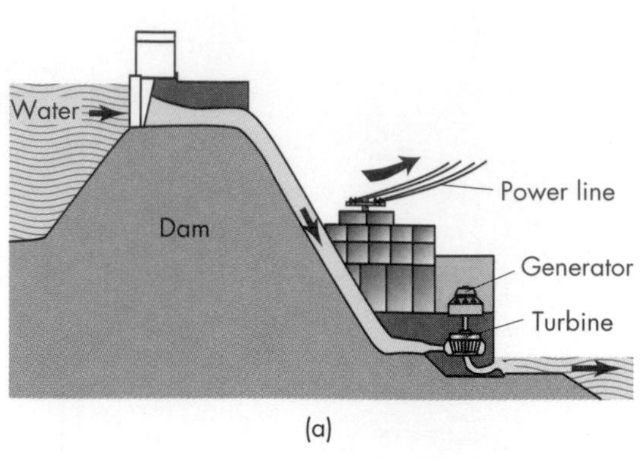

(a)

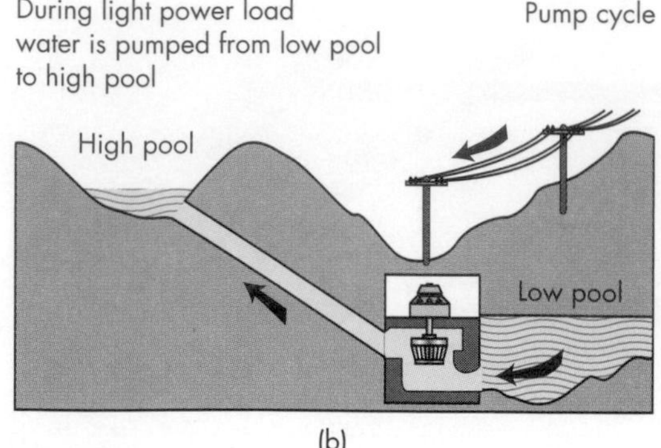

(b)

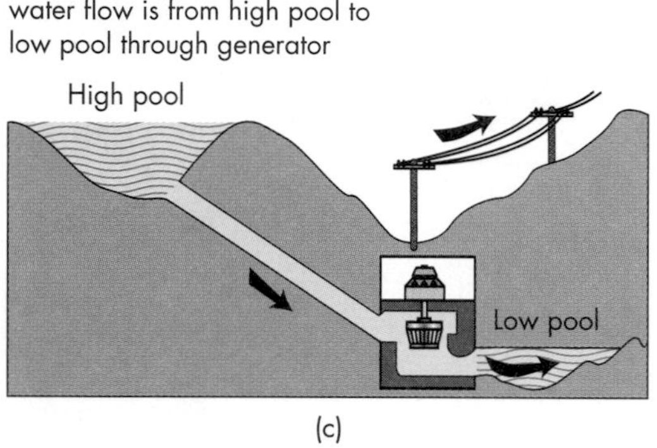

(c)

▲ **FIGURE 15.37** (a) Basic components of a hydroelectric power station. (b, c) How a pump storage system works. (*Modified from Council on Environmental Quality.* 1975. Energy alternatives: A comparative analysis. *Prepared by the Science and Public Policy Program, University of Oklahoma, Norman.*)

lies in the timing of energy production and use: The two units can be drawn during peak demand, and the three units are used to pump the water to the high pool when the demand is low.

Tidal Power Another form of water power might be derived from ocean tides in a few places where there is favorable topography, such as the Bay of Fundy region of the northeastern United States and Canada. The tides in the Bay of Fundy have a maximum rise of about 15 m. A minimum rise of about 8 m is necessary to even consider developing tidal power.

The principle of **tidal power** is to build dams across the entrance to bays, creating a basin on the landward side so as to create a difference in water level between the ocean and the basin. Then, as the water in the basin fills or empties, it can be used to turn hydraulic turbines that will produce electricity (46). A tidal power station on the river Rance near Saint Malo, France, produces more than 200,000 kilowatts of electricity from 24 power units across the dam.

Use of Water Power Water power is clean power. It requires no burning of fuel, does not pollute the atmosphere, produces no radioactive or other waste, and is efficient. There is an environmental price to pay, however. Water falling over high dams may pick up nitrogen gas, which enters the blood of fish, expands, and kills them. Nitrogen has killed many migrating game fish in the Pacific Northwest. Furthermore, dams trap sediment that would otherwise reach the sea and replenish the sand on beaches. In addition, for a variety of reasons, including displacement of people, loss of land, loss of wildlife (animals and fish), and adverse changes to river ecology and hydrology downstream, many people do not want to turn wild rivers into a series of lakes. In fact, in the United States, several dams have been removed and others are being considered for removal as a result of the adverse environmental impacts their presence is causing.

On the other hand, the world's largest dam is now being constructed in China. The Three Gorges Dam on the Yangtze River (Figure 15.38) will displace about 2 million people from their homes, drowning cities, farm fields, and highly scenic river gorges. Also lost will be habitat for endangered river dolphin. The dam will be about 185 m high and more than 1.6 km wide, and it will produce a reservoir nearly 600 km long. There is concern that the reservoir will become polluted by raw sewage and industrial pollutants currently disposed of in the river, turning the long narrow reservoir into an open sewer; drown valuable resources, including farmlands and archaeology sites; and degrade or eliminate deep-water shipping harbors at the upstream end of the reservoir, where sediments will most likely be deposited. The dam may also produce a false sense of security to people and cities downstream. If the dam and reservoir are unable to hold back floods in the future and if the dam

(a)

(b)

◀ **FIGURE 15.38** Three Gorges on China's Yangtze River. (a) Qutang Gorge, and (b) construction of temporary locks near the dam site on Xiling Gorge. (*Bob Sacha/Bob Sacha Photography*)

encourages further development in flood-prone areas, then loss of property and life may be greater than if the dam was never constructed. Finally, the dam is in a seismically active region where large landslides are common. If the dam fails, downstream cities, such as Wuhan with a population of 4 million people, would be submerged with catastrophic results (47). On the positive side, 26 giant turbines will produce about 18,000 mw of electricity (equivalent to 18 large nuclear or coal-burning power plants). However, opponents of the dam point out that a series of dams on tributaries to the Yangtze could have produced electrical power while not causing environmental damage to the main river (48).

While in the future the growth of large-scale water power is limited because of objections to building dams and the fact that many good sites for dams are already utilized, there seems to be increased interest in micro-hydropower (small dams) for supplying electricity or mechanical energy. The environmental impact of numerous micro-hydropower installations in an area may be considerable. The sites change the natural stream flow, affecting the stream biota and productivity. Small dams and reservoirs also tend to fill more quickly with sediment than larger installations, making their potential useful time period much shorter.

Because micro-hydropower development can adversely affect the stream environment, careful consideration must be given to its development over a wide region. A few such sites may cause little environmental degradation, but if the number becomes excessive, the impact over a wider region may be appreciable. This is a consideration that must be given to many forms of technology that involve small sites. The impact of a single site on a broad region may be nearly negligible, but as the number of sites increases, the total impact may become very significant.

Wind Power

Wind power, like direct solar power, has evolved over a long period of time, from early Chinese and Persian civilizations to the present. Wind has propelled ships and driven windmills for grinding grain or pumping water. More recently, wind has been used to generate electricity. The potential energy that might eventually be derived from the wind is tremendous, yet there will be problems because winds tend to be highly variable in time, place, and intensity (40).

Wind prospecting has become an important endeavor. On a national scale, regions with the greatest potential for development of wind energy are the Pacific Northwest coastal area, the coastal region of the northeastern United States, and a belt extending from northern Texas northward through the Rocky Mountain states and the Dakotas. For example, North Dakota alone has an enormous potential wind energy resource that is estimated at about 35 percent of the total electric power produced in the United States (49). There are also many other good sites, such as in mountain areas in North Carolina and the northern Coachella Valley in southern California.

At a particular site, the direction, velocity, and duration of the wind may be variable, depending on local topography and on the regional or local magnitude of temperature differences in the atmosphere (50). For example, wind velocity often increases over hilltops or mountains or when the wind

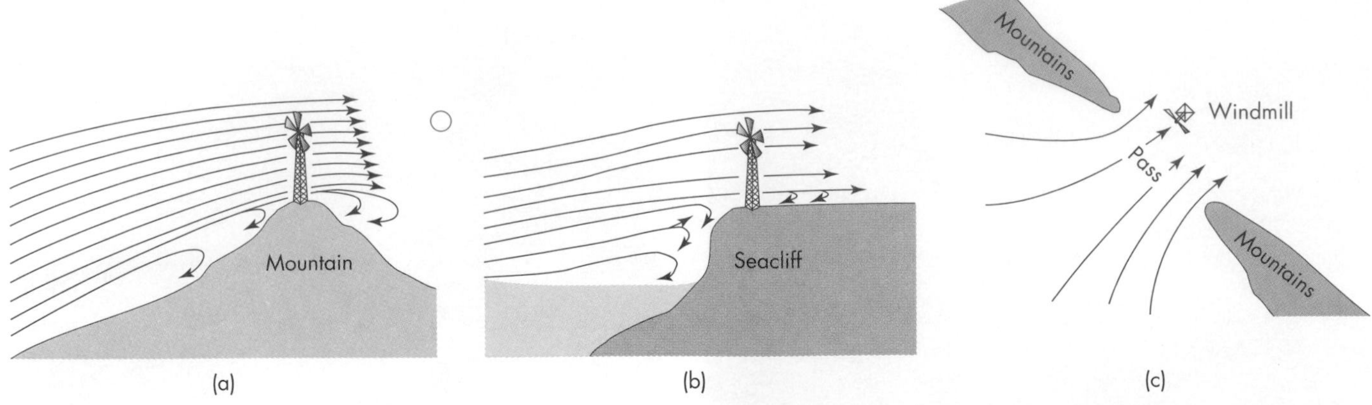

▲ **FIGURE 15.39** How wind may be converged (and velocity increased) vertically (a,b) or horizontally (c). Tall windmills are necessary on hilltops or on the top of a seacliff to avoid near-surface turbulence. (d) Wind farm in California. (*Glen Allison/Tony Stone Images*)

is funneled through a broad mountain pass (Figure 15.39). The increase in wind velocity over a mountain is caused by a vertical convergence of the wind, whereas through a pass, it is caused partly by horizontal convergence as well. Because the shape of a mountain or pass is often related to the local or regional geology, prospecting for wind energy is both a geologic and geographic and meteorologic problem.

Significant improvements in the size of windmills and the amount of power they produce occurred from the late 1800s through approximately 1950, when the United States and many countries in Europe became interested in larger-scale generators driven by the wind. In the United States, thousands of small wind-driven generators have been used on farms. Most of the small windmills generated about 1 kilowatt of power, which is much too small to be considered for central power generation needs.

Small-scale Producers Small-scale power production from windmills in the United States was made more feasible by passage of the Public Utility Regulatory Policy Act (PURPA) in 1978. This act requires utilities to interconnect with small-scale independent producers of power and pay fair market price for the electricity produced. The PURPA legislation initiated an entrepreneurism by providing a market to small-scale power producers.

Small-scale use of wind power is nowhere better exemplified than in California, where more than 17,000 windmills, with a generating capacity of about 1400 megawatts, have been installed. Individual windmills produce from 60 to 75 kilowatts of power and are arranged in wind farms consisting of clusters of windmills located in mountain passes. Electricity produced at these sites is connected to the general utility lines. These wind farms produced sufficient electricity to supply approximately 100,000 homes, making a significant contribution in the modern utility grid. Tax incentives have helped the wind-power industry become established, and improvements in technology have reduced the cost of making electricity by wind from 30 cents per kilowatt-hour in 1980 to 7 to 9 cents today. For comparison purposes, the average cost of producing electricity from fossil fuels is 6 to 7 cents per kilowatt-hour. In California it is expected that wind power may be the state's second least-expensive source of energy by the year 2012, second only to hydropower (51).

Use of Wind Power Wind power will not solve all our energy problems, but it is yet another alternative energy source that is renewable and can be used in particular sites to reduce our dependency on foreign oil. It is possible that wind power might eventually produce up to 20 percent of the electricity in the United States (49). Wind energy has a few detrimental qualities: First, demonstration projects suggest that vibrations from windmills may be a source of problems (noise and damage to the machinery); second, windmills may cause interference with radio and television broadcasting; third, windmills kill birds; and finally, windmills, particularly if there are many of them, may degrade the scenic resources of an area. Still, everything considered, wind energy has a relatively low environmental impact, and its continued use should be carefully evaluated.

Energy from Biomass

Biomass fuel is a new name for the oldest human fuel. **Biomass** is organic matter that can be burned directly as a fuel or converted to a more convenient form and then burned.

For example, we can burn wood in a stove or convert it to charcoal and then burn it. Biomass has provided a major source of energy for human beings throughout most of the history of civilization. When Europeans first settled in North America, there was more wood fuel than could be used. They often cleared sites for agriculture by girdling trees (cutting through the bark all the way around the trunk) to kill them and then burning the forest. Much less of the world is forested today, but more than 1 billion people still use wood as their primary source of energy for heat and cooking. Although firewood is the best known and most widely used biomass fuel, there are many others. In India and other countries, cattle dung is burned for cooking. In northern areas such as Scotland, where peat is abundant, it provides heating and cooking fuel.

Energy from biomass may take several routes: burning of biomass to heat water or air directly, burning of biomass to produce electricity, and distillation of biomass to produce alcohol for fuel. Today in the United States, various biomass sources supply more than 3 percent of the entire energy consumption (approximately 2.8 exajoules annually), primarily from the use of wood for industry and home heating (52,53).

The three primary sources of biomass fuels in North America are forest products, otherwise unused agricultural products, and urban waste. Biomass can be recycled, and today numerous facilities in the United States process urban waste for use in generating electricity or producing fuel. At processing plants such as those in Baltimore County, Maryland; Chicago, Illinois; Milwaukee, Wisconsin; Tacoma, Washington; and Akron, Ohio, municipal waste is burned and the heat energy is used to make steam for a variety of purposes ranging from space heating to industrial generation of electricity. The United States, owing to lack of public acceptance, has been slower than other nations to use urban waste as an energy source. In western Europe a number of countries now use between one-third to one-half of their municipal waste for energy production (38).

15.8 Conservation, Efficiency, and Cogeneration

We established earlier in this chapter that we will have to get used to living with a good deal of uncertainty concerning the availability, cost, and environmental effects of energy use. Furthermore, we can expect that serious shocks will continue to occur, disrupting the flow of energy to various parts of the world.

Both the supply and the demand for energy are difficult to predict because technical, economic, political, and social assumptions underlying projections are constantly changing. Large annual variations in energy consumption must also be considered: Energy consumption peaks during the winter months (heating), with a secondary peak in the summer (air conditioning). Future changes in population or intensive conservation measures may change this pattern, as might better design of buildings and more reliance on solar energy.

There has been a strong movement to change patterns of energy consumption through such measures as conservation, increased efficiency, and cogeneration. **Conservation** of energy refers to a moderation of our energy demand. Pragmatically, this means adjusting our energy uses to minimize the expenditure of energy necessary to accomplish a given task. **Efficiency** means designing and using equipment that yields more power from a given amount of energy, wasting less of the energy as heat (54). Finally, **cogeneration** refers to a number of processes that capture and use some of the waste heat rather than simply release it into the atmosphere or water as thermal pollution.

The three concepts of energy conservation, increased efficiency, and cogeneration are very much interrelated. For example, when electricity is produced at large coal-burning power stations, large amounts of heat may be emitted into the atmosphere. Production of electricity typically burns three units of fuel to produce one unit of electricity, an energy loss of about 67 percent. The use of a "unit of fuel" is arbitrary. It could, for example, be a barrel of oil or a ton of coal. Cogeneration, which involves recycling of that waste heat, can increase the efficiency of a typical power plant from 33 percent to as high as 75 percent. Put the other way around, cogeneration reduces energy loss from 67 percent to as little as 25 percent (54).

Cogeneration of electricity also involves the production of electricity as a by-product from industrial processes that normally produce and use steam as part of their regular operations. Optimistic energy forecasters estimate that we may eventually be able to meet approximately one-half of the electrical power needs of industry through cogeneration (55). In fact, it has been estimated that, by the year 2000, more than 10 percent of the power capacity of the United States could be provided through cogeneration. This would mean fuel savings of as much as 2 million barrels of oil daily! When we talk about savings, we are talking about conservation, showing that conservation and cogeneration are interrelated.

15.9 Energy Policy for the Future

Hard Path Versus Soft Path

Energy policy today is at a crossroads. One road leads to development of so-called *hard technologies*, which involve finding ever-greater amounts of fossil fuels and building larger centralized power plants. Following this **hard path** means continuing "business as usual." This is the more comfortable approach, in that it requires no new thinking or realignment of political, economic, or social conditions. It also involves little anticipation of the inevitable depletion of the fossil fuel resources on which the hard path is built.

Those in favor of the hard path argue that environmental problems have occurred in less-developed countries because people have had to utilize local resources such as wood for energy rather than, say, for land conservation and erosion control. In this view, the way to solve these problems is to provide less-developed countries with cheap energy that utilizes more heavy industrialization and technology. Furthermore, the United States and other nations with sizable resources of coal or petroleum should exploit these resources to prevent environmental degradation in their own countries. Proponents of this view maintain that giving the energy industry the freedom to develop available resources will ensure a steady supply of energy and less total environmental damage than if the government regulates the energy industry. They point to the recent increase in the burning of firewood across the United States as an early indicator of the effects of strong governmental controls on energy supplies. The eventual depletion of forest resources, they maintain, will have a detrimental effect on the environment, just as it has done in so many less-developed countries.

The other road is designated as the **soft path** (56). One of the champions of this choice has been Amory Lovins, who states that the soft path involves energy alternatives that are renewable, flexible, decentralized, and environmentally more benign than those of the hard path. As defined by Lovins, these alternatives have several characteristics:

1. They rely heavily on renewable energy resources such as solar, wind, and biomass.
2. They are diverse, tailored for maximum effectiveness under specific circumstances.
3. They are flexible and can be decentralized; that is, they are relatively low technologies that are accessible, understandable to many people, and can be used in a variety of locations and settings.
4. They are matched in both geographical distribution and scale to prominent end-use needs (the actual use of energy).
5. There is good agreement or matching between energy-grade and end-use. The term *energy-grade* refers to energy sources such as electricity (a high-grade energy) that can be used for many end-uses. An example of a lower-grade energy is solar-heated water to warm buildings.

The last point is of particular importance because, as Lovins points out, people are not particularly interested in having a certain amount of oil, natural gas, or electricity delivered to their homes; rather, they are interested in having comfortable homes, adequate lighting, food on the table, and energy for transportation (56). These latter uses are called *end-uses*, and only about 5 percent of end-uses really require high-grade energy such as electricity. Nevertheless, a lot of electricity is used to heat homes and water. Lovins points out the wastefulness of burning fossil fuels at high temperatures or using nuclear reactions at still higher temperatures to supply an end-use temperature increase of only a few tens of degrees.

The United States annually consumes about 90 exajoules of energy. Projections of U.S. energy consumption in the year 2030 suggest it may be as high as 120 exajoules or as low as 60. Why is there such a big discrepancy? If we stay on the hard path, the high value is probably appropriate. On the other hand, with intensive energy conservation and increased efficiency, annual consumption of energy could be cut in half. This low-energy scenario is the soft path. Actual energy consumption in the year 2030 will probably not be as low as 60 exajoules; it is hoped it will not exceed 90 to 100 exajoules. Achieving this level will require, with expected population increase, a substantial commitment to energy conservation and increased energy efficiency.

There is debate as to whether our energy planning should seek reduced energy consumption by following the soft path or to rely heavily on the fossil fuels, primarily coal. Both scenarios, however, speculate that nuclear energy will play a reduced role in the total energy picture by the year 2020. Certainly, there is considerable public criticism of nuclear energy in the United States and in other parts of the world (57). The energy option that can yield the greatest return in the near future is conservation. We have the technology now to save sufficient energy to eliminate or greatly reduce the need to import oil. Such a plan would reduce U.S. dependency on foreign sources of energy and improve our economy significantly.

A Sustainable Energy Policy

A major problem of energy planning for tomorrow is that we are now conscious that burning of fossil fuels is degrading our global environment. The soft path is called the *environmental path* because it would help reduce environmental degradation by reducing emissions of carbon dioxide and air pollutants. To be fair, advocates of the hard path propose increased energy conservation, increased efficiency, and cogeneration to reduce consumption and environmental problems associated with burning fossil fuels. Although there is sufficient coal to last hundreds of years, proponents of the soft path would prefer to use coal as a transitional rather than a long-term energy source. In their view, development of a **sustainable energy policy** means finding useful sources of energy that do not have the adverse environmental effects associated with the burning of fossil fuels.

A transition from the hard to the soft path would presumably involve continued utilization of fossil fuels. Electric power will still be essential for many purposes; in fact, with innovations like the fuel cell, electricity may help solve some of the problems of pollution and wasteful use of resources (see *A Closer Look: Fuel Cells*). The energy path we take must be one capable of supplying the energy we require for human activities without endangering the planet (58). This is the heart of the concept of sustainable energy policy.

Using less energy need not result in a lower quality of life. As Amory Lovins emphasizes, increased energy conservation and a more energy-efficient distribution of urban populations, agriculture, and industry will allow us to maintain our standard of living. Alternatives that will help achieve lower energy consumption include new settlement patterns that encourage "urban villages" with populations of about 100,000; constrained and minimized private transportation and maximum accessibility of services; agricultural practices that emphasize locally grown and consumed food consisting largely of

A CLOSER LOOK Fuel Cells

When we first started using fossil fuels in the internal combustion engine we were not particularly conscious of potential pollution problems from large numbers of automobiles. Since then we have built engines that burn fuel more efficiently, and we have added devices to reduce emissions from automobiles. Similarly, we have developed new technology to help reduce emissions from power plants that burn coal or oil. Despite all this good work, we still have serious environmental problems related to burning fossil fuels, which has led some people to conclude that we must seek an environmentally benign technology capable of generating electrical power efficiently (59). Fuel cells may be part of that technology.

Figure 15.F shows the basic components of a fuel cell. In the simplest case, the cell uses hydrogen as a fuel to which an oxidant (oxygen) is supplied. An electrolyte solution allows the flow of ions between the positive and negative electrodes. What happens is a chemical reaction in which hydrogen is combined with oxygen, just as if the hydrogen were burned. However, the reactants are separated, and in order for them to combine, they must be ionized and migrate as ions through the electrolyte solution. Electrons released in this process move from the negative to the positive electrode; if they were not diverted, their route would be through the solution. The function of the fuel cell is to reroute the electrons through an electric motor, thus supplying the current that keeps the motor running. Both the hydrogen fuel and oxygen are added as needed to produce electric power.

A fuel cell or a series of cells, depending on the amount of energy needed, may be used to provide electricity for a building complex or to run a vehicle such as a bus or automobile. Several fuel-cell-powered buses are now in operation at the Los Angeles International Airport, and several such buses are used by B.C. Transit in Vancouver, British Columbia. At a larger scale, fuel cells are being commercially produced to provide energy for building complexes, and at least one fuel-cell power plant with a capacity of approximately 2 megawatts is under construction (59).

If hydrogen is used as the fuel for a fuel cell, the only waste products are oxygen and water. Hydrogen can be generated from renewable energy sources such as solar or wind power that is used to split water into hydrogen and oxygen, but large amounts of hydrogen to be used as a fuel are not generally available at present. Fortunately, fuel cells can also use other fuels such as methane, which is a hydrogen-rich gas. Burning methane does produce some pollutants, but the amount is only about 1 percent of what would be produced by burning fossil fuels in a conventional power plant or internal combustion engine (59).

In the future we will have to make a transition from the fossil fuels to other energy resources. Fuel-cell technology is arriving at the right time to play an important role in that transition.

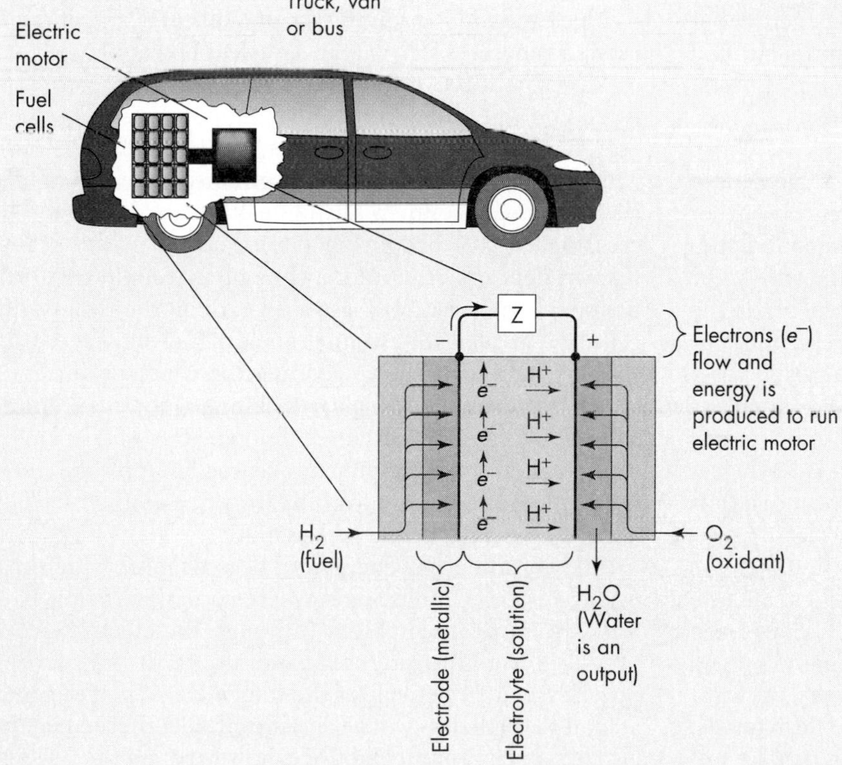

◀ **FIGURE 15.F** Idealized diagram showing a fuel cell.

grains, beans, potatoes, and vegetables; and new industrial guidelines designed to conserve energy and to reduce total energy demand by minimizing production and consumer waste (55). To an extent, these alternatives are already taking form: Highway construction is sometimes given a lower priority than development of mass transit systems, agricultural lands near urban centers are sometimes preserved, and industry is more receptive to recycling and to decreasing production of consumer wastes such as unnecessary packaging.

As long ago as 400,000 years, early humans used fire for warmth and cooking. Throughout most of human history, few people have cared much about the source of energy, but rather have been concerned with the services it provides. We want to cook our food, heat our homes, drive our automobiles, and take hot showers. However, in recent years we have become conscious of the fact that our use of some sources of energy may be causing global changes potentially detrimental to our environment. These include, among others, possible global warming, which will be discussed in Chapter 16. Therefore, the path we follow in energy management should be one that minimizes adverse local, regional, and global change (58).

SUMMARY

Both the ever-increasing population and appetite of the American people and the world for energy can be staggering. It is time to question seriously the need for and desirability of an increasing demand for electrical and other sources of energy in industrialized societies. Quality of life is not necessarily directly related to greater consumption of energy.

The fossil fuels—coal, oil, and natural gas—are essentially solar energy stored in organic material that has escaped destruction by oxidation. Although the geologic processes that formed these fuels are ongoing, they are too slow to be of use to us, so the fossil fuels are nonrenewable resources. The environmental disruption associated with exploration and development of these resources must be weighed against the benefits gained from the energy; but this development is not an either-or proposition, and sound conservation practices combined with pollution control and reclamation can help minimize the environmental disruption associated with fossil fuels.

Vast supplies of coal still exist in the world, 25 percent of them in the United States. The grade (carbon content) of coal determines its value as a fuel, but the sulfur content determines how much it pollutes the atmosphere with sulfur oxides. Increased coal mining worldwide in response to the diminishing of other energy sources may increase the supply of cleaner (lower-sulfur) coal. Coal mining is increasingly done by strip mining (open-pit mining), which has adverse environmental effects, particularly in humid regions where acid mine drainage is a problem. U.S. law now requires restoration of strip-mined land to (in most cases) its previous use.

The hydrocarbon fossil fuels (oil and natural gas) are found in large deposits (oil and/or gas fields), most of them located near tectonic plate boundaries. Oil fields have been extensively mined by means of oil wells that pump oil and natural gas to the surface, and barring discovery of major new fields, shortages of oil will occur in the near future. Peak production of oil will likely occur in about 2020, followed by reduced production and shortages.

Oil can be recovered from earth materials called *oil shale* and *tar sands*, but mining techniques that are economically and environmentally sound have not been fully developed. The potential for environmental disruption exists at every stage of oil development and use; disruptions include oil spills from tankers and air pollution from burning petroleum products in automobiles and power plants. Burning fossil fuels produces emissions of sulfur dioxide, which is responsible for acid rain.

Acid rain, which refers to both wet and dry acid deposition, is a serious environmental problem in many parts of the world today. Sulfur and nitrogen oxides emitted into the atmosphere undergo complex chemical processes that eventually produce sulfuric and nitric acid. Effects of acid rain include damage to soils and vegetation, to lake ecosystems, and to human artifacts such as buildings and monuments.

Nuclear fission produces vast amounts of heat that can be used to generate electricity in a nuclear power plant. It also produces radioactive wastes that must be disposed of safely. Fission will probably remain an important source of energy, but the growth of fission reactors in the United States will not be as rapid as originally projected because of concern about environmental and health hazards and the increasing cost of producing large nuclear power plants. Known deposits of uranium, the only naturally occurring fissionable material, will last until early in the twenty-first century. To avert an eventual nuclear fuel shortage, work should continue on the breeder reactor, which produces fissionable plutonium. Nuclear fusion is a potential energy source for the future. Although fusion reactors are still in the experimental stage, fusion may eventually supply a tremendous amount of energy from hydrogen, a readily available and nearly inexhaustible fuel supply.

Use of geothermal energy will become more widespread in the western United States where natural heat flow from the earth is relatively high. Although the electric energy produced from the internal heat of the earth will probably never exceed 10 percent of the total electric power generated, it nevertheless will be significant. Geothermal energy also has an environmental price, however, and the possibility of causing surface subsidence through withdrawal of fluids and heat, as well as possibly causing earthquakes from injection of hot wastewater back into the ground, must be considered.

Renewable sources of energy depend upon solar energy and can take a variety of forms, including direct solar, water, wind, and energy from biomass (including recycled biomass from urban waste). These energy sources, which are generally used to produce electric power, will not be depleted and are thus dependable in the long term. They have varying attributes, but generally cause little environmental disruption and, except for the burning of biomass, do not pollute the atmosphere. However, most of these sources are local and intermittent and some are still expensive to produce. Of particular importance in the next few years will be the continued growth in direct solar energy as well as continued growth in the recycling of urban waste to produce energy. Hydropower will undoubtedly continue to be an important source of electricity in the future, but it is not expected to grow much because of lack of potential sites and environmental considerations. Wind power and use of solar cells, along with various biomass alternatives, are still somewhat experimental, so it is difficult to predict their growth.

Energy for the future will continue to be troubled by uncertainties. However, it seems certain that we will continue to look more seriously at conservation, energy efficiency, and cogeneration. The most likely targets for energy efficiency and conservation as well as cogeneration are in the area of space heating for homes and offices, water heating, industrial processes that provide heat (mostly steam) for various manufacturing processes, and motor vehicles. These areas collectively account for approximately 60 percent of the total energy used in the United States today.

We may be at a crossroads today concerning energy policy. The choice is between the "hard path," characterized by centralized, high-technology energy sources, and the "soft path," characterized by decentralized, flexible, renewable energy sources. Perhaps the best path will be a mixture of the old and the new, ensuring a rational, smooth shift from depending so heavily on fossil fuels.

Finally, whatever path we take should have the goal of developing a sustainable energy plan that supplies the energy we need but does not harm the environment. Fuel cells are part of a potential technology that may help in the transition from hard to soft paths.

REFERENCES

1. **Averitt, P.** 1973. Coal. In *United States mineral resources,* eds. D. A. Brobst and W. P. Pratt, pp. 133–42. U.S. Geological Survey Professional Paper 820.
2. **British Petroleum Company.** 1998. *B.P. statistical review of world energy.*
3. **Rahn, P. H.** 1982. *Engineering geology: An environmental approach.* New York: Elsevier.
4. **Garbini, S., and Schweinfurth, S. P., eds.** 1986. *Symposium Proceedings: A national agenda for coal quality research, April 9–11, 1985.* U.S. Geological Survey Circular 979.
5. **U.S. Environmental Protection Agency.** 1973. *Processes, procedures and methods to control pollution from mining activities.* EPA-430/9-73-001.
6. **Committee on Environment and Public Planning.** 1974. Environmental impact of conversion from gas or oil to coal for fuel. *The Geologist: Newsletter of the Geological Society of America.* Supplement to vol. 9, no. 4.
7. **Nieman, T. J., and Meshako, D.** 1990. The Star Fire Mine reclamation experience. *Journal of Soil and Water Conservation* 45(5):29–32.
8. **Kesler, S. E.** 1994. *Mineral resources, economics, and the environment.* New York: Macmillan.
9. **McCabe, P. J., Gautier, D. L., Lewan, M. D., and Turner, C.** 1993. *The future of energy gases.* U.S. Geological Survey Circular 1115.
10. **McCulloh, T. H.** 1973. Oil and gas. In *United States mineral resources,* eds. D. A. Brobst and W. P. Pratt, pp. 477–96. U.S. Geological Survey Professional Paper 820.
11. **Culbertson, W. C., and Pitman, J. K.** 1973. Oil shale. In *United States mineral resources,* eds. D. A. Brobst and W. P. Pratt, pp. 497–503. U.S. Geological Survey Professional Paper 820.
12. **Duncan, D. C., and Swanson, V. E.** 1965. *Organic-rich shale of the United States and world land areas.* U.S. Geological Survey Circular 523.
13. **Committee on Environmental and Public Planning.** 1974. Development of oil shale in the Green River Formation. *The Geologist: Newsletter of the Geological Society of America.* Supplement to vol. 9, no. 4.
14. **Office of Oil and Gas, U.S. Department of the Interior.** 1968. *United States petroleum through 1980.* Washington, DC: U.S. Government Printing Office.
15. **Allen, A. R.** 1975. Coping with oil sands. In *Perspectives on energy,* eds. L. C. Ruedisili and M. W. Firebaugh, pp. 386–96. New York: Oxford University Press.
16. **Youngquist, W.** 1998. Spending our great inheritance. Then what? *Geotimes* 43(7):24–27.
17. **Canadian Department of Environment.** 1984. *The acid rain story.*
18. **Winkler, E. M.** 1998. The complexity of urban stone decay. *Geotimes* 43(9):25–29.
19. **U.S. Geological Survey.** 1996. *Trends in precipitation chemistry in the United States (1983–94): An analysis of the effects of Phase I of the Clean Air Act Amendments of 1990.* Title IV. Open File Report 96-0346.
20. **Edwards, J. D.** 1997. Crude oil and alternative energy production forecast for the twenty-first century: The end of the hydrocarbon era. *American Association of Petroleum Geologists Bulletin* 81(8):1292–1305.
21. **Finch, W. I., et al.** 1973. Nuclear fuels. In *United States mineral resources,* eds. D. A. Brobst and W. P. Pratt, pp. 455–76. U.S. Geological Survey Professional Paper 820.
22. **Duderstadt, J. J.** 1977. Nuclear power generation. In *Perspectives on energy,* eds. L. C. Ruedisili and M. W. Firebaugh, pp. 249–73. New York: Oxford University Press.

23. **Energy Research and Development Administration.** 1978. *Advanced nuclear reactors: An introduction.* ERDA-76-107.
24. **U.S. Geological Survey.** 1973. *Nuclear energy resources: A geologic perspective.* U.S.G.S. INF-73-14.
25. **U.S. Department of Energy.** 1980. *Magnetic fusion energy.* DOE/ER-0059.
26. **U.S. Department of Energy.** 1978. *The United States magnetic fusion energy program.* DOE/ET-0072.
27. **MacLeod, G. K.** 1981. Some public health lessons from Three Mile Island: A case study in chaos. *Ambio* 10:18–23.
28. **Anspaugh, L. R., Catlin, R. J., and Goldman, M.** 1988. The global impact of the Chernobyl reactor accident. *Science* 242:1513–18.
29. **Duffield, W. A., Sass, J. H., and Sorey, M. L.** 1994. *Tapping the earth's natural heat.* U.S. Geological Survey Circular 1125.
30. **Smith, R. L., and Shaw, H. R.** 1975. Igneous-related geothermal systems. In *Assessment of geothermal resources of the United States—1975,* eds. D. F. White and D. L. Williams, pp. 58–83. U.S. Geological Survey Circular 726.
31. **White, D. F.** 1969. *Natural steam for power.* U.S. Geological Survey Publication, GP0-19690-339-S36.
32. **White, D. F., and Williams, D. L., eds.** 1975. In *Assessment of geothermal resources of the United States—1975.* U.S. Geological Survey Circular 726.
33. **Muffler, L. J. P.** 1973. Geothermal resources. In *United States mineral resources,* eds. D. A. Brobst and W. P. Pratt, pp. 251–61. U.S. Geological Survey Professional Paper 820.
34. **Worthington, J. D.** 1975. *Geothermal development. Status Report—Energy Resources and Technology: A Report of the Ad Hoc Committee on Energy Resources and Technology.* Atomic Industrial Form, Inc.
35. **Tester, J. W., Brown, D. W., and Potter, R. M.** 1989. *Hot dry rock geothermal energy—A new energy agenda for the 21st century.* Los Alamos National Laboratory. LA-11514-MS, UC-251.
36. **Tenenbaum, D.** 1994. Deep heat. *Earth* 3(1):58–63.
37. **Papadopulus, S. S., et al.** 1975. Assessment of onshore geopressured-geothermal resources in the northern Gulf of Mexico basin. In *Assessment of geothermal resources of the United States—1975,* eds. D. F. White and D. L. Williams, pp. 125–46. U.S. Geological Survey Circular 726.
38. **Council on Environmental Quality.** 1979. Environmental quality. *Tenth Annual Report of the Council on Environmental Quality.* Washington, DC: U.S. Government Printing Office.
39. **Lovins, A. B.** 1979. Foreword. In *A golden thread,* eds. K. Butti and J. Perlin. Palo Alto, CA: Cheshire Books.
40. **Eaton, W. W.** 1976. Solar energy. In *Perspectives on energy,* 2nd ed., eds. L. C. Ruedisili and M. W. Firebaugh, pp. 418–50. New York: Oxford University Press.
41. **Stone, J. L.** 1993. Photovoltaics: Unlimited electrical energy from the Sun. *Physics Today* 64(9):22–29.
42. **Johnson, J. T.** 1990. The hot path to solar electricity. *Popular Science* 241(5):82–85.
43. **Becker, N. D.** 1992. The demise of LUZ: A case study. *Solar Today,* January–February, 24–26.
44. **DiChristina, M.** 1995. Sea power. *Popular Science* 246(5):70–73.
45. **Alward, R., Eisenbart, S., and Volkman, J.** 1979. *Microhydropower: Reviewing an old concept.* National Center for Appropriate Technology. U.S. Department of Energy.
46. **Committee on Resources and Man, National Academy of Science.** 1969. *Resources and man.* San Francisco: W. H. Freeman.
47. **Pearce, F.** 1995. The biggest dam in the world. *New Scientist,* January, pp. 25–29.
48. **Zich, R.** 1997. China's Three Gorges: Before the flood. *National Geographic* 192(3):2–33.
49. **Abelson, P. H.** 1993. Power from wind turbines. *Science* 261:1255.
50. **Nova Scotia Department of Mines and Energy.** 1981. *Wind power.*
51. **Weinberg, C. J., and Williams, R. H.** 1990. Energy from the sun. *Scientific American* 263(3):147–55.
52. **Energy Information Administration.** 1990. *Annual Energy Review 1989.*
53. **Office of Technology Assessment.** 1980. *Energy from biological processes,* OTA-E-124.
54. **Darmstadter, J., Landsberg, H. H., Morton, H. C., with Coda, M. J.** 1983. *Energy today and tomorrow.* Englewood Cliffs, NJ: Prentice-Hall.
55. **Steinhart, J. S., Hanson, M. E., Gates, R. W., Dewinkel, C. C., Briody, K., Thornsjo, M., and Kabala, S.** 1978. A low-energy scenario for the United States: 1975–2050. In *Perspectives on energy,* eds. L. C. Ruedisili and M. W. Firebaugh, pp. 553–88. New York: Oxford University Press.
56. **Lovins, A. B.** 1979. *Soft energy paths: Towards a durable peace.* New York: Harper & Row.
57. **Hafele, W.** 1990. Energy from nuclear power. *Scientific American* 263(3):137–44.
58. **Davis, G. R.** 1990. Energy for planet Earth. *Scientific American* 263(3):55–74.
59. **Kartha, S., and Grimes, P.** 1994. Fuel cells: Energy conversion for the next century. *Physics Today* 47(11):54–61.

KEY TERMS

coal (p. 402)
strip mining (p. 405)
oil (p. 407)
natural gas (p. 407)
oil shale (p. 412)
tar sands (p. 413)
acid rain (p. 414)
fission (p. 417)
burner reactors (p. 418)
breeder reactors (p. 420)
fusion (p. 421)
geothermal energy (p. 422)
renewable energy (p. 428)
solar energy (p. 428)
photovoltaics (p. 430)
water power (p. 433)
tidal power (p. 434)
wind power (p. 435)
biomass (energy source) (p. 436)
conservation (p. 437)
efficiency (p. 437)
cogeneration (p. 437)
hard path (p. 437)
soft path (p. 438)
sustainable energy policy (p. 438)

SOME QUESTIONS TO THINK ABOUT

1. When we first started using fossil fuels, particularly oil, we did not know much about potential environmental impacts of developing and burning oil, nor were we particularly concerned. Suppose at that time we had completed an environmental impact report concerning the use of oil and were able to predict consequences such as air pollution, toxicity, and acid rain. Do you think we would have developed the use of oil as fast as we did and become so dependent upon it? Justify your answer.
2. Some people have argued that the safety of nuclear power is not a technology question but a people question. That is, we can manufacture nuclear power plants that have an extremely low probability of an accident, but we cannot yet manufacture people who make no mistakes. As a consequence, it can be argued, we will never increase the safety of nuclear power unless we take safety issues completely out of the hands of individuals and leave it up to computers and sensors within the reactor. Do you agree with this? Explain your answer.
3. Sustainable energy development means developing an energy policy and energy sources that will provide the energy that society needs while not harming the environment. Do you think this is possible? Outline a plan of action to move the United States toward sustainable energy development.
4. Those who favor the hard path of energy development say we can solve environmental problems in less-developed countries by providing them with cheap energy, thus utilizing more heavy industrialization and technology, not less, as advocated by the soft path. They might further argue that intensive energy conservation, efficiency, and cogeneration will produce a higher-quality environment for people than will transforming our energy policy to follow the soft path. How would you respond to such a statement? Do you agree or disagree? Justify your answer.

PART 5

Global Change, Land Use, and Decision Making

CHAPTER 16
Global Change and Earth System Science

CHAPTER 17
Air Pollution

CHAPTER 18
Landscape Evaluation and Land Use

The concept of global change is not new: For more than 100 years, geologists have studied the changes in the earth that have occurred during its 4.6 billion years of history. The pace of change in the earth and its atmosphere has increased, however, as a result of human activity. Society today is very concerned with changes in the atmosphere that are taking place over a period of several decades. Foremost among the concerns are the increase in atmospheric carbon dioxide, which may already have brought about global warming, and reduction in the ozone layer, which may affect our health.

Although not all changes are beneficial, change is a natural characteristic of our environment, and wise management accepts this fact. Chapter 16 presents principles of global change and earth system science that foster better land use and decision making. Chapter 17 discusses ways of dealing with the specific problem of air pollution.

One controversial environmental issue today is determining the most appropriate uses of our land and resources. The controversy is a human issue that involves responsibility to future generations as well as compensation for present-day landowners. Before we can plan the use of land and resources, we must develop reliable methods to evaluate the land and develop a legal framework within which we can make sound environmental decisions. Although environmental law is not a branch of geology, there is a real need for applied earth scientists to know something about our judicial system and the laws and legal theory affecting the natural environment. Chapter 18 discusses several aspects of landscape evaluation such as land-use planning, site selection, landscape aesthetics, environmental impact, and environmental law.

16 Global Change and Earth System Science

Earth from space. (Courtesy of Lyndon B. Johnson Space Center/NASA)

Viewing the earth from space allows us to observe our home as a system. The portion of the earth shown here is from the Mediterranean Sea at the top to the Antarctic polar ice cap at the bottom. The large land mass is Africa. The swirling clouds reflect the dynamic nature of our atmosphere. Today we are trying to understand how earth's major systems (land, oceans, biosphere, and atmosphere) have evolved and are maintained. One of the most fascinating and important aspects is interactions between physical and biological processes that sustain life on earth.

16.1 Global Change and Earth System Science: An Overview

Preston Cloud—a famous earth scientist interested in the history of life on earth, human impact on the environment, and the use of resources—stated that two central goals of the earth sciences are (2):

- To understand how the earth works and how it has evolved from a landscape of barren rock to the complex landscape dominated by the life we see today.
- To apply that understanding to better manage our environment.

Cloud's comments emphasize that our planet is characterized by a complex evolutionary history. Interactions among the atmosphere, oceans, solid earth, and biosphere have resulted in development of a complex and abundant diversity of landforms—continents, ocean basins, mountains, lakes, plains, and slopes—as well as the abundant and diverse life-forms that inhabit a broad spectrum of habitats.

Early in this book we advanced the principle that the earth is a dynamic evolving system. We now return to that theme. To better understand global change we will discuss:

- The goals of earth system science and global change
- Earth's atmosphere and energy budget

LEARNING OBJECTIVES

It is often said that the only certainties in life are death and taxes, but to this we should also add change. We are all interested in changes that will affect us, our families, our society, and the world. The sorts of changes we might encounter are many—varying from gradual to accelerating, abrupt, chaotic, or surprising. Some changes will affect our local, regional, or global environment. Some changes will affect all our lives (1)! With this is mind, learning objectives for this chapter are:

- *To know what earth system science is and its two main goals.*
- *To know the major research priorities of earth system science and global change.*
- *To gain a modest appreciation for the composition of the atmosphere and earth's energy balance.*
- *To know what tools are used for studying global change.*
- *To become familiar with climate and climate change.*
- *To be able to discuss El Niño and La Niña events.*
- *To be able to discuss the issue of potential global warming and how it is related to increasing atmospheric carbon dioxide.*
- *To be familiar with basic issues related to stratospheric ozone depletion.*

Web Resources

Visit the "Environmental Geology" Web site at www.prenhall.com/keller to find additional resources for this chapter, including:

- Web Destinations
- On-line Quizzes
- On-line "Web Essay" Questions
- Search Engines
- Regional Updates

- The tools by which we may study change
- Climate change
- Increase in atmospheric carbon dioxide and the greenhouse effect
- Stratospheric ozone depletion
- Particulates in the atmosphere

Until recently it was generally thought that human activity caused only local, or at most, regional environmental change. It is now generally recognized that the effects of human activity on the earth are of such an extent that we are involved in unplanned planetary experiments. To understand and perhaps modify the changes we have initiated, we need to understand how the entire earth works as a system. The emerging discipline called **earth system science** seeks to further this understanding by learning how the various components of the system—the atmosphere, oceans, land, and biosphere—are linked on a global scale and interact to affect life on earth.

Figure 16.1 shows some of the major research areas of earth system science and global change, including solid earth, atmospheric chemistry, carbon cycle, hydrologic cycle, heat transport, and clouds and radiation. In many complex ways, these components interact with physical, chemical, and biological processes that cause global change. The primary goal of earth system science is understanding how the various components have evolved, are linked to each other, presently function, and can be expected to function and evolve in the future, at a time scale significant to people (3). A major secondary goal is prediction of global changes that are likely to occur within the next few decades to a century. Of particular importance is the distinction between human-induced changes and natural changes and interactions between the two (3).

Changes in the global system that can be observed within an individual's lifetime are called *first-order changes;* such changes occur over a period of up to a hundred years. *Second-order changes* are those that occur over the time span of the historic record, which in some areas is several thousand years. Third-order changes occur over periods of tens of thousands of years, corresponding, for example, to the phases of an ice age. Longer periods of change, such as fourth- or fifth-order variability, are related to the time necessary for an entire ice age or a major glacial event, which is on the order of approximately 2 million to 3 million years (4). First-order and second-order changes are the most significant to humans. These are the time ranges over which we must find a way to plan for and work with potential change.

To reach the goals of earth system science, it is necessary to establish research priorities. These have been summarized as (5):

- Establishment of worldwide measurement stations to better understand the physical, chemical, and biological processes significant for the evolution of our planet on a variety of time scales
- Documentation of global changes, especially those changes that are of a time period of particular interest to people

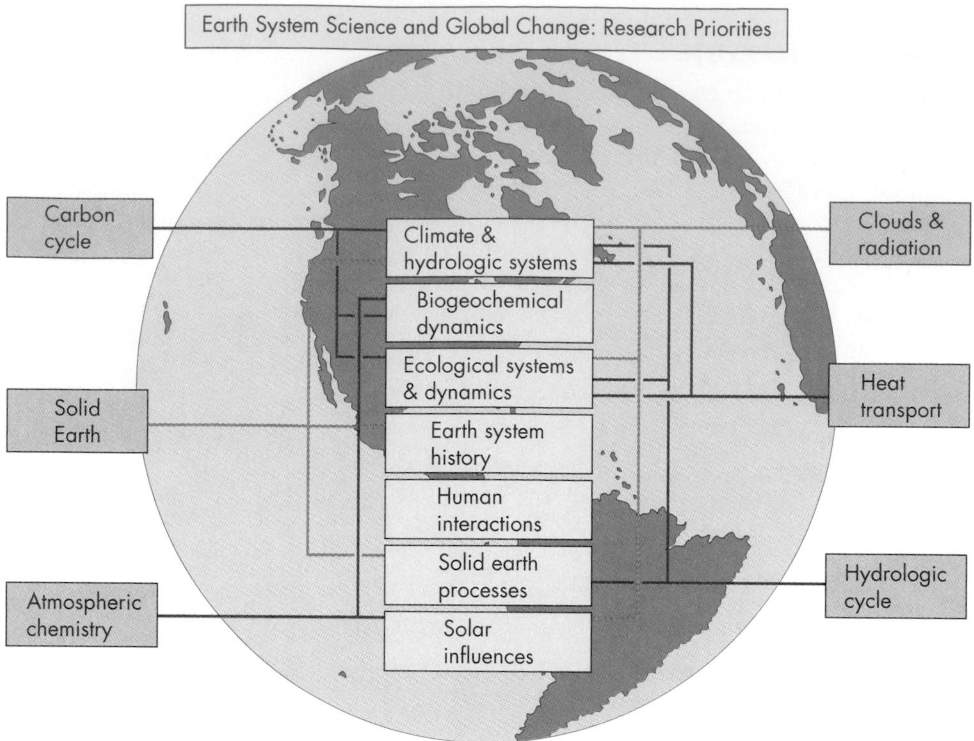

▶ **FIGURE 16.1** Research areas and priorities of earth system science. (*Source:* NASA. 1990. EOS: A *mission to planet earth.*)

- Development of quantitative models that may be used to predict and anticipate future global change
- Assistance in the gathering of essential information for decision making at the regional (national and international) and global levels

16.2 Tools for Studying Global Change

The Geologic Record

Sediments deposited on floodplains or in lakes, bogs, glaciers, or the ocean can be compared to pages of a history book. Organic material that is often deposited with sediment can be dated by a variety of methods to provide a chronology. In addition, the organic material itself can tell a story concerning the past climate, what lived in the area, and what changes have taken place.

Marine Sediments Studies of marine sediments have helped delineate the history of ocean-water temperature as well as biological and chemical changes occurring in the ocean basins over millions of years. Climatic information gleaned from ocean sediments has helped to establish the climatic change during the Quaternary (last 1.65 million years) in terms of the waxing and waning of continental ice sheets.

River Sediments Studying river sediments has helped to delineate past flood events and thus to determine the probability of future high-magnitude floods. Studies of river flooding have utilized the carbon-14 (radiocarbon) method. Carbon-14 (or ^{14}C) is an unstable isotope that undergoes radioactive decay at a known rate. When an organism dies, it stops collecting carbon, which starts the atomic clock ticking: We can determine the age of flood sediments from the proportion of ^{14}C in the organic matter buried with them. This method is good back to about 40,000 to 50,000 years before the present. It has been used, for example, at Pallett Creek in southern California, where earthquake history for the past 2000 years has been determined from organic material that was mixed with flood deposits disturbed by prehistoric earthquakes.

Lake and Floodplain Deposits Lake and floodplain deposits can tell us about changes in vegetation. A recent study (6) utilized analysis of pollen in floodplain deposits to establish the change in vegetation in northeastern Iowa during the past 12,500 years. That study used numerous ^{14}C dates to show that spruce forests present prior to 9100 years ago were followed by prairie vegetation from 5400 to 3500 years ago. The prairie vegetation was replaced about 3500 years ago by oak savanna, which was present until European settlement. This type of information is very important in establishing climatic conditions in the past, and thus regional and perhaps even global change.

Glacial Ice One of the more interesting uses of the geologic record has been the examination of glacial ice. This ice contains trapped air bubbles that may be analyzed to provide information concerning atmospheric carbon dioxide (CO_2) concentrations when the ice formed. The trapped air bubbles in a long core of glacial ice are time capsules of the atmosphere from the past. This method has been used to analyze the carbon dioxide content of air as old as 160,000 years (7). Glaciers also contain a record of heavy metals such as lead that settle out of the atmosphere (very fine particles of lead are picked up and carried by the wind) as well as a variety of other chemicals that can be used to study recent earth history.

Dendrochronology Another method of evaluating past earth history is **dendrochronology,** which is the study of tree rings. Many trees develop an annual growth ring, so counting and correlating the rings can establish a history of the area. The width of individual rings and other information can provide insights into the hydrologic and climatic conditions at the site. For example, careful analysis of the tree rings can reveal frequency of cycles of drought and wet years. This method (using prehistoric and fossil trees) has helped establish climatic changes during the past 12,000 years in many parts of the world.

Water Study of movement of water in the oceans and on land in lakes, rivers, and groundwater systems has potential for understanding linkages between land and water systems. For example, we may wish to predict the residence time of groundwater in a particular aquifer; estimate how long circulation of deep marine waters from the polar regions to the equator takes; or, based on groundwater chemistry, what the mean annual surface temperature was during the past 20,000 years. To answer these research questions requires investigation in use of environmental tracers or indicators. For example, some radioactive isotopes or dyes may be used to identify movement of water; human-made chemicals such as chlorofluorocarbons (CFCs), to be discussed shortly with respect to global warming and ozone depletion, are being used to estimate the age of young groundwaters; and dissolved gases in groundwater can provide information concerning paleotemperatures (prehistoric temperatures) from the mid-latitudes to the tropics (8).

Real-Time Monitoring

Monitoring is the regular collection of data for a specific purpose; real-time monitoring means collecting this data while a process is actually occurring. For example, we may monitor the flow of water in rivers to evaluate water resources or flood hazard. In a similar way, samples of atmospheric gases can help establish trends or changes in the composition of the atmosphere, and measurements of temperature composition of the ocean can be used to examine changes there. Gathering of real-time data is necessary for testing models and for calibrating the extended prehistoric record derived from geological data.

Methods of monitoring vary widely depending upon what is being measured. For example, deforestation can be monitored by evaluating remotely sensed data collected by satellite or high-altitude aerial photographs taken by cameras mounted on airplanes. Remote sensing utilizing satellite information has been particularly useful in monitoring at the global level. Nevertheless, the most reliable data are derived from monitoring that uses ground measurements to establish the validity of the airborne or satellite measurements.

Some of the most interesting evaluations concerning changes in ecosystems have resulted from long-term studies that gather specific data concerning interactions between physical and biological processes. For example, long-term studies of nutrient cycling in ecosystems of the eastern United States have helped establish principles of management for timber harvesting and other activities related to human use and interest in the land.

Mathematical Models

Mathematical models use numerical means to represent real-world phenomena and the linkages and interactions between the processes involved. Such models have been developed to predict flow of surface water and groundwater, erosion and deposition of sediment in river systems, ocean circulation, and atmospheric circulation.

The models of global change that have gained the most attention are the **Global Circulation Models** (GCM). The objective of these models is to predict atmospheric changes (circulation) at the global scale (9). Variables used in the model include, among others, temperature, relative humidity, and wind conditions. Many of these variables are estimated for the past based on proxies, such as the tree-ring records discussed earlier. Data used in the calculations are arranged into large cells that represent several degrees of latitude and longitude (Figure 16.2); typical cells represent an area about the size of Oregon or of Indiana and Ohio together. In addition, there are usually 6 to 20 levels of vertical data representing the lower atmosphere. Calculations

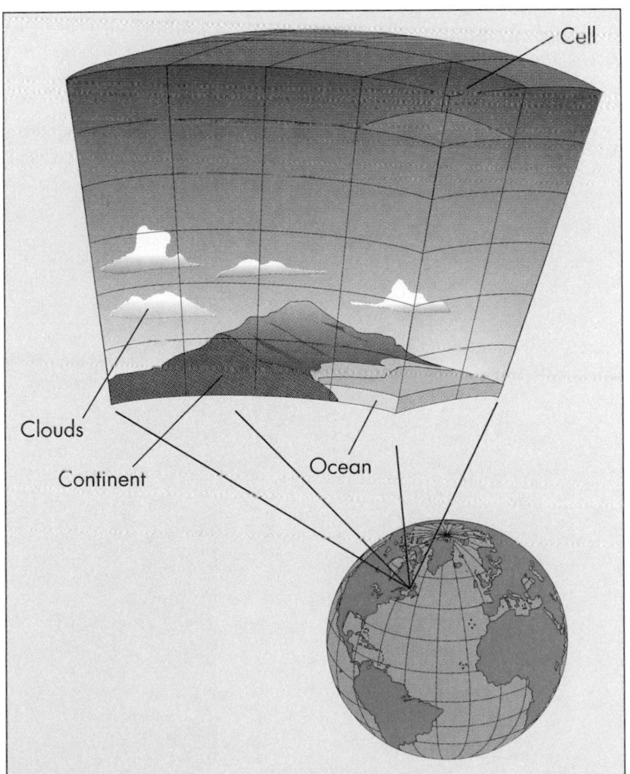

▲ **FIGURE 16.2** Idealized diagram illustrating the cells used in Global Circulation Models (GCM).

involving equations that represent the conservation of mass-energy and momentum are then used to make predictions.

Global Circulation Models are complex and require supercomputers for their operation (9). Unfortunately, the results are fairly crude and may not accurately represent future conditions (10). The models are inaccurate in part because methods used to calculate variables, such as temperature, relative humidity, and wind conditions, over a very large region (the cell) may have unrealistic assumptions. Also, other variables, such as cloud covers, are difficult to estimate. Global Circulation Models must be viewed as a first approach to solving very complex problems. In spite of their limitations, they provide information necessary for evaluating the earth as a system, and in pointing out which additional data are necessary to develop better models in the future. The models do predict which regions are likely to be wetter or drier (relative to other regions) if certain changes in the atmosphere occur, and these predictions are being taken seriously.

16.3 Earth's Atmosphere and Energy Balance

The study of global change is to a very great extent the study of changes in the atmosphere and the link between the atmosphere and the lithosphere, hydrosphere, and biosphere. Therefore, a modest acquaintance with earth's energy balance and solar radiation is critical in understanding global processes and environmental problems associated with the atmosphere, such as the greenhouse effect and ozone depletion.

The Atmosphere

Our **atmosphere** can be thought of as a complex chemical factory with many little-understood reactions taking place within it. Many of the reactions occurring are strongly influenced by sunlight and by the compounds produced by life. The air we breathe is a mixture of nitrogen, N_2 (78%); oxygen, O_2 (21%); argon, Ar (0.9%); carbon dioxide, CO_2 (0.03%); other trace elements (less than 0.07%); and compounds such as methane, ozone, carbon monoxide, oxides of nitrogen and sulfur, hydrogen sulfide, hydrocarbons, and various particulates. The most variable part of the atmosphere's composition is water vapor (H_2O), which can range from approximately 0 to 4 percent by volume in the troposphere (the lower 10 km of the atmosphere) (11).

Study of the earth's history suggests that the atmosphere has been changing since the earth first formed. Figure 16.3 shows an idealized cross section of the earth and its atmosphere as it may have been during early Precambrian times, more than 3 billion years ago. Volcanic outgassing produced water and carbon dioxide, and in the upper atmosphere pho-

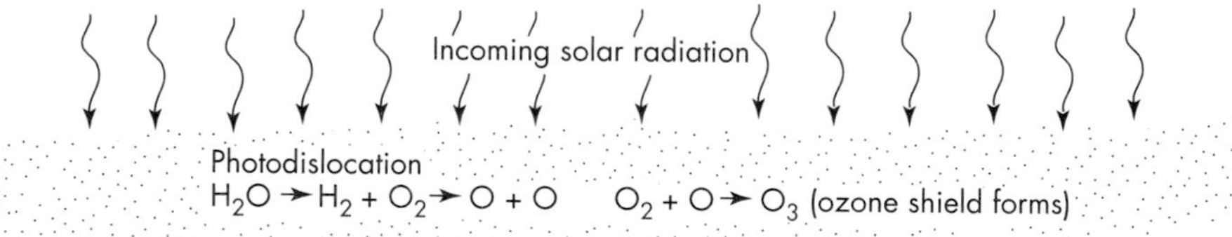

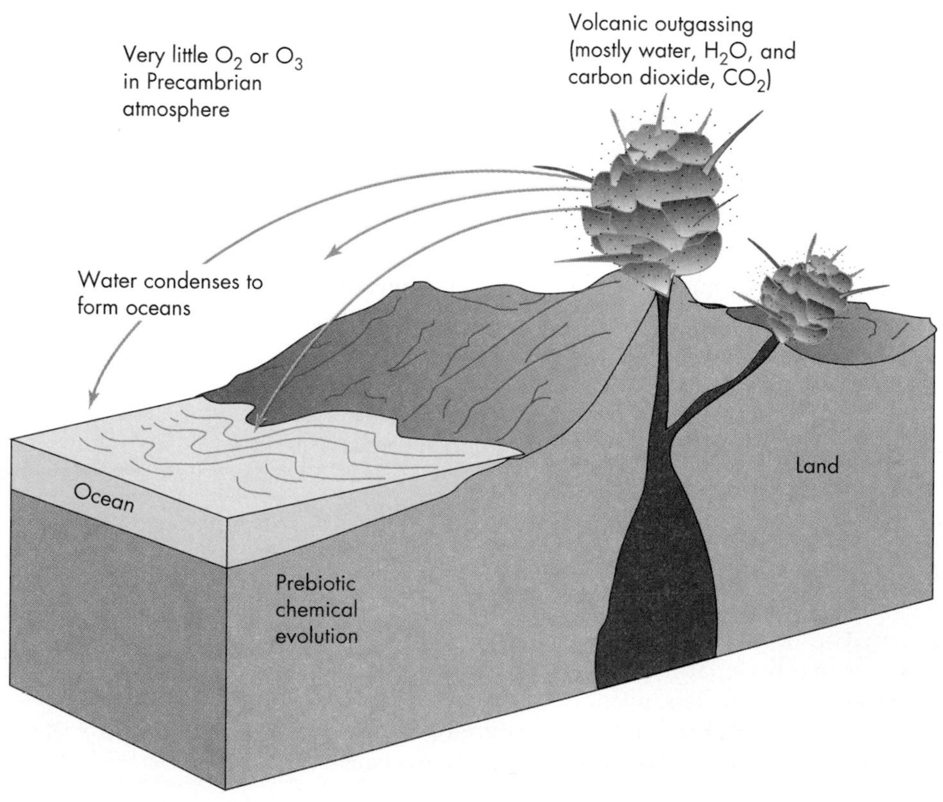

▶ **FIGURE 16.3** Idealized diagram showing selected global processes in the early Precambrian (about 3 billion years ago). (*Modified after* S. Kershaw. 1990. *Evolution of the earth's atmosphere and its geologic impact*, Geology Today, *March–April, pp.* 55–60. *Blackwell Scientific Publications Limited.*)

todissociation of water and molecular oxygen (O_2) produced an ozone (O_3) layer. At that time, however, the earth's atmosphere contained very little free molecular oxygen (O_2). Development of our oxygen-rich atmosphere was to wait about a billion years until photosynthesis by plants was widespread (12).

Solar Radiation and Earth's Energy Balance

At the planetary scale, the earth can be considered part of a larger solar energy system. The earth receives energy from the sun, and this energy affects the atmosphere, the oceans, the land, and living things before being radiated back into space. **Earth's energy balance** refers to the equilibrium between incoming and outgoing energy, which involves changes in the energy's form. Although the earth intercepts only a very tiny fraction of the total energy emitted by the sun, this intercepted energy sustains life on earth. It also drives many processes at or near the earth's surface, including the hydrologic cycle, ocean waves, and the circulation of air on a global scale.

Figure 16.4 shows some of the important components of the earth's energy budget. As you can see from this figure, nearly all of the energy that is available at the earth's surface comes from the sun. Geothermal heat generated from the interior of the planet is only a very small fraction of 1 percent of the earth's energy budget. Nevertheless, it is this internal heat that drives the great global tectonic cycle and moves the tectonic plates of the lithosphere, generating earthquakes and creating volcanoes.

Electromagnetic Energy Energy emitted from the sun is in the form of electromagnetic energy, or electromagnetic radiation, which travels through the vacuum of space at a speed of about 300,000 km per second (the speed of light). Different forms of electromagnetic radiation are distinguished by their wavelengths, and the collection of all possible wavelengths forms a continuous range known as the *electromagnetic spectrum* (Figure 16.5). The longer wavelengths (greater than 1 m) include radio waves, and the shortest waves are X rays and gamma rays. Electromagnetic radiation to which our eyes respond is known as visible electromagnetic radiation or light, and is only a very small fraction of the total spectrum. Other types of electromagnetic radiation with environmental significance include microwaves and ultraviolet (UV) radiation (which is sometimes called *UV light*, although

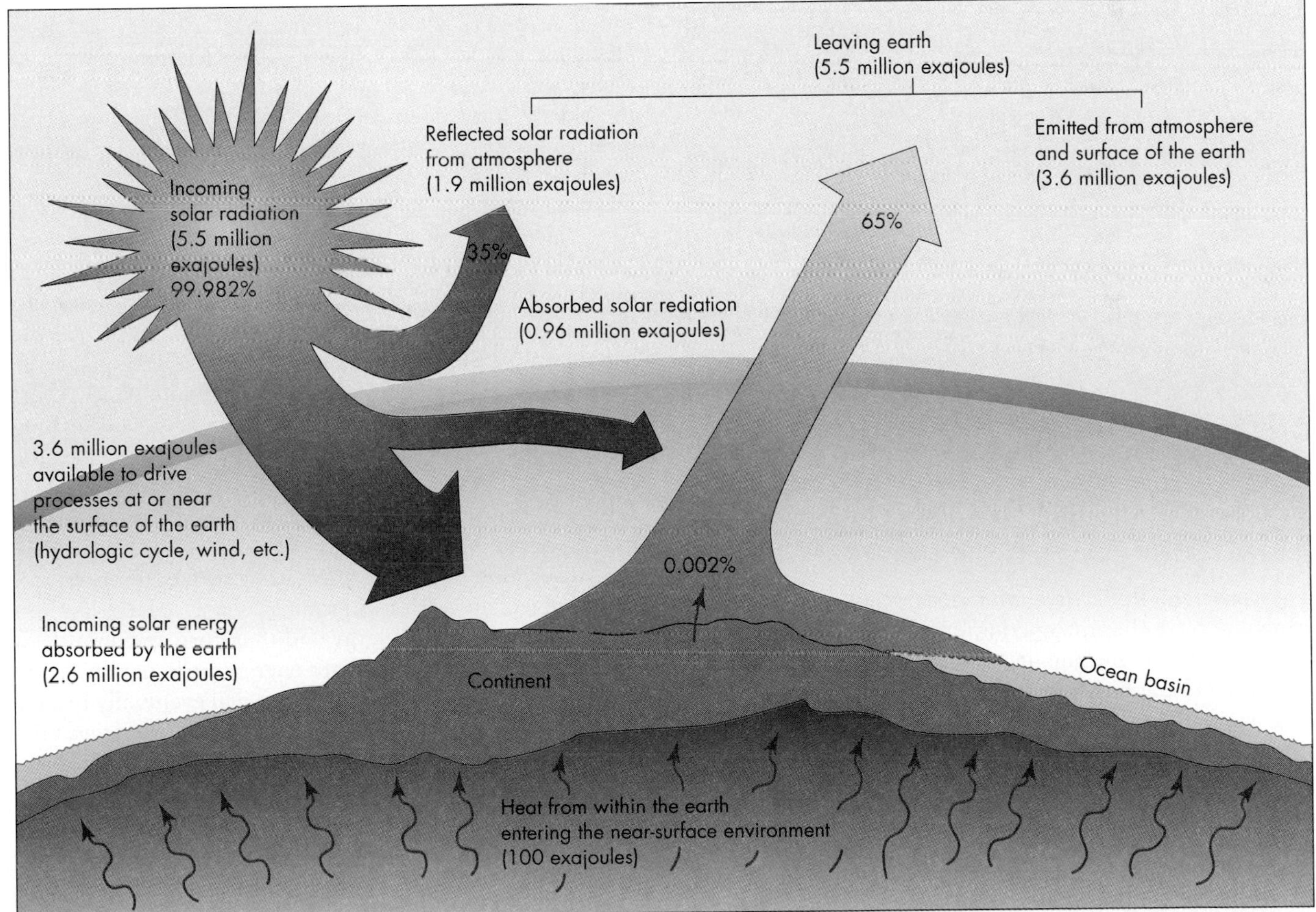

▲ **FIGURE 16.4** Annual energy flow to the earth from the sun. Also shown is the relatively small component of heat from the earth's interior to the near-surface environment. (*Modified after Marsh and Dozier,* Landscapes: An introduction to physical geography. *Copyright © 1981. John Wiley & Sons, Inc. Reprinted by permission of John Wiley & Sons, Inc.*)

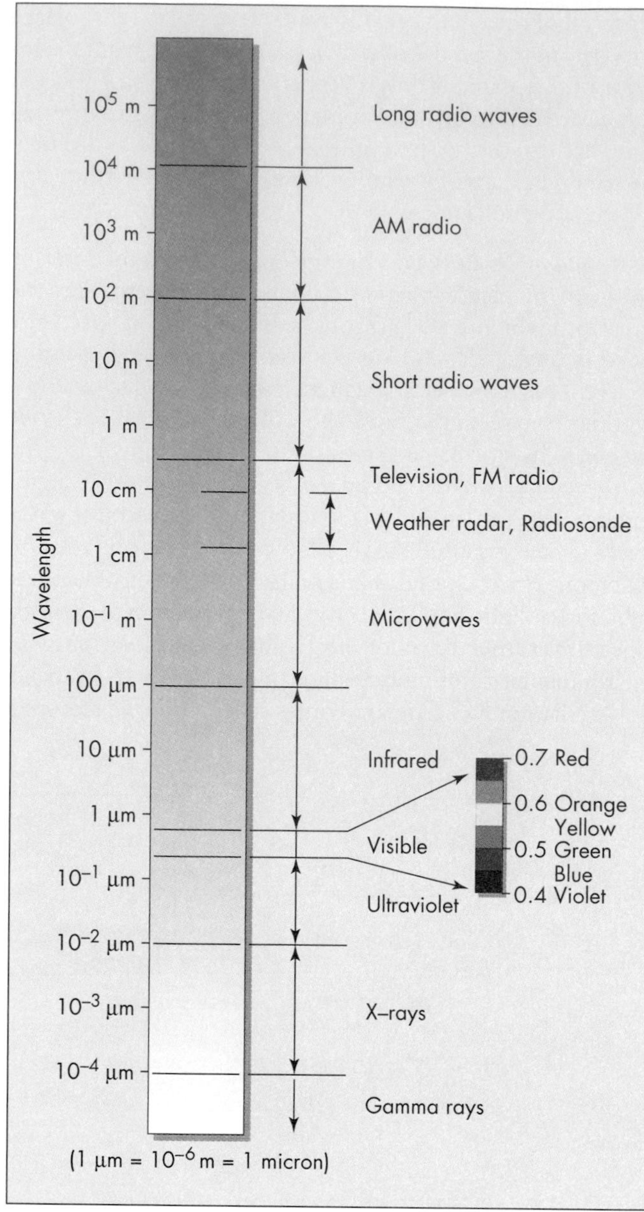

▲ **FIGURE 16.5** The electromagnetic spectrum.

it is not visible to humans). Microwaves have industrial uses, and UV radiation plays an important role in ozone depletion.

Transmission, Reflection, and Absorption When electromagnetic radiation encounters a material it may be transmitted (pass through it), reflected (bounce off of it), or absorbed (13). Which one of these processes occurs can have important environmental ramifications. Most visible radiation from the sun passes through the atmosphere without being absorbed, although some of it is reflected back into space by clouds. However, the absorption of a portion of the incoming UV radiation by ozone (O_3) in the upper atmosphere protects life at the surface of the planet.

It is not unusual for transmission, reflection, and absorption to all take place at the same time. For example, consider a skylight in a modern home. If you look at the skylight from outside you may see a sparkle of light coming from it, indicating that some of the incoming solar radiation is reflected. If you touch the glass, it may feel warm, indicating some of the energy has been absorbed. If you are inside and see the sky through the glass, you may conclude that some of the solar radiation has been transmitted through the skylight and entered your eyes. If you feel warmth below the skylight, you know that the air in the room has absorbed some of the radiation that was transmitted through the glass.

Thermal Energy In the example of the skylight, the glass and the air felt warm because the electromagnetic energy they absorbed was converted to thermal energy, which is the kinetic energy of vibrating molecules. (Thermal energy is closely related to heat, and for purposes of this discussion may be considered to be the same as heat.) When an object absorbs electromagnetic radiation, the vibration of its molecules increases and it becomes warmer. The vibrating molecules radiate some of the absorbed energy back into the surroundings as electromagnetic radiation, but of longer wavelengths than the radiation that was absorbed. As will be explained shortly, this reradiated energy is predominantly in the form of infrared radiation, which is invisible to our eyes.

Effect of Temperature on Radiation and Absorption The hotter an object—the sun, the earth, a lake, a rock—the more electromagnetic energy it emits. In fact, the amount of energy radiated each second varies with the fourth power of the object's surface temperature: If the surface temperature doubles, the radiated energy increases 16 times. This explains why the sun, with a surface temperature of 5800 °C, radiates so much more energy per unit area than the earth, which has an average surface temperature of 15 °C.

The temperature of a body also affects the type of radiation it emits, as shown in Figure 16.6. The hotter an object, the more rapidly it radiates energy, and the shorter the wavelength of the predominant radiation. This explains the fact that the sun's radiant energy has predominantly short wavelengths, whereas terrestrial radiation has relatively long wavelengths. The surface of the earth, which includes the surfaces of plants, clouds, water, rocks, plants, and animals, is so cool that energy is radiated predominantly in the infrared portion of the spectrum (13,14).

Absorption is likewise affected by surface temperature. A cold object on earth's surface may absorb a large amount of the incoming solar energy and therefore warm up. As it warms, however, it will radiate energy more rapidly. With a constant input of energy, an object will eventually reach a temperature that allows it to absorb and radiate energy at the same rate. The earth receives approximately 5.5 million exajoules from the sun, but it radiates the same amount of energy back into space (see Figure 16.4).

Reflectivity An object's color also affects its ability to absorb and radiate energy. Dark or black surfaces absorb and radiate electromagnetic energy readily. White surfaces, on the other hand, tend to reflect electromagnetic energy rather than absorb it. Ice reflects 80 to 95 percent of the

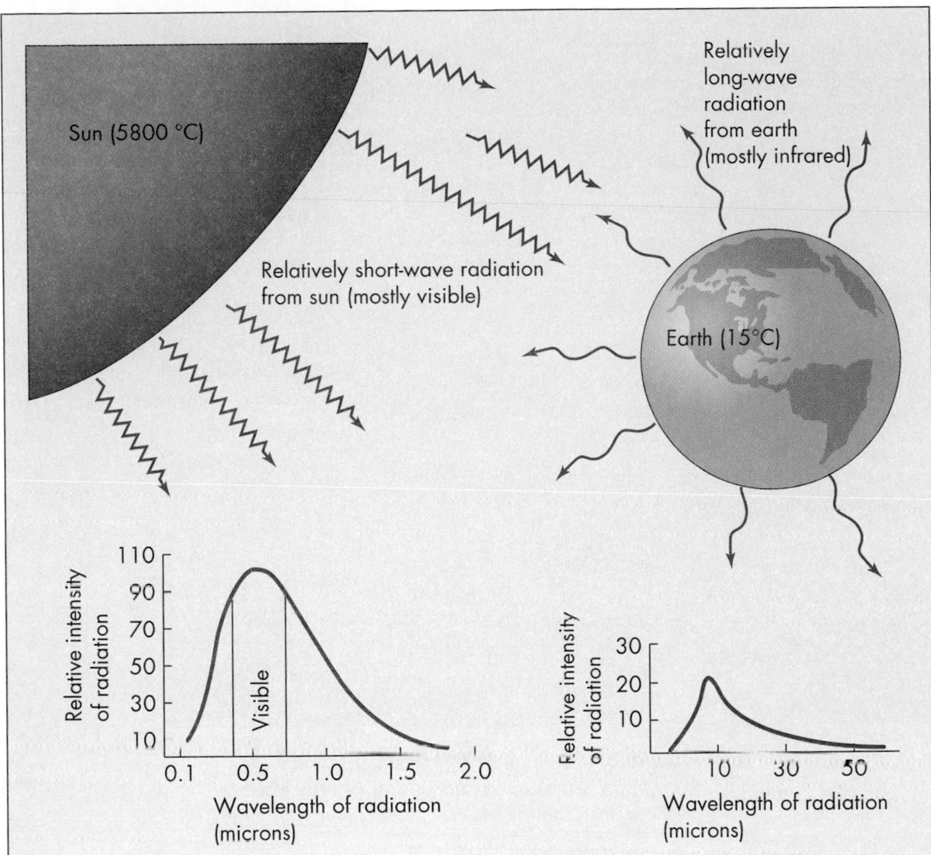

FIGURE 16.6 Idealized diagram comparing the emission of energy from the sun with that from the earth. Notice that the solar emissions have a relatively short wavelength, whereas those from the earth have a relatively long wavelength. (*Modified after Marsh and Dozier,* Landscapes: An introduction to physical geography. *Copyright © 1981. John Wiley & Sons, Inc. Reprinted by permission of John Wiley & Sons, Inc.*)

solar energy that falls on it, dry grassland 30 to 40 percent, and coniferous forest 5 to 15 percent. Ice that contains a good deal of rock and other dark material at the surface reflects much less than pure ice, because the dark objects absorb more of the incoming radiation. The dark objects grow warmer as a result, and they transmit their thermal energy to the ice, which may partially melt. This explains why darker glacial ice tends to be topographically lower than adjacent white ice.

Albedo is a measure of the reflectivity of a surface, expressed as the percentage of incoming light that is reflected. It has been estimated that, given present atmospheric conditions, a 1 percent change in the amount of sunlight reflected by the earth would cause an approximate 1.7 °C change in the surface temperature. Thus, the earth's surface temperature is very sensitive to a change in albedo. Actual changes in the absorption and reflection of solar energy are due to such factors as cloud cover, amount of ice at the earth's surface, type of plant cover, and presence of water (11,13).

16.4 Atmospheric Carbon Dioxide and Global Warming

The Greenhouse Effect

The temperature of the earth is determined for the most part by three factors, shown in highly idealized form in Figure 16.7: the amount of sunlight the earth receives, the amount of sunlight the earth reflects (and therefore does not absorb), and atmospheric retention of reradiated heat (9). Absorbed solar energy warms the earth's atmosphere and surface, which reradiate the energy as infrared radiation (9). Water vapor and several other atmospheric gases, including carbon dioxide, methane, and chlorofluorocarbons, tend to trap heat, which is to say they absorb some of the energy radiating from the earth's surface and are thereby warmed. For this reason, the planet is much warmer than it would be if all its radiation escaped into space without this intermediate absorption and warming. This effect is somewhat analogous to the trapping of heat by a greenhouse (although the retention of heat in a greenhouse is due mostly to reduced cooling by air circulation, with only a small amount due to trapped infrared radiation). The trapping of heat by the atmosphere is generally referred to as the **greenhouse effect** (Figure 16.7).

It is important to understand that the greenhouse effect is a natural phenomenon that has been occurring for millions of years, on earth and on other planets in our solar system. Were it not for the trapping of heat in the atmosphere, the earth would be approximately 33 °C cooler than it is now, and all surface water would be frozen (15). Most of the natural "greenhouse warming" is due to water vapor in the atmosphere. However, potential global warming due to human activity is related to carbon dioxide, methane, nitrous oxides, and chlorofluorocarbons. In recent years the atmospheric concentrations of these gases and others have been increasing due to human activities. These gases tend to absorb infrared radiation from the earth, and it has been

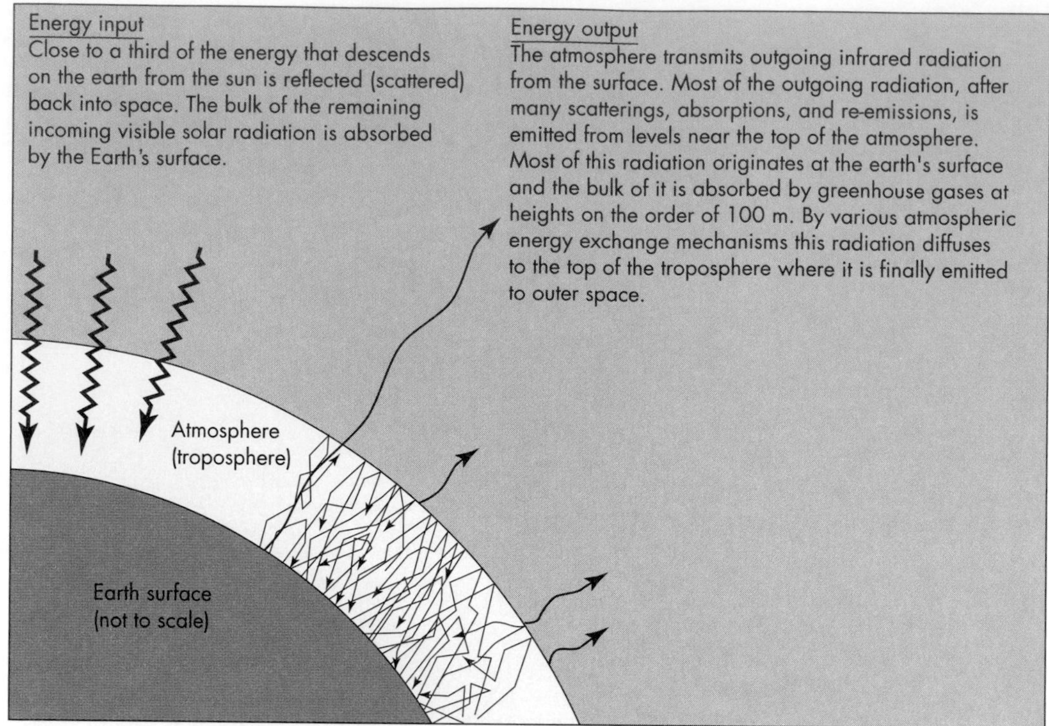

▲ **FIGURE 16.7** Idealized diagram showing the greenhouse effect. Incoming visible solar radiation is absorbed by the earth's surface to be reemitted in the infrared region of the electromagnetic spectrum. Most of this reemitted infrared radiation is absorbed by the atmosphere, maintaining the greenhouse effect. (*Developed by* M. S. *Manalis and* Edward A. *Keller,* 1990)

hypothesized that the planet may be warming due to increases in the amounts of "greenhouse gases." The major greenhouse gases, in terms of their contribution to the anthropogenic (human-induced) portion of the greenhouse effect, are shown in Table 16.1.

Changes in Greenhouse Gases

Carbon Dioxide As you can see from Table 16.1, 60 percent of the anthropogenic greenhouse effect is attributed to carbon dioxide (CO_2). Measurements of carbon dioxide trapped in air bubbles of the Antarctic ice sheet suggest that during the past 160,000 years the atmospheric concentration of carbon dioxide has varied from a little less than 200 ppm to about 300 ppm (7). The highest levels are recorded during interglacial periods that occurred about 125,000 years ago and at the present. About 130 years ago, at the beginning of the Industrial Revolution, the atmospheric concentration of carbon dioxide was approximately 280 ppm, a level that had been constant for at least the previous 700 years (7).

Since about 1860, there has been an exponential growth of the concentration of carbon dioxide in the atmosphere. The change from approximately the year 1500 to 1990 is shown in Figure 16.8. Data prior to the mid-twentieth century are from measurements made from air bubbles trapped in glacial ice. The concentration of carbon dioxide in the atmosphere today is nearly 350 ppm, and it is predicted to reach at least 450 ppm—more than 1.5 times the preindustrial level—by the year 2050 (16).

Table 16.1 Rate of increase and relative contribution of several gases to the anthropogenic greenhouse effect

	Rate of Increase (% per year)	Relative Contribution (%)
CO_2	0.5	60
CH_4	<1	15
N_2O	0.2	5
O_3*	0.5	8
CFC-11	4	4
CFC-12	4	8

*In the troposphere.

Source: Data from H. Rodhe, 1990. A comparison of the contribution of various gases to the greenhouse effect. *Science* 248, 1218, Table 2. Copyright 1990 by the AAAS.

The rate of increase of carbon emissions from burning fossil fuels has been approximately 4.3 percent per year since the Industrial Revolution began. Given this, it would be easy to assume that the increase in carbon dioxide in the atmosphere since then is all due to human-induced processes, such as burning of fossil fuels and deforestation (17). Unfortunately, proving this relationship has been difficult.

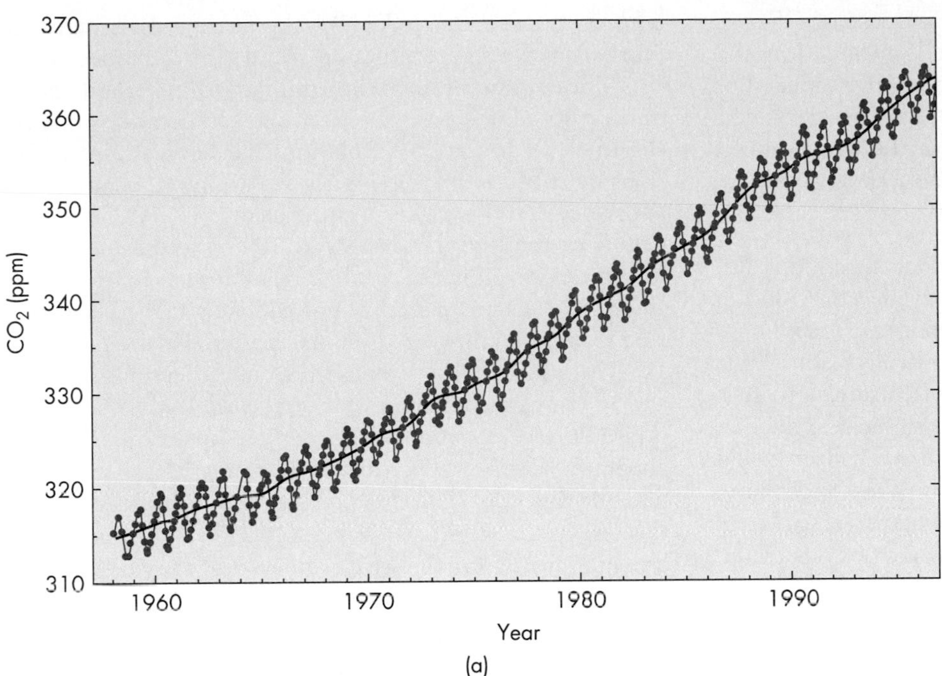

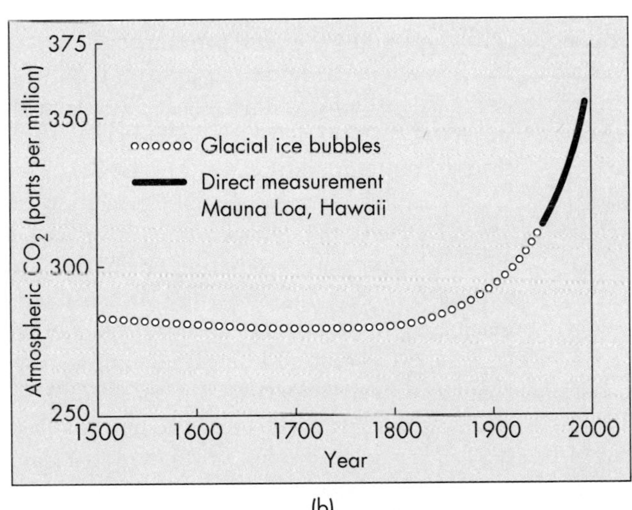

FIGURE 16.8 (a) Concentrations of atmospheric CO_2 from 1958 to 1997 at Mauna Loa, Hawaii. Annual cycles reflect changes in photosynthesis occurring in the Northern Hemisphere. High peaks of the cycles are winter months when many trees are dormant and not using CO_2. Low values are the summer months when photosynthesis is more vigorous. The apparent widening of the cycles with time probably is a result of greater volume of plants as CO_2 has increased (CO_2 is a plant nutrient). (C. D. *Keeling*. 1998. *Personal oral communication.*) The solid line is the average value of CO_2. (*Source: Scripps Institute of Oceanography and* NOAA. *http://mloserv.m/o.hawaii.90v/mlointo/programs/gasses/CO2graph.htm. Accessed* 2/11/99.) (b) Average concentration of atmospheric carbon dioxide, 1500–1995. (*Data in part from* W. M. *Post et al.* 1990. American Scientist 78[4]:210–26.)

The global carbon cycle is very complex, and the linkages and flow of carbon from the various natural sources and sinks are not well known. If all the carbon dioxide produced by human activities remained in the atmosphere, its concentration should be even higher than it is today. There must therefore be carbon dioxide sinks in the oceans or on land, but these are not well understood. Despite all this, it is clear that carbon dioxide concentration in the atmosphere has significantly increased since the Industrial Revolution, and this increase may contribute to global warming via the greenhouse effect.

Methane Methane is thought to contribute approximately 15 percent of the anthropogenic greenhouse effect. There is a controversy, however, concerning the sources and sinks of methane in the atmosphere, and much more work needs to be done in this area. Despite the uncertainties, it is fairly well understood that methane gas from anthropogenic sources is related to burial and decomposition of biomass (as, for example, in landfills), production of coal and natural gas, and agricultural practices such as growing rice and raising cattle and sheep. The methane related to rice production is associated with anaerobic activity in flooded lands. Methane from cattle and sheep results from digestive processes and expulsion of gas by the animals (4).

Chlorofluorocarbons Approximately 12 percent of the anthropogenic greenhouse effect may be related to chlorofluorocarbons (CFCs) in the atmosphere. These compounds have been and are still used in refrigeration units, i.e., Freon and the propellant in spray cans (18). Atmospheric concentrations of CFCs are increasing more rapidly than concentrations of carbon dioxide. Because CFCs are highly stable compounds, their residence time in the atmosphere is long. The rate of increase of CFCs is about 5 percent annually, and even if production and emissions of these chemicals were drastically reduced or completely eliminated, the elevated concentrations would be with us for many years, particularly in the stratosphere, where CFCs present there were probably produced 20 to 30 years ago (19).

Nitrous Oxide Nitrous oxide (N_2O) concentration is also increasing in the atmosphere and may be contributing approximately 5 percent to the anthropogenic greenhouse effect (18). Most of the anthropogenic nitrous oxide results from the application of fertilizers in agricultural activities; another contributor is the burning of fossil fuels. Reduction in burning of fossil fuels and reduced use of fertilizers could lower the emission of nitrous oxide into the atmosphere. However, nitrous oxide has a long residence time, so if rates of emission are stabilized, elevated concentrations of the gas could persist for at least several decades (4).

16.5 Global Temperature Change: Evidence for a Warming Trend

The Pleistocene ice ages began approximately 1.65 million years ago, and since then there have been numerous variations in earth's mean annual temperature and thus climate (11). It is important to distinguish between weather and climate. Weather is the day-to-day variation in factors such as surface temperature, rainfall, snowfall, and wind conditions. Climate is the characteristic atmospheric condition (weather) at a particular place or region over time periods of seasons, years, or decades. The climate at a particular location is more than average precipitation and temperature. It may be dependent on infrequent or extreme seasonal patterns, such as rain in the monsoon climate of parts of India and the southwestern U.S. Global change in temperature will have effects on the major climatic zones on earth (desert, humid, arctic, etc.; Figure 16.9). Causes of global climate change are complex and not well understood. We do know that significant changes may sometimes occur quickly, perhaps on a scale of decades! This has important implications for the future of humans (see *A Closer Look: Causes of Climatic Change*). Figure 16.10 shows the changes of the past million years on several time scales. The top scale shows the entire million years, during which there have been major climatic changes involving swings in mean temperature of several degrees Celsius. Periods of low temperature have coincided with major glacial events, and periods of high temperature have coincided with interglacial events. For the two scales that show changes over 150,000 and 18,000 years, respectively, interglacial and glacial events become increasingly prominent.

In the scale showing changes over 1500 years, several warming and cooling trends that have affected humans can be seen. For example, a major warming trend from about A.D. 800 to 1200 allowed the Vikings to colonize Iceland, Greenland, and northern North America. When glaciers made a minor advance about A.D. 1400, during a cold period known as the *Little Ice Age*, the Viking settlements in North America and parts of Greenland were abandoned (11).

Since about 1750, an apparent warming trend lasted until approximately the 1940s, when temperatures cooled slightly (11). In the time scale showing temperature variations over 140 years (see Figure 16.10), more changes are apparent, and the 1940s event is clearer (4). What is evident from the record from 1865 to 1998 (bottom-most scale in Figure 16.10) is that global mean annual temperature since 1890 has increased by approximately 0.5 °C. The period from 1986 through 1998 has been the warmest in the 139 years that global temperatures have been monitored, and 1998 was the warmest year on record. In the conterminous United States the average daily temperature has risen about 0.3 °C since 1900. Most of the increase has been in the past 30 years. The U.S. trend is consistent with estimates for the rest of the world as a whole (20,21). The trend of increasing global atmospheric temperature is evidence supporting **global warming.**

Additional support for global warming comes from observations on glaciers. Many more glaciers in the Northern Cascades, United States, Switzerland, and Italy are retreating (melting back at their terminus) today than advancing (growing), and the increase in the number of retreating glaciers is accelerating in the Cascades and Italy (Table 16.2) (26, 27). Evidently this is in response to a mean global temperature (1977–1994) that has averaged 0.4 °C above the long-term mean temperature. On Mount Baker in the Northern Cascades, for example, all eight glaciers were advancing in 1976. By 1990, all eight were retreating. In addition, 4 of 47 alpine glaciers observed in the Northern Cascades have disappeared since 1984. Although similar changes in Alaskan glaciers are not yet as apparent, the glaciers there are much larger and will require a longer lag time to respond to climate change (Pelto, M. S. 1998. Personal communication). The observations of recent increase in global temperature and retreat of glaciers are not proof of global warming resulting from humans burning fossil fuels. Such evidence is consistent with global warming and is a source of concern. In summary, with respect to global warming: (1) human activity is increasing the concentration of greenhouse gases in the atmosphere, (2) the mean temperature of earth increased by about 0.5 °C in the past 100 years, and (3) a significant portion of the observed increase in mean temperature of earth probably results from human activity.

Because of the complexities of the carbon cycle and the greenhouse effect, future global warming trends remain uncertain. A major source of complexity is the various possible positive and negative feedback mechanisms—for example, mechanisms related to the role of clouds in the atmosphere—that affect the response of the biosphere to change. As we do not know much about these mechanisms, our computer models are not as adequate for predicting future climatic change as we would like. Figure 16.11 shows potential range in predicted global warming derived from computer modeling. The models do predict fairly reliably that warming will occur and possibly accelerate in the com-

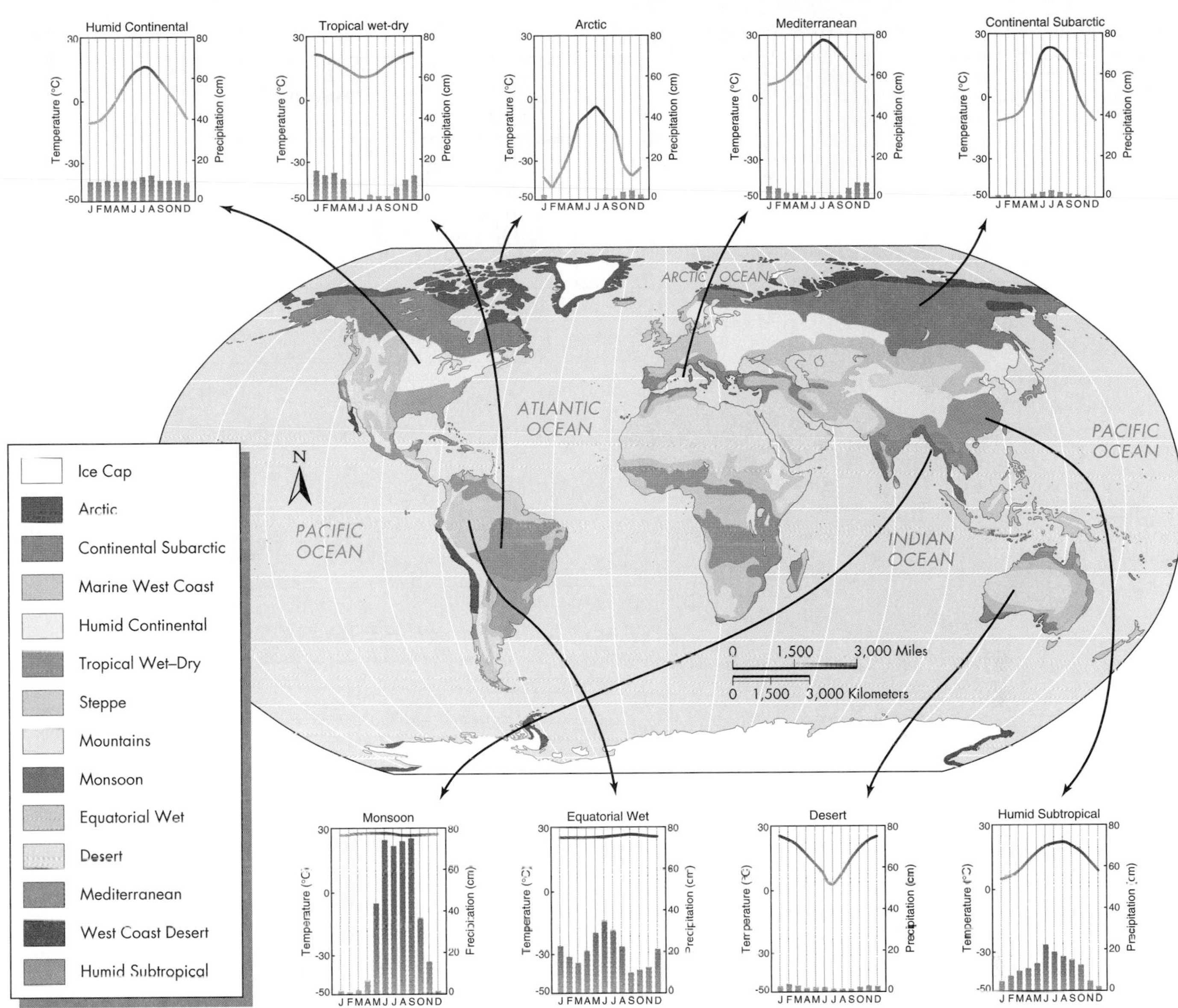

▲ **FIGURE 16.9** Climates of the world with characteristic temperature and precipitation conditions. (*Modified after* Marsh, W. M., *and* Dozier, J. Landscapes: An introduction to physical geography, *Copyright* © 1981. *John Wiley & Sons.*)

ing decades (21). Therefore, we need to examine carefully the potential effects of such a warming.

Role of Particulates in the Atmosphere

Human activity is adding **particulates** to the atmosphere. Agriculture, forestry, power production, and other industrial activities all involve burning, which produces particulate matter. Types and sizes of selected particulates are shown in Figure 16.12. Elevations that heavy particles may reach in the atmosphere depend on the specific atmospheric transport processes involved (for example, existence of upward-moving air masses). How particulate material in the atmosphere will affect global changes in the atmosphere's mean annual temperature is being debated.

Particulates have two important effects. First, they act as condensation nuclei and therefore cause an increase in precipitation or fog. Rain and fog form by condensation of water on small particles. Second, they affect the amount of sunlight reaching the earth. As the total amount of particulates in the atmosphere increases, a larger percentage of incoming solar radiation may be reflected away from the earth, causing mean annual temperature to decrease. On the other hand, some particles may absorb incoming solar radiation, causing a rise in the atmospheric and land-surface temperature. This second effect is observed in urban areas, where particulates in the atmosphere are concentrated.

Slight global cooling and spectacular sunsets for up to a year or so have followed volcanic eruptions that blast volcanic ash and other material into the atmosphere. The 1883 eruption of Krakatoa, a now small volcanic island in the East Indies, was one of the largest volcanic eruptions in historic times. The eruption blew about 2.6 km^3 of volcanic ash and

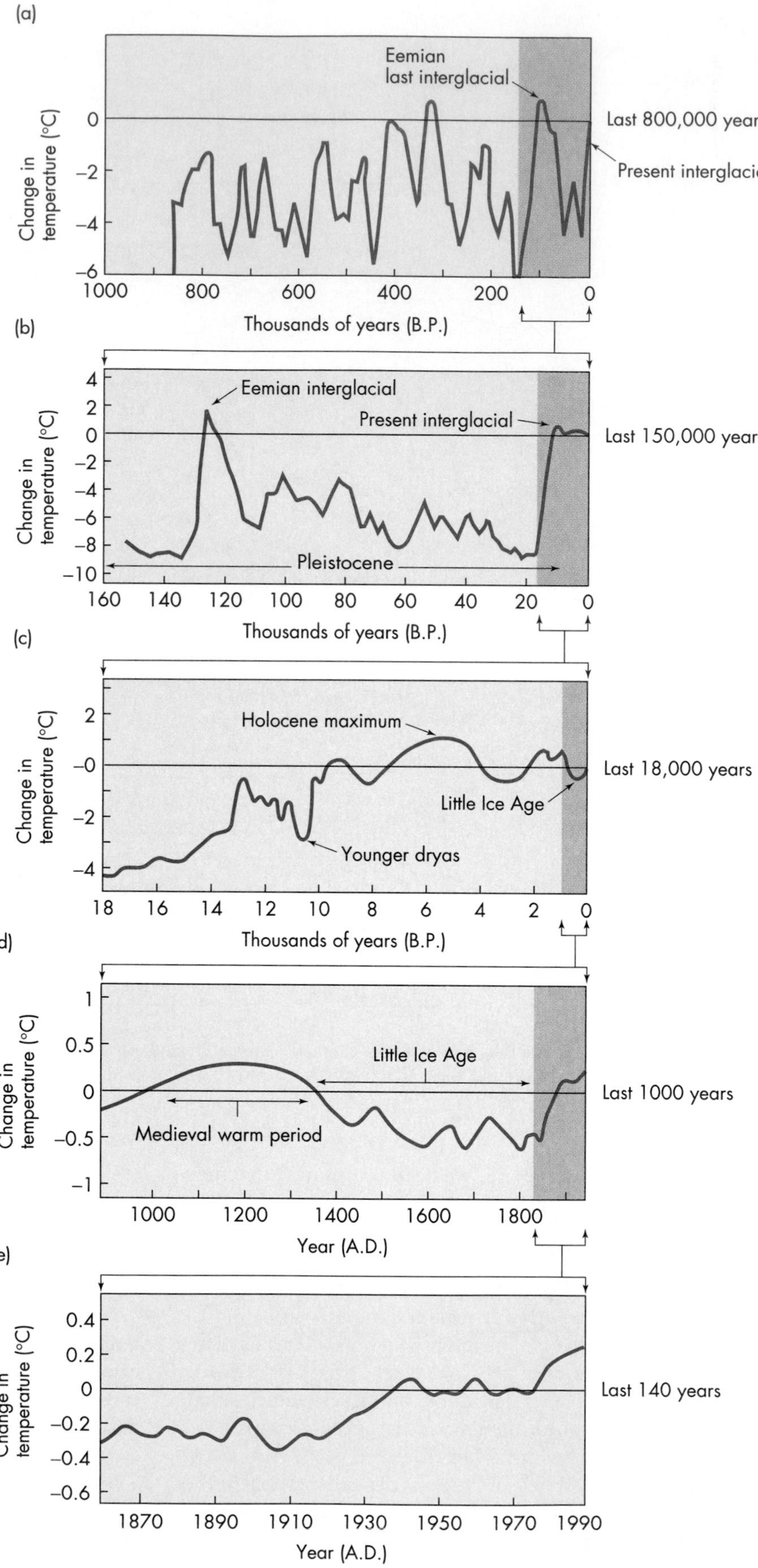

▶ **FIGURE 16.10** Change in temperature over different periods of time during the last million years. (*Modified after* UCAR [*University Corporation for Atmospheric Research*]. *Office for Interdisciplinary Studies*. 1991. *Science Capsule, Changes in time in the temperature of the earth*. EarthQuest, *vol.* 5, *no.* 1.)

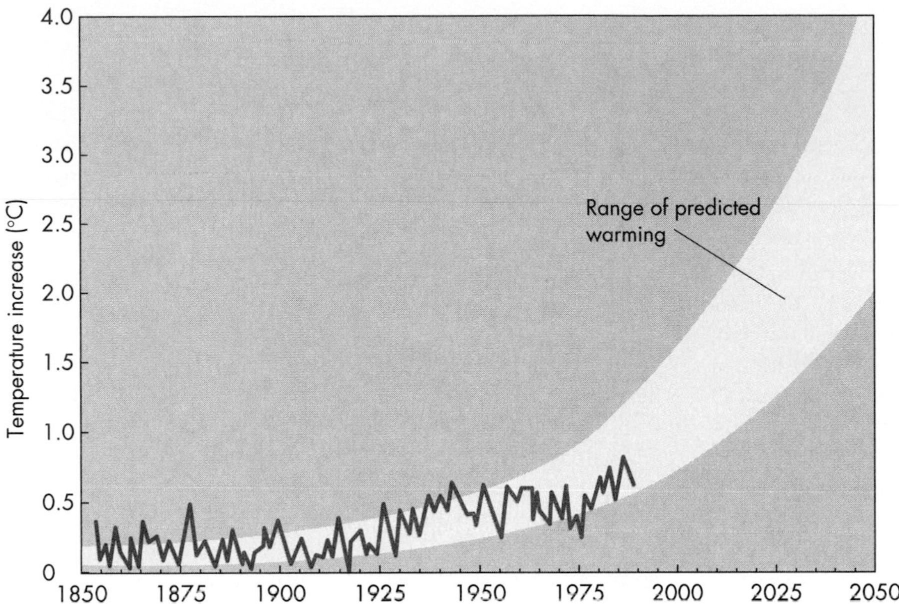

◀ **FIGURE 16.11** Historical trends and range of predicted warming from computer models. All models predict substantial warming. (*Adapted from* P. D. Jones *and* T. M. L. Wigley. 1990. [*Modified*] *by* John Deecken *from Global Warming Trends.* Scientific American 263[2]:84–91. *Copyright © 1990 by Scientific American, Inc. All rights reserved.*)

Table 16.2 Changes in glaciers, 1967–1995*

	Advancing	Stationary	Retreating
Northern Cascades			
1967	7	8	7
1974	9	0	2
1985	5	10	32
1990	1	5	41
1995	0	0	47
Switzerland			
1967	31	14	55
1975	70	3	28
1986	42	9	13
1989	19	3	83
1993	6	0	73
Italy			
1981	25	10	10
1985	25	6	14
1988	26	13	92
1990	9	9	123
1993	6	8	127

*The number of glaciers observed have varied during the period of observation.

Source: Data from Pelto, M. S. 1993. Changes in water supply in alpine regions due to glacier retreat. American Institute of Physics, Proceedings Vol. 277: *The World at Risk: Natural Hazards and Climate Change,* 61–67; and Pelto, M. S. 1996. Recent changes in glacier and alpine runoff in the North Cascades, Washington. *Hydrological Processes* 10:1173–1180.

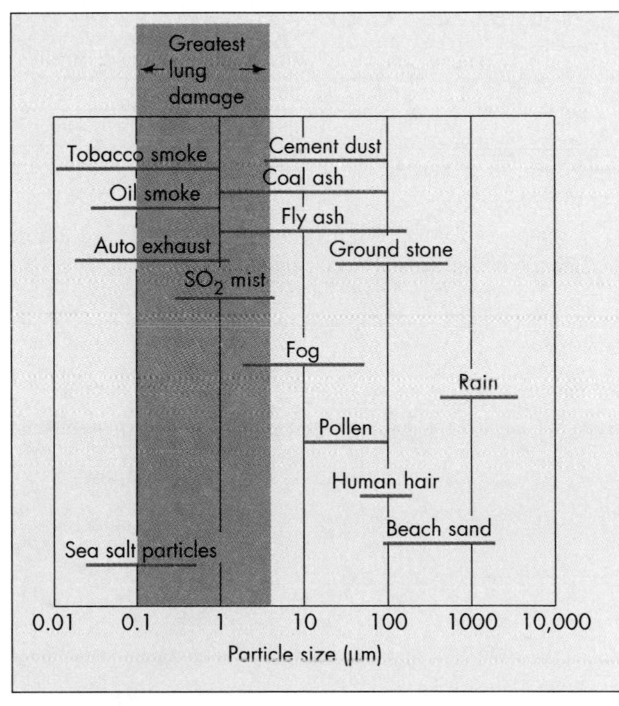

▲ **FIGURE 16.12** Types and sizes of selected particulates. (*Modified from* Giddings. 1973. Chemistry, man, and environmental change: An integrated approach. *Canfield Press; and Hidy and Brock.* 1971. Topics in current aerosol research. *Pergamon Press.*)

A CLOSER LOOK Causes of Climatic Change

Climate Cycles

Why does climate change occur? Examination of Figure 16.10 illustrates cycles of change with periods of about 100,000 years separated by shorter cycles of 20,000 years and 40,000 years in duration. These cycles were first identified by Milutin Milankovitch in the 1920s. He suggested, as a hypothesis, that the cycles produced climate change from glacial to interglacial periods. Milankovitch realized that the spinning earth, like a wobbling top, is not able to keep a constant position in relationship to the sun, and that periodic change in earth's position determines (in part) the amount of sunlight reaching and warming the earth. He suggested that variations in the earth's orbit around the sun follow an approximate 100,000-year cycle that correlates with the major glacial and innerglacial of Figure 16.10a. Cycles of approximately 40,000 and 20,000 years are the result of changes in the tilt of the earth's axis and wobble of the earth's axis, respectively. While Milankovitch cycles do reproduce most of the long-term climatic cycles, their effect on the amount of sunlight reaching the earth are insufficient by themselves to produce the large-scale climatic variations we have observed in the geologic record. Therefore, Milankovitch cycles can be looked at as natural forcing mechanisms that along with other processes may produce climatic change. Shorter cycles have also been suggested, and, in fact, one study suggests that during the past 4000 years, there has been an approximately 1500-year cycle, perhaps explaining the medieval warming period during the Little Ice Age as well as the present warming trend, which is predicted to continue naturally until approximately A.D. 2400 (22). If this is correct, then any warming caused by human activity would be superimposed on a system that is already naturally, slowly warming. However today the human component of warming may be greater than that warming not associated with human activity.

Ocean Conveyor Belt

It is becoming apparent that our climate system may be inherently unstable and capable of jumping quickly from one state (cold) to another (hot) over a time period as short as a few decades (23). Part of what may be driving the climate system and its potential to change is the so-called *ocean conveyor belt,* which is a global-scale circulation of ocean waters characterized by strong northward movement of surface waters in the Atlantic Ocean. These waters are approximately 12° to 13 °C when they arrive near Greenland, where they are cooled to 2° to 4°C (Figure 16.A) (23). As the water is cooled, it becomes more salty, also increasing the density and causing it to sink to the bottom. It then flows southward around Africa, adjoining the global pattern of ocean currents. The flow in this conveyor belt current is huge and about equal to that of 100 Amazon rivers (23). The Amazon River itself contains approximately 20 percent of the total river flow runoff of the earth! The amount of warm water and heat released is sufficient to keep northern Europe 5° to 10°C warmer than it would be if the conveyor belt was not present. If the conveyor belt were to "shut down," global cooling might result quick-

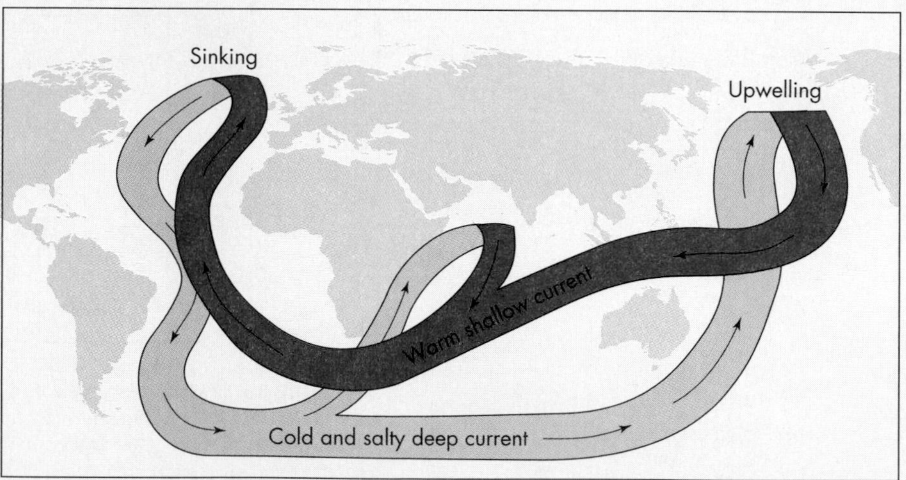

▲ **FIGURE 16.A** Idealized diagram of the oceanic conveyor belt. The natural system is more complex, but in general warm surface water (red) from the Pacific Ocean is transported to the southwest and northward, increasing in temperature due to solar heating and salinity due to evaporation. Near Iceland it cools from contact with cold Canadian air. As the water cools, it increases in density and sinks to the bottom, flowing south (blue) then east and north into the Pacific where upwelling occurs. The mass of sinking and upwelling waters are balanced, and total flow of the current rate is about 20 million m^3/sec. This is a tremendous volume of water, about equivalent to global annual precipitation. Heat released to the atmosphere from the warm surface water keeps northern Europe 5° to 10 °C warmer than if the conveyor were not present. (*After Broker,* W. 1997. *Will our ride into the greenhouse future be a smooth one?* Geology Today 7[5]:2–6.)

ly. As a result, northern Europe would become cooler and less habitable. If this happens over a period of years in the future when there are several billion people to feed, a global catastrophe might result (23)!

El Niño

The term *El Niño* refers to a child and, in particular, the Christ Child because an El Niño event often starts off South America near Christmas time. El Niño became a common term during the winter of 1997–98 when it was blamed for everything from tornadoes and thunderstorms in Florida to flooding and landslides in southern California, droughts and fires in Brazil and Australia, and catastrophic fires in Indonesia. For example, the 1997–98 El Niño is also thought to have been partly responsible for the warmer, more pleasant winter of both New York and Seattle (Figure 16.B). Major winter storms in 1997–98 in California produced widespread flooding and landsliding that statewide caused more than $550 million in property damage (24). Worldwide, the 1997–98 El Niño was partly responsible for the deaths of more than 2000 people while inflicting more than $30 billion in damages (25).

El Niño events are natural disruptions of the ocean–atmosphere system in the tropical Pacific, with important consequences for weather on a global scale. El Niño events are in part responsible for weather events causing billions of dollars in property damages and thousands of human lives lost. El Niño events occur at intervals of two to seven years, averaging about every three to four years between events, and they typically last for 12 to 18 months. El Niño events start with a weakening of the east-to-west trade winds and warming of eastern Pacific Ocean waters. This results in tropical rainfall shifting from Indonesia to South America, as shown ideally on Figure 16.C. During the more normal non-El-Niño conditions, the trade winds blow across the tropical Pacific toward the west. Warm surface water in the western Pacific increases in elevation, and the sea surface is as much as 0.5 m higher at Indonesia than at Peru (Figure 16.C).

In contrast, during El Niño, the trade winds weaken and may even reverse, and the eastern equatorial Pacific Ocean becomes anomalously warm. The westward-moving equatorial ocean current weakens or reverses, and the rise in the temperature of the sea surface waters off the South American coast has significant consequences. Because rainfall follows the warm water eastward, during El Niño years there are higher rates of precipitation and flooding in Peru, while droughts and fires are commonly observed in Australia and Indonesia. Because the warm ocean water provides an atmospheric heat source, the El Niño changes the global atmospheric circulation, which causes changes in weather in regions that are far removed from the tropical Pacific, as shown on Figure 16.C (24,25).

With respect to the 1997–98 El Niño, anomalously warm tropical ocean waters appeared in March and the global-scale temperature and precipitation changes were considerable, disrupting climates from the eastern United States to Australia.

An important point concerning El Niño is that it is a natural phenomena that is part of the dynamic system involving the coupling of the earth's atmosphere and the ocean. El Niño events temporarily alter the weather that we humans are used to, sometimes with devastating effects. El Niño events

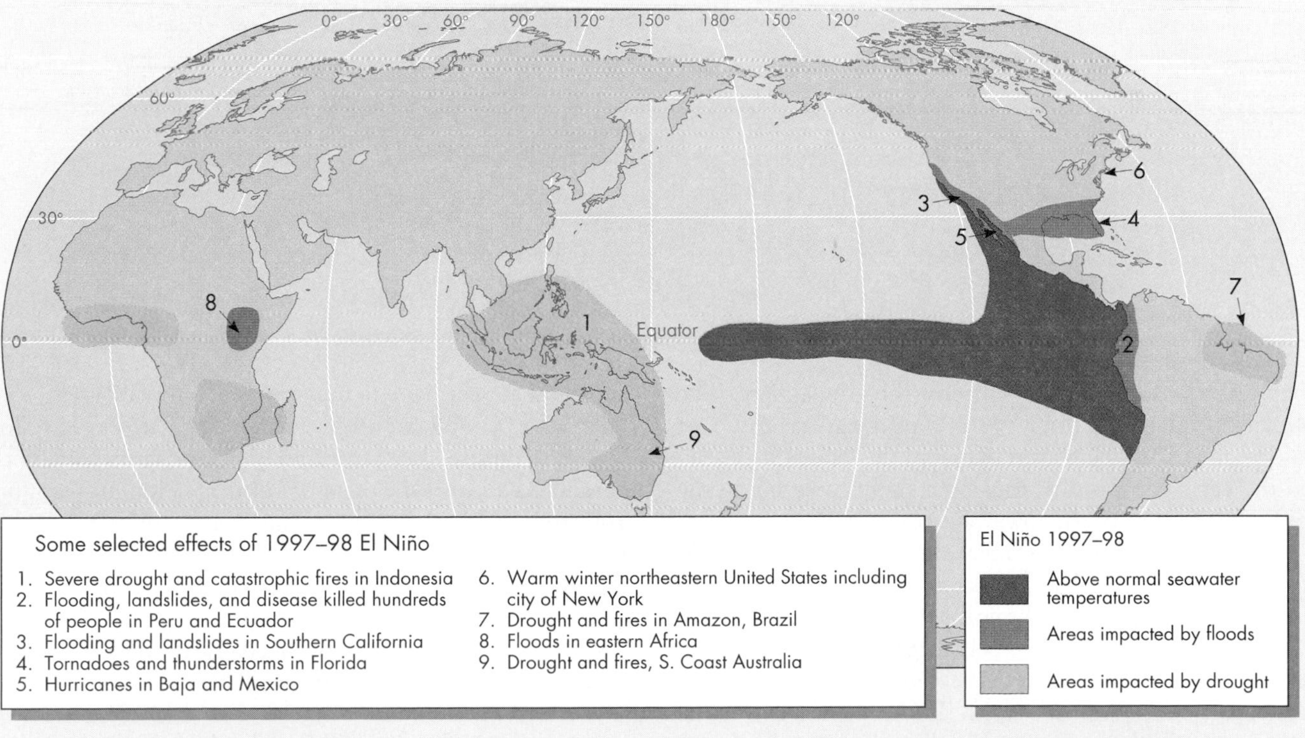

▲ **FIGURE 16.B** The 1997–98 El Niño event. Map shows the general extent of the El Niño effects and the regions damaged by floods, fires, or drought. (*Data from* NOAA, 1998)

can alternate with La Niña events, which are characterized by unusually cool ocean water temperatures, in marked contrast to the warm waters of the El Niño, the effects of which are the opposite. For example, during La Niña, winter temperatures are cooler than normal in the northwest (for example, the 1998–99 winter). La Niña events do not necessarily follow every El Niño event, but together they constitute a natural cycle of ocean-atmospheric disruption with global consequences (24,25).

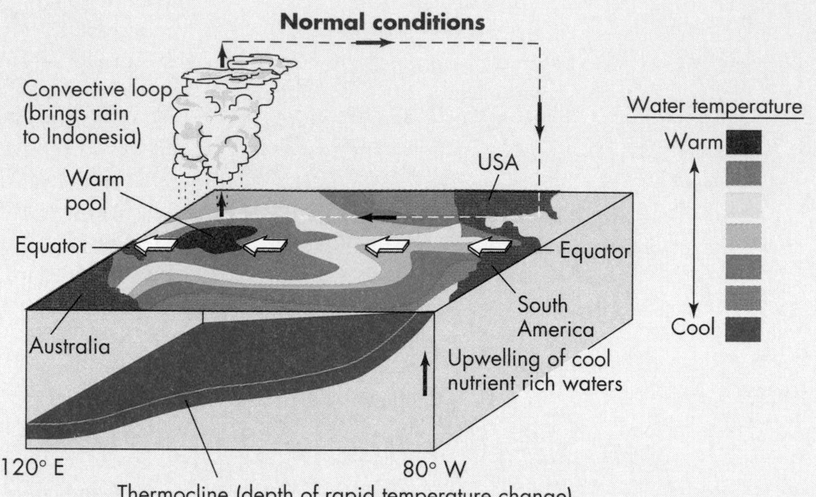

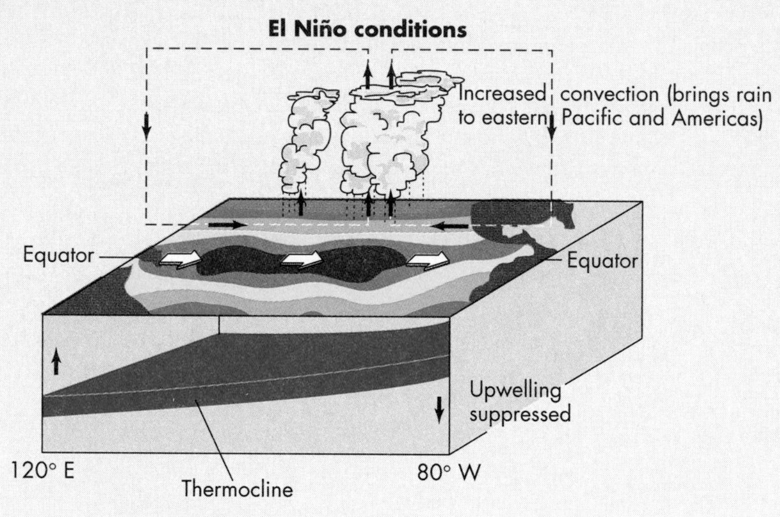

▶ **FIGURE 16.C** Idealized diagram contrasting normal conditions and processes (a) with those of El Niño (b). (*Modified after* NOAA. *Accessed 3/3/99 at http://www.elnino.noaa.gov/.*)

other material as high as 27 km into the atmosphere. The hole left in the ocean floor was 300 m deep, and the blast could be heard 5000 km away! Dust in the stratosphere circled the earth for months, causing a slight lowering in the mean temperature of the lower atmosphere.

Recent research suggests that the amount of ash reaching the stratosphere is less important than the composition of the ash in determining climatic effects. For example, the eruptions of Mount St. Helens in 1980 and El Chichón in southern Mexico in 1982 both ejected volcanic debris into the stratosphere (Figure 16.13). The two events were of similar magnitude and much smaller than the Krakatoa eruption. However, although Mount St. Helens caused little or no climatic disturbance, El Chichón was probably responsible for the drop in mean annual temperature of about 0.3 to 0.5 °C that occurred during a 3-year period following the eruption. The stratospheric cloud from El Chichón contained aerosol droplets of sulfuric acid and was about 100 times as dense as the cloud from Mount St. Helens, which contained little sulfur. While fine volcanic ash settles out in a few weeks, an aerosol of sulfuric acid droplets may circle the earth for several years.

In addition to its hypothesized effect on world temperatures, the El Chichón eruption also may have contributed to the formation of the strong 1982–83 El Niño (see *A Closer Look: Causes of Climatic Change*) by disrupting atmospheric circulation, which drives oceanic circulation (28). The chain of events is thought to be as follows: The sulfurous cloud ab-

▲ **FIGURE 16.13** Vertical eruption cloud from Mount St. Helens eruption in the mid-1980s. Volcanic eruptions can inject material such as ash and sulfur dioxide into the upper atmosphere and disturb global climate for a period of a year or so. (*Roger Werth/Woodfin Camp & Associates*)

sorbed incoming solar radiation in the stratosphere, resulting in warming of the stratosphere and cooling of the lower atmosphere. These processes reduced normal atmospheric circulation, making it easier for warm oceanic currents to move west near the equator off South America (rather than east as they normally do), and this shifting of the currents help produce the El Niño event.

The 1991 eruptions of Mt. Pinatubo in the Philippines were larger than the El Chicón events and also released tremendous quantities of sulfur dioxide that oxidized to sulfuric acid. The climatic effects of these eruptions are being measured and modeled. Yet it is not the particulate matter from volcanic eruptions that causes the most concern, but the increase in atmospheric particulates from human activities such as burning wood and coal, mining, and agriculture. We cannot yet say whether the cooling caused by these particulates will offset the warming trend caused by carbon dioxide from burning fossil fuels.

The fires deliberately set during the 1991 Persian Gulf War and the 1997 catastrophic fires in Indonesia, caused by controlled burning of rain forest that burned out of control, producing one of the world's most severe environmental disasters, provide dramatic examples of atmospheric pollution produced by human activity. The Iraqi military set several hundred Kuwaiti oil wells ablaze, burning millions of barrels of oil each day for months, and the Indonesian fires scorched more than 20,000 hectares of land. The fires and smoke produced an enormous amount of particulates in addition to carbon dioxide, carbon monoxide, and nitrogen oxides. These pollutants created significant environmental problems on a regional scale but have minor effect on the global environment. For example, the Gulf War fires contribute less than 2 percent of the annual global production of carbon dioxide.

Potential Effects of Global Warming

If the greenhouse gases double in the future, it is estimated that the average global temperature will rise about 1.2 °C (16), with significantly greater warming at the polar regions. Specific effects of this temperature rise are difficult to predict, but two of the main uncertainties being considered are change in global climate pattern and a rise in sea level due to thermal expansion of the warmer seawater and partial melting of glacial ice.

Climate Pattern Global rise in temperature might significantly change rainfall patterns, soil-moisture relationships, and other climatic factors important to agriculture. It has been predicted that some northern areas such as Canada and eastern Europe may become more productive, whereas lands to the south will become more arid. It is emphasized that such predictions are very difficult in light of the uncertainties surrounding global warming. Furthermore, even if optimal climatic growing zones move north, this does not necessarily mean that prime agricultural zones will move north, for maximum grain production also depends upon fertile soil conditions that may not be present in all new areas. For example, Canadian prairie soils tend to be thinner and less fertile than those of the U.S. Midwest. To some extent, it is this uncertainty that makes people nervous. Global agricultural activities that are stable or expanding are crucial to people throughout the world who depend upon the food grown in the major grain belts. Hydrologic variations associated with climatic change resulting from global warming might seriously affect food supplies at the global level.

Global warming might also change the frequency and intensity of violent storms, and this change may be more important than which areas become wetter, drier, hotter, or cooler. Warming oceans could feed more energy into high-magnitude storms such as hurricanes. More or larger hurricanes would increase the hazard of living in low-lying coastal areas, many of which are experiencing rapid growth of human population.

Although there is an emerging general consensus that global warming is occurring, studies are just beginning to verify if effects of warming can be recognized. Also, remember that consensus among scientists (scientific consensus) is not scientific proof. It was not too long ago that the consensus among scientists was that the continents were not moving or drifting apart (discovery of plate tectonics changed that consensus). The U.S. Geological Survey has initiated a pilot study of the Delaware River basin, with the major objective of determining what data are necessary for predicting impacts of climate change (9). The Delaware River is a good study site because the basin crosses four distinct physiographic provinces (Figure 16.14), providing a variety of environments to study. The plan is to complete investigations and environmental monitoring in terms of hydrologic response to existing management strategies related to processes associated with storage of water for New York City; maintaining stream flow requirements for a variety of water uses; and controlling migration of salt water into the Delaware estuary if the sea level changes. The Delaware River basin does not, however, encompass nearly the full range of hydrologic variability in the United States, and so other basins across the country will also be studied as potential indicators of climatic change (9).

Rise of Sea Level Rise of sea level is a potentially serious problem related to global warming. Estimates of the rise expected in the next century vary widely, from approximately 40 to 200 cm, and precise estimates are not possible at this time. However, a 40-cm rise in sea level would have significant environmental impacts. Such a rise could cause increased coastal erosion on open beaches of up to 80 m, rendering buildings and other structures more vulnerable to waves generated from high-magnitude storms. In areas characterized by coastal estuaries, a 40-cm sea level rise would cause a landward migration of existing estuaries, again putting pressure on human-built structures in the coastal zone. A sea level rise of 1 m would have very serious consequences; significant alterations would be needed to protect investments in the coastal zone. Communities would have to choose between making very substantial investments in controlling coastal erosion or allowing beaches and estuaries to migrate landward over wide areas (16).

Coastal erosion is a serious problem today in many parts of the world, and a substantial increase in sea level could double the current rate in some areas (16). It seems unavoidable that such a rise must lead to a major investment for protecting cities in the coastal zone. Building of dikes, seawalls, and other erosion-control structures will become common practice, and coastal erosion will continue to threaten urban property. In some areas where coastal development has been greatly restricted, erosion may proceed, with little consequence to people. Because coastal erosion is so difficult to deal with, it seems prudent to allow it to continue wherever feasible and to fight it only where absolutely necessary (see Chapter 9).

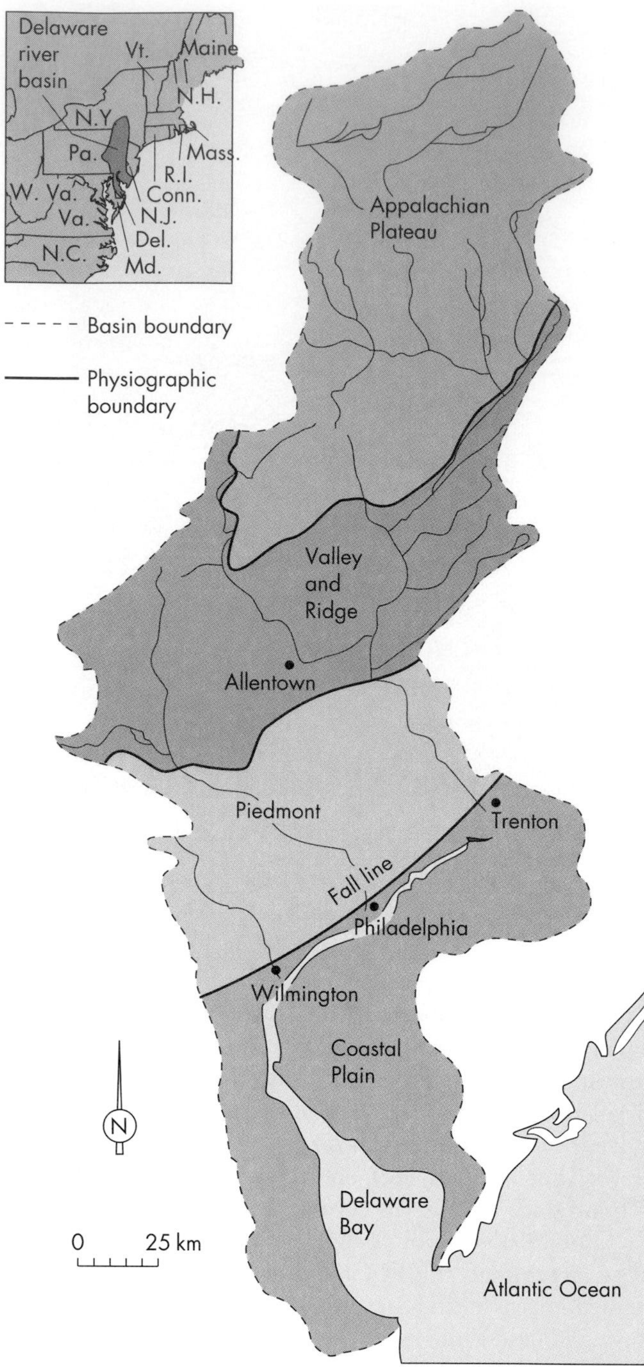

▲ **FIGURE 16.14** Delaware River basin with physiographic provinces shown: Appalachian Plateau (uplands), Valley and Ridge (alternating linear valleys and ridges), Piedmont (rolling hills), and Coastal Plain (low relief). (*From* M. E. Moss *and* H. F. Lins. U.S. *Geological Survey Circular* 1030.)

Strategies for Reducing the Impact of Global Warming

If it is true that global warming is due in part to the increased concentration of the greenhouse gases, reduction of these gases in the atmosphere must be a primary management strategy. Much of the carbon dioxide emission from anthropogenic sources is related to burning of fossil fuels; therefore, energy planning that relies more heavily on alternative energy sources such as wind power, solar, or geothermal will reduce emissions of carbon into the atmosphere. A change to greater use of nuclear energy would also lessen atmospheric carbon loading.

Another source of atmospheric carbon dioxide is related to deforestation. Burning of forest lands to convert them to agricultural use puts a lot of carbon dioxide into the atmosphere. Management plans to protect the world's forests would be another strategy to help reduce the potential threat of global warming.

In summary, better management of the production of "greenhouse gases" would ultimately lead to a lower probability of global warming. There are uncertainties as to whether global warming is occurring and, if so, what its magnitude and effects will be. Nevertheless, a conservative plan would involve attempting to reduce emissions of greenhouse gases as much as economically and politically feasible in the short range and preparing additional management strategies to further reduce emissions of the gases should the need arise. In light of the consensus that the global climate will warm further, and the significant adverse effects this warming will have on our environment, we need to determine when we should take steps to reduce those potential effects. Fortunately, recent model studies suggest that global warming, if it is occurring, is not an emergency and we have a decade or so to develop alternatives to the continued intensive burning of fossil fuels. However, it will be necessary to go through a transition from fossil fuel energy sources to alternative sources that produce much less CO_2 if we decide that we must stabilize the concentration of atmospheric CO_2 in the future. Stabilization will not be easy. To stabilize CO_2 concentrations at present-day levels would require about a 60 percent reduction of global carbon dioxide emissions. To address this problem, the United Nations Framework Convention on Climate Change (FCCC) was adopted at the Rio Earth Summit in 1992. This commits all signatories of the convention to produce a national program to reduce the amount of carbon dioxide produced to 1990 levels by the year 2000 and to develop methods for the protection of carbon dioxide sinks such as forests. The issue of reducing carbon dioxide emissions was revisited in 1997 at the Summit on Global Warming in Kyoto, Japan. The United States has made the issue an internal political debate. Congress has refused to acknowledge the problem and prohibited studying global warming unless developing countries also commit to reductions in carbon dioxide emissions. Congressional inactivity reduces the chances of real progress. However, the meetings did result in bringing the issue of global warming to the forefront for further debates, as had earlier environmental issues such as water pollution that eventually resulted in meaningful legislation.

16.6 Ozone Depletion

Ozone Formation and the Ozone Layer

Ozone (O_3) is a triatomic form of oxygen in which three atoms of oxygen (O) are bonded. The oxygen we breathe is diatomic oxygen (O_2), consisting of two oxygen atoms bonded together. Ozone is relatively unstable and releases an oxygen atom fairly easily. Ozone is therefore a strong oxidant and reacts with many different materials. Some of the reactions are useful; for example, ozone gas bubbling through water can purify it. In the lower atmosphere, ozone is a pollutant that may injure living organisms, including people.

Figure 16.15 shows the structure of the atmosphere. Ozone in the troposphere (the lower 10 km of the atmosphere, where weather, human activity, and urban air pollution occur) is produced in conjunction with photochemical reactions (light-induced chemical reactions) related to natural and anthropogenic emissions of nitrogen dioxide and hydrocarbons. Ozone in the stratosphere (the atmosphere from about 10 km to about 50 km above the earth) is produced by natural processes. Photodissociation of oxygen molecules (O_2) in the stratosphere produces single oxygen

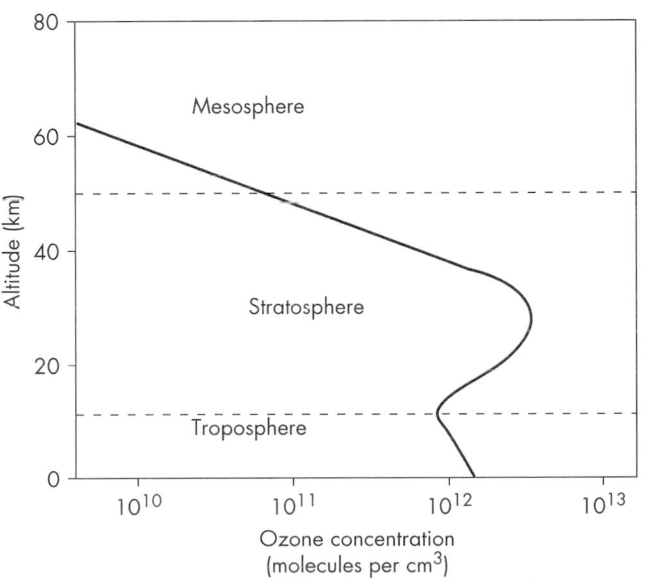

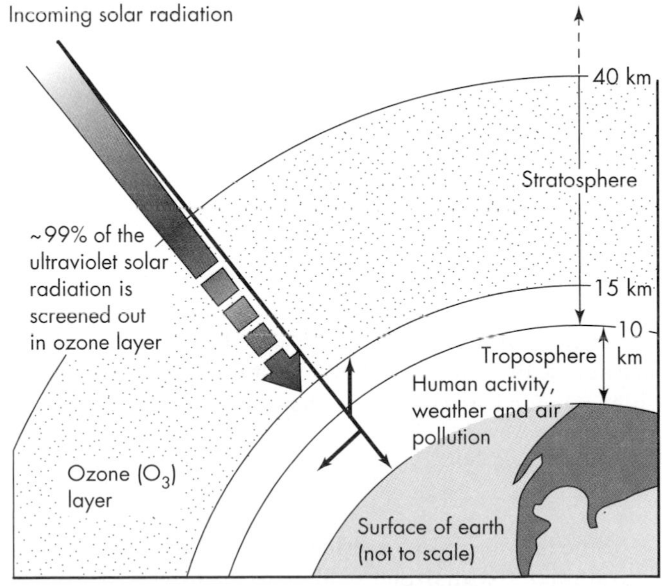

▲ **FIGURE 16.15** Structure of the atmosphere, ozone concentration, and reduction of ultraviolet radiation in the ozone layer. (*Ozone concentrations from* R. T. Watson. *Atmospheric Ozone, in* Effects of change in stratospheric ozone and global climate, *vol.* 1. *Overview, ed.* J. G. Titus, *p.* 70. U.S. *Environmental Protection Agency.*)

atoms (O). An oxygen atom may then combine with an oxygen molecule to produce ozone (O_3). The highest concentrations of stratospheric ozone occur from approximately 20 to 25 km above the surface of the earth, forming the ozone layer (19).

The ozone layer is sometimes referred to as earth's ozone shield or life shield, because it absorbs ultraviolet solar radiation, preventing transmission of harmful amounts of this radiation to the troposphere. The total amount of ozone in the stratosphere is quite small: If all of it were brought down to the surface of the earth, it would form a layer approximately 4 mm thick. Nevertheless, this small amount of ozone is responsible for maintaining the life shield.

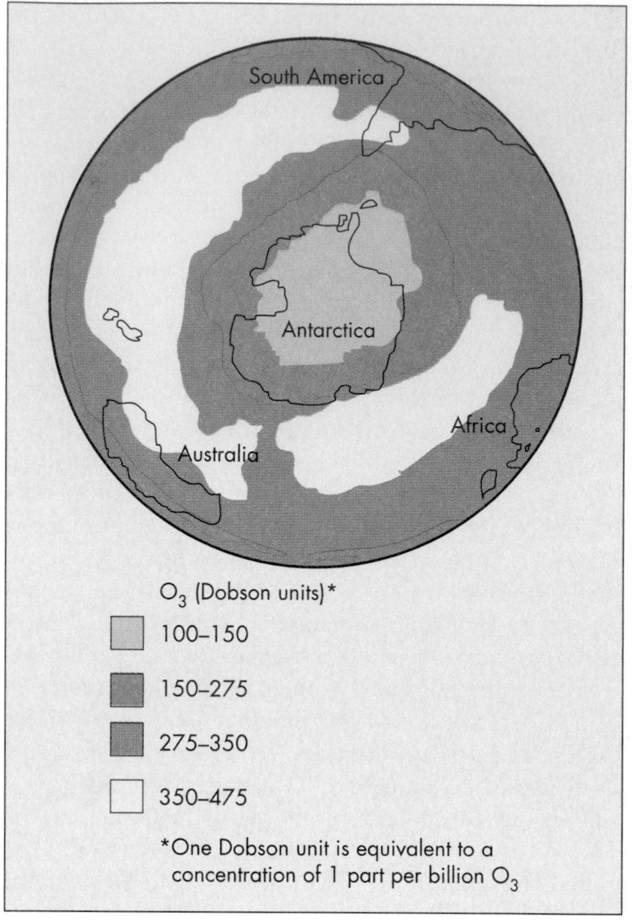

▲ **FIGURE 16.16** Antarctic "ozone hole" in early October 1994. (*Data from* NASA, *Goddard Space Flight Center*)

Removal of Stratospheric Ozone

The ozone layer in the stratosphere is naturally fragile. Natural processes are constantly forming, maintaining, and removing ozone; in fact, absorption of ultraviolet radiation in the ozone layer is a natural mechanism of ozone removal. Anything that changes the balance between ozone formation and renewal has the potential to disrupt the ozone layer.

It was first suggested in 1974 that human-made chemicals might be responsible for destroying the ozone layer (29). The hypothesis presented in 1974 was that release of chlorine by photodissociation of chlorofluorocarbon molecules (CFCs) is responsible for triggering a chain reaction that significantly depletes stratospheric ozone. The importance of the hypothesis was dramatically heightened in 1985 with the discovery of the so-called *Antarctic ozone hole* (Figure 16.16).

Chlorofluorocarbons are contained in many commercial products, including refrigerants, cleaning solvents, and aerosol propellants. The CFCs are extremely stable compounds; that is, they are insoluble and nonreactive. The two most commonly used CFCs are trichlorofluoromethane (CFC-11) and dichlorodifluoromethane (CFC-12), which have been released into the atmosphere for many years. Recently, annual total emissions of these materials have been in excess of 1 million tons.

When the stable CFCs are emitted into the atmosphere, they can remain there for a very long time, often for more than 100 years. During that period, they randomly diffuse through the atmosphere, gradually rising to higher altitudes. At the midstratospheric level they absorb ultraviolet radiation, which causes them to release chlorine atoms. The stratospheric chlorine may then descend back through the ozone layer and eventually be deposited as hydrochloric acid in rainfall. Prior to the formation of the acid, however, the chlorine may participate in a chain reaction (29):

$$Cl + O_3 \Rightarrow ClO + O_2$$

$$ClO + O \Rightarrow Cl + O_2$$

The first equation describes the combination of a chlorine atom (Cl) with ozone, leading to the formation of chlorine oxide (ClO) and a diatomic oxygen molecule (O_2). This reaction is the one that destroys ozone. In the second reaction the chlorine oxide combines with an oxygen atom to produce chlorine and another oxygen molecule; chlorine reappears, but ozone does not. The process becomes a chain reaction because the chlorine may combine with another ozone molecule, repeating the first reaction and causing further ozone destruction. It has been estimated that during the 1- to 2-year average residence time of one chlorine atom in the stratosphere, it might destroy approximately 100,000 molecules of ozone (30). Thus, the annual emission of approximately 1 million tons of CFCs may translate into a loss of ozone that could be about 100,000 times as great.

It is important to recognize that chlorine is not the only chemical capable of entering into reactions that destroy ozone. Other materials, such as nitrous oxide (N_2O) and bromine (Br), can enter into similar reactions. However, because CFCs are being released in great quantities and are increasing in concentration in the atmosphere, the focus has been on chlorine in the stratosphere.

The **ozone depletion** reactions just discussed may give the impression that the ozone depletion story is a simple one. This could not be further from the truth! These reactions greatly simplify what is happening in the Antarctic polar vortex, the mass of air that circulates counterclockwise over the South Pole. Complex chemical reactions occur on the edges of stratospheric clouds over the pole during the polar winter and early spring, resulting in the release of ozone-destroying chlorine from compounds it has formed with other atoms (19,31) (see *A Closer Look: Role of Stratospheric Clouds in Antarctic Ozone Depletion*).

The first reports of significant ozone depletion were from land-based data in Antarctica. Approximately 30 years of data suggest that there has been a significant springtime decrease in ozone each year, starting perhaps in the 1970s, which intensified during the mid-1980s (32). Satellite data confirm the ozone loss and the development of the Antarctic ozone hole. In 1987 the ozone decline above Antarctica was nearly 50 percent, although in 1988 the decline was closer to 15 percent (33). The lower decline in 1988 was thought to be a result of greater atmospheric mixing and a warmer stratosphere over Antarctica that produced fewer ice clouds to participate in ozone depletion reactions (31). Antarctic ozone depletion increased again between 1989 and 1994, with a record depletion of 70 percent in 1993, nearly matched in 1994 (Figure 16.16). Furthermore, the size of the Antarctic ozone hole is apparently increasing, from less than 2 million km^2 in the late 1970s to about 24 million km^2 in 1994 (34).

Environmental Effects of Ozone Depletion

Depletion of stratospheric ozone allows greater amounts of ultraviolet radiation to reach the surface of the earth. Potential effects on humans include increases in all types of skin cancers and greater severity of sunburns. Skin cancers in the United States have increased about 90 percent since the 1970s. This is an epidemic. However, it is controversial as to how much of the current epidemic of skin cancers in the United States is due to ozone depletion. Under a worst-case scenario, ozone depletion should only have resulted in an increase of 20 to 40 percent in skin cancers. Thus, other causes of skin cancer must play a role (35). There is concern also that UV radiation will affect productivity of food plants such as soybeans. In fact, it has been speculated that ozone depletion might adversely affect, through plant damage, part of the agriculture-based global food supply.

Increased UV radiation may also affect marine biologic processes. The radiation penetrates several meters of ocean water and could potentially cause damage to planktonic life in the ocean. Plankton form the base of the food chain in the marine environment, and so any disruption of plankton would cause potential problems throughout the entire food chain and eventually affect people who consume marine resources. Also, plankton are believed to be a major sink for carbon dioxide; hence, their disruption could increase atmospheric CO_2 concentrations. A recent study in the Antarctic Ocean beneath an air mass with depleted ozone suggested a reduction in planktonic life of 6 to 12 percent (36). More studies are needed to further test the hypothesis that increased UV radiation is destroying plankton.

Management Issues Related to Ozone Depletion

Ozone depletion is going to be with us for many decades. Even if all uses of CFCs were stopped today, millions of tons of these chemicals are already in the lower atmosphere, moving upward toward the stratosphere. This movement will cause an increase in their concentration in coming decades and a reduction in the concentrations of stratospheric ozone. As a result, reductions in CFC use will have only a minor effect on the ozone problem before the end of this century (30). Nevertheless, it is very prudent to move forward with management plans that greatly restrict the use of CFCs and other chemicals that cause ozone depletion. Concern for ozone depletion has led the United Nations' Environmental Program to develop a strategy to protect stratospheric ozone. The effect of what is known as the 1987 Montreal Protocol was a significant reduction in the emission of CFCs by the year 1999. In the United States, plans were put in place in 1992 to eventually eliminate all production of CFCs. As a result, there are programs to collect and reuse CFCs and to use substitute chemicals for CFCs.

16.7 Coupling of Global-Change Processes

The major global-change processes discussed in this chapter have interesting linkages. For example, the chlorofluorocarbons that cause ozone depletion when they reach the stratosphere may contribute to the greenhouse effect when they are released into the lower atmosphere. In fact, CFCs trap more heat than carbon dioxide because they absorb infrared radiation in an area of the spectrum in which carbon dioxide does not absorb. One study (29) concluded that CFCs, on a per-molecule basis, are 10,000 times more efficient in absorbing infrared radiation than is carbon dioxide. Nevertheless, carbon dioxide contributes much more to the total anthropogenic greenhouse effect than do CFCs because so much more carbon dioxide is being emitted.

The important point to be considered here is the coupling of the greenhouse and ozone problems via release of materials such as CFCs. Other couplings are related to processes such as burning of fossil fuels, which releases precursors to acid rain (see Chapter 15) as well as carbon dioxide and other greenhouse gases. As was pointed out in the last section, the burning of fossil fuels also produces particulates that result in atmospheric cooling, and we do not know whether the particulates or the carbon dioxide will have the predominant effect. Thus, we see the principle of environmental unity in action. Many aspects of one problem are related to other problems.

A CLOSER LOOK Role of Stratospheric Clouds in Antarctic Ozone Depletion

Polar stratospheric clouds form at altitudes of approximately 20 km above polar regions (Figure 16.D, part a). The clouds form during the polar winter when the polar air mass is isolated from the remainder of the atmosphere and in the Antarctic rotates counterclockwise in the polar vortex (Figure 16.D, part b). Chemical reactions and processes that ultimately result in Antarctic ozone depletion are shown on Figure 16.D, part c. Molecules of chlorine nitrate ($ClONO_2$) and hydrochloric acid (HCl) are sequestered on edges of the cloud particles. The chlorine nitrate and hydrochloric acid are the important sinks for chlorine in the stratosphere, and on the cloud edges they react to form dimolecular chlorine (Cl_2) and nitric acid (HNO_3). During the winter, denitrification occurs as nitric acid particles grow large in the intense cold and fall out of the stratosphere by gravitational settling. Thus, in the springtime, the dimolecular chlorine is available and is readily dissociated by the first sunlight into chlorine atoms that enter into ozone-depleting chain reactions as previously discussed. In the early spring this reaction may cause as much as 1 to 2 percent ozone depletion per day, ultimately producing the 70 percent reduction in ozone observed in 1993 and 1994 (19,31).

Before leaving this discussion it is prudent to discuss a little further the two natural sinks for chlorine in the atmosphere. The formation of chlorine nitrate ($ClONO_2$) occurs because chlorine oxide combines in the atmosphere with nitrogen dioxide. When this happens, ozone depletion is minimal. Hydrochloric acid is the other natural sink for chlorine. The reaction that occurs is that chlorine released from the CFCs combines with methane (CH_4) to form the hydrochloric acid that may then rain out of the atmosphere, effectively remov-

▶ **FIGURE 16.D** Polar stratospheric clouds (a). Clouds seen from NASA aircraft at approximately 12 km in elevation in the polar regions north of Datanger, Norway. Shown are two types of polar stratospheric clouds. The lower dark orange or brown clouds are called Type I clouds and consist of cloud particles comprising mostly nitrogen trihydrates. The pearly white clouds at the top are called Type II clouds and consist mostly of water molecules frozen as ice. (NASA *Headquarters*)

(a)

SUMMARY

For more than 100 years, geologists have studied the record of global change. Today people are very concerned about the accelerating changes taking place in our atmosphere, including an increase in carbon dioxide, which is enhancing or causing global warming; depletion of the ozone layer, which may affect our health; and increasing air pollution from particulates, which may influence climate changes.

The main goal of the emerging integrated study known as earth system science is to obtain a basic understanding of how our planet works and how its various components (such as the atmosphere, oceans, and solid earth) interact. An important secondary goal is to predict global changes that are likely to occur within a time frame of several decades, and thus are of particular importance to people.

Methods of studying global change include examination of the geologic record from lake sediments, glacial ice, tree rings, water, and other earth materials; gathering of real-time data from monitoring stations; and development of mathematical and computer models to predict change.

Earth's atmosphere is a mixture of gases in which complex and incompletely understood reactions take place. Many of these reactions are affected by sunlight and by coming the chlorine from the ozone-destroying chain reaction. The polar stratospheric clouds do not allow the natural sinks to operate fully and, as a result, in the springtime the chlorine is free to react and destroy ozone, as shown on Figure 16.D, part c.

Thus, we see that the chemistry of Antarctic ozone depletion that results in the ozone hole is really quite complicated, depending upon a number of factors, including the formation of the polar vortex and polar stratospheric clouds.

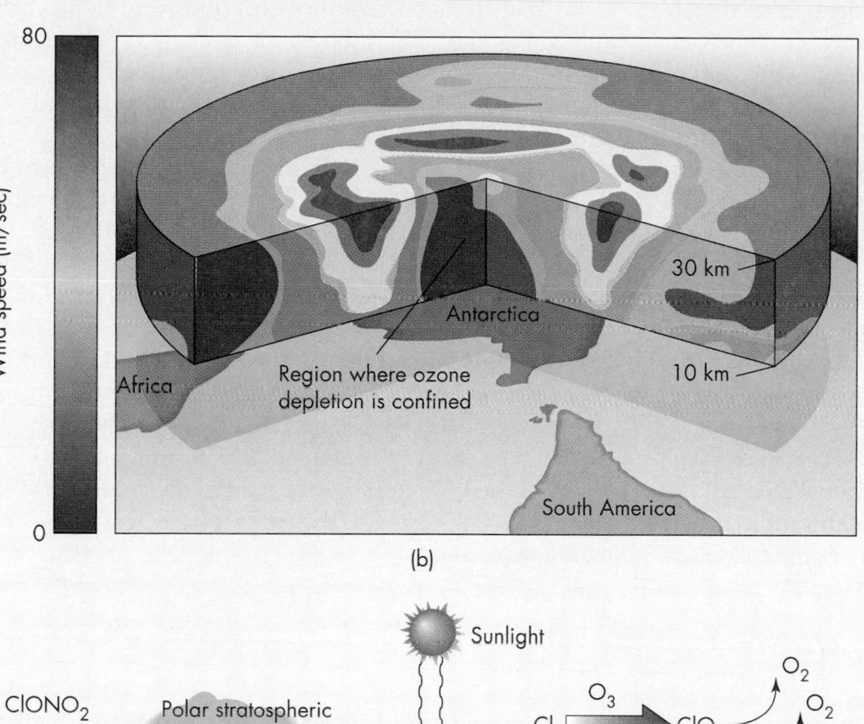

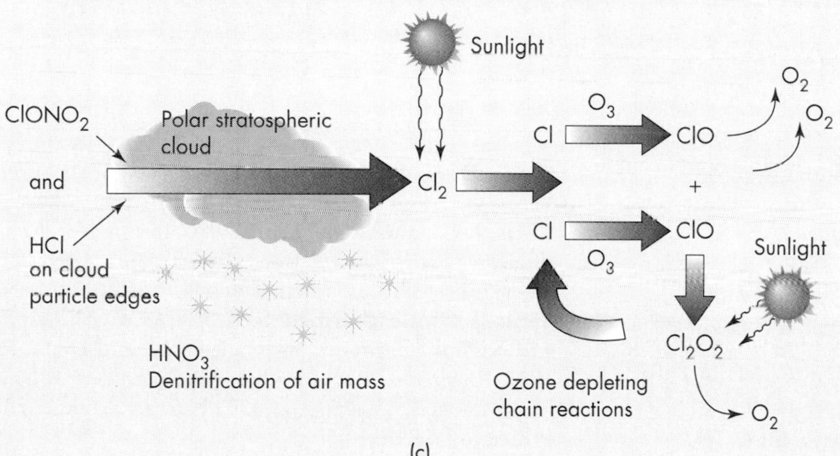

FIGURE 16.D (*continued*) The polar vortex (b) along with chemical processes and reactions associated with polar stratospheric clouds and Antarctic ozone depletion (c) emphasize the complex relations that occur associated with ozone depletion. (*[b] and [c] from* O. B. *Toon and* R. P. *Turco.* 1991. *Polar stratospheric clouds and ozone depletion.* Scientific American 246[6]: 68–74.)

pounds produced by living organisms. Most of the gaseous oxygen in today's oxygen-rich atmosphere has been produced by photosynthetic plants.

To understand changes in global processes in general and in atmospheric processes in particular, one must know something about earth's energy balance. The earth receives energy from the sun in the form of electromagnetic radiation; the visible wavelength of this radiation is called *light.* Solar radiation reaching the atmosphere may be reflected, transmitted, or absorbed; most visible radiation is transmitted through the atmosphere to the earth's surface, where it may be reflected or absorbed. Absorbed solar energy warms the earth's surface and is reradiated as infrared radiation, which causes warming of the atmosphere. The amount of energy the earth radiates back into space is equal to the amount it receives from the sun.

Global climate change is due in part to natural cycles (Milankovitch cycles) that change the amount of solar radiation reaching the surface of the earth. The so-called "conveyor belt" in the world's oceans are responsible for the transport of tremendous quantities of water and heat by way of near-surface and deep subsurface ocean currents, moderating the present climate and helping warm western Europe. El Niño and La Niña events in the Pacific provide important examples of natural earth processes that temporally alter weather patterns of much of the world.

The trapping of heat by the atmosphere is generally referred to as the *greenhouse effect.* Water vapor and several other gases (including carbon dioxide, methane, and chlorofluorocarbons) tend to trap heat and warm the earth, because they absorb some of the heat energy radiating from the earth. Since about 1860, there has been an exponential growth of the concentration of carbon dioxide in the atmosphere, which is responsible for approximately 60 percent of the anthropogenic (human-caused) portion of the greenhouse effect. In the past 100 years mean global temperature has increased by about 0.5 °C, and 1986 to 1998 has been the warmest period during the 139 years of global monitoring of temperature. Some model studies estimate that the average global temperature will rise about 1.2 °C in the next several decades owing to increases in the greenhouse gases and trapping of heat. Effects of such a temperature rise are difficult to predict; the more certain ones include a rise in global sea level and changes in rainfall patterns, high-magnitude storms (hurricanes), soil moisture, and other factors important to agriculture.

It was first hypothesized in 1974 that release of chlorine by photodissociation of chlorofluorocarbon (CFC) molecules caused a chain reaction that could significantly deplete stratospheric ozone. This hypothesis was dramatically confirmed in 1985 with the discovery of the Antarctic ozone hole. Ozone absorbs ultraviolet radiation from the sun, and depletion of stratospheric ozone allows more of this radiation to reach the surface of the earth. The potential effects include increase in all types of skin cancers and damage to plants and marine life. International cooperation is expected to lead to much tighter restrictions on the production and emission of CFCs. However, because CFCs are so stable in the environment, such reductions will have only a minor effect on the ozone problem before the end of the twentieth century.

Particulates are emitted into the atmosphere from both natural and human-induced sources. Eruptions of volcanic ash and aerosolic sulfuric acid into the stratosphere can cause slight global cooling for several years following a large eruption. Particulates emitted by human processes may help offset global warming caused by increase in atmospheric CO_2 from burning fossil fuels, but we do not understand these processes well enough to predict which will predominate.

Some global changes, such as ozone depletion and the greenhouse effect, are coupled. For example, the CFCs released in the lower atmosphere add to the greenhouse effect there before diffusing to the stratosphere, where they cause ozone depletion.

REFERENCES

1. **Myers, N.** 1990. *Future worlds.* London: Robertson McCarta.
2. **Cloud, P.** 1990. Personal written communication.
3. **NASA.** 1990. *EOS: A mission to planet earth.* Washington, DC.
4. **Council on Environmental Quality.** 1990. *Environmental Trends 1989.*
5. **Earth System Sciences Committee.** 1988. *Earth system science: A preview.* Boulder, CO: University Corporation for Atmospheric Research.
6. **Chumbley, C. A., Baker, R. G., and Bettis, E. A., III.** 1990. Midwestern Holocene paleoenvironments revealed by floodplain deposits in northeastern Iowa. *Science* 249:272–74.
7. **Post, W. M., Peng, T., Emanuel, W. R., King, A. W., Dale, V. H., and De Angelis, D. L.** 1990. The global carbon cycle. *American Scientist* 78(4):310–26.
8. **Clark, J.** 1995. Personal oral communication.
9. **Moss, M. E., and Lins, H. F.** 1989. *Water resources in the twenty-first century.* U.S. Geological Survey Circular 1030.
10. **NASA Goddard Space Flight Center; principal contributors: Rind, D., and Lebedeff, S.** 1984. *Potential climatic impacts of increasing atmospheric CO_2 with emphasis on water availability and hydrology in the United States.* Washington, DC: U.S. Environmental Protection Agency.
11. **Marsh, W. M., and Dozier, J.** 1981. *Landscape.* Reading, MA: Addison-Wesley.
12. **Kershaw, S.** 1990. Evolution of the earth's atmosphere and its geological impact. *Geology Today,* March–April, 55–60.
13. **Erlich, P. R., Erlich, A. H., and Holdren, J. P.** 1970. *Ecoscience.* San Francisco: W. H. Freeman.
14. **Gates, D. M.** 1980. *Biophysical ecology.* New York: Springer-Verlag.
15. **Titus, J. G., and Seidel, S. R.** 1986. In *Effects of changes in the stratospheric ozone and global climate,* vol. 1, ed. J. G. Titus, pp. 3–19. U.S. Environmental Protection Agency.
16. **Titus, J. G., Leatherman, S. P., Everts, C. H., Moffatt and Nichol Engineers, Kriebel, D. L., and Dean, R. G.** 1985. *Potential impacts of sea level rise on the beach at Ocean City, Maryland.* Washington, DC: U.S. Environmental Protection Agency.
17. **Council on Environmental Quality.** 1990. *Environmental trends, 1989.* Washington, DC: U.S. Government Printing Office.
18. **Rodhe, H.** 1990. A comparison of the contribution of various gases to the greenhouse effect. *Science* 248(6): 1217–19.
19. **Waters, J. W.** 1993. The chlorine threat to stratospheric ozone. *Engineering and Science* LVI(4):1–13.
20. **Karl, T. R.** 1995. Trends in U.S. climate during the twentieth century. *Consequences* 1(1):3–12.
21. **Jones, P. D., and Wigley, T. M. L.** 1990. Global warming trends. *Scientific American* 263(2):84–91.
22. **Campbell, I. D., Campbell, C., Apps, N. J., Rutter, N. W., and Bush, A. B. G.** 1998. Late Holocene approximately 1500-year climatic periodicies and their implications. *Geology* 26(5):471–73.
23. **Broker, W.** 1997. Will our ride into the greenhouse future be a smooth one? *Geology Today* 7(5):2–6.
24. **NOAA.** 1998. *What is El Niño?* Accessed 10/2/98 at http://www.elnino.noaa.gov.

25. **Suplee, C.** 1999. El Niño/La Niña. *National Geographic* 195(3):72–95.
26. **Pelto, M. S.** 1993. Changes in water supply in alpine regions due to glacier retreat. American Institute of Physics, Proceedings, volume 277: *The World at Risk: Natural Hazards and Climate Change*, 61–67.
27. **Pelto, M. S.** 1996. Recent changes in glacier and alpine runoff in the North Cascades, Washington. *Hydrological Processes* 10:1173–80.
28. **Rampino, M. R., and Self, S.** 1984. The atmospheric effects of El Chichón. *Scientific American* 250:48–57.
29. **Molina, M. J., and Rowland, F. S.** 1974. Stratospheric sink for chlorofluoromethanes: Chlorine atom-catalyzed destruction of ozone. *Nature* 249:810–12.
30. **Rowland, F. S.** 1989. Chlorofluorocarbons and the depletion of stratospheric ozone. *American Scientist* 77:36–45.
31. **Toon, O. B., and Turco, R. P.** 1991. Polar stratospheric clouds and ozone depletion. *Scientific American* 246(6):68–74.
32. **Farman, J. C., Gardiner, B. G., and Shanklin, J. D.** 1985. Large losses of total ozone in Antarctica reveal seasonal ClO_x/NO_x interaction. *Nature* 315:207–10.
33. **Kerr, R. A.** 1988. A shallower ozone hole, as expected. *Science* 28 (October):515.
34. **Kerr, R. A.** 1994. Antarctic ozone hole fails to recover. *Science* 266:217.
35. **Kane, R. P.** 1998. Ozone, depletion related to UVB changes, and increased skin cancer incidence. *International Journal of Climatology* 18(4):457–472.
36. **Smith, R. C., Prezelin, B. B., Baker, K. S., Bidigare, R. R., Boucher, N. P., Colen, T., Karentz, D., MacIntyre, S., Matlick, H. A., Menzies, D., Ondrusek, M., Wan, Z., and Waters, K. J.** 1992. Ozone depletion: Ultraviolet saturation and phytoplankton biology in Antarctic waters. *Science* 255:952–59.

KEY TERMS

earth system science (p. 447)
dendrochronology (p. 449)
monitoring (p. 449)
Global Circulation Models (p. 449)
atmosphere (p. 450)
earth's energy balance (p. 451)
albedo (p. 453)
greenhouse effect (p. 453)
global warming (p. 456)
particulates (p. 457)
ozone (p. 465)
ozone depletion (p. 466)

SOME QUESTIONS TO THINK ABOUT

1. Have a discussion with your parents or someone of similar age and write down the major changes that have occurred in their lifetime as well as yours. Characterize these changes as gradual, accelerating, surprising, chaotic, or other of your choice. Analyze these changes and discuss which ones were most important to you personally. Which of these affected our environment at the local, regional, or global level?
2. How do you think future change is likely to affect you, particularly considering the environmental problems we may be facing as a result of increased world population?
3. The subject of positive and negative feedback in systems was introduced in Chapter 2. Return to that topic now in light of potential global warming. Assume the following:
   - Algae populations will increase as global warming occurs and warms ocean waters.
   - A warming earth will cause increase in evaporation of water from the oceans, adding additional water vapor to the atmosphere.
   - Increasing the amount of carbon dioxide in the environment will stimulate plant growth, so that plants will collectively absorb more carbon dioxide from the atmosphere.
   - Global warming will cause more clouds to form.
   - Global warming will cause a reduction in the amount of snow and ice on earth.
   - Global warming will cause people to use additional air conditioning.

   Consider each of the above and decide which would be a positive and which a negative feedback mechanism for potential global warming. Justify your answers.
4. A friend of yours comments that environmental problems such as overpopulation, global warming, ozone depletion, and acid rain are simply being trumped up by overzealous environmentalists and that we should be more concerned with the pressing problems related to economics and a rising crime rate. How would you respond to your friend?

17 Air Pollution

Los Angeles, California. (*Ken Biggs/Tony Stone Images*)

City of the Angels, Los Angeles, California, was photographed (above) on a rare clear day. Although we have made progress in reducing automobile emissions, the city of Los Angeles still experiences many days when the air is unhealthy for people who live there. While the emissions from individual vehicles have dramatically decreased, especially for newer cars, the total number of automobiles in the Los Angeles metropolitan area continues to increase. Thus, the product of emissions per vehicle and the total number of vehicles results in production of pollutants that continue to cause air quality problems. Abatement of the air pollution problem in Los Angeles will not be achieved until we make dramatic changes in the vehicles and mode of transportation in the Los Angeles area. Positive signs include the increase in number of electric cars and the potential of use of fuels such as hydrogen that do not pollute the air.

17.1 Introduction to Air Pollution

Ever since life began on earth, the atmosphere has been an important resource for chemical elements and a medium for depositing wastes. The earliest photosynthetic plants dumped their waste oxygen into the atmosphere, and the oxygen-rich air that eventually resulted made possible the development and survival of higher life-forms.

As the fastest moving dynamic medium in the environment, the atmosphere has always been one of the most convenient places for people to dispose of unwanted materials. Ever since humans first used fire, the atmosphere has all too often been a sink for waste disposal. With the rapid urbanization and industrialization of the last two centuries, atmospheric circulation has often proved inadequate to dissipate human wastes, and urban air in particular has

LEARNING OBJECTIVES

Air pollution is one of our most serious environmental problems. It affects all of us in terms of the quality of the air we breathe and the public and private funds spent to control air pollution. Learning objectives of this chapter are:

- *To learn something about the history, general effects, and sources of air pollution.*
- *To become familiar with the common types of air pollution and their environmental effects.*
- *To know some of the important relationships among meteorology, topography, and air pollution.*
- *To understand basic components of the urban microclimate.*
- *To know what the two major types of smog are and how they are produced.*
- *To become familiar with some of the methods of controlling air pollution.*
- *To gain a modest acquaintance with U.S. air quality standards and how air quality is reported.*
- *To become familiar with the economic considerations associated with air pollution.*
- *To know some of the main provisions of the Clean Air Act Amendments of 1990.*

Web Resources

Visit the "Environmental Geology" Web site at www.prenhall.com/keller to find additional resources for this chapter, including:

- ▶ Web Destinations
- ▶ On-line Quizzes
- ▶ On-line "Web Essay" Questions
- ▶ Search Engines
- ▶ Regional Updates

become increasingly polluted (see *Case History: The London Smog Crisis of 1952*).

17.2 Pollution of the Atmosphere

Chemical pollutants can be thought of as compounds that are in the wrong place or in the wrong concentrations at the wrong time. As long as a chemical is transported away or degraded rapidly relative to its rate of production, there is no pollution problem. Pollutants that enter the atmosphere through natural or artificial emissions may be degraded by natural processes, not only within the atmosphere but also in the hydrologic and geochemical cycles. On the other hand, pollutants in the atmosphere may become pollutants in the hydrologic and geochemical cycles (as, for example, acid rain; see Chapter 15).

People have long recognized the existence of atmospheric pollutants, both natural pollutants and those induced by humans. Acid rain was first described in the seventeenth century, and by the eighteenth century it was known that smog and acid rain damaged plants in London. Beginning with the Industrial Revolution in the eighteenth century, air pollution became more noticeable; by the middle of the nineteenth century, particularly following the U.S. Civil War, concern with air pollution increased. The word *smog* was probably introduced by a physician at a public health conference in 1905 to denote poor air quality resulting from a mixture of smoke and fog.

Two major pollution events, one in the Meuse Valley in Belgium in 1930 and the other in Donora, Pennsylvania, in 1948, raised the level of scientific research on air pollution. The Meuse Valley event lasted approximately a week and caused 60 deaths and numerous illnesses. The Donora event caused 20 deaths and 14,000 illnesses. By the time of the Donora event, people recognized that meteorological conditions were an integral part of the production of dangerous smog events. This view was reinforced by the 1952 London smog crisis, after which regulations to control air quality began to be formulated.

General Effects of Air Pollution

Most of the adverse effects of air pollution are due to relatively low-level concentrations of toxins over a long period of time. The most serious of these effects are damage to green plants and aggravation of chronic illnesses in people. However, air pollution affects many aspects of our environment: soils, water quality, vegetation, animals, human health, natural and artificial structures, and visually aesthetic resources.

Effects on Soils and Water When pollutants from the air are deposited, soils and water may become toxic. Soils may also be leached of nutrients by pollutants that form acids.

Effects on Vegetation Effects of air pollution on vegetation include damage to leaf tissue, needles, or fruit; reduction in growth rates or suppression of growth; increased susceptibility to a variety of diseases, pests, and adverse weather; and the disruption of reproductive processes. Damage to vegetation can in turn damage entire terrestrial or aquatic ecosystems.

Effects on Animals Effects of air pollutants on vertebrate animals include impairment of the respiratory system; damage to eyes, teeth, and bones; increased susceptibility

CASE HISTORY The London Smog Crisis of 1952

In London, England, during the first week of December 1952, the air became stagnant and the cloud cover did not allow much of the incoming solar radiation to penetrate. The humidity climbed to 80 percent, and the temperature dropped rapidly until the noontime temperature was about 1 °C. A very thick fog developed, and the cold and dampness increased the demand for home heating. Because the primary fuel used in homes was coal, emissions of ash, sulfur oxides, and soot increased rapidly. The stagnant air became filled with pollutants, not only from home heating fuels but also from automobile exhaust. At the height of the crisis, visibility was greatly reduced and automobiles had to use their headlights at midday. Between December 4 and 10, an estimated 4000 people died from the pollution. Figure 17.A shows the increase in sulfur dioxide and smoke and the accompanying deaths during that period. The environment, not human activities, finally solved the problem: The siege of smog ended when the weather changed and the air pollution was dispersed.

Ever since the beginning of the Industrial Revolution and before, people had survived in London and other major cities despite the weather and pollution. What had finally gone wrong? During the London smog crisis, the stagnant weather conditions, together with the number of homes burning coal and of cars burning gasoline, exceeded the atmosphere's ability to remove or transform the pollutants; even the usually rapid natural mechanisms for removing sulfur dioxide were saturated. As a result, sulfur dioxide remained in the air and the fog became acid, adversely affecting people and other organisms, particularly vegetation. The health effects on people were especially destructive because small acid droplets became fixed on larger particulates, facilitating their being drawn deep into the lungs.

The 1952 London smog crisis was a landmark event. Finally, human activities had exceeded the natural abilities of the atmosphere to serve as a sink for the removal of wastes. The crisis was due in part to a positive, or reinforcing, feedback situation (see Chapter 1). Burning fossil fuels added particulates to the air, increasing the formation of fog and decreasing visibility and light transmission; the dense, smoggy layer increased the dampness and cold and accelerated the use of home heating fuels. The worse the weather and the pollution, the more people burned coal to keep warm, and this further worsened the weather and the pollution.

Before 1952, London was well known for its fog; what was relatively little known was the role of coal burning in intensifying fog conditions. Since 1952, London fogs have been greatly reduced because coal has been replaced to a large extent by much cleaner gas as the primary home heating fuel.

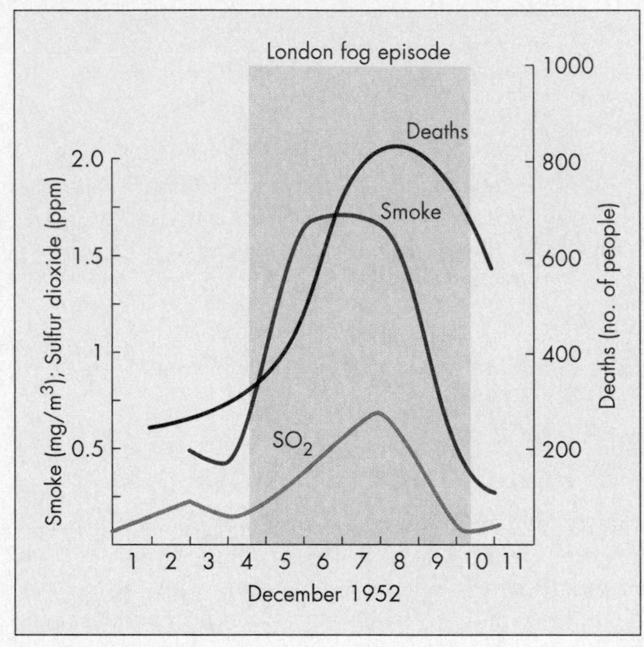

▲ **FIGURE 17.A** The relationship between the number of deaths and the London fog of 1952. (*Modified from* S. J. Williamson. 1973. Fundamentals of air pollution. *Addison-Wesley*.)

to disease, pests, or other stress-related environmental hazards; decrease in availability of food sources, such as vegetation impacted by air pollutants; and reduction in ability to reproduce (1).

Effects on Human Health Effects of air pollutants on human health include toxic poisoning, eye irritation, and irritation of the respiratory system. In urban areas, people suffering from respiratory diseases are more likely to be affected by air pollutants, whereas healthy people tend to acclimate to pollutants relatively quickly. Nevertheless, urban air can cause serious health problems. Many pollutants have synergistic (mutually reinforcing) effects; for example, sulfate and nitrate can attach to particles in the air, thus facilitating their inhalation deep into lung tissue.

Effects on Human Artifacts Effects of air pollution on buildings and monuments include discoloration, erosion, and decomposition of construction materials, as discussed in the section on acid rain in Chapter 15.

Aesthetic Effects Air pollutants affect visual resources by discoloring the atmosphere and reducing visual range and atmospheric clarity. We cannot see as far in polluted air, and what we do see has less color contrast. Once limited to cities, these effects now extend even to the wide open spaces of the United States. For example, emissions from the Four Corners fossil-fuel-burning power plant, near the junction of New Mexico, Arizona, Colorado, and Utah, are altering visibility in an area where previously one could see 80 km from a mountaintop on a clear day (1).

Sources of Air Pollution

Many pollutants in our atmosphere have natural origins. Natural emissions of air pollutants include release of gases such as sulfur dioxide by volcanic eruptions; release of hydrogen sulfide by geyser and hot-spring activity and by biological decay in bogs and marshes; increased concentration of ozone in the lower atmosphere as a result of violent thunderstorms and other unstable meteorological conditions; and emission of a variety of particles from wildfires and windstorms (1).

Major natural and human-produced air pollutants and sources are shown in Table 17.1. These data suggest that, with the exception of sulfur and nitrogen oxides, natural emissions of air pollutants exceed human-produced input. Nevertheless, the human component is most abundant in urban areas and leads to the most severe air pollution events for human health.

The two major kinds of air pollution sources are stationary and mobile. **Stationary sources,** those with a relatively fixed location, include point sources, fugitive sources, and area sources. **Point sources** emit air pollutants from one or more discrete controllable sites, such as smokestacks of industrial power plants. **Fugitive sources** generate air pollutants from open areas exposed to wind processes. Examples include dirt roads, construction sites, farmlands, storage piles, and surface mines. Figure 17.1 shows a large dust cloud on the island of Maui (Hawaiian Islands) generated in an area near Kihei where sugarcane fields were burned a few days earlier. The dust from recently burned fields is an example of a fugitive source. *Area sources* are locations that emit air pollutants from several sources within a well-defined area, as, for example, small urban communities or areas of intense industrialization within an urban complex. *Mobile sources* move from place to place while yielding emissions and include automobiles, aircraft, ships, and trains (1).

Air Pollutants*

The major air pollutants occur either in a gaseous form or as particulate matter. The **gaseous pollutants** include sulfur dioxide (SO_2), nitrogen oxides (NO_x), carbon monoxide (CO), ozone (O_3), volatile organic compounds (VOC), hydrogen sulfide (H_2S), and hydrogen fluoride (HF). **Particulate-matter pollutants** (PM-10) are particles of solid or liquid substances less than 10 μm in diameter and may be either organic or inorganic.

Air pollutants are also classified as primary and secondary according to their origin. **Primary pollutants** are emitted directly into the air and include particulate matter, sulfur oxides, carbon monoxide, nitrogen oxides, and hydrocarbons. **Secondary pollutants** are produced when primary pollutants react with normal atmospheric compounds. As an example, ozone forms over urban areas through reactions among primary pollutants, sunlight, and natural atmospheric gases. Thus, ozone becomes a serious pollution problem on bright, sunny days in areas with much primary pollution. Although particularly well documented for southern California cities such as Los Angeles, this occurs worldwide under appropriate conditions.

The primary pollutants that account for nearly all air pollution problems are particulates, hydrocarbons, carbon monoxide, nitrogen oxides, and sulfur oxides. Each year well over a billion metric tons of these materials enter the atmosphere from human-related processes. About half of this is carbon monoxide, and the other four pollutants account for a few percent each. At first glance this quantity of pollutants appears very large. However, if uniformly distributed in the atmosphere, this would amount to only a few parts per million by weight. Unfortunately, pollutants are not uniformly distributed but tend to be released, produced, and concentrated locally or regionally. For example, over large cities weather and climatic conditions combine with urbanization-industrialization to produce local air pollution problems.

Sulfur Dioxide **Sulfur dioxide** (SO_2) is a colorless and odorless gas under normal conditions at the earth's surface. One significant aspect of SO_2 is that once emitted into the atmosphere it may be converted through complex reactions to fine particulate sulfate (SO_4) and removed from the atmosphere by wet and dry deposition. This is the fate of about 30 percent of atmospheric SO_2. Another major process is the oxidation of SO_2 to form H_2SO_4 (sulfuric acid) (2). The major source for the anthropogenic component of sulfur dioxide is burning of fossil fuels, mostly coal in power plants. Industrial processes, ranging from refining of petroleum to production of paper, cement, and aluminum, are another major source.

Destructive effects associated with sulfur dioxide include corrosion of paint and metals and injury or death to plants and animals, especially to crops such as alfalfa, cotton, and barley. Sulfur dioxide, particularly in the sulfate form, can cause severe damage to the lungs of humans and other animals.

Nitrogen Oxides **Nitrogen oxides** (NO_x) are emitted in several forms, the most important of which is nitrogen dioxide (NO_2), a light yellow-brown to reddish-brown gas with an irritating odor. It is toxic and extremely corrosive (2). Nitrogen dioxide may be converted by complex reactions in the atmosphere to fine particulate nitrate (NO_3^-) and then HNO_3 (nitric acid). Additionally, nitrogen dioxide is one of the main pollutants contributing to the development of photochemical smog. Nearly all nitrogen dioxide is emitted from anthropogenic sources; the two major contributors are automobiles and power plants that burn fossil fuels such as coal and oil.

Environmental effects of nitrogen oxides are variable but include irritation of eyes, nose, and throat; increased susceptibility of animals and humans to infections; suppression

*This discussion is summarized in part from *Air Resources and Management Manual*, National Park Service, 1984.

Table 17.1 Major natural and human-produced components of air pollutants (U.S.) 1996

Air Pollutant	Total 1996 Emissions (million metric tons)	Emissions (% of total)		Major Sources of Human-Produced Component	% of Total Emissions
		Natural	Human-produced		
Particulates (PM-10), less than 10 μm	28.4	17	83	Fugitive dust (roads, construction)	55
				Agriculture	29
				Combustion of fuels (stationary sources)	4
Sulfur dioxide	17.3		nearly all	Combustion of fuels (stationary sources, mostly coal)	88
				Industrial processes	8
Carbon monoxide (CO)	80.6	2	98	Transportation (mostly automobiles)	60
				Non-road engines and vehicles	19
Nitrogen dioxide (NO_2)	21.2		nearly all	Vehicles and engines (mostly lawn and garden, commercial, and boats)	50
				Combustion of fuels (stationary sources, mostly natural gas and coal)	40
Ozone (O_3)	N/A	A secondary pollutant derived from reactions with sunlight, NO_2, and oxygen (O_2)		Concentration that is present depends on reaction in lower atmosphere involving hydrocarbons and thus automobile exhaust	
Volatile organic compounds	17.3	3	97	Hydrocarbons (vehicles)	41
				Solvent use (industrial and non-industrial)	33

Source: Environmental Protection Agency. *National air pollution trends report 1900-1996.* Accessed 3/1/99 at http://www.epa.gov/ttn/chief/trends97/emtrnd.html

(a)

(b)

▲ **FIGURE 17.1** Burning sugarcane fields near Kihei on the island of Maui, Hawaii, 1994 (a); and dust rising from the area a few days later (b). (*Edward A. Keller*)

of plant growth and damage to leaf tissue; and impaired visibility when the oxides are converted to their nitrate form in the atmosphere. On the other hand, when nitrate is deposited on the soil, it can promote plant growth.

Carbon Monoxide **Carbon monoxide** (CO) is a colorless, odorless gas that at very low concentrations is extremely toxic to humans and other animals. This toxicity results from a striking physiological effect—namely that carbon monoxide and hemoglobin attract each other. Hemoglobin in our blood takes up carbon monoxide nearly 250 times more rapidly than it takes up oxygen. Therefore, if any carbon monoxide is in the vicinity, a person will take it in very readily. Many people (about 200 annually in the United States) have been accidentally asphyxiated by carbon monoxide produced from incomplete combustion of fuels in campers, tents, and houses. Detectors (similar to smoke detectors) are now available to warn people if CO in a building becomes concentrated at a toxic level. Actual effects of CO toxicity can range from dizziness and headaches to death. Carbon monoxide is particularly hazardous to people with known heart disease, anemia, or respiratory disease. Finally, the effects of carbon monoxide tend to be worse in higher altitudes, where oxygen levels are naturally lower.

Approximately 90 percent of the carbon monoxide in the atmosphere comes from natural sources and the other 10 percent comes mainly from fires, automobiles, and other sources of incomplete burning of organic compounds. Local concentrations of carbon monoxide can build up and cause serious health effects.

Ozone and Other Photochemical Oxidants Photochemical oxidants result from the atmospheric interaction of sunlight with pollutants such as nitrogen dioxide. The most common photochemical oxidant is **ozone** (O_3), a colorless, unstable gas with a slightly sweet odor. The major sources of photochemical oxidants and particularly of ozone are automobiles, burning of fossil fuels, and industrial processes that produce nitrogen dioxide. The effects of ozone and other oxidants are well known and include (among others) damage to rubber, paint, and textiles. Biological effects include damage to plants and animals.

The effects of ozone on plants can be subtle. At very low concentrations, ozone can reduce growth rates while not producing any visible injury. At higher concentrations, ozone kills leaf tissue, eventually killing entire leaves and, if the pollutant levels remain high, whole plants. The death of white pine trees planted along highways in New England is believed to be due in part to ozone related to automobiles. Ozone also causes harm to humans and other animals, especially to the eyes and the respiratory system.

Volatile Organic Compounds Volatile organic compounds (VOCs) include a wide variety of organic compounds used as solvents in industrial processes, such as dry cleaning, degreasing, graphic arts and adhesives. **Hydrocarbons** (a group of VOCs) are compounds composed of hydrogen and carbon. Thousands of such compounds exist, including natural gas, or methane (CH_4), butane (C_4H_{10}), and propane (C_3H_8). Analysis of urban air has identified many different hydrocarbons, some of which react with sunlight to produce photochemical smog. Potential adverse effects of hydrocarbons are numerous because many are toxic to plants and animals or may be converted to harmful compounds through complex chemical changes that occur in the atmosphere. On a global basis only about 5 percent of hydrocarbon emissions (which are primary pollutants) are anthropogenic. However, nearly half of hydrocarbons entering the atmosphere in the United States are emitted from anthropogenic sources, the most important of which is the automobile (see Table 17.1). Anthropogenic sources are particularly abundant in urban regions. Nevertheless, in some southern American cities such as Atlanta, Georgia, natural emissions probably exceed those from automobiles and other human sources (2).

Hydrogen Sulfide Hydrogen sulfide (H_2S) is a highly toxic corrosive gas easily identified by its rotten egg odor. Hydrogen sulfide is produced from natural sources such as geysers, swamps, and bogs and from human sources such as industrial plants that produce petroleum or that smelt metals. Potential effects of hydrogen sulfide include functional damage to plants and health problems ranging from toxicity to death for humans and other animals.

Hydrogen Fluoride Hydrogen fluoride (HF) is a gaseous pollutant released primarily by industrial activities such as production of aluminum, coal gasification, and burning of coal in power plants. Hydrogen fluoride is very toxic, and even a small concentration (as low as 1 ppb) may cause problems for plants and animals. It is particularly dangerous to grazing animals because some forage plants can become toxic.

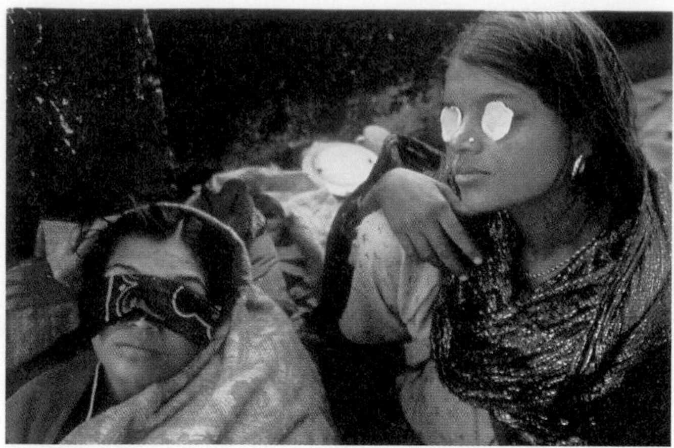

▲ **FIGURE 17.2** Many victims in the Bhopal, India, leak of toxic gas from a pesticide plant suffered eye damage. This resulted because the gas burns tissue on contact. (*Pablo Bartholomew/Liaison Agency, Inc.*)

Other Hazardous Gases It is a rare month when the newspapers do not carry a story of a truck or train accident that releases toxic chemicals in a gaseous form into the atmosphere. People are often evacuated until the leak is stopped or the gas has dispersed to a nontoxic level. Chlorine gases or a variety of other materials used in chemical and agricultural processes may be involved.

Another source of gaseous air pollution is sewage treatment plants. Urban areas deliver a tremendous variety of organic chemicals, including paint thinner, industrial solvents, chloroform, and methyl chloride to treatment plants by way of sewers. These materials are not removed in the treatment plants; in fact, the treatment processes facilitate the evaporation of the chemicals into the atmosphere, where they may be inhaled. Many of the chemicals are toxic or are suspected carcinogens. It is a cruel twist of fate that the treatment plants designed to control water pollution are becoming sources of air pollution. Although some pollutants can be moved from one location to another and can even change form, as from liquid to gas, we really cannot get rid of them as easily as we once thought.

Some chemicals are so toxic that extreme care must be taken to ensure they do not enter the environment. This was tragically demonstrated on December 3, 1984, when a toxic gas (stored in liquid form) from a pesticide plant leaked, vaporized, and formed a deadly cloud that settled over a 64-km^2 area of Bhopal, India. The gas leak lasted less than an hour, yet more than 2000 people were killed and more than 15,000 were injured. The colorless gas, methyl isocyanate, causes severe irritation (burns on contact) to eyes, nose, throat, and lungs (Figure 17.2). Breathing the gas in concentrations of only a few parts per million (ppm) causes violent coughing, swelling of the lungs, bleeding, and death. Less exposure can cause a variety of problems, including loss of sight.

Methyl isocyanate is an ingredient of a common pesticide known in the United States as Sevin and of at least two other insecticides used in India. A plant in West Virginia also makes the chemical, and small leaks evidently occurred there both before and after the catastrophic accident in Bhopal. The accident clearly shows that chemicals that can cause catastrophic injuries and death should not be stored close to large population centers. Furthermore, chemical plants need to have more reliable accident-prevention equipment and personnel trained to control leaks or other problems.

***Particulate Matter* (PM-10)** **Particulate matter** encompasses the small particles of solid or liquid substances less than 10 μm in diameter (10 millionths of a meter) that are released into the atmosphere by many natural processes and human activities. Desertification, volcanic eruptions, fires, and modern farming all add considerable amounts of particulate matter into the air. Nearly all industrial processes release particulates into the atmosphere, as does the burning of fossil fuels. Much particulate matter is easily visible as smoke, soot, or dust, but some particulate matter is not easily visible. Particulates include airborne asbestos particles and small particles of heavy metals, such as arsenic, copper, lead, and zinc, which are usually emitted from industrial facilities such as smelters.

Of particular importance are the very fine particulate pollutants, with particles less than 2.5 μm in diameter, which can cause the greatest damage to the lungs. Among the most significant of the fine particulate pollutants are sulfates and nitrates, which are mostly secondary pollutants produced in the atmosphere through chemical reactions between sulfur dioxide or nitrogen oxides and normal atmospheric constituents. These reactions are especially important in the formation of sulfuric and nitric acids in the atmosphere, which are precipitated as acid rain (1). When measured, particulate matter is often referred to as total suspended particulates (TSP).

Particulates affect human health, ecosystems, and the biosphere. Particulates that enter the lungs may lodge there and have chronic effects on respiration; asbestos is particularly dangerous in this way. Dust raised by road building and plowing and deposited on the surfaces of green plants may interfere with their absorption of carbon dioxide and oxygen and their release of water; heavy dust may affect the breathing of animals. Particulates associated with large construction projects may therefore kill organisms and damage large areas, changing species composition, altering food chains, and generally affecting ecosystems. In addition, modern industrial processes have greatly increased the total suspended particulates in the atmosphere. Particulates block sunlight and thus cause changes in climate, which can have lasting effects on the biosphere.

Asbestos Asbestos particles have only recently been recognized as a significant hazard. In the past, asbestos was treated rather casually, and people working in asbestos plants were not protected from dust. Asbestos was used in building insulation and in brake pads for automobiles. As a result, asbestos fibers are found throughout industrialized countries, especially in Europe and North America and especially in urban environments. In one case, asbestos products were sold in burlap bags that eventually were reused in plant nurseries and other secondary businesses, thus further spreading the pollutant. Some asbestos particles are believed to be carcinogenic, or to carry with them carcinogenic materials, and so must be carefully controlled (see Chapter 13).

Lead Lead is an important constituent of automobile batteries and other industrial products. When lead is added to gasoline, automobile engines run more evenly. The lead is emitted into the environment via automobile exhaust, and in this way lead has been spread widely around the world, reaching high levels in soils and waters along roadways. Lead is no longer used in gasoline in the United States, but it is still used elsewhere. In the United States lead emissions have decreased by 98 percent since 1970 in direct response to reductions and elimination of lead in gasoline.

Once released, lead can be transported through the air as particulates to be taken up by plants through the soil or deposited directly on plant leaves. Thus, it enters terrestrial food chains. When lead is carried by streams and rivers, deposited in quiet waters, or transported to the ocean or lakes, it is taken up by aquatic organisms and thus enters aquatic food chains.

Cadmium Some of the cadmium in the environment comes from coal ash, which is spread widely from smokestacks and chimneys. Cadmium exists as a trace element in the coal, in the very low concentration of 0.05 ppm. When the coal ash falls on plants, the cadmium is incorporated into plant tissue and concentrated three to seven times. As cadmium moves up through the food chain, each higher level concentrates it approximately three times more. Herbivores have approximately three times the concentration of green plants, and carnivores have approximately three times the concentration of herbivores. This process is known as *biomagnification* or *biological concentration.*

17.3 Urban Air Pollution

Air pollution is not distributed uniformly around the world. Much of it is concentrated in and around urban areas, where automobiles and heavy industry emit an enormous amount of waste into the environment. The visible air pollution known as smog is present in nearly all urbanized areas, although it is much worse in some places than in others. In this section we will consider the factors contributing to urban air pollution and discuss the composition and formation of smog.

Influence of Meteorology and Topography

The extent to which air pollution occurs in an urban area depends largely on the emission rates, topography, and weather conditions because these factors determine the rate at which pollutants are concentrated and transported away from their sources and converted to harmless compounds in the air. When the rate of production exceeds the rates of transport and chemical transformation, dangerous conditions can develop.

Meteorological conditions can determine whether air pollution is only a nuisance or is a major health problem. Most pollution periods in the Los Angeles Basin and other smoggy areas do not cause large numbers of deaths. However, serious pollution events can develop over a period of days and lead to an increase in deaths and illnesses.

Restricted circulation in the lower atmosphere due to the formation of an inversion layer may lead to a pollution event. An **atmospheric inversion** occurs when warmer air is found above cooler air, and this is particularly a problem when there is a stagnated air mass. Figure 17.3 shows two types of developing inversions that may worsen air pollution problems. In the upper diagram, which is somewhat analogous to the situation in the Los Angeles area, descending warm air from inland warm arid areas forms a semipermanent inversion layer. Because the mountains act as a barrier to the pollution, polluted air moving in response to the sea breeze and other processes tends to move up canyons, where it is trapped. The air pollution that develops occurs primarily during the summer and fall, when warm inland air comes over the mountains and overlies cooler coastal air, forming the inversion.

The lower part of Figure 17.3 shows a valley with relatively cool air overlain by warm air, a situation that can occur in several ways. When cloud cover associated with a stagnant air mass develops over an urban area, the incoming solar radiation is blocked by the clouds, which absorb some of the energy and thus heat up. On or near the ground, the air cools. If the humidity is high, a thick fog may form as the air cools. Because the air is cool, people living in the city burn more fuel to heat their homes and factories, thus delivering more pollutants into the atmosphere. As long as the stagnant conditions exist, the pollutants will build up.

Evaluating meteorologic conditions can be extremely helpful in predicting which areas will have potential smog

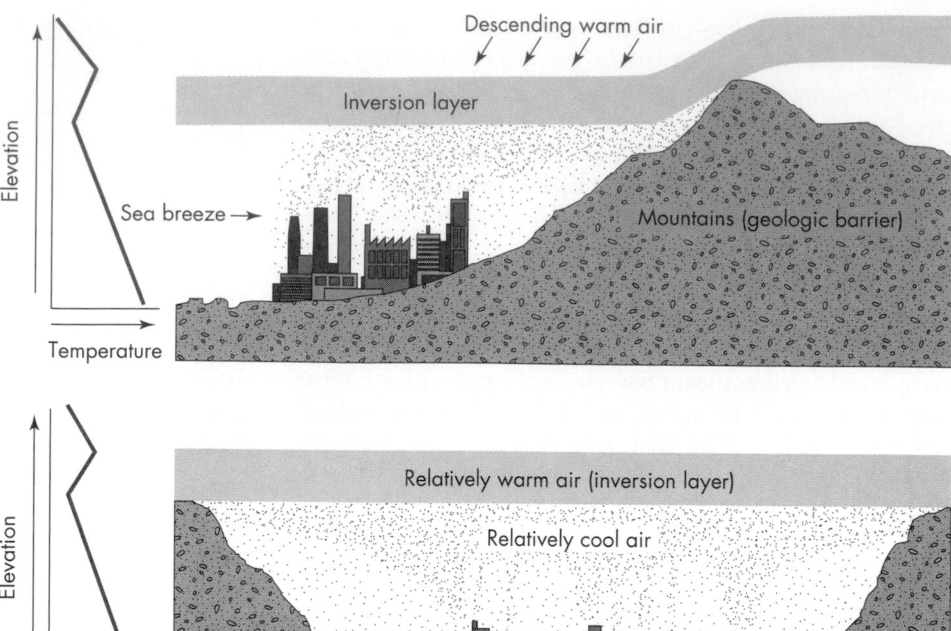

▶ **FIGURE 17.3** Two causes for the development of a temperature inversion, which may aggravate air pollution problems.

problems. Figure 17.4 shows the number of days in a 5-year period for which conditions were favorable for reduced dispersion of air pollution for at least a 48-hour period. This illustration clearly shows that most of the problems are located in the western United States (where the greatest number of days of reduced dispersion is present). This helps explain why cities such as Los Angeles and San Diego, in California, and Phoenix, Arizona, have air quality problems from time to time.

Cities situated in a topographic "bowl" surrounded by mountains are more susceptible to smog problems than are cities in open plains. Cities where certain kinds of weather conditions, such as temperature inversions, occur are also particularly susceptible. Both the surrounding mountains and the temperature inversions prevent the transportation (dispersion) of pollutants by the winds and weather systems.

Potential for Urban Air Pollution

The potential for air pollution in urban areas is determined by the following factors, as illustrated in Figure 17.5: the rate of emission of pollutants per unit area; the distance downwind of the pollution sources that a mass (column) of air may move through an urban area; the average speed of the wind; and, finally, the height of the mixing layer, that is, the height to which potential pollutants may be thoroughly mixed in the lower atmosphere (3). Assuming a constant rate of emission of air pollutants, the column of air will collect more and more pollutants as it moves through the urban area. Thus, the concentration of pollutants in the air is directly proportional to the first two factors: As either the emission rate or the downwind travel distance increases, so will the concentration of pollutants. On the other hand, city air pollution decreases with increases in the wind velocity and the height of mixing. The stronger the wind and the higher the mixing layer, the lower the pollution.

If an inversion layer is present, it acts as a lid for the pollutants, but near a geologic barrier such as a mountain a "chimney effect" may let the pollutants spill over the top of the barrier (see Figure 17.5). This effect has been noticed particularly in the Los Angeles Basin, where pollutants can climb several thousand meters, damaging mountain pine trees and other vegetation and spoiling the air of mountain valleys.

The Urban Microclimate

The very presence of a city affects the local climate, and as the city changes, so does its climate. Although air quality in urban areas is in part a function of the amount of pollutants present or produced, it is affected also by the city's ability to ventilate and thus flush out pollutants. The amount of ventilation depends on several aspects of the urban microclimate.

Cities are warmer than surrounding areas. The observed increase in temperature in urban areas is approximately 1° to 2 °C in the winter and 0.5° to 1.0 °C in the summer for midlatitude areas. The temperature increase results from both increased heat production in the area and a decreased rate of heat loss. The additional heat is emitted from burning fossil fuels and other industrial, commercial, and residential sources. The decreased heat loss is due to the dust in the urban air, which traps and reflects back into the city longwave (infrared) radiation emitted from city surfaces.

In the winter, in cold-weather regions, space heating from the city is primarily responsible for heating the local air environment. In Manhattan, for example, the input of

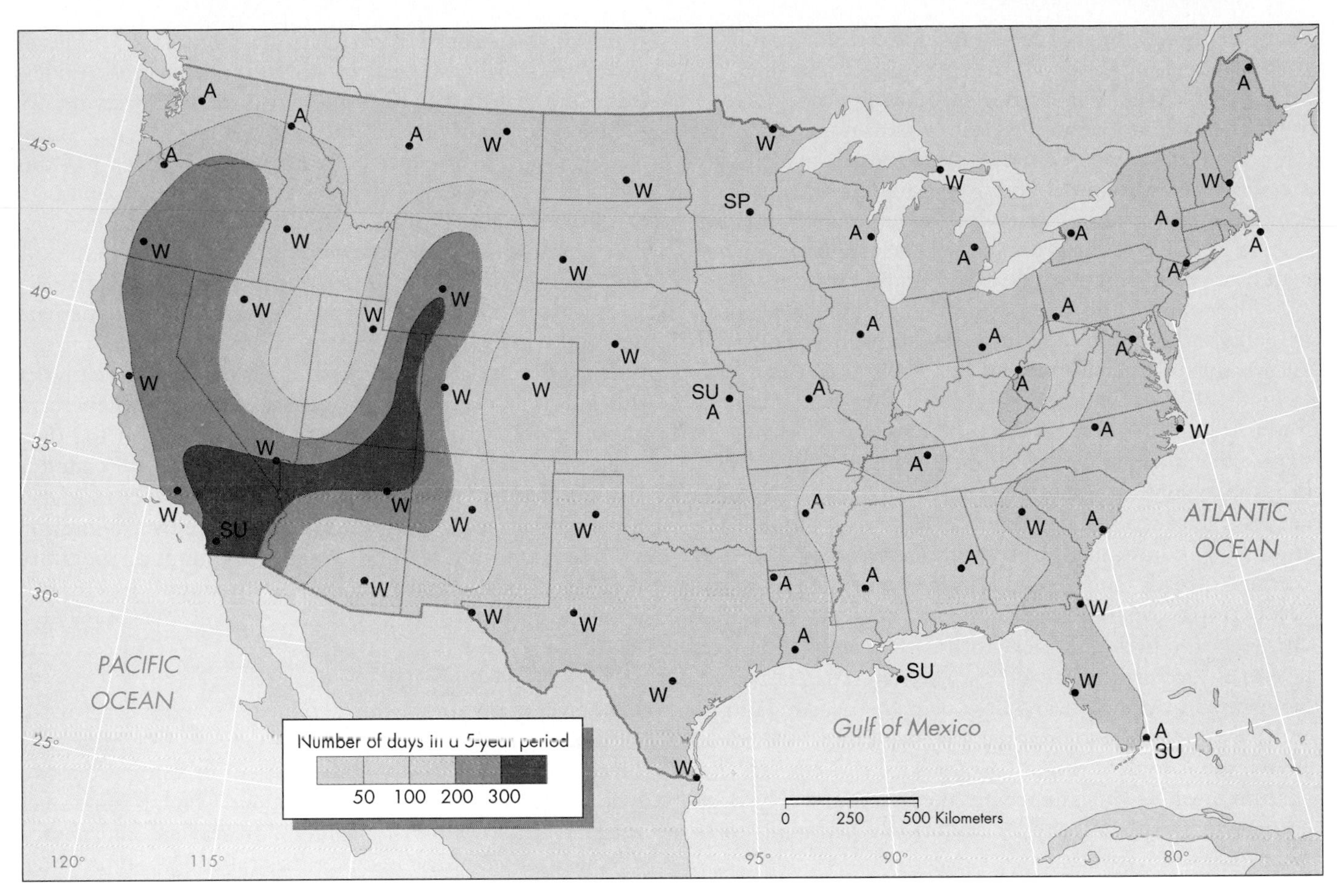

▲ **FIGURE 17.4** Number of days in a 5-year period characterized by conditions favorable for reduced dispersion and local buildup of air pollutants that existed for at least a 48-hour (2-day) period. The time of year when most of these periods occurred is shown by the season: SP (spring), A (autumn), W (winter), or SU (summer). (*Modified after Holzworth, as presented in Neiburger, Edinger, and Bonner.* 1973. Understanding our atmospheric environment. W. H. *Freeman.*)

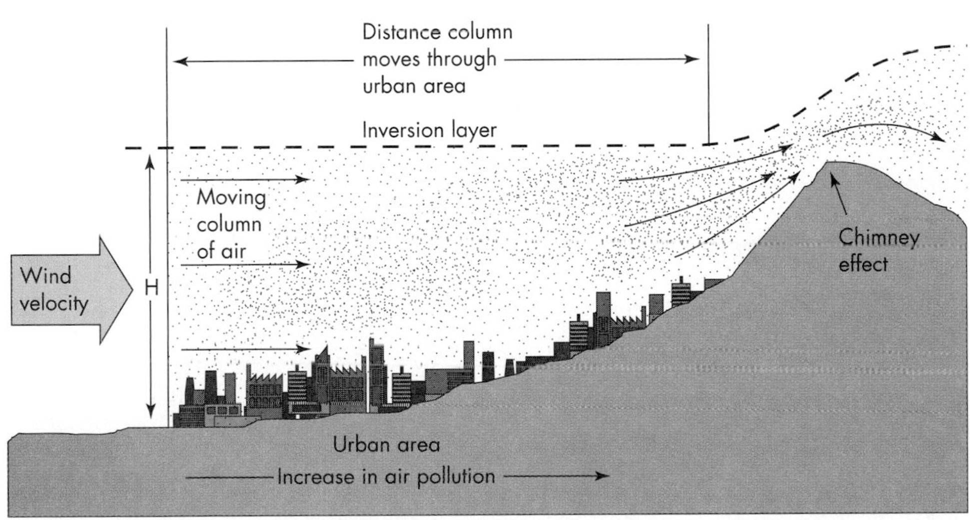

◀ **FIGURE 17.5** The higher the wind velocity and the thicker the mixing layer (shown here as H), the less is the potential air pollution. The greater the emission rate and the longer the downwind length of the city, the greater the air pollution. The "chimney effect" allows polluted air to move over a mountain down into an adjacent valley.

heat from industrial, commercial, and residential space heating in the winter is about 2.5 times the solar energy input, whereas the annual average is about 33 percent of the solar input. In large urban areas characterized by warmer winters, the heat input from artificial sources is much less (4). Concrete, asphalt, and roofs tend to act as solar collectors and quickly emit heat, helping to increase the sensible (felt) heat in cities (5).

Particulates in the atmosphere over a city are often 10 times or more as abundant as in surrounding areas. Although the particulates tend to reduce incoming solar radiation by up to 30 percent and thus cool the city, this effect is small relative to the effect of urban heat-producing processes (6).

Cities are in general less windy than nonurban areas. Air over cities tends to move more slowly than in surrounding areas because buildings and other structures obstruct the flow of air. Thus, wind velocities are frequently reduced by 20 to 30 percent, and urban areas have 20 percent more calm days than do nearby rural areas (6).

The combination of lingering air and abundance of particulates and other pollutants in the air produces the well-known **urban dust dome** and **heat-island effect,** shown in Figure 17.6. This figure also shows the general circulation pattern of air moving from the rural or suburban areas toward the inner city, where it flows up and then laterally out near the top of the dust dome. This circulation pattern often occurs when a strong heat island develops over the city. The rising hot air draws in air from surrounding areas. As it rises, it cools, spreads, and descends away from the heat island. For example, when a dust dome and heat island develop during a calm period in New York City, there is an upward flow of air over the heavily developed Manhattan Island, accompanied by a downward flow over the nearby Hudson and East rivers, which are greenbelts and thus have cooler air temperatures (5). Figure 17.6 also shows the air-temperature profile, which delineates the heat island. The dust-dome effect explains why air pollution often tends to be most intense at city centers.

Particulates in a dust dome provide condensation nuclei, and thus urban areas experience 5 to 10 percent more precipitation and considerably more cloud cover and fog than do surrounding areas. The formation of fog is particularly troublesome in the winter and may impede air traffic. If the pollution dome moves downwind, increased precipitation may occur outside the urban area. For example, in the mid-1960s, effluent from the southern Chicago–northern Indiana industrial complex apparently caused a 30 percent increase in precipitation at La Porte, Indiana, 48 km downwind to the south. La Porte also has almost 2.5 times as many hailstorms, 38 percent more thunderstorms, and less sunshine than the countryside not directly downwind (7). The La Porte case is extreme, but it is not unique. Atmospheric particulate matter from urban sources has undoubtedly altered local weather at numerous locations.

In summary, cities are cloudier, warmer, rainier, and less humid than their surrounding areas. Cities in middle latitudes receive about 15 percent less sunshine and 5 percent less ultraviolet light during the summer and 30 percent less ultraviolet light during the winter than do nonurban areas. They are 10 percent rainier and 10 percent cloudier and have a 25 percent lower average wind speed, 30 percent more summer fog, and 100 percent more winter fog than nonurban areas. Average relative humidity is 6 percent less in cities, partly because they have large impervious surfaces and little surface or soil water to exchange by evaporation with the atmosphere. The average maximum temperature difference between a city and its surrounding region is about 3 °C (8).

Smog Production

Wherever many sources are producing air pollutants over a wide area—as, for example, automobiles in Los Angeles—there is potential for the development of smog. The two major types of smog are sulfurous smog, which is sometimes referred to as London-type smog, or gray air, and photochemical smog, which is sometimes called L.A.-type smog, or brown air. Sulfurous smog is produced primarily by burning coal or oil at large power plants. Under certain meteorological conditions, the sulfur oxides and particulates produced by this burning combine to produce a concentrated **sulfurous smog** (Figure 17.7).

The reactions that produce **photochemical smog** are complex, involving nitrogen oxides (NO_x), organic compounds (hydrocarbons), and solar radiation (Figure 17.8). Development of photochemical smog is directly related to automobile use. In southern California, for example, when

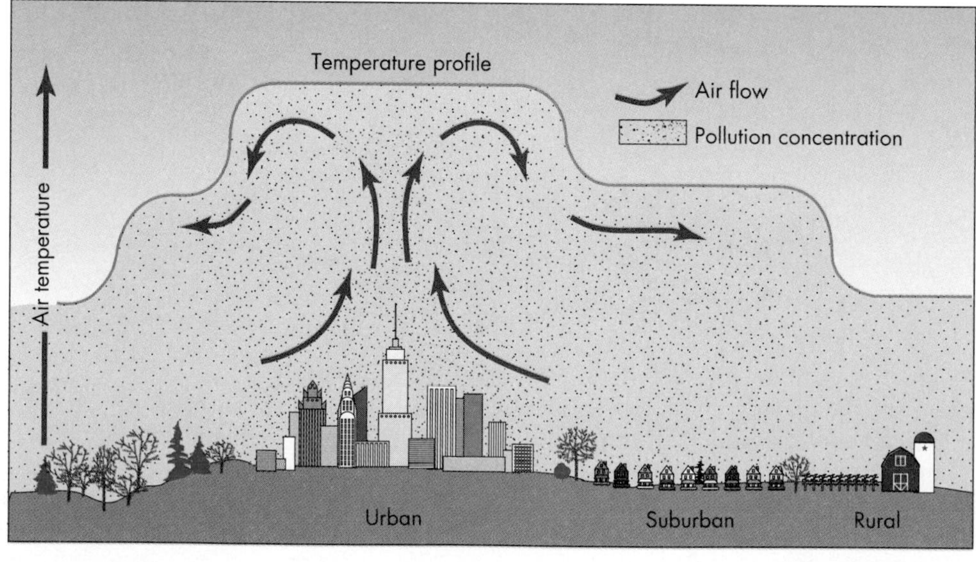

▶ **FIGURE 17.6** Lingering air, an abundance of particulates, and the flow of air over heavily built-up areas create an urban dust dome and a heat island.

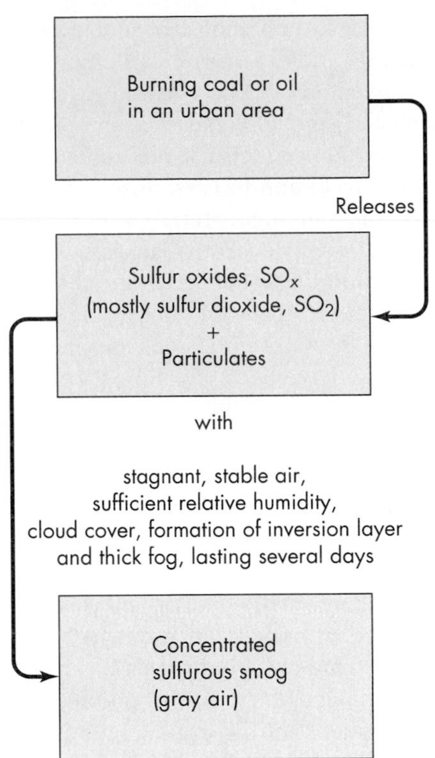

▲ **FIGURE 17.7** How concentrated sulfurous smog and smoke might develop.

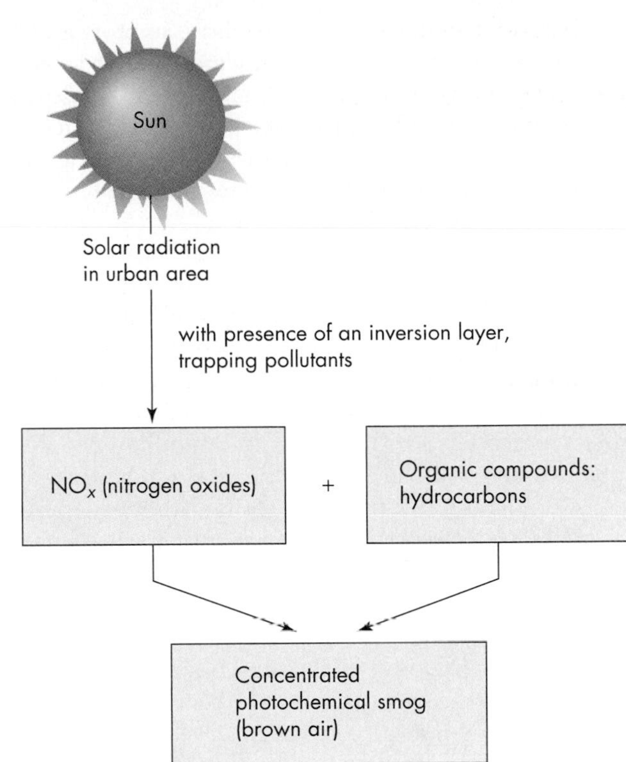

▲ **FIGURE 17.8** How photochemical smog might be produced.

commuter traffic begins to build up early in the morning, the concentrations of nitrogen oxide (NO) and hydrocarbons begin to increase. At the same time, the amount of nitrogen dioxide (NO_2) may decrease owing to the sunlight's action on NO_2 to produce NO plus atomic oxygen (O). The atomic oxygen is then free to combine with molecular oxygen to form ozone, so after sunrise the concentration of ozone also increases. Shortly thereafter, oxidized hydrocarbons react with NO to increase the concentration of NO_2 by midmorning. This reaction causes the NO concentration to decrease and allows ozone to build up, producing a midday peak in ozone and a minimum in NO. As the smog matures, visibility may be greatly reduced (Figure 17.9) owing to light scattering by aerosols.

The Future of Air Pollution in Urban Areas

Air pollution levels in many cities in developed countries have a mixed but positive record. Data from major U.S. metropolitan areas in recent years show a decline in the total number of days characterized as unhealthful and very unhealthful,

(a)

(b)

▲ **FIGURE 17.9** City of Los Angeles, California, on a clear day (a); and a smoggy day (b). (*Kathleen Campbell/ Liaison Agency, Inc.*)

which suggests that the nation's air quality is improving. This is despite the fact that there are many more motor vehicles. Improvements have resulted from building automobiles that burn less fuel more efficiently, have smog-control devices, and burn cleaner (improved) fuel. However, most major urban areas such as New York and Los Angeles still have unhealthful air much of the time (9).

Cities in less-developed countries with burgeoning populations are particularly susceptible to air pollution now and in the future; they do not have the financial base necessary to fight air pollution. People in these countries tend to be more concerned with basic survival and finding ways to house and feed their growing urban populations. For example, Mexico City, with more than 24 million people, is one of the largest urban areas in the world today. Industry and power plants in Mexico City, along with the numerous cars, trucks, and buses (many of which are aging and in poor condition), emit hundreds of thousands of tons of particulates, sulfur dioxide, nitrogen oxides, and hydrocarbons into the atmosphere each year. The city is at an elevation of about 7400 feet (2255 m) in a natural basin surrounded by mountains—a perfect situation for a severe air pollution problem. The mountains can rarely be seen from Mexico City now. Physicians report a steady increase in respiratory diseases, while headaches, irritated eyes, and sore throats are common.

The encouraging news is that emissions of some major air pollutants in the United States are decreasing (Figure 17.10). For example, since 1970 SO_2 emissions have declined significantly as a result of burning less coal, using low-sulfur coal and treating effluent gases from power plants before release into the environment. Emissions of volatile organic compounds have also decreased since 1970 to levels not recorded since the 1940s. A decrease of about 38 percent is a result in part of successful control of automobile emissions and substitutions of water-based components for VOC components in products such as asphalt.

17.4 Indoor Air Pollution

In recent years buildings have been constructed more tightly to save energy. As a result, air is filtered through rather extensive systems in many structures. Unless filters are maintained properly, indoor air can become polluted with a variety of substances, including smoke, chemicals, disease-carrying organisms, and radon, a naturally occurring radioactive gas suspected of causing lung and other cancers (see Chapter 13). It was shown that the virus responsible for the respiratory infection known as *Legionnaires' disease* multiplies and is transported through buildings in air filters and ventilation systems.

Modern urban structures are built of many substances, some of which release minute amounts of chemicals and other materials into the nearby air. In some buildings asbestos fibers are slowly released from insulation and other fixtures; people exposed to a particular type of asbestos fiber may develop a rare form of lung cancer. The poisonous gases carbon monoxide and nitrogen dioxide may be released in homes from unvented or poorly vented gas stoves, furnaces, or water heaters. Formaldehyde, present in some insulation materials and wood products used in home construction, is known to cause irritation to ears, nose, and throat. Radon is present in some building materials, such as concrete block and bricks, if they are made from materials with a high radon concentration. Buildings that lack a good system to recirculate the air with clean air are likely to have indoor pollution problems. Recommendations on how to improve indoor air quality have been developed—most are related to improving circulation of clean air and reducing emissions of pollutants.

Interestingly, people who lived centuries ago also suffered from indoor air pollution. In 1972 the body of a fourth-century Eskimo woman was discovered on St. Lawrence Island in the Bering Sea. The woman evidently was killed during an earthquake or landslide and her body froze soon after death. Detailed autopsies showed that the woman suffered from black lung disease, which occasionally afflicts coal miners today. Anthropologists and medical personnel concluded that the woman breathed very polluted air for a number of years. They speculate that the air she breathed included hazardous fumes from lamps that burned seal and whale blubber, causing the black lung disease (10).

17.5 Control of Air Pollution

Reducing air pollution requires a variety of strategies tailored to specific sources and type of pollutants. From an environmental perspective, the best strategy is to reduce emissions through conservation and efficiency measures—that is, to burn less fossil fuel. For both stationary and mobile sources, strategies have been to collect, capture, or retain pollutants before they enter the atmosphere. Other strategies may only make the problem worse, as happened at smelters in Sudbury, Ontario, Canada. The ores contain a high percentage of sulfur, and the smelter stacks emit large amounts of sulfur dioxide as well as particulates containing nickel, copper, and other toxic metals. Attempts to minimize the pollution problem close to the smelters by increasing the height of the smokestacks backfired by spreading the pollution over a larger area.

Pollution problems vary in different regions of the world and even within the United States. In the Los Angeles Basin, for example, nitrogen oxides and hydrocarbons are particularly troublesome because they combine in the presence of sunlight to form photochemical smog. Furthermore, in Los Angeles most of the nitrogen oxides and hydrocarbons are emitted from automobiles—a nonpoint source (7). In other urban areas, such as in Ohio and the Great Lakes region in general, air quality problems result primarily from emissions of sulfur dioxide and particulates from industry and coal-burning power plants, which are point sources.

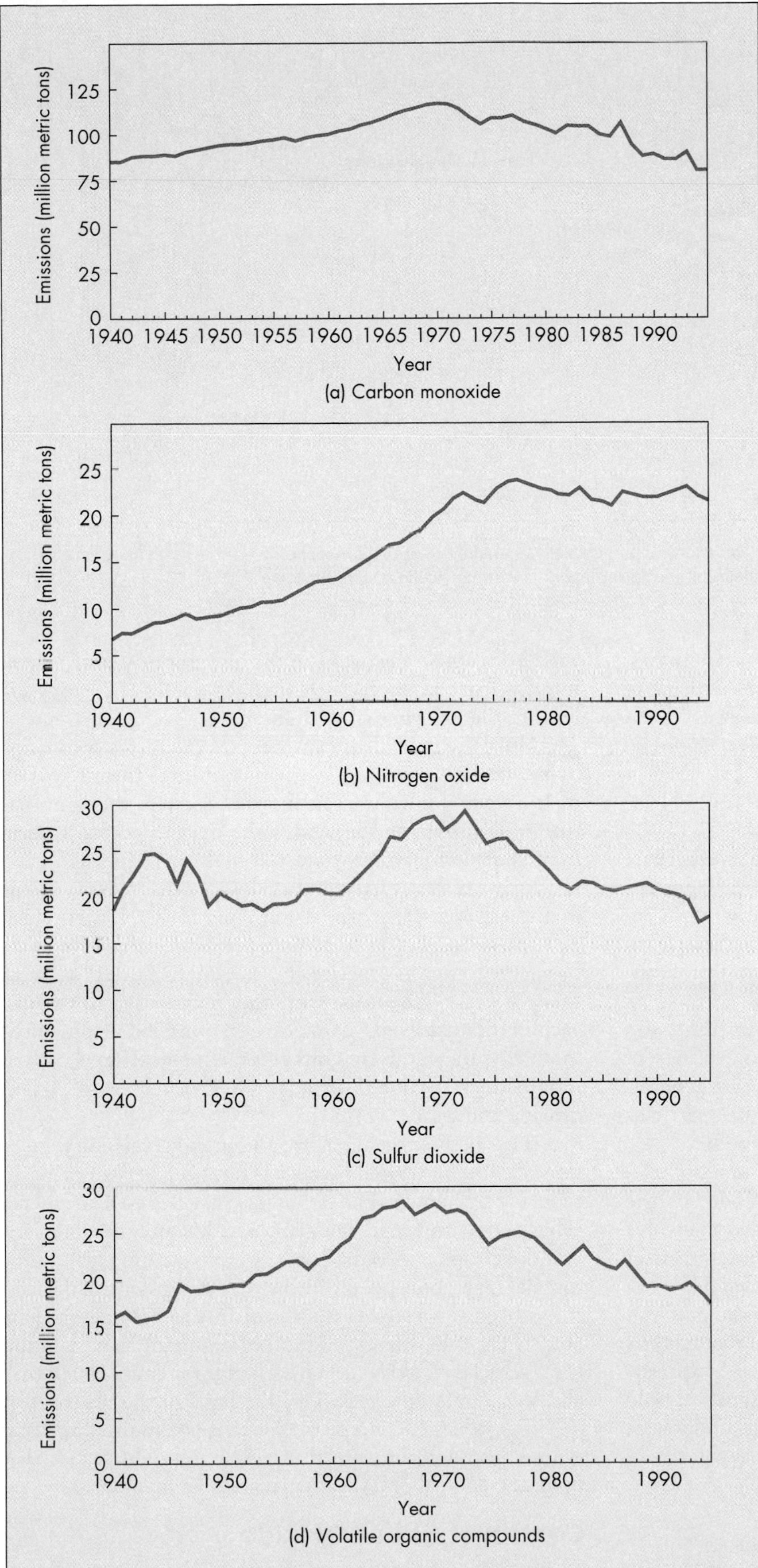

◀ **FIGURE 17.10** Trends in U.S. emissions of selected air pollutants: carbon monoxide (a); nitrogen oxide (b); sulfur dioxide (c); and volatile organic compounds (d). (*Source: Environmental Protection Agency*. National air pollution trends report 1900–1996. *Accessed 3/1/99 at http://www.epa.gov/ttn/chief/trends97/emtrnd.html*)

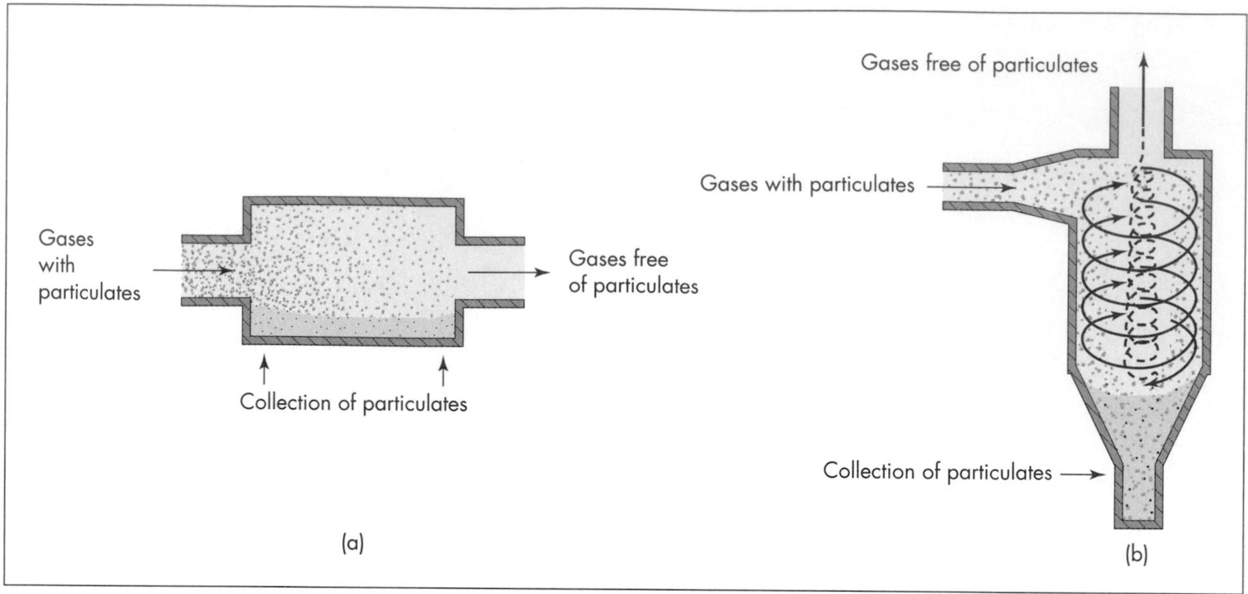

▲ **FIGURE 17.11** Devices being used to control emissions of particulates before they enter the atmosphere: (a) a simple settling chamber that collects particulates by gravity settling; (b) centrifugal collector, in which particulates are forced to the outside of the chamber by centrifugal force and then fall to the collection site.

Because the problems vary greatly from country to country and from region to region, it is often difficult to obtain both the money and the political consensus for effective control. Nevertheless, effective control strategies do exist for many air pollution problems.

Control of Particulates

Particulates emitted from stationary sources—fugitive, point, or area sources—are easier to control than are the very small particulates of primary or secondary origin released from mobile sources such as automobiles. As we learn more about these very small particles, new control methods will have to be devised.

Control of coarse particulates from power plants and industrial sites (point or area sources) utilizes a variety of settling chambers or collectors, some of which are generalized in Figure 17.11. These methods settle out particles where they may be collected for disposal in landfills.

Particulates from fugitive sources (such as a waste pile) must be controlled on site before they are eroded by the wind and carried into the atmosphere. Measures can involve protecting open areas, dust control, or reducing the effect of wind. For example, waste piles can be protected by covering them with plastic or other material, and soil piles can be protected by planting them with vegetation that inhibits wind erosion. Dust on a road can be controlled by spreading water or a combination of water and chemicals to hold it down; structures or vegetation may be used as windbreaks, that is, placed to lessen wind velocity near the ground.

Control of Automobile Pollution

Carbon monoxide, nitrogen oxides, and hydrocarbons in urban areas are best controlled by regulating automobile exhaust, the source of most of the anthropogenic portion of these pollutants. Furthermore, control of these materials will also regulate the ozone in the lower atmosphere, where it forms from reactions with nitrogen oxides and hydrocarbons in the presence of sunlight.

Nitrogen oxide from automobile exhaust is controlled by recirculating exhaust gas, which dilutes the air-to-fuel mixture being burned. The dilution reduces the temperature of combustion, decreases the oxygen concentration in the burning mixture (that is, it makes a richer fuel), and produces fewer nitrogen oxides. Unfortunately, this method increases hydrocarbon emissions. Nevertheless, exhaust recirculation to reduce nitrogen oxide emissions has been common practice in the United States for more than 20 years. The most common device used to remove carbon monoxide and hydrocarbon emissions from automobiles is the **catalytic converter,** which converts carbon monoxide to carbon dioxide and hydrocarbons to carbon dioxide and water (11).

It has been argued that the automobile emission regulation plan in the United States has not been very effective in reducing pollutants. The pollutants may be reduced when a vehicle is relatively new, but many people simply do not maintain their emission control devices over the life of the automobile. Also, some people disconnect smog-control devices. It has been suggested that effluent fees replace automobile controls as the primary method of regulating air pollution (12). Vehicles would be tested each year for emission control, and fees would be assessed on the basis of the test results. The fees would encourage the purchase of automobiles that pollute less, and the annual inspections would ensure that pollution control devices are properly maintained.

Control of Sulfur Dioxide

Sulfur dioxide emissions can be reduced by abatement measures performed before, during, or after combustion. The technology to clean up coal so that it will burn cleanly is already available, although the cost of removing the sulfur does make the fuel more expensive.

Changing from high-sulfur coal to low-sulfur coal seems an obvious solution to reducing sulfur dioxide emissions. Unfortunately, most low-sulfur coal is located in the western part of the United States, whereas most burning of coal occurs in the eastern part. Therefore, using low-sulfur coal is a solution only where it is economically feasible to transport the coal over long distances. Another possibility is cleaning relatively high-sulfur coal by washing it. By washing finely ground coal with water, the iron sulfide (the mineral pyrite) settles out because it is relatively more dense than coal. Although washing removes some of the sulfur, it is expensive. Another option is coal gasification, which converts relatively high-sulfur coal to a gas and removes the sulfur in the process. The gas obtained from coal is quite clean and can be transported relatively easily, augmenting supplies of natural gas. The synthetic gas produced from coal is now fairly expensive compared to gas from other sources, but may become more competitive in the future (4).

During combustion of coal, a process known as **fluidized-bed combustion** can eliminate sulfur oxides. The process involves mixing finely ground limestone with coal and burning it in suspension. Sulfur can also be removed by a process known as *limestone injection*, which involves injecting ground limestone into a special burner. In both processes, the sulfur oxides combine with the calcium in the limestone to form calcium sulfides and sulfates, which can be collected.

Sulfur oxide emissions from stationary sources such as power plants can also be reduced by removing the oxides from the gases in the stack before they reach the atmosphere. Perhaps the most highly developed technology for the cleaning of gases in tall stacks is **scrubbing** (Figure 17.12). In this method the gases produced during burning (including SO_2) are treated with a slurry of lime (calcium oxide, CaO) or limestone (calcium carbonate, $CaCO_3$). The sulfur oxides react with the calcium to form insoluble calcium sulfides and sulfates, which are collected. However, the residue containing the calcium sulfides and sulfates is a sludge that must be disposed of at a land-disposal site. Because the sludge can cause serious water pollution if it interacts with the hydrologic cycle, it must be treated carefully (4). Furthermore, scrubbers are expensive and add significantly (10% to 15% or more) to the total cost of electricity produced by burning coal.

Finally, an innovative approach has been taken at a large coal-burning power plant near Mannheim, Germany. Gases (smoke) from combustion are treated with liquid ammonia (NH_3), which reacts with the sulfur to produce ammonium sulfate. Sulfur-contaminated gases entering the pollution-control system are cooled by outgoing sulfur-depleted (clean) gases to a temperature that favors the reaction; then outgoing clean gases are heated by incoming sulfur-rich (dirty) gases to force it out of a vent. Waste heat from the cooling towers also heats nearby buildings, and the plant sells the ammonium sulfate to farmers as fertilizer in a solid granular form. The plant, finished in 1984, was built in response to tough pollution-control regulations enacted to help abate acid precipitation, thought to contribute to pollution that is killing forests in Germany.

Pollution abatement has been successful in Japan. Japan had what was considered the most severe sulfur pollution problem in the world, and the health of the Japanese people was being directly affected. It was not uncommon to see residents of Yokohama and other areas wearing masks over their mouths when out on the streets. About 30 years ago the

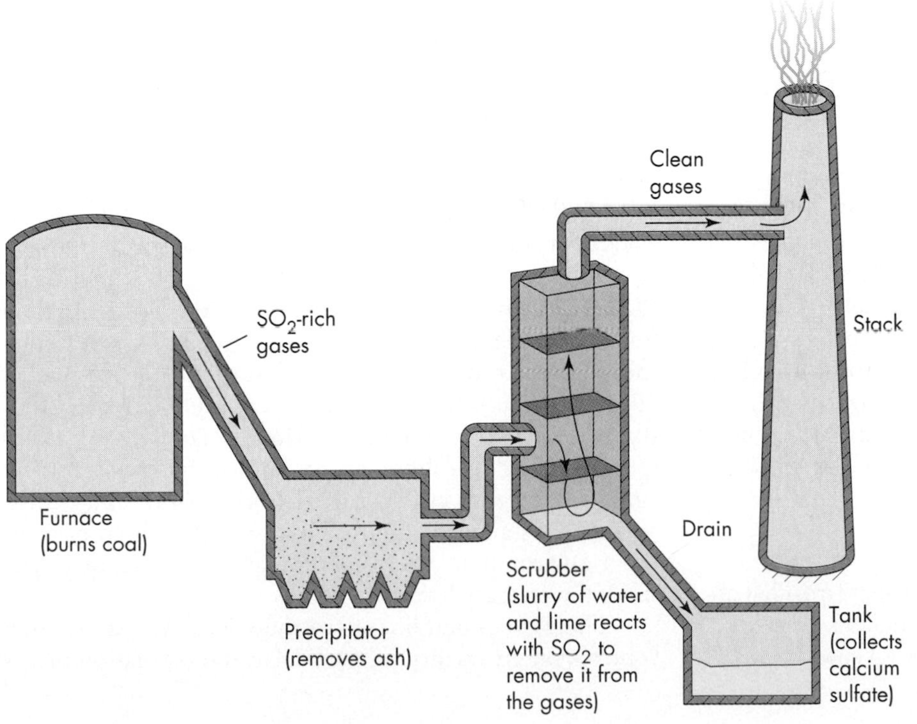

FIGURE 17.12 A scrubber that is used to remove sulfur oxides from the gases emitted by tall smokestacks.

Table 17.2 Pollutants Standards Index[a]

PSI Value	Air Quality Level	Cautionary Statements	Health Effect Label	TSP (particulates (24-hour) μg/m³
500	Significant harm	All persons should remain indoors, keeping windows and doors closed. All persons should minimize physical exertion and avoid traffic.		1000
400	Emergency	Elderly and persons with diseases should stay indoors and avoid physical exertion. General population should avoid outdoor activity.	Hazardous (PSI > 300)	875
300	Warning	Elderly and persons with existing heart and lung disease should stay indoors and reduce physical activity.	Very unhealthful (PSI = 200 to 300)	625
200	Alert	Persons with existing heart or respiratory ailments should reduce physical exertion and outdoor activity.	Unhealthful (PSI = 100 to 200)	375
100	NAAQS[b]		Moderate (PSI = 50 to 100)	260
50	50% of NAAQS[b]		Good (PSI = 0 to 50)	75[c]

[a]One measure of air quality is the Pollutant Standards Index. It is a highly summarized health-related index based on five of the criteria pollutants: total suspended particulates, sulfur dioxide, carbon monoxide, ozone, and nitrogen dioxide. The PSI for one day will rise above 100 in a Standard Metropolitan Statistical Area when one of the five pollutants at one station reaches a level judged to have adverse short-term effects on human health.

[b]NAAQS = National Ambient Air Quality Standard.

Japanese government issued control standards, and in a 5-year period following initiation of the program, the sulfur dioxide level was reduced by 50 percent while the level of energy consumption more than doubled. New goals have been set and emission levels have steadily decreased. The Japanese have also begun to control nitrogen oxides. Power plants have met requirements for the most part by installing scrubbers known as *flue gas desulfurization systems*, which can remove more than 95 percent of the sulfur from smokestacks. More than 1000 of these are in use in Japan today (13).

Air Quality Standards

Air quality standards are tied to emission standards that attempt to control air pollution. Emission standards enacted by various countries set a maximum emission of specific pollutants, but these maximums vary greatly from one country to another, reflecting a lack of agreement concerning the concentrations of pollutants that cause environmental problems (12).

Clean Air Legislation in the United States In the United States, the Clean Air Act (with the objective of improving the nation's air quality) was passed in 1970 and amended in 1977 and 1990. The 1977 amendments define two levels or types of air quality standards. Primary standards are levels set to protect the health of people but which may not protect against damaging effects to structures, paint, and plants. Secondary standards, although designed to help prevent other environmental degradation, are the same, or nearly so, as the primary levels (12).

The 1990 amendments to the Clean Air Act tighten controls on air quality. In particular, the legislation places

Pollutant Level				
SO_2 (24-hour) μg/m³	CO (8-hour) μg/m³	O_3 (1-hour) μg/m³	NO_2 (1-hour) μg/m³	General Health Effects
2620	57,500	1200	3750	Premature death of ill and elderly. Healthy people will experience adverse symptoms that affect their normal activity.
2100	46,000	1000	3000	Premature onset of some diseases in addition to significant aggravation of symptoms and decreased exercise tolerance in healthy persons.
1600	34,000	800	2260	Significant aggravation of symptoms and decreased exercise tolerance in persons with heart or lung disease, with widespread symptoms in the healthy population.
800	17,000	400	1130	Mild aggravation of symptoms in susceptible persons, with irritation symptoms in the healthy population.
365	10,000	240	(c)	
80[d]	5,000	120	(c)	

[c]There are no index values reported at concentrations below those specified by Alert criteria.

[d]Annual primary NAAQS.

Source: Council on Environmental Quality

more stringent controls on emissions of sulfur dioxide (SO_2) produced from coal-burning power plants, requiring these emissions to be reduced by approximately 50 percent by the year 2000. To reach this goal, utility companies will have to practice a variety of pollution abatement strategies as discussed earlier. The legislation also attempts to provide for economic incentives to reduce pollution. This is being attempted through a system of allowances (credits) to emit SO_2. Utility companies are issued a certain number of these allowances, which they may buy and sell on the open market. Those utility companies with low emissions may sell their credits to other utilities that have higher emission rates. Of course, environmentalists may also purchase the allowances, which would effectively keep the allowances from the utility companies, forcing them to employ tighter pollution abatement technology to their operation.

The 1990 amendments also require reductions in the emissions of nitrogen oxides. This will be more difficult to achieve than reduction in sulfur dioxide emissions, as nitrogen oxide emissions are generally associated with automobiles rather than stationary coal-burning power plants (14). Finally, the 1990 amendments also address emissions of toxic materials into the atmosphere that are believed to have the most damaging effect on human health, including causing cancer. The goal is to reduce these emissions by as much as 90 percent.

Monitoring Air Pollution In the United States, air quality in urban areas is often reported as good, moderate, unhealthy, very unhealthy, or hazardous (Table 17.2). These levels are derived from monitoring the concentration of five major pollutants: total suspended particulates, sulfur dioxide, carbon monoxide, ozone, and nitrogen dioxide. During a pollution episode in Los Angeles, hourly ozone levels are reported, and a first-stage smog episode begins if the primary National Ambient Air Quality Standard (NAAQS) of 0.12 ppm (240 μg/m^3) is exceeded. This corresponds to unhealthy air with a Pollutant Standards Index (PSI) between 100 and 300. A second-stage smog episode is declared if the PSI exceeds 300, a point at which the air quality is hazardous to all people. As the air quality decreases during a pollution episode, people are asked to remain indoors, minimize physical exertion, and avoid driving automobiles. Industry also may be asked to reduce emissions to a minimum during the episode.

The Cost of Controls

In the United States today a great deal of money (upward of $10 billion and more) is spent annually trying to control air pollutants. This is to reduce emissions from stationary sources (power plants, factories, etc.) and motor vehicles and to develop efficiency and conservation measures. The Clean Air Act Amendments of 1990 are very broad, and the impact of their pollution abatement measures will be widely felt throughout many sectors of American society and industry. This impact will include increases in the cost of many industrial products and services, including higher costs of autos and gasoline.

The cost and benefits of air pollution control are controversial subjects. Some argue that the present system of setting air quality standards is inefficient and unfair. They maintain that regulations are tougher for new sources of pollution than for existing ones, and that even if the benefits of pollution control exceed total costs, the cost of air pollution control varies widely from one industry to another.

Another economic consideration is that as the degree of control of a pollutant increases, eventually a point is reached where the cost of incremental control (reducing additional pollution) is greater than the additional benefits. Because of this and other economic factors, it has been argued that fees and taxes for emitting pollutants might be preferable to attempting to evaluate uncertain costs and benefits, and that it makes more economic sense to enforce fees rather than standards.

Some variables that must be considered in the economics of air pollution are shown in Figure 17.13. With increasing air pollution controls, the capital cost to control air pollution increases, and as the controls for air pollution increase, the loss from pollution damages decreases. The total cost is thus the sum of these two items, and the minimum is well defined in terms of a particular average pollution level. Figure 17.13 indicates that if the desired pollution level is lower than that at which the minimum total cost occurs, then additional costs will be necessary. This type of diagram, while valuable for considering some of the major variables, does not consider adequately all of the loss from pollution damages. For example, long-term exposure to air pollution may aggravate or lead to chronic diseases in human beings, with a very high cost. How do we determine what portion of the cost is due to air pollution? Despite these drawbacks, it seems worthwhile to reduce the air pollution level below some particular standard. Thus, in the United States, ambient air quality standards have been developed as a minimum acceptable pollution level.

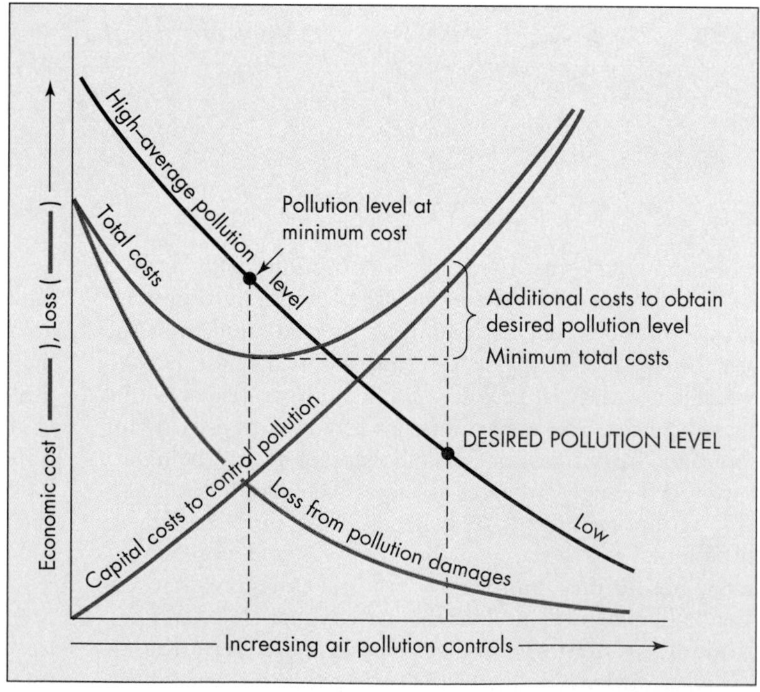

▶ **FIGURE 17.13** Some of the relationships between economic cost and increasing air pollution controls. (*Modified after* S. J. *Williamson*. 1973. Fundamentals of air pollution. *Addison-Wesley*.)

SUMMARY

The atmosphere is a fast-flowing and convenient medium for disposal of gaseous wastes. Every year several hundred million metric tons of pollutants enter the atmosphere above the United States from processes related to human activity. Considering the enormous volume of the atmosphere, this is a relatively small amount of material. If it were distributed uniformly, there would be fewer problems with air pollution. Unfortunately, the pollutants are not generally evenly distributed, but rather are concentrated in urban areas, or in other areas where the air naturally lingers.

The two major types of air pollution sources are stationary and mobile. Stationary sources have a relatively fixed position, and include point sources, fugitive sources (open places exposed to wind), and area sources. Air pollutants are classified as gaseous and particulate, or as primary and secondary. Primary pollutants are those emitted directly into the air: particulates, sulfur oxides, carbon monoxide, nitrogen oxides, and hydrocarbons. Secondary pollutants are those produced through reactions among primary pollutants and other atmospheric compounds. A good example of a secondary pollutant is ozone, which forms over urban areas through photochemical reactions among primary pollutants and natural atmospheric gases. Be careful not to confuse ozone as a pollutant in the lower atmosphere with ozone depletion in the stratosphere (see Chapter 16).

Meteorological and topographic conditions greatly affect whether polluted air is a problem in a particular urban area. In particular, restricted lower-atmosphere circulation associated with temperature inversion layers may lead to pollution events, especially in cities lying in bowls surrounded by mountains. Air pollution over an urban area is directly proportional to the emission rate of pollution and the downwind travel distance of pollutants over the city; pollution is inversely proportional to the wind velocity and the height of the mixing layer of the atmosphere over the urban area.

A city creates an environment different from that of the surrounding region. Cities change the local climate and, in general, are cloudier, warmer, rainier, less humid, and less windy. The combination of lingering air and abundance of pollutants produces the phenomena called the *urban dust dome* and the *heat-island effect*, which concentrate pollution in the center of a city.

The two major types of smog are sulfurous (gray) and photochemical (brown). Sulfurous smog is produced mainly by industrial burning of fossil fuels. Photochemical smog is produced by photochemical reactions of the nitrogen oxides and hydrocarbons in automobile exhaust. Each type causes particular environmental problems that vary with geographic region, time of year, and local urban conditions.

A growing problem in urban areas is indoor air pollution, caused by tight insulation of modern buildings, emission of toxic gases from building materials, and circulation of toxic gases, smoke, and disease-causing organisms through ventilation systems.

Methods to control air pollution are tailored to specific sources and types of pollutants. These methods vary from settling chambers for particulates to catalytic converters to remove carbon monoxide and hydrocarbons from automobile exhaust to scrubbers or combustion processors that use lime to remove sulfur before it enters the atmosphere.

Air quality in urban areas is usually reported in terms of whether the quality is good, moderate, unhealthy, very unhealthy, or hazardous. These levels are defined in terms of the Pollutant Standards Index (PSI) and National Ambient Air Quality Standard (NAAQS). The nation's air quality seems to have improved in recent years; however, in numerous areas urban air quality is still unhealthful a good deal of the year.

The relationships between emission control and environmental cost are complex. The minimum total cost is a compromise between capital costs to control pollutants and losses or damages resulting from such pollution. If additional controls are used to lower the pollution to a more acceptable level, then additional costs are incurred, which can increase rapidly.

REFERENCES

1. **National Park Service.** 1984. *Air resources management manual.* Washington, DC: U.S. Government Printing Office.
2. **Godish, T.** 1991. *Air quality*, 2nd ed. Chelsea, MI: Lewis Publishers.
3. **Pittock, A. B., Frakes, L. A., Jenssen, D., Peterson, J. A., and Zillman, J. W., eds.** 1978. *Climatic change and variability: A southern perspective.* (Based on a conference at Monash University, Australia, December 7–12, 1975.) New York: Cambridge University Press.
4. **Anthes, R. A., Cahir, J. J., Fraser, A. B., and Panofsky, H. A.** 1981. *The atmosphere*, 3rd ed. Columbus, OH: Charles E. Merrill.
5. **Marsh, W. M., and Dozier, J.** 1981. *Landscape.* Reading, MA: Addison-Wesley.
6. **Detwyler, T. R., and Marcus, M. G., eds.** 1972. *Urbanization and the environment.* North Scituate, MA: Duxbury Press.
7. **Gates, D. M.** 1972. *Man and his environment: Climate.* New York: Harper & Row.
8. **Lynn, D. A.** 1976. *Air pollution—Threat and response.* Reading, MA: Addison-Wesley.
9. **Council on Environmental Quality and the Department of State.** 1980. *The global 2000 report to the President: Entering the twenty-first century.* Washington, DC: U.S. Government Printing Office.

10. **Zimmerman, M. R.** 1985. Pathology in Alaskan mummies. *American Scientist* 73:20–5.
11. **Stoker, H. S., and Seager, S. L.** 1976. *Environmental chemistry: Air and water pollution*, 2nd ed. Glenview, IL: Scott, Foresman.
12. **Stern, A. C., Boubel, R. T., Turner, D. B., and Fox, D. L.** 1984. *Fundamentals of air pollution*, 2nd ed. Orlando, FL: Academic Press.
13. **Anonymous.** 1981. How many more lakes have to die? *Canada Today* 12(2):1–11.
14. **Molina, B. F.** 1991. Washington report. *GSA Today* 1(2):33.

KEY TERMS

stationary sources (p. 475)
point sources (p. 475)
fugitive sources (p. 475)
gaseous pollutants (p. 475)
particulate-matter pollutants (p. 475)
primary pollutants (p. 475)
secondary pollutants (p. 475)
sulfur dioxide (p. 475)
nitrogen oxides (p. 475)
carbon monoxide (p. 477)
ozone (p. 477)
hydrocarbons (p. 477)
particulate matter (p. 478)
atmospheric inversion (p. 479)
urban dust dome (p. 482)
heat-island effect (p. 482)
sulfurous smog (p. 482)
photochemical smog (p. 482)
catalytic converter (p. 486)
fluidized-bed combustion (p. 487)
scrubbing (p. 487)
air quality standards (p. 488)

SOME QUESTIONS TO THINK ABOUT

1. Consider the Case History of the London smog crisis in 1952 as well as the Meuse Valley incident in Belgium and the Donora, Pennsylvania, incident in 1948. What lessons have been learned from these pollution events? Do you think they are relevant to the air pollution problems being faced in the world today? Where might such events occur? Why?
2. From an environmental perspective, the best way to lower emissions of air pollutants is not to emit them in the first place. For your community, region, or state, develop a series of actions that could use principles of conservation or efficiency to help reduce real or potential air pollution problems.
3. One of the measures for helping reduce emissions of sulfur dioxide from burning fossil fuels is allowing utilities to purchase allowances or credits for emissions. These may then be bought and sold on the open market. As only a certain number of credits are available, this limits the pollution emissions to the desired level. Some environmentalists do not like this because they believe it allows utilities to buy their way out of pollution problems. What do you think about this idea? Make a list of advantages and disadvantages of this innovative system to try to help control emissions of sulfur dioxide.

18 Landscape Evaluation and Land Use

Chankanab Lagoon National Park, Cozumel, Mexico. (*Edward A. Keller*)

Chankanab Lagoon National Park is located on the island of Cozumel, in the Caribbean Sea, near Cancun, Mexico. Managing the park is difficult because there are several potentially conflicting objectives, including restoration of the lagoon and maintaining a tourist attraction. Shown here is an artificial sand beach constructed to cover the coarse limestone coastal rock shown in the foreground. The lagoon itself is located back a few meters from the beach, and a series of caves allow water in the lagoon to communicate with the waters of the Caribbean. Restoration work has the objective of keeping the caves open so that water continues to circulate and to remove sand and other debris from the lagoon. Swimming is no longer allowed in the lagoon, but the construction of the artificial beach has drawn many tourists to view the abundant marine resources on the Caribbean side. This management strategy has had problems, because during large storms, particularly the 1989 hurricane, sand was transported into the shallow lagoon, nearly filling it. Burial of coral by sediment kills it! As a result, the use of artificial sand near a coral reef lagoon is questionable. An alternative might be to build wooden decks and walkways over the coarse limestone, allowing entrance into the clear waters of the Caribbean. Hurricanes would tear up the wood decks and walks, but floating wood in the lagoon could easily be retrieved and sand would not enter the fragile environment.

18.1 Landscape Evaluation, Land Use, and the Earth Scientist

Evaluating the landscape for such purposes as land-use planning, site selection, construction, and environmental impact is common practice. A primary role of earth scientists in landscape evaluation is to provide geologic information and analysis before planning, design, and construction begin. As a member of the landscape evaluation team, earth scientists commonly work with geographers, civil engineers, architects, planners, lawyers, and public administrators. Specific

LEARNING OBJECTIVES

The study of landscape evaluation and environmental law requires integration of much of what we have learned throughout this book. In this sense, the chapter is a capstone and fitting conclusion to our discussion of environmental geology. Learning objectives of the chapter are:

- *To understand the role of the earth scientist in landscape evaluation.*
- *To know what an environmental geology map is and how it might be used.*
- *To become familiar with the basic concept and processes of the Geographic Information System.*
- *To know how the methods of cost-benefits analysis and physiographic determinism relate to site selection and evaluation of land for specific purposes.*
- *To become familiar with environmental impact analysis, including the major components of the environmental impact statement and the processes of scoping and mitigation.*
- *To become familiar with the major processes involved in land-use planning.*
- *To gain a modest acquaintance with the processes of law and particularly the emerging process of mediation and negotiation.*

Web Resources

Visit the "Environmental Geology" Web site at www.prenhall.com/keller to find additional resources for this chapter, including:

- Web Destinations
- On-line Quizzes
- On-line "Web Essay" Questions
- Search Engines
- Regional Updates

information that earth scientists supply as part of the landscape evaluation process varies from area to area, and from project to project, but often centers on factors such as:

- Former and present land use (that is, site history)
- Physical and chemical properties of earth materials, including soil and rock type, presence of pollutants or contaminants, depth to bedrock, and engineering properties of soil and rock
- Extent of natural hazards present, including seismic risk, slope instability, volcanic activity, flooding, erosion potential, and others
- Depth to water table and groundwater flow characteristics, as well as any water pollution problems that might be present

An important contribution of earth scientists to landscape evaluation is emphasizing that not all land is the same, and that particular physical and chemical characteristics of the land may be more important to society than geographic location. Earth scientists recognize that there is a limit to our supply of land, and that we must strive to plan, so that suitable land is available for specific uses for this generation and those that follow (1).

18.2 Environmental Geology Mapping

The primary goal of environmental geology mapping is to help the planner and other decision makers (as, for example, city council members). Therefore, results of mapping must be presented in a way that is easy for a nongeologist to interpret (2). A good environmental geology map presents geologic and hydrologic information in terms of the engineering properties of materials, expressed in nontechnical language.

Interpretive Environmental Geology Maps

An **environmental geology map** might indicate the suitability of different areas for a specific land use, as soil maps do (see Chapter 3). The idea is to produce a series of maps for planners, one for each possible land use. These interpretive environmental geology maps might include a color code: green for "go" (favorable conditions); yellow for "caution"; and red for "stop" (unfavorable conditions or problem area).

Interpretive maps are sometimes prepared for siting of a specific facility such as a landfill for solid-waste disposal. Such a map was prepared for De Kalb County, Illinois, by the Illinois Geological Survey. The evaluation considered topography, type, and thickness of unconsolidated material; present and potential sources of surface water and groundwater; and type of bedrock. Sites with a minimum of problems for solid-waste disposal are colored green, whereas less favorable areas are yellow. Acceptable disposal sites can be found in yellow areas but require careful preliminary investigation. Areas colored red on the maps are the least favorable, but even within the red areas there may be sites that would make satisfactory landfills following detailed evaluation and engineering to improve the site (3). It is emphasized that interpretive maps are seldom drawn to appropriate scale for detailed evaluation. They serve as general guidelines to assist individuals and other parties interested in purchasing land.

Interpretive maps may also be prepared to show specific geologic hazards in an area, such as land subject to flooding, landslide susceptibility, potential for ground rupture and seismic shaking from earthquakes, and soil conditions such as potential for liquefaction, potential to expand and contract, and potential for production of radon gas. Ideally, a master hazard map could be prepared. A prospective buyer or developer could consult the map and quickly and easily determine whether the property in question has potential problems.

Regional maps that delineate an area's relative vulnerability to environmental disruption are also valuable in planning future development and land use. Figure 18.1, for example, shows major existing sources of potential groundwater contamination and the relative vulnerability of aquifers to pollution from surface discharge in New Mexico. Relative vulnerability to groundwater pollution was predicted by considering aquifer type, depth to the water table, permeability, and presence or absence of natural barriers to migration of pollutants (4). The maps thus present geologic and hydrologic data as well as sources of potential groundwater pollution in a format useful to regional planners.

Although environmental geology maps such as those in the Illinois and New Mexico examples are valuable to planners, larger-scale, more detailed maps that show accurate geologic boundaries are preferable. However, it is not always possible to produce such a map—the scale of an environmental geology map is based on the accuracy and amount of available geologic data, and natural geologic boundaries are often gradual rather than sharp (3).

Environmental Resource Units (ERUs)

Another interesting approach to environmental mapping is the concept of an **environmental resource unit (ERU).** This method uses a multidisciplinary team approach to define and analyze the total natural environment—geologic, hydrologic, and biologic. A portion of the environment with a similar set of physical and biological characteristics is the environmental resource unit. Ideally, every ERU is a natural division characterized by specific assemblages of structural components, such as rocks, soils, water, wildlife, and vegetation, and of natural processes, such as erosion, runoff, and soil formation (2).

A study site of 10.4 km^2 near Morrison, Colorado, illustrates the concept of defining and working with environmental resource units. Figure 18.2 shows the topography and the area, and Figure 18.3 shows the four ERUs. The first is the mountain forest ERU—ridge and ravine topography supporting a pine and Douglas fir forest. Running-water and mass-wasting processes are important in modifying the land, and the dominant ecological control is the variation of moisture with elevation and exposure. The second is the floodplain forest ERU—recent alluvial (stream-deposited) material deposited by channel and overbank stream flow. This ERU supports a cottonwood-willow forest that needs abundant water. The third is the hogback wood and grassland ERU—a series of tilted sedimentary rock units with steep scarp slopes, gentle dip slopes, and gentle debris slopes or level bottoms. Mass-wasting processes are controlled by slope and exposure. The scarp slopes and bottomlands support grass, and the dip slopes support junipers. The fourth is the Pleistocene grasslands ERU—a complex of older alluvial and mass-wasting material deposited by processes that are no longer operating. The natural vegetation of this type of land is grass, but in the study area much of the ERU is used for agricultural purposes (2).

Environmental resource units are used to establish patterns of structure and process and are valuable in establishing land capabilities and suitability. The ERU concept is most valuable as part of a systems approach involving the entire environment. These units can then be analyzed separately or collectively for a specific land use. The ultimate value of ERUs is realized when multidisciplinary teams, including geologists, soil scientists, biologists, landscape architects, engineers, and planners, use remote sensing along with field survey techniques to identify different units and then analyze the data for planning purposes by means of computer-drawn maps for rapid data analysis and display.

Urbanizing areas in the United States, from California to Indiana to the District of Columbia, offers opportunities to integrate environmental resource units with earth science maps to plan land use that is sensitive to the natural setting and hazards. The synthesis resulting in the final land-use capability map is an example of computer composite mapping using input of topographic, geologic, and hydrologic data (Figure 18.4). Maps such as this may have a significant impact on decisions to rezone existing land or allow new development in urban areas.

The Geographic Information System

A computer composite map such as a land-use capability map (Figure 18.4) is an example of the application of an information system to environmental work. In recent years the concept has evolved to what is known as the **Geographic Information System (GIS).** The GIS is a form of technology capable of storing, retrieving, transforming, and displaying spatial environmental data (5). The spatial data used in the GIS may be points, lines, or areas. Following are examples of the GIS spatial data: water wells and intersections of roads are points; rivers, roads, and faults are lines; and rock type or soil type define areas. It is important to recognize that the Geographic Information System is more than simply a way to categorize and store data. The system has the ability to manipulate data and thus create a new product. That product may be a geologic hazard or land-use capability map, as shown in Figure 18.4. It may also be a sophisticated analysis of environmental variables for future land-use planning.

The Geographic Information System has four essential elements (6):

- **Data acquisition:** Collecting aerial photographs, maps, satellite imagery, census data, and so forth
- **Data processing and management:** Recording data from the maps and other sources in a systematic way.

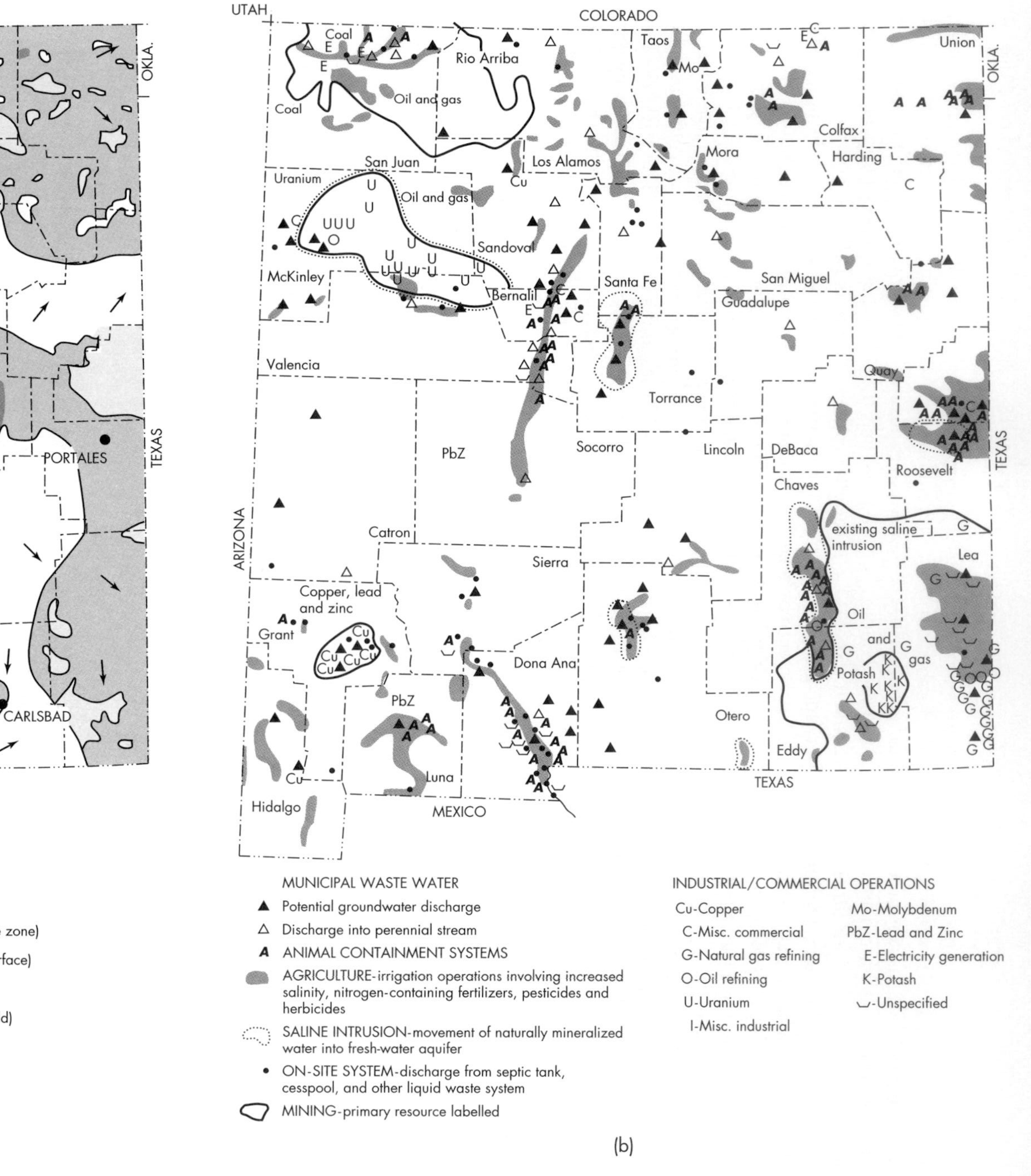

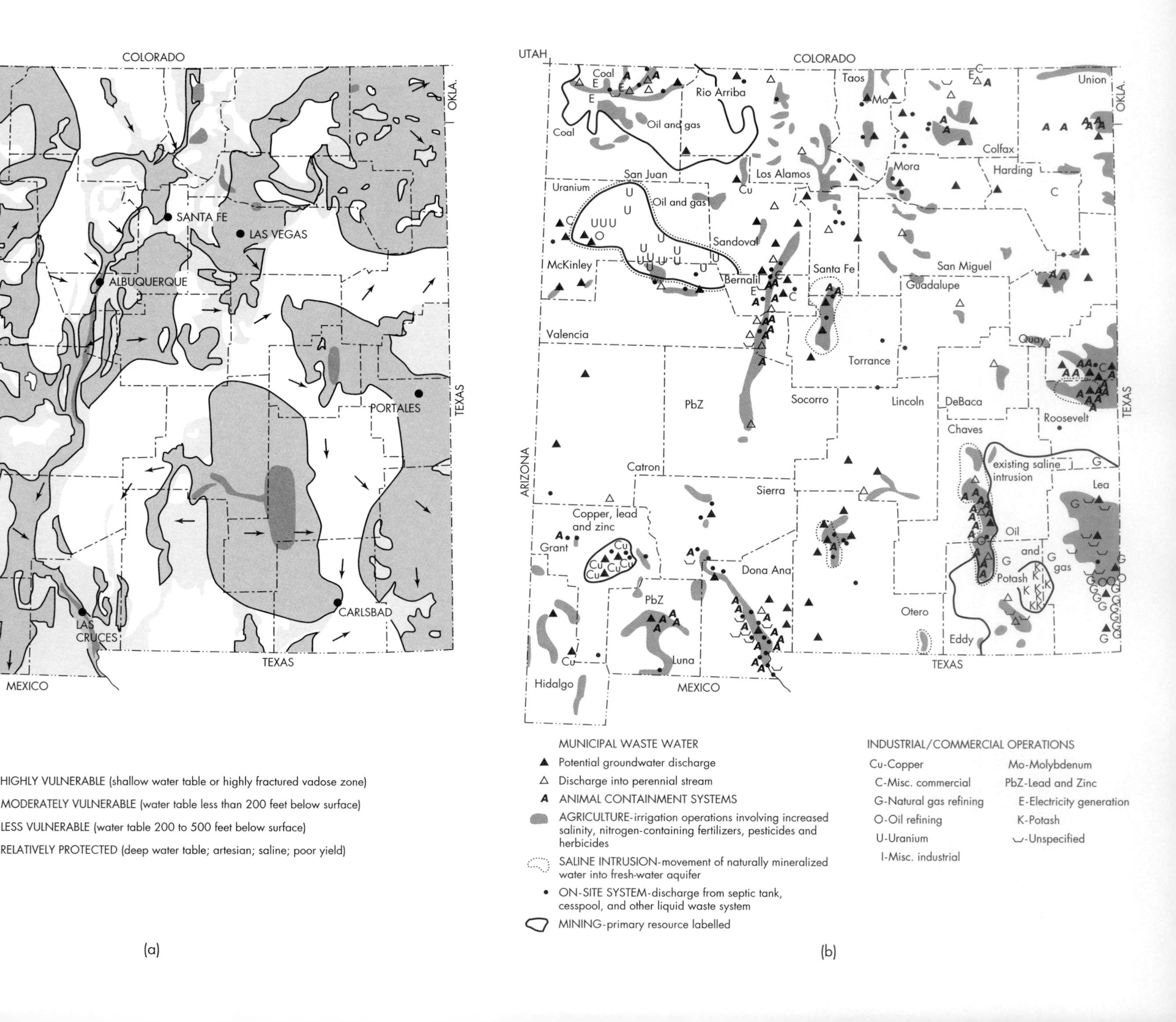

UTAH
COLORADO
OKLA.
TEXAS
MEXICO
ARIZONA
GALLUP
SANTA FE
LAS VEGAS
ALBUQUERQUE
PORTALES
CARLSBAD
LAS CRUCES
HIGHLY VULNERABLE (shallow water table or highly fractured vadose zone)
MODERATELY VULNERABLE (water table less than 200 feet below surface)
LESS VULNERABLE (water table 200 to 500 feet below surface)
RELATIVELY PROTECTED (deep water table; artesian; saline; poor yield)
(a)
UTAH
COLORADO
OKLA.
TEXAS
MEXICO
ARIZONA
Coal
Rio Arriba
Oil and gas
San Juan
Uranium
Los Alamos
Taos
Mora
Colfax
Union
Harding
Sandoval
McKinley
Bernalil
Santa Fe
San Miguel
Guadalupe
Quay
Valencia
Torrance
PbZ
Socorro
Lincoln
DeBaca
Roosevelt
Chaves
existing saline intrusion
Lea
Catron
Sierra
Copper, lead and zinc
Grant
Dona Ana
Oil and gas
Potash
Otero
Eddy
Luna
Hidalgo
MUNICIPAL WASTE WATER
Potential groundwater discharge
Discharge into perennial stream
ANIMAL CONTAINMENT SYSTEMS
AGRICULTURE-irrigation operations involving increased salinity, nitrogen-containing fertilizers, pesticides and herbicides
SALINE INTRUSION-movement of naturally mineralized water into fresh-water aquifer
ON-SITE SYSTEM-discharge from septic tank, cesspool, and other liquid waste system
MINING-primary resource labelled
INDUSTRIAL/COMMERCIAL OPERATIONS
Cu-Copper
C-Misc. commercial
G-Natural gas refining
O-Oil refining
U-Uranium
I-Misc. industrial
Mo-Molybdenum
PbZ-Lead and Zinc
E-Electricity generation
K-Potash
-Unspecified
(b)

◀ **FIGURE 18.1 (*previous page*)** (a) Relative vulnerability of aquifer contamination and (b) major sources of potential groundwater contamination in New Mexico. Arrows in (a) indicate general groundwater flow direction. (*After* L. *Wilson*, 1981, *New Mexico Geological Society Spec. Pub.* No. 10, *pp.* 47–54.)

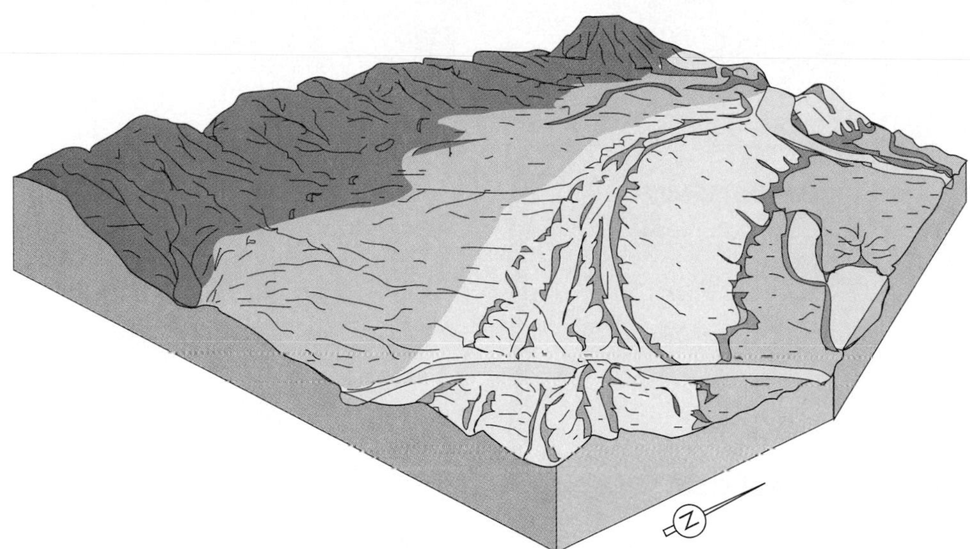

◀ **FIGURE 18.2** Physiographic diagram of the Morrison (Colorado) Test Site. (*From Turner and Coffman*. 1973. *Geology for Planning: A Review of Environmental Geology*, Quarterly of the Colorado School of Mines 68[3].)

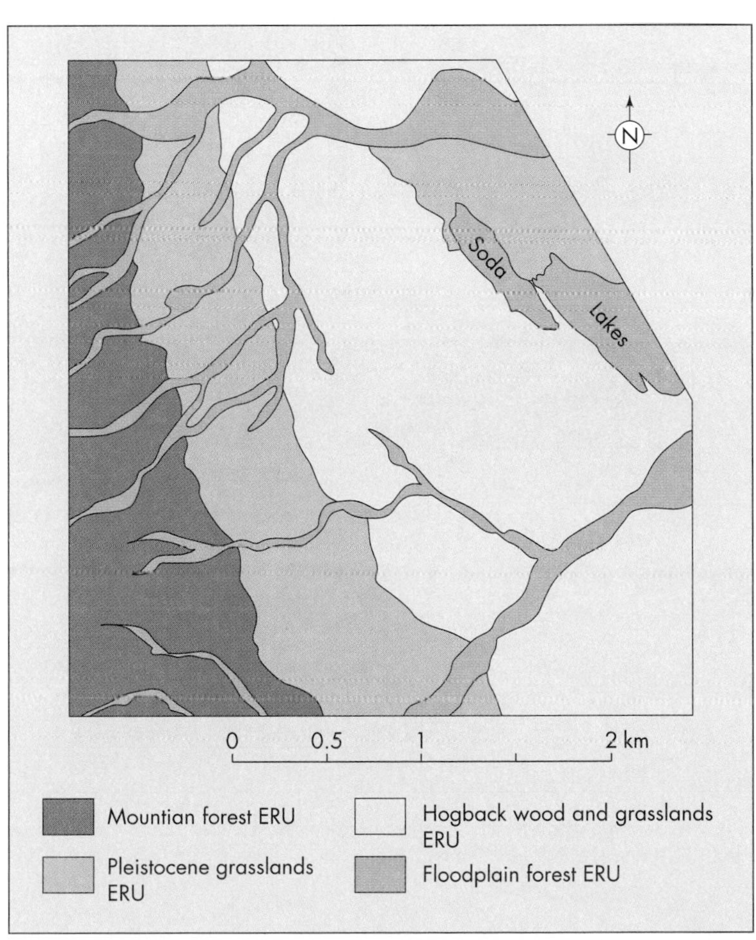

▲ **FIGURE 18.3** Environmental resource units for the Morrison (Colorado) Test Site. (*After Turner and Coffman*. 1973. *Geology for Planning: A Review of Environmental Geology*, Quarterly of the Colorado School of Mines 68[3].)

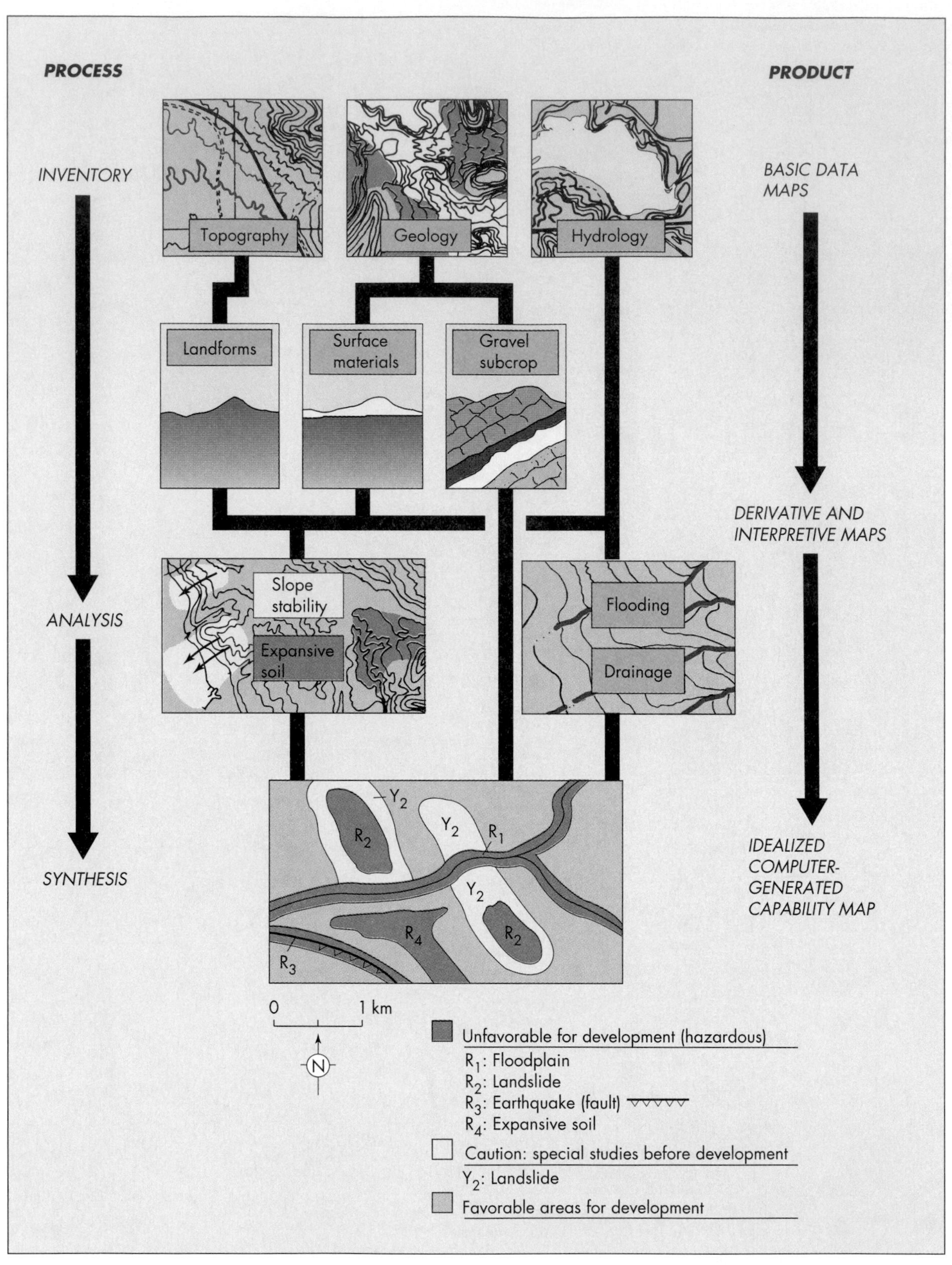

▲ **FIGURE 18.4** Flowchart and idealized diagrams showing the use of earth science information in a Geographic Information System (GIS) for natural hazards and land use.

This process involves the creation of a computer database where all the data are stored and accurately identified and located. An important feature here is developing consistent ways that data may be entered, deleted, updated, and retrieved.

- **Data manipulation and analysis:** This is the part of the GIS that can develop new information. For example, as shown earlier (Figure 18.4), topographic, geologic, and hydrologic data may be combined and overlain with other information, such as data on landforms, to evaluate slope stability.
- **Generation of products:** This is the stage in GIS work that produces the desired output. This might include land capability maps, statistical analysis of environmental variables, or natural hazard maps.

Geographic Information Systems are fast becoming the dominant method of data collection, storage, and retrieval for landscape evaluation. Therefore, it behooves the serious earth scientist to become familiar with this exciting field of geographical research and application.

18.3 Site Selection

Site selection is the process of evaluating a physical environment that will support human activities. It is a task shared by professionals in engineering, landscape architecture, planning, earth science, social science, and economics, and thus involves a multidimensional approach to landscape evaluation.

The goal of site evaluation for a particular land use is to ensure that site development is compatible with both the possibilities and limitations of the natural environment. Although it is obviously advantageous to know the possibilities and limitations of a site before development begins, site evaluation is often overlooked. People still purchase land for various activities without considering whether the land use they have in mind is compatible with the site they have chosen. There are well-known examples of poor siting that have resulted in increased expense, limited production, or even abandonment of partially completed construction:

- Construction of a West Coast nuclear power facility was terminated when fractures in the rock (active faults) were discovered and possible serious foundation problems arose.
- The productivity of a large chicken farm in the Southeast was greatly curtailed because the property was purchased before it was determined whether there was sufficient groundwater to meet the projected needs.
- A housing developer in northern Indiana purchased land and built country homes in one of the few isolated areas where bedrock (shale) is at the surface. Thus, plans for septic-tank systems had to be abandoned, and a surface sewage treatment facility had to be built. The rock also made it much more expensive to excavate for basements and foundations.

Role of the Earth Scientist in Site Selection

The process of site selection begins with a thorough understanding of the general purposes and specific requirements of the designed activity. This step also includes knowing for whom the site is being selected. The next step is an analysis of possible alternative sites in terms of physical limitations and cost. The final step is to recommend which site or sites are most advantageous for the desired land use (7).

The earth scientist plays a significant role in the evaluation process by providing crucial geologic information. This information includes soil and rock types, rock structure (especially fractures), drainage characteristics, groundwater characteristics, landform information, and estimates of possible hazardous earth events and processes, such as floods, landslides, earthquakes, and volcanic activity. The engineering geologist also takes samples, makes tests, and predicts the engineering properties of the earth materials.

The geologic aspect of the evaluation process requires these steps:

- Collection of existing data, such as geologic reports, topographic maps, aerial photographs, soil maps, water surveys, engineering reports, and climate records
- Reconnaissance of likely sites to note potential problems and possibilities
- Collection of necessary new data, including soil, rock, and water analyses
- Determination of the magnitude and importance of geologic limitations and possibilities of the site or sites

Methods of Site Selection

Although specific methods of determining the suitability of a site vary with the intended use, a few methods are common for evaluating the desirability of a particular site for relatively large-scale land uses, such as construction of reservoirs, highways, and canals and development of parks and other recreational areas. Two common methods are cost-benefits analysis and physiographic determinism. The first approach is strictly economic, whereas the second balances economics with less tangible aspects, such as aesthetics.

Cost-benefits Analysis **Cost-benefits analysis** is a way to assess the long-range desirability of the particular project, such as construction of a highway or development of a recreational site. The idea is to develop the benefits-to-cost ratio; that is, the total benefits in dollars over a period of time are compared to the total cost of the project, and the most desirable projects are those for which the benefits-to-cost ratio is greater than one (8).

Cost-benefits analysis assumes that all relevant costs and benefits can be determined. There are three difficult aspects of this determination: which costs and which benefits are to be evaluated, how they are to be evaluated, and how intangible costs and benefits such as aesthetic degradation or improvement are to be evaluated (8). Cost-benefits analysis is common for three general types of projects: water resource projects, including irrigation, flood control, and hydroelectric power systems (as well as multipurpose systems, such as reservoirs used for flood control, power generation, and recreation); transportation projects such as highways, roads, railways, and river and canal work to facilitate navigation; and land-use projects, such as urban renewal and recreation (8).

Examining some of the costs and benefits for a hypothetical flood-control project will further illustrate the method. The initial tangible costs of a flood-control project are relatively straightforward, but evaluation of repair and maintenance costs and intangible costs over the life of the project is much more difficult. The main benefit from flood control is averting loss of property. (*Loss* here refers primarily to different types of assets, such as property, furnishings, and crops.) The general principle is to evaluate the annual flood loss and damage based on the most probable flood levels for several recurrence intervals, say, the 10-year and 25-year floods. This estimate of annual loss and damage is then taken as the maximum yearly amount that people would be willing to pay for flood control. Other benefits to consider include reducing the number of deaths by drowning and eliminating recurring costs of evacuating flood victims, emergency sandbag work, and incidence of sanitary failure (8).

Physiographic Determinism An interesting method of site selection is **physiographic determinism,** which involves application of ecological principles to planning. The method considers physical, social, and aesthetic data, and the objective is to maximize social benefit while minimizing social cost. The philosophy of physiographic determinism is to "design with nature" (9). The idea is that site selection is determined by the characteristics of the site—we might almost say that the site selects itself. That is, rather than laying down an arbitrary design or plan for an area, it may be more advantageous to find a plan that nature has already provided (10). For example, if the soils of an area vary considerably in their ability to corrode, then the best corridor for an underground pipeline might not be the shortest route. Rather, the route might be designed so that the greatest possible length lies in the soils with the least corrosive potential. Although such a route is slightly longer and requires a higher initial cost, it may, by taking advantage of nature, produce tremendous savings in maintenance and replacement costs (11).

The actual methodology of physiographic determinism in selecting a site or a route for a highway, regional development plan, or other land use involves selecting factors that reflect physical, biological, and social values (including economic). For each factor, data in the form of three or more grades of the value (for example, low, medium, and high) are input and mapped using a Geographic Information System. The lower values indicate areas of conflict, where physical conditions are not appropriate or social values conflict with the desired land use. The data are then physically or mathematically superimposed on one another. Areas representing those places with the greatest physical limitations, construction costs, and social costs can be distinguished from places where these factors are at a minimum. By this method, the best sites may be selected. No matter how poor the choices may look, there will always be relatively good ones (9).

The idea of determining the lowest social cost in site selection is innovative in addressing social values as well as traditional physical-economic values, but it is not above criticism. It is extremely subjective, in that the person who is evaluating chooses the factors and thus essentially builds the formula used in the evaluation. For example, in evaluating the siting of a highway route, the evaluator may decide that property values are important and that a relatively high negative rating should be assigned to high property values. This decision essentially prescribes that the highway should be routed away from people, and most of all from wealthy people. This is a point some planners and others would (based on social justice issues) vigorously argue against (10).

Another issue is that the method gives all factors the same weight. Critics point out that a geologic factor such as suitability of soil for foundation material or susceptibility to erosion should not necessarily be weighted equally with social values such as scenic or recreational value. Regardless of these shortcomings, the method of physiographic determinism in evaluating both physical-economic factors and social values is a step in the right direction (10).

Site Evaluation for Engineering Purposes

The geologist has a definite role in the planning stages of site development for engineering projects, such as construction of dams, highways, airports, tunnels, and large buildings.

Dams Construction of either masonry (concrete) dams or earth dams (Figure 18.5) requires careful and complete geologic mapping and testing early in the planning process. Testing should include:

- Evaluation of the valley and the dam site to determine the present and expected stability of the slopes
- Identification of possible problems, such as active faults and fracture zones in rocks or adverse soil conditions where the dam foundation would be in contact with rock or soil
- Prediction of the rate of sediment accumulation
- Assessment of the availability of building material for the construction of the dam

Continuous geologic mapping during construction of dams is important because the geologic history of a river valley often involves several periods of alternating erosion and deposition, producing irregular or unexpected deposits that

▲ **FIGURE 18.5** Flaming Gorge Dam on the Green River of Utah. Environmental impact analysis associated with dams and reservoirs is an important endeavor. (*Michael Collier*)

can cause problems in construction, operation, and maintenance of the dam and reservoir. We emphasize, however, that major irregularities or problems that might affect the dam or reservoir should be recognized by direct and indirect geologic investigation before the construction phase of the project.

Foundations for dams vary according to what rock types are encountered. Generally:

- Igneous rocks, such as granite and some pyroclastic rocks, are satisfactory foundations for a dam site. Leakage, if any, will occur along fractures, and fractures can be grouted, that is, filled with a mixture of cement, sediment, and water.
- Metamorphic rocks may also be good foundation material, and the best sites are those in which the foliation of the rocks is parallel to the axis of the dam.
- Sedimentary rocks such as limestone and particularly compaction shale can be troublesome. Limestone may have large underground cavities, and compaction shales are noted for problems stemming from their tendency to deform and settle when loaded (12,13). Particularly important is the stability of sedimentary rocks following wetting and drying or freezing and thawing. Failure of the St. Francis Dam in California (see Chapter 2) was caused in part by the deterioration upon wetting of the sedimentary rocks in the foundation.

Highways Site evaluation for highways requires considerable geologic input, including mapping the geology along the proposed route and isolating problems of slope stability, topography, flooding, or weak earth materials that would cause foundation problems. It is also helpful to locate and estimate the amount of possible construction materials in close proximity to the proposed route (13).

Airports There are four important considerations in site selection for airports:

- Topography as it relates to necessary grading, drainage, and surfacing
- Soils, the best of which are coarse-grained with high bearing capacities and good drainage
- Availability of construction materials
- Good surface drainage—areas subject to flooding should be avoided (12)

Tunnels Construction of tunnels requires detailed knowledge of geology, and evaluation of geologic information is basic to selection of the route and determination of proper construction methods. Geology is a major factor determining the economic feasibility of the project; therefore, a geologic profile along the center line of the proposed tunnel route is essential before tunneling. This profile is developed from mapping the surface geology and using subsurface data whenever possible (12,13). As with dam investigations, it is important to continue the geologic mapping as construction proceeds so as to ensure that unexpected hazards do not produce problems.

Two distinct types of tunnel conditions are defined by whether the tunnel is in solid rock or in soft ground—that is, in cohesionless (plastic) earth—that may tend to flow and fill the tunnel (13). Both conditions may be present in one tunnel, and each presents its own particular problems.

Potential problems in rock tunnels are determined primarily by the nature and abundance of fracturing or other partings in the rock. Fracturing greatly affects how much overbreak (excess rock that falls into the tunnel) is present. In general, the closer the fractures and other partings, the greater the overbreak (12). Tunnels in granite or horizontally layered rocks generally develop a symmetrical, naturally arch-shaped overbreak, whereas inclined rocks produce an irregular (asymmetric) roof with potentially more unstable blocks (Figure 18.6). When possible, it is advisable to select tunnel routes through rock that has a minimum of fractures.

Rock structure, particularly folds and faults, is also significant in locating and orienting tunnels. When tunnels are built through synclines (a type of fold in rocks that is often in the shape of a basin or U), water problems are likely, because the natural inclination of the rocks facilitates drainage into the tunnel (Figure 18.7). In addition, rock fractures in synclines tend to form inverted keystones, and thus the roof of the tunnel may require substantial support. On the other hand, tunnels through anticlines (a type of fold that is often in the shape of an arch, like the McDonald's® sign) are less likely to have water problems, and the fractures tend to form normal keystones that are more nearly self-supporting (12). Faults in tunnels usually cause water problems because water often migrates along open fractures associated with fault systems. Therefore, whenever possible, tunnels should be driven at right angles to faults to minimize the length of tunnel that will be in contact with the fault (12).

Tunneling in soft ground is generally done at shallower depths, thus facilitating the collection of more detailed subsurface information before construction than is possible

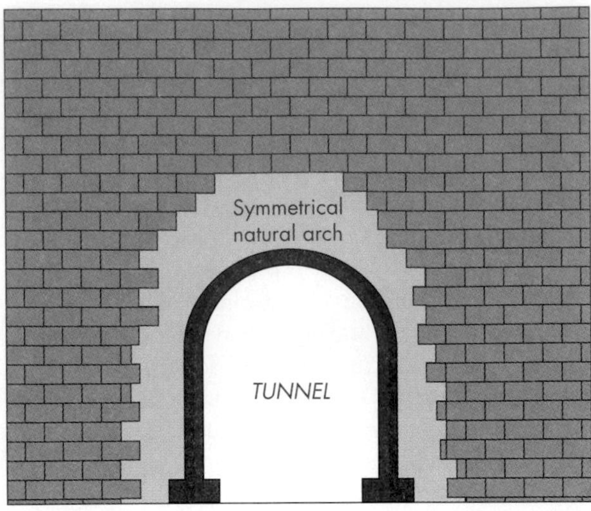

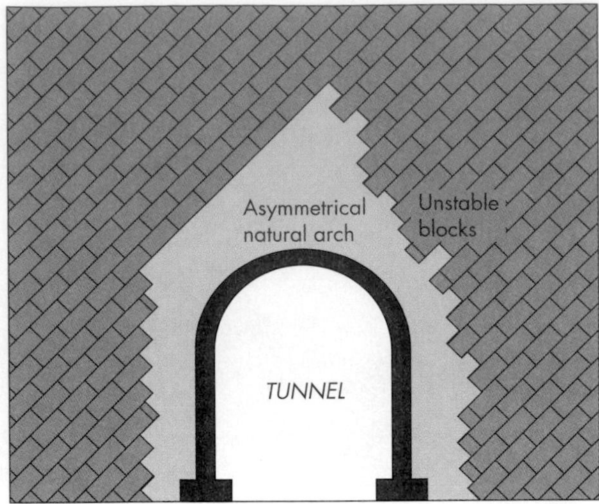

▲ **FIGURE 18.6** Overbreak that develops in (a) horizontally layered rocks and (b) inclined rocks.

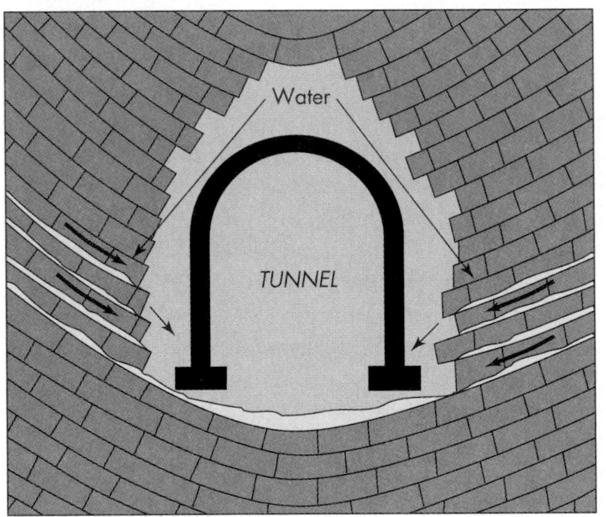

▲ **FIGURE 18.7** Possible groundwater hazard associated with tunnels driven through synclines.

in deep rock tunnels (12,13). Most problems in earth tunneling are compounded when groundwater saturates the material through which the tunnel is being driven. When it is saturated, sandy material may slowly flow and fill a tunnel or suddenly develop a hazardous condition known as a "blowout or boil," in which the material quickly fills the tunnel. Thorough evaluation of soil and hydrogeology is necessary when tunneling in soft ground (12,13).

***Large* Buildings** The geologist's role in evaluating a site for a large building is primarily to provide geologic information that can be used in designing the building's foundation. The site investigation includes drilling for soil or rock samples (Figure 18.8) to be evaluated for engineering properties, to help engineers anticipate geologic problems that might arise during construction of the foundation. Engineers also design buildings to withstand predicted shaking of the ground from earthquakes. The fact that field investigation is in most cases simple does not mean that detailed investigation can be routinely shortened (12,13).

18.4 Environmental Impact Analysis

The probable effects of human use of the land are generally referred to as the *environmental impact.* This term became popular in 1969 when the National Environmental Policy Act (NEPA) required that all major federal actions that could possibly affect the quality of the human environment be preceded by an evaluation of the project and its impact on the environment.

Environmental Impact Statements

To carry out both the letter and the spirit of NEPA, the Council on Environmental Quality set forth guidelines to help in preparing an **environmental impact statement (EIS).** The major components of the statement, according to the revised guidelines issued in 1979, are (14):

- A summary of the EIS
- A statement concerning the purpose and need for the project
- A rigorous comparison of the reasonable alternatives
- A succinct description of the environment of the area to be affected by the proposed project
- A discussion of the environmental consequences of the proposed project and of the alternatives. This must include direct and indirect effects; energy requirements and conservation potential; possible depletion of resources; impact on urban quality and cultural and/or historical resources; possible conflicts with state or local land-use plans, policies, and controls; and mitigation measures.

In addition, a list of the names and qualifications of the persons primarily responsible for preparing the environmental impact statement, a list of agencies to which the statement has been sent, and an index are also required.

Scoping The environmental impact statement process was criticized during the first 10 years under NEPA because it initiated a tremendous volume of paperwork by requiring

(a)

(b)

▲ **FIGURE 18.8** Drilling large-diameter holes as part of constructing a foundation for a large building at the University of California, Santa Barbara (a), and drilling to obtain soil and groundwater samples at the site of a leaking underground gasoline storage tank (b). (*Edward A. Keller*)

detailed reports that tended to obscure important issues. In response, the revised regulations (1979) introduced the idea of **scoping,** the process of identifying important environmental issues requiring detailed evaluation early in the planning of a proposed project. As part of this process, federal, state, and local agencies, as well as citizen groups and individuals, are asked to participate by identifying issues and alternatives that should be addressed as part of the environmental analysis.

Mitigation Another concept of importance to environmental impact analysis is that of **mitigation.** By definition, mitigation involves identification of actions that will avoid, lessen, or compensate for anticipated adverse environmental impacts of a particular project. For example, if a project involves filling in wetlands, a possible mitigation might be enhancement or creation of wetlands at another site.

Although NEPA requires agencies to consider the potential environmental impacts associated with a project, it does not necessarily require mitigation measures. As a result, an agency might simply ignore alternatives and mitigation measures recommended by an environmental impact statement and pursue actions that might cause environmental degradation (15). However, the potential for an agency to ignore the advice of an EIS is diminished by the fact that draft EISs are reviewed by other public agencies and by private environmental organizations, such as the Sierra Club and Audubon Society.

Mitigation is becoming a common feature of many environmental impact statements and state environmental impact reports. Unfortunately, it might be being overused. Sometimes no mitigations are possible for a particular environmental disruption. Furthermore, we often do not know enough about restoration and creation of habitats and environments such as wetlands to do a proper job. Requiring mitigation procedures may be useful in many instances, but must not be considered an across-the-board acceptable way to circumvent problems of adverse environmental impacts associated with a particular project.

State Environmental Impact Legislation

Following the passage of NEPA, it was apparent that there was a need for state and local governments to better control activities not covered by the federal legislation. To date, approximately one-half of the states have enacted some sort of environmental review process. Some states, including Hawaii and California, have enacted State Environmental Policy Acts (SEPAs), patterned after the federal legislation (15).

The California Environmental Quality Act (CEQA) of 1970 requires the completion of an environmental impact report (EIR) for a wide variety of projects that may affect the environment. In 1972 requirements for environmental review were extended to include all private as well as public projects in California. The California law provides more protection for the environment than does the federal legislation. This can even be noted to some extent in the terms used to describe the acts. The federal legislation considers "environmental policy," whereas the California state law is concerned directly with "environmental quality." The CEQA goes further than NEPA in that it may prohibit agencies from approving projects that will cause significant adverse environmental impacts when alternatives are not feasible and mitigation is not possible (16).

Negative and Mitigated Negative Declarations

Negative declarations are filed by an agency that determines that a particular project does not have a significant adverse impact on the environment. The negative declaration requires a statement that includes a description of the project and detailed information supporting the contention that the project will not have a significant effect on the environment. A negative declaration does not have to consider a wide variety of alternatives to the project, but it should be a complete and comprehensive statement concerning potential environmental problems (16).

A **mitigated negative declaration** may be filed when the initial study of a project suggests that significant environmental problems will probably occur but that the project could be modified in such a way as to reduce those effects to being nearly insignificant (15). An important step in the process of mitigated negative declaration is the assurance that the appropriate agency will implement the necessary mitigation measures. This involves the initiation of a monitoring program to ensure compliance to the agreed-upon mitigation measures (16).

Negative declarations and mitigated negative declarations are an important addition to the environmental review process. They provide a mechanism whereby projects that do not cause environmental disruption may avoid preparation of a lengthy environmental impact report. However, they do not avoid the important steps of public review and comment prior to adoption of the statement and approval of the project. Although both the language and the law associated with the concept of the negative declaration are different at the federal and state levels, the negative declaration is an important component of environmental assessment.

Methodology and Achievements of Environmental Impact Analysis

There is no accepted methodology for assessing the environmental impact of a particular project or action. Furthermore, because of the wide variety of topics, such as hunting of migratory birds or construction of a large reservoir, no one method of impact assessment is appropriate for all situations. What is important is that those responsible for preparing the statement strive to minimize personal bias and maximize objectivity. The analysis must be scientifically, technically, and legally defensible and prepared according to highly objective scientific inquiry standards (17).

Environmental impact analysis for major projects or actions generally benefits from the combined effort of a task force or a team of investigators, each of whom is assigned specific resource topics or disciplines. It is important to remember that the specific function of the task force is to evaluate the issue, not to decide it. The work of a task force is to provide information to enable those with authority to make equitable decisions (17).

The spirit of NEPA and SEPAs arose in response to the recognized need to consider the environmental consequences of an action before its implementation. However, environmental impact analysis has also been used to develop management plans for areas that developed problems before there were environmental regulations. The objective is to recognize potential conflicts and problem areas as early as possible, so as to minimize regrettable and expensive environmental deterioration. In general, the NEPA and SEPA processes have:

- Focused attention on potential environmental degradation
- Provided a framework for evaluation of environmental consequences of a project
- Greatly increased efforts to protect the environment in the United States

18.5 Case Histories: Uses of Environmental Impact Analysis

The case histories of Cape Hatteras National Seashore in North Carolina and the San Joaquin Valley in California demonstrate how environmental impact analysis has been used to develop corrective management policies for areas already experiencing adverse effects of development.

Cape Hatteras National Seashore

The Outer Banks of North Carolina has for generations been inhabited by people living and working in a marine-dominated environment. Until recently, their way of life depended on raising livestock, fishing, hunting, boat building, and other marine pursuits (18).

The landscape of the Outer Banks can change in a very short time in response to major storms such as hurricanes and northeasters that periodically strike the islands. Of the two types of storms, the more frequent northeasters probably cause the most erosion. On the other hand, infrequent hurricanes can trigger major changes, including extensive overwash and formation of new inlets. Historically, the people of the Outer Banks have philosophically lived with and adjusted to a changing landscape. In recent times, however, this philosophy has changed because of economic pressure to develop coastal property. A new philosophy of coastal protection has arisen in an attempt to stabilize the coastal environment. Stabilization, or a constant-appearing landscape, is a prerequisite to commercial development.

In 1937 Congress approved an act authorizing the Cape Hatteras National Seashore as the first national seashore. The seashore was finally established in 1953. The park consists of 115 km^2 along a 120-km portion of the Outer Banks (Figure 18.9) and includes portions of three islands of the more than 240 km of the Barrier Island system, which bounds and protects the largely undeveloped coastal plains lowland of North Carolina (18).

Eight unincorporated villages (Figure 18.10) are bounded by the Cape Hatteras National Seashore and are spaced along nearly the entire length of the shore. Legislation authorizing the park provided for the continued existence of these villages, including their beaches on the ocean side of the islands. This legislation has been interpreted by many as an obligation of the federal government to maintain and stabilize the beach access and frontage of these villages. This philosophy of stabilizing areas subject to change is contrary to the spirit of the early development of the islands, which was mainly on the protected inland side. However, following construction of the first artificial dune systems in the 1930s and opening of the road link in the 1950s, the villages began to spread toward the ocean, and this migration has grown at an ever-increasing pace (18). In 1953 there were only 75 oceanside subdivision lots, compared to more than 4000 today. Unfortunately, stabilization of the beaches has been difficult, and in some villages less than 60 m of beach remains because of recent coastal erosion (Figure 18.11).

It was once assumed that much of the Outer Banks had been heavily forested and that logging and overgrazing had

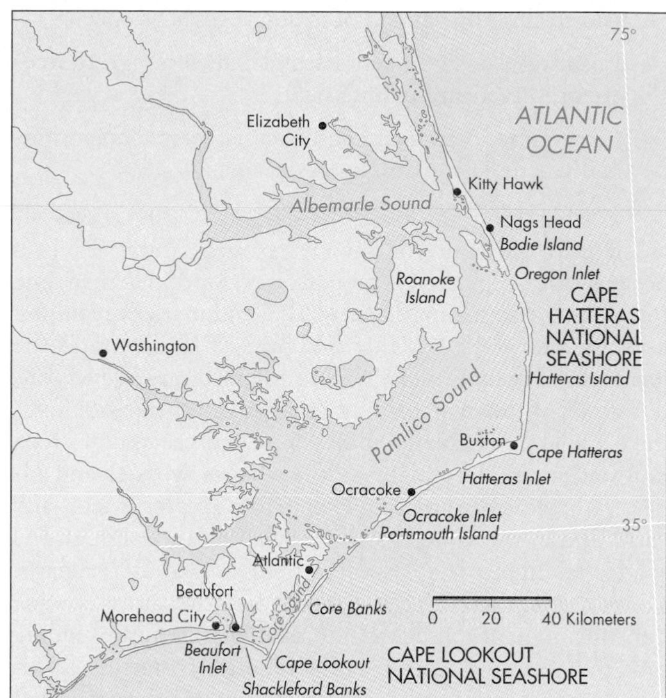

▲ **FIGURE 18.9** Map showing the Outer Banks of North Carolina and the Cape Hatteras National Seashore. (*After Godfrey and Godfrey.* 1971. *In* D. R. *Coates, ed.*, Publications in geomorphology. *Binghamton: State University of New York.*)

▲ **FIGURE 18.10** Some of these homes at Nags Head, North Carolina, are on the beach and obviously have a high potential for suffering from coastal erosion. Many of the homes on the beach at Nags Head were originally constructed with a considerable setback. In some instances homes now on the beach were in fact constructed as the second row back from the shoreline. This brings up an important point—even if a good setback is established at the time of construction, eventually erosion will catch up and the buildings may still be endangered by coastal erosion. (*Henley and Savage/Tony Stone Images*)

▲ **FIGURE 18.11** Rodanthe, North Carolina, during the passage of Hurricane Felix off the coast, August 17, 1995. Some of the houses at this location on the shoreline are virtually in the water. (E. *Robert Thieler, Duke University, Program for the Study of Developed Shorelines*)

destroyed the forests. Evidence for this is controversial, and extensive forests may have existed on only a few islands with an orientation that protected the forests from storms. The wooded areas were protected by a natural dune barrier, so the logical conclusion was to construct an artificial dune system to inhibit beach erosion and help the area return to a natural state. Of course, the dune line would also protect the roads and communities. An artificial dune system now extends along the entire Cape Hatteras National Seashore at an average dune height of 5 m and an average distance of 100 m from the ocean (18). This erosion-control measure, along with programs to artificially keep sand on the beach, could cost about $1 million annually if continued.

The expense of dune maintenance, along with a geological argument as to whether artificial dune lines will jeopardize the future of the Barrier Islands, has led to a controversy about how to best manage the park in order to maintain the natural environment for the use and enjoyment of future generations. The geologic controversy is about how dynamic earth processes create and maintain barrier islands. Although there are several known ways that coastal processes can develop barrier islands, debate on the processes needed to maintain them is heated. The Cape Hatteras Barrier Islands may have developed about 5000 years ago in response to complex wind and water processes associated with a slowly rising sea level (19). Although there is debate about this hypothesis, most investigators agree that the rise going on today is closely associated with tremendous changes in the character of the islands due to erosion of the coastline. This is a real and immediate hazard to future coastal development, which demands a static shoreline.

Debate about geologic processes that maintain the "naturally changing" Barrier Islands centers on the importance of periodic overwash of the frontal dune system by storm waves. If the overwash is important in maintaining the islands, building a dune line is contrary to natural processes, and the final result over a period of years may be deterioration of the islands as a natural system. This would be contrary to the National Park Service's stated objective of preserving the islands in a natural state. Some geologists have argued that barrier islands are maintained in natural landward migration by periodic overwash of frontal dunes or by inlet breaching that allows landward transportation of sediment. This argument is unconvincing, however. Most of the energy expended by storm waves is concentrated in the surf zone rather than splashing over low dunes. In fact, one recent study concluded that the present rise in sea level is causing erosion on both the seaward and landward sides of many U.S. Atlantic Coast barrier islands. Rather than migrating toward the land, these barrier islands are becoming narrower in response to rising sea level (20).

Frequent overwash may be atypical and occur only where islands are already in an unnatural state from overgrazing and subsequent lowering of dunes resulting from loss of protective vegetation. Nevertheless, overwash is a natural process that gives islands a more "natural appearing" coast and is probably significant in the migration history of some barrier islands. This does not mean that it is always bad to selectively control rapid coastal erosion by means of a protective frontal dune system. Proper placement of dunes and subsequent sound conservation practice in maintaining them remain a feasible solution to selected short-range coastal erosion problems, particularly near settlements and critical communication lines.

Faced with the problem of selecting a management policy for the Cape Hatteras National Seashore, the Park Service in 1974 presented five possible alternatives (Table 18.1) ranging from essentially no control of natural coastline processes (Alternative 1) to attempting complete protection (Alternative 5). Each alternative was analyzed to determine the entire spectrum of possible impacts. Few people want complete protection or complete lack of control of the seashore (18); the position of the Park Service is to attempt to strike a compromise in which the Barrier Islands for the most part may be preserved in a natural state while maintaining a transportation link with the mainland. Thus, residents in the island communities will have to live with and adjust to the dynamic high-risk environment in which they choose to live, as the people of the Outer Banks historically have done. In contrast to prevailing trends in coastal development that assume a more stable environment is desirable, this philosophy acknowledges that natural processes play a significant role in preserving the natural environment (18).

A general management plan for the Cape Hatteras National Seashore was completed in the 1980s after several years of careful environmental impact work and public review. The plan proposed several actions (21):

- Allowing natural seashore processes and dynamics to occur, except in instances when life, health, or significant cultural resources on the major transportation link are jeopardized
- Controlling use of off-road vehicles
- Expanding and improving access to recreation sites on the beach and sound
- Controlling the spread of exotic vegetation species
- Ensuring that significant natural and cultural resources are preserved and maintained
- Cooperating with state and local governments in mutually beneficial planning endeavors

For planning purposes, the national seashore was divided into four environmental resource units (ERUs): ocean/beach, vegetated sand flats, interior dunes/maritime forest, and marsh/sound. Table 18.2 summarizes planning objectives for each ERU, and Figure 18.12 illustrates the principle of management zoning for Hatteras Island. Implications of the management program are threefold: first, the Cape Hatteras National Seashore is preserved in a natural state; second, residents of the villages live with and adjust to the effects of natural events to a greater extent, and third, life in the villages will change, because though the road is maintained, it is subject to more periodic damage. For example, the Park Service no longer performs routine dune maintenance or beach nourishment (artifically adding sand to beaches). Dunes are only rebuilt in some cases, following significant erosion that threatens private property.

Ever since the management plan was implemented, it has been controversial. Coastal erosion has continued to be a serious problem in several areas in the 1990s. The primary reason for continued shoreline stabilization is to maintain the highway that extends along the entire national seashore. In 1995 the National Park Service began a 5-year multiagency investigation to explore and evaluate alternatives to long-term maintenance of the highway. Finally, the conflict between preserving land and allowing natural processes is at odds with protection of private property and public facilities, such as the famous Cape Hatteras Lighthouse, which was moved nearly 0.5 km inland during the summer of 1999 (see Chapter 9).

The San Joaquin Valley

The San Joaquin Valley in central California is one of the richest farm regions in the world. Farming on the western side of the valley has been made possible through extensive irrigation from deep wells and water transported by canals from the Sierra Nevada range to the agricultural area.

On the western side of the valley (Figure 18.13), soils have developed on alluvial fan deposits shed from the Coast Ranges. With an annual average precipitation of less than 25 cm, irrigation is necessary for agricultural productivity. A drainage problem exists on the western side because at shallow depths of 3 to 15 m there are often clay barriers that restrict the downward migration of water and salts. When irrigation waters are applied, these clay barriers produce a local zone of saturation near the surface that is well above the regional groundwater level (a condition known as *perched groundwater*). As more and more irrigation water is applied, the saturated zone may rise to the crop-root zone and drown the plants (22). To avoid this hazard, a subsurface drainage system may be installed to remove the excess water (Figure 18.14).

Table 18.1 Possible alternative management plans for the Cape Hatteras National Seashore

Alternative 1	
Management objectives:	To manage Cape Hatteras National Seashore as a primitive wilderness so as not to impede natural processes.
Proposed action:	Let nature take its course, and oppose any attempts to control natural forces or stabilize the shoreline.
Legislative changes:	Classify Cape Hatteras as a recreation area, and officially designate the Seashore a wilderness area by administrative decree.
Major impacts:	A natural environment would be restored; the existing highway could not be maintained; property within the villages would be lost; historic structures would be destroyed unless relocated; and the economic base and economic development of the village enclaves would decline.
Cost of alternative:	No direct federal costs except possible relocation of Cape Hatteras Lighthouse and other historic structures. State costs of maintaining the road would be excessive.
Alternative 2	
Management objectives:	To manage Cape Hatteras National Seashore in accordance with the policies established for natural areas of the National Park Service, and to preserve the area in a natural state.
Proposed action:	Develop an ongoing resource management program, revegetate major overwash areas following destructuve storms, and investigate alternative transportation modes.
Legislative changes:	Classify Cape Hatteras as a recreation area, and officially designate the Seashore a natural area by administrative decree.
Major impacts:	The Seashore would be preserved in a natural state, and residents of the villages would have to live with and adjust to natural events. Some property would be lost, and economic development of the villages would be retarded. Historic structures would be relocated or abandoned. The road would be damaged or washed out in places but could be maintained over the next several years.
Cost of alternative:	Estimated at $125,000 annually, or $3,125,000 over a 25-year period.
Alternative 3	
Management objectives:	Continue present shore erosion-control practices.
Proposed action:	Management practices of the past—dune building, dune repair, and beach nourishment would continue.
Legislative changes:	No changes would be required.
Major impacts:	The Seashore environment would become increasingly artificial as efforts were concentrated on protecting private development. The village enclaves would continue to grow and the highway would have to be widened to serve these areas.
Cost of alternative:	Estimated at $1 million annually, or $25 million over a 25-year period.
Alternative 4	
Management objectives:	To acquire threatened property within the village enclaves to preserve a public beach fronting the villages and avoid having to protect such property.
Proposed action:	Identify threatened private property, determine fair market value, and purchase the property.
Legislative changes:	Authorize legislation for Cape Hatteras to allow for the acquisition of additional land.
Major impacts:	The economic base of the village enclaves would be reduced and, as the shoreline continues to recede, future acquisition would be required at much greater cost. This alternative would establish the precedent of compensating unwise development.
Cost of alternative:	Very high—more than $25 million for Kinnakeet Township alone (1973 prices). The land involved is the costliest on the Seashore. Loss of tax revenue and revenue derived from visitor expenditures is not included.
Alternative 5	
Management objectives:	To protect the development taking place within the village enclaves and protect the highway running through the Seashore.
Proposed action:	Stabilize the shoreline fronting threatened property and protect exposed sections of highway by relocating or elevating the road or by stabilizing shoreline structures or building dunes.
Legislative changes:	Amend the authorizing legislation for Cape Hatteras, deleting the section pertaining to managing the area as a primitive wilderness.
Major impacts:	The natural setting of the Seashore would be destroyed and the resource base would be seriously degraded. The village enclaves would develop into urban complexes and dominate the landscape. This alternative would acknowledge a federal responsibility to subsidize unwise economic development.
Cost of alternative:	Overall initial cost of construction would be $40 million to $56 million with annual maintenance of $3.2 million to $6.4 million.

Source: National Park Service, 1974

Table 18.2 Planning objectives for Cape Hatteras National Seashore Environmental Resource Units (ERUs)

ERU	Characteristics	Planning Objectives
Ocean/Beach	Shifting sands, frequent overwash, limited vegetation on dunes	Allow natural processes to continue unhampered; allow for wide range of unstructured recreational activities by visitors; no construction allowed
Vegetated sand flats	Located between dune line and edge of saltwater marsh	Continue use as transportation corridor; allow development necessary to support visitor activities and resource protection
Interior dunes/ Maritime forest	Found in relatively few locations; variable topography, remoteness, dense vegetation	Maintain in natural state; allow passive recreation; design any construction to minimize impact on natural processes and systems
Marsh/Sound	Includes the sound, sound shore, and associated marshes	Maintain in natural state; provide limited access to the sound; allow limited development to support passive recreational activities

Source: National Park Service, 1984

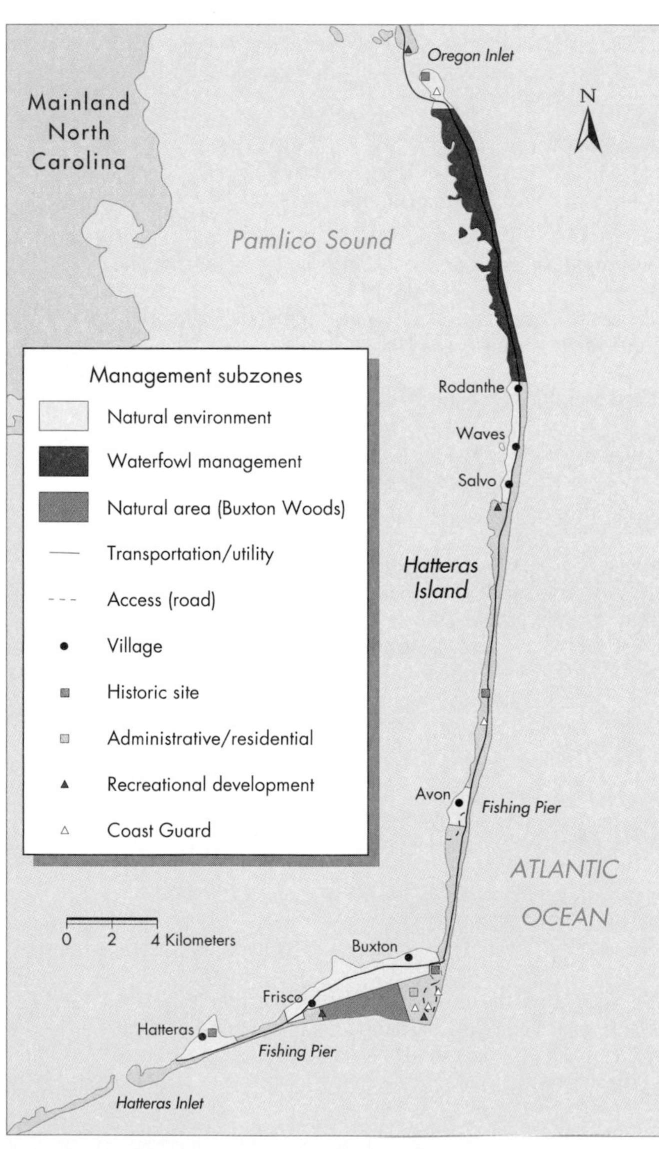

▲ **FIGURE 18.12** Proposed management subzones for part of the Cape Hatteras National Seashore. (*After National Park Service*, 1984).

Irrigation water contains a variable amount of dissolved salts, and as soil water evaporates and plants remove water through transpiration, the salts are left behind in the soil. Periodic leaching by applying large amounts of fresh water will remove the salts, but subsurface drains are necessary to remove the salty water (Figure 18.14). Tile drains (pipes with holes in them to let in the water) beneath each field connect to larger drains (ditches or canals) that eventually join a master drainage system, usually a canal.

The San Luis Drain (a 135-km-long concrete-lined canal) was constructed between 1968 and 1975 to convey the agricultural drainage water northward toward the San Francisco Bay Area. The drain to the bay was never completed because of limited funding and uncertainty over the potential adverse environmental effects of discharging drain water into the bay. As a result, Kesterson Reservoir (a series of 12 ponds with average depth of 1.3 m and total surface area of about 486 hectares) is the terminal point for the drainage water (Figure 18.15). The reservoir is part of the much larger Kesterson Wildlife Refuge.

During the first few years following construction of Kesterson Reservoir, the flow in the drain was primarily fresh water that was purchased to provide water for the Kesterson Wildlife Refuge, established primarily for waterfowl. The reservoir started receiving agricultural drainage in 1978, and by 1981 this was its main source of water. The canal and reservoir seemed like a good system to get rid of the wastewater from farming activities, but unfortunately, as the water infiltrated the soil, it picked up not only salts and agricultural chemicals but also the heavy metal selenium. The selenium is derived from weathering of sedimentary rock in the Coast Ranges on the western side of the valley. The sediments, including the selenium, are carried by streams to the San Joaquin Valley and deposited there.

In 1982 a United States Fish and Wildlife study revealed unusually high selenium levels in fish from Kesterson Reservoir. Dead and deformed chicks from water birds

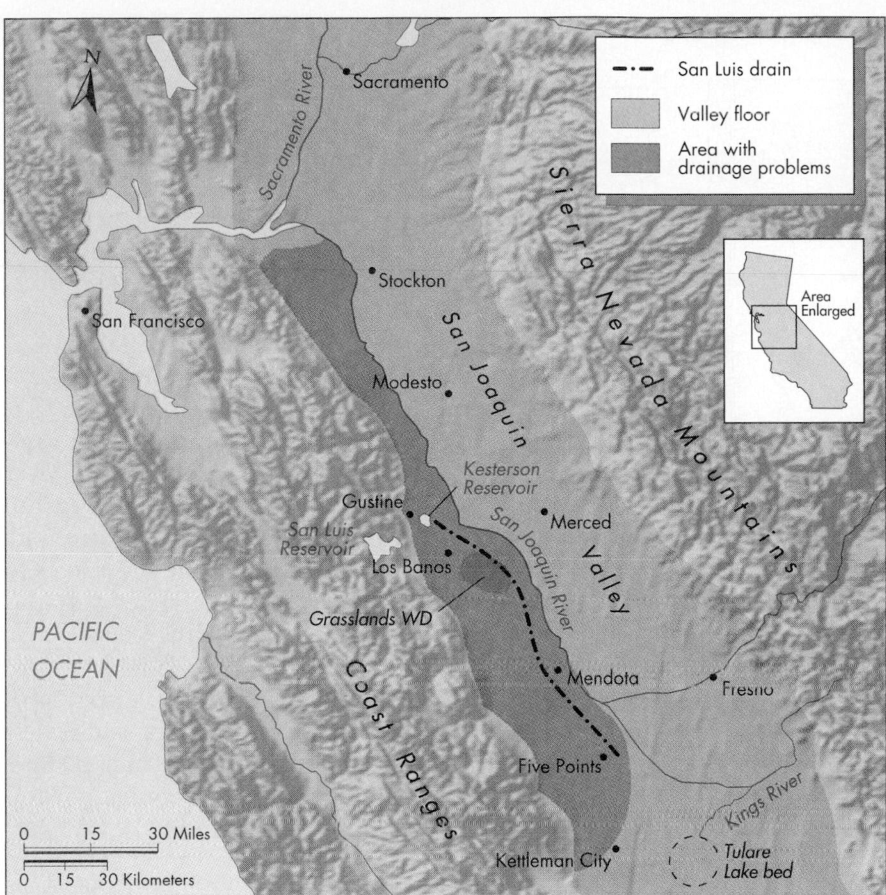

◀ **FIGURE 18.13** The San Joaquin Valley and San Luis Drain, which terminates at Kesterson Reservoir. (*Modified from K. Tanji, A. Lauchli, and J. Meyer.* 1986. *Selenium in the San Joaquin Valley.* Environment 28[6].)

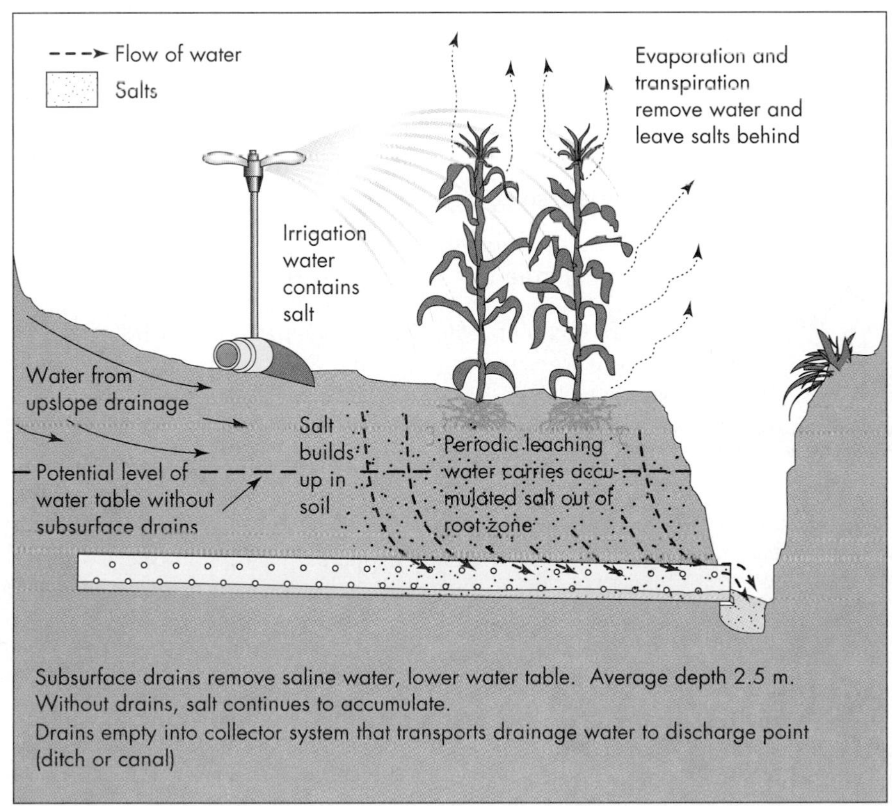

◀ **FIGURE 18.14** Diagram indicating the need for subsurface drains on the western side of the San Joaquin Valley. (*Modified after U.S. Department of the Interior, Bureau of Reclamation.* 1984. Drainage and salt disposal, Information Bulletin 1, *San Luis Unit, Central Valley Project, California.*)

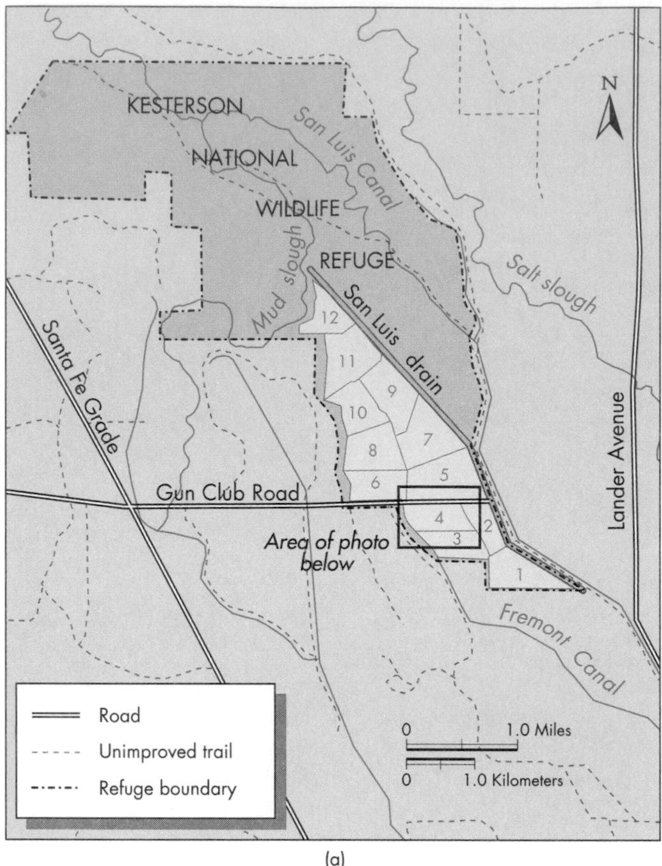

(a)

(b)

▲ **FIGURE 18.15** Map of Kesterson Reservoir ponds (a). (*After* U.S. *Department of* Interior. 1987. Kesterson program, *Fact Sheet* No. 4.) Aerial photograph of the site (b). Pond #4 is in the foreground, and at the time this photograph was taken in 1983, was filled with seleniferous agriculture drainage water. (*San Luis National Wildlife Refuge Complex*)

and other wildfowl were reported from 1983 to 1985. Analysis of water samples revealed selenium in the ponds ranged from 60 to 390 parts per billion (ppb). Selenium concentrations as high as 4000 ppb were measured in the agricultural drain waters. These concentrations of selenium are many times the drinking water standard of 10 ppb set by the Environmental Protection Agency. As a result of the studies it was inferred that selenium toxicity was the probable cause of the deaths and deformities of the wildfowl. The selenium was clearly shown to come from the San Luis Drain, and as the water spread out through the Kesterson Reservoir, the heavy metal was incorporated in increasingly higher concentrations from the water to the sediment, to the vegetation, and eventually to the wildlife (a process known as *biomagnification*).

In 1985 Kesterson wastewater was classified as hazardous and a threat to public health. At that time the California Water Resources Control Board ordered the U.S. Bureau of Reclamation, which was running the facility, to alleviate the hazardous condition at the reservoir. They were given 5 months to develop a cleanup plan and 3 years to comply with it (22). The immediate strategy was threefold: first, initiate a monitoring program; second, take steps to ensure that wildfowl would no longer use the wetlands (this was done by scaring off the birds); and third, reduce public exposure to the hazard.

The Kesterson case history is interesting because it provides insight into the environmental planning and review process. Primary environmental concerns and impacts were identified by the scoping process, which identified the following issues:

- Disposition of Kesterson Reservoir and the San Luis Drain following cleanup
- Environmental, social, and economic impacts of closing the Kesterson Reservoir to agricultural drain water discharge
- Potential public health impacts of selenium and other contaminants in the drain water
- Consideration of alternative methods to clean up the water, soils, sediment, and vegetation
- Potential for migration of contaminated groundwater away from the Kesterson area

In October 1986 the U.S. Department of Interior, Bureau of Reclamation, issued a final environmental impact statement that addressed the impacts of alternative methods to clean up Kesterson Reservoir and the San Luis Drain (23). In late March 1987 the California Water Resources Control Board ordered the Bureau of Reclamation to excavate all contaminated vegetation and soil and dispose of it on site in a monitored, lined, and capped landfill. The Board also required the Bureau to mitigate the loss of Kesterson Wildlife Refuge by providing alternative wetlands for wildlife. The disposal plan is estimated to cost as much as $50 million. Although the disposal plan may help solve the Kesterson problem, if the selenium problem is widespread in the western San Joaquin Valley (as it probably is) then a more general long-term solution is needed. People in the valley cannot afford to construct and maintain a series of toxic waste dumps for contaminated soil and vegetation. Research to evaluate the selenium problem is continuing, so the initial plans may change with time as more information is obtained.

At Kesterson the capping of ponds has been successful in reducing impacts on wildfowl. Monitoring of the site has been an ongoing project since the ponds were filled. Wetlands created through the process of mitigation are providing valuable wildlife habitat in the 1990s in an area where historically wetlands have been lost to other land uses. Selenium-laden agriculture drain water from the 17,000 ha of irrigated agriculture land in the area has been eliminated, but at a significant cost to farmers. Groundwater has risen in some fields as drainage was eliminated, threatening their productivity. Finally, the selenium problem in the broader San Joaquin Valley hasn't disappeared. Selenium-contaminated water from agriculture drains for fields not entering Kesterson continues to cause concern.

18.6 Land Use and Planning

Land use in the conterminous United States is dominated by agriculture and forestry, with only a small portion of the land (about 3%) used for urban purposes. Conversion of rural land to nonagricultural uses is currently several thousand square kilometers each year. About one-half of the conversion is for wilderness areas, parks, recreation areas, and wildlife refuges. The other half is for urban development, transportation networks, and facilities. Seen on a national scale, urbanization of rural lands may appear to be proceeding slowly. However, it often occurs in rapidly growing urban areas, where it may be viewed as destroying agricultural and natural lands and intensifying existing urban environmental problems. Urbanization in more remote areas with high scenic and recreation value is often viewed as potentially damaging to important ecosystems.

Scenic Resources

Scenery in the United States has been recognized as a natural resource since 1864, when the first state park, Yosemite Valley in California, was established. The early recognition of scenic resources was primarily concerned with outdoor recreation, focusing on preservation and management of discrete parcels of unique scenic landscapes. Public awareness of and concern for the scenic value of the "everyday" nonurban landscape is relatively recent. Society now fully recognizes scenery—even when it is not spectacular—as a valuable resource. There is a growing awareness that landscapes have varying degrees of scenic value, just as more tangible resources have varying degrees of economic value (24). Earth scientists, as members of a team evaluating the entire environment, assist in characterization of the landscape and its resources, including scenery.

Sequential Land Use

The need to use land near urban areas for a variety of human activities has in some instances led to application of the concept of sequential land use rather than permanent, exclusive use. The concept of sequential use of the land is consistent with the fundamental principle we discussed earlier: that effects of land use are cumulative, and therefore we have a responsibility to future generations. The basic idea is that after a particular activity (perhaps mining, or a landfill operation) has been completed, the land is reclaimed for another purpose (see *Case History: Butchart Gardens*).

There are several examples of sequential land use. Sanitary landfill sites have been planned so that when the site is completed, the land will be used for recreational purposes, such as a golf course. The city of Denver used abandoned sand and gravel pits once used for sanitary landfill as sites for a parking lot and the Denver Coliseum (Figure 18.16). Bay Harbor, Michigan, is a modern upscale community and world-class resort in Little Traverse Bay, near the north end of Lake Michigan. The harbor and other development (Figure 18.17) resulted from restoration of an abandoned shale quarry and cement plant. The restoration transformed a site that was a source of cement dust (an air pollutant) into a high-value landscape where people wish to visit. Finally, enormous underground limestone mines in the Missouri cities of Kansas City, Springfield, and Neosho

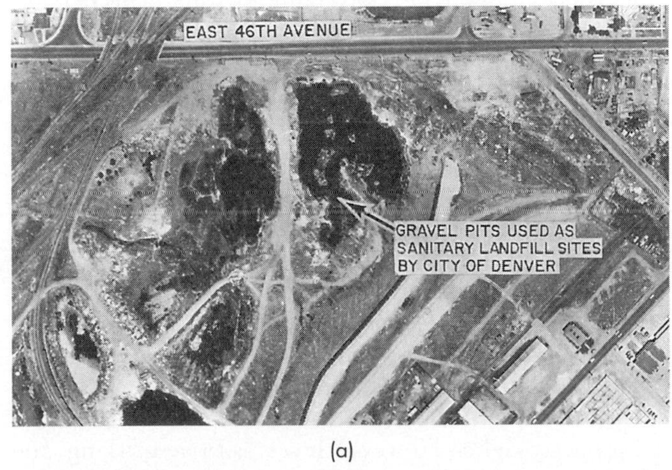

(a)

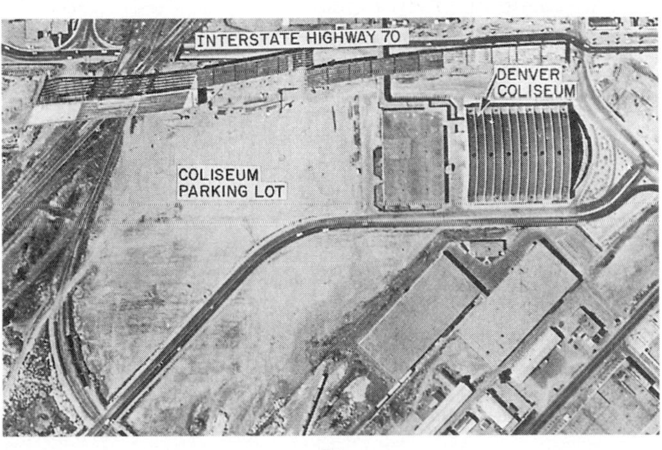

(b)

▲ **FIGURE 18.16** Example of sequential land use in Denver, Colorado. Gravel pits were used as a sanitary landfill site as long ago as 1948 (a). Today the Denver Coliseum and parking lots cover the landfill site (b). (*Courtesy of Missouri Geological Survey*)

(a)

(b)

▲ **FIGURE 18.17** Sequential land use at Little Traverse Bay, Lake Michigan, near Petoskey, Michigan, transformed an abandoned quarry and cement plant (a) (note quarry wall behind the plant and small building right of the access road) into a world-class resort (b). ((*a*) *Courtesy of Ned Tanner; and* (*b*) *courtesy of* Bob Fell)

have been profitably converted to warehousing and cold storage sites, offices, and manufacturing plants. Other possibilities also exist, such as use of abandoned surface mines for parking below shopping centers, chemical storage, and petroleum storage (25).

Land-Use Planning

Land-use planning is an important environmental issue. Sound land-use planning is essential for sound economic development, for avoiding conflicts between land uses, and for maintaining a high quality of life in our communities. It has been pointed out that when a business manages its capital and resources in an efficient manner, we call that good business. When a city or county efficiently manages its land and resources, we call that good planning (27). From an earth science perspective, the basic philosophy of good land-use planning is to avoid hazards, conserve natural resources, and generally protect the environment through the use of sound ecological principles. This involves understanding earth systems, as outlined in Chapter 2, and planning for an ever-changing environment.

The land-use planning process shown in Figure 18.18 includes several steps (28):

- Identify and define issues, problems, goals, and objectives
- Collect, analyze, and interpret data (including inventory of environmental resources and hazards)
- Develop and test alternatives
- Formulate land-use plans
- Review and adopt plans
- Implement plans
- Revise and amend plans

The role of the earth scientist in the planning process is most significant in the data-collection and analysis stage. Depending on the specific task or plan, the earth scientist will use available earth science information, collect necessary new data, prepare pertinent technical information, such as interpretive maps and texts, and assist in the preparation of land-capability maps that, ideally, should match the natural capability of a land unit to specific potential uses (29).

Comprehensive Planning

The **comprehensive plan** is an official document adopted by local government that states general and long-range policies on how the community will deal with future development (27). A comprehensive plan may sometimes be referred to by other, potentially misleading, names such as the older terms *master plan* or *general plan*. However, a comprehensive plan has a unique feature that the older types of official plan did not: the development of an environmental inventory (including geologic resources and hazards) as a basis for planning and zoning by local governments.

Only a few states in the Union have developed statewide planning programs requiring local comprehensive plans to be approved at the state level. Oregon has one of the more rigorous statewide programs. Oregon state law requires cities and counties to develop comprehensive plans that implement a set of 19 statewide goals developed by the Oregon Land Conservation and Development Commission (LCDC). One of these goals is concerned with citizen involvement in planning; another is to protect against natural hazards; another addresses the quality of air, water, and land resources. The Oregon Department of Land Conservation and Development (DLCD) reviews and approves local plans. All land must be planned and zoned on the basis of inventories of environmental resources and hazards, and cities and counties must prepare their plans and issue permits for development according to state goals and guidelines (27). State-mandated planning in Oregon has been labeled as innovative but controversial concerning individual property rights. However, it seems to be working, and other states such as Florida have adopted similar programs.

Some states, including California, have developed a system of recommending but not requiring comprehensive

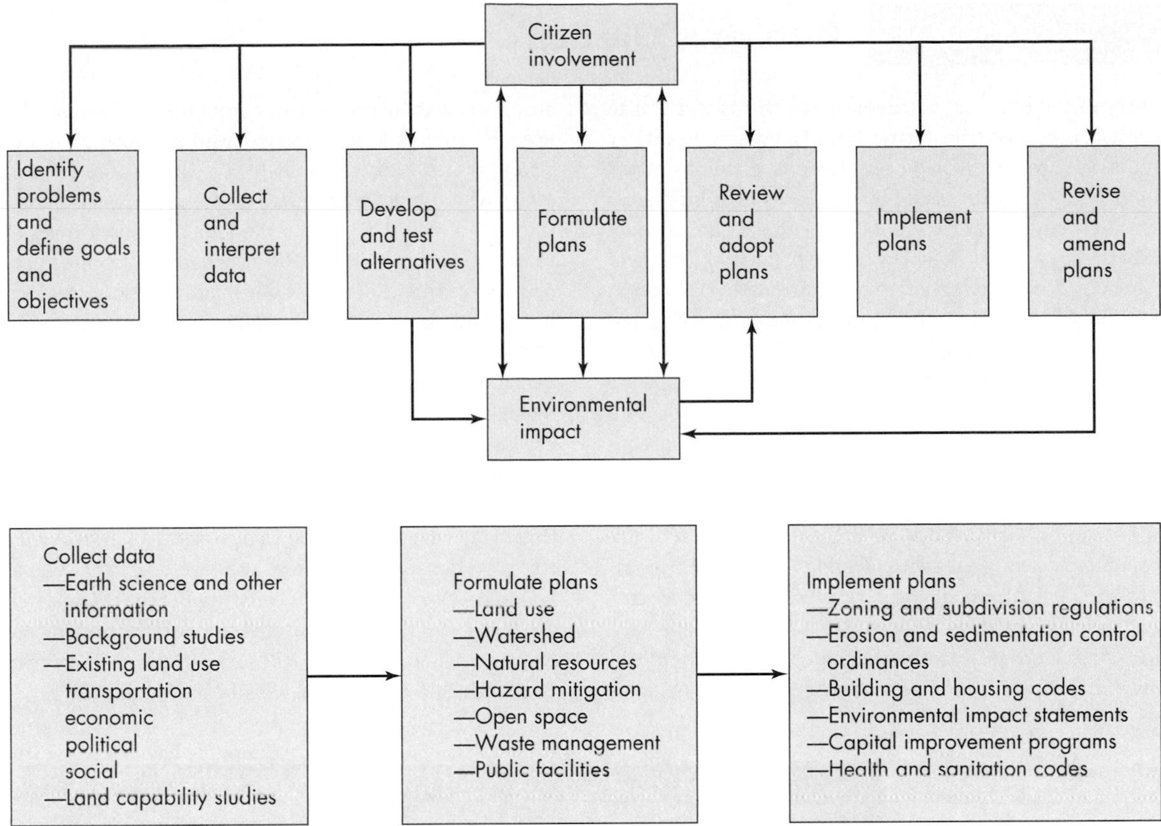

▲ **FIGURE 18.18** The land-use planning process. (*Modified after* U.S. *Geological Survey Circular* 721, 1976)

planning. Comprehensive plans for a city or county in California must address several elements identified by state law as being important (28). These required elements include land use, housing, conservation, open space, and safety (natural hazards). Cities and counties in California may add elements that they feel are relevant to their specific comprehensive plan. California also has a State Environmental Quality Act, and environmental review is an integral part of the planning process. In fact, California has recommended that local governments should consider combining the comprehensive plan and the environmental impact report on the plan to save both time and cost (28). This is a major difference from comprehensive planning in Oregon, where a commission approves the plans developed by cities and counties.

The three elements of most interest to earth scientists that must, by state law, be included in California comprehensive plans are land use, conservation, and safety. The land-use element is the heart of the plan. It serves as a statement of land-use policy that:

- Offers a guide for public investment in land, while protecting environmental resources and alerting planners and the public to environmental hazards
- Provides a useful framework for budgeting and planning construction of facilities, such as schools, roads, and sewer and water systems
- Helps coordinate regulatory policies and decisions

The land-use element contains maps that designate intensity and uses of the land for purposes such as housing, business, industry, open space, recreation, public facilities, waste-disposal sites, and other uses. What is shown on the land-use maps reflects the goals, objectives, and policies of the other elements in the comprehensive plan. Thus, the land classification maps of the land-use element are a very visible part of the comprehensive plan and often are the most-used part (28).

The conservation element of the comprehensive plan is intended to set goals and policies concerning development and use of natural resources such as water, soils, wildlife, forests, and mineral deposits (28). Because this element addresses so many different resources, it is crucial that it be carefully coordinated with a number of local, state, and federal agencies that deal with resources and their use.

The safety element is of particular interest to the earth scientist. The purpose is to establish goals, policies, and programs to protect people from natural hazards, such as earthquakes, landslides, floods, and wildfires (28). Included with this element are discussions of the various hazards, as well as maps that help delineate the hazards and areas where particular development should be avoided. This would include maps that show active faults, land subject to liquefaction from seismic shaking, landslides, and floodplains. The safety element also establishes evacuation routes, emergency procedures, and other issues related to disaster preparedness. This

CASE HISTORY Butchart Gardens

The story of the Butchart Gardens, Vancouver Island, British Columbia, is one of transformation of a limestone quarry that was an eyesore into world-class gardens visited by about one million people a year today. The story is also intimately related to the cement industry, and in particular the production of Portland cement. The process of producing Portland cement was developed in England in the year 1824. The process involves a fine grinding of a precise mixture of limestone and shale, which has been heated in a rotary kiln to near-fusion temperature. The resulting material is fine-ground again, placed in sacks and sold to consumers as Portland cement. The process of making cement and adding aggregate to make concrete is an integral part of building our urban environment. Robert Pim Butchart learned of the process of making Portland cement on his honeymoon in England in 1884. In 1902 Butchart became aware of the availability of limestone approximately 20 km north of the city of Victoria and determined the site was ideal for establishing a cement plant. As a result, Tod Inlet Cement Plant was started in 1904, and the first sacks of Portland cement were shipped in 1905. The limestone used to produce the cement was exhausted in 1908, leaving behind a large open excavation about 20 m deep, consisting of near-vertical walls of gray limestone. Butchart's wife Jenny, who initially had little experience with gardens, became fascinated with the site and the idea of transforming the quarry into a sunken garden. With the support of her husband and many workmen, she imported massive amounts of topsoil by horse and cart to form the beds for the garden on the floor of the quarry. The garden was completed in 1921 and quickly became a tourist attraction. The garden today (Figure 18.A) is much larger, including fountains, lakes, rose gardens, and a Japanese garden, among others, in addition to the sunken garden.

The history and story of Butchart Gardens is an interesting one because it shows the perseverance of one person (Jenny Butchart), who had a vision to transform an exhausted mine pit into a garden. Today the term we would use for such a project would be *mine reclamation*. Thus, the Butchart Gardens story is an early example that illustrates the potential of environmental restoration and sequential land use—in this case, transformation of a limestone quarry to a world-class garden (26).

FIGURE 18.A Butchart Gardens, Victoria Island, British Columbia, Canada, an example of sequential land-use and mine reclamation. (*Edward A. Keller*)

element may be expanded when necessary to include issues such as disposal of hazardous waste, transport of hazardous materials, and failure of utility services (28).

Land Management

Land management and land-use planning are related to one another in that management often follows development under the guidance of land-use plans. That is, once land has been designated for a particular use—as, for example, urban development or recreation—then management of that land follows. Before developing a management plan, we should first consider what actually needs to be managed. We can usually identify several kinds of environmental impact that will need to be monitored and managed, including:

- Impact of natural processes (for example, flooding, landslides, erosion)
- Impact of human use and interest on physical, chemical, and biological processes and on natural resources, such as soil, water, wildlife, and vegetation

We also need to define the goals of management. For example, if we are interested in urbanizing a watershed, management goals might include:

- Protection of human lives and property
- Protection of water quality and supplies
- Protection of wildlife resources
- Ecosystem protection
- An increase of public access and recreation potential to natural portions of the drainage basin

The overlying management approach for many areas is to try to understand the physical, biological, and hydrologic processes of the ecosystem. An important principle is that we must learn to manage complex systems that change frequently as a result of human-induced and natural disturbance. For example, if we are considering an urbanizing drainage basin, as discussed above, then some of the management approaches might be:

- Delineating floodplains, landslide-prone areas, and areas susceptible to excessive erosion
- Delineating water processes and sediment processes in the drainage basin
- Identifying sources of potential soil and water pollution
- Identifying useful functions of the drainage basins
- Identifying sensitive environmental areas
- Delineating upstream processes
- Understanding the role of extreme events

Integration of land-management goals with appropriate management approaches must include an understanding of the system being managed, linkages among physical, chemical, and biological processes, and potential impacts attributable to human use and interests in the area. For our urbanizing watershed, the management plan might include such elements as:

- Floodplain avoidance
- Control of sediment and storm-water runoff
- Development of pollution-abatement measures
- Protection of natural habitats
- Development of guidelines to allow for restoration following wildfires, high-magnitude storms, and other disturbances

Emergency Planning

In recent years, two types of local and regional planning have emerged. The first is planning of long-term projects in which engineering design and environmental impact are an integral consideration; this type includes land-use planning and comprehensive planning, which we have already discussed. The second type is **emergency planning** of short-term projects following catastrophic events such as hurricanes, floods, wildfires, or volcanic eruptions that cause widespread damage.

Emergency planning is in response to pressing needs and an influx of emergency money. Too often, the authorized work is either overzealous and beyond what is necessary, or is not carefully thought out. As a result, emergency projects can trigger further environmental disruptions instead of help. Acknowledgment of these problems does not imply that all emergency work is poor or unnecessary; nevertheless, when millions of dollars of emergency funds arrive, care must be taken to ensure that it is used for the best possible purposes. Emergency planning is important, and even if there is not enough time to prepare an environmental impact statement, all projects should be considered carefully to determine if they are really necessary and will not cause future problems. Another course of action is to have emergency plans in place before an emergency happens. That is, develop what actions should ensue following a wildfire, flood, or earthquake before it happens. Then when an emergency occurs, the plans are quickly implemented.

18.7 Environmental Law

Environmental planning requires implementation and enforcement, and **environmental law** is therefore becoming an important part of our jurisprudence. The many works devoted to environmental law attest to the fact that the subject is becoming increasingly popular among lawyers and environmental law societies, and environmental courses have even been established at law schools.

The Process of Law

It is beyond the scope of our discussion to consider in detail the process of law as it relates to the environment. Suffice it to say that law is a technique for the ordered accomplishment of economic, social, and political purposes, and the most desirable legal technique generally is one that most quickly allows ends to be reached. Law serves primarily the major interest that dominates the culture, and in our sophisticated society the major concerns are wealth and power. To these might also be added the quality of life—with the understanding that standard of living and quality of life are not the same and should not be confused. These concerns stress society's ability to use the resource base to produce goods and services, and the legal system provides the vehicle to ensure that productivity (30).

Some environmental lawyers today believe that the process of law as it has generally been practiced is not working satisfactorily when environmental issues are concerned. In general, when two views conflict, adversarial confrontation occurs. Emotional levels are often high, and it may be difficult for disagreeing parties to see positions other than their own. An emerging view in environmental law is one that stresses problem solving. This may take the form of **mediation** through **negotiation.** For example, in the 1970s the U.S. Environmental Protection Agency (EPA) often announced new environmental regulations that were thrust upon various sectors of society without warning. Not surprisingly, many of these regulations ended up in lawsuits and lengthy litigation. In the 1980s the EPA began a practice of consulting interested parties prior to regulation. Consultation, negotiation, and mediation may well prove much more successful than earlier strategies that produced unproductive adversarial reactions.

One of the major problems with mediation and negotiation is how to get all the important players in a given case to sit down and talk about the issues in a meaningful way. Whichever side seems to have the upper hand may try to make the negotiations particularly difficult for opposing views. Increasingly, however, both parties in environmental issues are recognizing that it is advantageous to work together toward a solution that is satisfactory for everyone. This is basically a collaborative process that seeks solutions favoring the environment while allowing activities and projects to go forward.

It is important to recognize that collaboration is different from and broader than compromise, which often requires giving something up to get something else. Collaboration is more comprehensive in that it includes—in fact, necessitates—the parties working together to create opportunities for mutual gain. It creates a climate of joint problem solving. What is required for negotiation and mediation to work is for both sides clearly and honestly to state their positions and then work to see where common ground might be found. Relationships built upon mutual trust are then developed. It should come as no surprise that almost all issues are negotiable, and alternatives can often be worked out that avoid or at least minimize costly litigation and lengthy delays.

Some Legal Case Histories

A discussion of particular case histories is useful for understanding some legal aspects of environmental problems. Our selection of cases does not imply a judgment of any particular activity but rather indicates the considerable variability and possibilities in environmental law. The cases we will discuss include areas of air pollution, aesthetic pollution, and land use.

The Ducktown Case It is fitting to return in this final chapter to the story of Ducktown, Tennessee, which we described in some detail in Chapter 1. The Ducktown story began in 1843 with what was thought to be a gold rush in Tennessee that turned out to be a copper rush. By 1855, approximately 30 companies in the Ducktown area were transporting copper ore by mule to a place called Copper Basin. Trees were cut from the surrounding area to fuel giant open-pit ovens—up to 200 m wide and 30 m deep—that were used to separate the copper from zinc, iron, and sulfur. Eventually an area of more than 100 km^2 was deforested to fuel these giant smelters. Deposition of the sulfur, combined with a wet climate, resulted in massive acid precipitation and acid dust deposition, killing the vegetation surrounding the smelters. This plus the deforestation led to massive soil erosion (see Figure 1.12). Adding insult to injury, grazing cattle consumed the plants that managed to survive (31).

Not surprisingly, people in the surrounding area were not pleased by the mining operations and pollution, which prevented them from using their farms and homes as they did before. In what was an early environmental case in 1904 (*Madison v. Ducktown Sulphur, Copper, and Iron Co.)*, the court chose to balance the equities in favor of the mining industry and declined to grant injunctive relief (an order to stop operations). The court reasoned that, because it was not possible for the mining industry to process the ores in a different way or move to a remote location, forcing companies to curtail air pollution would compel them to stop operating their plants. Such an order would cause 10,000 people to lose their jobs, destroy the tax base of the county, and make plant properties practically worthless (32). Although damages were awarded, mining operations were allowed to continue.

Another legal action began in 1907, when the state of Georgia sued the Tennessee Copper Company in an effort to halt the pollution from drifting across the nearby border and damaging Georgia's land. In 1915 the copper smelter was ordered to limit sulfur emissions to 20 tons per day during the summer months. This was a very significant reduction and might have caused the industry to flounder. During the case, the famous U.S. Supreme Court Justice Oliver Wendell Holmes argued that Georgia was well within its rights to decide whether degradation of its land and its air by pollution was acceptable. Limiting the emissions of sulfur dioxide led to development of a new technology that converted the sulfur in the smelting operations to a useful product (sulfuric acid). Ironically, the sulfuric acid became the main product that Tennessee Copper sold (31,33).

The legacy of mining with open-pit smelting in Tennessee was an area of approximately 140 km^2 that was essentially an eastern desert. Import of low-cost, high-grade copper ore from countries such as Chile, Peru, and Zambia put pressure on the mines, and when they finally closed in 1987, the economic future of the area was in trouble. Revegetation of the area devastated by the earlier smelting and refining activities started in the 1930s, and today approximately two-thirds of the area has some vegetation. Some people, however, want to leave part of the area as a desert. Because the landscape is unique in the eastern United States, there is hope of developing tourism as an industry (31).

The legal issues and court battles over copper mining, smelting, and refining in the Ducktown area have important historical meaning for the practice of environmental law. The cases involved states' rights, and they helped establish important aspects of the Balancing Doctrine, which asserts that the public benefit from, and the importance of, a particular action should be balanced against potential injury to certain individuals. The Supreme Court case in 1907 was one of the nation's first environmental-rights decisions, and the state of Georgia came out the winner (31,33).

The important lessons from the Ducktown experience will, it is hoped, alleviate or lessen similar activities in other parts of the world. Certainly, Ducktown is not unique. Refineries and smelters in the Sudbury, Ontario, region have damaged an area much larger than the Copper Basin. Similar activities are operating today in the Third World. Destruction of vegetation in countries such as Brazil and Mexico is occurring where environmental regulations are far behind industrial activities (31).

The Storm King Mountain Case The Storm King Mountain dispute is a classic example of conflict between a utility company and conservationists. In 1962 the Consolidated Edison Company of New York announced plans for a hydroelectric project approximately 64 km north of New York City in the Hudson River highlands, an area in the Appalachian Mountains considered by many to have unique aesthetic value. Nowhere else in the eastern United States is there a major river eroded through mountains at sea level, giving the effect of a fjord (34).

Early plans called for construction of an aboveground powerhouse, which would have required a deep cut into Storm King Mountain. The project was redesigned to site the powerhouse entirely underground, eliminating the cut on the mountain (Figure 18.19). Regardless, conservationists continued to oppose the project, and the issues broadened to include possible damage to fishery industries. The argument was that the high rate of water intake from the river, 883 m^3/sec, would draw many fish larvae into the plant, where they would be destroyed by turbulence and abrasion. The most valuable sport fish in the river is the striped bass, and one study showed that 25 to 75 percent of the annual bass hatch might be destroyed if the plant were operating. The fish return from the ocean to tidal water to spawn, and because the Hudson River is the only estuary north of the Chesapeake Bay where the striped bass spawn, concern for the safety of the fisheries was justified. The problem was even more severe because the proposed plant was to be located near the lower end of the 13-km reach in which the fish spawn (34).

The Storm King Mountain controversy is interesting because it emphasizes the difficulty of making decisions

(a)

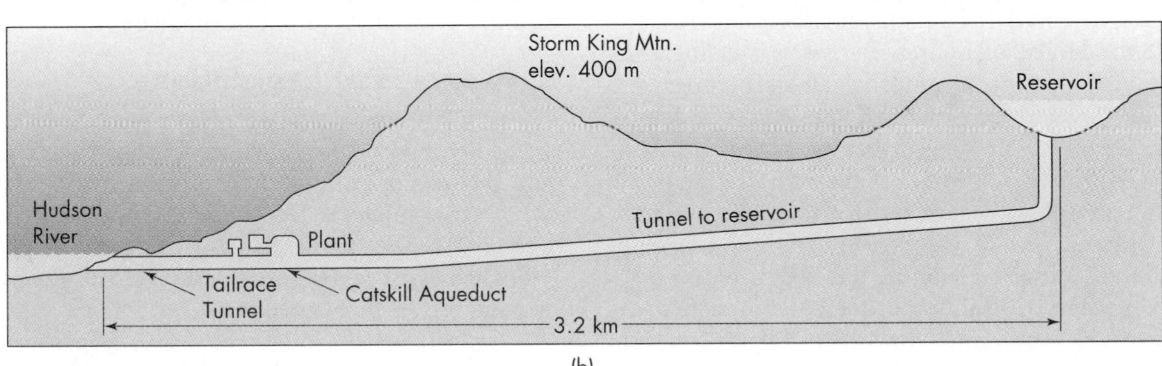

(b)

▲ **FIGURE 18.19** (a) Storm King Mountain and the Hudson River highlands, New York. (*Foto by Joe Deutsch*) (b) Diagram showing how the entire Storm King Mountain hydroelectric project might be placed underground. (*After L. J. Carter,* Science 184 (*June 1974*), *pp.* 1353–58. *Copyright* 1974, *American Association for the Advancement of Science.*)

about multidimensional issues. On the one hand, a utility company was trying to survive in New York City, where unbelievably high peak-power demands are accompanied by high labor and maintenance costs. On the other hand, conservationists were fighting to preserve a beautiful landscape and fishery resources. Both sides had legitimate arguments, but in light of their special interests, it was difficult to resolve the conflict. Existing law and procedures were sufficient to resolve the issues, but trade-offs were necessary. Ultimately, an economic and environmental price must be paid for any decision, a price that reflects our desired life-style and standard of living.

The first lawsuit in the Storm King dispute was filed in 1965, and following 16 years of intense courtroom battles, the dispute was settled in 1981. The total paper trail exceeded 20,000 pages, and in the end the various parties used an outside mediator to help them settle their differences. This famous case has been cited as a major victory for environmentalists, but the real question is: Could the outcome have been decided much earlier? Had the various sides been able to sit down and talk about the issues openly, then the process of negotiation and mediation might have been completed much earlier and at much less cost to the individual parties and society in general (34). This case emphasizes that environmental law issues may best be approached from a problem-solving viewpoint rather than by the sides taking adversarial positions.

▲ **FIGURE 18.20** Florissant Fossil Beds National Monument, Colorado. The fossils shown are beach leaves approximately 35 million years old. Fossils such as these provide evidence of what life may have been like during that important period that occurred 30 million years before humans appeared on earth. (*Fred Mang, Jr./Wind Cave National Park*)

The Florissant Fossil Beds Case A Colorado case emphasizes the significance of the Trust Doctrine and the Ninth Amendment as they pertain to land management (36). The conflict surrounded the use of 7.3 km^2 of land near Colorado Springs, which is part of the Florissant Fossil Beds, where insect bodies, seeds, leaves, and plants were deposited in an ancient lake bed about 30 million years ago (Figure 18.20). Today these fossils are remarkably preserved in thin layers of volcanic shale. They are delicate, however, and unless they are protected, they tend to disintegrate when exposed. Many people consider them unique and irreplaceable. At the time of the controversy, a bill had been introduced in Congress to establish a Florissant Fossil Beds National Monument. The bill passed the Senate, but the House of Representatives had not yet acted on it (36).

While the House of Representatives was deliberating the bill, a land development company that had contracted to purchase and develop recreational homesites on 7.3 km^2 of the ancient lake bed announced that it was going to bulldoze a road through a portion of the proposed national monument site to gain access to the property it wished to develop. A citizens' group was formed to fight the development until the House acted on the bill. The group tried to obtain a temporary restraining order, which was first denied because no law prevented the owner of a property from using that land in any way he wished, provided that existing laws were upheld.

The conservationists then argued before an appeals court that the fossils were subject to protection under the Trust Doctrine (the idea that land may be held in trust for the public and future generations) and the Ninth Amendment of the U.S. Constitution. Their argument was that protection of an irreplaceable, unique fossil resource was an unwritten right retained by the people under the Ninth Amendment, and that furthermore, since the property had tremendous public interest, it was also protected by the Trust Doctrine. An analogy used by the plaintiffs was that if a property owner were to find the Constitution of the United States buried on the land and wanted to use it to mop the floor, certainly that person would be restrained. After several more hearings on the case, the court issued a restraining order to halt development; shortly thereafter, legislation to establish a national monument was passed by Congress and signed by the president (36).

The court order prohibiting destruction of the fossil beds may have deprived a landowner of making the most profitable use of the property, but it does not prohibit all uses consistent with protecting the fossils. For instance, the property owners are free to develop the land for tourism or scientific research. Although this might not result in the largest possible return on the property owner's investment, it probably would return a reasonable profit (35).

Environmental Legislation

The ultimate goal of those who are concerned with how we treat and use our natural environment and resources is to ensure that ecologically sound, responsible, socially acceptable, economically possible, and politically feasible legislation is passed (36). To attain this, we need professional people to assist legislators at all levels of government in drafting the needed laws and regulations.

Environmental legislation has already had a tremendous impact on the industrial community. New standards in regulations limiting the discharge of possible pollutants into the environment have placed restrictions on industrial activity. Furthermore, any new activity that directly or indirectly involves the federal government must be preceded by an evaluation of the environmental impact of the proposed activity. Beyond this, federal legislation has set an example that many states have followed in passing environmental protection legislation.

We have already mentioned the most significant piece of federal environmental legislation—the **National Environmental Policy Act (NEPA),** which was passed in 1969 and amended in 1979. The philosophical purposes of NEPA are:

- To declare a national policy that will encourage harmony with the physical environment
- To promote efforts preventing or eliminating environmental degradation, thereby stimulating human health and welfare
- To improve our understanding of relations between ecological systems and important natural resources

The act establishes the Council on Environmental Quality, whose function is to promote interest and research concerning these goals, and to confer the necessary authority for achieving them. The council is in the Executive Office of the President and is responsible for preparation of a yearly Environmental Quality Report to the Nation. It also provides advice and assistance to the President on environmental policies.

The most significant aspect of the act is that it requires an environmental impact statement (EIS) before major federal actions are taken that could significantly affect the quality of the human environment. (We discussed the contents of EISs in Section 18.4, Environmental Impact Analysis.) This requirement extends to such activities as construction of nuclear facilities, airports, federally assisted highways, electric power plants, and bridges; release of pollutants into navigable waters and their tributaries; and resource development on federal lands, including mining leases, drilling permits, and other uses. Ever since enactment of NEPA, many thousands of environmental impact statements have been prepared.

Land-Use Planning and Law

Few environmental topics are as controversial as land-use planning legislation. The controversy results from several factors. First, unlike air and water pollution that can be measured, evaluated, and possibly corrected, it is very difficult to determine the "highest and best use" of the natural environment as opposed to the "most profitable use." Second, landowners fear that land-use planning will take away their right to decide what to do with their property; that is, individual ownership of property could be converted to a social property ownership in which the individual would have a caretaker role.

Americans greatly value private ownership of land, so a law that requires rural landowners to use their properties in very restricted ways is not always popular. Some people argue that private ownership of property is a sacred right and that only in countries where private ownership is permitted are there also other personal rights. Everything considered, it is extremely unlikely that private ownership in the United States will be abolished. On the other hand, it is increasingly apparent that private ownership of land does not mean the owner has the right to deliberately or inadvertently degrade the environment. Sustainable development requires that we maintain a quality environment for future generations. As a result, we need to realize that we never really own the land but, rather, have custodial rights for varying periods of time. Finally, to be effective, land-use planning should leave the property owner (custodian) alternatives of land use from which to choose freely.

It seems inevitable that some form of federal land-use planning act will eventually become law, with the purpose of assuring that all the land in the nation will be used in a way that facilitates harmonious existence with our natural environment, so that environmental, economic, social, and other requirements of present or future generations are not degraded. The argument in favor of a federal land-use planning act stresses that such planning is urgently needed to minimize degradation of land resources of statewide or national importance. Those in favor of such an act point out that land use is the most important aspect of environmental quality control that remains to be stated as national policy (37).

Many states have implemented some form of land-use legislation. The most common programs include comprehensive planning, coastal zone management, wetlands management, siting of power plants, surface mining regulation, and floodplain management. However, only a few states have enacted mandatory local planning.

SUMMARY

A primary role of the earth scientist in landscape evaluation is to provide information and analysis before planning, design, and construction. An important contribution of earth scientists is emphasizing that all land is not the same, and that particular physical and chemical characteristics of the land can greatly affect appropriate land use.

Environmental geology mapping often involves the preparation of interpretive maps that show suitability for a particular land use or the nature and extent of geologic hazards, such as flooding, landslide, and seismic risk. Environmental resource units (ERUs), portions of the environment with shared structural and biological characteristics, are used as the basis for environmental maps that include geologic, hydrologic, and biologic information. The Geographic Information System (GIS) is an emerging technology capable of storing, retrieving, transforming, and displaying environmental data. An important aspect of the GIS is that it has the ability to manipulate data and create new products, such as a land capability map or hazard map. Geographic Information Systems are being widely used in a variety of fields related to land-use planning.

Site selection is the process of evaluating the physical environment to determine its capability of supporting human activities, as well as the possible effects of the proposed activities on the environment. Two approaches to site selection are cost-benefits analysis and physiographic determinism. Cost-benefits analysis assesses the proposed use of a site by comparing the monetary value of the benefits to the cost of the project. Physiographic determinism, which attempts to let natural characteristics determine the choice of a site, is innovative in considering social and aesthetic as well as economic factors but has been criticized as being subjective in determining which factors are objectionable or desirable. Site evaluation for engineering purposes—such as construction of dams, highways, airports, tunnels, and large buildings—requires careful geologic study before planning and designing the project. The role of the geologist is to work with engineers and indicate possible adverse or advantageous geologic conditions that might affect the project.

Probable effects of human use of the land are termed *environmental impact.* The National Environmental Policy Act (NEPA) requires preparation of an environmental impact statement (EIS) for all major federal actions that could have a significant impact on the quality of the human environment. Major components of the EIS are a declaration of the purpose and need for the project, a comparison of reasonable alternatives, a description of the environment to be affected, and discussion of the environmental consequences of the project and of potential alternatives. Scoping and mitigation are important processes in the analysis outlined by the EPA. Scoping is early identification of important issues that will require detailed evaluation in the planning of a project. Mitigation is identification of actions that will avoid, lessen, or compensate for adverse environmental impacts of the project. Negative and mitigated negative declarations are issued for projects having no negative impact and projects whose negative impact can be eliminated by mitigation measures. Some states have additional requirements or recommendations for environmental impact analysis.

Because no one methodology can be established for the diverse situations subject to environmental impact analysis, it is important to recognize the intent of the federal and state guidelines, which is to identify potential problems of an activity early so as to minimize adverse environmental consequences. Environmental impact analysis can also be used to mitigate the adverse effects of existing situations, as demonstrated by the history of Cape Hatteras National Seashore and of California's San Joaquin Valley. The federal and state guidelines have focused public attention on environmental problems, provided a framework for project evaluation, and greatly increased efforts to protect the environment.

Land use is an important environmental issue. Urbanization of rural areas may destroy agricultural land, intensify existing urban environmental problems, or damage ecosystems with high scenic or recreational values. Scenery is increasingly recognized as a valuable resource. The limited land available for expansion of urban areas has led to the concept of sequential land use rather than permanent consignment of land to a single use. Land-use planning is essential for sound economic development and maintaining a high quality of life. The basic philosophy of good land-use planning, from an earth science perspective, is to plan to avoid hazards, conserve natural resources, and generally protect the environment through the use of sound ecological principles. The land-use planning process includes several steps: identification of issues, problems, goals, and objectives; collection and analysis of data; development of alternatives; formulation of land-use plans; review and adoption of plans; implementation of plans; and procedures to revise and amend plans.

A comprehensive plan is one designed for long-range local development based on an environmental inventory of resources and hazards. A few states have statewide planning programs that require communities to have comprehensive plans in accord with state environmental policies. Land management and land-use planning are related to one another, because management often follows development of land-use plans. The overlying management approach in many instances is to attempt to understand the physical, biological, and hydrologic processes of the ecosystem to identify the environmental impacts that need to be managed. An important principle is that we must learn to manage complex systems that change frequently as a result of natural disturbance and human processes.

Emergency planning differs from other local and regional planning in that it follows catastrophic events that cause widespread damage. Emergency planning is an important endeavor, but its projects must be considered carefully to determine whether they are necessary and to ensure that they will not cause future problems.

The term *environmental law* has gained common usage, and the subject is an important part of law. Topics of particular concern to individuals and society are degradation of air, water, scenery, and other natural resources. The process of environmental law is undergoing some important changes. There is a move toward problem solving and mediation rather than confrontation and adversarial position when dealing with environmental matters. When negotiation replaces inflexibility, progress in solving problems is enhanced. Early in this century, court cases established the Balancing Doctrine, which asserts that public benefit of an action must be weighed against its potential to injure certain individuals; this was a milestone in the establishment of environmental rights. Other cases have established that environmental resources are subject to protection under the Trust Doctrine, which states that land may be held in trust for the public and future generations.

Finally, environmental legislation is having a tremendous impact on society. Perhaps the most controversial and potentially significant legislation is associated with land-use planning. Conflicts arise because it is often difficult to determine the best use of land and because landowners fear that land-use planning will cause them to lose control over their property. The argument in favor of land-use planning legislation stresses that planning is urgently needed to control degradation of the land and that land use is the one remaining important aspect of environmental quality control that has not yet been included in national policy statements.

REFERENCES

1. **Bartelli, L. J., Klingebiel, A. A., Baird, J. V., and Heddleson, M. R., eds.** 1966. *Soil surveys and land-use planning.* Soil Science Society of America and American Society of Agronomy.
2. **Turner, A. K., and Coffman, D. M.** 1973. Geology for planning: A review of environmental geology. *Quarterly of the Colorado School of Mines* 68.
3. **Gross, D. L.** 1970. Geology for planning in DeKalb County, Illinois. *Environmental Geology Notes No. 33.* Springfield, IL: Illinois State Geological Survey.
4. **Wilson, L.** 1981. Potential for ground-water pollution in New Mexico. In *Environmental geology and hydrology in New Mexico,* eds. S. G. Wells and W. Lambert. New Mexico Geological Society Special Publication No. 10, pp. 47–54.
5. **Parker, H. D.** 1987. What is a geographic information system? In *GIS '87: Second annual international conference, exhibits and workshops on Geographic Information Systems.* American Society for Photogrammetry and Remote Sensing, pp. 72–79.
6. **Star, J., and Estes, J.** 1990. *Geographic Information Systems.* Englewood Cliffs, NJ: Prentice-Hall.
7. **Lynch, K.** 1962. *Site planning.* Cambridge, MA: MIT Press.
8. **Prest, A. R., and Turvey, R.** 1965. Cost-benefit analysis: A survey. *The Economic Journal* 75:683–735.
9. **McHarg, I. L.** 1971. *Design with nature.* Garden City, NY: Doubleday.
10. **Whyte, W. H.** 1968. *The last landscape.* Garden City, NY: Doubleday.
11. **Flawn, P. T.** 1970. *Environmental geology.* New York: Harper & Row.
12. **Schultz, J. R., and Cleaves, A. B.** 1955. *Geology in engineering.* New York: John Wiley & Sons.
13. **Krynine, D. P., and Judd, W. R.** 1957. *Principles of engineering geology and geotechniques.* New York: McGraw-Hill.
14. **Council on Environmental Quality.** 1979. *Environmental quality.* Annual Report. Washington, DC: U.S. Government Printing Office.
15. **Callies, D. L.** 1984. *Regulating paradise.* Honolulu: University of Hawaii Press.
16. **Remy, M. H., Thomas, T. A., and Moose, J. G.** 1991. *Guide to the California Environmental Quality Act,* 5th ed. Point Arena, CA: Solano Press Books.
17. **Brew, D. A.** 1974. *Environmental impact analysis: The example of the proposed Trans-Alaska Pipeline.* U.S. Geological Survey Circular 695.
18. **National Park Service.** 1974. *Cape Hatteras Shoreline Erosion Policy Statement.* Washington, DC: U.S. Department of the Interior.
19. **Hoyt, J. H., and Henry, V. J., Jr.** 1971. Origin of capes and shoals along the southeastern coast of the United States. *Geological Society of America Bulletin* 82:59–66.
20. **Leatherman, S. P.** 1983. Barrier dynamics and landward migration with Holocene sea-level rise. *Nature* 301(5899): 415–18.
21. **National Park Service.** 1984. *General management plan, development concept plan, and amended environmental assessment, Cape Hatteras National Seashore.* Washington, DC: U.S. Government Printing Office.
22. **Tanji, K., Lauchli, A., and Meyer, J.** 1986. Selenium in the San Joaquin Valley. *Environment* 28(6):6–11, 34–39.
23. **U.S. Department of the Interior, Bureau of Reclamation.** 1987. *Kesterson Program.* Fact Sheet No. 4.
24. **Zube, E. H.** 1973. Scenery as a natural resource. *Landscape Architecture* 63:126–32.
25. **Hayes, W. C., and Vineyard, J. D.** 1969. *Environmental geology in town and country.* Missouri Geological Survey and Water Resources, Educational Series No. 2.
26. **Anonymous.** 1998. *The Butchart Gardens.* Victoria, British Columbia: The Butchart Gardens Ltd.
27. **Rohse, M.** 1987. *Land-use planning in Oregon.* Corvalis, OR: Oregon State University Press.
28. **Curtin, D. J., Jr.** 1991. *California land-use and planning law,* 11th ed. Point Arena, CA: Solano Press Books.
29. **William Spangle and Associates; F. Beach Leighton and Associates; and Baxter, McDonald and Company.** 1976. *Earth-science information in land-use planning—Guidelines for earth scientists and planners.* U.S. Geological Survey Circular 721.
30. **Murphy, E. F.** 1971. *Man and his environment: Law.* New York: Harper & Row.
31. **Barnhardt, W.** 1987. The death of Ducktown. *Discover,* October, 35–43.
32. **Juergensmeyer, J. C.** 1970. Control of air pollution through the assertion of private rights. *Environmental Law,* pp. 17–46. Greenvale, NY: Research and Documentation Corp.
33. **Rodgers, W. H., Jr.** 1977. *Handbook on environmental law.* St. Paul, MN: West Publishing.
34. **Carter, L. J.** 1974. Con Edison: Endless Storm King dispute adds to its troubles. *Science* 184:1353–58.
35. **Bacow, L. S., and Wheeler, M.** 1984. *Environmental dispute resolution.* New York: Plenum Press.
36. **Yannacone, V. J., Jr., Cohen, B. S., and Davison, S. G.** 1972. *Environmental rights and remedies 1.* San Francisco: Bancroft-Whitney.
37. **Healy, M. R.** 1974. National land use proposal: Land use legislation of landmark environmental significance. *Environmental Affairs* 3:355–95.

KEY TERMS

environmental geology map (p. 494)
environmental resource unit (ERU) (p.495)
Geographic Information System (GIS) (p. 495)
site selection (p. 499)
cost-benefits analysis (p. 499)
physiographic determinism (p. 500)
environmental impact statement (EIS) (p. 502)
scoping (p. 503)
mitigation (p. 503)
negative declaration (p. 503)
mitigated negative declaration (p. 504)
land-use planning (p. 512)
comprehensive plan (p. 512)
emergency planning (p. 515)
environmental law (p. 515)
mediation (p. 515)
negotiation (p. 515)
National Environmental Policy Act (NEPA) (p. 519)

SOME QUESTIONS TO THINK ABOUT

1. From your geology department or library, examine geologic and topographic maps of the county in which you live. What environmental resource units (ERUs) are found in your county? How could you incorporate the mapping of these environmental resource units into producing environmental geology maps, as, for example, locating sites for waste disposal or mapping hazards such as flooding or landslides?
2. The college or university you are attending is trying to find a site of approximately one acre on or adjacent to the campus for future development of an academic center. Complete a survey of your campus and surrounding area and make a recommendation as to where the new buildings should go. What criteria did you use in your decision-making process? What values are involved in your decision? Could you use a Geographic Information System to help in the analysis?
3. Keeping to our theme of adding the additional buildings to your campus, consider the process of scoping as part of the environmental impact analysis. That is, identify the important environmental issues that will require detailed evaluation early in the planning of the proposed project. Could some of the potential adverse environmental impacts be mitigated? If so, how?
4. Let us continue to push this idea of expanding your campus. Get together with a couple of fellow students and do some role-playing. One of you might play the role of the developer or the people in favor of the development while another takes the role of an environmentalist opposed to the project. Still, a third person could assume the role of a lawyer or mediator trying to find common ground for negotiation. Each of you take a few minutes to present your case and then try to come to some sort of negotiated settlement. What factors are necessary for such a settlement to occur?

APPENDIX A • World Wide Web: An Introductory Statement

Today is the information age! The World Wide Web (WWW) contains a large number of Web sites from which information may be obtained. You may be writing a term paper for your geology course, or you may be interested in obtaining further information on a particular subject such as earthquakes or climate change and wish to use information taken from the Web. Also, there is a trend for scientists to release the results of their research on the WWW. It is important to keep in mind that some of the documents that are in electronic format from the Web may be a from published paper and will remain unchanged in the future, whereas others may be periodically modified. Some information on the WWW is of an ephemeral nature—about a "hot topic" or event that has occurred—whereas other information may be data from continuous monitoring of activities, such as the flow of water in rivers or gases emitted from a volcanic area. An important question is: How reliable is data taken from the WWW? This of course depends upon the source of the data, and I would recommend that you primarily use sites from major government agencies, whose task it is to report information to the public, or from other sources that have a long history of providing reliable information, such as PBS or National Geographic.

Another common question that arises when using the WWW is: How do you cite Web site information in your report or paper? This subject was recently discussed by the U.S. Geological Survey's Water Resources Division in Memorandum no. 98.31, transmitted on July 21, 1998. Suggested formats for citing documents on the WWW are similar to those used to cite reports and other published data. However, reports and data do not yet have a standard format for the information they contain. In the list that follows, some of the information is in brackets and this information is to be provided in the cited reference if it is stated on the report or data you are using. At any rate, in the References section of your report, you should provide the following information for a WWW reference: name of information providers (authors); [position]; [publication date]; [title]; [publication series and number]; access date; and Uniform Resource Locator (URL). Two examples from the Water Resources Division are listed below:

- Lanfear, K. J., 1986. Digital map of landfill locations, 1986: U.S. Geological Survey data available on the World Wide Web, accessed January 10, 1998, at URL http://water.usgs.gov/lookup/getspatial/landfill
- Boughton, C. J., Rowe, T. G., Allander, K. K., and Robledo, A. R., 1997. Stream and Ground-Water Monitoring Program, Lake Tahoe Basin, Nevada and California: U.S. Geological Survey Fact Sheet FS-100-97, accessed March 10, 1998, at URL http://water.usgs.gov/lookup/get:fs10097

Citing WWW references in your report is similar to citing any other literature. For example, for the Boughton citation above, you might write in your report, "During the uplift of the Sierra Nevada, approximately 2–3 million years ago, the Lake Tahoe Basin was formed by block faulting (Boughton and others, 1997)." Of course you may choose another referencing system, such as that used in *Environmental Geology*, which provides a numbered list of references cited and then includes those numbers, in parentheses, in the text.

WWW SITES FOR PART 1

The first three chapters in *Environmental Geology* that comprise Part 1 are related to some of the important fundamental concepts of environmental geology; the internal structure of the earth and plate tectonics; properties of minerals and rocks with respect to environmental concerns; and soils. The WWW sites below reflect these subjects and will provide an introduction as to where to find further information:

- PBS's Savage Earth: **http://www.pbs.org/wnet/savageearth/** An excellent Web site with background information about the internal structure of the earth and plate tectonics, including animations and a plate boundary map. You may also access information about earthquakes, volcanoes, and tsunamis, and how they relate to plate tectonics.
- NASA's Observatorium: **http://observe.ivv.nasa.gov/nasa/earth/earth_index.shtml** Includes much information about the earth system and how it changes. You may access links to plate tectonics information, as well as general information about the planet, ranging from tsunamis to El Niño. In fact, this site will be useful as a supplement to many chapters in this book.
- U.S. Geological Survey Online edition of *This Dynamic Earth: the Story of Plate Tectonics:* **http://pubs.usgs.gov/publications/text/dynamic.html** An online version of *This Dynamic Earth: the Story of Plate Tectonics* by Jacquelyne Kious and Robert I. Tilling. Gives background about the history and development of the theory of plate tectonics, and includes sections explaining plate motion, hot spots, unanswered questions, and how plate tectonics relates to human lives.

- Population Reference Bureau United States Population Data: **http://www.prb.org/prb/pubs/usds98.htm#usdata** This site has demographic data for the United States as well as a number of other interesting aspects concerning environmental indicators. The site also has links that discuss U.S. population growth and world population growth.
- Population Reference Bureau: **http://www.prb.org** This site contains information on both the U.S. and international population statistics and trends in changes in human population.
- World Watch Institute: **http://www.worldwatch.org/** This site lists a number of interesting environmental information, as well as a list of the publications of World Watch.

WWW Sites for Part 2

The subjects discussed in Part 2 are related to surficial earth processes and natural hazards, including earthquakes, volcanic activity, rivers and flooding, slope processes, and coastal processes. Of particular importance to environmental geology is the recognition of these processes, how they operate, and how we might minimize the hazards they present. An important issue related to people and natural hazards are the recognition and perception of the hazards. A major objective of studying environmental geology is to gain the necessary knowledge to make better decisions as to where we build our homes and other facilities necessary for our complex urban and rural life. For example, we don't want to build or buy a home on a floodplain or a slope that is likely to fail, and we would like to know what sorts of hazards are present at a variety of locations, such as national parks, airports, and industrial facilities, so that we might apply better principles of land-use planning to avoid potential problems related to natural processes. Listed below are some sites to get you started in this endeavor and provide additional information:

- Southern California Earthquake Center (SCEC): **http://www.scecdc.scec.org/** The SCEC Data Center is the primary archive for seismological data recorded by the South California Seismic Network. The center archives the set of historical records for seismicity as well as current seismicity and other interesting data sources about earthquakes in southern California.
- Federal Emergency Management Agency: **http://www.fema.gov/** This site contains information on current emergencies, including storms, fires, floods, and earthquakes, as well as provides information about FEMA's role in dealing with these hazards.
- Smithsonian Institution Global Volcanism Program: **http://www.volcano.si.edu/gvp/** Access lists of Holocene volcanoes, notices of current volcanic activity, and links to other volcano Web sites.
- National Weather Service Hydrologic Information Center: **http://www.nws.noaa.gov/oh/hic/index.html** This site contains updated flood information, current stream conditions, snow conditions, and weather reports.
- U.S. Geological Survey National Landslide Information Center: **http://gldage.cr.usgs.gov/html_files/nlicsun.html** At this site you'll find background information about landslides, facts about the hazard posed by landslides, and state information about landslides in your area. You can also access images of landslides.
- U.S. Geological Survey, Pasadena Office Earthquake Information: **http://www-socal.wr.usgs.gov/** This site has a good deal of information concerning past, present, and potential future seismicity in southern California, as well as products such as ground motion maps, shaking maps, and crustal deformation maps for southern California.
- California Division of Mines and Geology: **http://www.consrv.ca.gov/dmg/** This site has a lot of information for a variety of hazards in California, from earthquakes to landslides with geology and hazards mapping. Also covered are areas such as mineral resources and water resources.
- National Hurricane Center (NHC): **http://www.nhc.noaa.gov/** This site maintains a continuous world watch of storm activity for cyclones and hurricanes over the Atlantic, Caribbean, Gulf of Mexico, and eastern Pacific. It also contains information concerning past storms and other interesting information concerning hazards related to hurricanes.
- U.S. Geological Survey Geologic Hazards Team: **http://geohazards.cr.usgs.gov/welcome.html** This site contains interesting information concerning a variety of hazards, including earthquakes and landslides. It also has additional linkages to interactive maps of earthquake faults and landforms.
- U.S. Geological Survey Education (exploring maps): **http://www.usgs.gov/education/learnweb/Maps.html** This site contains an interdisciplinary set of materials that helps you understand basic mapping and map reading skills.
- Seismological Laboratory, California Institute of Technology: **http://www.gps.caltech.edu/seismo/seismo.page.html** This site has a variety of information concerning the earthquake hazard in California, including ongoing research and list of publications.
- Southern California Earthquake Center: **http://www.scec.org/** This site has a good deal of in-

formation concerning the earthquake hazard of southern California as well as a list of earthquake information resources and other interesting information.

- U.S. Geological Survey Volcanoes, Long Valley: **http://quake.wr.usgs.gov/VOLCANOES/LongValley/Response.html** This site has information concerning the volcanic hazard of Long Valley Caldera and the response plan.
- U.S. Geological Survey Cascades Volcano Observatory: **http://vulcan.wr.usgs.gov/** This site contains various interesting features related to the Cascade volcanoes and, in particular, Mount St. Helens.

WWW Sites for Parts 3 and 4

Several important environmental topics are introduced in Parts 3 and 4, the most important being a discussion of our resources necessary for living on earth and the processes that lead to pollution of those resources or as a result of resource utilization. The major resources discussed are water, minerals, energy, and air. WWW sites that will provide enrichment of your study of environmental geology, as well as data and other information for resources in pollution, are listed below:

- U.S. Environmental Protection Agency Office of Water: **http://www.epa.gov/watrhome/** This site provides considerable information, including programs and activities for protecting our drinking water, groundwater, and other resources; water quality; environmental indicators of water quality; and information about the Clean Water Act.
- U.S. Geological Survey Water Resources Division: **http://h2o.er.usgs.gov/** See data, graphs, and maps of water use, river flows, and reservoir levels throughout the United States as well as access real time data and information about topics such as acid rain.
- U.S. Department of Energy: **http://www.doe.gov/** The Department of Energy home page is a great place to start when looking for information about all aspects of energy in the United States.
- U.S. Geological Survey Minerals Information: **http://minerals.er.usgs.gov/minerals/** Browse the most recent information on the worldwide supply, demand, and flow of minerals, the importance of minerals to the U.S. economy, and statistics on where the United States obtains most of its mineral resources.
- U.S. Department of the Interior Office of Surface Mining: **http://www.osmre.gov/osm.htm** Read about the latest efforts in environmental protection and restoration, as well as new technology pertaining to mining in the United States.
- Earth Communications Office: **http://www.earthcomm.org/welcome.html** This nonprofit organization's Web page contains information about ocean pollution, the Clean Water Act, ozone depletion, and recent environmental news.
- U.S. Geological Survey Atmospheric Deposition and Precipitation Chemistry: **http://btdqs.usgs.gov/acidrain/** This site discusses the role of the USGS as a lead agency in the monitoring of wet atmospheric deposition in the United States. The site contains a number of maps and a variety of data related and important to acid rain.
- U.S. Geological Survey Minerals Information, Commodities Statistics and Information: **http://minerals.er.usgs.gov/minerals/pubs/commodity/** This site contains a variety of information concerning mineral resources used in the United States. The main site has a list of the various minerals and includes links to the minerals yearbook and information concerning use of a particular resource as well as recycling endeavors.
- Environmental Protection Agency Emission Trends: **http://www.epa.gov/ttn/chief/trends97/emtrnd.html** This site contains reports with tables and graphs concerning emissions of major air pollutants from 1900 to 1996.
- British Petroleum Statistical Review of World Energy 1998: **http://www.bp.com//bpstats/index.htm** This site contains voluminous information concerning use of fossil fuels, nuclear energy, and hydroelectricity. Topics include distribution of reserves as well as amounts of proven reserves at the end of 1997 in terms of tables and maps for each energy source.
- U.S. Geological Survey Mineral Resources Program: **http://minerals.er.usgs.gov/** This site provides a lot of information concerning the quality, quantity, and availability of different mineral resources. It also has linkages to a variety of other information.
- U.S. Geological Survey Water Resources of the United States: **http://water.usgs.gov/** This site has a lot of information concerning our water resources, and lists a variety of reports that may be downloaded and reviewed concerning water resources.
- Environmental Protection Agency Acid Rain Program: **http://www.epa.gov/docs/acidrain/overview.html** This site has excellent information concerning the overall goal of the acid rain program as well as data concerning emissions of sulfur dioxide and significant features of the acid rain program.

WWW Sites for Part 5

The capstone for *Environmental Geology* is the information provided in Part 5 that deals with environmental management, global change, and specific examples of relationships between geology and society that transcend those we have discussed earlier related to hazards and resources. In particular, the subject of a potential climate change and global warming as a result of activities of people is a "hot" (pun intended) topic. The WWW sites listed below will help you better understand some of the information related to global change, environmental management, and how people and geology interact:

- U.S. Environmental Protection Agency's Global Warming Page: **http://www.epa.gov/globalwarming/index.html** An excellent source of information pertaining to global warming and its causes and impacts and the actions being taken to lessen the expected effects of continued global warming.
- National Climatic Data Center: **http://www.ncdc.noaa.gov/** Access information about El Niño; temperature, precipitation, and drought data; historical climate data; and extreme weather events. Also includes radar and satellite resources.
- United Nations Environment Programme's Fact Sheets: **http://www.unep.ch/iucc/factcont.html** Browse through fact sheets pertaining to greenhouse gases, their sources, predicted climate changes, and ways to limit greenhouse gas emissions.
- U.S. Global Change Research Program: **http://www.usgcrp.gov/** This interagency program conducts research to monitor and understand factors contributing to a changing climate. Check out findings from the Intergovernmental Panel of Climate Change, and see what data is being collected. You'll also find many good links to other climate change Web sites.
- Global Climate Information Programme: **http://www.doc.mmu.ac.uk/aric/gcc/gcciphm.html** This program of the Atmospheric Research and Information Centre at Manchester Metropolitan University in Great Britain includes many fact sheets about global warming, climate change, and the greenhouse effect.
- Goddard Distributed Active Archive Center: **http://daac.gsfc.nasa.gov/** From the Goddard Space Flight Center at NASA, this site has substantial resources. You can access data and satellite images, and find interesting facts on general atmospheric dynamics and chemistry.
- U.S. Environmental Protection Agency Office of Solid Waste: **http://www.epa.gov/swerrims/** Access information about hazardous waste and the volume of solid waste generated and recycled in the United States. Check out information about the Superfund project and see its progress. Also, try this link for information specific to recycling in the United States: **http://www.epa.gov/epaoswer/non-hw/recycle/**
- National Park Service, Cape Hatteras National Seashore: **http://www.nps.gov/caha/** This site discusses interesting information concerning the Barrier Islands and Cape Hatteras National Seashore.
- U.S. Department of Commerce, NOAA El Niño Page: **http://www.elnino.noaa.gov/** This site defines El Niño and La Niña and includes characteristic maps of the events as well as updates on what is happening at a particular time.

APPENDIX B • *Maps and Related Topics*

TOPOGRAPHIC MAPS

As the name suggests, topographic maps illustrate the topography at the surface of the earth. The topography is represented by contour lines, which are imaginary lines every point of which on a particular contour line is at the same elevation (generally relative to mean sea level). Individual contour lines on a topographic map are a fixed interval of elevation apart, and this interval is known as the *contour interval.* Common contour intervals are 5, 10, 20, 40, 80, or 100 m or ft. The actual contour interval of a particular map depends upon the topography being represented as well as the scale of the map. If the topography has a relatively low relief (difference in elevation between the highest point and lowest point of a map), then the contour interval may be relatively small, but if the relief is large, then larger contour intervals are necessary or the lines will become so bunched up as to be impossible to read. This brings up an important point: Where contour lines are close together, the surface of the earth at that point has a relatively steep slope compared to other areas where the contour lines are further apart.

The scale of a topographic map (or for any map, for that matter) may be delineated in several ways. First, a scale may be stated as a ratio, say 1 to 24,000 (1:24,000), which means that 1 in. on a map is equal to exactly 24,000 in., or 2000 ft on the ground. Second, a topographic map may have a bar scale often found in the lower margin of the map useful for measuring distances. Finally, scales of some maps are stated in terms of specific units of length on the map; for example, it might be stated on the map that 1 in. equals 200 ft. This means that 1 in. on the map is equivalent to 200 ft (which is 2400 in.) on the ground. In this example we could also state that the scale is 1 to 2400. The most common scale used by the U.S. Geological Survey Topographic Maps is 1:24,000, but scales of 1:125,000 or smaller are also used. Remember 1÷24,000 is a larger number than 1÷125,000, so 1:125,000 is the smaller scale. The smaller the scale of maps of similar physical map size, the more area is shown.

In addition to contour lines, topographic maps also show a number of cultural features, such as roads, houses, and other buildings. Features such as streams and rivers are often shown in blue. In fact, a whole series of symbols are commonly used on topographic maps. These symbols are shown in Figure B.1.

Reading Topographic Maps

Reading or interpreting topographic maps is as much an art as it is a science. After you have looked at many topographic maps that represent the variety of landforms and features found at the surface of the earth, you begin to recognize these forms by the shapes of the contours. This is a process that takes a fair amount of time and experience in looking at a variety of maps. However, there are some rules of thumb that do help in reading topographic maps:

- Valleys containing rivers or even small streams have contours that form "Vs" that point in the upstream direction. This is sometimes known as the *rule of Vs.* Thus, if you are trying to draw the drainage pattern that shows all the streams, you should continue the stream in the upstream direction as long as the contours are still forming a "V" pattern. Near a drainage divide, the "Vs" will no longer be noticeable.
- Where contour lines are spaced close together, the slope is relatively steep, and where contour lines are spaced relatively far apart, the slope is relatively low. Where contour lines that are spaced relatively far apart change to become closer together, we say a "break in slope" has occurred. This is commonly observed at the foot of a mountain or where valley side slopes change to a floodplain environment.
- Contours near the upper parts of hills or mountains usually show a closure. These may be relatively oval or round in shape for a conical peak or longer and narrower for a ridge. Remember the actual elevation of a peak is higher than the last contour shown and may be estimated by taking the half contour interval. That is, if the highest contour on a peak is 1000 m with a contour interval of 50 m, then you would assume the top of the mountain is 1025 m.
- Topographic depressions that form closed contours have hachure marks on the contours and are useful in indicating the presence of such a depression.
- Sometimes the topography on a slope is hummocky and anomalous to the general topography of the area. Such hummocky topography is often suggestive of the existence of mass wasting processes and landslide deposits.

In summary, after observing a variety of topographic maps and working with them for some time, you will begin to see the pattern of contours as an actual landscape consisting of hills and valleys and other features.

Locating Yourself on a Map

The first time you take a topographic map into the field with you, you may have some difficulty locating where you are. Determining where you are is absolutely crucial in trying to

Control data and monuments

Vertical control

Third order or better, with tablet	BM×16.3
Third order or better, recoverable mark	×120.0
Bench mark at found section corner	BM 18.3
Spot elevation	×5.3

Contours

Topographic

- Intermediate
- Index
- Supplementary
- Depression
- Cut; fill

Bathymetric

- Intermediate
- Index
- Primary
- Index primary
- Supplementary

Boundaries

- National
- State or territorial
- County or equivalent
- Civil township or equivalent
- Incorporated city or equivalent
- Park, reservation, or monument

Surface features

Levee	Levee
Sand or mud area, dunes, or shifting sand	Sand
Intricate surface area	Strip mine
Gravel beach or glacial moraine	Gravel
Tailings pond	Tailings pond

Mines and caves

Quarry or open pit mine	
Gravel, sand, clay, or borrow pit	
Mine dump	Mine dump
Tailings	Tailings

Vegetation

Woods	
Scrub	
Orchard	
Vineyard	
Mangrove	Mangrove

Glaciers and permanent snowfields

- Contours and limits
- Form lines

Marine shoreline

Topographic maps

- Approximate mean high water
- Indefinite or unsurveyed

Topographic-bathymetric maps

- Mean high water
- Apparent (edge of vegetation)

Coastal features

Foreshore flat	Mud
Rock or coral reef	Reef
Rock bare or awash	
Group of rocks bare or awash	
Exposed wreck	
Depth curve; sounding	3
Breakwater, pier, jetty, or wharf	
Seawall	

Rivers, lakes, and canals

- Intermittent stream
- Intermittent river
- Disappearing stream
- Perennial stream
- Perennial river
- Small falls; small rapids
- Large falls; large rapids
- Masonry dam
- Dam with lock
- Dam carrying road

Perennial lake; Intermittent lake or pond	
Dry lake	Dry lake
Narrow wash	
Wide wash	Wide wash
Canal, flume, or aquaduct with lock	
Well or spring; spring or seep	

Submerged areas and bogs

Marsh or swamp	
Submerged marsh or swamp	
Wooded marsh or swamp	
Submerged wooded marsh or swamp	
Rice field	Rice
Land subject to inundation	Max pool 431

Buildings and related features

Building	
School; church	
Built-up area	
Racetrack	
Airport	
Landing strip	
Well (other than water); windmill	
Tanks	
Covered reservoir	
Gaging station	
Landmark object (feature as labeled)	
Campground; picnic area	
Cemetery: small; large	Cem

Roads and related features

Roads on Provisional edition maps are not classified as primary, secondary, or light duty. They are all symbolized as light duty roads.

- Primary highway
- Secondary highway
- Light duty road
- Unimproved road
- Trail
- Dual highway
- Dual highway with median strip

Railroads and related features

- Standard gauge single track; station
- Standard gauge multiple track
- Abandoned

Transmission lines and pipelines

Power transmission line; pole; tower	
Telephone line	Telephone
Aboveground oil or gas pipeline	
Underground oil or gas pipeline	Pipeline

▲ **FIGURE B.1** Some of the common symbols used on topographic maps prepared by the U.S. Geological Survey. (From U.S. *Geological Survey*)

prepare maps to show particular features, for example, the locations of floodplains, landslides, or other features. One way to locate yourself on a map is to recognize certain features such as mountain peaks, intersections of roads, a prominent bend in a road or river, or some other readily recognized feature. Then work it out from there with a compass. That is, you might take a bearing (compass direction) to several prominent features and draw these bearings on the map; your location is where they intersect. Today, however, we have more modern technology for locating ourselves and working with maps. Global Positioning Systems (GPS) are readily available at a very modest price and assist in locating a position on the surface of the earth. Hand-held GPS systems are readily available today and can generally locate your position on the ground with an accuracy of about 100 m. The GPS receivers work by receiving signals from three or four satellites and measuring the distance from the satellite to your location. This is done by measuring the time it takes for the signal from your receiver to reach the satellite and return. The accuracy of defining your position can be reduced to approximately 6 m by utilizing a reference receiver on the ground that communicates with the satellites and then working out your position relative to the reference receiver, as shown in Figure B.2.

Global Positioning Systems are commonly linked to computers, so when a position is known, it may be plotted directly on a map viewed on a computer screen. GPS technology is revolutionizing the way we do our mapping, is becoming widely available, and is a valuable research tool in the field.

A word of information here concerning the term *in the field*. When geologists use the term *in the field*, we mean "outside on the surface on earth" and are not referring to the "field of geology." For example, I am now about to go out and study the coastline of California following the El Niño storms. I would say to my colleagues, "I am going into the field."

An Example from a Coastal Landscape

A coastal landscape is shown in Figure B.3. As illustrated in (a), which is an oblique view of the topography, the area is characterized by two hills with an intervening valley. A coastal seacliff is shown on the east (right) part of the diagram

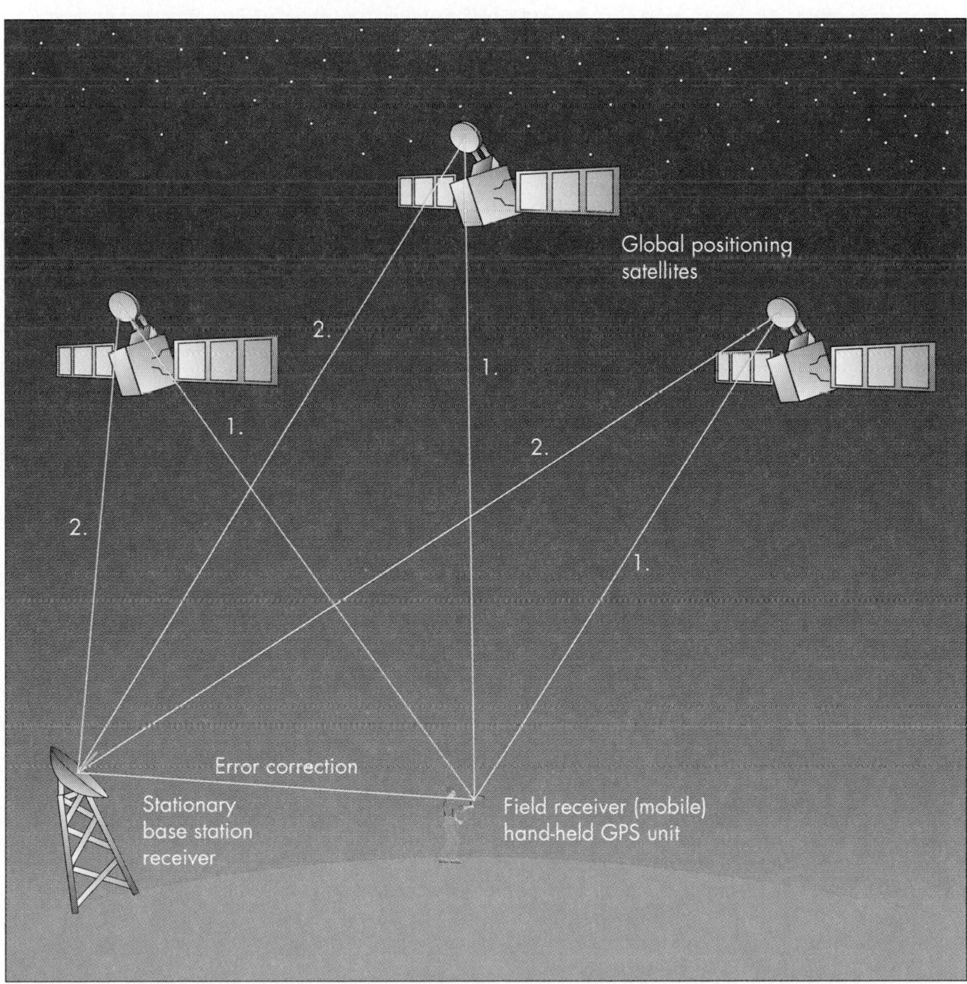

FIGURE B.2 Idealized diagram showing how GPS systems work.

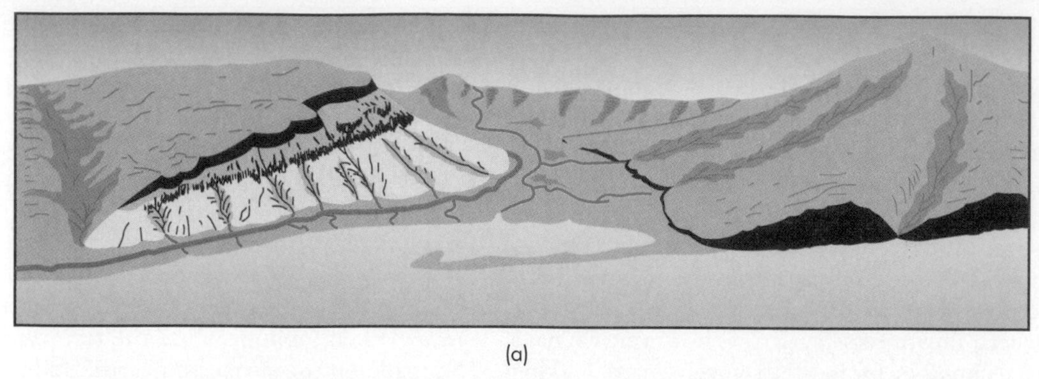

(a)

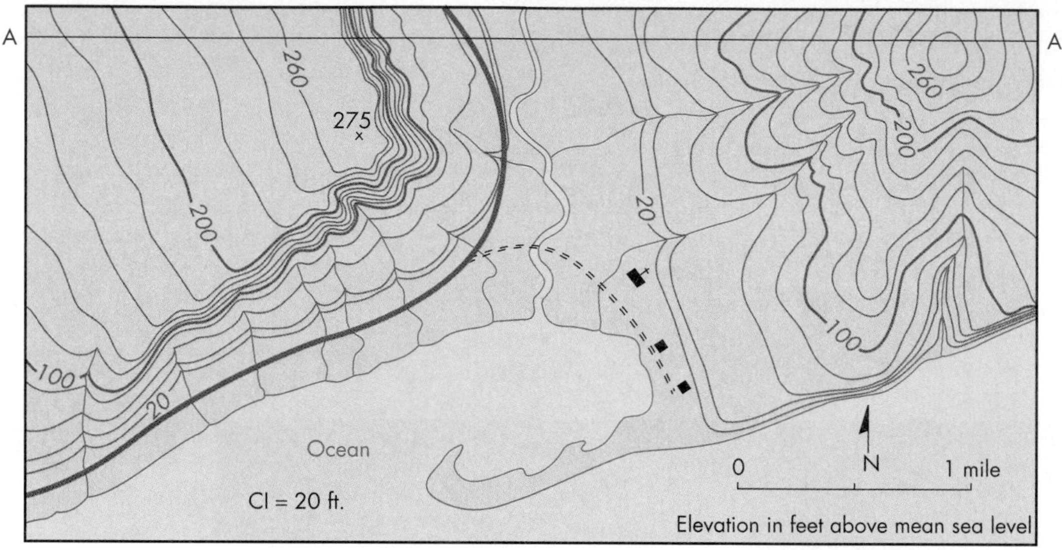

(b)

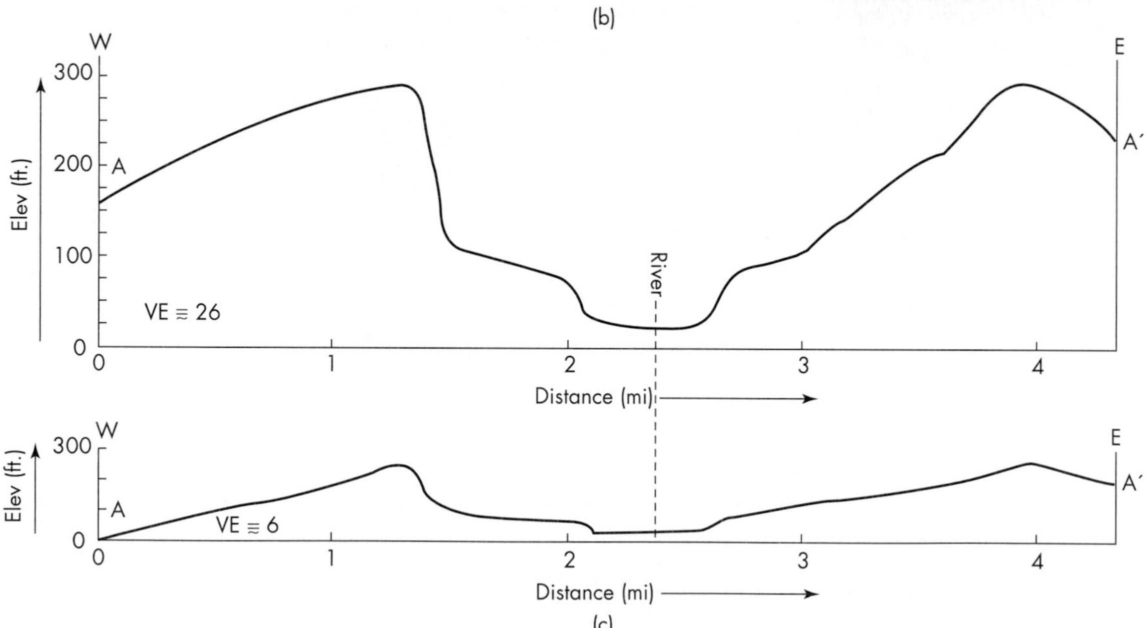

(c)

▲ **FIGURE B.3** (a) Set of diagrams illustrating topographic maps. Drawing of a landscape along a coastal area. (b) Topographic map for the same area with a contour interval of 20 ft. (c) Topographic profiles along line A–A′ of the topographic map shown with vertical exaggeration of approximately 26 times and 6 times. The vertical exaggeration is the ratio of the vertical to horizontal scale of the topographic profile. The ratio for the upper profile is approximately 26, so we say that the vertical exaggeration is approximately 26 times. For the lower profile, the vertical exaggeration is about 6 times. In the real world, of course, there is no vertical exaggeration (the vertical and horizontal scales are the same). As an experiment, you might try to make a topographic profile along line A–A′ with no vertical exaggeration. What do you conclude from this? (*From U.S. Geological Survey*)

and a sandspit produced by longshore coastal processes (see Chapter 8) with a hook on the end suggests the direction of sand transport in the surf zone and beach is from the east (right) to the west (left) along this coastal area. A topographic map for the area is shown in (b). Notice that the contour interval is 20 ft. The elevation at the top of the highest hill on the east side of the diagram would be approximately 290 ft because the last contour is 280 and so the hill cannot be higher than 300, so we split the difference. Several streams flow into the main valley; notice that the contours "V" in the upstream direction particularly toward the peak of elevation of approximately 270 ft. Other information that may be "read" from the topographic map include the following:

- The landform on the western portion of the map (left side) is a hill with elevation 275 ft and a gentle slope to the west, and a steep slope to the east (right) and toward the ocean. The eastern slope is particularly steep near the top of the mountain where the contours are very close together.
- The center part of the map shows a river flowing into a bay protected by a hooked sand bar. The relatively flat land in the vicinity of the river is a narrow floodplain with a light-duty road (see Figure B.3) delineated by two closely spaced lines extending along the western side. A second unimproved road crosses the river and extends out to the head of the sandspit, providing access to a church and two other buildings. The floodplain is delineated on both sides of the river by the 20-ft contour, above which to the 40-ft contour is a break in slope at the edge of the valley.
- The eastern and southern slopes of the hill on the west (left) side of the map with elevation 275 ft has a number of small streams flowing mostly southward into the ocean. These streams may be defined as relatively narrow, steep gullies that are rapidly dissecting (eroding) the hill.
- In contrast to the hill on the western part of the map, the hill on the eastern part with elevation of approximately 290 ft has more gentle slopes on the western flank with streams spaced relatively far apart flowing toward the river and the ocean. Thus, we could state that the streams flowing to the main river valley from the east have dissected (eroded) the landscape less than those streams flowing into the main valley from the west.
- In particular, note the stream channel just south of the hill with elevation 275 ft on the west side of the map. That stream is actively eroding headward, forming a narrow, steep gully denoted by the pattern of closely spaced contours that form a concave indentation into the more gentle topography that forms the surface that is sloping to the west from the top of the mountain.

To continue a study of this area, we might construct topographic profiles across the area from east to west and also construct profiles of some of the streams flowing into the main valley. The next task might be to obtain aerial photographs and geologic maps to draw conclusions concerning the topography and geology.

Geologic Maps

Geologists are interested in the types of rocks found in a particular location and in their spatial distribution. To understand the geology of an area, a very basic step is to produce a geologic map. Geologic maps are the fundamental database from which to interpret the geology of an area. The first step in preparing a geologic map is to obtain good base maps (usually topographic) or aerial photographs on which the geologic information may be transferred. The geologist then goes into the field and makes observations at outcrops where the rocks may be observed. Different rock types, units, or formations are mapped and the contacts between varying rock types are mapped as lines on the map. The attitude of sedimentary rock units is designated by a T-shaped strike and dip symbol. The strike is the compass direction of the intersection of the rock layer with a horizontal plain, and the dip is the maximum angle the rock layer makes with the horizontal (Figure B.4). Figure B.5a shows a very simple geologic map with three rock types—sandstone, shale, and limestone—over an area of approximately 1350 km^2. The arrangement of the strike and dip symbols suggest that the major structure is an anticline. A geologic cross section constructed across the profile along the line *E–E′* is shown in Figure B.5b. Geologists often make a series of cross sections to try to better understand the geology of a particular area. Geologic maps at a variety of scales from 1 to 250,000 down to 1 to 24,000 are generally available from a variety of sources, including the U.S. Geological Survey.

Digital Elevation Models

Topographic data for many areas in the United States and other parts of the world are now available on computer disk. The arrays of elevation values which may include, for example, the elevation of the ground on a 30 m × 30 m ground spacing grid are the basic topographic data. Computer

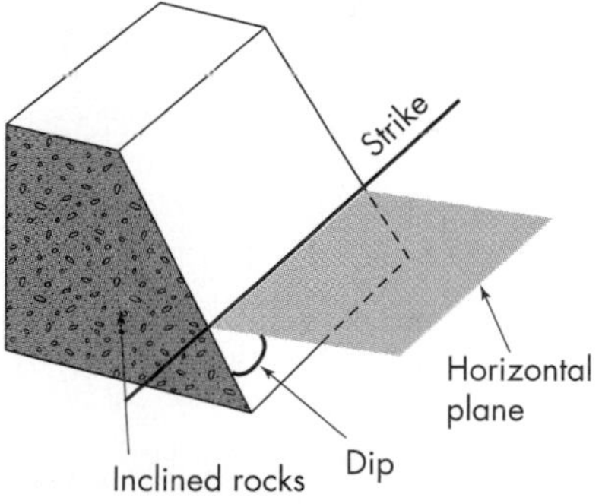

▲ **FIGURE B.4** Idealized block diagram showing the strike and dip of inclined sedimentary rocks.

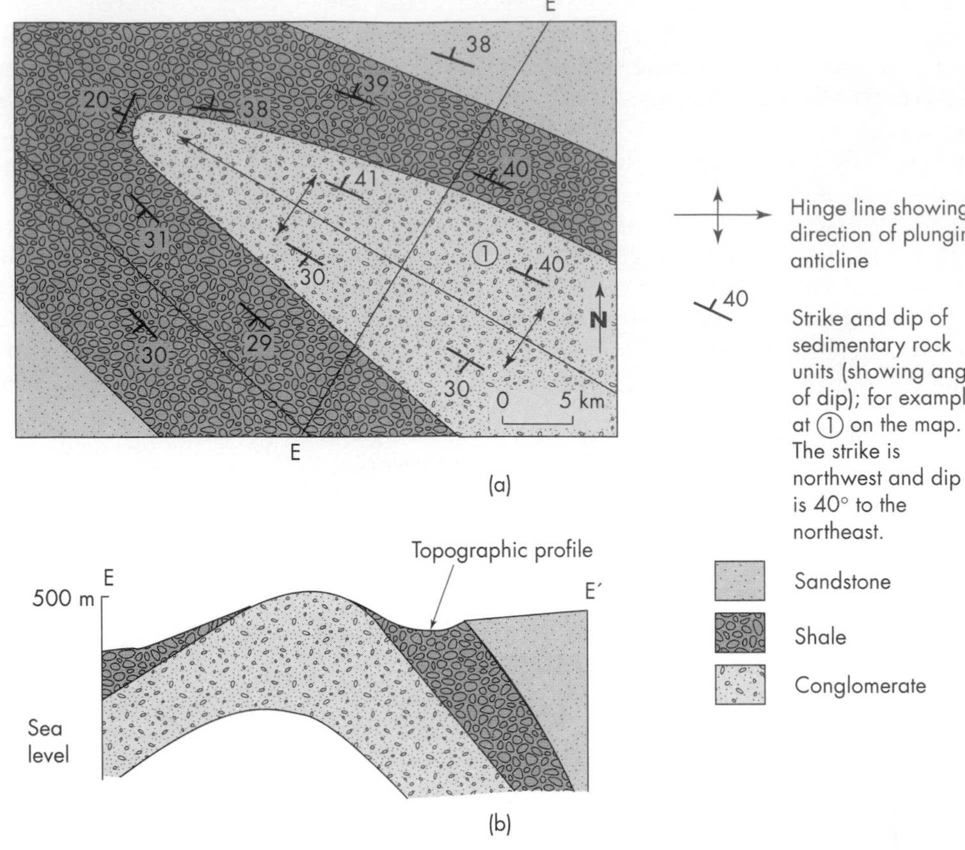

▶ **FIGURE B.5** (a) Idealized diagram showing a very simple geologic map with three rock types, and (b) geologic cross section showing the topography and structure of the anticline.

programs are then used to view the data, and color shading may be provided to better show the topography, which is the Digital Elevation Model (DEM). DEMs provide a visual representation of the surface of earth and can be "viewed" from a variety of different angles. That is, you may decide you would like to view the topography obliquely from the south, north, east, or west. The vertical dimension may also be exaggerated so that minor topographic differences may become more apparent. Figure B.6 shows a digital elevation model for the Los Angeles Basin. This clearly illustrates that Los Angeles is nearly surrounded by mountains and hills, all of which are related to recent tectonic activity. DEMs are becoming important research tools in evaluating the topography of an area. They may be used to delineate and map such features as fault scarps, drainages, marine terraces, floodplains, and many other features of the landscape.

SUMMARY

Our discussion above establishes that there are several types of maps that are useful in evaluating a particular area and its geology. Of particular importance are topographic maps and geologic maps. Digital Elevation Models may be constructed from topographic data, and a variety of other special purpose maps are also available. Examples include maps of recent landslides, maps of floodplains, and engineering geology maps that show engineering properties of earth materials.

▲ **FIGURE B.6** Digital elevation model (DEM) of the Los Angeles area with a fair amount of vertical exaggeration. The flat area in the central part of the image is the Los Angeles Basin. The smaller basin just to the left is the San Fernando Valley, which is separated from the Los Angeles Basin by the Santa Monica Mountains. (*Courtesy of Robert Crippin*, NASA, *Jet Propulsion Laboratory*)

GLOSSARY

A soil horizon Uppermost soil horizon, sometimes referred to as the *zone of leaching*.

Absorption The process of taking up, incorporating, or assimilating a material.

Acid rain Rain made artificially acidic by pollutants, particularly oxides of sulfur and nitrogen. (Natural rainwater is slightly acidic owing to the effect of carbon dioxide dissolved in the water.)

Active fault There are a variety of definitions. One is that of a fault along which there has been displacement in the past 10,000 years. Another definition is a fault where there has been multiple displacement in the past 35,000 years.

Adsorption Process in which molecules of gas or molecules in solution attach to the surface of solid materials with which they come in contact.

Advanced treatment Wastewater treatment that includes a variety of processes, such as use of chemicals, sand filters, or carbon filters, to further purify and remove contaminants from wastewater.

Aerobic Characterized by the presence of free oxygen.

Aesthetics Originally a branch of philosophy, defined today by artists and art critics.

Age element Elements that characteristically accumulate in tissue with age.

Aggregate Any hard material, such as crushed rock, sand, gravel, or other material, that is added to cement to make concrete.

Air quality standards Concentrations of selected air pollutants set to protect the health of people as well as prevent environmental degradation from air pollution.

Albedo A measure of reflectivity or the amount (by decimal or percent) of electromagnetic radiation reflected by a material or surface.

Alkaline soil Soil found in arid regions that contains a large amount of soluble mineral salts (primarily sodium) that in the dry season may appear on the surface as a crust or powder.

Alluvium Unconsolidated sediments, including sand, gravel, and silt, deposited by streams.

Alpha particles Type of nuclear radiation consisting of two protons and two neutrons emitted during radioactive decay.

Anaerobic Characterized by the absence of free oxygen.

Andesite A type of volcanic rock consisting of the minerals feldspar and ferromagnesium with an intermediate silica content of about 60 percent. Andesite lava flows form some of the most beautiful stratified volcanoes in the world. An example is Mount Fuji in Japan.

Angle of repose The maximum angle that loose material will sustain.

Anhydrite Evaporite mineral ($CaSO_4$) calcium sulfate.

Anthracite A type of coal characterized by a high percentage of carbon and low percentage of volatiles, providing a high heat value. Anthracite often forms as a result of metamorphism of bituminous coal.

Anticline Type of fold characterized by an upfold or arch. The oldest rocks are found in the center of the fold.

Appropriation doctrine, water law Holds that prior usage of water is a significant factor in determining water rights. The first to use the water for beneficial purposes is prior in right.

Aquifer Earth material containing sufficient groundwater that the water can be pumped out. Highly fractured rocks and unconsolidated sands and gravels make good aquifers.

Aquitard Earth material that retards the flow of groundwater.

Area (strip) mining Type of strip mining practiced on relatively flat areas.

Argillic B soil horizon Designated as *Bt*, a soil horizon enriched in clay minerals that have been translocated downward by soil-forming processes.

Artesian Term for a groundwater system in which the groundwater is isolated from the surface by a confining layer and the water is under pressure. Groundwater that is under sufficient pressure will flow freely at the surface from a spring or well.

Asbestos Fibrous mineral material used as insulation. It is suspected of being either a true carcinogen or a carrier of carcinogenic trace elements.

Ash, volcanic Unconsolidated volcanic debris, less than 4 mm in diameter, physically blown out of a volcano during an eruption.

Ash fall Volcanic ash eruption that blows up into the atmosphere and then rains down on the landscape.

Ash flow Mixture of volcanic ash, hot gases, and fragments of rock and glass that flows rapidly down the flank of a volcano. May be an extremely hazardous event.

Atmosphere Layer of gases surrounding the earth.

Atmospheric inversion or inversion layer Restricted circulation in the lower atmosphere as a result of a warmer air mass overlying a cooler one.

Avalanche A type of landslide involving a large mass of snow, ice, and rock debris that slides, flows, or falls rapidly down a mountainside.

Average residence time The amount of time it takes for the total stock or supply of material in a system to be cycled through the system.

Azonal soil Recent surface materials such as floodplain deposits that do not have distinctive soil layering.

B soil horizon Intermediate soil horizon; sometimes known as the *zone of accumulation.*

Balance of nature The idea that nature undisturbed by human activity will reach a balance or state of equilibrium, thought by some to be an antiquated idea about a state that probably never has existed in nature.

Balancing doctrine Asserts that public benefits and importance of a particular action should be balanced against potential injury to certain individuals. The method of balancing is changing, and courts are considering possible long-range injury caused by certain activities to large numbers of citizens other than the immediate complainants.

Barrier island Island separated from the mainland by a salt marsh. It generally consists of a multiple system of beach ridges and is separated from other barrier islands by inlets that allow exchange of seawater with lagoon water.

Basalt The most common type of volcanic rock, consisting of the minerals feldspar and ferromagnesium with relatively low silica content of about 50 percent; forms shield volcanoes.

Basaltic Engineering geology term for all fine-grained igneous rocks.

Base flow The low flow discharge of a stream or river that is produced by groundwater seeping into the channel.

Bauxite A rock composed almost entirely of hydrous aluminum oxides. It is a common ore of aluminum.

Beach Accumulation of whatever loose material (commonly sand, gravel, or bits of shell and so forth) that accumulate on a shoreline as the result of wave action.

Beach budget Inventory of sources and losses of sediment to a particular stretch of coastline.

Beach nourishment Artificial process of adding sediment (sand) to a beach for recreational and aesthetic purposes as well as to provide a buffer to coastal erosion.

Bed material Sediment transported and deposited along the bed of a stream channel.

Bedding plane The plane that delineates the layers of sedimentary rocks.

Bentonite A type of clay that is extremely unstable; upon wetting, it expands to many times its original volume.

Berms The relatively flat part of a beach profile produced by deposition from waves. It is the part of the beach where people sunbathe.

Beta particles Type of nuclear radiation consisting of electrons emitted during radioactive decay.

Biochemical oxygen demand (BOD) A measure of the amount of oxygen necessary to decompose organic materials in a unit volume of water. As the amount of organic waste in water increases, more oxygen is used, resulting in a higher BOD.

Biogeochemical cycle Movement of a chemical element or compound through the various earth systems, including atmosphere, lithosphere, biosphere, and hydrosphere.

Biomass Organic matter. As a fuel, biomass can be burned directly (as wood) or converted to a more convenient form (such as charcoal or alcohol) and then burned.

Biopersistence A measure of how long a particular material will remain in the biosphere.

Biosphere The zone adjacent to the surface of the earth that includes all living organisms.

Biotechnology With respect to resource management, refers to use of organisms to assist in mining of ores or cleaning up of waste from mining activities.

Biotite A common ferromagnesian mineral; a member of the mica family.

Bituminous coal A common type of coal characterized by relatively high carbon content and low volatiles; sometimes called *soft coal.*

Bk soil horizon Soil horizon characterized by accumulation of calcium carbonate that may coat individual soil particles and fill some pore spaces but does not dominate the morphology of the horizon.

Blowout Failure of an oil, gas, or disposal well resulting from adverse pressures that can physically blow part of the well casing upward. May be associated with leaks of oil, gas, or, in the case of disposal wells, harmful chemicals.

Braided river A river channel characterized by an abundance of islands that continually divide and subdivide the flow of the river.

Breaker zone That part of the beach and nearshore environment where incoming waves peak up, become unstable, and break.

Breakwater A structure (such as a wall), which may be attached to a beach or located offshore, designed to protect a beach or harbor from the force of the waves.

Breccia A rock or zone within a rock composed of angular fragments. Sedimentary, volcanic, and tectonic breccias are recognized.

Breeder reactor A type of nuclear reactor that actually produces more fissionable (fuel) material than it uses.

Brine Water that has a high concentration of salt.

British thermal unit (Btu) A unit of heat defined as the heat required to raise the temperature of one pound of water one degree Fahrenheit.

Brittle Material that ruptures before any plastic deformation.

Bulk element Any of the common elements that make up the bulk of living material.

Burner reactor Type of nuclear reactor used to generate electricity that consumes more fissionable material than it produces.

C soil horizon Lowest soil horizon, sometimes known as the *zone of partially altered parent material.*

Cadmium A metallic element with an atomic number of 48 and an atomic weight of 112.4. As a trace element, it has been associated with serious health problems.

Calcite Calcium carbonate ($CaCO_3$); common carbonate mineral that is the major constituent of the rock limestone. It weathers readily by solutional processes, and large cavities and open weathered fractures are common in rocks that contain the mineral calcite.

Caldera Giant volcanic crater produced by very rare but extremely violent volcanic eruption or by collapse of the summit area of a shield volcano following eruption.

Caldera eruption Relatively infrequent large volcanic eruption that is associated with a catastrophic explosion that may produce a very large volcanic crater several tens of kilometers in diameter.

Caliche A white-to-gray irregular accumulation of calcium carbonate in soils of arid regions.

Calorie The quantity of heat required to raise the temperature of one gram of water from 14.5° to 15.5 °C.

Cambic soil horizon Soil horizon diagnostic of incipient soil profile development, characterized by a slightly redder color than the other horizons.

Capable fault Defined by the U.S. Nuclear Regulatory Commission as a fault that has exhibited movement at least once in the last 50,000 years or multiple movements in the 500,000 years.

Capacity A measure of the total load a river may carry.

Capillary action The rise of water along narrow passages, facilitated and caused by surface tension.

Capillary fringe The zone or layer above the water table in which water is drawn up by capillary action.

Capillary water Water that is held in the soil through capillarity.

Carbon cycle One of the earth's major biogeochemical cycles that involves the movement of carbon in the atmosphere, biosphere, lithosphere, and hydrosphere.

Carbon monoxide (CO) A colorless, odorless gas that at low concentration is very toxic to humans and other animals. Anthropogenic component of emissions is only about 10 percent and comes mainly from fires, automobiles, and other sources of incomplete burning of organic compounds. Many people have been accidentally asphyxiated by carbon monoxide from poorly ventilated campers, tents, and homes.

Carbonate A compound or mineral containing the radical (CO_3). The common carbonate is calcite.

Carcinogen Any material known to produce cancer in humans or other animals.

Catalytic converter A device utilized in gasoline-burning vehicles that reduces emissions by converting carbon monoxide and hydrocarbons to carbon dioxide and water.

Catastrophe An event or situation causing sufficient damage to people, property, or society in general that recovery and/or rehabilitation is long and involved. Natural processes most likely to produce a catastrophe include floods, hurricanes, tornadoes, tsunamis, volcanoes, and large fires.

Cesium-137 A fission product with a half-life of 33 years produced from nuclear reactors.

Channel pattern The shape of a river channel as viewed from above in a "bird's-eye view." Patterns include straight, meandering, and braided.

Channel restoration The process of restoring stream channels and adjacent areas to a more natural state.

Channelization An engineering technique to straighten, widen, deepen, or otherwise modify a natural stream channel.

Circum-Pacific belt One of the three major zones where earthquakes occur. This belt is essentially the border of the Pacific plate. It is also known as the *ring of fire*, as many active volcanoes are found on the edge of the Pacific plate.

Clay May refer to a mineral family or to a very fine-grained sediment. It is associated with many environmental problems, such as shrinking and swelling of soils and sediment pollution.

Clay skins Oriented plates of clay minerals surrounding soil grains and filling pore spaces between grains.

Closed system A system with boundaries that restrict the flow of energy and matter. For example, with respect to mineral resources, earth can be considered a closed system.

Coal A sedimentary rock formed from plant material that has been buried, compressed, and changed.

Coastal erosion Erosion of a coastline caused by a number of processes, including wave action, landsliding, wind, and runoff from the land.

Cogeneration With respect to energy resources, the recycling of waste heat to increase the efficiency of a typical power plant or factory. May involve production of electricity as a by-product from industrial processes.

Colluvium Mixture of weathered rock, soil, and other, usually angular, material on a slope.

Columnar jointing System of fractures (joints) that break rock into polygons of typically five or six sides; the polygons form columns. This type of fracturing is common in basalt and most likely is caused by shrinking during cooling of the lava.

Common excavation Excavation that can be accomplished with an earth mover, backhoe, or dragline.

Competency A measure of the largest-sized particle that a river may transport.

Complex response Mechanism of operation of a system in which changes occur at a variety of scales and time without input of multiple perturbations from outside the system.

Composite volcano Steep-sided volcanic cone produced by alternating layers of pyroclastic debris and lava flows.

Composting A biochemical process in which organic materials are decomposed to a humus-like material by aerobic organisms.

Comprehensive plan Planning document adopted by local government that states general and long-range policies on how the community plans to deal with future development.

Compressibility (soil) Measure of a soil's tendency to decrease in volume.

Conchoidal fracture A shell-like or fan-shaped fracture characteristic of the mineral quartz and natural glass.

Cone of depression A cone-shaped depression in the water table caused by withdrawal of water at rates greater than those at which the water can be replenished by natural groundwater flow.

Confined aquifer An aquifer that is overlain by a confining layer (aquitard).

Conglomerate A detrital sedimentary rock composed of rounded fragments, 10 percent of which are larger than 2 mm in diameter.

Connate water Water that is no longer in circulation or in contact with the present water cycle; generally, saline water trapped during deposition of sediments.

Conservation Policy for resources such as water and energy that moderates or adjusts demands in order to minimize expenditure of the resource. May mean getting by with less through improved technology to provide just the amount of the resource necessary for a given task.

Consumptive use A type of offstream use in which the water does not return to the stream or groundwater resource following use. The water evaporates, is incorporated into crops or products, or is consumed by animals or humans.

Contact metamorphism Type of metamorphism produced when rocks are in close contact with a cooling body of magma below the surface of the earth.

Continental drift Movement of continents in response to seafloor spreading. The most recent episode of continental drift supposedly began about 200 million years ago with the breakup of the supercontinent Pangaea.

Continental shelf Relatively shallow ocean area between the shoreline and the continental slope that extends to approximately a 600-foot water depth surrounding a continent.

Contour (strip) mining Type of strip mining used in hilly terrain.

Convection Transfer of heat involving movement of particles; for example, the boiling of water in which hot water rises to the surface and displaces cooler water that moves toward the bottom.

Convergent plate boundary Boundary between two lithospheric plates in which one plate descends below the other (subduction).

Corrosion A slow chemical weathering or chemical decomposition that proceeds inward from the surface. Objects such as pipes experience corrosion when buried in soil.

Corrosion potential Potential of a particular soil to cause corrosion of buried iron pipes as a result of soil chemistry.

Cost-benefits analysis A type of site selection in which benefits and costs of a particular project are compared. The most desirable projects are those for which the benefits-to-cost ratio is greater than 1.

Creep A type of downslope movement characterized by slow flowing, sliding, or slipping of soil and other earth materials.

Crystal settling A sinking of previously formed crystals to the bottom of a magma chamber.

Crystalline A material with a definite internal structure such that the atoms are in an orderly, repeating arrangement.

Crystallization Processes of crystal formation.

Cultural eutrophication Rapid increase in the abundance of plant life, particularly algae in freshwater or marine environments resulting from input of nutrients from human sources to the water.

Darcy's law Empirical relationship that states that the volumetric flow rate, such as cubic meters per day, is a product of hydraulic conductivity, hydraulic gradient, and cross-sectional area of flow. Developed by Henry Darcy in 1856.

Darcy flux The product of the hydraulic conductivity and hydraulic gradient, and thus has the units of a velocity.

Debris flow Rapid downslope movement of earth material often involving saturated, unconsolidated material that has become unstable because of torrential rainfall.

Deep-well disposal Method of waste disposal that involves pumping waste into subsurface disposal sites, such as fractured or otherwise porous rocks.

Dendrochronology Study of tree rings.

Desalination Engineering processes and technology that reduces salinity of water to a level that it may be consumed by people or used in agriculture.

Desertification Conversion of land from a more productive state to one more nearly resembling a desert.

Design basis ground motion In the state of California, this is defined as the ground motion that has a 10 percent chance of being exceeded in a 50-year period as determined by site-specific seismic hazard analysis or from a hazard map drawn specifically for ground motion.

Design response spectrum A graph representing the dynamic effects of the design basis ground motion, calculated from simple equations if the seismic zone factor is known as well as the soil type and seismic source type.

Detrital Mineral and rock fragments derived from preexisting rocks.

Diamond Very hard mineral composed of the element carbon.

Dilatancy of rocks Inelastic increase in volume of a rock that begins after stress on the rock has reached one-half the rock's breaking strength.

Dilatancy-diffusion model A model to explain the occurrence of earthquakes based upon a variety of precursor stages including seismic velocity, earth movements, and number of seismic events. Stages include buildup of elastic strain, influx of water and microfracturing, and sudden drop in stress during an earthquake that is followed by aftershocks.

Directivity Refers to the fact that during some moderate to large earthquakes the rupture of the fault is in a particular direction and the intensity of seismic shaking is greater in that direction.

Discharge The quantity of water flowing past a particular point on a stream in a given amount of time, usually measured in cubic feet per second (cfs) or cubic meters per second (cms).

Disease From an environmental viewpoint, disease may be considered as an imbalance that results in part from a poor adjustment between an individual and his or her environment.

Disseminated mineral deposit Mineral deposit in which ore is scattered throughout the rocks; examples are diamonds in kimberlite and many copper deposits.

Disturbance From an ecological viewpoint, refers to an event that disrupts a system. Examples include wildfire or hurricanes that cause considerable environmental change.

Divergent plate boundary Boundary between lithospheric plates characterized by production of new lithosphere; found along oceanic ridges.

Dose dependency Refers to the effects of a certain trace element on a particular organism being dependent upon the dose or concentration of the element.

Dose-response curves A graph showing relationship between response and dose of a particular trace element on a particular population of organisms.

Doubling time The time necessary for a quantity of whatever is being measured to double.

Downstream floods Floods produced by storms of long duration that saturate the soil and produce increased runoff over a relatively wide area. Often are of regional extent.

Drainage basin Area that contributes surface water to a particular stream network.

Drainage control Refers to the development of surface and subsurface drains to increase the stability of a slope.

Drainage net System of stream channels that coalesce to form a stream system.

Dredge spoils Solid material, such as sand, silt, clay, or rock deposited from industrial and municipal discharges, that is removed from the bottom of a water body to improve navigation.

Driving forces Those forces that tend to make earth material slide.

Ductile Material that ruptures following elastic and plastic deformation.

E soil horizon A light-colored horizon underlying the *A* horizon that is leached of iron-bearing compounds.

Earthquake Natural shaking or vibrating of the earth in response to the breaking of rocks along faults. The earthquake zones of the earth generally correlate with lithospheric plate boundaries.

Earthquake cycle A hypothesis to explain periodic occurrence of earthquakes based upon drop in elastic strain following an earthquake and reaccumulation of strain prior to the next event.

Earth system science The study of the earth as a system.

Earth's energy balance Refers to the balance between incoming solar radiation and outgoing radiation from the earth. Involves consideration of changes in the energy's form as it moves through the atmosphere, oceans, and land, as well as living things, before being radiated back into space.

Ease of excavation (soil) Measure of how easily a soil may be removed by human operators and equipment.

Ecology Branch of biology that treats relationships between organisms and their environments.

Economic geology Application of geology to locating and evaluating mineral materials.

Effective dose (ED) Dose or concentration of a particular toxin necessary to cause a particular response to a population. For example, ED-50 is the concentration or effective dose at which 50 percent of the population shows a response such as a symptom.

Effluent Any material that flows outward from something; examples include wastewater from hydroelectric plants and water discharged into streams from waste-disposal sites.

Effluent stream Stream in which flow is maintained during the dry season by groundwater seepage into the channel.

El Niño An event during which trade winds weaken or even reverse and the eastern equatorial Pacific Ocean becomes anomalously warm. The westward-moving equatorial current weakens or reverses.

Elastic deformation Type of deformation in which the material returns to its original shape after the stress is removed.

Emergency planning Planning for projects following catastrophic events such as hurricanes, floods, or other events.

Engineering geology Application of geologic information to engineering problems.

Environment That which surrounds an individual or a community; both physical and cultural surroundings. Environment also sometimes denotes a certain set of circumstances surrounding a particular occurrence; for example, environments of deposition.

Environmental crisis Refers to the possible convergence of resources and people. Also refers to the hypothesis that environmental degradation has reached a crisis point as a result of human use of the environment.

Environmental geology Application of geologic information to environmental problems.

Environmental geology map A map that combines geologic and hydrologic data expressed in nontechnical terms to facilitate general understanding by a large audience.

Environmental impact statement A written statement that assesses and explores the possible impacts of a particular project that may affect the human environment. The statement is required by the National Environmental Policy Act of 1969.

Environmental law A field of law concerning the conservation and use of natural resources and the control of pollution.

Environmental resource unit (ERU) A portion of the environment with a similar set of physical and biological characteristics, a supposedly natural division characterized by specific patterns or assemblages of structural components (such as rocks, soils, and vegetation) and natural processes (such as erosion, runoff, and soil processes).

Environmental unity A principle of environmental studies that states that everything is connected to everything else. Garrett Harden would further state that environmental unity means you cannot do just one thing without affecting something else.

Ephemeral Temporary or very short-lived. Characteristic of beaches, lakes, and some stream channels that change rapidly (geologically).

Epicenter The point on the surface of the earth directly above the hypocenter (area of first motion) of an earthquake.

Erodibility (soil) Measure of how easily a soil may erode.

Evaporite Sediments deposited from water as a result of extensive evaporation of seawater or lake water; dissolved materials left behind following evaporation.

Exponential growth A type of compound growth in which a total amount or number increases at a certain percentage per year, and each year's rate of growth is added to the total from the previous year; characteristically stated in terms of a particular doubling time, that is, the time in years it will take the original number to double. Commonly used in reference to population growth.

Extrusive igneous rock Igneous rock that forms when magma reaches the surface of the earth; a volcanic rock.

Factor of safety With respect to landsliding, refers to the ratio of resisting to driving forces. A safety factor of greater than 1 suggests a slope is stable.

Fault A fracture or fracture system that has experienced movement along opposite sides of the fracture.

Fault gouge A clay zone formed by pulverized rock during an earthquake, which may create a groundwater barrier.

Fault segmentation A concept recognizing that faults may be divided into specific segments depending upon their geometry, structure, and earthquake history.

Fecal coliform bacteria A type of bacteria commonly found in the gut of humans and other animals. Usually harmless but can cause some disease. Commonly used as a measure of biological pollution.

Feedback The response of a system in which output from the system serves as an input back into the system, causing change.

Feldspar The most abundant family of minerals in the crust of the earth; silicates of calcium, sodium, and potassium.

Ferromagnesian mineral Minerals containing iron and magnesium, characteristically dark in color.

Fertile material Material such as uranium-238, which is not naturally fissionable but upon bombardment by neutrons is converted to plutonium-239, which is fissionable.

Fission The splitting of an atom into smaller fragments with the release of energy.

Floodplain Flat topography adjacent to a stream in a river valley, produced by the combination of overbank flow and lateral migration of meander bends.

Floodplain regulation A process of delineating floodplains and regulating land uses on them.

Floodway district That portion of a channel and floodplain of a stream designated to provide passage of the 100-year regulatory flood without increasing elevation of the flood by more than one foot.

Floodway fringe district Land located between the floodway district and the maximum elevation subject to flooding by the 100-year regulatory flood.

Fluidized-bed combustion A treatment process during the combustion of coal that removes sulfur and thus reduces emissions of sulfur dioxide into the atmosphere.

Fluorine Important trace element, essential for nutrition.

Fluvial Concerning or pertaining to rivers.

Fly ash Very fine particles (ash) resulting from the burning of fuels such as coal.

Focus The point or location in the earth where earthquake energy is first released. During an earthquake event, seismic energy radiates out from the focus.

Fold Bend that develops in stratified rocks because of tectonic forces.

Foliation Property of metamorphic rock characterized by parallel alignment of the platy or elongated mineral grains; environmentally important because it can affect the strength and hydrologic properties of rock.

Formation Any rock unit that can be mapped.

Fossil fuels Fuels such as coal, oil, and gas formed by the alteration and decomposition of plants and animals from a previous geologic time.

Fracture zone A fracture system that may or may not be active and may or may not have an alteration zone along the fracture planes. Fracture zones are environmentally important because they greatly affect the strength of rocks.

Fugitive sources Those air pollutants emitted from open areas exposed to wind processes.

Fumarole A natural vent from which fumes or vapors are emitted, such as the geysers and hot springs characteristic of volcanic areas.

Fusion, nuclear Combining of light elements to form heavy elements with the release of energy.

Gaging station Location at a stream channel where discharge of water is measured.

Gaia hypothesis A series of hypotheses to explain how the earth as a system may operate with respect to life. Metaphorically, the earth is viewed as a giant organism consisting of various interactive systems with distinct feedback and thresholds that result in producing an environment beneficial to the many life forms on earth. Furthermore, life is an important ingredient in producing that environment.

Gamma radiation Type of nuclear radiation consisting of energetic and penetrating rays similar to X rays emitted during radioactive decay.

Gasification Method of producing gas from coal.

Geochemical cycle Migratory paths of elements during geologic changes and processes.

Geographic Information System (GIS) Technology capable of storing, retrieving, transforming, and displaying spatial environmental data.

Geologic cycle A group of interrelated cycles known as the *hydrologic*, *rock*, *tectonic*, and *geochemical cycles*.

Geomorphology The study of landforms and surface processes.

Geopressured system Type of geothermal energy system resulting from trapping the normal heat flow from the earth by impermeable layers such as shale rock.

Geothermal energy The useful conversion of natural heat from the interior of the earth.

Glacial surge A sudden or quick advance of a glacier.

Glacier A landbound mass of moving ice.

Global circulation models Refers to computer models used to predict global change such as increase in mean temperature, precipitation, or some other climatic variable.

Global warming Refers to the hypothesis that the mean annual temperature of the lower atmosphere is increasing as a result of burning fossil fuels and emitting greenhouse gases into the atmosphere.

Gneiss A coarse-grained, foliated metamorphic rock in which there is banding of light and dark minerals.

Grading of slopes Cut-and-fill activities designed to increase the stability of a slope.

Gravel Unconsolidated, generally rounded fragments of rocks and minerals greater than 2 mm in diameter.

Gravitational water Water that occurs in pore spaces of a soil and is free to drain from the soil mass under the influence of gravity.

Greenhouse effect Trapping of heat in the atmosphere by water vapor, carbon dioxide, methane, and CFCs.

Groin A structure designed to protect shorelines and trap sediment in the zone of littoral drift, generally constructed perpendicular to the shoreline.

Groundwater Water found beneath the surface of the earth within the zone of saturation.

Groundwater flow Movement of water in the subsurface below the groundwater table.

Grout A mixture of cement and sediment that is sufficiently fluid to be pumped into open fissures or cracks in rocks, thereby increasing the strength of a foundation for an engineering structure.

Growth rate A rate usually measured as a percentage by which something is changing. For example, if you earn 5 percent interest in a bank account per year, then the growth rate is 5 percent per year.

Gypsum An evaporite mineral, $CaSO_4 \cdot 2H_2O$.

Half-life The amount of time necessary for one-half of the atoms of a particular radioactive element to decay.

Halite A common mineral, NaCl (salt).

Hardpan soil horizon Hard, compacted or cemented soil horizon, most often composed of clay but sometimes cemented with calcium carbonate, iron oxide, or silica. This horizon is nearly impermeable and often restricts the downward movement of soil water.

Hard path From an environmental energy perspective, refers to use of large centralized power plants. May be coupled to energy conservation and cogeneration.

Heat island effect Mass of warmer air over urban centers that results from rising hot air from the centers that pull in air from surrounding areas. As the warmer air rises, it cools and spreads, descending away from heat island.

Hematite An important ore of iron, a mineral (Fe_2O_3).

High-value resource Materials such as diamonds, copper, gold, and aluminum. These materials are extracted wherever they are found and transported around the world to numerous markets.

Hot igneous system Type of geothermal energy system in which heat is supplied by the presence of magma.

Hot spot Assumed stationary heat source located below the lithosphere that feeds volcanic processes near the earth's surface.

Hot springs and geysers Features at the surface of the earth where hot water and steam are quietly released or may be explosively erupted.

Humus Black organic material in soil.

Hurricane Tropical cyclone characterized by circulating winds of 100 km/hr or greater generated over an area of about 160 km in diameter. Also known as typhoons (in the Pacific Ocean).

Hydraulic conductivity Measure of the ability of a particular material to allow water to move through it. Units are length per time such as meters per day.

Hydrocarbon Organic compounds consisting of carbon and hydrogen.

Hydroconsolidation Consolidation of earth materials upon wetting.

Hydrofracturing Pumping of water under high pressure into subsurface rocks to fracture the rocks and thereby increase their permeability.

Hydrogeology Discipline that studies relationships between groundwater, surface water, and geology.

Hydrograph A graph of the discharge of a stream over time.

Hydrologic cycle Circulation of water from the oceans to the atmosphere and back to the oceans by way of precipitation, evaporation, runoff from streams and rivers, and groundwater flow.

Hydrologic gradient The driving force for both saturated and unsaturated flow of groundwater. Quantitatively, it is the slope or rate of change of the hydraulic head, which at the point of measurement is the algebraic sum of the elevation head and pressure head.

Hydrology The study of surface and subsurface water.

Hydrosphere The water environment in and on earth and in the atmosphere.

Hydrothermal convection system Geothermal energy system characterized by the circulation of hot water. May be dominated by water vapor or hot water.

Hydrothermal ore deposit A mineral deposit derived from hot water solutions of magmatic origin.

Hygroscopic water Refers to water absorbed and retained on fine-grained soil particles. This water may be held tenaciously.

Hypocenter The point in the earth where an earthquake originates; also known as the *focus*.

Hypothesis A statement intended to be a possible answer to a scientific question. The best hypothesis may be tested. Often multiple hypotheses are developed to answer a particular question.

Igneous rocks Rocks formed from solidification of magma; they are extrusive if they crystallize on the surface of the earth and intrusive if they crystallize beneath the surface.

Impermeable Earth materials that greatly retard or prevent movement of fluids through them.

Incineration Reduction of combustible waste to inert residue (ash) by burning at high temperatures.

Infiltration Movement of surface water into rocks or soil.

Influent stream Stream that is everywhere above the groundwater table and flows in direct response to precipitation. Water from the channel moves down to the water table, forming a recharge mound.

Input-output analysis A type of systems analysis in which rates of input and output are calculated and compared.

Instream use Water that is used but not withdrawn from its source. For example, water used to generate hydroelectric power.

Integrated waste management A complex set of management alternatives for waste management, including source reduction, recycling, composting, landfill, and incineration.

Intrusive igneous rock Igneous rock that forms when magma solidifies below the surface of the earth; a volcanic rock.

Iodine A nonmetallic element needed in trace amounts by the human body. Iodine is necessary for the normal functioning of the thyroid gland; a shortage can hinder growth or cause goiter.

Island arc A curved group of volcanic islands associated with a deep-oceanic trench and subduction zone (convergent plate boundary).

Isotopes Forms of the same element having a variable atomic weight.

Itaiitai disease Extremely painful disease that attacks bones, causing them to become very brittle so that they break easily. Associated with heavy metals, especially cadmium, in concentrations of a few parts per million in the soil or in food consumed by victims of the disease.

Jetty Often constructed in pairs at the mouth of a river or inlet to a lagoon, estuary, or bay, a jetty is designed to stabilize a channel, control deposition of sediment, and deflect large waves.

Juvenile water Water derived from the interior of the earth that has not previously existed as atmospheric or surface water.

K soil horizon A calcium-carbonate-rich horizon in which the carbonate often forms laminar layers parallel to the surface. Carbonate completely fills the pore spaces between soil particles.

Karst topography A type of topography characterized by the presence of sinkholes, caverns, and diversion of surface water to subterranean routes.

Kimberlite pipe An igneous intrusive body that may contain diamond crystals disseminated (scattered) throughout the rock type.

Land application Alternative for disposal of certain types of hazardous chemical waste in which the waste is applied to the soil and degraded by natural biological activity in the soil.

Land ethic Ethic that affirms the right of all resources, including plants, animals, and earth materials, to continued existence and, at least in some locations, continued existence in a natural state.

Landslide Specifically, rapid downslope movement of rock and/or soil; also a general term for all types of downslope movement.

Land-use planning Complex and extremely controversial process involving development of a land-use plan to include a statement of land-use issues, goals, and objectives; summary of data collection and analysis; land-classification map; and report describing and indicating appropriate development in areas of special environmental concern.

Lateral blast Type of volcanic eruption characterized by explosive activity that is more or less parallel to the surface of the earth. Lateral blast may occur when catastrophic failure of the side of a volcano occurs.

Laterite Soil formed from intense chemical weathering in tropical or savanna regions.

Lava Molten material produced from a volcanic eruption, or rock that forms from solidification of molten material.

Lava flow Eruption of magma at the surface of the earth that generally flows downslope from volcanic vents.

Lava tube A natural conduit or tunnel through which magma moves from a volcanic event downslope sometimes many kilometers to where the magma may again emerge at the surface. Following the volcanic eruption, the tubes are often left as open voids and are a type of cave.

Leachate Noxious liquid material capable of carrying bacteria, produced when surface water or groundwater comes into contact with solid waste.

Leaching Process of dissolving, washing, or draining earth materials by percolation of groundwater or other liquids.

Lethal dose (LD) Concentration or dose of a toxin that kills a particular percentage of a population. For example, LD-50 refers to the concentration or dose of a toxin necessary to kill 50 percent of the population exposed.

Lignite A type of low-grade coal.

Limestone A sedimentary rock composed almost entirely of the mineral calcite.

Limonite Rust; hydrated iron oxide.

Liquefaction Transformation of water-saturated granular material from the solid state to a liquid state.

Lithosphere Outer layer of the earth approximately 100 kilometers thick of which the plates that contain the ocean basins and continents are composed.

Littoral Pertaining to the nearshore and beach environments.

Littoral cell Segment of coastline that includes an entire cycle of sediment delivery to the coast, longshore littoral transport, and eventual loss of sediment from the nearshore environment.

Loess Angular deposits of windblown silt.

Longshore bar and trough Elongated depression and adjacent ridge of sand roughly parallel to shore produced by wave action.

Low-level radioactive waste Materials that contain only small amounts of radioactive substances.

Low-value resource Resources such as sand and gravel that have primarily a place value, economically extracted because they are located close to where they are to be used.

Magma A naturally occurring silica melt, much of which is in a liquid state.

Magma tap Attempt to recover geothermal heat directly from magma. Feasibility of such heat extraction is unknown.

Magnetite A mineral and important ore of iron, Fe_3O_4.

Magnitude-frequency concept The concept that states that the magnitude of an event is inversely proportional to its frequency.

Manganese oxide nodules Nodules of manganese, iron with secondary copper, nickel, and cobalt, that cover vast areas of the deep-ocean floor.

Marble Metamorphosed limestone.

Marl Unconsolidated clays, silts, sands, or mixtures of these materials that contain a variable content of calcareous material.

Maximum credible earthquake The largest earthquake that may reasonably be assumed to occur at a particular area in light of the tectonic environment, historic earthquakes, and paleoseismicity.

Meanders Bends in a stream channel that migrate back and forth across the floodplain, depositing sediment on the inside of the bends, forming point bars, and eroding outsides of bends.

Mediation Process of working toward a solution to a conflict concerning the environment that is advantageous for everyone; a collaborative process that seeks a solution that favors the environment while allowing activities and projects to go forward.

Metamorphic rock A rock formed from preexisting rock by the effects of heat, pressure, and chemically active fluids beneath the earth's surface. In foliated metamorphic rocks, the mineral grains have a preferential parallel alignment or segregation of minerals; nonfoliated metamorphic rocks have neither.

Meteoric water Water derived from the atmosphere.

Methane A gas, CH_4; the major constituent of natural gas.

Mica A common rock-forming silicate mineral.

Mineral A naturally occurring, solid, crystalline substance with physical and chemical properties that vary within known limits.

Mitigated negative declaration Environmental statement filed when the initial study of a project suggests that any significant environmental problems that will occur from a project could be modified to mitigate those problems.

Mitigation The identification of actions that will avoid, lessen, or compensate for anticipated adverse environmental impacts.

Modified Mercalli Scale A scale with twelve divisions that subdivide the amount and severity of shaking and damage from an earthquake.

Moment magnitude The magnitude of an earthquake based on its seismic moment, which is the product of the average amount of slip on the fault that produced the earthquake, the area that actually ruptured, and the shear modulus of the rocks that failed.

Monitoring Periodic or continuous gathering of samples of soil, vegetation, vadose zone water, and groundwaters in and near waste management facilities, such as landfills or hazardous waste disposal facilities.

Mudflow A mixture of unconsolidated materials and water that flows rapidly downslope or down a channel.

Myth of superabundance The myth that land and water resources are inexhaustible and management of resources is therefore unnecessary.

National Environmental Policy Act of 1969 (NEPA) Act declaring a national policy that harmony between man and his physical environment be encouraged. Established the Council on Environmental Quality, and established requirements that an environmental impact statement be completed prior to major federal actions that significantly affect the quality of the human environment.

Natural gas Also referred to as *natural energy gas*, hydrocarbons that include ethane, propane, butane, and hydrogen.

Negative declaration The filing of a statement that declares that there are no significant effects on the environment from a particular project or plan.

Negative feedback A type of feedback in which the outcome moderates or decreases the process, often leading to a steady-state system or a system in quasi-equilibrium (is self-regulating).

Neutron A subatomic particle having no electric charge, found in the nuclei of atoms. Neutrons are crucial in sustaining nuclear fission in a reactor.

Nitrogen oxides (NO_x) Include compounds such as nitrogen dioxide (NO_2), which is a light yellow-brown to reddish-brown gas that is a main pollutant contributing to the development of photochemical smog. Emitted as a result of burning fossil fuels from automobiles and power plants.

Non-point sources Diffused and intermittent sources of air or water pollutants.

Nonrenewable resource A resource cycled so slowly by natural earth processes that, once used, will be essentially unavailable during any useful time frame.

Nuclear reactor Device in which controlled nuclear fission is maintained; the major component of a nuclear power plant.

O soil horizon Soil horizon that contains plant litter and other organic material. Found above the *A* soil horizon.

Ocean pollution Pollution of the oceans of the world due to direct or indirect (deliberate or not) injection of contaminants to the marine environment. Often results from the process of ocean dumping, but there are many sources of ocean pollution.

Offstream use Water removed or diverted from its primary source for a particular use.

Oil May also be known as petroleum or crude oil; a liquid hydrocarbon generally extracted from wells.

Oil shale Organic-rich shale containing substantial quantities of oil that can be extracted by conventional methods of destructive distillation.

Open system A type of system in which there is a constant flow of energy and matter across the borders of the system.

Ore Earth material from which a useful commodity can be extracted profitably.

Osteoporosis Disease characterized by reduction in bone mass.

Outcrop A naturally occurring or human-caused exposure of rock at the surface of the earth.

Overburden Earth materials (spoil) that overlie an ore deposit, particularly material overlying or extracted from a surface (strip) mine.

Overland flow Flow of water on the surface of the earth not confined to channels; results because the intensity of precipitation is greater than the rate at which rainwater infiltrates into the ground.

Oxidation Chemical process of combining with oxygen.

Ozone Triatomic oxygen (O_3).

Ozone depletion Refers to stratospheric loss of ozone generally at the South Pole related to release of chlorofluorocarbons (CFCs) into the atmosphere.

P wave One of the seismic waves produced by an earthquake; the fastest of the seismic waves, it can move through liquid and solid materials.

Particulate matter Small particles of solid or liquid substances that are released into the atmosphere by natural processes and human activities. Examples include smoke, soot, or dust as well as particles of heavy metals, such as copper, lead, and zinc.

Pathogen Any material that can cause disease; for example, microorganisms, including bacteria and fungi.

Pebble A rock fragment between 4 and 64 mm in diameter.

Ped An aggregate of soil particles. Peds are classified by their shapes as spheroidal, blocky, prismatic, etc.

Pedology The study of soils.

Pegmatite A coarse-grained igneous rock that may contain rare minerals rich in elements such as lithium, boron, fluorine, uranium, and others.

Perched water table A water table of relatively limited extent that is found at a higher elevation than the more regional water table.

Percolation test A standard test for determining the rate at which water will infiltrate into the soil. Primarily used to determine feasibility of a septic-tank disposal system.

Permafrost Permanently frozen ground.

Permeability A measure of the ability of an earth material to transmit fluids such as water or oil. *See* Hydraulic conductivity.

Petrology Study of rocks and minerals.

Photochemical smog Sometimes called *L.A.-type smog* or *brown air*; smog resulting from complex chemical reactions involving pollutants such as nitrogen oxides, and hydrocarbons that react with solar radiation.

Photovoltaics Type of solar technology that converts sunlight directly to electricity.

Physiographic determinism Site selection based on the philosophy of designing with nature.

Physiographic province Region characterized by a particular assemblage of landforms, climate, and geomorphic history.

Placer deposit Ore deposit found in material transported and deposited by such agents as running water, ice, or wind; for example, gold and diamonds found in stream deposits.

Plastic deformation Deformation that involves a permanent change of shape without rupture.

Plate tectonics A model of global tectonics that suggests that the outer layer of the earth known as the *lithosphere* is composed of several large plates that move relative to one another; continents and ocean basins are passive riders on these plates.

Plunging breaker A type of wave or breaker that strikes a shoreline with a relatively steep beach profile. Plunging breakers tend to be associated with beach erosion.

Plutonium-239 A radioactive element produced in a nuclear reactor; has a half-life of approximately 24,000 years.

Point bar Accumulation of sand and other sediments on the inside of meander bends in stream channels.

Point sources Usually discrete and confined sources of air or water pollutants, such as pipes, that enter into a stream or river or stacks emitting waste from factories or other facilities into the atmosphere.

Pollution Any substance, biological or chemical, in which an identified excess is known to be detrimental to desirable living organisms.

Pool Common bed form produced by scour in meandering and straight stream channels with relatively low channel slope; characterized at low flow by slow-moving, deep water. Generally, but not exclusively, found on the outside of meander bends.

Porosity The percentage of void (empty space) in earth material such as soil or rock.

Positive feedback A type of system in which the output amplifies the input, leading to what some call a vicious cycle. Another way of looking at this concept: "The more you have, the more you get."

Potable water Water that may be drunk safely.

Primary pollutants Refers to those air pollutants emitted directly into the atmosphere, including particulates, sulfur oxides, carbon monoxide, nitrogen oxides, and hydrocarbons.

Primary treatment Wastewater treatment that includes screening and removal of grit and sedimentation of larger particles from the waste stream.

Pyrite Iron sulfide, a mineral commonly known as *fool's gold;* environmentally important because, in contact with oxygen-rich water, it produces a weak acid that can pollute water or dissolve other minerals.

Pyroclastic activity Type of volcanic activity characterized by eruptive or explosive activity in which all types of volcanic debris, from ash to very large particles, are physically blown from a volcanic vent.

Pyroclastic flow Rapid subaerial flowage of eruptive material consisting of volcanic gases, ash, and other materials that move rapidly down the flank of a volcano. Often form as the result of the collapse of an eruption column. May also be known as *ash flows, fiery clouds,* or *nuée ardentes.*

Quartz Silicon oxide, a common rock-forming mineral.

Quartzite Metamorphosed sandstone.

Quick clay Type of clay that, when disturbed, as by seismic shaking, may experience spontaneous liquefaction and lose all shear strength.

R soil horizon Consolidated bedrock that underlies the soil.

Radioactive waste Type of waste produced in the nuclear fuel cycle, generally classified as high-level or low-level.

Radioisotope A form of a chemical element that spontaneously undergoes radioactive decay, changing from one isotope to another and emitting radiation in the process.

Radon A colorless, radioactive, gaseous element.

Rapid drawdown Rapid decrease in the elevation of the water table at a particular location due to a variety of processes, including pumping of groundwater, receding floodwaters, or lowering of the water in a reservoir.

Reclaimed water Water that has been treated by wastewater handling facilities and may be used for other purposes on discharge, such as irrigation of golf courses or croplands.

Reclamation, mining Restoring land used for mining to other useful purposes, such as agriculture or recreation, after mining operations are concluded.

Record of decision A concise statement by an agency planning a proposed project as to which alternatives were considered and, specifically, which alternatives are environmentally preferable. The record of decision is becoming an important part of environmental impact work.

Recycling The reuse of resources reclaimed from waste.

Reduce, recycle, and reuse The three Rs of integrated waste management that have the objective of reducing the amount of waste that must be disposed of in landfills or other facilities.

Regional metamorphism Wide-scale metamorphism of deeply buried rocks by regional stress accompanied by elevated temperatures and pressures.

Relative profile development Refers to soils that may be weakly, moderately, or strongly developed, depending upon specific soil properties.

Renewable energy Energy sources that are replenished quickly enough to maintain a constant supply if they are not overused. Examples include solar energy, water power, and wind energy.

Renewable resource A resource such as timber, water, or air that is naturally recycled or recycled by human-induced processes within a useful time frame.

Reserves Known and identified deposits of earth materials from which useful materials can be extracted profitably with existing technology under present economic and legal conditions.

Resisting forces Forces that tend to oppose downslope movement of earth materials.

Resistivity A measure of an earth material's ability to retard the flow of electricity; the opposite of conductivity.

Resources Includes reserves plus other deposits of useful earth materials that may eventually become available.

Rhyolite A type of volcanic rock consisting of feldspar, ferromagnesian, and quartz minerals with a relatively high silica content of about 70 percent. Rhyolite is associated with volcanic events that may be very explosive.

Richter magnitude A measure of the amount of energy released by an earthquake, determined by converting the largest amplitude of the shear wave to a logarithmic scale in which, for example, 2 indicates the smallest earthquake that can be felt and 8.5 indicates a devastating earthquake.

Riffle A section of stream channel characterized at low flow by fast, shallow flow; generally contains relatively coarse bed-load particles.

Riparian doctrine Part of our prevailing water law; restricts water rights, for the most part, to owners of land adjoining a stream or body of standing water.

Riparian rights, water law Right of the landowner to make reasonable use of water on his or her land, provided the water is returned to the natural stream channel before it leaves his or her property; the property owner has the right to receive the full flow of the stream undiminished in quantity and quality.

Rippable excavation Type of excavation that requires breaking up soil before it can be removed.

Riprap Layer or assemblage of broken stones placed to protect an embankment against erosion by running water or breaking waves.

Risk From an environmental viewpoint, risk may be considered the product of the probability of an event times the consequences.

Risk assessment The process of determining potential adverse environmental health effects following exposure to a particular toxic material.

Risk management The process of integrating risk assessment with legal, social, political, economical, and technical issues to develop a plan of action for a particular toxin.

Riverine environment Land area adjacent to and influenced by a river.

Rock *Geologic:* An aggregate of a mineral or minerals. *Engineering:* Any earth material that must be blasted to be removed.

Rock cycle Group of processes that produce igneous, metamorphic, and sedimentary rocks.

Rock salt Rock composed of the mineral halite.

Rotational landslide Type of landslide that develops in homogeneous material; movement is likely to be rotational along a potential slide plane.

S wave Secondary wave, one of the waves produced by earthquakes.

Safety factor *See* Factor of safety.

Saline Salty; characterized by high salinity.

Salinity A measure of the total amount of dissolved salts in water.

Salt dome A structure produced by upward movement of a mass of salt; frequently associated with oil and gas deposits on the flanks of a dome.

Sand Grains of sediment with a size between 1/16 and 2 mm in diameter. Often, sediment composed of quartz particles of this size.

Sand dune Ridge or hill of sand formed by wind action.

Sandstone Detrital sedimentary rock composed of sand grains that have been cemented together.

Sanitary landfill Method of solid-waste disposal that does not produce a public health problem or nuisance; confines and compresses waste and covers it at the end of each day with a layer of compacted, relatively impermeable material, such as clay.

Saturated flow A type of subsurface or groundwater flow in which all the pore spaces are filled with water.

Scarp Steep slope or cliff commonly associated with landslides or earthquakes.

Scenic resources The visual portion of an aesthetic experience; scenery is now recognized as a natural resource with varying values.

Schist Coarse-grained metamorphic rock characterized by foliated texture of the platy or elongated mineral grains.

Schistosomiasis Snail fever, a debilitating and sometimes fatal tropical disease.

Scientific method The method by which scientists work, starting with asking a question concerning a problem, followed by development and testing of hypotheses.

Scoping Process of identifying important environmental issues that require detailed evaluation early in the planning of a proposed project. Scoping is an important part of environmental impact analysis.

Scrubbing Process of removing sulfur dioxide from gases emitted from burning coal in power plants producing electricity.

Seacliff Steep (commonly, near-vertical) bluff adjacent and adjoining a beach or coastal environment. The seacliff is produced by a combination of erosional processes, including wave activity and subaerial processes, such as weathering, landsliding, and runoff of surface water from the land.

Seawall Engineering structure constructed at the water's edge to minimize coastal erosion by wave activity.

Secondary enrichment Weathering process of sulfide ore deposits that may concentrate the desired minerals.

Secondary pollutants Refers to pollutants produced when primary pollutants react with normal atmospheric compounds. An example is ozone, which forms through reactions among primary pollutants, sunlight, and natural atmospheric gases.

Secondary treatment Wastewater treatment that includes aerobic and anaerobic digestion of waste in the wastewater stream, primarily by bacterial breakdown. The final stage is disinfecting of treated water, usually with chlorine.

Secure landfill Type of landfill designed to contain and dispose of hazardous chemical waste. Many of these facilities have been shut down because containment of the hazardous waste has been impossible to maintain.

Sedimentary rock A rock formed when sediments are transported, deposited, and then lithified by natural cement, compression, or other mechanism. Detrital sedimentary rock is formed from broken parts of previously existing rock; chemical sedimentary rock is formed by chemical or biochemical processes removing material carried in chemical solution.

Sediment pollution Pollution of some part of the environment either on land or in a body of water by sediment that has been transported into that environment by wind or water. An example includes turbidity of a water supply (muddy water).

Sediment yield Volume or mass of sediment per unit time produced from a particular area.

Sedimentology Study of environments of deposition of sediments.

Seismic Refers to vibrations in the earth produced by earthquakes.

Seismic gaps Areas along active fault zones that are capable of producing large earthquakes but have not produced one recently.

Seismograph Instrument that records earthquakes.

Selenium Important nonmetallic trace element with an atomic number of 34.

Sensitivity (soil) Measure of loss of soil strength due to disturbances such as human excavation and remolding.

Septic tank Tank that receives and temporarily holds solid and liquid waste. Anaerobic bacterial activity breaks down the waste, solid wastes are separated out, and liquid waste from the tank overflows into a drainage system.

Sequential land use Development of land previously used as a site for the burial of waste. The specific reuse must be carefully selected.

Serpentine A family of ferromagnesian minerals; environmentally important because they form very weak rocks.

Sewage sludge Solid material that remains after municipal wastewater treatment.

Shale Sedimentary rock composed of silt- and clay-sized particles; the most common sedimentary rock.

Shield volcano A broad, convex volcano built up by successive lava flows; the largest of the volcanoes.

Shrink-swell potential (soil) Measure of a soil's tendency to increase and decrease in volume as water content changes.

Silicate minerals The most important group of rock-forming minerals.

Silt Sediment between 1/16 and 1/256 mm in diameter.

Sinkhole Surface depression formed by solution of limestone or collapse over a subterranean void such as a cave.

Sinuous channel Type of stream channel (not braided).

Slate A fine-grained, foliated metamorphic rock.

Slip rate Long-term rate of slip (displacement) along a fault. Usually measured in millimeters or centimeters per year.

Slump Type of landslide characterized by downward slip of a mass of rock, generally along a curved slide plane.

Snow avalanche Rapid downslope movement of snow, ice, and rock.

Soft path Refers to development of an energy policy that involves alternatives that are renewable, flexible, decentralized and (to some people) more benign from an environmental viewpoint than the hard path.

Soil *Soil science:* Earth material so modified by biological, chemical, and physical processes that the material will support rooted plants. *Engineering:* Earth material that can be removed without blasting.

Soil chronosequence A series of soils arranged in terms of relative soil profile development from youngest to oldest.

Soil fertility Capacity of a soil to supply nutrients (such as nitrogen, phosphorus, and potassium) needed for plant growth when other factors are favorable.

Soil horizons Layers in soil (*A*, *B*, *C*, etc.) that differ from one another in chemical, physical, and biological properties.

Soil profile development Weathering of earth materials that, along with biological activity and time, among other factors, produces a soil that contains several horizons distinct from the parent material from which the soil formed.

Soil sensitivity Refers to soils that when disturbed and remolded and have only a portion of their original strength. These soils are said to be sensitive.

Soil strength The shear strength of a soil, usually in terms of cohesive and frictional forces.

Soil survey A survey consisting of a detailed soil map and descriptions of soils and land-use limitations; usually prepared by the Soil Conservation Service in cooperation with local government.

Solar energy Energy that is collected directly from the sun. Broadly, also includes energy that is collected indirectly.

Solid waste Material such as refuse, garbage, and trash.

Specific risk The product of the elements at risk, the probability that a specific event will occur, and the vulnerability defined as the proportion of elements at risk.

Spilling breaker A type of wave associated with a shoreline of relatively low slope. Spilling breakers tend to be associated with deposition of sand on a beach.

Spoils, mining Banks or piles that are accumulations of overburden removed during mining processes and discarded on the surface.

Stationary sources Those sources of air pollution that are relatively fixed in location.

Steady-state system A system in which the input is approximately equal to the output, and so a rough equilibrium is established.

Storm surge Wind-driven oceanic waves.

Strain Change in shape or size of a material as a result of applied stress; the result of stress.

Stream power The product of the discharge of a river and its energy slope (can use channel slope as approximation).

Strength Ability of a rock or soil to resist deformation. Results from cohesive and frictional forces.

Stress Force per unit area; may be compressive, tensile, or shear.

Strip mining A method of surface mining.

Subduction Process in which one lithospheric plate descends beneath another.

Subsidence Sinking, settling, or other lowering of parts of the crust of the earth.

Subsurface water All the waters within the lithosphere.

Sulfur dioxide (SO_2) Colorless and odorless gas whose anthropogenic component in the atmosphere results primarily from burning of fossil fuels.

Sulfurous smog Sometimes referred to as *London-type smog* or *gray air*; produced primarily by burning coal or oil at large power plants where sulfur oxides and particulates produced by the burning produce a concentrated smog.

Surf zone That part of the beach and nearshore environment characterized by borelike waves of translation after waves break.

Surface impoundment Excavated or natural topographic depressions used to hold hazardous liquid waste. Although impoundments are often lined, they have been criticized because they are especially prone to seepage and pollution of soil and groundwaters.

Surface water Waters above the surface of the lithosphere.

Surface wave One type of wave produced by earthquakes; these waves generally cause most of the damage to structures on the surface of the earth.

Suspended load Sediment in a stream or river carried off the bottom by the fluid.

Sustainability A difficult term to define but generally refers to development or use of resources in such a way that future generations will have a fair share of the earth's resources and inherit a quality environment. In other words, *sustainability* refers to types of development that are economically viable, do not damage the environment, and are socially just.

Sustainable energy policy Development of energy policy that finds useful sources of energy that do not have adverse environmental effects, or minimizes those effects, such that future generations will have access to energy resources and a quality environment.

Sustainable global economy A global economic development that will not harm the environment, will provide for future generations, and is socially just.

Syncline Fold in which younger rocks are found in the core of the fold; rocks in the limbs of the fold dip inward toward a common axis.

System Any part of the universe that is isolated in thought or in fact for the purpose of studying or observing changes that occur under various imposed conditions.

Tar sand Naturally occurring sand, sandstone, or limestone that contains an extremely viscous petroleum.

Tectonic Referring to rock deformation.

Tectonic creep Slow, more or less continuous movement along a fault.

Tectonic cycle Part of the geologic cycle. At the global scale it is the cycle of plate tectonics that produces ocean basins and mountain ranges.

Tephra Any material ejected and physically blown out of a volcano; mostly ash.

Texture, rock The size, shape, and arrangement of mineral grains in rocks.

Theory A strong scientific statement. A hypothesis may become a theory after it has been tested many times and has not been rejected.

Threshold A point of change where something happens. For example, a stream bank may erode when the water has sufficient force to dislodge particles. The point at which the erosion starts is the threshold.

Throughflow Downslope shallow subsurface flow above the groundwater table.

Tidal energy Electricity generated by tidal power.

Tidal flood A type of flood that occurs in estuaries or coastal rivers as the result of interactions between high tides and storm waves.

Till Unstratified, heterogeneous material deposited directly by glacial ice.

Toxic Harmful, deadly, or poisonous.

Transform fault Type of fault associated with oceanic ridges; may form a plate boundary, such as the San Andreas Fault in California.

Translation (slab) landslide Type of landslide in which movement takes place along a definite fracture plane, such as a weak clay layer or bedding plane.

Tropical cyclone Severe storm generated from a tropical disturbance; called *typhoons* in most of the Pacific Ocean and *hurricanes* in the Western Hemisphere.

Tsunami Seismic sea wave generated by submarine volcanic or earthquake activity; characteristically has very long wave length and moves rapidly in the open sea; incorrectly referred to as *tidal wave*.

Tuff Volcanic ash that is compacted, cemented, or welded together.

Unconfined aquifer Aquifer in which there is no impermeable layer restricting the upper surface of the zone of saturation.

Unconformity A buried surface of erosion representing a time of nondeposition; a gap in the geologic record.

Unified soil classification system Classification of soils, widely used in engineering practice, based on amount of coarse particles, fine particles, or organic material.

Uniformitarianism Concept that the present is the key to the past; that is, we can read the geologic record by studying present processes.

Unsaturated flow Type of groundwater flow that occurs when only a portion of the pores is filled with water.

Upstream floods Floods that occur in the upper part of drainage basins as the result of local storms. May be produced by intense rainfall of short duration over a relatively small area.

Urban dust dome Mass of dirty air that often accumulates over urban areas.

Urban ore Refers to the fact that in some communities the sewage sludge from waste disposal facilities contains sufficient metal deposits to be considered an ore.

Vadose zone Zone or layer above the water table in which some water may be suspended or moving in a downward migration toward the water table or laterally toward a discharge point.

Volcanic breccia, agglomerate Large rock fragments mixed with ash and other volcanic materials cemented together.

Volcanic dome Type of volcano characterized by very viscous magma with high silica content; activity is generally explosive.

vx (actual velocity of groundwater flow through pores) The ratio of the product of hydraulic conductivity and hydraulic gradient to porosity.

Wastewaters renovation and conservation cycle A process of recycling liquid waste that includes return of treated wastewater to crops or irrigation and continued renovation through recharge of groundwater. The reused part involves pumping out of the groundwater for municipal, industrial, or other purposes.

Water budget Analysis of sources, sinks, and storage sites for water in a particular area.

Water conservation Practices taken to use water more efficiently and to reduce withdrawal and consumption of water.

Water cycle *See* Hydrologic cycle.

Water management Practice of managing our water resources.

Water pollution Degradation of water quality as measured by biological, chemical, or physical criteria.

Water power Use of flowing water, such as in a reservoir, to produce electrical power.

Water table Surface that divides the vadose zone from the zone of saturation; the surface below which all the pore space in rocks is saturated with water.

Watershed Land area that contributes water to a particular stream system. *See* Drainage basin.

Wave climate Statistical characterization on an annual basis of wave height period and direction for a particular site.

Weathering Changes that take place in rocks and minerals at or near the surface of the earth in response to physical, chemical, and biological changes; the physical, chemical, and biological breakdown of rocks and minerals.

Wetlands Landscape features such as swamps, marshes, bogs, or prairie potholes that are frequently or continuously inundated by water.

Wind power Technology (mostly windmills) used to extract electrical energy from the wind.

Zinc An important trace element necessary in life processes.

Zone of saturation Zone or layer below the water table in which all the pore space of rock or soil is saturated.

INDEX

Other Conversion Factors

1 ft^3/sec = .0283 m^3/sec = 7.48 gal/sec = 28.32 liters/sec
1 acre-foot = 43,560 ft^3 = 1,233 m^3 = 325,829 gal
1 m^3/sec = 35.32 ft^3/sec
1 ft^3/sec for one day = 1.98 acre-feet
1 m/sec = 3.6 km/hr = 2.24 mi/hr
1 ft/sec = 0.682 mi/hr = 1.097 km/hr
1 billion gallons per day (bgd) = 3.785 million m^3 per day
1 atmosphere = 1.013×10^5 N/m^2 = approximately 1 bar
1 bar = approx. 10^5 N/m^2 = 10^5 pascal (Pa)

Strength of Common Rock Types

	Rock Type	Range of Compressive Strength (10^6 N/m^2)	Comments
Igneous	Granite	100 to 280	Finer-grained granites with few fractures are the strongest. Granite is generally suitable for most engineering purposes.
	Basalt	50 to greater than 280	Brecciated zones, open tubes, or fractures reduce the strength.
Metamorphic	Marble	100 to 125	Solutional openings and fractures weaken the rock.
	Gneiss	160 to 190	Generally suitable for most engineering purposes.
	Quartzite	150 to 600	Very strong rock.
Sedimentary	Shale	Less than 2 to 215	May be a very weak rock for engineering purposes; careful evaluation is necessary.
	Limestone	50 to 60	May have clay partings, solution openings, or fractures that weaken the rock.
	Sandstone	40 to 110	Strength varies with degree and type of cementing material, mineralogy, and nature and extent of fractures.

SOURCE: Data primarily from *Handbook of Tables for Applied Engineering Science,* ed. R. E. Bolz and G. L. Tuve (Cleveland, OH: CRC Press, 1973).

Commonly Used Multiples of 10

Prefix (Symbol)	Amount	Prefix (Symbol)	Amount
exa (E)	10^{18} (million trillion)	centi (c)	10^{-2} (one-hundredth)
peta (P)	10^{15} (thousand trillion)	milli (m)	10^{-3} (one-thousandth)
tera (T)	10^{12} (trillion)	micro (μ)	10^{-6} (one-millionth)
giga (G)	10^9 (billion)	nano (n)	10^{-9} (one-billionth)
mega (M)	10^6 (million)	pico (p)	10^{-12} (one-trillionth)
kilo (k)	10^3 (thousand)		